32nd International DAAAM Symposium on Intelligent Manufacturing and Automation 2021

Vienna, Austria
28-29 October 2021

Editor:

B. Katalinic

ISBN: 978-1-7138-4139-5

Printed from e-media with permission by:

Curran Associates, Inc.
57 Morehouse Lane
Red Hook, NY 12571

Some format issues inherent in the e-media version may also appear in this print version.

DAAAM International Vienna
Bahnallee 10
A-2120 Wolkersdorf
Austria, EU

Phone: +436646424398

http://www.daaam.info

Additional copies of this publication are available from:

Curran Associates, Inc.
57 Morehouse Lane
Red Hook, NY 12571 USA
Phone: 845-758-0400
Fax: 845-758-2633
Email: curran@proceedings.com
Web: www.proceedings.com

32nd International DAAAM Symposium on Intelligent Manufacturing and Automation 2021

Vienna, Austria
28-29 October 2021

ISSN 1726-9679

DANUBE ADRIA ASSOCIATION FOR AUTOMATION & MANUFACTURING

DAAAM INTERNATIONAL VIENNA

www.daaam.info

PROCEEDINGS

OF THE 32ND INTERNATIONAL DAAAM SYMPOSIUM
"INTELLIGENT MANUFACTURING & AUTOMATION"
28-29TH OCTOBER 2021
HOSTED FROM VIENNA UNIVERSITY OF TECHNOLOGY, VIENNA, AUSTRIA, EU

ORGANIZED BY:
DAAAM INTERNATIONAL VIENNA & UNIVERSITY OF MOSTAR

IN COOPERATION WITH:
ÖIAV 1848, VIENNA UNIVERSITY OF TECHNOLOGY, INTERNATIONAL ACADEMY OF ENGINEERING, AND UNIVERSITY OF APPLIED SCIENCES – TECHNIKUM WIEN

UNDER THE AUSPICES OF THE DANUBE RECTORS' CONFERENCE &
RECTORS' AND PRESIDENTS' HONOR COMMITTEE OF DAAAM INTERNATIONAL

EDITOR: B.[RANKO] KATALINIC

IMPRESSUM 2021

Proceedings of the 32nd International DAAAM Symposium "Intelligent Manufacturing & Automation" 28-29th October 2021, hosted from Vienna University of Technology, Vienna, Austria, EU, Volume 32, No.1,
Abbreviated key title: ***Ann. DAAAM 2021***

ISSN 1726-9679

Ed. B. Katalinic,

Published by DAAAM International, Vienna, Austria, EU, 2021
Layout & Design: B[ranko] Katalinic
Technical Editors: I[lija] Zec / L[uka] Zec

Publisher:

DAAAM International Vienna
TU Wien, Getreidemarkt 9, 1060 Wien, Austria
E-mails: president@daaam.info / branko.katalinic@tuwien.ac.at
DAAAM homepage: http://www.daaam.info

CONTENTS

AUTHOR INDEX

ORGANIZING COMMITTEE FOR 2021

KATALINIC, BRANKO – PRESIDENT OF DAAAM INTERNATIONAL
COVIC, DRAGAN – HONOR PRESIDENT OF ORGANIZING COMMITTEE
GUSEV, BORIS – PRESIDENT OF INTERNATIONAL ACADEMY OF ENGINEERING
ZEC, ILIJA - SECRETARY

ABURIA, MOHAMED
BEVANDA, IVAN
BIELIC, TONI
BOSNJAK, IGOR
BREIDO, IOSSIF
BUCHMEISTER, BORUT
COLAK, IVO
COLAK, STANKO
COSIC, ILIJA
CVETKOVSKA, ROBERTA
DURAKBASA, NUMAN
FUERST, LAURENZ
GODEC, DAMIR
HOELLER, LIANE
KALENIC, MIODRAG
KNEZOVIC, NIKOLA
KOSTOLANI, DAVID
KUKUSHKIN, ILYA
MARIC, PETAR
NAUMBURGER, JOERG
PANFILOV, PETER
PARK, HONG-SEOK
PRYANICHNIKOV, VALENTIN
RAKIC, KRESIMIR
SARAVANJA, LUKA
SCHMOELLEBECK, FRITZ
SEREMET, ZELJKO
STAZHKOV, SERGEY
STAROVESKI, TOMISLAV
STOJKIC, INJA
STOJKIC, ZELJKO
STUJA, KEMAJL
TEKIC, ZELJKO
TOMIC, ZORAN
UDILJAK, TOMA
UGLESIC, ANTE
VLADOV, MAREK
ZEC, ANTE
ZEC, LUKA

RECTORS' AND PRESIDENTS' HONOR COMMITTEE

ABDRASILOV, BOLATBEK	RECTOR, KARAGANDA STATE INDUSTRIAL UNIVERSITY
ABRUDAN, IOAN VASILE	RECTOR, TRANSILVANIA UNIVERSITY OF BRASOV
ALEXANDROV, ANATOLIY	RECTOR, BAUMAN MOSCOW STATE TECHNICAL UNIVERSITY
BAAIJENS, FRANK	RECTOR, EINDHOVEN UNIVERSITY OF TECHNOLOGY
BIAŁKIEWICZ, ANDRZEJ ANTONI	RECTOR, CRACOW UNIVERSITY OF TECHNOLOGY
BONDREA, IOAN	PRESIDENT, LUCIAN BLAGA UNIVERSITY OF SIBIU
BORAS, DAMIR	RECTOR, UNIVERSITY OF ZAGREB
BOROVIKOV, YURY	RECTOR, THE YAROSLAV-THE-WISE NOVGOROD STATE UNIVERSITY
CHEON, YEON	PRESIDENT, UNIVERSITY OF ULSAN
CISNEROS PÉREZ, GUILLERMO	RECTOR, UNIVERSIDAD POLITECNICA DE MADRID
COSTOIU, MIHNEA	RECTOR, UNIVERSITY POLITEHNICA OF BUCHAREST
CZIGANY, TIBOR	RECTOR, BUDAPEST UNIV. OF TECHNOLOGY AND ECONOMICS
DAM, HENRIK	DIRECTOR, UNIVERSITY OF SOUTHERN DENMARK
DANIELS, MITCHELL	PRESIDENT, PURDUE UNIVERSITY
DI SCIASCIO, EUGENIO	RECTOR, POLYTECHNIC OF BARI
DRĂGAN, FLORIN	RECTOR, POLYTECHNIC UNIVERSITY OF TIMISOARA
G. BOUDOUVIS, ANDREAS	RECTOR, NATIONAL TECHNICAL UNIVERSITY OF ATHENS
GALIJEV, ILJHAM	PRESIDENT, OMSK STATE TRANSPORT UNIVERSITY
GARCÍA MARÍN, JUAN FRANCISCO	RECTOR, UNIVERSITY OF LEON
GARCIA, EVA FERREIRA	RECTOR, UNIVERSITY OF THE BASQUE COUNTRY
GELENCSER, ANDRAS	RECTOR, UNIVERSITY OF PANNONIA
GEYER, SYLVIA	RECTOR, UNIV. OF APPLIED SCIENCES TECHNIKUM, VIENNA
GONOKAMI, MAKOTO	PRESIDENT EMERITUS, UNIVERSITY OF TOKYO
GUBERAC, VLADO	RECTOR, STROSSMAYER UNIVERSITY OF OSIJEK

International Scientific Committee

International Program Committee

BLEICHER, FRIEDRICH
BLEY, HELMUT
BOTEZATU, CORNELIA PAULINA
BOTEZATU, MIHAI
BRAJLIH, TOMAZ
BREZAK, DANKO
BREZOCNIK, MIRAN
BUCHMEISTER, BORUT
BUCURESTEANU, ANCA
CAMPBELL, R. IAN
CEKIC, AHMET
CELAR, STIPE
CHAN, CHEE SENG
CHAUS, ALEXANDER
CHRAMCOV, BRONISLAV
COHODAR, MAIDA
COLEMAN, J. J.
COSIC, ILIJA PETAR
COSMA, SMARANDA
COTET, COSTEL EMIL
CUBONOVA, NADEZDA
CURTU, IOAN
DAMIC, VJEKOSLAV
DENKENA, BEREND
DEVAHASTIN, SAKAMON
DOLGA, LIA
DOLGA, VALER
DOLINAY, JAN
DOLINAY, VILIAM
DOSTAL, PETR
DOSTALEK, PETR
DRAGCEVIC, ZVONKO
DRAGICEVIC, MARIJA
DRSTVENSEK, IGOR
DUHOVNIK, JOZE
EJUB DZAFEROVIC,
FIRST ROGALE, SNJEZANA
FUJII, NORIO
GHEORGHIU, HORIA
GODEC, DAMIR
GOMEZ, ANTONIO
GORDIN, DOINA MARGARETA
HLAVA, JAROSLAV
HOFREITER, MILAN
HRUBINA, KAMIL
HRUSKA, FRANTISEK
HRUSKOVA, ERIKA
IBISHI, ISMET M.
JANACOVA, DAGMAR
JANSSENS, THIERRY
JARVIS, CHRISTINE W.
JAVOROVA, ANGELA
JERBIC, BOJAN
JURAGA, IVAN
JURUM-KIPKE, JASNA
KARAULOVA, TATYANA
KATALINIC, BRANKO
KAWALEC, ANDRZEJ
KHANNA, ASHUTOSH
KHOMCHENKO, VASILY
KIRK, CHRIS. S.
KISS, IMRE
KITTL, BURKHARD
KLIMENKO, S. A.
KNEGO, NIKOLA
KNEZEVIC, PETAR
LUJIC, ROBERTO
LUKIC, IVANA
LUPEA, IULIAN
LUPU, CIPRIAN
LUPULESCU, NOURAS BARBU
MACHEDON PISU, TEODOR
MACKU, LUBOMIR
MACUROVA, ANNA
MAGHETI, IOAN
MAJETIC, DUBRAVKO
MALISA, VIKTORIO
MALOPOLSKI, WALDEMAR
MAN, MARIANA
MANDIC, LIDIJA
MANDRU, DAN
MANESCU, TIBERIU STEFAN
MARASCU KLEIN, VLADIMIR
MARCOS, MARIANO
MARCU, MARIUS
MAREK, KUBALCIK
MARIAN TOLNAY
MARIN, DUMITRU
MARIN, MIHNEA
MARINESCU, NICULAE ION
MARN, FERDINAND
MARONEK, MILAN
MARSLAND, DAVID
MARTA, CONSTANTIN
MARTINEZ, SUSANA
MASTACAN, LUCIAN
MATUSOVA, MIRIAM
MATUSU, RADEK
MEHTA, NARINDER KUMAR
MEIER, HORST
MENZ, PETER
MICHAL, BLIZNAK
MICIETA, BRANISLAV
MICLOSI, VIOREL
MOHORA, CRISTINA
MOK, H.
MORARU, SORIN AUREL
MORIWAKI, T.
MORTOIU, DOINA
MOTOMANCEA, ADRIAN
MUNTEAN, MIHAELA
MUNTEAN, SEBASTIAN
MUNTEANU, ANCA
MUSCA, GAVRIL
MUSTAFIZUL, KARIM
MYSHLYAVTSEV, ALEXANDR
NAGI, MIHAI
NAGLA, K. S.
NAKAMOTO, NOBUYUKI
NEUMANN, PETR
NICA COTET, GABRIELA BEATRICE
NIKOLIC, GOJKO
NKUMBWA, RODGER
NOH, SANG DO
NOVAC, OVIDIU
NOVOSELOVIC, DANIEL
NURUL AMIN A.K.M.
OANCEA, GHEORGHE I.
OKAMOTO, J.
OLARU, ADRIAN, DUMITRU
OLARU, ONISIFOR
OLEKSIK, VALENTIN
PJERO, ELENICA
PLESA, ANGELA
PODURAEV, JURY
POKORNY, PAVEL
POMAZAN, VALENTINA
POPA, CICERONE LAURENTIU
POPPEOVA, VIERA
PRODAN, DAN
PROKOP, ROMAN
PRYANICHNIKOV, VALENTIN
PYTKO, STANISLAW
QUINN, ROGER D.
RAMAJ, VEHBI XH.
RAVAS, RUDOLF
RAY, T. S.
RIKALOVIC, ALKSANDAR
RIZESCU, SABIN VASILE
ROGALE, DUBRAVKO
ROHDE, ULRICH L.
ROMERO, FERNANDO
ROS, OLIMPIA
ROSADO, V.O.G.
RUDAS, IMRE J.
RUGESCU, DRAGOS RADU DAN
RUJA, IOAN
RUSU ANGHEL, NICOLAE
RUSU, LIGIA
RUSU-CASANDRA, AURELIA
RUTEMI, GANI BRAHIM
RUZIC, DRAGO
SAKULSKI, DUSAN
SALIHU, AVDI HAJDIN
SAMEK, DAVID
SCHREIBER, PETER
SELVAKUMAR, ARUL LAWRENCE
SEONG, POONG HYUN
SERCER, MLADEN
SHIBASAKA, TOSHIROH
SHIH, MING CHANG
SIEMIATKOWSKI, M.
SIHN, WILFRIED
SINAY, JURAJ
SISEK, BORIS
SITUM, ZELJKO
SKOCZYNSKI, WACLAW
SKOLUD, BOZENA
SLUGA, ALOJZ
SMIRNOV, ALEXANDER
STANKOVSKI, STEVAN
STAROVEŠKI, TOMISLAV
STEFANIC, NEDELJKO
STOPPER, MARKUS
STRGAR KURECIC, MAJA
SUBRAMANIAN, ARUNKUMAR
SUGAR, PETER
SYROVA, LIVIA
SYSALA, TOMAS
TADAMURA, KATSUMI
TAKALA, JOSU
TAMASILA, MATEI
TAN, KOK KIONG
THIERHEIMER, WALTER WILHELM
TORIMS, TOMS
ULRICH, GUENTHER
VAARIO, JARI
VASEK, VLADIMIR

REVIEWING COMMITTEE

32ND DAAAM INTERNATIONAL SYMPOSIUM ON INTELLIGENT MANUFACTURING AND AUTOMATION

Preface

The 32nd DAAAM International Symposium on Intelligent Manufacturing and Automation was organised as virtual online conference hosted by the Vienna University of Technology, Vienna, Austria, between the 27th and 30th October 2021, during the DAAAM International Week. The Symposium was organized by DAAAM International Vienna in cooperation with ÖIAV 1848, Vienna University of Technology, International Academy of Engineering, University of Mostar and University of Applied Sciences – Technikum Wien and Under the Auspices of the Danube Rectors' Conference & Rectors' and Presidents' Honor Committee of DAAAM International for 2021. The Symposium was hosted from Vienna, Austria. This year's symposium aimed at continuing the success of the previous years, focusing on the five-fold traditional objectives of the symposium: the presentation of the most recent high-quality results, support of development of young scientists and researchers, organization of international (summer) doctoral school, inauguration of new members of Central European Branch of International Academy of Engineering and the provision of the necessary setting for stimulating discussions, brainstorming and networking among European and international researchers coming both from the academia government agencies and industry

DAAAM International Symposium has been established as the leading Central European conference devoted to the cutting-edge advancements in the field of modern production and automation. Symposium provides an international forum for engineers, scientists and industrial managers to present and discuss the current status and impact of advanced manufacturing and automation. The international series of Symposium was initiated by the first meeting in Vienna (1990) and continued in Štrbske Pleso (1991), Budapest (1992), Brno (1993), Maribor (1994), Cracow (1995), Vienna (1996), Dubrovnik (1997), Cluj - Napoca (1998), Vienna (1999), Opatija (2000), Jena (2001), Vienna (2002), Sarajevo (2003), Vienna (2004), Opatija (2005), Vienna (2006), Zadar (2007), Trnava (2008), Vienna (2009), Zadar (2010), Vienna (2011), Zadar (2012), Zadar (2013), Vienna (2014), Zadar (2015), Mostar (2016), Zadar (2017), Zadar (2018), Zadar (2019), Mostar (2020), Vienna (2021).

Regular organization of the annual Symposium is one of the central activities of the DAAAM International association. The association was founded on 5. November 1990 in Vienna during a meeting organised by Professor B. Katalinic in order to celebrate the 175 Years of Vienna University of Technology. The main idea of the association was to establish a network for a new kind of international scientific and academic cooperation in the fields of intelligent manufacturing and automation. Our primary focus is on: all aspects of production of technical products made out of parts and corresponding automation and knowledge, trans-disciplinary application of methods and all kinds of education.

DAAAM International activities, during the last thirty-two years, have been supported by scientists and experts from over fifty countries all over the world. In the course of the last decade, DAAAM International has focused its activities on the regional and global long-term co-operation between (technical) universities. The result of this initiative is the founding of the University Network of DAAAM International Vienna and DAAAM International Network for Advanced Technologies which was founded on 2002-10-25 in Vienna.

The main activities of DAAAM are:

- Organisation of annual DAAAM International Symposium.
- To maintain and improve the University Network of DAAAM International Vienna as a strategic alliance and permanent international open platform for long-term co-operation and partnership in all fields of technology, research and education.
- Organisation of long-term projects, initially known as Inter-University Doctoral Studies.
- Organisation of long-term projects, initially known as the Virtual University.
- Publishing of international journals, books, and proceedings.
- Promotion of the exchange of people, scientific, and technical information.
- Promotion of international co-operation among scientists and engineers.
- Promotion of R & D, education, and industrial application.
- To co-operate with other national and international organisations in matters of common interest.
- Organisation of International Doctoral School as a platform for the support of the next generation of engineers at the highest level.

This proceeding contains 103 refereed papers from 278 authors presented at the 32nd DAAAM 2021. In DAAAM proceedings, since the first symposium in 1990, we published papers from 9462 authors.

All the papers included in these proceedings are reviewed by experts in double blind peer review way, and finally revised by the authors for the improvement according to referee's reports and advices.

The papers discuss many aspects of modern manufacturing and automation such as: Algorithms, Artificial Intelligence, CAX, Computer Integration, Control, Cutting Tools, Design, FEM, Invited Lectures, Knowledge, Management, Manufacturing System, Mechatronics, Methodology, Methods, Modelling, Optimization, Robotics, Simulation, Technical Solutions, Technology and Trends.

Appreciation is forwarded to reviewers who contributed to improve the quality of papers reviewing the papers, providing constructive critical comments, improvements, and corrections to the authors. We also recognize the support of reviewers, who were in charge of acceptance/rejection process. We wish to thank the Scientific, Program and Organizing Committees, Session Chairs, student helpers and administrative assistants, for contributing to the success of this Conference. They did an excellent job.

Finally, we wish to thank all the authors who submitted papers, making this conference possible and the authors of accepted papers for making appropriate modifications and corrections with care and patience and for preparing the final manuscript. It is the quality of their presentations and their passion to communicate with other participants that really makes this conference series a success.

Special thanks to Festo company for support of our activities and support of young researchers and scientists in the form of Festo Prize and Festo scholarships for the participants of the 10th DAAAM International Doctoral School, which was held parallel to the symposium and was positively integrated.

We would like to thank all members of the International Academy of Engineering – Central European Branch, Rector of the Vienna University of Technology Professor Dr Sabine Seidler, and Rector of the University of Mostar Professor Dr. Sc. Zoran Tomić and his team, management and all those nice colleagues for their excellent support in the organization of 32nd DAAAM International Symposium and the 10th DAAAM International Doctoral School.

Vienna, 2021-12-03
Editor
Chair of the 32nd DAAAM 2021

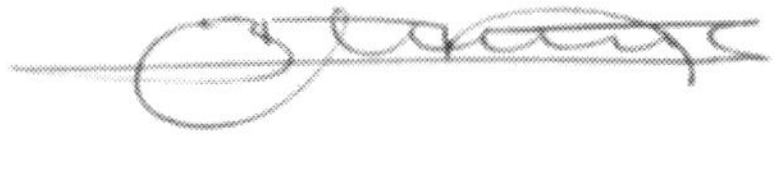

Branko Katalinic

Professor rtd Dr.sc. Dr.mult.h.c.Prof.h.c. Branko Katalinic
President of DAAAM International
TU Wien, IFLT-IMS, Getreidemarkt 9
1060 Wien, Austria, Europe
URL: http://www.daaam.info
E-mail: president@daaam.info
branko.katalinic@tuwien.ac.at

DOI: 10.2507/32nd.daaam.proceedings.001

MATHEMATICAL MODEL FOR HUMAN RESOURCES PLANNING IN THE PRODUCTION PROCESS

Aleksandra Kostic, Bogdan Maric, Melisa Kustura & Valentina Timotic

This Publication has to be referred as: Kostic, A[leksandra]; Maric, B[ogdan]; Kustura, M[elisa] & Timotic, V[alentina] (2021). Mathematical Model for Human Resources Planning in the Production Process, Proceedings of the 32nd DAAAM International Symposium, pp.0004-0009, B. Katalinic (Ed.), Published by DAAAM International, ISBN 978-3-902734-33-4, ISSN 1726-9679, Vienna, Austria
DOI: 10.2507/32nd.daaam.proceedings.001

Abstract

Problems of planning, management and optimal use of available human resources in the production process are very important for rational business management. The training of workers to service more jobs or vast number of machines with the aim of reducing the number of workers in the production process, thus balancing the workload, is an economic imperative. With the COVID 19 pandemic, the problem of reducing the number of workers in the production plant became even more notable. This paper presents a mathematical model for human resource planning. The application of the model was examined on the work potential, in terms of the interdisciplinarity of individuals, in the production system.

Keywords: Management; planning; human potential; mathematical model.

1. Introduction

The world of the new millennium is characterized by globalization, liberalization, technological development and lifelong learning. In such conditions, modern production systems were, until two years ago, consumer-oriented, which were key factors in planning the work process in production systems. Consumers were in a position to set high criteria in terms of product price, quality and time to market, because they had a large selection of products from different manufacturers at their disposal. We should not forget the fact that in the last decade, the world economy has been shaken by the economic crisis and recession, which is especially reflected in production processes and equally affects small and large companies. In the last two years, we have witnessed the COVID 19 pandemic, which significantly affects production processes, which were affected by lockdown, epidemiological measures and the requirement to carry out the production process with as few workers as possible in production halls. In such conditions, great demands are placed on the systems of production process management, especially human resources, with the aim of more rational business. The COVID 19 pandemic has once again put human resources on an almost equal footing with consumers in terms of the orientation of production systems.

During the last years of the last century, one became aware of all the shortcomings of the classical system of production process management and in response to this situation, the Lean concept was developed as a source of improving efficiency and effectiveness in work processes. The development of this approach was initiated by Lean Production.

This represented a breakthrough in more significant process analyses in order to improve production and the factory environment in which products are created. Lean production has changed the opinion and behaviour of participants in work processes and created an environment in which work is focused on: work process, elimination of losses in the work process, permanent training of employees at all levels and in all functions in order to achieve shorter delivery times and cheaper products. In this way, an ambience was created in the factories, which represents the Lean concept or doctrine in the way of managing and organizing production plants. Many authors state the efficiency in the development of organizational and managerial structures of production or industrial systems called "efficiency of the third dimension" which has a basic role in the Lean concept. The basic characteristics of managers who lead the work process in the Lean concept are dynamic behaviour, quick decision-making and teamwork focused on products, customers and the market, and training and maintaining employees to respond to environmental conditions. More about the Lean concept can be found in [1]. Due to the great advantages provided by the Lean concept, it is still most often used today to solve various problems in production, which can be seen from the works [3], [6], [7], [8].

Workers are seen as an economic factor of the production process and therefore the problem of planning, management and optimal use of available human resources plays an important role in the production economy. To manage work potential, the tool of the Lean concept - Nagara System can be applied, which in human resources imperatively requires greater interdisciplinarity of the individual in the interest of overall efficiency. Experimental research Lean concepts - Nagara System on the distribution and duration of the overhaul cycle in the process of repairing a turbojet engine (TMM), the redistribution of jobs to workers trained for more jobs is shown in [4], [5].

Experiences with the pandemic have again put the worker in an important place in terms of human resource management. Namely, previous experiences show that such emergencies are possible and that the production system must adapt to new challenges such as: reducing the number of workers in one shift, shortening the duration of the shift, enabling social distance between workers and the fact that for each job there must be at least two workers trained to perform the same task. Due to the unpredictability of viral infections as well as their duration, production must not be allowed to depend on a single worker who may be incapacitated by the disease to perform tasks. Therefore, it is necessary to develop a simpler mathematical model, without complex diagrams, which will effectively deploy workers in order to increase production efficiency and overcome production problems in emergency conditions such as a pandemic. This is also the subject of our paper.

The work is organized as follows: Section 2 discusses the position of workers in the new organizational concept of production systems as well as the conditions of the pandemic, the demands of management to the worker and the worker to management. The advantages of lifelong learning as well as the impact of the new organizational concept on employee motivation and effectiveness are also considered. The mathematical model of worker redistribution in terms of achieving higher productivity and compliance with epidemiological measures will be discussed in Section 3. The proposed mathematical model is characterized by ease of application and clear dynamic, without the use of complex graphs. In section 4, the mathematical model is applied and tested on a specific production (overhaul) process. Section 5 provides a critical consideration of the application of the mathematical model and guidelines for further research.

2. Human resources

The quality and quantity of the workforce is changing significantly. The requirements of the employee towards the employer as well as the requirements of the management towards the employee are also changing. Therefore, it is necessary to analyse the demands of workers and management. Worker requirements are usually the following: guaranteed wages, more favourable working conditions, greater responsibility and opportunities for decision-making in the work process, protection of workers at work, permanent training and education and implementation of epidemiological measures in order to reduce COVID 19 infection. On the other hand, the requirements of the employer are: more efficient production, interdisciplinary work potential that is able to perform a variety of activities with the aim of increasing the level of utilization of working time to meet market demands. We will give one simple example where the production bottleneck can be easily overcome by redirecting workers to different working operation. Workers and the duration of the operation are marked in series production with W_1, W_2, W_3 and W_4. The graph shows the sequence of their operations and indicates the time of their execution. First, a situation is shown where each worker is trained to perform only operations that are performed recursively one after the other.

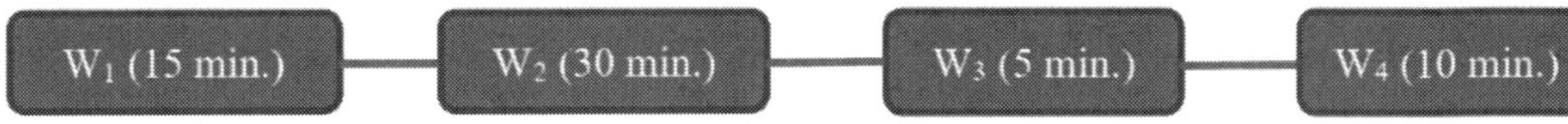

Fig. 1. Diagram 1

It is clear that the bottleneck in production has occurred among W_2 workers who, due to the length of the operation, reduce the capacity of the entire chain with their limited production capacity.

If now workers W_2 and W_3 are both trained for both operations, and if they perform a fusion of their activities we get the following diagram 2.

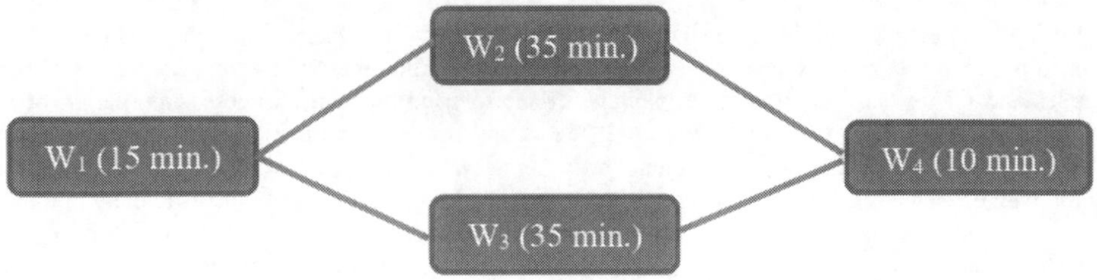

Fig. 2. Diagram 2

This means that after the same waiting, the W_4 worker would receive 2 semi-finished products for further processing and not one, and that the production bottleneck was partially overcome. From this simple example, it can be seen that interdisciplinarity and redistribution of workers can significantly affect the increase of production efficiency.

By applying the Lean concept of the Nagara system in one overhaul-production system, the number of workers decreased by 51.5%, while the overhaul time increased by 3.1%, which is satisfactory in this field of research. There is also a reduction in the entropy of the system that is created due to the presence of incompetent workers who refuse or prolong the execution of jobs due to constant requests for help in diagnosing the condition. More details on this use of human resources in the overhaul system can be found in [4], [5].

As we have seen, it is possible to significantly reduce the number of workers in a production plant without notable increase in time production. This is very important because in the conditions of a pandemic, production must be performed with a smaller number of workers in the plant or with reduced working hours. There are two options. The first option is to divide the number of workers of one production line into two sub-lines, each of which will work every other day. Another option is to work in four shifts of 6 hours each, while one quarter of the workers from each shift will make up that fourth line. In a pandemic, it is better to have more workers trained to perform more operations because production must not be allowed to stop due to the sudden illness of individuals. However, it should be noted that there is a fear among workers that they will be fired due to improved productivity. Reassurance must be given to workers that a higher level of individual training will lead to more paid workplaces and better distribution of workers to work tasks in terms of balancing norms. Worker needs to be reassured that he will not be laid off to improve productivity. Workers should be actively involved in improving performance, developing a climate of openness, respect and participation through KAIZEN meetings and thus creating mutual trust and respect between workers and management. Workers should be responsible for product quality. It is clear that the orientation of the organization of production towards worker is stronger. However, the management method cannot be completely simplified and reduced to the use of human resources, because the use of human resources is related to other components of Lean production and concept, as pointed out in the study [6]. In the end, even in the Lean concept, when it is necessary to reduce production costs, costs of worker training are reduced first. Thus, there is a real need to find a simple mathematical model, which will better allocate interdisciplinary workers with the aim of better productivity and which is easy to apply and modify in accordance with the existing situation. Such model will be presented in the next section.

3. Mathematical model

In mathematical modelling, we assume that all workers have accepted the Lean concept, which means awareness of the need for interdisciplinarity and redistribution to different work operations in order to increase work efficiency and reduce the number of workers on one production line. This concept assumes that workers are paid not only by experience but by ability to perform multiple different operations. Here we will identify work operations with the workplace and measure the complexity of the operation with a norm. More complex operations have a lower norm and less complex operations have a greater one. We will also assume that they want to redistribute workers without buying new machines. In the redistribution of workers, we will start from the redistribution of workers who do not already work on CNC machines, because great efficiency of production has already been achieved on these machines. There are production lines where a large number of operations cannot be performed on CNC machines. Such is, for example, the confection of the sewing line. We will denote workers as before with W_1, W_2,..., W_n, work operations with X_1, X_2,..., X_k and the corresponding norm by N_1, N_2,..., N_k. Each worker must be trained for at least two independent operations or for at least two work stations. We will assume that it is optimal for workers who do not work on CNC machines to be trained for three operations, provided that we can combine operations on the same machines or operations on the same type of machines. There are two strategies for redirecting human resources:

1. Redistribution was preceded by a serious analysis of the production process, the observation of production bottlenecks caused by the poor distribution of human resources and according to that, workers are trained for certain operations and after the training a new distribution is made.
2. Workers are already trained for certain operations and based on the available potential, redistribution is made.

We have already stated that in the case of reducing production costs, even in the Lean concept, the training of workers should be abolished first. Therefore, it is necessary in the case we already have training funds that are less than the funds for the purchase of highly automated machines to approach the problem in a planned way and thus prefer the first strategy while the second is done in case of suspension of funds for training workers. By redistributing workers, we want to solve the following problem:

a) Solving bottlenecks in production caused by poor distribution of labour potential. But here we should be extremely careful because sometimes the problem is easily solved under the assumption that we have the option of introducing a new machine or purchase more expensive automated one, which is often not the case.
b) Balancing the norm or workload. It happens that some operations have a low norm due to their complexity, while others have a high norm due to the simplicity of the operation, which workers cannot achieve in real work time.
c) Increasing the efficiency and quality of production.
d) Reduction in the number of workers in the plant due to the COVID 19 pandemic.

We will analyse both strategies and offer appropriate mathematical models with the aim of solving problems a) - d). The mathematical model is required to be simple, adaptable and dynamic and thus easy to use. Since the set of workers is a discrete set from a mathematical point of view there is no need for excessive model complexity. Let us now consider the first strategy.
Suppose that all workers are already assigned to their jobs, and we want to train them further for the l operations. Then we have

$$\binom{k-1}{l} = \frac{(k-1)!}{(k-1-l)!\,l!} \tag{1}$$

possibilities of choosing these additional operations. In our case

$$\binom{k-1}{2} = \frac{(k-1)(k-2)}{2} \tag{2}$$

which is objectively a lot. Therefore, we will first locate the worker with the min norm W_{min} and the worker with the maximum norm W_{max} and mark their norms and work operations appropriately. Here we must be careful because N_{min} can refer to the minimal norm in the whole line or to one group of operations performed on the same type of machine. The same goes for the maximal norm. Please note that it is not desirable to send the same worker to several different machines. However, it is quite acceptable to train a worker who does manual work to work on at least one machine. An example of important manual work is quality control. A worker who works e.g. on the machine that performs the operation that is at the beginning of the production process, we can easily send to the product control in the last third of his working time. So the proposal is to perform a fusion of operations with the highest and lowest norm and add one of the adjacent operations. This is the reason for training workers for performing new tasks and operations. This way the problem of the production bottleneck is solved. We increase the input to the next operation not too much to overwhelm the worker, but again enough to speed up the production process. Therefore, we reduced the number of workers because we sent a worker away from the machine to one of the manual operations, in one part of the working hours, preferably at the beginning of production.
The new norm of workers in the designation N' where the bottleneck of production is calculated by formulae

$$N'_j = a \cdot N_{max} + b \cdot N_{min} + c \cdot N_{adjacent}, \; 0 \le a \le 1, 0 \le b \le 1, 0 \le c \le 1. \tag{3}$$

The exact value of a, b and c can be determined by a technologist or manager based on a recording of the production process or can be determined by applying fuzzy logic. More information about fuzzy logic can be obtained in [2]. In place of one worker at the bottleneck of production we introduce 2 workers W_{min} and $W_{adjectant}$ so that the worker $W_{adjectant} + 1$ which usually has a higher norm gets instead of one semi-finished product two semi-finished products for processing. Usually $b = c$ is taken and a is then easily calculated by $a = 1 - 2b$. It is possible to integrate similar operations which are too simple and are done on the same machine, meaning they have high norms, into one. This way we obtain new norm for the workplace

$$N'_m = a \cdot N_{m_1} + b \cdot N_{m_2} + c \cdot N_{m_3}, \quad 0 \le a \le 1, \; 0 \le b \le 1, \; 0 \le c \le 1. \tag{4}$$

Norm of the production line depends greatly from the first operation in line. To be exact, first operation defines it in some way because the number A cannot be exceeded and is obtained as

$$A = p_1 \cdot N_1 \tag{5}$$

where p_1 is number of workers on the first operation. This is why it makes sense to combine work operations at the end of the production cycle with the first operation in the cycle, with the purpose of increasing the input of the semi products so we can get

$$N'_1 = a \cdot N_1 + b \cdot N_{e_1} + c \cdot N_{e_2}, \quad 0 \leq a \leq 1,\ 0 \leq b \leq 1,\ 0 \leq c \leq 1. \tag{6}$$

Index e denotes end operation. Since we can define wanted input we get

$$A' = p_1' \cdot N_1' \tag{7}$$

where we define A' ourselves.

When calculating a new norm, we multiply the existing norm by the numbers a, b and c. Since the norms must be natural numbers, we round the number of the norm to the first larger natural number.

4. The Mathematical Model Application in Textile Line Production

For our mathematical model application, we have chosen textile industry trousers production line. We observed process of trousers sewing because in this process CNC machine use is minimal. This means that in this process the interdisciplinarity of workers can be achieved and their redistribution to different work operations can bring significant results. The factory in which the production line is located has a fully automated tailoring department in which the located machines are Gerber Paragon LX Cutting System cutters, GERBER XLs50 Spreaders, GERBER MP+ Series Plotters, Aesus Labelers. For this paper we were more interested in sewing machines Veit automated Pants Finishers 8741, Veit Fusing Machines AX 450 C, Brother Industries NEXIO KE-430HS Electronic Direct Drive Lockstitch Bar Tackers, Brother Industries RH-9820 Eyelet Buttonhole, Brother Industries NEXIO T-8752D Twin Needle Direct Drive Split Needle Bar Lock Stitcher, Brother Industries NEXIO HE-800C Electronic Direct Drive Lockstitch Button Holer. In the production of trousers, the following operations are used, which take place according to the following schedule:

Type of operation	Norm for 1 worker	Number of workers
X_1 (processing)	65 pieces	2
X_2 (non-sewing operation)	300 pieces	1
X_3 (sewing operation)	70 pieces	2
X_4(sewing operation)	75 pieces	2
X_5 (non-sewing operation)	150 pieces	1
X_6(sewing operation)	110 pieces	1
X_7(sewing operation)	150 pieces	1
X_8(sewing operation)	120 pieces	1
X_9(sewing operation)	150 pieces	1
X_{10}(sewing operation)	220 pieces	1
X_{11} (non-sewing operation)	55 pieces	2
X_{12}(sewing operation)	180 pieces	1
X_{13}(sewing operation)	220 pieces	1
X_{14}(sewing operation)	180 pieces	1
X_{15}(sewing operation)	45 pieces	3
X_{16}(sewing operation)	170 pieces	1
X_{17}(sewing operation)	120 pieces	1
X_{18} (non-sewing operation)	320 pieces	1
X_{19}(sewing operation)	140 pieces	1
X_{20} (non-sewing operation)	65 pieces	2
X_{21} (non-sewing operation)	300 pieces	1
X_{22} (quality control operation)	500 pieces	1
X_{23} (non-sewing operation)	700 pieces	1

Table 1. Number of operations needed for trousers production and norm of workers

Total line production norm is 125 pieces done by 30 workers. Production bottlenecks occur in the production of welted pockets (X_3, X_4), as well as in operations related to the placement of zippers (X_{11}, X_{12}, X_{13}). The first bottleneck in production occurs when setting up welted pockets. We can overcome this by procuring pocket welting machines, however if we do not have enough funds to purchase a new machine, we can integrate operations, X_3 and X_4 with quality control X_{22} as follows

$$N_3' = N_4' = N_{22}' = 0{,}5 \cdot N_3 + 0{,}4 \cdot N_4 + 0{,}1 \cdot N_{22} \tag{8}$$

Workers perform product control in the last third of their working time.

To solve problems related to the placement of the zippers we will integrate operations X_{11}, X_{12} i X_{13}. This way four workers will be sawing zippers instead of two. Their norms will be described as follows

$$N_{11}' = N_{12}' = N_{13}' = 0{,}6 \cdot N_{11} + 0{,}2 \cdot N_{12} + 0{,}2 \cdot N_{13} \quad (9)$$

We can integrate many operations which are not related to sewing to further decrease number of workers. We can integrate the sewing operations if we need smaller number of workers on the production line because of the pandemic COVID 19.

5. Conclusion

Due to the problems of modern society such as the economic crisis and the COVID 19 pandemic, the distribution of human resources is once again coming into the focus of interest. On the other hand, production must be oriented towards the demands of consumers and the market. The way to address these conflicting requirements is to accept the Lean concept of thinking, which implies openness and trust between employees and management, as well as trust between employees themselves. The lean concept also implies the acceptance of additional training of workers to perform multiple work operations. It would be desirable for the training of workers to be planned, because with unplanned training, conditions of difficult production with delays can be created. An example of a problem that can occur in unplanned training of workers is reflected in the ability to train redundant workers for some work operations, while for other work operations there will be a shortage of manpower. It should also be considered that all funds for the planned training of workers should be used rationally. If production costs are reduced, the cost of training workers will be reduced first. A mathematical model is needed for the planned training of workers. Based on the mathematical model, a plan for training workers would be made, and later their redistribution. The goal is to provide efficient production, to enable production with a reduced number of workers in the event of a pandemic, whereby the worker would have the opportunity to make decisions in the production process.

Therefore, in this paper we have presented the mathematical model for redistribution of workers to different working places and operations. The presented mathematical model corresponds to the requirements of simplicity and dynamism and is not based on complex diagrams. We have seen that this way we can overcome bottled neck production and decrease the number of workers. For calculating coefficients a, b and c in our formulas one can use either the experience of technologist, engineers, workers and managers or apply fuzzy logic. At a specific plant, we showed how a quality result can be obtained by applying the proposed model. Further research will go in the direction of using the least squares method which is a mathematical method typical for solving engineering problems.

6. References

[1] Bozickovic, R. & Maric, B. (2013). Lean Concept – a Challenge to Managers for the Better Future of the Company, Chapter 27 in DAAAM International Scientific Book 2013, Katalinic, B & Tekic, Z. (Eds.), page numbers (491-510), DAAAM International, ISBN 978-3-901509-94-0, Viena, Austria

[2] Chen, G, & Pham T. T. (2002). Introduction to Fuzzy Sets, Fuzzy Logic and Fuzzy Control Systems, CRC Press LLC, ISBN 0-8493-1658, New York

[3] Cvetkovic, N.; Moraca S.; Jovanovic, M.; Medojevic, M. & Lalic, B. (2017). Enhancing the Agility and Performances of a Project with Lean Manufacturing Practices, Proceedings of the 28th DAAAM International Symposium, 08-11th November 2017, ISBN 978-3-902734-11-2, Katalinic, B. (Ed.), pp.0661-0670, DAAAM International Viena, Austria, DOI 10.2507/28 th.daaam.proceedings093

[4] Maric, B.; Bozickovic, R. & Cosic, I. (2010). Nagara System for Planning Human Potentials in the Repair Process of Technical Systems, Process of Technical Systems, Proceedings of Conference Maintenance 2010, 10-13 June 2010, ISSN 1986-583x, Brdarevic, S.& Jasarevic, S. (Eds.) pp. 23-28, University of Zenica, Faculty of Mechanical Engineering in Zenica, Zenica Bosnia and Herzegovina

[5] Maric, B.; & Bozickovic, R. (2014). Maintenance of Technical Systems & Lean concept, Faculty of Mechanical Engineering, East Sarajevo, University of East Sarajevo, ISBN 978-99976-623-0-9, East Sarajevo

[6] Olegehe, O. & Salonitis, K. Hybrid simulation modelling of the human-production process interface in lean manufacturing systems. International Journal of Lean Six Sigma, Vol., 10, No., 2, (2019)., page numbers (665-690), IJLSS-01-208-0004

[7] Softic, A.; Basic, H.; Lulic, H. & Nakic, A. (2020). Application of the Lean Concept for Analysis and Optimization of the Automobile Filter Production Company, Proceedings of the 31th DAAAM International Symposium, 21-24th October 2020, ISBN 978-3-902734-29-7, Katalinic, B. (Ed.), pp.0708-0716, DAAAM International Viena, Austria, DOI 10.2507/31st.daaam.proceedings098

[8] Vujica Herzog, N. & Buchmeister, B. (2012). Lean Manufacturing in Slovenian Companies, Chapter 11 in DAAAM International Scientific Book 2012, Katalinic, B. (Ed.), page numbers (117-128), DAAAM International, ISBN 978-3-901509-86-5, Viena, Austria

DOI: 10.2507/32nd.daaam.proceedings.002

INTRODUCTION AND MAINTAINING PUBLIC RELATIONS OF SMALL ENTERPRISES IN VIRTUAL ENVIRONMENT

Berislav Andrlic, Kresimir Lackovic & Martin Stimac

This Publication has to be referred as: Andrlic, B[erislav]; Lackovic, K[resimir] & Stimac, M[artin] (2021). Introduction and Maintaining Public Relations of Small Enterprises in Virtual Environment, Proceedings of the 32nd DAAAM International Symposium, pp.0009-0015, B. Katalinic (Ed.), Published by DAAAM International, ISBN 978-3-902734-33-4, ISSN 1726-9679, Vienna, Austria
DOI: 10.2507/32nd.daaam.proceedings.002

Abstract

In this paper, the object of consideration is a small company and as a subject and goal is to explore an effective model of introducing and maintaining public relations. In this sense, the methods of analysis and synthesis, ie induction and deduction, as well as abstraction and modeling, have been applied. The researched data indicate a large participation of small enterprises in terms of employment and social product creation, but also their participation in the dynamization of the national economy. Special attention is paid to public relations and it starts from the scientific assumption that in small companies it is not only a marketing but also a managerial activity. In addition, the goal is to research the model of introducing and maintaining public relations and defining the process of activities at the lowest possible cost. Therefore, the second hypothesis implies that the development of information technologies enables the setting up of a public relations process via the Internet with the aim of cost and time efficiency. From this approach, a model is proposed as the final solution for the effective introduction and maintenance of public relations in small businesses.

Keywords: Public Relations; Small Business; Internet; Information Technology

1. Introduction

Every company is part of the wider community, so in addition to the market, it must respect and maintain exemplary and mutually beneficial relations with the rest of the public. According to the already established definitions, public relations or English Public Relations (hereinafter also PR) are part of the promotion tool as an element of the marketing mix. The object of consideration are small enterprises, which play a major role, especially in developed economies. For example, in the U.S. economy, there are more than 27 million small businesses that generate about 50% of U.S. domestic product [4].

In modern conditions, and especially in small companies, PR is becoming an increasingly important managerial activity because all managerial and creative work in most cases is performed by one person or owner-manager. [2] Therefore, in the mutual interest, it should maintain appropriate relations with the relevant public, such as: customers and suppliers, local self-government and financial institutions, the media, and existing databases and centers of excellence.

This is extremely important because the company is an integral part of the social and economic being of the community, whether local or national, so it is obliged to conduct its business in a way that improves the social and economic progress of the community, and not to hinder. [5]. In addition, the PR of every small business must deal with the overall image and reputation of the company as an entity in its environment. [6].

Given the limited costs, but also time, a small entrepreneur does not have great opportunities to establish and maintain relationships with the relevant public. Information technology can help him here, especially the recently ubiquitous social networks. [1] The development of information technologies provides the entrepreneur with the opportunity to achieve all PR goals with a minimum of time and costs. In this sense, the hypothesis is set that by systematic introduction in stages with the aim of establishing and maintaining PR, all business and especially marketing goals can be achieved. Given that the object of this paper is a small company and as a subject and goal to research an effective model of introducing and maintaining public relations, appropriate scientific research methods have been applied. In order to determine the basic characteristics and role of small enterprises in national economies, but also the function of managers, methods of analysis and synthesis, ie induction and deduction, were applied. These methods have also been used in researching and identifying opportunities offered by Internet technologies as a public relations support. By applying the methods of abstraction and modeling, a phased approach to the introduction and maintenance of public relations in small enterprises is defined.

As part of the consideration and research in the text, attention is paid to the special features of a small business as an object of research. The research related to the development of public relations has covered a part of historical development. From a corporate perspective, the first PR department was established in 1889 at Westinghouse for the purpose of promoting a revolutionary AC system. In addition, at Jale University in 1899, the secretary's office was successfully converted into an office for former student and public relations officials. [7].

With the development of information technology, public relations activities have increasingly taken place on the Internet to communicate with potential and current customers in order to provide desirable information about the company, which greatly reduces unwanted comments. Public relations can accelerate and expand the distribution of press releases especially in today's global and highly dynamic market. In addition, interactive two-way and multi-way communication as well as communication mediation can be realized via the Internet. In most cases, Internet PR aimed to influence the external image of a company or brand with the following tasks: [8].

- Acquisition and retention of new customers
- Communication and information
- Increasing public attention
- Creating a high cost-benefit ratio
- Improving reputation
- Measuring and managing success
- Achieving a high degree of topicality

The main feature of these beginnings of the Internet approach to PR is the asynchronous flow of communications because communication did not take place in real time. Teaching development went in the direction of modeling with the aim of increasing mutual synergy between companies and the public. The analysis of the development of possibilities of information technologies shows that different approaches are possible, and on this occasion the systematic introduction of PR by phases was chosen. Such an approach ensures the gradual introduction of the public relations function on the one hand and the education of entrepreneurs on the other. The discussion with concluding remarks has the purpose of explaining the research and warning about the proposed possibilities arising from the proposed PR model and introduction by stages. From the above considerations, it is clear that PR is a strong managerial activity in a small business that can be performed in an online environment with minimal loss of time and negligible costs. The continuation of this research is aimed at monitoring the development of public relations and especially information technologies and continuous education with the aim of effective application.

2. Elementary characteristics of small business

In modern conditions, most, especially developed economic systems, are made up of a large number of small enterprises, but most employees are also in small and medium enterprises. It has long been known that data on small business appeared 4,000 years ago. According to data on the U.S. economy from 30 years ago, 88.9% of employees worked in companies with up to 10 employees and 91.3% in companies with up to 20 employees. To this can be added the indicative data that 99.1% of employees worked in companies that had less than 100 employees. Of course, not all companies are successful because there is data that in the US out of 600,000 new companies established annually, only half will live 18 months, and only a fifth ten years, but a similar number of new small businesses will be reactivated next year. From the previous data, it can be concluded how dynamic such an economy is. Therefore, the success of any economic system is based on the performance and volume of small businesses.

In modern conditions in the Republic of Croatia, according to the Act on the Promotion of Small and Medium-Sized Entrepreneurship, the sphere of small business has been formed, which consists of micro, small and medium-sized enterprises. They vary in size based on the following characteristics:[8]:

Micro small business entities (natural and legal persons)

- employ less than 10 workers (annual average).
- generate annual operating income in the amount of the equivalent of up to EUR 2,000,000.00, or have total assets if they are taxpayers of profit tax, or have fixed assets if they are taxpayers of income tax, in the amount of the equivalent of up to EUR 2,000,000.00.

Small business entities (natural and legal persons)

- have less than 50 employees (annual average)
- earn annual operating income in the amount of the equivalent of up to EUR 10,000,000.00, or have total assets if they are taxpayers of profit tax, or have fixed assets if they are taxpayers of income tax, in the amount of the equivalent of up to EUR 10,000,000.00

Medium-sized small businesses (individuals and legal entities)

- have less than 250 employees (annual average)
- generate annual operating income in the amount of the equivalent of up to EUR 50,000,000.00, or have total assets if they are taxpayers of profit tax, or have fixed assets if they are taxpayers of income tax, in the amount of the equivalent of up to EUR 43,000,000.00.

According to the Report on Small and Medium-Sized Enterprises in Croatia for 2018, the importance of small and medium-sized enterprises is visible. Namely, in the period from 2013 to 2017, the numerical department of micro and small enterprises in the total number of economic entities in their years was 98%. If small and micro enterprises are added to medium, then their percentage is 99.7%. This means that special attention should be paid in the research of smaller economic entities in order to find various innovations whose application can enable the stability of these entities. [3] This is especially important due to the fact that even in developed economies, many small start-ups do not last. In addition to this information, the fact that small businesses have advantages and certain disadvantages should be taken into account. [9]. Small businesses have the following advantages:

- In a small business, each employee feels part of a shared process.
- There is a greater opportunity to highlight personal particularly professional ambitions.
- Every employee feels an appropriate share of responsibility.
- A small business will be more productive and with a greater degree of discipline.
- Almost all interested employees can be involved in development programs.

In addition to the obvious advantages, the following disadvantages should be borne in mind:

- Insufficient knowledge, especially in the field of finance and informatics-
- Patriarchal leadership.
- Insufficient level of group decision making.
- Great reliance on improvisation and intuition.
- Overloaded with a large number of functions.
- Insufficient use of the achievements of the development of technique and technology.

These advantages, and especially the disadvantages of small businesses, suggest that there is considerable scope for researching the way businesses do business and imposing proposals to increase the efficiency of small businesses. As already mentioned in small businesses, the owner is in most cases a manager, so he is responsible for almost all creative and management functions and thus marketing. [4] As part of the marketing mix or promotion as part of sales promotion are public relations. Further research shows that especially in small businesses, public relations is becoming an increasingly important managerial activity. Therefore, in modern online conditions, in order to reduce costs, public relations should be performed with the help of Internet technologies.

3. Public relations in modern conditions- virtual environment

Since all marketing activities are usually performed by one person or owner-manager in a small business, this must be supported by information technology. [5] Namely, in a large number of cases, all production and marketing activities are contained in one person or manager who is also the owner of a small business. In this sense, it is necessary to research the market, which means to find a group of consumers who will be similar in their reactions to the products offered and to identify a group that will behave similarly in buying and consuming, which means to determine the market segment. Similar procedures are carried out in the procurement as well as in the sales market. After determining the market segment of purchasing and sales, one constant marketing activity or promotion is needed. Promotional activities can be done in different ways, in the form of advertising, personal sales and sales promotion. Special attention should be paid to public relations activity. There are several definitions of this activity and in this case we have the following: "Public relations is a decision-making practice that has the task of building relationships and interests between organizations and their publics by providing information using communication methods that are ethical and trustworthy."

In terms of publicity in the company's environment are: the media, customers and suppliers, financial institutions, local government and databases. In a situation where usually one person needs to perform all management and creative activities, an extremely important factor is cost and time. In addition to limiting costs, there are research results where 60% of respondents-entrepreneurs believe that PR, along with operations, is an important managerial strategic activity. [10]. Therefore, it is proposed in the organizational structure of a small business that PR be, in addition to tools to improve sales, one of the important activities of managers or owners of small businesses.

4. Introducing and maintaining public relations in an online environment

After the necessary IT education, a small entrepreneur can gradually approach the introduction of PR, and the Internet is becoming an increasingly important tool. According to Figure 1, it is recommended that this be done in stages, because in this way the process of PR development and IT education of entrepreneurs will take place in parallel, ie how to use new information technologies and different information platforms for their own public relations.

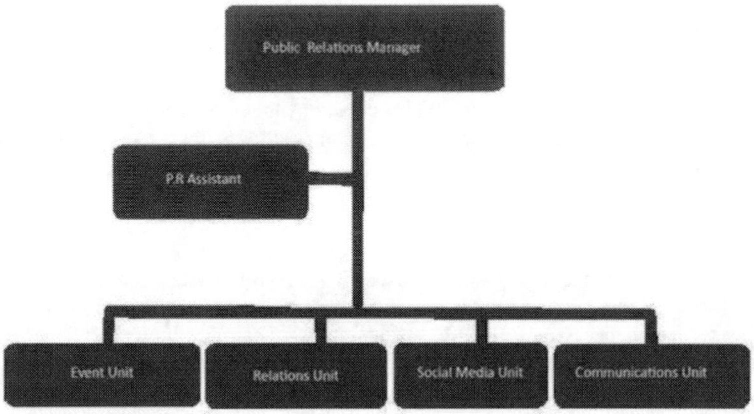

Fig. 1. Proposal of PR organizational structure of a small business

The above process in Figure 1 should be correlated with the development of information hardware and software. It all started when a networked hypertext system was developed for CERN called the World Wild Web (www for short) that included HTML language and HTTP protocol, so after installing it on a computer, the first web server was created. After that, the possibility of connecting the www network with other computers was developed and Web 1.0 was developed, as well as connecting commercial and non-commercial networks, which became cheaper. [6]

Opening a web portal or profile on the most famous social networks is the minimum that every new entrepreneur should do. With the help of this, in a short time it will determine the possible boundaries of the market segment and even the group of the most important customers and suppliers. After determining the market segment and the grouping of the most important customers and suppliers, the entrepreneur can move on to the next phase, which is a personalized web portal. The main goal of personalization is to deliver information that is important to an individual or group of individuals among customers and suppliers in a specific form, schedule and time intervals. [11].

At this stage of development, the PR entrepreneur already addresses his public directly, which means that he communicates asynchronously because he can get answers and comments after his information. In this way, conditions were created for other sections of the public to be involved in order to participate in the discussion. This type of discussion is known as a forum which is actually the technological evolution of bulletin boards. From a technological point of view, forums or web boards are applications that manage user-generated content. Early online forums could be described as a web version of an electronic list or newsgroup; allowing people to post messages and comment on other messages. Subsequent developments have mimicked different newsgroups or individual lists, providing more than one forum, dedicated to a particular topic. The team has the opportunity to provide additional information about their own products and services and to reconcile common interests. [7]

The development of information technologies on the one hand and the entrepreneur's education in the next phase enables synchronous communication through video conferencing, which is organized by agreement from time to time with all interested sections of the public. Videoconferencing is the connection of individuals and groups through telecommunications networks and video technology so that participants have the impression of simultaneously attending meetings in the form of a live video link on a computer screen. This type of synchronous discussion provides an opportunity for all participants to inform each other, but also to draw conclusions of mutual interest. After the establishment of PR with all participants in the previous stages in the coming period should ensure the sustainability and effectiveness of established relationships.

Additional features that maintain and refresh PR can be social networks, which in principle can be defined as a web-based service that allows an individual to create a (semi) public profile and within a limited system, articulates a list of other users with whom he shares connections and sees and compares his list. contacts with those made by other users within the system. [12].

There are many ways in which a social network can be used, such as: creating your own profile, areas of interest and preferences, searching for and finding old or making new friends, exchanging pictures and videos, but you can become a member of certain interest groups. There are several types of social networks, and on this occasion, Facebook, Instagram and Linkedin will be especially highlighted. The concept of Facebook was designed to connect people who already know each other in real life via the Internet. The main feature of Facebook are numerous applications because, for example, it is possible to have virtual pets, receive virtual gifts and greeting cards, solve quizzes and play video games. Facebook offers the possibility of privacy control, so the user can hide their profile and photos and protect themselves from unknown visitors. Instagram is a popular free app for processing and sharing photos via mobile phones on social networks. Linkedin is a business-oriented social networking service and is mainly used for professional networking especially in recruitment and sales. All these social networks provide the opportunity to personalize each other's information after synchronous discussions. This means that the entrepreneur creates even closer relationships with customers and suppliers, but also with the rest of the public. In that way, a long-term connection is realized with minimal costs and loss of time.

In order to successfully apply the proposed phased approach, positive attitudes and suggestions were made to pay special attention to the following: [12].

- Providing information on the website on aspects such as investor relations, organizational programs and policies, community involvement and employee relations. This could open the door for public participation and debate.
- Creating a forum for comments and suggestions. This forum could function as a chat room where a visitor can enter a message and receive an immediate response from a company representative or another visitor. Such a forum could function like a 24-hour, 7-day focus group. [8]
- Intuitiveness and ease of display is required along with a hierarchical relationship between graphics and text in order to promote efficiency and download speed, as too many graphics can bother users who are short of time. Websites should not be designed to be accessible only to users who have the latest hardware or software but should be interesting, informative and contain valuable information to all visitors. Graphics and sounds must be useful tools, but content should take precedence over aesthetic considerations. Dialogic public relations must create lasting and valuable relationships with your public and therefore the website must not be used just as propaganda, marketing or an advertising tool.
- In order to obtain a return of visits, the first important element is the need for useful and timely information for various publics.
- Websites that provide limited or immutable information will not generate repeat visits because after one visit they are no longer useful and therefore not suitable for dialogue communications. Websites that are regularly updated and contain valuable information because it seems credible to the relevant public and gives the impression that the organization is responsible, thus helping to create the preconditions for dialogical relationships. [9]
- When it comes to retaining visitors, website designers should be careful not to do this by engaging in links from their websites to other useful sites engaging in links from their websites. Once visitors leave the site using the link, they may never return to the original site.
- Generating return visits and retaining visitors and the likelihood of return visits from different publics is an important element in building long-term relationships through organizational websites.

Previous analyzes of the researched possibilities of Internet technologies show that there are numerous possibilities for effective introduction and maintenance of public relations, especially when this activity is contained in one person or manager of a small company. After the establishment of the aforementioned process of establishing public relations, a small entrepreneur should continue to hold synchronous discussions in order to give innovations to all interested parties in one go, and further maintain this through personalization through social networks. In any case, a particularly small entrepreneur at the beginning of the business should organize the process step by step. In conclusion, it can be stated that PR is not only a marketing tool but also one of the important management activities that can be most effectively carried out using Internet technologies.

5. Discussion

Analyzing the previous content, one can notice several peculiarities contained in the object of research, which is a small business entity. New theories, new techniques, cultural changes and technological advances all combine to create a dynamic environment within which marketers try to ensure that their messages get through to their target audiences. [13].The basic observation lies in the fact that small enterprises are by all parameters an essential component of all, and especially developed economic systems. The situation is similar in the Croatian economy and thus in the economy of Slavonia and Baranja. This is why small businesses deserve special attention. As a subject and the main goal of the research are public relations or PR. The researched and presented data show that the development of this activity followed at the end of the 19th and the beginning of the 20th century.

Public relations have gradually changed, especially with the development of new information technologies, so PR has increasingly taken place in terms of synergy of common interests of the company and its public. Research shows that PR is still an important promotional tool in marketing, but it is gaining in importance and is becoming one of the essential management activities especially in a small business. As the essence of the research and confirmation of the hypothesis are the possibilities provided by the development of IT hardware and software. Namely, information technologies provide the possibility of effective public relations with minimal costs and little waste of time. The proposal of the PR application is approached in stages. The goal of such an approach is aimed at the gradual and effective introduction of PR via the Internet.

In addition, the small entrepreneur is enabled to perform permanent education only with a minimum of time and costs. In this sense, the exchange of information in synergy with the environment simply leads to new knowledge in various spheres of interest. The introduction and maintenance of PR is gradually developed towards the goal, which is an all-synchronous discussion that enables mutual communication and information and even education. Teaching relations through social networks can help maintain public relations, because it personalizes everything and thus strengthens the contacts of entrepreneurs with the relevant environment. Given that a small entrepreneur is usually the owner and manager, it is clear that he will personally perform all PR activities himself. The effectiveness of this approach can be seen from the proposed model but also from several of the views cited in the relevant sources.

6. Conclusion

More can be deduced from the content analysis and discussion. First of all, small businesses are an important factor in any, especially developed economic system. Problem is that small businesses are not using Internet technology enough in promotion. This is shown by data obtained from relevant foreign and domestic sources. In addition, PR activities should be given special attention because it is part of marketing but also a managerial activity in small businesses. In particular, it should be borne in mind that all creative and management tasks in a small business are performed by one person. Therefore, it is proposed to use Internet opportunities in the introduction and maintenance of public relations in a small business. After analyzing and defining PR as one of the managerial activities, the phased introduction of PR with the help of the Internet is proposed. In that sense, the steps are listed and their functions are explained.

In scientific terms, it has been confirmed that PR in small businesses should also be performed online and introduced in stages. This is confirmed by the analysis of several relevant sources as the fact that PR, especially in small businesses, is not only a marketing tool but also an important managerial activity. In terms of development, PR can be maintained and increase quality in accordance with the development of information technology. This approach can be effective in the application in small enterprises of the Republic of Croatia and the wider region, especially the part that is even at a slightly lower level of economic development. As a direction and goal of future research and scientific plans on the development of public relations, it can be research on the greater role of centers of excellence and especially scientific research educational institutions.

7. References

[1] Chen, P. (1976). The Entity-Relationship Model – Toward a Unified View of Data, ACM Transactions and Database Systems, Vol. 1., No. 1. pp 9-36..

[2] Havaš, L., Medved, V., Skočir, Z. (2015). Application of Mobile Technologies in the Preparations for Long Distance Running, Springer International Publishing, pp 85-99.

[3] Office of Advocacy, U.S. Small Business Administration (2010). The Small Business Economy: A Report to the President, Appendix A

[4] Osredečki, E. (1995). Public relations, Edo-Biblioteka poslovnih priručnika, Zagreb, pp.142.

[5] Jonston, J., Zawawi, C. (2004). Public Relatons: Theory and Practice, 2nd.Ed. Allen & Unwin Croes Nest, Australia, 2004, pp. 390.

[6] Tomić, Z. (2008). Public relations - theory and practice, Synopsis, Zagreb-Sarajevo, pp. 34.

[7] Nowsu,I.E .(2000), Information Technology as an Indispensable Tool for Public Relations in the New Millennium, Public Relations Journal, Vol.2., No.1,pp. 18

[8] Lentz, I. (2000), 10 commandments for the success of small and medium enterprises, HUP,Zagreb, pp.55.

[9] Ružić, D. (2003) e-marketing, Economic faculty Osijek, pp. 65.

[10] Kim, W., Personalization: Definition, Status and Challenges ahead, Journal of Object Technology, No.1.

[11] Alfred C. W., Morrison, B., B., (2008), Social Networking, IEEE Computer 41(2), pp. 98.

[12] Cooley, T. (1999). Interactive communication – Public relations on the Web. Public Relations Quarterly, Vol.44, No.2, pp. 41-42.

[13] Markova, P, Prajova, V, Salgovicova, J. (2011). "The internet and integrated marketing communications." Annals of DAAAM & Proceedings, p. 271.

DOI: 10.2507/32nd.daaam.proceedings.003

ON THE EFFECTIVE SUBSTITUTION OF TURNING BY PERIPHERAL MILLING

Miroslav Piska & Stepan Kolomy

This Publication has to be referred as: Piska, M[iroslav] & Kolomy, S[tepan] (2021). On the Effective Substitution of Turning by Perifheral Milling, Proceedings of the 32nd DAAAM International Symposium, pp.0016-0021, B. Katalinic (Ed.), Published by DAAAM International, ISBN 978-3-902734-33-4, ISSN 1726-9679, Vienna, Austria
DOI: 10.2507/32nd.daaam.proceedings.003

Abstract

Turning operation seems to by prevailing for a production of rotational parts with standard demands on quality on machined surfaces. However, there are some applications that prevent effective use of the technology due to chip removal, entwining, clogging that prevents safe cooling of the turning tools and it ends frequently in a tool fracture or damage of the machined surface. Milling technology can suppress many of these problems, because the machining action is interrupted, chip is short and the cutting tool can machine even more surfaces in parallel. A difficult to machine welded assembly of stainless steel materials (the austenitic EN 1.4307 and the ferritic EN 1.4511) of profiled workpiece shape was selected for a substitution of the standard turning using several cutting tools by a milling and just one PVD coated monolithic cemented carbide tool. The series production of the part was selected and all quality parameters were first predicted, monitored and optimized. Finally, the economics of the production was assessed in terms of productivity and costs. All results confirmed a very stabilized production of good surface quality and long endurance of the milling tool (machined more then 30,000 pieces with one milling cutter).

Keywords: turning; milling; chip; loading;

1. Introduction

Modern technologies deal with many topics today how to enhance the productivity, to reduce the production cost, energy consumption, etc. Sometimes the researches investigate how to suppress vibrations when machining [1], the limits of the machine tools [2] and how to increase the cutting conditions [3]. There are advances in cutting materials [4], optimization of cutting geometry [5] or the principle of technology itself [6]. Modern technologies are focused on not only high speed or high feed machining of big parts, providing big material removals, but also to high quality of difficult to machine parts with small dimensions. The benefits of the productions stem frequently in several times higher added values, compared to the standard parts, and the production is more "green" and gentle to the environment. However, the orientation to the small dimensions reflects in special machinery, close tolerances and high demands on integrity of the produces surfaces.

The selected part for the study was made as a welded assembly of two materials – the austenitic stainless steel 1.4307 and a ferritic stainless steel 1.4511 (X3CrNb17 [7]), and has a very special function in many applications. Anyway, this work deals only with machining the ferritic part only, which is difficult to machine due to big toughness and high tendency to production of built-up-edges, that result in many instabilities in the production (unacceptable dimensions, poor quality of surface, irregular wear of cutting tools, chipping, fractures etc. The first selected technology for production in the past was a classical turning using several cutting tools. However, the chip formation made the technology very risky. The second technology was a circular inner milling with the use of special indexable inserts. However, it failed also due to poor fastening of the inserts. So finally, a peripheral milling with a whole carbide tool was offered as a perspective alternative. The details of the technology undergo special *know-how*, so a limited information can be shared. Nevertheless, the solution was successful and used in daily practise and some information can be published today.

2. Theoretical background, analysis of the technology

The very basic information about the studied technology can be found in Fig. 1. The first problem is a machining of all surfaces in parallel in the up-milling circular interpolation. From the physical point of view, it is a loading of a beam by torque, shear and bending. The acting forces are generated from the material removal and wear mostly.

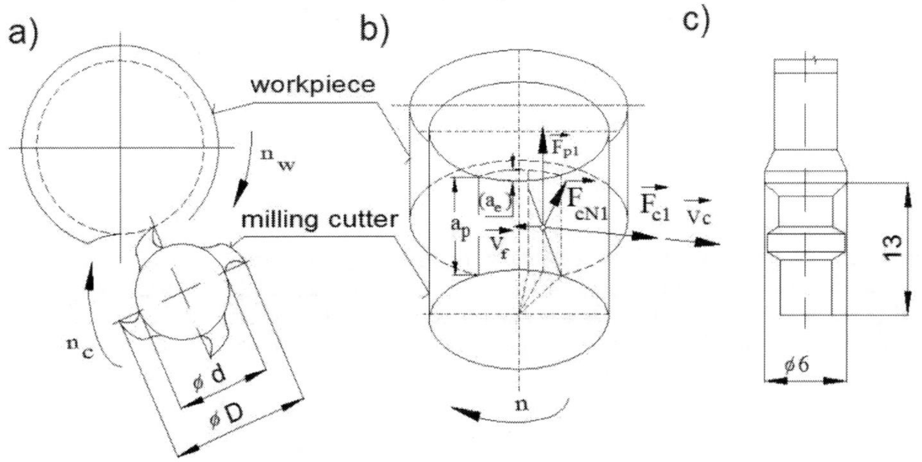

Fig. 1. Description of the machining: (a) principle of the peripheral milling, (b) the expected force loading, (c) a piece of the part machined (simplified)

The sculptured design of the cutting tool or workpiece were complicated so an exact solution of the stresses and deformations by FEA might be more complicated than the works for simple face milling cutters in the past [8]. However, the expected deformations are needed for a CNC tool compensations [9], reflecting various rake geometry [10]. Some simplified solutions were assumed for the two solid cylinders corresponding to the outer diameter (maximal diameter of the beam) and inner diameter (upon the bottom of the flutes) as the first calculative iterations – Fig. 2.

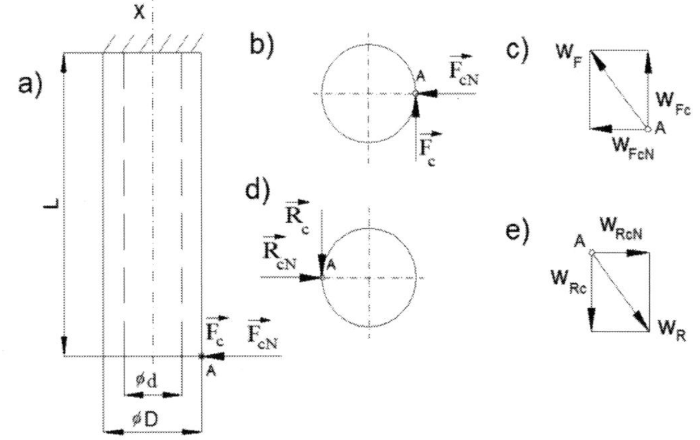

Fig. 2. Force loading and deformations at the machining: (a) a simplified model of the tool and the workpiece, (b,c) the force loadings and deformations of the tool, (d,e) the force loadings and deformations of the workpiece

The tool and workpiece deflections can be calculated analytically as deformations of the fixed beams, loaded by the force F (or its reactions R). The clamping end was regarded as fixed. In general, the beam is defined by the inertia moment J, Young`s elasticity modulus E and the length L. The translational deformations of the cylindrical beams (either the milling tool or the workpiece) can be derived in general from the accumulated elastic energy (1):

$$w_f = \frac{\partial W}{\partial F} = \frac{1}{E}.\frac{M_y(x)}{J_y(x)}.\frac{\partial M_y(x)}{\partial F}.dx + \frac{\beta}{6}.\int_0^l \frac{T_z(x)}{G_{(x)}}.\frac{\partial T_z(x)}{\partial F}.dx + \frac{1}{G}.\int_0^l \frac{M_k(x)}{J_p(x)}.\frac{\partial M_k(x)}{\partial F}dx \quad (1)$$

and the angular deformations (2):

$$\varphi_M = \frac{\partial W}{\partial M} = \frac{1}{E}.\frac{M_y(x)}{J_y(x)}.\frac{\partial M_y(x)}{\partial M}.dx + \frac{\beta}{6}.\int_0^l \frac{T_z(x)}{G_{(x)}}.\frac{\partial T_z(x)}{\partial M}.dx + \frac{1}{G}.\int_0^l \frac{M_k(x)}{J_p(x)}.\frac{\partial M_k(x)}{\partial M}dx \quad (2)$$

where W is total stored elastic energy of the beam, M, Myx and Mk are bending and torque moments, Tz is a shear force (equel to force F), Jy and Jp are inertia and polar inertia moments, G is a shear modulus of elasticity, β is a geometrical constant (1.66 for the cylindrical shape of a beam of L in length). The values of elasticity constants can be found at the material standards, but the values unfortunately vary. In general, the cemented carbides increase the rigidity of the beams approximately three times more than the steel materials, due to high values of E (about $6.86.10^{11}$ Pa vs $2.17.10^{11}$ Pa). The cutting force for the teeth in a spiral was predicted according to the Fig. 1b and the equation (3):

$$F_c = \int_{\varphi_1}^{\varphi_2} dF_c = \int_{\varphi_1}^{\varphi_2} k_{c1}.dA_D = k_{c1}.\int_{\varphi_1}^{\varphi_2} \sqrt{R^2 + \left(\frac{p_{scr}}{2\pi}\right)^2}.fz^{1-mc}.\sin\varphi^{1-mc}.d\varphi = c_o.\int_{\varphi_1}^{\varphi_2} \sin\varphi^{1-mc}.d\varphi, \quad (3)$$

where A_D is the undeformed chip cross-section, k_{c1} is the specific cutting force acting on one cutting edge, R is the radius of the tool, p_{scr} is the pitch of the spiral, and ϕ_1, ϕ_2 the angle tool engagements, mc is the material coefficient and co is a multiplicative constant. The next problem is the predicted roughness of the machined surface – Fig. 3. The residual inequality Rzt depends on the diameters of the tool and workpiece, and the feed per teeth. So the angle ω should be computed according to the desired roughness as a programmed increment.

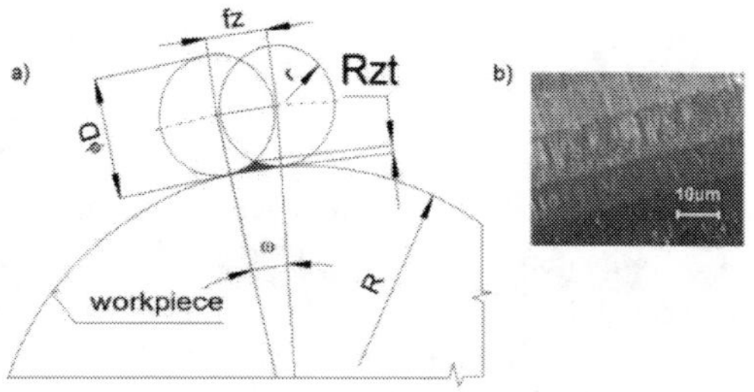

Fig. 3. A prediction of surface roughness as a function of tool geometry and cutting conditions (a), the real surface (b).

A theoretical value of arithmetical mean deviation of the assessed profile Ra for the peripheral milling can be predicted approximately according to the equation (4) [11]:

$$Ra = \frac{1000.f_z^2}{18\sqrt{3}.r}. \quad (4)$$

3. Experimental part

The exact material composition and mechanical properties of the workpiece can be found in the Tables 1 and 2. The diameter of the blank material was 6mm, the length 60 mm, so it was easy to clamp it into a precise collet and chuck of a special milling machine (both developed just for the production). A special block of the ferritic stainless material with dimensions 50x100-120mm was also tested.

C	Si	Mn	P	S	Cr	Nb	Fe
max. 0.05	max. 1.0	max. 1.0	max. 0.04	max. 0.015	16.0 – 18.0	0.6 – 1.0	balance

Table 1. Chemical composition (in weight %) of steel X3CrNb17 (1.4511): EN 10088-2-2005

Tensile strength Rm [MPa]	Proof strength $R_{p0.2}$ [MPa]	Min. elongation at fracture A_5 [%]
420-620	200-230	20-23

Table 2. Mechanical properties of steel X3CrNb17 (1.4511)

The cutting tool of 6mm in diameter, with four cutting flutes in helix 30°, was made of ultra fine grained (0.5-0.9 μm) tungsten carbides ISO M20 and protected with extra smooth and uniform (Al,Ti)N coating (thickness 2.5-3.0 μm, the PVD HIPIMS technology of CemeCon). The cutting conditions can be seen in the Table 3. The machining routine consisted from two passes - roughing and finishing, following without any break. The run-ins and run-outs were made via the circular interpolation and six soft approaching points, so no tool marks after the machining could be detected from the entering and leaving of the cutter from the workpiece.

Operation	Average adial depth of cut a_e [mm]	Number of tool rotations n_c [min^{-1}]	Number of workpiece rotations n_w [min^{-1}]	Average cutting speed v_f [mm/min]	Feed speed v_f [mm/min]	Axial depth of cut a_p [mm]
Roughing	0.9	7,000	400	132	125	13
Finishing	0.1	7,000	400	132	80	13
Cooling: Oil emulsion 12% with EP additives, 6 outer nozzles, cooling pressure 20 bars, total cooling intensity 0.8 l/s.						

Table 3. Cutting conditions for the milling operation

The cutting conditions were taken into consideration for the effective cutting rake and flank geometry. The abrasive mode of wear showed to be convenient and the machining exhibited a self-sharpening effect on the cutting flutes. Those synergetic effects resulted in low cutting forces, low deflections tool and workpiece and a stabilized machining.

The cutting forces were measured with the apparatus Kistler 9272, fully controlled with a PC. The data acquisition at the frequency 16 kHz in every channel was set, the DynoWare v. 3 and Microsoft Excel 2021 were used for data processing – Fig. 4. The measurements were organized in two groups – the first one with machining of solid block of workpiece materials 50x100-120mm (to analyse the deflection of the tools) and then second one with the real workpieces).

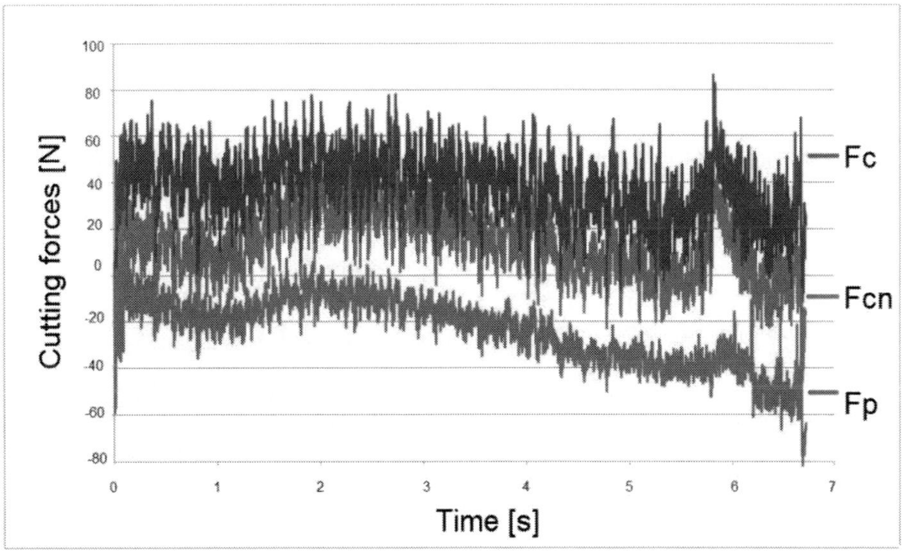

Fig. 4. An exhibition of cutting force measurement during one machining pass.

The predictions of the deformations and real measured values are plotted in the Fig. 5. The statistics is presented with the mean values of the deflections and the standard deviations (10 times repeated measurements with Alicona IF G5)

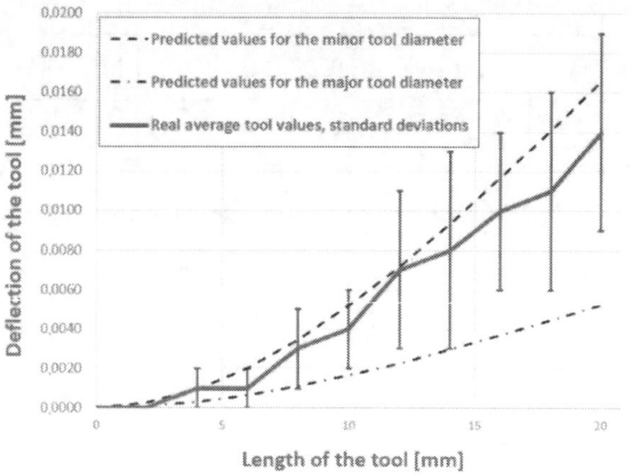

Fig. 5. Total deflections of the tool and workpiece when machining – predicted and measured values

The study of surface quality expressed by parameters Ra and Rz can be seen in Fig. 6. It can be said that the interval for lower feed speeds did not make good conditions for an effective chip production and many passes were not followed with a material removal as a chip. Moreover, the higher feed rates suppressed not only the built-up-edge production, but helped to acquire a better economy of the technology also.

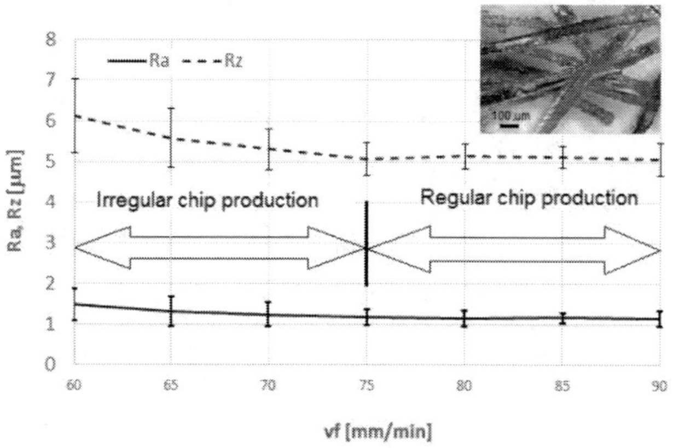

Fig. 6. The roughness parameters as a function of feed speed

A study devoted to the wear confirmed a very smooth abrasive mode of wear, stabilized in the whole tool life - Fig. 7. The cutting conditions altered in every cutting pass and the local places of the cutter underwent slightly different cutting conditions, but the endurance of the tool cutter was excellent and no big problem with chip removal has been watched.

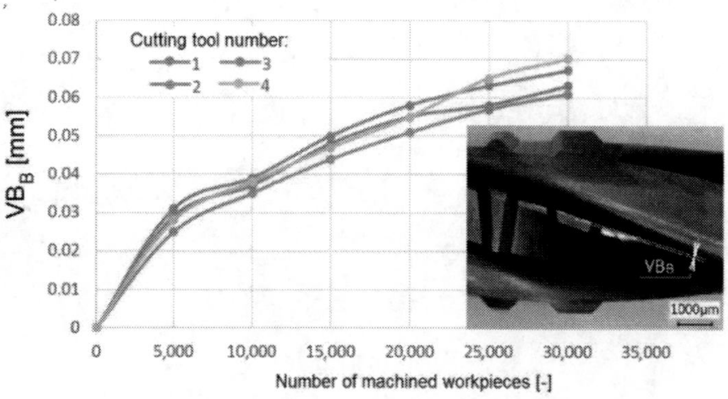

Fig. 7. Tool life as a function of finished machined workpieces

The cutting performance – 30,000 pieces made by one milling cutter without any catastrophic failure was a very good record and there are some ways how to improve it more.

4. Conclusions

The change of technology from turning to peripheral milling means a break point in the production technology and showed an efficient technology how to machine ferritic stainless steel. The substitution of the turning operation by milling was successful in all monitored parameters:

- the milling produced a very convenient chip formation and regular short chips,
- the milling cutter with a sharp geometry, protected with HIPIMS PVD technology provided excellent cutting performance even for the slim milling cutter and the overhang from the tool holder,
- the average cutting speed 132 m/min and feed speed 90 mm/min were verified as recommended values,
- the cutting conditions can be recommended for similar milling technologies of the material,
- the precision of the all machined dimensions was within IT10-11,
- the surface parameters Ra and Rz were below the allowed limits (2 μm for Ra, and 6 μm for Rz),
- the cooling of the milling was very effective and beneficial for the tool life,
- the outstanding cutting performance of the cutters was reached (30,000 parts per one milling tool),
- very low production costs were reached,
- the production has been running stabilized and smoothly so far for several months.

The new research and plans are devoted to a new tool design for 4-axis machining of the parts, taking into account the tool and workpiece deformations to reach the dimensional precision of IT 8-9.

5. Acknowledgments

These research works were kindly supported by the Brno University of Technology, FME, Specific research 2016, FSI S-16-3717 and the grant „Application of the New Surface Treatment Technologies in Metal Packaging Industry“, PID: FV40313, Ministry of Industry and Trade, The Czech Republic.

6. References

[1] Olt, J. & Maksarov, V. (2016). Development of Chatter-Resistant System of Cutting Tool, Proceedings of the 26th DAAAM International Symposium, Katalinic, B. (Ed.), pp. 0223-0228, Published by DAAAM International, ISBN 978-3-902734-07-5, ISSN 1726-9679, Vienna, Austria DOI:10.2507/26th.daaam.proceedings.031

[2] Olt, J. & Maksarov, V. (2016). Cutting process simulation on the basis of rheological properties of metals. Proceedings of the 26th International DAAAM Symposium, Katalinic, B. (Ed.), pp. 229–237. Published by DAAAM International, ISBN 978-3-902734-07-5, ISSN 1726-9679, Vienna, Austria. DOI:10.2507/26th.daaam.proceedings.032

[3] Kulenovic, M.; Begic, D. & Cekic, A. (2007). Experimental investigation of carbon steel in high speed cutting. Annals of 18th DAAAM and Proceedings of the International DAAAM Symposium. Katalinic, B. (Ed.). pp. 411-412, Published by DAAAM International, ISBN: 3-901509-58- 5. Vienna, Austria

[4] Hronek, O. & Zetek, M. (2017) The influences of cutting edge radius on surface roughness when milling nickel alloy. Proceedings of the International DAAAM Symposium. pp. 1037-1043. Katalinic, (Ed.). ISSN 1726-9679, ISBN 978-3-902734-08-2. Vienna, Austria. DOI:10.2507/28th.daaam.proceedings.144

[5] Belan, M. & Michalik, P. (2012). The durability and wear of coated mills from HSS-PM in stainless steel cutting without cutting fluid. Proceedings of the 23rd International DAAAM Symposium, pp. 1031-1034. ISSN 2304-1382, ISBN 978-3-901509-91-9. Vienna, Austria

[6] Brenci, L.M., Cismaru, I., Cosereanu, C., Curtu, I., Lica, D. & Fotin, A. (2009). The influence of the profiled milling parameters to the quality of the surface. Annals of the 19th DAAAM and Proceedings of the International DAAAM Symposium. Katalinic, B. (Ed.). pp. 413-414. ISSN 1726-9687, ISBN: 978-3-901509-69-8. Vienna, Austria

[7] ACERINOX. Https://www.acerinox.com/ [online]. [cit. 2021-09-25]. Available at: https://www.acerinox.com/opencms901/export/sites/acerinox/.content//galerias/galeria-descargas/galeria documentos-producto/ACX525-low.pdf

[8] Píška, M.; Zouhar, J. & Polzer, A. (2009). Stress- strain analysis of end milling HSS cutters reflecting the wear process. CIRP Annual Meeting 2009. 2009-01-29/5. 3. Paris, France: CIRP, pp. 1-6.

[9] Píška, M.; Zouhar, J. & Polzer, A. (2009). On the precise shoulder milling with an optimized 4- axis CNC machining. In Modern production technologies for the 21st century. Brno: CERM, 2009. pp. 43-47. ISBN: 978-80-214-3914-6.

[10] Zouhar, J. & Píška, M. (2008). Modeling of the Orthogonal Machining Process using Cutting Tools with Different Geometry. In MATAR PRAHA 2008, Part 2: Testing, Technology. 1. Prague: SMT - Society for Machine Tools, pp. 81-86. ISBN: 978-80-904077-0-1.

[11] Davim, J. P. (2010). Surface Integrity in Machining. London: Springer. ISBN 978-1-84882-873-5.

DOI: 10.2507/32nd.daaam.proceedings.004

APPLICATION OF GAME-BASED LEARNING IN MATLAB USING OBJECT-ORIENTED PROGRAMMING

Karel Perutka & Michal Vymazal

This Publication has to be referred as: Perutka, K[arel] & Vymazal, M[ichal] (2021). Application of Game-based Learning in MATLAB using Object-oriented Programming, Proceedings of the 32nd DAAAM International Symposium, pp.0022-0032, B. Katalinic (Ed.), Published by DAAAM International, ISBN 978-3-902734-33-4, ISSN 1726-9679, Vienna, Austria
DOI: 10.2507/32nd.daaam.proceedings.004

Abstract

This paper focuses on game-based learning and educational digital games. The first theoretical part summarizes the basic concepts and follows the problems with the introduction of game-based learning in practice. The work focuses on game-based learning as a modern method of education, which is used by many experts from practice and confirms the benefits of its use in teaching. The next part is devoted to specific types of players, then presents the characteristics of digital educational games used in practice and tries to map the use of this method of education in teaching abroad and in the Czech Republic. Furthermore, examples of the use of digital games in teaching and in other industries are presented. The practical part deals with the design of the original game, designed primarily to teach MATLAB, the description of the created application from the user's point of view, and its verification by students during teaching. The game was created using object-oriented programming in MATLAB, which is a very rare and new approach.

Keywords: Game-based learning; computer game; education; object-oriented programming; MATLAB.

1. Introduction

The issue of game-based learning is currently a frequently discussed phenomenon presented in electronic media [3], [4]. The first technical elements in the process of education and learning include the use of pictures and symbols during lessons followed by a spoken word that complements the printed text. Over time, in the field of education technology development occurs great progress when computers came in the 20th century. It is a time of more significant change, a time of great progress, with books and television being replaced by the Internet and digital games [5], [6].

The advent and constant expansion of digital technology mean a major impact on the whole society [1], [2]. People in Europe, the USA, Japan, and China are most affected by this technology. Digital technology enters several areas in which it fulfills the function of education through play. First, the method of John Amos Comenius can be mentioned as a "school by play" whose main idea is “Teaching is fun”, and it is still used today. Today's world, in an environment connected with technique, offers an educational method called "Game-based learning". The concept is possible to conceive as “digital game-based education or digital learning games”. This method of education permeates the entire paper.

The goal of game-based learning is to find out how and where it is possible to incorporate the simpler, more effective, and more interesting method of education involving digital games. Implementation using object-oriented programming is quite a new approach [7], [8], [9]. But it is popular to use games for teaching, there we created nice applications in this field [11], [12], [13]. This paper contains a brief overview of educational games with the potential to educate and also own educational games based on game-based learning [10].

2. Theoretical background

2.1. Definitions

Game - Basically, there is no clear definition that defines the very concept of the game. Because it is this work focuses on the issue of game-based learning, we will use the definition of one of the most respected experts in this field, Marc Prensky, whose goal is to make the game engaging. According to his publication, the following structural elements serve this purpose:

- Goals
- Rules
- Interactivity
- Entertainment
- Adaptability
- Winning states
- Outputs and feedback
- Conflict/rivalry/challenge/objection
- Interaction
- Problem-solving
- Representation and story

From the above, it is clear that there is no uniform definition of the game, not even within one scientific field. Because with many definitions, the game is characterized by key elements, we assume that they are necessary for its functioning. However, not every game has to identify with all key elements. Key elements are:

- Rivalry - The goal is to achieve a better result than others can on students act on two factors, motivational and demotivating. There can also be rivalry affected by competition
- Illusion - The existence of an environment that can be explored, commonly used in the simulation.
- Interaction - is an important element of education, it helps students better understand the topic substance, group teaching is recommended, in the form of discussion or counseling, (action changes the state of the game and triggers feedback
- People - more people are recommended (students, teachers)
- Challenge - tasks require effort and are not very trivial, so they become challenging for him and difficult
- Rules - should be understood as recommendations rather than regulations, and thus lead the student to successful education
- Exploration - The existence of an environment that can be explored
- Security - this is a quiet place, set for the learning process and verification knowledge of students.
- Results - specific results are important for evaluation (e.g. scoring)
- Goals - explicit goals and objectives.
- Research - is important, looking for new interesting information, offers other questions, researches the more detailed topic

Digital game - The word digital means the use of some kind of electronic device that can, however, we don't have to be connected to the internet. These devices include, for example, laptops, pocket gaming consoles, classic desktop computers, mobile phones, music players, etc. These are actually about a device with a microchip that allows you to play the game.

Video game - Some people have no problem imagining a digital game under this name. It can be said that They are right because video games are the equivalent of a digital game, but the display of the game is the screen that is required. The main purpose of video games is to entertain the players.

Computer game - Computer games are run on a computer or laptop. It is an interactive program that allows you to communicate with the user using sounds, 2D and 3D graphics, and text. The form of tension and fun fills and diversifies the free time of all players.

2.2. Types of players

Bartl's taxonomy - Richard Bartle (1960) is a British writer, professor, and researcher in the online gaming industry according to the structure, there are multiplayer games. For example, in collaboration with Roy Trubshaw, was created game MUD (Multi-User-Dungeon) was. It is a game based on the virtual world, specifically on text. Bartl took advantage of the game to create a survey to find out what games players enjoy, as different players may not have the fun same game. According to the behavior of players in the community (Multi-User-Dungeon), it was found that players are divided into four categories: Achiever, explorer, socializer, killer [10].

Achiever - The main goal of the player is to get as many awards or rewards in the game world as possible, for example, badges, points, money, levels, and more. The main motivation for players is to achieve goals and win. He adapts the tactics to obtain as many material things as possible. The player is accompanied constantly by the idea of comparing with other players.

Explorer - This type of player focuses on individual elements of the game world. First, he maps out what rules the virtual world works with. Then he looks for shortcomings. The usual characteristic of A player of this type is that he admires secret places and other unexplored locations. It does not accept restrictions games and trying to overcome them. Once he finds nothing new, the game becomes unattractive for him.

Socializer - This type of player is primarily interested in socializing, so communication with people is at stake most important to him. The player strives to create a strong relationship in the sharing community thoughts, opinions, stories. The game itself, for him, is not so important. Games for one player only lays when this game attracts a lot of attention from the surroundings. Just because to talk to other players about the game. His biggest weakness is the fear of rejection.

Killer - A killer is the type of player he kills to defeat an opponent. The player can play games from sports to action, the most important element is the opponent. From the psychological point of view, this type of person exploits his aggression by bringing it into play.

2.3. Game-based learning

The term game-based learning can be translated as "learning through digital games or education based on digital games." It is a combination of digital games with form education. Today's students are accustomed to digital technology from an early age, and therefore they want to combine it with teaching. The primary goal and premise of game-based learning (GBL) is teaching that makes students more enjoyable and based on which it is possible to achieve the same or even better results than the classic form of education, which is already boring for students.

Certain procedures and teaching methods have been provided for the older generation of students.
Some teachers still think that these methods could work today. This idea has already been refuted. "Today's students are no longer the people for whom our education was system designed". There is a problem for both teachers and students. The problem of teachers is engaged with their teaching method and effectively educate students. Students can't concentrate and usually tend to be bored and inattentive. Therefore, it is appropriate to change the teaching, to match current education. By introducing a more interesting and fun form of teaching we achieve educational games.

Educational games are seen as the opposite of fun games, otherwise translated as "serious games". Sarou de Freitas defines educational games as: "Applications using characteristics of video games and computer games to create a catchy learning experience to transfer of identified learning objectives, outcomes, and experiences ". There is still debate among educators on the issue of digital games in the field of education, so-called digital game-based learning.

The game with educational content works as a means of education mainly in fulfilling these reasons:

- By including the content of education in the game, some obligation arises that may happen interesting for students, aversion to learning.
- Depending on the educational goals, it is possible to apply different methods of the interactive educational process.
- The grouping of the interactive educational process and the integration of the educational one content into the context of the game. It can be achieved in several ways, but the best solution will follow according to the needs of the teacher and the student.

A game that meets these requirements can become completely ineffective, if not the educational content does not correspond to the abilities and interests of the student.

Practical use - Digital games serve not only as a very important and indispensable element in teaching but as well as preparation for various crises. Examples are natural disasters or terrorist attacks requiring rapid decision-making under time pressure. Their use can also be used in training for driving cars, controlling aircraft, to create models of solutions to problems in different areas, or serve as a tool for team building and team training cooperation that promotes goal-oriented coherence.

It also allows providing information on social networks, and especially its importance as a team is needed within companies and in their cooperation. Digital educational games are a success used in many sectors, such as insurance, law, medicine, pharmaceutical industries, banking, etc. Subsequently, I deal with these in more detail
industries in which game-based learning is used - education, employment, military, economy.

3. Selected computer games created in MATLAB

In the following chapter, we will find examples of games in MATLAB. The criterion for choosing these games was the fact to what extent they were fun. How they improve logical thinking and motivation to the game. All of the following games can be found on the server of MathWorks.

3.1. Chess

This is one of the oldest games and the first mentions coming from the western world are dated to the 6th century. The game comes from India, the date of origin cannot be determined exactly. Los Almos is the first program developed to play chess and dates back to 1956. According to available Paul Stein and Mark Well for the MANIAC I computer was to take part in the development of the information. The basis of the game is a chessboard measuring 8x8. On the board, you can see that individual the squares of the chessboard alternate between the two colours for better orientation. For complete clarity as well used to name individual rows and columns. The player can play for either a white or black group of figures. The first version of Chess (Chess Master) in MATLAB, programmed by Brian Moore was created in 2014.

3.2. 2048

The original game was programmed in JavaScript. This is not a particularly old game, the game dates back to 2014 and the author of the game was Gabriele Ciruili from Italy. The playing field is about the size of a 4x4 grid (16 squares). When using the arrows, the boxes move according to the selected direction. The game works on the principle of connecting two cells, with the same number. This way we get to the number 2048. This is a game created in 2014 in MATLAB by an employee of MathWorks, nicknamed Jiro.

3.3. MATLABTETRIS

Tetris is the most successful computer game in modern history. This is a game programmed in MATLAB, from 1985. The main author of this game is Russian scientist Alexei Pažitov. As the name suggests, each shape consists of four squares. Everybody shape falls from the top down and our goal is to place it in the lower part of the playing area so that it will be filled in its entirety line so that it can then be deleted. The game automatically increases the difficulty the longer it is played time, thereby increasing the speed of the falling body. The game itself is controlled with the arrow keys. The up arrow has the function of turning the page and the down arrow has the function of setting body fall velocity. In the game environment, the individual shapes that follow the existing ones are displayed. It is too available information table, in which it is stated: number of points obtained, current level and the number of rows deleted. Matt Fig reprogrammed the game into MATLAB, releasing a total of two versions in the same year 2012.

Fig. 1. MATLABTETRIS main window

3.4. Pac-Man

Toru Iwatani is the author of this famous computer game, which he named Pac-Man. Toru Iwatani began to develop it because there was no other similar game on the market. Most games in the given tend to come more from the action game industry. The aim of the game is for Pacman to collect all the balls. But a player who plays for Pacman must be careful not to be caught by ghosts. There is also the possibility of temporary control over ghosts. To gain this power, the player will have to eat a ball that is larger than the other. This would put Pacman in a phase where he could hunt ghosts and get some for it points, but this moment is usually very short. Once Pacman collects all the balls, gets to the next level, where the ghosts gain higher speeds. Athanasios Kehagias rewrote this game to MATLAB, working specifically with the version R2016a. Now the current version is from 2017.

4. A new application of game-based learning in MATLAB

TBU Quiz is a new 2D educational game focused primarily on testing your knowledge players. In general, he can verify knowledge in various areas, because it is enough to exchange question and answer database files. And all in a fun way. The game is programmed in general. It is currently focused on knowledge of MATLAB, but in the future, it can be used to test any other knowledge. The list of questions for the quiz is not strictly limited to MATLAB and is freely distributable.

Fig. 2. Opening window of new game

4.1. Game breakdown

The game is divided into a total of four parts, which are configuration, help, questionnaire, and the quiz itself. Each part is designed to be fully managed and controlled by a graphical interface created using MATLAB. Players and any administrators thus are not burdened with the knowledge of the structures of the files used and their interconnection. Everything is possible conveniently set up with an intuitive interface.

4.2. Configuration

Most games, whether simple or complex, allow a variety of settings a potential player can adjust the environment according to their preferences. This is not the case with TBU Quiz either otherwise. This is not a configuration in the true sense of the word, but rather a simpler derivative. It does not allow any customization of the interface, but only the basic configuration. Specifically, it goes to manages the quiz and questionnaire questions on which the whole game is based. That's why it's their management is an integral part of the game.

4.3. Help

Help is another invaluable helper in every right game. That's why it's not missing either here. The game itself is quite intuitive, so it wouldn't even need help, but only for the completeness of the solution deserves it.

4.4. Questionnaire

The questionnaire part is the only parallel to the main playing part. You could say it works about a minigame designed to analyze the type of respondent, in this case, the player, based on provided answers. It may seem that such a questionnaire does not completely fit into the strategy quiz games, especially if it focuses on knowledge of MATLAB syntax and semantics. On the other hand, it is very valuable information that the game provides, without any special knowledge. Interestingly, there is no wrong answer. Each so can shape the character of the player.

4.5. Quiz

The quiz is the main part of the application. The part that every player looks forward to when running any games. Because it is a variation on a similar, well-known television quiz that it is intended for two players, and here it is no different. As with the TV quiz, it is the alpha and omega everything a list of predefined questions hidden behind the field that the player chooses. Unlike the TV version of the game offers a list of potentially correct answers, when correct is always only one. The second difference is the way individual fields are assigned to players. In the TV quiz, the selected field, to which the player answers incorrectly, is neutralized. TBU Quiz as part of simplification and optimization assigns it to opponents. The third and also the last main difference is the evaluation of both players after completing the quiz. The game at its end evaluates both the knowledge of the winner and the opponent. They will both find out how they are doing with their knowledge.

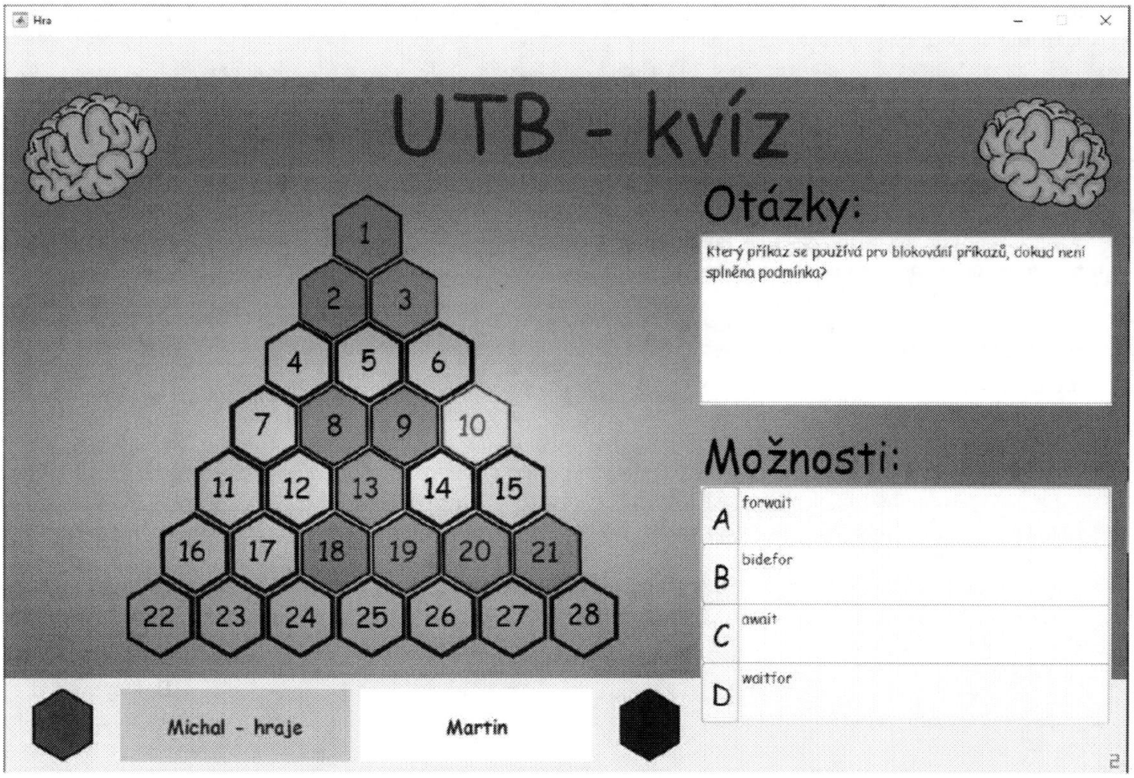

Fig. 3. The main window of new game

4.6. Quiz rules

The quiz is intended for two players and is not limited by any number of rounds or time limits. The game ends when one of the players manages to connect all three sides with their fields playing pyramids. The fields defining the winning path through the pyramid are marked for clarity in light green colour. The game statistics of both players are displayed. Each round always determines only one active player. They take turns. Active the player must always choose one free field hiding the question. After making a selection, it is displayed the question and the active player must answer it by clicking on one of the buttons "A", "B", "C" or "D" that precedes the answer. If the answer is evaluated as correct, the field is assigned to the active player. Otherwise, the field is assigned to the opponent. In both cases, however, the field cannot be re-selected during the game.

The individual fields of the pyramid may be in different states, but never simultaneously in more. The first state is an unoccupied field. It is only framed in black. In such fields, players can vote in each round. Another state is the active field, which contains in addition to a black frame also light yellow background. If the field is highlighted in red or blue, signals that it belongs to one of the players. The last condition is a light green coloration. This state is used in case of a win and only the fields that connect all of them are marked three sides of a playing pyramid.

4.7. The technical solution of the new application in MATLAB

The whole game application is designed using the MATLAB development environment in the version R2020a. The individual files are divided into classes that contain data logic that MATLAB indicates the extension ".m" and files containing the graphic design of the application. They are marked with the extension ".mlapp."

- The Question.m class is a data model that combines information about specific questions and theirs answers. It is used mainly in configuration, then also in the game part. It is designed so to be able to completely take the data from the configuration file questions, and subsequently in the case provided them for back-serialization to a file.
- The QuestionAnswer.m class is a data model that combines information about a specific answer question. It is insignificant in itself. Therefore, it is generally used in conjunction with the class Question.m, where individual answer objects are linked to a specific question using reference relationship 1: N.
- The GameField.m class is a data model that associates data with a specific field of a game pyramid. In principle, only a reference to an object of class Question.ma contains its designation, which is the whole positive number from 1 to 28 depending on where it is in the pyramid.
- The GamePlayer.m class is a data model that associates data about a specific player. It preserves information about his nickname and also statistical data. Specifically, erroneous statistics and correct answers are divided according to individual groups of questions.
- The GameResult.m class is a data model that stores a reference to the player who won and at the same time to the player who lost.
- The MenuWindow.mlapp class contains a graphical design of the application's main menu. At the same time provides an interface for transitions between different parts of the application.
- The QuestionnaireWindow.mlapp class contains a graphical design of a questionnaire. It also provides the evaluation logic of the answers that the player chooses.
- The PlayersWindow.mlapp class contains a graphical design of a form that allows settings nicknames of individual players. It does not contain any additional logic. It's one from simpler application forms.
- The GameWindow.mlapp class contains a graphical design of the main game form application. It also contains most of the game logic of the quiz. This is the most complex form whole application.
- The MessageWindow.mlapp class contains the graphical design of the form that allows passing information to users. Used to notify various setting errors alert or to view quiz winning statistics.
- The InputWindow.mlapp class contains a graphical design for a takeover from user input. This is almost certainly the simplest form in the entire application. Used only to accept the user's password when entering the configuration.
- The ConfigWindow.mlapp class contains a graphical design of the configuration submenu form. It thus provides a guide between configuring the questionnaire and the quiz.
- The QuestionaireConfigWindow.mlapp class contains a graphical form design enabling the management of questionnaire questions. It also provides features for saving them to the configuration file.
- The QuestionConfigWindow.mlapp class contains a graphical design of the form quiz questions management. It also provides functions for storing them in the configuration file.

Below is shown the sample code of the created application using object-oriented programming in MATLAB, the class for game field

```
classdef GameField

    properties(Access = private)
        Number
        Question
    end

    methods
        function        obj        =
GameField(question,number)
            obj.Question = question;
            obj.Number = number;
        end

            function        results        =
        getNumber(obj)
                results = obj.Number;
            end

            function        results        =
        getQuestion(obj)
                results = obj.Question;
            end
        end
    end
```

4.8. Data storage

The game itself, in essence, places only minimal demands on any data storage, possibly the need to use a database system. It requires only two specific files that store information about questionnaires and quiz questions. To store this data, just a simple file system will suffice. Both of these files must be in the directory Config, which is located relatively at the same level as all other files' applications.

- Questionnaires.xml is a file data structure that stores all questions and the relevant answers used in the questionnaire. The file is for easy editing in xml. Despite this fact, it is not recommended to edit it outside the configuration panel itself application.
- Questions.xml is a file data structure that stores all questions and relevant answers used in the quiz. The file is for easy editing in xml format. Despite this, however, it is not recommended to edit it outside the configuration panel of the custom application.

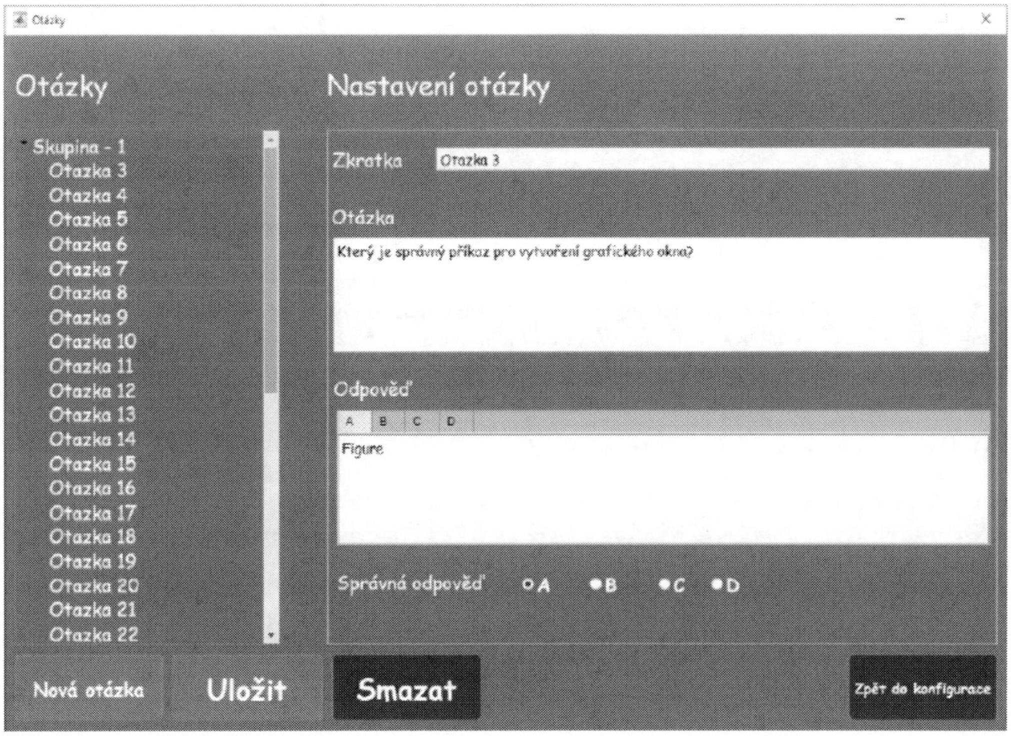

Fig. 4. The window of quiz data configurator

4.9. Main algorithm

Matlab brings with it various limitations in the development of form applications. For an application containing only one form, there is no need to deal with any additional complexities and eventuality. However, in case there is a need to use more than one form on the fly application, a few major obstacles need to be addressed. An example is a modal form, to ensure the risk of closing a form other than the one currently open. Next An issue that needs to be addressed in a multi-form application is passing parameters between forms. Last but not least, it is also a challenge to deal with various generic lists, such example, queues, dictionaries, or data sheets only. The main application form is MenuWindow.mlapp. It is only after its completion that it ends the whole application. For this purpose, the game uses two principles that ensure the impossibility of his closing before all other forms are closed. The first principle is the implementation of top bridge technology.

This informs the operating system of the highest display priority the form so marked. Any other form is then displayed after this, namely including forms from other applications. But this alone is not enough for the desired behaviour, because it is still possible to close another form than currently displayed, thus getting the application into a non-deterministic state. Therefore, the application still internally implements an internal state in the forms, which controls whether it is possible to close or not. The state is represented by a variable containing the logical value 0 or 1. If any other form is displayed, this variable is set to a logical value of 0, making it impossible to close the form from which the new modal was displayed form. In its close callback, it checks the status of the variable and if it is set to logical 0, cancels it and does not close.

4.10. Algorithm evaluating pyramid questions

One of the main algorithms of the whole application is undoubtedly the evaluation algorithm game pyramid issues. It runs in each round of the game quiz. The algorithm begins by switching the current player to an opponent. As for the first round, set to the current player always red. It continues by loading user input in the form of the selected field. If when the field is occupied, it returns to the game field selection. Otherwise, it loads the question also answer and display them on the board. In the next step, it reloads the user input in the form of selected answers. He has to evaluate it. If the answer is correct, it assigns the selected field active player. Otherwise, it is assigned by the opponent. In both cases, however, it continues by evaluating the completeness of the user's path. And be an active player if it is the right one answer or the opponent in the case of an incorrect answer. The player's path is evaluated using graph theory. Specifically, the width-through algorithm. This is also the case of a signature algorithm, which is also the most complex. If the path of the player, if any opponent is incomplete and does not connect all three sides of the pyramid, the algorithm returns to the beginning. Here the player switches to the opponent. The algorithm is executed until one of the players reaches the completeness of the journey and thus wins.

4.11. Algorithm evaluating the completeness of the path

TBU Quiz uses graph theory to evaluate a player's path in the game pyramid. Specifically, searching the graph in width. This algorithm requires a queue at the beginning of processed items. Matlab offers AsyncBuffer for this purpose. AsyncBuffer allows you to read elements exactly in the order in which they were written to it, which is a great advantage in processing. After creating the queue, the algorithm will add it directly to the field currently being processed. It also creates an auxiliary list for the fields that have already been processed need to remember, otherwise there would be a loop. Here Matlab provides a function containers.Map, which allows the definition of the dictionary structure.

When the algorithm has a queue and an auxiliary list ready, it starts a cyclic pass the element queue, which it executes as long as the queue contains something. For each individual, the field within the cycle reads all neighbouring ones and also tests them cyclically. If given the field belongs to the player and is not in the help list, it adds it to the queue for processing and continues to the next until he passes all the neighbours. The moment the queue is empty, he checks the algorithm the last condition that is required to end the game by winning, and that is connecting all three sides of the game pyramid. This part is already simple. Here the algorithm only tests whether the auxiliary queue of processed elements always contains at least one field on each side of the game pyramid and if so, ends the game by winning the player. Otherwise, the game continues for another round.

5. Application verification

5.1. Self-testing

Before it is possible to pass the application containing the game to the agreed testers, it must go through its testing process. If this were not the case, possible errors would delay testers and distracted them from the very essence of game testing. The result of such testing should not be a list of bugs, but suggestions for improving or modifying existing algorithms.

Internal testing takes place in three steps. The first is to test the baseline scenarios application control. The second is playability testing according to the documentation and the third is random, one could say that even a chaotic click through the whole application. In general, development should never be tested by a programmer. Or more precisely, to test certainly had, but such output can not be considered as a proven functionality, because the programmer knows exactly how he wrote and is affected by this fact during testing. He doesn't always succeed reveal everything. For this reason, the game will go through both internal testing and pre-testing agreed on testers.

5.2. Testing baseline scenarios

Testing basic scenarios deals with the issue of basic application control procedures. It focuses mainly on the detection of shortcomings and unfinished business resulting from various additional modifications to algorithms that cause the application to crash unexpectedly and untreated.

Thanks to this testing, it was possible to detect semantic indexing errors of various kinds of collections that are used in the application. The first versions of the game have suffered from this issue since they are built on collections and cycles that go through them using indexes. Generally speaking, that the custom is to index the collection from the number 0. Thus, the first element in the collection is at index 0. There are two different approaches in MATLAB. One is classically based on indexing from 0, the second, on the contrary, is from 1. And that was the source of many problems in the early development of this game. Get used to it Take a dual approach and in each specific case, thinking which one to choose was enough demanding, and this generated a lot of unpredictable situations. An example may be a vector containing randomly shuffled indexes of quiz questions. Such vectors are indexed from 1. But that's not all. If it is cyclical, it is necessary to think even that the vector ends with a length index and is therefore not reduced by one.

The opposite approach but provides, for example, collections that publish individual XML child elements. This collection is indexed from 0, so its last element is at the specified index the length of the collection reduced by one. Another problem resulting from this testing was in the object access to data. The program itself contains several classes, where their instances are used in different places application. The class itself is a very useful thing. But the problem arises when it is necessary for its instance, not only to keep the data static but also dynamic. That is if the class must also allow data to change. Specifically, the GamePlayer class stores statistics for a given player. These change after each round that the player plays. And that was exactly the problem. The values of the statistics appear when the instance is created sets the class to 0. Then, in each player's round, the value 1 is added to the specific one object properties. However, the changed value lasted only in the context of the method in which it was changed. Even if the object was stored in a global variable. As if accessing a variable made local copies. This scandal was removed by inheriting the class from an ancestor named handle.

5.3. Gameplay testing

When the application is debugged for unexpected crashes, it is necessary to proceed with testing game algorithms. These were written almost flawlessly. The only problem was the presentation of game statistics of individual players. The achieved percentages were displayed in exponential shapes, or formatted to a lot of decimal places. This problem was more complicated than it may seem. The data is written to the form using the sprintf method. She takes it as parameters a formatted string and an array of values to print in the string. As a solution to a problem, the use of the Round function was offered. But it did not work as expected and the value was still did she misspelled. The solution was provided by changing the format string itself. Originally the value is printed using the wildcard % d to be used for integer values. New the % 0.2f symbol is used, which ensures correct formatting to only two decimal places.

5.4. Final random testing

Random testing includes a random element in the process. So it tries to get the application to a non-deterministic state by simply randomly clicking on any buttons that the application offers. Usually, this activity is suitable for a person who does not know the application at all and does not know what he is allowed to do and what he is not allowed to click on. By this testing, as in the previous section, more or less only one problem was revealed. But in more places. This problem was the absence of code addressing the logic of the dialog form. More precisely, some forms did not contain the DialogResult property, so they could not be responded to from forms superiors. But that was a problem that caused the game not to be turned off in any way since she did not allow her termination. She was still waiting for the result of the dialogue. The problem was fixed quite simply. The affected forms provided the implementation of the required interface and everything began to work as expected.

6. Evaluation

This chapter evaluates what exists, what is new in this paper, and the plans for the future. There exist games in MATLAB and applications of game-based learning written in different programming languages than MATLAB, mostly in western European countries and the United States. But there are only a few of them created or used in the Czech Republic. The same is with the object-oriented programming applications in MATLAB, there is a minimum of them worldwide. This paper combines object-oriented programming in MATLAB and game-based learning with implementation in testing the knowledge of MATLAB, which is a new approach. In the future, a database of questions and answers will be enlarged together with the implementation of another taught subject, discrete control.

7. Limitations and plans

Although the paper has reached its aims, there were some limitations. Because the created application is new, it was verified only by a small group of students who were attending the classes. Therefore, it is necessary to use it for some years to generalize the result and to find what to improve and enlarge. Future work is divided into several parts. Firstly, it is necessary to use the materials for more years in purpose to evaluate them from the users´ point of view. Lately, there will be realized necessary correction based on the experience with these materials.

8. Conclusion

The paper dealt with implementing an important method of education based on digital games. Many authors of professional articles are convinced of the use of games not only directly intended for teaching but also games with quality and useful educational goals, i.e. commercial games. Although it does not exist a unified interpretation of a digital game with educational value, according to a literary survey, several experts agree that sophisticated digital games bring players faster decision-making, develop logical and strategic thinking, and while maintaining thoughtful educational content serves as a full-fledged educational tool.

The current generation has a huge advantage in using digital educational games and education, as it has an innate ability to take an interest in digital technology and discover the whole digital world. The introduction of game-based learning in schools, in companies, and in other collective activities contributes to maintaining mutual relations and stronger teamwork.

Examples of game-based learning research prove that the world is on this topic. Many professionals have been dedicated to education for many years. Based on the statistics discussed above in comparison with the world, game-based learning is not so widespread in the Czech Republic. But it is already to see a growing interest in a method of education based on digital games. The potential for teaching is huge. This medium can be used to teach various subjects. Scientists and teachers almost agree that games work positively, but at the same time, it is necessary to give a big one and to be careful not to create a gambling addiction to the player, in which about one is seen as the main negative influence.

The current trend is the technology that pupils and students cannot avoid. After leaving school, whether at work or in everyday life, will need and use it. The theme of game-based learning is very likely to appear in a large number of games focused on learning through play, thus realizing a fun form of learning for as many as possible students.

These procedures are implemented in the designed educational game TBU-quiz in MATLAB. This game is for two players. Its task is to connect the cells that are arranged in a pyramid shape. Beneath each cell is a randomly generated question with four answer options. The questions are focused on the MATLAB program. Specifically, it is about three areas of questions. Each topic is represented by a specific cell. To whom first manages to unite all three sides of the pyramid, wins. The resulting window offers a percentage success of players in individual categories. The goal is to use this game for teaching or dissemination of knowledge.

To sum up, in general, what was the problem? The problem was to implement the modern technique of education in the field of MATLAB teaching. How was it solved? It was solved by creating the game using object-oriented programming in MATLAB. What are the solutions? The solution is the created application in MATLAB as teaching What are plans? In the future, we will focus on the enlargement of questions and solution databases.

9. References

[1] Sugahara, S. & Cilloni, A. (2021). Mediation effect of students' perception of accounting on the relationship between game-based learning and learning approaches. Journal of Accounting Education, 56, 100730

[2] Dudzinski, M.; Rozgic, M. & Stiemer, M. (2018). oFEM: An object oriented finite element package for Matlab. Applied Mathematics and Computation, 334, 117-140

[3] Xu, Y.; Lau, Y.; Cheng, L. J. & Lau, S. T. (2021). Learning experiences of game-based educational intervention in nursing students: A systematic mixed-studies review. Nurse Education Today, 107, 105139

[4] Kao, Ch.-W. (2020). The effect of a digital game-based learning task on the acquisition of the English Article System. System, 95, 102373

[5] Greipl, S.; Klein, E.; Lindstedt, A.; Kiili, K.; Moeller, K.; Karnath, H.-O.; Bahnmueller, J.; Bloechle, J. & Ninaus M. (2021). When the brain comes into play: Neurofunctional correlates of emotions and reward in game-based learning, Computers in Human Behavior, 125, 106949

[6] Krath, J.; Schürmann, L. & von Korflesch, H. F. O. (2021). Revealing the theoretical basis of gamification: A systematic review and analysis of theory in research on gamification, serious games and game-based learning, Computers in Human Behavior, 125, 106963

[7] Richard, B.; Rastiello, G.; Giry, C.; Riccardi, F.; Paredes, R.; Zafati, E., Kakarla, S. & Lejouad, Ch. (2019). CastLab: an object-oriented finite element toolbox within the Matlab environment for educational and research purposes in computational solid mechanics, 128, 136-151

[8] Berger, R.; Bruns, M.; Ehrmann, A., Haldar, A., Häfele, J.; Hofmeister, B.; Hübler, C. & Rolfes, R. (2021). EngiO – Object-oriented framework for engineering optimization, Advances in Engineering Software, 153, 102959

[9] Noroozi, O.; Dehghanzadeh, H. & Talaee, E. (2020). Entertainment Computing, 35, 100369

[10] Vymazal, M. (2021). Application of Game-based Learning in MATLAB, Bachelor Thesis, Faculty of Applied Informatics, Tomas Bata University in Zlin, Zlin, Czech Republic

[11] Karuovic, D.; Radosav, D. & Glusac, D. (2010). Educational Game Model for Pre-School Children, Chapter 12 in DAAAM International Scientific Book 2010, pp. 107-116, B. Katalinic (Ed.), Published by DAAAM International, ISBN 978-3-901509-74-2, ISSN 1726- 9687, Vienna, Austria DOI: 10.2507/daaam.scibook.2010.12

[12] Polcar, J.; Gregor, M/; Horejsi, P. & Kopecek, P. (2016). Methodology for Designing Virtual Reality Applications, Proceedings of the 26th DAAAM International Symposium, pp.0768-0774, B. Katalinic (Ed.), Published by DAAAM International, ISBN 978-3-902734-07-5, ISSN 1726-9679, Vienna, Austria DOI:10.2507/26th.daaam.proceedings.107

[13] Musalek, M. & Adamek, M. (2020). The Use of Virtual Reality Games as a Tool to Overcome Acrophobia, Proceedings of the 31st DAAAM International Symposium, pp.0485-0489, B. Katalinic (Ed.), Published by DAAAM International, ISBN 978-3-902734-29-7, ISSN 1726-9679, Vienna, Austria DOI: 10.2507/31st.daaam.proceedings.068

DOI: 10.2507/32nd.daaam.proceedings.005

DESIGN OF A ROBOT APPLICATION WITH REGARD TO ENERGY EFFICIENCY

Kemajl Stuja, Branko Katalinic, Mohamed Aburaia & Arbnor Pajaziti

This Publication has to be referred as: Stuja, K[emajl]; Katalinic, B[ranko]; Aburaia, M[ohamed] & Pajaziti, A[rbnor] (2021). Design of a Robot Application with Regard to Energy Efficiency, Proceedings of the 32nd DAAAM International Symposium, pp.0033-0039, B. Katalinic (Ed.), Published by DAAAM International, ISBN 978-3-902734-33-4, ISSN 1726-9679, Vienna, Austria
DOI: 10.2507/32nd.daaam.proceedings.005

Abstract

Design of robotics applications based on energy efficiency is not a new idea or new requirement of industry. Furthermore, the new circumstances of the global warming of the earth's climate system oblige the robotic application designers and system integrators to improve their systems about energy efficiency. Especially robotic systems for palletizing applications, when the Robot moves the parts very quickly, should be improved for the energy efficiency. Inopportunely the implementation of this requirement during design phase often happens partially. In this regard, the designers see the lack of appropriate tools as the cause of this problem as well as missing of physical components during this early project stage. In this context, this works highlights the energy efficiency improvement during design stage using simulation methods/tools for virtual robots. Even in this early stage of the design process, these tools support energy evaluation of different robot moving patterns for different layout concepts. To demonstrate this, a robotic pick-and-place (palletising) application for compost bags is used as a case study. The results of this investigation (energy diagrams) for each concept and scenario are graphically presented and in details discussed. Further research will focus on developing tools for automatic energy check of robotic trajectories within the software user interface.

Keywords: Energy efficiency; robot simulation; robot applications; palletizing; design methods.

1. Introduction

To remain competitive, robot manufacturer continuously improves their products for power consumptions. Both the electronics and the mechanical system are optimized for an area of application, working range and for an array of payloads. The idea for energy-efficient robotic systems has as well been widely used in the academic world [1]. In this comprehensive publication, almost 100 papers from 1993 to 2017 on the topic of energy consumption have been reviewed and discussed in detail. Two works [2] and [3] are familiar to our publication. Against this, our research is concerned with the design methodology for robotic cells, as well as the design of the flow of materials within the robotic cell, rather than the design of the robot / manipulator itself. The reason is simple: it is not possible for robot manufacturers to produce step less robot sizes or to offer customer-specific solutions as in our investigation [4] but instead of that, they should provide appropriate software tools to improve planning/integration of the robot applications. The design of the robot cell and scheduling of material flow has great potential for energy savings. In this context, it is very important selection of the correct manipulator (structure, payload and working range), design of end-effector and periphery components. Amalgamate of all components gives the total energy consumption, which must be improved. Regarding to that, dynamic applications such as palletizing of heavy workpieces are particularly affected, as the total mass moved (including the robot mass) and the workpiece throughput strongly influence the energy consumption. Well-known robot manufacturers ABB, KUKA, Fanuc etc. already offer software simulation tools, which abridge planning, designing, and optimizing of robot systems for energy efficiency. RobotStudio [5] is software tool of ABB and aids programming of robot paths in 3D Environment respectively checking the energy consumptions of robot movements/programmes. This wide-ranging task should be carried out by system integrators or customers. To support this large group of engineers, this paper briefly explains the design methodology for designing palletizing robot applications based on energy efficiency.

2. Problem Statement- Materials and Methods

ISO Standard 50001 [6] defines the requirements respectively energy performance only for an entire company. As there is no standard defining explicitly the energy performance of robotic applications, often not all options and possibilities are exhausted in the design of the robotic cell, which can have an impact on energy consumption. Especially small companies because of the lack of corresponding designing departments attempts to find one "quick" solution, without evaluating possible better alternatives in term of energy efficiency. To express the importance of energy performance evaluation a palletizing application for compost bags will be used. This repetitive dynamic robot task can be provided during the design stage of robot cell. The design process and their stages are fully explained in [7]. Safety and Security relationship was discussed at [8] and will be implemented during the design process. For simulation during design process a very powerful software of ABB (RobotStudio) will be used.

2.1. Requirements of robot application

The Requirements is the first and the most important task during the design process. In this stage the system integrators should establish the list of customer requirements, initial situations and to define the outgoing functions of the robot applications that must be fulfilled. In our case-study the requirements of the customer are shown below in the table 1. The table in this work is very short list of customer requirements for the automatizing of bag-application. In the column "demands" are the requirements, which must be fulfilled. In the column of the "wishes" are the features "nice to have", which can be considered, if the engineering costs allows. Work piece is earth composition for flowers. The bag has big volume with small mass (vacuum suction gripper can be used for palletizing/de-palletizing too).

Requirements			**Demands**	**Wishes**
Technical Req.				
	Workpiece	Weight	Bags 12,5kg	Bags 25kg
		Content	Compost for flowers	
		Dimension	500mm x 300mm x 120 mm	
		Flow Rate	<3,5 Second / Bag	3
	Pallets	Form	Euro	
		Layer pattern	See the figure 4	Different patterns
		Loading/Unloading	Manual by operator	
	Machine	Material Workflow	Discontinuously	Continuously
Safety Req.				
	Mechanical		National guidelines/ ISO	
	Electrical		National guidelines/ ISO	
Secure Req.	**Software**	Authorization	OPC UA	
	Hardware			
Cost Req.			<200 000€	

Table 1. Short list of customer requirements for palletizing composite bags.

2.2. Conceptual design – searching for solutions

The conceptual design stage [7] is a process for generating of ideas, which can be implemented to the system being designed. The entire system can be imaged like black box with inputs/outputs of material, energy and information's (required/wished by customer). Complex systems/tasks will be divided into subsystems/small tasks in order to simplify the design process and to generate many solutions or concepts as possible. The sub-solutions can be placed into the Zwicky "morphological box". Putting the ideas into "morphological box" stimulates creativity and finding of new solutions. After combining the possible solutions, the concept candidates are selected for evaluation [7]. However, the selection process requires a lot of experience and should consider all customer requirements, compliance with security guidelines and technical/economic requirements. Table 2 shows the possible solution candidates.

		Sub-Solutions for the Sub-Functions			
	Sub-Functions of Robot cell	**Sub-Solution 1**	**Sub-Solution 2**	**....**	**Sub-Solution n**
Conceptual design for a robotic palletizing	**Handling**	6 axis Robot	4 axis Robot	....	Cartesian Robot
	Picking	claw gripper	vacuum gripper single	....	Vacuum gripper double
	Parts Moving	conveyor	AGV	....	Mobile Robot
	Part checking	presence bag sensors conveyor	Presence sensor gripper	....	Vision systems
	Safety components	Safety Gates	Light Grids	....	Optical monitoring
	....	....	....	.....	....
	Authorization	OPC UA	Windows embedded	....	Linux
	Solution Candidate (SC) - Variants	SC1	SC2		SCn

Table 2. Simplified and shortened Zwicky - "Morphological box" for creating of solutions -variants [7].

For palletizing of bags, engineers use traditionally the claw grippers. This sub-solution has more technical/economic advantages differing to vacuum suction gripper. For heavy bags the claw-gripper is necessary solution. But thoughts of energy efficiency push the engineers to choose other solutions like a modular vacuum suction gripper, which in contrast to claw-gripper allows maybe gripping of more work pieces at once. If the bags are generally lightweight and offer large permeable suction surfaces a modular vacuum suction gripper can be used for de-palletizing.

On the other hand, palletizing is very high dynamic application. Robot moves continuously the parts very fast from one pick-position to the place-position. Therefore, energy efficiency of the system depends on moving mass. For that reason, it is very significant to find a solution with "lean moving components" and reducing number of movement paths. Thus, idea for using of double suction gripper for picking/placing of two parts at once, reduces for more than 40% the energy consumption of the entire system. Like in [7] the best solution will be selected considering all requirements of the customer list and will be prepared for the next design stage – preliminary design.

2.3. Preliminary design

In this stage of the design the chosen concept from 1.2 will be drafted and digitally modelled for simulation purposes. Here and now is the final moment for risk/vulnerability analysis. As discussed in [7] this analysis affects the costs and technical features of the entire system. On the other hand, different work piece flow scenarios or different grippers cause different safety concepts. In the previous example, the vacuum suction gripper was compared to the claw gripper ([5] models used from RobotStudio library). Using the claw gripper and both filled pallets will be exchanged by the operator, safety concept shown in Fig. 2 can be used. On the other hand, depending on the working environment, this concept cannot be used if the vacuum suction gripper (Fig. 1. b.) has been chosen for the final layout. Should the necessary pneumatic pressure be lost during the rapid robot movement to the "placing"-station, the released bags can fly unhindered through the light grids. This means that the safety concept as in Fig. 2 is not sufficient for the vacuum gripper and must be modified by the safety gates. In this case, replacing the pallets costs more time.

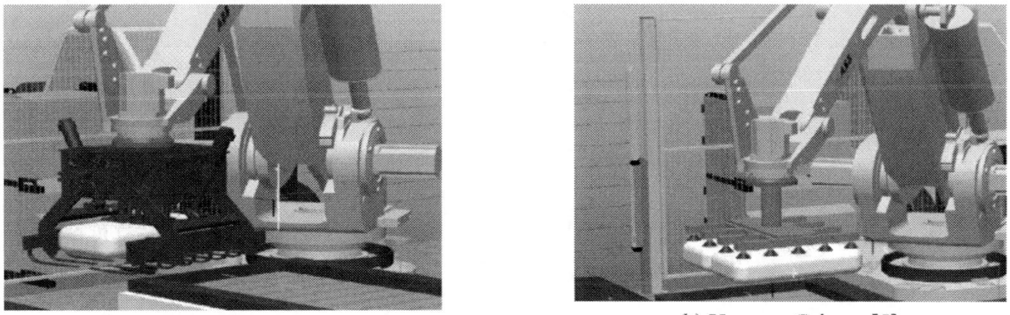

a) Claw gripper [5]   b) Vacuum Gripper [5]

Fig. 1. Possible end-effectors for composite bags.

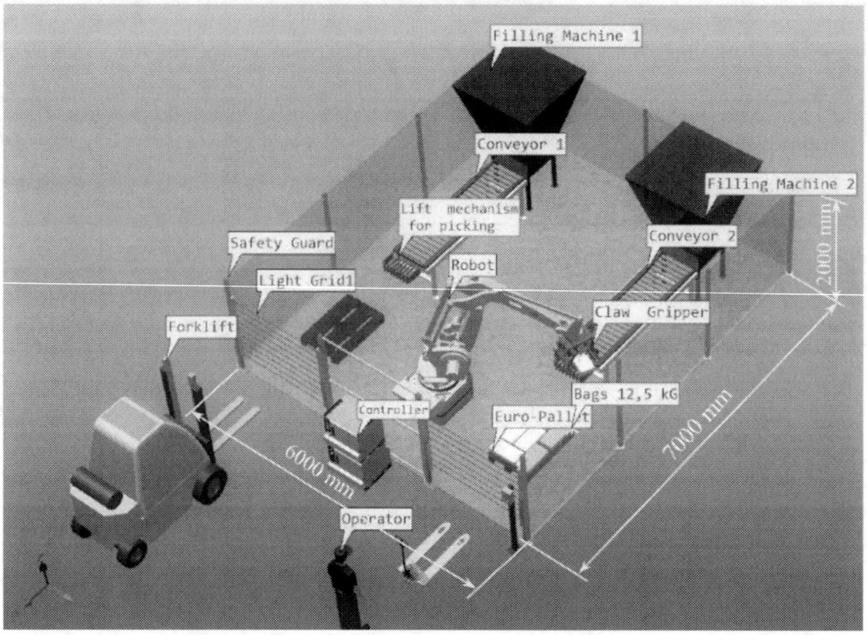

Fig. 2. Layout and simple safety concept for robot cell using claw gripper and discontinuously work piece flow.

2.4. Final design-layout

The last task of the system-design is the final design, where the preliminary design should be optimised and all documents, digital models, diagrams, manuals, etc. should be created. In our case study, the simulation layout serves as an output only to answer the customer's requests. Figures 2 and 3 show two similar layouts. The similar part of the layout consists of two filling machines and linear conveyors up to the robot. Depending on the gripper used, an additional lifting/release mechanism is needed to pick up/grip the bags. The first layout (Fig. 2) was designed for the case, when the claw gripper is used. The second layout (Fig. 3) consists of a vacuum suction pad that allows handling of two bags at once. However, there is less safety if the bag falls out from the suction pad. This means that the robotic cell must be enclosed (safety guard) to achieve the required safety level. In both cases, the entire robot cell must be stopped, to exchange pallets. This step can be automated with the exit conveyors for pallets, but this was not what the customer required.

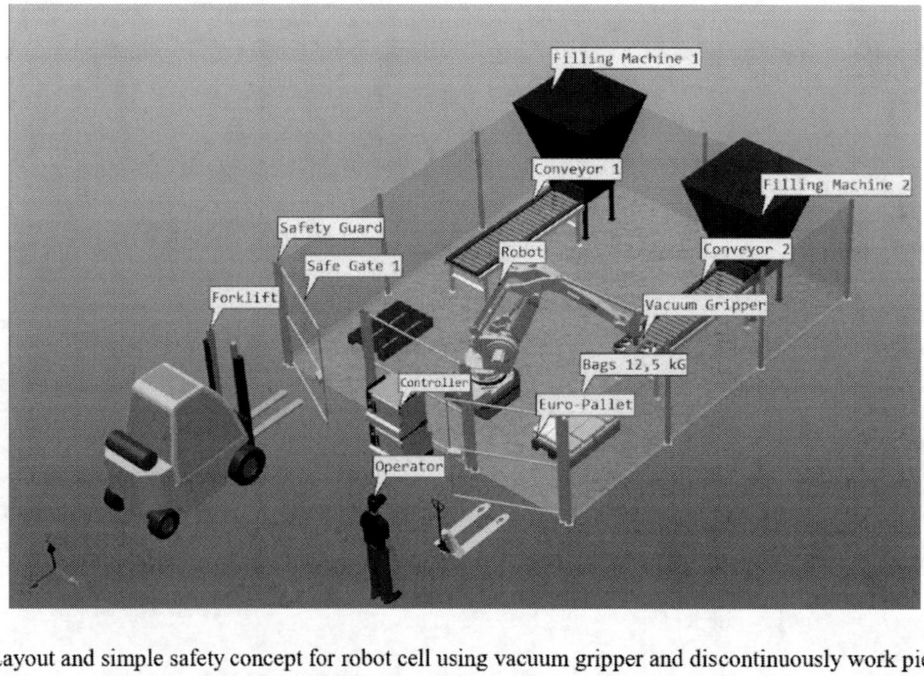

Fig. 3. Layout and simple safety concept for robot cell using vacuum gripper and discontinuously work piece flow.

3. Design of experiments, results and discussion

For the preliminary and final design the simulation results are very important. These results can be used to evaluate the variants and work piece flow scenarios for the preliminary design as well as to optimize the design – final design. Since there are many robot movement parameters that need to be considered, the experimental design can be very extensive. The major moving parameters of robot (for example ABB industrial Robot) are:

- Moving Type (Linear -MoveL, Point to Point – MoveJ etc.)
- Moving velocity and acceleration (TCP - Tool Center Point relative to world or user coordinate system)
- Approach parameters (flying by point or stop point)
- Waiting times (for external events, for vacuum gripping, etc.).
- Robot configurations (Axis positions - where are more than one possible configuration for reaching the target).

Table 3 and Figure 4 show the result of the variant with vacuum suction gripper for a single machine. That means when the first filling machine is working the second will be prepared for the production. In addition, the simulation of the claw gripper is obsolete, since with this end effector only one bag is handled and thus the energy consumption comparing to previous case can be approximately doubled. Table 3 is divided to three columns. The first column shows the scenarios S from 1 to 16. The second column are input parameters (variables). The third column represents the major results (output parameters). The flow rate of the work pieces is constant 1 Bag/3 seconds for single machine and was used from the customer requirements (table 1), respectively acts as time constraint for the simulation. For parallel working machines and continuously work piece flow, robot must move alternately from one machine to another. This solution would be inefficient. The second column is moving type between pick and place stations. Linear movement is not efficient (S16). Third column is tool-center-point (TCP) velocity relative to work-object-coordinate system. Fourth column is approach parameter (flying by point) for "air" targets. That means by z100, robot-TCP 100 mm before reaching the programmed target, moves toward next programmed target (flying around the point to truncate the distance respectively to spare brakes, to reduce energy consumption's, etc.). This parameter can be applied for the "air" targets between stations. Pick and Place targets remains to "fine" zone. They are stop targets. The last two columns are two output parameters: cycle time including steady state and waiting times. The last column (from left) shows the energy consumption for different input parameters. Scenario 1 to 5 (S1 to S5) take too much time respectively they are inefficient since fine zone (stop points) are used for all targets. Scenarios 6 and 11 are disqualified, then they take too much time and cycle time required by customer cannot be achieved. Scenario 7 and 12 is possible with risk tendency: queues (conveyors) to become overflowed. Scenario 8 or 13 can be used for the final programming. By remaining scenarios robot must wait for work pieces and thus consumes more energy.

Scenario	Input Parameters				Output Parameters	
(S)	Flow Rate [Bag / s]	Moving Type	TCP Velocity [mm/s]	TCP Zone [mm]	Cycle Time including Waiting [s]	Energy Consumption (Joule)
S1	3	PTP*	1000	fine	Not possible	
S2	3	PTP	2000	fine	02:24	98209,2
S3	3	PTP	3000	fine	02:07	128484,2
S4	3	PTP	4000	fine	02:07	149044,5
S5	3	PTP	5000	fine	02:07	159723,2
S6	3	PTP	1000	z100**	Not possible	
S7	3	PTP	2000	z100	02:12	93439,3
S8	3	PTP	3000	z100	01:57	122288,0
S9	3	PTP	4000	z100	01:57	142517,5
S10	3	PTP	5000	z100	01:57	151792,5
S11	3	PTP	1000	z200**	Not possible	
S12	3	PTP	2000	z200	02:05	93407,24
S13	3	PTP	3000	z200	01:57	122167,8
S14	3	PTP	4000	z200	01:57	142412,0
S15	3	PTP	5000	z200	01:57	151799,2
S16	3	Linear	5000	z200	01:57	188989,2

*) Robot moves between two stations (long distances) with MoveJ (PTP-Point to point) and flying by. End-positions "Pick" or "Place" are linear. **) End-positions "Pick" or "Place" remains with "fine" zone.

Table 3. Results of energy consumption for Layout (vacuum gripper) moving Robot with different parameters.

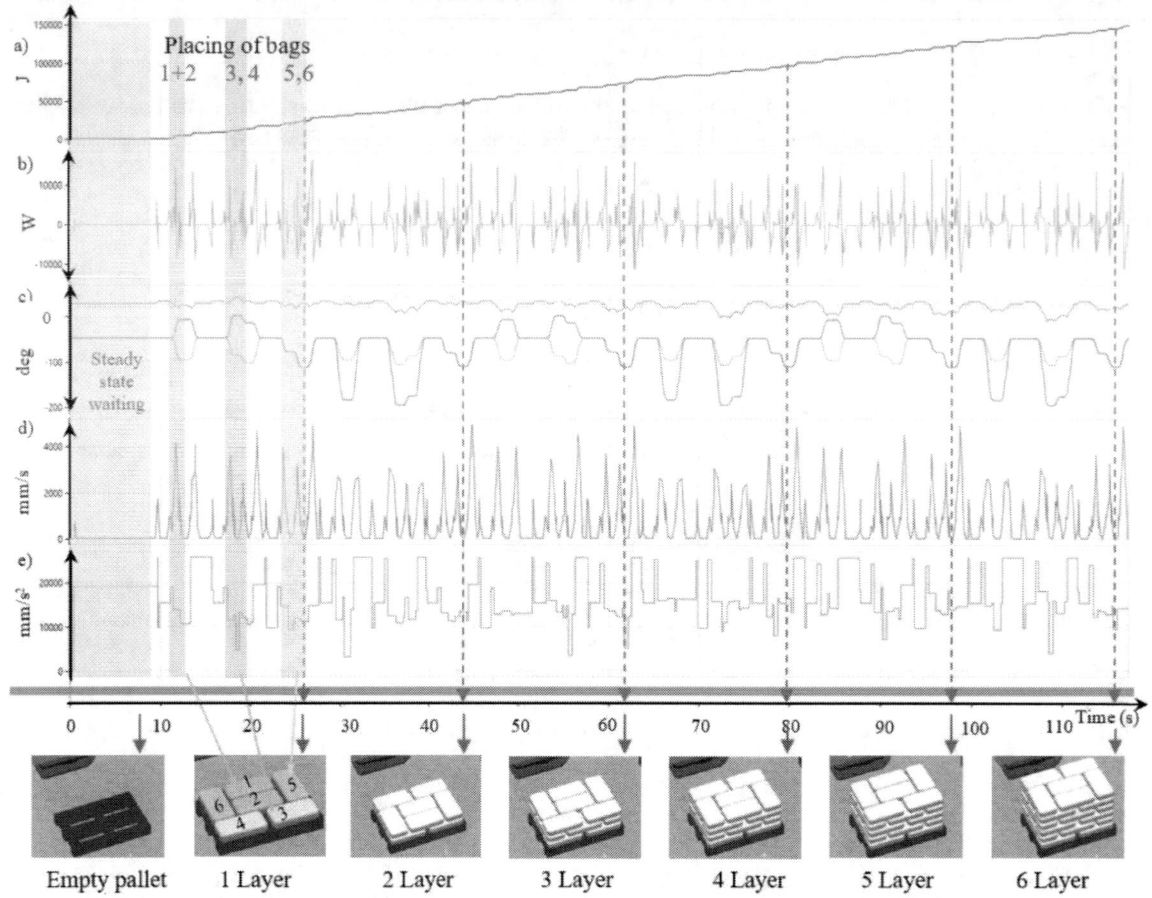

Fig. 4. Energy consumption of virtual robot for scenario S10-Table 3 and palletizing patterns.

Fig. 4 shows the energy consumption for scenario S10 (see Table 3 for more information). For each scenario there is a different graph. Abscise shows the time needed to palletize 6 layers. There are 36 bags in total = 6 layers x 6 bags. The even patterns 1, 3, 5 or the odd patterns 2, 4, 6 are the same depending on the customer's request. Grey pillar in diagram shows steady state – waiting time of Robot for arriving of first two bags (can be omitted). The blue, red and green columns show the time needed to place two bags, with the blue column being narrower as two bags (1 and 2) are placed at the same time. Furthermore, Fig. 4 also shows the following: Fig. 4 a) shows the cumulative energy consumption of virtual robot during the time (expressed in Joule - J). Fig. 4 b) shows the work needed for different robot motion (expressed in Watt -W). Fig. 4 c) shows the Joint (J) angles during robot motion, where colours mean: ■ J1, ■ J2, ■ J3 and ■ J4 (expressed in degrees). The second and third Joint moves very smoothly. The first and fourth Joints are more dynamically during the time. Fig. 4 d) shows the velocity as well Fig. 4 e) shows the acceleration of tool centre point (TCP) relating to work object coordinate system. To find the optimal velocity for the cycle time this diagram is very important. In diagram are many speed profiles like trapezoid, triangle and truncated triangle. Trapezoid speed profile means that the distance between distance AD (Fig. 5 a) programmed speed can be reached. Triangle profile (Fig. 5 b) shows that the programmed speed can be reached only at the point B. Truncated triangle profile (Fig. 5 b yellow shape) shows that the programmed speed cannot be reached, since distance AD is too small. Furthermore, this programming style, would mechanically strains the structure of manipulator and should therefore be avoided.

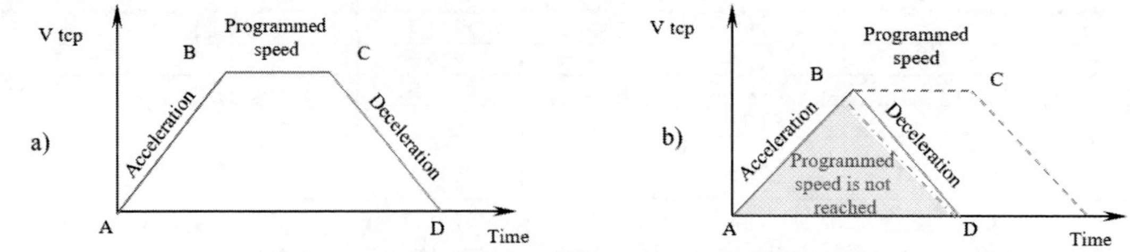

Fig. 5. Velocity (speed) profiles of the Robot TCP: a) Trapezoid, b) Triangle resp. truncated triangle (yellow shape).

4. Summary, Conclusion and outlook

Design of robot applications is a comprehensive engineering work. To find and work out an optimal solution for these kinds of applications, interdisciplinary knowledge is necessary. Especially for the design of highly dynamic material flow applications, the software tools are necessary. A methodical approach is very significant for finding a creative as well as the optimal solutions, in term of energy efficiency. In this work, a combination of two methodological approaches is used: VDI-2221 for mechanical design and VDI-2206 for mechatronic design. It has been shown (headings 1.1 to 1.4) that the use of these approaches makes efficient and shortens engineering work. Moreover, these methods allow to find and evaluate creative solutions (claw- and vacuum -gripper). On the other hand, the simulation tool supported the design of various robot cell layouts with different material flow scenarios using virtual models at a very early stage of the project. By simulating of 16 scenarios in very shortly time, this simulation tool has proven its worth. Different concepts for the work piece flow in the robotic cell result in different energy profiles.

According to the customer conditions, a concept of two parallel filling machines with robot in the centre was chosen. Switching on the machines one after the other when the pallet is full is the best variant in terms of energy consumption by discontinuously work piece flow. Rearrangements of machines and conveyor belts to reduce / optimize the throughput time of the work pieces (bags) are not desired by the customer and for that reason are not investigated. Among others, the simulation results have shown the rocky relationship between safety and the design process. For Example: depending on used gripper (claw or vacuum suction) the safety concept will be different, causing possibly additional cost. Therefore, during the designing process is very important the consideration of safety/security as well as energy efficiency of robot application. The last part significantly affects the final design, base cost for the robot cell and running cost. According to the simulation results, it was confirmed once again that the shortest robot paths (linear paths) are not less energy consuming (Scenario S16, Table 3). Hence, linear movement was used only between offset and final targets (Pick /Place targets). Using Point-to-Point movement type and variation approach parameters (fly by point) was possible to reduce energy consumption rep. to reduce the total cycle time (Scenario S6 to S15, Table 3). Another advantage of simulation was the option for analysing of speed profiles (shown in fig 4). It has been shown that the cause of the after-swinging of mechanical structure (manipulator) is the truncated triangle profile. Although this phenomenon cannot be observed in the animation, the diagram clearly shows that the programmed speed of the TCP cannot be reached. That means, for short distances, the high speeds are not useful and should be avoided.

Further research should focus on add-ins for the simulation software to automatically analyse robot trajectories/trajectories. Highlighting robot trajectories with different colours (e.g. green for trajectories with lower energy consumption) directly in the simulation software user interface would help designers and reduce development time.

5. Acknowledgments

We would like to thank the company ABB Robotics in Austria for providing the software and for equipping the laboratory at the Vienna University of Applied Sciences with robots. We would like to express our special thanks of gratitude to our sponsor MA23 Vienna and to "TÜV –Austria Gruppe" for the technical support.

6. References

[1] Carabin, G; Wehrle, E. & Vidoni, R. (2017). A Review on Energy-Saving Optimization Methods for Robotic and Automatic Systems, Robotics 2017, 6, 39; doi:10.3390/robotics6040039 pp. 1-21. www.mdpi.com/journal/robotics.

[2] Paes, K; Dewulf, W.; Vander Elst, K.; Kellens, K. & Slaets P. (2014). Energy Efficient Trajectories for an Industrial ABB Robot. 21st CIRP Conference on Life Cycle Engineering, Volume 15, pp105-110 ISSN 2212-8271, DOI: 10.1016/j.procir.2014.06.043.

[3] Liu, A; Liu, H.; Yao, B.; Xu, W. & Yang, M. (2018). Energy consumption modeling of industrial robot based on simulated power data and parameter identification. SAGE Publications, Advances in Mechanical Engineering, Vol. 10(5), pp 1–11 https://doi.org/10.1177/1687814018773852.

[4] Stuja K., Bruqi, M., Markl E. & Aburaia, M. (2016). Lightweight 4 -Axis Scara Robot for Education and Research, Proceedings of the 27th DAAAM International Symposium, pp. 0102-0108, B. Katalinic (Ed.), Published by DAAAM International, ISBN 978-3-902734-08-2, ISSN 1726-9679, Vienna, Austria.

[5] https://new.abb.com/products/robotics/robotstudio. ABB AG. Accessed on: 2021-08-21.

[6] ISO 50001(2018). Energy management systems - Requirements with guidance for use (ISO 50001:2018).

[7] Stuja, K.; Poszvek, G.; Walter, W. & Markl, E. (2018). Integrated Method for The Design and Evaluation of Safety & Secure Manufacturing Systems, Proceedings of the 29th DAAAM International Symposium, pp.157-163, B. Katalinic (Ed.), Published by DAAAM International Vienna, ISBN 978-3-902734-20-4, ISSN 1726-9679, Vienna, Austria DOI: 10.2507/29th.daaam.proceedings.022.

[8] TÜV Austria Gruppe (2018). Sicherheit in der Mensch-Roboter-Kollaboration, Fraunhofer Austria Research GmbH, pp. 23-25, White Paper, No.3, March 2018.

DOI: 10.2507/32nd.daaam.proceedings.006

MONITORING OF INDOOR AIR QUALITY IN CLASSROOM

Halima Hadziahmetovic, Nurin Zecevic & Amina Neimarlija

This Publication has to be referred as: Hadziahmetovic, H[alima]; Zecevic, N[urin] & Neimarlija, A[mina] (2021). Monitoring of Indoor Air Quality in Classroom, Proceedings of the 32nd DAAAM International Symposium, pp.0040-0047, B. Katalinic (Ed.), Published by DAAAM International, ISBN 978-3-902734-33-4, ISSN 1726-9679, Vienna, Austria

DOI: 10.2507/32nd.daaam.proceedings.006

Abstract

Measurement of Indoor Air Quality parameters (IAQ), as CO_2 concentration, temperature, and relative humidity, was performed in one classroom at the Faculty of Mechanical Engineering in Sarajevo, with the aim to assess ventilation efficiency. In 2011, several energy efficiency measurements as part of the pilot project, were implemented on the building of the Faculty of Mechanical Engineering in Sarajevo, such as thermal insulation of the building and replacement of existing boiler, but without adequate system for ventilation and cooling. Faculty still continues to use natural ventilation. With implementation of mentioned energy efficiency measures, savings on the heating energy were achieved for about 43 %, but at the same time due to the lack of appropriate ventilation system, there is a significant impact on indoor air quality, especially during the heating season. Beside achieved airtightness of the building, IAQ is affected by other factors, such as the volume of the room, number of users and ventilation frequency. Measurement has lasted for ten days and results were compared with the standard EN16798-1:2019. The measured values showed that in the classroom, during the winter period, the average indoor air temperature was 21 °C, relative humidity 35 % and measured CO_2 concentrations were up to almost 3 times higher than the recommended by the standard.

Keywords: building; indoor air quality; thermal comfort; measurement; CO_2 concentration.

1. Introduction

Buildings of educational institutions represent a significant factor in every social community, characterized by their specific role and requirements. From the aspect of energy consumption in buildings, they belong to the sector of public buildings, which is more heterogeneous when compared to residential, differing in structure, number of users and energy consumption. The importance of their role is reflected in the fact that none institution has as much influence on young people as educational institutions, where they gain their knowledge and findings, and where young people spend most of their time at the stage of life when there is rapid development of their intellectual, social and emotional capabilities [1].

High expectations of these institutions should not imply only well-educated employees and fulfillment of their primary educational objectives, but also indoor environmental quality (IEQ) has great importance for all users [1]. However, numerous studies and energy audits provide a different picture, indicating that in most cases there is an imbalance between meeting needs for energy and achieving the IEQ requirements. This is reflected in a several cases.

One of the reasons is that many school buildings were built in the 1960s and 1970s to solve two problems: the high number of "baby boomer" children coming into the public school system, and the increasing cost of energy needed to heat or cool old school buildings. Therefore, many of the schools built during those years used "modern" energy-saving features like windows that could not be opened. While "closing the envelope" reduced the cost to heat and cool the buildings, ironically, it resulted in "sealed" classrooms that trapped poor air quality, contaminants, and drove up CO_2 levels past acceptable limits [2]. On the other side, recent and emerging global standards in buildings dictate tight building envelopes, purposely limiting air infiltration with the aim to achieve energy savings and reduction of carbon footprint. These measures also have the aim to reduce inside air pollutants, which is contradictory in the situation when implementation of energy efficiency measures deteriorates health and productivity of users [3].

Unfortunately, studies have shown that many universities, classrooms, and indoor learning centers, do not have the proper ventilation rates, to begin with [2]. Most state educational institutions because of limited budget, often put the need for improvement of the facility on the second place or implement partial energy efficiency measures, which later leads to higher costs for upgrading the system. Usually it refers to the fact that most of the students still have classes in buildings where the ventilation is only ensured by natural ventilation [2].

The main causes of indoor air pollution are insufficient and inadequate ventilation frequency, emissions from cleaning products, and the chemicals emitted by building materials or furnishings [4]. Existing ventilation standards are usually minimum recommendations and optimal air quality mostly cannot be achieved in that manner, especially in the case when there is a high occupancy [5]. The most common way of expressing adequate ventilation rate is through CO_2 concentration. It is directly linked to the room characteristics, the number of persons present, outdoor air quality, and ventilation rate [5]. However, a wide study of schools located in a Mediterranean area with natural ventilation system showed that the measured CO_2 concentration exceeded the 1000 ppm even when the windows were open, with CO_2 concentration falling below 1000 ppm in only 28% of case studies with open windows studies [4]. In a recent study it was found that as the CO_2 concentration in the classroom increases, the performance test results decrease, and it can lead to the lower attendance levels. The cause is not directly the CO_2 concentration, but it can serve as an indicator of a bad IAQ which can cause more illnesses between the pupils [4]. Children are particularly sensitive to air pollution, because they inhale a large volume of air relative to their body weight and consequently the concentration of pollutants can result in a greater load on the children's body than in adults [6]. This age group is more susceptible to the influence of Sick Building Syndrome [7]. The impact of unhealthy environment and poor air quality in schools is reflected through child-illness such as asthma and allergies, respiratory diseases (fever, flue, legionellosis and sinus infections), neurological problems (headache, difficulties in concentration, of productivity, uninteresting in learning, poorer test results) and increased absences and treatment costs [6], [8]. Some studies show that ill-advised energy conservation measures can reduce children's performance of schoolwork by as much as 30%. This is a significant problem since students due to their age can be more susceptible to long-term health damage caused by poor IAQ in school buildings [4]. The ideal indoor environment is one with lots of outside, fresh air replacing the stale, contaminated air inside. The available research related to the improved student performance with increased classroom ventilation rates has shown that 8 out of 11 studies reported statistically significant improvements in at least some measures of student performance with increased ventilation rates or lower carbon dioxide concentrations [2].

Especially, the importance of adequate ventilation and IAQ came to the fore in a recent year, when humanity has been confronted with pandemic and COVID-19. Numerous studies today are doing research on how to reduce the spread of COVID virus in classrooms. For example, SARSCoV- 2 can also be transmitted through aerosols, particles smaller than droplets that remain in the air over time and distance, especially in a poorly ventilated room, with the potential to lead to super spreading events. Current evidence shows that it takes about 4 min for the number of small droplets in the air to be halved with no ventilation; whereas with only mechanical ventilation turned on in a room, the number of respiratory particles is halved in 1.4 min. In a room that also has a door and window open, the number is halved after 30 s; substantially faster than in poorly ventilated and unventilated rooms. An important approach to lowering the concentrations of indoor air pollutants or contaminants, including any viruses that may be in the air, is to increase ventilation. Increasing air flow by ventilation, whether by window and door opening or mechanical systems – which could mix outdoor air with indoor air or air filtration and cleaning devices, reduces the risk of infection by diluting concentrations of respiratory particles and removing them in the ambient air [17]. Educational buildings should be designed adhering to the principle of sustainability and taking into account that measurements for energy savings should not impact on conditions in the classrooms and negatively affect the productivity of students, as this would mean that the implementation of energy efficiency measures has a counterproductive effect [1].Simply said, energy declaration without declaration related to the inside environmental quality does not make sense [3].

2. Recommended values of IAQ parameters

Regulatory restrictions on CO_2 concentration in buildings have been established to assess the indoor air (IAQ). Ventilation and continuous air renewal in the building should ensure that the introduced air should, as far as possible, be free from any impurity that may affect the health of users; occupants should not be exposed to uncomfortable draughts and it is recommended, where necessary, to provide a mechanical ventilation systems as an addition to natural ventilation, if it is not possible to ensure a sufficient number of air changes naturally with regard to climatic conditions and purpose of the room [1], [9].

The concentration limits for CO_2 can be described according to European Standard EN 16 798-1: 2019 [16], corresponding to three indoor air quality categories (Table 1). EN 16 798-1: 2019 replaces EN 15 251 and it focusses on parameters at category I, II, III level – sometimes also level IV (level I is best, III / IV is worst) for thermal environment, indoor air quality, lighting and acoustics [10]. Note that this guideline is not designed to limit the amount of CO_2, but rather to indicate that a proper level of clean air (15-20 CFM/person) is being distributed in a classroom [2].

According to ASHRAE, the recommended CO_2 level in buildings should be no more than 700 parts per million (ppm) above outdoor air. Since outdoor air is approximately 400 ppm, indoor CO_2 levels should be no more than 1100 ppm. Building Bulletin 101, 2016, recommends for naturally ventilated educational buildings, that the CO_2 concentration should be less than 1500 ppm [9].

	1st category	**2nd category**	**3rd category**
CO_2 concentration (above outdoor concentration) (ppm)	550	800	1350
CO_2 concentration (absolute values; outdoor CO_2 concentration 480 ppm) (ppm)	1030	1280	1830
Temperature (winter) (°C)	21-23	20-24	19-25
Temperature (summer) (°C)	23,5-25,5	23-26	22-27

Table 1. CO_2 concentration and temperature in schools according to EN 16798-1: 2019 [16]

The microclimate in educational buildings needs to maintain a thermal balance between students, ensuring the thermal comfort of the environment, which would lead to increased attention and concentration during the learning period [11]. It can be achieved with air temperature, relative humidity and air flow, whose values at a specific level are called "comfort zone". Relative humidity is determined based on temperature values. ASHRAE in a separate document (ASHRAE Standard 62), recommends that the relative humidity should be maintained between 30 % and 60 % [12], and according to [13] relative humidity in educational buildings should be between 45-55%.

Digital equipment for carbon dioxide measurement is a relatively economical and efficient way to determine carbon dioxide concentration as indicator of IAQ and efficiency of ventilation system and it can detect the problem before users are aware of it [14], [15]. With long-term carbon dioxide monitoring in combination with measurement of air temperature and relative humidity, building management can have better insight in supporting decision making about ventilation control and improving HVAC system, which can lead to greater energy savings, improved overall air quality and raising social awareness, necessary for successful achievement of high environmental standards [5].

3. Aim and methodology of research

In 2011, at the Faculty of Mechanical Engineering Sarajevo (MEF) was implemented a pilot project of energy efficiency, which included the rehabilitation of facade, through thermal insulation, replacement of old windows and upgrading the heating system with replacement of an existing boiler. In the first year after the implementation of energy efficiency measures, about 43 % of annual energy savings were achieved. After one month of monitoring the average indoor air temperature in the building, an increase of temperature was recorded from 19 °C to 21 °C. However, the Faculty did not have mechanical ventilation at all, except of split systems in some rooms and it mainly relied on the natural ventilation [2]. Due to the partial implemented measures of energy efficiency, not including thermal insulation of a flat roof and adequate system for ventilation and cooling, in the first years after the implementation of the project, it was noticeable that these measures had impact of on the users' thermal comfort in classrooms and offices and increased cooling needs during the warmer period of the year.

The paper analyses the measurement of IAQ parameters, i.e. CO_2 concentration, temperature and relative humidity, in the classroom 201. This classroom was chosen because of the usable area, as one of the larger classrooms at the Faculty, frequency of the use and orientation. Classroom 201 (Figure 1) is located on the second floor of the Western building of the Mechanical Engineering Faculty. The dimensions of the room are 15,9 x 9,32 m, the classroom area is 148,62 m^2, and the room volume is 498,84 m^3. The orientation of window surfaces of the longer side is to the South and the shorter side to the East. The classroom has sun protection in the form of mechanized panels that can be moved and adjusted as needed. These is no mechanical ventilation, but it relies on natural ventilation. Heating at the MEF is automatically regulated throughout the whole week and starts at 06:00 a.m.

Fig. 1. Classroom 201 (left); location of data logger during measurement (right)

The device used to measure IAQ parameters is OPUS 20-TCO data logger, manufactured by Lufft (Figure 1). It is consisted of three internal sensors that measure values: air temperature (in the range from -20 to +50 °C), relative humidity (in the range 10-95 %) and CO_2 concentration (0-5000 ppm). The device is placed 1m above the floor, at the professor's desk, due to the distance from the entrance door to the classroom, but also duo to security reasons that may affect the accuracy of measurement data, because any movement and touching of data logger affects the reliability of the data. Air should circulate freely within the minimum zone, at a distance of 0,5 m around the sensor. The measurement range is set to measure every 10 seconds, and the value reading is displayed every 30 minutes.

4. Analysis of measured IAQ parameters in the classroom 201

The measurements were performed in the winter semester and consisted of two measurement periods. First measurement period lasted from 27th November to 1st December of 2019, and second measurement period lasted from 1st December to 6th December of 2019. The heating season was chosen for the measurement because this period is characterized by a greater differences between indoor and outdoor temperatures, but also due to the user's behavior in terms of habits and awareness of the necessary frequent ventilation of the room. Usually during the winter period, due to the lower outside temperature values, windows are less open for ventilation because it affects the decrease of the internal temperature and creating discomfort for the users.

The number of students in the classroom 201 in one class ranged in average from 21 to 52, depending on the subject. According to the schedule, one hour of lectures lasts 45 minutes, while the break between classes lasts 15 minutes. During the measurement, the usual behavior of users in the classroom was monitored, i.e. the environment was not controlled. The schedule of classes was not 100 % reliable, because there were deviations between the actual situation and the predicted schedule. Some classes were held based on an agreement between professor and students. In the following subsections, the measured values of IAQ parameters in the classroom 201 are analyzed.

5. Results and discussion

5.1. First measurement period (27th November – 1st December 2019)

Figure 2 shows the measured values of IAQ parameters for the first measurement period in the classroom 201.

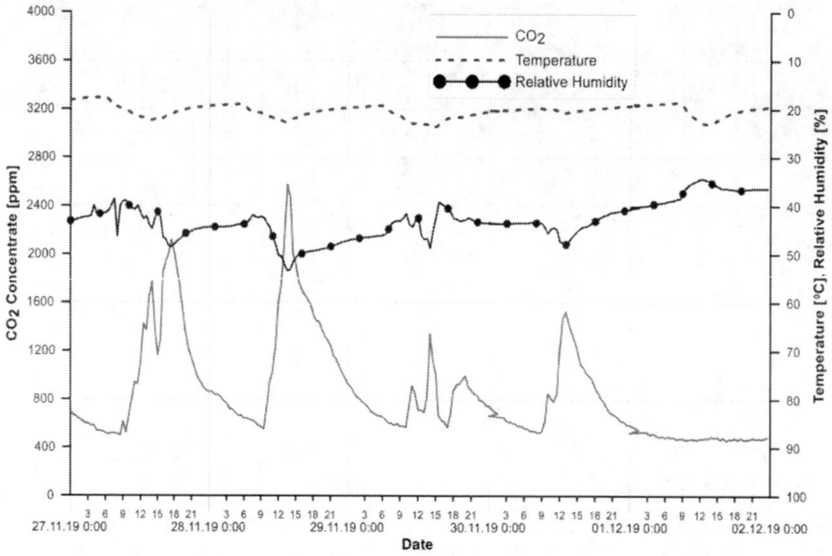

Fig. 2. Measured values of IAQ parameters for first measurement period in the classroom 201 (27th November- 1st December 2019)

With the start of lecture on the first measurement day, at 9:15 a.m., CO_2 concentration was low, 499 ppm. The lectures lasted according to the schedule from 9:15 a.m. to 15:00 p.m., and an additional lecture was held from 15:15 p.m. to 17:30 p.m. During the first lecture (9:15-11:15), which was attended by about 21 students, CO_2 concentration increased from 499 ppm to 942 ppm, at the end of class.

During the second lecture (11:15-15:00) with about 51 students attended, CO_2 concentration ranged from 927 ppm at 11:15 a.m., to 1165 ppm at 15:00 p.m. From the beginning of the lecture, there was a continuous increase in the concentration, with a peak value for that period, 1771 ppm at 14:00 p.m. After 14:00 p.m., the value began to decline, indicating that class has finished earlier, and students left the classroom. At 15:15 p.m., the CO_2 concentration starts to rise again from 1165 ppm to 2118 ppm, measured at 17:30 p.m., which is the maximum concentration of CO_2 for that day. It took about 2 hours and 30 minutes for the CO_2 concentration to drop from the maximum measured value of that day, i.e. from 2118 ppm at 14:30 p.m. to 985 ppm at 22:00 p.m.

CO_2 concentration decreased slightly during the breaks, which is shown in the example when at the beginning of the break at 11:00 the concentration was 942 ppm, and at 11:30 a.m. with the start of class, the measured concentration was 972 ppm. Usually, during the winter period, students rarely go out because the hallways are much colder than the classroom.

On Thursday, 28.11.2019, lectures lasted from 9:15 a.m. to 14:00 p.m., with about 50 students attended. Measured value of CO_2 concentration before the lecture was 552 ppm, and after the first class it was 727 ppm, with the continues growth. Half an hour before the end of classes, the maximum value of 2574 ppm for that day was measured, and after 14:00 p.m. it decreased to 2470 ppm. Average increase of CO_2 concentration every half hour from the beginning of the lecture till the end at 14:00 p.m., was 253 ppm. The total increase of concentration at the end of classes compared to the beginning of the lecture was 2 022 ppm. It took 9 hours for the CO_2 value to drop below 1000 ppm, i.e. by 23:00 p.m. concentration had dropped by 1489 ppm to 981 ppm.

Although scheduled lectures for Friday were supposed to be held in the morning, they were still held from 13:00 p.m. to 15:00 p.m. The average number of students present was 53. At the beginning of the lecture, the measured CO_2 concentration was 695 ppm, while at the end of classes, the concentration was 1011 ppm, which is the maximum value for that day. On Saturday, the classroom was used for additional classes, in the period from 11:00 a.m. to 14:00 p.m., which can be seen on Figure 2 through increase in CO_2 concentration. Concentration of CO_2 in that period ranged from 728 ppm at 11:00 a.m. to 1353 ppm at 14:00 p.m. After the class, it took about 3 hours to CO_2 concentration decrease below 1 000 ppm. On Sunday, when it is non-working day, the concentration ranged from 458 ppm to 585 ppm.

On Friday, the highest temperature during the classes was measured for the first measurement period of 23,8 °C, with a gradual decrease after the end of classes. The average temperature during classes was 21,9 °C, with a minimum value of 20,5 °C and a maximal 23,5 °C. During the weekend, an increase of air temperature was observed with the heating turned on, when temperature rose from 18,8 °C at 06:00 a.m. to 22,8 °C at 14:00 p.m. After 14:00 p.m., when the heating was switched off, the temperature values gradually decreased.

Average humidity during classes was around 45 %, with a minimum of 43 % and a maximum of 53,4 %. The highest relative humidity of 54 % for the first measurement period, was measured on Thursday at the time when the highest CO_2 concentration of 2574 ppm was measured. On the other days, the highest values of relative humidity were measured at the time when the highest concentration of CO_2 was measured.

5.2. Second measurement period (2nd December – 6th December 2019)

Figure 3 shows the measured values of IAQ parameters for the second measurement period in the classroom 201. On Monday, 02.12.2019, lecture started at 11:00 a.m. with a total number of 57 of present students. CO_2 concentration measured at the beginning of lecture was 595 ppm, and at 13:30 p.m., it reached maximum weekly value of 3543 ppm, which is almost three times higher than recommended standard. In half an hour, by about 14:00 p.m., the value decreased by 290 ppm, and by 15:00 p.m. total CO_2 drop was by 2495 ppm, i.e. the value dropped to 1048 ppm. After 15:00 p.m., CO_2 values were below 1000 ppm.

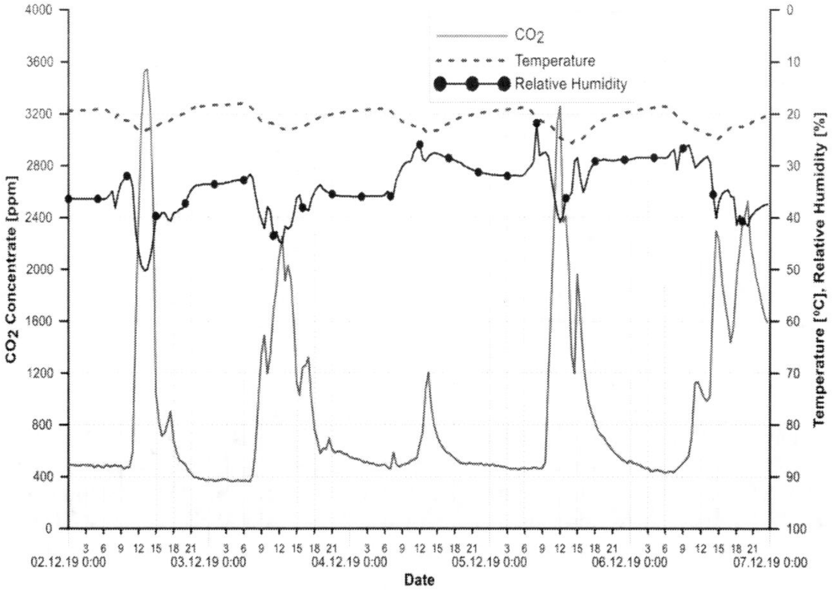

Fig. 3. Measured values of IAQ parameters for first measurement period in the classroom 201 (2nd December-6th December 2019)

On Tuesday, CO_2 concentration during the first lecture (8:15-14:00) ranged from 612- 2252 ppm, and during the second held lecture (15:00-17:00), it ranged from 1027-1322 ppm. During the break between these two classes, the value dropped by 793 ppm, i.e. from 1923 to 1130 ppm. After the end of classes at 17:00 p.m., it took about 30 minutes for the CO_2 value to drop by 325 ppm to below 1000 ppm (997 ppm). On Wednesday, average CO_2 concentration ranged from 652-1 207 ppm.

Second maximum CO_2 value for the second measurement period, was measured on Thursday, 05.12.2019. According to the schedule for the classroom 201, lecture was held in the period from 9:15 a.m. to 14:00 p.m., attended by 50 students. CO_2 concentration measured at 9:00 a.m. was 466 ppm. At 12:00 a.m., it reached a value of 3256 ppm, which is also the maximum daily value of CO_2. The lecture lasted until 13:00 p.m. and at the end of lectures, measured CO_2 concentration has reached the value of 2410 ppm. It took 4 hours from maximum measured value, i.e. after 17:00 p.m. for the concentration to decrease by about 1449 ppm, to a value below 1000 ppm (961 ppm).

On Friday, a characteristic increase in CO_2 concentration was measured throughout the day, as shown on Figure 3. The first increase of concentration was observed from 10:00 a.m., with the value of 561 ppm, and within the hour, the concentration doubled, to 1124 ppm. From 12:00 a.m., the value decreases from 1021 to 985 ppm at 13:00 p.m. From 13:00 p.m., the value doubled again, i.e. at 14:30 p.m., the value of 2 298 ppm was measured. By 18:30 p.m., CO_2 concentration decreased by 370 ppm to 1928 ppm, and after 18:30 p.m., there was again an increase of 601 ppm. At 20:00 p.m. the value of 2529 ppm was measured, and at 23:30 p.m. the concentration of CO_2 was 1593 ppm, which is a higher value for that period of time. These values are an indicator of how important is to ventilate the space after the end of classes because higher values will continue to be present until the beginning of the new working week.

The average temperature during classes was 23 °C, with a minimum of 19,8 °C and a maximal 25 °C. Temperature values followed the increase of CO_2 concentration during the classes. After the end of classes, with the decrease of CO_2 concentrations, the room temperature also decreased. On Thursday when the maximum CO_2 concentration of 3 256 ppm was measured, the air temperature was 24,7 °C and it continued to rise by one degree till the end of class, after which there was decrease along with the decrease of CO_2. Average humidity during classes was around 36 %, with a minimum of 26 % and a maximum of 50,3 %, which is the highest value measured on Monday, at the time when the highest CO_2 concentration of 3251 ppm was measured during the second measurement period.
Table 2 shows the overall measured values of IAQ parameters during the whole day and during the lectures, for the first and second measurement period in the classroom 201.

Measured period	I measurement period		II measurement period	
Parameters	All day	Lectures	All day	Lectures
Maximum CO_2 concentration (ppm)	2574	2574	3543	3543
Average CO_2 concentration (ppm)	853	1106	891	1596
Average air temperature (°C)	20	21,5	20,9	23
Average relative humidity (%)	43	44,1	33,9	35,8

Table 2. Measured IAQ values for the first and second measurement period in the classroom 201

Based on the measurement results, it can be concluded that the maximum values of CO_2 exceed 2-3 time more than the recommended values according to the standard EN 16798-1:2019. The mean CO_2 concentration during the day was in the recommended values, while the mean value during the lectures exceeded the recommended value by 36 %. Temperature values during the first measurement period, for the whole day and during the lectures were in the recommended values according to the EN 13790, while during the second measurement period, a slightly higher value was measured by about 2 °C, compared to the recommended 21°C. The mean relative humidity was in the recommended range. On Thursday of the first measurement period, it took about 10 h to decrease the maximum CO_2 concentration of 2 574 ppm, measured at 13:30 p.m., to 981 ppm at 23:30 p.m. While in the second measurement period, on Monday it took about 2 hours for the maximum value of 3543 ppm, measured at 13:30 p.m., to decrease to 831 ppm at 15:30 p.m. It is possible to conclude based on the shorter time required to decrease the CO_2 value on Monday of the second measurement period, that the classroom was ventilated by opening the windows after the end of class, thus enabling a faster decrease of CO_2 concentration, which again emphasize that after the end of classes, it is very important to ventilate the classroom, to change indoor stale air and with introduced fresh air reduce the concentration of carbon dioxide, lower the temperature, and humidify the existing air. By ventilation of the room, the rate of decrease of CO_2 concentration increased almost by 3 times. Measured high values of CO_2 concentration during the lectures indicate that the high values are not only caused by airtightness of the building, but also there is insufficient ventilation of the spaces and the user's attitude towards this issue.

6. Conclusion

The paper analyzed the measurements of IAQ parameters in the classroom 201, at Faculty of Mechanical Engineering in Sarajevo, to estimate ventilation efficiency from the aspect of measured CO_2 concentration in the chosen classroom. In introduction part, the importance and impact of the indoor environmental parameters on the health and well-being of users are described, as well as the recommended values according to the different standards. Standard EN 16798-1:2019, that replaced EN 15251, was used to assess the air quality in the classroom. Experimental research was performed in the classroom 201, where the parameters of IAQ (CO_2 concentration, temperature, and relative humidity) were measured. The results of measurement showed the presence of higher CO_2 concentrations during the heating season, where individual values exceeded the recommended value by up to two or three times. Since the most common way of expressing adequate ventilation rate is through CO_2 concentration, high measured CO_2 concentration indicates low or inefficient ventilation, which is often the case in the buildings that are naturally ventilated. Some of the factors that affect the higher values of CO_2 are airtightness of the buildings, which is the case at the Faculty of Mechanical Engineering Sarajevo where energy efficiency measures have been implemented, but without adequate ventilation system. In addition, the higher concentration is also affected by the room spatial characteristics, the class schedule, number of users, and the frequency of ventilation of the room. The choice of the most appropriate ventilation system for improvement of IAQ in the classrooms depends on many factors, and very often there are spatial limitations or financial restrictions, especially if it is a case of a State institutions and limited budget. Accordingly, it is recommended to apply, as much as possible, simple measures as passive ventilation (night and early morning), more frequent ventilation of the classroom before, during and after the end of class, installation of CO_2 concentration sensor as recommendation for schools because it indicates high concentration and the need for ventilation. If energy efficiency measures do not include the measure of providing adequate ventilation system, users should partly take an obligation to maintain a quality indoor environment through more frequent ventilation. This is where the importance of user's awareness about the indoor air quality and their role in achieving and maintaining it comes to the fore, as the factor that should not be neglected. This model of behavior should not only be applied to, in this case, educational institutions, but also to the wider community and households.

After analyzing the research of the impact of implemented energy efficiency project on the building of Faculty, through insulation of the building but without adequate ventilation system, future research plans will be concentrated on two areas. First one refers to the measurement of IAQ parameters in the building of primary school, since many research have shown that younger children are more susceptible to the influence of poor indoor air.

The frequency of ventilation in classes with younger children compared to students, depends on the teacher's attitude, and because children are not capable and aware to such an extent that it is necessary to ventilate the space. Two primary schools will be selected, a school building which is thermal insulated and a school that has not been renovated since the construction period. The aim is to compare measurement results in dependence of the structural characteristics of the building. The second area relates to the analysis of IAQ measurements in the classroom when air-purifier is being used. Since the air in Sarajevo is more polluted every year, most institutions use indoor air purifiers. The IAQ parameters will be measured and analyzed for the period when the air purifier is not used in the classroom and for the period when it is used, in order to compare measurement IAQ results.

7. References

[1] Zečević, N. (2019). Development of a model for determining the optimal combination of energy efficiency measures and renewable energy technologies on the buildings of higher education institutions in Bosnia and Herzegovina, Ph.D. Dissertation, University of Sarajevo, Faculty of Mechanical Engineering Sarajevo, Bosnia and Herzegovina.

[2] CO2 meter. Indoor Air Quality in Classrooms and COVID19. (2021). Available from: http:// www.co2meter.com/blogs/news/7334762-indoor-air-quality-in-the-classroom, Accessed: 2021-06-10.

[3] Olesen, W.B. (2012). Revision of EN 15251: Indoor Environmental criteria, International Centre for Indoor Environment and Energy, REHVA Journal, Department of Civil Engineering, Technical University of Denmark. Available from: https://www.rehva.eu/rehva-journal/chapter/revision-of-en-15251-indoor-environmental-criteria, Accessed: 2021-06-15.

[4] Zemitis, J., Bogdanovics, R, Bogdanovica, S. (2021). The study od CO2 concentration in the classroom during the COVID-19 safety measures, E3S Web of Conferences 246, 01004, Cold Climate HVAC & Energy 2021. Available from: https://www.e3s- conferences.org/articles/e3sconf/pdf/2021/22/e3sconf_hvac2021_01004.pdf, Accessed: 2021-06-24, DOI: 10.1051/e3sconf/202124601004.

[5] Zecevic, N., Husika, A., Dzaferovic, E. (2018). Impact of energy efficiency measures on indoor air quality in building of Mechanical Engineering Faculty Sarajevo, Proceedings of the 29th DAAAM International Symposium, pp. 0197-0201, B. Katalinic (Ed.), Published by DAAAM International, ISBN 978-3-902734-20-4, ISSN 1726-9679, Vienna, Austria. DOI: 10.2507/29th.daaam.proceedings.028

[6] Mendell, M.J., Heath, G.A. (2005). Do Indoor pollutants and thermal conditions in schools influence student performance? A critical review of the literature, Indoor Air 2005, pp. 27-52. DOI:10.1111/j.1600-0668.2004.00320.x

[7] Shendell, D.G., Pril, R., Fisk, R., Apte, M.G, Blake, D., Faulkner, D. (2004). Association between classroom CO2 concentrations and student attendance, Indoor Air 2004; 14 (5): 333-41.

[8] Kats, G. (2006). Greening America's Schools: Costs and benefits, A Capital E. Reports, Available from: https://www.usgbc.org/resources/greening-america039s-schools-costs-and-benefits, Accessed: 2021-06-24.

[9] Building Bulletin 101. (2016). Guidelines on ventilation, thermal comfort and indoor air quality in schools, Department for education. Information Policy Team, The National Archives, Kew, London, TW9 4DU. Available from:https://consult.education.gov.uk/capital/bb101-school-design-iaq-comfort-and-ventilation/supporting_documents/DfE%20Ventilation%20guide%20consultation%20draft%2029%2006%202016.pdf, Accessed: 2021-06-26.

[10] Bogdanovica, S., Zemitis, J., Bogdanovics, R. (2020). Effects of CO2 concentration on children's well-being during the process of learning, Energies 2020, 13, 6099, Available from: https://www.mdpi.com/1996-1073/13/22/6099/pdf, Accessed: 2021-06-28.

[11] United States Environmental Protection Agency (EPA). (2020). Why Indoor Air Quality is important to Schools. Available from: https://www.epa.gov/iaq-schools/why-indoor-air-quality-important-schools,Accessed: 2021-06-28.

[12] School Service Health and Safety. (2010). Code of practice and guidelines for schools, nurseries and children centres on thermal comfort, Ealing Education Service.

[13] Auf-Franić, H., Oluić, V., Žarnić, T., Bertina, M., Korlaet, L., Rister, V., Roth-Čerina, M. (2003). Osnovne škole – programiranje, planiranje i projektovanje, Arhitektonski fakultet Sveučilišta u Zagrebu.

[14] Lowitz, G. (2015). Managing Carbon-Dioxide Risk: What You Should Know, ONSET HOBO, Available from: https://www.microdaq.com/downloads/dl/file/id/917/product/3090/managing_ carbon_dioxide_risk.pdf , Accessed: 2021-07-20.

[15] AirTest, (n.d). CO2 Control in School Classrooms. Available from: http://www.airtest.com /support/reference/CO2&SchoolClassrooms.pdf , Accessed: 2021-07-05.

[16] EN 16 798-1:2019. (2019). Energy performance of buildings- Ventilation for buildings- Part 1: Indoor environmental input parameters for design and addressing indoor air quality, thermal environment, lighting and acoustics- Module M1-6.

[17] Asanati, K., Voden, L., Majeed, A. (2021). Healthier schools during the COVID-19 pandemic: ventilation, testing and vaccination, Journal of the Royal Society of Medicine; 2021, Vol. 114 (4), pp.160-163. DOI:10.1177/0141076821992449

DOI: 10.2507/32nd.daaam.proceedings.007

THE CONCEPT OF CLOUD OF THINGS FOR OVERCOMING THE SHORTCOMINGS OF EXISTING TECHNOLOGIES

Kresimir Rakic & Zeljko Seremet

This Publication has to be referred as: Rakic, K[resimir] & Seremet, Z[eljko] (2021). The Concept of Cloud of Things for Overcoming the Shortcomings of Existing Technologies, Proceedings of the 32nd DAAAM International Symposium, pp.0048-0051, B. Katalinic (Ed.), Published by DAAAM International, ISBN 978-3-902734-33-4, ISSN 1726-9679, Vienna, Austria
DOI: 10.2507/32nd.daaam.proceedings.007

Abstract

This paper describes and explains the need to integrate state-of-the-art computing technologies, the Internet of Things and cloud computing. The synergy of these technologies, the Cloud of Things, represents a new aspect of the use of everyday objects connected to the Internet and with each other and gives a new dimension to the possibilities of building smart resources.

Keywords: Information Society; Internet of Things; Cloud Computing; Cloud of Things.

1. Introduction

The birth of the personal computer in the 1970s, the graphical user interface in the 1980s, the World Wide Web in the 1990s and today the Internet of Things have launched our global society into the so-called information age. For the vast majority of people, digital change happens outside of their conscious perception. They are faced with the ever-changing requirements and needs posed to them. They need digital identification cards to communicate with government institutions; they feel the need to own a smartphone or maintain a page on a social network to stay in touch with their family and friends. At the same time, the technologies and processes behind these changes have become smaller, more complex, and more hidden in our everyday physical environment [1].

The amount of information today, practically every day, exceeds the situation from previous days. This fact is reflected in the explosion of media channels and products, the accelerated informatization of society, the development of education and the growing interest in information-related occupations. Living in the information age, we have become an information society.

The information society is a term for a society in which the creation, distribution, use, dissemination, integration and management of information have become the most important cultural, economic and political activity. The goal of the information society is to achieve a global competitive advantage, using information technology in a creative and useful way [2]. What is implicit in this term, in the way it is more widely used, is the pervasive nature of some digital information technologies, as well as our personal and professional dependence on computers and the Internet. When talking about the information society, people are likely talk of the incredibly fast ways in which digital communication has developed in the recent decades, and on the impact it has had on their lives and habits [3].

In 2020, the amount of data on the Internet hit 40 zettabytes. A zettabyte is about a trillion gigabytes [4]. Most of this data is first captured and created by humans - by typing, pressing a capture key, taking a digital picture, or scanning a barcode. However, people have limited time, attention and accuracy - all of which means that they are not very good at collecting data about things in the real world. People, like their environment, are physical. Our society is based on physical things, not on ideas or information. Ideas and information are important, but things are more important.

Yet today's information technology depends so much on data coming from humans that our computers know more about ideas than things. If we could train computers to know about things, based on data they collected independently, directly from those things, without our help, we could track and count everything, greatly reducing waste, losses and costs. We would know when to replace, repair, or recall things and, in general, manage them better and more efficiently [5]. But, in fact, we can. The solution to this problem is the Internet of Things.

2. Concept of Internet of Things

The basic concept of the Internet of Things (IoT) is the ability to upgrade everyday objects with identification, sensory, network and process capabilities that will enable them to communicate with each other, as well as communicate with other devices and services over the Internet, to achieve its purpose [6]. These objects are usually called "smart objects" because they require minimal human intervention to generate and exchange data, and they have capabilities for data collection, analysis and management [7].

The IoT represents the evolution of the use of existing technologies: Radio Frequency Identification (RFID) technologies, sensor networks, device communication, etc., in terms of the number and type of devices, as well as connecting the network of these devices via the Internet [8]. Over the past decade, the IoT has garnered significant attention from both academia and industry. The main reasons behind this interest are the opportunities that the IoT will offer. The IoT describes a world in which all objects, i.e., smart objects, are connected to the Internet and communicate with each other with minimal human intervention. The ultimate goal is to create a "better world for all people", in which the objects around us will know what we love, what we want, and what we need, and will act accordingly without explicit instructions. In its periodic report, Cisco predicts that in the next 10 years, the IoT will generate up to 500 billion devices connected to the Internet and to each other. This development will contribute to higher productivity, create new business models and generate new revenue streams [9]. Areas of application of IoT technologies are numerous and diverse, and IoT solutions are expanding every day and today find their place in almost all areas of human activity [10][11].

The most prominent areas of application include, for example, the smart industry, in which the IoT is a communication channel for the collaboration of smart manufacturing systems, with each other and with people. This application is referred to as Industry 4.0 [12]. In the field of smart homes and buildings, security systems, intelligent thermostats and other IoT technology solutions are receiving a lot of attention. Research in the field of smart energy consumption focuses on smart electricity, gas and water meters. Smart transportation solutions include, among other things, fleet tracking and mobile ticket sales [13]. In the field of smart health, applications on patient surveillance and chronic disease management stand out [14]. In the context of smart city projects, solutions such as monitoring the availability of parking spaces in real-time, intelligent street lighting, etc. are being explored [15].

The number of devices connected to the IoT is rapidly increasing, and with this increase grows the amount of data that needs to be collected, processed and stored. The need for storage space will soon not be able to be met locally because the devices at the IoT end are, as a rule, of the low vein and modest capabilities. The importance of the collected data and their efficient use goes beyond simple processing intending to form information and knowledge but should make them a means of wisdom for the user. Nor can this be accomplished by the limited capabilities of devices on the IoT side. A possible solution is the lease of storage space and processing power that can be achieved with the technologies of cloud computing [16].

3. Cloud Computing

Cloud computing represents the delivery of various services, i.e., tools and applications, via the Internet, including data storage, servers, databases, networking and the necessary software. As an alternative to storing and processing data on a proprietary hard drive or local storage device, cloud storage allows you to save it in a remote database, access that data, and software applications to work with it [17].

Cloud computing is called such because the information it accesses is located remotely, in virtual space, or the cloud. The user does not have to be in a specific place to access and be able to work with the data, which allows him to work remotely. This approach to storing and processing data takes away the demanding work from the device you are carrying or sitting at it and work. All business moves into computer clusters far away in cyberspace. The Internet is becoming a cloud, our data, business and applications are available from any device you can connect to the Internet, anywhere in the world.

The cloud computing system is not a unique technology, but consists of three basic services [18]:

- Software-as-a-Service (SaaS) includes licensing software applications to users. Licenses are usually issued through a pay-as-you-go or on-demand model.
- Infrastructure-as-a-service (IaaS) includes a method for delivering everything from operating systems to servers and storage via IP-based connectivity, as part of an on-demand service. Clients can avoid the need to purchase software or servers and instead acquire these resources in an external on-demand service.
- Platform as a Service (PaaS) is considered to be the most complex of the three layers of cloud computing. PaaS shares some similarities with SaaS, and the primary difference is that instead of delivering software online, it is actually a platform for creating software that is delivered over the Internet.

The IaaS service is an appropriate solution to the problem posed before. Users of this service can obtain storage, processing, networks and other basic computer resources, and use them by deploying and running arbitrary software [19]. By using cloud computing resources and services, even modest devices with limited computing capabilities, such as things in the IoT, become able to perform the complex calculations needed to effectively perform the assigned task. Everything they need is their equipment with the appropriate sensors and actuators, and their decision-making capabilities can be facilitated by the almost endless computing capabilities of the cloud. The synergistic effect of IoT and cloud computing, which emphasizes the positive aspects of these technologies, and by interacting reduces or even cancels their Cloud of Things not only enables the discovery and integration of things related to the Internet, but also their offer as a resource. In doing so, things can not only be discovered and combined, but also provided as a service to meet the agreed requirements of users

4. Cloud of Things

Cloud of Things (CoT) is a high-performance cloud based IoT application platform that allows connecting our IoT enabled devices and remotely monitor, manage and control them. It represents the next step in processing and analysing the data and creating more usefulness from the data generated by IoT by allowing the development of more advanced smart applications.

As an upgrade to basic cloud computing services for building, storing, creating and using data and applications, i.e., SaaS, IaaS and PaaS, different models and forms of CoT services are interconnected by different network environments:

- Sensing-as-a-service,
- Big-Data-Analytics-as-a-Service,
- Database-as-a-service,
- Identity-and-Policy-Management-as-a-service,
- Video-Monitoring-as-a-service,
- Data-as-a-service,
- Sensor-as-a-service,
- etc.

CoT not only enables the discovery and integration of things related to the Internet, but also their offer as a resource. In that way, things can not only be discovered and combined, but also provided as a service to meet the agreed requirements of users. This approach envisions new scenarios and innovative, ubiquitous value-added applications, revealing the world of CoT to every user, customer, and potential service provider, thus enabling an open market of things [21].

5. Conclusion

The Internet of Things, realized at the primitive level of connected things with poor computing capabilities, has brought many possibilities that make our everyday lives smart. It's positioning on a robust, cloud platform, which could serve as an effective backbone for achieving a network of sensors and actuators that can be used to improve the performance of connected devices, would bring added value to these already effective technologies.

Cloud of Things represents an integration of the Cloud platform, Cloud infrastructure and IoT middleware. Cloud platforms are the components responsible for providing IoT with the necessary and current requirements, such as real-time processing, scalable storage and global access, as well as expanding towards other capabilities such as machine learning. The cloud infrastructure provides the storage, networking, and computing resources needed by IoT and Cloud platforms. Finally, IoT middleware provides a layer of abstraction for basic IoT devices, as well as mechanisms for interacting with Cloud Computing. However, given that this is an emerging technology, further work is needed to mitigate the challenges of the practical implementation of IoT in the cloud.

6. References

[1] https://waag.org/en/article/we-live-information-age-what-does-actually-mean, (2013). Zandbergen, D. We live in an Information Age: What does that actually mean?, Accessed on: 2021-10-02

[2] Zadeh, S. K., Veisi, A. G., Zadeh, M. K. (2013). Do We Live in an Information Society? Does It Matter? International Journal of Advanced Research, Vol.1, No. 3, pp. 362-366. ISSN NO 2320-5407.

[3] Ou, M. S. (2016). Do we live in an "information society"?. University of London.

[4] https://healthit.com.au/how-big-is-the-internet-and-how-do-we-measure-it, (2020). How Big is the Internet, and How Do We Measure It, Accessed on: 2021-10-15

[5] Ashton, K. (2009). That "Internet of things" Thing. RFID Journal, Vol. 22, No. 7, pp. 97-114.

[6] Gubbi, J., Buyya, R., Marušić, S., & Palaniswami, M. (2013). Internet of Things (IoT): A Vision, Architectural Elements, and Future Directions. Future Generation Computer Systems, Vol. 29, No. 7, pp. 1645-1660.

[7] Stankovski, S., Ostojic, G., Laslo, T., Stanojevic, M., & Babic, M. (2019). Challenges of IoT Payments in Smart Services, Proceedings of the 30th DAAAM International Symposium, pp.0004-0009, B. Katalinic (Ed.), Published by DAAAM International, ISBN 978-3-902734-22-8, ISSN 1726-9679, Vienna, Austria

[8] Majstorovic, V[lado] & Rakic, K[resimir] (2017). Internet of Things and Social Media: Tools of a Successful Information Organization, Proceedings of the 28th DAAAM International Symposium, pp. 0295-0298, B. Katalinic (Ed.), Published by DAAAM International, ISBN 978-3-902734-11-2, ISSN 1726-9679, Vienna, Austria

[9] https://www.cisco.com/c/r/en/us/internet-of-everything-ioe/internet-of-things-iot/index.html, (2017). Cisco Internet of Things, Accessed on: 2017-08-27

[10] Atzori, L., Iera, A., & Morabito, G. (2010). The Internet of Things: A Survey. Computer Networks, Vol. 54, No. 15, pp. 2787-2805.

[11] Efremov, S., Pilipenko, N., & Voskov, L. (2015). An Integrated Approach to Common Problems in the Internet of Things. Procedia Engineering, 100, 1215-1223.

[12] Shrouf, F., Ordieres, J., & Miragliotta, G. (2014). Smart Factories in Industry 4.0: A Review of the Concept and of Energy Management Approached in Production Based on the Internet of Things Paradigm. Proceedings of the 2014 IEEE International Conference on Industrial Engineering and Engineering Management, pp 697-701. IEEE.

[13] Vermesan, O., et al. (2014) Internet of Things Strategic Research and Innovation Agenda. In: Internet of Things – From Research and Innovation to Market Deployment, Vermesan, O., Friess, P. (Eds), pp. 7–142, River Publishers, Aalborg,

[14] Fleisch, E. (2010). What is the internet of things? An economic perspective. Economics, Management, and Financial Markets, Vol. 5, No. 2, pp. 125-157.

[15] Wortmann, F., & Flüchter, K. (2015). Internet of things. Business & Information Systems Engineering, Vol. 57, No. 3, pp. 221-224.

[16] Aazam, M., Khan, I., Alsaffar, A. A., & Huh, E. N. (2014). Cloud of Things: Integrating Internet of Things and Cloud Computing and the Issues Involved. In Proceedings of 2014 11th International Bhurban Conference on Applied Sciences & Technology (IBCAST) Islamabad, Pakistan, 14th-18th January 2014, pp. 414-419. IEEE.

[17] https://www.investopedia.com/terms/c/cloud-computing.asp, (2020). Frankenfield, J., Cloud Computing. Accessed on: 2021-10-10

[18] Sunyaev, A. (2020). Cloud computing. In Internet computing (pp. 195-236). Springer, Cham, Switzerland

[19] Mell, P., & Grance, T. (2011). The NIST definition of cloud computing, Oracle.

[20] Parwekar, P. (2011). From Internet of Things towards Cloud of Things. In 2011 2nd International Conference on Computer and Communication Technology (ICCCT-2011) (pp. 329-333). IEEE.

[21] Distefano, S., Merlino, G., & Puliafito, A. (2012, July). Enabling the Cloud of Things. In 2012 Sixth International Conference on Innovative mobile and Internet Services in Ubiquitous Computing (pp. 858-863). IEEE.

DOI: 10.2507/32nd.daaam.proceedings.008

SMART TEMPORARY TRAFFIC LIGHTS

Radek Votrubec

This Publication has to be referred as: Votrubec, R[adek] (2021). Smart Temporary Traffic Lights, Proceedings of the 32nd DAAAM International Symposium, pp.0052-0057, B. Katalinic (Ed.), Published by DAAAM International, ISBN 978-3-902734-33-4, ISSN 1726-9679, Vienna, Austria
DOI: 10.2507/32nd.daaam.proceedings.008

Abstract

The aim of this paper is to introduce a control system for a traffic light. In this case, these are not fixed traffic lights at a light-controlled intersection, but temporary traffic signs at an ordinary intersection, for example during road repairs. It is necessary to install four portable traffic lights, two pairs, in this place. These traffic lights cannot be connected with cables. It is therefore proposed to control them using independent microprocessors communicating with each other via wifi. Various control strategies are proposed for this automaton-based system.

Keywords: Traffic lights; Microprocessor; Wifi; Arduino; Internet of Things.

1. Introduction

At our department we created several production systems or models of production system and home automation systems at in the last years. Production systems and their models are suitable for teaching the modern principles of Industry 4.0. [6], [8], [4], [5]. Manufacturing factories increase in product quality like the main competitive criterion today [9], [7]. All our systems and models consist of elements with their own control systems and communicates with the other elements of the systems via Internet of Things. There are several examples like a model of a smart factory with the production of a colourful bracelet [1], a production system of the little plate with name with milling machine [2] or two models of intelligent houses with smart home automation systems [3]. We have now developed a model of temporary portable traffic lights based on demand from engineering network construction and management companies.

The control system used on the model is suitable for commercial use at real portable traffic lights. In this case, these are not fixed traffic lights at a light-controlled intersection, but temporary traffic signs at an ordinary intersection, for example during road repairs, Fig.1. It is necessary to install four portable traffic lights masts, two pairs, in this place. By default, there are only a few traffic lights for a straight section. These traffic lights do not normally communicate with each other. They cannot be connected using cables because of long distance between them. Their synchronization is based on the use of precise time, which is synchronized by a radio signal from Frankfurt am Main. In our solution, four traffic lights cannot be connected using cables. It was therefore proposed to control them using independent microprocessors communicating with each other via wifi. Various control strategies for this automata-based system are proposed for these traffic lights.

Fig. 1. Portable traffic lights

2. Control unit for portable traffic lights

There are several strategies to control the lights and there are two situations. The first is the central control system with one Arduino only, Fig.2. This is an intermediate stage of development. All six colours R1, O1, G1, R2, O2 and G2 are connected with wires to this Arduino.

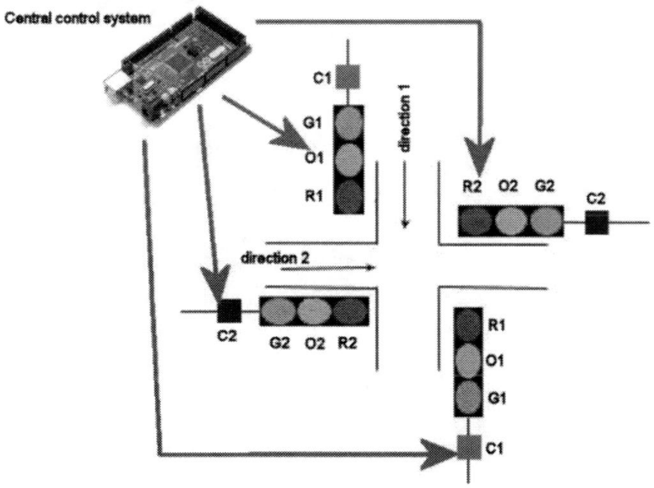

Fig. 2. Central control system

The final solution consists of four alone control systems with four Arduinos, because it is not possible to use wires on the real crossroad. There are four traffic lights masts. First mast is a Master. He does the timing and sends synchronization signal, current state, to other masts. There are two kinds of Slaves. Slave1 is in the same direction like Master and two Slaves2 are in second direction, Fig 3. The photograph of the solution is in Fig.4.

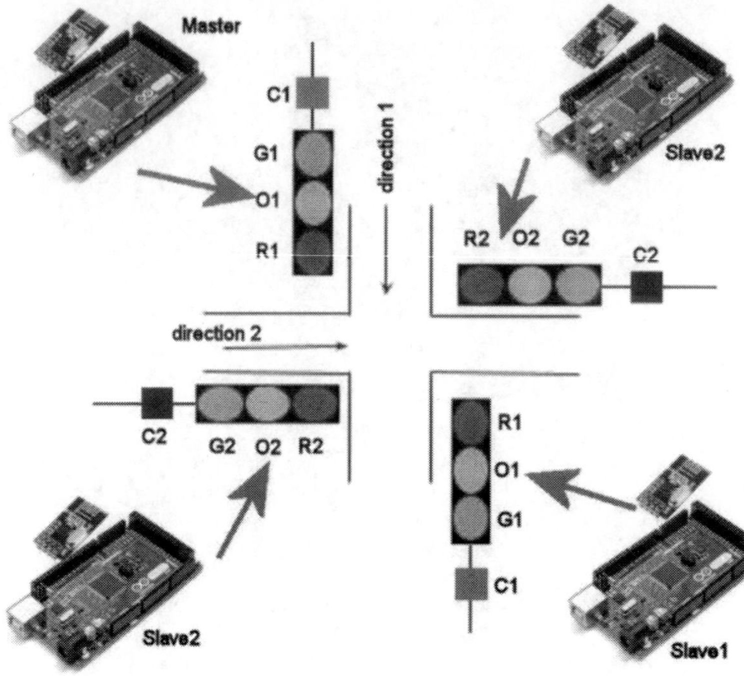

Fig. 3. Independent components IoT

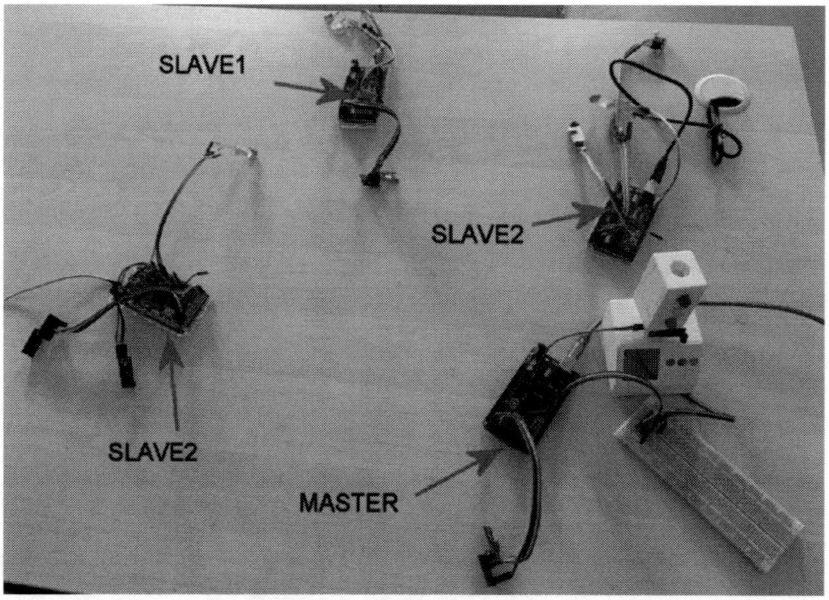

Fig. 4. Four control units

Four strategies are designed:

- Strategy 1 - Traffic lights are alternating with predefined time sequence only. (30s for green colour)
- Strategy 2 – Sensors of cars added. Arriving car changes these 30s to 10s and gets green sooner.
- Strategy 3 – One direction is the main and has green. Second direction has sensor of arriving car. When car arrives it gets green and after a while (30s) the green returns to main direction.
- Strategy 4 – Both direction have the same priority. Green is taken by sensors of arriving cars in both directions.

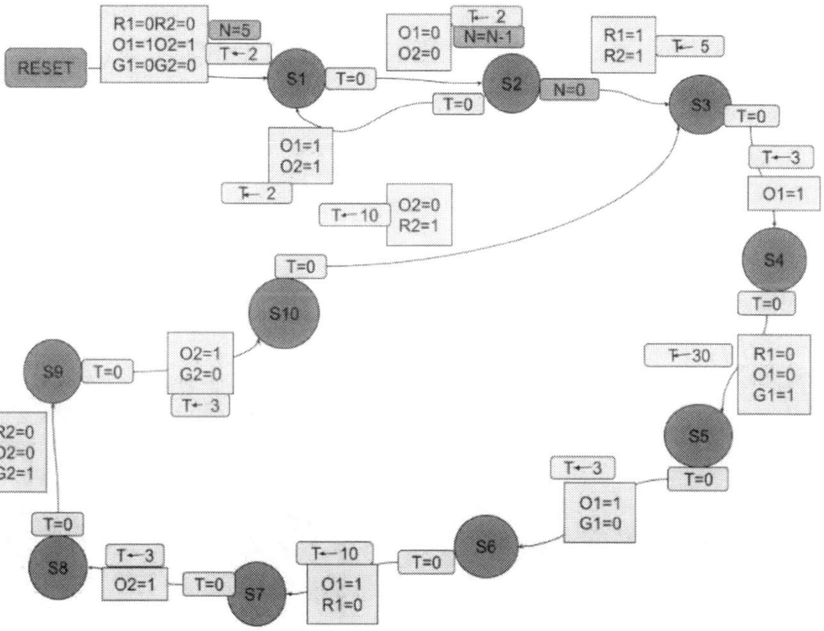

Fig. 5. The automaton of strategy 1 – central control

The control algorithm is based on the automaton graph, Fig.5. The automatons and their components are described in the article [10]. In the Fig.5 there is a strategy 1 automaton for central control unit. It is necessary to split this automaton to three new automata for Master, Slave1 and Slave2 control units of the final solution with four masts. These automatons are in Fig.6, 7. And 8.

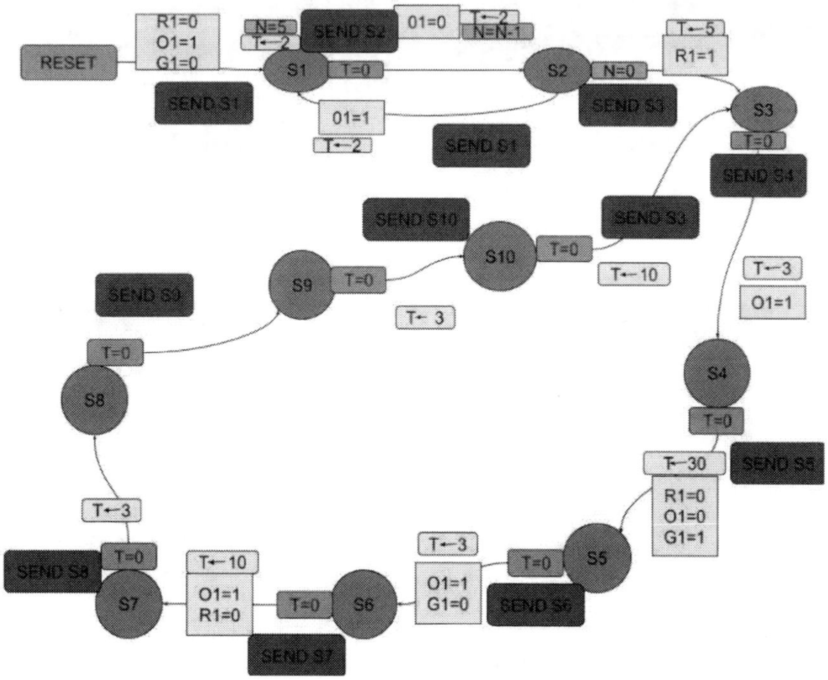

Fig. 6. The automaton of strategy 1 – Master unit

The automaton for Master, Fig.6. is created by modifying the automaton in Fig.5. All action interventions with lights with index 2 are removed and the transmission of a synchronization signal with the current position in the automaton graph is added. The automations of Slave units receive this signals and changes their current position.

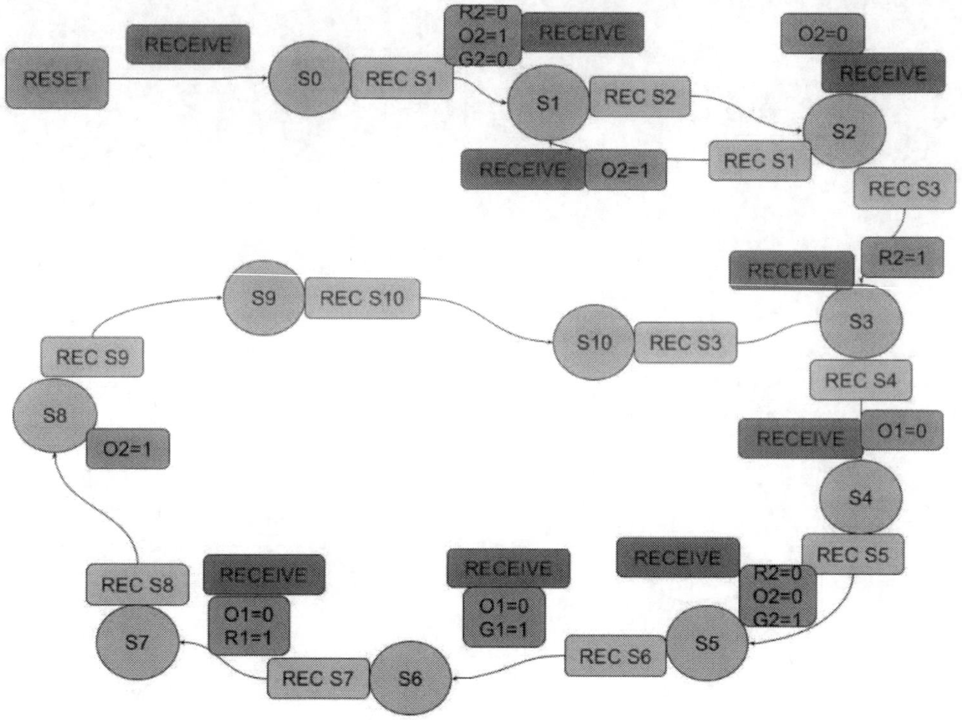

Fig. 7. The automaton of strategy 1 – Slave1 unit

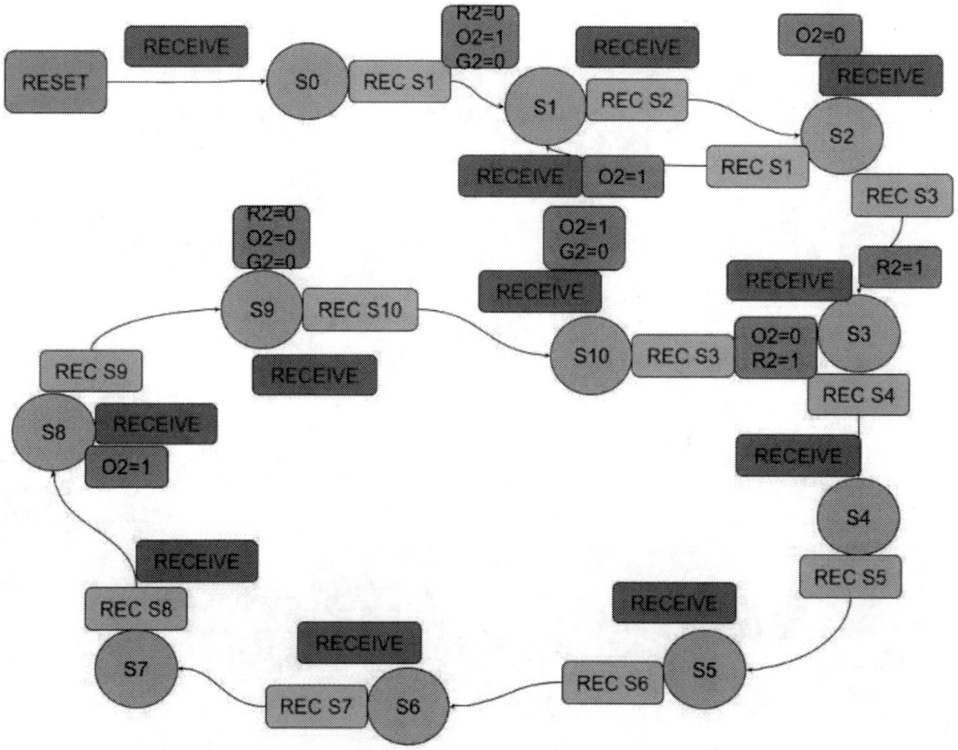

Fig. 8. The automaton of strategy 1 – Slave2 unit

3. Conclusion

The teaching model of the crossroad with traffic lights controlled by one central Arduino or four independent Arduinos was created. This control system is suitable for industrial using at real crossroad. The led diodes on the model are replaced with amplifiers and bigger LED lights. This was tested on real traffic lights. Several strategies were designed. First strategy was timing of lights only, when each direction has green for adjusted time. This strategy was improved using sensors of the car. In this case arriving car can decrease the time to green light – strategy 2. Next two strategies replace adjusted time to green forever of till the car in red direction appears. In this moment it takes green for itself and in strategy 3 it returns to previous direction after adjusted time or in strategy 4 the green stays forever or to next car at this road.

There were two control systems designed and created. One system is one Arduino which controls all lights, all four masts. It is not usable in real situation. Then the second control system was developed with four independent arduinos communication via wifi messages. The education model was created when one sample of the mast was printed on 3D printer. The next steps will be to improve the algorithms with regard to transmission security and to create a cover for the control unit.

4. Acknowledgments

This publication was written at the Technical University of Liberec as part of the project "(21278) – Optimization of manufacturing systems, 3D technologies and automation" with the support of the Specific University Research Grant, as provided by the Ministry of Education, Youth and Sports of the Czech Republic in the year 2019"

5. References

[1] Votrubec, R & Koblasa, F. (2019). Control System of Vehicle for Smart Factory Model with Principles of Industry 4.0, Proceedings of the 30th DAAAM International Symposium, pp.0261-0267, B. Katalinic (Ed.), Published by DAAAM International, ISBN 978-3-902734-22-8, ISSN 1726-9679, Vienna, Austria

[2] Sevic, M. (2017). Design of small CNC machine for the model of production system with principles Industry 4.0. Master Thesis, Department of production systems and automation, Technical University of Liberec, Liberec, Czech Republic.

[3] Koblasa, F.; Votrubec, R.; Lelek, L. & Koudelka, J. (2019). Project oriented teaching in Industry 4.0 – smart home. Industry engineering 2019: International student scientific conference. Pilsen, Czech Republic. ISBN 9788026108948.

[4] Bartodziej, Ch. J., (2016), The concept industry 4.0. New York, NY: Springer Berlin Heidelberg, ISBN 978-3-658-165017.

[5] Thanes, L. & Schaefer, D. (2017). Cybersecurity for Industry 4.0: Analysis for Design and Manufacturing. Switzerland: Springer International Publishing, ISBN 978-3-319-50659-3.

[6] Takakuwa, S.; Veza, I. & Celar, S. (2018). Industry 4.0 in Europe and East Asia, Proceedings of the 29th DAAAM International Symposium, B. Katalinic (Ed.), Published by DAAAM International, ISBN 978-3-902734-20-4, ISSN 1726-9679, Vienna, Austria.

[7] Palcic, I.; Ojstersek, R. & Buchmeister, B. (2018). Creating Value in Manufacturing Companies, Proceedings of the 29th DAAAM International Symposium, B. Katalinic (Ed.), Published by DAAAM International, ISBN 978-3-902734-20-4, ISSN 1726-9679, Vienna, Austria.

[8] Medojevic, M.; Diaz Villar, P.; Cosic, I.; Rikalovic A.; Sremcev, N. & Lazarevic, M. (2018). Energy Management in Industry 4.0 Ecosystem: a Review on Possibilities and Concerns, Proceedings of the 29th DAAAM International Symposium, B. Katalinic (Ed.), Published by DAAAM International, ISBN 978-3-902734-20-4, ISSN 1726-9679, Vienna, Austria.

[9] Ojstersek, R.; Acko, B. & Buchmeister, B. (2020). Simulation study of a flexible manufacturing system regarding sustainability, International Journal of Simulation Modelling, Vol. 19, No. 1, 65-76, doi:10.2507/IJSIMM19-1-502

[10] Votrubec, R. (2020). Communication Layer between Control Units within the Internet of Things, Proceedings of the 31st DAAAM International Symposium, pp.0072-0079, B. Katalinic (Ed.), Published by DAAAM International, ISBN 978-3-902734-29-7, ISSN 1726-9679, Vienna, Austria, DOI: 10.2507/31st.daaam.proceedings.010

DOI: 10.2507/32nd.daaam.proceedings.009

CONTROL UNIT FOR SEAT WITH VARIABLE STIFFNESS

Radek Votrubec, David Cirkl & Michal Sivcak

This Publication has to be referred as: Votrubec, R[adek]; Cirkl, D[avid] & Sivcak, M[ichal] (2021). Control Unit for Seat with Variable Stiffness, Proceedings of the 32nd DAAAM International Symposium, pp.0058-0066, B. Katalinic (Ed.), Published by DAAAM International, ISBN 978-3-902734-33-4, ISSN 1726-9679, Vienna, Austria
DOI: 10.2507/32nd.daaam.proceedings.009

Abstract

This paper describes the control system of car seat with variable stiffness. The pressure in pneumatic element inside the seat is controlled. It improves the comfort of sitting person and it has an influence against the fatigue. The control system is based on microprocessor using the Arduino board. There are several control strategies using two-state and proportional pneumatic valves. Central system with air reservoir or own compressor is used.

Keywords: Vibroisolation; Car Seat; Arduino; Pneumatic System; Active Damping.

1. Introduction

At our Technical University in Liberec, we have been dealing with vibration isolation systems for many years [2], [7]. This issue is very current [3], [4], [5], [7] We are now developing seats with variable seat stiffness, Fig.1. One or more pneumatic elements are inserted into the seat cushion.

Fig. 1. The seat with variable stiffness

The control of the pressure in the pneumatic element affects the stiffness of the seat. At zero pressure, the support is provided by a foam filling [1], [6]. The seat with variable stiffness brings several benefits. Adjusting a softer or harder seat improves the driver's driving comfort. Dynamic changes in stiffness contribute to reducing the feeling of fatigue.

The use of seats with variable stiffness is expected in the areas of cars and trucks, work machines and means of transport, both at the driver's seat and for passengers. Evaluating of comfort is difficult. Parameters such as pressure or temperature measured in the contact zone between the seat and the body can be used to objectively assess the improvement, fig.2.

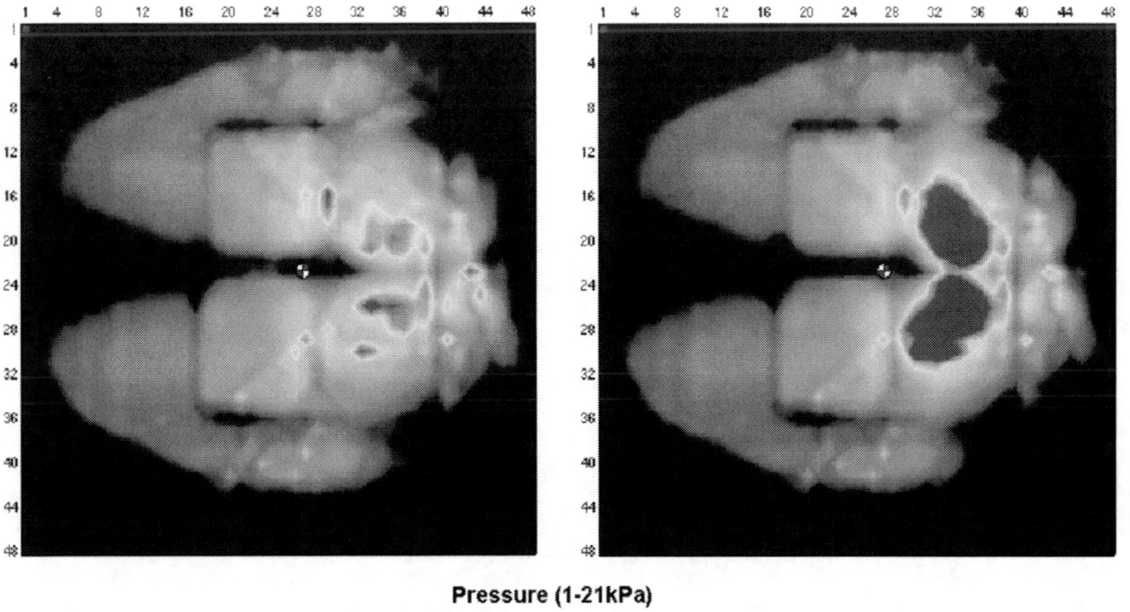

Fig. 2. Pressure measurement in contact zone between the seat and body.

As part of the solution of this project, a requirement arose to create a control unit for this seat with adjustable stiffness. In previous years, the seat was controlled using a measuring card and a control computer with Labview. This variant is only suitable for development purposes. The project has now moved towards commercial use and it was necessary to replace this solution with a separate microprocessor-based control unit. Similar control units have been developed in the past [8], [9]. This task was further divided into two variants.

The first version of the control unit is used for laboratory use. It enables the operation of the seat in all control modes using various types of sensors, valves, etc. It has the possibility of setting a large number of parameters and it is possible to perform laboratory measurements on it, export the measured data for further processing. The second version consists of only the most necessary components and switches only between the two driving modes. It is optimized in terms of size, price, reliability, heat dissipation and electromagnetic compatibility.

2. Variants of the source of compressed air

The pneumatic circuit consists of a compressor C, pneumatic valves Vin, Vout, pressure sensors PS, PR and a pneumatic element of the seat S. Depending on the source of compressed air used, the solution can be divided into three variants. The simplest version is shown in Fig. 3. The source of compressed air here is only a controlled compressor. The compressor can be controlled either two-state, on-off, or continuously using pulse width modulation.

For this connection, it is necessary to use a compressor that seals and maintains pressure when switched off. Or it is possible to connect a non-return valve behind it. Schematically, the control circuit is indicated in blue with the control unit R. In this case, the compressor C and the outlet valve Vout are controlled on the basis of the pressure measured by the sensor PS.

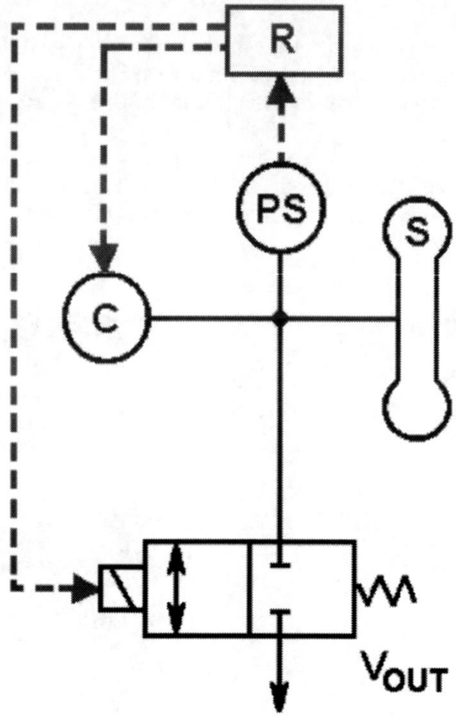

Fig. 3. Pneumatic circuit with own local compressor only

The second variant is shown in Fig.4, where an inlet valve is connected in series to the compressor. The control of this valve is synchronized with the control of the compressor.

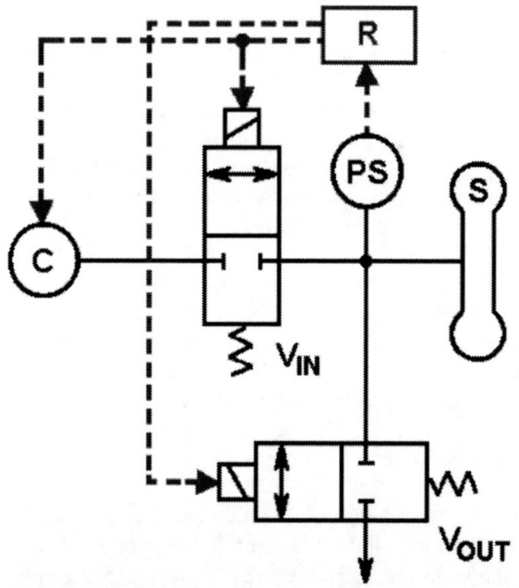

Fig. 4. Pneumatic circuit with own compressor and input valve

The last configuration is shown in Fig. 5. A central compressor C is used as a source of compressed air together with an air reservoir AR. The pressure in this air tank is measured by a PR sensor and is independently controlled by an auxiliary control system R1. The main control circuit then controls the inlet valve Vin if necessary to increase the pressure in the seat element or the valve Vout if necessary to decrease the pressure in the seat.

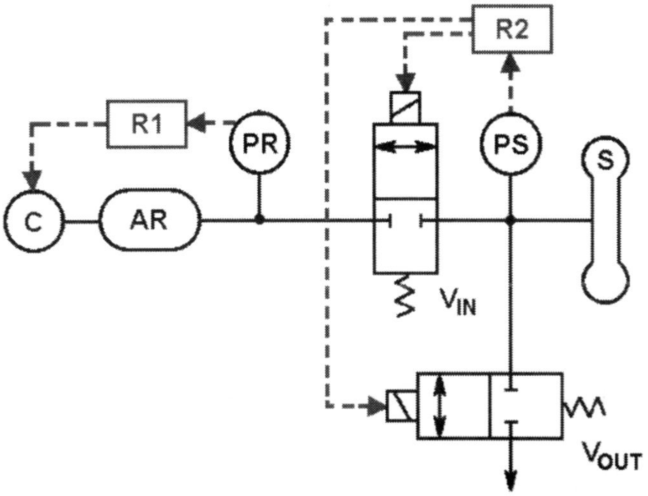

Fig. 5. Pneumatic circuit with independent compressor, air reservoir and auxiliary control circuit

3. Pneumatic valve configuration variants

Figure 6 shows the overall pneumatic circuit with all components used for the laboratory version of the seat. Compressor C is connected to the analog output of the control unit and can be used either as a two-state element, ie with states off - on, or its rotation speed can be continuously controlled by PWM. There is a non-return valve at its outlet. A two-state valve Vin and a proportional valve Vain are connected in parallel at the air supply to the pneumatic element S. A pair of two-state and proportional valves Vout and Vaout are similarly connected in parallel in the discharge pipe. Manually adjustable throttle valves are added in both of these paths. The pressure in element S is measured by a pressure sensor PS. Exhaust air from the pneumatic element can alternatively be realized via the discharge valve Vout1.

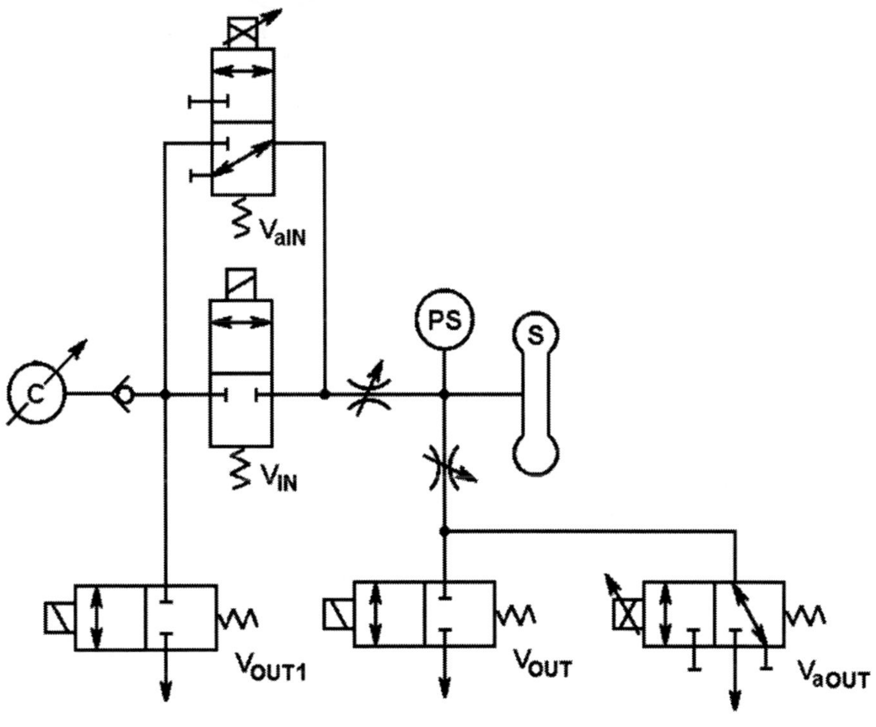

Fig. 6. Pneumatic circuit with all the components

The controls in this pneumatic circuit can all be used simultaneously, reducing the time required to inflate and deflate the element by opening parallel paths. Or part of the members can be discarded and the pneumatic circuit is simplified. Different circuit variants are suitable for different control methods. On the standard seat, it is necessary to take into account the complexity and cost of the overall solution. Of the sixteen variants tested, the most important ones are selected in Fig.7. In variant A) two two-state valves are used and it is one of the simplest variants. For measurement purposes, throttle valves with the possibility of manual flow adjustment are located on version B). These two options are suitable for three-position seat pressure control. In variant C) a pair of analog valves is used. This variant allows continuous adjustment of the valve flow and is thus suitable for finer control using a PI controller. Variant D) is then the most economical variant suitable for three-position control or purely manual control. The inlet valve is removed and is only replaced by switching the compressor with a non-return valve. Analog proportional valves are the most expensive components used. Therefore, variants E) and F) have been proposed, where it is also a proportional control, but using only one analog valve.

In variant E), PWM compressor speed control is used as a continuous action variable on the supply line. In variant F) one of the analog valves is replaced by a cheaper discrete valve. To switch on the air supply to the seat, it is necessary to switch on the compressor C, open the control valve Va and close the outlet valve Vout1. To release the air to the atmosphere, switch off compressor C and open control valve Va and outlet valve Vout1.

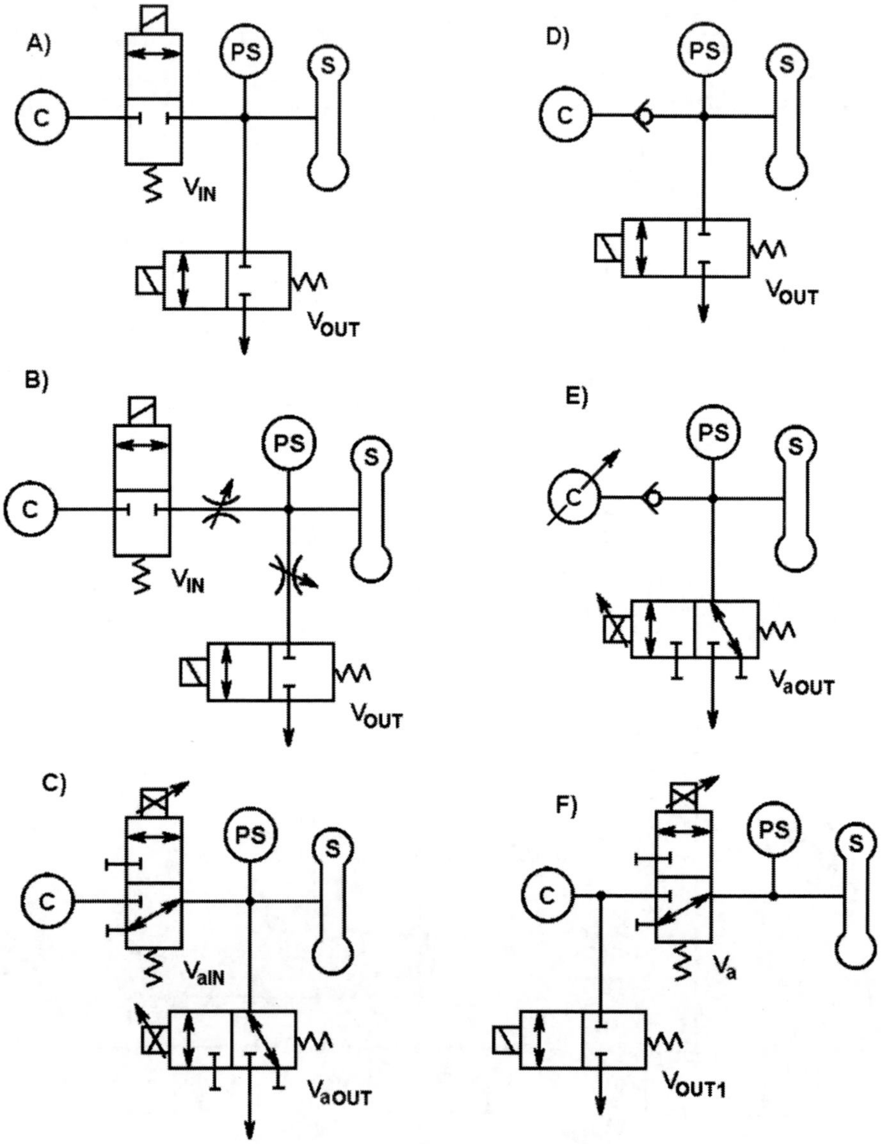

Fig. 7. Pneumatic circuits with reduced components

4. Control unit for the seat

The control unit for laboratory use was designed with an Arduino microprocessor development board, Fig. 8.

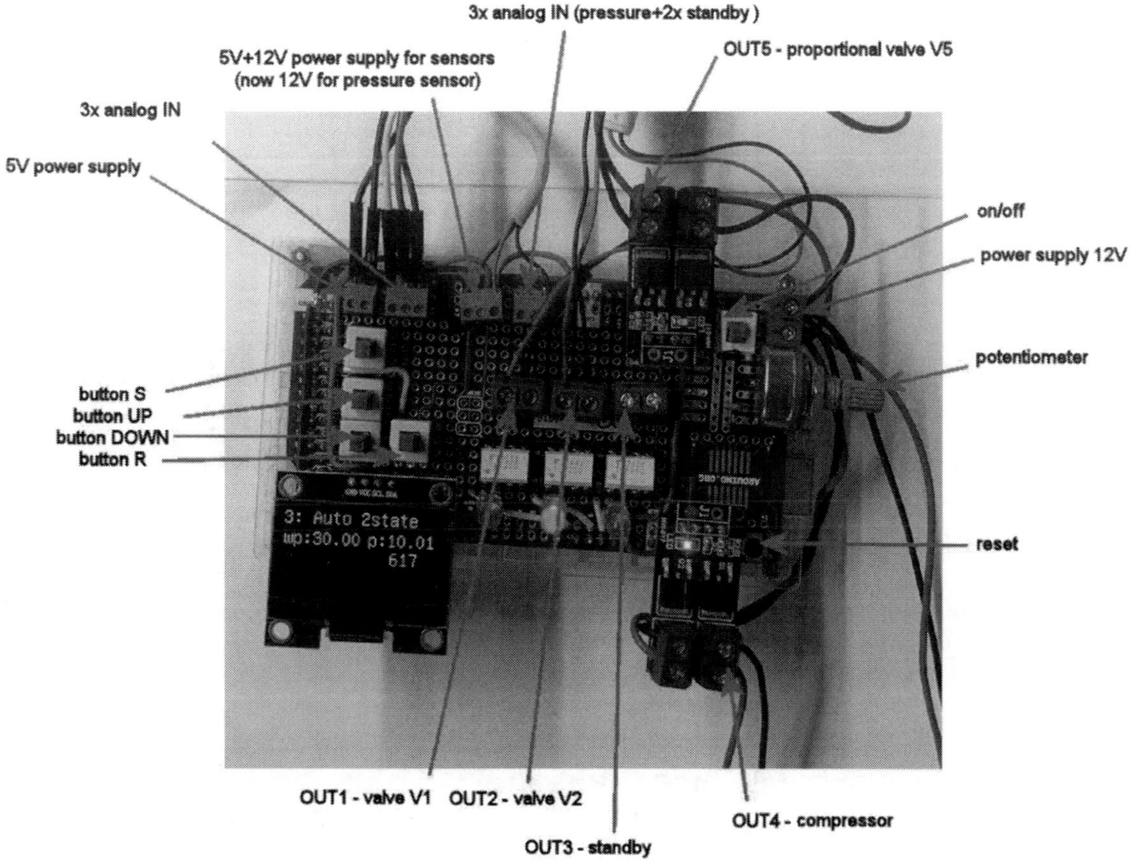

Fig. 8. Control Unit for the seat

The control unit is designed to be able to control all the components described above in chapters 2 and 3, ie three digital valves, two analog valves, a compressor and two pressure sensors can be connected to it. Optical separation and amplifiers are designed for both discrete and analog actuators.

The control unit also contains controls for setting control modes and parameters such as pressure desired value. These elements are the buttons and the potentiometer. The individual modes and the required parameters are shown on the OLED display. Measured data can be sent online via a serial line for further processing on a PC. The block diagram of the control unit is shown in Fig.9.

There are several regimes to test various functions and control algorithms on the board. Regimes of the control unit are:

- Hand control of all actuators (valves and compressor) and reading of all sensors (pressure sensors, buttons, potentiometer)
- Hand control of pressure inside the pneumatic element
- Automatic control of pressure using three-state control algorithm, Fig.10
- Automatic control of pressure using PI control algorithm, Fig.11

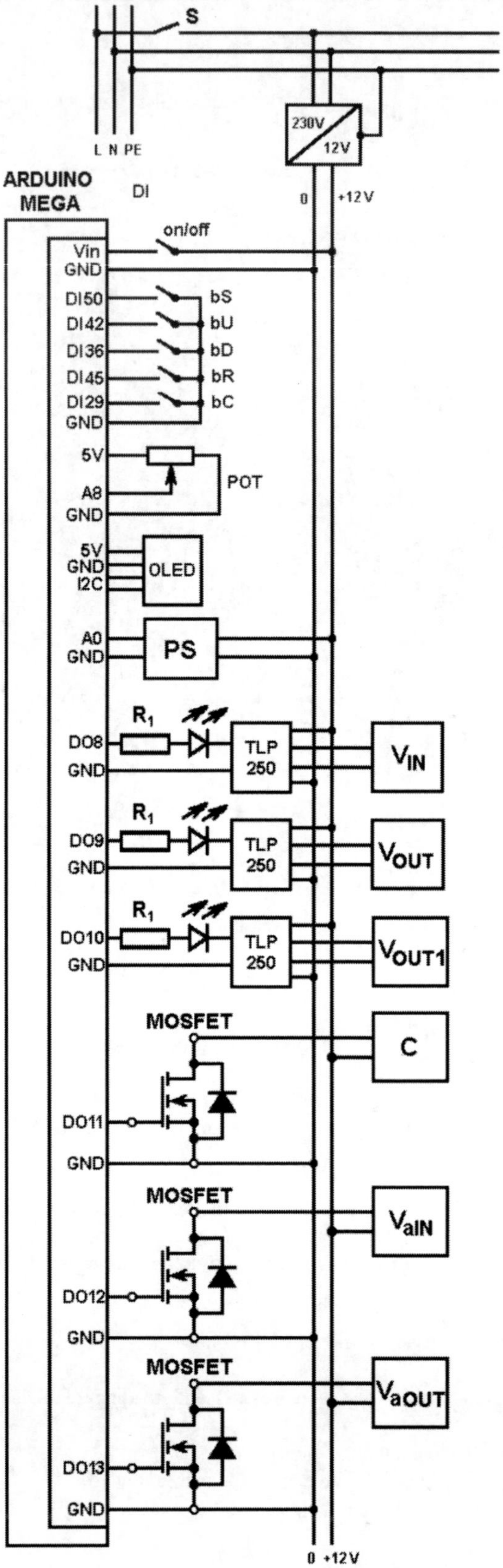

Fig. 9. Scheme of control unit

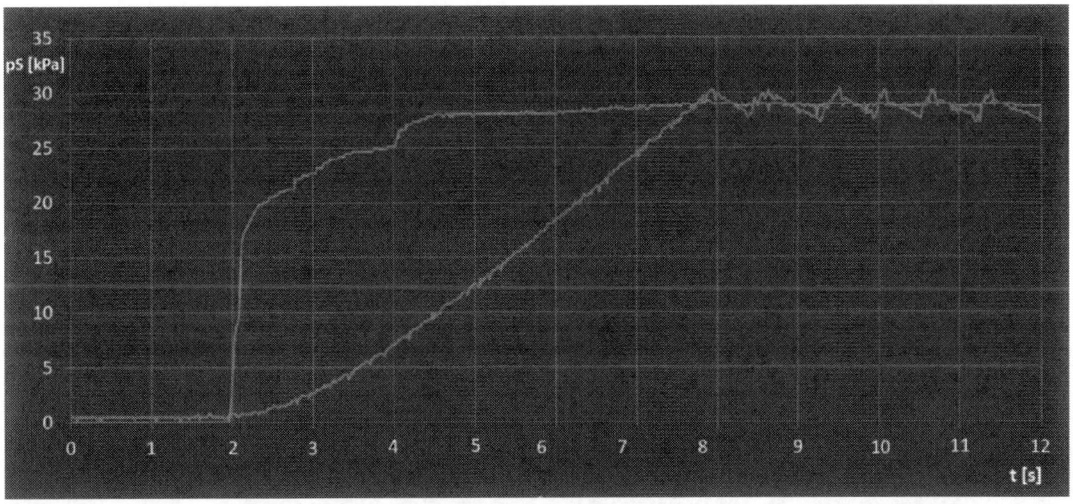

Fig. 10. Three state control

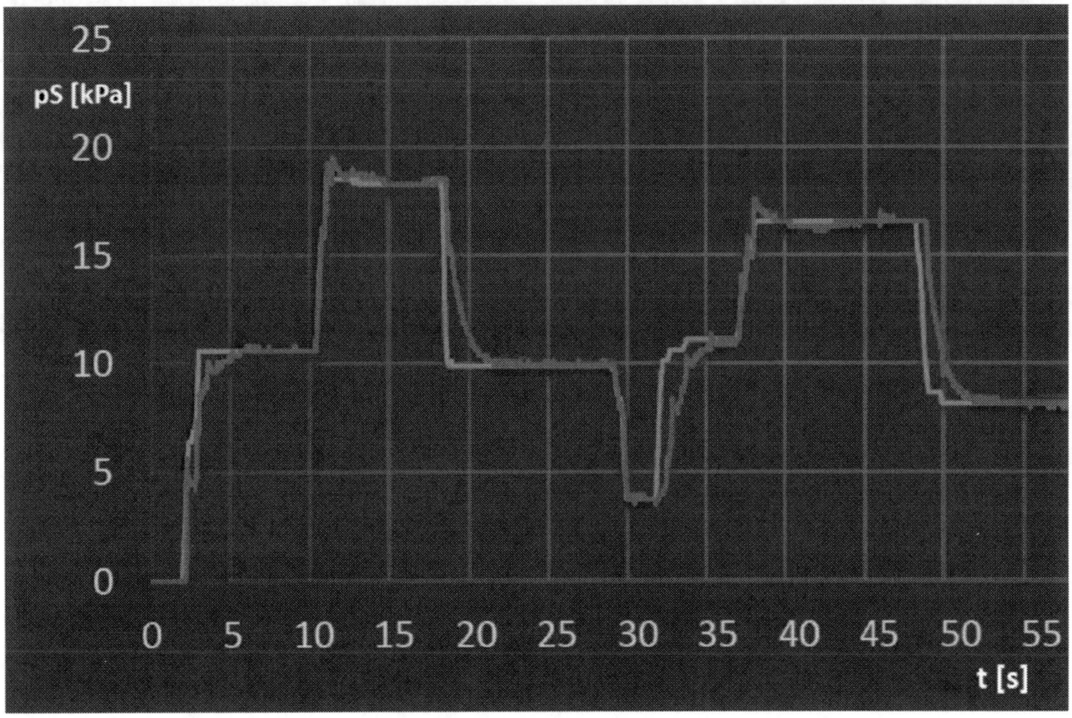

Fig. 11. PI control algorithm

There are examples of control algorithms in Fig. 10 and Fig 11. The desired value is displayed in orange and is entered using a potentiometer. The pressure course in the seat is shown in blue.

5. Conclusion

This article describes the development and variants of the pneumatic circuit and its control unit for a seat with adjustable seat pressure. The control unit has been designed for laboratory purposes to be able to operate in many test modes and in many pneumatic circuit configurations. The unit was implemented in a test seat and has been used for laboratory measurements. Next step in development of control unit will be transfer of presented solution from laboratory scale to industrial standard in cooperation with producer of electronic devices. Also final solution of pneumatic circuit will be selected considering the simplicity of constructional arrangement. Implemented working modes of the seat with variable stiffness must be designed with regard to estimated consumption of compressed air and capability of compressor.

6. Acknowledgments

This work was supported by project „European Smarter Seats for Global Innovation in the Automotive Sector“ (ESSGIAS, no. LTE 120004). The project is co-funded by Czech Ministry of Education, Youth and Sports within the INTER-EUREKA (LTE) INTER-EXCELENCE program, supported by EURIPIDES[2] and monitored by AENEAS.

7. References

[1] Cirkl, D., Modelling of damping force of dynamically loaded specimen of polyurethane foam in dependence on its density. Shock and Vibration 17 (2010), ISSN: 1070-9622, 373-381

[2] Votrubec, R. (2011). Control of Active Sanitary Coach, Annals of DAAAM for 2011 & Proceedings of the 22nd International DAAAM Symposium, 23-26th November 2011, Vienna, Austria, Volume 22, No. 1, ISSN 1726-9679, ISBN 978-3-901509-83-4, Katalinic, B. (Ed.), pp. 0617-0618, Published by DAAAM International Vienna, Vienna.

[3] Saravanja, D. & Grbesic, M. (2019). Application of Vibration Analysis in Journal Bearing Problems Diagnostics, Proceedings of the 30th DAAAM International Symposium, pp.0100-0106, B. Katalinic (Ed.), Published by DAAAM International, ISBN 978-3-902734-22-8, ISSN 1726-9679, Vienna,Austria, DOI: 10.2507/30th.daaam.proceedings.013

[4] Cekic, A; Muhamedagic, K.; Cohodar, M.; Begic-Hajdarevic,D. & Biogradlija, S. (2019). Experimental Investigation of Effect of Overhang Tool Length on Tool Vibration and Surface Roughness, Proceedings of the 30th DAAAM International Symposium, pp.0184-0191, B. Katalinic (Ed.), Published by DAAAM International, ISBN 978-3-902734-22-8, ISSN 1726-9679, Vienna, Austria, DOI: 10.2507/30th.daaam.proceedings.024

[5] Wrzochal, M.; Adamczak, S.; Lomza, H. & Luszczak,R. (2018). Testing the Susceptibility of Rolling Bearing Vibration Measurement Systems, Proceedings of the 29th DAAAM International Symposium, pp.0860-0865, B. Katalinic (Ed.), Published by DAAAM International, ISBN 978-3-902734-20-4, ISSN 1726-9679, Vienna, Austria DOI: 10.2507/29th.daaam.proceedings.124

[6] Herman, A.; Vaclav Stourac, F.; Kyncl, J. & Beranek, L. (2018). New Design of Polyurethane Foam Moulds for Vibrational Casted Refractory Materials, Proceedings of the 29th DAAAM International Symposium, pp.1231-1238, B. Katalinic (Ed.), Published by DAAAM International, ISBN 978-3-902734-20-4, ISSN 1726-9679, Vienna, Austria DOI: 10.2507/29th.daaam.proceedings.178

[7] Kharuzin, S.; Ivanov, A. & Shmakov, O. (2017). Autonomous Control System for a Vehicle with Actively Articulated Frame, Proceedings of the 28th DAAAM International Symposium, pp.0759-0766, B. Katalinic (Ed.), Published by DAAAM International, ISBN 978-3-902734-11-2, ISSN 1726-9679, Vienna, Austria DOI: 10.2507/28th.daaam.proceedings.107

[8] Votrubec, R. (2020). Communication Layer between Control Units within the Internet of Things, Proceedings of the 31st DAAAM International Symposium, pp.0072-0079, B. Katalinic (Ed.), Published by DAAAM International, ISBN 978-3-902734-29-7, ISSN 1726-9679, Vienna, Austria DOI: 10.2507/31st.daaam.proceedings.010

[9] Votrubec, R. & Buchta, J. (2019). Air Flow Control of Centrifugal Fans for Personal Protection Equipment, Proceedings of the 30th DAAAM International Symposium, pp.0268-0275, B. Katalinic (Ed.), Published by DAAAM International, ISBN 978-3-902734-22-8, ISSN 1726-9679, Vienna, Austria DOI: 10.2507/30th.daaam.proceedings.035

DOI: 10.2507/32nd.daaam.proceedings.010

DIAGNOSIS OF ACOUSTIC PROPERTIES OF RECYCLED MATERIALS FROM VEHICLES

Miroslav Badida, Lydia Sobotova, Kristian Pastor, Alzbeta Novakova, Miroslava Badidova & Veronika Gumanova

This Publication has to be referred as: Badida, M[iroslav]; Sobotova, L[ydia]; Pastor, K[ristian]; Novakova, A[lzbeta]; Badidova, M[iroslava] & Gumanova, V[eronika] (2021). Diagnosis of Acoustic Properties of Recycled Materials from Vehicles, Proceedings of the 32nd DAAAM International Symposium, pp.0067-0074, B. Katalinic (Ed.), Published by DAAAM International, ISBN 978-3-902734-33-4, ISSN 1726-9679, Vienna, Austria
DOI: 10.2507/32nd.daaam.proceedings.010

Abstract

The aim of this paper is to present the results of the diagnosis of acoustic properties of selected single-layer and double-layer, three-layer sandwich materials. The analysis was performed based on measurements using an impedance tube and the analysed acoustic properties were the sound absorption coefficient α (-) and the transmission attenuation (dB). Measurements were performed in the frequency range from 100 Hz to 6300 Hz for sample thicknesses from 10 mm up to 60 mm. The results of the measurements were analysed and compared in terms of changes in the acoustic properties of the examined materials according to the changing thickness of these materials and the frequency band of the noise.

Keywords: Impedance tube; Sound absorption coefficient; Transmission attenuation; Recycled materials

1. Introduction

The acoustic properties of materials are playing an increasingly important role today. This is the reason why a number of new acoustically suitable materials with different sound insulation properties appear and develop almost every day. Recycled components from automobiles over their lifetime are often used to develop such materials. These materials are used for passive sound control in the design of car interiors, factories, workshops, residential buildings, noise barriers, etc. to prevent unwanted noise. The determination of acoustic properties is therefore extremely important.[1][2][4]

End-of-life vehicles represent a whole range of components (waste) of various material composition which should be further recovered. Focus is mainly on problematic materials from components, such as tyres, car seats, carpets, upholstery, airbags, glass, etc., and on the development of technologies and techniques for waste recovery into acoustic, thermal insulation, and other products.[4] The products from used car recycling that can be used for the production of sound absorbing elements are products from recycled tyres, car seats, seat covers, carpets and airbags. [4]

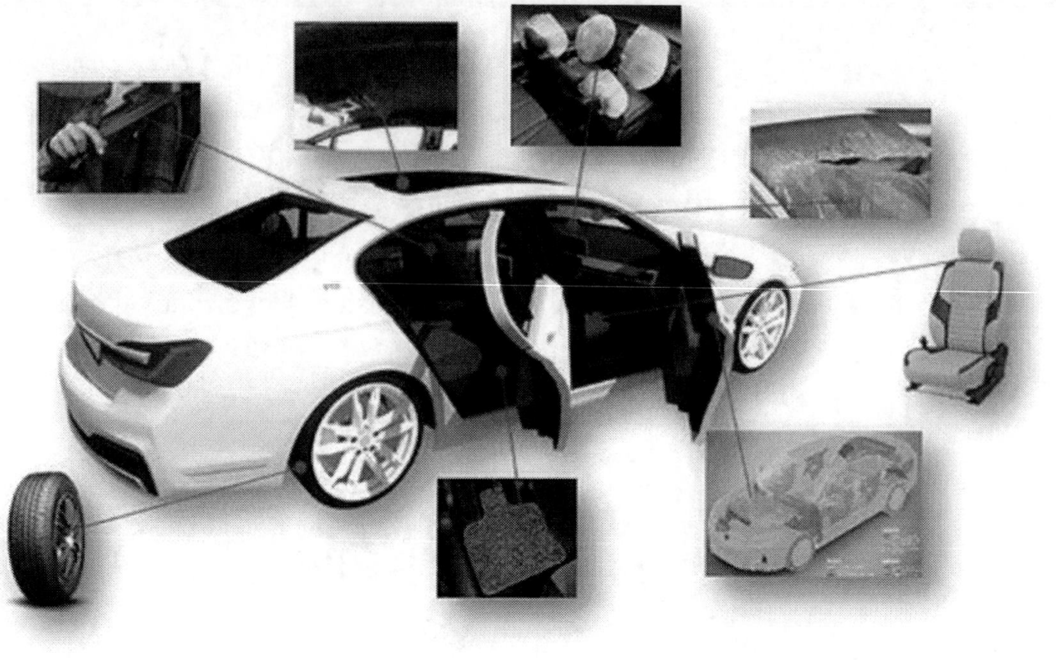

Fig. 1. Materials suitable for recycling

Research in this area seeks to control and minimize the sounds that occur in machines, cars, homes and other places. There are a large number of studies and research experiments that are trying to develop, discover new materials that can be used in various acoustic projects. For the analysis of these materials, sound absorption is one of the most important physical phenomena.[3]

Various synthetic materials have excellent acoustic properties, but their price and harmful effects on the environment must be taken into account. Bhingare N.H. describes the properties of natural fibres and waste (recycled) material. Natural materials are an effective alternative to synthetic ones due to their low cost and lower environmental impact. Waste material present in the environment, such as rubber crumbs from tires, waste wood etc., can be used with acoustic applications. This is emphasized in current research of natural fibrous materials, such as jute, kenaf, kapok and coconut fibres, fibres from palm and sugar cane, corn husk fibre and their acoustic performance is being researched. It was concluded that the properties of these natural materials can be effectively used for acoustic applications. The possibilities of using recycled natural or synthetic polymers from waste rubber in the production of acoustic materials for its good acoustic properties was studied by Xinwu X.[5][6]

This paper deals with the analysis of acoustic properties, namely the sound absorption coefficient α and the attenuation index R, for selected materials using an impedance tube. Several recycled materials from automobiles were selected, which were evaluated in terms of analysed acoustic properties in different thicknesses as separate single-layer materials, but also as sandwich two-layer and three-layer materials.

2. Description of the measurement method and sample preparation

The measurements were performed using a BSWA SW466 impedance tube in a configuration with two and four microphones using the transfer-function method according to the ISO 10534-2:1998 standard. The impedance tube has an advantage over other measurement methods due to its compactness, low cost and fast result. Compared to the reverberation chamber, the impedance tube requires significantly smaller samples of the tested material. A sound-generating speaker is located on one side of the impedance tube and a removable sample holder with a fixed end is located at the other end. Microphones are placed between the speaker and the sample. Depending on the number of microphones used in the measurement, there are two, three or four microphone methods (Fig. 2).[1][2][7]

The sound absorption coefficient (α) is a dimensionless number ranging from <0, 1>. The closer the value of the coefficient is to the number 1, the better the sound absorption of the material. An impedance tube in a two-microphone configuration is used to measure the sound absorption factor of selected materials [8]. At one end of the tube is placed the sample to be examined and at the other end is a loudspeaker which is powered by a signal generator. The signal is then amplified in an amplifier. Two microphones of the same type are placed on the tube to measure acoustic pressures. Other components of the measuring technology are the MC3242 4-channel analyser for data collection, the PA50 measuring power amplifier from BSWA TECH and a computer equipped with VA-Lab4 software for information evaluation and tube control (Fig.3).

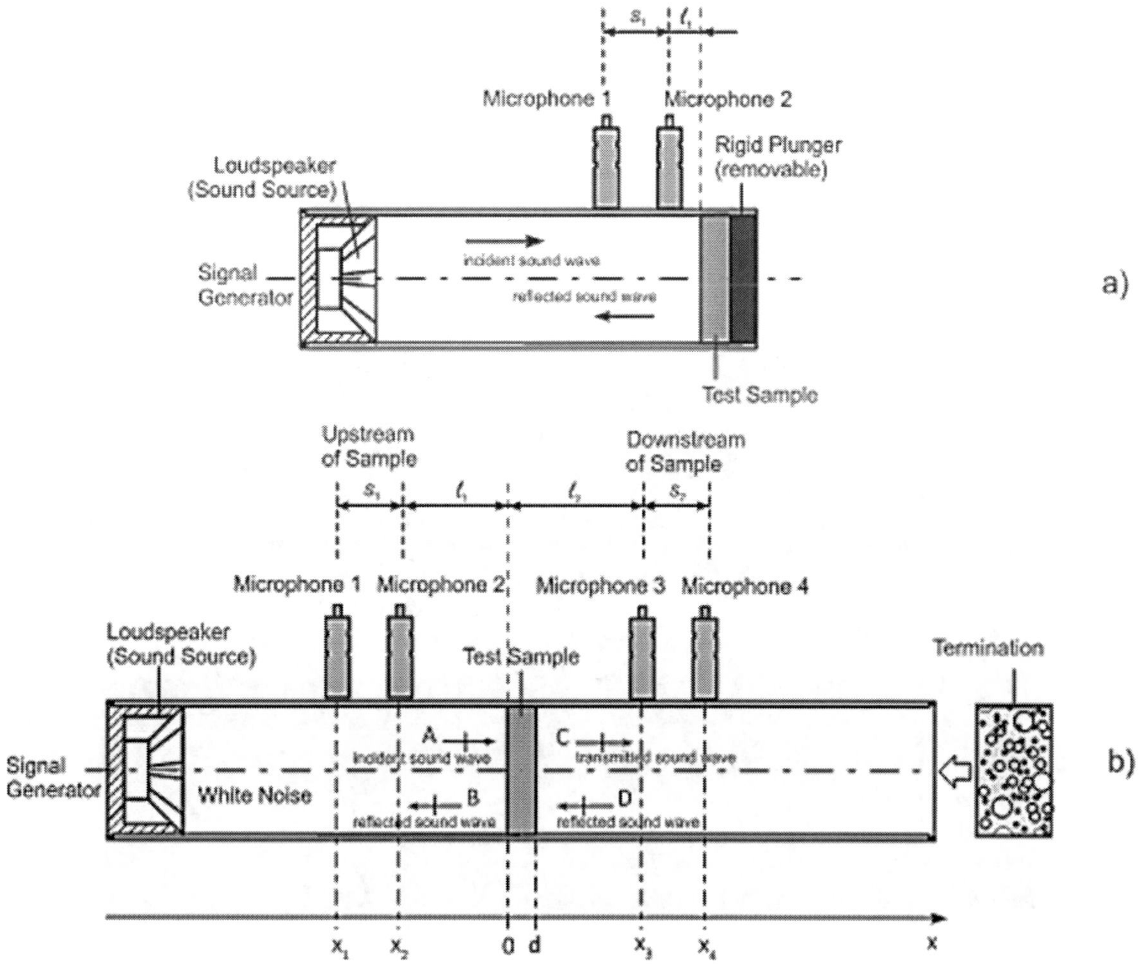

Fig. 2. Schematic of the SW466 impedance tube, a) method with two microphones, b) method with four microphones

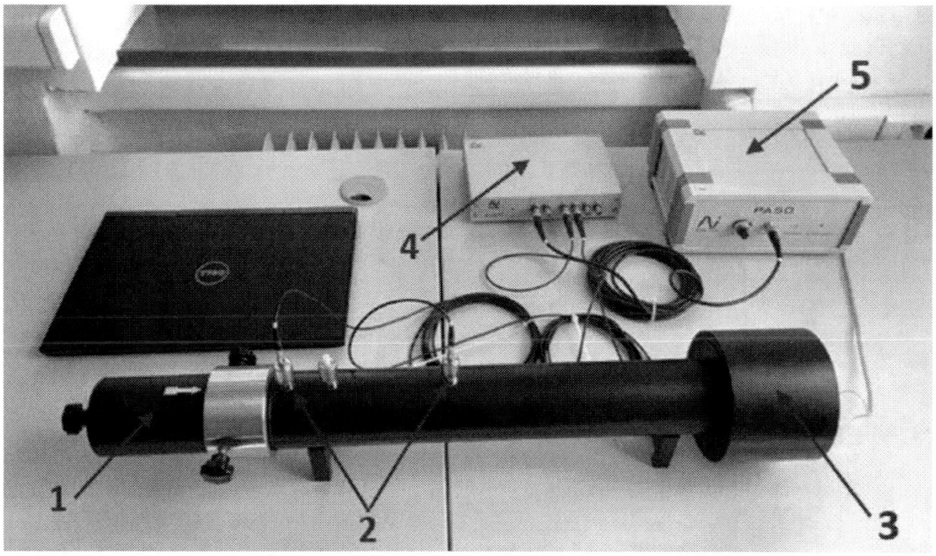

1 - Sample holder for measuring the sound absorption coefficient, 2 - Two microphones for measuring the sound absorption coefficient, 3 - Built-in speaker, 4 - Four-channel analyser, 5 - Power amplifier

Fig. 3. Measuring apparatus for measuring the sound absorption coefficient α (-)

The attenuation index R is a value given in decibels (dB) based on the ratio of the wave of sound incident on the front of the sound-absorbing material to the sound waves transmitted back. The attenuation index represents the damping properties of the material, which means that the higher the value of the attenuation index, the more attenuated the sound and the better the damping properties of the material. For the measurement of the attenuation index, either two reverberation rooms with a sample of material in the opening between the rooms or the impedance tube method are traditionally used.[9]

An impedance tube in a configuration with four microphones was used to measure the attenuation index of the examined materials. At one end of the tube there is a loudspeaker which is powered by a signal generator and the signal is subsequently amplified in an amplifier and at the other end of the tube in this case the sample holder is not connected as in sound absorption measurement but an extension tube is connected. The sample is located in the middle. Other components of the measuring technology are the MC3242 4-channel analyser for data collection, the PA50 measuring power amplifier from BSWA TECH and a computer equipped with VA-Lab4 software for information evaluation and tube control. (Fig.4)

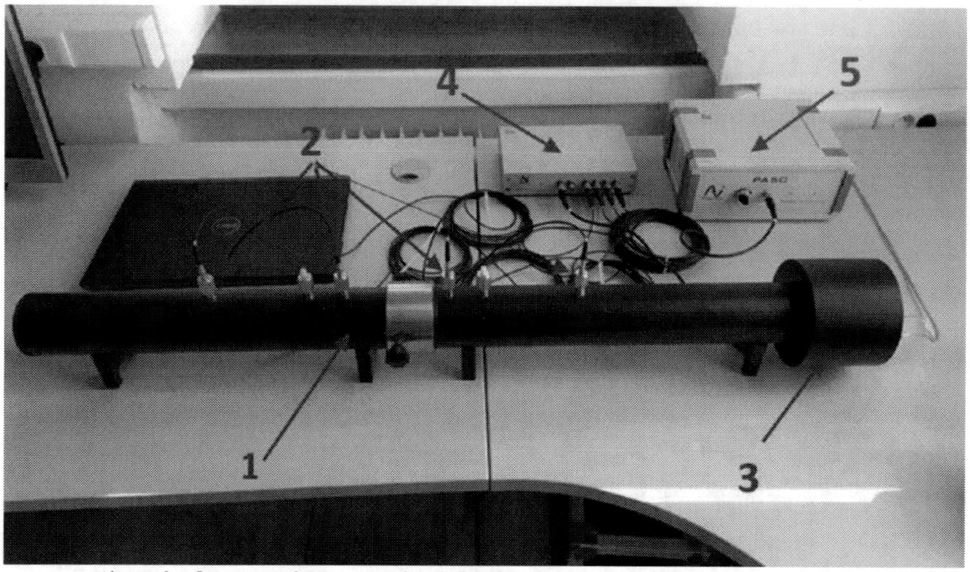

1 – Large extension tube for measuring attenuation index, 2 - Four microphones for measuring attenuation index, 3 - Built-in speaker, 4 - Four-channel analyzer, 5 - Power amplifier

Fig. 4. Measuring apparatus for measuring The attenuation index R (dB)

Table 1 shows samples of materials used for the experiment. All samples had the same diameter, 30 mm or 60 mm, which corresponds to the inner diameter of the tube. The thickness of the samples varied from 10mm to 60 mm.
The materials that were the subject of the research were:

- Nobasil
- Recycled rubber
- Ecomolitan

Materiál	Photo of samples	Thickness of material (mm)	Material diameter (mm)	Layer arrangement
Nobasil	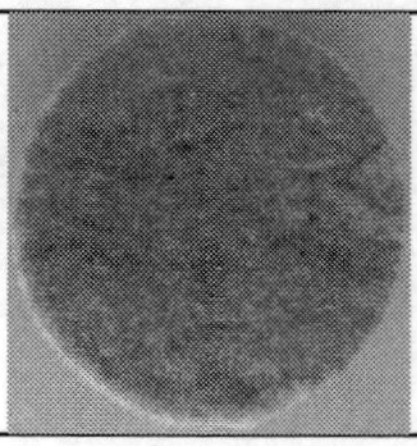	10	60/ 30 mm	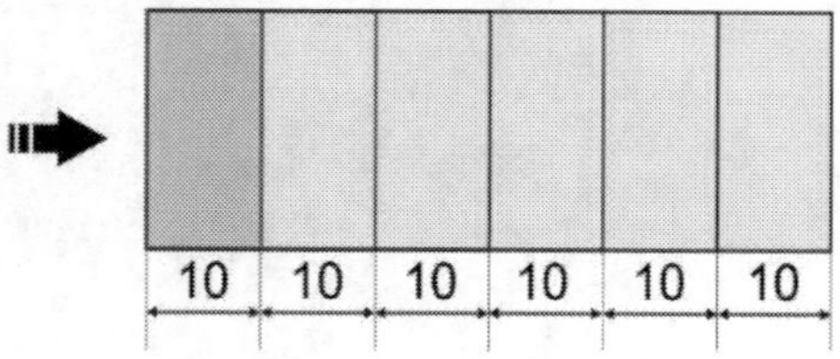
		20		
		30		
		40		
		50		
		60		
Ecomolitan		20		
		22		
		30		
		40		

	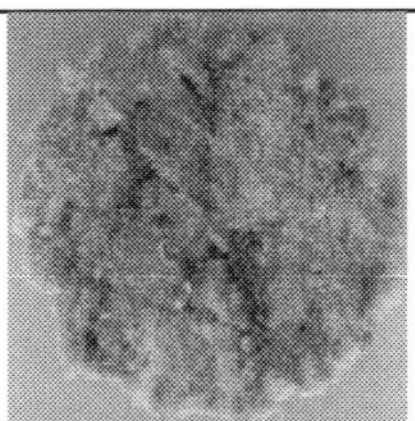	50 60		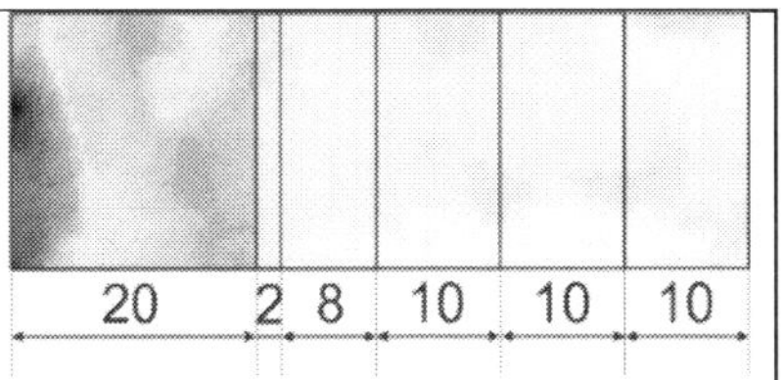
Recycled rubber		20 30 40 50 60		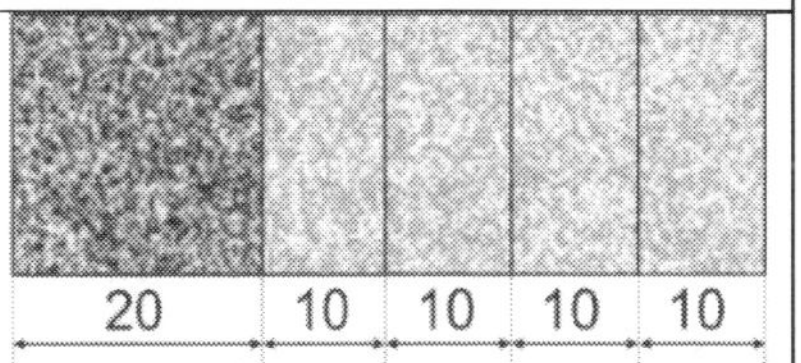

Table 1. Samples of the materials that were the subject of the research

3. Processing of measurement results, comparison and evaluation

The results (Fig. 5) show that at the sound absorption coefficient α, the worst values are generally reached by the investigated materials at low frequencies in the range of approximately 100 Hz to 350 Hz. The best values range in the wider frequency spectrum from 630Hz up to 6300 Hz. The best values are achieved for materials with thicknesses: Ecomolitan 30 mm, Recycled rubber 20 mm, Nobasil 30 mm and 40 mm. It follows that thickness does not play a significant role in sound absorption, and that increasing the sound absorption coefficient is not directly proportional to the increasing thickness of the material.

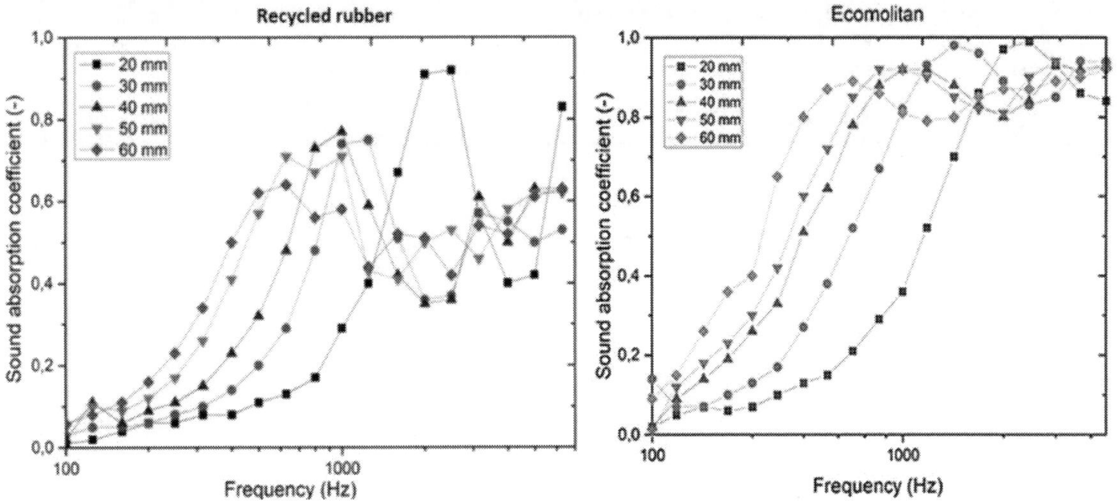

Fig. 5. Sound absorption coefficient values α for single layer materials

As with single-layer also with double-layer materials, it has been shown that the worst values of sound absorption are achieved by the investigated materials at lower frequencies, from 100 Hz to 125 Hz. The best values, however, reach mainly in the middle frequency band from 315 Hz to 800 Hz.

It can also be seen from the results that the values of the sound absorption coefficient are influenced not only by the composition of the sandwich material, i.e. what materials it is composed of, but also by the arrangement of these materials with respect to the direction of sound propagation. This can be seen, for example, in the sandwich material in the composition Recycled rubber + Nobasil (Fig. 6). In such an arrangement, when the Recycled Rubber is the first layer as opposed to the direction of sound propagation, the highest value of the sound absorption coefficient achieved was 0.75 at a frequency of 630 Hz. In the reverse order of these materials, Nobasil + Recycled Rubber had the highest value of 0.86 at a frequency of 6300 Hz.

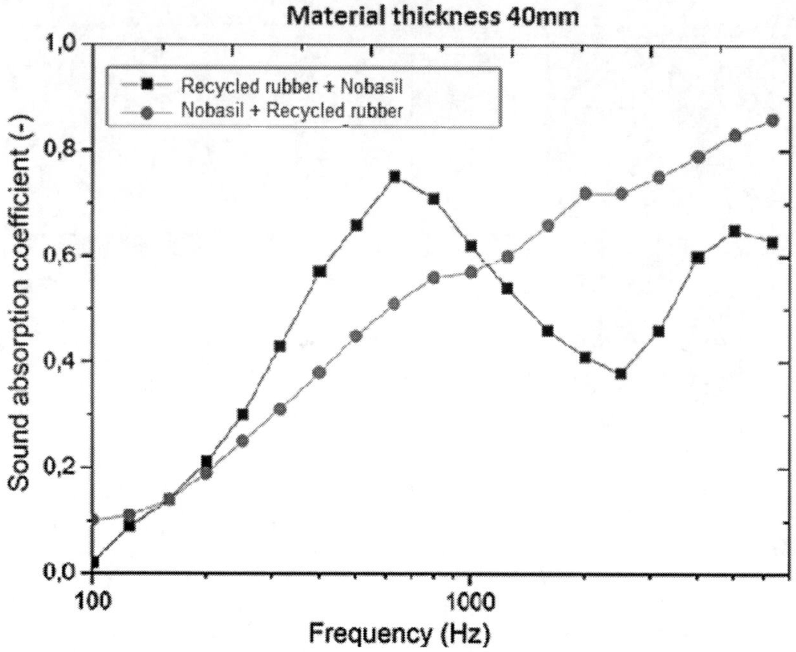

Fig. 6. Sound absorption coefficient values α for sandwich material Recycled rubber + Nobasil

As with the sound absorption coefficient α (-), the results of the Attenuation Index (dB) show that the investigated materials reach their best values at high sound frequencies, namely 5000 - 6300 Hz. The results show that, in contrast to the sound absorption coefficient, the thickness of the material plays an important role in the attenuation index. Most materials achieved the best values at the largest sample thickness, i.e. 60 mm. The graphs (Fig. 7) show the growth of the values, which is directly proportional to the increasing thickness of the material sample.

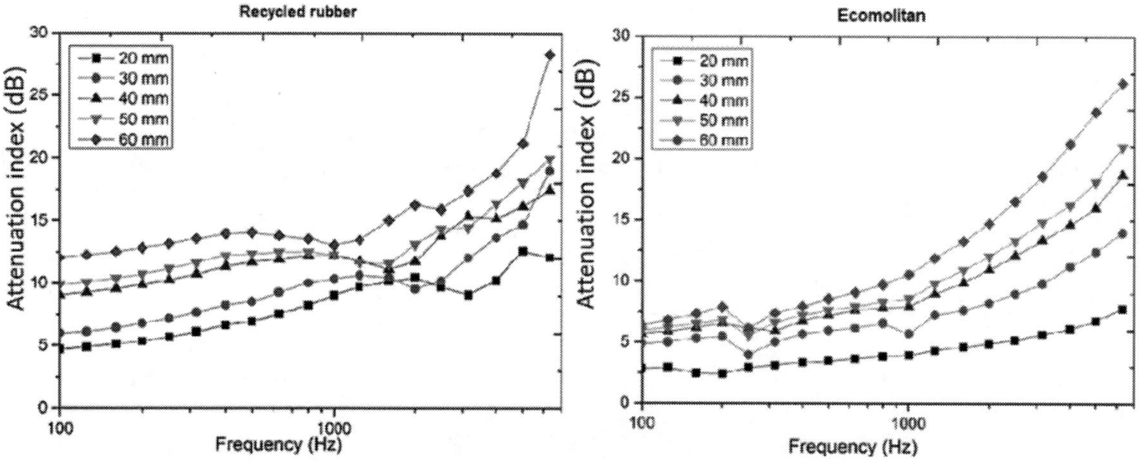

Fig. 7. Attenuation index values for single layer materials

It can also be seen in the graphs (Fig. 8) that, in contrast to the sound absorption in the attenuation index, the arrangement of the materials in the layers plays essentially no significant role. For the pair Recycled Rubber + Nobasil, slight differences in the frequency band from approximately 2000 Hz to 5000 Hz can be seen on the graph. For the other pairs, however, no difference in values is seen depending on the arrangement of the materials. It can also be seen that the two-layer material Recycled Rubber + Ecomolitan achieves in its two configuration variants the values of the attenuation index 17.98 and 17.56 dB at a sound frequency of 6300 Hz and a common thickness of 40 mm. However, at the same thickness and frequency, these materials as single-layer values reach 17.49 dB for Recycled Rubber and even higher than the two-layer material for Ecomolitan at 18.71 dB.

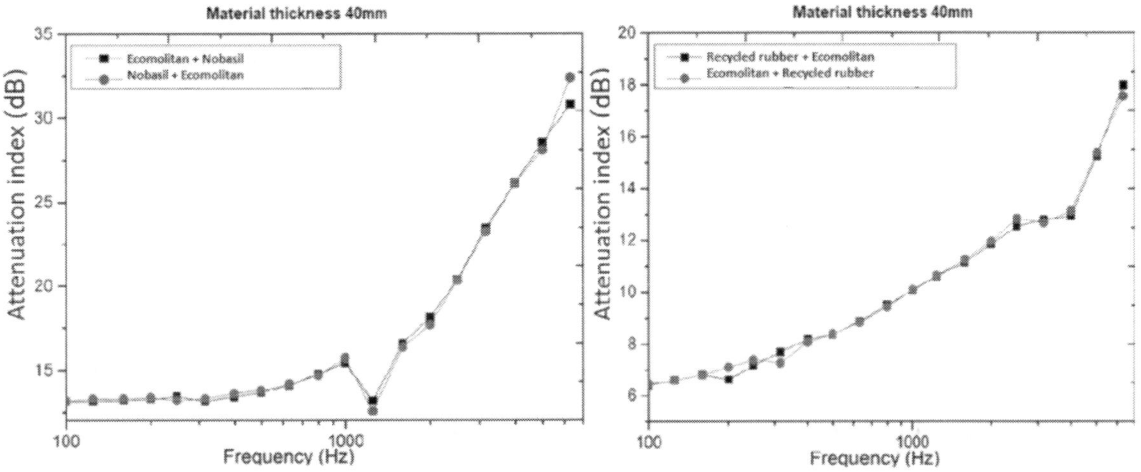

Fig. 8. Attenuation index values for sandwich materials

It can therefore be concluded that the arrangement of the materials in the case of two-layer sandwich materials does not play a major role in terms of transmission attenuation. Three-layer sandwich materials generally show better TL properties than two-layer materials. This, however, according to previous results, can be attributed more to the increasing thickness of the material than to its composition and the fact that it has three layers.

4. Conclusion

The aim of this research was to assess recycled materials from cars in terms of their acoustic properties. According to the measured results, it can be stated for the sound absorption coefficient α that the thickness of the analysed materials had no major effect on the value of the coefficient. The formation of sandwich materials, however, significantly affected the value of the coefficient and the properties of materials in different frequency bands. For sandwich materials, the arrangement of the materials with respect to the direction of noise propagation also played a role.

In the attenuation index r, in contrast to the sound absorption coefficient α, the thickness of the material had a significant effect. The tables and graphs showed an increase in the value of the attenuation index, which was directly proportional to the increasing thickness of the material. Also, the formation of sandwich materials did not show any significant advantages in terms of attenuation index. In several cases, even when comparing sandwich and single-layer materials of the same thickness, single-layer materials showed higher transmission attenuation values than sandwich materials.

Based on these results, we can conclude that recycled materials from vehicles can be used as a sound insulation material after their lifetime. Also, by creating sandwich materials and their suitable distribution, we can very well manipulate the properties of these materials at different frequencies in terms of sound absorption and thus obtain a suitable insulating material for specific tasks. In the future, we would like to further investigate recycled materials from vehicles and assess their acoustic properties, assess their possible use in practice and also include other recycled materials in research.

5. Acknowledgments

This work was supported by the project titled KEGA No. KEGA 009TUKE-4/2021: Implementation of the latest knowledge from recycling technologies for material recovery of product components at the end of their life cycle

6. References

[1] Niresh, J.; et al. (2015). Review of acoustic characteristics of materials using impedance tube, ARPN Journal of Engineering and Applied Sciences, Vol. 10, No. 8, 2015, pp. 3319-3326, ISSN 1819-6608, Available from: http://www.arpnjournals.com/jeas/research_papers/rp_2015/jeas_0515_1928.pdf Accessed: 2021-09-07

[2] Arunkumar, B.; Jeyanthi, S. (2017). Design and analysis of impedance tube for sound absorption measurement, ARPN Journal of Engineering and Applied Sciences, Vol. 12, No. 5, 2017, pp. 1400-1405, ISSN 1819-6608, Available from: http://www.arpnjournals.org/jeas/research_papers/rp_2017/jeas_0317_5774.pdf Accessed: 2021-09-07

[3] Rosely, C.V.M.; Silva, R.H.A.; Linhares, T.M.; Rocha, C.M. (2012). The impedance tube construction and acoustic absorption study applied materials in side vehicles, 41st International Congress and Exposition on Noise Control Engineering 2012, Vol. 2, No.1, 2012, pp. 872-888, ISBN: 978-1-62748-560-9, Available from:

https://www.researchgate.net/publication/289289454_The_impedance_tube_construction_and_acoustic_absorption_study_applied_materials_inside_vehicles Accessed: 2021-09-07

[4] Badida, M.; et al. (2021). Progresívne technológie zhodnocovania odpadov v automobilovom priemysle, Spektrum STU Bratislava, ISBN 978-80-553-3867-5

[5] Bhingare, N. H.; Prakash, S.; Jatti, V. S. (2019). A review on natural and waste material composite as acoustic material, Polym. Test., doi: 10.1016/j.polymertesting.2019.106142.

[6] Xu, X.; Wang, H.; Sun, Y.; Han, J.; Huang, R. (2018). Sound absorbing properties of perforated composite panels of recycled rubber, fiberboard sawdust, and high density polyethylene, Journal of Cleaner Production, vol. 187, 2018, doi: 10.1016/j.jclepro.2018.03.174.

[7] Labašová, E.; Ďuriš, R. (2019). Measurement of the acoustic absorption coefficient by impedance tube, Research Papers Faculty of Materials Science and Technology Slovak University of Technology, Vol. 27, No. 45, 2019, pp. 94-101, ISSN 1338-0532, Available from: https://www.researchgate.net/publication/338031592_Measurement_of_the_Acoustic_Absorption_Coefficient_by_Impedance_Tube Accessed: 2021-09-07

[8] Měření zvukové pohltivosti materiálů, Laboratorní cvičení z Environmentální fyziky Fakulta technologická, UTB ve Zlíně, p. 8, Available from: http://ufmi.ft.utb.cz/texty/env_fyzika/EF_lab_02.pdf Accessed: 2021-09-07

[9] Badida, M.; Lumnitzer. E.; Bartko, L. (2011). Možnosti znižovania dopravného hluku. Elfa,s.r.o, Košice, 2011. ISBN 978-80-8086-181–0

DOI: 10.2507/32nd.daaam.proceedings.011

OPTIMIZING KUBERNETES PERFORMANCE, EFFICIENCY AND ENERGY FOOTPRINT IN HETEROGENOUS HPC ENVIRONMENTS

Vedran Dakic, Mario Kovac & Jasmin Redzepagic

This Publication has to be referred as: Dakic, V[edran]; Kovac, M[ario] & Redzepagic, J[asmin] (2021). Optimizing Kubernetes Performance, Efficiency and Energy Footprint In Heterogenous HPC Environments, Proceedings of the 32nd DAAAM International Symposium, pp.0075-0083, B. Katalinic (Ed.), Published by DAAAM International, ISBN 978-3-902734-33-4, ISSN 1726-9679, Vienna, Austria
DOI: 10.2507/32nd.daaam.proceedings.011

Abstract

This paper describes the most common complexities in heterogenous HPC (High Performance Computing) environments, the most common problems facing heterogenous HPC environments arising from these complexities, the most commonly used solutions to these problems and the potential solutions to these problems by using new methodologies. One of these methodologies could be using containers and container orchestration methodologies instead of virtual machines to lower compute overhead when dealing when large-scale, distributed computing concepts like HPC. This will require significant efforts in different phases of the HPC process - workload preparation, producing different types of images and layers for various HPC applications, potential changes in HPC schedulers etc. In the era of green computing, that will also require significant improvements in energy management process, that needs to be an integral part of the overall design.

Keywords: High Performance Computing; Kubernetes; containers; virtual machines; automation; orchestration.

1. Introduction

Ever since the invention of supercomputers in the 1960s, science world has been one of the key customers on that market. As technology, programming and computational models improved, another type of supercomputer was introduced, and its subsequent application was called HPC (High Performance Computing). High Performance Computing is the application of supercomputers to computational problems that are either too large for standard computers or would take too long.

In the past decade, HPC environments have been extensively using virtual machines/clouds because of the need for computational power, agility, and manageability. But there are other methodologies for running workloads, like containers. Containers offer a broad array of benefits, including a consistent lightweight runtime environment through OS-level virtualization, as well as low overhead to maintain and scale applications with high efficiency [1]. This makes them a good candidate to apply to HPC environments, especially as HPC environments become more and more heterogenous. IT industry in general became more and more aware of the fact that significant gains in computational power can be achieved by moving towards containers, years ago. Further operational advantages can be achieved by implementing automation and orchestration principles as they can help us do a lot more while working manually a lot less. The question that we need to ask is - can the same idea be used in the area of HPC?

What are the real obstacles on the way of using containers as compute resources for HPC? Are there different ways of overcoming those issues and tapping into the idea of green computing at the same time? We think that the correct way forward is to implement tighter hardware and software integration, and to create programmable, custom Kubernetes schedulers for HPC workloads. And this is precisely the subject of our paper.

2. Theoretical background

By definition, heterogenous computing means mixing different types of cores within the limits of a single node. We can use this approach in a variety of different ways, like:

- by using CPU and GPU in the same node (for example, x86 CPU and NVIDIA Tesla GPU).
- by using CPU and ASIC in the same node (for example, x86 CPU coupled with Intel eASIC or Xilinx UltraScale).
- by using different CPU architectures in the same node (for example, AMD SkyBridge).

The most widely defined goal of heterogenous computing is to maximize performance and to provide energy efficiency. We can add additional performance by utilizing other hardware devices in the same nodes, by using concepts like VFs (Virtual Functions) [2]. As an example, we can partition a PCI Express device to multiple sub-devices that can be independently configured (for performance, cost, Quality of Service or other reasons). We can use specialized PCI Express cards to partition them across different compute objects. Typical example of that methodology is SR-IOV (Single Root Input/Output Virtualization), that enables us to use Virtual and Physical Functions to manage access to physical, underlying specialized hardware. Virtual functions (VFs) are logical PCI Express instances of Physical Functions (PF) on a specialized PCI Express device that can be virtualized by using a Hypervisor. For example, if we have a SR-IOV compatible physical network card that is placed in a server's PCI Express slot that supports SR-IOV, we can present that physical network card as multiple virtual network cards. We can then connect these virtual network cards to compute objects like virtual machines and containers, in essence giving them much more direct approach to the underlying specialized hardware, which means less overhead, more bandwidth and less latency. Having in mind that HPC applications are very sensitive to latency and bandwidth, using specialized devices like these can have a positive impact on the performance of HPC environment.

If we were to use this kind of functionality inside a container, other criteria need to be met. For example, hardware vendor needs to have a driver that's able to expose these functionalities to a container as it's a completely different methodology to using a hypervisor. For example, Mellanox has a plugin called k8s-rdma-sriov, which can be used in such a scenario if we have a Mellanox InfiniBand adapter and want to assign it to multiple different containers. There are multiple different technologies that are prominent - on desktop, server and HPC markets. Some common examples include:

- GPGPU (General-Purpose Graphics Processing Unit) technologies like CUDA (Compute Unified Device Architecture) and OpenCL (Open Computing Language) on various hardware platforms like NVIDIA/AMD GPUs paired with x86 processors.
- Intel Xeon Phi and similar accelerators, that can take advantage of concepts like OpenCL, OpenMP, C/C++, Fortran etc.
- Xilinx and similar FPGA/ASICs, that can also take advantage of various concepts like OpenCL, HLS etc.

All these solutions can be mixed and matched and used side-by-side with other solutions, like Open MPI (Open Message Passing Interface), Intel MPI, MPICH (Message-Passing Interface Chameleon) for distributed parallel computing, which fits well with the concept of HPC. Idea behind these technologies is quite straightforward – to run code on multiple processors and use one of these solutions via library calls. Other aspects of modern computing – Big Data, AI, HPC - are just further proof to the following point – the present and future of computing are based on heterogenous model. There are different reasons for that, such as:

- x86-based ecosystem has been well-developed over the past 40+ years, which means that a lot of workloads can be handled by it.
- ARM-based ecosystem has been focused more on non HPC markets in the past However, the newly developed 64-bit architecture along with their respective SoCs will have a large impact on the CPU performance and memory bandwidth of these ARM processors in HPC domain [3].
- RISC-V based ecosystem has been focused on open-sourcing CPU technology via open instruction set and could be the "Linux of CPU design" if everything goes well.
- GPUs are very fast for specific types of workloads.
- ASICs are even faster for specific types of workloads.

Combining these different technologies in one or multiple nodes could yield the best possible design and architecture for a modern HPC data centre. Of course, there are various geopolitical and strategical reasons why current state of the industry in which x86-based environments are prominent might not be feasible for the long-term feature. EU has been investing a lot of resources into EPI (European Processor Initiative), showing its intent to break free from non-EU owned technology.

3. Hardware and software integration

Let's now discuss the software aspect. For the most part, workloads that we're running today can use either virtual machines (VMs) or containers (Docker, Podman). Integrating these two types of solutions offers the best of both worlds, and it's also a valid approach for HPC workloads. But that also means we need to pay close attention to how we're going to integrate these two concepts. Simple brute-force integration just won't do in the 21st century - randomly placing virtual machines or containers on various nodes isn't efficient nor is it a good policy. Nodes differ in capabilities, compute power, GPU/ASIC availability, storage subsystem performance, performance per Watt figures etc. Connecting performance per Watt figures with node capabilities, automatization and orchestration technologies like Kubernetes can be a good approach for intelligent workload placement, given a specialized, fully programmatic scheduler. There are multiple questions that need answers if we were to use the Kubernetes approach as a valid alternative to virtual machines. Questions like:

- Are containers a better solution than virtual machines?
- Are containers a solution that can be used for any workload?
- How can using containers improve execution of HPC workloads?
- If containers offer a possibility for improving HPC workload execution, can we somehow connect that to the idea of green computing?

Let's start by discussing differences between virtual machines and containers and how these differences can be used to improve HPC workload efficiency. Key architectural difference between a virtual machine and a container is the size of those two concepts. Virtual machines are much bigger objects than containers, as they require more resources to work. First and foremost, they require a virtual disk with installed operating system, for every virtual machine installed on a node. If we multiply that by X virtual machines, that overhead can easily mean terabytes of disk space being used to store virtual machine elements that are needed for it to boot, let alone do some kind of usable work. There are cloud and VDI (Virtual Desktop Infrastructure) concepts that we could use to alleviate this problem, but they also bring additional performance and management overhead. For example, we could use linked clones - a concept that uses virtual machine virtual disk (usually referred to as "Golden image") assigned to multiple virtual machines at the same time.

As it's impossible to assign the same virtual disk to multiple virtual machines in read-write mode at the same time, a hypervisor would need to create *delta* virtual disk per virtual machine so that we can store data to all virtual machines linked to the same image. But, at the same time, multiple virtual machines using the same image to boot operating system and work can lead significant performance degradation. Booting multiple virtual machines using the same image at the same time leads to a phenomenon called *boot storms*. Result of this phenomenon is a significant delay in booting virtual machines that can lead to hours of boot processes if it happens in a big environment. Furthermore, real-life usage of linked-clone virtual machines is also slower than using regular virtual machines. As there are a lot of HPC applications that are very sensitive to storage I/O and latency, this might have a significant influence on HPC application performance. Containers use a completely different approach, as they use container images that are made up out of different layers unique to specific container and application that's containerized. Generally speaking, for everything else, containers use resources from the underlying operating system, therefor not wasting disk space on content that's used X times for X containers.

Furthermore, as far as compute size goes (compute size equals CPU resources plus memory resources), virtual machines are a much bigger concept. If we want virtual machines to run properly, we need to reserve resources for them, which increases their context size. And that's on top of the fact that virtual machines need a lot of compute power just to run the operating system. And again, we can multiply that by X number of virtual machines. In contrast to that, containers are regular processes running directly on the underlying operating system. We can manipulate container resources by using well-known ideas like *cgroups*, kernel limits (by using *pam_limits* module), *ulimit*, *cpulimit*, etc. We can also easily manipulate container execution priorities, before we start them or after we start them - by using concepts like *nice* and *renice*. We can't implement these concepts on virtual machines easily, and even when we do, similar concepts in virtualization work by manipulating behavior of virtual machine scheduler, not virtual machines *before* they undergo their scheduling process.

These two key architectural differences can be summed up from our perspective to say that virtual machines need a lot of resources and time (measured in minutes) to be spooled up so that they can do the work. In contrast, containers use less resources and take seconds to spool up. All of this leads us to the following conclusion - given the right set of circumstances, containers are easily the better solution to run almost any workload. "Almost any workload" is an important part of that conclusion, as it helps us to answer the next two questions. It's impossible to state that containers are the better solution for any kind of workload. There are various capabilities of virtual machines that containers don't have at all or have partial virtual machine capabilities. For example, working with advanced PCI Express devices like ASICs, GPUs and FPGAs is sometimes possible *only* if we're using virtual machines. This will depend solely on the ASIC/GPU/FPGA vendor, if they're willing to offer software support that enables us to use all their products' capabilities inside containers. We mentioned *k8s-rdma-sriov* project, which is a good example of what can be done with containers if a manufacturer develops deep integration between a hardware device and its corresponding software stack. So, the answer to our next two questions is simple - containers can not necessarily be used for *any* type of workload, but given the right set of circumstances, they *can* be used.

The answer to the last question seems logical - if we have a large infrastructure and compatible workloads, using containers can offer huge strides towards the idea of green computing. Containers offer the capability to use our infrastructure in a much more efficient manner, without added overheads of running virtual machines. That means that we can use our nodes to run more containers, which leads to an environment where compute power is used to do more computing, less general, unnecessary work. So, generally speaking, because of improved efficiency, we're either given a chance to shrink our infrastructure to suit our computational requirements which means less power usage, or we can do more work given the same infrastructure. Both of these scenarios are completely in line with the idea of green computing. We can further this point by saying that introducing various, more efficient CPU architectures will also have a huge impact on performance-per-Watt calculation for any infrastructure.

There are other ways in which using containers can lead to efficiency improvements. We mentioned container images and layers. Container images are objects that consist of sub-objects - layers. These layers can be used to create new images, that can be customized to create new container images. For example, let's consider the following scenario:

- we have a layer that contains all the necessary binaries and libraries for application 1.
- we have a layer that contains all the necessary binaries and libraries for application 2.
- we have a layer that contains all the necessary binaries and libraries for application 3.
- we have a layer that contains all the necessary binaries and libraries for application 4.
- .....
- we have a layer that contains all the necessary binaries and libraries for application n

At any given moment, we can use layer with application 1 and layer with application 2 to create a new image that will have both of these applications inside. Or, we can use layer with application 3 and layer with application n to create a new image that will have those two applications. We just need to create a file that describes which layers need to be combined and create a new image. The resulting image will be small in size, and constantly re-usable in our environment, which is a very good approach from both operational and CapEx standpoint.

If we were to try to do the same thing on a virtual machine, we'd soon realize that it's almost impossible to do this in environments that don't have additional capabilities not native to HPC and regular virtualized environments. There are concepts in VDI that can make this approach happen, but this is a completely different use case with different architectural requirements and prohibitive licensing cost. If we exclude VDI principles as non-viable for our HPC use cases, the only thing left to do for us in HPC environments would be to create X number of virtual machines with different application combinations. That would just help us confirm what we already discussed - that from overhead standpoint, virtual machines are a demanding concept that's not necessarily the best for every use case we might need them for. Having the capability to do compute operations with smaller objects is the way to go, if possible.

This is the reason why, as a concept, Kubernetes is going to have a major role to play in our work. Using containers in manual fashion would mean a lot of additional administrative work, which we want to avoid at all costs. We also don't want to use built-in methods for container placement (concepts like Docker Swarm), as they're too limiting in their capabilities and not aimed at use cases that involve large infrastructures with hundreds or thousands of nodes.

Kubernetes is there to do all that work for us, as its primary objective is to *manage* and *schedule* containers to run assigned workloads on Kubernetes nodes. We want these workloads to be placed on Kubernetes nodes in an intelligent way, that doesn't just use node CPU and memory availability as the only way to determine container placement. We are going to use Kubernetes to provide us with storage orchestration capabilities, as we need permanent storage mounted to our containers, but we also want that done in a secure way. Then, we are going to use Kubernetes for its clustering and self-healing capabilities, as it can re-spool containers that fail (for example, when a Kubernetes node fails).

We are going to focus on problems that are facing our use cases – heterogenous HPC environments based on all technologies - x86, ARM, RISC-V, ASIC, and GPU or any combination thereof. We will make sure to have virtual machine-based data so that we have basis to compare to, which we will later use to work on improving efficiency of our Kubernetes scheduler. Our goal will be to try to answer to the following question - how to combine performance, workload priority, efficiency, and energy footprint to get the best possible "green" result on an infrastructure/data centre level, by using all available technologies. From the green computing perspective, it is crucial to be able to estimate and predict the power and consequently the energy of a data centre [4].

4. Related research

Katenbrink et al. presented the idea of Dynamic Scheduling for Seamless computing for Kubernetes or DYSCO [5], a methodology that uses Kubernetes to schedule applications between different nodes at runtime [5]. Their application of this methodology aims to solve the problem of movement of IoT devices, as Kubernetes itself doesn't really consider the mobility of nodes [5] in multi-layer environments that have Edge, Fog and Cloud infrastructure layers. The reality is that not all the HPC environments have capability to do a fallback to cloud solution. Furthermore, there might be privacy problems involved, especially if we're working in fields like molecular biology, stem cell research, AI-assisted facial or speech recognition, motion estimation for video surveillance etc. Load balancing workloads across multiple nodes is a very important aspect of container-based workloads. We can look at this problem from two different levels – how to load balance workloads across nodes, and how to load-balance traffic that's coming to those nodes.

As we mentioned, Kubernetes' default workload load-balancing schedulers are based on Filtering and Scoring. These mechanisms are decent when we have more generic environments running similar types of workloads – for example, many containers running Apache web server, or PHP solution. If we're talking about traffic load balancing when using Kubernetes in larger environments, we will have to investigate additional solutions, ranging from HAproxy and nginx to hardware load balancers. Qingyang et al. explain the reason in their paper about how to design a multi-metric load balancer for Kubernetes, by stating that the built-in load balancing component has a few functions, only supports static load balancing strategies, and cannot adapt to complex needs [6]. They propose a solution that checks additional I/O metrics for network IO, CPU usage, memory usage and disk IO in real time with ability to check application running state, which is the basic idea behind load balancing on OSI Layer 4 or 7, depending on scenario and application used.

Scalability of Kubernetes-based solutions is very important for any kind of enterprise-level deployment. Previously, large-scale deployments went through a phase where workloads were installed on physical servers (in HPC space, often on supercomputers), and virtualized servers, by using an underlying hypervisor. Dewi at al. concluded that virtualizing using containers enables much flexibility for capacity management in a server [7]. Furthermore, their research indicates that implementation of scalability to the containers (on multiple nodes) reduces the CPU usage in a pod due to the distribution of loads [7], by running them across multiple nodes.

Flexibility and effectiveness of Kubernetes-based environment plays a key role in designing a Kubernetes-based data centre. Tesliuk et al. discussed the possibility of using MPI, HPC and data parallelism for HPC applications. Deployment of virtual MPI cluster allows us to run MPI software without any modification of the code [8]. They also note that detailed analysis of the performance of this approach is a subject of future research [8]. By using kube-openmpi [9], we have a standardized way of integrating Kubernetes with Open MPI, which is a good starting point for our research.

When designing HPC data centre and thinking about using container-based methodology, performance of Docker Swarm and Kubernetes-based solutions for HPC workloads is the one of the key aspects to evaluate. Cloud-based methodologies at scale (when using massive number of nodes) are problematic at best – we need to have enough resources for these nodes, which means multiple regions or cloud datacentres. That, in turn, means increase in latency, along with additional infrastructure complications. Let's use Microsoft Azure as an example. If we wanted to deploy one thousand nodes in Azure across five or six different regions, we would first need to connect those regions by using VPN gateways, software services that aren't necessarily consistent in bandwidth and latency. That's a big administrative overhead, as well, and if we need to move those workloads to other regions, it's only going to get worse, even without discussing the variable cost and other problems across different regions. Which is why local, high-speed networking in local datacentres (local infrastructure) is the best way to get low-latency, high-bandwidth network backend. Beltre at al. note that Kubernetes and Docker Swarm can achieve near bare metal performance over RDMA communication when high performance transports are enabled [10]. They also note that using Kubernetes presents overheads for several HPC applications over TCP/IP protocol [10]. But, when Infiniband is used as network backend, performance in within 1% of the bare metal performance both for Docker Swarm and Kubernetes in low complexity applications like HPL, while HPCG overhead drops from 13.15% to 10.31% [10].

Fairness of workload distribution is also an important characteristic of Kubernetes scheduler. By design, default Kubernetes schedulers don't have any available options here, if we discount node/pod affinity and anti-affinity capabilities. These enable us to do a static selection which nodes are going to be selected by Kubernetes scheduler. As this approach decreases efficiency and agility, using it might mean that we can't schedule a workload to be run on Kubernetes pod. According to Hamzeh et al., fairness is a missing point in Kubernetes [11]. They also note that some applications may require intensive resources such as CPU and memory that should be maximized to satisfy them [11], which can be done by using different fair allocation policies based on algorithms – Hamzeh et al. MLF-DRS [12], Hamzeh et al. FFMRA [13], and Ghodsi et al. DRF [14]. All very important points when designing a Kubernetes scheduler. Efficiency and power usage for a large HPC datacenter are very important from the design standpoint, as we live in the era of green computing. From datacenter standpoint, green computing principles can be used on multiple levels:

- Approach to building a data centre from architectural standpoint – for example, materials used, energy certifications etc.
- Approach to energy recovery, waste management and utilization – for example, are we only using grid power or have solar or wind-based power generation.
- Efficient design from the hardware standpoint – for example, trying to get either as much performance from a given power envelope or customizing our hardware design to fit a specific set of workloads to be as efficient as possible for those specific workloads.
- Data centre location – is it inland, underwater, or in higher altitude, as this will have significant impact on data centre design. For example, Microsoft conducted a series of tests using Northern Isles data centre underwater (Project Natick), which might define Microsoft's future data centre strategy.

Our approach will be to be as efficient as possible in terms of workload placement to be done in an environmentally responsible way. That means that we will not be searching just for outright performance – we will try to place workloads to be done with as less energy usage as possible. Guzek et al. proposed a holistic model for hypervisors in this respect, by dividing a system into three layers: physical node (server), container (VM), and task (application) [4]. Their research yielded two approaches – multi-stage linear regression approach and neural network approach.

By using the first approach, they were able to accurately estimate the power consumption in a virtualized system with acceptable temporary and cumulative errors [4]. Second approach introduced a lot more errors. This stems from the usage of neural networks, as they require a lot of time for training. As they state, research should be continued by adding more elements to their model, and to be used on a larger set of applications and machines [4].

The other aspect of achieving high efficiency comes from the platforms used. Maqbool et al. noted that ARM-based CPUs performed four times better in tests of the performance to power ration on a single core and 2.6 times better for tests on multiple cores [15]. In terms or memory performance, as expected, Intel CPUs performed significantly better in massively parallel applications, while ARM CPUs performed better in other I/O bound applications [15]. But one of the most important conclusions is related to other parts of the HPC system used for testing revealed the impact of network bandwidth, workload type, and messaging overhead on scalability, floating point performance, the performance of large-scale applications and energy-efficiency of the cluster [15]. This was also emphasized by Nam et al. They state that improving the QoS of latency-sensitive workloads is crucial to the users at a data centre [16].

5. Application of programmable, custom Kubernetes scheduler on HPC workloads

From a container workload perspective, we need to consider multiple different scenarios. In the following cases we will assume that GPU can in general case be replaces by other type of HW accelerator. Let's go through them:

- If the workload that we need to execute cannot be GPU-accelerated, we need to have capability to place those workloads on node(s) without GPUs. We don't want to waste CPU on a GPU-enabled node on workloads that cannot use the GPU.
- If the workload we need to execute can be GPU-accelerated, it needs to be placed on a GPU-enabled node to be done as fast and as efficiently as possible.
- If we have nodes with multiple different GPUs, we need to find the best way of using them (in terms of power, speed, and overall efficiency).
- If the workload that needs to be done is bound to a certain CPU architecture (for example, HPC application that's only available for ARM), it needs to be placed on an adequate host that meets the demands of that specific workload.
- If the workload that needs to be done isn't bound to a specific CPU architecture (for example, HPC application that's available both on x86 and ARM CPU architectures), it needs to place the workload on the node that's going to go through that workload in an efficient manner from an overall environment and efficiency perspective.
- If there's a chance for further optimization on the host-level post container deployment, migrating a container to another node could offer better efficiency.
- If there's a node failure, workload needs to be migrated to another node and either continued or restarted (depending on the HPC application that we're using).
- It needs to take power per kWh pricing as a variable when making decisions on workload placement.
- For ease of use, it would be best to use multi-architecture container images, a Docker image feature that allows the specification of the same container image for multiple CPU architectures [5].
- Therefore, a fair, efficiency-focused custom scheduler for Kubernetes needs to take all these scenarios into consideration and have the following characteristics:
- It needs to take the actual workload into account (whether it can be accelerated by using a GPU or other type of accelerator, if these devices support Docker or Podman containerization). Therefore, it needs to be able to distribute workloads based not only on capacity or CPU speed, but on node capability.
- It needs to have insert capability, so that we can insert workloads into the existing workload queue at will for some priority workloads.
- It needs to be mindful of the energy footprint as one of the key priorities.
- It needs to be able to learn about workload placement either directly or by using external service, and, if possible, interact with the underlying OS platform to redistribute workloads on node level.

This would give us „the best of both worlds“ type of result – we will have the capability to insert any high-priority workload to the fastest nodes, but at the same time, be mindful of every other non-priority workload that can be placed on the best „performance-per-Watt“ node selected by our scheduler. Kubernetes scheduler component is responsible for deploying pods and services in the suitable nodes [11]. Fastest nodes aren't just nodes with fastest CPU's – it's a terminology that we use to describe nodes where the workload should be placed to be fastest in terms of execution time. We mentioned the idea of placing workloads based on capabilities – let's call that capability-based placement. For example, if we have a HPC workload that can take extensive usage of NVIDIA GPU, then it makes sense to put that workload on a node that has a NVIDIA GPU and assign it to that specific workload.

6. Initial performance measurement and baselining

We must establish performance baselines so that we can make sure that our programmatic approach to Kubernetes scheduler gives real-life benefits. First step in that process is going to be introducing general and HPC-workload simulation benchmarks to measure baseline performance.

After that, next step will be cross-referencing these results with power usage data points and develop methodology that will take both data sets as inputs to decision-making process for Kubernetes scheduler. Generic Kubernetes schedulers use a two-step process to select a node for running a Kubernetes pod:

- Filtering – does a node have enough compute resources to handle the workload.
- Scoring – after filtering, scoring system is applied to remaining nodes to find the best fit for any given workload.

There are some attributes and methods that we could configure in generic scheduler by using Extension Points, Scheduling Profiles and Policies, but these are nowhere near fine enough to manage big environments like HPC-based data centres. By extension, that also means that we can't get finely grained workload-to-node assignments which can lead to much greater energy consumption, especially on a HPC data centre scale. Number of cores in HPC data centres are often in the millions, and numbers of servers running those cores are often in the hundreds of thousands. This represents a significant chance for optimization. These synthetic and HPC tests will include various cross-platform benchmarks like:

- sysbench tests for CPU (single and multi-core), and memory performance – a good general measurement of all-around CPU and memory I/O, by using prime number calculation (64-bit integers). Memory tests are sequential read/write tests.
- stress-ng tests for CPU (single and multi-core), and memory performance.
- HPCC benchmarks – a collection of benchmarking applications that consists of multiple tests Specifically, the suite is composed of several well-known computational kernels (STREAM, HPL, matrix multiply – DGEMM, parallel matrix transpose – PTRANS, FFT, RandomAccess, and bandwidth/latency tests – b_eff) that attempt to span high and low spatial and temporal locality space [1]. This is a standard HPC benchmark suite that has been used for years.
- HPCG (High Performance Conjugate Gradient) benchmarks – a second standard HPC benchmark suite that consists of multiple tests that are much more memory-bound than HPL/HPCC stack of benchmarks. HPCG is composed of computations and data-access patterns commonly found in scientific applications [2].

By running these benchmarks on our Kubernetes nodes, we can get a really good idea about the general level of performance we can expect from our nodes. We could also run all these tests in Docker/Podman containers, to get information how the physical performance any given Kubernetes node translates to container-based performance. That will also give us a good insight into overall compute overhead of the underlying OS. We can also use information from servers' onboard management hardware to get a rough estimate of how much power it uses, which we will turn into a more reliable power measurement method with time and additional data. We're going to use a shell script to run through all these tests on physical Kubernetes nodes. Output from that script is going to be CSV file which we can then use to do our analysis. Running these benchmarks in a container requires external storage for results, which we will achieve by mounting a NFS storage from network attached storage device. We're going to use the same basic shell script principle to get the same type of output to make our results directly comparable.

7. Preliminary test results

We created a small test environment with different x86 and ARM-based nodes. Purposefully, we chose different x86 nodes in terms of capabilities so that we can gather as much data as possible, as shown in the following table:

Hardware	Kubernetes node specifications			
	Node 1	*Node 2*	*Node 3*	*Node 4*
CPU type	2xXeon E5-2630L	2xXeon E5-2450L	2xXeon E5-2680	1xCortex A72 (ARM v8)
Core number /SMT core number	12/24	16/32	16/32	4/4
Base frequency	2 GHz	1.8 GHz	2.7 GHz	1.8 GHz
L1 cache (per core)	64 KB	64 KB	64 KB	80 KB
L2 cache	256 KB per core	256 KB per core	256 KB per core	1 MB shared (with GPU)
L3 cache (shared)	15 MB	20 MB	20 MB	-
Memory	96 GB	96 GB	96 GB	4 GB
Memory frequency	1333 MHz	1333 MHz	1333 MHz	3200 MHz
Average power consumption	220 W	230 W	250 W	13W

Table 1. Nodes used for preliminary testing

Let's now contrast that to testing results, using previously mentioned benchmarking tools:

Hardware	Kubernetes node performance			
	Node 1	*Node 2*	*Node 3*	*Node 4*
Sysbench single core (events per second)	643.89	593.16	815.52	1787.74
Sysbench multicore (events per second)	1299.85	1196.87	1630.47	3574.08
Sysbench memory (total time)	9.55366	10.0014	7.57572	10.0002
Stress-ng single core (bogo ops)	12258.4	11274.4	14836.4	1253
Stress-ng multi core (bogo ops)	24286.8	22398.0	29626.8	2134
HPL (Gflops)	2.01437	1.76892	2.47621	10.9701
RandomAccess (GUP/s)	0.0622563	0.0471568	0.0416779	0.0075283
FFT (Gflops)	1.99787	1.74458	2.06731	0.186364
STREAM (GB/s)	6.19918	5.35783	6.1504	0.537999
PTRANS (GB/s)	1.59295	1.35349	1.70072	0.130684
DGEMM (Gflops)	1.52678	1.32662	1.58446	4.19122
HPL time (s)	331.03	376.96	269.29	60.7851
HPCG (Gflops)	1.48279	1.37273	1.85292	0.100408

Table 2. Node performance

We used the same settings for all tests and limited the amount of available CPU cores for all tests on all servers to 4, so that we can have basis for per-CPU-core analysis. We also turned off all of the non-used CPUs (we only used cores from 0-3 on x86 platforms), and HyperThreading so that we can get a good estimate on power usage and performance. There are other factors that can influence efficiency and speed of HPC environments - we already mentioned low-latency and high-bandwidth networking as a must. HPC is increasingly faced with a QoS-sensitive computing demand, coming from that class of applications whose correctness depends on both performance and timing requirements, and the failure to meet either of them is critical [17]. Assigning network adapters in a direct fashion (PCI Express forwarding) or via SR-IOV could be the step in the right direction, which is the next step in Linux kernel development as not all of these features are supported if we're using containers as compute objects.

8. Conclusion and future work

If we take a deeper look at the preliminary test results, we can easily notice some trends. Although x86 CPU architecture has been the de-facto standard for decades now, its performance per Watt leaves a lot to be desired for. And although ARM's primary target isn't necessarily outright performance, even an inexpensive platform based on ARM A72 can give it a run for its money in certain use cases. Scaling that idea up to many nodes, coupled with other CPU architectures like RISC-V seems like a viable option for a high-performance, yet energy efficient HPC solution. Of course, we cannot discount x86 as its maturity and the scope of available hardware and software solutions still make it a platform to be used. Combining our preliminary test results with all the related scientific work that was previously done with relation to Kubernetes and HPC makes us confident that there are a lot of unexplored areas yet to be analysed. Therefore, focus of our future work should be on using all the measured performance parameters and expand the amount of performance testing to have as many data sets as possible for creating our much more efficient Kubernetes scheduler. We will also add more platforms and nodes to our environment – more ARM-based hardware and some RISC-V-based hardware, as well. This will help us gain even more understanding about performance scaling as we add more nodes to our environments. We will use all these tests to make scheduler as efficient as possible from the energy usage standpoint while keeping an eye on workload priority and distribution fairness. We will also have to introduce a methodology to do continuous power usage measurement. We can use the following data mechanisms for that:

- Remote management mechanisms like iLO (Integrated Lights Out), IPMI (Integrated Platform Management Interface) and others to get continuous stream of data from Out-of-band management technologies at our disposal.
- Software-based management utilities via hypervisor or OS-based tools (if available).
- PDU (Power Distribution Unit) remote management, which can give us much more insight and information about power usage over time.

This overall concept will require a fully orchestrated, programmatic approach to delivering containers. This is specifically what our Kubernetes scheduler needs to do for us. That also means that we must prepare a container image library for every HPC application and workload type, which requires time. Some of the latest research done by Zhou et al. clearly states that there are ways of integrating HPC schedulers and containers, which is why they developed a tool named Torque-Operator that bridges Kubernetes and TORQUE, one of the most commonly used HPC schedulers [18].

We can even go to greater lengths there and combine multiple container layers to create unified containers with multiple applications – such is the power of containers. This will help overcome major challenges implied by the HPC: performance, the total property cost (the cost of electric energy), programming, scalability and reliability [19]. Time involved in creating these customized containers will make it a lot easier for us to run all our HPC workloads in the future. So, it should be time well spent.

9. References

[1] https://www.researchgate.net/publication/238654737 (2005). University of Tennessee Knoxville, Oak Ridge National Laboratory, MITRE, High Performance Computing Center (HLRS), Information Sciences Institute, University of South California, MIT Lincoln Lab, IBM Austin, Lawrence Berkeley National Laboratory, University of Tsukuba, Introduction to the HPC Challenge Benchmark Suite, Accessed on: 2020-10-10

[2] Dogarra, J.; Heroux A. M. & Luszczek, P. (2016). High-performance conjugate-gradient benchmark: A new metric for ranking high-performance computing systems, The International Journal of High Performance Computing Applications Vol.30(I)3-10, DOI: 10.1177/1094342015593158

[3] Smith, J. & Hamilton, A. (2015). Massive affordable computing using ARM processors in high energy physics, Journal of Physics: Conference Series vol.608, IOP Publishing, DOI:10.1088/1742-6596/608/1/012001

[4] Guzek, M.; Varrette, S.; Plugaru, V.; Pecero, E.J. & Bouvry, P. (2013). A Holistic Model of the Performance and the Energy-Efficiency of Hypervisors in an HPC Environment, COST ICO804 European Conference on Energy Efficiency in Large Scale Distributed Systems, Volume 8046, Pages 133-152, DOI: 10.1007/978-3-642-40517-4_13

[5] Tatenbrink, A.; Seitz, A.; Mittermeimer, L; Mueller, H. & Bruegge, B. (2018). Dynamic Scheduling for Seamless Computing, IEEE 8th International Symposium on Cloud and Service Computing (SC2), IEEE, ISBN: 978-1-7281-0236-8, DOI: 10.1109/SC2.2018.00013

[6] Liu, Q.; Haihong, E. & Song, M. (2020). The Design of Multi-Metric Load Balancer for Kubernetes, Proceedings of the Fifth International Conference on Inventive Computation Technologies (ICICT-2020), IEEE, ISBN: 978-1-7281-4685-0 , DOI: 10.1109/ICICT48043.2020.9112373

[7] Dewi, L.P.; Noertjahyana, A.; Palit, H.N. & Yedutun, K. (2019). Server Scalability Using Kubernetes, The 2019 Technology Innovation Management and Engineering Science International Conference (TIMEs-iCON2019), ISBN:978-1-7281-3755-1, DOI: 10.1109/TIMES-iCON47539.2019.9024501

[8] Tesliuk, A.; Bobkov, S.; Ilyin, A.; Novikov, A.; Poyda, A. & Velikhov, V. (2019). Kubernetes container orchestration as a framework for flexible and effective scientific data analysis, 2019 Ivannikov Ispras Open Conference (ISPRAS), 67-71, ISBN:978-1-7281-6055-9, DOI: 10.1109/ISPRAS47671.2019.00016

[9] https://github.com/everpeace/kube-openmpi (2017). Omura, S., kube-openmpi [Accessed: 04-Jan-2021].

[10] Beltre, A.; Saha, P.; Govindaraju, M.; Younge, A.J. & Grant, R.E. (2019). Enabling HPC workloads on Cloud Infrastructure using Kubernetes Container Orchestration Mechanisms, 2019 IEEE/ACM Workshop on Containers and New Orchestration Paradigms for Isolated Environments in HPC (CANOPIE-HPC), 11-20, ISBN:978-1-7281-6028-3, DOI: 10.1109/CANOPIE-HPC49598.2019.00007

[11] Hamzeh, H.; Meacham, S. & Khan, K. (2019). A New Approach to Calculate Resource Limits with Fairness in Kubernetes, 2019 First International Conference on Digital Data Processing (DDP), 51-58., ISBN:978-1-7281-5363-6, DOI: 10.1109/DDP.2019.00020

[12] Hamzeh, H.; Meacham, S.; Virginas B. & Khan, K. (2019). MLF-DRS: A Multi-level Fair Resource Allocation Algorithm in Heterogeneous Cloud Computing Systems, 2019. IEEE 4th International Conference on Computers, Communications and Systems (ICCCS), ISBN:978-1-7281-1322-7, DOI: 10.1109/CCOMS.2019.8821774

[13] Hamzeh, H.; Meacham, S.; Khan, K.; Phalp, P. & Stefanidis, A. (2019). FFMRA: A Fully Fair Multi-Resource Allocation Algorithm in Cloud Environments, 2019 IEEE SmartWorld, Ubiquitous Intelligence & Computing, Advanced & Trusted Computing, Scalable Computing & Communications, Cloud & Big Data Computing, Internet of People and Smart City Innovation, 279-286, ISBN:978-1-7281-4034-6, DOI: 10.1109/SmartWorld-UIC-ATC-SCALCOM-IOP-SCI.2019.00091

[14] Ghodsi, A.; Zaharia, M.; Hindman, B.; Konwinski, A.; Shenker, S. & Stoica, I. (2011). Dominant Resource Fairness: Fair Allocation of Multiple Resource Types, Proceedings of the 8th USENIX conference on Networked systems design and implementation, 323-336.

[15] Maqbool, J.; Oh, S. & Fox, G.C. (2015). Evaluating ARM HPC Clusters for Scientific Workloads, Concurrency and Computation: Practice & Experience, Wiley, ISSN: 1532-0634.

[16] Nam, Y.; Kang, M.; Sung, H.; Kim, J. & Eom, H. (2016). Workload-aware resource management for software-defined compute , Cluster Computing, Vol. 19, Issue 3, 1555-1570., DOI: 10.1007/s10586-016-0613-6

[17] Fitch, J.; Agosta, G.; Ampletzer, P.; Alonso, D.A.; Cilardo, A. ; Fornaciari, W. ; Kovač, M.; Roudet, F. & Zoni, D. (2015). The MANGO FET-HPC Project: An Overview, 18th IEEE International Conference on Computational Science and Engineering, ISBN:978-1-4673-8297-7, DOI: 10.1109/CSE.2015.57

[18] Zhou, N.; Georgiou, Y.; Pospieszny, M.; Zhong, L.; Zhou, H.; Niethammer, C.; Pejak, B.; Marko, O.; Hoppe, D. (2021). Container orchestration on HPC systems through Kubernetes, Journal of Cloud Computing: Advances, Systems and Applications, DOI: 10.1186/s13677-021-00231-z

[19] Novacesku, F. (2011). Trends in the field of High Performance Computing and influence on HPC field, Annals of DAAM for 2011 & Proceedings of the 22nd International DAAAM Symposium, Volume 22, No.1, ISSN 1726-9679.

DOI: 10.2507/32nd.daaam.proceedings.012

DISTRIBUTED STORAGE PERFORMANCE MEASUREMENT IN A VIRTUALIZED ENVIRONMENT

Vedran Dakic, Davor Baksa & Jasmin Rezepagic

This Publication has to be referred as: Dakic, V[edran]; Baksa, D[avor] & Redzepagic, J[asmin] (2021). Distributed Storage Performance Measurement in a Virtualized Environment, Proceedings of the 32nd DAAAM International Symposium, pp.0084-0089, B. Katalinic (Ed.), Published by DAAAM International, ISBN 978-3-902734-33-4, ISSN 1726-9679, Vienna, Austria
DOI: 10.2507/32nd.daaam.proceedings.012

Abstract

This paper describes a very simple, yet efficient methodology to measure storage performance in a virtualized environment. By "virtualized environment", we are talking about an environment that's based on a hypervisor and virtual machines, commonly used compute objects in IT industry today. Having in mind that virtual machines need a significant amount of storage in terms of capacity, lowest latency possible, and best possible throughput in terms of outright performance, having a capability to measure storage performance in a distributed environment is a welcome addition to any datacenter. If nothing else, it could give us a good performance baseline that we can use for future storage performance tuning. That's why - as a result of this paper - we want to use a virtual machine-based methodology to implement a BASH (Bourne Again SHell) shell script-based set of automated test procedures that are easy to use and don't require a lot of time to be interpreted.

Keywords: hypervisor; virtual machine; storage performance; open-source; multipathing

1. Introduction

Measuring storage performance in virtualized environments - our first goal - has always been a difficult issue to tackle. There are different problems in the way - which tests to use, how to simulate workloads, when to start workloads, which workloads to use... All of that is very important to the second subject of this paper, which is all about functionality and performance. The last step - which is fundamental to this paper - is actually about tuning storage and hypervisor settings to achieve better storage performance for that environment. The caveat being that every environment is different - different workloads, different servers, different switches, different cabling... The aim of this paper is primarily to introduce a testing methodology for storage performance measurement. After that, it's also about applying that methodology to a specific environment in all directions - top-down, bottom-up, left-right.

As a basis for this paper will use OSNEXUS Quantastor open-source storage system through various connectivity scenarios and across multiple platforms to determine whether such a system is suitable for use in a variety of business environments. Companies often require dedicated storage systems that they use in conjunction with virtualization technology, and the paper will try to simulate loads when using such systems and measure performance metrics with keeping track of all of the results, possible difficulties or benefits, conclude on the cost-effectiveness of such a system, and tuning specific hardware or software adapters and attributes with respect to the measured results.

2. Testing environment description

Our testing environment aims to simulate a virtualized environment that uses some of the most commonly used solutions within the context of virtual machines and the performance of used disk systems. For this purpose, two HP Proliant DL380p Gen8 servers were prepared, one as a compute node, the other for storage. Server1 (compute node) took on the role of virtual machine hypervisor (VMware ESXi, a well-known hypervisor solution), while OSNEXUS system was installed on Server2 (storage system), which enables all capabilities of a SDS (Software Defined Storage) platform.

For testing purposes, we used Linux virtual machines, configured by using VMware paravirtual storage controller. VMware presents the Paravirtual virtual storage controller as the one to use when we need lower latency, better throughput, and lower CPU overhead from the host perspective [1]. In the Linux variant, sixteen (16) virtual machines based on the CentOS operating system were prepared, and LAMP (Linux-Apache-MySQL-PHP, Hypertext Preprocessor) was installed on each instance. Each virtual machine had a PHP-based website that uses the MySQL database and running on the Apache 2.4 platform. These virtual machines were using DD (Copy and Convert), IOPING, and FIO (Flexible I/O tester) tools to provide metrics such as bandwidth, latency, and IOPS for each environment used.

As for the storage system (Server2), 8 SSD (Solid State Disk) SATA (Serial Advanced Technology Attachment) disks and 1 NVMe (Non-Volatile Memory express) disk were used in the most commonly used RAID (Redundant Array of Inexpensive Disks) array configurations (RAID 5 and RAID 10 for SATA disks) and direct access for the NVMe disk. NVMe SSDs, as PCIe (PCI Express) devices, employ the high-speed PCIe bus to communicate with the host [2]. Also, NVMe is designed from the ground up for non-volatile memory technology such as NAND and 3D XPoint over PCIe [3]. Reasons behind our choices of SSDs and NVMe devices are:

- RAID 5 is often used in systems where attempts are made to ensure maximum capacity with guaranteed minimum redundancy (single drive redundancy - in a field with 8 disks, the usable capacity is 7). In this variant, more focus is added to capacity versus speed and redundancy.
- RAID 10 is also a commonly used RAID system configuration - especially in virtualized environments - where 50% utilization of available disk capacity is ensured, but 50% of disk space is redundant and better performance versus RAID 5 array is ensured.

Both variants use 8K and 64K block sizes on a disk system presented to virtual machines to measure performance metrics depending on the purpose. For example, the 64K variant is recommended for the general storage scenario while the 8K variant is recommended for usage scenarios where the IOPS (Input/output Operations Per Second) metric versus raw bandwidth is more important - a good example for this usage model would be databases.

Also, two different connections of Server1 and Server2 were used - Ethernet and Fibre Channel (FC), in single-uplink (10 Gbit/s Ethernet or 8 Gbit/s FC) and four links using multipathing (40 Gbit/s Ethernet or 32 Gbit/s FC) configurations to be able to measure scalability. For each iteration of the test, we also changed the amount of available working memory on the storage server (32 GB / 64 GB) since OSNEXUS is designed to use the available working memory for the cache where possible. We can also tune the amount of memory that goes to caching out of the overall server memory amount, as it's one of the available options.

IOPS usage in idle state was tested for 24 hours, so we set up a network connection to a Synology NAS, installed Zabbix in a Docker container on Synology NAS (Network Attached Storage), and Zabbix agent on all of the VMs (Virtual Machines) to set everything up for IOPS monitoring. We initially started with running all of the tests manually (FIO, DD, IOPING) and typed in the table, but as we soon realized it didn't make sense because the manual process was slowing us down significantly. That's the reason why we created a script to do the work for us, as we will describe in "Testing methodology and storage performance measurement complexity".

While developing our script, we started with one virtual machine, tested everything, added a couple of virtual machines more, re-tested and, as the number of virtual machines grew, we also moved to the next step in our testing process, which was to use a simple methodology to start all of the tests in sync, on all of the virtual machines at the same time. We did this so that we can test the absolute worst case - when all of the virtual machines are overloading the storage system with I/O requests. There was one more reason - experience teaches us that we can't use all of the available bandwidth by using just a single object accessing storage to extract the absolute maximum that our storage device can provide.

For this purpose, we used mobaXterm's MultiExec mode to do multi-VM testing and had to overcome a few issues along the way. Specifically, as some of the tests are simple to use within our testing context - we can just start them to run for a specific amount of time. *dd* utility, on the other hand, doesn't do its job that way - it runs until the test is done, which means that it will run for a variable amount of time on multiple virtual machines. This will be further compounded by the fact that we are running that test on multiple virtual machines at the same time, which will add more latency to the process (summary of latencies plus a bit more random latency), and all of that will have direct negative implications on bandwidth. If we wanted to avoid that, there are VMware-based QoS (Quality of Service) methodologies to do that - via the usage of Storage I/O Control and by creating Virtual Machine Storage Policies. This would be a logical thing to do at the *end* of our process when we find out the actual performance of our storage device. This is where implementing a semaphore-like functionality and sleep logic (for synchronization) in our script was hugely useful for us as we can sync all of the tests to start at the same time by using it. We included a link to our script in the " Testing methodology and storage performance measurement complexity" section below.

3. Testing methodology and storage performance measurement complexity

We believe that our script clearly shows how involved the programming work behind the virtual machine and storage performance testing premise can be. Our script is available at https://github.com/vEddYcro/BASH/test-dd.sh, feel free to download it, work with it, modify it, comment on it as it's too long to be directly copy-pasted in this paper. When we dissect our script, there's a clear thought process behind it. Let's explain in detail our design criteria behind the script:

- All local temporary (temp) files are removed from each machine because this is the default place to store the temporary results of any previous tests.
- A DD write test with 64K disk block size (disk mapped to /64k-storage) is performed and for each machine, the final result is written to share (/fiologs/log) in the form $ host-dd-64k-write-final.txt
- DD read test with 64K disk block size (disk mapped to /64k-storage) is performed and for each machine, the final result is written to share (/fiologs/log) in the form $ host-dd-64k-read-final.txt
- A DD write test with 8K disk block size (disk mapped to /8k-storage) is performed and for each machine, the final result is written to share (/fiologs/log) in the form $ host-dd-8k-write-final.txt
- A DD read test with 8K disk block size (disk mapped to /8k-storage) is performed and for each machine, the final result is written to share (/fiologs/log) in the form $ host-dd-8k-read-final.txt
- FIO read test with 8K block size is performed on storage with 8K block size and final results are written to share (/fiologs/log) in the form $ host-read-bw-8-8.txt for bandwidth and $ host-read -IOPS-8-8.txt for IOPS results
- FIO read test with 8K block size on storage with 64K block size is performed and the final results are written to share (/fiologs/log) in the form $ host-read-bw-8-64.txt for bandwidth and $ host-read -IOPS-8-64.txt for IOPS results
- FIO read test with 64K block size in storage with 8K block size is performed and the final results are written to share (/fiologs/log) in the form $ host-read-bw-64-8.txt for bandwidth and $ host-read -IOPS-64-8.txt for IOPS results
- FIO read test with 64K block size is performed on storage with 64K block size and final results are written to share (/fiologs/log) in the form $ host-read-bw-64-64.txt for bandwidth and $ host-read -IOPS-64-64.txt for IOPS results
- FIO test of writing with 8K block size on storage with 8K block size is performed and the final results are written to share (/fiologs/log) in the form $ host-write-bw-8-8.txt for bandwidth and $ host-write -IOPS-8-8.txt for IOPS results
- FIO test of writing with 8K block size on storage with 64K block size is performed and the final results are written to share (/fiologs/log) in the form $ host-write-bw-8-64.txt for bandwidth and $ host-write -IOPS-8-64.txt for IOPS results
- FIO 64K block size write test is performed on 8K block size storage and final results are written to share (/fiologs/log) in the form $ host-write-bw-64-8.txt for bandwidth and $ host-write -IOPS-64-8.txt for IOPS results
- A FIO 64K block size write test is performed on a 64K block size storage and the final results are written to share (/fiologs/log) in the form $ host-write-bw-64-64.txt for bandwidth and $ host-write -IOPS-64-64.txt for IOPS results
- Direct (non-cached) latency check is performed with IOPING tool on 8K storage and the result is written to share in the form $ host-IOPING-DIRECT-8K.txt
- Cached latency check with IOPING tool on 8K storage is performed and the result is written to share in the form $ host-IOPING-CACHED-8K.txt
- Direct (non-cached) latency check is performed with the IOPING tool on 64K storage and the result is written to share in the form $ host-IOPING-DIRECT-64K.txt
- Cached latency check with IOPING tool on 64K storage is performed and the result is written to share in the form $ host-IOPING-CACHED-64K.txt
- Summarize the results obtained for all virtual machines into a single result for all variations of tests and save to share (/fiologs/log/TOTAL)

The script also implements a semaphore system that ensures that the next test step does not move until all virtual machines report that they have completed the current test.

4. Improving storage performance

By using our script, we got a wealth of data about storage performance. The next step in our optimization process is to use that data to modify hypervisor and storage settings to try to improve virtual machine storage performance. Therefore, this initial dataset can be used as a storage performance baseline that we are going to use to check new data against after we implement some configuration changes to our environment. As with any other technology used for any purpose, there is always a way to optimize the delivery process of a given service, which is true especially in the case of storage systems. Starting from the lowest layer - the hardware used to various software optimizations, the user can almost always expect at least some positive changes in terms of the resulting system performance.

4.1. Improving storage performance on hardware level

As various technologies have been used for this paper, it is important to be familiar with how all of them work to find the optimal method of using all of them together in all variants of use.

Specifically, in this case - the already mentioned optimizations at the lowest layer would be the active use of both server power supplies to ensure that the speed of the PCIe bus is not limited and thus create a "bottleneck" in the data flow at the lowest layer. A lot of servers will drop their PCI Express performance if we have only one power supply installed. In addition, detailed knowledge of the equipment used is crucial, so it is extremely important to get acquainted with its operation and the best way is to study this is the manufacturer's documentation in detail. An example of this would be something seemingly banal like the arrangement of memory modules in the slots on the server - if the memory modules are not set according to the default method defined by the manufacturer it will not be possible to use all the capabilities - in our case on a storage server using 2 processors is to fill the memory slots on the server motherboard in a specific order, because otherwise, you can not activate Advanced ECC (Error Correction Code) memory profiles that ensure optimal performance - effectively working memory runs at a lower clock speed and with less bandwidth and thus directly affects the performance of the storage server. Also, separating PCI Express devices across CPUs by using multiple PCI Express riser cards is absolutely crucial if we want to achieve maximum performance from the hardware level.

4.2. Improving storage performance on the storage network level

As both FC and Ethernet connections were used for performance metrics testing, it is important to distinguish how both technologies connect to the storage server itself. FC is more of a native plug and play variant when using a direct connection between servers does not require additional configuration nor is there room for additional optimizations on that layer. If multiple hosts were used, i.e. if there was an FC switch between the servers, there would be additional optimization possibilities in terms of the smart definition of FC zones and careful planning of connection methods to maximally mitigate negative scenarios in case of unwanted outages. An example of this would be that, for example, when using multiple FC cards and switches, we don't connect individual FC connectors on a single card to one switch, but that a certain number of connectors of one FC card is connected to one FC switch and thus to another set of connectors of other FC cards and switches. This ensures the maximum available bandwidth in scenarios when a single FC port fails, the card itself, or the switch during which time high availability is constantly ensured in the background. Although performance degradation will occur in this scenario (because some of the uplinks are no longer physically connected) storage service will not be interrupted.

When using Ethernet - iSCSI connections - there are a several settings that can be adjusted to optimize performance - primarily because iSCSI as a protocol relies on Ethernet which is relatively universal and its primary purpose is not, as in the case of FC, to provide super-fast, super low-latency method communications with the storage server. These ideas came much later when Ethernet standards were further developed and concepts like RDMA/ROCE were introduced.

Primarily, it is recommended to use network cards that are declared as hardware HBA (Host Bus Adapter), i.e. those that can configure iSCSI initiators in firmware to establish a connection to the server (iSCSI target) before booting the operating system - this way all the work on target-initiator relationship takes place on such a card. That means that no server resources are used and thus better performance is achieved. In our case, the software HBA (iSCSI initiator) was used because such a scenario covers a much higher percentage of usage in real conditions because most companies use this model. If the performance of your hardware bus adapters (HBAs) is unsatisfactory or your SAN storage processors are overutilized, you can adjust your ESXi hosts' maximum queue depth value [4]. Hardware HBA cards are significantly more expensive, so they are often not the first choice. That being said, iSCSI connection is then created via software HBA and the difference in the price is often invested in the storage capacity itself.

Furthermore, in VMware systems, the default queue depth for iSCSI connections is 128 - as this is not a fixed variable (variable name *iscsivmk_LunQDepth*), it was changed to 256 to make the most of the available storage system. As it is not possible to make a change from the ESXi GUI itself, but only through the terminal by issuing the following command and restarting the system:

esxcli system module parameters set -m iscsi_vmk -p iscsivmk_LunQDepth = 256

The following command can be used to check if a new value has changed:

esxcli system module parameters list -m iscsi_vmk

In the advanced iSCSI settings on the initiator side (ESXi server) additional things can be set such as:

- Size of each data segment sent via iSCSI (default 128 bytes)
- Parallelism of iSCSI process - the default value is 1, but as we have a total of 4 used uplinks, this value was set to 4.

The maximum value can be set to 8 because VMware's hypervisor supports a maximum of 8 iSCSI initiators.

Then, we can move to additional multipathing settings, as this is also something that can be easily changed on VMware's hypervisor. Specifically, we need to configure a *Path Selection Policy* called *Round Robin*. Round Robin policy provides load balancing by cycling I/O requests through all Active paths, sending a fixed (but configurable) number of I/O requests through each one in turn [5].

That means that hypervisor is going to use all available paths to access a block device on our storage system that we exposed. Then, it's about deciding *how often* will the path between hypervisor and storage change on the *I/O* or *bytes* level [6]. For iSCSI devices, it's good to start with IOPS close to 1, while for FC devices, it's best to start around 1000 or 1500. That will depend on the storage device used, but to know more about those details about a specific storage device, we'd need to check the manufacturer's documentation as it's different from manufacturer to manufacturer. For us to be able to configure these settings, we need to know what storage device ID is on the hypervisor level, which is usually its EUI or NAA ID. Then, we would do something similar to this procedure for iSCSI block device with EUI ID *eui.3435353637303632* - to set both PSP and IOPS settings:

```
esxcli storage nmp device set --device eui.3435353637303632 --psp VMW_PSP_RR
esxcli storage nmp psp roundrobin deviceconfig set --type=iops --iops=1 --device= eui.3435353637303632
esxcli storage core claimrule load
esxcli storage nmp device list
```

Also, we'd do something like this on FC block device with EUI ID *eui.3438316432326162*, to set both PSP and IOPS settings:

```
esxcli storage nmp device set --device eui.3438316432326162 --psp VMW_PSP_RR
esxcli storage nmp psp roundrobin deviceconfig set --type=iops --iops=1500 --device= eui.3438316432326162
esxcli storage core claimrule load
esxcli storage nmp device list
```

We used these two EUI IDs as these were the ones generated by the storage system, and these IDs will vary per environment.

The final thing we can optimize on the network layer for iSCSI is the MTU (Maximum Transmission Unit). The default value is 1500, but support for jumbo frames can be turned on and the MTU value can be effectively set to 9000. This enables the transmission packets that are six times the size on the network layer and thus reduces losses that would be caused by encapsulation of the packet itself. thus less usable bandwidth is lost to traffic encapsulation, which translates to overhead and latency. To achieve this, it is important that all devices (physical and virtual) must be set to use an MTU with a value of 9000. This refers to:

- VMkernel port (ESXi)
- vSwitch (ESXi)
- physical switch ports
- any network card that serves as an iSCSI target (OSNEXUS)

Some of these changes aren't trivial and require a bit of time and effort. But in the end, we are usually going to be rewarded with much better performance.

4.3. Improving storage performance on the storage level

Regardless of the connection medium used (FC or Ethernet), certain things can also be optimized on the storage server itself, such as default profiles for individual disk volumes. The default IO profile in OSNEXUS is "Virtualization", which is adequate for most uses in providing storage systems to any type of hypervisor (Hyper-V, ESXi,...), but when we check the exact specifications of this profile, we see that in this profile *max queue depth* variable is set to 128. Since we use a queue depth of 256 in our tests to make the most of the available potential of the IOPS value (which would be desirable for data storage systems for databases, etc.), this value must be changed.

Reviewing other available profiles, we find that the profile "SSD (Default)" uses the default value of max queue depth of 256 and that other variables of this profile mostly reflect our scenario, so this profile was selected for all tests. In case there is no already predefined profile that would suit us, you can always create a new one, with parameters specified by user input.

5. Conclusion

We proposed a two-step, iterative process of measuring storage performance. First, we set up our environment, start our script and let it complete its run, and use acquired data as a baseline. Then, we start introducing changes to environment configuration so that we can extract more storage performance out of our storage system and iterate over this step.

There are two harsh realities to be aware of here:

- This procedure will take time and it is a bit complicated, but it will give us much, much more performance. By using this methodology, we've seen storage bandwidth increase from 700-800 MB/s to 2-2.5 GB/s. We've also noticed significant increases in terms of IOPS, but not as much. Bandwidth and IOPS are often in the inversely proportional relationship, depending on the workload.
- OSNEXUS, as a product, is cheap, and extracting these levels of performance from it is stunning. Comparing our storage solution in terms of pricing and performance vs some kind of commercial storage device capable of the same numbers is something that usually puts commercial storage devices to shame, as they're either slower, or much more expensive. We can confirm this from the past five years of experience working with it.

Implementing these changes had a significant impact on how our storage device performed - performance was significantly better. There's a bit of trial and error involved here to find the correct combination, but it usually doesn't take a lot of time. Also, it enables us to have quite a bit of freedom to tune our storage performance towards more bandwidth (MB/s or GB/s), IOPS, or both. That means that we can create multiple block devices, tune them differently for different purposes, which is a rather cheap way of having application-specific storage devices, tuned for specific applications. At the end of the day, this is where the storage industry has been going for the past couple of years. And, to top it off, the raw performance is excellent, and on par with commercial storage systems 10x the price, or more.

Next research steps could include making all of this completely automated and orchestrated by using Ansible to do everything from copying our script to virtual machines to starting tests. Next, we could also make modifications to our script to use SSH keys to automatically connect to all VMs and be completely automated, without having to use mobaXterm's MultiExec mode. After that, we could chase even more performance out of our storage system by implementing the same system in containers and use Docker or Podman to spool up containers for testing instead of virtual machines. There are reasons why containerization has been a key cloud computing platform that can perfectly take advantage of this parallelism by allowing multiple isolated instances (i.e., containers) of the storage resources [8]. That would mean significantly less overhead as containers have less compute and storage overhead than virtual machines. If we wanted to, we could also implement all of the storage tests as functions, and then create some kind of algorithm to randomize these tests so that different tests are started at the same time to create a completely random scenario. So, there are plenty of opportunities for further research and development of the principle.

6. References

[1] Dakic, V. (2019). Influence of virtual storage controller type on Microsoft SQL Server 2019 performance, Proceedings of DAAAM., at Zadar, Croatia, ISSN 1726-9679, doi:10.2507/30th.daaam.proceedings.050

[2] Liu, J.; Wang, F.; Feng, D.; (2019). CostPI: Cost-Effective Performance Isolation for Shared NVMe SSDs, ICPP 2019, Kyoto, Japan, doi:10.1145/3337821.3337879

[3] Dennerman, F.; Hagoort, N. (2017). VMware vSphere 6.5 Host Resources Deep Dive, Frank Dennerman and Niels Hagoort, ISBN 9781540873064

[4] Elder, K.; Kusek, Cristopher; Sarkar, P. (2017). vSphere High Performance Cookbook, Packt Publishing, ISBN: 978-1-78646-462-0

[5] VMware (2015). Performance Best Practices for VMware vSphere 6.0, VMware

[6] VMware (2021.) Adjusting Round Robin IOPS limit from default 1000 to 1 (2069356), URL: https://kb.vmware.com/s/article/2069356

[7] Gastermann, B.; Stopper, M.; Kossik A.; Katalinic, B. (2014). Secure Implementation of an On-Premises Cloud Storage Service for Small and Medium-Sized Enterprises, 25th DAAAM International Symposium on Intelligent Manufacturing and Automation, Procedia Engineering 100, doi: 10.1016/j.proeng.2015.01.407

[8] Yang, Z.; Hoseinzadeh, M.; Wong, P.; Artoux, J.; Mayers, C.; Evans, D.T.; Bolt, R.T.; Bhimani, J.; Mi, N.; Swanson, S. (2017). H-NVMe: A Hybrid Framework of NVMe-based Storage System in Cloud Computing Environment, 36th International Performance Computing and Communications Conference (IPCCC), ISSN: 2374-9628, DOI: 10.1109/PCCC.2017.8280431.

DOI: 10.2507/32nd.daaam.proceedings.013

SECURING COMMUNICATION IN CHATTER APPLICATION

Zlatan Morić, Jure Pinjuh & Sanjin Kurelic

This Publication has to be referred as: Moric, Z[latan]; Pinjuh, J[ure] & Kurelic, S[anjin] (2021). Securing Communication in Chatter Application, Proceedings of the 32nd DAAAM International Symposium, pp.0090-0097, B. Katalinic (Ed.), Published by DAAAM International, ISBN 978-3-902734-33-4, ISSN 1726-9679, Vienna, Austria
DOI: 10.2507/32nd.daaam.proceedings.013

Abstract

Chatter is an Angular web application that offers registered users' ability to chat with other registered users. Users can chat with multiple users in a secure fashion. The main problem in designing a good backend for the Chatter application is dealing with message exchange security. In this paper, we will present a few security protocols and suggestions for exchanging secure messages. From single substitution ciphers (*Caesar* and *Atbash*) and multi-alphabet substitution ciphers (Vigenère and Playfair) to more complex like symmetric (XOR stream cipher and DES block ciphers) and asymmetric algorithms (using private and public key mechanisms with Diffie-Hellman key exchange and RSA algorithm with hash functions - SHA). We will present the pros and cons of using each algorithm by using man-in-the-middle attack and choose a secure asymmetric algorithm as the main security for exchanging messages at the end.

Keywords: secure message exchange; substitution cipher; asymmetric algorithm; hash; man-in-the-middle attack.

1. Introduction

Globally, the number of internet users increased from only 413 million in 2000 to over 3.4 billion in 2016. The one billion barriers were crossed in 2005. Every day over the past five years, an average of 640,000 people went online for the first time [1]. As more and more people got online more of them are becoming technically educated and more of them care about their personal data. Personal data can be defined as any information relating to an identified or identifiable natural person ('data subject'); an identifiable natural person can be identified, directly or indirectly, in particular by reference to an identifier such as a name, an identification number, location data, an online identifier or to one or more factors specific to the physical, physiological, genetic, mental, economic, cultural or social identity of that natural person [2].

People want to protect their personal data and activity on the Internet. In recent times more and more people are protesting big tech companies like Facebook, Google, Microsoft, etc. They are moving from Facebook Messenger and WhatsApp because of the new usage policy to more secure platforms like Telegram or Signal. Secure communication between two entities become the most important feature of chatting application – protecting both entities from the hacker (man-in-the-middle) and from the company itself. Companies that are offering chatting applications should not install a backdoor in communication nor they should be man-in-the-middle by offering secure communication only between user and server. Choosing the right security mechanism for message exchange can be tricky if it's not done properly.

Man-in-the-middle attack is one of the most popular and challenging threats in network security and there is a large body of research dedicated to the detection and analysis of different forms of these. A man-in-the-middle attack is defined as an attack in which the intruder can read and write messages communicated between two parties without either party being conscious of this fact [3]. Chatter is a chat web application built in Angular that aims to provide secure communication between registered users. Security is implemented by using the Forge framework for communication and the Keypair framework for generating RSA keys.

2. Classic cryptography

Securing message communication between two entities has been in the focus of people for more than 4000 years. During the war or trade, people used hidden and secret writing so enemy or competition doesn't reveal their plans. In past, cryptography was almost exclusively used by militaries and governments. Cryptography probably began in or around 2000 B.C. in Egypt, where hieroglyphics were used to decorate the tombs of deceased rulers and kings. These hieroglyphics told the story of the life of the king and proclaimed the great acts of his life. As time went by, these writings became more and more complicated, and eventually, the people lost interest in deciphering them [4].

In past, the most popular cipher was the substitution algorithm. Substitution is a cipher in which each occurrence of a plaintext symbol is replaced by a corresponding ciphertext symbol to generate the ciphertext. The key for such a cipher is a table of the correspondence or a function from which the correspondence is computed. Gary Knight uses the following classification for substitution algorithms: mono-alphabetic (replacing one character with ciphertext), polyalphabetic (each plaintext letter has defined multiple substitutions), poly-graphic (substitution in groups of text), and fractionating (ciphertext depends on multiple plaintext characters).

2.1. Mono-alphabetic substitution

Caesar cipher is probably the most famous mono-alphabetic substitution cipher, used by Julius Caesar. Every letter is shifted by a fixed number of spaces to the left or the right in the alphabet. In this cipher key is the number of characters that is shifting. For example, the word "Chatter" shifted by 4 is: "Glexxiv".

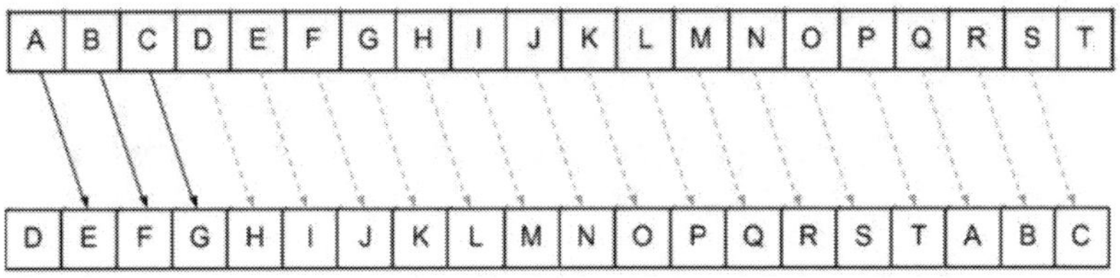

Fig. 1. Caesar cipher shifted by 4 characters

Caesar cipher can easily be implemented in JavaScript by using pure ASCII codes:

```
// Go through each character
for (var i = 0; i < str.length; i++) {
  // Get the character we'll be appending
  var c = str[i];

  // If it's a letter...
  if (c.match(/[a-z]/i)) {
    // Get its code
    var code = str.charCodeAt(i);

    // Uppercase letters
    if (code >= 65 && code <= 90) {
      c = String.fromCharCode(((code - 65 + amount) % 26) + 65);
    }

    // Lowercase letters
    else if (code >= 97 && code <= 122) {
      c = String.fromCharCode(((code - 97 + amount) % 26) + 97);
    }
}
```

The problem with Caesar cipher is that it can be decrypted easily. A simple decrypt algorithm could shift all letters until it gets valid words. This can also be done interactively with an attacker who is checking the validity of sentences. The problem with Caesar cipher is also how to exchange the key (number of characters that is shifting). Hardcoding it in the application is even less secure because that information compromises all chats on the platform.

Atbash is a variation of the mono-alphabetic substitution algorithm which supplement the first letter of the alphabet to the last, the second letter to the second last, etc (Fig. 2). Using this cipher, we get a less obvious cipher technique, but once the attacker detects the algorithm, it's not possible to fix it (without changing the encryption algorithm). This is the opposite Caesar cipher with hardcoded key and security is only in hiding algorithm that application uses. Word "Chatter" using this algorithm would be: "xszggvi".

A	B	C	D	E	F	G	H	I	J	K	L	M	N	O	P	Q	R	S	T	U	V	W	X	Y	Z
Z	Y	X	W	V	U	T	S	R	Q	P	O	N	M	L	K	J	I	H	G	F	E	D	C	B	A

Fig. 2. Atbash cipher for the English alphabet

2.2. Poly-alphabet substitution

The statistical attack on the mono-alphabetic substitution cipher can be carried out because the key defines a fixed mapping that is applied letter-by-letter to the plain text. Such an attack could be thwarted by using polyalphabetic substitution cipher [5]. The most famous polyalphabetic substitution algorithm is Vigenère cipher. Vigenère cipher also known as a polyalphabetic shift cipher works by applying several independent instances of the shift cipher in sequence (Caesar cipher) [5]. For this cipher user must define the key that is used for encryption of the message. If a key is shorter than the length of the message, then the key is repeated (example: "key" becomes "keykeykey" for plaintext that has a length of nine characters). Vigenère cipher use Tabula Recta (Fig. 3) for encryption/decryption. Word "Chatter" with key "Chatter" by using Vigenère cipher becomes "eoammii".

	A	B	C	D	E	F	G	H	I	J	K	L	M	N	O	P	Q	R	S	T	U	V	W	X	Y	Z
A	A	B	C	D	E	F	G	H	I	J	K	L	M	N	O	P	Q	R	S	T	U	V	W	X	Y	Z
B	B	C	D	E	F	G	H	I	J	K	L	M	N	O	P	Q	R	S	T	U	V	W	X	Y	Z	A
C	C	D	E	F	G	H	I	J	K	L	M	N	O	P	Q	R	S	T	U	V	W	X	Y	Z	A	B
D	D	E	F	G	H	I	J	K	L	M	N	O	P	Q	R	S	T	U	V	W	X	Y	Z	A	B	C
E	E	F	G	H	I	J	K	L	M	N	O	P	Q	R	S	T	U	V	W	X	Y	Z	A	B	C	D
F	F	G	H	I	J	K	L	M	N	O	P	Q	R	S	T	U	V	W	X	Y	Z	A	B	C	D	E
G	G	H	I	J	K	L	M	N	O	P	Q	R	S	T	U	V	W	X	Y	Z	A	B	C	D	E	F
H	H	I	J	K	L	M	N	O	P	Q	R	S	T	U	V	W	X	Y	Z	A	B	C	D	E	F	G
I	I	J	K	L	M	N	O	P	Q	R	S	T	U	V	W	X	Y	Z	A	B	C	D	E	F	G	H
J	J	K	L	M	N	O	P	Q	R	S	T	U	V	W	X	Y	Z	A	B	C	D	E	F	G	H	I
K	K	L	M	N	O	P	Q	R	S	T	U	V	W	X	Y	Z	A	B	C	D	E	F	G	H	I	J
L	L	M	N	O	P	Q	R	S	T	U	V	W	X	Y	Z	A	B	C	D	E	F	G	H	I	J	K
M	M	N	O	P	Q	R	S	T	U	V	W	X	Y	Z	A	B	C	D	E	F	G	H	I	J	K	L
N	N	O	P	Q	R	S	T	U	V	W	X	Y	Z	A	B	C	D	E	F	G	H	I	J	K	L	M
O	O	P	Q	R	S	T	U	V	W	X	Y	Z	A	B	C	D	E	F	G	H	I	J	K	L	M	N
P	P	Q	R	S	T	U	V	W	X	Y	Z	A	B	C	D	E	F	G	H	I	J	K	L	M	N	O
Q	Q	R	S	T	U	V	W	X	Y	Z	A	B	C	D	E	F	G	H	I	J	K	L	M	N	O	P
R	R	S	T	U	V	W	X	Y	Z	A	B	C	D	E	F	G	H	I	J	K	L	M	N	O	P	Q
S	S	T	U	V	W	X	Y	Z	A	B	C	D	E	F	G	H	I	J	K	L	M	N	O	P	Q	R
T	T	U	V	W	X	Y	Z	A	B	C	D	E	F	G	H	I	J	K	L	M	N	O	P	Q	R	S
U	U	V	W	X	Y	Z	A	B	C	D	E	F	G	H	I	J	K	L	M	N	O	P	Q	R	S	T
V	V	W	X	Y	Z	A	B	C	D	E	F	G	H	I	J	K	L	M	N	O	P	Q	R	S	T	U
W	W	X	Y	Z	A	B	C	D	E	F	G	H	I	J	K	L	M	N	O	P	Q	R	S	T	U	V
X	X	Y	Z	A	B	C	D	E	F	G	H	I	J	K	L	M	N	O	P	Q	R	S	T	U	V	W
Y	Y	Z	A	B	C	D	E	F	G	H	I	J	K	L	M	N	O	P	Q	R	S	T	U	V	W	X
Z	Z	A	B	C	D	E	F	G	H	I	J	K	L	M	N	O	P	Q	R	S	T	U	V	W	X	Y

Fig. 3. Tabula Recta with annotated letters for encryption/decryption of word "Chatter"

The problem with Vigenère cipher is the length of the key. If the attacker finds the length of the key then she/he could easily decrypt text by using Tabula Recta and common attacks on shift cipher algorithms. The length of the key could be found by observing patterns in the ciphertext. In chat applications, people usually start a conversation with "Hi". Smaller messages could also contain "Ok", "See you", "Bye" and similar. Taking this into consideration, the attacker could easily find the start of the key, but not the length of it.

At first, this could be good for chatting applications, but users who like to type more words on chat or for example students who copy/paste a chunk of text from some literature could lead to breaking of this algorithm. By observing common words, like "the", the attacker can find the length of key, and break the algorithm by using techniques from mono-alphabetic substitution attacks.

2.3. Poly-graphic and fractionating substitution

Poly-graphic substitution adds more complexity to polyalphabetic substitution by doing encryption/decryption on blocks of text. One of the examples of poly-graphic substitution is Playfair cipher that encrypts pairs of letters instead of single letters. The frequency analysis of bigrams is possible, but more difficult with 600 possible bigrams rather than the 26 possible monograms. Fractionating substitution is a method of splitting letters so that each plaintext letter is represented by two or more symbols. This makes the message harder to break, however, the attacker could break it just like mono-alphabetic substitution by grouping letters in pairs (or diagraphs).

Plaintext Alphabet	a	b	c	d	e	f	g	h	i	j	k	l	m	n	o	p	q	r	s	t	u	v	w	x	y	z
Ciphertext Alphabet	01	02	03	04	05	06	07	08	09	10	11	12	13	14	15	16	17	18	19	20	21	22	23	24	25	26

Table 1. Example of fractionating substitution

Both these types of substitution make more complexity in breaking the message, but they are not enough for modern computers. By observing repetition patterns in the ciphertext and changing some letters and observing changes in cipher text, a hacker can easily break these algorithms and find cipher key or plain text value.

3. Modern cryptography

None of the mentioned security algorithms satisfies the premise of secure communication in Chatter application. Unlike classical cryptography, modern cryptography is based on known mathematical problems which means that even if the attacker knows the algorithm/code it could not compromise the information (Table 1). The only way attackers can compromise the system is by knowing the private key, unlike for example Atbash cipher where knowing the cipher algorithm can compromise the data. With these characteristics, Chatter application can be an open-source application and everyone can check the implemented security mechanism.

Classic Cryptography	**Modern Cryptography**
Manipulates traditional characters (letters and digits directly)	Operates on binary bit sequences.
Mainly based on "security through obscurity". The techniques employed for coding were kept secret and only the parties involved in communication knew about them	Relies on publicly known mathematical algorithms for coding the information. Secrecy is obtained through a secrete key which is used as the seed for the algorithms. The computational difficulty of algorithms, absence of the secret key, etc., make it impossible for an attacker to obtain the original information even if he knows the algorithm used for coding
Requires the entire cryptosystem for communicating confidentially.	Modern cryptography requires parties interested in secure communication to possess the secret key only
Manipulates traditional characters (letters and digits directly)	Operates on binary bit sequences.

Table 2. Comparison between classic and modern cryptography

In modern cryptography, the encryption of plain text can be done in one of two ways. The first way is a stream cipher, and the second way is a block cipher. In stream cipher, every bit in the plaintext will combine with pseudorandom key as a way to encrypt the plaintext bit by bit. In block cipher, many bits in the plaintext will be combined to be processed with a single key to produce the same size of block as a ciphertext [6]. Key distribution can be symmetric (same key for encryption and decryption) or asymmetric (different keys for encryption and decryption).

3.1. Symmetric algorithms

Symmetric key cryptography means that the same key is used to encrypt the message and to decrypt it, making the keys "symmetric". Symmetric algorithms are always faster than asymmetric but do have the issue of exchanging the keys. Symmetric cipher can be easily done using exclusive OR (XOR) function which encrypts text as a stream cipher. XOR is a logical operation that outputs true only when inputs differ (Table 2).

Input		**Output**
A	B	C
0	0	0
0	1	1
1	0	1
1	1	0

Table 3. The truth table for XOR operation

The most important feature of XOR operation is its reversibility. If plain text is "01101" and the private key is "01010" then XOR operation will produce ciphertext: "00111". If ciphertext "00111" is again put through XOR operation output will be the same plain text as in beginning "01101". More complex symmetric algorithms are block ciphers that use blocks of data. The most popular algorithms in that area are DES and AES. AES is an asymmetric algorithm based on a design principle known as a substitution-permutation network and is efficient in both software and hardware [7]. It has a fixed block size of 128 bits and a key size of 128, 192, or 256 bits. AES operates on a 4x4 matrix called state. AES consists of four rounds which have several processing steps:

1. Key expansion – round keys are derived from the cipher key using key schedule
2. Initial round:
    a. Add round key – each byte of the state is combined with the round key using bitwise XOR
3. Rounds:
    a. Sub bytes – a non-linear substitution step where each byte is replaced with another according to a lookup table
    b. Shift rows – a transportation step where each row of the state is shifted cyclically a certain number of steps
    c. Mix columns – a mixing operation that operates on the column of the state, combining the four bytes in each column
    d. Add round key
4. Final round:
    a. Sub bytes
    b. Shift rows
    c. Add round key

Even though symmetric algorithms like AES are currently impossible to break, their usage in chatting applications is problematic because sending the key over the network can be insecure (intercepted) and also usage of the application can get more complicated. If the application requires that the user exchanges a key in some other way (example: by SMS), usage of it gets complicated and the user will probably use short passwords which are prone to dictionary attacks.

3.2. Asymmetric algorithms

Asymmetric key cryptography means that there are two keys, one for encryption, one for decryption. One of the most popular asymmetric key algorithms is RSA. Using two separate keys in cryptography gives us more flexibility in exchanging keys. One of the most popular protocols for exchanging keys over an insecure channel is Diffie-Hellman algorithm. Diffie-Hellman key exchange establishes a shared secret between two parties that can be used for secret communication for exchanging data over a public network.

For a better understanding of the concept, we usually use Alice and Bob analogy. Alice and Bob want to exchange a secret key over an insecure network:
1. Alice and Bob publicly agree to use a modulus p and base q $(0 < q < p, n = g^k \ mod \ p)$
2. Alice chooses secret integer a, then sends Bob $A = g^a \ mod \ p$
3. Bob chooses a secret integer b, then sends Alice $B = g^b \ mod \ p$
4. Alice computes $s = B^a \ mod \ p$
5. Bob computes $s = A^b \ mod \ p$
6. s from Alice computation and s from Bob computation are equal s=s, which means they successfully exchange data s

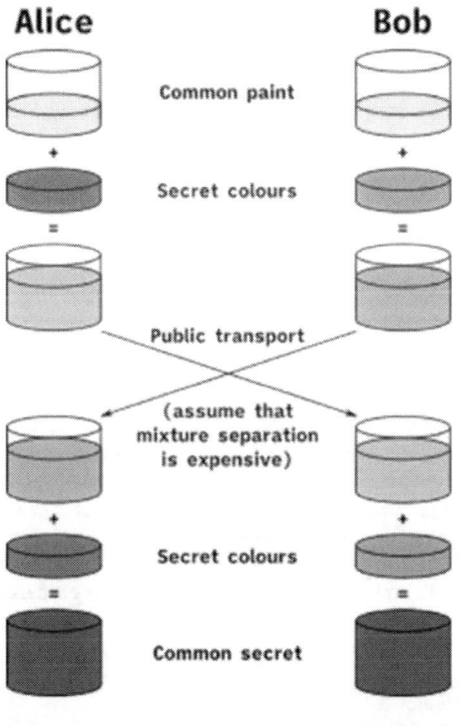

Fig. 4. Alice and Bob communicate by using Diffie-Hellman algorithm
(https://www.comparitech.com/blog/information-security/diffie-hellman-key-exchange/)

The algorithm of Alice and Bob communication can be shown by using colors as analogies (Fig. 4). The problem with using Diffie-Hellman protocol is the man-in-the-middle attack. An attacker could intercept the channel and behave as a middle point – communicate with Alice with some private-public pair and communicate with Bob with other private-public key pair. Alice does not know about Bob, and Bob does not know about Alice, so an attacker could easily become a middle point.

Man-in-the-middle attacks could be avoided using public key infrastructure (PKI). The PKI is a set of roles, policies, hardware, software, and procedures needed to create, manage, distribute, use, store and revoke digital certificates and manage public-key encryption. PKI can be viewed as a contract that connects the public key with a specific user and that contract is defined in central authority (CA). Modern protocols use this technique to identify users. This leads us to the next requirement for our Chatter application - it should use HTTPS for communication.

The most famous of the public key cryptosystem is RSA which is named after its three developers Ron Rivest, Adi Shamir, and Leonard Adleman. At the time of the algorithm's development (1977), the three were researchers at the MIT Laboratory for Computer Science [8]. Security of the RSA lies in multiplying two prime numbers (ex. 15 = 3 x 5). Multiplying two or more numbers (factors) is an easy task but finding factors of one number is complicated - especially if we're discussing two large prime numbers.

The RSA involves four steps: key generation, key distribution, encryption and decryption. The keys for the RSA algorithm are generated in the following way:

1. Choose two distinct prime numbers p and q
2. Compute $n = pq$
3. Compute $m = (p-1)(q-1)$
4. Choose a small number e that doesn't have the same factors
   Find d for which this equation is true: $de \,\%\, m = 1$

Public keys are n and e, and private key is d. Numbers p, q and m can all be discarded after d is computed. Chatter application stores public and private key (generated on user login) on local storage:

```
generateKeyPair() {
    var keyPair = keypair();
    localStorage.setItem("pk", keyPair.private);
    return keyPair.public;
}
```

Message is encrypted with public key and decrypted with private key, which is done like this:

```
encryptMessage(message: string, publickey: string) {
  var publicKey = forge.pki.publicKeyFromPem(atob(publickey));
  var encrypted = publicKey.encrypt(forge.util.encodeUtf8(message));
  return encrypted;
}

decryptMessage(message: string) {
  var privatekey = forge.pki.privateKeyFromPem(localStorage.getItem("pk"));
  var decrypted=privatekey.decrypt(message);
  return forge.util.decodeUtf8(decrypted);
}
```

Communication can be verified using hash functions. A hash function is a function that takes a variable-size input and returns a fixed-size string. A hash function is easy to compute, one-way, and collision-free. Hash functions are hard to invert which means that it's computationally not feasible to find the original text from the hash value. To make dictionary attacks more difficult with hash functions, we usually use salting. Salt is a random set of bits that are used as one of the inputs to the hash. The most widely used hash algorithm today is SHA, specifically the SHA-2 algorithm. SHA-2 consists of two similar hash functions - SHA-256 and SHA-512 - that have different block sizes, 256 and 512 bits.

When padding is important, mask generation functions are used. Padding is adding nonsense data to the beginning, middle, or end of a message before encryption. In that way, the same plaintext can have multiple different ciphertexts. A mask generation function (MGF) is a cryptographic primitive similar to a cryptographic hash function except that while a hash function's output is a fixed size, an MGF supports variable-length output. Chatter application is using RSASSA-PSS signatures (Fig. 5) for verifying messages. Probabilistic Signature Scheme (PSS) is a cryptographic signature scheme, and in Chatter application it's used as SHA 512 hash function with MGF1 masking function and a 20-byte salt. Code for verification and generating signature is:

```
createSignature(message) {
  var md = forge.md.sha512.create();
  var pss = forge.pss.create({
    md: forge.md.sha512.create(),
    mgf: forge.mgf.mgf1.create(forge.md.sha512.create()),
    saltLength: 20
  });
  var privatekey = forge.pki.privateKeyFromPem(localStorage.getItem("pk"));
  md.update(message,'utf-8');
  return privatekey.sign(md,pss);
}

verifySignature(signature,message,publickey){
  publickey=atob(publickey);
  var publicKey=forge.pki.publicKeyFromPem(publickey);
  var md = forge.md.sha512.create();
  var pss = forge.pss.create({
    md: forge.md.sha512.create(),
    mgf: forge.mgf.mgf1.create(forge.md.sha512.create()),
    saltLength: 20
  });
  md.update(message,'utf-8');
  return publicKey.verify(md.digest().getBytes(),signature,pss);
}
```

There are other methods that we could use to further improve the security of our application. In the future, when available, we could look into using quantum cryptography, as it's the logical next step to be used in secure communication. Chatter application could use this idea as well, to further increase the security level as end users want to have as much privacy as possible, and secure communication is one of the key parameters in that process. After all, it is necessary to strike a balance between the freedoms of individuals on the Internet; their rights should not be violated [9]. Not all methods need to be just encryption-based, they could also be context-based. We could look into neural networks to classify individual data packets so that we can implement some sort of classification-based mechanism to reject packages aimed at our Chatter application that are coming from unknown or insecure sources - for example, sources that are not involved in the chat session[10].

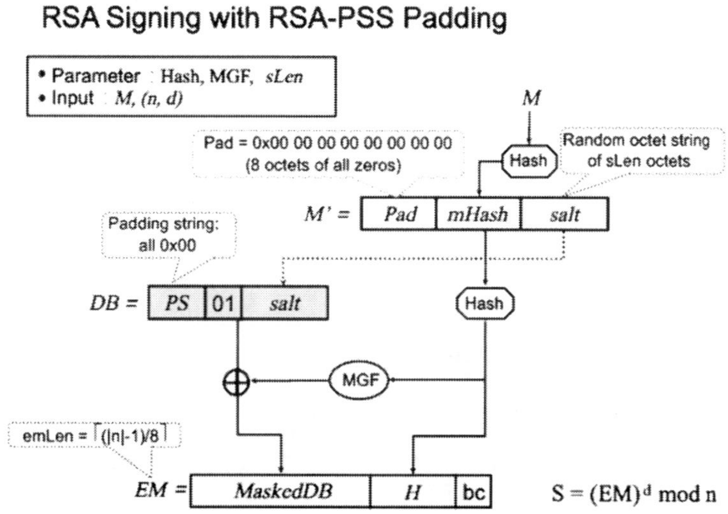

Fig. 5. Diagram of RSA signing with RSA-PSS padding
(https://slidetodoc.com/presentation_image/a6b6804dc2d0898c8a752ad3552d4ea3/image-53.jpg)

This would also fit well with the idea of layered security, where we don't use just one technology or component to deliver a secure solution - we use multiple different solutions on different levels. We also see these two concepts as potential areas for future study and improvement for our application.

4. Conclusion

Implementing a secure chatting application on modern computers is impossible with traditional cryptography. Modern cryptographic algorithms can be divided into symmetric and asymmetric ciphers. Symmetric ciphers are faster to compute but have a problem in exchanging keys over insecure networks. Communication can be verified by using hash functions with extra padding. This also adds extra obfuscation to the data, so it makes breaking it even harder. By using the session approach, each user gets a new key with a new login, older communications could not be compromised. Our research indicates that, by using a hybrid approach, the asymmetric algorithm could be used to transfer symmetric keys over an insecure network, and then use symmetric cipher for securing the data. A popular algorithm for implementing this hybrid approach is RSA, which is used in our Chatter application. We find this methodology to be the best in terms of the price/performance ratio of computational power used vs time-based overhead that gets introduced into the communication process.

5. References

[1] Roser, M., Ritchie, H. & Ortiz-Ospina, E. (2015). Our world in data, Available from: https://ourworldindata.org/internet#growth-of-the-internet, Accessed: 2021-01-31
[2] European Parliament and Council of European Union (2016). Regulation (EU) 2016/679. Available at: https://gdpr-info.eu/art-4-gdpr/, Accessed: 2021-01-31
[3] Aziz, B. & Hamilton, g. (2009). Detecting man-in-the-Middle Attacks by Precise Timing, Third International Conference on Emerging Security Information, Systems and Technologies
[4] Cohen, F. (1987). Introductory Information Protection, ASIN: B000715AKW
[5] Jonathan K. & Yehuda L. (2014). Introduction to Modern Cryptography, Chapman and Hall/CRC, 978-1466570269, New York
[6] Bokhari, M. U. & Shallal, Q. M. (2016). A Review on Symmetric Key Encryption Techniques in Cryptography, International Journal of Computer Applications, Volume 147 - Number 10, 0975 – 8887
[7] Bruce S. et al. (2000). The Twofish Team's Final Comments on AES Selection, unpublished
[8] Christensen, C. (2006), Introduction to RSA and to Authentication, unpublished
[9] Zharova, A., Elin, V. & Panfilov, P. (2018). Technological and legal issues of identifying a person on the internet to ensure information security, Annals of DAAAM and Proceedings of the International DAAAM Symposium, 29(1), pp. 0471–0478. doi: 10.2507/29th.daaam.proceedings.069.
[10] Igor, H. et al. (2014). Application of neural networks in computer security, 24th DAAAM International Symposium on Intelligent Manufacturing and Automation, 2013. doi: 10.1016/j.proeng.2014.03.111.

DOI: 10.2507/32nd.daaam.proceedings.014

ENTERPRISE TOOLS FOR DATA FORENSICS

Zlatan Moric, Jasmin Redzepagic & Frano Gatti

This Publication has to be referred as: Moric, Z[latan]; Redzepagic, J[asmin] & Gatti, F[rano] (2021). Enterprise Tools for Data Forensics, Proceedings of the 32nd DAAAM International Symposium, pp.0098-0105, B. Katalinic (Ed.), Published by DAAAM International, ISBN 978-3-902734-33-4, ISSN 1726-9679, Vienna, Austria
DOI: 10.2507/32nd.daaam.proceedings.014

Abstract

Companies today face different types of cyber-crime-related attacks. There were loads of security-related problems discovered in the past couple of years - Poodle, Heartbleed, Spectre, Meltdown, Ransomware, Microsoft Exchange security problems - these are just some of the more common ones from the past couple of years. As these attacks become more prevalent, it is only a matter of time before the incident will require digital forensics investigation. Tools for data imaging, hashing, analysis, and reporting must be a part of the administrator toolkit. Many tools provide modules or toolkits. These can be applied for different functions in the forensic processes. Competencies and knowledge are often outsourced in most organizations. There will always be a demand for data forensics procedures, and it will only increase as IT develops.

Keywords: Digital Forensic; Computer Forensics Tool; RAM Forensics; Disk Forensics; Network Forensics.

1. Introduction

Digital forensics and tools that make it possible are changing quickly as the entire field is changing. [1] At the same time, it is important to have the right tool at the right time since your digital toolkit must be compatible with whatever is on the market today. This means that we must do two things at once - we need to keep track of all the things happening in the user area, both on the software and the hardware side and at the same time we need to keep track of tools that provide all the features necessary for working with said user resources. Just as an example, if we wanted to re-configure our Microsoft IIS (Internet Information Services) due to SSL-related security problems, we can use a tool like IIS Crypto that can make it much easier for us to do that. If we had issues with Ransomware, we could probably deal with that by having a correctly set up backup and disaster recovery solution. That doesn't mean that we can solve all the security problems easily, and for most of them there are precautionary steps that we can take before something happens. After a security issue happens, all that's left to do is to do forensic analysis, and that is the core subject of this paper - which tools to use to do a proper forensic analysis.

2. Use cases for digital forensic tools

Organizations face all kinds of IT challenges, from planning infrastructure to configuring stable and secure services and applications. But what is the action plan once the administrator discovers a major security issue has happened e.g., data leakage like for example a former employee downloading the source code and uploading it to the internet.

Digital Forensics is a time-consuming art of preservation, acquisition, documentation, analysis, and interpretation of evidence from different storage media types found in the organizations that reported the incident. So, in this example, an image from all disks of the suspected employee should be captured in a tamper-proof state. This information can be provided as evidence in the court or as an indicator of vulnerabilities that must be patched. If digital forensic procedures are not conducted a system can be left in a vulnerable state because of the absence of information on what caused the incidents. Also, if Digital Forensics is not performed in a criminal case, law authorities can make wrong decisions. They will make these decisions because of the misleading data or incorrect shape and form of evidence that is presented. So, ignoring the need for forensic cases in the organization can seriously damage the reputation of the business and stakeholders. Companies need to have forensic readiness. In other words, they need to have proactive procedures, people, tools, and prerequisites defined and configured. Nowadays most of the organization don't have the required skills and resources so InfoSec and supports teams are responsible for all security-related incidents. In the incident case in most organizations, the responsiveness would be reactive, not proactive. By default, Windows systems don't log the deletion or manipulation of files, this audit needs to be configured manually. Free and native features such as logging with required details need to be utilized. Administrators need to know these features exist and must configure them. Retention of logs must be defined and edited if necessary. It is a good practice to have a centralized logs system running in an environment. Types of forensic incidents can vary but data recovery is the most common use case.

Criminals always try to hide the track, so investigators try to identify all data that has been deleted, changed to different extensions, or hidden in any other way. In most cases, digital forensic include the process of discovery and/or recovery of data (evidence) using the various tools and methods available to the investigator. Also, malware investigation is a big part of digital forensic because ransomware spread by trojans and worms across a network is a major threat to individuals, companies, and military organizations. A digital forensic investigation can include tracking down identity theft of any type e.g., stolen credit card usage or a fake social media profile. Network and internet investigation are processes where investigators try to figure out sources of the network-based incident such as Denial-of-Service (regular and distributed). Also, email investigation is a big source of IT security incidents because of spam, phishing attacks, and uneducated users that open every attachment without checking the validity of the sender. Forensic investigators must identify the source IP and geo-location of the attacker. Investigator will perform investigations of malicious attachments.

Corporate espionage can also be a case for digital forensic investigations, companies nowadays need to secure their assets so that sensitive information cannot be accessed or transmitted. Lastly, forensic investigators can face global problems such as drug trafficking and child abuse that are widely exploited within the Deep Web and the internet. Highly skilled forensic analysts can bring down these dark realities by analysing internet traffic, transaction and browser history, email records, and images. To solve these kinds of incidents, investigators need to be well educated and driven, it is an intense job that requires punctuality and professionalism. Most organizations will outsource these problems to other firms that are competent in digital forensic investigation. Meaning these skills are of high value and demand for digital investigation will increase in the future.

3. Tools used in forensic processes

Any IT process needs to be well documented, so proper procedures are a key tool in any forensic investigation. Procedures ensure the integrity of the investigation. The National Institute of Standards and Technology (NIST) has implemented the Computer Forensics Tool Testing (CFTT) program that's main purpose is to test and make reports for the community. Most of the tools that will be mentioned have reports available for the public can be found on the CIFTT website (https://www.dhs.gov/science-and-technology/nist-cftt-reports#). Also, tools need to be tested in the lab environment.

This can be done using a Kali Linux (or some other tool such) virtual machine as a Forensic workstation and a client virtual PC or external media (hard drive or USB with some dummy files) to image and analyse. Kali is the best operating system for information security since it is natively equipped with all sorts of tools (and can be upgraded with additional) [2]. It comes suited with many forensic tools such as Autopsy and Guymager. Other tools can be run as a live operating system as a virtual machine (or as a USB live boot). Budget is always an issue and some solutions cost thousands of dollars but are robust, accurate, and have GUI. These tools can be open source e.g., CAINE, or commercial e.g., Paladin. Either way, forensic people need to spend a lot of time testing tools and have a skill set to adapt to case requirements. [3]

3.1. Identification and precollection

Firstly, a new case arrives and needs to be categorized accordingly [4]. A case can first be recognized as an incident alert that is acquired by email or a security operations centre system (e.g., Darktrace). In the first stage, the incident needs to be discovered, no onsite action is needed at this time. The goal is to gather as much information about the problem and the types of systems involved. Interview with the client should be done and verified by some form of the written communication (such as email) so, that the initial incident information is retained permanently. To focus efforts a digital investigation plan is created, this is a roadmap to a successful investigation. It is important to define the background of the incident (circumstances), the scope of investigation, goals, and plans of actions in terms of evidence gathering. Then the pre-collection stage needs to be done.

The stage consists of processes such as organizing the acquisition kit and documentation. The data acquisition kit contains a card reader, various adapters (various types of USB and USB to SATA), device cables (power, SATA, HDMI, VGA), networking cables (straight-through, crossover, console), and a write blocker. Write blockers are used in the investigation to protect and preserve the original evidence from being modified. If a hardware write blocker isn't available, a software tool should be used. [5] The advantage to the software write blocker is that it is installed on the investigator's workstation, one less device in the acquisition kit. Paladin is an Ubuntu-based forensic toolbox. It is a live operating system (with a small footprint) that has a range of forensic features including write blocking. Currently, there is no free version available for commercial use and the minimal price is 25 USD. CAINE (Computer Aided Investigation Environment) is a GUI-based open-source Linux distribution that integrates forensic tools as modules. It also provides features such as read-only access to drives by filtering any IO command sent to the interface. Software write blockers operate by the policy definitions that can be changed to enable writing on the disk.

On the other hand, it is easier to run daily imaging operations with the hardware write blocker, it works independently providing indicators (lights or output on the mini screen) that the computer is not tempering with the drive which contains evidence. When investing in hardware write blockers different features such as connection types, writing capability, and data transfer rate should be considered. Hard drives have increased sector size to 4096 bytes so it's a write blocker should have Advanced Format compatibilities. That format defines disk sectors that exceed the size of 512, 520, or 528 bytes. A larger sector size ensures data integrity with stronger error correction maintenance algorithms. Coolgear hardware write blocker adapter costs 65 USD and has support for Windows, Linux, and Mac operating systems. For more intense workloads Tableau Tk35u USB 3.0 Forensic Ide/SATA bridge kit but this product costs 487 USD in a brand-new state. [6]

3.2. *Acquisition*

Acquisition of evidence is the first most important step in any digital forensic case. It is described as collecting data for further investigation. Accountable staff retrieves a copy of disks and volatile memory from the suspected digital asset. The goal of this phase is to collect and protect the data. All data from the suspect's computers need to be collected without harming the integrity of evidence. In this procedure evidence integrity methods such as write blocker, hash generating techniques, and chain of custody must be used.

The acquisition type and requirements need to be defined. Getting approvals for the action and rehearsing the imaging process (it saves time). When a technician arrives on the scene, he will prepare to create a forensic image which is a tamper-evident container that contains files, file systems, media storage, and device data. To simplify it is a copy of a drive that is compressed into a single file. He will then photograph the evidence and start the chain of custody. The chain of custody is a form that legally ensures the integrity of the evidence as the evidence is an exchange between individuals. It also provides accountability because personal information is matched with the date and time of evidence transfer. So, in this way it is well known who had the evidence at a certain time. A document created by the Scientific Working Group on Digital Evidence (SWGDE) defines a list of aspects to consider when executing a live system acquisition procedure. The list is defined from most to least volatile and crucial. When initiating the process of imaging the device's power state needs to be checked because it is important to store data that is in RAM and paging file with as little modification as possible. Working memory often contains more evidence than hard disk data because the running process, open documents, and network connections can be identified using different tools. There is only one opportunity to collect volatile data. RAM and running processes must be captured first then system settings and storage media. 4-Magnet is a free RAM acquisition tool. It can be used to capture the current state of RAM and running processes on Windows 10 system. It allows investigators to capture the contents of physical memory into a dump file that can be analysed. 7-RAMcapture (developed by Belkasoft) is a portable tool compatible with Windows operating system. It is used for RAM capture. It exports the contents of working memory to a raw file that can be analysed by other tools.

Active network connections information such as active sockets and processes that are correlated with than can be useful for the case. Xplico is an open-source GUI based Network Forensic Analyst Tool that main goal is to extract information from network and internet captures. It can automatically decode packet capture files (.pcp) files using an IP decoder and decoder management components. It can use analyse files generated by other packet capture software such as WireShark and Fiddler. Packet captures often contain useful information such as websites browsed, emails, chats, printed files, and VOIP (Voice over Internet Protocol) or RTP (Real-time Transport Protocol) data. Xplico cant view traffic encrypted by SSL (Secure Sockers Layer).

Volatility is Python open-source forensic software that can be used to analyse RAM on most used operating systems (Windows, Linux, Mac, and Android). It can identify and export network information from the hosts such as PID, Port, Protocol, Address, the creation time of the connection. So, it is a cross-platform tool that creates a snapshot of memory also known as a memory dump. Profile of the operating system has to be chosen firstly; different systems store information in different locations. Additional plugins can be added to the existing framework. It can identify and analyse the hidden running processes that can be useful for malware analysis. Volatility can also capture command line history and a list of timestamped events, which is a must in every forensic investigation. For Microsoft profiles, there is an option for a DLL (Dynamic Link Libraries) analysis [7].

Additionally, system settings from the browser and registry can be exported and analysed. This is important because the information from users, setting, operating systems, and even hashed passwords can be found in the registry. Digital evidence and forensics toolkit Linux is an Ubuntu desktop-based tool that can be used in situations when shutting down the host is not possible. It has compatibility for on-the-fly RAM and swap file analysis. It is also equipped with tools for antimalware, data recovery, imaging, hashing, mobile forensics, and reporting. Browser history capturer by Foxton is used to extract and view internet history. It extracts all useful user data from the browser (history, sessions, searches, bookmarks, passwords, etc.) into a single file.

The next step is to collect data from storage media using a write blocker tool. Which is known as a forensic disk controller that permits read-only access to the data storage device without compromising the integrity of the data. Evidence must be preserved in a persistent state, no changes to the evidence are allowed. A user data user must be saved in a dedicated external hard drive(s) and nowhere else. A laptop is just a tool to save data on external hard disks. For this part of the process, the hard drive needs to be at least the size of the data for investigation, but it's recommended to have the double amount of storage that is required for investigation. The process of imaging can be described as extracting or copying data either as a file, partition, entire storage media or a drive [8].

- Dc3dd (Department of Defense Cyber Crime Center dd) is an upgraded version of the dd (data dump) which is a copying and conversion tool originated on 1970s Unix systems that copied all contents of a storage device in a single file. Dc3dd is maintained parallel to dd, so when the dd tool gets updated dc3dd gets an update too. Dc3dd has many enhancements over dd. It copies data from one location to another while performing data conversion. It also performs hashing on the fly (MD5, SHA-1, SHA-256, and SHA-512). It can create multiple image file copies to different devices (in different formats) and split files into multiple parts. It can verify and wipe media and has a progress indicator.
- DCFLDD (Defense Computer Forensics Labs dd) is an enhanced version of dd that is maintained and supported by the Defense Computer Forensic Lab.
- Guymager is an open-source imaging tool that's not command-line, meaning its GUI based. It has hashing and verification operations and supports files using EnCase, raw, and AFF formats. EnCase format allows investigators to conduct a detailed analysis of user files (browser data, images, documents, and registry) that are collected. Raw data is unprocessed data (common extensions are .bin, .dd, .img, .iso, .raw), sectors are copied in order. Source and target will have the same value. No additional information is stored in a raw file. Raw images are not forensic files but can be used in an investigation.
- AFF (Advanced Forensics Format) format is used to capture digital evidence for legal proceedings.
- Libewf is an open-source library that is used to create Encase evidence image files. It can be used to convert, acquire EnCase files outside of EnCase Forensic.
- Afflib can create, convert, and verify AFF image files. So, this tool can convert raw images created by guymager, dd, and other tools to AFF format.
- Forensic Toolkit (aka. FTK) is free software developed by AccessData that is used on Windows operating systems. It includes a standalone imaging program (FTK imager). It can calculate MD5 has to confirm the integrity of data that is collected. Confirmation of the validity of evidence is important for presenting data that is important for the investigation case.
- Helix 3 Pro is the enterprise commercial toolkit for digital forensics. It has compatibility for live system acquisition and disk imaging. Along with reporting features for data discovery and digital analysis.

3.3. Examination

When collecting data investigator needs to maintain the integrity of evidence. Hash values should be provided before, during, and after an acquisition. If the copy is edited, the hash values will be quite different meaning the evidence integrity is ruined. To verify that the data set has not been altered Digital Forensic analysts use tools to calculate and verify cryptographic hashing. Cryptographic hashing is a type of mathematical function that creates a unique, fixed-size mathematical value from an arbitrary set of input values. [9]

The hash value is a unique integer that is a digital fingerprint of a file or a disk. Items in a collection of digital evidence can be identified by their hash value rather than relying on the less accurate and changeable indicators such as timestamps, filename, sizes, or by direct viewing. Files can be whitelisted or blacklisted by their hash values or hash sets. Hash values are calculated by the one-way hash functions and are not reversible (this doesn't mean they are secure). Hashing is not encryption. Hashing is used to determine the integrity of content by detecting modifications to the data that consequently change the hash output. Encryption is a two-way function that encodes the data to maintain data confidentiality so only authorized entities can read it (they pose the decryption key). Hashing is irreversible while encryption is reversible. The hash algorithm will always produce the same size hash value. Typical hash sizes are 128-bits, 160-bits, and 256-bits. Larger hashes have greater reliability in representing digital information than hashes, more bits more unique items. Hash collision, mathematically impossible to prevent but the probability of two files with the same hash can be minimized by increasing the bit length of the hash. Hash length is increased by choosing the hashing algorithm that can provide larger length hashes. Hashing unambiguously identifies digital information by detecting water the digital information has changed. Hashed made from items of evidence that are physical-digital storage devices.

Hashes are part of the meta-information describing an item of evidence. Hash also identifies known information present on the storage device and is used to verify that the item of evidence has not changed. Forensic investigators store hashes in hash sets that are described in the examination report in the digital forensics' cases [10].

- Message-Digest (MD2, MD4, MD5. MD5) is used in digital forensics. It creates 128-bit message digests and is free to use.
- Secure hash algorithm (SHA-0. SHA-1, SHA-224, SHA-256, SHA-384, SHA-512). SHA-1 and SHA-256 are used in digital forensics. SHA-1 is 160-bit and SHA-256 is 256-bits. It is a US standard and free to use.
- RIPMEND is a cryptographic hash algorithm designed by Hans Dobbertin, it has a length of 160 bits.
- WHIRLPOOL is a hashing algorithm created by Vincent Rijmen and Paulo Barreto that produces the 512-bit message digest.
- TIGER algorithm is used by modern computing because of its speed and efficiency
- Sumtools (summary tools) calculate a hash value using MD5 and SHA hashing algorithms for files stream or a file system. It is also used to compare files to pre-calculated hash values.
- Hash my files is a portable free tool that is used to calculate MD5 and SHA1 hash values of files. It can generate HTML reports with information useful for the investigation such as size, timelines of data creation and modification.
- Rahash2 is a part of the radare 2 toolkit, which supports MD and SHA hashing capabilities but also and CRC, Xor, and many other algorithms that are used for unpacking and decoding all sorts of binary files. So, it can also be used for advanced work with malware files in terms of decompiling, reverse engineering, and testing malware programs in a sandbox environment. In digital forensic, it is mostly used to check the identity and verify the integrity of files.
- The Hashdeep is installed by default on the current Kali version, and it provides tools to provide forensic audit mode, block hashing (MD5, SHA1, SHA256, TIGER, WHIRLPOOL), matching using hash sets, hash disk recursively in a file tree.

After the hash, values have been checked and the integrity of the evidence integrity has been verified the next phase is to analyse the data in the search for concrete evidence.

3.4. Analysis

The goal of the analysis phase is to examine the data, identify the evidence data that is in the scope of the case and extract useful evidence information. In this phase, the recovered data (artifacts, metadata, and initial facts) is heavily researched. The investigator tries to find data that is relevant to the case and answers important questions (such as who, what, when, where why, and how) about the crime or an incident. Correlating the evidence is important to determine what happened so timestamps of events are crucial. For example, if the users deny the execution of the malware. Investigators would need to prove this by identifying the program execution artifacts and user logon events on the affected computer. Then this information would be collated with timestamps that can provide more detailed conclusions. Investigators keep in mind that an analysis is the interpretation of evidence, not absolute proof. In this case, we also need to verify that the suspected user was using his account on the computer and no one else. This can be verified by correlating the timestamps with the user's entry to the building, most companies now track the working hours with the entrance ID cards that employees need to use to enter the building. In some severe cases, video camera footage should be also considered to correlate and verify the evidence timestamps. In Digital Forensics timeline analysis is important because it arranges the events from multiple data sources in the correct order, to determine when and what happened on the system and add context to the analysis.

The Sleuth Kit (TSK) is a set of tools that can be used to analyse disk images and recover deleted files from them. Additional modules can be added to the existing framework. Autopsy runs The Sleuth Kit in the background. It identifies and lists all allocated or unallocated (deleted) files. It can identify files hidden by rootkits. It can display metadata information. It categorizes data by type and provides timeline information.

The Autopsy (installed by default on Kali) is a GUI-based tool that is used by the military, law enforcement, and corporate users to investigate cybercrime. It is free and provides features like other tools such as web artifact analysis and registry analysis. It also provides features for hash filtering, file system analysis, keyword searching and extraction geolocations, and camera information from JPEG files. It can group the files by their type. For media files, it can display images as thumbnails for a quick view and videos without the need for an external viewer. For mail fails it can parse MBOX format messages.

Encase (acquired by OpenText) is created by Guidance Software. It is used for acquisition, deep forensic analysis, and reporting. It provides functionalities for several processes in Digital Forensics including acquisition, analysis, and reporting. This tool has all the required features for conducting a digital forensic investigation such as in-depth analysis of user files and data (e.g., internet history) and data recovery. It has a scripting facility called EnScipt. All features are integrated into one tool.

3.5. Reporting

In the post-investigation reporting phase, investigators prepare the presentation of the analysis conclusions. The report needs to be prepared in a way that all stakeholders can understand. This phase requires technical and legal evaluation. Most of the law enforcement (if they do not have the technical background) doesn't have the required knowledge to determine the facts from the evidence. It is the Digital investigator's job to present the evidence and explain the techniques and methodologies used in the data analysis phase. So, the main goal is to present the evidence adequately and acceptably, defined by the court of law.

There is also a corporate type of investigation report. This report differs from law enforcement case reporting. The most important part is to present the evidence in the way in which a client who outsourced the incident can understand the cause, consequences, and responsibilities. All forensic reports should have a transparent view of the investigation process. EnCase tool is an all-purpose forensic tool. It has an integrated reporting module. It can create reports that can be easily shared. It also provides the ability to create report templates. Helix framework has a reporting component to pull concise reports on audits or ad-hoc reports based on different criteria. Processes such as collection and preservation should be explained.

Details of the analysis should be described and supported with evidence material. All crucial conclusions should be verified using at least two different techniques or tools to increase the validity and present the evidence as tested facts. The snippets from all tools used should be presented. The best approach is to avoid absolute phrases and long sentences. Also, it is a good practice to prepare a summary section for non-technical readers of the report. Presentation tools such as Sway and PowerPoint could be great tools. Short presentations that provide all important evidence, their validity, and a graphical workflow with events of the incident could be created to ensure that all participates understand the nature and cause of the incident that has occurred.

4. Conclusion

Data forensic tools must be used by an experienced forensic expert. This is important because these tools can harm the data on disks and consequently do damage to the investigation (keep in mind that dd is known as data dump but also as data destroyer). To mitigate risks, investigations must be done systematically, defined by procedures. Well-defined procedures are a crucial tool. Also, some tools like EnCase have all the required features for conducting these procedures. It is better to gather evidence with multiple tools to increase the validity of evidence. Also, some tools can perform better for different use cases. Investigators need to have experience with all tools mentioned. They need to know the best toolkits for different platforms to adapt to all scenarios that are possible. The forensic procedure can be run using only open-source tools.

First, the problem or incident needs to be defined in detail. The most important step of the investigation is acquisition. Before executing acquisition is executed a plan with evidence gathering goals should be created (to avoid time-wasting). If the acquisition is not done correctly no further investigation can be done. There is no point to analyse the data if the data was partially wiped or changed (the hash values). The experienced technician needs to execute the acquisition process, he is responsible for creating the chain of custody document, photographing the incident site and assets, and checking the power state of the affected computer (or any other IT asset). If the power is on the procedure for the live acquisition should be executed. RAM data should be exported in the memory dump file using the Volatility and other capture tools to create multiple files for verification. Volatility is a great tool because of cross-platform support and features such as active network connection and process identification and analysis with timestamps. Xplico should be used for packet capture and inspection. Browser history capturer should be used to gather information from the browser. To verify the result tools such as Fiddler, Wireshark, and command line should be used. After all the necessary that has been gathered the acquisition process should continue to the hard drives.

There are types of data alteration prevention devices. Hardware or software write blockers. Depending on the case the technician will either use PALADIN live OS as a software write blocker or connect the hard drive to the Tableau hardware write blocker and use the acquisition tool in the Kali operating system. On Kali tools such as DC3DD and Guymager will be used. The image needs to be copied to several external hard drives. One hard drive will stay as the initial state of the evidence and others will be working versions. Backup of working versions of the discs should be created. The evidence images and files will be transported to the investigation facilities. The chain of cost should be updated after every transfer or major manipulation of evidence. All the steps should be documented. When the evidence arrives for forensic analysis tools such as Hashdeep and Sum tools should be used to check the integrity of data. Afterward, the investigation of generated images, memory dumps, user settings, and data will begin. This part of the investigation will reveal the crucial evidence for the case. Tools such as The Sleuth Kit and Autopsy will be used to determine the data manipulation, deleted files, malware execution artifacts, registry manipulation, and user settings on the mounted image. Other tools such as Volatility and Xplico should be used to determine the active processes and sockets from the generated dumps. Browser data should be analysed to gather the history of internet activity and useful information such as saved passwords. In this part of an investigation, case tools are only media for evidence presentation. The investigator needs to know what to find and where to look for. The future of the case relies on the investigator's skills and knowledge. The biggest focus is on the events and data created in the incident time range. A forensic examiner can be in communication with the client's internal IT team for additional information.

The goal is to gather the evidence and answer the important questions provided by the client (such as who, what, when, where why, and how). An event timestamp collaboration will be identified. The whole process of gathering and analysing the information that led to the conclusion will be documented. The reports will be created using EnCase and Helix 3 Pro tools and consolidated into a single document along with snippets from different tools. A different version of the report document will be available. For internal use, a document with technical details will be created. A PDF summary and conclusion report of the event will be created for the client. On the investigation closer meeting or in the court of law, the fact will be presented understandably, coherently, and concisely, along with an explanation of the procedures and verifications of the evidence integrity. Graphical representation of workflow with the timestamps of the crucial events will be presented in a PowerPoint presentation. Digital forensics is a long process that's why only fast tools that provide reliability and accuracy need to be used. Criminals can make life hard and stressful for investigators. They try to cover their tracks using various anti-forensics methods (encryption, obfuscation, and cloaking techniques) to trick investigators. Cybercrime is a lucrative business and response from law enforcement agencies needs to be equally advanced. They need to continually develop intelligent solutions and educate people to use them. That is the only way to put up with a never-ending battle in the digital world. This is the resulting table from our research, with different tools suggested for different types of forensic analysis:

Tool Type	Primary	Secondary
Write Blocker	Tableau Tk35u, Tableau Tk35e	Coolgear, PALADIN
Acquisition (Hard drives)	EnCase, Guymager	EnCase, PALADIN
Acquisition (RAM)	Volatility, FTK	4-Magnet, 7-RAMcapture
Verification	Sumtools, The Hashdeep	Hash my files, Rahash2
Analysis	The Sleuth Kit, The Autopsy	Encase, Xplico
Reporting	The Sleuth Kit, EnCase, Helix 3 Pro	Powerpoint, Sway
Additional tools	PowerShell, Bash shell	Wireshark, Fiddler

Table 1. Suggested tools

To maintain the integrity of the investigation multiple tools, need to be used. This is a good practice because multiple tools can ensure the verification and backup of evidence. So, if the primary tool is not accurate, we can rely on the secondary tool which is ready. All tools (hardware and software) need to be prepared and maintained. Because of the advancements in technology tools used in digital forensic procedures must be updated regularly. Technology has come a long way and digital forensic incidents are nowadays related to solid-state drives, cloud storage, the Internet of things, fiber-optic connections with gigabit speeds, and SD cards for storage. These new challenges also promote advancements. Understanding core technological concepts such as cryptographic hashing and hard drives are important, the core foundation is still the same.

The future of forensic tools and procedures will depend on the changes in storage demands and storage systems (petabytes and later exabytes of data). The changes will also be influenced by new laws (changes to the existing laws) and practices. The malware and anti-forensic techniques will also improve so the tools used by investigators need to evolve to meet the compatibilities and sophistication of the bad guy's tools. I think that most admins do not think that these tools will ever be useful for their day-to-day tasks, so when the incident happens, they will be forced to outsource the case to some external firm and pay a lot of money for their services. A new investigator that wants to invest time in this field of information technology needs to continually learn to use cross-platform tools in an automated fashion to improve efficiency for the growing demand. Automation of forensics is a topic that needs more research because the incident and systems on which they accrue are not homogeneous, meaning incidents that happen to one client most likely will not happen the same way to others. This is a never-ending, time-consuming process and requires a lot of effort and strong nerves because a single wrong move can jeopardize the whole investigation. The logging and monitoring configurations in the company's systems are of the most value when the incident happens. This system can alert admins and help in reconstructions of incidents and evidence gathering. I think most of the IT admin crowds don't know how to improve the audit and logging settings and leave everything to the default values (why should I change it if it works). This can be problematic when they get hacked or some major incident happens. From a forensic standpoint, we shouldn't be limited to whatever evidence is available, we are responsible for the audit and logging configuration.

Machine-learning-based systems can learn and know the baseline of services and can proactively prevent incidents when the anomaly event happens. These tools will be able to send detailed reports to admins providing facts and evidence in the PDF document that can be automatically sent to the email. These tools such as Darktrace are the best for IT security assurance but very expensive on the other hand. Scientific papers suggest that many tools are impractical and do not meet the industry changes. There is also a lack of standardization of policies, ethics, development, and tools. Currently, there is no single best tool and framework that can cover every process of the investigation. Multiple tools are needed to complete the necessary actions and to verify the results. To provide accountability and retrieval of the lost data state of the art tools need to be used. But a tool is as good as a person using it. Knowledge, experience, and well-defined procedures are the keys to a successful investigation.

5. References

[1] Cisar, P.; Maravic, C.S. & Bosnjak, S. (2014). Cybercrime and digital forensics – Technologies and Approaches, DAAAM international scientific book, pp. 525–542.

[2] Parasram, S.V.N. (2020), Digital forensics with Kali Linux, Packt Publishing, ISBN 978-1838640804

[3] Murry, J. D. (2020). Digital forensic tools in Kali Linux, Pluralsight

[4] Arshad, S. (2020). Digita forensic Getting started, Pluralsight

[5] Kessler, G.C. & Carlton, G.H. (2014) A study of forensic imaging in the absence of write-blockers, Journal of Digital Forensics, Security and Law

[6] Best hardware write blockers for digital investigators. Available at: https://linuxhint.com/best_hardware_write_blockers/ Accessed: 2020-02-11

[7] Montenegro, M. (2020). Starting using Volatility, Pluralsight

[8] Dweikat, M., Eleyan, D. and Eleyan, A., Digital forensic tools used in analysing cybercrime, Journal of University of Shanghai for Science and Technology, ISSN: 1007-6735

[9] The difference between encryption, hashing and salting. Available at: https://www.thesslstore.com/blog/difference-encryption-hashing-salting/ Accessed: 2020-01-10

[10] Kumar, K. at all. (2012), Significance of hash value generation in digital forensic, International Journal of Engineering Research and Development

DOI: 10.2507/32nd.daaam.proceedings.015

ISSUE OF CRITICAL PARTS MARKING IN THE FIELD OF PRECISE METROLOGY

Martin Melichar, Jan Kutlwašer & Nikola Skřivanová

This Publication has to be referred as: Melichar, M[artin]; Kutlwaser, J[an] & Skrivanova, N[ikola] (2021). Issue of Critical Parts Marking in the Field of Precise Metrology, Proceedings of the 32nd DAAAM International Symposium, pp.0106-0109, B. Katalinic (Ed.), Published by DAAAM International, ISBN 978-3-902734-33-4, ISSN 1726-9679, Vienna, Austria
DOI: 10.2507/32nd.daaam.proceedings.015

Abstract

In the field of higher quality management systems, such as automotive IATF 16949, aeronautics EN9100 or railway standards IRIS, unambiguous identification throughout the production process is one of the fundamental principles on which production philosophies are based. This paper deals with an attempt to influence the measurement results of a reference piece using different marking methods. Parameters were evaluated under the same environmental conditions of an accredited laboratory for the 4 tracers, and are thus suitable for statistical comparison. To understand the process and its limitations, an analysis of the measurement system (Cg/Cgk) was performed and a clear limit was established beyond which single-label methods are no longer applicable.

Keywords: Metrology; Traceability; Identification; Marking

1. Introduction

Regarding higher quality management systems, such as the automotive IATF 16949, aviation EN9100 or railway ISO/TS 22163:2017 standards, unambiguous identification throughout the production process is one of the basic principles on which production philosophies are based. For example, aviation standard EN9100:2018 in clause 8.4.2 Identification and traceability literally states that "The organisation shall control the unique identification of the outputs when traceability is a requirement, and shall retain the documented information necessary to enable traceability." Finished products are most often identified by a code, enabling traceability often back to the specific melt of the semi-finished material and material inputs. For obvious reasons, this code is only assigned to the product after the final release of the product towards the customer and after verification that the product meets the specifications defined in the technical documentation. [1], [2]. This presents manufacturing companies with a difficult dilemma - how to appropriately identify and temporarily identify parts during production. [3]

a) Products with loss of identification can be treated as non-conforming products according to higher standards with all the consequences until the contrary is clearly proven. This is a requirement that the marking be of sufficient quality and durability
b) The marking must not affect or distort the characteristics of the products and intermediate products

2. Experimental analysis

Within the accredited metrology laboratory L-1718 of the Regional Institute of Technology at the Faculty of Mechanical Engineering, University of West Bohemia in Pilsen, a comprehensive analysis was carried out to investigate the influence of commonly used markers and markers on the outputs of high-precision shape using a Taylor Hobson Talyrond 585 L device. All measurements were subject to DIN ISO 1101 environmental conditions, i.e. a temperature of 20 °C ± 1 °C in a vibration-neutral environment, a filter of 15 waves/circumference LSC or 2.5 mm LSS, 5 rpm or 5 mm/s and a standard scanning arm with a 3 mm diameter ball. [7], [8]

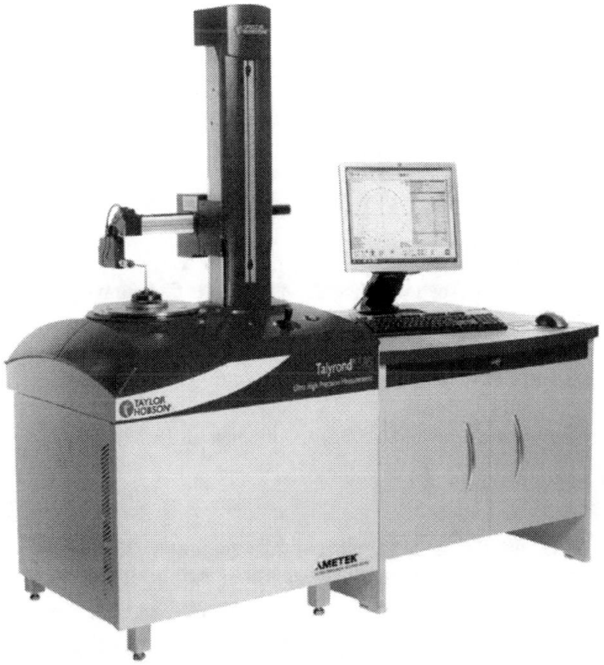

Fig. 1. Talyrond 585 Lt

A precision polished standard a cylinder with a diameter and height of 50mm was chosen as the test body. This etalon was first tempered in the laboratory to ambient temperature for 12 hours and then carefully cleaned using an ultrasonic cleaner by machine and then manually cleaned using isopropyl alcohol. [5], [6]

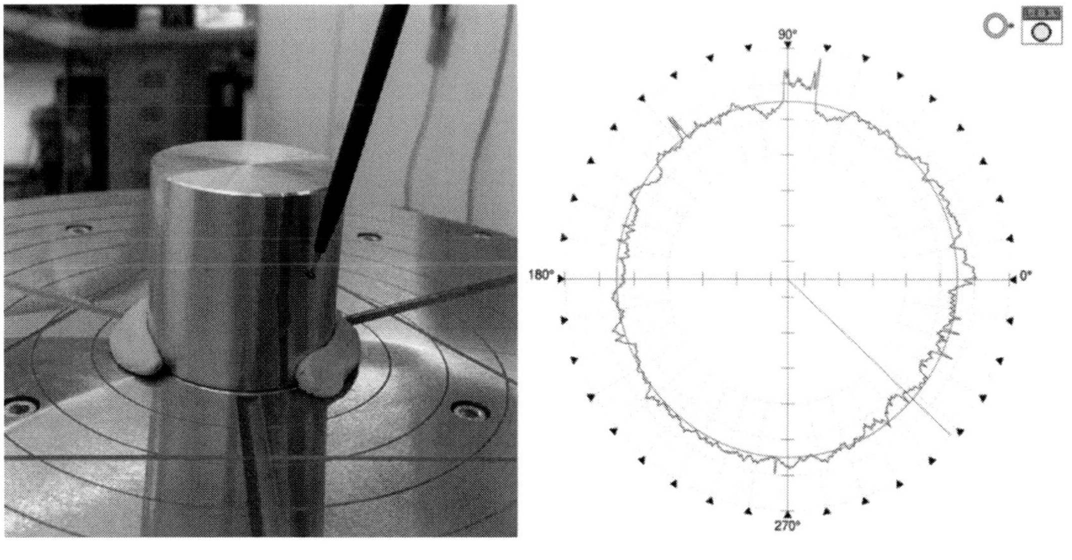

Fig. 2 Marker line evaluation

3. Marker layer analysis results

As part of the statistical verification of the results, the individual experiments were repeated at least 25 times and the resulting data were used to verify the normality of the results and to determine the mean or standard deviation of the results. [4]

- Green colour alcohol marker for DVD KORES => 1μm
- Industrial lacquer MARKER 9100 by Centropen => 3μm
- Thin lacquer marker Paint marker => 2μm
- Red WHITEBOARD marker Centropen => beyond the machine's resolution

a) Following the standard definition of Cg/Cgk meter eligibility according to IATF 16949, namely that a gauge or etalon is only eligible if it is 20 times more accurate than the identified parameter, it is then conventionally correct and acceptable within the supply chain to use verified marking methods in downstream cases:
Prescribed technical specification values:

- Green colour alcohol marker for DVD KORES => 0,02mm
- Industrial lacquer MARKER 9100 by Centropen => 0,06mm
- Thin lacquer marker Paint marker => 0,04mm
- Red WHITEBOARD marker Centropen => an inconclusive result

b) Following the standard definition of the so-called "golden rule of metrology", namely that a gauge or etalon is only eligible if it is 10 times more accurate than the identified parameter, it is then conventionally correct and acceptable within the supply chain to use verified marking methods in downstream cases:

- Green colour alcohol marker for DVD KORES => 0,01mm
- Industrial lacquer MARKER 9100 by Centropen => 0,03mm
- Thin lacquer Paint marker => 0,02mm
- Red WHITEBOARD marker Centropen => beyond the machine's resolution

4. Result confirmation

Finally, due to the focus of this article on the aerospace and automotive industry, the VDA5 methodology was chosen as the evaluation methodology. Considering the lengthy calculations and the need for a considerable amount of data (MSA I + II analysis was also performed as part of the validation calculations), the supporting tool Yarvyn was used.

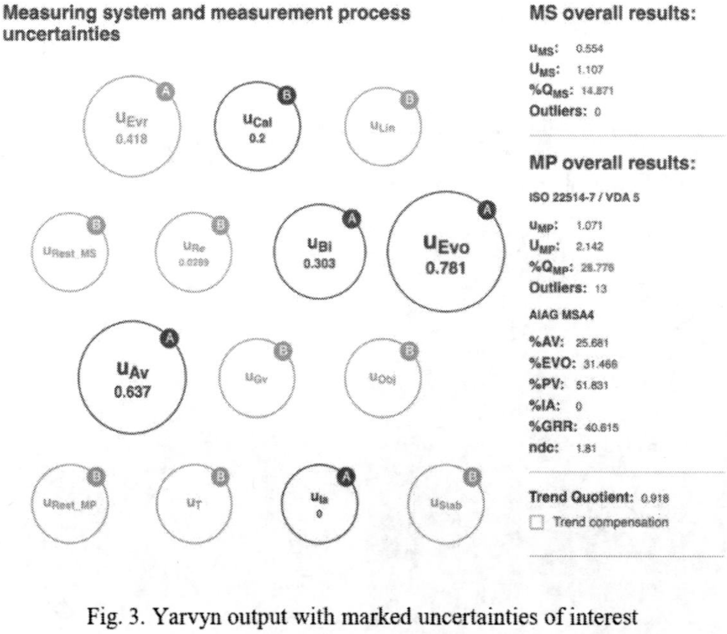

Fig. 3. Yarvyn output with marked uncertainties of interest

In the case of the variants: green alcohol marker / MARKER 9100 / thin varnish marker - the influence and experimentally verified substantial increase of the A-type uncertainty components, namely: uEvo, uBias, uCal, uAv, has been unequivocally proven, thus leading to the degradation of the entire measuring system, including its accuracy, which drops by up to 3 orders of magnitude. [10], [9]

5. Conclusion

The motivation for this document was to consider the effectiveness of the growing trend of 100% identification in relation to measurement accuracy. Researchers in an accredited laboratory have addressed the question of how the commonly used method of marking products can affect the reliability of dimensional inspection results and thus the inspection itself. A standard 50 mm diameter standard cylinder was chosen as the test object. To obtain reliable results, it was first necessary to obtain data from a real inspection. Statistically processed data from repeated roundness measurements on a high precision Taylor Hobson roundness tester were used. The parameters were measured under the same environmental conditions for 4 different commonly used markers. Due to the identical input conditions, it was also possible to compare the data afterwards. In order to understand the process and its limitations, an analysis of the measurement system was carried out according to the VDA5 methodology and an evaluation of the individual effects on the uB type uncertainty and thus on the overall uc uncertainty. As can be seen from the results at a glance, an inappropriate way of marking a part before inspection can degrade the reliability of the measurement by hundreds of percent. This was most evident in the Marker 9100, where the measurement uncertainty increased to an extreme. This article is part of a comprehensive study of the effect of the human factor on the accuracy and reliability of results. The individual sub-experiments performed in 2021 and planned for 2022 will be published in a comprehensive method manual on shop-floor metrology.

6. Acknowledgments

The article contribution has been prepared under project SGS-2019-008

7. References

[1] Svingerova, M. & Melichar, (2017). Evaluation of Process Risks in Industry 4.0 Environment, Proceedings of the 28th DAAAM International Symposium, pp.1021-1029, B. Katalinic (Ed.), Published by DAAAM International, ISBN 978-3-902734-11-2, ISSN 1726-9679, Vienna, Austria DOI: 10.2507/28th.daaam.proceedings.142

[2] Banduka, N.; Macuzic, I.; Stojkic, Z.; Bosnjak, I. & Peronja, I. (2016). Using 80/20 Principle to Improve Decision Making at PFMEA, Proceedings of the 27th DAAAM International Symposium, pp.0487-0492, B. Katalinic (Ed.), Published by DAAAM International, ISBN 978-3-902734- 08-2, ISSN 1726-9679, Vienna, Austria DOI: 10.2507/27th.daaam.proceedings.073

[3] Markova, P.; Prajova, V.; Homokyova, M. & Horvathova, M. (2019). Human Factor in Industry 4.0 in Point of View Ergonomics in Slovak Republic, Proceedings of the 30th DAAAM International Symposium, pp.0284-0289, B. Katalinic (Ed.), Published by DAAAM International, ISBN 978-3-902734-22-8, ISSN 1726-9679, Vienna, Au

[4] Skrivanova, N. & Melichar, M. (2019). The Comparison of Destructive and Non-Destructive Forms of Measurement in the Automotive Industry, Proceedings of the 30th DAAAM International Symposium, pp.0995-1002, B. Katalinic (Ed.), Published by DAAAM International, ISBN 978-3-902734- 22-8, ISSN 1726-9679, Vienna, Austria DOI: 10.2507/30th.daaam.proceedings.138

[5] Poor, P. & Basl, J. (2018). Czech Republic and Processes of Industry 4.0 Implementation, Proceedings of the 29th DAAAM International Symposium, pp.0454-0459, B. Katalinic (Ed.), Published by DAAAM International, ISBN 978

[6] Brkljac, M. & Sudarevic, T. (2018). Sharing Economy and „Industry 4.0" as the Business Environment of Millennial Generation - a Marketing Perspective, Proceedings of the 29th DAAAM International Symposium, pp.1092-1101, B. Katalinic (Ed.), Published by DAAAM International, ISBN 978- 3-902734-20-4, ISSN 1726-9679, Vienna, Austria DOI: 10.2507/29th.daaam.proceedings.156

[7] Clares Rodriguez, J. M.; Vazquez Martinez, J. M.; Gomez-Parra, A.; Puerta Morales, F. J. & Marcos, M. (2016). Experimental Methodology for Evaluating Workpieces Surface Integrity in Dry Turning of Aerospace Alloys, Proceedings of the 26th DAAAM International Symposium, pp.0849-0855, B. Katalinic (Ed.), Published by DAAAM International, ISBN 978-3-902734- 07-5, ISSN 1726-9679, Vienna, Austria DOI:10.2507/26th.daaam.proceedings.118

[8] Kuts, V.; Sarkans, M.; Otto, T. & Tahemaa, T. (2017). Collaborative Work Between Human And Industrial Robot In Manufacturing By Advanced Safety Monitoring System, Proceedings of the 28th DAAAM International Symposium, pp.0996-1001, B. Katalinic (Ed.), Published by DAAAM International, ISBN 978-3-902734-11-2, ISSN 1726-9679, Vienna, Austria DOI: 10.2507/28th.daaam.proceedings.138

[9] Bicova, K[aterina]; Kutlwaser, J[an] & Sklenicka, J[osef] (2016). Issue of High Precision Manufacturing Analysis in Automotive Industry, Proceedings of the 27th DAAAM International Symposium, pp.0353-0357, B. Katalinic (Ed.), Published by DAAAM International, ISBN 978-3-902734-08-2, ISSN 1726-9679, Vienna, Austria DOI: 10.2507/27th.daaam.proceedings.052

[10] Kubatova, D[ana] & Melichar, M[artin] (2018). Effect of Scanning Speed when Measuring Surface Roughness, Proceedings of the 29th DAAAM International Symposium, pp.1249-1255, B. Katalinic (Ed.), Published by DAAAM International, ISBN 978-3-902734-20-4, ISSN 1726-9679, Vienna, Austria DOI: 10.2507/29th.daaam.proceedings.180

DOI: 10.2507/32nd.daaam.proceedings.016

ENERGY EFFICIENCY OF WORM GEAR DRIVES

Antoaneta Dobreva & Petyr Pavlov

This Publication has to be referred as: Dobreva, A[ntoaneta] & Pavlov, P[etyr] (2021). Energy Efficiency of Worm Gear Drives, Proceedings of the 32nd DAAAM International Symposium, pp.0110-0116, B. Katalinic (Ed.), Published by DAAAM International, ISBN 978-3-902734-33-4, ISSN 1726-9679, Vienna, Austria
DOI: 10.2507/32nd.daaam.proceedings.016

Abstract

A theoretical investigation of cylindrical worm gear drives in the field of energy efficiency is presented. The research objective is to analyse and determine the possibilities for a selection of worm gear drives with maximum energy efficiency at the initial stage of the design process within the technical and design requirements set up. An optimisation study of worm gear drives has been implemented, based upon several groups of criteria: tribological, strength, kinematic and geometry parameters. The main research criterion determined is the maximum gearing efficiency. Scientific methods have been applied, which consistently determine the influence of different groups of criteria mentioned. A database of the obtained calculation results is structured. Different graphical relationships show the dependencies between the relevant tribological, kinematic, strength and geometric parameters of the worm gear drives. The analysis of the results indicates the most advantageous areas of application for worm gear drives according to the main optimisation criterion selected. Important relationships and recommendations are suggested.

Keywords: energy efficiency; worm gear drives; gearing efficiency coefficients; tribology approach; optimisation research.

1. Introduction

The research of worm gear drives is a particularly important issue. Its solution is directly related to the improvement of the strategy for the modern development of mechanical engineering in order to increase the competitiveness of the sector. Improving the energy efficiency and the functional capability of worm gear drives are significant research problems for the scientific and academic staff. The research problem of improving the performance of the worm gear drives is also an interdisciplinary problem. Its solution is possible through the application of modern methods and approaches.

The calculation of the efficiency of a system requires the knowledge of either the power input and power loss, or the power input and power output. With regard to gearboxes, the overall power loss can be described as the sum of partial power losses of the gearbox components. They are caused mainly by the gears, bearings and by contacting seals. Gear losses generally cause a significant proportion of the overall power loss. Friction within the contact of two tooth flanks relates to the applied load of the tooth system and results in load-dependent gear losses. Churning losses, squeezing losses, impulse losses and ventilation losses are related to the oil flow in the gearbox [1]. They are referred to as no-load gear losses as they are almost independent from the applied load.

2. Theoretical background

Worm gear drives are usually required at torque conversions with high gear ratio, at compact installation space and 90-degree axis-crossing angle. Due to their high-power density and sliding speeds within the tooth contact, frictional heat and thermal stresses are relatively high [1]. Therefore, the prediction of energy efficiency coefficients of gearboxes containing one or more worm gear stages is very important, especially during the design phase. However, determining the optimal geometry parameters is a complex problem, as they have to meet all the constraints, practical limitations and rules defined in the corresponding design case, [2], [3] and [4]. Some important transmission components have been successfully analysed aiming to improve their functional parameters with the help of modern methods and approaches. Important basic influencing parameters have also been defined, [5], [6] and [7]. Significant contributions concerning mathematical modelling of the tooth contact of worm gears have been developed by Niemann [8] and Weber [9]. Important tribological research activities have been performed by Wilkesmann [10], concerning elasto-hydrodynamic lubrication (EHL) calculations for different worm tooth geometries and by Predki [11], concerning significant parameter studies. Considerable parts of these investigations formed the basis of DIN 3996: 2019-09, [12].

Bouché [13] formulated a physics-based model for the calculation of the coefficient of friction for worm gears. The USA standard [14] recommends the important analytical relationships for estimating the coefficient of friction in the meshing for a hardened steel worm (58 HRC minimum), smoothly ground, or polished, or rolled, or with an equivalent finish, operating on a bronze worm gear. The approach for selecting the values of friction coefficient is explained in details in [15]. It depends on the sliding velocity in the worm gear drive.

Magyar [16] has made some research contributions, concerning the dynamics of worm gears and has derived a tribological calculation model for the calculation of the coefficient of friction, which is applied for the calculation of worm gear efficiency in [17]. Monz [18] and Mautner et al. [19] have made interesting investigations about the load capacity and efficiency of worm gears lubricated by consistent grease. They have used a specific Thermal Network Method for heat balance calculations, which correspond closely to the measurements.

Since worm gears show different gear losses, depending on the direction of the power flow, the calculation of the meshing efficiency must be considered separately. When the worm shaft is driving, according to DIN 3996:2019-09 [12], the meshing efficiency coefficient depends upon the reference lead (pitch) angle and the tooth friction angle. Beside geometrical and operational data such as gear ratio u, worm wheel torque T_2, worm shaft drive speed n_1 and reference pitch (lead) angle of the worm, the calculation of the load-dependent gear losses depends on the mean coefficient of friction. When the worm wheel is driving, the efficiency is generally lower. Furthermore, a self-locking effect can occur in this operation mode if the meshing efficiency coefficient is less than 0.5, according to [12].

The mean coefficient of friction represents the complex friction characteristic of meshing tooth flanks by one single mean value. In the area of worm gear drives, there are currently several different approaches and calculation models available. DIN 3996:2012-09 [20] describes a simpler, empirical model, while Oehler et al. [21] & [22] present a more detailed and complicated method. The latter was standardized in DIN 3996:2019-09 [12], replacing the simpler approach in DIN 3996:2012-09 [20] very recently.

As already discussed, the efficiency calculations for worm drives are standardized in AGMA 6034-B92 [14], BS 721 [23] and DIN 3996 [12]. The author's team completed successfully some research tasks in the field of energy efficiency in the area of gear drives and machine science. The results of these investigations are described in details in the following publications: [24], [25], [26], [27] and [28]. Based upon the literature publications studied, the authors' team focused on the following research objective: to analyse and determine the possibilities for selecting worm gear drives with maximum meshing efficiency coefficient at the initial stage of the design process within the technical requirements set up.

3. Research methods

A scientific methodology is applied, in which an optimisation investigation is successively performed within the following several stages:

The first stage includes the definition of the groups of criteria, according to which the optimisation investigation will be performed: tribological (gearing efficiency coefficient, tooth friction coefficient) and strength: determination of the nominal bending stresses in the gear profile of the worm wheel. The gearing friction coefficient is determined by the method described in details in Mott [15] and AGMA [14]. Bending strength calculations are determined by the methods of Niemann [1], [8] and [10].

The second stage of the research methodology includes tribological and strength calculations based upon different values of the kinematic parameters (gear ratio, worm shaft drive speed and sliding speed) and different values of the geometry parameters (module, center distance).

The third stage involves structuring a database of the calculation results obtained. Different graphical dependences are represented, showing the dependences between the relevant tribological, kinematic, strength and geometrical parameters of the worm gear drives.

The fourth stage of the research methodology aims to determine the main criterion of the optimisation investigation, which, according to the authors' team, is the gearing efficiency coefficient. The analysis of the results indicates the optimal areas of application for worm gear drives, according to the main optimisation criterion selected.

Significant relationships and recommendations are to be suggested.

4. Research results and discussion

The optimisation research has been implemented on the criteria of sliding speed and gearing efficiency coefficient. It turns out that the influence of the worm gear module is particularly significant. Fig. 1 shows the variation in the sliding speed and the gearing efficiency when the module of the worm gear increases. The alterations of the indicated parameters at different numbers of starts of the worm $z_1 = 4$, $z_1 = 2$ and $z_1 = 1$ are also shown. The results described graphically in the indicated figures are obtained at center distance a = 200 mm and worm shaft drive speed $n_1 = 700$ rpm.

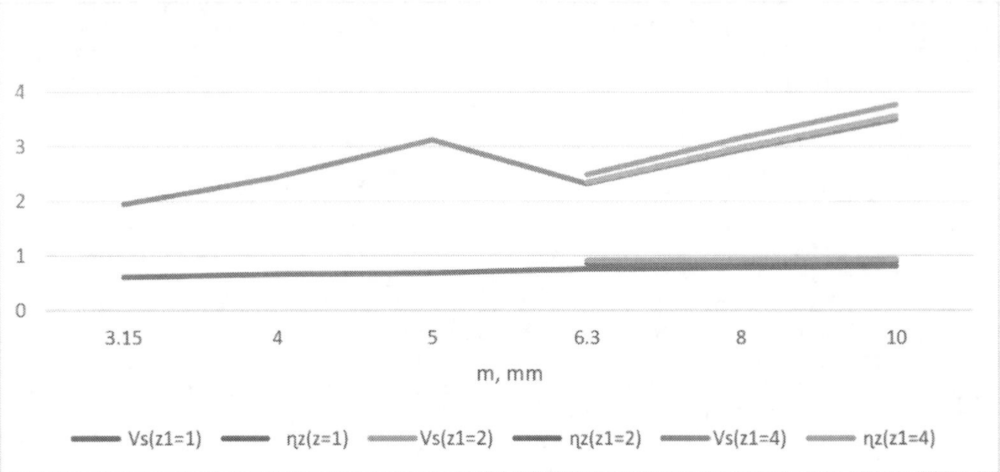

Fig. 1. Variation of sliding speed and gearing efficiency coefficient depending on the module

For a specific centre distance, the best gearing efficiency coefficient values are achieved at modules between the following boundaries: 6.3 and 10 mm. With a larger number of worm starts, there is a slight increase in the gearing efficiency coefficient when the other geometry and kinematic parameters do not change.

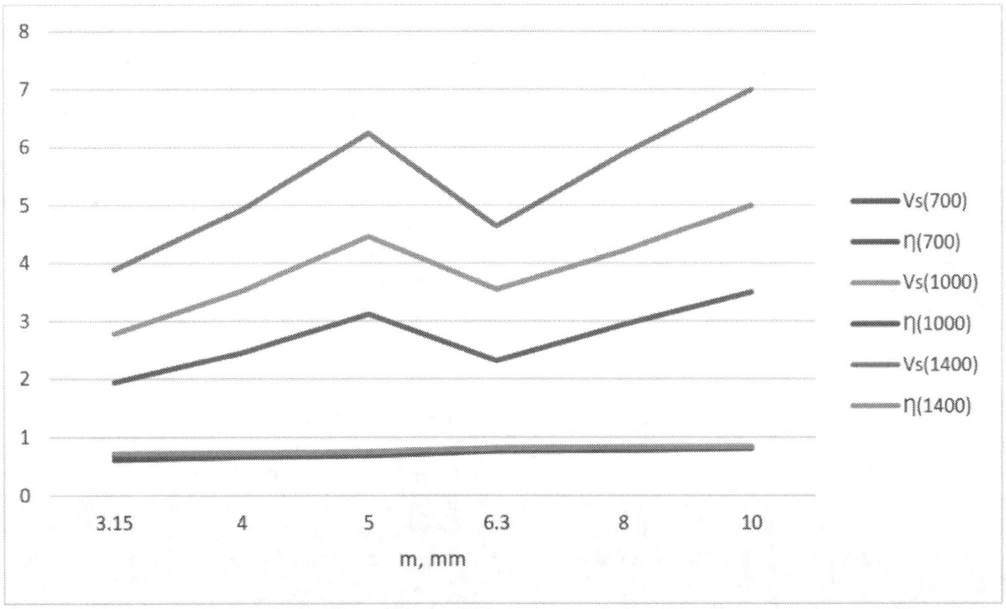

Fig. 2. Variation of sliding speed and gearing efficiency depending on the module and worm shaft drive speed

The graphical representations on Fig. 2 show the variations of the gearing efficiency coefficient and the sliding speed as functions of the values of the module and the worm shaft drive speed. It can be concluded that the gearing efficiency increases - although insignificantly - with increasing the worm shaft drive speed at equal geometry parameters of the worm gear drives. The influence of the worm shaft drive speed upon the sliding speed is much more clearly represented, as the speed of sliding increases with the increase of the rotational speed of the worm shaft.

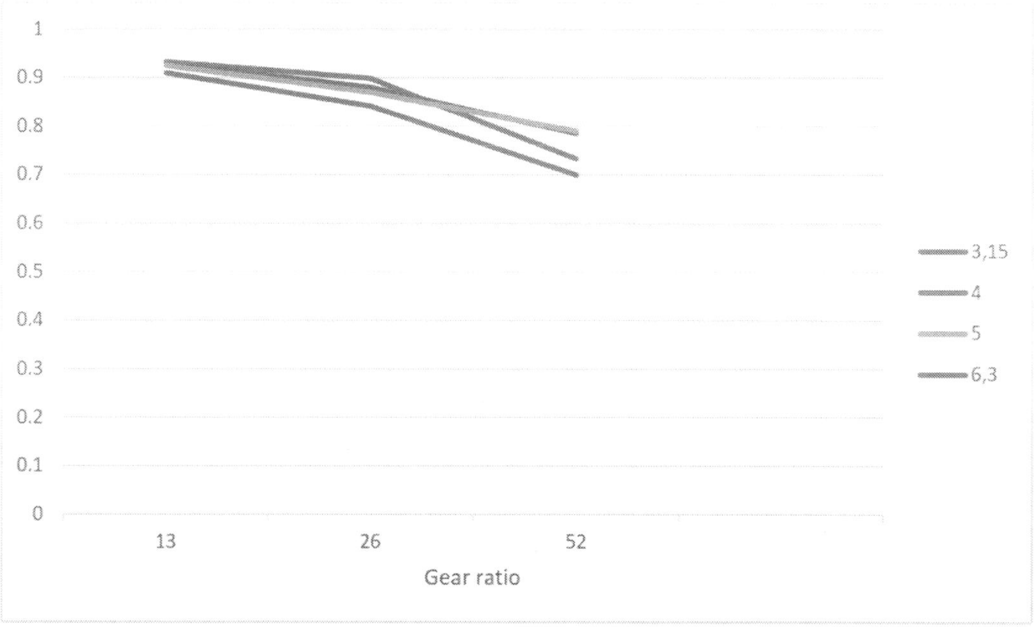

Fig. 3. Variation of gearing efficiency coefficient as a function of the module and the gear ratio

The graphical curves on Fig. 3 show the changes in the gearing efficiency coefficient depending on the values of the module and the values of the gear ratio at a centre distance a = 125 mm and at worm shaft drive speed n = 1500 rpm. It can be observed that the influence of the ratio on the gearing efficiency is much more significant in comparison with the influence of the factor "worm shaft drive speed". The application of lower gear ratios (up to i = 26) leads to optimal values of gearing efficiency coefficients.

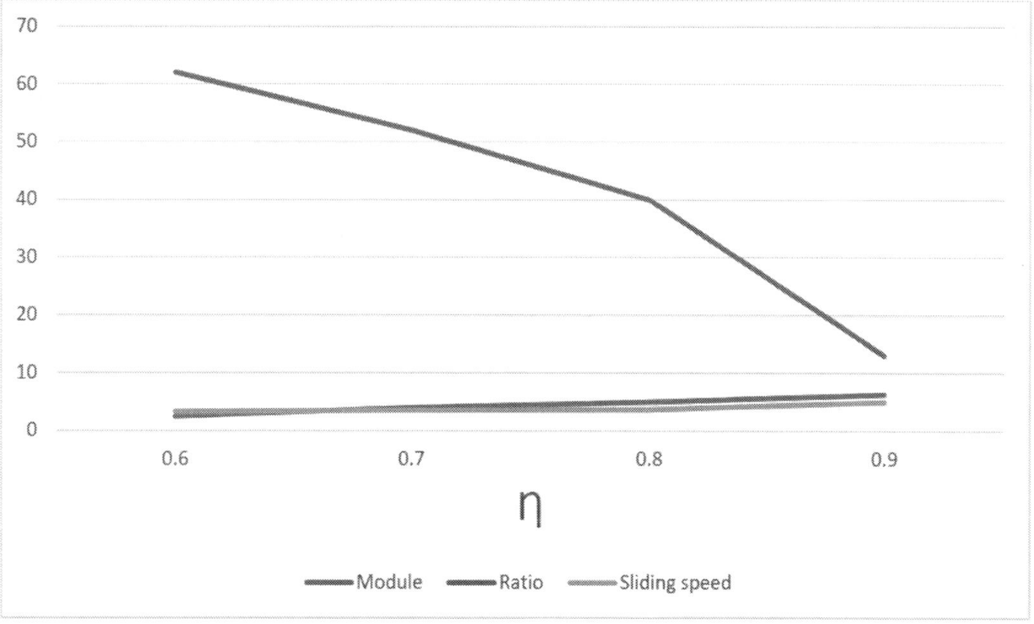

Fig. 4. Influence of the module, the gear ratio and the sliding speed

The representations on Fig. 4 show the simultaneous influence of the module, the gear ratio and the sliding speed to the gearing efficiency coefficient values at the center distance a = 125 mm and at the worm shaft drive speed n = 1500 rpm. The tendency for an abrupt increase of the gearing efficiency coefficient with a decrease in the gear ratio is especially clear.

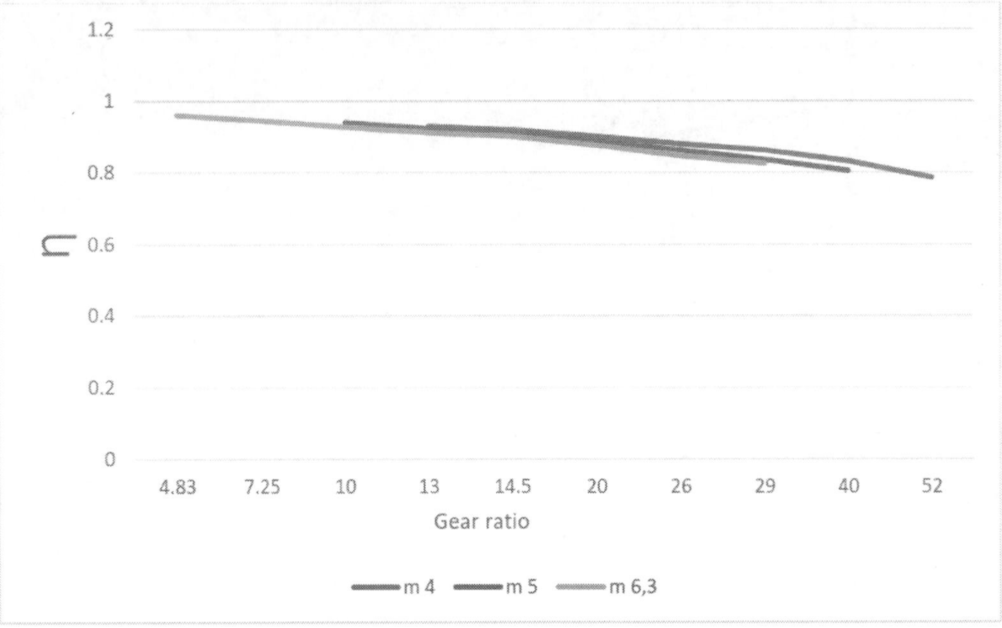

Fig. 5. Influence of the module and the gear ratio on the gearing efficiency

Fig. 5 shows the simultaneous influence of the module, the gear ratio and the sliding speed on the gearing efficiency coefficients at a centre distance a = 125 mm and at a worm shaft drive speed n = 1500 rpm. The analysis of the results shows that the most diverse kinematic possibilities are obtained when using gear ratios in the range from 10 to 26 while maintaining very good values of gearing efficiency, higher than 0.8.

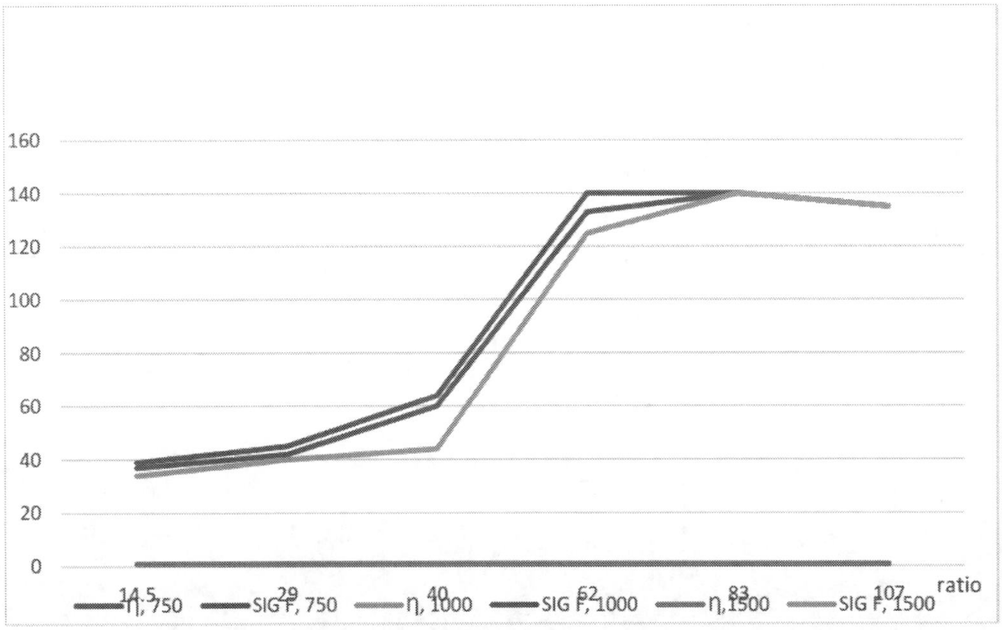

Fig. 6. Influence of kinematic parameters on gearing efficiency and bending stresses

The graphical curves in Fig. 6 show the change of the calculated maximum bending stresses of the gear profile of the worm wheel at nominal torque, taking into account the influence of the gear ratio, the rotational speed of the worm shaft on the gearing efficiency at centre distance a = 125 mm and at worm shaft speeds n_1 = 750, 1000 and 1500 rpm. The analysis of the results shows that the optimal values of bending stresses and gearing efficiency are obtained when using gear ratios in the range up to 40.

5. Conclusions

The worm gear efficiency coefficient is highly dependent and its ways of optimisation are the problem to be solved in this paper. An optimisation study of worm gear drives is presented. Several groups of criteria have been applied: tribological, kinematic and load capacity. Calculations and analyses of the results have been performed. The influence of the gear ratio, module and the rotational speed of the worm shaft on the meshing efficiency in the worm gearing and the variation of the values of the bending stresses have been established.

The graphical representations show the variations of the gearing efficiency coefficient and the sliding speed as functions of the values of the module and the worm shaft drive speed. It can be concluded that the gearing efficiency increases - although insignificantly - with increasing the worm shaft drive speed at equal geometry parameters of the worm gear drives. The results show the changes in the gearing efficiency coefficient, depending on the values of the module and the values of the gear ratio. It can be observed that the influence of the ratio on the gearing efficiency is much more significant in comparison with the influence of the factor "worm shaft drive speed". The application of lower gear ratios (up to $i = 26$) leads to optimal values of gearing efficiency coefficients. The tendency for an abrupt increase of the gearing efficiency coefficient with a decrease in the gear ratio is especially clear. The analysis of the results shows that the most diverse kinematic possibilities are obtained when using gear ratios in the range from 10 to 26 while maintaining very good values of gearing efficiency, higher than 0.8. The results show the change of the calculated maximum bending stresses of the gear profile of the worm wheel at nominal torque, taking into account the influence of the gear ratio and the rotational speed of the worm shaft on the gearing efficiency. The analysis of the results shows that the optimal values of bending stresses and gearing efficiency are obtained when using gear ratios in the range up to 40.

The influence of the worm shaft drive speed upon the sliding speed is much more clearly represented. The speed of sliding increases with the increase of the rotational speed of the worm shaft. The results obtained can be used at an early stage of a design process of worm gear drives, allowing the designer to choose the optimal parameters in terms of maximum gearing efficiency coefficients in worm gear drives. Due to the small number of criteria used, this paper gives only the main dependencies of the worm gear set. The approach is to be optimised as well, showing how all criteria together influence the efficiency.

The future work on this topic is to involve more criteria into the research in order to determine in higher detail their influence on the maximisation of the gear efficiency coefficient. Three-dimensional graphical representations are to be created. These diagrams will contribute to facilitating the decision-making process, concerning the selection of the optimal worm gear drive in view of the energy efficiency within the relevant design assignment.

6. Acknowledgments

The authors acknowledge the financial support from the Scientific Research Funds of the University of Ruse, Bulgaria. The research work done has been realised in the framework of several scientific seminars at the Department of "Machine Science, Machine elements, Engineering Graphics and Physics" at the Transport Faculty at the University of Ruse. The support is gratefully acknowledged.

7. References

[1] Niemann, G. & Winter, H. (1983). Helical, bevel, worm, chain, belt, friction gear, clutches, brakes, freewheels, 2nd edn. Machine elements, vol 3. Springer, Berlin, Heidelberg, (Original title: Schraubrad-, Kegelrad-, Schnecken-, Ketten-, Riemen- Reibradgetriebe, Kupplungen, Bremsen, Freiläufe).

[2] Raspudic, V. (2020). Optimal Design of Laterally Unrestrained I-Beams Using Genetic Algorithm, Proceedings of the 31st DAAAM International Symposium, pp.0683-0691, B. Katalinic (Ed.), Published by DAAAM International, ISBN 978-3-902734-29-7, ISSN 1726-9679, Vienna, Austria DOI: 10.2507/31st.daaam.proceedings.095.

[3] Hrubina, K.; Katalinic, B.; Jadlovska, A.; Wessely, E.; Macurova, A. & Majercak, J. (2010). Modeling and Computing Methods for Solving Optimization Problems, Chapter 42 in DAAAM International Scientific Book 2010, pp. 471-488, B. Katalinic (Ed.), Published by DAAAM International, ISBN 978-3-901509-74-2, ISSN 1726-9687, Vienna, Austria, DOI: 10.2507/daaam.scibook. 2010.42.

[4] Valjan, V. & Raspudic, V. (2020). Stress and Fatigue Analysis of a Single Cylinder Engine Crankshaft, Proceedings of the 31st DAAAM International Symposium, pp.0692-0700, B. Katalinic (Ed.), Published by DAAAM International, ISBN 978-3-902734-29-7, ISSN 1726-9679, Vienna, Austria DOI: 10.2507/31st.daaam.proceedings.096.

[5] Trobradovic, M.; Blazevic, A.; Pikula, B. & Bibic, D. (2020). Development of Simulation Model for Dry Clutch Engagement Analysis, Proceedings of the 31st DAAAM International Symposium, pp.0819-0825, B. Katalinic (Ed.), Published by DAAAM International, ISBN 978-3-902734- 29-7, ISSN 1726-9679, Vienna, Austria DOI: 10.2507/31st.daaam.proceedings.114.

[6] Likaj, R.; Bruqi, M.; Shala, A. & Bajrami, X. (2016). Optimal Design and Analysis of Quarter Vehicle Suspension System by Using Matlab, Proceedings of the 27th DAAAM International Symposium, pp.0082-0090, B. Katalinic (Ed.), Published by DAAAM International, ISBN 978-3-902734-08-2, ISSN 1726-9679, Vienna, Austria, DOI: 10.2507/27th.daaam.proceedings.012.

[7] Beles, H.; Rus, A.; Dragomir, G.; Mitran, T.; Trusca, D.& Tolea, B. (2016). Researches Regarding the Continuous Improvement of the ABS (Anti-Lock Braking System) Operation for the Passenger Cars, Proceedings of the 26th DAAAM International Symposium, pp.0196-0205, B. Katalinic (Ed.), Published by DAAAM International, ISBN 978-3-902734-07-5, ISSN 1726-9679, Vienna, Austria, DOI:10.2507/26th.daaam.proceedings.027.

[8] Niemann, G. (1942). Worm gear with liquid friction - dependence of the transferable power and the coefficient of friction on tooth shape, dimensions, speed and lubricating viscosity. VDI publishing house, Berlin (Original title: Schneckentriebe mit flüssiger Reibung – Abhängigkeit der übertragbaren Leistung und des Reibwertes von Zahnform, Abmessung, Drehzahl und Schmierzähigkeit)

[9] Weber, C.; Maushake, W. & Niemann, G. (1956). Investigation of cylindrical worm drives with axes crossing at right angles - Report 125 of the Research Center for Gears and Gear Construction Technical University of Munich. Vieweg + Teubner Verlag, Wiesbaden. (Original title: Untersuchung von Zylinderschneckentrieben mit rechtwinklig sich kreuzenden Achsen)

[10] Wilkesmann, H. (1974). Calculation of worm gears with different tooth profile shapes - load capacity and power loss for hollow circle, involute and straight line profile. Dissertation, Technical University of Munich, Munich. (Original title: Berechnung von Schneckengetrieben mit unterschiedlichen Zahnprofilformen – Tragfähigkeit und Verlustleistung für Hohlkreis-, Evolventen- und Geradlinienprofil)

[11] Predki, W. (1982). Hertzian pressures, lubrication gap heights and efficiency of worm drives. Dissertation, Ruhr University Bochum, Bochum. (Original title: Hertzsche Drücke, Schmierspalthöhen und Wirkungsgrade von Schneckentrieben)

[12] DIN 3996:2019-09: (2019). Load capacity calculation of cylindrical worm gears with axes crossing at 90 degree angles. (Original title: Tragfähigkeitsberechnung von Zylinder-Schneckengetrieben mit sich rechtwinklig kreuzenden Achsen).

[13] Bouché, B. (1991). Coefficients of friction of worm gear gears in the mixed friction area. Dissertation, Ruhr University Bochum, Bochum. (Original title: Reibungszahlen von Schneckengetriebeverzahnungen im Mischreibungsgebiet).

[14] AGMA 6034-B92: (2010). Practice for Enclosed Cylindrical Worm gear Speed Reducers and Gearmotors, AGMA, Alexandria, USA.

[15] Mott, R. L.; Vavrek, E. M. & Wang, J. (2018). Machine Elements in Mechanical Design. 6th Ed. ISBN 10: 0-13-444118-4 & ISBN 13: 978-0-13-444118-4. Pearson Education, Inc., New York.

[16] Magyar, B. (2012). Tribo-dynamic investigations of cylindrical worm gears. Dissertation, Technical University of Kaiserslautern, Kaiserslautern. (Original title: Tribo-dynamische Untersuchungen von Zylinderschneckengetrieben).

[17] Oehler, M.; Magyar, B. & Sauer, B. (2017). A new, standardizable calculation approach for the efficiency of worm gears. Research in Engineering 81 (2–3): 145–151. https://doi.org/10. 1007 / s10010-017-0225-1.(Original title: Ein neuer, normungsfähiger Berechnungsansatz für denWirkungsgrad von Schneckengetrieben. Forschung im Ingenieurwesen).

[18] Monz, A. (2012). Load capacity and efficiency of worm gears when lubricated with consistent gear greases. Dissertation, Technical University of Munich, Garching (Original title: Tragfähigkeit und Wirkungsgrad von Schneckengetrieben bei Schmierung mit konsistenten Getriebefetten).

[19] Mautner, E. M.; Sigmund, W.; Stemplinger J.P. & Stahl, K. (2016). Efficiency of worm gearboxes. Proc Inst Mech Eng Part C: JMech Eng Sci 230(16):2952–2956.

[20] DIN 3996:2012-09: (2012). Load capacity calculation of cylindrical worm gears with axes crossing at 90 degree angles. (Original title: Tragfähigkeitsberechnung von Zylinder-Schneckengetrieben mit sich rechtwinklig kreuzenden Achsen).

[21] Oehler, M.; Magyar, B. & Sauer, B. (2018). Coupled thermal and tribological analysis of worm gear. Tribol lubrication technology 65(1):54–60. (Original title: Gekoppelte thermische und tribologische Analyse von Schneckengetrieben).

[22] Oehler, M.; Magyar, B. & Sauer, B. (2016). IGF-Nr. 18275N, FVANr. 729 I, Worm gear efficiency - worm gear efficiency. Final report, Frankfurt am Main. (Original title: Schneckengetriebewirkungsgrade – Schneckengetriebewirkungsgrade).

[23] BS 721-1: (1963). Specification for Worm Gearing — Imperial Units, BSI, London, UK.

[24] Dobreva, A. (2013): Theoretical Investigation of the Energy Efficiency of Planetary Gear Trains. Mechanisms and Machine Science, No 13, pp 289-298.

[25] Dobreva, A. (2013). Methods for Improving the Geometry Parameters and the Energy Efficiency of Gear Trains with Internal Meshing. VDI – Berichte, No 2199.2, pp 1291 – 1302, (2013).

[26] Dobreva, A. & Dobrev, V. (1993). Improving the Tribological Characteristics of Heavy Loaded Gear Boxes. Proceedings of the First Balkan Conference on Tribology "Balkantrib'93", Volume 2.3, Sofia, pp 166-170.

[27] Dobreva, A. & Dobrev, V. (2018). Innovative Methodology for Decreasing Mechanical Losses in Vehicles. Proceedings of the 4th International Congress of Automotive and Transport Engineering (AMMA 2018), Springer Verlag, pp 234 – 242.

[28] Dobreva, A. & Stoyanov, S. (2012). Optimization Research of Gear Trains with Internal Meshing. Ruse, University Publishing Centre, pp 144.

DOI: 10.2507/32nd.daaam.proceedings.017

OPEN SOURCE LOW-COST APPROACH TO TERRAIN MAPPING USING DRONE WITH LIDAR

Adnan Masic, Boran Pikula, Dzevad Bibic & Faruk Razic

This Publication has to be referred as: Masic, A[dnan]; Pikula, B[oran]; Bibic, D[zevad] & Razic, F[aruk] (2021). Open Source Low-cost Approach to Terrain Mapping Using Drone with LiDAR, Proceedings of the 32nd DAAAM International Symposium, pp.0117-0122, B. Katalinic (Ed.), Published by DAAAM International, ISBN 978-3-902734-33-4, ISSN 1726-9679, Vienna, Austria
DOI: 10.2507/32nd.daaam.proceedings.017

Abstract

A novel method for terrain mapping is presented in this paper. It is based on low-cost open source components, which brings this innovative technique to a much wider group of researchers than ever before. The in-house developed drone was used together with a commercially available LiDAR sensor to produce a terrain mapping platform. This system was tested in two field locations with good results. In some aspects of performance, the drone surpassed commercially available solutions that are much more expensive and closed.

Keywords: LiDAR; drone; terrain mapping; inertial measurement unit; global navigation satellite system.

1. Introduction

In the context of the development of automation and robotics, the use of LiDAR (light detection and ranging) is an increasingly popular technological trend. For example, the use of LiDAR in the fields of autonomous vehicles, aerial surveying, power line inspections, forest research, etc. is being intensively investigated. There are numerous examples of papers in the literature. Hopkinson investigated laser pulse return intensity and canopy frequency distribution [1], while Wulder analyzed LiDAR plots [2]. Lipovsky used LiDAR to study rock and ice avalanches [3], and Hopkinson [4] assessed the influence of glacier downwasting on water resources. Marselis evaluated the potential of full-waveform LiDAR for mapping pan-tropical tree species richness [5]. Kettridge used airborne LiDAR in [6]. Chasmer investigated laser pulse penetration through a conifer canopy by integrating airborne and terrestrial LiDAR [7]. Yeong highlighted some of the challenges in sensor fusion technology (including LiDAR) in autonomous vehicles [8]. Tang analyzed the LiDAR sensor performance on detecting pedestrians under different weather conditions [9]. Chen performed research on control and dynamic property of autonomous vehicle adaptive LiDAR system [10]. Elhousni surveyed 3D LiDAR localization for autonomous vehicles [11].

However, what is common to all these references is the use of sophisticated and expensive equipment, which is generally inaccessible to researchers who do not have a large budget. In the case of aerial measurements, costs increase further due to the need to use an aircraft capable of carrying sophisticated (and relatively heavy) measuring equipment. The goal of our work is to develop an open and accessible drone-based platform, the price of which is one or two orders of magnitude lower than commercially available solutions.

2. The drone-based platform

Unmanned aerial vehicles (drones) are used for a variety of purposes, including data acquisition [12], temperature inversion measurements [13], particulate matter (PM) concentration measurements [14], and simultaneous measurements of temperature and PM concentration above urban area [15]. Our drone platform was developed in-house, based on Ardupilot open-source autopilot and commercially available electronic components [16]. This platform has a modular design and can carry various sensors and equipment, including electrochemical sensors for toxic gases [17], [18], and the abovementioned PM sensors. In our previous work we discussed quality assurance (QA) procedures that were applied to ensure reliable data [19], [20]. For this research, LiDAR with 100 meters range, LW20/C (LightWare, USA) was configured and installed on the drone (Figure 1). For terrain mapping the measurements from LiDAR are not sufficient for the calculation of terrain height, because the drone moves during the flight. Thus, it is necessary to combine raw readings from the LiDAR with the drone Attitude and Heading Reference System (AHRS) to calculate terrain height at given geographical coordinates. During the flight of the drone, which is in our case limited to 30 minutes, a discrete set of points is obtained. Spatial interpolation methods were employed afterward, using our previously explained methodology [21].

The following steps are performed to calculate terrain height:

1. Choose an area to survey and create an automatic mission for the drone.
2. Execute mission and record data from both lidar and drone AHRS.
3. Calculate terrain height in the geographic coordinate system.
4. Based on the density of measured points, create a grid for interpolation and calculate interpolated values at cell centers. In this research Inverse distance weighting (IDW) and Kriging algorithm were used for interpolation.
5. Plot 3D surface map of the terrain.

Fig. 1. The in-house developed drone based on open-source flight controller.

3. Field Tests

Two field tests in Sarajevo Canton were selected for this research (Figure 2): Trebević mountain (test 1) and Nišići plateau (test 2). In test 1 we have surveyed an area of 113531 m^2 to map the terrain which has a complex topology (Figure 3). The second test was the measurement of elevation profile over 1.5 km long line and to compare that profile with available data from Google Earth.

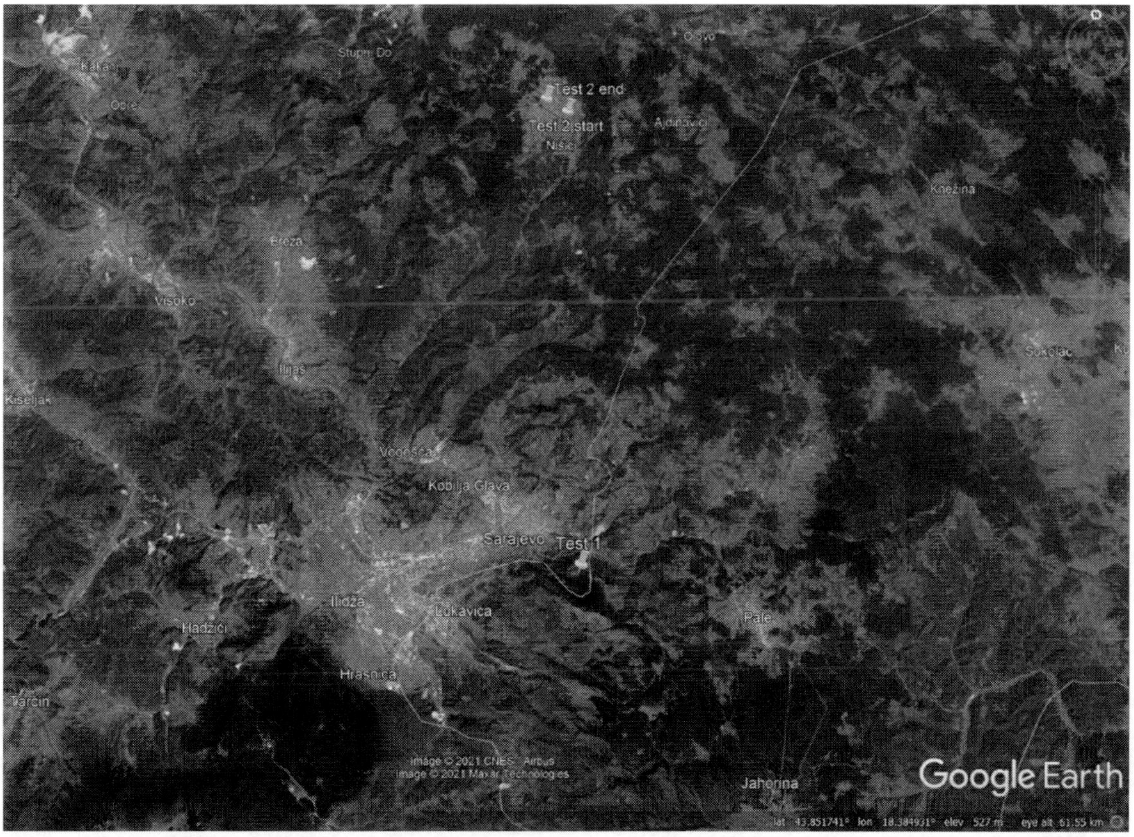

Fig. 2. Test locations.

Fig. 3. Clockwise from top left: image of the flying drone taken from above, mapping of terrain with buildings, forest mapping, and elevation profile measurement.

4. Results

Figure 3 shows the terrain model computed using the 5 steps explained in section 2. During the drone flight we have collected 23987 individual LiDAR measured points, over an area of 113531 m^2, which gives the point density of 0.211 points per m^2. We would like to note that the aim of this research is proof of concept, rather than high point density (which will be increased in future work). For IDW calculations power exponent 2 was used.

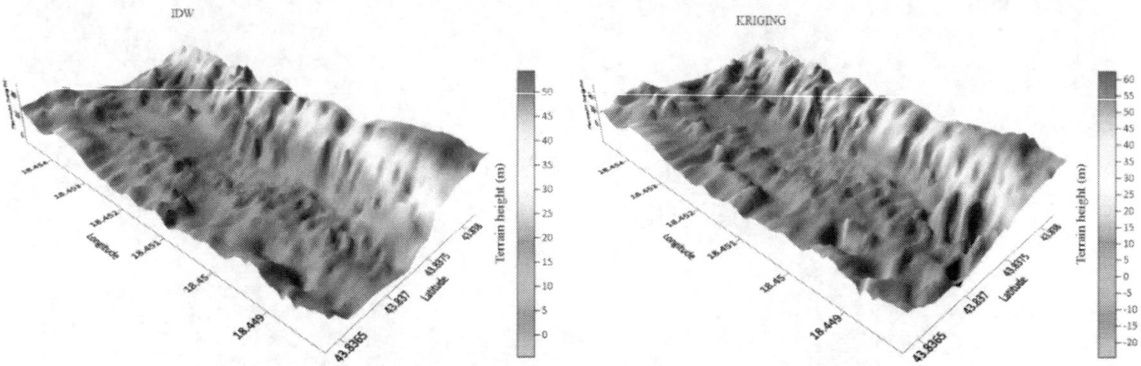

Fig. 4. Terrain height was calculated using IDW and Kriging algorithm.

Elevation profile measurement (Figure 5) was straightforward, giving the interesting result: the profile is very similar to one obtained from Google Earth, but it gives more details and is generally more accurate.

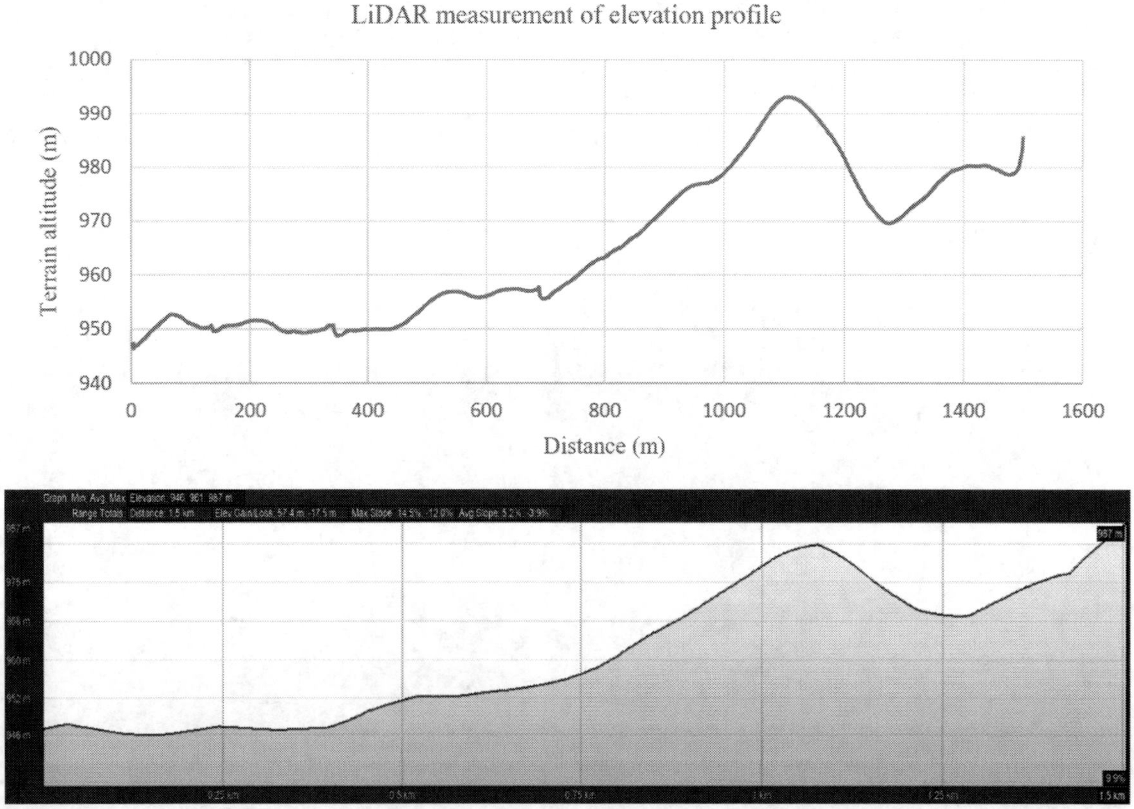

Fig. 5. Elevation profile at Nišići. Top: actual measurement, bottom: screenshot from Google Earth profile of the same terrain.

These results show that terrain mapping and elevation profile measurement are suitable test cases for the proposed concept of a low-cost aerial LiDAR platform. Autonomous flight mode was used with the predefined mission, so the measurements could be repeated if necessary.

5. Conclusion

The concept of a low-cost and open-source drone platform with LiDAR for terrain mapping was successfully demonstrated using two field tests. The survey area of 113531 m^2 at Trebević mountain was covered in single drone flight and terrain height was calculated using IDW and Kriging algorithm. The second test produced an 1.5 km long elevation profile at Nišići plateau, which agrees well with Google Earth profile but gives more details.

The first results are a positive and encouraging step towards massive adoption of this advanced technology in scientific research. In future work we will increase the capabilities of our platform, including the design and construction of a more powerful drone, installation of a more capable LiDAR sensor, and better data acquisition system to open new possibilities for low-cost drone investigations. These may include: high-resolution terrain mapping, research of forest, mapping of buildings and wider urban area, etc.

6. References

[1] Hopkinson, C. (2007). The influence of flying altitude, beam divergence, and pulse repetition frequency on laser pulse return intensity and canopy frequency distribution, Canadian Journal of Remote Sensing, 33:4, 312-324, DOI: 10.5589/m07-029

[2] Wulder, M. A.; White, J. C.; Bater, C. W.; Coops, N. C.; Hopkinson, C. & Chen, G. (2012). Lidar plots — a new large-area data collection option: context, concepts, and case study, Canadian Journal of Remote Sensing, 38:5, 600-618, DOI: 10.5589/m12-049

[3] Lipovsky, P. S.; Evans, S. G.; Clague, J. J. et al. (2008). The July 2007 rock and ice avalanches at Mount Steele, St. Elias Mountains, Yukon, Canada. Landslides 5, 445–455 (2008). DOI: 10.1007/s10346-008-0133-4

[4] Hopkinson, C. & Demuth, M. N. (2006). Using airborne lidar to assess the influence of glacier downwasting on water resources in the Canadian Rocky Mountains, Canadian Journal of Remote Sensing, 32:2, 212-222, DOI: 10.5589/m06-012

[5] Marselis, S. M.; Abernethy, K.; Alonso, A. et al. (2020). Evaluating the potential of full-waveform lidar for mapping pan-tropical tree species richness. Global Ecol Biogeogr. 2020; 29: 1799– 1816. DOI: 10.1111/geb.13158

[6] Kettridge, N.; Lukenbach, M. C.; Hokanson, K. J.; Hopkinson, C.; Devito, K. J.; Petrone, R. M.; Mendoza, C. A. & Waddington, J. M. (2017). Low evapotranspiration enhances the resilience of peatland carbon stocks to fire. Geophysical Research Letters, 44, 9341– 9349. DOI: 10.1002/2017GL074186

[7] Chasmer, L.; Hopkinson, C. & Treitz, P. (2006). Investigating laser pulse penetration through a conifer canopy by integrating airborne and terrestrial lidar, Canadian Journal of Remote Sensing, 32:2, 116-125, DOI: 10.5589/m06-011

[8] Yeong, D. J.; Velasco-Hernandez, G.; Barry, J. & Walsh, J. (2021). Sensor and Sensor Fusion Technology in Autonomous Vehicles: A Review. Sensors 2021, 21, 2140. DOI: 10.3390/s21062140

[9] Tang, L.; Shi, Y.; He, Q.; Sadek, A. W. & Qiao, C. (2020). Performance Test of Autonomous Vehicle Lidar Sensors Under Different Weather Conditions. Transportation Research Record. 2674(1):319-329. DOI:10.1177/0361198120901681

[10] Chen, J.; Zhang, H.; Lu, Y. & Zhang, Q. (2020). The Research on Control and Dynamic Property of Autonomous Vehicle Adaptive Lidar System, 2020 International Conferences on Internet of Things (iThings) and IEEE Green Computing and Communications (GreenCom) and IEEE Cyber, Physical and Social Computing (CPSCom) and IEEE Smart Data (SmartData) and IEEE Congress on Cybermatics (Cybermatics), 2020, pp. 462-468, DOI: 10.1109/iThings-GreenCom-CPSCom-SmartData-Cybermatics50389.2020.00086.

[11] Elhousni, M. & Huang, X. (2020). A Survey on 3D LiDAR Localization for Autonomous Vehicles, 2020 IEEE Intelligent Vehicles Symposium (IV), 2020, pp. 1879-1884, DOI: 10.1109/IV47402.2020.9304812.

[12] Masic, A. (2015). Unmanned Aerial Vehicle as Data Acquisition System. Journal of Trends in the Development of Machinery and Associated Technology. Vol. 19, No. 1, 2015, ISSN 2303-4009 (online), p.p. 181-184.

[13] Masic, A.; Musemic, R. & Dzaferovic-Masic, E. (2016). Temperature Inversion Measurements in Sarajevo Valley Using Unmanned Aerial Vehicles, Proceedings of the 27th DAAAM International Symposium, pp.0423-0427, B. Katalinic (Ed.), Published by DAAAM International, ISBN 978-3-902734-08-2, ISSN 1726-9679, Vienna, Austria, DOI: 10.2507/27th.daaam.proceedings.062

[14] Masic, A.; Pikula, B. & Bibic, D. (2017). Mobile Measurements of Particulate Matter Concentrations in Urban Area, Proceedings of the 28th DAAAM International Symposium, pp.0452-0456, B. Katalinic (Ed.), Published by DAAAM International, ISBN 978-3-902734-11-2, ISSN 1726-9679, Vienna, Austria DOI: 10.2507/28th.daaam.proceedings.063

[15] Masic. A.; Bibic, D.; Pikula, B.; Dzaferovic-Masic, E. & Musemic, R. (2019). Experimental study of temperature inversions above urban area using unmanned aerial vehicle. Thermal Science, Vol. 23, No. 6A, pp. 3327-3338, DOI: 10.2298/TSCI180227250M

[16] Saric I.; Masic A. & Delic, M. (2021). Hexacopter Design and Analysis. In: Karabegović I. (eds) New Technologies, Development and Application IV. NT 2021. Lecture Notes in Networks and Systems, vol 233. Springer, Cham. DOI: 10.1007/978-3-030-75275-0_9

[17] Masic, A.; Bibic, D.; Pikula, B. & Razic, F. (2018). New approach of measuring toxic gases concentrations: principle of operation, Proceedings of the 29th DAAAM International Symposium, pp.0882-0887, B. Katalinic (Ed.), Published by DAAAM International, ISBN 978-3-902734-20-4, ISSN 1726-9679, Vienna, Austria, DOI: 10.2507/29th.daaam.proceedings.127

[18] Bibic, D.; Pikula, B.; Masic, A. & Razic, F. (2018). New approach of measuring toxic gases concentrations: application examples, Proceedings of the 29th DAAAM International Symposium, pp.0876-0881, B. Katalinic (Ed.), Published by DAAAM International, ISBN 978-3-902734-20-4, ISSN 1726-9679, Vienna, Austria, DOI: 10.2507/29th.daaam.proceedings.126

[19] Masic, A.; Bibic, D. & Pikula, B. (2019). On the applicability of low-cost sensors for measurements of aerosol concentrations, Proceedings of the 30th DAAAM International Symposium, pp.0452-0456, B. Katalinic (Ed.), Published by DAAAM International, ISBN 978-3-902734-22-8, ISSN 1726-9679, Vienna, Austria DOI: 10.2507/30th.daaam.proceedings.060

[20] Masic, A.; Pikula, B.; Bibic, D.; Musemic, R. & Halac, A. (2018). Calibration and Assessment of Low-cost Dust sensors, Proceedings of the 29th DAAAM International Symposium, pp. 523-528, B. Katalinic (Ed.), Published by DAAAM International, ISBN 978-3-902734-xx-x, ISSN 1726-9679, Vienna, Austria, DOI: 10.2507/29th.daaam.proceedings.075

[21] Masic, A.; Bibic, D.; Pikula, B.; Dzaferovic-Masic, E. & Razic, F. (2018). Visualization of Data from Network of Sensors: Appropriate Spatial Interpolation Method, Proceedings of the 29th DAAAM International Symposium, pp.0529-0533, B. Katalinic (Ed.), Published by DAAAM International, ISBN 978-3-902734-20-4, ISSN 1726-9679, Vienna, Austria DOI: 10.2507/29th.daaam.proceedings.076

DOI: 10.2507/32nd.daaam.proceedings.018

DRONE MEASUREMENTS OF TEMPERATURE INVERSION CHARACTERISTICS AND PARTICULATE MATTER VERTICAL PROFILES IN URBAN ENVIRONMENTS

Adnan Masic, Boran Pikula, Dzevad Bibic,
Vahidin Hadziabdic & Almir Blazevic

This Publication has to be referred as: Masic, A[dnan]; Pikula, B[oran]; Bibic, D[zevad]; Hadziabdic, V[ahidin] & Blazevic, A[lmir] (2021). Drone Measurements of Temperature Inversion Characteristics and Particulate Matter Vertical Profiles in Urban Environments, Proceedings of the 32nd DAAAM International Symposium, pp.0123-0126, B. Katalinic (Ed.), Published by DAAAM International, ISBN 978-3-902734-33-4, ISSN 1726-9679, Vienna, Austria
DOI: 10.2507/32nd.daaam.proceedings.018

Abstract

A new approach for measurements of vertical profiles of temperature and particulate matter (PM) concentrations is developed and tested. Special purpose unmanned aerial vehicle with data acquisition system has been developed in-house. The temperature sensor is based on a fast response thermistor, while the PM sensor uses optical detection of particles. All sensors were tested in the laboratory and calibrated against reference instruments. Two urban locations were selected for field tests. This research produced a very useful asset that characterizes the relation between temperature inversion and air pollution.

Keywords: temperature inversion; particulate matter; drone; boundary layer.

1. Introduction

Temperature inversion is a natural phenomenon in the atmospheric boundary layer which has strong effects on local climate and air pollution. Understanding temperature inversion plays important role in air quality management. If the temperature inversion is strong enough, natural convection is prevented, and if this happens over an urban area, the concentration of air pollutants rises quickly. Thus, quantitative characteristics of temperature inversion should be regularly investigated to understand and predict air quality in any area prone to temperature inversions, particularly urban valleys.

Quantitative characteristics of the temperature inversion layer may be obtained using various field measurements and observations. Some of the most widely used methods include satellite observations [1], microwave radiometer [2], radiosonde [4]-[7], and drone base measurements [8]-[13]. The novelty in this research is a combination of instruments and conditions: simultaneous measurements of vertical profiles of temperature and particulate matter (PM) concentration in urban environments using an unmanned aerial vehicle (drone).

2. The drone and sensors

An in-house developed drone [14] named MEF-707 was used in this research. It is a medium-sized hexacopter, with a diagonal distance between motors of 707 mm, powered by open-source hardware and software (Figure 1). Major components of MEF-707 are:

- Arducopter open-source flight stack,
- Pixhawk flight controller,
- DJI E800 motors,
- DragonLink long-range radio control system,
- Carbon fiber frame.

Data acquisition system [15] is also in-house developed, and it is called MAQS (Mobile Air Quality System) [16], [17]. It is a modular platform for measurements of various parameters in the atmosphere. The primary application of MAQS is air pollution measurement [18]-[21], but it is expandable with other sensors, including pressure, temperature, relative humidity, carbon dioxide, wind speed, and GNSS coordinates. For the measurements of altitude above ground level (AGL), a barometer was used. A thermistor with a negative coefficient (NTC) was used for temperature measurements. All sensors were calibrated against reference instruments. MAQS system was installed atop of the UAV, in a spot with minimal air turbulence.

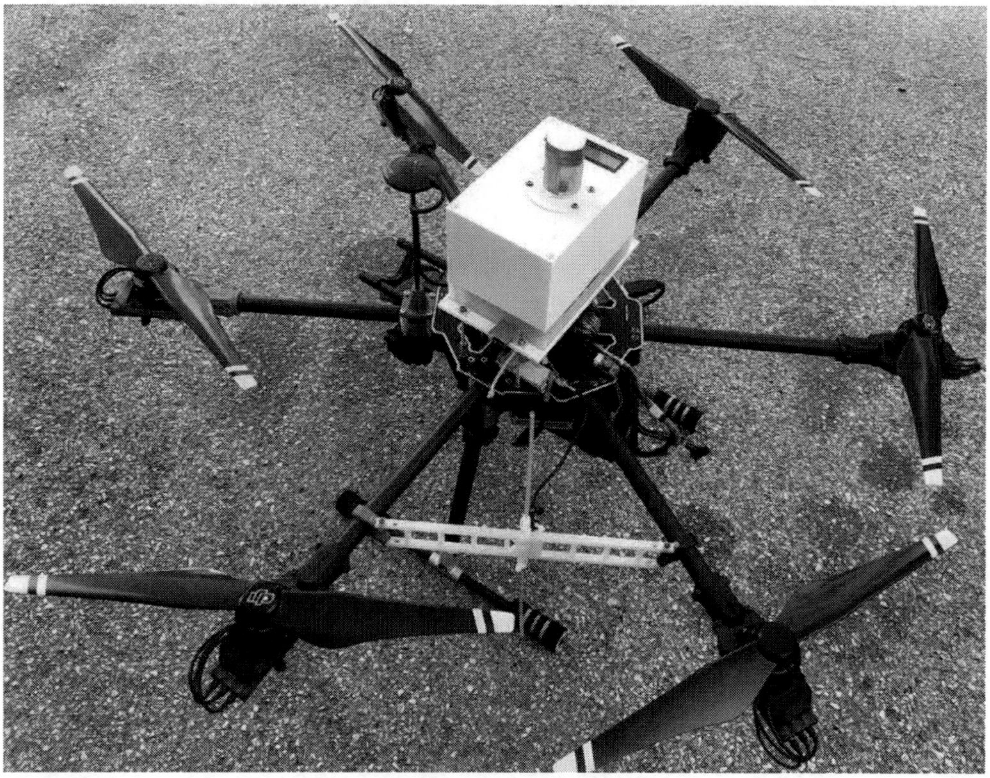

Fig. 1. The MEF-707 drone.

3. Vertical profiles

By taking into account the results of our previous campaigns [9] and [13] we were looking for representative test locations for vertical measurements up to 500m above ground level. The following locations were selected:

- Hadžići: 43.8210000(N), 18.1961210(E), 562 m ASL,
- Vogošća: 43.8946190(N), 18.3736600(E), 553 m ASL.

Figure 2 shows vertical profiles in Vogošća on 19.01.2021. with strong temperature inversion starting from the ground and very low air temperature. As a consequence, alarmingly high concentrations of $PM_{2.5}$ were measured on the ground. Temperature inversion appeared at 180m above ground in Hadžići on 02.02.2021. (Figure 3), but the temperature gradient in the inversion layer was moderate and the air pollution level wasn't too high.

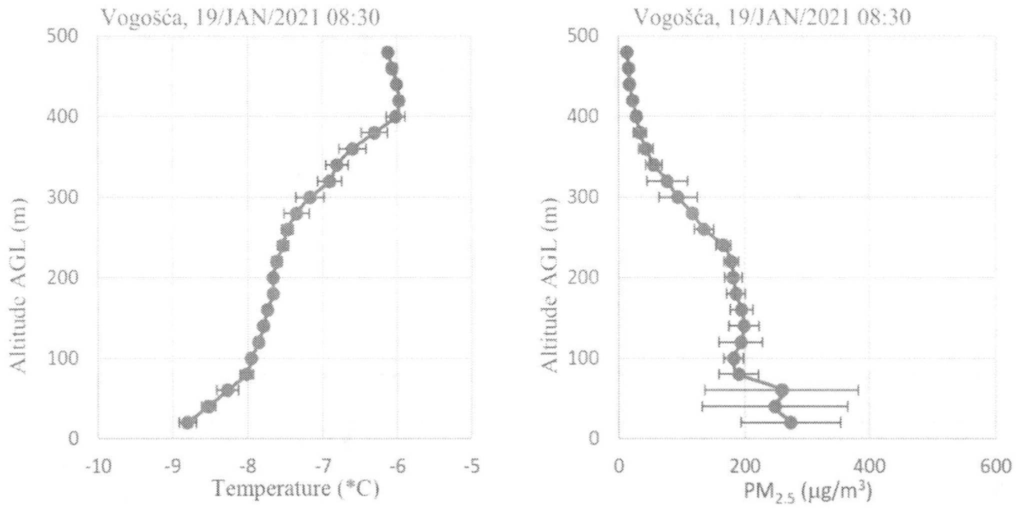

Fig. 2. Vertical profiles in Vogošća.

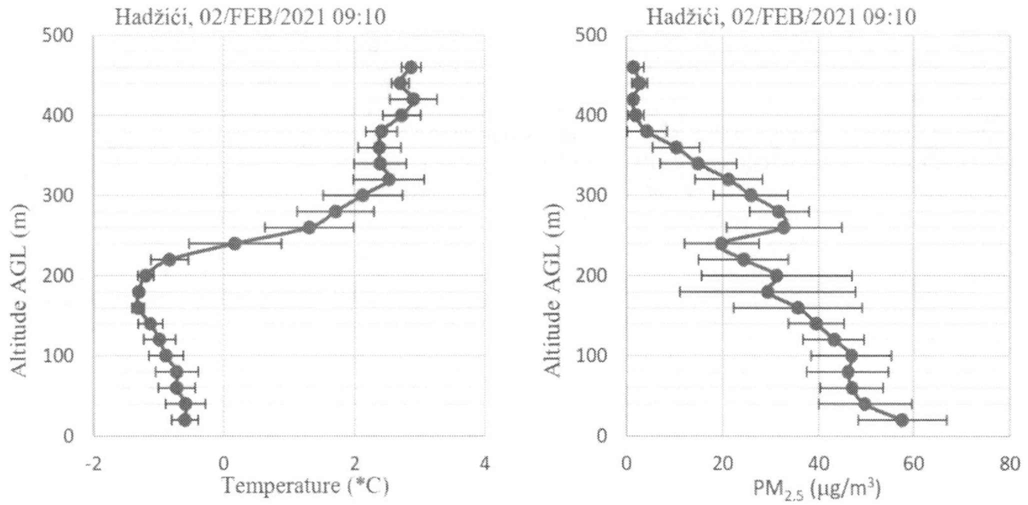

Fig. 3. Vertical profiles in Hadžići.

Figures 2 and 3 show the complexity of the relationship between temperature inversion and air quality. Not every temperature inversion leads to a high concentration of air pollutants. If the gradient in the inversion layer is mild, then natural convection is not prevented completely. But if the temperature inversion gradient is strong and close to the ground, leaving only a shallow atmospheric layer below, this may lead to a dramatic concentration of particulate matter, especially if the ambient air temperature is very low.

4. Conclusion

Simultaneous measurements of temperature and $PM_{2.5}$ were performed using a drone at the different field locations. The results provided evidence of the strong influence of temperature inversion on air pollution in urban environments. An innovative technique based on an in-house developed drone and data acquisition system was used. This platform offers key advantages over other techniques for vertical profiles measurements: low cost per single measurement and full control of flight trajectory and experimental setup in both vertical directions.

Two quantitative characteristics of the temperature inversion layer are particularly important: the height above ground level at which the inversion starts, and the temperature gradient in the inversion layer. As a recommendation for future work, it would be useful to include measurements of concentrations of toxic gasses, such as carbon monoxide and sulfur dioxide. Long-term speaking, the drone-based platform for atmospheric observation and measurements has the potential to change the paradigm of research in this field.

5. References

[1] Boylan, P., Wang, J., Cohn, S. A., Hultberg, T. & August, T. (2016). Identification and intercomparison of surface-based inversions over Antarctica from IASI, ERA-Interim, and Concordiasi dropsonde data, Journal of Geophysical Research: Atmospheres, 121, 9089–9104, DOI: 10.1002/2015JD024724

[2] Pietroni, I., Argentini, S. & Petenko, I. (2014). One Year of Surface-Based Temperature Inversions at Dome C, Antarctica, Boundary-Layer Meteorology, 150, 131–151, DOI: 10.1007/s10546-013-9861-7

[3] Hudson, S. R. & Brandt, R. E. (2005). A Look at the Surface-Based Temperature Inversion on the Antarctic Plateau, Journal of Climate, 18, 1673–1696, DOI: 10.1175/JCLI3360.1

[4] Bradley, R. S., Keimig, F. T. & Diaz, H. F. (1993). Recent changes in the North American Arctic boundary layer in winter, Journal of Geophysical Research: Atmospheres, 98, 8851–8858, https://doi.org/10.1029/93JD00311

[5] Lesins, G., Duck, T. J. & Drummond, J. R. (2010). Climate trends at Eureka in the Canadian high arctic, Atmosphere-Ocean, 48, 59–80, DOI: 10.3137/AO1103

[6] Zhang, Y., Seidel, D. J., Golaz, J.-C., Deser, C., & Tomas, R. A. (2011). Climatological Characteristics of Arctic and Antarctic Surface-Based Inversions, Journal of Climate, 24, 5167 – 5186, DOI: 10.1175/2011JCLI4004.1

[7] Lesins, G., Duck, T. J. & Drummond, J. R. (2012). Surface Energy Balance Framework for Arctic Amplification of Climate Change, Journal of Climate, 25, 8277–8288, DOI: 10.1175/JCLI-D-11-00711.1

[8] Cassano, J. J. (2014). Observations of atmospheric boundary layer temperature profiles with a small unmanned aerial vehicle, Antarctic Science, 26, 205–213, DOI: 10.1017/S0954102013000539

[9] Masic, A.; Musemic, R. & Dzaferovic-Masic, E. (2016). Temperature Inversion Measurements in Sarajevo Valley Using Unmanned Aerial Vehicles, Proceedings of the 27th DAAAM International Symposium, pp.0423-0427, B. Katalinic (Ed.), Published by DAAAM International, ISBN 978-3-902734-08-2, ISSN 1726-9679, Vienna, Austria, DOI: 10.2507/27th.daaam.proceedings.062

[10] Cassano, J. J., Seefeldt, M.W., Palo, S., Knuth, S. L., Bradley, A. C., Herrman, P. D., Kernebone, P. A. & Logan, N. J. (2016). Observations of the atmosphere and surface state over Terra Nova Bay, Antarctica, using unmanned aerial systems, Earth System Science Data, 8, 115–126, DOI: 10.5194/essd-8-115-2016

[11] Chabot, D. & Bird, D. M. (2015). Wildlife research and management methods in the 21st century: Where do unmanned aircraft fit in? Journal of Unmanned Vehicle Systems, 3, 137–155, DOI: 10.1139/juvs-2015-0021

[12] Gaffey, C. & Bhardwaj, A. (2020). Applications of Unmanned Aerial Vehicles in Cryosphere: Latest Advances and Prospects. Remote Sensing, 12, 948, DOI: 10.3390/rs12060948,

[13] Masic. A.; Bibic, D.; Pikula, B.; Dzaferovic-Masic, E. & Musemic, R. (2019). Experimental study of temperature inversions above urban area using unmanned aerial vehicle. Thermal Science, Vol. 23, No. 6A, pp. 3327-3338, DOI: 10.2298/TSCI180227250M

[14] Saric I.; Masic A. & Delic, M. (2021). Hexacopter Design and Analysis. In: Karabegović I. (eds) New Technologies, Development and Application IV. NT 2021. Lecture Notes in Networks and Systems, vol 233. Springer, Cham. DOI: 10.1007/978-3-030-75275-0_9

[15] Masic, A. (2015). Unmanned Aerial Vehicle as Data Acquisition System. Journal of Trends in the Development of Machinery and Associated Technology. Vol. 19, No. 1, 2015, ISSN 2303-4009 (online), p.p. 181-184.

[16] Masic. A.; Bibic, D.; Pikula, B.; Blazevic, A.; Huremovic, J. & Zero, S. (2020). Evaluation of optical particulate matter sensors under realistic conditions of strong and mild urban pollution. Atmospheric Measurement Techniques, 13, 6427–6443, DOI: 10.5194/amt-13-6427-2020

[17] Masic, A.; Kepnik, G.; Bektesevic, J.; Mehuljic, M.; Saric, I. & Hadziabdic, V. (2020). The Network of Smart Sensors for Indoor Air Quality Monitoring, Proceedings of the 31st DAAAM International Symposium, pp.0232-0235, B. Katalinic (Ed.), Published by DAAAM International, ISBN 978-3-902734-29-7, ISSN 1726-9679, Vienna, Austria, DOI: 10.2507/31st.daaam.proceedings.032

[18] Masic, A.; Pikula, B. & Bibic, D. (2017). Mobile Measurements of Particulate Matter Concentrations in Urban Area, Proceedings of the 28th DAAAM International Symposium, pp.0452-0456, B. Katalinic (Ed.), Published by DAAAM International, ISBN 978-3-902734-11-2, ISSN 1726-9679, Vienna, Austria DOI: 10.2507/28th.daaam.proceedings.063

[19] Masic, A.; Bibic, D.; Pikula, B. & Razic, F. (2018). New approach of measuring toxic gases concentrations: principle of operation, Proceedings of the 29th DAAAM International Symposium, pp.0882-0887, B. Katalinic (Ed.), Published by DAAAM International, ISBN 978-3-902734-20-4, ISSN 1726-9679, Vienna, Austria, DOI: 10.2507/29th.daaam.proceedings.127

[20] Masic, A.; Bibic, D. & Pikula, B. (2019). On the applicability of low-cost sensors for measurements of aerosol concentrations, Proceedings of the 30th DAAAM International Symposium, pp.0452-0456, B. Katalinic (Ed.), Published by DAAAM International, ISBN 978-3-902734-22-8, ISSN 1726-9679, Vienna, Austria DOI: 10.2507/30th.daaam.proceedings.060

[21] Masic, A.; Pikula, B.; Bibic, D.; Musemic, R. & Halac, A. (2018). Calibration and Assessment of Low-cost Dust sensors, Proceedings of the 29th DAAAM International Symposium, pp. 523-528, B. Katalinic (Ed.), Published by DAAAM International, ISBN 978-3-902734-xx-x, ISSN 1726-9679, Vienna, Austria, DOI: 10.2507/29th.daaam.proceedings.075

DOI: 10.2507/32nd.daaam.proceedings.019

REFERENCE MEASUREMENTS OF PM_{10} AND $PM_{2.5}$ PARTICULATE MATTER CONCENTRATIONS

Adnan Masic, Emir Hanic, Emir Bibic,
Dzevad Bibic & Boran Pikula

This Publication has to be referred as: Masic, A[dnan]; Hanic, E[mir]; Bibic, E[mir]; Bibic, D[zevad] & Pikula, B[oran] (2021). Reference Measurements of PM10 and PM2.5 Particulate Matter Concentrations, Proceedings of the 32nd DAAAM International Symposium, pp.0127-0130, B. Katalinic (Ed.), Published by DAAAM International, ISBN 978-3-902734-33-4, ISSN 1726-9679, Vienna, Austria
DOI: 10.2507/32nd.daaam.proceedings.019

Abstract

To provide the reference dataset for particulate matter concentrations, a long-term campaign of continuous gravimetric measurements of PM_{10} and $PM_{2.5}$ concentrations was performed. Strict implementation of standard EN12341:2014 was fulfilled. Daily and monthly concentrations of both PM_{10} and $PM_{2.5}$ were measured with the highest possible accuracy. Ratio $PM_{2.5}/PM_{10}$ was determined and analyzed. Different distribution of particles was observed during the winter period, where most of the particulate matter originates from burning fossil fuel, and during an episode of Sahara dust, occasionally observed in spring and summer.

Keywords: PM_{10}; $PM_{2.5}$; gravimetric measurement; Sahara dust.

1. Introduction

Air pollution by particulate matter (PM) is a global problem. Measurement techniques for PM concentrations have evolved over the past decades. Tapered element oscillating microbalance (TEOM) is one of the traditional methods [1], [2]. One of the most widely used methods is the beta attenuation mass monitor (BAM) [2]. Many governmental institutions use it because they are robust, relatively accurate, and easy to maintain. Then, laser scattering optical instruments are increasingly more popular nowadays [3], especially the low-cost variants [4]-[11]. They are less accurate but they have better time resolution and give information of several fractions of PM at once. There are also remote sensing techniques such as measurements from satellites [12]. What is common for all the abovementioned methods is significant measurement uncertainty. Thus, the reference method is necessary to verify all other measurements under different environmental conditions. The reference method for measurements of PM_{10} (all particulate matter smaller than 10 μm in diameter) and $PM_{2.5}$ (all PM smaller than 2.5 μm) concentrations is a gravimetric procedure, based on collecting the particles on filters, and weighing the filters before and after the sampling in the controlled environment. However, this method is expensive and requires a lot of manual work and it is rarely used in long-term campaigns. In this research we present a valuable asset of reference measurements of PM_{10} and $PM_{2.5}$ in Sarajevo, Bosnia Hercegovina, from 27.10.2020. till 31.5.2021.

2. The laboratory for gravimetric measurements

The laboratory for gravimetric measurements (Figure 1) is part of a larger laboratory for physical measurements, where we calibrate all instruments used in research [13]-[21]. A specific procedure for reference measurements of PM is defined in standard EN12341:2014, which is strictly implemented in our laboratory.

The following equipment was used:

- Dual channel (PM2.5 and PM10) air sampler Gemini, Dadolab, Italy,
- Quartz filters Munktell Ahlstrom 47 mm,
- Laboratory with controlled temperature and humidity,
- Analytical balance ABT 101-5NM, Kern, Germany (Mettler Toledo XSR105DU in the final stage).

The sampling begins at 12:00 each day and ends at 11:58 the next day. Two minutes are reserved for filter exchange (automatic process). Each filter is conditioned before and after sampling using the following procedure: filters are stored in the room with a controlled temperature of (20±1) ^{0}C and relative humidity (47,5±2,5) % two days. After 24 hours we take the first measurement of mass, and after 48 hours we take the second measurement. If the difference between the two measurements is not greater than 40 μg, the filter is ready for sampling, otherwise, the procedure is repeated until two successive mass measurements satisfy the given criterion. The same procedure is performed after sampling. Airflow for sampling is 2.3 m^3/h (55.2 m^3/day).

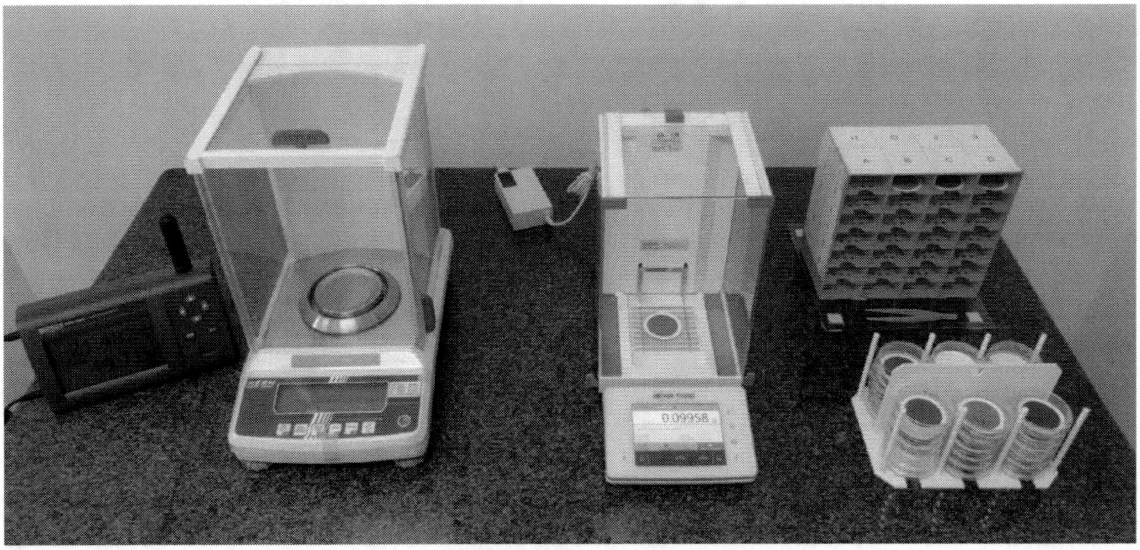

Fig. 1. The gravimetric laboratory.

3. Results

Figure 2 shows daily average values of $PM_{2.5}$ and PM_{10} concentration, while Table 1 gives the monthly average values and ratio $PM_{2.5}/PM_{10}$ which was on average 84.4%. The primary source of air pollution in this period was the burning of fossil fuels for domestic heating and from vehicles' emissions. However, additional sampling was performed from 24.6.2021. until 26.6.2021. when Sahara dust episode was detected by satellites. Table 2 shows a very different ratio $PM_{2.5}/PM_{10}$ of around 30% in the case when Sahara dust was the primary source of pollution.

	Dec 2020	Jan 2021	Feb 2021	Mar 2021	Apr 2021	May 2021
$PM_{2.5}$ (μg/m^3)	76.3	45.6	41.3	26.4	16.2	8.0
PM_{10} (μg/m^3)	91.3	49.1	50.4	31.1	20.7	13.1
$PM_{2.5}/PM_{10}$	83.5%	92.9%	81.9%	85.0%	78.2%	61.2%

Table 1. Monthly average values of $PM_{2.5}$ and PM_{10}.

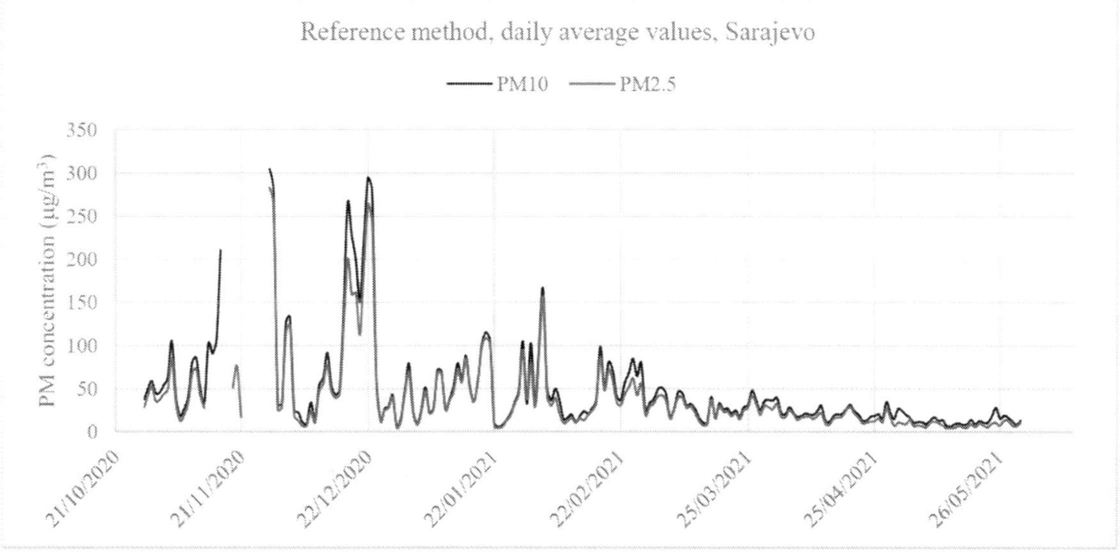

Fig. 2. Daily average values of $PM_{2.5}$ and PM_{10}.

Start of sampling	24.6.2021. 12:00	25.6.2021 12:00
$PM_{2.5}$	21.08	14.18
PM_{10}	73.24	39.62
$PM_{2.5}/PM_{10}$	28.8%	35.8%

Table 2. Sampling during the presence of Sahara dust.

Excellent accuracy of gravimetric instruments enables precise determination of $PM_{2.5}/PM_{10}$ ratio, which is a challenging task for all other techniques [3]. This is especially important in the atypical composition of particulate matter, such as the Sahara dust episode.

4. Conclusion

The very valuable asset of reference measurements of $PM_{2.5}$ and PM_{10} concentrations was produced in this research. It can be used for benchmarking all other measurement techniques, such as the research-grade equivalent methods or even the low-cost sensors. The average ratio $PM_{2.5}/PM_{10}$ was 84.4% during the winter period when the primary source of air pollution is the burning of fossil fuels. High daily average values up to 300 μg/m³ of $PM_{2.5}$ were observed occasionally, which is extremely dangerous for public health. However, at the beginning of Summer, an episode of Sahara dust was detected. When the Sahara dust was the primary source of PM pollution, the ratio $PM_{2.5}/PM_{10}$ was only around 30% which is very interesting. This should be investigated more in future work.

5. References

[1] Allen, G.; Sioutas, C.; Koutrakis, P.; Reiss, R.; Lurmann, F. W. & Roberts, P. T. (1997). Evaluation of the TEOM Method for Measurement of Ambient Particulate Mass in Urban Areas, Journal of the Air & Waste Management Association, 47:6, 682-689, DOI: 10.1080/10473289.1997.10463923

[2] Schweizer, D.; Cisneros, R. & Shaw, G. (2016). A comparative analysis of temporary and permanent beta attenuation monitors: The importance of understanding data and equipment limitations when creating PM2.5 air quality health advisories, Atmospheric Pollution Research, Volume 7, Issue 5, 2016, Pages 865-875, ISSN 1309-1042, DOI: 10.1016/j.apr.2016.02.003.

[3] Masic. A.; Bibic, D.; Pikula, B.; Blazevic, A.; Huremovic, J. & Zero, S. (2020). Evaluation of optical particulate matter sensors under realistic conditions of strong and mild urban pollution. Atmospheric Measurement Techniques, 13, 6427–6443, DOI: 10.5194/amt-13-6427-2020

[4] Zheng, T.; Bergin, M. H.; Johnson, K. K.; Tripathi, S. N.; Shirodkar, S.; Landis, M. S.; Sutaria, R. & Carlson, D. E. (2018). Field evaluation of low-cost particulate matter sensors in high- and low-concentration environments, Atmos. Meas. Tech., 11, 4823–4846, DOI: 10.5194/amt-11-4823-2018

[5] Zheng, T.; Bergin, M. H.; Sutaria, R.; Tripathi, S. N.; Caldow, R. & Carlson, D. E. (2019). Gaussian process regression model for dynamically calibrating and surveilling a wireless low-cost particulate matter sensor network in Delhi, Atmos. Meas. Tech., 12, 5161–5181, DOI: 10.5194/amt-12-5161-2019

[6] Mukherjee, A.; Stanton, L. G.; Graham, A. R. & Roberts, P. T. (2017). Assessing the utility of low-cost particulate matter sensors over a 12-week period in the Cuyama Valley of California, Sensors, 17, 1805, DOI: 10.3390/s17081805

[7] Mukherjee, A.; Brown, S. G.; McCarthy, M. C.; Pavlovic, N. R.; Stanton, L. G.; Snyder, J. L.; D'Andrea, S. & Hafner, H. R. (2019). Measuring Spatial and Temporal PM2:5 Variations in Sacramento, California, Communities Using a Network of Low-Cost Sensors, Sensors, 19, 4701, DOI: 10.3390/s19214701

[8] Morawska, L.; Thai, P. K.; Liu, X.; Asumadu-Sakyi, A.; Ayoko, G.; Bartonova, A.; Bedini, A.; Chai, F.; Christensen, B.; Dunbabin, M.; Gao, J.; Hagler, G. S. W.; Jayaratne, R.; Kumar, P.; Lau, A. K. H.; Louie, P. K. K.; Mazaheri, M.; Ning, Z.; Motta, N.; Mullins, B.; Rahman, M. M.; Ristovski, Z.; Shafiei, M.; Tjondronegoro, D.; Westerdahl, D. & Williams, R. (2018). Applications of low-cost sensing technologies for air quality monitoring and exposure assessment: How far have they gone?, Environ. Int., 116, 286–299, DOI: 10.1016/j.envint.2018.04.018

[9] Sayahi, T.; Butterfield, A. & Kelly, K. E. (2018). Long-term field evaluation of the Plantower PMS low-cost particulatematter sensors, Environ. Poll., 245, 932–940, DOI: 10.1016/j.envpol.2018.11.065

[10] Tryner, J.; L'Orange, C.; Mehaffy, J.; Miller-Lionberg, D.; Hofstetter, J. C.; Wilson, A. & Volckens, J. (2019). Laboratory evaluation of low-cost PurpleAir PMmonitors and in-field correction using collocated portable filter samplers, Atmos. Environ., 220, 117067, DOI: 10.1016/j.atmosenv.2019.117067

[11] Malings, C.; Tanzer, R.; Hauryliuk, A.; Saha, P. K.; Robinson, A. L.; Presto, A. A. & Subramanian, R. (2020). Fine particle mass monitoring with low-cost sensors: Corrections and longterm performance evaluation, Aerosol Sci. Tech., 54, 160–174, DOI: 10.1080/02786826.2019.1623863

[12] Christopher, S. A. & Gupta, P. (2010). Satellite Remote Sensing of Particulate Matter Air Quality: The Cloud-Cover Problem, Journal of the Air & Waste Management Association, 60:5, 596-602, DOI: 10.3155/1047-3289.60.5.596

[13] Masic, A.; Pikula, B. & Bibic, D. (2017). Mobile Measurements of Particulate Matter Concentrations in Urban Area, Proceedings of the 28th DAAAM International Symposium, pp.0452-0456, B. Katalinic (Ed.), Published by DAAAM International, ISBN 978-3-902734-11-2, ISSN 1726-9679, Vienna, Austria DOI: 10.2507/28th.daaam.proceedings.063

[14] Masic, A.; Bibic, D.; Pikula, B. & Razic, F. (2018). New approach of measuring toxic gases concentrations: principle of operation, Proceedings of the 29th DAAAM International Symposium, pp.0882-0887, B. Katalinic (Ed.), Published by DAAAM International, ISBN 978-3-902734-20-4, ISSN 1726-9679, Vienna, Austria, DOI: 10.2507/29th.daaam.proceedings.127

[15] Bibic, D.; Pikula, B.; Masic, A. & Razic, F. (2018). New approach of measuring toxic gases concentrations: application examples, Proceedings of the 29th DAAAM International Symposium, pp.0876-0881, B. Katalinic (Ed.), Published by DAAAM International, ISBN 978-3-902734-20-4, ISSN 1726-9679, Vienna, Austria, DOI: 10.2507/29th.daaam.proceedings.126

[16] Masic, A.; Bibic, D. & Pikula, B. (2019). On the applicability of low-cost sensors for measurements of aerosol concentrations, Proceedings of the 30th DAAAM International Symposium, pp.0452-0456, B. Katalinic (Ed.), Published by DAAAM International, ISBN 978-3-902734-22-8, ISSN 1726-9679, Vienna, Austria DOI: 10.2507/30th.daaam.proceedings.060

[17] Masic, A.; Pikula, B.; Bibic, D.; Musemic, R. & Halac, A. (2018). Calibration and Assessment of Low-cost Dust sensors, Proceedings of the 29th DAAAM International Symposium, pp. 523-528, B. Katalinic (Ed.), Published by DAAAM International, ISBN 978-3-902734-xx-x, ISSN 1726-9679, Vienna, Austria, DOI: 10.2507/29th.daaam.proceedings.075

[18] Masic, A.; Bibic, D.; Kamenica, I.; Bektesevic, J.; Haziabdic, V. & Mehuljic, M. (2020). Mobile Measurements of Black Carbon for Source Apportionment Studies, Proceedings of the 31st DAAAM International Symposium, pp.0227-0231, B. Katalinic (Ed.), Published by DAAAM International, ISBN 978-3-902734-29-7, ISSN 1726-9679, Vienna, Austria, DOI: 10.2507/31st.daaam.proceedings.031

[19] Masic, A.; Bibic, D.; Pikula, B.; Dzaferovic-Masic, E. & Razic, F. (2018). Visualization of Data from Network of Sensors: Appropriate Spatial Interpolation Method, Proceedings of the 29th DAAAM International Symposium, pp.0529-0533, B. Katalinic (Ed.), Published by DAAAM International, ISBN 978-3-902734-20-4, ISSN 1726-9679, Vienna, Austria DOI: 10.2507/29th.daaam.proceedings.076

[20] Masic, A.; Kepnik, G.; Bektesevic, J.; Mehuljic, M.; Saric, I. & Haziabdic, V. (2020). The Network of Smart Sensors for Indoor Air Quality Monitoring, Proceedings of the 31st DAAAM International Symposium, pp.0232-0235, B. Katalinic (Ed.), Published by DAAAM International, ISBN 978-3-902734-29-7, ISSN 1726-9679, Vienna, Austria DOI: 10.2507/31st.daaam.proceedings.032

[21] Masic. A.; Bibic, D.; Pikula, B.; Dzaferovic-Masic, E. & Musemic, R. (2019). Experimental study of temperature inversions above urban area using unmanned aerial vehicle. Thermal Science, Vol. 23, No. 6A, pp. 3327-3338, DOI: 10.2298/TSCI180227250M

DOI: 10.2507/32nd.daaam.proceedings.020

LONG-TERM MEASUREMENT OF BLACK CARBON CONCENTRATION AND SOURCE APPORTIONMENT ANALYSIS

Adnan Masic, Goran Kepnik, Dzevad Bibic, Elmedin Mesic & Boran Pikula

This Publication has to be referred as: Masic, A[dnan]; Kepnik, G[oran]; Bibic, D[zevad]; Mesic, E[lmedin] & Pikula, B[oran] (2021). Long-term Measurement of Black Carbon Concentration and Source Apportionment Analysis, Proceedings of the 32nd DAAAM International Symposium, pp.0131-0135, B. Katalinic (Ed.), Published by DAAAM International, ISBN 978-3-902734-33-4, ISSN 1726-9679, Vienna, Austria
DOI: 10.2507/32nd.daaam.proceedings.020

Abstract

Continuous measurement of black carbon (BC) concentrations in ambient air was performed using the optical-based instrument called aethalometer. The attenuation of light by BC was measured at 7 different wavelengths. This gives not only the pure concentration of BC in the ambient air but also dependence of attenuation on the wavelength of light, from which the Angstrom exponent can be derived. Based on the Angstrom exponent, we can estimate the percentage of BC that originates from solid fuel (wood and coal burning) or liquid fuel (diesel). Thus, a simple source apportionment analysis is presented.

Keywords: black carbon; source apportionment; fossil fuels; biomass burning.

1. Introduction

Aerosol black carbon (BC) is a very important component of air pollution. It originates from incomplete combustion of fossil fuels and has deep effects on climate. It is argued that BC is the second largest anthropogenic climate change factor [1], [2] (carbon dioxide being first). The negative health effects of black carbon are well documented [3], [4]. Thus, the research on black carbon is a very significant step towards decarbonization. Previous research articles include Evaluation with a Photoacoustic Instrument [5], method comparisons [6], vertical profiles over Milan [7], diesel exhaust [8], wood combustion particles [9], emissions from fresh biomass burning [10], and source apportionment [11]. The novelty in this article is the long-term campaign of continuous BC measurement with source apportionment analysis.

2. The aethalometer

The instrument for the measurement of BC concentration is called aethalometer (the word comes from the classical Greek verb *aethaloun* which means "to blacken with soot"). The AE33 aethalometer, Magee Scientific, USA (Figure 1) is used in this research. The main specifications of AE33 are:

- 7 measurement wavelengths: 370, 470, 520, 590, 660, 880 and 950 nm,
- Airflow: set-points of 2, 3, 4, and 5 LPM,
- Max power consumption: 90 W. Typical power consumption: 25 W.

Figure 2 shows the flow diagram of AE33. The inlet cyclone, at the beginning of the air sampling line, separates particles greater than 2.5 mm in diameter. BC particles are collected on filter tape and the light absorption method is used to calculate BC concentration at 880 nm of wavelength, as described in our previous work [12]. The black carbon facility is one component of a larger project of our research group, where different aspects of air pollution are observed and compared, including toxic gases [13], [14], and particulate matter [15]-[21].

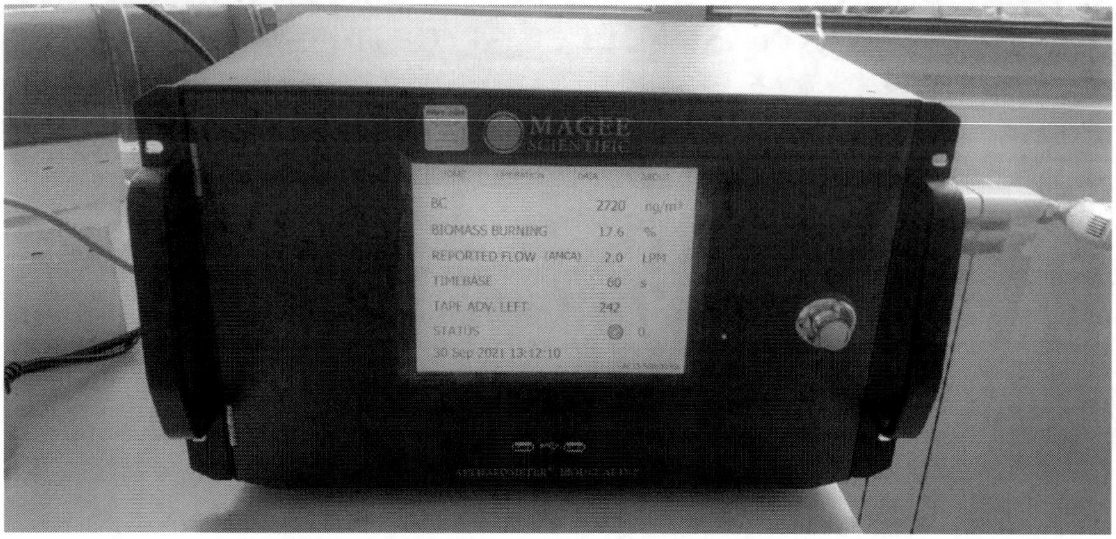

Fig. 1. The aethalometer.

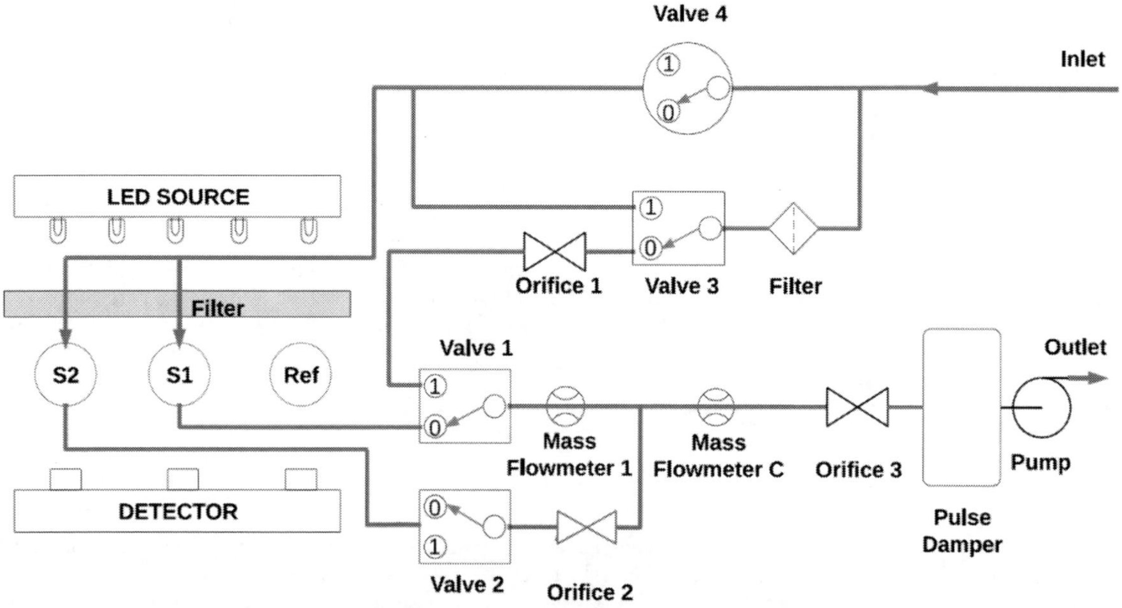

Fig. 2. AE33 flow diagram (image source: AE33 user manual)

3. Measurements of black carbon in Sarajevo

The test facility is located at the Faculty of Mechanical Engineering in Sarajevo (Figure 3). The average concentration of BC in the period 27.11.2020 - 31.5.2021 was 5,67 μg/m³, or 14,82% of total $PM_{2.5}$ mass. Figure 4 shows the daily average values of BC concentrations. Since we have measurements from all 7 wavelengths of light, we can calculate the Angstrom exponent [12]. If we assume that value 1 corresponds to liquid fuel (diesel) and value 2 of Angstrom exponent corresponds to biomass burning, we can estimate the biomass burning percentage in total BC mass (Figure 5). From Figure 5 we can see that up to 70% of BC comes from burning solid fuels in the winter (domestic heating season), and only around 20% in May, when the majority of BC comes from diesel engines.

Fig. 3. The location of measurements.

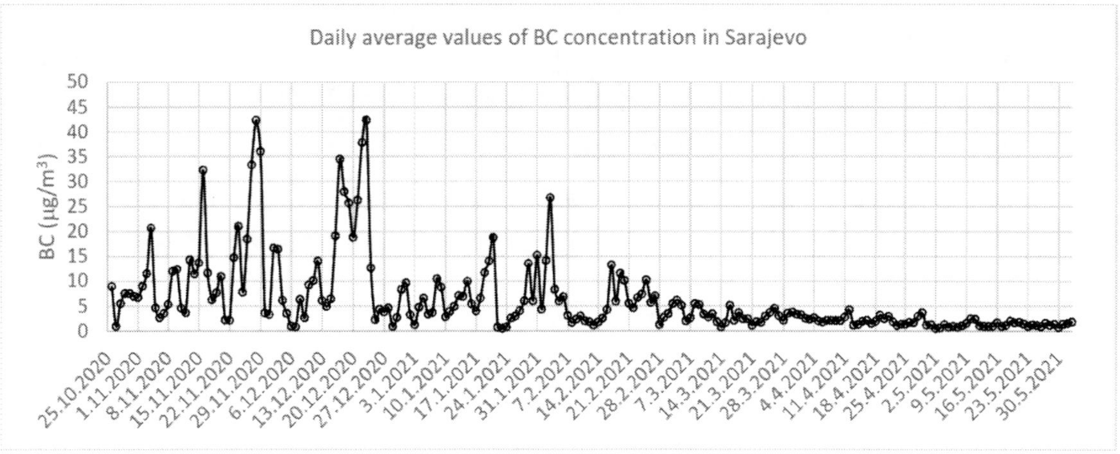

Fig. 4. Daily average values.

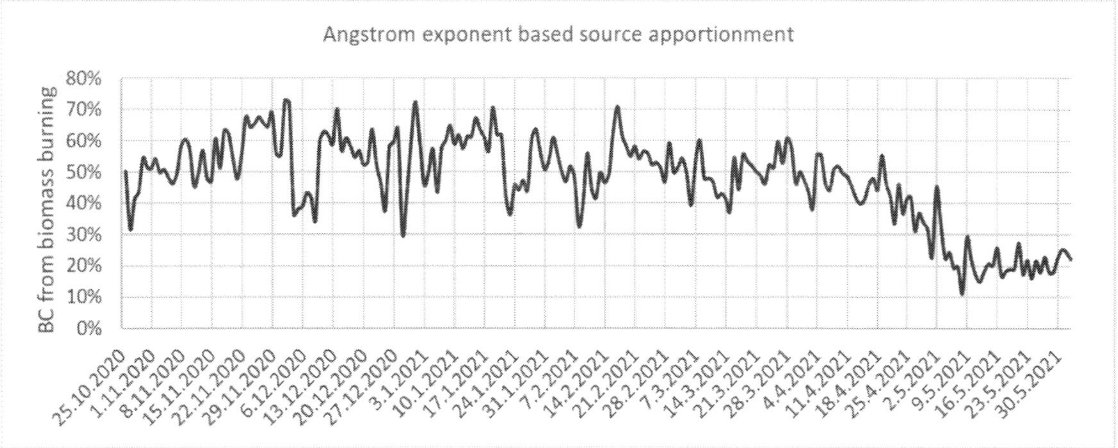

Fig. 5. Biomass burning percentage.

Figure 5 also shows significant noise in the results, which should be investigated in more details. But even though this is a rough estimate, this method is a very promising tool for source apportionment of BC.

4. Conclusion

The AE33 aethalometer was successfully used for continuous measurement of BC concentration in Sarajevo from 27.11.2020 - 31.5.2021. The average value of BC concentration was 5,67 $\mu g/m^3$, or 14,82% of total $PM_{2.5}$ mass. Angstrom exponent based source apportionment analysis shows that during the domestic heating season most of the BC originates from solid fuels, as expected. However, during May, only 20% of solid fuels and 80% of liquid fuels contributed to the total BC mass. The Angstrom exponent based source apportionment of BC is a very promising tool and it should be investigated more in future work.

5. References

[1] Ramanathan, V. & Carmichael, Gregory. (2008). Global and Regional Climate Changes Due to Black Carbon. Nat Geosci. 1. DOI: 10.1038/ngeo156.

[2] Bond, T. C. et al. (2013), Bounding the role of black carbon in the climate system: A scientific assessment, J. Geophys. Res. Atmos., 118, 5380– 5552, DOI: 10.1002/jgrd.50171.

[3] Bové, H.; Bongaerts, E.; Slenders, E. et al. (2019). Ambient black carbon particles reach the fetal side of human placenta. Nat Commun 10, 3866 (2019). DOI: 10.1038/s41467-019-11654-3

[4] Janssen, N. A.; Hoek, G.; Simic-Lawson. M.; Fischer, P.; van Bree, L.; ten Brink, H.; Keuken, M.; Atkinson, R. W.; Anderson, H. R.; Brunekreef, B. & Cassee, F. R. (2011). Black carbon as an additional indicator of the adverse health effects of airborne particles compared with PM10 and PM2.5. Environ Health Perspect. 119(12):1691-9. DOI: 10.1289/ehp.1003369

[5] Arnott, W. P.; Hamasha, K.; Moosmüller, H.; Sheridan, P. J. & Ogren, J. A. (2005). Towards Aerosol Light-Absorption Measurements with a 7-Wavelength Aethalometer: Evaluation with a Photoacoustic Instrument and 3-Wavelength Nephelometer, Aerosol Science and Technology, 39:1, 17-29, DOI: 10.1080/027868290901972

[6] Babich, P.; Davey, M.; Allen, G. & Koutrakis, P. (2000). Method Comparisons for Particulate Nitrate, Elemental Carbon, and PM2.5 Mass in Seven U.S. Cities, Journal of the Air & Waste Management Association, 50:7, 1095-1105, DOI: 10.1080/10473289.2000.10464152

[7] Ferrero, L.; Mocnik, G.; Ferrini, B.S.; Perrone, M.G.; Sangiorgi, G. & Bolzacchini, E. (2011). Vertical profiles of aerosol absorption coefficient from micro-Aethalometer data and Mie calculation over Milan, Science of The Total Environment, Volume 409, Issue 14, 2011, Pages 2824-2837, ISSN 0048-9697, DOI: 10.1016/j.scitotenv.2011.04.022.

[8] Ning, Z.; Chan, K.L.; Wong, K.C.;.Westerdahl, D.; Močnik, G.; Zhou, J..H. & Cheung, C. S. (2013). Black carbon mass size distributions of diesel exhaust and urban aerosols measured using differential mobility analyzer in tandem with Aethalometer, Atmospheric Environment, Volume 80, 2013, Pages 31-40, ISSN 1352-2310, DOI: 10.1016/j.atmosenv.2013.07.037.

[9] Wang, Y.; Hopke, P. K.; Rattigan, O. V.; Xia, X.; Chalupa, D. C. & Utell, M. J. (2011). Characterization of Residential Wood Combustion Particles Using the Two-Wavelength Aethalometer Environmental Science & Technology 2011 45 (17), 7387-7393, DOI: 10.1021/es2013984

[10] Pan, X.; Kanaya, Y.; Taketani, F.; Miyakawa, T.; Inomata, S.; Komazaki, Y.; Tanimoto, H.; Wang, Z.; Uno, I. & Wang, Z. (2017). Emission characteristics of refractory black carbon aerosols from fresh biomass burning: a perspective from laboratory experiments, Atmos. Chem. Phys., 17, 13001–13016, DOI: 10.5194/acp-17-13001-2017

[11] Zotter, P.; Herich, H.; Gysel, M.; El-Haddad, I.; Zhang, Y.; Močnik, G.; Hüglin, C.; Baltensperger, U.; Szidat, S. & Prévôt, A. S. H. (2017). Evaluation of the absorption Ångström exponents for traffic and wood burning in the Aethalometer-based source apportionment using radiocarbon measurements of ambient aerosol, Atmos. Chem. Phys., 17, 4229–4249, DOI: 10.5194/acp-17-4229-2017

[12] Masic, A.; Bibic, D.; Kamenica, I.; Bektesevic, J.; Haziabdic, V. & Mehuljic, M. (2020). Mobile Measurements of Black Carbon for Source Apportionment Studies, Proceedings of the 31st DAAAM International Symposium, pp.0227-0231, B. Katalinic (Ed.), Published by DAAAM International, ISBN 978-3-902734-29-7, ISSN 1726-9679, Vienna, Austria DOI: 10.2507/31st.daaam.proceedings.031

[13] Masic, A.; Bibic, D.; Pikula, B. & Razic, F. (2018). New approach of measuring toxic gases concentrations: principle of operation, Proceedings of the 29th DAAAM International Symposium, pp.0882-0887, B. Katalinic (Ed.), Published by DAAAM International, ISBN 978-3-902734-20-4, ISSN 1726-9679, Vienna, Austria, DOI: 10.2507/29th.daaam.proceedings.127

[14] Bibic, D.; Pikula, B.; Masic, A. & Razic, F. (2018). New approach of measuring toxic gases concentrations: application examples, Proceedings of the 29th DAAAM International Symposium, pp.0876-0881, B. Katalinic (Ed.), Published by DAAAM International, ISBN 978-3-902734-20-4, ISSN 1726-9679, Vienna, Austria, DOI: 10.2507/29th.daaam.proceedings.126

[15] Masic, A.; Pikula, B.; Bibic, D.; Musemic, R. & Halac, A. (2018). Calibration and Assessment of Low-cost Dust sensors, Proceedings of the 29th DAAAM International Symposium, pp. 523-528, B. Katalinic (Ed.), Published by DAAAM International, ISBN 978-3-902734-xx-x, ISSN 1726-9679, Vienna, Austria, DOI: 10.2507/29th.daaam.proceedings.075

[16] Masic, A.; Bibic, D. & Pikula, B. (2019). On the applicability of low-cost sensors for measurements of aerosol concentrations, Proceedings of the 30th DAAAM International Symposium, pp.0452-0456, B. Katalinic (Ed.), Published by DAAAM International, ISBN 978-3-902734-22-8, ISSN 1726-9679, Vienna, Austria DOI: 10.2507/30th.daaam.proceedings.060

[17] Masic, A.; Pikula, B. & Bibic, D. (2017). Mobile Measurements of Particulate Matter Concentrations in Urban Area, Proceedings of the 28th DAAAM International Symposium, pp.0452-0456, B. Katalinic (Ed.), Published by DAAAM International, ISBN 978-3-902734-11-2, ISSN 1726-9679, Vienna, Austria DOI: 10.2507/28th.daaam.proceedings.063

[18] Masic, A.; Bibic, D.; Pikula, B.; Dzaferovic-Masic, E. & Razic, F. (2018). Visualization of Data from Network of Sensors: Appropriate Spatial Interpolation Method, Proceedings of the 29th DAAAM International Symposium, pp.0529-0533, B. Katalinic (Ed.), Published by DAAAM International, ISBN 978-3-902734-20-4, ISSN 1726-9679, Vienna, Austria DOI: 10.2507/29th.daaam.proceedings.076

[19] Masic, A.; Bibic, D.; Pikula, B.; Blazevic, A.; Huremovic, J. & Zero, S. (2020). Evaluation of optical particulate matter sensors under realistic conditions of strong and mild urban pollution. Atmospheric Measurement Techniques, 13, 6427–6443, DOI: 10.5194/amt-13-6427-2020

[20] Masic, A.; Pikula, B.; Haziabdic, V.; Mehuljic, M. & Bektesevic, J. (2020). Evaluation of Face Mask Filtering Efficiency Using High-End Scanning Mobility Particle Sizer, Proceedings of the 31st DAAAM International Symposium, pp.0224-0226, B. Katalinic (Ed.), Published by DAAAM International, ISBN 978-3-902734-29-7, ISSN 1726-9679, Vienna, Austria DOI: 10.2507/31st.daaam.proceedings.030

[21] Masic, A.; Kepnik, G.; Bektesevic, J.; Mehuljic, M.; Saric, I. & Haziabdic, V. (2020). The Network of Smart Sensors for Indoor Air Quality Monitoring, Proceedings of the 31st DAAAM International Symposium, pp.0232-0235, B. Katalinic (Ed.), Published by DAAAM International, ISBN 978-3-902734-29-7, ISSN 1726-9679, Vienna, Austria DOI: 10.2507/31st.daaam.proceedings.032

DOI: 10.2507/32nd.daaam.proceedings.021

MASS SPECTROMETRY OF NANOPARTICLES

Adnan Masic, Jasmin Bektesevic, Dzevad Bibic, Midhat Mehuljic & Boran Pikula

This Publication has to be referred as: Masic, A[dnan]; Bektesevic, J[asmin]; Bibic, D[zevad]; Mehuljic, M[idhat] & Pikula, B[oran] (2021). Mass Spectrometry of Nanoparticles, Proceedings of the 32nd DAAAM International Symposium, pp.0136-0139, B. Katalinic (Ed.), Published by DAAAM International, ISBN 978-3-902734-33-4, ISSN 1726-9679, Vienna, Austria
DOI: 10.2507/32nd.daaam.proceedings.021

Abstract

Nanoparticles with diameters from 10 nm to 1000 nm were analyzed using the scanning mobility particle sizer (SMPS) continuously from October 2020 till May 2021. Mass spectrometry was performed in 129 channels. Measurements of some other air pollution parameters were used for comparison and analysis. A large number of nanoparticles was detected in the period of strong air pollution, as expected. However, a surprisingly large number of nanoparticles was detected in spring and summer, where other parameters of air pollution were good.

Keywords: SMPS; nanoparticles; air pollution; mass spectrometry.

1. Introduction

Although there are several definitions of nanoparticles, we can accept the definition of the European Commission: "A natural, incidental or manufactured material containing particles, in an unbound state or as an aggregate or as an agglomerate and where, for 50 % or more of the particles in the number size distribution, one or more external dimensions is in the size range 1 nm - 100 nm." [1]. Similarly, ultrafine particles (UFP) are defined as aerosols with an aerodynamic diameter of 100 nm or less. In the context of air pollution by particulate matter, nanoparticles represent smallest of all particles. Detection of nanoparticles is difficult due to their physical properties.

But it is very important to analyze nanoparticles because they may have a serious effect on health [2], may even be used as drugs [3], or may have a role in COVID-19 transmission [4]-[6]. The importance of research of nanoparticles is recognized in the latest World Health Organization's Global Air Quality Guidelines [7]. Some of the previous studies include a comparison of nanoparticle monitoring instruments [8], [9], characterization of nanoparticles in aerosolized photocatalytic and regular cement [10], systems of monitoring airborne nanoparticles [11], and development of a new nanoparticle sizer [12]. Novelty in this research is long-term continuous monitoring of nanoparticles in an urban environment with periods of strong and mild air pollution.

2. Scanning Mobility Particle Sizer

A mass spectrometer of nanoparticles called SMPS (Scanning Mobility Particle Sizer) was used at the Laboratory for physical measurements of the Faculty of Mechanical Engineering in Sarajevo, which is focused on tropospheric and air quality research [13]-[21]. Figure 1 shows three major components of SMPS:

- X-ray charge conditioner,
- differential mobility analyzer (DMA),
- condensation particle counter (CPC).

The system is configured with long DMA (10 nm – 1000 nm) and counts the nanoparticles in 129 channels using alternating upscans and downscans. Each scan takes about 6 minutes [20]. Before entering the charge conditioner, ambient air is dried using an in-house developed diffusion dryer.

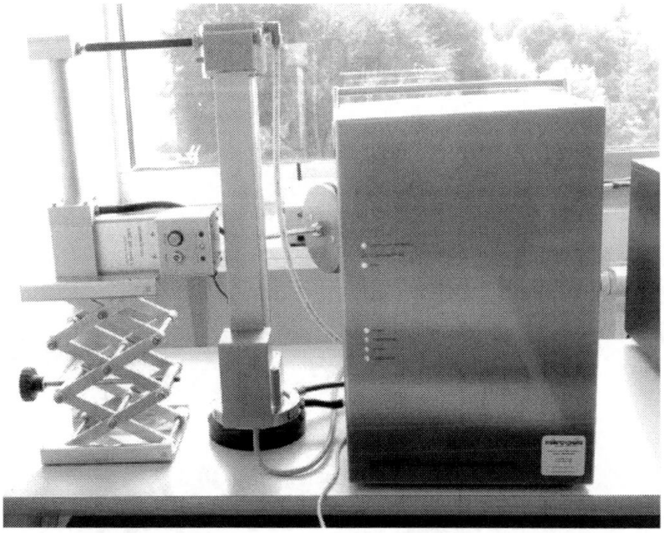

Fig. 1. The Grimm SMPS.

3. Results

Daily average concentrations of nanoparticles in the period 26.10.2020. - 28.5.2021. are shown in Figure 2.

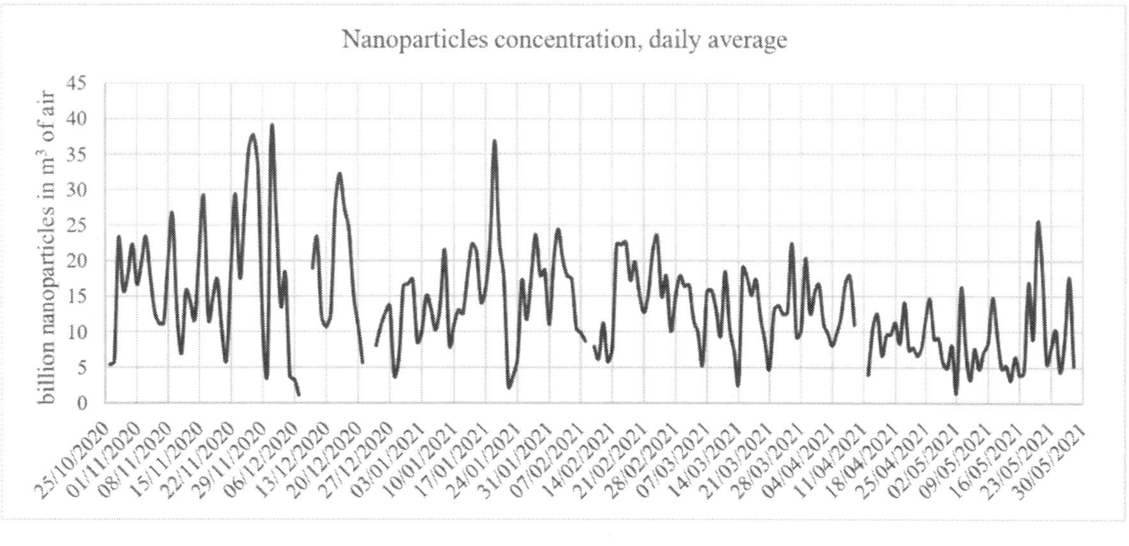

Fig. 2. Concentration of nanoparticles.

A high concentration of nanoparticles during the domestic heating season is expected. Measurement for three months period from December 2020 until February 2021 gave the average value of 15.62 billion nanoparticles per m^3 of air. However, a relatively high concentration was recorded even in May 2021: 8.91 billion nanoparticles per m^3 of air. We should note that the air in May 2021 was 'clean' in Sarajevo, based on $PM_{2.5}$ measurements, with the monthly average concentration of $PM_{2.5}$ 8.0 μg/m^3. This may be explained using different sources of nanoparticles, some of which are frictional brakes on vehicles, but it must be investigated further before conclusions.

Figure 3 shows mass spectrometry of nanoparticles, where we can see a peak around 100 nm, but a large number of particles is present in the entire 10 nm – 100 nm range. The time evolution of nanoparticles is given in Figure 4. Hourly average values of particle concentrations were calculated and plotted against time and diameter (on a logarithmic scale). Local peaks are usually around 100 nm, but sometimes they appear for even smaller nanoparticles.

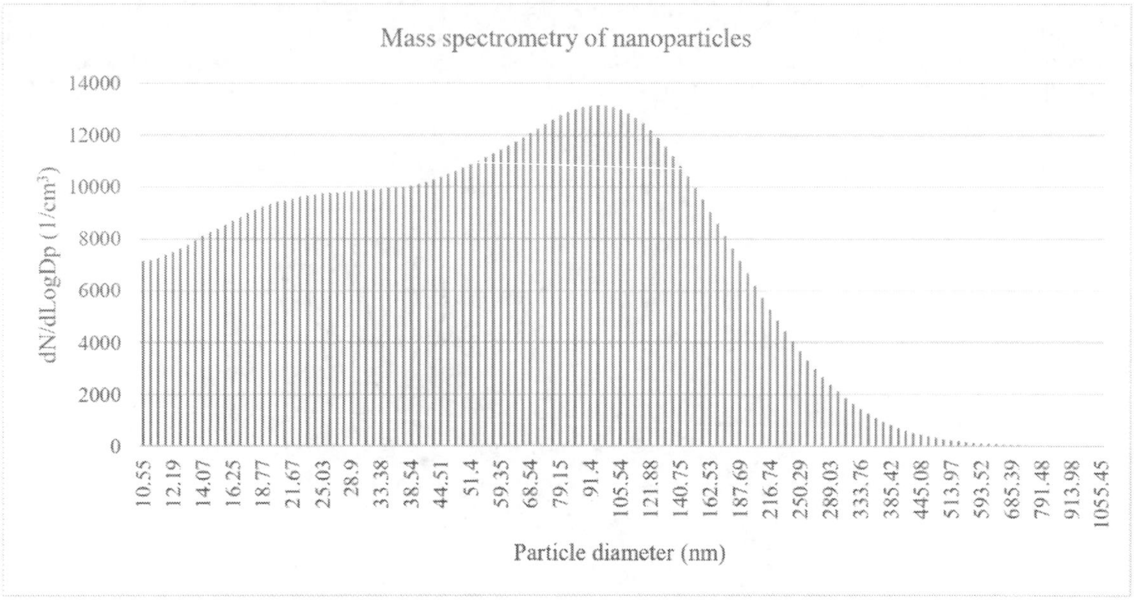

Fig. 3. Mass spectrometry of nanoparticles for entire period.

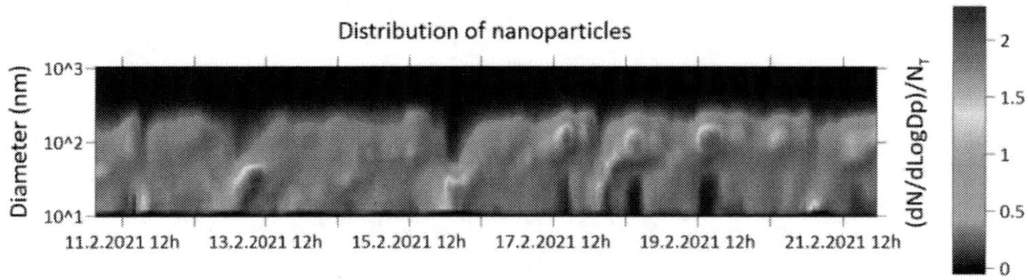

Fig. 4. Distribution of nanoparticles from 11.2.2021 until 21.02.2021.

4. Conclusion

The long-term campaign of mass spectrometry of nanoparticles from 10 nm to 1000 nm was performed. The largest number concentration belongs to particles around 100 nm in diameter, although a significant number of smaller particles (10 nm – 100 nm) exist most of the time. During the period of strong air pollution (December 2020 - February 2021) the number of nanoparticles was high, on average 15.62 billion nanoparticles per m^3 of air. However, a surprisingly large number of nanoparticles was measured in May 2021, 8.91 billion nanoparticles per m^3 of air, even though the air was considered clean, based on $PM_{2.5}$ concentrations (only 8 mg/m3). This discovery, that a large number of nanoparticles may be present even at low pollution levels, is very important and should be clarified in the future.

5. References

[1] https://ec.europa.eu/environment/chemicals/nanotech/faq/definition_en.htm (2021), Definition of a nanomaterial, Accessed on: 2021-09-30

[2] von Mikecz, A. & Schikowski, T. (2020). Effects of Airborne Nanoparticles on the Nervous System: Amyloid Protein Aggregation, Neurodegeneration and Neurodegenerative Diseases. Nanomaterials (Basel, Switzerland), 10(7), 1349. DOI: 10.3390/nano10071349

[3] Leszek, J.; Md Ashraf, G.; Tse, W. H.; Zhang, J.; Gasiorowski, K.; Avila-Rodriguez, M. F.; Tarasov, V. V.; Barreto, G. E.; Klochkov, S. G.; Bachurin, S. O. & Aliev, G. (2017). Nanotechnology for Alzheimer Disease. Curr Alzheimer Res. 2017;14(11):1182-1189. DOI: 10.2174/1567205014666170203125008

[4] Shen, X.; Sun, J.; Yu, F.; Wang, Y.; Zhong, J.; Zhang, Y.; Hu, X.; Xia, C.; Zhang, S. & Zhang, X. (2021). Enhancement of nanoparticle formation and growth during the COVID-19 lockdown period in urban Beijing, Atmos. Chem. Phys., 21, 7039–7052, DOI: 10.5194/acp-21-7039-2021

[5] Almanza-Reyes, H.; Moreno, S.; Plascencia-López, I.; Alvarado-Vera, M.; Patrón-Romero, L.; Borrego, B. et al. (2021). Evaluation of silver nanoparticles for the prevention of SARS-CoV-2 infection in health workers: In vitro and in vivo. PLoS ONE 16(8): e0256401. DOI: 10.1371/journal.pone.0256401

[6] Masic, A.; Pikula, B.; Haziabdic, V.; Mehuljic, M. & Bektesevic, J. (2020). Evaluation of Face Mask Filtering Efficiency Using High-End Scanning Mobility Particle Sizer, Proceedings of the 31st DAAAM International Symposium, pp.0224-0226, B. Katalinic (Ed.), Published by DAAAM International, ISBN 978-3-902734-29-7, ISSN 1726-9679, Vienna, Austria DOI: 10.2507/31st.daaam.proceedings.030

[7] https://www.who.int/news/item/22-09-2021-new-who-global-air-quality-guidelines-aim-to-save-millions-of-lives-from-air-pollution (2021), New WHO Global Air Quality Guidelines aim to save millions of lives from air pollution, Accessed on: 2021-09-30

[8] Jørgensen, R. B. (2019). Comparison of four nanoparticle monitoring instruments relevant for occupational hygiene applications. J Occup Med Toxicol 14, 28 (2019). DOI: 10.1186/s12995-019-0247-8

[9] Ham, S.; Lee, N.; Eom, I.; Lee, B.; Tsai, P.-J.; Lee, K. & Yoon, C. (2016). Comparison of Real Time Nanoparticle Monitoring Instruments in the Workplaces, Safety and Health at Work, Volume 7, Issue 4, 2016, Pages 381-388, ISSN 2093-7911, DOI: 10.1016/j.shaw.2016.08.001

[10] Batsungnoen, K.; Hopf, N. B.; Suárez, G. & Riediker, M. (2019). Characterization of nanoparticles in aerosolized photocatalytic and regular cement, Aerosol Science and Technology, 53:5, 540-548, DOI: 10.1080/02786826.2019.1578334

[11] Chiu, Y. J.; Leong, K. S. & Chang, L. M. (2011). CPC and SMPS Systems of Monitoring Airborne Nanoparticles-Theory and Experiment. Applied Mechanics and Materials, 101–102, 60–66. DOI: 10.4028/www.scientific.net/amm.101-102.60

[12] Lee, H. K.; Lee, H. & Ahn, K.-H. (2020). Development of a new nanoparticle sizer equipped with a 12-channel multi-port differential mobility analyzer and multi-condensation particle counters, Atmos. Meas. Tech., 13, 1551–1562, DOI: 10.5194/amt-13-1551-2020

[13] Masic, A.; Bibic, D.; Kamenica, I.; Bektesevic, J.; Haziabdic, V. & Mehuljic, M. (2020). Mobile Measurements of Black Carbon for Source Apportionment Studies, Proceedings of the 31st DAAAM International Symposium, pp.0227-0231, B. Katalinic (Ed.), Published by DAAAM International, ISBN 978-3-902734-29-7, ISSN 1726-9679, Vienna, Austria DOI: 10.2507/31st.daaam.proceedings.031

[14] Masic, A.; Bibic, D.; Pikula, B. & Razic, F. (2018). New approach of measuring toxic gases concentrations: principle of operation, Proceedings of the 29th DAAAM International Symposium, pp.0882-0887, B. Katalinic (Ed.), Published by DAAAM International, ISBN 978-3-902734-20-4, ISSN 1726-9679, Vienna, Austria, DOI: 10.2507/29th.daaam.proceedings.127

[15] Bibic, D.; Pikula, B.; Masic, A. & Razic, F. (2018). New approach of measuring toxic gases concentrations: application examples, Proceedings of the 29th DAAAM International Symposium, pp.0876-0881, B. Katalinic (Ed.), Published by DAAAM International, ISBN 978-3-902734-20-4, ISSN 1726-9679, Vienna, Austria, DOI: 10.2507/29th.daaam.proceedings.126

[16] Masic, A.; Pikula, B.; Bibic, D.; Musemic, R. & Halac, A. (2018). Calibration and Assessment of Low-cost Dust sensors, Proceedings of the 29th DAAAM International Symposium, pp. 523-528, B. Katalinic (Ed.), Published by DAAAM International, ISBN 978-3-902734-xx-x, ISSN 1726-9679, Vienna, Austria, DOI: 10.2507/29th.daaam.proceedings.075

[17] Masic, A.; Bibic, D. & Pikula, B. (2019). On the applicability of low-cost sensors for measurements of aerosol concentrations, Proceedings of the 30th DAAAM International Symposium, pp.0452-0456, B. Katalinic (Ed.), Published by DAAAM International, ISBN 978-3-902734-22-8, ISSN 1726-9679, Vienna, Austria DOI: 10.2507/30th.daaam.proceedings.060

[18] Masic, A.; Pikula, B. & Bibic, D. (2017). Mobile Measurements of Particulate Matter Concentrations in Urban Area, Proceedings of the 28th DAAAM International Symposium, pp.0452-0456, B. Katalinic (Ed.), Published by DAAAM International, ISBN 978-3-902734-11-2, ISSN 1726-9679, Vienna, Austria DOI: 10.2507/28th.daaam.proceedings.063

[19] Masic, A.; Bibic, D.; Pikula, B.; Dzaferovic-Masic, E. & Razic, F. (2018). Visualization of Data from Network of Sensors: Appropriate Spatial Interpolation Method, Proceedings of the 29th DAAAM International Symposium, pp.0529-0533, B. Katalinic (Ed.), Published by DAAAM International, ISBN 978-3-902734-20-4, ISSN 1726-9679, Vienna, Austria DOI: 10.2507/29th.daaam.proceedings.076

[20] Masic. A.; Bibic, D.; Pikula, B.; Blazevic, A.; Huremovic, J. & Zero, S. (2020). Evaluation of optical particulate matter sensors under realistic conditions of strong and mild urban pollution. Atmospheric Measurement Techniques, 13, 6427–6443, DOI: 10.5194/amt-13-6427-2020

[21] Masic, A.; Kepnik, G.; Bektesevic, J.; Mehuljic, M.; Saric, I. & Haziabdic, V. (2020). The Network of Smart Sensors for Indoor Air Quality Monitoring, Proceedings of the 31st DAAAM International Symposium, pp.0232-0235, B. Katalinic (Ed.), Published by DAAAM International, ISBN 978-3-902734-29-7, ISSN 1726-9679, Vienna, Austria DOI: 10.2507/31st.daaam.proceedings.032

DOI: 10.2507/32nd.daaam.proceedings.022

3D SCANNING - ACCURACY OF A VOLUME MODEL TRANSFORMED FROM SCANNED DATA

Dana Kubatova & Aneta Kaufnerova

This Publication has to be referred as: Kubatova, D[ana] & Kaufnerova, A[neta] (2021). 3D Scanning - Accuracy of a Volume Model Transformed from Scanned Data, Proceedings of the 32nd DAAAM International Symposium, pp.0140-0144, B. Katalinic (Ed.), Published by DAAAM International, ISBN 978-3-902734-33-4, ISSN 1726-9679, Vienna, Austria

DOI: 10.2507/32nd.daaam.proceedings.022

Abstract

Quality control is one of the most important areas in the production process. Quick inspection requires the most efficient techniques for comparing and inspecting manufactured parts with CAD design. One of the methods for evaluating product quality is to compare the CAD model with the scanned data. However, it may also happen that the customer wants to compare two production technologies for the same component. This issue is the topic of this article. The resulting deviations are circa 1 μm. This finding is satisfactory, and it is possible to use the converted model from the scanned data as reference data and thus compare products produced by another technology or apply it to another problem. Today, 3D scanning is a commonly used method for inspecting products. This work is based on a real situation that can help other solvers. The result of this work is to determine the deviation of the conversion of scanned data to a volume model in Geomagic control X.

Keywords: Scanning; Geomegic control X; CMM; Volume model.

1. Introduction

Product accuracy and quality control are an integral part of the production process. Until recently, the inspection of components, i.e. the measurement of dimensional and shape accuracies, was previously performed only by conventional methods, such as the contact method on a coordinate measuring machine (CMM). The data obtained by contact measurement are very accurate, but in some cases this method of measurement is insufficient, e.g. measurement of complex surfaces, soft or flexible products. The detailed surface of the part is measured by a sufficient density of points, which are obtained with difficulty using the contact method and it is very time consuming. For this reason, the measurement of complex surfaces using the contactless method, with 3D scanners, came to the fore. This modern measurement method is very fast, and, thanks to the overall description of the measured component, they enable us to perform complex and illustrative analyses easily. The transfer of real shapes to virtual 3D models is increasingly appearing in classical engineering due to the greater availability of scanning technologies. [1,7] One of the biggest advantages of 3D scanning is the fast acquisition of an accurate three-dimensional model. The model created by scanning can then be used for production using, for example, stereolithography, rapid prototyping or conventional CNC machining. [2,6]

One of the methods for evaluating product quality is to compare the CAD model with the scanned data. However, it may also happen that the customer wants to compare two production technologies of the same component. This issue is the topic of this article.

2. Methodology

The required result from the measurement is to compare two products made by different technologies. One is made by unconventional technology and the other is made by conventional technology. [7], [8] This requirement can be met by comparing the scanned data with the "3D colour map" function, but this is only possible when the model is selected as the reference data. There is a problem here as to the exact conversion from the scanned data to the volume model in Geomagic control X 2020. The measurement was performed on a coordinate measuring machine at the Regional Institute of Technology at the University of West Bohemia in Pilsen. The measured object is a sample (Fig.1) that was printed from PLA material on the Prusa MK3S 3D printer (Fig. 2).

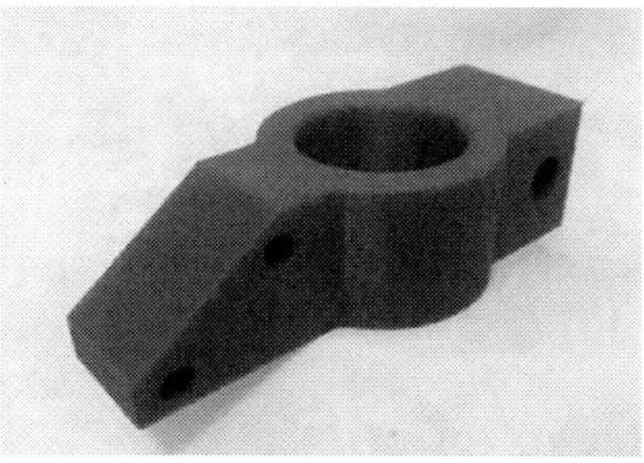

Fig. 1. The sample

2.1. The equipment

The sample was printed on a Prusa MK3S (Fig. 2) developed by Prusa Research. The filament used was Prusament PLA Azure Blue.

Fig. 2. PRUSA MK3S

The Laboratory of Workshop Metrology, which belongs to the Regional Institute of Technology, has a CMM from Zeiss, Prismo 7 Navigator (Fig.3). This machine is a bridge type CMM. This type is ideal for all measuring tasks. It accurately measures the geometry of a work piece along X, Y and Z axes using a contact probing system and optical probing system. The Prismo 7 Navigator measures lengths with a maximum permissible error of 0.9 + L / 350 µm.

Fig. 3. Prismo 7 Navigator

The machine has two heads, the RDS rotary head and the VAST XXT fixed head. It is possible to clamp both the head for contact measurement and the sensing system for LineScan (Fig.4) in the rotating head. LineScan is an optical laser line scanner designed for high-speed data digitization. It records 250 000 points per second and is compatible with Calypso software. It is also possible to monitor the recorded data in the WBscan software over time.

Fig. 4. LineScan

Software used:

- Calypso version 2020
- Geomagic Design X
- Geomagic Control X
- WBscan

2.2. Procedure

2.2.1. Data scaning

The scanner was calibrated before the measurement. The sample was then set on a cube for better scanner accessibility. The scan was divided into 3 measurements. [9] QSP (Qualified Surface Points) and RSL (Raw Scan Lines/Points) data were exported for further editing of the scanned data.

2.2.2. Work with data

Subsequently, the data were edited in Geomagic Design X software. Editing data in Geomagic Design X:

- Filter out noise from scanned points
- Smooth points to the middle level
- Convert points to a fine mesh
- Fill the holes
- Create the 3D model
- Export the mesh in stl format and the 3D model in step format

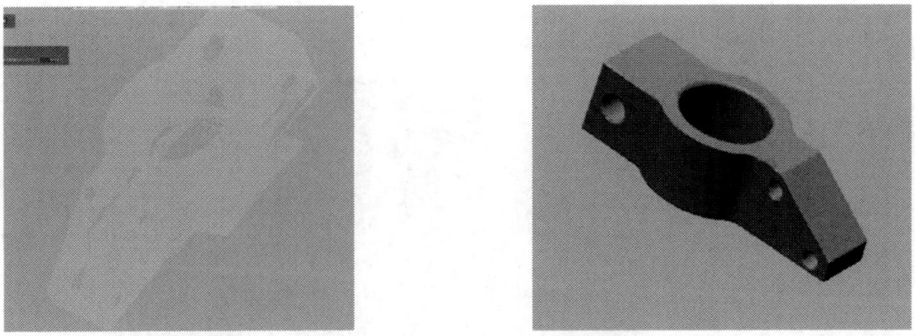

Fig. 5. A point cloud on the left and a polygon mesh on the right

2.2.3. *Data analysis*

Individual meshes and 3D models were compared by 3D comparison in Geomagic Control X. The procedure in Geomagic Control X is:

- Import reference data (model volume)
- Import polygon mesh
- Align imported data
- Adjust the colour scale range and accuracy of the 3D scanner
- 3D analysis

3. Evaluation in Geomagic Control X

First, the 3D model and mesh, which were created from RSL point clouds, were compared. As shown in the figure below, the selected colour scale range was ± 0.5 and the specific tolerance of the 3D scanner was ± 0.025. According to the colour scale, the conversion of the RSL mesh to the model is high quality (in green). 95.075% of values are in a tolerance of ± 0.025 mm. The lowest point is -0.2006 mm and the highest point is +0.2031mm. These points are out of tolerance. This can cause defects in the measured object, insufficient mesh conversion, or scanning of RSL points that represent all scanned points.

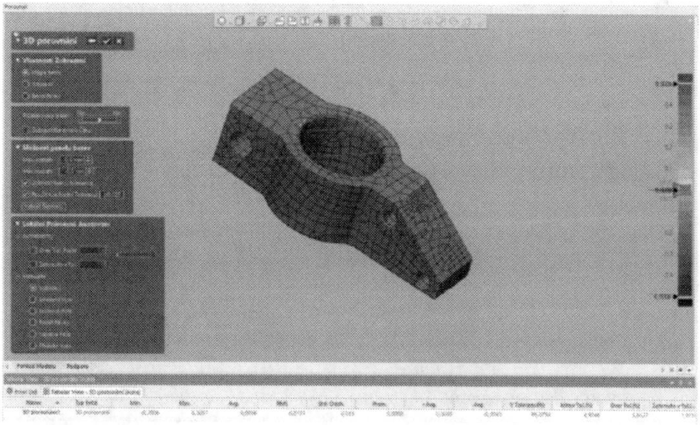

Fig. 6. The comparison - RSL volume model with RSL polygon mesh

According to the calculated results, the average value is 0.0016 mm. This value expresses how good the conversion of the RSL mesh to the model is. Second, the mesh from QSP data and the model transferred from this mesh were compared. As shown in the figure below, the colour scale range was ± 0.5 and the specific tolerance of the 3D scanner was ± 0.025 mm. 96.0748% of values are in the tolerance of ± 0.025 mm . The lowest point is located at -0.0828 mm and the highest point is at +0.083 mm. According to the calculated results, the average value is 0.0014 mm. The average value is smaller than for RSL points because QSP points are calculated by a mathematical operation and RSL points are real scanned points.

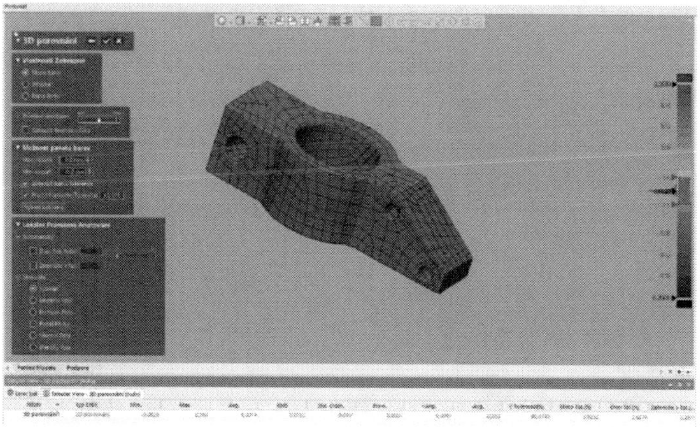

Fig. 7. The comparison - QSP volume model with QSP polygon mesh

The resulting values of the converted models differ according to the selected scanned data (QSP, RSL). The resulting deviations are circa 1 μm. This finding is satisfactory, and it is possible to use the converted model from the scanned data as reference data and thus compare products produced by different technologies or apply it to another problem. Slightly better results are obtained from QSP, namely 0.0002 mm.

4. Conclusion

Product accuracy and quality control is an integral part of the production process. Until recently, the inspection of components, i.e. the measurement of dimensional and shape accuracies, was previously performed by conventional methods, such as contact method on a CMM. The data obtained by contact measurement are very accurate, but in some cases this method is insufficient, e.g. measurement of complex surfaces, soft or flexible products. These parts are more often measured by noncontact methods. One of these methods is 3D scanning. Today, 3D scanning is commonly used for inspecting products. This work is based on a real situation that can help other designers. The measured object is a printed sample on a 3D printer that was scanned using a CMM. The scanned data was further edited and converted to mesh and 3D models. These were further compared and analysed.

The aim of this work is to determine the deviation of the conversion of scanned data to a 3D model in Geomagic control X. The resulting deviations are circa 1 μm. From the resulting values, it can be concluded that the conversion of the scanned data does not affect the resulting measurement. Thanks to these results, this step can be applied to compare the same parts produced by different technologies (3D metal printing, milling, etc.). The article describes only a part of the research that is carried out in this area within the University of West Bohemia in Pilsen. This research will help members of the metrology laboratory and perhaps not only check parts faster and easier that cannot be measured by contact methods.

5. Acknowledgments

This paper was created under the project GA UWB in Pilsen: SGS-2019-008 "Research and development for innovation in the field of manufacturing processes – Technology of metal cutting III".

6. References

[1] Sze S. M., and Leciejewicz J.; (1969) Physics of Semiconductor Devices (New York: Wiley–Interscience)

[2] Caplar R., and Kulisic P.; (1973) Density matrix theory of coherent ultrafast dynamics Proc. Int. Conf. on Nuclear Physics (Munich) vol 1 (Amsterdam: North-Holland/American Elsevier) p 517

[3] Lomakin V. O., Kuleshovav M. S., and Bozh'eva S. M. (2016) Numerical Modeling of Liquid Flow in a Pump Station Power Technology and Engineering 49 (5) pp 324–327

[4] Lomakin V. O., Kuleshova M. S., and Kraeva E. A.; (2015) Fluid Flow in the Throttle Channel in the Presence of Cavitation Procedia Engineering 106 p 27

[5] Christie B.; (2014) Quality Online Learning and Teaching (QOLT) Evaluation Instrument Sections and Objectives (California State University) (http://qolt.sfsu.edu/sites/default/files/QOLT2-pgnumberslogoCore24CClicense111715.pdf) accessed 20 November 2016

[6] Stojkic, Z[eljko]; Culjak, E[va] & Saravanja, L[uka] (2020). 3D Measurement - Comparison of CMM and 3D Scanner, Proceedings of the 31st DAAAM International Symposium, pp.0780-0787, B. Katalinic (Ed.), Published by DAAAM International, ISBN 978-3-902734-29-7, ISSN 1726-9679, Vienna, Austria DOI: 10.2507/31st.daaam.proceedings.108

[7] Sykora, J[indrich] (2020). Additive Manufacturing of Unmanned Aerial Vehicle, Proceedings of the 31st DAAAM International Symposium, pp.0836-0841, B. Katalinic (Ed.), Published by DAAAM International, ISBN 978-3-902734-29-7, ISSN 1726-9679, Vienna, Austria DOI: 10.2507/31st.daaam.proceedings.117

[8] Habrman, M[artin] (2019). Influence of the Infill on the Tensile Strength and the Economic Factors of 3D Printing, Proceedings of the 30th DAAAM International Symposium, pp.1056-1060, B. Katalinic (Ed.), Published by DAAAM International, ISBN 978-3-902734-22-8, ISSN 1726-9679, Vienna, Austria DOI: 10.2507/30th.daaam.proceedings.147

[9] Tatarevic, M[ichal]; Gapinski, B[artosz] & Swojak, N[atalia] (2019). The Use of Optical Scanner for Analysis of Surface Defects, Proceedings of the 30th DAAAM International Symposium, pp.0076- 0085, B. Katalinic (Ed.), Published by DAAAM International, ISBN 978-3-902734-22-8, ISSN 1726-9679, Vienna, Austria DOI: 10.2507/30th.daaam.proceedings.010

DOI: 10.2507/32nd.daaam.proceedings.023

THE REAL APPLICATION OF THE SOFTWARE FILTER SELECTION METHODOLOGY ON PRACTICAL EXAMPLES

Dana Kubatova

This Publication has to be referred as: Kubatova, D[ana] (2021). The Real Application of the Software Filter Selection Methodology on Practical Examples, Proceedings of the 32nd DAAAM International Symposium, pp.0145-0150, B. Katalinic (Ed.), Published by DAAAM International, ISBN 978-3-902734-33-4, ISSN 1726-9679, Vienna, Austria
DOI: 10.2507/32nd.daaam.proceedings.023

Abstract

The functional surfaces of components are usually finished by chip and chipless machining methods. The designer prescribes the basic properties of the surface with respect to its intended function. The task of the technology is to implement this prescription. Ideally, in the case of completely error-free production, the finished part is directly ready for functional use. In normal practice, however, the conditions of the production process do not remain completely constant, and it is, therefore, necessary to include inspection operations to detect any deviations that may affect the function of the component (e.g. deviations in dimensions, shape, surface quality, hardness of the component material, etc.). Changes in the manufacturing process affect the surface structure in various ways. Importantly, the measurement of the surface structure not only leads to the determination of the actual surface condition but also provides a basis for the control of the manufacturing process and the prediction of the functional properties of the surface. In the area described in this article, the area of software filter selection in real examples, this is not always the case. And the main reason for this is the low awareness and the high complexity of choosing the appropriate combination of possible procedures and techniques for setting up software filters. In this paper, possible differences in not following the correct measurement procedure, and especially the evaluation of surface roughness, are described. The article does not focus on one particular area of surface roughness measurement but describes the possible differences for both periodic and aperiodic surfaces, depending on the measurement method used (contact or non-contact method).

Keywords: Roughness; Methodology of selection of software filter; Aperiodic and periodic surface; Contact and non-contact measuring.

1. Introduction

Measurement and evaluation of surface roughness have made significant qualitative progress in recent years. Leading manufacturers of measuring instruments (Hommel, Carl Zeiss, and others) actively responded to the new requirements. This is mainly due to the fact that the surface structure is often checked with the help of single-purpose measuring devices. For this reason, the manufacturers of these machines are also the main creators of pressure to develop standards related to the evaluation of the quality of measurement of machined surfaces. [2] This has ultimately led to improvements in the technologies of existing tools for measuring and evaluating surface structure, as well as to better methods, measuring systems, and surface structure evaluation systems that are still under development. [6; 7]

The surface structure evaluation system is defined by a set of standards that describe the marking, measurement, and evaluation of the surface structure, calibration of measuring instruments, and other aspects. These are GPS (Geometrical Product Specification) standards [5]. In general, the measurement and evaluation of surface structure is a separate area of metrology. Where using special techniques, the data needed to characterize the surface quality can be obtained.

The functional surfaces of the parts are usually finished by methods of chip and chipless machining. The designer prescribes the basic properties of the surface with respect to its intended function. The task of technology is to implement this regulation. In the ideal case of completely flawless production, the finished part is directly ready for functional use. In common practice, however, the conditions of the production process do not remain completely unchanged, and therefore it is necessary to include control operations to detect all deviations that may affect the function of the part (eg deviations in dimensions, shape, surface quality, hardness of part material, etc.). Changes in the production process affect the surface structure in various ways, e.g. the effect of tool wear, the stress in the material, the use of unsuitable cutting conditions, etc. [3], [4] It is important that measuring the surface structure not only determines the actual surface condition, but at the same time, it will also provide data for the regulation of the production process and the prediction of the functional properties of the surface.

Very important properties depend on the quality (quality) of the surface, such as efficiency, reliability, functional properties, external appearance, and finally, the cost of the entire production process. And if there is an inaccurate assessment when evaluating the integrity of the surface, especially when assessing the surface roughness, this can lead to fatal problems in the actual use of the inspected part in practice. To meet all the requirements of surface structure evaluation, it is necessary to keep up with the times and keep under control new possibilities and knowledge in the field of measuring and evaluating surface integrity. And implement this knowledge into real conditions of use.

In the area described in this article, the area of choice of software filters in real examples, this is not always the case. And the main reason is low awareness and great complexity of choosing a suitable combination of possible procedures and techniques for setting software filters. Therefore, a comprehensive methodology for the selection of software filters for both periodic surfaces and subsequently for aperiodic surfaces described in Article 1 was created.

2. Surface roughness measuring systems

In today's practice, we find two basic methods of evaluating the surface structure. The older method of surface evaluation is the contact method. It is the most common surface evaluation method that measures the surface profile using high-sensitivity instruments. [8] The contact device means that a specially modified tip scans the coordinates of the evaluated surface, which are then computer-processed. Their disadvantage is that due to the size of the tip 2; 5 or 10 μm there is mechanical filtering of the obtained data. (See Figure 1) [9]

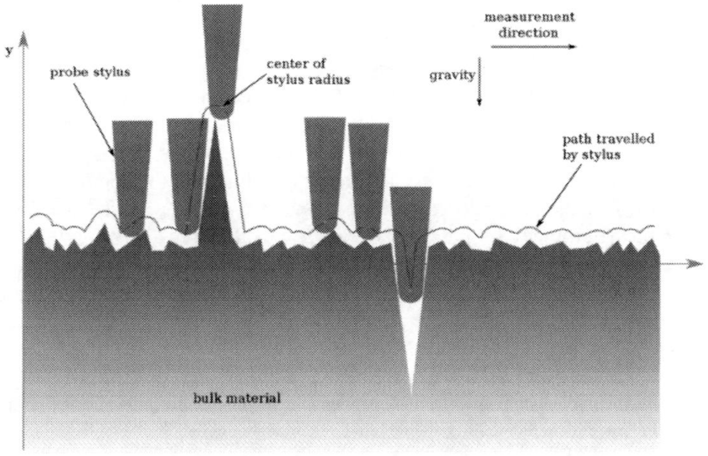

Fig. 1. Contact measurement [9]

This shortcoming has been eliminated by creating surface roughness measurements by non-contact methods. The principle of the non-contact method is based on the use of a light beam, which is projected on a rough surface. The irregularities on the inspected surface created after machining are scattered by the light beam and the light scattering are determined. While with an ideally smooth surface, there is no light scattering, and the beam is reflected as it came. The greater the roughness, the lower the mirror component and the more the scattering components increase. If the height of the irregularities is less than the wavelength of the scattered light, the diffraction vector and the scalar Beckmann theory are used to determine the surface roughness. Only the Gaussian normal distribution is used for the parameters, where precise electromagnetic boundary conditions are obtained due to the difficulty in obtaining the analytical value of the roughness from the light scattering. (see Figure 2)

Fig. 2. Non-contact measurement [9]

3. Brief characteristics of the methodology for the selection of software filters [2]

In both cases of surface evaluation, however, it is necessary to ensure that the measured data are correctly evaluated and processed. This is much easier thanks to the software filter selection methodology. The software filter selection methodology has 4 (periodic surface) and 5 (aperiodic surface) steps, respectively, thanks to which the user of the profilometer, whether contact or contactless, gets to the correct setting of the device. Furthermore, thanks to the use of the methodology, it should be ensured that two people measure the same or at least in the order of hundredths of the same results (values), which currently does not work completely.

The first step in the methodology of selecting a software filter for a periodic surface is to select the optimal stylus tip size. After summarizing and final evaluation of the tests, it was decided that to achieve optimal results, it is appropriate to use a tip of 2 µm for roughness values up to Ra = 6.3 µm. In the second step of selecting the software filter for the periodic surface, it is necessary to select the optimal distance between the scanned points. The procedure here is that a compromise must be found between the small variance of the measured value and the value of the distance of the measured value from the standard value.

In the third step, the stylus feed rate is selected when scanning the primary profile. Based on the findings of this test, information emerged that the scan rate has no effect on the Rsm parameter. The results for the parameters Ra, Rv, and Rz using scanning speeds up to 1.07 mm/s are almost identical. At higher speeds, the inaccuracy of the measured data increases for all three parameters and increases the variance of values. Therefore, it is not recommended to use a feed rate value greater than 1.07 mm/s. The fourth (last) step of the methodology is the choice of software filter. Here again, it is important to find a compromise between the smallest variance of values and the maximum measured value.

When creating the methodology for choosing a software filter for an aperiodic surface, it was decided on the basis of the tests performed and described in Articles 2, only to supplement the existing methodology. The methodology was supplemented by one whole step and several other important findings for the evaluation of aperiodic surfaces. The most important thing is to add a step - choose the scanning direction. At this point, the need to measure aperiodic surfaces in several directions per surface trace, at different locations of the test sample, is pointed out. As described in Article [3]. Another important finding for measuring aperiodic surfaces is in the step of choosing a software filter, the finding that compared to the periodic surface came out as the most suitable filter, the filter EN ISO 16610-31 (Gaussian robust filter).

4. Description of the experiment

Within the performed experiment, the task was to verify the correct function of the method of software filter selection in the application on selected contact and non-contact measuring devices.

4.1. Testing samples

The tested samples were from various applications for the automotive, aerospace, and medical industries. For each tested sample, the optimal conditions for measurement were determined according to the software filter selection methodology described in Article 1, supplemented by new points of the methodology for the aperiodic surface. See Table. 1.

It was always based on four basic areas of software filter selection methodology. So in the first step, the size of the touch was chosen. In the second step, the number of points to be scanned and the scanning direction in which the measurement is to be performed were selected. Which plays a major role for the aperiodic surface. In the third step, the scanning speed and the measured length were selected. Subsequently, the used software filter was selected. In the last step, the parameters suitable for the given application of the part were designed. The choice of the settings described above was always based on the expected surface roughness of the test sample and its intended use.

Parameter - recommended setting	Aperiodic standard	Flap Longitudinally	Flap Longitudinally	Engine	Form	Spraying before	Spraying after	Mirror surface
Expected roughness	0.2 μm	to Ra=15 μm	to Ra=10 μm	Ra=1 μm	to Ra=0,8 μm	to Ra=3,2 μm	to Ra=1 μm	DtoRa=0,2 μm
Tip size	2 μm	2 μm	2 μm	2 μm	2 μm	2 μm	2 μm	2 μm
Scanning direction	different directions	different directions	One direction	One direction	different directions	different directions	different directions	different directions
Number of points	48000	6858	4572	48000	16000	4800	4800	48000
Measured length	4.8 mm	4.8 mm	3.2 mm	4.8 mm	4.8 mm	4.8 mm	4.8 mm	4.8 mm
Feed rate	0.5 mm/s	0.5 mm/s	0.5 mm/s	0.3 mm/s	0.5 mm/s	0.5 mm/s	0.5 mm/s	0.5 mm/s
Cuf-off	0.25 mm	2..5 mm	0.8 mm	0.25 mm	0.8 mm	0.8 mm	0.25 mm	0.25 mm
Software filter	16610-31	4768	4768	16610-31	16610-21	4768	16610-31	16610-31
Evaluated parameters	Pa, Pz, Rk, Rvk, Rpk, Mr1, Mr2,Ra, Rz, Rv	Pa, Pz, Rk, Rvk, Rpk, Mr1, Mr2,Ra, Rz, Rv	Pa, Pz, Rvk, Rpk, Mr1, Mr2,Ra, Rz, Rv	Ra, Rz, Rt, Rp, Rvk, Rpk, Mr1, Mr2,	Ra, Rv, Rp, Rk, Rvk, Rpk, Mr1, Mr2,	Ra, Rv, Rp, Rk, Rvk, Rpk, Mr1, Mr2,	Ra, Rv, Rp, Rk, Rvk, Rpk, Mr1, Mr2,	Ra, Rq, Rk, Rvk, Rpk, Mr1, Mr2,

Table 1. Recommended parameters for measurement based on the methodology of software filter selection

4.2. Tested devices

1) Contact devices - from the area of contact surface roughness measurement, both measurements on laboratory devices and on devices intended for workshop measurements were considered. Each of these devices has its pros and cons that enter into the measurement, and it is necessary to take them into account.

   a. Hommel Etamic T8000 Roughness Tester - is one of the most widespread profilometers/roughness gauges in the world. I use them both for common surface roughness measurements and it is very widely used for calibration of bound roughness standards. Its undeniable advantage is its variability of settings. Whether in the field of software filters or in the field of the possibility of evaluating an inexhaustible range of roughness parameters up to the possibility of exchanging a stylus of various sizes or shapes.
   b. Workshop roughness meter Mitutoyo SURFTEST SJ-410 - is a workshop roughness tester, which due to the impossibility of replacing the probe has only limited use but in the case of a suitably selected tip when buying the device based on the product portfolio, for which it will be intended laboratory instruments. [10]

2) Non-contact instruments - for the test described in this article, consideration was given to instruments that are used primarily to measure surface roughness and measure using laser surface sensing. Thus, instruments that can also be used for measuring dimensions and sensing the surface are performed by illuminating with light in a plan view.

   a. Optical-scanning microscope IFM G4 (Alicona) - the device works on the principle of scanning the surface with a laser beam through objectives, which performs a magnification of the surface so that it can be evaluated. The disadvantage of the instrument is that it allows measurement with only one present software filter. [11]
   b. Kyence VHX-6000 - the instrument is a compact measuring microscope which is used not only to measure the dimensions but is also a good alternative to instruments used exclusively for measuring surface roughness.
   c. Olympus LEXT OLS5000 laser microscope - the instrument falls into the field of confocal microscopes. Thanks to the appropriately selected objectives, working with the microscope is very simple and fast.

5. Test results

The actual evaluation of the functionality of the methodology was started by an experiment in which the measured data from different devices were compared but using the same conditions determined according to the methodologies. For these purposes, measurements were performed on a standard piece for which the value of selected parameters is known. The measured data can be seen in Table 2. This table shows the achieved values for the tested standard with an aperiodic surface. The nominal value of the standard is Ra = 0.2 μm. This value was chosen quite intentionally, given that it is the limit value for some selected devices for the test.

	Hommel	Mitutoyo	IFM	LEXT	VHX-600
Pa [µm]	0.2	0.22	0.25	0.2	0.21
Pz [µm]	0.49	0.48	0.53	0.54	0.5
Rk [µm]	0.23	0.25	0.27	0.28	0.25
Rvk [µm]	0.06	0.059	0.062	0.062	0.061
Rpk [µm]	0.06	0.056	0.064	0.064	0.063
Mr1 [%]	13.713	13.701	13.816	13.796	13.805
Mr2 [%]	93.426	93.469	93.526	93.396	93.416
Ra[µm]	0.2	0.22	0.25	0.2	0.21
Rz [µm]	0.49	0.48	0.53	0.54	0.5
Rv [µm]	0.24	0.25	0.27	0.28	0.25

Table 2. Measured values from the standard

Already the initial measurement pointed out some small differences in the measured results, but based on the difference between the meters, the differences were expected. Given that the differences between the plots were up to 5% of the difference, it was decided to continue the test. The evaluation of the test was performed based on a comparison of the achieved (measured) results for individual measuring systems, their use, and the character of the required surface. The disadvantage is, as already described above, that not all devices are able to measure according to all software filters. But even so, it was possible to build the possibilities of use and the expected results achieved.

1) Contact roughness meter

   a. Roughness meter Mahr MarSurf LD 120 - allows using the methodology in the whole range when achieving results with a deviation of up to 5% (the size of measurement uncertainty) in the resulting measured value. This was found by cross comparing the results with the collaborating laboratory.
   b. Mitutoyo SURFTEST SJ-410 workshop roughness tester - the device allows the use of only part of the methodology because the device has only a limited number of software filters, a fixed measurement speed, including the distance between points and a fixed one specific tip. But in cases where it was possible to make a comparison based on the recommendations, values with a maximum deviation of up to 5% were again achieved.

2) Non-contact roughness meter
   a. Optical-scanning microscope IFM G4 (Alicona) - the device does not allow changing the software filter based on the user's request, only changing parameters. The device has a fixed filter 16610-21, based on this it was not possible to verify whether the methodology works on this device.
   b. Laser microscope OLS5000 - LEXT - allows to use the methodology in the whole range to achieve results with a deviation of up to 10% (usually higher roughness values, which is due to the laser scanning system, which has a diameter of 0.2 µm in its entire length compared to touch sensing where the tip has a radius of 2 µm) in the resulting measured value. This was found by cross comparing the results with the collaborating laboratory.
   c. Kyence VHX-6000 - allows using the methodology in the whole range when achieving results with a deviation of up to 10% (usually higher values of roughness, which is caused by a laser beam scanning system, which has a diameter of 0.2 µm in its entire length compared to touch sensing where the tip has a radius of 2 µm) in the resulting measured value. This was found by cross comparing the results with the collaborating laboratory.

6. Conclusion

We define texture, or in other words the structure of a surface, as an area with a huge set of grains, which are located between the surroundings and the surface of the material, ie a fixed part. The surface is an extremely complex system, where there are repeated (periodic) or random (aperiodic) deviations from the ideal geometric shape of the surface. These deviations can be expressed in several ways, the oldest is the line profile, and the most common nowadays is probably the expression of the three-dimensional topography of the surface. According to the size of the pitch of the respective irregularities, the surface structure is divided into three components. The basic profile determines the component with the largest pitch of inequalities and is called the P-profile.

The component with the smallest pitch is determined by the surface roughness, which in turn is referred to as the R-profile. And the component between the largest and smallest spacing is called the surface ripple, W-profile. [3], [9]. The composition of these components on the surface generally directly affects the functional properties of the surface. The often-expressed view of the surface - "the smoother, the better" - may not always be correct, as there are other factors. E.g., to ensure lubrication, depressions must be left on the surface to hold the lubricating medium.

In all cases, it is necessary to assess the economic side of things - the increase in the cost of producing a very "smooth" surface may not always correspond to the functional benefit. Therefore, functional use is decisive for the properties of the surface. The requirements for surfaces that will move relative to each other will be different from the requirements for surfaces forming sealing surfaces or for surfaces intended for subsequent painting, etc. It follows from the above that measuring and evaluating the surface structure is very important from a practical point of view.

And ensuring the correct measurement procedure is very demanding and complicated. Mainly due to the very large possible variation of possible settings that the control device can allow. However, the article aims to reduce at least a small part of this possible variability and, thanks to the described methodology and demonstration of the achieved results, outline the recommended roughness setting in evaluating selected typical situations in evaluating surface roughness in the automotive, aerospace, or medical industries.

As can be read from the article, based on the tests performed while adhering to the methodology of software filter selection, whether the periodic or non-periodic surface is already measured, it is possible to achieve almost identical results despite the different instruments used in the measurement. For contact devices, the difference between the measured values is up to 5%, for optical devices, this value of the difference in the measured values is up to 10%. Which are extreme values and in most of the performed tests the differences were in the order of 2 to 3%.

7. Acknowledgments

The paper was created with the support of the Motivation System of the University of West Bohemia in Pilsen, part of POSTDOC. And it was co-financed by project GA UWB in Pilsen: SGS-2019-008 "Research and development for innovation in the field of manufacturing processes – Technology of metal cutting III".

8. References

[1] Kubátová, D.; & Melichar, M. (2018) Post Processing of Roughness Raw Data Solid State Phenomena 278 15–22 https://doi.org/10.4028/www.scientific.net/ssp.278.15

[2] Kubatova, D. (2018) Software surface roughness filters V Plzni Disertační práce (Ph.D.). Západočeská univerzita v Plzni, Fakulta strojní.

[3] Kubátová D (2020) Problems of defining and subsequently measuring surface roughness in aperiodic surfaces Proceedings of the 31st DAAAM International Symposium pp. xxxx-xxxx, B. Katalinic (Ed.), Published by DAAAM International, ISBN 978-3-902734-xx-x, ISSN 1726-9679, Vienna, Austria, DOI: 10.2507/31st.daaam.proceedings.xxx

[4] Cekic, A.; and Muhamedagic, K.; and Cohodar. M.; and Begic-Hajdarevic, D.; and Biogradlija, S. (2019) Experimental Investigation of Effect of Overhang Tool Length on Tool Vibration and Surface Roughness Proceedings of the 30th DAAAM International Symposium pp.0184-0191 B. Katalinic (Ed.), Published by DAAAM International, ISBN 978-3-902734-22-8, ISSN 1726-9679, Vienna, Austria DOI: 10.2507/30th.daaam.proceedings.024

[5] Kubatova, D.; and Melichar, M. (2019) Roughness Evaluation Using Abbott-Firestone Curve Parameters, Proceedings of the 30th DAAAM International Symposium, pp.0467-0475, B. Katalinic (Ed.), Published by DAAAM International, ISBN 978-3-902734-22-8, ISSN 1726-9679, Vienna, Austria DOI: 10.2507/30th.daaam.proceedings.063

[6] EN ISO 4287. Geometrical product specifications (GPS) - Surface structure: Profile method Terms, definitions and parameters of surface structure. 1. Brusel: CEN, 1999. [7] EN ISO 3274. Geometrical product specifications (GPS) - Surface structure: Profile method - Nominal characteristics of tactile devices. 1. Brusel: CEN, 1999

[7] Chen, S.; and Feng, R.; and Zhang, Ch.; and Zhang, Z.; Surface roughness measurement method based on multi-parameter modeling learning. Measurement [online]. 2018, 129, 664-676 [cit. 2020-09-09]. DOI: 10.1016/j.measurement.2018.07.071. ISSN 02632241. Dostupné z: https://linkinghub.elsevier.com/retrieve/pii/S0263224118307103

[8] Purkait, B.; and Majumdar, D, D.; and Mazumder, R.; (2017) Bed roughness and grain sorting- an experimental study over fine to medium sand beds; International Journal of Sediment Research; Volume 32; Issue 3; Pages 384-400 ISSN 1001-6279 https://doi.org/10.1016/j.ijsrc.2017.06.002.

[9] Chen, S.; Feng, R.; Zhang, Ch.; and Zhang, Y. (2018) Surface roughness measurement method based on multi-parameter modeling learning. Measurement [online]. 129, 664-676 [cit. 2020-09-09]. DOI: 10.1016/j.measurement.2018.07.071. ISSN 02632241.

[10] Purkait, B.; Majumdar, D. D.; Mazumder, R.; (2017) Bed roughness and grain sorting- an experimental study over fine to medium sand beds, International Journal of Sediment Research, Volume 32, Issue 3, Pages 384-400, ISSN 1001-6279, https://doi.org/10.1016/j.ijsrc.2017.06.002.

[11] Zatloukal, T.; Gombar, M.; Fulemova, J.; Rehor, J. & Povolny, M. (2019). Optimization of the Thermal Spraying Process Depending on the Surface Roughness, Proceedings of the 30th DAAAM International Symposium, pp.1026-1033, B. Katalinic (Ed.), Published by DAAAM International, ISBN 978-3-902734-22-8, ISSN 1726-9679, Vienna, Austria DOI: 10.2507/30th.daaam.proceedings.143

DOI: 10.2507/32nd.daaam.proceedings.024

PROBLEMS OF CHOOSING AN INTERMODULE INFORMATION INTERACTION PROTOCOL FOR MOBILE ROBOTS WITH MODULAR CONTROL SYSTEM ARCHITECTURE

Victor Andreev & Pavel Pletenev

This Publication has to be referred as: Andreev, V[ictor] & Pletenev, P[avel] (2021). Problems of Choosing an Intermodule Information Interaction Protocol for Mobile Robots with Modular Control System Architecture, Proceedings of the 32nd DAAAM International Symposium, pp.0151-0157, B. Katalinic (Ed.), Published by DAAAM International, ISBN 978-3-902734-33-4, ISSN 1726-9679, Vienna, Austria
DOI: 10.2507/32nd.daaam.proceedings.024

Abstract

In the modern world, there is an increasing need to combine the results of the development of various engineers, vendors and research groups to quickly create and efficiently operate robotic systems (RTS) for various purposes. The creation of common ways of interaction between parts of the software and hardware of mechatronic devices and robotic systems (modules) is one of the important tasks facing robotics as a science and a field of technology. The article proposes a criterion for assessing the applicability of existing protocols for creating methods of intermodule communication in RTS with a modular control system architecture. The criterion includes 12 requirements, the presence or absence of which determines the degree of suitability for the implementation of intermodular information interaction in the modular structure of a robotic system. As a result of the analysis of existing systems, the main sources of problems of organizing such interaction are identified and the applicability of the protocol chosen based on this criterion for the multilevel hierarchical (pyramidal) network topology of the RTS control system is justified. The article proposes a new concept of a generalized method of intermodule interaction implemented in the form of a specification.

Keywords: modular robot; communication interface; protocols; internet of things.

1. Introduction

When using robotic systems (RTS) in extreme scenarios, it is often impossible to predict the conditions in which this RTS will work; therefore, it is impossible to determine its functional composition in advance. There are times when the robot must be fully or partially functionally modified at the site of the upcoming work (or even during the work itself) – that is, the robot must have the ability to perform operational reconfiguration. The works [1], [2], [3] brought attention to this problem. Control systems with a mono-computer do not allow for operational reconfiguration of the RTS; to ensure the reconfigurability of a mobile robot (MR) with a centralized control system (CS) in practice means creating a new robot with a given new functionality. As noted in [4], the solution to these problems is seen in the creation of a modular architecture of the control system of a mobile robot.

Operational reconfiguration is a change in the configuration and/or technical characteristics of a mobile robot when adding new functional units or replacing them without the need to reconfigure the software (software) of the control system of the entire robot. The requirement of efficiency leads to the need for such an organization of the CS, in which it is automatically set to a new configuration – plug-and-play mode. Since such RTSs operate in difficult and dangerous conditions, the requirements for autonomy, and, consequently, for the speed of reaction to a changing situation in such RTSs are increasing. This entails ensuring the real-time operation of the RTS for all components of the robot.

This article is a continuation of the research [5], [6], [7] to develop the principles of creating a network infrastructure for modular mobile robots with a hierarchical pyramidal structure of the control system. This topic is part of a general new direction related to the creation of mobile robots based on modular architectures. Due to its novelty, this direction has not yet been clearly defined. Even though such systems consist of modules and are capable of reconfiguration, they do not belong to either homogeneous or heterogeneous robots. This is due to differences in the very concept of construction: in both homogeneous and heterogeneous robots, the implementation of reconfiguration is solved by changing the mechanical configuration – a kinematic scheme consisting of a limited set of modules, due to their rearrangement and mutual displacement. In an MR with a modular architecture, the target task is solved via the correct selection of functional nodes-modules and their automatic software integration into the robot's control system as a whole without any changes in the software aspects of both the control systems of the modules themselves and the general control system. Distributed computing is implemented in the modular architecture of the control system, which eliminates the problem of an infinite increase in the computing power of the control computer to provide real-time mode.

There are reasons to believe that the modular construction of the control system makes it possible to improve the functionality of a robot as a whole. This can be achieved by reducing energy consumption and heat dissipation of the entire control system due to the use of energy-efficient embedded systems – microcontrollers of low performance in the modules. The possibility of using such computing devices in modules is provided by reducing the complexity of algorithmic solutions of the functionality of a separate module. Also, an increase in the overall functionality of the RTS can be achieved through the division of engineering labor during the development of the robot - the designer of a separate module can find more effective and/or intelligent solutions than the designer of the robot as a whole, forced to deal with all design problems simultaneously. As soon as modules of sufficiently general purpose are created by *different manufacturers* for any of the fields of robotics, the use of such modules will be economically advantageous due to their mass production and competition between manufacturers. Therefore, it is necessary to create such means of intermodule interaction so that modules with the same functionality from different manufacturers are interchangeable.

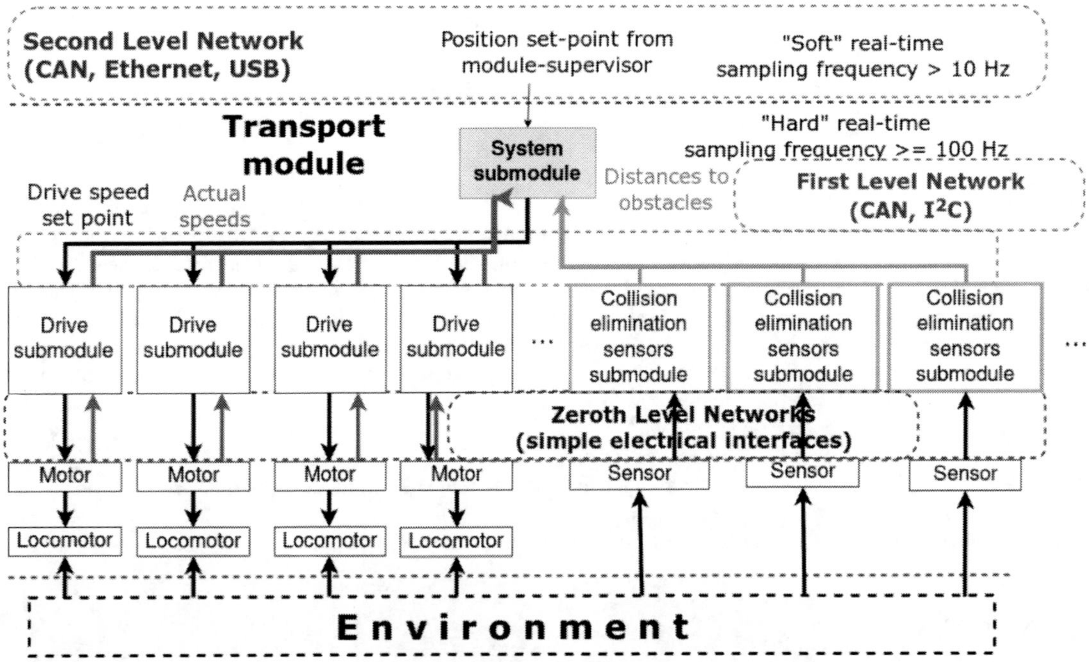

Fig. 1. Transport module's structure.

As a theoretical justification in [8], it was shown that the multilevel hierarchical (pyramidal) network topology of the MR control system can be represented as a multi-agent system in which "agent-modules" must interact with each other when performing their specific tasks. In a mobile robot with a modular architecture, considered in [6] and [7], interaction between different modules and submodules occurs at different levels of the hierarchy of agent-modules through different environments.

Such an environment is made up of service software that implements various communication protocols at different levels of the OSI model [9]. At the same time, in an MR with a modular hierarchical CS structure, there may be many such interaction environments. In our previous works, at least two levels of subordination of interaction environments were identified – communication networks of the first and second levels (an example of such a structure of a transport module is shown in Figure 1). In these works, studies were conducted on the applicability of various transport protocols for the operation of a transport module in a modular robot. It has been shown that the CAN bus and the UDP/IP protocol from the Ethernet technology stack have sufficient speeds for use in modular robots. These two protocols are well-known and popular among software developers for different purposes, but they require the introduction of a data exchange discipline – that is, network protocols of a higher (in the sense of OSI) level - the application level. Such a protocol will be called the intermodule communication method (IMCM). IMCM can be built on one or more transport layer network protocols (CAN, UDP/IP, TCP/IP, etc.), which we will call "IMCM transports" for short.

The purpose of this work is to develop requirements for methods of intermodule communication of agents-modules, on the basis of which to assess the applicability of existing protocols for creating IMCMs in the hierarchical structure of mobile robots with modular architecture, to identify their strengths and weaknesses.

2. Requirements for the method of intermodule interaction

Based on the results of previous studies [5], [6], [7] it is possible to formulate the following requirements for IMCM in the hierarchical structure of the control system of a modular mobile robot:

R1. IMCM's transports must ensure the data transfer speed, sufficient for the functioning of CS of modules in real time, and, if necessary, to provide deterministic message delivery.
R2. IMCM must be defined as an open standard or an open source specification.
R3. IMCM must have one or more portable (multi-platform), and preferably open implementations.
R4. It is desirable that IMCM can be implemented on top of different transports (see definition above).
R5. IMCM should be implementable on relatively simple (low performance) and cheap microcontrollers and occupy a minimum amount of both their program memory and RAM.
R6. IMCM must support the connection and disconnection of modules during operation (hotplug).
R7. Messages in IMCM must be typed. Message exchange in IMCM should be built to exclude the reception and transmission of corrupted or incorrect typed messages.
R8. IMCM transports should not have an explicit main node for data transmission, i.e. the communication network should ensure the creation of direct horizontal links.
R9. IMCM with transports that are able to provide the node with information about the topology of the connection of other nodes in the network should have priority.
R10. There should be a unified way of transmitting messages between different IMCM transports.
R11. As many levels as possible (in the sense of OSI) of IMCM transport protocols should be implemented at the hardware level in embedded systems (microcontrollers and embedded computers).
R12. The physical and/or channel layer of IMCM transports must be protected from interference and/or have error detection and correction mechanisms for transmitting information at module-specific distances.

The R1 and R12 requirements are necessary so that the modules can reliably maintain real-time operation even in the presence of external interference. It should be noted that the requirements of R2 and R3 are key since they form the requirement of interoperability (in the sense of the standard [10]). In case of non-compliance, the use of IMCM to combine modules from different manufacturers and different teams will be extremely difficult due to the lack of a unified approach to implementation. In order to achieve the stated goals of reducing energy consumption and minimizing cost, the requirements R5 and R11 have been introduced. The requirements R4 and R10 are necessary in order to be able to select the same IMCMs for different levels of the hierarchy. The R6 requirement is important for implementing operational reconfiguration. The R7 requirement is important for creating easy-to-understand and error-resistant IMCMs.

Since the modules in a modular robot must work in real time, a robot should not have a central point of failure of all systems - the R8 requirement is responsible for this. Since changing the structure of a modular robot (reconfiguration) changes various characteristics (its mass, moment of inertia, etc.), it is necessary to rebuild and adjust the control systems of robot modules. Therefore, in order for the mechanisms to work correctly during the automatic reconfiguration of modules and submodules at all levels, it is necessary to automatically update the knowledge of modules about the relative location of modules and submodules, their size, weight, and so on. Thus, IMCM must provide modules with information about their mutual topology, i.e. about where exactly the next module is connected or disconnected from - requirement R9.

It should be noted that R8 and R9, in fact, contradict each other. Therefore, any choice of IMCM will be a compromise between the possibility of creating horizontal connections and obtaining information about the topology and connection points of nodes. For each of the IMCM under consideration, we will evaluate compliance with the corresponding requirement with the values of 1 (met) or 0 (not met) points. Then we add up the points obtained by IMCM to form an overall assessment of the applicability of the IMCM for mobile robots with a modular hierarchical control system architecture. The maximum score is 12 points (all requirements are met).

3. Assessment of the requirements in various networks

We will conduct a study of the use of the infrastructure of various intermodule communication methods for integration possibilities, the use of ready-made program blocks and for the organization of intermodule interaction. Preliminary analysis of publications [7], [11], [12], [13], [14], [15], [16], [17], [18], [19] has shown that in many works there is a "unifying" approach to intermodule interaction - it is proposed to combine several "transport" communication networks with one common protocol of intermodule interaction. "Transport" networks are understood here as communication networks whose details of implementation and operation are hidden from the programmer by a generalized application programming interface (API).

Such "unifying" is, for example, IMCM EmSBoT [11]. The authors propose to create a common abstraction from various operating systems, both conventional – Linux and Windows, and real-time, such as Chibi/RTOS. It is also proposed to abstract from the way messages are transmitted between modules. The speed of the transports offered by EmSBoT is sufficient for service robots - Ethernet and CAN are used, which is also implemented in our previous work [7] - the R1 requirement is met. Article [11], unfortunately, is the only source of information about EmSBoT, i.e. there is no open specification, as there is no available implementation of this technology – the requirements of R2 and R3 are not met. Because of this, it is difficult to assess the remaining requirements – we will consider them unfulfilled due to the lack of information provided by the authors. EmSBoT gets 1 point out of 12.

IMCM R2P [12] is designed for various mechatronic and robotic devices, starting with toy robots and ending with mobile service platforms, unmanned aerial vehicles and ground vehicles. The authors of R2P do not provide unified mechanical components, but focus only on the electronic and software components necessary for robotic systems, as well as on tools for developing embedded robot software. For intermodule communication in R2P, the author's RTCAN real-time protocol is used [13]. RTCAN is primarily aimed at real-time communication, and not at fault tolerance: erroneous messages are only marked as damaged, and higher-level protocols (above RTCAN) are responsible for all retransmissions. The authors use only the CAN bus, which has sufficient bandwidth. Conclusion - requirement R1 is met, but requirement R4 is not met. The authors published the source code for intermodule interaction, but not the specification - the requirements of R3, R5 are met, and R2 is not. IMCM R2P is based only on the CAN network - interaction with more complex modules based on Linux OS is carried out through the transmission of messages to IMCM ROS (about which below) - the requirements of R4 and R9 are not met, and the requirements of R8, R10, R12 are met. CAN can support disconnecting individual modules from the network, but it is not clear from the article whether there is support for dynamic connection and disconnection in the protocol – the R6 requirement is not met. All types of transmitted data are declared in common header files – the R7 requirement is met. Since the CAN network is used, which is usually implemented at the hardware level in microcontrollers, the R11 requirement is met. The total score for R2P is 10 out of 12 points.

The Robotic Operating System (ROS), its successor is ROS2 [14], its low-level implementations of H-ROS (Hardware ROS) [15] and microROS [16] are IMCMs and an ecosystem of code for robotic applications. In this article, we will not consider the old version - ROS1 due to its missing implementation on microcontrollers and H-ROS because further implementation of this project has stopped. The authors of ROS and ROS2 tend to use modern technologies that are standards for software development. Both versions use XML format that is difficult to read and write for configuration, and strictly Ethernet is used for network interaction. The first version of ROS used a synchronous connection based on its own protocol based on the TCP/IP protocol stack. The second version uses the DDS – Data Distribution Service technology, which is quite difficult to implement and understand by simple developers. Direct use of this technology on microcontrollers is impossible due to the huge (by the standards of microcontrollers) sizes of the resulting code. To cope with this complexity, the developers of DDS - Object Management Group - have developed an alternative specification - XRCE-DDS (eXtremely Resource Constrained Environment). This partially solves the problem, but there is still a need for a central node that performs arbitration of all messages and communication with the rest of the ROS2 network. Due to the protocol features, it is not possible to establish direct horizontal connections at the lower level. Also, the amount of code for XRCE-DDS to work is still quite large. The Ethernet network speeds are enough to work in soft real time - the R1 requirement is met. All protocols included in ROS2 and microROS are implemented in the form of specifications and there are different both open and closed DDS implementations (see [16]– - the requirements of R2 and R3 are met. Theoretically, ROS2 can be built on top of any transports, since the specifics of a particular transport or its software implementation are hidden in separate ROS Middleware interface libraries, therefore, in theory, the R4 requirement is met. Due to the peculiarities of microROS, the requirements of R4, R5 and R10 are met. Thanks to DDS, ROS2 provides automatic detection and reconnection of elements – the R6 requirement is met. In ROS, all messages are typed – the R7 requirement is met. Due to XRCE-DDS, the requirements of R8 and R9 are not met.

The R11 requirement is not met, since all microROS integration methods require the installation of additional peripheral devices – a rare microcontroller can be connected via Ethernet without an additional module. The Ethernet network is sufficiently protected from interference, but the UART interface, which is the basis for the interaction of the microROS module with the central broker module, is not resistant to interference, therefore, the requirement – R12 is not met. The total score for ROS2 and microROS is 8 out of 12 points.

Our approach, described in the previous work [7], to the creation of IMCM MODROB-1 is based on the design methods of ZeroMQ libraries – C4 and offers a common specification for all levels of interaction and its specification for implementation based on a different "transport" communication networks. Just like IMCM EmSBoT, MODROB-1 proposes abstraction from specific communication networks. As it was shown in the noted work, the speeds of the selected transports are sufficient for data transmission, therefore the requirement R1 is fulfilled. MODROB-1 was originally designed as an open specification and the authors had plans to open access to the source code of the implementations - then, from the authors' point of view, the requirements of R2 and R3 are met. MODROB-1 was originally designed to work only via the CAN network, but then it was expanded to work on other communication networks, such as Ethernet or UART - the requirements of R4, R11 and R12 are met.

The authors have an implementation of MODROB-1 for different microcontrollers and computers. The implementation of this specification requires a fairly small amount of RAM and program memory – the R5 requirement is met. IMCM MODROB supports hot-plugging and disconnecting individual modules – other modules "connected" to this module will simply stop receiving new variable values from the disconnected module. There is also no need for special configuration to "connect" the module, since all modules constantly publish their values and the introduction of a new module into the network occurs automatically. But the mechanisms of (de)centralized module detection and top-level management mechanisms that should set a new configuration have not yet been described, i.e. there is no mechanism that should automatically enable it when a new module (submodule) is connected. Thus, the R6 requirement is not fully met. Messages in MODROB-1 have only 1 type – a single-length floating-point real number. Other types are not supported by the specification. This means that the R7 requirement has not been met. IMCM MODROB-1 assumes the use of one global bus on top of different transports - the requirement of R8 is met, and R9 is not. The message type is the same for all the subject communication networks – the requirements of R10 are met. IMCM MODROB-1 scores 9 out of 12 points.

IMCM UAVCAN (Uncomplicated Application-level Vehicular Communication And Networking) [18], [19] is a new open standard developing according to the open specification model (requirement R2) with open implementations in several programming languages for different embedded systems (requirements R3, R5). The author of the standard shows the simplicity of implementing UAVCAN on new platforms – only 1500 lines of code are enough [19]. The zero version of the standard assumed the use of only the CAN bus, but its further development to version 1 led to the inclusion in the "orbit" of implementations for other networks, such as UDP over Ethernet, UART and even the IEEE-802.15.4 wireless channel (requirements R1, R4, R11 and R12). This flexibility is achieved primarily due to the static typing of all messages (R7 requirement) and a single and extremely efficient binary encoding format for information to be transmitted through transports. The network was built initially to create direct horizontal links between modules without a central broker and the need to explicitly subscribe to certain publications (the requirements of R8 are met, and R9 is not). Also, the authors of the specification have released quite a lot of additional programs based on this specification. For example, this specification does not assume dynamic addresses for modules, but with the help of additional software, they can also be implemented. Also, additional software allows you to implement dynamic reprogramming of embedded systems (requirement R6). Since the binary interface is the same for all modules, it is possible to use any of the transports and transmit messages between networks (R10 requirements). The UAVCAN framework scores 11 out of 12 points.

Req.	EmSBoT	R2P	ROS2	MODROB-1	UAVCAN
R1	V	V	V	V	V
R2	X	X	V	V	V
R3	X	V	V	V	V
R4	X	X	V	V	V
R5	X	V	V	V	V
R6	X	X	V	X	V
R7	X	V	V	X	V
R8	X	V	X	V	V
R9	X	X	X	X	X
R10	X	V	V	V	V
R11	X	V	X	V	V
R12	X	V	X	V	V
Sum	1	8	8	9	11

Table 1. Comparison of various MMK (V – the requirement is fulfilled, X – the requirement is not fulfilled)

The results of the study are summarized in Table 1. The letter V indicates fulfilled requirements, X – unfulfilled. As can be seen from the table, the most suitable for implementing the method of intermodule interaction are MODROB-1 and UAVCAN. These methods most fully implement the necessary requirements, except for the R9 requirement for determining the topology.

4. Conclusion

Our earlier research aimed at developing a protocol for intermodule interaction in the network structure of the control system of mobile robots with modular architecture was only partially successful. Nevertheless, these studies made it possible to develop requirements for such protocols and analyze existing protocols developed by other authors in order to assess the possibility of their use in intermodule communication methods (IMCM) in real-time systems. The analysis given in this paper showed that the UAVCAN protocol meets the listed requirements the most and differs from our developments (IMCM MODROB-1) in that it is already supported by the community of developers of unmanned aerial vehicles (UAVs). Thus, unlike IMCM MODROB-1, this protocol has a great prospect for use. There is not as much ready-made software written for UAVCAN as for ROS/ROS2, but proposals have already been made in the developer community to replace the DDS networking layer inside ROS2 with UAVCAN. Also, the software of a "bridge" between ROS and UAVCAN has already been written, which will allow reusing many of the developments in the field of software from ROS. However, its direct use in our hierarchical architecture of the control system of mobile modular robots is impossible. Although the UAV is also a mobile robot, its operating conditions and functional purpose are significantly different. In this regard, it is necessary to develop add-ons for the UAVCAN protocol – unified protocols for different types of modules and submodules and a unified set of software libraries designed to develop their software aspects. It is also necessary to experimentally test the throughput of intermodule communication channels with different protocols. It should also be checked whether UAVCAN is suitable for use both for basic control on microcontrollers and for more complex interaction at higher levels of the hierarchy, since hard or soft real-time mode is required at different levels of the hierarchy.

We believe that our further actions should be aimed at an experimental study of the use of the UAVCAN protocol in the communication network of a modular robot with a multi-level (hierarchical) control system architecture. It is also necessary to investigate the applicability of this protocol for the implementation of dynamic reprogramming of computing devices of control systems of modules and submodules.

5. Acknowledgments

Research is supported by the Russian Foundation for Basic Research: Grant 19-07-00892a.

6. References

[1] Babich, A.V.; Baranov, A.G.; Kalabin, I.V. & others (1982). Industrial robotics[Промышленная робототехника], Mashinostroenie[Машиностроение], Moscow, 1982.

[2] Vorobyov E.I.; Kozyrev, Yu.G. & Tsarenko, V.I. (1988) Industrial robots of aggregate-modular type [Промышленные роботы агрегатно-модульного типа], Mashinostroenie[Машиностроение], ISBN: 5-217-00166-6, Moscow.

[3] Lopota A.V. & Yurevich E.I. (2013) Stages and prospects of development of the modular principle of building robotic systems [Этапы и перспективы развития модульного принципа построения робототехнических систем], Scientific and technical bulletin of SPbSPU [Научно-технические ведомости СПбГПУ], Vol 1 2013 Computer science. Telecommunications. Management.[Информатика. Телекоммуникации. Управление.], pp. 98 - 103, ISSN 1994-2354

[4] Yurevich E.I. (2003) Robots of the Central Research Institute of RTK at the Chernobyl Nuclear Power Plant and the development of extreme robotics [Роботы ЦНИИ РТК на Чернобыльской АЭС и развитие экстремальной робототехники], Polytechnic University Publishing House [Издательство Политехнического университета], ISBN 5-7422-0481-7, Saint Petersburg

[5] Andreev, V. & Pletenev, P. (2017) Organizing Intermodular Communication for Heterogeneous Modular Mobile Robot. pp. 0474–0480. https://doi.org/10.2507/28th.daaam.proceedings.066

[6] Andreev, V.P. & Pletenev, P. (2019) Organizing Network Interaction in Modular Robot with Multilevel Control System, in: Proceedings of the 30th DAAAM International Symposium, DAAAM International, Vienna, Austria, pp. 0485–0492. https://doi.org/10.2507/30th.daaam.proceedings.065

[7] Andreev, V.; Pletenev, P. & Eprikov, S., (2020). Development of a Communication Protocol Specification for The Modrob-1 Modular Mobile Robot. pp. 0244–0253. https://doi.org/10.2507/31st.daaam.proceedings.034

[8] Andreev V. (2019). Control System Mobile Robots with Modular Architecture as a Multi-Agent System with a Hierarchical Topology, Proceedings of the 30th DAAAM International Symposium, B. Katalinic (Ed.), Published by DAAAM International, ISBN 978-3-902734-22-8, ISSN 1726-9679, Vienna, Austria, 2019, pp.0010-0019

[9] GOST R ISO/MEK 7498-1-99. — « Information technology. Open systems interconnection. Basic reference model. Part 1. The basic model». — OKS: 35.100.70. — In use since 01.01.2000. — 62c.

[10] ISO Central Secretary, (2021). Robotics – Modularity for service robots – Part 1: General requirements (Standard No. ISO/IEC TR 22166-1:2021). International Organization for Standardization, Geneva, CH.

[11] Peng, L.; Guan, F.; Perneel, L.; Fayyad-Kazan, H. & Timmerman M. (2016) EmSBoT: A lightweight modular software framework for networked robotic systems. 2016 3rd International Conference on Advances in Computational Tools for Engineering Applications (ACTEA), Beirut, 2016, pp.216-221. doi: 10.1109/ACTEA.2016.7560142.

[12] Bonarini, A.; Matteucci, M.; Migliavacca, M .& Rizzi, D. (2014). R2P: An open source hardware and software modular approach to robot prototyping. Robotics and Autonomous Systems. – 2014. – No.62. – pp.1073-1084.

[13] Migliavacca, M.; Bonarini, A. & Matteucci, M., (2013). RTCAN - A Real-time CAN-bus Protocol for Robotic Applications, in: ICINCO. https://doi.org/10.5220/0004484303530360

[14] Thomas, D.; Woodall, W. & Fernandez, E., 2014. Next-generation ROS: Building on DDS, in: ROSCon Chicago 2014. Open Robotics, Mountain View, CA. https://doi.org/10.36288/ROSCon2014-900183

[15] Mayoral, V.; Hernandez, A.; Kojcev, R.; Muguruza, I. & others. (2017). The shift in the robotics paradigm – the Hardware Robot Operating System (H-ROS); an infrastructure to create interoperable robot components, NASA/ESA Conference on Adaptive Hardware and Systems (AHS), Pasadena, CA. – 2017. – pp.229-236.

[16] micro-ROS URL https://micro.ros.org/ (accessed 29.09.21).

[17] Kronauer, T., Pohlmann, J., Matthe, M., Smejkal, T., Fettweis, G., 2021. Latency Analysis of ROS2 Multi-Node Systems. arXiv:2101.02074 [cs].

[18] Kirienko, P.; Dixon, S. & others (2014) UAVCAN: an open lightweight protocol for reliable intravehicular communication in aerospace and robotic applications over CAN bus, Ethernet, and other robust transports. Available from: https://uavcan.org/specification/UAVCAN_Specification_v1.0-beta.pdf (accessed 29.09. 21).

[19] Kirienko, P. (2019). Publisher/subscriber for distributed fault-tolerant real-time onboard systems in 1500 lines of code [Издатель/подписчик для распределённых отказоустойчивых бортовых систем реального времени в 1500 строк кода] URL https://habr.com/ru/post/512570/ (accessed 29.09.21)

DOI: 10.2507/32nd.daaam.proceedings.025

ANALYSIS OF EMOTIONS USING EEG DATA AND MACHINE LEARNING

Vyacheslav V. Potekhin & Ogul Unal

Peter the Great St. Petersburg Polytechnic University

This Publication has to be referred as: Potekhin, V[yacheslav] & Unal, O[gul] (2021). Analysis of Emotions using EEG Data and Machine Learning, Proceedings of the 32nd DAAAM International Symposium, pp.0158-0167, B. Katalinic (Ed.), Published by DAAAM International, ISBN 978-3-902734-33-4, ISSN 1726-9679, Vienna, Austria
DOI: 10.2507/32nd.daaam.proceedings.025

Abstract

Emotions might be one of the most significant differences between a machine and a human being. Currently, Machine Learning can learn and make predictions using Data in many different fields such as Medicine. In this paper, the goal is to evaluate how human emotions are being executed then test the performance of ML in terms of Emotion analysis. DREAMER Data was used to perform all the tasks. The classification of rankings for 23 subjects' was done. The reactions of subjects and EEG data were recorded while 23 participants were watching 18 different short movies. Each candidate has a ranking, between one to five, based on their affective state after each stimuli in terms of valence, arousal, and dominance. Afterward, Using Generative Adversarial Network (GAN), synthetic data will be created and analysed using supervised and unsupervised learning algorithms. Finally, all results will be compared. The research can help to evaluate the basic human emotions for robots or devices in Medicine. Robotics or cyber-physical machines in healthcare are already growing every day therefore, the quality of surgeries, treatments, or medical assistance can be improved.

Keywords: Machine Learning; Emotion Analysis; Electroencephalography; Brain Computer Interface; Generative Adversarial Networks

1. Introduction

Electroencephalography, according to the American Heritage dictionary, is a neurological test that works with electronic monitoring equipment to estimate and document the electrical activity in the brain. EEG can monitor and record the electrical activity of the human brain. Therefore it is possible to use a Medicine Dataset for Machine Learning (ML) algorithms. Today, many kinds of research are aiming to solve how the human brain controls our emotions and actions. Several examples are; Facial expression recognition or Emotion analysis using a Human-machine interface.

Today, the human brain is not fully understood. It is likely to get the unique language occurring in the human brain with the help of signal processing technology and instruments and transform them into machine understandable language. Therefore in this project, it was possible to use EEG Affect Recognition Dataset (DREAMER) based on the signals. Emotion theories and brain structure have been studied. Afterward, neurons and neurotransmitters were analysed to understand the electrochemical structure of the brain.

The Supervised, Unsupervised, and Generative Adversarial Networks in the Machine Learning field have been used to perform experiments to understand the dataset in detail. The classification tests had significantly high accuracy results therefore, it is possible to use the Machine Learning approach in affect recognition systems. The results might be useful to provide more information to new technical solutions related to human emotions in intelligent systems.

Due to the importance of the health industry in the world and the overpopulation, robots might help in the healthcare systems thus, intelligent machines can assist people. This research aims to study human emotions that are unknown and difficult to understand mathematically and explain emotions to computers using mathematics, EEG, and machine learning. Brain structure is not fully understood. It is aimed that this research will contribute to the development of the connection between the human brain and computers and help patients in hospitals.

2. Literature Review

2.1 The Human Brain

Approximately one and a half kilogram organ that organizes every function to manage the body itself. The brain is also responsible to analyse the information as well as controlling emotions, intelligence, and memory. The main parts of the brain are Cerebrum, Cerebellum, and Brainstem [1]. The five senses can receive by the brain as tasting, touching, hearing, smelling, and sight simultaneously. It can assemble any message that is meaningful for us because the brain can collect and interpret the memory. Instincts, desires, pain happiness, and any other emotional condition is controlled by the brain.

2.2 Emotions in Human Brain

The limbic system consists of parts such as the thalamus, hypothalamus, hippocampus, and amygdala. The limbic system is also called the emotional brain because it is the control center for emotions. The limbic system is associated with emotions [7]. Thus, the limbic system governs behaviours, exciting events, and behaviours necessary for the survival of the individual. The other parts of the brain and the limbic system work together to regulate purposeful behaviours according to the information coming from the inner and outer environment of the person.

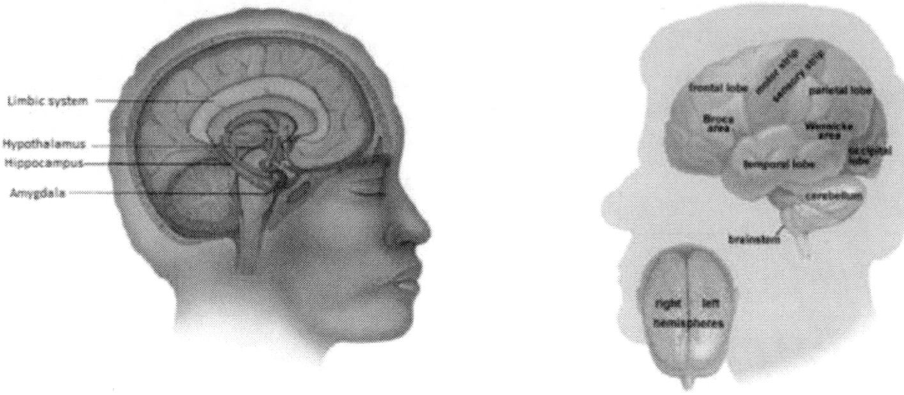

Fig. 1. Human brain structure (left) and the Limbic System (right)
Source: www.macalester.edu/psychology

2.3 Neurons and Nervous System

The cells that make up the nervous system are called neurons. The neuron consists of three main parts. Those are the cell body (soma), dendrite, and axon. Dendrites are numerous and usually short extensions [7]. These extensions receive stimuli from receptors and other nerves. The action potential is the brief electrical event that is generated in the axon which signals the neuron as "active". An action potential travels as the length of the axon and causes the release of neurotransmitters into the synapse. The action potential and consequent transmitter release allow the neuron to contact other neurons. The nervous system is the organized group of cells specialized for the conduction of electrochemical stimuli from sensory receptors through a network to the site at which a response occurs. [18].

Electrophysiology of a neuron; The general idea of electrical signals in a neuron is; there is a voltage difference between the interior of the neuron relative to the space surrounding the neuron, the extracellular space. This voltage is known as the membrane potential which is a constant of about -70 mV [8]. When a neuron is stimulated, the membrane voltage may indicate both positive and negative variants. These voltage variations could carry signals across the body of a neuron, and also contribute to signals that are broadcasted from one to another across the synapse [8].

2.4 Neurotransmitters

Neurons are communicating through the body and they communicate with one another to transmit signals. Although, neurons are not simply connected physically. Each neuron's end is a small gap that is a synapse and to communicate with the next cell, the signal has to be able to cross the space. This process is known as neurotransmission. There are excitatory and inhibitory neurotransmitters. Excitatory will increase the likelihood that the neuron will fire an action potential while inhibitory will decrease it. The functions of the neurotransmitters are explained and shown below.

Neurotransmitters and their functions are explained as [6]:

- Adrenaline (excitatory): Also called Epinephrine, it increases blood flow and heart rate.
- Norepinephrine: (NE)*:* also known as noradrenaline (excitatory) Increases blood flow and attention.
- Dopamine (both excitatory and inhibitory)*:* is mostly responsible for feelings of reward and pleasure. Abnormal dopamine level is common in specific disorders such as schizophrenia [2].
- Serotonin (inhibitory): feelings of happiness and well-being, stable mood, sleep cycle, and digestive system.
- Gamma-aminobutyric acid (GABA) (inhibitory): Inhibits neuron firing in the Central Nervous System also, high levels improve focus and low levels causes anxiety, contributes to motor control and vision.
- Acetylcholine(excitatory): learning, memory, muscle contraction, awakening
- Glutamate (excitatory): learning, memory, and creating new nerve pathways.
- Endorphins*:* natural pain killer, excitement, and exercise.

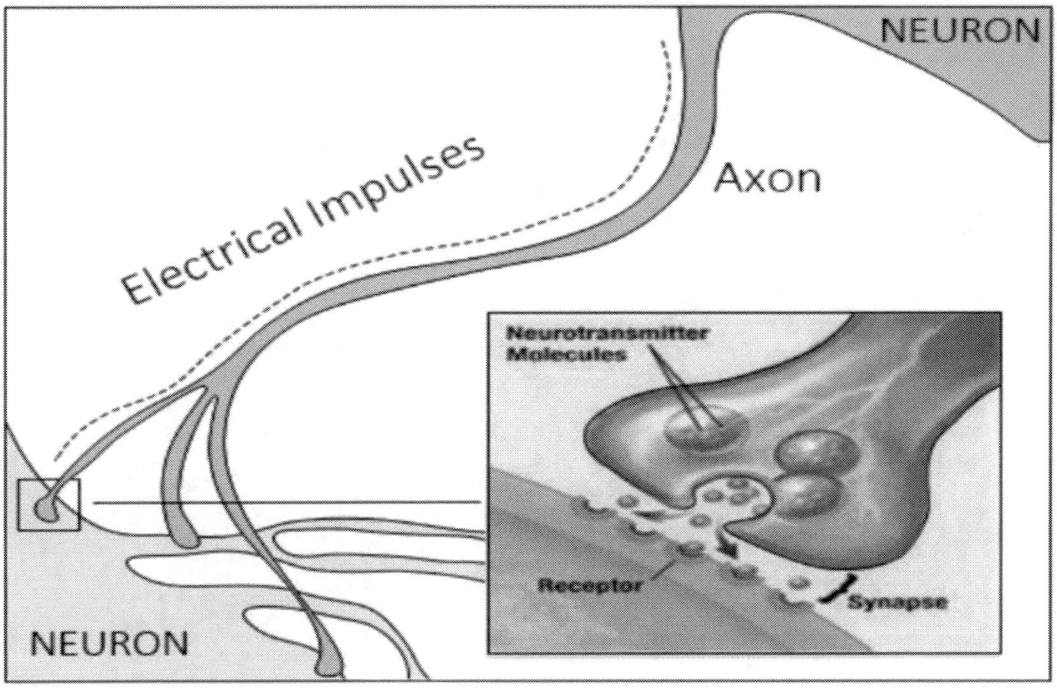

Fig. 2. Illustration of the major elements in chemical synaptic transmission

2.5 Emotions and Neurotransmitters

Neurotransmitters and have different physiological signatures while transmitting signals throughout the brain. Currently, there are many unknown areas about complex emotions in the human brain, basic emotions might be detected in the responsible areas [10]. Basic emotions such as fear, anger, pleasure (when combined) create another layer for creating different types of secondary emotions, therefore, primary emotions can be combined with the experience of other, higher-order emotions that are more affected by specific learning and experience [11]. Using this approach, primary emotions are observable in evolutionarily diverse organisms, which allows us to functionally “dissect” the mechanisms of the presumed associated internal emotional states and their externally manifest behaviours [11]. The first important theory about emotions is Basic Emotion Theory which is evolved to handle fundamental life tasks, e.g., fear and anger can aid survival. The second theory is the Dimensional Theory of Emotion that has three types of emotions those are: pleasant-unpleasant, tension-relaxation, and excitation-calm [11].

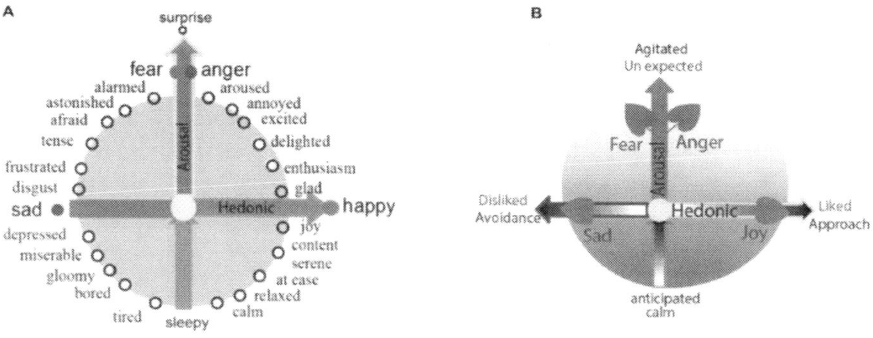

Fig. 3. Dimensional Theory of Emotion

2.6 EEG

The brain has hundreds of billions of neurons that are performing electrical transmissions with densely interconnected via of the synapses. The nerve cells in our brains are active they create electrical signals that rotate through the surface of the scalp. Thus by using sensors placed on the scalp it is viable to measure those changes in Voltage while performing an activity or solving a task [3]. The measurement of EEG divides into two parts that are considered as Non-Invasive and Invasive techniques. In this research, all tasks were under Non-Invasive systems.

It is possible to say that the source of EEG signals is the pyramidal cells in the cortex. All cells in the corresponding area are perpendicular and parallel to each other. Each pyramidal cell has approximately 30000 exhibitory and 1700 inhibitory inputs [4]. Thus the EEG is created by the electrical flow of neurons that are placed in extracellular space. The parallel configuration is; the Pyramidal neurons are spatially aligned perpendicular to the articular surface. Thus it creates a dipole layer in the cortex. Therefore EEG represents the mainly postsynaptic potentials of those pyramidal neurons which are close to the recording electrode [5].

3. Methods & Materials

3.1 EEG and DREAMER Database

EEG nodes, Participant number, and Age columns were used in the dataset for Machine Learning. Gender 0 is Male, Gender 1 represents Female subjects. It is not fully understood differences of emotions for different genders therefore in this research, Genders have been separated and tested individually. From EEG nodes near to limbic to were used.

DREAMER is a multi-modal database containing electroencephalogram (EEG) and electrocardiogram (ECG) signals recorded. Signals from 23 participants were recorded along with the participants' self-assessment of their affective state after each stimulus, in terms of dominance, valence, and arousal. All the signals were captured using wearable, portable and wireless equipment. [20]. The Emotiv-EPOC wireless EEG headset was used for EEG The database contains the participant ratings and physiological recordings of an experiment where 23 volunteers have watched 18 film clips [19]. EEG and ECG signals were recorded and each participant rated their emotion by reporting the felt arousal, valence, and dominance on five-point scales. The peak of the main wave of AI is about to hit industries, governments, and societies accelerating the pace of product and service innovation and putting data in the focus of the business interest [9]. Using AI in Medicine is also another challenge nowadays. Advanced distributed coordination and decision-making techniques allow more effective operation and availability therefore it is possible to combine AI, Medicine, and a cyber-physical system to build a human-computer interface [28].

In this work, only the EEG Stimuli part of the Database was used to perform Machine Learning experiments. Dataset was created based on 23 participants that watched 18 films. Out of 18 films, nine different film clips were chosen. Each of these nine films targets emotions as shown in the figure below. Each prepared dataset contains one film for 23 subjects. Each subject's results were appended to the next one respectively. During the ML experiment datasets were shuffled, validated and Random State was used to avoid overfitting.

Out[4]:

	F7	F3	FC5	T7	T8	FC6	F4	F8	Participant	Gender	Age	ScoreValence	ScoreArousal	ScoreDominance
0	4110.769231	4080.000000	4336.923077	4277.948718	4232.820513	3873.846154	4414.871795	4288.205128	1	0	22	5	4	4
1	4101.538462	4106.153846	4392.820513	4341.025641	4309.743590	3919.487179	4425.641026	4334.871795	1	0	22	5	4	4
2	4097.948718	4106.666667	4398.974359	4346.153846	4330.256410	3929.230769	4424.615385	4350.256410	1	0	22	5	4	4
3	4110.256410	4089.230769	4356.923077	4299.487179	4272.307692	3896.410256	4420.000000	4319.487179	1	0	22	5	4	4
4	4121.538462	4092.820513	4346.153846	4295.897436	4251.282051	3882.051282	4428.205128	4315.384615	1	0	22	5	4	4

Fig. 4. The first 5 lines of each of 9 Supervised Dataset in python

ID	Film clip	Target emotion	Valence	Arousal	Dominance
1	Searching for Bobby Fischer	calmness	3.17 ± 0.72	2.26 ± 0.75	2.09 ± 0.73
2	D.O.A.	surprise	3.04 ± 0.88	3.00 ± 1.00	2.70 ± 0.88
3	The Hangover	amusement	4.57 ± 0.73	3.83 ± 0.83	3.83 ± 0.72
4	The Ring	fear	2.04 ± 1.02	4.26 ± 0.69	4.13 ± 0.87
5	300	excitement	3.22 ± 1.17	3.70 ± 0.70	3.52 ± 0.95
6	National Lampoon's VanWilder	disgust	2.70 ± 1.55	3.83 ± 0.83	4.04 ± 0.98
7	Wall-E	happiness	4.52 ± 0.59	3.17 ± 0.98	3.57 ± 0.99
8	Crash	anger	1.35 ± 0.65	3.96 ± 0.77	4.35 ± 0.65
9	My Girl	sadness	1.39 ± 0.66	3.00 ± 1.09	3.48 ± 0.95
10	The Fly	disgust	2.17 ± 1.15	3.30 ± 1.02	3.61 ± 0.89
11	Pride and Prejudice	calmness	3.96 ± 0.64	1.96 ± 0.82	2.61 ± 0.89
12	Modern Times	amusement	3.96 ± 0.56	2.61 ± 0.89	2.70 ± 0.82
13	Remember the Titans	happiness	4.39 ± 0.66	3.70 ± 0.97	3.74 ± 0.96
14	Gentlemans Agreement	anger	2.35 ± 0.65	2.22 ± 0.85	2.39 ± 0.72
15	Psycho	fear	2.48 ± 0.85	3.09 ± 1.00	3.22 ± 0.9
16	The Bourne Identity	excitement	3.65 ± 0.65	3.35 ± 1.07	3.26 ± 1.14
17	The Shawshank Redemption	sadness	1.52 ± 0.59	3.00 ± 0.74	3.96 ± 0.77
18	The Departed	surprise	2.65 ± 0.78	3.91 ± 0.85	3.57 ± 1.04

Fig. 5. DREAMER database films and expected emotions

The length of the film clips was between 65 to 393 s (M = 199 s), which is considered sufficient since, according to psychologists, video stimuli between 1 to 10 min is capable of detecting single emotions [20]. The target emotions after the experiment were Valence, Arousal, or Dominance. The database has results based on levels of subjects after watching the corresponding films. The Scale is between 1 (lowest) to 5 (highest) reactions for each Valence, Arousal, and Dominance measurements. The database provides information about Affect Recognition therefore it was possible to use it in further Machine Learning experiments. The dataset of film clips: 1, 2, 4, 5, 6, 7, 9, 12, 14 were chosen.

The dimensional approach proposes that every emotion has different amounts of hedonic and arousal value. In the DREAMER database, there are three different emotions based on different film clips. This means in the Database, there are measurements of levels for each emotion after different film clips.

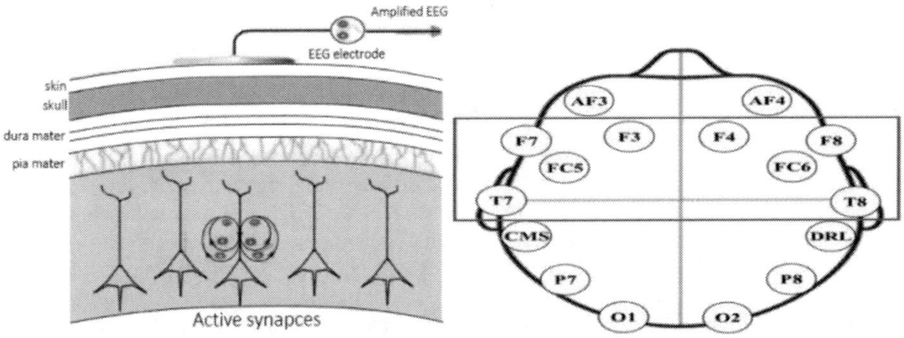

Fig. 6. The source of EEG signals (left) and tThe corresponding EEG nodes in prepared Dataset

3.2 Valence, Arousal and Dominance

Wundt (1912/1924) was among the first psychologists to argue that affective experiences involve at least two properties: valence (ranging from feeling pleasant to unpleasant) and arousal (ranging from feeling quiet to active) [24]. In Psychology, emotional valence describes the measurement to which an emotion is negative or positive, and arousal refers to its intensity. Dominance on the other hand represents the control and influence over circumstances [25]. It is explained in the section on Emotions and Neurotransmitters and it is shown in figure 6. The emotions have levels as explained in Dimensional Theory of Emotion [11], therefore approach in DREAMER Database using 1 to 5 to determine the levels of each emotion suits the experiment.

3.3 Algorithms

In this project, SVM, Logistic Regression, KNN, XGBoost, and Random forest (RF) were used. RF is a bagging technique and XGBoost is a boosting technique. RF gave the highest accuracy. The algorithm takes votes of the similar or same implementations, XGBoost on the other hand, improves the weights of the model [23]. Each model, in this case, is influenced by the previous implementation. Neural Networks may give acceptable results when the Data-set is big enough. A Dense Neural Network means (DNN) all the nodes in each layer are connected. Before a neural network can be used for the required problem or task, it must be trained.

The learning is based on given rules and data. After determined (100 epochs and 500 batch size and 0.01 learning rate in this project) epochs, the NN weights the connections of the neurons afterward, it develops patterns, relations which also means intelligence. The learning rules depend on how the learning material changes the neural network iteratively. Thus the Neural Network learns how to connect the weights and relations according to its neurons properly.

The RELU (Rectified Linear Units) activation function is used in input and hidden layers. The advantages of using RELU are: it converges faster than traditional activation functions (tan-h, sigmoid), and more particularly; RELU is non-saturation of its gradient, which means it accelerates the convergence of stochastic gradient descent [21]. Therefore RELU is used as the main activation function of neurons. The Soft-max gives at least a minimal amount of probability to all elements in the output vector [22]. The Soft-Max activation function was used for multi-class classification. Data was scaled using Min-Max before the training. Neural network modelling requires knowledge of historical data on supply volume large enough as to provide a good training of the network [29].

The shape of the neural network might be related to the number of features, in this case; the training data has nine features(8 EEG columns and age column) for the model evaluation and training, NN has 9 input neurons, two hidden layers with 18 hidden layer units (nodes), and 9 output units shown in the figure below.

Layer (type)	Output Shape	Param #
dense (Dense)	(None, 9)	90
dense_1 (Dense)	(None, 18)	180
dense_2 (Dense)	(None, 18)	342
dense_3 (Dense)	(None, 9)	171

Fig. 7. Structure of Neural Network visualization in python

3.4 K-Means Algorithm

K-Means is an Unsupervised Learning algorithm. Unsupervised algorithms make inferences from datasets using only input vectors without referring to known, or labelled, outcomes. This means for the K-Means algorithm the synthetic Data (generated using GAN) and original Data were used without labels.

3.5 Generative Adversarial Networks

A Generative Model is an unsupervised approach in ML that automatically discovers and learns the patterns in input data therefore the model can be used to generate or new data that could have been drawn from the original dataset [26]. It is possible to use GAN to generate Tabular data [27], in this research Conditional GAN for a Tabular data model was used (CTGAN). CTGAN outperforms all methods to date and surpasses Bayesian networks on at least 87.5% of tested datasets [27]. And conditional generator in GTGAN and training-by-sampling can overcome the imbalanced training data issue [27]. DREAMER database is a big and not balanced tabular database therefore this method was chosen.

	F7	F3	FC5	T7	T8	FC6	F4	F8	Participant	Age
0	4203.552046	4156.127160	4382.279428	3864.216931	4279.113371	3833.542173	4463.083137	4311.492870	7	24
1	4307.395474	4410.086663	4412.589617	4060.456142	4000.183823	4073.934100	4498.636255	4269.469152	17	24
2	3995.793459	4169.346434	4401.219550	4286.014600	4256.597210	3984.843198	3930.496456	4303.901104	17	23
3	4090.673105	4386.967971	4179.603535	4083.889777	4380.213973	3906.740936	4095.513307	4584.518766	2	28
4	4097.154870	4168.952474	4116.434074	4336.664542	4381.233043	4537.257813	3960.494053	4499.348790	1	23

Fig. 8. Synthetic Data (GAN) in tabular format is shown

4. Results

4.1 Random Forest Algorithm using Original Dataset

The first method was RF (supervised learning) for multi-class classification. After training and testing, (70% train, 30% test data) models were able to predict with 98% accuracy. There were five different classes for three of these groups in the Dataset. The classification groups are individual which means they do not affect each other. One more important issue here was not all the classification groups have five different levels. In this case, Arousal for film 4 had only 3, 4, and 5 while the Valence group has 1,2,3,4, and 5 as classes. RF tests were completed for 9 different films as explained in the previous chapter and accuracies were significantly close to each other around %98 for both female and male groups.

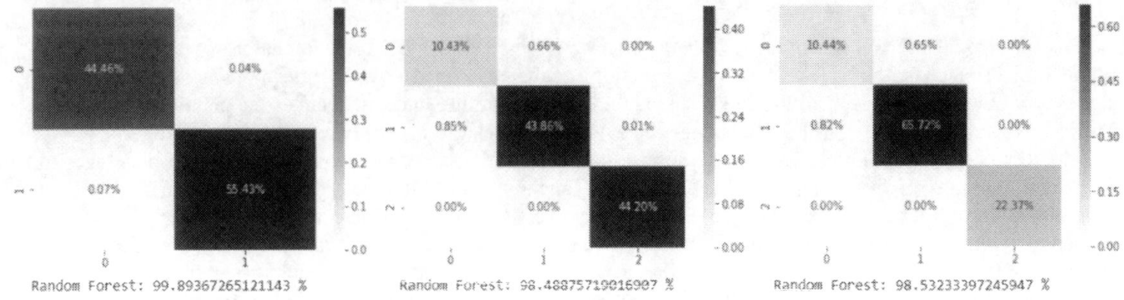

Fig. 9. Random Forest results of arousal, dominance and valence shown for Film 4

4.2 Dense Neural Network Using Original Dataset

DNN (Dense Neural Network) was used and tested for 9 films. Data was shuffled and used the same Random state (42) and cross-validation techniques to prevent overfitting. The accuracy for DNN was 91.41% for male candidates for film 4 to classify Dominance. The accuracy was 92.44 for female candidates for film 4 to classify Dominance. After testing Valence and Arousal the results were similar between 90% - 95% accuracy for males and females. There were no overfitting (Train-Val was close to each other) during the training process. Therefore the model gave high accuracy to detect emotion levels (1 to 5) as a multi-class classification for Valence, Arousal, and Dominance Separately.

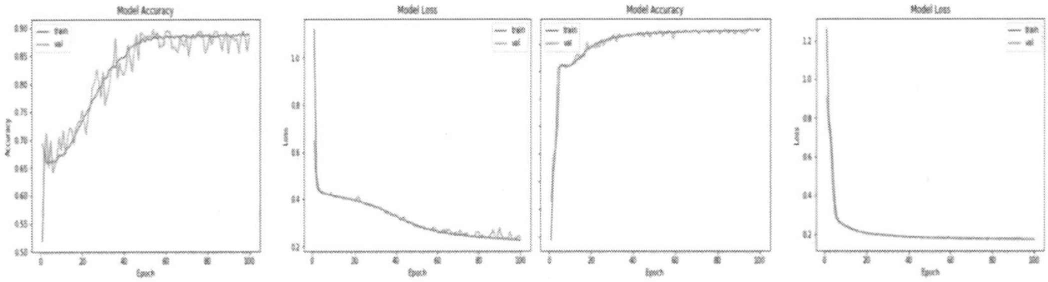

Fig. 10. DNN implementation results for dominance, male (left) and for female (right)

4.3 K-Means using Original Dataset

The unsupervised learning method K-Means was applied to the original Dataset. For film14 Dataset. The results of dataset film14-Females were clustered properly on the other hand the results of dataset film14-Male were not reliable shown in the figures below. The purpose of this test was to detect the possible clusters related to Dominance, Valence, or Arousal using an unsupervised learning approach. Elbow method using the edge on graph was used that selects the optimal number of clusters by fitting the model with a range of values for K.

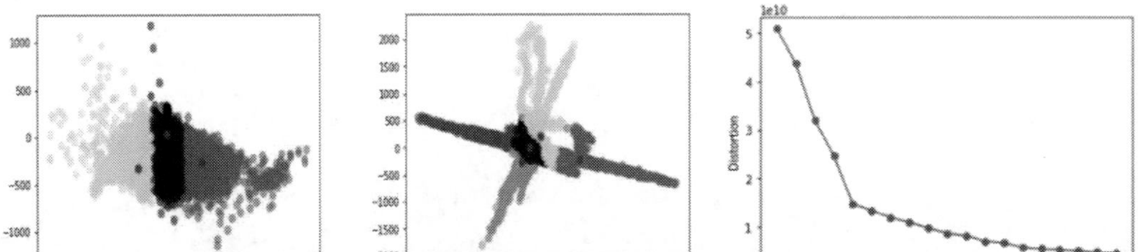

Fig. 11. K-Means and Elbow results for film14 Female (left) and Male (Right) respectively.

4.4 Generative Adversarial Networks for Synthetic Data Generation

To create Tabular Data using GAN there were 300 epochs for generator and discriminator. To observe the effect of quantity of GAN data; three different sample numbers of Data as 1000, 10000, and the original number of subjects, for instance, there were 25472 samples for females in film-1 or 21760 for females in the film-12 dataset.

The correlation results between original and Synthetic data are shown in the figure below. To test the generated data, the same supervised (RF and DNN) and unsupervised (K-Means) algorithms were used.

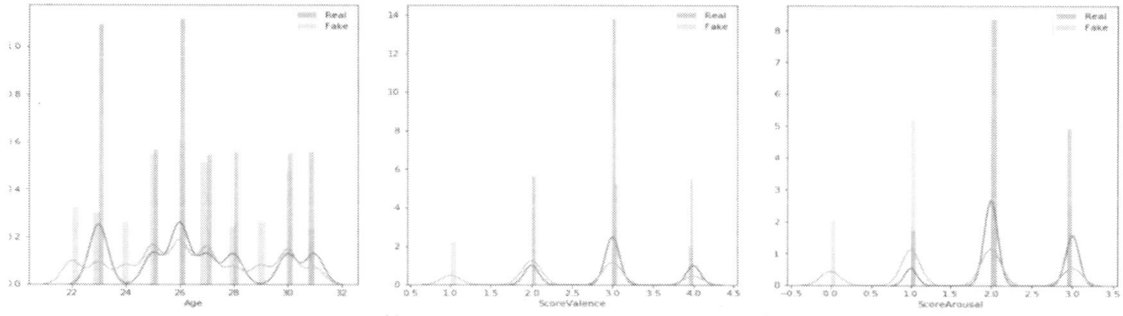

Fig. 12. GAN results of attributes(correlations) for Synthetic Data generation

4.5 Random Forest and DNN using Synthetic Data

Synthetic Data (GAN) was tested for all 9 films using 1000, 10000, and original dataset size (chosen). Dataset size did not affect accuracy. The best results were Female Film1 Dominance 51.67%, Female Film14 Valence 51.3%, Male Film12 Valence 42%, Male-Film14 Dominance 41.67%. Arousal gave the worst results around 30% in every dataset. An extra algorithm SVM (Support Vector Machine classifier) was used and it slightly performed better than RF in Female-Film1 Dominance. The highest accuracies were in films 1, 12, and 14, Dominance and arousal data for both genders and synthetic data of Female subjects performed better up to 10% accuracy difference.

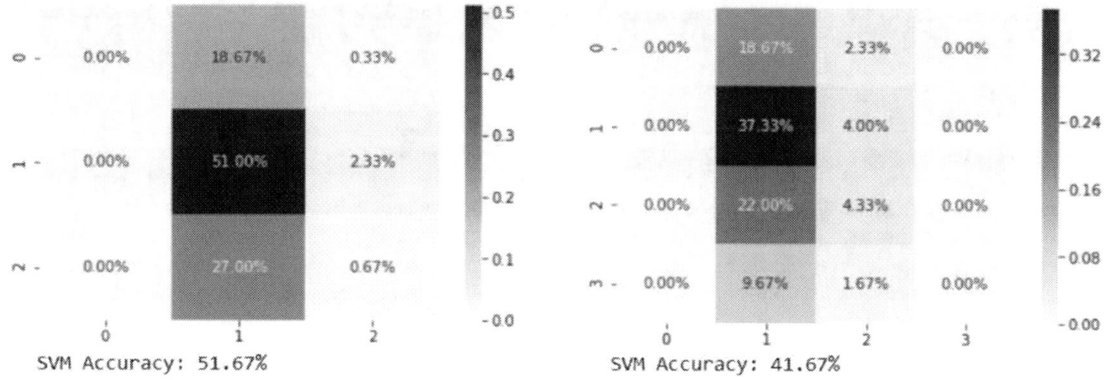

Fig. 13. Dominance -Film1 Female and film 14 Male Supervised learning results for Synthetic Dataset

4.6 K-Means Using Synthetic Data

The same approach above was implemented using synthetic (GAN) data. The results were similar to the original data tests. In each unsupervised test, algorithms have failed to determine the exact numbers of clusters although the optimal number of clusters was 5 according to the Elbow Method. The results of K-Means were inconsistent.

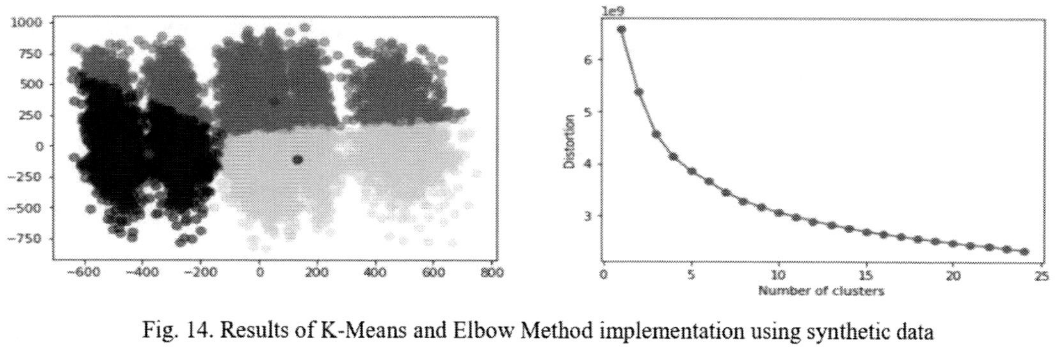

Fig. 14. Results of K-Means and Elbow Method implementation using synthetic data

5. Discussion

In this paper, the structure of the brain, limbic system, neurotransmitters, and emotion theories were explained. Based on two different emotional theories, it is possible to say that emotions, limbic system work together with other parts of the brain such as decision making and memory. Grossberg and Levine's theory proposed that emotions are the neural signals that connect the instinctual sensor with the conscious brain, to make the instinctual needs to be consciously known to the brain, to make the instinct conceptually recognized-understood [12]. The basic emotions are evoked by sensing the basic bodily instinctual needs, while the "complex" emotions, including love, aesthetic emotions are evoked by higher cognitive needs of a human being [13]. In Maslow's Hierarchy of Needs, physiological needs and safety needs might be directly related to basic emotions, while the other needs, such as the need for love, esteem, and self-actualization, are related to feelings such as aesthetic emotions [13].

There are two different generalizations for emotions. Firstly basic emotions are related to our instincts. The second one is emotions by experience that are complex to detect and analyse. Experience learning is related to personal behaviour, this means a subject might have arousal or valence based on her desires, wishes, or experiences. Thus to analyse complex emotions using this database was not possible. The tests were related to the dimensional theory of emotion using basic emotions as valence, arousal, and dominance in the DREAMER dataset. The research was able to classify rankings of emotions although it was not possible to specify which exact emotion is occurring in the brain during the DREAMER data experiments. The neurotransmitter dopamine (DA), is known to mediate unconditioned pleasure and reward from food, sex, and drugs, but recent findings suggest that DA is also involved in the anticipatory, preparatory, approach, and coping phases of reward behaviour [14]. Alterations in one neurotransmitter affect other neurotransmitters. Numerous studies have pointed to an important role for neuromodulators (e.g., DA, 5-HT, and NE) in the emotional process [15]. There are relationships between higher levels of arousal and higher relation of dopamine [16] [17]. Based on this research, we can say that neurotransmitters control our emotions and EEG might detect signal alterations in the human brain. Thus using EEG signals, it was possible to analyse rankings of basic emotions.

The supervised learning classifications for original data were successful with 95-98% accuracy. Thus, it is possible to say that using EEG data rankings of basic emotions can be classified. On the other hand, unsupervised learning tests were inconsistent to create clusters analysing EEG data. GAN has been tested to generate synthetic EEG data to widen the experiments on the database. The highest accuracy was 51% on the female subject using GAN data for supervised learning algorithms. GAN results were not high enough as in the original dataset although it is possible to say that GAN was able to generate synthetic EEG data with around 50% accuracy.

6. Conclusion

In conclusion, emotions in the human brain were analysed. The limbic system not only controls emotions but also works with other parts of the brain. Neurons are the tiniest units in the brain that are synchronously communicating together. Neurotransmitters allow the communication process electrochemically. Because of the relation between emotions, neurotransmitters, and EEG signals, it is possible to detect alterations in neurotransmitter levels and using those signals occurring in the brain to perform Affect Recognition. The multi-output multi-class accuracies were high enough to use the model in real-time equipment. In this research some of the results were acceptable. Using different methods and models, not all models gave high accuracies and there were unstable results for several subjects. Emotions are still unknown in theory and practice and this issue was the biggest limitation of this research because different subjects can give different results for ML systems which will cause unstable models. Not only neurotransmitters but experience, memory, and expectations can affect emotions. It is not yet possible to generalize emotions based on experiments. The goal is to help patients in hospitals by understanding basic emotions. Unfortunately, using such systems might have dangerous outcomes related to privacy. For future work, the Machine Learning model will be tested using Cyber-Physical and control systems. Finally, the EEG signals can be used to develop an Affect Recognition system to analyse or detect basic emotions. Therefore human-computer interface can be used in the healthcare industry to help patients in hospitals or to improve the quality.

7. Acknowledgments

Firstly, I would like to thank my father Dr. Turgay Ünal for his help in EEG studies and would like to thank Associate professor Vyacheslav Potekhin for all his help during my studies. Finally, I thank my mom Dr. Seher Gülçin Ünal for her assistance in giving all the support in my life. This research is being conducted at the Peter the Great St. Petersburg Polytechnic University and partially funded by the Ministry of Science and Higher Education of the Russian Federation as part of World-class Research Center programme: Advanced Digital Technologies (contract No. 075-15-2020-934 dated 17.11.2020).

8. References

[1] Marieb, H. & Hoehn, K. (2009). Brain Anatomy Adapted from Human Anatomy & Physiology 9th edition, Pearson, pp. 335, ISBN-13: 9780321743268

[2] Brisch, R.; Saniotis, A. & Rainer, W. (2014). The role of dopamine in schizophrenia from neurobiological and evolutionary perspective: old fashioned, but still in vogue, Frontiers in Psychiatry, Volume 5, pp. 1-12 doi: 10.3389/fpsyt.2014.00047

[3] John, A. & Putman, A. (2012). Signal Processing Techniques, EEG Info, USA, pp. 3-15, doi:1s10.1061/ (ASCE) CP.1943-5487.0000719

[4] Asli, A. (2015). EEG Elektroensephalogram, Bio-physics YDU Medicine Faculty Biophysics, pp.3-11, doi: 10.5350/BTDMJB201107303

[5] Bright Brain Centre. (2021).zEEG- a Window into the Real Time Neural Processing, Available from: cwww.brightbraincentre.co.uk/electroencephalogram-eeg-brainwaves Accessed: 2021-07-23

[6] Marianne K.; Daniella S.; Ariel R.; David N.; Jackson, C. & Ricardo xG. (2019). Dopamine: Functions, Signaling, and Association with Neurological Diseases, Cellular and Molecular Neurobiology, pp.1-30, doi: 10.1007/s10571-x x018-0632-3

[7] Engin U. (2015). The Impact of Neuroanatomical and Neurochemical Structure of Brain on Personality and Behaviour, International Journal of Alanya Faculty of Business, ResearchGate, pp. 193-219

[8] Potekhin, V.; Muhammed, E. & Ogul, U. (2018). Development of Artificial Neural Network To Determine Hand Activity Using Real Motor Activity From Electroencephalography Signals, Apple Space Engineering, Technologies and Exploration, pp.214-224., ISBN 978-3-00-060580-2

[9] Zeljko, T.; Cosic, I. & Katalinic B. (2019). Manufacturing And The Rise Of Artificial Intelligence, Innovation Challenges Innovation Challenges, Proceedings of the 30th DAAAM International Symposium, pp. 192-x196.Published by DAAAM International, ISBN 978-3-902734-22-8, doi:10.2507/30th.daaam.proceedings.025

[10] Fushun, W.; Yang, J.; Fang, P.; Roger, C. & Jason H. H. (2020). Neurotransmitters and Emotions, Frontiers in Psychology Emotion Science, pp. 1-5, doi: 10.3389/fpsyg.2020.00021

[11] Simeng, G.; Fushun, W.; Nitesh, P.; Patel, J. & Jason H. H. (2019). A Model for Basic Emotions Using Observations of Behavior in Drosophila, Frontiers in Psychology Emotion Science, pp. 1-10, doi: 10.3389/fpsyg.2019.00781

[12] Yi, L.; Zhao, J. & Wenbin, G. (2018). Emotional Roles of Mono-Aminergic Neurotransmitters in Major Depressive Disorder and Anxiety Disorders, Frontiers in Psychology Emotion Science, pp.x1-9, doi: 10.3389/fpsyg.2018.02201

[13] Zhang, K.; Guo, J.; Peng, Y.; Xi, W. & Guo A. (2007). Dopamine-mushroom body circuit regulates saliency-based decision-making in Drosophila. Science 316, pp. 1901–1904, doi: 10.1126/science.1137357

[14] Schultz, W.; Dayan, P. & Montague, P. R. (1997). A neural substrate of prediction. Science 275,pp. 1593–1599, doi: 10.1126/science.275.5306.1593

[15] Pereira, T. & Murthy, M. (2017). To fight or not to fight. Neuron 95, pp.986–988, doi: 10.1016/j.neuron.2017.08.029

[16] David, C.; Vandana P.; Deborah, S. & Don, B. (1987) The relationship of affiliative arousal to dopamine release, Springer-Motivation and Emotion volume 11, pp. 51-66, doi: 10.1007/BF00992213

[17] Jan, W. (2018). Effects of positive emotion, extraversion, and dopamine on cognitive stability-flexibility and frontal EEG asymmetry, Pubmed Psychophysiology, pp. 1-10, doi: 10.1111/psyp.12727

[18] Solomon D. (2020). nervous system. Encyclopedia Britannica, Available from: www.britannica.com/science/nervous-system Accessed: 2021-08-13

[19] Gabert-Quillen, C.; Bartolini, E. & Abravanel, B. (2015). Ratings for emotion film clips, Behavior Research Methods, vol. 47, no. 3, pp. 773–787, doi: 10.3758/s13428-014-0500-0

[20] Katsigiannis, S. & Ramzan, N. (2017). DREAMER: A Database for Emotion Recognition Through EEG and ECG Signals from Wireless Low-cost Off-the-Shelf Devices, IEEE Journal of Biomedical and Health Informatics, pp.1-10, doi: 10.1109/JBHI.2017.2688239

[21] Krizhevsky, A.; Ilya, S. & Geoffrey H. (2012). ImageNet Classification with Deep Convolutional Neural Networks, Advances in neural information processing systems, pp. 1-9., ISSN 1049-5258, doi: 10.1145/3065386

[22] Nielsen, M. (2019). Neural Networks and Deep Learning, Determination Press, Unpublished

[23] Blockeel, H. & Raedt, L. (1997). Top-down induction of first-order logical decision, Information Technologies, pp. 285-297, doi: 10.1007/978-3-662-04599-2_6

[24] Kuppens, P.; Tuerlinck F.; Russell, J. & Feldman B. (2012). The Relation Between Valence and Arousal in Subjective Experience, Psychological Bulletin 139, pp.1-10, doi: 10.1037/a0030811

[25] Sheri, L.; Liane, J.; & Luma, M. (2012). The Dominance Behavioral System and Psychopathology: Evidence from Self-Report, Observational, and Biological Studies, NCBI, pp. 1-12, doi: 10.1037/a0027503

[26] Ian, J.; Jean, P; Mehdi, M.; Bing, X.; Warde, D.; Ozairy, S.; Courville, A. & Bengio, Y. (2014). Generative Adversarial Nets, Montreal University, pp. 1-9

[27] Lei, X. & Kalyan, V.(2018). Synthesizing Tabular Data using Generative Adversarial Networks, MIT, pp. 1-12

[28] Katalinić, B.; Kostenko, D.; Onufriev, V. & Potekhin, V. (2020). Cyber-Physical Systems in Complex Technologies and Process Control, pp. 40-54., Lecture Notes in Networks and Systems, doi: 10.1007/978-3-030-34983-7-5

[29] Ciupan, E. (2013). A Study Regarding the Possibility of Optimizing the Supply Batch using Artificial Neural Networks, 24th DAAAM International Symposium on Intelligent Manufacturing and Automation, pp. 141-149, doi: 10.1016/j.proeng.2014.02.214

DOI: 10.2507/32nd.daaam.proceedings.026

ENGLISH AS MEDIUM OF INSTRUCTION: LANGUAGE AND CONTENT COMBINATION IN DIGITIZED HIGHER ENGINEERING EDUCATION

Oleg Ipatov, Sofia Saramaha, Darina Barinova,
Anna Rubtsova & Maria Odinokaya

Peter the Great St.Petersburg Polytechnic University

This Publication has to be referred as: Ipatov, O[leg]; Saramaha, S[ofia]; Barinova, D[arina]; Rubtsova, A[nna] & Odinokaya, M[aria] (2021). English as Medium of Instruction: Language and Content Combination in Digitized Higher Engineering Education, Proceedings of the 32nd DAAAM International Symposium, pp.0168-0176, B. Katalinic (Ed.), Published by DAAAM International, ISBN 978-3-902734-33-4, ISSN 1726-9679, Vienna, Austria
DOI: 10.2507/32nd.daaam.proceedings.026

Abstract

The study was intended to explore the use of digital media and demand for EMI (English as Medium of Instruction) contents in higher education engineering sector. EMI is being increasingly implemented via top-down policies, sometimes with little attention to the educational implications that learning through a second language can have for the millions of students affected. Thus, this research aims to fill this void in its exploration of the effects of EMI on language learning, content learning, teaching delivery, quality of education, inequalities of access, language flexibility and hybridity, the competencies needed to be a successful EMI teacher, and other multi-faceted aspects of EMI. The research provides the results of the study connected with the investigation of implementing EMI within the Master of Engineering program in Peter the Great Saint-Petersburg Polytechnic University. The results obtained are compared with the ones presented by the EMI Oxford research group having conducted a profound study on the analysis of expanding implementation of EMI on a world scale. The article serves the interest for those who research digital trends in education as well as those who are interested in introducing EMI into the educational process.

Keywords: English as Medium of Instruction (EMI); Master program; Engineering; Foreign language; Digitized contents; ICT (information and communication technology) university curriculum.

1. Introduction

Currently, in all countries of the world, including Russia, there is an increasing trend of using English as a means of teaching academic subjects. This approach to the use of English is called EMI (English as Medium of Instruction) – "the use of English for teaching academic subjects in countries or jurisdictions where English is not the first language (L1) for the majority of the population".

Despite the urgency of this problem, it is necessary to state the almost complete lack of serious research on this topic in Russia. In 2014, representatives of Oxford EMI, with the assistance of the British Council, conducted the first large-scale study aimed at confirming the growing popularity of the use of English for teaching academic subjects in non-English-speaking countries of the world, as well as identifying the main problems that arise with this approach to teaching [11]. The results of the study were presented in the report by Julie Dearden [19]. The problem of teaching academic subjects in English attracts the attention of researchers, developers of relevant programs and directly teachers who implement this approach.

The purpose of this article is to present the results of a study aimed at identifying a number of features of the EMI approach implementation in training Master of Engineering students and to compare these results with global trends in EMI training. To solve the problems of the study, the literature review and the analysis of the program documents related to the experience of implementing the EMI approach in different countries and identifying problems associated with this process was carried out.

The problem of using English as a means of teaching academic subjects (EMI) has been widely covered in scientific publications of researchers around the world [1], [2], [4], [6], [7], [8], [10], [12], [23], [24]. At the University of Oxford, a special center for research and development has been established to study the problem of EMI (Oxford EMI: Center for Research and Development on English Medium Instruction) [11]. It was the representatives of this center, together with members of the British Council, who conducted a world-wide study from October 2013 to March 2014 (representatives of 55 countries took part in the study), which aimed to get a detailed picture of the scale of the use of English as a teaching tool, as well as to identify the features of this process in different countries. The findings were presented in the report "English as a medium of instruction: a growing global phenomenon" by Julie Dearden [19]. It is worth noting that this report does not contain data on Russia. The report includes a description of the methodology of the study (two stages: primary and main); the main conclusions concerning the confirmation of the growing popularity of the implementation of this approach in the educational process; the review of the policies of individual states in connection with the adoption or rejection of this approach; the prospects for the implementation of EMI in different countries; the public point of view on this process; problems arising in the process of teaching and learning within EMI; the problem of internationalization of higher education; as well as a call for the academic community to continue research in the field of EMI implementation. The report itself is a 34-page text with graphic information, as well as tables; in addition, photocopies of the EMI seminars are attached to the text of the report. From our point of view, due to the scale of the project and the absolute significance of both quantitative and qualitative information, this particular document is of the greatest interest to researchers of the EMI implementation problem. The opportunities offered by the digital education are relevant to the goals of reforming modern education systems. Thanks to these technologies new forms of education have emerged - distant and blended, the unified information and education space of universities are being created.

This paper presents only some of the conclusions of the report that are directly relevant to our research. In particular, the report notes that some survey participants believe that, from a social point of view, the implementation of EMI is controversial, since teaching in English may limit access to education for students from low-income socio-economic groups and/or because of concerns that this may negatively affect the use of their native language and national identification. It is also noted that the educational infrastructure in many countries does not guarantee high-quality education within the framework of EMI: there is a shortage of teachers with good linguistic training; there is no established level of English proficiency required for training under EMI; there is an insufficient number of organizational and pedagogical guidelines of a recommendatory nature that contribute to effective teaching based on EMI. Educational programs of teachers' professional development practically do not address the problems of training within the framework of EMI. One of the issues raised in the report is the issue of knowledge control. What is being evaluated: the English language or the content of the subject? (What is being assessed: English or the subject content?). Should an EMI teacher take on the responsibility of teaching not only the subject, but also the English language? The report notes that the teachers who took part in the preliminary survey were all opposed to being required to teach in English. In the future, many noted that in the implementation of knowledge control, they evaluated only the content side and ignored the way of presenting this material, that is, the language design [19], [20]. Another problem highlighted in the report is the establishment of a certain level of language proficiency for both students and teachers in the digitized education. Digitized education has peculiar characteristics of learning tools that should be considered as their natural qualities that can be used with didactic purposes. General didactic properties of such form of education are based on their two key functions - informative and communicative - and are used in any educational system. These properties are manifested at all stages of the learning process - from the presentation of educational information to its consolidation and control.

The main didactic property of digitized education is multimedia. Multimedia gives mobile and computer devices some signs of animation: it is possible to show videos, full-color images, speak and understand individual phrases, listen to music, play animation, etc. All this, in its turn, contributes to the increase of learning motivation, speed of learning and development of skills. Multimedia allows the formation of skills that are impossible to form with the help of other technologies, for example, pronunciation in the study of a foreign language. Research by psychologists in this field shows that the enrichment of sensory experience formed in the process of perception of reality is a necessary condition for the mental development of a person.

The initial level of human mental and intellectual development is the sensory-perceptual level. Modern multimedia can provide the creation of a special system for the development of sensory-perceptual abilities of a person, make it possible to demonstrate those processes and phenomena that a person cannot directly perceive due to the limitations of the senses. In multimedia learning systems, along with training and error analysis, there is mainly the following type of learning: visually presented communicative situation - inclusion in it - achievement of some result. The use of such technologies in the process of teaching foreign languages leads to some changes in the structure of learning activities. Spatial parameters of subject-subject interaction, communicative components of activity, its needs-motivational regulation, goal-setting, etc. are transformed. In the process of applying multimedia programs and resources, a number of important methodological problems are solved which sometimes cannot be solved in traditional (without the use of digital technology) instruction:

- adaptation to the authentic language environment;
- formation of a vivid visual image of the country and society of the language being studied;
- simulation of the linguistic and structural environment.

Thus, the use of multimedia programs and resources in foreign language teaching has a number of undoubted advantages, which consist in the multidimensional presentation of educational information and create a more natural atmosphere for language learning, as students can simultaneously see, hear and speak, in the ability to combine all four types of speech activity in one task. The report states that respondents from different countries mention a different desired level for students: from "good enough" to B2 (CEF) level. With regard to teachers, it is noted that not all highly qualified teachers of special subjects speak English at a sufficient level to conduct classes within the framework of EMI, while younger teachers who have a better command of the language do not have sufficient professional knowledge and experience in teaching specific professional subjects. On the other hand, when implementing the EMI approach, priority may be given to foreigners, native speakers, as a result of which local teachers, highly qualified specialists, may lose their jobs. Another problematic issue is who will contribute to helping students master English if teachers in the EMI approach do not consider it their responsibility. At the same time, EMI teachers generally believe that students generally improve their English proficiency in their classes, at least in reading and understanding professional texts in English, and it is the students themselves who should be responsible for the progress in mastering English.

Among the approaches to solving the problems mentioned above, it is possible to distinguish the suggestion of der Walt and Kidd to use both English and local languages in some educational situations, for example, to create instructions for tests [23]. We must admit that the length of this article does not allow us to cover in detail all the problems discussed in the Oxford EMI report, as well as other researchers of EMI implementation; At the same time, it is worth noting that all of the provisions related to EMI listed in the report are also considered by other researchers to some extent, and the Julie Dearden report is the most comprehensive, revealing the essence of the EMI phenomenon, which is why this part of the article was based mainly on the provisions of this report.

Many universities, both foreign and domestic, for example, Peter the Great St. Petersburg Polytechnic University (SPbPU), with the goal of integrating into the world space and attracting foreign students, include courses in their training programs that involve teaching academic subjects in English. For example, according to data from foreign researchers, there were more than 64,000 EMI programs in Europe in 2013 [13], [17], [18], [20], [23]. SPbPU currently has more than 200 courses taught in English [14]. The multidimensional nature and significance of the findings contained in this report prompted us to conduct our own research and analyze the features of implementing the EMI approach [3], [5], [7], [9], [15], [22]. The university offers a wide range of Master Degree Programs in English in various areas of expertise.

To prove the title of a true 'cosmopolitan university' SPbPU has launched a wide range of international educational programs and cutting-edge projects, expanded international and English-speaking faculty, increased enormously a number of out-bound and inbound talented students [14]. It allows students to acquire international experience in a pertinent area of expertise, broaden scientific horizons and streamline their skills [15], [16], [21], [22]. The university offers a wide range of Master of Engineering Degree Programs in English in the following areas:

1. Civil Engineering;
2. Energy Efficient and Sustainable Building;
3. Environmental Engineering in Urban Construction;
4. Emergency Preparedness and Response;
5. Energy Technology;
6. Power Plant Engineering;
7. Electrical Engineering;
8. Nuclear Power Engineering;
9. New Materials and Additive Technologies;
10. Laser and Fiber Optic Systems;
11. Microelectronics of Telecommunication Systems;
12. Continuum Mechanics: Fundamentals and Applications;
13. Mechanics and Mathematical modeling;
14. Intelligent Systems.

All the students (both foreign and native speakers) gain practical skills in up-to-date and state-of-the-art labs at the university and perform academic and research activity using a plenty of helpful resources. With the aim to deliver pertinent knowledge and develop practical skills in the most efficient way the following teaching methods are applied:

- implementation of problem-based teaching methods;
- organization of research projects;
- development of students' motivation;
- simulation of learning activity (i.e., creation of a virtual linguistic space);
- organization of self-evaluation and group evaluation.

Game applications and computer games, programs for 3D communication with the help of avatars - Second Life, Machinima - are used both in distance and mixed forms of foreign language teaching. Quite often elements of gamification are used in open educational resources, in tests. For example, the web-based test creation platform Quizlet https://quizlet.com has recently launched two templates to create interactive tests of Gravity and Scatter format, based on maximum visualization of educational material. Software developers offer very easy-to-use game builders or game engines that language teachers can use to design their own game scenarios and game assignment formats.

Scientific research activities and publishing. These activities imply working in re-search labs as well as writing scientific articles under a professor's supervision. Students might get their research and scientific articles published on global citation web-sites. Students gain practical skills in the highly equipped and modern laboratories of the university and carry out research activities using many resources of SPbPU. Studying in English in the Master program is a step that guarantees future master students significant achievements in the professional field and the beginning of a successful career.

2. Methods

Theoretical foundations for the study are the methodological principles of teaching a foreign language in the format of mixed and online learning (by means of ICT), principles of professionally-oriented teaching of a foreign language, as well as the basic principles of the communicative approach to teaching a foreign approach to foreign language teaching. The material of the study served as theoretical works devoted to creation and evaluation of courses in the format of blended and online learning. As the main methods of research were questionnaire survey, analysis of scientific literature.

3. Experiment

With the advent of digitized education, which provides vast amounts of knowledge and information at any time and place, the need for special competencies in education and learning has arisen. Today, no one questions the fact that it is not just possible, but necessary to introduce digital technologies into the pedagogical process but, unfortunately, often the use of innovative methods based on such technologies do not facilitate and optimize learning activities, but rather overloads and complicates them, leading to frustrations and failures.

In collaboration with the Graduate School of Engineering Pedagogy, Psychology and Applied Linguistics, the main research method was chosen to conduct interviews with the creators of the course "History and Methodology of Science" (taught in English), Civil Engineering Institute for the relevant areas of training. The course was developed with the use of e-learning technologies and placed in the electronic information and educational environment on the basis of SPbPU MOOK platform. Foreign Languages Department This course is mandatory for all the first year Master students. The course examines the historical and methodological aspects of Engineering as a field of science. Its task is to demonstrate that despite the historical variability, there are certain criteria of scientific character, in particular, the requirement to implement a certain scientific methodology, which, with all the rigor and objectivity of scientific knowledge, cannot do without the subjectivity of the researcher. The course is taught in the frames of integration of digital technology in the educational process. An anonymous survey of students who completed this course was also conducted, and a number of personal interviews with students were organized.

The study was conducted at Peter the Great St. Petersburg Polytechnic University in the frames of Master of Engineering programs (Civil Engineering Institute). The students under review were only non-native English speakers. What distinguished this course from similar courses taught at SPbPU was the fact that this course was taught to the 1st-year postgraduate students without their preliminary taking the course of English for Specific Purposes (ESP). The absence of the ESP course in the Master program curriculum was explained by the peculiarities of the English language training program in 2017-2019, which involved teaching English only within the framework of the Foreign language in professional communication course. Thus, the course "History and methodology of science" was actually the introduction to the majoring field of studies, equal to the similar course in Russian.

4. Results

The survey was attended by 5 course developers, 20 lecturers (10 institutes) who give lectures and practical classes, and 119 students having attended this course. The response score range is from 0 to 10, where 0 is the lowest score. The most important information provided by the course developers when analyzing their experience of teaching the previously mentioned course is presented in Table 1, Figure 1:

Main difficulties	Evaluated aspects	Result in %
Language proficiency level	Lack of general knowledge of English	30
	Lack of professional vocabulary	47
Self-motivation evaluation	High level	23
	Average level	44
	Low level	33
Students' analytical skills	High level	16
	Average level	22
	Low level	62
Students' self-responsibility	High level	45
	Average level	31
	Low level	24
Knowledge control	Language proficiency level	10
	Course content	90
Academic writing skills	High level	29
	Average level	32
	Low level	49

Table 1. Results of Course developers' opinion poll

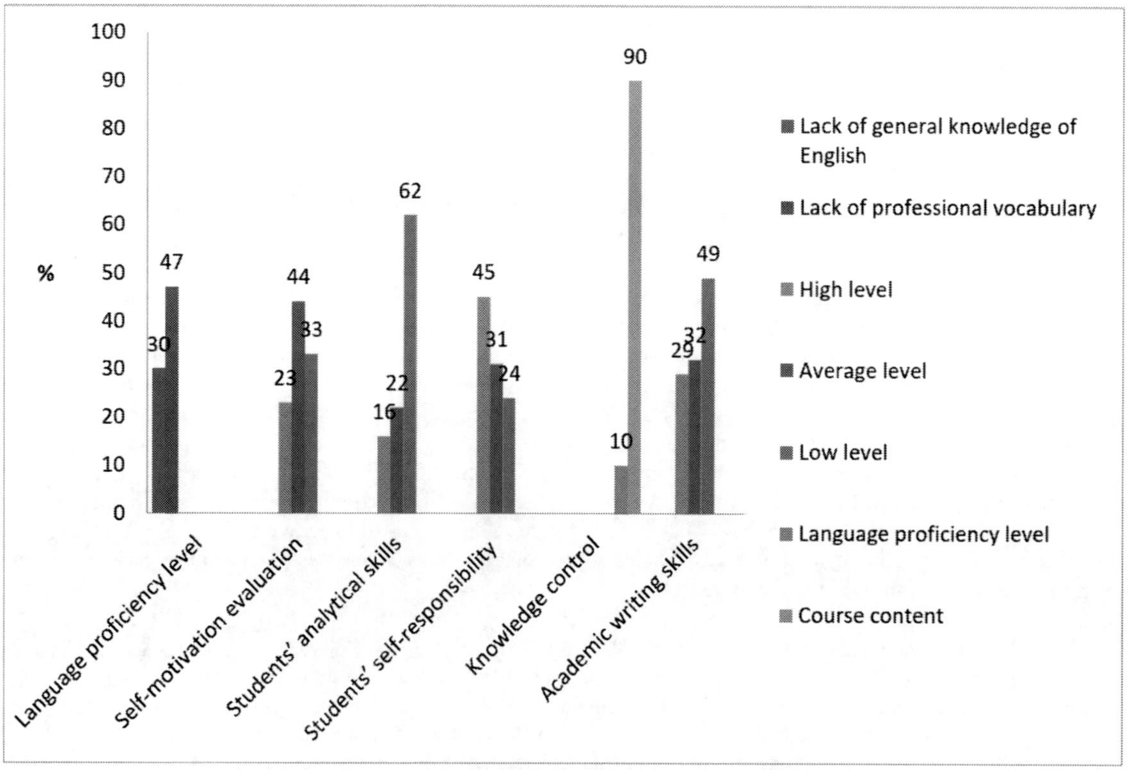

Fig. 1. The analysis of the experience of teaching the course

While teaching, the lecturers had to adjust the course to the level of the English language proficiency of the students, since the students had a rather low level of proficiency in general vocabulary, which made it difficult to master the main course material, including the introduction of professional vocabulary. One third of students do not have the knowledge of academic vocabulary.

Mastering the course was complicated by the lack of serious academic work skills, as well as general skills related to the analysis and the ability to correctly formulate the thoughts in a foreign language. When performing control tasks, some of the students' works were almost plagiarism, which was explained both by the inability to work independently with volumes of information and the lack of academic skills in general, and by the difficulty of expressing their ideas in English. Both the irresponsibility of students and the desire to facilitate work by cheating were noted, which, according to the authors of the article, sharply distinguished Russian students from foreign ones.

As the lecturers noted, the control of knowledge was carried out without taking into account the level of the English language proficiency, the emphasis was on checking the content of the material offered for assimilation (the lecturers do not consider themselves obliged, in addition to teaching special students, to teach English as well).

Despite the complexity of the course, as noted by the students, some students did an excellent job with the course tasks. The success was due not so much to the level of the English language proficiency required for the course, but to the responsibility when working in the classroom and completing assignments.

During the interview, the course developers first stressed that the main thing for students to master the course in English was to have a sufficiently high level of General English proficiency, and it was possible to learn professional vocabulary during the course. However, in the future, the lecturers expressed a wish related to the fact that before passing the course, it was desirable for students to master the professional academic vocabulary. There were also suggestions for strengthening the requirements for the initial level of the students' English language proficiency when entering programs taught in a foreign language, up to the introduction of the internal exam in English. At the moment, the English language exam is being included in the tests required for the admission to these areas of training.

After completing the course, students were given an anonymous questionnaire. Of the 124 students who formally took this course, 119 students took part in the survey. For most students, this course was the first course taught in English, with knowledge of the subject being assessed in English. The results of the student's survey are presented in Table 2, Figure 2:

Main difficulties	Evaluated aspects	Result in %
Language proficiency level	Lack of general knowledge of English	25
	Lack of professional vocabulary	45
Self-motivation evaluation	High level	10
	Average level	70
	Low level	20
Students' analytical skills	High level	25
	Average level	34
	Low level	41
Students' self-responsibility	High level	3
	Average level	6
	Low level	91
Knowledge control	Language proficiency level	10
	Course content	90
Academic writing skills	High level	29
	Average level	32
	Low level	49

Table 2. Results of Master students' opinion poll

Approximately two-thirds of the students who answered the questionnaire noted that they had difficulties with both general and professional English vocabulary. The majority also noted the difficulties associated with preparing an abstract in English (mainly, ignorance of the structural features of writing such academic works in English). Assessing their motivation in mastering the course, most noted an average level. Among the explanations for the low motivation, the students noted the previously mentioned difficulties, as well as the lack of awareness of the importance of studying majors in English, as well as problems associated with "too voluminous materials offered for assimilation" and "the monotony of the classes' design" (from the comments of students). When asked whether the course improved the level of the foreign language proficiency, the overwhelming majority (85%) answered positively.

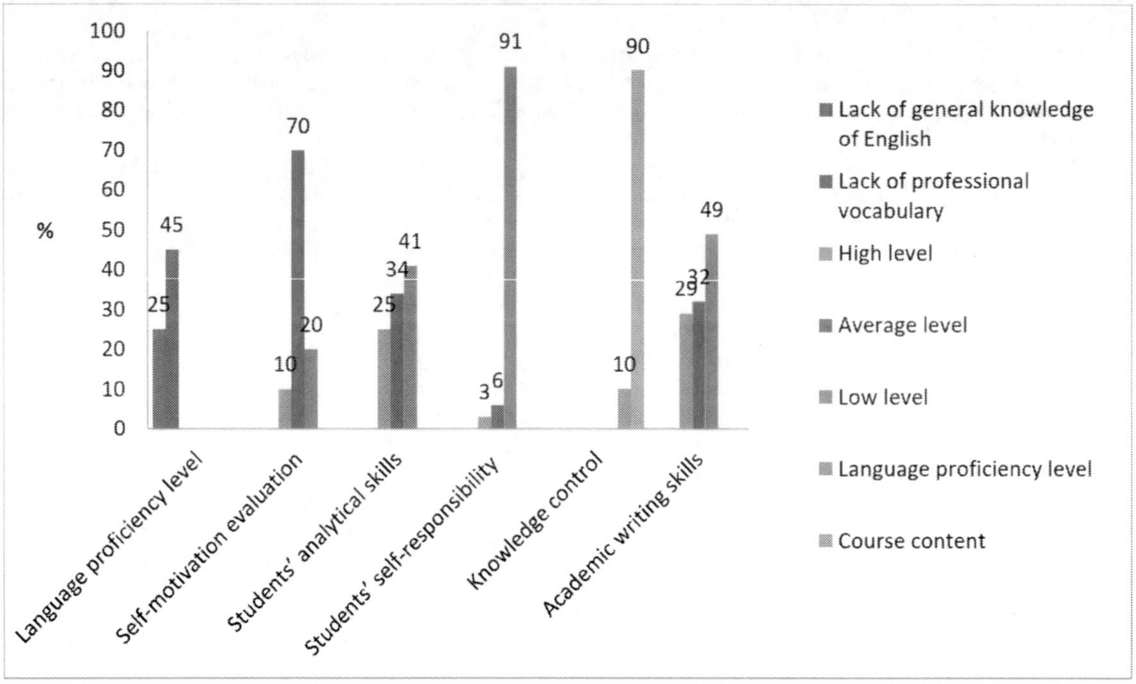

Fig. 2. The analysis of the results of the student's survey

The majority of respondents (96%) supported the possibility of parallel teaching of their majors in English by the lecturers in majors and conducting ESP classes aimed at familiarizing with the specific vocabulary and features of the professional communication in English in oral and written forms. 25% noted the opportunity to do without an introductory course in their majoring field of study and to give lectures and conduct seminars in English at once. In general, all students noted the need to include professionally-oriented courses in English in the university curriculum. At the same time, 80% considered that these courses should be included in the mandatory program, and 20% – as an elective.

As a result of the personal interviews with students, the interviewers identified a dual attitude of students to passing a professionally-oriented course in English in their first year. On the one hand, there was a clear interest of students in this course and awareness of the seriousness of the approach to teaching the majors, on the other hand there were doubts about the validity of the introduction of this course when students have not mastered the basics of their major.

The analysis of the results obtained in the study on the implementation of the EMI approach in the frames of Master of Engineering programs allowed stating that the main problems identified in the study are similar to the problems raised in the Oxford EMI report.

The main common problems are the lack of requirements for the level of the English language proficiency on the part of students and the conduct of knowledge control, in which the knowledge of professional realities was subjected to control, while the language design was absolutely not evaluated, without which it is impossible to talk about the formation of the foreign-language professional competence.

The positive aspects of the introduction of the course "History and methodology of science" are the high professional level of the lecturers chosen for the development and implementation of this course. The fact of the introduction of this course within the first year of the Master program without having mastered the basics of the academic and professional vocabulary in English can be considered a drawback. However, it should be noted that this drawback was taken into account and subsequently a solution was found to overcome it.

The only significant drawback noted by the authors is the fact that the high-quality education provided by SPbPU to those who do not speak English is not available. Even if the L2 of the students before admission was a foreign language other than English, it is English that becomes a mandatory language for mastering. A lot of majors are taught in English and are not duplicated in Russian. A low level of the English language proficiency or the lack of knowledge of it is a serious obstacle for the students to enter the University and to be further trained. These facts indicate the existence of social problems associated with social inequality, and may cause discontent among some segments of the population.

5. Conclusion

In conclusion, it should be noted that the conducted research may be of interest to those who are engaged in the study of new trends in education, as well as those who directly consider the problem of implementing EMI into the educational process both in higher educational institutions, and in secondary educational schools and vocational educational institutions.

The advantage of the courses conducted in the English language is a large selection of materials, as the number of professional academic texts, textbooks and anthologies in English is much greater than in similar publications in Russian. The main obstacles to achieving the objectives of the course lie far beyond the language itself. To optimize the use of innovation technologies methodologically sound and expedient application of them should take into account and be based on the basic conditions for their successful integration into the educational process, namely:

1) provision of technical equipment of a university or school and computer security of the learning process;
2) formation of information and learning environment of higher education institution or school, which is a unified system of software, databases, electronic teaching and methodological resources and other elements that implement information and educational processes;
3) sustainable motivation of a teacher and a student;
4) the level of development of information and communication competence of teachers and students.

In general, the problem of implementing EMI-based e-learning in Russia is quite undeveloped and therefore presents great prospects for further research.

6. References

[1] Alenkina, T. (2020). English medium instruction in the Russian university. Vysshee Obrazovanie v Rossii, Vol. 29, No. 1, pp. 98–103., DOI:10.31992/0869-3617-2020-29-1-98-103

[2] Almazova, N., Barinova, D., Ipatov, O. (2018). Forming of information culture with tools of electronic didactic materials. In: Annals of DAAAM and Proceedings of the International DAAAM Symposium, Vol. 29, No.1, pp. 0587–059., DOI:10.2507/29th.daaam.proceedings.085

[3] Almazova, N., Krylova, E., Rubtsova, A., Odinokaya, M. (2020). Challenges and opportunities for Russian higher education amid covid-19: Teachers' perspective. Education Sciences, Vol. 10, No.12, pp. 1–11

[4] Anderson, K. D. (2004): The Nature of Teacher Leadership in Schools as Reciprocal Influences Between Teacher Leaders and Principals. School Effectiveness and School Improvement, No. 15, pp. 97–113., DOI:10.1076/sesi.15.1.97.27489

[5] Bekirogullari, Z., Bylieva, D., Lobatyuk, V., Nam, T. (2021). How Virtual Personal Assistants Influence Children's Communication. Lecture Notes in Networks and Systems, No. 184, pp. 112–124., DOI:10.1007/978-3-030-65857-1_12

[6] Botha, W. (2013): English-medium instruction at a university in Macau: Policy and realities. World Englishes, No. 32, pp. 461–475., DOI:10.1111/weng.12053

[7] Bylieva, D., Lobatyuk, V., Fedyukovsky, A. (2019). Ways of sociotechnical integration of scientists and volunteers in citizen science. IOP Conference Series Materials Science and Engineering, DOI:10.1088/1757-899X/940/1/012150

[8] Chappel, J. (2015). Teaching in English Is Not Necessarily the Teaching of English. International Education Studies, Vo. 8, No. 3, DOI:10.5539/ies.v8n3p1

[9] Chernysheva, M.A., Kozlova, A.O., Donova, E.V. (2019). Obuchenie akademicheskogo pis'ma na angliiskom yazyke russkoyazychnykh nauchno-pedagogicheskikh rabotnikov (iz opyta raboty). Vysshee obrazovanie v Rossii, Vol. 28, No .10, pp. 50-57., DOI:10.31992/0869-3617-2019-28-10-50-57

[10] Doiz, A., Lasagabaster D., Sierra, J.M. (Eds.). (2013). English-medium instruction at universities: Global challenges. Clevedon: Multilingual Matter, DOI:10.1093/elt/cct045

[11] http://www.education.ox.ac.uk/crdemi-oxford, (2021). EMI Oxford: Centre for Research and Development on English Medium Instruction, Accessed on: 2021-03-21.

[12] Hasan, Al-W. (2016). Role of Using English as medium of Instruction in Reshaping Bahraini Senior Teachers' Perceptions of Their Roles as Middle Leadership in their Schools. Journal of Language and Education No. 1, pp. 6-15., DOI:10.17323/2411-7390-2016-2-1-6-15

[13] Hollenbeck, A., Kalchman, M. (2012). Professional development for conceptual change: Extending the paradigm to teaching reading comprehension in US schools. Professional Development in Education, Vol. 39, No. 5, pp. 638-655., DOI:10.1080/19415257.2012.728535

[14] https://www.spbstu.ru/international-cooperation/international-educational-programs/international-graduate-programs/graduate-programs-in-english, (2021). International Master's Degree Programs, Accessed on: 2021-03-21

[15] Karpovich, I., Odinokaya, M. (2020). Interactive technology of pedagogical assistance as a means of adaptation of foreign first-year students. IOP Conference Series Materials Science and Engineering, DOI:940. 10.1088/1757-899X/940/1/012130

[16] Klochkova E., Volgina A., Dementyev S., Klochkov Y. (2016). Human FactorinQualityFunctionDeployment.SecondInternationalSymposiumon Stochastic Models in Reliability Engineering, Life Science and Operations Management (SMRLO), IEEE, 466-468, DOI:10.1109/SMRLO.2016.81

[17] Lukacs, K.S., Galluzzo, G.R. (2014). Beyond empty vessels and bridges: Toward defining teachers as the agents of school change. Teacher Development: An international journal of teachers' professional development, Vol. 18, No. 1, pp. 100-106., DOI:10.1080/13664530.2013.856338

[18] Lo, Y.Y. (2017). Development of the beliefs and language awareness of content subject teachers in CLIL: Does professional development help? International Journal of Bilingual Education and Bilingualism. Advance online publication, DOI:10.1080/13670050.2017.1318821.

[19] Macaro, E., Curle, S., Pun, J., An, J., Dearden, J. (2018). A systematic review of English medium instruction in higher education. Language Teaching, Vol. 51, No. 1, pp. 36–76., https:// https://dx.doi.org/10.1017/S0261444817000350

[20] O'Dowd, R. (2018). The training and accreditation of teachers for English medium instruction: an overview of practice in European universities. International Journal of Bilingual Education and Bilingualism Vol.21, No. 5, pp. 553–563., Doi:10.1080/13670050.2018.1491945

[21] Rubtsova, A., Odinokaya, M., Krylova, E., Smolskaia, N. (2021). Problems of Mastering and Using Digital Learning Technology in the Context of a Pandemic. Lecture Notes in Networks and Systems, No. 184, pp. 324–337.

[22] Shipunova, O., Berezovskaya, I., Smolskaia, N. (2019). The Role of Student's Self-actualization in Adapting to the e-Learning Environment. TEEM'19: Proceedings of the Seventh International Conference on Technological Ecosystems for Enhancing Multiculturality, DOI:745-750. 10.1145/3362789.3362884.

[23] Wiseman, A., Odell, A. (2021). Should non-English-speaking countries teach in English? Available from: https://www.britishcouncil.org/voices-magazine/should-nonenglish-speaking-countries-teach-in-english, Accessed: 2021-03-21.

DOI: 10.2507/32nd.daaam.proceedings.027

HYBRID- AND NANOCOMPOSITES BASED ON THERMOELASTOPLASTICS FILLED IN BASALT SCALES AND NANO METALS AND ITS OXIDES

Vera Myasoedova, Elena Zakharova & Ilja Vasiljev

This Publication has to be referred as: Myasoedova, V[era]; Zakharova, E[lena] & Vasiljev, I[lja] (2021). Hybrid- and Nanocomposites Based on Thermoelastoplastics Filled in Basalt Scales and nano Metals and Its Oxides, Proceedings of the 32nd DAAAM International Symposium, pp.0177-0184, B. Katalinic (Ed.), Published by DAAAM International, ISBN 978-3-902734-33-4, ISSN 1726-9679, Vienna, Austria
DOI: 10.2507/32nd.daaam.proceedings.027

Abstract

This article aims to develop innovations in the area of obtaining the hybrid- and polymer nanocomposites and investigation of its structural, thermodynamics' and physico-mechanical properties. As promising and extraordinary materials, nano metals, its oxides and hydroxides have been emerging in many areas of technology. The excellent properties of nanofiller makes this polymer nanocomposite material an ideal for manufacture of the new goods obtained by extrusion and improve some properties of polymer matrices. The success of the process depends on the obtainment of a homogeneous dispersion of nanoparticles without exfoliation. Investigation focuses on the melting process by extrusion and properties of innovation hybrid- and nanocomposite and material. Results open the possibility to evaluate the influence of different filler and polymer matrix chemical nature, sizes and content on increasing the thermostability and elasticity of the new hybrid- and nanocomposites. It is aiming the moulding mentioned above hybrid- and nanocomposites in form of plates, films, irrigation hoses for industrial application.

Keywords: Nano metals; polymer; nanocomposites; elastic properties.

1. Introduction

The use of hybrid composites and nanocomposites based on filled in micron- and nanosized particles of synthetic and natural thermoplastic and thermoelastic polymer materials as an alternative to replace traditional materials is of great of industrial, economic and environmental interest [1], [2],[3],[4]. A hybrid- and nanocomposite material can be defined as a multiphase composite material, containing at least two components, with an interface or interphase separating them. Choice of components should apply predetermined chemical, physical, mechanical (such as flexibility and elasticity or stiffness, or thermal stability of hybrid- and nanocomposites. The components are divided into two main phases: the matrix phase, which is a continuous phase and corresponds to the component with the highest content in the composition (> 50 wt%), and the dispersed phase, also called reinforcement or filler that is a discontinuous phase of the component in less quantity.

The use of hybrid composites and nanocomposites based on filled in micron- and nanosized particles of synthetic and natural thermoplastic and thermoelastic polymer materials as an alternative to replace traditional materials is of great industrial, economic and environmental interest. A hybrid- and nanocomposite material can be defined as a multiphase composite material, containing at least two components, with an interface or interphase separating them. The methods for nano metals synthesis described by [5], [6]. Historically, the first reference of the study of nanocomposites was in the 1950s, and studies of polyamide nanocomposites appeared in the 1970s. However, it was only at the end of the 1980s that there was a deepening of the subject with research carried out by Toyota's Central Research and Development Laboratories for the development of nanocomposites composed of polymers and nanoceramics, causing research on nanocomposites to evolve for an industrial scale [7].

The nanocomposites are classified according to the filler's nanometric dimensions (0 to 100 nm) [8]. Nanometric fillers (nanofillers) have specific characteristics and induce different behaviours in the matrix when compared to micro and macro fillers, so, these materials can be considered as promising alternatives for use as materials that overcome the limitations of regular fillers. Some mechanical and physical properties which cannot be achieved in polymer composites made with regular fillers can be improved with the use of nanofillers. Nanofillers have a high surface/volume ratio, which allows the addition of smaller amounts (1-5% v/v), when compared to micrometric fillers (15-40% v/v), obtaining similar results of mechanical behaviour. Also, the surface/volume ratio of the nanofillers can induce better interaction between filler and matrix, which can be improved by the functionalization of the nanofiller Choice of components should apply predetermined chemical, physical, mechanical (such as flexibility and elasticity or stiffness, or thermal stability of hybrid- and nanocomposites. The components are divided into two main phases: the matrix phase, which is a continuous phase and corresponds to the component with the highest content in the composition (> 50 wt%), and the dispersed phase, also called reinforcement or filler that is a discontinuous phase of the component in less quantity [2], [3]. The concern in its use is based on the fact that the variation between the components and their contents in the composition allows to obtain a range of materials with unique properties, many times different from the properties of their precursor materials. If the combination of these materials creates synergy between them, the properties of the hybrid- and nanocomposites may even outgrow the properties of the precursors. The composite materials can be classified based in: a) Matrix constituent: ceramic-matrix composites (CMC), metal-matrix composites and organic-matrix composites (OMC), that can be divided in carbon-matrix composites (carbon-carbon composites) and polymer-matrix composites (PMC); b) Type of reinforcement: scales, wickers, fibbers and particles; c) Size of fillers: macro-composites, micro- composites and nanocomposites. Hybrid- and nanocomposites are multiphase materials, as well as composite materials, however, at least one of its phases must be present on a nanometric scale, which may be one, two, or three nanometric dimensions [6]. The nanometric phase must also present one of its dimensions in the order of 1 and 100 nm, obtaining a nanocomposite with excellent properties. However, nanofillers can have some difficulties related to their high surface/volume ratio, which can generate clusters (or agglomerates) or, mainly in high contents, and it can be formed during the processing of the nanocomposite or even in the production of the nanoparticle. For this reason, it is necessary to study the best methodology for dispersing the nanofillers (particles or fibres) to create a more homogeneous microstructure, otherwise, these aggregations or agglomerations could lead the nanocomposite to a heterogeneous dispersion, producing an unbalanced material and with loss of the interfacial area, presenting lower than expected properties. There are several ways to minimize agglomerates of powder particles of fillers is to choose a suitable methodology to develop a homogeneous dispersion of the fillers in the matrix. There are several ways to produce hybrid- and nanocomposites from existing (In situ polymerization: solution mixing (blending); melt mixing (melt blending): the micro- or nanofiller dispersion occurs directly in the polymer. In the last method process going in mixers, moulding injectors, or extruders. The mixing is carried out between the molten or rubbery state polymer and the micro- and nanofillers. Other techniques: there is also the possibility of choosing the best nanocomposite production technique for your application and adjust it in the way to produce new techniques, such as layering, hand lay-up, spray lay-up, sol-gel process, electrospinning, centrifugal casting, additive manufacturing, among others [7].

As founded by us the methodology chosen to extrusion process for the hybrid- and nanocomposites depends on several factors: the choice of the polymer matrix (individual synthetic or natural polymer, or their blends, containing thermoplastics (including, for example, polyolefins: polyethylene, polypropylene) and thermo elasto plastics (styrene-butadiene-styrene (SBS), styrene-ethylene-butadiene-styrene (SEBS), copolymer ethylene-octene (POE Solumer) and the micro- and nanofillers (rubber powder, active dyes, pigments dyestuff, as published by one of the authors of this paper [8,10]. Also the novel technical solutions and results are in the development of temperatures and concentrations regimes of extrusion process, the geometry and of the final product in form of films, strands, the application area - for containers and flexible eyeliners and hoses, among others. We received the new experimental results in the area of injection moulding process for the new composites [4] for polymer composites based on engineering plastics/thermoplastics/pigment. Polymer composites was investigated by structural and thermodynamics methods. The planning of the production process must take all these variables into account to obtain a hybrid- and nanocomposites with excellent properties, with low relative cost, and to ensure to fulfil the requirements of the project research and development and approach in industry for the new additive technologies (3-D printing, for example) [3]. We focus our attention on possibilities to develop the innovation hybrid- and nanocomposite materials with predetermined properties based on synthetic polymers with high level elongation at break (at rupture) based our preliminary experience [10] of extrusion of filled in nano-size powder of devulcanized rubber polymer blends (polyolefins / styrene-ethylene-butadiene-styrene, polyolefins / copolymer ethylene-octene) and hybrid composites based on copolymer ethylene-octene filled in basalt scales.

2. Materials and Methods

Researches are conducted with polymers of thermo elastic type of co-polymer of ultra-low density - copolymer ethylene-octene grade POE SOLUMER 875L obtained by Nexlene technology brand SK Global, density 0,87 g/sm^3 ; melt flow rate 5,0 g/sm^3, tensile strength 6,1 MPa, elongation at break 1000%; basalt scales (sizes 0-80 mkm, 80-160mkm, 160-300 mkm); nano Nickel; nano Nickel Oxide, nano metals synthesis describe as following.

In the course of the study, $NiSO_4$ and KOH reagents were used for nano nickel. Precipitation of nickel hydroxide was carried out in a solution containing nickel sulphate and alkali. Production of nickel hydroxide was separated on the Buchner funnel using a vacuum pump. After receiving $Ni(OH)_2$ dried at 60 ^{0}C, ground and sifted, washed from soluble salts, and dried again. Nickel nano powder was obtained by reducing nickel hydroxide ($Ni(OH)_2$) with subsequent passivation in the atmosphere of technical nitrogen, which contains a small amount of oxygen. The recovery process was carried out at a speed of 10 ^{0}C/min to 280 ^{0}C, followed by exposure for 5 minutes and a decrease in speed by 5 ^{0}C /min to a temperature of 300 ^{0}C with an exposure of 1 hour. Qualitative X-ray phase analysis of the sample was carried out on a diffractometer "Diffray–401", the survey was carried out according to the Bregg-Brentano scheme. The study of the sample surfaces was carried out using a scanning electron microscope (SEM) "Tescan" "VEGA 3SB". With the help of differential scanning calorimetric and thermogravimetric analyses on the installation of Simultaneous DSC-TGA series Q Series TM SDT Q600, the nature of the recovery of nickel nanopowder was determined. Identification of polymer spent by Fourie-IR-spectroscopy method by VERTEX (Bruker Corporation, USA) in area 350-7500 sm-1 in the oxygen atmosphere of air. Thermostability and heat flow of source samples of the components and obtained composites investigated by methods: thermal analysis (TGA) and differential scanning calorimetry (DSC) and properties of solid samples of Nickel particles, POE SOLUMER 875L and nanocomposites (obtained by melt extrusion) investigated by equipment NETZSCH Erich Netzsch GmbH & Co. Holding KG, (Germany,) on platinum crucible in area of temperature 0 - 2400°C, velocity of the heating of the samples 10°C /min. Temperature scale and enthalpy of melting was calibrated by standard sample of In (temperature of melting T_{melt} = 156,6 °C, specific heat of melting ΔH = 28,44 Joule/g). Results were applied for the degree of crystallinity determination. Preparation of hybrids-and nanocomposites was by extrusion and injection moulding method. Temperature regime shown in following Table 1.

Thermo Elastoplast	Temperature in cylinder zones of extruder, °C				
	Zones of extruder				
	1 zone	2 zone	3 zone	4 zone	5 zone
POE Solumer 875L	160	175	185	190	190

Table 1. Temperature zones of extruder for strand obtaining from granules of copolymer ethylene-octene

3. Results and discussion

3.1. The structural characterization of the synthesized nickel nanopowder

The characterization of the synthesized nickel nano powder (nanoNi), which was later used to produce a nanocomposite, is shown on Figure 1. The study of samples by scanning electron microscopy showed that nanoNi powder is aggregates consisting of densely aggregated spherical particles. Figure 4 shows a micrograph of a nanoNi sample, the average size of which is 29 nm. Based on the obtained microphotographs, a histogram of the particle size distribution is constructed. It can be seen that the histogram of the particle size distribution of the sample obeys the logarithmic-normal law. According to the results of qualitative X-ray phase analysis, it was found that only two phases are present in the nanoNi sample: Ni, NiO. In this regard, it can be assumed that the nickel nano powder began to oxidize due to high surface activity, which is confirmed by the presence of oxygen of the element according to chemical analysis.

3.2. POE Solumer 875L and filled in composites thermodynamic properties investigation

Endothermic peak is pronounced on the DSC the curve of the sample POE SOLUMER 875L on Figure 2 in area of temperatures 30-80 °C. with significance of the value of the enthalpy of formation - 41,33 J/g [8] responds to the melting process (it is known that the temperature of melting of investigating copolymer POE SOLUMER 875L respond 61 °C.

Strand of nanocomposite (POE SOLUMER 875 L/0,5 % mass. nano Ni and strands of hybrid composites (POE SOLUMER 875 L/Basalt scales) obtained by mould extrusion method. Further DSC the curve obtained for several samples of composites based on nanocomposite 0.5% mass..nano Ni /POE Solumer 875L and composites filled in basalt scales of different dispersion (0-80-160-300 mkm) and content 2,5% mass.; 5,0% mass.and 10 % mass. in POE SOLUMER 875L. Some of results of DSC analysis - shown further (on Figures 8 and 9) - demonstrate that extremum of melting process shifted and smoothed to area more high temperatures (about 90-100 °C.), that testifies of shifting of melting process to higher temperature the new obtained nanocomposite containing nano nickel and composites based on POE SOLUMER 875L filled in basalt scales

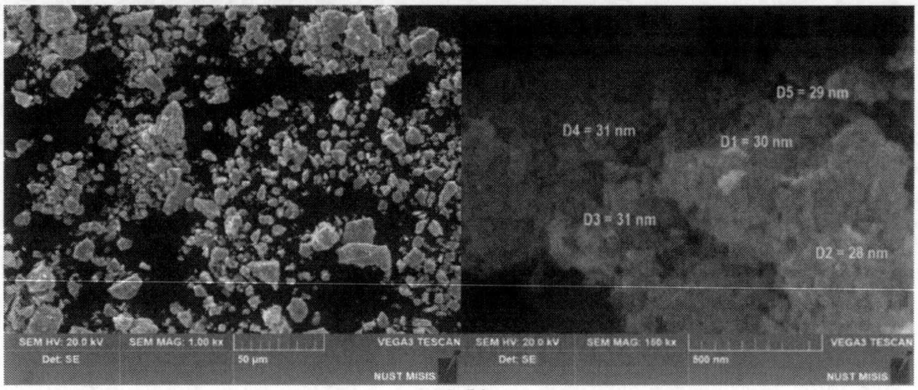

(a) (b)

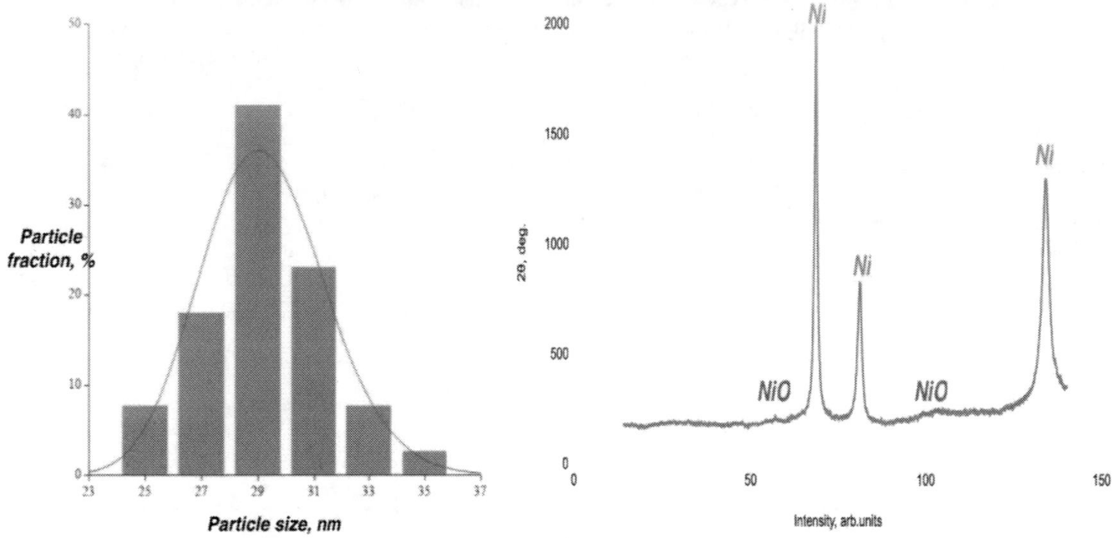

(c)(d)

Fig. 1. Characteristics of nickel nanopowder:.(a,b) Microphoto of nanoNi;(c) Histogram of the particle size distribution of the nanoNi, (d) XRD patterns of NanoNi

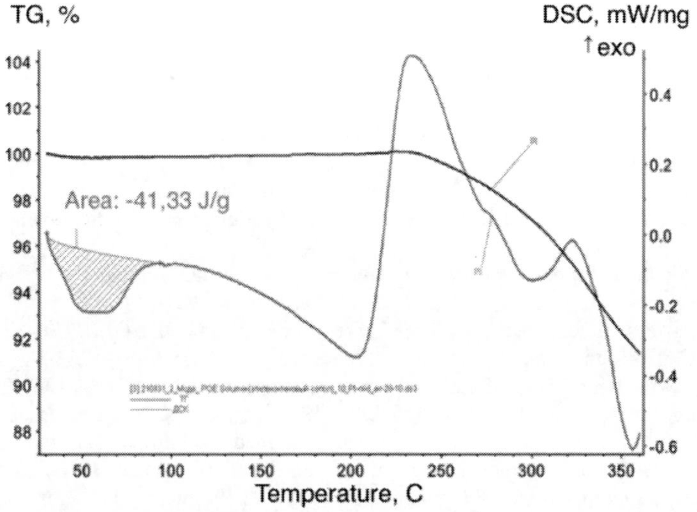

Fig. 2.TGA curve of the ethylene-octene copolymer in the temperature range 20-350 °C (black, 2), DSC curve of the ethylene-octene copolymer in the temperature range 20-350 °C (red,1)

3.3. Nanocomposites based on POE Solumer/0.5% mass.. nanoNi and hybrid-composites POE Solumer/Basalt scales:structure, thermostability and elasticity

Strands obtained by melt extrusion of filled in nano nickel and basalt scales of sizes mentioned above in part 2 Materials and methods [11] shown below on photo (Figure 3 and Figure 4). We investigated structural, thermodynamics, and physico-mechanical properties of the samples of obtained composites in film form and strands by several experimental methods described above in Part 1.

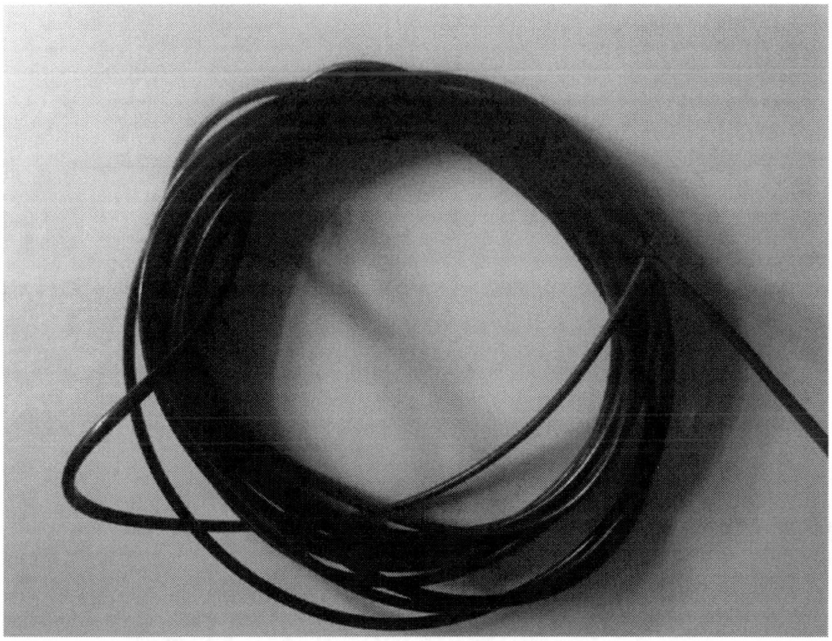

Fig. 3. Foto of strand nanocomposite (POE SOLUMER 875 L/0,5 %mass nano Ni) obtained by melt extrusion method

Fig. 4. Foto of strands hybrid composite (POE SOLUMER 875 L/Basalt scales) obtained by extrusion method

Qualitative X-ray diffraction analysis (Figure 5) showed that two phases are present in nanocomposite samples: nickel phase and C-H, which corresponds to this copolymer, POE SOLUMER 875L.

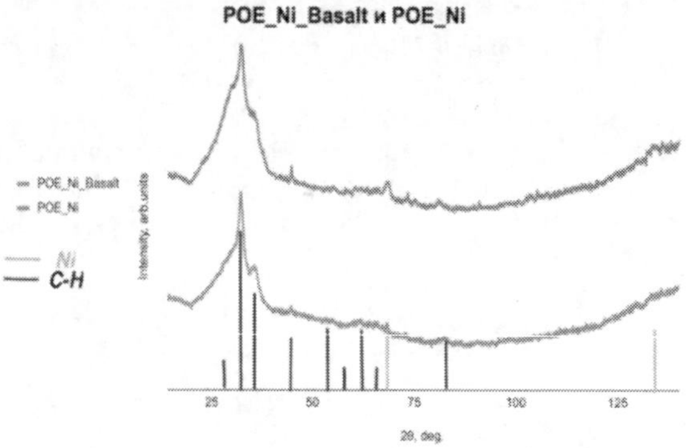

Fig. 5. XRD patterns of POE Solumer 875L filled in blend of nano Ni and Basalt Scales and POE Solumer filled in nano Ni powde

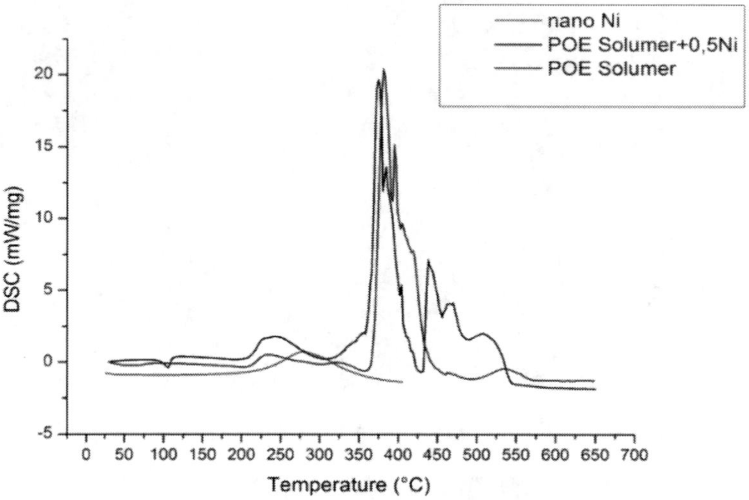

Fig.6 DSC analysis of samples of a nano Ni, POE Solumer 875L, nanocomposite 0.5% wt. nano Ni /POE Solumer 875L

DSC analysis of the samples of samples of a nano Ni, POE Solumer 875L, nanocomposite 0.5% wt. nano Ni /POE Solumer 875L represented on Figure 6. Figure 7 and 8, 9 show the results of comparative investigation of the loss of mass of the samples thermo elastoplast POE SOLUMER 875, nanocomposite 0.5% wt. nano Ni /POE Solumer 875L and hybrid composites with different content of basalt scales. It is found that the thermostability of nanocomposite is higher than thermostability of polymer.

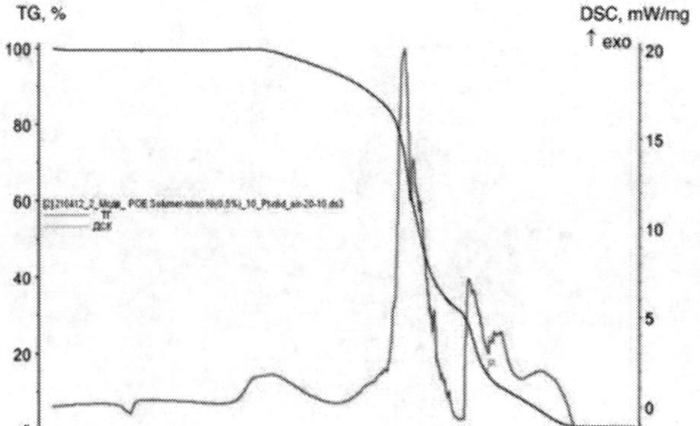

Fig. 7. TGA/DSC analysis of the sample of nanocomposite 0.5% wt. nano Ni /POE Solumer 875L

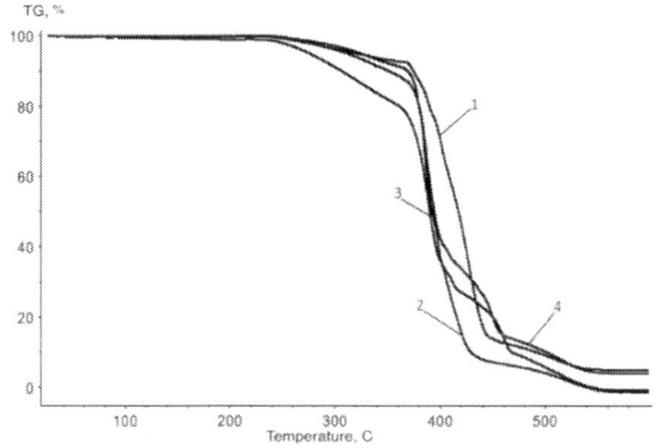

Fig. 8. TGA curves of a sample POE SOLUMER 875L (2) and hybrid composites filled in basalt scales sizes 80-160 mkm: 10% mass.(1), 2,5 % mass.(3), 5% mass.(4)

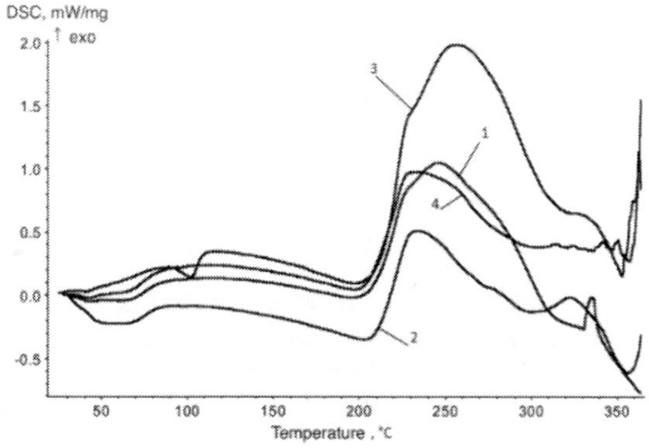

Fig. 9. DSC curves of a sample POE SOLUMER 875L (2) and hybrid composites filled in basalt scales sizes 80-160 mkm: 10% mass.(1), 2,5 % mass..(3) , 5% mass/(4)

The most important result of physico-mechanical properties investigation (by methods described above, in part 1 of this paper) is increasing tensile strength of elasticity and of the new composites: tensile strength value (when breaking) of the sample POE SOLUMER 875, as known, equal to 6,1 MPa) and the tensile strength hybrid-composites samples of copolymer containing basalt scales growing to7 MPa: tensile strength of the samples of nanocomposite nano Ni|/POE Solumer 875L became 10 MPa; elongation of break of the samples of nanocomposites (from 1000% for POE SOLUMER 875L to 1240% for nanocomposite sample nano Ni|/POE Solumer 875L).

4. Conclusion

Propositions of the paper as well as the author's views of the practical implications of the results for increasing elasticity of the new hybrid- and nanocomposites based on polymers type thermo elastoplasts filled in basalt scales and nano powder of metals. The most important results of presented investigations is obtained by reinforcement of investigated thermal elastoplast, (ethylene-octene copolymer) by coextrusion with nanoparticles powder (30 nm) of metals and its oxides. Determination of improve the open possibilities to obtain the new hybrid- and nanocomposite materials based on ethylene-octene copolymer filled in basalt scales and nanomaterials and metal oxides with predetermined properties for industrial approach for the new goods manufacture in form films, fibres, strands, hosepipes and the other forms with higher elasticity than initial polymer. Conducted research and new results open the possibility of application of the new hybrid- and nanocomposites based on thermoplastics filled in basalt scales, nano metals powder and their blends for application for increase of tensile strength and elongation of break. in additive technologies such as 3D-printing.

5. Acknowledgments

Research and development of presented paper supported by Programme 35 of the Presidium Russian Academy of Sciences. A Scientific Basis to develop New Functional Materials state order No 0082-20019-0008 Developing "Methods to make Nanostructured Polymer, Biopolymer and Composite Materials and their Target Modification Aimed at Wide Range of Practical Applications".

An experimental investigation was conducted with the experimental support of y Grachev A.(Senior researcher FRC ChPh NN Semenov RAS) and Golobokov D. (student of Polytech University, Moscow) CCP №5006694 and UNU № 144073 of the Federal Research Centre of Chemical Physics Semenov NN of the Russian Academy of Sciences.

6. References

[1] Hodzik et al.(2020) Strain Rate Influence on Mechanical Characteristics of FDM 3D Printed. Materials. proceedings of 32th DAAAM International Symposium, B.Katalinich (Ed.), Vienna, Austria, ISBN 978-3902734 http://dx.doi.org/10.2507/31st.daaam.proceedings.02

[2] Myasoedova V.V. and Taran I.A. (2020) Chemistry of Energy-Effective Polymer Nanocomposites (ID Academia Zhukovskogo, Moscow, 212 p.(in Russian), ISBN 978-5-907275-12

[3] Fedorov L.Yu.(2017) DSC/TGA Analysis of composite material based on highest density polyethylene and nanoparticles CuO, POLYCOM TRIB-2017, Gomel, Belorussia, June, 27-30, pp.-20-23

[4] Svoboda P.,TherraValappil R., Svobodova D. et al. (2010) Elastic properties of polypropylene/ethylene-octene copolymer blends.Polymer Testing, Vol.129, Issue 6, September, pp.742-748

[5] Myasoedova V.V., Taran I.A., Lushkova A.V. Physical and Chemical Properties of Composites Produced on Disperse-filled Blends of Polymers , Polymer Science, Series D, 2019, Vol.12, No 2, pp.203-206, ISSN 1995-4212, DOI:10.1134/S1995421219020151

[6] Keihan A.N. Veisi H., Veasi H (2017) Green synthesis and characterization of spherical copper nanoparticles as organometallic antibacterial agent, Appl.Organomet.Chem , https://doi.org/10.1002/aoc.3642

[6] Dahoumane S.A. et al (2017). Biosynthesis of inorganic nanoparticles: A fresh look at the control of shape, size and composition, Bioengineering. https://doi.org/10.3390/bioengineering 4010014

[7] Drexler R.T. (1981) Molecular engineering: an approach to the development of general capabilities for molecular manipulation. Proc Natl lAb afSci. USA78, p.5275

[8] Myasoedova V.V. (2018) Recycling of different chemical nature polymers for approach as composites (Plenary lecture) //The Book of the Fourth International Conference on Reuse and Recycling of Materials, ICRM-2018, March 09-11, pp.12-15

[9] Svoboda P.,TherraValappil R., Svobodova D. et al. (2010) Elastic properties of polypropylene/ethylene-octene copolymer blends.Polymer Testing, Vol.129, Issue 6, September, pp.742-748

[10] Myasoedova V.V. Lushkova A.V.(2020) Eco solution: composites based on recycled polypropylene modified by thermoelasoplasts and filled in disperse filler. CHEMICAL SAFETY SCIENCE, No 4(2), pp.265-281

[11] Myasoedova V.V.(2021) Graded Recycled Polymer Modified by Thermoelastoplasts: Composite Materials to be Used for Injection Molding, Pol. Sci., Series D, Vol.14, No 3, pp. 462-466, ISSN 1995-4212, DOI:10.1134/s1995421221030217

DOI: 10.2507/32nd.daaam.proceedings.028

AIR GUN PRESSURE INFLUENCE ON THE PROJECTILE PENETRATION OF A 2MM MILD STEEL PLATE

Faruk Razic, Zahida Ademovic, Jasmin Terzic,
Sabina Serdarevic-Kadic & Miralem Burek

This Publication has to be referred as: Razic, F[aruk]; Ademovic, Z[ahida]; Terzic, J[asmin]; Serdarevic-Kadic, S[abina] & Burek, M[iralem] (2021). Air Gun Pressure Influence on the Projectile Penetration of a 2mm Mild Steel Plate, Proceedings of the 32nd DAAAM International Symposium, pp.0185-0190, B. Katalinic (Ed.), Published by DAAAM International, ISBN 978-3-902734-33-4, ISSN 1726-9679, Vienna, Austria
DOI: 10.2507/32nd.daaam.proceedings.028

Abstract

Guns that operate on compressed air (Air Guns) are widely used in entertainment business. They are very common in Paintball and Airsoft competitive team sports and in performance riffles. Projectiles fired from these types of guns can reach more than 400 m/s. A Projectile weighing 5 g at this velocity has a kinetic energy of 400 J upon impact. This energy is sufficient for skin penetration and infliction of serious life threating injuries. This paper deals with projectile velocities, achieved with compressed air, needed for full penetration of the mild steel plate of 2 mm thickness. Projectile firing has been performed with custom made Air Gun capable of reaching compressed air levels up to 10 MPa and projectile velocities up to 160 m/s. Velocity of the fired projectiles has been recorded with High Speed camera after which it was possible to determine kinetic energy upon the impact and energy density as well. Experiments presented in this paper show that compressed air operated guns can deliver significant kinetic energy to the target, capable of inflicting serious injuries

Keywords: Velocity; Air Gun; Pressure; Energy; Target.

1. Introduction

Paintball and Airsoft guns, more often called Markers, and performance rifles are capable of firing projectiles at very high velocities. Playground facilities for Paintball and Airsoft have limited levels of projectile velocities, 91 m/s for Paintball and 150 m/s for Airsoft markers [1] [2]. Performance riffles on the other hand are capable for even higher projectile velocities, more than 400 m/s. One of the fastest pellet gun on the market today is probably a Magnum Riffle by Gamo which can reach 470 m/s (1542 fps) with 4.5 mm pellet [3].

Kinetic energy levels at the muzzle of a Paintball markers are quite low, around 12 J [4]. This energy level is achieved with 17.3 mm ball whose mass is 3 g, flying at 91 m/s velocity. Airsoft balls are quite lower in mass compared to paintball balls. At 0.4 g and 150 m/s they achieve only 4.5 J of kinetic energy at the muzzle. On the other hand, performance rifles can achieve 5-10 times higher kinetic energy because of the much higher projectile velocity. Also, these types of Air Guns are capable of penetration of a human skin which is not the case for previous mentioned markers.

Skin penetration can be observed as a first step in prediction of possible injury of a human body. Energy levels and energy densities at which it is possible to penetrate human skin are usually investigated using porcine body parts. Human cadavers on the other hand are rarely used for this kind of experiments and use of human cadavers is often considered unethical. In this work, penetration of a 2 mm thick mild steel plate was tested. Unquestionably steel plate is more resilient than human skin. Eventual penetration of a steel plate using compressed air for projectile propelling will provide visual confirmation of air gun penetration possibilities and its potential. When velocity and mass of the fired projectiles were taken into account it is possible to calculate kinetic energy and energy density at which penetration occurs.

Above mentioned kinetic energy levels are calculated for the projectile muzzle velocity. It is expected that these energy levels decrease with the shooting distance since projectile is losing velocity on its trajectory to the target. This leads to the linear decrease in wounding potential with increased distance to the target [5]. There are also other weapons used in entertainment for firing projectiles, like slingshots or blowpipes. However, energy levels achieved with these types of weapons are not sufficient for skin penetration nor they can cause serious injuries to the muscles, bones or skin [6].

Main goal of this paper is to provide data about kinetic energy and energy density at which it is possible to provide full penetration of the steel target using compressed air. There are many different reports and papers that are describing wounds gained from Air Guns. Some of those injuries turn out to be fatal for the person being hit [7]. However, data about full potential and capability of such type of guns are rather scarce.

All Air Guns are capable of firing projectiles at even higher velocities than the ones designated on the playground. This is achieved by pressure increasing in the gun tank. To which levels this pressure can be increased depends on the gun/marker properties. With increased projectile velocity, the kinetic energy and energy density will also increase. In that case there is a higher possibility for a serious injury of a human body.

As it was already stated, mass of the projectiles used in paintball and airsoft markers are quite low. Considering their low mass, velocity of the projectiles should be increased significantly to provide desired penetration energy levels. Performance rifles on the other hand are capable of delivering very high pellet velocities. However, mass of the pellet is also very low which decreases penetration abilities. Custom made Air Gun was designed for this study. This Air Gun was capable of firing heavy steel projectiles at high velocities.

2. Case study

Penetration capability of a compressed air driven projectiles has been examined through experiment. Custom made Air Gun, Fig. 1. has been assembled in order to provide higher levels of compression, up to 10 MPa (1450 Psi). Achieved air pressure level was monitored with high accuracy manometer connected to the Air Gun tank. Air Guns are in general of simple construction. Main parts of an Air Gun are: Tank, Valves and Barrel. By assembling these three components one can custom made an Air Gun and use it to fire projectiles. Hence, it is not important what kind of device operated on compressed air will be used for firing. Main goal is to provide sufficient velocity for the projectile which will as a result produce desired level of kinetic energy upon impact.

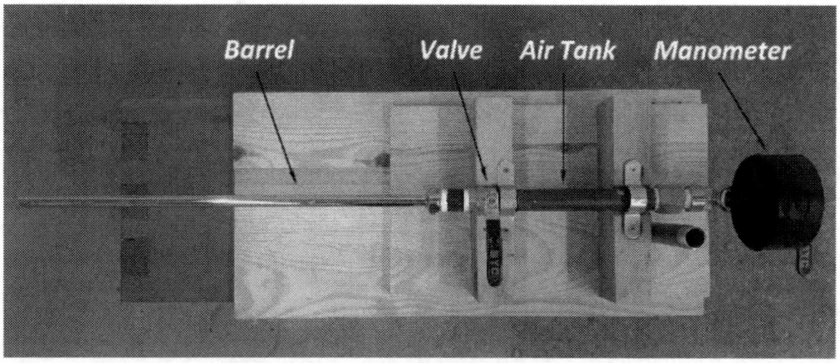

Fig. 1. Custom made Air Gun

Projectiles used in this test were 5.5 mm mild steel projectiles machined in to a shape that will provide high sectional density (mass to projected surface area ratio). These projectiles were fired at different pressure levels into a steel plate of 2 mm thickness, Fig. 2. Dimensions of the plate were 500 x 500 mm. Penetration of this target visually presented capabilities of impact energy provided with compressed air.

Three shooting series are performed at three different air pressure levels. Velocity was measured for each projectile fired, and with known mass of the projectile, kinetic energy and energy density are determined. In the third shooting series, piezo electric pressure sensor was installed. This sensor was capable of measuring pressure inside the Air Gun barrel. Using proper data acquisition system, pressure vs time curve was obtained. This curve describes the pressure vs time integral where it can be seen how the maximum pressure was achieved. Pressure loss was also evident.

Fig. 2. Blunt and Sharp tip projectiles (left), 2 mm steel plate target (right)

Two different types of projectiles are used in penetration test. One with blunt and the other with sharp tip. Pressure levels at which projectiles were fired at the target, placed 500 mm in front of the muzzle, were 2, 3 and 5 MPa providing velocities from 80 to 160 m/s. In order to provide flight stability and gas sealing during projectile motion through the gun barrel, plastic sabot with fins was designed. Mass of the steel penetrator without sabot was ranging from 10.00 – 10.86 g. When sabot was added to the steel penetrator it formed a projectile of total mass around 12 g.

Both, sharp and blunt projectiles are fired five times into the target at three different pressure levels and at different velocities. Deformation of the steel plate was observed and analysed after the impact. Four possible penetration outcomes are observed: Full penetration "fp" (projectile went through the target), Partial penetration "pp" (projectile made a hole in the target, but did not went through), projectile did not penetrate the target "dnp" (projectile bounced off leaving indentation in the target plate and p-stuck "ps" (projectile perforated target and stayed in the target protruding at the backside of the plate). For each penetration outcome, velocity and energy density were defined and compared to the skin penetration criteria. These data, combined with the damage on the steel plate, will clearly present energy levels of compressed air-based weapons and its possibilities.

3. Results and Discussion

Penetration test results are presented in the tables bellow. Velocity and energy density were determined for each projectile fired and penetration outcome is registered. Kinetic energy of the projectiles upon the impact was calculated according to:

$$E_k = \frac{mv^2}{2} \tag{1}$$

Where "E_k" is kinetic energy of the projectile, "m" is mass of the assembled projectile (penetrator + sabot) and "v" is the velocity of the projectile. Velocity was recorder with high speed camera and calculated for the distance projectile has travelled. Velocity was measured only at the muzzle and kinetic energy and energy density were calculated also for muzzle velocity. Because target was placed only 500 mm in front of the barrel, there was no significant change in velocity prior to impact.

Table 1. presents penetration data for projectiles fired at 2 MPa tank pressure. Both, blunt and sharp tip projectiles are fired at the target plate.

Projectile No.	Chamber pressure 2 MPa											
	Projectile diameter 5.5 mm											
	Sharp Projectile						Blunt Projectile					
	Mass [g]	Vel. [m/s]	Energy [J]	Energy density [J/cm^2]	Sectional density [g/mm^2]	Penet.	Mass [g]	Vel. [m/s]	Energy [J]	Energy density [J/cm^2]	Sectional density [g/mm^2]	Penet.
1	12.30	95	55.39	233.13	0.518	dnp	12.56	105	69.24	291.42	0.529	dnp
2	12.23	107	69.40	292.10	0.515	pp	12.49	104	67.55	284.30	0.526	dnp
3	12.32	97	57.78	243.20	0.519	dnp	12.58	107	72.01	303.11	0.529	dnp
4	12.34	102	63.82	268.60	0.519	pp	12.62	111	77.75	327.24	0.531	pp
5	12.36	116	83.47	351.35	0.503	pp	12.61	114	81.94	344.89	0.531	pp

Table 1. Penetration outcome for 2 MPa chamber pressure

Based on the penetration outcome, it can be seen that partial penetration is possible at velocities generally above 100 m/s. Projectiles that were able to perforate steel plate (made a small hole in the target) had higher energy than the ones which left only indentations on the target (dnp). Energy density was much higher than skin penetration limit which is ranging from 23.99 J/cm^2 up to 52.74 J/cm^2, depending on the place of impact on a human body [8].

Projectile No.	Chamber pressure 3 MPa											
	Projectile diameter 5.5 mm											
	Sharp Projectile						Blunt Projectile					
	Mass [g]	Vel. [m/s]	Energy [J]	Energy density [J/cm^2]	Sectional density [g/mm^2]	Penet.	Mass [g]	Vel. [m/s]	Energy [J]	Energy density [J/cm^2]	Sectional density [g/mm^2]	Penet.
1	12.66	135	115.36	485.57	0.533	ps	12.56	124	95.56	406.43	0.529	ps
2	12.59	125	98.36	414.00	0.530	ps	12.64	110	76.47	321.87	0.532	ps
3	12.68	118	88.28	371.57	0.534	ps	12.55	120	90.36	380.33	0.528	ps
4	12.70	126	100.81	424.33	0.535	ps	12.66	108	73.83	310.77	0.533	ps
5	12.72	120	91.58	384.48	0.535	ps	12.68	118	88.28	371.57	0.534	ps

Table 2. Penetration outcome for 3 MPa chamber pressure

All of the projectiles fired at 3 MPa chamber pressure, experienced p-stuck penetration outcome "ps", Table 2. Projectiles perforated steel plate and protruded on the other side staying in the target. Velocities at which impact has occurred were higher in some cases for more than 30 % in comparison to 2 MPa chamber pressure.

Projectile No.	Chamber pressure 5 MPa											
	Projectile diameter 5.5 mm											
	Sharp Projectile						Blunt Projectile					
	Mass [g]	Vel. [m/s]	Energy [J]	Energy density [J/cm^2]	Sectional density [g/mm^2]	Penet.	Mass [g]	Vel. [m/s]	Energy [J]	Energy density [J/cm^2]	Sectional density [g/mm^2]	Penet.
1	12.13	148	132.85	559.16	0.511	ps	12.50	145	131.41	553.10	0.526	fp
2	12.06	156	146.75	617.66	0.508	ps	12.58	150	141.53	595.69	0.529	fp
3	12.15	146	129.49	545.05	0.511	ps	12.49	141	124.16	522.58	0.526	fp
4	12.17	137	114.21	480.71	0.512	ps	12.61	135	114.91	483.66	0.531	fp
5	12.19	145	128.15	539.38	0.513	ps	12.48	149	138.53	583.10	0.525	fp

Table 3. Penetration outcome for 5 MPa chamber pressure

At 5 MPa chamber pressure, Table 3, only blunt projectiles were capable of full penetration of the steel plate while sharp projectiles expressed same behaviour as it was the case in 3 MPa tests, that is the projectiles were stuck in the target. Sectional density was higher for the blunt projectiles because of their higher mass compared to the sharp ones. Since caliber of the projectile was identical for all shooting series, mass has the major influence on sectional density in this case.

In case of a 5 MPa Chamber pressure conditions, piezo electric pressure sensor Kistler 6215 was added in front of the trigger valve in order to measure pressure inside the barrel after triggering. Recorded Pressure vs Time curves, Figure 3, resembles to those curves obtained in internal ballistics of firearms. Pressure drop is evident from these curves.

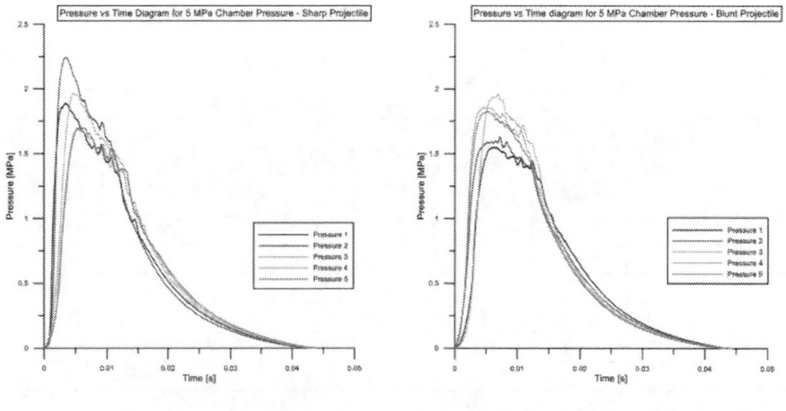

Fig. 3. Pressure vs Time diagrams for sharp and blunt projectiles fired at 5 MPa

Fig. 4. presents achieved penetration outcomes of the 2 mm steel plate. First image presents target damage in case of a projectile that bounced of the target leaving indentation in the target material (dnp). Second image is the result of the partial penetration (pp). Projectile did make a hole in the plate, but it was not capable of full penetration through the target body. In this case projectile also bounced of the target. The results of full penetration are presented in the third image (fp). Projectile went through the whole target continuing its motion after penetration. Fourth image presents effects of the projectile that has stuck in the target material (ps). This projectile was capable of penetrating the target making a hole size of a projectile caliber but it did not have enough kinetic energy to continue its motion further. Markings of the tool used to remove projectile from the target are visible on the fourth image from left to right (ps).

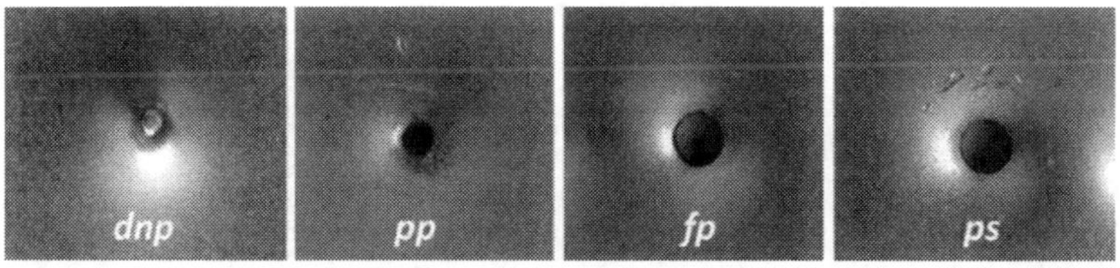

Fig. 4. Penetration outcomes

Table 4. shows energy levels and other parameters calculated for projectiles fired from Air Guns used in Paintball, Airsoft and for Performance riffles. Velocity and projectile mass were acquired from riffle/marker manufacturers and safety measurements provided from playground sites that use these kinds of guns according to regulations [1] [2] [3] [9].

Projectile parameters		Paintball Marker	Airsoft Marker	Air Riffle
Mass	[g]	3.0	0.4	0.490
Diameter	[mm]	17.270	6.000	4.500
Projected Area	[cm^2]	2.342	0.283	0.159
Velocity	[m/s]	91.0	150.0	470.0
Energy	[J]	12.422	4.500	54.121
Sectional density	[g/mm^2]	0.013	0.014	0.031
Energy density	[J/cm^2]	5.303	15.915	340.288

Table 4. Parameters of projectiles used in various Air Guns

Energy density at which 50% risk of skin penetration is possible is ranging between 23.99 J/cm^2 up to 52.74 J/cm^2, depending on the place where human body was hit [8]. According to velocity levels regulated by the safety measures, energy density achieved from Paintball and Airsoft markers is not sufficient for skin penetration. Energy and Sectional density are too low to produce any significant damage to the skin, bones or muscles. However, it is still possible to cause serious eye injury with these types of Air Guns [10] [11]. When performance riffle characteristics are observed, it is evident that skin penetration energy density limit is exceeded more than 10 times. Even though energy density of the projectiles fired from custom made air gun could be compared with performance rifles, sectional density was significantly lower in performance rifle projectiles. Gun used in this test was of much higher power than markers/riffles used in entertainment.

4. Conclusion

Air guns used in entertainment generally have low velocities and low operational pressure. Excluding performance rifles which can achieve velocity up to couple of hundred meters per second, other entertainment guns should not be able to inflict serious injuries to bones, skin, muscles etc.

Experiment described in this paper is not solely intended to compare air guns and its injury capabilities. Main premise was to present energy levels of compressed air in projectile propelling and its possibilities in penetration of the steel target. Energies achieved in custom made air gun clearly describes possibilities of compressed air by visual interpretation of the plate perforations using steel projectiles.

Even though these levels of compression are not present in the entertainment guns/markers, it is possible to achieve them by some sort of marker modification. It is important to emphasize that these modifications are against the regulations. Modified markers must not be used in playgrounds or any sort of activity where someone could be injured. Custom made Air gun used in these experiments was operated by the trained personnel in safe laboratory conditions with protective plates against ricochet. Safety measures are also taken in case of air tank burst.

Further research on the subject of penetration abilities of the projectiles should be performed on organic targets such as porcine meat with skin and bones. These results will provide possible comparison of energies needed for target penetration similar to a human body.

5. References

[1] http://lvlupsports.com, (2021). LVL UP Sports, Sports Safety Rules, Accessed on: 20-07-21.

[2] http://paintball-command.com, (2021). Paintball Command, Airsoft Equipment Rules, Accessed on: 20-07-21.

[3] http://www.gamo.co.za, (2021). Gamo Precision Air Guns, Gamo G magnum, Accessed on: 07-07-21.

[4] http://www.floridaairsoft.com, (2021). FA, Airsoft and Paintball Kinetic Energies, Accessed on: 08-08-21.

[5] Ficek, M.; Malanik, Z.; Mikulicova, M. & Gracla, M., (2019). Influence of the Shooting Distance on the Depth of Penetration of the Bullet into the Replacement Material for Air Gun Weapons, Proceedings of the 30th DAAAM International Symposium, pp.0663-0672, B. Katalinic (Ed.), Published by DAAAM International, ISBN 978-3-902734-22-8, ISSN 1726-9679, Vienna, Austria DOI: 10.2507/30th.daaam.proceedings.091.

[6] Ficek, M.; Juricek, L.; Gracla, M.; Malanik, Z. & Mikulicova, M. (2020). Determination of Wounding Potential of Guns Known from Childhood - Blowpipe and Slingshot, Chapter 19 in DAAAM International Scientific Book 2020, pp.227-236, B. Katalinic (Ed.), Published by DAAAM International, ISBN 978-3-902734-27-3, ISSN 1726- 9687, Vienna, Austria DOI: 10.2507/daaam.scibook.2020.19.

[7] Dumencic, B.; Rajc, J.; Pavokovic, D. & Damjanovic, T. (2020). "Fatal injury by air gun: a case report," Egyptian Journal of Forensic Sciences, Vol. 10, No. 7., DOI:https://doi.org/10.1186/s41935-020-00182-7.

[8] Bir, C. A., Stewart, S. J. & Wilhelm, M. (2005). "Skin Penetration Assessment of Less Lethal Kinetic Energy Munitions", Journal of Forensic Sciences, Vol. 50, No. 6, pp. 1426-1429., Paper ID JFS2004551.

[9] http://www.liveabout.com, (2018). LiveAbout, Choosing the Right Paintball Caliber, Accessed on: 23-08-21.

[10] Farr, A. K. & Fekrat, S. (1998). "Eye injuries associated with paintball guns", The International Journal of Clinical Ophthalmology and Visual Sciences, Vol. 22, pp. 169-173., DOI: 10.1023/a:1006216125152.

[11] Kennedy, E. A., P Ng, T. & Duma, S. (2006). "Evaluating eye injury risk of Airsoft pellet guns by parametric risk functions", Biomedical Sciences Instrumentation, Vol. 42, pp. 7-12., PMID: 16817577.

DOI: 10.2507/32nd.daaam.proceedings.029

INFLUENCE OF INFILL DESIGN ON COMPRESSIVE AND FLEXURAL MECHANICAL PROPERTIES OF FDM PRINTED PLA MATERIAL

Damir Hodzic & Adi Pandzic

This Publication has to be referred as: Hodzic, D[amir] & Pandzic, A[di] (2021). Influence of Infill Design on Compressive and Flexural Mechanical Properties of FDM Printed PLA Material, Proceedings of the 32nd DAAAM International Symposium, pp.0191-0199, B. Katalinic (Ed.), Published by DAAAM International, ISBN 978-3-902734-33-4, ISSN 1726-9679, Vienna, Austria
DOI: 10.2507/32nd.daaam.proceedings.029

Abstract

Today, one of the fastest growing manufacturing technologies is certainly additive manufacturing, where there are several different product manufacturing processes, and one of the most famous is FDM 3D printing process. From literature review, it can be concluded that many parameters affect the final mechanical properties of FDM printed materials. Also, one of the great advantages provided by this technology is the production of product materials with an infill structure, which brings several advantages such as: reduced time of production, reduced material consumption, lower product weight and generally more economical production. The aim of this study is to analyse the influence of infill design on the compressive and flexural properties of FDM printed material. In particular, the influence of infill pattern and density for PLA FDM printed material was analysed. Based on the results of the statistical analyses, it can be concluded that the infill pattern and density affect the compressive and flexural strength and modulus of PLA FDM printed material.

Keywords: PLA material; FDM; Infill design; Material properties; 3D printing

1. Introduction

In general, additive manufacturing (AM), known as 3D printing, is defined as the novel technology used to produce the 3D/4D parts from a CAD model by adding layers of material bottom-up of various types of materials such as polymers, ceramics and composites. Fused deposition modelling (FDM) is one of AM technology that has experienced a great increase since it was presented on the market [1], [2], [3]. Also, one of the most well-known AM technologies is FDM, and it can be said, mostly due to the relatively low cost of the devices and materials used for production. In addition to the relatively low cost, FDM has other advantages presented below (Fig. 1) [4], [5], [6].

This technology finds its applications in education, research and many fields of engineering and industry, such as automotive products, aircraft, medical implants, dental and more. Also it is very popular in maker and start-up hubs, and changed the standard way of developing prototypes and products [7].

FDM technology advantages
Complex geometry fabrication
Product material with infill structure
Easy aplication
A large number of different materials
Fast production of small series products
Rapid prototyping
Low cost

Fig. 1. FDM technology advantages over traditional manufacturing production

FDM technology belongs to the group of AM processes that, in general, uses thermoplastic polymers, but in today's market there are FDM machines that also makes products from metallic materials [8]. In this process, the material is melted in a liquid state in a printer head (extruder) and then deposited through a nozzle that traces the parts cross sectional geometry to create 3D object (product, part) on printer build platform directly from a CAD model (Fig. 2) [7].

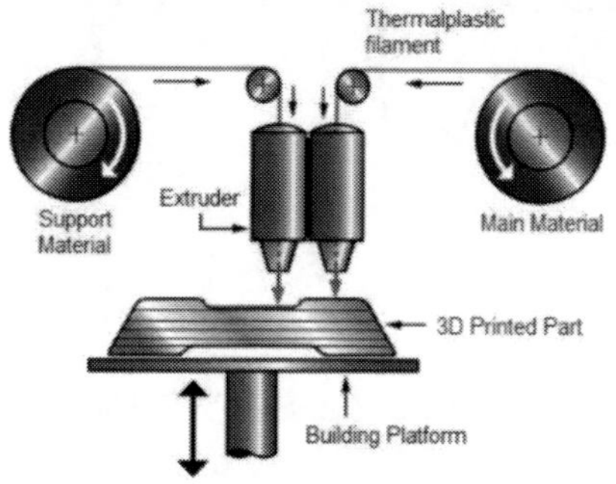

Fig. 2. Basics of FDM 3D printing process [9]

FDM 3D printer uses a CAD model for rapid prototyping process. The 3D model from CAD software, is converted to stereo lithography (.STL) format and after that, using "slicer" software, user is creating G-code for 3D printer [10]. Despite the number of advantages offered by these technologies, designers, when designing new products, must adapt the design of their product in order to take full advantage of these technologies. In addition to product design, the properties of FDM-produced parts significantly depend on the processing parameters. These parameters include layer height, nozzle temperature, nozzle diameter, air gap, build direction, print orientation, extrusion width, raster orientation, raster angle, infill pattern, infill density, infill angle, raster overlap and more [11], [12]. One of the benefits of FDM technology is the ability to fabricate the product material with an infill structure. Product material with infill structure provides several advantages (Fig. 3) such as reduced time of production, reduced material consumption, lower product weight and generally more economical production [1], [13], [14].

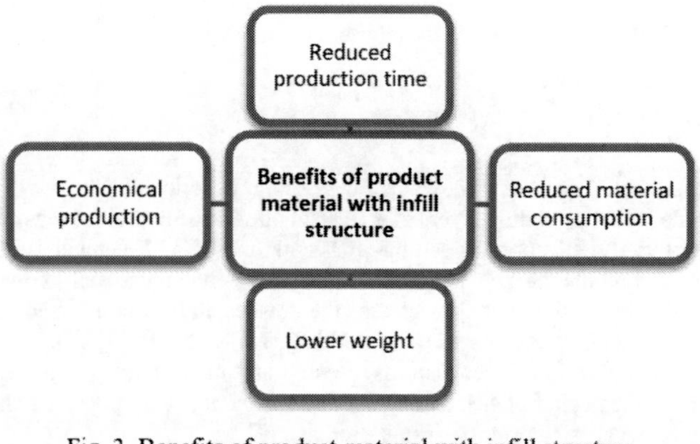

Fig. 3. Benefits of product material with infill structure

A review of the above mentioned literature, as well as the study [15], shows the direct impact of infill design on the mechanical properties of FDM printed material. The focus of this research paper was to study the effect of infill pattern and density on compressive and flexural mechanical properties of PLA FDM printed material. Five different infill patterns, with variation of infill density, were analysed and compared with samples with 100% infill density. All results are analysed using statistical methods and presented in further text. Also, the literature review shows that the mechanical properties of printed materials are influenced by the material colour [3], the manufacturer (brand) of the material [4] as well as the model of the 3D printer [16]. In order to reduce the influencing factors on the results, for each test the samples were made on the same 3D printer, from materials of one manufacturer as well as the same colour.

2. Research methodology

Methodology of the experimental part of this research is presented on Fig. 4, where the goal was to analyse the influence of the infill pattern and density on the compressive and flexural properties of FDM printed PLA material.

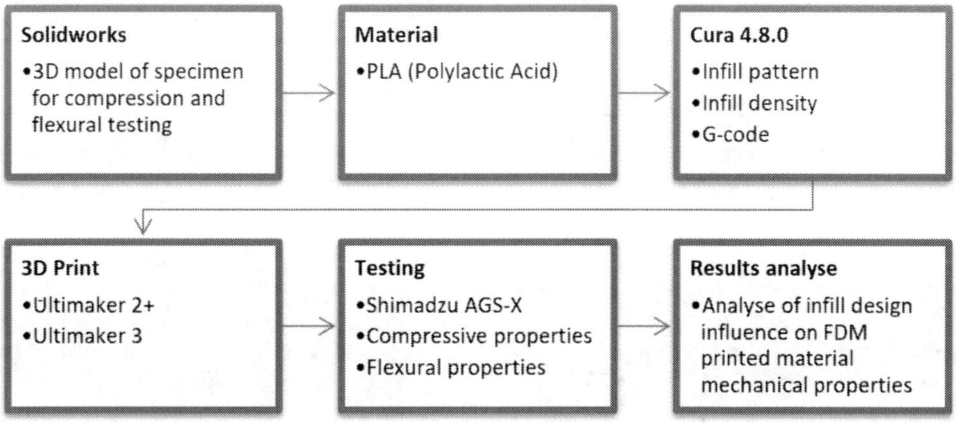

Fig. 4. Methodology of research experiment

Using Solidwors CAD software, 3D models of testing specimens were created, where the specimen for compression testing were designed according to ASTM D695 and for the flexural testing according to the EN ISO 178 standard (Fig. 5). As there are still no standards for FDM materials testing, but also for materials in additive technologies in general, material testing in this study is based on these standards, as can be seen in other research papers.

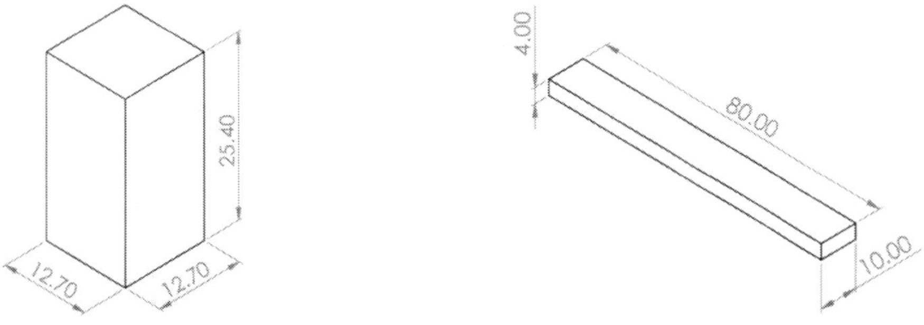

Fig. 5. Compressive (ASTM D695) and flexural (EN ISO 178) testing specimen

In this study, significantly researched and widely used PLA (supplied by 3D Republika) material was examined. PLA is from group of aliphatic polyesters, nontoxic and biodegradable material. A biopolymer prepared from corn, sugarcane, with relatively good mechanical properties and its major drawbacks include brittleness and low thermal stability [4], [6], [17]. As meintoined before, in general, FDM printed materials properties depends on 3D printing parameters and for this experiment all specimens are printed with same 3D printing parameters defined by default Ultimaker printing profile "Normal 0,15mm". Specimens and 3D printing parameters are prepared for printing using Cura 4.8.0 slicer and all specimens are printed with Ultimaker FDM 3D printers.

In slicer Cura, it is possible to choose 13 different infill patterns, but for this study, five different patterns were analysed (Fig. 6.), as follows: concentric (CO), grid (GR), gyroid (GY), octet (OC) and triangle (TR), and for each pattern the density was varied (20%, 40%, 60%, 80% and 100%).

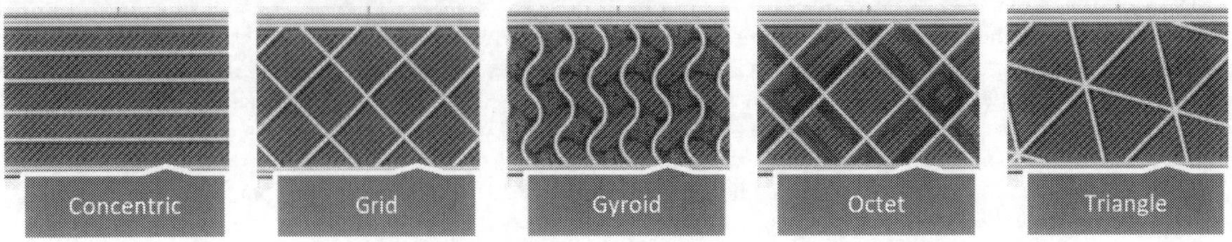

Fig. 6. Five different infill patterns used for testing

From slice Cura, g-code with printing details and parameters is transferred to 3D printers. All specimens for compressive testing are printed with Ultimaker 2+ and for flexural testing with Ultimaker 3. Compressive (ASTM D695) and flexural (EN ISO 178) testing of printed samples were performed on a Shimadzu AGS-X universal testing machine (Fig. 7), where strain rate was 5mm/min. For every test, five specimens are tested. The results and Hooks diagrams are recorded and collected with Shimadzu Trapezium X software, and in further text statistical analyses are presented.

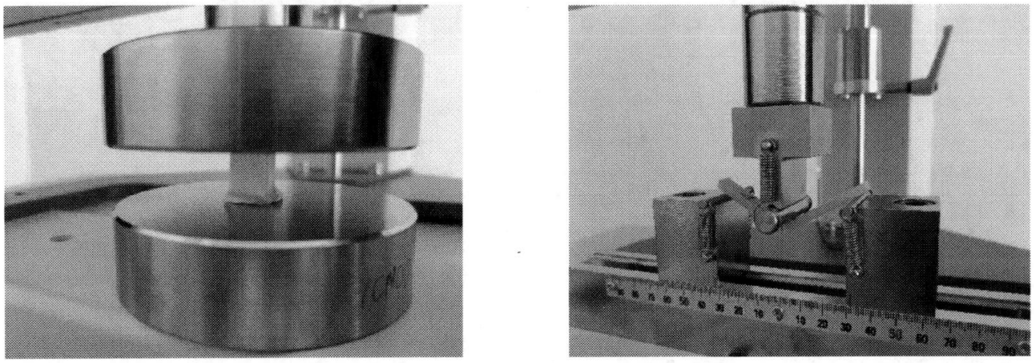

Fig. 7. Compressive (left) and flexural (right) testing on Shimadzu AGS-X testing machine

3. Results and discussion

After analysing the obtained data of compressive and flexural material testing, the results with discussion are presented below. What has been analysed is whether the infill pattern and density affect the compressive and flexural properties of the FDM printed PLA material with infill structure. Compressive and flexural strength as well as young modulus were analysed from the material properties. A total of five different infill patterns were analysed, where different infill density was varied for each pattern (20%, 40%, 60% and 80%). What should be noted is that the strength results obtained by the test are values for the "full" cross section of the material, but as the aim of the research is to see percentage differences in mechanical properties, so this will not affect the final results.

Compressive and flexural stress-strain diagrams for PLA FDM 3D printed material with "Triangle" infill structure with different infill density are presented on Fig. 8.

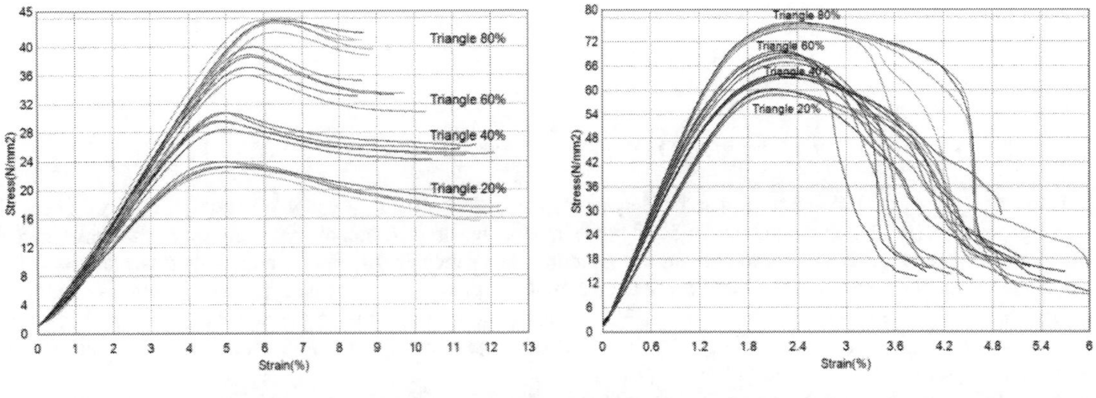

Fig. 8. Compressive (left) and flexural (right) stress – strain diagrams for "Triangle" infill pattern with different infill density

From Fig. 9. it can be seen that infill density have influence on compressive strength for every tested infill pattern, and for each pattern the compressive strength increases linearly with increasing infill density. Also, as expected, the highest values of compressive strength were obtained for an infill density of 80%.

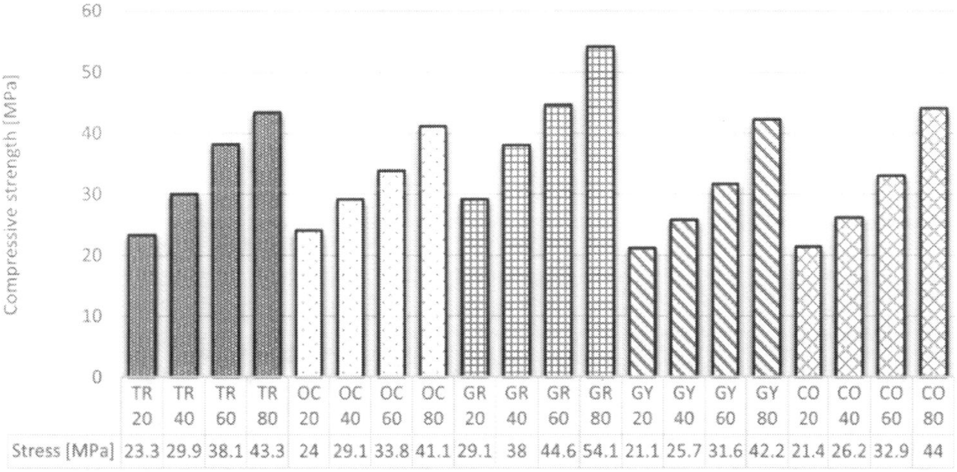

Fig. 9. Influence of infill density for five different infill patterns on compressive strength

The diagram below (Fig. 10) shows that in addition to the infill density, the infill pattern also has an effect on the compressive strength. In general, it can be observed that from five tested infill patterns, "Grid" pattern gives the highest values of compressive strength for all tested infill densities. It is interesting to see that the "Grid" pattern with density of 20% shows higher compressive strength than "GY" and "CO" patterns with density of 40%, and the same value as the "OC" pattern. Also, the "Grid" pattern with 60% density gives a higher value of compressive strength than the "TR", "OC", "GY" and "CO" patterns with a density of 80%.

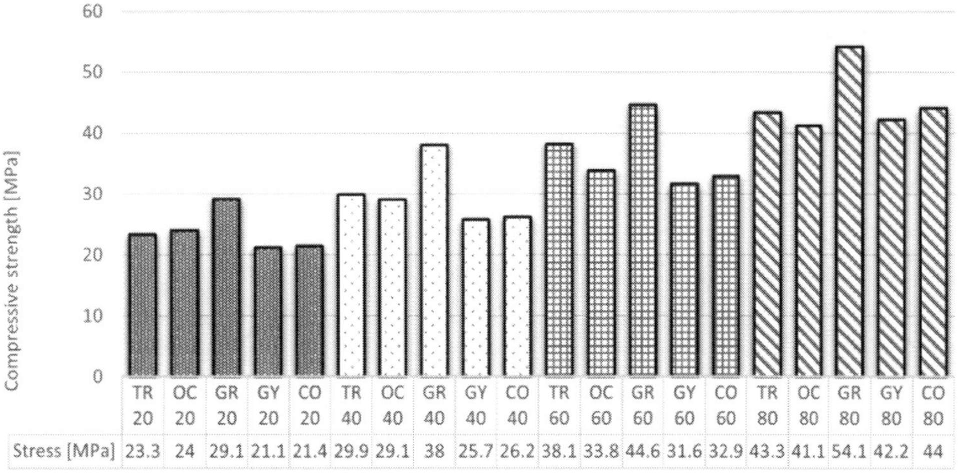

Fig. 10. Influence of infill pattern with different infill density on compressive strength

The diagram in Fig. 11. shows that the infill density also affects the compressive modulus, and generally for all five tested infill patterns a linear relationship between the infill density and the compressive modulus can be seen. In the case of "Triangle" and "Grid" patterns, it can be noticed that for infill densities of 60% and 80%, relatively equal values of compression modulus are obtained, where the difference for "Triangle" pattern is about 3% and for "Grid" pattern about 1,5%.

Influence of infill pattern with different infill density on compressive modulus is presented on diagram bellow (Fig, 12). From the diagram it can be concluded that the infill pattern affects the compressive modulus, where for infill density of 20%, 40% and 60%, depending on the infill pattern, compressive modulus differs up to 15%. While for a infill density of 80%, the compressive modulus for the "Gyroid" pattern is lower by 7% compared to other patterns that have relatively the same compressive modulus value.

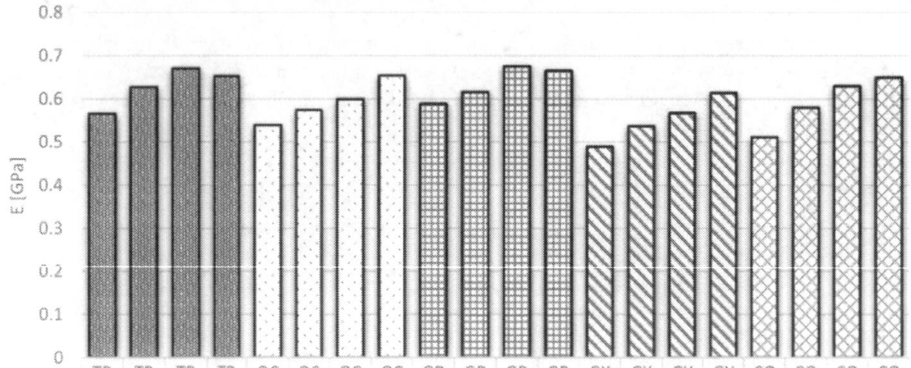

	TR 20	TR 40	TR 60	TR 80	OC 20	OC 40	OC 60	OC 80	GR 20	GR 40	GR 60	GR 80	GY 20	GY 40	GY 60	GY 80	CO 20	CO 40	CO 60	CO 80
E [GPa]	0.57	0.63	0.67	0.65	0.54	0.57	0.6	0.65	0.59	0.61	0.67	0.66	0.49	0.54	0.57	0.61	0.51	0.58	0.63	0.65

Fig. 11. Influence of infill density for five different infill patterns on compressive modulus, E [GPa]

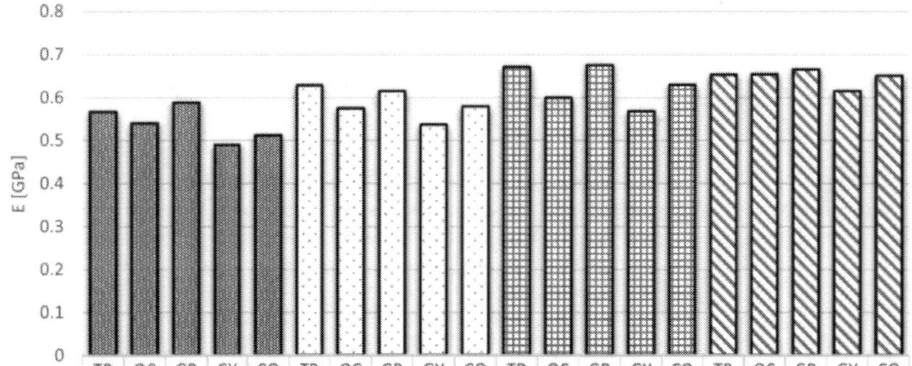

	TR 20	OC 20	GR 20	GY 20	CO 20	TR 40	OC 40	GR 40	GY 40	CO 40	TR 60	OC 60	GR 60	GY 60	CO 60	TR 80	OC 80	GR 80	GY 80	CO 80
E [GPa]	0.57	0.54	0.59	0.49	0.51	0.63	0.57	0.61	0.54	0.58	0.67	0.6	0.67	0.57	0.63	0.65	0.65	0.66	0.61	0.65

Fig. 12. Influence of infill pattern with different infill density on compressive modulus, E [GPa]

The obtained results of compressive mechanical properties of FDM 3D printed PLA material with infill structure could not be compared with the results of samples with 100% infill density, because the testing machine has a maximum force of 10 kN, and samples of PLA material with 100% infill density require higher force at compressive tests.

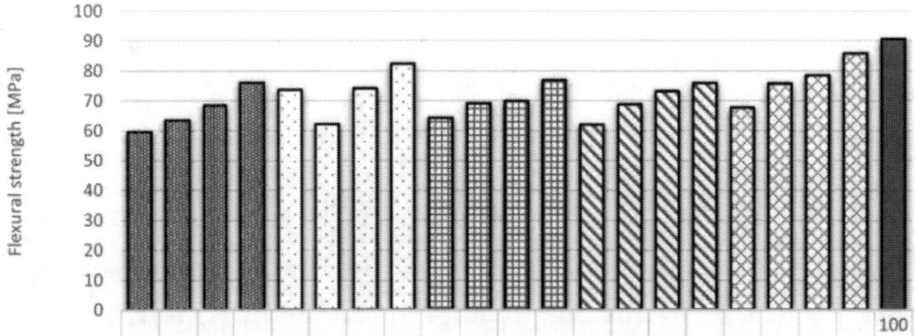

	TR 20	TR 40	TR 60	TR 80	OC 20	OC 40	OC 60	OC 80	GR 20	GR 40	GR 60	GR 80	GY 20	GY 40	GY 60	GY 80	CO 20	CO 40	CO 60	CO 80	100 % Infill
Stress [MPa]	59.6	63.5	68.5	76	73.7	62.2	74.1	82.4	64.2	69.1	69.8	76.8	62.1	68.7	73.1	75.8	67.6	75.6	78.5	85.7	90.5

Fig. 13. Influence of infill density for five different infill patterns on flexural strength

From diagram above (Fig. 13) it can be seen that infill density have influence on flexural strength for every tested infill pattern, and for each pattern the flexural strength increases linearly with increasing infill density. Also, with the "Octet" infill pattern, it can be seen that an infill density of 20% gives a higher value of flexural strength, wich is not expected, so these measurements should certainly be repeated.

As expected, the highest values of flexural strength were obtained for an infill density of 80%, and when they are compared with the flexural strength values for infill density of 100% the flexural strength varies from 5% to 17% depending on the shape of the infill.

The diagram from Fig. 14. shows that the infill pattern influences the flexural strength. If we exclude the results for the "Octet" pattern with infill density of 20% due to the previously mentioned reasons, it can be concluded that the "Concentric" pattern gives the highest values of flexural strength. What also can be concluded from the diagram is that "Concentric" pattern with density of 20% shows higher flexural strength than "TR" and "OC" patterns with density of 40%. Also, the "Concentric" pattern with 60% density gives a higher value of flexural strength than the "TR", "GR" and "GY" patterns with a density of 80%.

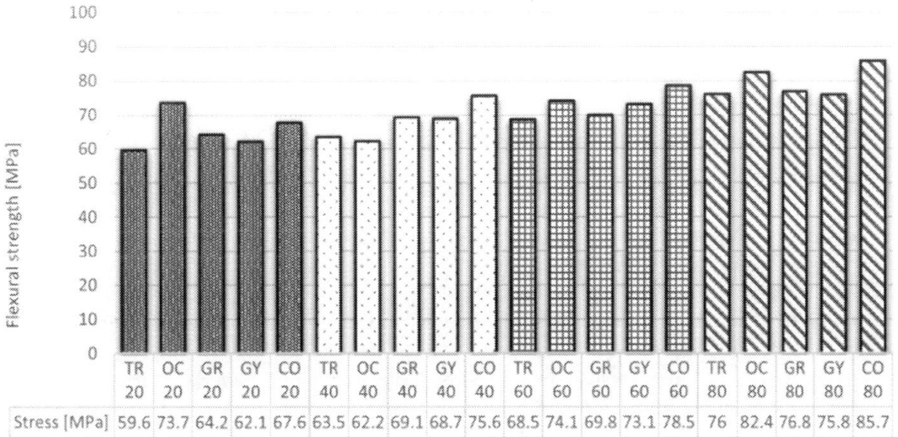

Fig. 14. Influence of infill pattern with different infill density on flexural strength

The diagram in Fig. 15. confirms the influence of infill density on flexural modulus, where the linear dependence of flexural modulus and infill density are also confirmed. It can be noticed that for the "Octet" and "Concentric" infill pattern with an infill density of 80%, the flexural modulus value differs by 4% (for Octet) and 0,7% (for Concentric) compared to the test sample with 100% infill density.

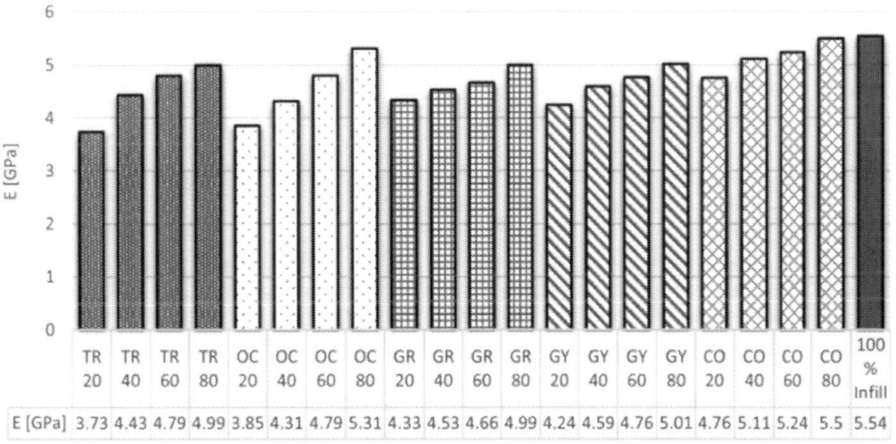

Fig. 15. Influence of infill density for five different infill patterns on flexural modulus, E [GPa]

The results of influence of infill pattern with different infill density on flexural modulus are presented on diagram bellow (Fig, 16). From the diagram it can be concluded that the infill pattern affects the flexural modulus, and depending on the infill pattern, for infill density of 20% differs up to 15%, for infill density of 40% differs up to 17%, for infill density of 60% differs up to 12% and for infill density of 80% differs up to 10%.

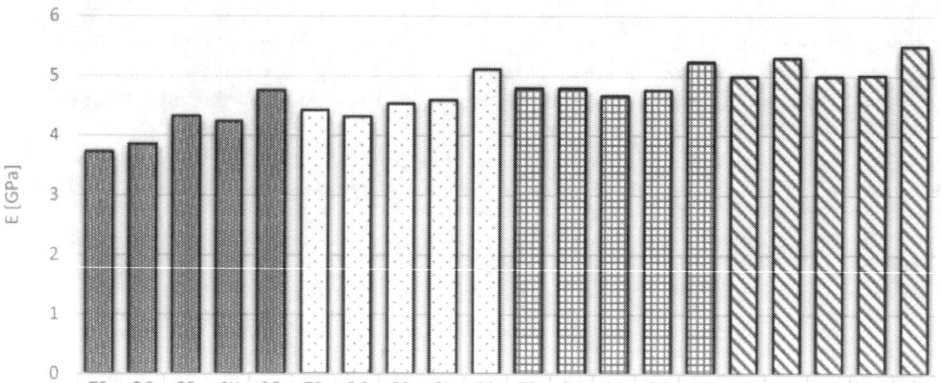

	TR 20	OC 20	GR 20	GY 20	CO 20	TR 40	OC 40	GR 40	GY 40	CO 40	TR 60	OC 60	GR 60	GY 60	CO 60	TR 80	OC 80	GR 80	GY 80	CO 80
E [GPa]	3.73	3.85	4.33	4.24	4.76	4.43	4.31	4.53	4.59	5.11	4.79	4.79	4.66	4.76	5.24	4.99	5.31	4.99	5.01	5.5

Fig. 16. Influence of infill pattern with different infill density on flexural modulus, E [GPa]

4. Conclusion

The experimental study presented in this research paper aims to present the influence of infill pattern and density on compressive and flexural mechanical properties of FDM printed PLA material. Five different infill patterns (Triangle, Octet, Grid, Gyroid and Concentric) with different infill densities (20%, 40%, 60% and 80%) were analysed. Of the mechanical properties, compressive and flexural strength as well as modulus of elasticity were analysed, and results obtained from this research conclude the followings:

- Infill density have influence on compressive strength for every tested infill pattern, and compressive strength increases linearly with infill density increase.
- Infill pattern influences the compressive strength. From five tested infill patterns, "Grid" pattern gives the highest values of compressive strength for all tested infill densities. The "Grid" pattern with density of 20% shows higher compressive strength than "GY" and "CO" patterns with density of 40%, and the same value as the "OC" pattern. Also, the "Grid" pattern with 60% density gives a higher value of compressive strength than the "TR", "OC", "GY" and "CO" patterns with a density of 80%.
- Infill density affects the compressive modulus, where for all five examined infill patterns a linear relationship between the infill density and the compressive modulus is shown. For "Triangle" and "Grid" patterns, with infill densities of 60% and 80%, relatively equal values of compression modulus are obtained, where the difference for "Triangle" pattern is about 3% and for "Grid" pattern about 1,5%.
- Infill pattern affects the compressive modulus, where for infill density of 20%, 40% and 60%, depending on the infill pattern, compressive modulus differs up to 15%. While for a infill density of 80%, the compressive modulus for the "Gyroid" pattern is lower by 7% compared to other patterns that have relatively the same compressive modulus value.
- Infill density have influence on flexural strength for every tested infill pattern, and for each pattern the flexural strength increases linearly with increasing infill density. Also, with the "Octet" infill pattern, an infill density of 20% gives a higher value of flexural strength, which is not expected, so these measurements should certainly be repeated.
- Infill pattern influences the flexural strength, and it can be concluded that the "Concentric" pattern gives the highest values of flexural strength. What also can be concluded from the results is that "Concentric" pattern with density of 20% shows higher flexural strength than "Triangle" and "Octet" patterns with density of 40%. Also, the "Concentric" pattern with 60% density gives a higher value of flexural strength than the "Triangle", "Grid" and "Gyroid" patterns with a density of 80%.
- Results confirms the linear influence of infill density on flexural modulus. It can be noticed that for the "Octet" and "Concentric" infill pattern with an infill density of 80%, the flexural modulus value differs by 4% (for Octet) and 0,7% (for Concentric) compared to the test sample with 100% infill density.
- Infill pattern affects the flexural modulus, and depending on the infill pattern it differs from 10% to 17%.

In general, the printed materials with an infill structure are the future of production, which provides a number of advantages. In this research, only one segment of the influence of infill design on the compressive and flexural mechanical properties of PLA material was performed. In order to better understand and describe the behaviour of printed materials with an infill structure, similar tests should be performed on other materials too in future research. Also, the influence of infill design should be examined in the other two build orientation, because in this paper only one build orientation was analysed. What else should be also done on the topic of material characterization, where the results of real strength values would be obtained, because in the case of materials with infill structure, the cross section is not solid. In future, research on obtaining models of materials for FEM/FEA analyses and more should be done.

5. Acknowledgments

The authors would like to acknowledge the Ultimaker and 3D Republika companies for supporting this research with 3D printers and materials.

6. References

[1] Pandzic, A.; Hodzic, D. & Milovanovic, A. (2019). Effect of Infill Type and Density on Tensile Properties of PLA Material for FDM Process, Proceedings of the 30th DAAAM Interational Symposium, pp.0545-0554, B. Katalinic (Ed.), Published by DAAAM International, ISBN 978-3-902734-22-8, ISSSN 1726-9679, Vienna, Austria, DOI: 10.2507/30th.daaam.proceedings.074

[2] Rahmatabadi, D.; Aminzadeh, A.; Aberoumand, M. & Moradi M. (2021). Mechanical Characterization of Fused Deposition Modeling (FDM) 3D Printed Parts, In: Dave H.K., Davim J.P. (eds) Fused Deposition Modeling Based 3D Printing. Materials Forming, Machining and Tribology. Springer, Cham. DOI: 10.1007/978-3-030-68024-4_7

[3] Pandzic, A.; Hodzic, D. & Milovanovic, A. (2019). Influence of Material Colour on Mechanical Properties of PLA Material in FDM Tecgnology, Proceedings of the 30th DAAAM Interational Symposium, pp.0555-0561, B. Katalinic (Ed.), Published by DAAAM International, ISBN 978-3-902734-22-8, ISSSN 1726-9679, Vienna, Austria, DOI:10.2507/30th.daaam.proceedings.075

[4] Hodzic, D.; Pandzic, A.; Hajro, I. & Tasic, P. (2020). Strength Comparison of FDM 3D Printed PLA made By Different Manufacturers, TEM Journal, Vol. 9, No. 3, 2020, pp. 966-970., ISSN 2217-8390, DOI:10.18421/TEM93-18

[5] Nuha, H. J. Al Hasan. (2020). The influence of an Infill Density, Percent on the Flexural Strength of the 3D, Key Engineering Materials, Vol. 870, pp. 73-80., ISSN 1662-9795, DOI: 10.4028/www.scientific.net/KEM.870.73

[6] Dey, A.; Hoffman, D. & Yodo, N. (2019). Optimizing Multiple Process Parameters in Fused Deposition Modeling With Particle Swarm Optimization, International Journal on Interactive Design and Manufacturing, Vol. 14, pp. 393-405, DOI: 10.1007/s12008-019-00637-9

[7] Farhad M. O.; TahseenFadhil A. & Hind B. A. (2018). Influence of Process Parameters on Mechanical Properties and Printing Time of FDM PLA Printed Parts Using Design of Experiment. Journal of Engineering Research and Applications, Vol. 8, No. 7, (2018) pp. 65-69, ISSN: 2248-9622

[8] Hodzic, D.; Pandzic, A.; Hajro, I. & Tasic, P. (2020). Strain Rate Influence on Mechanical Characteristics of FDM 3D printed Materials, Proceedings of the 31st DAAAM Interational Symposium, pp.0168-0175, B. Katalinic (Ed.), Published by DAAAM International, ISBN 978-3-902734-29-7, ISSN 1726-9679, Vienna, Austria, DOI:10.2507/31st.daaam.proceedings.022

[9] Huu, N. H.; Phuoe, D. P.; Huu, T. N. & Thu, H. T. T. (2019). Optimization of the FDM Parameters to Improve The Compressive Strength of The PLA-Copper Based Products. IOP Conference Series: Materials Science and Engineering, 530 012001. DOI:10.1088/1757-899X/530/1/012001

[10] Shabana, S.; Santosh, R. V. N.; Vikram, K. A. & Lakshmi, V. V. K. (2019). Evaluating the Mechanical Properties of Commonly Used 3D Printed ABS and PLA Polymers With Multi Layered Polymers. International Journal of Engineering and Advanced Technology, Vol. 8, No. 6, 2019, ISSN: 2249-8958, DOI: 10.35940/ijeat.F8646.088619

[11] Gebisa, A. W. & Lemu, H. G. (2018). Investigation Effects of Fused-Deposition Modeling (FDM) Processing Parametes on Flexural Properties of ULTEM 9085 using Designed Experiment. Materials, Vol. 11, No. 4, DOI:10.3390/ma11040500

[12] Akhoundi, B. & Behravesh, A. H. (2019). Experimental Mechanics, Vol. 59, pp. 883-897, ISSN 0014-4851, DOI: 10.1007/s11340-018-00467-y

[13] Habrman, M. (2019). Influence of the Infill on the Tensile Strength and the Economic Factors of 3D Printing, Proceedings of the 30th DAAAM Interational Symposium, pp.1056-1060, B. Katalinic (Ed.), Published by DAAAM International, ISBN 978-3-902734-22-8, ISSSN 1726-9679, Vienna, Austria, DOI: 10.2507/30th.daaam.proceedings.174

[14] Pandzic, A., Hodzic, D. & Kadric, E. (2021). Experimental Investigation on Influence of Infill Density on Tensile Mechanical Properties of Different FDM 3D Printed Materials, TEM Journal, Vol. 10, No. 3, DOI: 10.18421/TEM103-25

[15] Cabreria, V. & Santana R. M. C. (2020). Effect of Infill Pattern in Fused Filament Fabrication (FFF) 3D Printing on Materials Performance, Revista Materia, Vol. 25, No. 3, DOI: 10.1590/S1517-707620200003.1126

[16] Pandzic, A.; Hodzic, D.; Hajro, I. & Tasic, P. (2020). Strength Properties of PLA Material Obtained By Different Models of FDM 3D Printer, Proceedings of the 31st DAAAM Interational Symposium, pp.0315-0322, B. Katalinic (Ed.), Published by DAAAM International, ISBN 978-3-902734-29-7, ISSN 1726-9679, Vienna, Austria, DOI: 10.2507/31st.daaam.proceedings.044

[17] Valerga, A. P.; Batista, M.; Salguero, J., & Girot, F. (2018). Influence of PLA filament conditions on characteristics of FDM parts. Materials, Vol. 11, No. 8: 1322, DOI:10.3390/ma11081322

DOI: 10.2507/32nd.daaam.proceedings.030

INFLUENCE OF LAYER HEIGHT, BUILD ORIENTATION AND POST CURING ON TENSILE MECHANICAL PROPERTIES OF SLA 3D PRINTED MATERIAL

Adi Pandzic

This Publication has to be referred as: Pandzic, A[di] (2021). Influence of Layer Height, Build Orientation and Post Curing on Tensile Mechanical Properties of SLA 3D Printed Material, Proceedings of the 32nd DAAAM International Symposium, pp.0200-0208, B. Katalinic (Ed.), Published by DAAAM International, ISBN 978-3-902734-33-4, ISSN 1726-9679, Vienna, Austria
DOI: 10.2507/32nd.daaam.proceedings.030

Abstract

In this research, an experimental analysis of the influence of layer height, build orientation and post curing on the tensile mechanical properties of SLA 3D printed material was performed. The material "Grey Resin" from the manufacturer Fromlabs was specifically analysed. When it comes to layer height, four different layer heights were analysed, with 160, 100, 50, and 25 microns. Specimens were manufactured in three different build orientations (vertical, side and flat). All specimens were tested with and without post curing (30 min, 60 °C). The results in further text showed that the layer height, build orientation as well as the post curing have an impact on the tensile mechanical properties of the SLA "Grey Resin" material. Also, in addition to the effect on mechanical properties, the paper presents how these factors affect the time of 3D printing.

Keywords: SLA; Layer thickness; Build orientation; Curing; Tensile testing

1. Introduction

The technologies that are slowly changing the way products are made today are certainly additive technologies. Production of very complex physical models geometries with good quality and in a wide range of different materials is possible with additive manufacturing (AM) technologies. This technology is used in many fields such as engineering, medicine (prosthetics, bionics, orthotics), automotive and aircraft industries, dental restorations, prototyping, education and more [1], [2] [3]. Today it can be seen the continuous development of new AM technologies, materials and applications. In terminology additive manufacturing is also known as rapid prototyping (RP). On the market there are several different groups (families) of AM technologies (according to ASTM F2792 standard) such as VAT Photopolymerization, Powder Bed Fusion (PBF), Binder Jetting, Material Jetting, Sheet Lamination, Material Extrusion, Directed Energy Deposition (DED) and Hybrid, also all the listed technologies can be found in literature under different names. In this research study, stereolithography (SLA) AM technology was used, which in addition to DLP (Digital Light Processing), 3SP (Scan, Spin and Selectively Photocure) and CLIP (Continuous Liquid Interface Production) belongs to VAT Photopolymerization group (Fig. 1) [4], [5], [6].

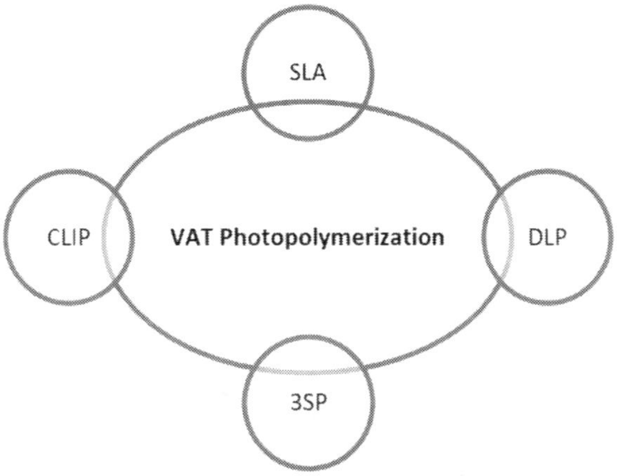

Fig. 1. VAT Photopolymerization AM "family" according to ASTM F2792 standard

With AM technologies, the first step is creating a 3D CAD model of the product using some of CAD software such as Solidworks, Catia, NX or others. The 3D CAD model is then converted to STL or another similar format, and using "Slicer" software, g-code with all printing parameters is generated which is further used by 3D printer to 3D print the product. In general, slicer cuts a 3D model into 2D cross-sectional layers, and based on that the 3D printer makes the final product layer by layer. [7], [8].

Stereolithography 3D printing technology is known for being able to produce products of simple but also complex geometries with high resolution (up to 25 microns) and accuracy, watertight and isotropic characteristics with smooth surface finish and fine features. SLA is technology that basically uses polymeric materials, known as resins, but today it can be seen that some companies developing also composites (Formlabs Ceramic Resin) [9], [10], [11]. There are a number of different materials (resins) on the market as well as the manufacturers themselves, and often the resins are divided by purpose as shown on Fig. 2.

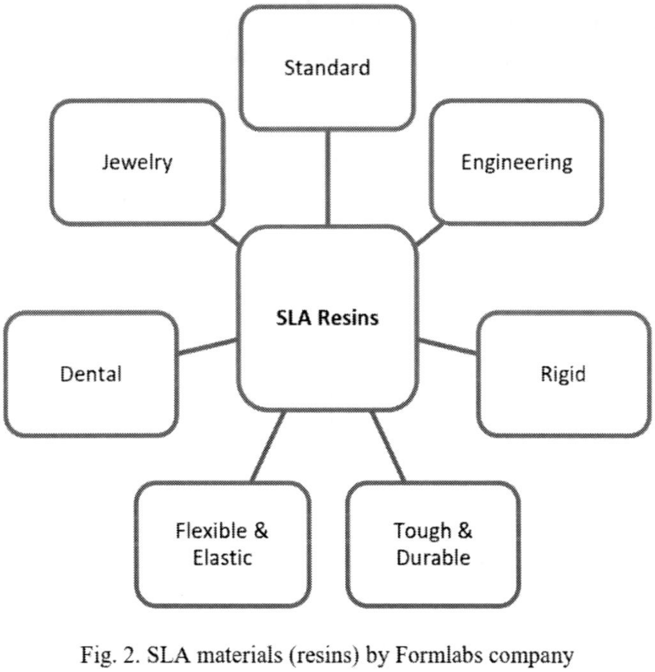

Fig. 2. SLA materials (resins) by Formlabs company

Working principle of SLA 3D printer is based on hardening (curing) a liquid photopolymer resin layer by layer using electronically controlled UV laser beam. The parts are 3D printed on a moving build platform located above fixed resin tank where the material (resin) is stored. Generally, final parts are created bottom-up, and when each new layer is made, they are immersed in the liquid resin. And this procedure is repeated until final part is completely 3D printed [12], [13]. Also, schematic representation of the basic mechanics and parts of SLA 3D printer (Formlabs Form 3) is presented on Fig. 3.

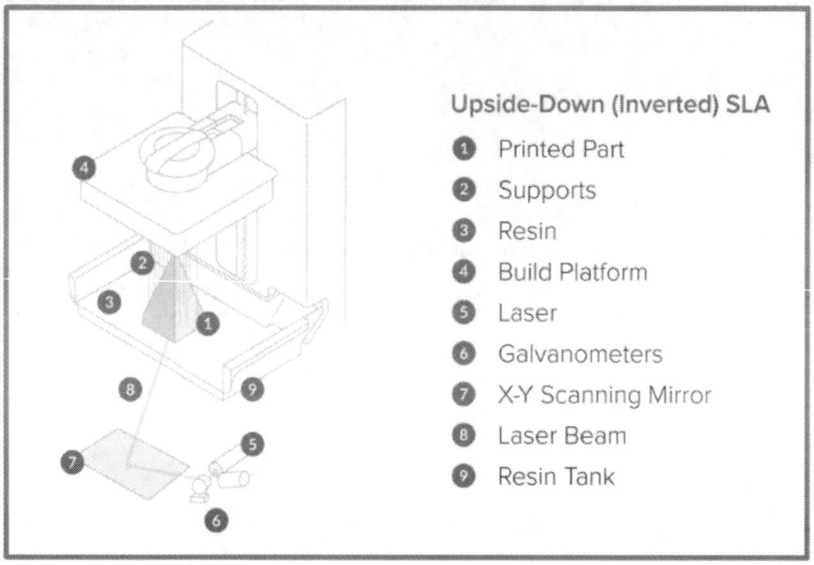

Fig. 3. Schematic representation of the basic mechanics of SLA 3D printer (figure taken from formlabs.com)

Once the final product is 3D printed and material solidifies, the build platform of the SLA printer is lifted out and the excess liquid drains away. After this, printed part usually goes in bath with IPA (isopropanol) alcohol to be washed. Washing can be manually with hands or automated using for example "Form Wash" where time can be controlled, and better cleaning can be achieved due to the circulation of IPA alcohol during the washing. After washing and drying printed product can be used or it can be additionally post cured, with different time and temperature which depends on type of material. Post curing helps 3D printed parts achieve their highest possible strength and stability [14], [15], [16].

Through a literature review, it can be concluded that mechanical properties of printed material in SLA technology are influenced by various factors such as: layer height, build orientation, supports, washing time, post curing time and temperature (Fig. 4). Also, if SLA technology is compared to FDM technology, here is a significant smaller number of influencing factors on the final characteristics of the printed product that user can change [17], [18], [19].

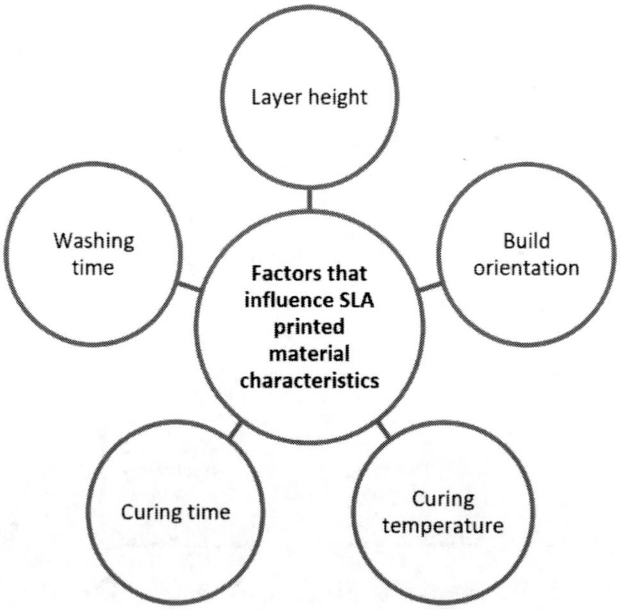

Fig. 4. Influencing factors and printing parameters on SLA printed material characteristics

Based on this, the idea of this paper is to analyse, through experimental research and statistical analysis of the obtained data, the influence of layer height, build orientation and post curing on the tensile mechanical properties of SLA printed material and the printing time.

2. Experimental details

The experiment of this study is divided into several phases shown in the flow diagram below (Fig. 5). In the first stage, the CAD 3D model of tensile testing specimens has been designed and in STL format prepared. Than, in slicer software g-code with printing parameters was prepared and specimens were SLA 3D printed. All specimens are washed after 3D printing in IPA alcohol and half of the total amount of specimens were post cured. In the end all specimens were tensile tested, collected data was analysed and results are presented.

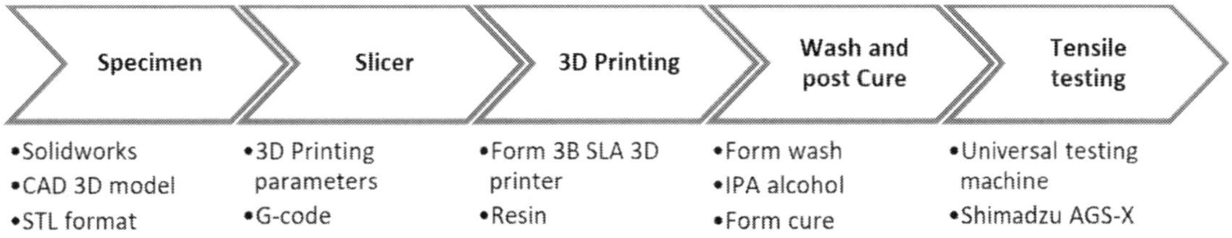

Fig. 5. Experiment phases flow diagram

CAD 3D model of tensile testing specimen was designed in Solidworks 2021 software, and STL format was prepared. Tensile testing specimen dimensions are according to ISO 527-2, and technical drawing is presented in Fig. 6.

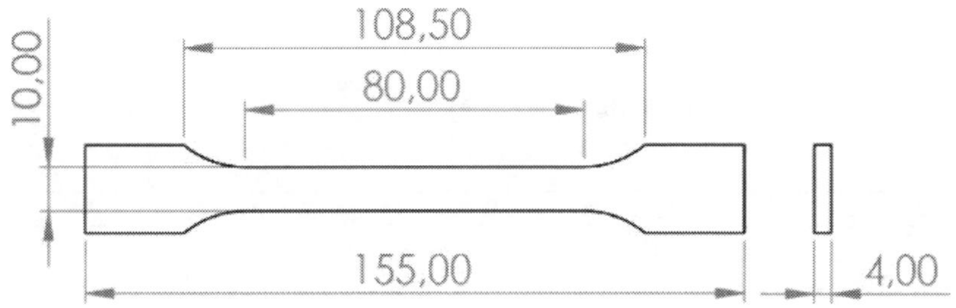

Fig. 6. Tensile testing specimen dimensions according to ISO 527-2

Specimen 3D model was prepared for printing in slicer software PreForm (FormLabs, 3.16.0 version), where printing parameters are also defined and presented in Table 1.

3D Printer	Resin (Material)	Layer height [microns]	Build orientation
Form 3B	Grey, V4	25	Vertical (V) Flat (F) Side (S)
		50	
		100	
		160	

Table 1. Printing parameters defined in PreForm slicer

Different build orientations are presented on Fig. 7. Also, the tensile test specimens in the PreForm slicer as well as after they were manufactured are shown in the Fig. 8.

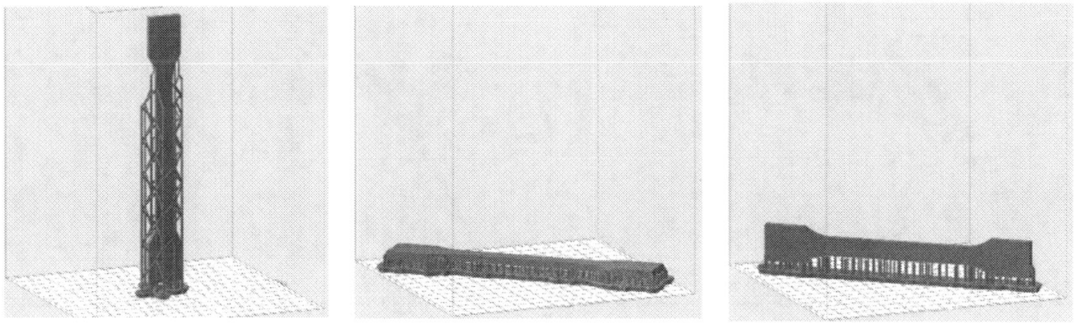

Fig. 7. Different specimen printing build orientations (from left: vertical, flat and side)

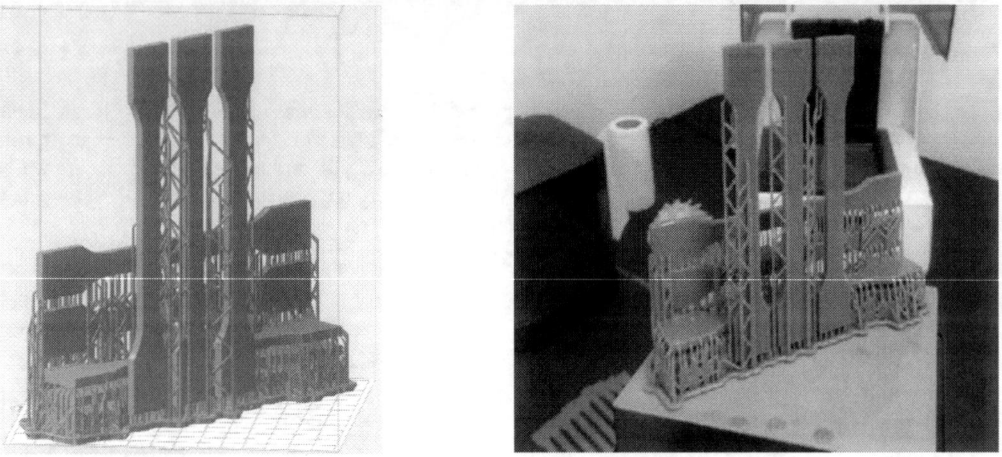

Fig. 8. Prepared tensile specimens in PreForm slicer (left) and printed specimens (right)

Grey resin (SLA material), from FormLabs manufacturer, was analysed in this paper. It is from group of standard resins for 3D printing strong, precise concept models and prototypes. With a matte surface finish, opaque appearance, and precise details, prints can be used right of the printer but also can be post cured. Material properties data by manufacturer is presented in Table 2.

Material properties	Air dried without curing [1]	Post-Cured [2]	Method
Ultimate tensile strength	38 MPa	65 MPa	ASTM D 638-10
Tensile modulus	1,6 GPa	2,8 GPa	ASTM D 638-10
Elongation at failure	12 %	6,2 %	ASTM D 638-10
Flexural modulus	1,25 GPa	2,2 GPa	ASTM D 790-10
Notched IZOD	16 J/m	25 J/m	ASTM D 256-10

[1] Data was obtained from parts, printed using Form 2, 100 microns, Clear settings, washed and air dried without post cure.

[2] Data was obtained from parts printed using Form 2, 100 microns, Clear setings, and post cured with 1,25 mW/cm^2 of 405 nm LED light for 60 minutes at 60 °C.

Table 2. Grey resin material properties data by manufacturer

After setting all parameters in slicer, g-code was created, and all specimens were printed on Form 3B SLA 3D printer. Then, specimens were washed using Form Wash for 10 min and air dried. Post curing settings were 30 min with 60 °C.

Tensile testing was performed according to ISO 527-2 and using universal testing machine Shimadzu AGS-X with maximum load of 10 kN. Testing speed (strain rate) was 5 mm/min, and all specimens were tested under laboratory conditions. Data monitoring during tensile testing was performed using Shimadzu Trapezium X software, and overall stress-strain diagrams for specimens with and without post curing are presented on Fig. 9.

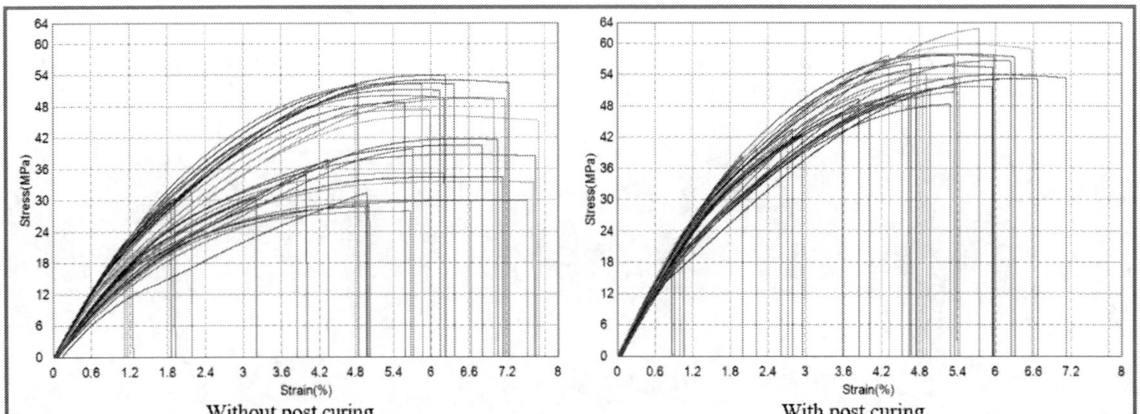

Fig. 9. Overall Stress-Strain diagrams for tensile tested specimens with and without post curing

3. Results and discussion

In this section tensile testing results are presented and discussed, where influence of layer height (LH), build orientation (BO) and post curing (PC) on tensile strength and elastic modulus were analysed. Also, influence of those parameters on printing time was analysed. In general, the results are presented through diagrams where they are analysed and commented.

Influence of layer height and post curing on tensile strength and elastic modulus for specimens printed in flat (F) build orientation are presented on Fig. 10.

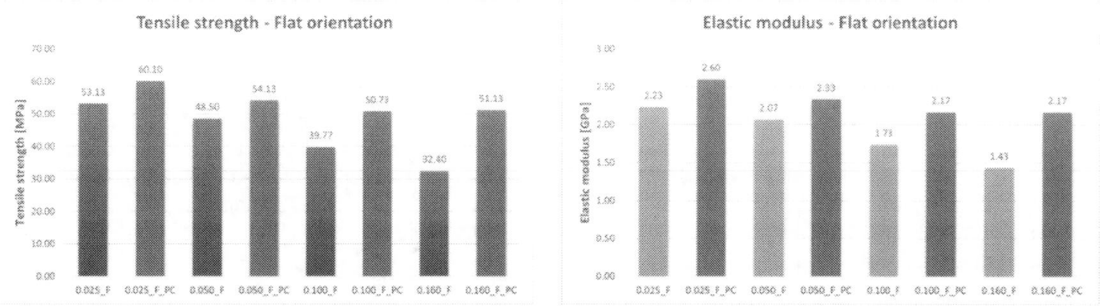

Fig. 10. Influence of LH and PC on tensile strength and elastic modulus for specimens printed in flat BO

Diagrams from Fig. 10 are showing that LH have influence on tensile strength and elastic modulus for specimens printed in flat BO. Depending on LH tensile strength varies up to 48% for specimens without PC, and up to 17% for specimens with PC. The value of the elastic modulus varies up to 43% for specimens without PC, and up to 18% for specimens with PC. When analysing the PC effect, it can be seen that in general the value of the tensile strength and elastic modulus has increased. Depending on the LH, the PC caused an increase in tensile strength up to 44%, and for elastic modulus up to 41%.

Influence of layer height and post curing on tensile strength and elastic modulus for specimens printed in side (S) build orientation are presented on Fig. 11.

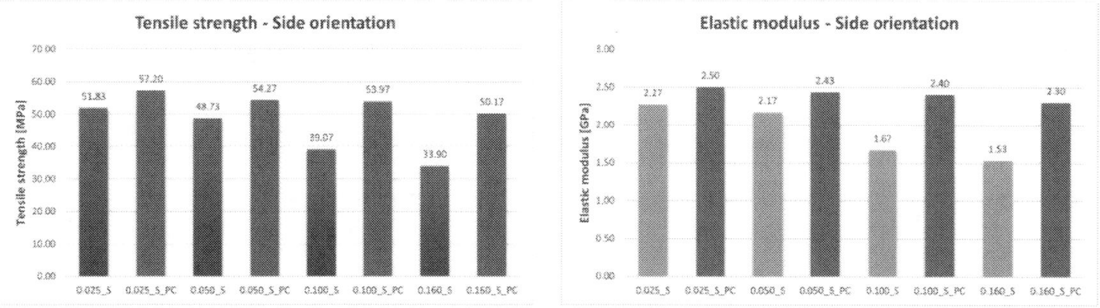

Fig. 11. Influence of LH and PC on tensile strength and elastic modulus for specimens printed in side BO

Diagrams from Fig. 11 are showing that LH have influence on tensile strength and elastic modulus for specimens printed in side BO. Depending on LH tensile strength varies up to 41% for specimens without PC, and up to 13% for specimens with PC. The value of the elastic modulus varies up to 39% for specimens without PC, and up to 8% for specimens with PC. When analysing the PC effect, it can be seen that the value of the tensile strength and elastic modulus has increased. Depending on the LH, the PC caused an increase in tensile strength up to 38%, and for elastic modulus up to 40%.

Influence of layer height and post curing on tensile strength and elastic modulus for specimens printed in vertical (S) build orientation are presented on Fig. 12, and diagrams are showing that LH have influence on tensile strength and elastic modulus for specimens printed in vertical BO. Depending on LH tensile strength varies up to 37% for specimens without PC, and up to 71% for specimens with PC. The value of the elastic modulus varies up to 46% for specimens without PC, and up to 7% for specimens with PC. When analysing the PC effect, in general the value of the tensile strength and elastic modulus has increased. Depending on the LH, the PC caused an increase in tensile strength up to 48%, and for elastic modulus up to 47%.

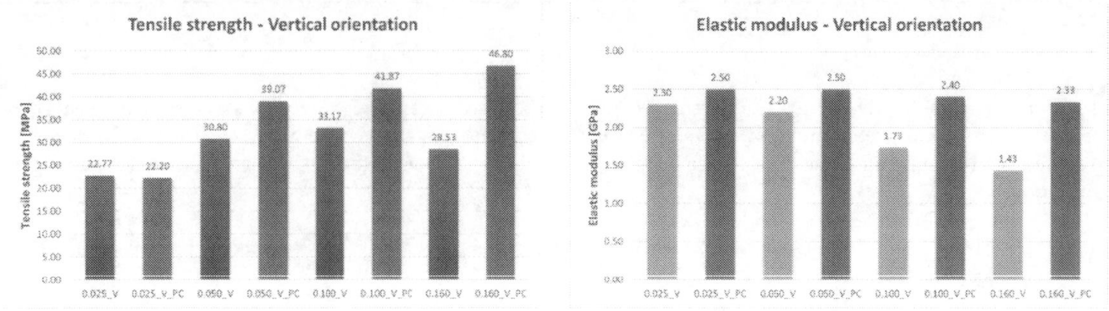

Fig. 12. Influence of LH and PC on tensile strength and elastic modulus for specimens printed in vertical BO

Influence of build orientation on tensile strength and elastic modulus is presented on diagrams below (Fig. 13). As expected, and based on these diagrams, it can be concluded, that the build orientation during SLA printing has an impact on the tensile mechanical properties. Results also showing that depending on the layer height, the impact of the build orientation changes, and specimens with lower layer height showed higher impact of build orientation on tensile strength. Analysing all samples, with and without post curing, for samples SLA printed in flat and side build orientation, the difference in tensile strength does not exceed 6%. For elastic modulus, in general, vertical build orientation is showing highest elastic modulus as expected.

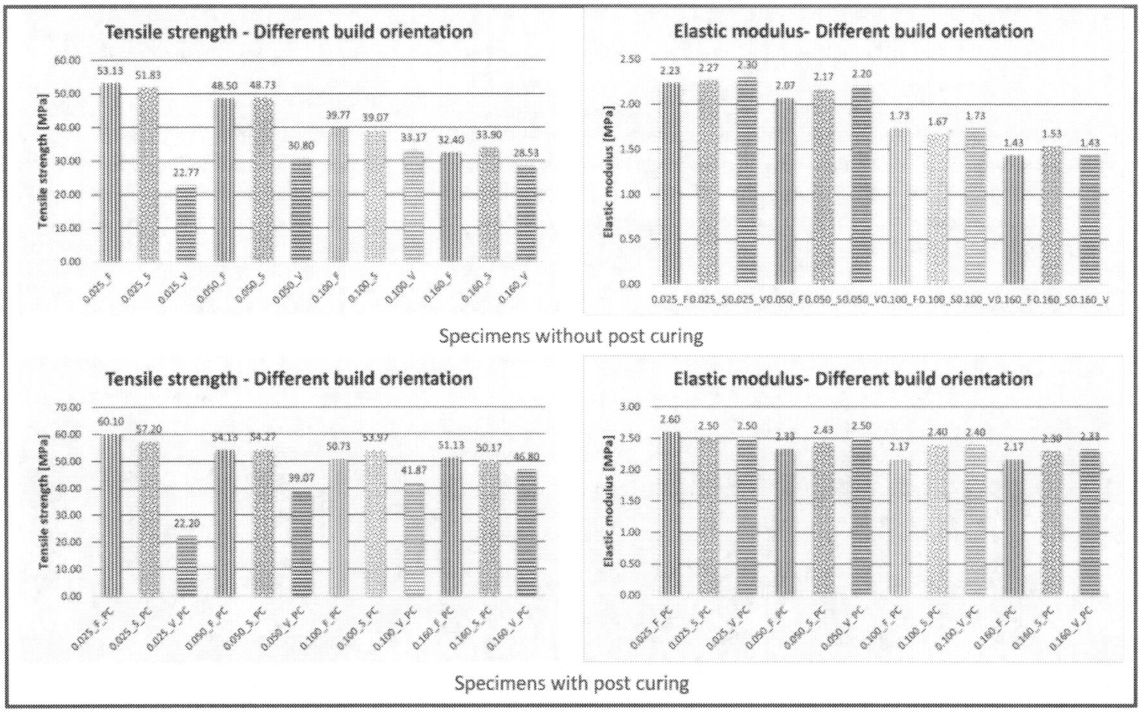

Fig. 13. Influence of build orientation on tensile strength and elastic modulus for tested specimens with and without post curing

It is known that the layer height and build orientation affect the time of product manufacturing by 3D printing, which can be seen from diagrams in Fig. 14. As expected, when the layer height decreases, the printing time increases. Analysing build orientation, the maximum printing time is achieved with vertical, then with the side and flat build orientation. Vertical build orientation increases printing time up to 6 times compared to side and flat build orientation. Post cure generally increased tensile strength and modulus of elasticity, which could be inferred from previous results as well.

When printing time is analysed, which is important in most cases because it directly influence on final product cost, from diagrams it can be seen that with post curing we can print products with higher layer height and achieved mechanical properties similar or higher than specimens with lower layer height, and with that, save the printing time and final product cost. In this research, post curing was 30 min which are added on overall manufacturing time (printing time + post curing time) and presented in diagrams (Fig. 14). So based on the results from diagrams, it can be concluded that with post curing after SLA printing, mechanical properties can be achieved with saving overall manufacturing time.

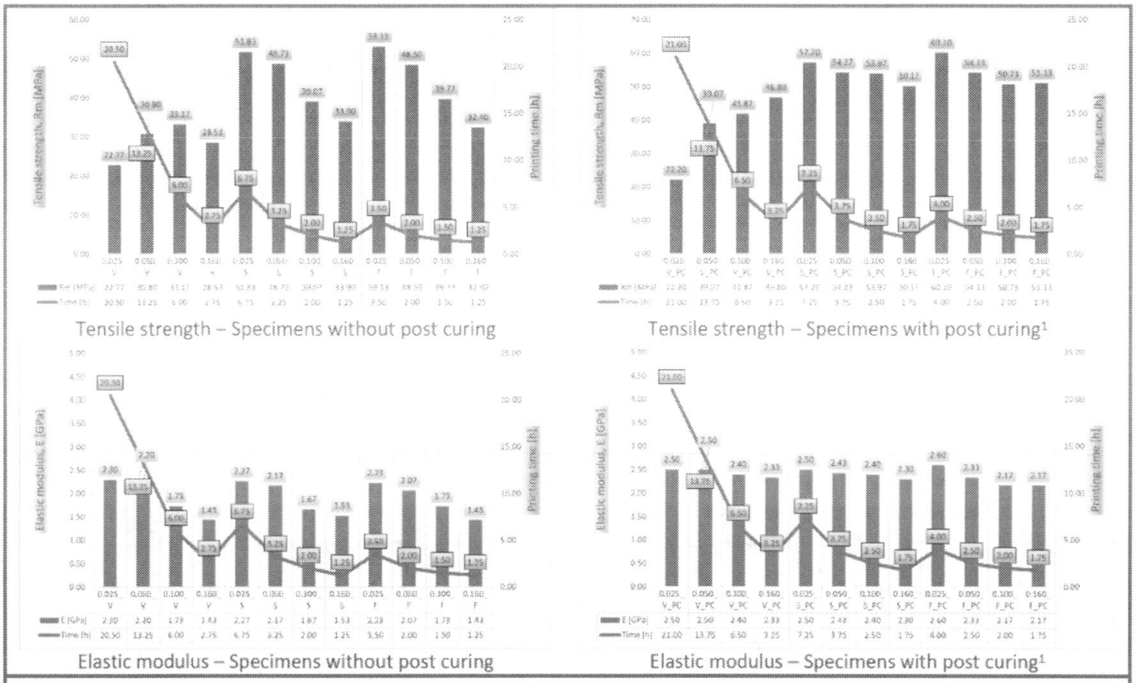

[1] Post cure time was 30min, so on diagrams with post cure 30 min was added on printing time.

Fig. 14. Influence of layer height, build orientation and post cure on printing time for tested specimens with and without post curing

4. Conclusion

Through this experimental research in this study, the influence of layer height, build orientation and post cure on tensile mechanical properties of "Grey resin" (by Formlabs) material SLA 3D printed on a Form 3B device was analysed. The results showed the importance of all three parameters for manufacturing SLA products (in this work tensile specimens) with the aim of obtaining the appropriate tensile strength at an acceptable cost. Three replicas were printed for each combination of parameters, so that a total of 72 tensile specimens were examined according to ISO 527-2, and with strain rate of 5 mm/min. Based on the results from this study, the following conclusions were drawn:

- Layer height have influence on tensile strength and elastic modulus. For samples SLA printed in flat and side build orientations, a linear increase in tensile strength and elastic modulus was shown with a decrease in layer height. While for samples printed in vertical build orientation, a lower layer height showed lower tensile strength.
- Build orientation also affects the tensile strength and elastic modulus, but the intensity of the impact depends on the layer height. Vertical build orientation increases printing time up to 6 times compared to side and flat build orientation. What can also be concluded is that flat and side build orientation show small differences in tensile strength and elastic modulus. For layer height of 160 microns difference in tensile strength for specimens without post curing is less than 17 %, and after post curing the difference is less than 9 %.
- Post curing according to the manufacturer's recommendations for this "Grey resin" material, 30 min at 60 °C, really had the effect of increasing the tensile mechanical properties, where the increase in tensile strength went up to 48 % and in elastic modulus up to 47 %, depending on layer height and build orientation. Also, post curing reduced the difference of tensile strength and elastic modulus for samples made with different layer height.
- Analysing printing time, it can be concluded that with a decrease in layer height, the production time increases, as well as with a different build orientation (increases from flat, side to vertical).

In addition to the above conclusions, if the goal is to reduce the printing time and to achieve adequate strength, this can be done using post curing after printing, so for future research it would be good to do optimization of these parameters. It is also recommended to analyse the influence of different post curing parameters (curing time and temperature), and to examined similar experiment but for other material mechanical properties such as compressive, flexural, etc. And detailed analysis of the impact of various factors and printing parameters on other SLA materials (resins) is of great importance in order to be able to make the most of the advantages of this technology.

5. Acknowledgments

The author would like to acknowledge the "3D Republika" company for supporting this research.

6. References

[1] Pandzic, A.; Hodzic, D.; Hajro, I. & Tasic, P. (2020). Strength Properties of PLA Material Obtained by Different Models of FDM 3D Printer, Proceedings of the 31st DAAAM International Symposium on Intelligent Manufacturing and Automation, Mostar, ISSN: 1726-9679, ISBN 978-3-902734-29-7, Katalinic, B. (Ed.), pp. 0315-0322, Published by DAAAM International, Vienna, DOI: 10.2507/31st.daaam.proceedings.044

[2] Cekic, A.; Begic-Hajdarevic, Dz.; Cohodar, M.; Muhamedagic, K. & Osmanlic, M. (2019). Optimization of Stereolitography and Fused Deposition Modeling Process Parameters, Proceedings of the 30th DAAAM International Symposium on Intelligent Manufacturing and Automation, Zadar, ISSN: 1726-9679, ISBN 978-3-902734-22-8, Katalinic, B. (Ed.), pp. 0681-0687, Published by DAAAM International, Vienna, DOI: 10.2507/30th.daaam.proceedings.093

[3] Pandzic, A.; Hodzic, D. & Milovanovic, A. (2019). Effect of Infill Type and Density on Tensile Properties of PLA Material for FDM Process, Proceedings of the 30th DAAAM International Symposium on Intelligent Manufacturing and Automation, Mostar, ISSN: 1726-9679, ISBN 978-3-902734-22-8, Katalinic, B. (Ed.), pp. 0545-0554, Published by DAAAM International, Vienna, DOI: 10.2507/30th.daaam.proceedings.074

[4] Castro-Sastre, A.; Fernandez-Abia, A. I.; Rodriguez-Gonzalez, P.; Martinez-Pellitero, S. & Barreiro, J. (2018). Characterization of materials Used in 3D-Printing Technology With Different Analysis Techniques, Proceedings of the 29th DAAAM International Symposium on Intelligent Manufacturing and Automation, Zadar, ISSN: 1726-9679, ISBN 978-3-902734-20-4, Katalinic, B. (Ed.), pp. 0947-0954, Published by DAAAM International, Vienna, DOI: 10.2507/29th.daaam.proceedings.136

[5] Pandzic, A.; Hodzic, D. & Kadric, E. (2021). Experimental Investigation on Influence of Infill Density on Tensile Mechanical Properties of Different FDM 3D Printed Materials, TEM Journal, Vol. 10, No. 3, 2021, pp. 1195-1201, ISSN 2217-8309, DOI: 10.18421/TEM103-25

[6] Dana, M.; Zetkova, I.; Hanzl, P. & Hronek, J. (2017). Accuracy of Holes Created by 3D Printing (DMLS), Proceedings of the 28th DAAAM International Symposium on Intelligent Manufacturing and Automation, Zadar, ISSN: 1726-9679, ISBN 978-3-902734-11-2, Katalinic, B. (Ed.), pp. 0467-0473, Published by DAAAM International, Vienna, DOI: 10.2507/28th.daaam.proceedings.065

[7] Walcawik-Macura, E. & Krol, M. (2021). Influence of Post-Processing and the Type of Filling on Strength Properties of Elements Printed by Stereolithography Technology, Solid State Phenomena, Vol. 236, pp. 51-60, Trans Tech Publications Ltd, Switzerland

[8] Yun, J. S.; Tae-Wan, P.; Jeong, Y. H. & Cho, J. H. (2016). Development of Ceramic-Reinforced Photopolymers for SLA 3D Printing Technology, Applied Physics A 122, 629, DOI: 10.1007/s00339-016-0157-x

[9] Dizon, J.R.C.; Gache, C.C.L.; Cascolan, H.M.S.; Cancino, L.T. & Advincula, R.C. (2021). Post-Processing of 3D-Printed Polymers, Technologies, Vol. 9, No. 3, DOI: 10.3390/technologies9030061

[10] Saini, J.S.; Dowling, L.; Kennedy, J. & Trimble, D. (2020). Investigations of the Mechanical Properties on Different Print Orientations in SLA 3D Printed Resin, Proceedings of the Institution of Mechanical Engineers, Part C: Journal of Mechanical Engineering Science, Vol. 234, No. 11, pp. 2278-2293, DOI: 10.1177/0954406220904106

[11] Xiaoyan, X.; Goyanes, A.; Trenfield, S.J.; Diaz-Gomez, L.; Alvarez-Lorenzo, C.; Gaisford, S. & Basit, A. W. (2020). Stereolithography (SLA) 3D Printing of a Bladder Device for Intravesical Drug Delivery, Materials Science & Engineering C, Vol. 120, 10.1016/j.msec.2020.111773

[12] Cosmi, F. & Dal Maso, A. (2020). A Mechanical Characterization of SLA 3D-Printed Specimens for Low-Budget Applications, Materials Today: Proceedings, Vol. 32, No. 2, pp. 194-201, DOI: 10.1016/j.matpr.2020.04.602

[13] Taneva, I. & Uzunov, T. (2020). Influence of Post-Polymerization Processing on the Mechanical Characteristics of 3D-Printed Occlusal Splints, Journal of Physics: Conference Series, 1492, DOI: 10.1088/1742-6596/1492/1/012018

[14] Espino, M. T.; Tuazon, B. J.; Robles, G. S. & Dizon, J. R. C. (2020). Application of Taguchi Methodology in Evaluating the Rockwell Hardness of SLA 3D Printed Polymers, Materials Science Forum, Vol. 1005, pp. 166-173, DOI: 10.4028/www.scientific.net/MSF.1005.166

[15] Dall-Argine, C.; Hochwallner, A.; Klikovits, N.; Liska, R.; Stampf, J. & Sangermano, M. (2020). Hot-Lithography SLA-3D Printing of Epoxy Resin, Macromolecular Materials and Engineering, Vol. 305, No. 10, DOI: 10.1002/mame.202000325

[16] Miedzinska, D.; Gieleta, R. & Ma-Lek, E. (2019). Experimental Study of Strength Properties of SLA Resins Under Low and High Strain Rates, Mechanics of Materials, Vol. 141, DOI: 10.1016/j.mechmat.2019.103245

[17] Shim, J. S.; Kim, J. E.; Jeong, S. H.; Choi, Y. J. & Ryu, J. J. (2020). Printing Accuracy, Mechanical Properties, Surface Characteristics, and Microbial Adhesion of 3D-Printed Resins With Various Printing Orientation, The Journal of Prosthetic Dentistry, Vol. 124, No. 4, pp. 468-475, DOI: 10.1016/j.prosdent.2019.05.034

[18] Bryant, S. J. & Mcleod, R. R. (2018). Understanding and Improving Mechanical Properties in 3D Printed Parts Using a Dual-Cure Acrylate-Based Resin for Stereolithography, Advanced Engineering Materials, Vol. 20, No. 12, DOI: 10.1002/adem.201800876

[19] Watters, M. P. & Bernhardt, M. L. (2018). Curing Parameters to Improve the Mechanical Properties of Stereolithographic Printed Specimens, Rapid Prototyping Journal, Vol. 24, No. 1, DOI: 10.1108/RPJ-11-2016-0180

DOI: 10.2507/32nd.daaam.proceedings.031

DESIGN AND OPTIMIZATION OF PRODUCTION TECHNOLOGY OF THIN-WALLED VANADIUM PARTS

Lubos Kroft & Jindrich Sykora

This Publication has to be referred as: Kroft, L[ubos] & Sykora, J[indrich] (2021). Design and Optimization of Production Technology of Thin-Walled Vanadium Parts, Proceedings of the 32nd DAAAM International Symposium, pp.0209-0214, B. Katalinic (Ed.), Published by DAAAM International, ISBN 978-3-902734-33-4, ISSN 1726-9679, Vienna, Austria
DOI: 10.2507/32nd.daaam.proceedings.031

Abstract

This article deals with one of the research directions solved at the Department of Machining Technology of the University of West Bohemia in the field of thin-walled parts machining technology. The main objective of this research was to find suitable conditions for machining parts made of nearly pure vanadium. Materials such as vanadium are used more and more frequently in the industry and their application, for example in medicine, is nowadays much more common. One of the most critical issues in machining vanadium parts is its relatively high toughness combined with low thermal conductivity, which makes its machining characteristics close to materials like titanium. The combination of these material properties together with a thin-walled part and the requirement for very high precision of the product adds up and creates a rather complex problem that needed to be solved so that the whole manufacturing process could be subsequently deployed in a real production environment.

Keywords: Machining; vanadium; thin-walled part; manufacturing technology; high precision

1. Introduction

The present time brings the use of various non-traditional solutions and their deployment in practice, especially with regard to the continuous development of technology and the desire to achieve higher utility values, reduce economic requirements and also reduce the environmental burden. For this reason, the use of materials with specific mechanical and chemical properties that were previously little used is now quite common. There are many areas dealing with this development, such as the use of hot spraying, superalloys and others. [1] One such material is vanadium. Most people in engineering practice are familiar with vanadium as an alloying element in various alloys. However, its use as a stand-alone material for the manufacture of machine parts with special properties is still very uncommon today. [5]

The properties of vanadium are very close to those of hard machinable materials [5] In combination with a part with a very thin wall, the whole issue can be considered very complicated. Research on machining vanadium parts has been divided into investigating the properties of vanadium itself and its machining properties. Furthermore, the research focused on machining thin-walled components. Based on this research, an experimental part was carried out where the machining strategy was designed and optimized and the stability of the whole process was tested.

2. Properties of vanadium and its machining

It is a silvery metal that, along with niobium and tantalum, is in the 5th group of metals in the periodic table of elements. Vanadium has a cubic centred crystal lattice. These metals are characterized by their high melting points. Vanadium has come into technical use mainly as an alloying element in the preparation of high-quality steels. In steel, vanadium carbide forms V4C3 and thus increases wear resistance. Other uses are found in titanium and aluminium alloys where it increases the mechanical resistance of the alloys.

The use of vanadium as a main material is even nowadays uncommon, the main disadvantage is, as with almost all pure metals, especially the low hardness. However, thanks to a certain amount of alloying elements it is possible to increase the mechanical properties and the alloy thus achieves particularly high strength, especially at high temperatures, toughness, corrosion resistance and ductility while maintaining a relatively low density. These are properties that make the use of vanadium particularly suitable for certain industries such as the medical, energy or military. The chemical composition and properties of the material used are given in Tables 1 and 2 [1] ,[3], [4]

	Oxygen	Nitrogen	Carbon	Hydrogen	Total
ppm by weight	200	60	50	2	312

Table 1. Interstitial impurity content of the commercially pure vanadium sample used (ppm by weight) [5]

Properties	Column A
Tensile strength	800 MPa
Yield strength	776 MPa
Elastic modulus	125.5 GPa
Poisson's ratio	0,36
Elongation at break (in 50 mm)	2%
Hardness, Brinell	165
Hardness, Rockwell A	52
Thermal expansion co-efficient (20-100°C/68-212°F)	8.33 µm/m°C
Thermal conductivity	31 W/mK

Table 2. Mechanical and thermal properties of vanadium [4]

As the main objective of the problem addressed in this research was the machining of this material. Due to its chemical composition and its mechanical and physical properties, it can be expected that during machining the material will stick to the cutting edge and overall machining will be problematic. From a machining point of view, vanadium can be characterised as a material approaching the harder to machine stainless steels of ISO M. In these materials, the high plasticity and toughness of the material results in the formation of a chip which is very difficult to cut. This fact results in complicated setting of cutting conditions with regard to chip leaving the cutting point and also possible damage to the workpiece or tool. [1], [2], [7]

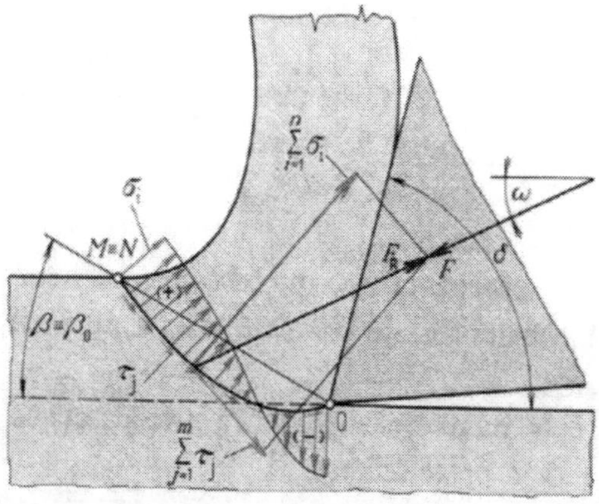

Fig. 1. Stress history in machining of tough materials [2]

Furthermore, the material generates a considerable amount of heat. There are two reasons for the increased amount of heat in these types of material. The main reason is the large amount of heat generated in the area of primary plastic deformation. This heat is generated by the low willingness of the material to separate, i.e. to reach a state of upsetting deformation. This state is due to the high toughness.

This makes it necessary to supply a much greater amount of energy to separate the individual grains. The formation of deformations in the primary plastic deformation zone is shown in Figure 1. The second factor is the low conductivity of vanadium, where almost all of the heat generated must be absorbed by the workpiece and the tool. This results in significant thermal stress and thus increased tool wear. The difference in heat concentration with respect to thermal conductivity is shown in Figure 2. [3], [8]

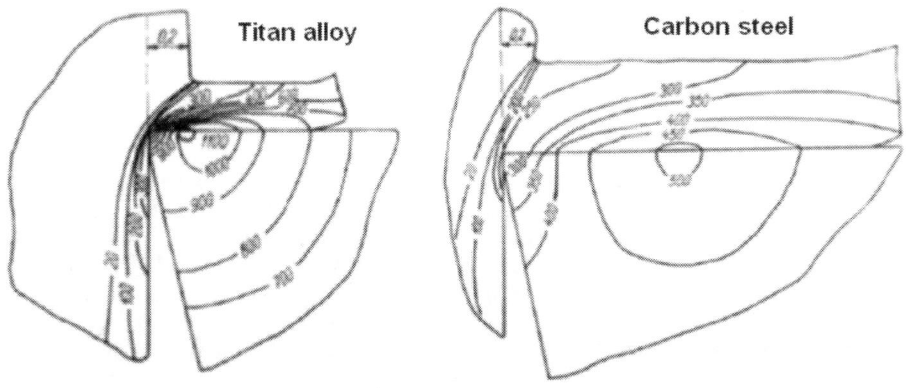

Fig. 2. Schematic representation of turning temperatures in titanium alloy and carbon steel [3]

Compared to stainless steels, however, there is no hardening of the material and the hardness is usually lower. Due to these properties, it is the closest to machining ISO S group materials such as titanium. From all these characteristics it follows that in general vanadium machining as such is possible but has certain specificities that need to be taken into account when designing the technology. [1]

3. Machining of thin-walled components

The shape of the part is always crucial when designing a machining strategy. Its geometric dimensions, tolerances and in combination with the material to be machined has a fundamental impact on the design of the machining technology. The desire to reduce the weight of each part as much as possible is logical in terms of saving energy and resources. Most companies today, thanks to advanced modelling algorithms, are trying to optimise the distribution of material on each part to produce parts with very thin walls. While these parts are very suitable from an environmental and economic point of view, they are not so suitable from a machining point of view. The machining of thin-walled parts involves a number of major complications and if the part is made of, for example, stainless steel or one of the ISO S materials such as titanium or Inconel it is often a very complex issue.

Machining thin-walled components involves a number of problematic factors that must be taken into account when designing the technology. One of the fundamental problems arising on very thin wall parts is vibration. The generation of vibration is fundamentally linked to the machining process itself and cannot be avoided to some extent, but it can be minimised. However, if the vibration exceeds a certain value, it is a spontaneous vibration of the whole assembly, and this is a phenomenon in machining in a second of cases caused by the instability of the process and is sought to be avoided. If vibrations occur, it is necessary to adjust the cutting conditions, the machining strategy, change the cutting tool or intervene in other ways. In most cases, vibration has very negative consequences such as damage to the part, the tool or the machine. In the case of thin-walled parts, the occurrence of vibration is probably the most common problem associated with their machining and therefore considerable attention must be paid to the setup of the entire process.

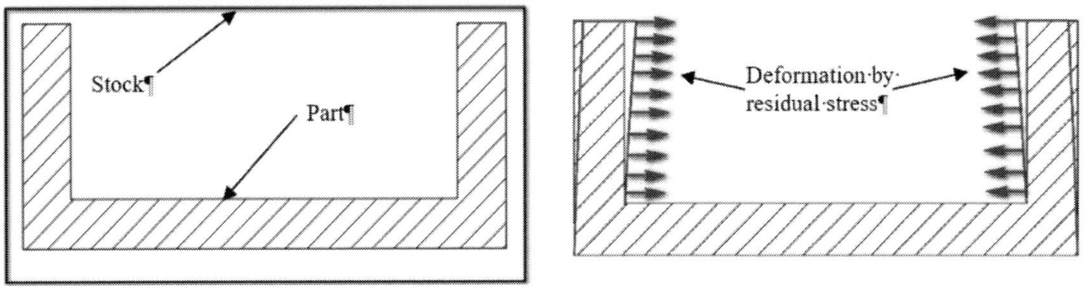

Fig. 3. Possible deformation of the part at the base of uneven material removal

Another very common phenomenon occurring when machining thin-walled parts is shape deformation or dimensional instability. This is due in most cases to internal stresses in the part which are caused by the removal of the blank material. The removal of large amounts of material results in an uneven distribution of stresses and these forces can then deform the part. There are a number of methods to try to avoid this problem. One of these is the appropriate choice of machining strategy, not only with regard to manufacturability itself, but specifically to internal stresses. Thus, it is common for thin-walled parts to be machined in multiple clamping operations precisely to compensate for the stresses induced by material removal. Another option is to include suitable heat treatment to reduce the internal stress, obviously taking into account whether the rest of the part allows it. One final option, used for example in high-volume production, is to adapt the machined shape to the previously calculated deformations. Here, it is possible to design a modification of the shape of the machined part so that, after accounting for the deformations, the workpiece will conform to the desired shape. This method is very efficient, but requires very precise calculations and also a considerable amount of experience to find a suitable procedure for the whole process. In spite of the methods mentioned, the production of a thin-walled ring in which 95 % of the material inside and only 5 % outside has to be removed will always be very complicated, despite the fact that circular profiles have the most suitable shape with regard to deformation.

There are a number of problems in machining thin-walled parts, but the last one, which is also connected with parts made of materials that generate a lot of heat during machining, is the removal and possible consequences of excessive heat accumulation in the workpiece. When machining materials with low thermal conductivity, it is necessary to take into account that the heat generated during machining is not dissipated as it is for example in carbon steels, but the distribution of heat dissipation is completely different. By default, the greatest amount of heat is dissipated through the chip where a significant amount of heat is also generated. The problem of low conductivity is that the workpiece material does not absorb as much heat as would be required and the heat accumulates in the area of primary plastic deformation. Here it is then distributed to other parts that are in direct contact with the region of origin. Thus, apart from the tool and the environment, much more heat is concentrated in the workpiece. Due to the low conductivity, the heat is absorbed slowly, but due to the low wall thickness, there is much less heat to induce thermal deformation. This fact then manifests itself again in shape deformation or dimensional instability. [6]

In engineering practice, all of the above factors usually occur to some extent simultaneously and it is therefore always necessary to find the appropriate combination of cutting conditions, machining strategy, cutting tool and a range of other parameters to achieve the required quality and dimensional accuracy of the part. In addition to those mentioned above, a number of factors such as tool wear, cutting forces, clamping method and many others intervene in the process. However, these are not the dominant effect of machining thin-walled parts or parts made of difficult to machine materials and their analysis will therefore not be dealt with further in this article.

4. Test subject and experimental methods

A part for medical development was selected as the test sample. It is a special capsule designed for further experiments. Unfortunately, further details cannot be disclosed due to cooperation with the test sponsor. The tested samples are shown in Figure 4. The larger part is called the capsule and the smaller cap. These are two parts that are supposed to fit together. After inserting the experimental sample into the capsule, the capsule is closed with the cap and then the two parts are welded together using an electron beam.

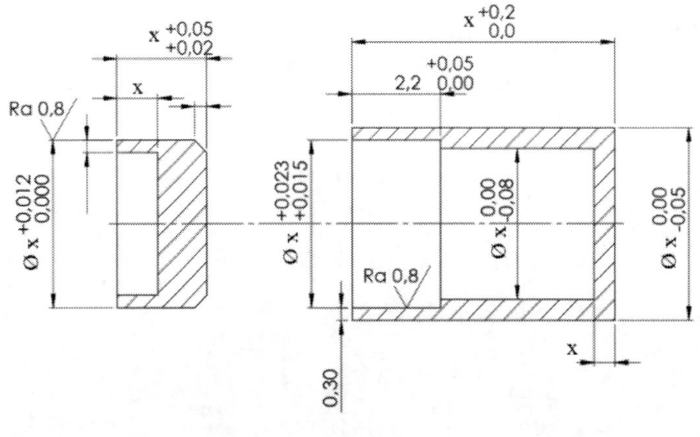

Fig. 4. Tested part, with respect to secret data, is a drawing without basic dimensions only with tolerances

From a machining perspective, these are very precise parts, and in particular the two diameters that fit together have very close tolerances, down to 0.008 mm. This tolerance is very thin with a wall length of 2.2 mm and a wall thickness of 0.3 mm and problems were expected. Such tight tolerances are there with respect to electron beam welding so that the least possible layer is affected. Other problem areas can be found in the corners where there was a requirement for minimum corner radii.

All these constraints had to be taken into account when designing the machining strategy. The experiments were conducted in two phases. In the first phase, the appropriate tools and their sequence were sought with respect to the deformation of the part and the achievement of geometric dimensions. In the second phase, the production of the part was carried out with respect to the compliance with the overall tolerances and to establish the stability of the process. After each of the tests, the experimental sample was taken to the metrology laboratory and all the necessary dimensions were measured. Based on these results, the whole process was then optimised and further testing was carried out.

5. Experiments and their evaluation

With regard to the shape of the part, turning was chosen as the initial machining. The part was manufactured from a solid round bar with a diameter of 6 mm. As mentioned above the first exception was the choice of tooling. The tooling assembly was chosen mainly to achieve the required dimensions and tolerances, but also according to the tooling of the machine used. The machine used was an EMCO MAXXTURN 25 turning centre.

The first operation was the alignment of the face. Here, a standard corner turning tool with a replaceable insert was used. The same tool was then used for the longitudinal external turning. Given the size of the parts and the requirement for minimum radius dimensions at the corner, a monolithic carbide tool with a tip radius of r = 0.1 mm was chosen for the internal turning. Before finishing, the hole was drilled with a monolithic drill bit. Allowances for all finishing operations ranged from 0.08 mm to 0.15 mm depending on the tool radius used.

Despite considerable efforts to find suitable cutting conditions for the above strategy, it had to be stated that it was not possible to achieve the required dimensions in a repeated production of only a few pieces. The main problematic element would be the dimension with a tolerance of 0.008 mm, where the turning tool showed considerable instability with regard to uneven chip formation and chip evacuation. Therefore, based on experience, finishing was carried out using a milling operation. This was a three-axis spiral milling of a circular pocket. A monolithic milling cutter with a very small corner radius and also a very small edge rounding was used as a tool, thus ensuring that tool deburring did not occur. The milling strategy also greatly improved chip evacuation by forming an articulated chip that was much easier to remove from the cut point.

Unfortunately, even after this improvement, the result was not satisfactory. The required surface roughness of the inner diameter was problematic. After several test passes it was found that vibration was occurring. This vibration was caused by the surface waviness after the drilling operation. The final optimization step was therefore to include a milling operation before the finishing operation. Also, the size of the drill bit was reduced so that there was sufficient allowance and the surface waviness could be safely removed. This optimised strategy already met the requirements, both for accuracy and surface quality. The whole course of optimization of the machining strategy is shown in Table 3. Table 4 shows the cutting conditions for each tool used.

Strategy I	Tools used	Strategy II	Tools used	Strategy III	Tools used
Face turning	CCMT 09T308 FM	Face turning	CCMT 09T308 FM	Face turning	CCMT 09T308 FM
External turning	DCGT 11T304 SF3	External turning	DCGT 11T304 SF3	External turning	DCGT 11T304 SF3
Drilling	Drill – 3xD -IK	Drilling	Drill – 3xD -IK	Drilling	Drill – 3xD -IK
Internal turning	PICCO L 050.3-10	Finishing milling	Helix solid carbide endmills	Semi-finishing milling	Helix solid carbide endmills
				Finishing milling	Helix solid carbide endmills

Table 3. Development of a machining strategy for the test piece

Tools used	Cutting speed v_c [m/min]	Feed rate f_n [mm/ot]	Deep of cut a_p [mm]
CCMT 09T308 FM	100	0,1	0,1
DCGT 11T304 SF3	100	0,15	0,5
Drill – 3xD -IK	60	0,06	-
PICCO L 050.3-10	35	0,08	0,15
Helix solid carbide endmills	80	0,12	0,25

Table 4. Cutting conditions for individual tools

The last point of the optimization was the verification of the chosen strategy during the production of the test batch. It consisted of 8 pieces of the whole chapel assembly, i.e. the cap and the capsule body. This series was used to test both the tool life and the reliability of the process. The lifetime results were satisfactory. The tools selected after the test series showed only minimal wear and no significant change of operation was required. The only minor problem was the chip material sticking to the cutting edges of the cutter used. This problem was eliminated by using a tool with a different type of coating. In terms of process reliability, 100% reliability was achieved after the inclusion of the pre-finishing operation. After production debugging, all tolerances on all parts were in the middle of the tolerance range. Based on these results, it was possible to conclude that the desired goal was achieved. The entire machining strategy will then be deployed at the customer's site and only after testing it will it be decided whether it is necessary to proceed with further investigation.

6. Conclusion

The whole problematic dealt with in this article was related to the machining of very thin parts made of difficult to machine materials. In the course of the investigation of the whole problem, the material to be machined was described, which was almost pure vanadium. This material is not a very common material and so it was necessary to describe the expected behaviour of the material before the actual tests could begin. It was then necessary to focus on the main problematic features of the part. This was the very weak walls and also the very tight tolerances of some dimensions. The machining strategy used was also chosen with this in mind.

In the course of the experiments it was found that for a given combination of the material to be machined and the shape and tolerances of the part, milling is a preferable strategy to turning despite the rotational shape of the part. After analysis of the results obtained, this was attributed to poor chip evacuation caused by heat build-up and increased plasticity of the material. After using the milling strategy and with the inclusion of a pre-finishing operation, the desired results were achieved. The overall stability of the process was verified by producing a test batch.

The results presented within this research serve both for finding a suitable technology for the production of the given component, but also as a basis for further research in the field of machining of thin-walled parts made of difficult to machine materials, which are currently given considerable attention.

7. Acknowledgments

This article was created under the project SGS-2019-008: Research and Development for Innovation in the Field of Manufacturing Technology - Machining Technology III.

8. References

[1] Zatloukal, T., Rehor, J., Fulemova, J. & Povolny, M. (2020). Optimization of the Machining Process of NiCrBSi Thermal Spraying. Proceedings of the 31st International DAAAM Symposium 2020. DAAAM International Vienna, 2020, 0632-0637. DAAAM Proceedings. ISBN 9783902734297. Available from: doi:10.2507/31st.daaam.proceedings.087

[2] Strnad, T. (2011). Machinability of Austenitic Steel. Pilsen, Internal documents -DMT - UWB. University of West Bohemia.

[3] Mikovec, M. & collective. (1963). Machining of Difficult to Machine Materials; Praha: State Publishing House of Technical Literature. SNTL, 1963, 316 s. ISBN: 04-220-63.

[4] Vanadium - Mechanical Properties and Material Applications [online]. 2012 [cit. 2021-10-18]. Available from: https://www.azom.com/article.aspx?ArticleID=7643

[5] Henshall, G. A. a S. G. (1993). Torres. Tensile Properties of Commercially Pure Vanadium from Room Temperature to 1200 °C. USA: Livermore National Laboratory, , 38.

[6] Masoudi, S., Amini, S., Saeidi, E. & Eslami-Chalander, H. (2015). Effect of Machining-Induced Residual Stress on the Distortion of Thin-Walled Parts. The International Journal of Advanced Manufacturing Technology. 2015, 76(1-4), 597-608. ISSN 0268-3768. Available from: doi:10.1007/s00170-014-6281-x

[7] Baksa, T., Schornik, T., Adamek, P. & Miroslav Zetek. (2016). Machining of Inconel 718 Using Uncoated Cutting Tools with Different Cutting Edge Quality. Proceedings of the 27th International DAAAM Symposium 2016. DAAAM International Vienna, 2016, 0441-0446. DAAAM Proceedings. ISBN 9783902734082. Available from: doi:10.2507/27th.daaam.proceedings.065

[8] https://www.sandvik.coromant.com/cs-cz/knowledge/materials/pages/workpiece-materials.aspx, (2018). Machined Materials, Accessed: 2021-10-10

DOI: 10.2507/32nd.daaam.proceedings.032

FORMAL SPECIFICATION OF A MOBILE ROBOT WITH MODULAR ARCHITECTURE FOR NAVIGATION IN DYNAMIC ENVIRONMENT

Victor Andreev & Valerii Kim

This Publication has to be referred as: Andreev, V[ictor] & Kim, V[alerii] (2021). Formal Specification of a Mobile Robot with Modular Architecture for Navigation in Dynamic Environment, Proceedings of the 32nd DAAAM International Symposium, pp.0215-0227, B. Katalinic (Ed.), Published by DAAAM International, ISBN 978-3-902734-33-4, ISSN 1726-9679, Vienna, Austria
DOI: 10.2507/32nd.daaam.proceedings.032

Abstract

This article discusses the organization of inter-modular information interaction in a mobile robot with a hierarchical architecture of the control system. The use of the principle of the full functionality of the modules allows real-time reconfiguration of such mobile robots and the distribution of the computational load between the modules to provide a real-time mode. The purpose of the work is to present a formal description, mathematical and computer models of the control system of a modular mobile robot, and inter-modular information interaction in the process of solving the general problem of a mobile robot. The two-level hierarchical architecture of the control system is considered, which consists of an intelligent control module, transport and sensor modules. The described robot control system provides navigation in an environment with dynamic obstacles. The functionality of the main modules of the robot and the requirements for their inter-module interfaces have been determined. An analysis of well-known methods for avoiding moving obstacles is given, which made it possible to establish the requirements for the functionality of the transport and sensor modules and their inter-module interface. The article presents the results of a computer simulation of the movement of a modular mobile robot in a dynamic environment. The specificity of the work of a modular system with a hierarchical organization of the control system is shown. A modular mobile robot is proposed to be represented as a network of intelligent agents in a multi-agent environment, the mathematical model of which can be formally described using the theory of finite automata.

Keywords: modular robot; finite-state machine; multi-agent system; collision avoidance; computer networks.

1. Introduction

This article is a continuation of the works [1], [2], [3], aimed at the implementation of a hierarchical modular architecture of the control system (CS) of mobile robots (MR). A review of similar solutions and the rationale for the proposed approach are considered in [3]. The hierarchical modular architecture of CS is based on the principle of distributed computing by analogy with multi-agent systems, which allows distributing the general MR task between modules. This approach leads to a simplification of the algorithms for the operation of the control systems of the modules, which, in turn, makes it possible to use relatively simple and cheap computing devices in the control systems of the modules - the so-called embedded systems (microprocessors and single-board computers).

The use of embedded systems allows reducing the total cost of robotic systems (RTS), energy consumption, and their dimensions. When organizing distributed computing, the general computational control process is parallelized, which makes it possible to provide a real-time mode even with an increase in the RTS functionality.

According to our approach, the control system of a mobile robot consists of 6 main modules, while the functional solution of each module is determined by the principle of full functionality formulated in our work [1]: *each robot module must be capable of performing its target function in any way convenient for it, using only its means to execute commands from an external control system*. In other words, each module must be fully functional. **A fully functional module** is a module that is capable of independently performing the task assigned to it by the supervisor (external control module). To obtain a formal description of such a modular architecture, it is necessary to solve a number of the following problems. First of all, it is necessary to limit ourselves to a specific task that a mobile robot must perform. Then one should define what specific functions each module is responsible for. Based on this, the requirements for inter-modular interfaces and inter-modular interaction protocols are formed.

Thus, the **purpose of the work** is to provide a formal description of the information interaction of modules, a mathematical and computer model of a control system for a mobile robot, consisting of an intelligent control module, transport, and sensor modules. We restrict the general MR task to navigation in an environment with static and dynamic obstacles.

The article is organized as follows: Section 2 provides a formal description of the information interaction of the mobile robot modules in the process of solving the task assigned to the MR. Section 3 provides a brief overview of various algorithms for avoiding moving obstacles. This overview is provided to reasonably distribute the computational load between the transport and sensor modules when avoiding obstacles. Section 4 provides a mathematical model of the transport module and intelligent control module, section 5 - a computer model of the system and the results of the simulation. At the end of the article, conclusions are given, including the direction of the following research.

2. Formal description of information interaction between mobile robot modules

The key feature of the architecture under consideration is the use of the principle of full functionality of modules. Let a modular robot move on a flat underlying surface in some partially known environment. The robot can be surrounded by static and dynamic (movable) obstacles. The robot receives a general task from an external supervisor (usually a human operator). The general task is to move the MR from the current position to the target position with coordinates X_G, Y_G relative to some world coordinate system. In the problem under consideration, the orientation angle at the target position is not taken into account.

The intelligent control module (ICM), the transport module (TM), and the short-range sensor module (SRSM) in the process of information interaction cooperatively solve the problem of moving the MR with the division of responsibility (Table 1).

Module	Main functions
ICM	Global planning and task formation for other modules
TM	Local planning and movement
SRSM	Collection and processing of information about the environment in the near zone

Table 1. Main functions of the modules

Basic assumptions when solving the problem:

- the TM path does not have to be optimal in terms of the distance traveled and the trajectory;
- the maximum permissible time of arrival (deadline) of the TM at a given position must be set;
- if possible, the TM should avoid all obstacles (including moving ones); in the event of a collision with an obstacle (or with a high probability of collision), the TM must report this to the ICM and wait for further instructions;
- all obstacles are solid bodies with dimensions that allow them to be detected by the sensor system;
- all obstacles fall into the scanning plane of the SRSM range finders;
- the number of objects in the field of view of the SRSM should not exceed some critical value (for example, no more than 10).

The general scheme of information interaction of fully functional modules is shown in fig. 1.

The ICM receives a general task from an external supervisor, which he divides into tasks for each module. Such a task for the TM is to move from an initial position to a target position $[X_i, Y_i]^T$ with some acceptable error in coordinates. In general, this target position is one of the many points of the route planned by the ICM, the final position of which has coordinates X_G, Y_G. According to the principle of full functionality, the TM must independently reach a given position using the control methods embedded in its control system and information received from its sensory system.

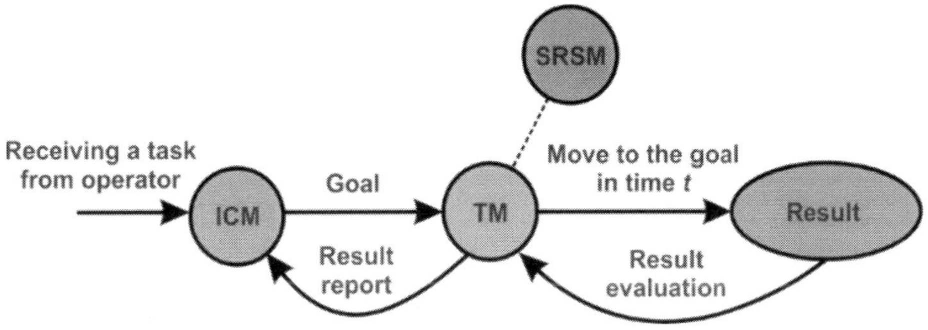

Fig. 1. The general scheme of information interaction of modules

It is assumed that the TM sensor system should have a minimum functionality, sufficient only for avoiding static obstacles. For avoiding dynamic and static obstacles with complex shape, the data received by the SRSM should be used.

Consider the information interaction between the ICM and the TM.

The ICM forms a path from the starting position to the target position using data about environment presented in the form of a global map. Usually the map contains information about static obstacles, the data about dynamic obstacles is directly received from the sensor module while moving. There are various methods of global planning on the map to find the optimal path. These include well-known graph search algorithms: A* [4], D* [5], random sampling algorithms: RRT [6], BIT* [7], etc. The choice of a specific search algorithm depends on the formal description of the map.

As a result of a path planning, the ICM must form a set of sub-goal points of the desired path that will be reached by the TM. In this case, the trajectory of the TM to the next point does not have to be optimal. The interaction between the ICM and the TM can be schematically represented as follows (fig. 2).

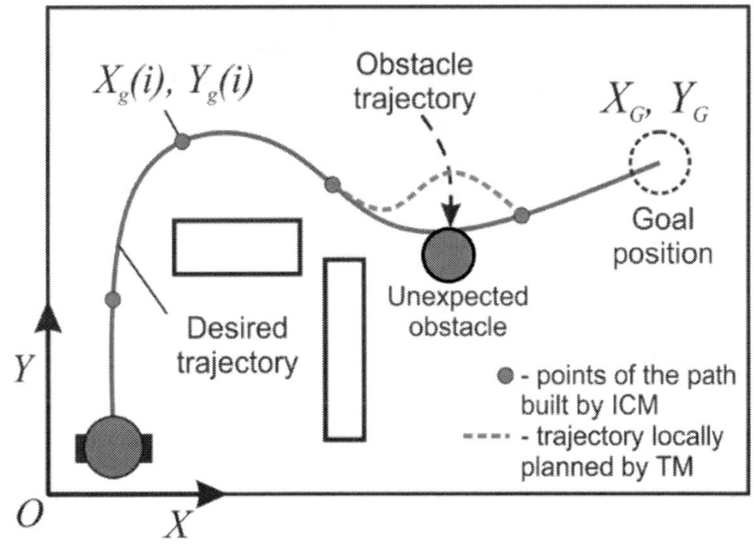

Fig. 2. Scheme of a modular robot movement in an environment with a moving obstacle known to the ICM

Let the intelligent control module construct a path from the initial position to the target one. The way of constructing the path can be any, while the path must be a finite set of points $\{(X_i, Y_i)\}$, $i = 1, 2, ..., n$ on the estimated trajectory. The distances between the points must be at least a certain value (for example, 3 m is a range of the SRSM reliable work), otherwise the interaction between the ICM and the TM will approach continuous, which negates the advantages of the principle of full functionality.

Figure 2 shows that in the event of a moving obstacle, the TM should try to avoid it and get to the next target point using the data of the SRSM and its local path planner. In the general case, the procedure for interaction between the TM and the ICM is described by the following algorithm (fig. 3).

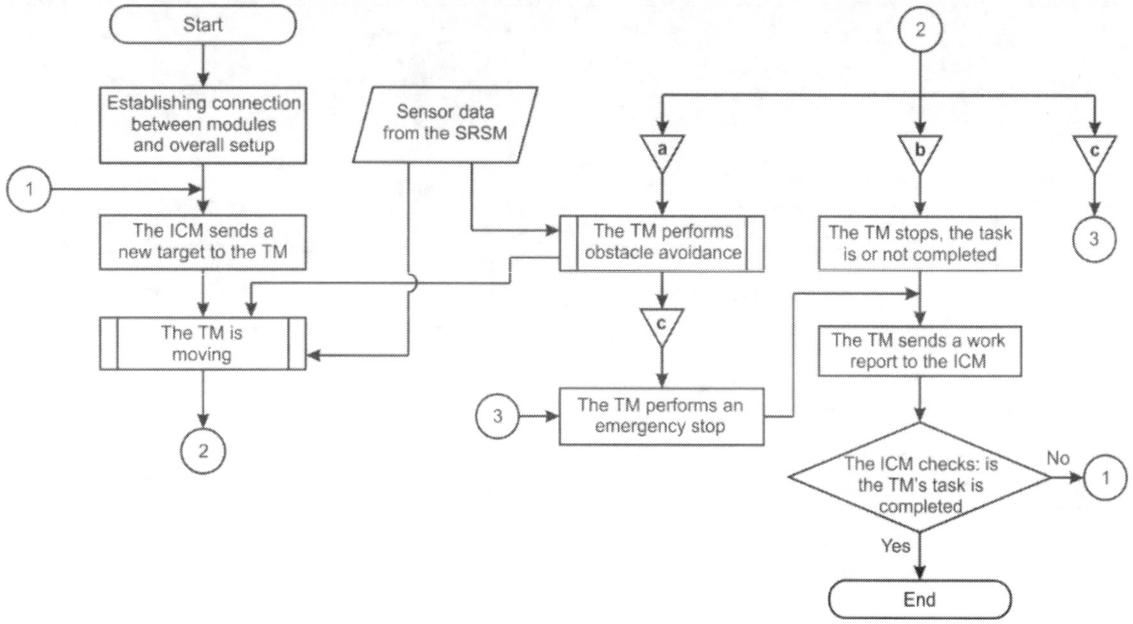

Fig. 3. General algorithm of interaction of ICM, TM and SRSM when the modular mobile robot moves

In this block diagram, the triangles with the symbols a, b, c represent the transfer of control from one process to another:

a – transition to the process of avoiding an obstacle by the transport module in the event of a static or dynamic obstacle on the way. If the obstacle is successfully passed, then there is a return to the process of normal movement to the next point.

b – transition to the stop process, when the TM considers that it has completed the task according to the data of its control system.

c – transition to the emergency stop process, when there was a collision with an obstacle (or a high possibility of a collision) or a failure of the TM internal systems. The block diagram (fig. 3) shows that this interruption can occur when the TM is moving towards the target and the path is clear, or the TM is avoiding an obstacle.

A movement task sent by the ICM to the TM is represented by a data structure with the following parameters:
1. target point coordinates: X_i and Y_i, m;
2. coordinates of the current initial point where the robot is located: X_0 and Y_0, m;
3. admissible error in a target position, m;
4. arrival time limit (deadline), s.

The TM sends a work report to the ICM for any outcome, as can be seen in the block diagram above. The report is presented as a data structure with parameters:
1. target points coordinates: X_i and Y_i, m;
2. elapsed time, s;
3. own assessment of the task (completed or not, emergency stop).

It should be noted that the target and current positions of the robot are set relative to the world coordinate system associated, for example, with the room. At the same time, the ICM knows the location of the fixed base coordinate system, and the coordinate system of the robot itself. The block diagram (fig. 3) shows that the execution of the movement to the target and the obstacle avoidance are carried out while receiving information from the SRSM. This information should allow the TM to avoid both static and dynamic obstacles.

It should be noted that, according to the principle of full functionality, the division of the computational load between the SRSM and the TM when avoiding obstacles can be implemented in many ways. For example, the SRSM can scan the environment in the near zone, detect obstacles and issue a motion vector for the TM, which in turn must only move in this direction. On the other hand, the SRSM can only be responsible for scanning and periodically send an array of points to the TM for further processing. It is necessary to divide the task of avoiding obstacles so that the computational load between the SRSM and the TM is divided approximately equally. To solve this problem, it is first necessary to analyze various algorithms for avoiding moving obstacles.

3. Algorithms for avoiding moving obstacles and analysis of the interaction of the transport and sensor module

The review includes mainly algorithms designed for a flat environment: [8], [9], [10], [11], [12], [13], [14], [15], [16], [17], [18], [19], [20] and [21]. In our case, the following characteristics of the algorithms are of interest:

- the need for a global map of the environment,
- the path planner used,
- representation of the local environment,
- the shape and size of the robot and obstacles,
- required information about moving/static obstacles,
- kinematic model of the robot,
- dynamic characteristics of the robot, taken into account in the work.

The analysis of considered algorithms showed that a global map of the environment is rarely used since in a changing environment with dynamic obstacles, local scanning is important for operational decision making. The path planner is used in those algorithms where it is necessary to construct an optimal path on a global occupancy grid map or a state graph. The A* algorithm and various optimization methods are used as the path planner.

The local environment around the robot can be defined in various ways. For example, in the classical method of velocity obstacles [15], the local environment is represented as the region of attainable robot velocities, built on the basis of its accelerations and "velocity cones" of moving obstacles. In papers [8], [11], the local environment is presented as an occupancy grid map, which is convenient for localizing both static and dynamic obstacles. At the same time, grid maps work well together with rangefinder sensors [22].

In the overwhelming majority of the considered algorithms, the robot is represented as a moving circle, less often as a polygon [13]. Dynamic obstacles are also represented as circles. This is due to the fact that the circles allow the most simple calculation of the nearest distance from the robot to the obstacle, which reduces the requirements for the computing resources of the robot. If a grid map is used, then obstacles are displayed as a set of occupied map cells [8], [11].

The analysis of considered algorithms showed that to avoid moving obstacles, the local planner of the robot must know the coordinates of the obstacles, the vectors of their velocities (projections of the velocities on the axes of the coordinate system) and the radii of the overall circles. This is usually enough to avoid a collision and get to the target position. There are methods that require a set of points on the trajectory of the obstacle [18]. However, there are ways to predict the trajectory of an obstacle based on its current speed, position and kinematic model.

Modern algorithms for avoiding obstacles take into account the kinematic constraints of robots (nonholonomicity), since this can narrow the range of permissible movements of the robot. Also, most of the methods take into account the dynamic characteristics of robots, in particular, acceleration. Taking acceleration and deceleration into account limits the permissible range of motion, but reduces a collision probability. So, to solve the problem of avoiding obstacles in a partially known environment, the following division of the computational load between TM and SRSM is proposed. SRSM should perform the following main functions:

1. Scanning the environment for obstacle detection and elimination of "false" scan points, i.e. points that are not obstacles (points filtering). The field of view (FOV) of the sensor system must be at least 180° and cover the front of the robot.
2. Solution of the SLAM problem.
3. Tracing moving obstacles. Obstacles should be presented as moving circles.

The SLAM problem should be solved by the SRSM in conjunction with the ICM, which stores a global map of the environment, as well as when using data received from a long-range sensor module (LRSM). The LRSM uses range sensors such as video camera or lidars.

The TM cannot perform the localization task, since its sensors do not have the required accuracy. Moving obstacles can only be traced by the SRSM, since it is equipped with appropriate sensors and performs the localization of the robot, without which it is impossible to distinguish a moving obstacle from a stationary one, especially if the robot is moving.

TM, in turn, must perform the following functions:

1. Local planning within the near zone of the SRSM.
2. Obstacle avoidance.
3. Control of actuators.
4. Sampling of low-level sensors (limit switches, simple rangefinders).

The near zone of the SRSM can have different sizes and shapes depending on the assigned task, the size of the robot and the size of the terrain. In this work, it is assumed that the near zone is limited by a circle with a radius of 3 m centered at the characteristic point of the robot. The robot itself is circular.

The algorithm for avoiding obstacles can theoretically be any, if the information received from the SRSM is sufficient to execute the algorithm. It should be taken into account that the time to avoid an obstacle is limited from above by the maximum arrival time. As part of inter-modular interaction, the SRSM sends a data structure with the following parameters to the TM:

1. coordinates of the robot current position with respect to world coordinate system: X_a and Y_a, m;
2. set of the static obstacles points: $\{\mathbf{x}_1, \mathbf{x}_2, ..., \mathbf{x}_n\}$, where $\mathbf{x}_i = (x_i, y_i)^T$ – coordinates of the i-th point in the world coordinate system, n – the number of scanning points in the near field after filtering points in the SRSM and setting a limit on the maximum number of points (see below);
3. moving obstacles dataset: $\{\mathbf{O}_1, \mathbf{O}_2, ..., \mathbf{O}_m\}$, where $\mathbf{O}_i = (x_{O_i}, y_{O_i}, v_{x_i}, v_{y_i}, r_{O_i})$ – tuple of the i-th obstacle's parameters, m – number of obstacles within the near zone of the robot, x_{O_i}, y_{O_i} – coordinates of the center of the overall circle of the obstacle with a radius r_{O_i} in the world coordinate system, v_{x_i}, v_{y_i} – the projections of the obstacle velocity vector on the axes of the robot coordinate system.

The number of points obtained as a result of scanning obstacles can be large: for example, a laser scanner with an angular resolution of 0.1 ° and a full all-round view can produce up to 3600 points at a frequency of 10-20 Hz. In this case, some of these points will be discarded after filtering. Not all scan points should be sent to TM, but only those n points that are within the near zone (up to 3 m).

Since the requirements for the accuracy of scanning the surrounding space in mobile robotics strongly depend on the specific task, conditions and the robot itself (a mechanical design, software, electronic components), in this work it is proposed to set the limit for the number of scan points sent by the SRSM to the TM. Modern mobile robots use high-resolution laser rangefinders for navigation, which can detect both static and dynamic obstacles (table 2).

Paper	Laser scanner	FOV	Resolution	Maximum number of points
[14]	TIM551	270°	1°	270
[11]	SICK LD-MRS	110°	0.125°	880
[8]	Hokuyo URG-04LX	240°	0.36°	667
[10]	SICK 2D LRF	270°	0.33°	818

Table 2. The main characteristics of typical laser scanners in various mobile robots for the task of motion in a dynamic environment

In all the articles indicated in the table, mobile robots successfully reach the target position by avoiding both static and dynamic obstacles. The number of scan points in these studies rarely exceeds 900. Therefore, in this work, we accept with a small margin that the maximum number of points of static obstacles transmitted to the TM should not exceed 1000 pcs. If the SRSM sensor system is capable of receiving a scan with a large number of points, then the SRSM should select only 1000 of them, excluding the rest.

The frequency with which the SRSM sends data to the TM primarily depends on the frequency of the module's rangefinders and the processing time. For laser rangefinders, the typical scanning frequency is 10-30 Hz [23]. The SRSM should send data as often as its sensors allow. TM sends information about its position and speed, obtained using odometry, to the SRSM. This is necessary to improve the accuracy of the SLAM algorithms. The sending frequency must be at least the odometer polling rate.

The structure of the sent data has the following parameters:

1. coordinates of the robot current position with respect to world coordinate system: X_a and Y_a, m;
2. the projections of the TM velocity vector on the axes of the robot coordinate system, m/s.

Sending information about the speed to the SRSM can be useful for adjusting the settings of the sensors of the sensor module. For example, if the speed of the TM is below a predetermined threshold, then the frequency of the range finders polling should be reduced to decrease the traffic of inter-module interaction and the power consumption of the SRSM.

4. Mathematical model of the mobile robot with modular architecture

The mathematical model of a modular mobile robot should take into account the interaction of modules and their functioning as independent intelligent agents. In this paper, it is proposed to use the theory of finite-state automata to formally describe some of the main parts of the robot control system. The transport module of the robot can be described by the finite state machine of the form [24]:

$A_T = (S_T, \Sigma_T, \delta_T, s_{T0}, F_T)$, where

the finite non-empty set of the TM states: $S_T = \{S_{T0}, S_{T1}, S_{T2}, S_{T3}\}$,
the input alphabet: $\Sigma_T = \{a_{T0}, a_{T1}, a_{T2}\}$,
$s_{T0} = S_{T0}$ - is an initial state,
δ_T - the state-transition function: $\delta_T : S_T \times \Sigma_T \to S_T$,
F_T - the set of final states, $F_T \subseteq S_T$, $F_T = \{S_{T2}, S_{T3}\}$.
The state diagram (a directed graph of the automata A_T) is shown in the fig. 4.

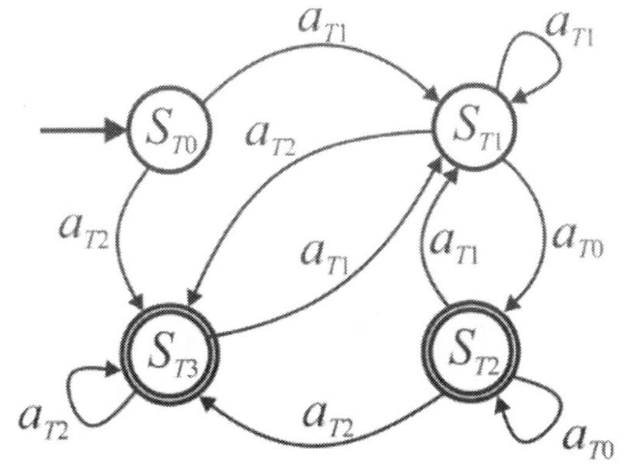

a

S \ a	a_{T0}	a_{T1}	a_{T2}
S_{T0}	-	S_{T1}	S_{T3}
S_{T1}	S_{T2}	S_{T1}	S_{T3}
S_{T2}	S_{T2}	S_{T1}	S_{T3}
S_{T3}	-	S_{T1}	S_{T3}

b

Fig. 4. The state diagram of the TM (a), state transition table (b)

The transport module can be in four main states: S_{T0} – the TM waits for a new target (position), S_{T1} – the TM moves to a new target, S_{T2} – the TM has reached the target, S_{T3} – an emergency stop caused by a collision, expiration of time to move to a target, etc. The initial state of the TM is the expectation of a new goal, therefore $s_{T0} = S_{T0}$. In this case, there are two final states: S_{T2} and S_{T3}, in which the TM is waiting for new instructions from the ICM. For the transport module the S_{T2} state is the same both in the case of arriving at an intermediate point and the final one.

The input alphabet describes the events (symbols) that relate to the work of the TM: a_{T0} – the next goal is achieved, a_{T1} – the new goal is received, a_{T2} – an emergency situation has arisen: a collision with an obstacle (or a high probability of a collision), the time of arrival at a new position has expired.

The state-transition function δ_T is most easily represented in the form of a table (fig. 4, b). Although the function δ_T is not defined for the combination (S_{T0}, a_{T0}) and (S_{T3}, a_{T0}), the presented state machine can be classified as deterministic (DFA), since for each combination of a state and input there is only one unique next state.

Consider the finite-state machine for ICM - $A_I = (S_I, \Sigma_I, \delta_I, s_{I0}, F_I)$. This module is a rather complex device, but for the task under consideration it is possible to reduce its operation to several states (fig. 5, a): $S_I = \{S_{I0}, S_{I1}, S_{I2}\}$, where S_{I0} – the robot moves along the planned path, S_{I1} – creating a new path, S_{I2} – task completed or not completed, stop; the task may not be completed if movement along the original path is impossible and a new one cannot be created. The initial state of the automata is: $s_{I0} = S_{I0}$, i.e. the machine starts working by moving along the planned path. The final state of the automata is: $F_I = S_{I2}$.

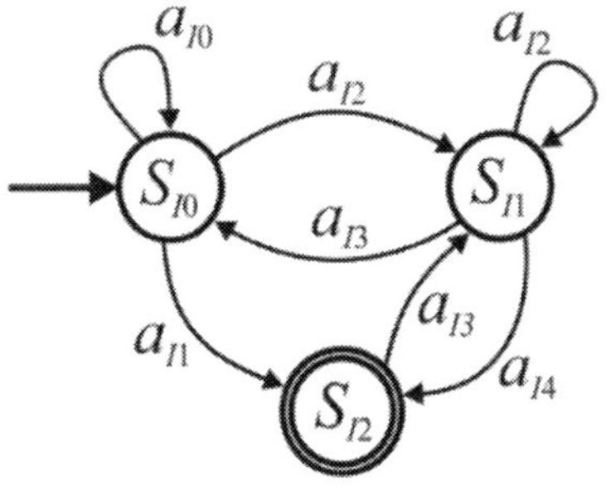

a

S \ a	a_{I0}	a_{I1}	a_{I2}	a_{I3}	a_{I4}
S_{I0}	S_{I0}	S_{I2}	S_{I1}	-	-
S_{I1}	-	-	S_{I1}	S_{I0}	S_{I2}
S_{I2}	-	-	-	S_{I1}	-

b

Fig. 5. The state diagram of the ICM (a), state transition table (b)

The input alphabet $\Sigma_I = \{a_{I0}, a_{I1}, a_{I2}, a_{I3}\}$ consists of the next symbols: a_{I0} – the current target is reached, a_{I1} – the last target (end position) is reached, a_{I2} – the transport module reports an emergency, a_{I3} – continue the task (movement), a_{I4} – impossible to create a new path.

The state-transition function δ_I is described by the table (fig. 5, b). The transition function is also not defined for all combinations of states and inputs, however the state machine A_I can be considered deterministic.

The implementation of the two presented automata in the computer model of the robot is described in the next section.

According to the scheme of information interaction between TM and ICM, the transport module must reach the next target point with a given accuracy in a certain time (state S_{T1} from finite automata). When an unexpected obstacle appears on the robot's path, TM must try to avoid it. In this work, we assume that the TM is a wheeled platform with an omnidirectional drive. Then it is possible to control the module by adjusting the speed vector of the wheeled platform – $\mathbf{v}_r = (v_x, v_y)^T$.

To solve the problem of moving to the target point, we use the simplest linear control law:

$$\begin{cases} v_x = KL\cos(\gamma - \varphi), \\ v_y = KL\sin(\gamma - \varphi), \end{cases} \tag{1}$$

where K – controller coefficient, L – distance from the current robot position to the target, γ – the angle between the OX axis of the world coordinate system and the line segment connecting the current position of the robot and the target point.

To eliminate the oscillatory movements of the robot near the target point, the TM should stop when the position error becomes less than a certain predetermined threshold.

In this work, the classic Velocity Obstacle (VO) method [15] was chosen to avoid moving obstacles. This is one of the most well-known algorithms that provides motion in a dynamic environment and has a large number of modifications [9], [13], [19], [21]. We will not give a description of the algorithm here, but note only that to avoid moving obstacles $B_1, B_2, ..., B_m$ the robot's new speed at the current time interval must be chosen outside the union of the VO sets:

$$VO_{A|B_1,B_2,...,B_m} = \bigcup_{i=1}^{m} VO_{A|B_i},$$

where $VO_{A|B_i}$ - velocity obstacle - area of forbidden robot speeds relative to a moving obstacle B_i.

The motion of the transport module is controlled at each time interval Δt, which is set based on the computational capabilities of the module and the frequency of obtaining new sensor data. The magnitude of the speed that the TM can reach through this time interval is determined by the maximum possible acceleration of the module [15], [16]. The robot's speed $\mathbf{v}_r(t + \Delta t)$ on the next interval Δt must satisfy the condition:

$$v_x(t+\Delta t) \in \left[v_x(t) - \dot{v}_x\Delta t, v_x(t) + \dot{v}_x\Delta t\right],$$
$$v_y(t+\Delta t) \in \left[v_y(t) - \dot{v}_y\Delta t, v_y(t) + \dot{v}_y\Delta t\right],$$

where $\dot{v}_x$ and $\dot{v}_y$ – accelerations along axes of the coordinate system fixed to the TM.

This condition can be graphically represented by a rectangular area of reachable velocities (RV). In general this area can intersect with cones $VO_{A|B_i}$. The portion of the reachable zone that does not belong to VO is the set of reachable speeds to avoid moving obstacles. In the article [15] this set is denoted as RAV:

$$RAV(t+\Delta t) = RV(t+\Delta t) \setminus VO_{A|B_1,B_2,...,B_n}(t)$$

Any RAV speed will allow the robot to avoid an obstacle, however the TM must select a speed that is close to the preferred speed $\mathbf{v}_r^{pref}$. This is the speed at which the TM should move towards the target point if there are no obstacles. For example, it can be estimated according to formula (1).

The choice of speed at the next time step is a typical optimization problem, which can be written as follows [25]:

$$\mathbf{v}_r^{new} = \underset{\mathbf{v}_r \in RAV(t+\Delta t)}{\arg\min} \left\| \mathbf{v}_r - \mathbf{v}_r^{pref} \right\| \tag{2}$$

This condition takes into account only dynamic obstacles, however, as noted above, static obstacles can be in the path of the robot. Usually, VO regions are also built for static obstacles, but it is difficult to do this for arbitrary-shaped obstacles. Therefore, in this work, all those speeds that will lead to a collision with a static obstacle at a time $t+\Delta t$ are excluded from the RAV set. Excluding all collision-related velocities from the RAV set, we get the RAV* set. Then finally condition (4) can be rewritten:

$$\mathbf{v}_r^{new} = \underset{\mathbf{v}_r \in RAV^*(t+\Delta t)}{\arg\min} \left\| \mathbf{v}_r - \mathbf{v}_r^{pref} \right\|$$

Thus, the algorithm of motion in a dynamic environment using the VO method is formulated as follows:

1. **Input parameters of the algorithm**: list of moving obstacles $\{\mathbf{O}_1, \mathbf{O}_2, ..., \mathbf{O}_m\}$, points of static obstacles $\{\mathbf{x}_1, \mathbf{x}_2, ..., \mathbf{x}_n\}$.
2. Calculate the preferred velocity $\mathbf{v}_r^{pref}$.
3. Construct the set of reachable velocities *RV*.
4. For every $\mathbf{O}_i$ from $\{\mathbf{O}_1, \mathbf{O}_2, ..., \mathbf{O}_m\}$:
5.   Construct $VO_{A|\mathbf{O}_i}$.
6.   Construct *RAV*: exclude $VO_{A|\mathbf{O}_i}$ from *RV*.
7. Construct *RAV** (using known points of obstacles).
8. If set *RAV** is empty, then
9.   Stop TM: $\mathbf{v}_r^{new} = (0,0)^T$.
10.   The TM informs the ICM that the task has not been completed (failure).
11. else
12.   Calculate new velocity $\mathbf{v}_r^{new} \in RAV^*$, which is close to $\mathbf{v}_r^{pref}$

Algorithm 1. Algorithm of TM motion in a dynamic environment

In this algorithm lines 9 and 10 indicate the transition to the S_{T3} state of the TM state machine. If there are no dynamic obstacles, then steps 4-6 are not performed, and the new speed is selected from the subset of RV*, which eliminates collision with static obstacles. In fact, the ICM route built for the robot must be safe (collisions with static obstacles are excluded), so this check is needed for additional safety.

5. Computer model of the mobile robot with modular architecture

To assess the performance of the proposed modular architecture, computer simulation was carried out in the MATLAB (Simulink). Simulation of the movement of a robot consisting of the TM, ICM and SRSM was carried out in a flat environment with moving and static obstacles. The block diagram of the model is shown in fig. 6.

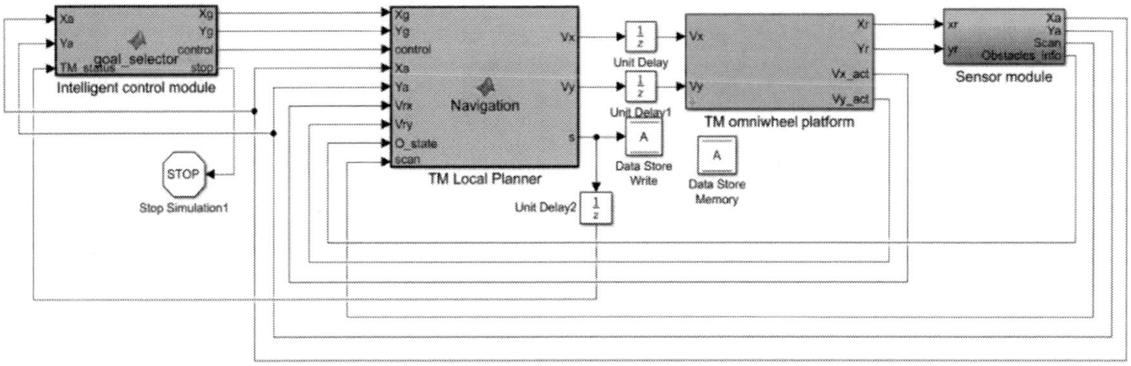

Fig. 6. Computer model of the mobile robot consisting of three main modules

The main blocks of the model:

1. Intelligent control module – a block that performs part of the ICM functions: stores data about the path of the robot, compares the position of the robot received from the TM with the given one. The model assumes that the TM receives information about its position using odometry. If the next target position is reached with an acceptable error (no more than 5 cm in this model), the ICM sends the coordinates of the next point to the TM.

The block partially implements the automaton shown in fig. 5, a. In the model, a new path is not built after a transition to a state S_{I1}; instead, an alternative path is used for a known map configuration (this is acceptable since the ICM must store a map of the environment).

2. TM Local Planner – block of TM local planning which implements the obstacle avoidance algorithm considered above (Algorithm №1) and the law of motion given by the formula (1). The block accepts the coordinates of the target position as input – Xg, Yg; coordinates of the current robot position – Xa, Ya; the TM speed projections – Vrx, Vrx; obstacles data presented in the form of a matrix with dimensions $m \times 5$, where m is the number of obstacles in a near zone of the TM and the points of the scan of the SRSM. The model assumes that the SRSM consists of a sensor system with 36 idealized rangefinders.

The TM Local Planner block outputs the projection of the linear speed, which ensures the avoidance of obstacles and is the closest to the reference linear speed, which is determined by the formula (1).

The block implements the state machine shown in fig. 4, a. The state of the TM is described by the "s" variable, which is issued to the block output for interaction with the ICM:

- s = 1 – the TM moves to a next target (the S_{T1} state),
- s = 2 – the TM has reached the target (the S_{T2} state),
- s = 3 – an emergency situation (the S_{T3} state): a collision has occurred, the RAV* set is empty or the time of arrival at the target position has expired.

Here, a collision is understood as a situation when the distance between the robot and the obstacle becomes less than a predetermined threshold. There is no S_{T0} state in the model since at the beginning of the simulation the robot immediately has a point to which it must arrive. When s = 3 the ICM starts issuing points of the new path.

3. TM omniwheel platform – a block containing a dynamic model of the TM with an omnidirectional drive. The block receives the projection of the given speed as input and outputs the actual speeds, as well as the coordinates of the platform position.

4. Sensor module – the block that scans an area around the robot using ideal rangefinders. The block returns the scan of obstacles obtained by determining the intersection points of virtual beams of rangefinders with polygons that are models of static obstacles. Inside the block there are models of moving obstacles, information about which is grouped into a matrix.

Consider the movement of the robot in the environment in the presence of both dynamic and static obstacles (Fig. 7). At the initial moment, the robot is in the position with coordinates (-0.5, -0.5). Let the ICM planned a path consisting of three points with the following coordinates: (1.5, 2), (4.5, 2) and (6, -0.75). TM must consistently pass these positions. The TM receives the next target point from the ICM after the distance between the actual position of the robot and the target becomes less than 0.05 m. There is one moving obstacle in the environment that moves to the left from the position with coordinates (5, 2). It starts moving 6 seconds after starting the robot and stops at point (2, 2).

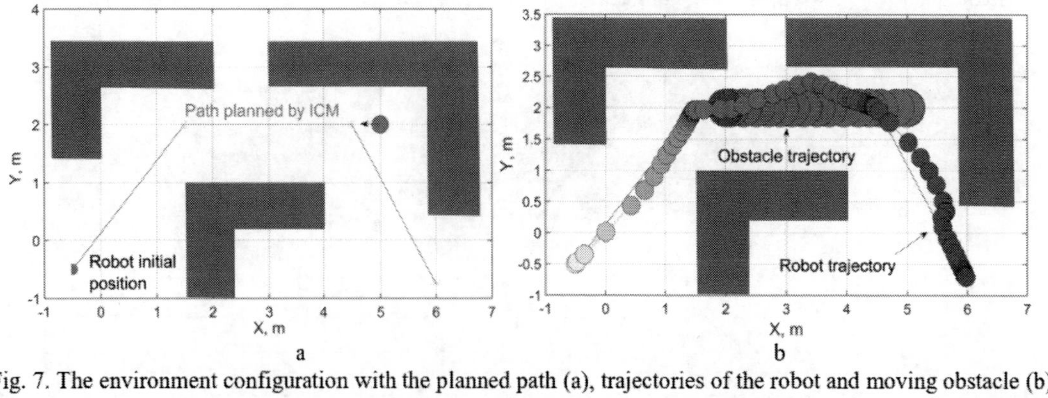

Fig. 7. The environment configuration with the planned path (a), trajectories of the robot and moving obstacle (b)

The trajectory of the modular robot and obstacle after the computer simulation is shown in Fig. 7, b. It can be seen from the graphs in the figure that the robot successfully avoided a moving obstacle and avoided collision with walls. Consider the following situation. The robot is in a room with one rectangular obstacle in the center. In the environment map, this obstacle was taken into account and the ICM planned two identical optimal paths (fig. 8).

Let the ICM choose path №1 with the coordinates of the target points (1, 3.5), (4, 3.5) and (5, 2). These points must be visited by the TM in sequence. At the same time, an obstacle moves towards it from the position (4, 3.5), which could not be taken into account at the stage of path planning.

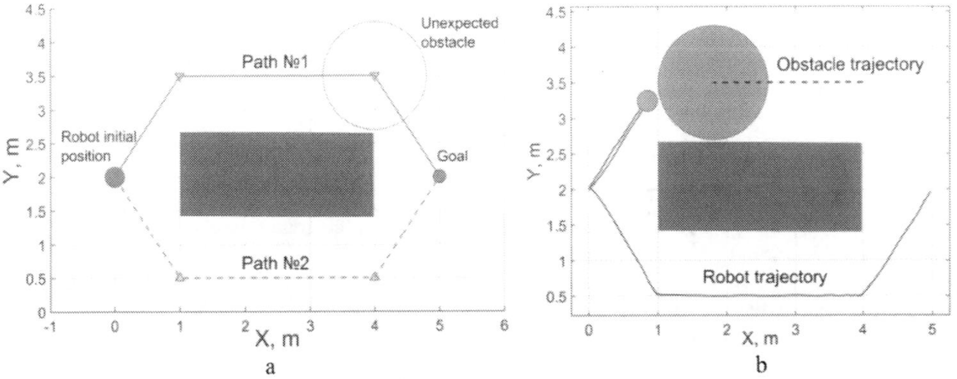

Fig. 8. The model of the environment with one static obstacle and one unknown moving obstacle (a), trajectories of the robot and moving obstacle after simulation (b)

At first the TM moved to the point (1, 3.5) in the normal mode (S_{T1} state), but a large obstacle moving towards it did not allow reaching the first point and a collision with the robot occurred. Due to the collision the TM changes its state to S_{T3} and reports this to the ICM. The ICM also changes its state and chooses the second path, which is similar to the first, and starts sending the coordinates of the target points of the second path to the TM. In this case, the first point to be visited by the TM is the initial position of the robot before starting a simulation. Figure 8 (b) shows the trajectory of the robot, which shows that the robot first tried to get to position (1, 3.5), but after colliding with an obstacle, it returned to its original position and began moving along path №2.

Consider a similar situation: there are two corridors on a map and the ICM builds two identical paths to the goal (Fig. 9, a). At the same time, one of the passages is blocked by obstacles that the ICM does not know about. This can happen if these obstacles appeared after the map was built. Let the ICM choose the path №1 and the TM begins to consistently visit the points of the path.

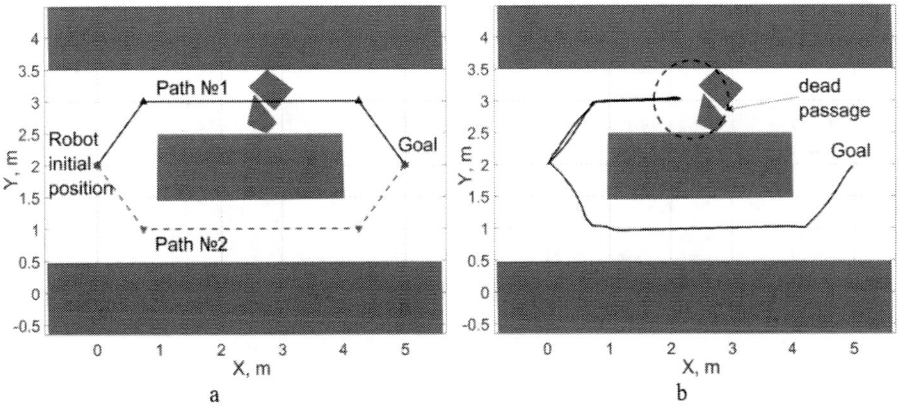

Fig. 9. The model of the environment with two corridors, one of which is blocked (a), the trajectory of the robot after simulation (b)

Figure 9 (b) shows the trajectory of the robot after the simulation. The TM visited only one point of the first path. Further in the corridor, obstacles did not allow TM to pass to the next target, and he began to make small movements in the dead passage and almost didn't move trying to go forward. Before the start of the simulation, a deadline was set to move to the next point (15 sec). When the time to reach the next point exceeded the deadline, the TM switched to the state S_{T3} (s = 3). This switch was captured by the ICM, the state of which changed from S_{I0} to S_{I1}. The ICM switched to performing a movement along path №2, and began sending points of this path to TM. After that, the TM has successfully reached the final position with coordinates (5, 2).

The results of the computer simulation showed the efficiency of the mathematical model of the modular robot. The use of the apparatus of the theory of finite automata makes it possible to describe the robot modules as independent agents. In this case, the state machine model can be easily implemented in the software of a particular module.

6. Conclusions

This article presents a developed formal description of the information interaction of the ICM, TM and SRSM modules, as well as a mathematical and computer model of a mobile robot consisting of these modules. The analysis of information inter-modular interaction was carried out based on the principle of full functionality for the problem of navigation in a flat environment. The division of functions between the modules has been substantiated and a general algorithm for the interaction of the ICM, TM and SRSM has been developed. The data structures that modules must send to each other are defined, which is necessary for drafting a specification for inter-modular interfaces of modules. To formalize the information interaction between TM and SRSM, it was necessary to conduct an analytical review of the known algorithms for avoiding moving obstacles. The analysis of known works showed that many algorithms use the same information about moving obstacles, which made it possible to formulate requirements for the TM-SRSM interface.

It is proposed to consider the control systems of the TM and ICM as finite automata. This approach allows studying modules as hardware and software agents in a multi-agent environment, changing their state and behavior depending on external and internal conditions. To avoid moving obstacles, it was decided to use the VO method in the TM control system. Analysis of the results of computer modeling allows us to conclude that the proposed mathematical model of the control system reflects the specifics of the modular hierarchical architecture. On the one hand, the modules work independently: the TM avoids obstacles; the ICM chooses the path on the map. On the other hand, the state of one module affects the operation of another, which was shown in the results of computer simulation.

7. Future work

In the course of further research, it is necessary to expand the mathematical model of a mobile robot with a modular architecture, namely: to investigate the behavior of the system under various internal and external conditions, to complicate the TM model, taking into account the sub-modules that are part of it, i.e. introduce the 3rd level of the hierarchy. This will increase the performance of the TM control system by further parallelizing the computational process. A big task is to add elements to the robot's computer model that are responsible for simulating the operation of communication channels to assess inter-module traffic in various situations.

8. Acknowledgments

The Russian Foundation for Basic Research supports this research: Grant 19-07-00892a.

9. References

[1] Andreev, V.; Kim, V. & Pletenev, P. (2017). The Principle of Full Functionality – the Basis for Rapid Reconfiguration in Heterogeneous Modular Mobile Robots, Proceedings of the 28th DAAAM International Symposium, 08-11th November 2017, Zadar, Croatia, ISSN 1726-9679, ISBN 978-3-902734-11-2, B. Katalinic (Ed.), pp.0023-0028, Published by DAAAM International, Vienna, Austria, DOI: 10.2507/28th.daaam.proceedings.003.

[2] Andreev V. (2019). Control System Mobile Robots with Modular Architecture as a Multi-Agent System with a Hierarchical Topology, Proceedings of the 30th DAAAM International Symposium, 23-26th October 2019, Zadar, Croatia, ISSN 1726-9679, ISBN 978-3-902734-22-8, B. Katalinic (Ed.), pp.0010-0019, Published by DAAAM International, Vienna, Austria, DOI: 10.2507/30th.daaam.proceedings.002.

[3] Andreev, V.; Kim, V., & Eprikov S. (2020). ModRob: the hardware-software framework for modular mobile robots prototyping, Proceedings of the 31st DAAAM International Symposium, 21-24th October 2020, Hosted in Mostar, BiH, ISSN 1726-9679, ISBN 978-3-902734-29-7, B. Katalinic (Ed.), pp. 0391–0402, Published by DAAAM International, Vienna, Austria, DOI: 10.2507/31st.daaam.proceedings.054.

[4] Hart, P. E.; Nilsson, N. J. & Raphael, B. (1968), "A Formal Basis for the Heuristic Determination of Minimum Cost Paths", IEEE Transactions on Systems Science and Cybernetics, Vol. 4, No. 2, 1968, pp. 100-107.

[5] Stentz, A. (1997). Optimal and efficient path planning for partially known environments, In: Intelligent Unmanned Ground Vehicles, Martial H., Charles E., Anthony Stentz, (Ed.), pp. 203–220, Springer US, ISBN 978-0-7923-9833-2, Germany.

[6] LaValle, S.M. (1998). Rapidly-Exploring Random Trees: A new tool for path planning, Available from: http://lavalle.pl/papers/Lav98c.pdf Accessed: 2020-10-09.

[7] Gammell, J. D.; Srinivasa S. S. & Barfoot, T. D. (2014). Batch informed trees (bit*): sampling-based optimal planning via the heuristically guided search of implicit random geometric graphs, arXiv preprint arXiv, vol. 1405, pp. 5848.

[8] Gerasimov, V.N. (2015). Sistema navigatsii servisnogo robota v srede s dinamicheskimi prepyatstviyami [service robot navigation system in an environment with dynamic obstacles], Ph.D. Dissertation, Bauman Moscow State Technical University, Moscow, Russian Federation.

[9] Berg, J. van den; Lin M., & Manocha, D. (2008). Reciprocal Velocity Obstacles for real-time multi-agent navigation, in 2008 IEEE International Conference on Robotics and Automation, 19-23 May 2008, Pasadena, CA, USA, Print ISSN: 1070-9932, pp. 1928–1935, IEEE, DOI: 10.1109/ROBOT.2008.4543489.

[10] Chen, Y. & Liu M. (2018). RRT* Combined with GVO for Real-time Nonholonomic Robot Navigation in Dynamic Environment, Available from: http://arxiv.org/abs/1710.07102 Accessed: 2021-10-01.

[11] Cherubini, A.; Spindler, F. & Chaumette, F. (2014). Autonomous visual navigation and laser-based moving obstacle avoidance, IEEE Transactions on Intelligent Transportation Systems, Vol. 15, No. 5, 2014, pp. 2101–2110, Print ISSN: 1524-9050, DOI: 10.1109/TITS.2014.2308977.

[12] Guy S.J. et al. (2009). ClearPath: highly parallel collision avoidance for multi-agent simulation, in Proceedings of the 2009 ACM SIGGRAPH/Eurographics Symposium on Computer Animation, 1-2 August 2009, New Orleans, Louisiana, USA, ISBN 978-1-60558-610-6, pp. 177-187, Association for Computing Machinery, New York, USA, DOI: 10.1145/1599470.1599494.

[13] Claes, D.; Hennes, D.; Tuyls, K. & Meeussen, W. (2012). Collision avoidance under bounded localization uncertainty, in 2012 IEEE/RSJ International Conference on Intelligent Robots and Systems, 7-12 Oct. 2012, Vilamoura-Algarve, Portugal, Print ISBN:978-1-4673-1737-5, Print ISSN: 2153-0858, pp. 1192–1198, IEEE, DOI: 10.1109/IROS.2012.6386125.

[14] Chen, Y. F.; Liu, M.; Everett, M. & How, J. P. (2017). Decentralized non-communicating multiagent collision avoidance with deep reinforcement learning, IEEE International Conference on Robotics and Automation (ICRA), 29 May – 3 June 2017, Singapore, Electronic ISBN: 978-1-5090-4633-1, pp. 285–292, IEEE, DOI: 10.1109/ICRA.2017.7989037.

[15] Fiorini, P. & Shiller, Z. (1998). Motion planning in dynamic environments using velocity obstacles, The International Journal of Robotics Research, Vol. 17, No. 7, 1998, pp. 760–772, ISSN 0278-3649, DOI: 10.1177/027836499801700706.

[16] Fox, D.; Burgard, W. & Thrun, S. (1997). The dynamic window approach to collision avoidance, IEEE Robotics & Automation Magazine, Vol. 4, No. 1, 1997, pp. 23–33, ISSN 10709932, DOI: 10.1109/100.580977.

[17] Shahriari, M.; Svogor, I.; St-Onge, D. & Beltrame G. (2018). Lightweight collision avoidance for resource-constrained robots, in 2018 IEEE/RSJ International Conference on Intelligent Robots and Systems (IROS), 1-5 Oct. 2018, Madrid, Spain, Electronic ISBN: 978-1-5386-8094-0, Electronic ISSN: 2153-0866, pp. 1–9, IEEE, DOI: 10.1109/IROS.2018.8593841.

[18] Narayanan, V.; Phillips, M. and Likhachev, M. (2012). Anytime Safe Interval Path Planning for dynamic environments, in 2012 IEEE/RSJ International Conference on Intelligent Robots and Systems, Oct. 2012, Vilamoura-Algarve, Portugal, Print ISBN: 978-1-4673-1737-5, Print ISSN: 2153-0858, pp. 4708–4715, IEEE, DOI: 10.1109/IROS.2012.6386191.

[19] Alonso-Mora, J.; Breitenmoser, A.; Rufli, M.; Beardsley, P. and Siegwart, R. (2013). Optimal Reciprocal Collision Avoidance for Multiple Non-Holonomic Robots, Distributed Autonomous Robotic Systems, Vol. 83, 2013, pp. 203–216, 2013, Print ISBN 978-3-642-32722-3, DOI: 10.1007/978-3-642-32723-0_15.

[20] Savkin, A. & Wang, C. (2013). A simple biologically inspired algorithm for collision-free navigation of a unicycle-like robot in dynamic environments with moving obstacles, Robotica, Vol. 31, No. 6, 2013, pp. 993–1001, ISSN 0263-5747, DOI: 10.1017/S0263574713000313.

[21] Snape, J.; Berg, J. van den; Guy, S.J. & Manocha, D. (2011). The Hybrid Reciprocal Velocity Obstacle, IEEE Transactions on Robotics, Vol. 27, No. 4, 2011, pp. 696–706, ISSN 1552-3098, DOI: 10.1109/TRO.2011.2120810.

[22] Siegwart, R.; Nourbakhsh, I. and Scaramuzza, D. (2011). Introduction to Autonomous Mobile Robots, Second edition. The MIT Press, ISBN-13: 978-0262015356.

[23] Wang, D.; Watkins, C. and Xie, H. (2020). MEMS Mirrors for LiDAR: A Review, Micromachines, Vol. 11, No. 5, p. 456-480, ISSN 2072-666X, DOI: 10.3390/mi11050456.

[24] Balogh, R. & Obdržálek, D. (2019). Using Finite State Machines in Introductory Robotics, in Robotics in Education, vol. 829, W. Lepuschitz, M. Merdan, G. Koppensteiner, R. Balogh, and D. Obdržálek, Eds., pp. 85–91, Springer International Publishing, DOI: 10.1007/978-3-319-97085-1_9.

[25] Snape, J.; Berg, J van den; Guy, S. J. and Manocha, D. (2009). Independent navigation of multiple mobile robots with hybrid reciprocal velocity obstacles, in 2009 IEEE/RSJ International Conference on Intelligent Robots and Systems, 10-15 Oct. 2009, St. Louis, MO, USA, pp. 5917–5922, DOI: 10.1109/IROS.2009.5354821.

DOI: 10.2507/32nd.daaam.proceedings.033

PRODUCTION SCHEDULING USING TWK AND SLK METHODS

Robert Ojstersek, Iztok Palcic & Borut Buchmeister

This Publication has to be referred as: Ojstersek, R[obert]; Palcic, I[ztok] & Buchmeister, B[orut] (2021). Production Scheduling Using TWK and SLK Methods, Proceedings of the 32nd DAAAM International Symposium, pp.0228-0234, B. Katalinic (Ed.), Published by DAAAM International, ISBN 978-3-902734-33-4, ISSN 1726-9679, Vienna, Austria DOI: 10.2507/32nd.daaam.proceedings.033

Abstract

The manuscript presents the importance of integrating mathematical methods of due date determination for the maturity optimization in evolutionary computation (EC) methods in the flexible job shop problem (FJSSP). The use of mathematical modelling methods for due date determination, the Slack Method (SLK) for low-dimensional and the Total Work Content Method (TWK) for medium- and high-dimensional problems was used. The results of FJSSP optimization of makespan, machine utilisation and due date scheduling using Multi-Objective Heuristic Kalman Algorithm (MOHKA) algorithm were compared with two existing multi-objective particle swarm algorithms. The obtained results of the tested benchmark datasets proved the high level of FJSSP importance in scheduling. The presented results demonstrate the importance of the due date optimization parameter in achieving a low numbers of tardy jobs, short job tardiness and potentially a lower tardy jobs costs in relation to a short makespan. The main results show that the TWK and SLK method is the key to achieving an on-time and financially justified production system.

Keywords: SLK method; TWK method; due date; production scheduling; evolutionary computation

1. Introduction

Researchers restricted themselves to optimization parameters that define the due date of jobs [1], assuming independence from other optimization parameters that significantly affect production flexibility [2]. The dynamic changes in FJSSP production due to dynamic customer demand and high-mix low-volume production make Pareto-based optimization approaches a suitable optimization approach [3]. The use of fuzzy approaches that satisfactorily solve the optimization problem of FJSSP production usually treats the problem only at a single level of the primary optimization criteria and restricts the multi-level structure of the FJSSP problem [4]. Heuristic methods, which allow a detailed estimation of the optimization approach and the satisfactory optimization method, are usually limited by the transfer of the optimization results to a real or simulated environment [5]. The need for an efficient optimization method to plan and schedule the FJSSP problem with the optimization parameter of due dates is the key to achieve a comprehensive optimization approach [6]. However, the research results must allow the devaluation of both test and real datasets to adequately integrate the proposed methods into the real production environment [7].

The present research work is organised as follows: In section two, the problem description of the FJSSP is presented through the mathematical formulation of the objective functions and constraints. Section three describes the TWK and SLK methods and proposes a methodology to apply them to the FJSSP. Section four presents the experiments used to evaluate the efficiency and robustness of MOHKA [8] compared to two similar algorithms from EC. Section five presents the optimization results numerically and graphically. The final sixth section describes conclusions and future research steps.

2. Problem description

Flexible job shop production can define flexibility by a four-stage architectural model that treats production flexibility as a multidimensional optimization problem [9]. The degree of production flexibility depends on the characteristics of each production system. Manufacturing systems can define production flexibility in an active or proactive manner [10]. The active mode represents the active use of flexibility to compensate for unknown uncertainties in the production system and deals with both flexibility inside and outside the production process (flexibility of production according to demand and competition policies) [11]. The active approach defines production flexibility as the current ability of the manufacturer to adapt and efficiently use the production system [12]. A proactive approach to production flexibility allows a company to achieve global competitiveness by increasing customer expectations (new orders) and increasing the uncertainty of competing companies [13]. In a proactive approach, production flexibility defines the dynamic change of environmental, social and economic changes [14]. From the point of achieving orders dynamic due dates, it is necessary to adjust production according to customer demand, consider the availability of flexibility of existing products and thus meet customer needs [1]. In the flexible job production optimization problem (FJSSP), the three most common optimization parameters (makespan, total workload of all machine and the workload of the most loaded machine) are defined by Eq. 1, Eq. 2, and Eq. 3 [15]:

- Makespan

$$MC = \max\ \{C_j \mid j = 1, \ldots, n\}, \tag{1}$$

- total workload of all machine

$$\boldsymbol{TW} = \sum_{i=1}^{n}\sum_{j=1}^{n_i}\sum_{k=1}^{m} \boldsymbol{p_{ijk}x_{ijk}},\ \boldsymbol{k} = \boldsymbol{1, 2, \ldots, m} \text{ and} \tag{2}$$

- the workload of the most loaded machine

$$\boldsymbol{MW} = \boldsymbol{max}\sum_{i=1}^{n}\sum_{j=1}^{n_i} \boldsymbol{p_{ijk}x_{ijk}},\ \boldsymbol{k} = \boldsymbol{1, 2, \ldots, m}, \tag{3}$$

where Cj is the completion time of job j, n represents the number of jobs, and m represents the number of machines. These three optimization parameters refer to the time for completing the orders and achieving the highest possible utilization of the machines. In highly flexible job shop production, a very important parameter is the due date of the orders dj, which can be very tight to satisfy market demands. Fig. 1 shows the order entry sequence of four jobs, where each job must be executed with a different due date in the production system defined by different machines suitable for perform individual operations. The optimization of the jobs process sequence must be done considering the multi-objective nature of the FJSSP, where three parameters must be minimized (makespan, machine workload and due date).

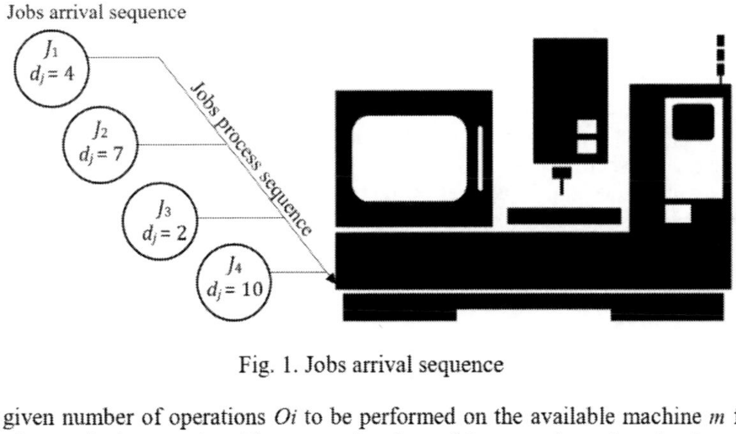

Fig. 1. Jobs arrival sequence

Each job j has a given number of operations Oi to be performed on the available machine m from the given set of machines that can perform a single operation. The processing time of the operation pjk varies depending on the assigned machine that can execute the single operation Oi. The optimization algorithm must arrange the job process sequence in such a way that it handles all four optimization parameters appropriately. For example, using the conventional Earliest Due Date (EDD) priority rule [16], the job process execution sequence would be J_3, J_1, J_2, and J_4. This job process execution sequence handles only one optimization parameter dj, which is defined as follows: the due date of job j is the estimated dispatch date of job j (dispatch date promised to the customer or scheduled in the company). Completing order j after the due date is allowed but represents an additional financial or time penalty to the company. When considering the optimization parameter dj, important associated goals must be established. The primary objective is to reduce the maximum lateness, which is defined as in Eq. 4:

$$L_{max} = max\ (L_j, \ldots, L_n), \tag{4}$$

where the lateness of an individual job j is defined by the Eq. 5:

$$L_j = C_j - d_j, \tag{5}$$

depends on the completion time of job j and the assumed delivery time of job dj. The temporal L_{max} is easier to define with the parameter number of tardy jobs. This optimization parameter only defines whether the individual job j has exceeded the estimated delivery time or not. Tardiness of job j is defined as in Eq. 6:

$$T_j = max\ (C_j - d_j, 0), \tag{6}$$

and the corresponding target function defined by Eq. 7.

$$\sum_{j=1}^{n} Tj. \tag{7}$$

Due to the shortcomings of the above optimization function, which relates to some very tardy jobs, it is useful to determine the importance weights of jobs j by w_j parameter. The higher the weight, the more important the job j is. The assumption in this research work is that the given weight w_j refers only to the importance of a single job j, which can be weighted directly by the scheduling team of the manufacturing system. The importance of multicriteria decision making between the four optimization parameters (MC, TW, MW and dj) does not determine the importance of the correlation between them.

3. TWK and SLK methods formulation

To model the due dates of jobs, a random FJSSP data sets are generated and divided into three groups in terms of their complexity dimensions [17]:

- Low dimensional optimization problems, the dataset of J5, M11, O66 has been configured to evaluate the capabilities of MOHKA with respect to the optimization results of conventional priority rules.
- Middle dimensional optimization problem is represented by the dataset of J10, M11, O122.
- High dimensional optimization problem is represented by dataset of J15, M11, O176.

TWK [18] method due date optimization parameter modelling is performed according to the Eq. 8.

$$\boldsymbol{d_j} = \boldsymbol{at_j} + \boldsymbol{K_x} * \sum_{i \in o_j} \boldsymbol{p_{i,j,k}} \tag{8}$$

The tightness allowance factor of the due date K_x determines the tightness of individual orders. Numerical values in the literature, range it as $3 \leq K_x \leq 5$, for the TWK method. The smaller the value of the tightness allowance factor, the tighter the due date of order j is. The experiments in the manuscript use the value of $K_x = 3$. The due date modelled by the TWK method depends on the arrival time of order j (a_{tj}), the total time of processing all operations (p_{ijk}), and the described tightness allowance factor. The MOHKA algorithm schedules the orders according to four optimization criteria, including the due date dj. The comparison of the adequacy of the proposed MOHKA method is performed using two comparison algorithms: Multi-objective particle swarm optimization (MOPSO) [19] and Bare-bones multi-objective particle swarm optimization (BBMOPSO) [20]. These two algorithms do not use an integrated mathematical decision model to schedule jobs according to the due date criterion, this criterion is numerically calculated in the experiment at the end of the optimization results. As stated in the initial research question, the due date parameter in the FJSSP optimization problem is not well studied, especially when it comes to using the EC method to obtain optimal solutions. All the algorithms in the experiment use the same initialization parameters: Population size ($N_s = 300$), maximum number of archived non-dominant solutions ($N_a = 100$), and maximum number of algorithm iterations ($MaxIter = 300$).

The optimization parameter for scheduling jobs by due date is analysed using three criteria: Number of tardy jobs, average job tardiness and tardy jobs cost. The tardy jobs costs is modelled as shown in Eq. 9, where the initial job cost (J_{cost}) is multiplied by the constant value of three, divided by the value constant K_s, and multiplied by subtracting the completion time C_j and the due date d_j.

$$L_{cost} = \begin{cases} \dfrac{3 * J_{cost}}{K_S} * (C_j - d_j), \text{if } C_j > d_j \\ 0, \text{if } C_j \leq d_j \end{cases} \tag{9}$$

A constant numerical value determines three times the cost of tardy jobs compared to the cost of orders completed on time (presenting penalty costs). The value K_S is automatically determined time variable, presenting individual order makespan. The parameter J_{cost} is the machines cost processing an individual order on a specific machine.

The modelling of the due dates and the achievement of the other three optimization parameters were performed using conventional methods (priority rules) and the heuristic GSBR (General Shifting Bottleneck Routine) method to evaluate the efficiency of the conventional scheduling methods compared to the proposed MOHKA EC method. The comparison of priority rules and GSBR method is performed in the Lekin software environment. Since the Lekin software environment only allows the optimization of datasets with up to one hundred operations in the FJSSP optimization problem, the evaluation is performed with a randomly generated benchmark dataset classified as a low-dimensional optimization problem J_5, M_{11}, O_{66}. A randomly generated benchmark dataset does not contain data where two or more operations are performed on the same machine within a single job. This limitation is made according to the Lekin decision making strategies. In contrast to larger benchmark datasets where the TWK method is used to model due dates, the SLK (Slack) [21] method is recommended for smaller datasets. For low-dimensional dataset and SLK method, the due date parameter is modelled by the Eq. 10.

$$d_j = at_j + \sum_{i \in o_j} p_{i,j,k} + K_y \quad (10)$$

In the SLK method the determination of the time reserve constant K_y determines the looseness-tightness of due dates. Reserve constant values given by the literature are $4 \leq K_y \leq 16$. In the presented research work the middle value of $K_y = 8$ is used.

4. Experiment design

To test the performance of the MOHKA algorithm for job scheduling (according to the due date parameter) four randomly generated benchmark datasets were used, describing FJSSP optimization problem. The datasets were randomly created using the dependency function between different optimization parameters. We divided these benchmark datasets into three groups according to the complexity of the optimization problem (low-, middle- and high-dimensional). Due to the classification of the used datasets into different groups according to their complexity we have been able to evaluate individual benchmark and algorithm capabilities. Use of TWK and SLK method for different optimization problems were made as: TWK method for medium and high dimensional optimization problems and SLK method for low dimensional optimization problems. With the presented classification approach, the complexity of the optimization problems can be evaluated more accurately to determine the advantages and limitations of each method. The proposed MOHKA algorithm performed the optimization of data sets with four parameters as described in section three: makespan (MC), total workload of all machines (TW), maximum workload of a single machine (MW) and added due date parameter (dj). The obtained optimization results were compared with two particle swarm based multi-objective optimization algorithms (MOPSO and BBMOPSO). The experiments were performed on a personal computer with Intel i7 processor and 16 GB internal memory.

5. Due date scheduling results

The results in Table 1 show the high reliability of scheduling jobs with the TWK method considering due dates with the MOHKA optimization algorithm. The method is successful in scheduling jobs with tight due dates, low average jobs tardiness, potentially low tardy jobs cost and short orders makespan. The medium sized data set J_{10}, M_{11}, O_{122} did not cause any problems for all three evolutionary computation algorithms in scheduling orders with tight due dates of the TWK method with the tightness coefficient of $K_x = 3$. No job has missed the scheduled due date, which in turn resulted in no additional tardy jobs costs in the production system. Since only the referential MOHKA optimization algorithm considers the mathematical architecture of the TWK method, we see that the results of multi-objective optimization have a positive impact on the achievement of the minimum orders makespan. The MOHKA algorithm achieves the shortest makespan with up to 226 h, in contrast to MOPSO and BBMOPSO, where the makespan is 272 h and 247 h, respectively.

When the dimension of the datasets increases from medium to high dimensional optimization problems, the difference between the optimization results of the comparison and reference algorithms becomes more obvious, as shown in Fig. 2 (green graph lines). In dataset J_{15}, M_{11}, O_{176}, the MOHKA algorithm terminates orders such that four orders are late by the expected due date, with an average job tardiness of 22.8 h, resulting in a tardy jobs costs of 2,738.6 EUR. The MOPSO algorithm finishes orders such that only two orders miss the expected due date, but with higher average job tardiness of 54.6 h, which is 139% higher average job tardiness than the MOHKA algorithm. Longer average job tardiness leads to higher tardy jobs costs, which is 3,559.8 EUR for the MOPSO algorithm. The BBMOPSO algorithm had the most inefficient job schedule, six orders missing the scheduled due date, with an average job tardiness of 24.72 h and a high tardy jobs costs of 4,270.2 EUR. With that results BBMOPSO represents a 56.6% increase in the tardy jobs cost compared to the MOHKA algorithm. The results show that the MOHKA algorithm is most successful even with an order makespan parameter of 347 h, which is shorter than 407 h for the BBMOPSO algorithm and 356 for the MOPSO algorithm. Based on the above optimization results of the due date FJSSP optimization, the complexity of solving higher dimensional optimization problems with the TWK and SLK methods increases. The initial research question is confirmed for the high-dimensional dataset of J_{20}, M_{11}, O_{240}, where the reference MOHKA algorithm dominates over the results of the two comparison algorithms shown in Fig. 2 (blue graph lines). The reference MOHKA algorithm has the number of tardy jobs equal to eight, with an average job tardiness of 37.32 h compared to 53.28 h and 56.16 h for MOPSO and BBMOPSO, respectively, which corresponds the higher number of tardy jobs of twelve for MOPSO and twenty two for BBMOPSO. Given the higher number of tardy jobs and the longer average job tardiness, the cost of tardy jobs is also higher for the two comparison algorithms is also higher than for the reference algorithm MOHKA, amounting to 7,318.1 EUR.

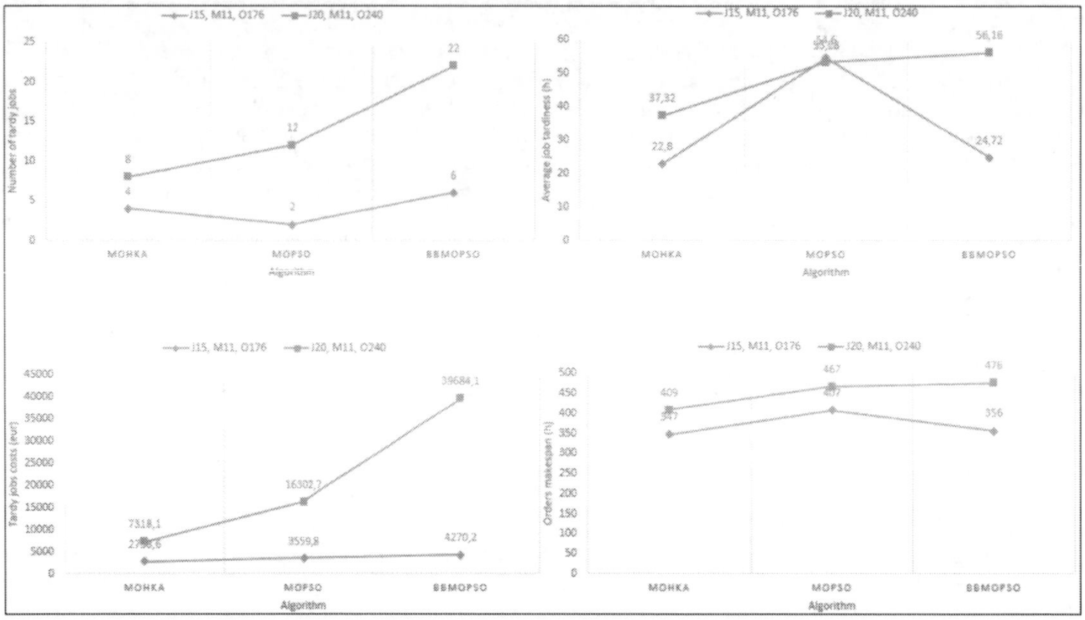

Fig. 2. Optimization results of datasets J_{15}, M_{11}, O_{176} and J_{20}, M_{11}, O_{240}.

Algorithm	Dataset	Number of tardy jobs	Average job tardiness (h)	Tardy jobs costs (EUR)	Orders makespan (h)
MOHKA	J_{10}, M_{11}, O_{122}	0	0	0	226
	J_{15}, M_{11}, O_{176}	4	22.8	2,738.6	347
	J_{20}, M_{11}, O_{240}	8	37.32	7,318.1	409
MOPSO	J_{10}, M_{11}, O_{122}	0	0	0	272
	J_{15}, M_{11}, O_{176}	2	54.6	3,559.8	407
	J_{20}, M_{11}, O_{240}	12	53.28	16,302.7	467
BBMOPSO	J_{10}, M_{11}, O_{122}	0	0	0	247
	J_{15}, M_{11}, O_{176}	6	24.72	4,270.2	356
	J_{20}, M_{11}, O_{240}	22	56.16	39,684.1	476

Table 1. Algorithm's optimization results

With the MOPSO algorithm, the tardy jobs costs amount to 16,302.7 EUR, while with BBMOPSO it is 39,684.1 EUR. The presented results prove the high importance of mathematical modelling with the parameters of due date optimization, as they have a crucial impact on the makespan and financial justification of highly dynamic manufacturing systems [22]. A suitable mathematical model of the multi-objective optimization problem is also reflected in the achievement of a short order makespan. For the high-dimensional data set of J_{20}, M_{11}, O_{240}, the reference MOHKA algorithm achieved a makespan of 409 h and the two comparison algorithms 467 h and 476 h, respectively.

5.1 Results of the mathematical modelling with SLK method

With the aim of comparing the solutions of MOHKA algorithm and the solutions of conventional priority rules, a comparison of the results of MOHKA optimization with the results of job scheduling of J_5, M_{11}, O_{66} in the Lekin software environment for the low-dimensional optimization problem was performed. Table 2 shows the optimization results of a J_5, M_{11}, O_{66} data set and comparisons between MOHKA and priority rules optimization results, using SLK method. The optimization was performed using MOHKA optimization algorithm in MATLAB software environment and seven optimization approaches in Lekin software environment. Of the seven optimization approaches, six are conventional priority rules and one is a heuristic algorithm called General Shifting Bottleneck Routine (GSBR). The SLK method with a time reserve constant of $K_y = 8$ was used to optimise the due dates. The results show high reliability of scheduling production jobs scheduling with MOHKA optimization algorithm. In the considered dataset, MOHKA schedules orders in such a way that two orders miss the scheduled due dates with an average job tardiness of 11.4 h and a tardy jobs costs of 3,181 EUR. Among the six priority rules, we see that the five priority rules, except for the priority rule SPT, terminate orders in a way that all five orders miss the scheduled due date. The average job tardiness is higher than the MOHKA algorithm, ranging from 35.3 h for the priority rule CR to 57.6 h for the priority rule LPT. Moreover, the cost of delayed orders is significantly higher. The only algorithm that partially approximates the results of the MOHKA algorithm is the heuristic GSBR algorithm, where four orders are tardy with an average job tardiness of 11.3. Another significant difference is that the MOHKA algorithm finishes jobs so that they are completed in a makespan of 118.8 h, while all other algorithms finish jobs with makespan between 249.6 h, by GSBR algorithm, and 307.2 h by CR priority rule.

Algorithm	MOHKA	EDD	MS	FCFS	LPT	SPT	CR	GSBR
Number of tardy jobs	2	5	5	5	5	4	5	4
Average job tardiness (h)	11.4	41.8	57.1	49.9	57.6	38.9	35.3	11.3
Tardy jobs costs (EUR)	3,181	13,590	17,618	15,984	17,323	12,684	10,686	4,447
Orders makespan (h)	118.8	253.2	265.2	258	260.4	279.6	307.2	249.6

Table 2. SLK optimization results of MOHKA algorithm vs. priority rules due date scheduling

The presented optimization results prove the high term fidelity of the MOHKA algorithm to achieve tight due dates from low-dimensional optimization cases (with SLK method) up to medium and high-dimensional optimization cases (with TWK method) in comparison to optimization solutions according to MOSPO, BBMOPSO and priority rules.

6. Conclusion

Scheduling multi-objective FJSSP optimization problem is defined as a non-deterministic polynomial-time hard optimization problem. The original research question of scheduling FJSSP production with the optimization parameter of due date importance and considering the standard optimization parameters related to the makespan of orders and machine utilisation was evaluated in the presented research using MOHKA optimization method and SLK and TWK methods for modelling due dates. As the number of optimization parameters increases, the computational complexity of the optimization algorithm increases. The presented research presents the integration of the mathematical structure of SLK (for low-dimensional optimization problems) and TWK method (for medium- and high-dimensional optimization problems) into the MOHKA optimization algorithm. The proposed MOHKA algorithm was used to schedule benchmark data sets, focusing on achieving an order due date, corelated with due date tightness coefficient. The optimization results were compared with the results of two EC algorithms: MOPSO and BBMOPSO optimization algorithms. The disadvantage of the comparative optimization methods (MOPSO and BBMOPSO) becomes obvious when dealing with medium and high dimensional optimization problems in scheduling FJSSP. The limited scheduling capabilities of the MOPSO and BBMOPSO algorithms are reflected in the limited mathematical structure, considering due date, which do not consider the SLK and TWK methods as decision parameters in obtaining optimally scheduled orders. The optimization results of the reference MOHKA algorithm prove the high importance of the due date optimization parameter, since the proposed method optimises the order scheduling with respect to the two comparison algorithms for low, medium, and high dimensional optimization problems. Since it involves multi-objective decision making and finding trade-offs between different (even conflicting) optimization parameters. The results of MOHKA algorithm prove its high ability to achieve all four optimization parameters (MC, TW, MW and d_j) equally and efficiently. The short order makespan, tight due dates, low average order tardiness and associated low associated job tardiness costs are achieved. The answer to the question about the efficiency of EC methods in multicriteria decision making compared to the conventional optimization approach of priority rules was given by the presented study, in which the optimization results of MOHKA algorithm are compared with the optimization results of six priority rules and an integrated heuristic method in the Lekin software environment. The obtained results demonstrate the high dominance of the optimization results of the EC method MOHKA, which completes the FJSSP production most efficiently according to the used low-dimensional dataset for all optimization parameters. Randomly generated benchmark datasets were the basis for conducting the validation of the applicability of the proposed method in real manufacturing systems, and the satisfactory optimization results were demonstrated in the experiment. Since the presented research work only deals with FJSSP, which is the main part of the research problem of multi-objective optimization of job shop production, it is necessary to further investigate the importance of scheduling in dynamic job shop production (DJSSP). The main features are dynamic job changes during the execution of the algorithm (in the initialization phase, the entire job record is unknown), machine failures during the execution of operations, and determining the importance of different jobs characteristics to be studied. Further research on the DJSSP research problem would remove the limitations of current research, in which the FJSSP optimization problem assumes of an initially known order data set, an initially empty production system, uniform importance of orders, and known production capacities that do not change during operations. Further jobs shop due date scheduling usability can be transferred to mass production and other types of manufacturing systems.

7. Acknowledgments

This research was funded by the Slovenian Research Agency (ARRS), Research Core Grant P2-0190.

8. References

[1] Saha, C.; Aqlan, F.; Lam, S. S. & Boldrin, W. (2016). A decision support system for real-time order management in a heterogeneous production environment, Expert Systems with Applications, Vol. 60, 16-26, doi:10.1016/j.eswa.2016.04.035

[2] Ojstersek, R. & Buchmeister, B. (2020). The impact of manufacturing flexibility and multi-criteria optimization on the sustainability of manufacturing systems, Symmetry, Vol. 12, No. 1, Paper 157, 22 pages, doi:10.3390/sym12010157

[3] Wisittipanich, W. (2013). An efficient pso algorithm for finding pareto-frontier in multi-objective job shop scheduling problems, Vol. 12, No. 2, 151-160, 10.7232/iems.2013.12.2.151

[4] Koblasa, F.; Kralikovà, R. & Votrubec, R. (2020). Influence of EA control parameters to optimization process of FJSSP problem, International Journal of Simulation Modelling, Vol. 19, No. 3, 387-398, doi:10.2507/ijsimm19-3-519

[5] Ojstersek, R.; Palcic, I.; Buchmeister, B. (2020). Simulation modelling of environmental and economic aspects in manufacturing systems, Proceedings of 31st DAAAM International Symposium on Inteligent Manufacturing and Automation, Katalinic, B. (Ed.), pp. 436-443, DAAAM International, Zadar, doi: 10.2507/31st.daaam.proceedings.061

[6] Xu, Y.; Wang, L.; Wang, S. & Liu, M. (2013). An effective shuffled frog-leaping algorithm for solving the hybrid flow-shop scheduling problem with identical parallel machines, Engineering Optimization, Vol. 45, No. 12, 1409-1430, doi:10.1080/0305215X.2012.737784

[7] Ojstersek, R.; Lalic, D. & Buchmeister, B. (2019). A new method for mathematical and simulation modelling interactivity: A case study in flexible job shop scheduling, Advances in Production Engineering And Management, Vol. 14, No. 4, doi:10.14743/apem2019.4.339

[8] Ojstersek, R.; Zhang, H.; Shifeng, L. & Buchmeister, B. (2018). Improved heuristic kalman algorithm for solving multi-objective flexible job shop scheduling problem, Procedia Manufacturing, Vol. 17, 985-902, doi:10.1016/j.promfg.2018.10.142

[9] Ripon, K. S. N.; Siddique, N. H. & Torresen, J. (2011). Improved precedence preservation crossover for multi-objective job shop scheduling problem, Evolving Systems, Vol. 2, No. 2, 119-129, doi:10.1007/s12530-010-9022-x

[10] Yang, W. & Takakuwa, S. (2017). Simulation-based dynamic shop floor scheduling for a flexible manufacturing system in the industry 4.0 environment, 2017 Winter Simulation Conference, pp. 3908-3916, IEEE, Las Vegas, doi:10.1109/WSC.2017.8248101

[11] Istokovic, D.; Perinic, M.; Vlatkovic, M. & Brezocnik, M. (2020). Minimizing total production cost in a hybrid flow shop: a simulation-optimization approach, International Journal of Simulation Modelling, Vol. 19, No. 4, 559-570, doi:10.2507/ijsimm19-4-525

[12] Ojstersek, R. & Buchmeister, B. (2017). Use of simulation software environments for the purpose of production optimization, Proceedings of the 28th DAAAM International Symposium, pp. 750-758, B. Katalinic (Ed.), DAAAM International, Zadar, doi:10.2507/28th.daaam.proceedings.106

[13] Zhang, H.; Buchmeister, B.; Li, X. & Ojstersek, R. (2021). Advanced metaheuristic method for decision-making in a dynamic job shop scheduling environment, Mathematics, Vol. 9, No. 8, Paper 909, 22 pages, doi:10.3390/math9080909

[14] Frutos, M.; Tohmé, F.; Delbianco, F. & Miguel, F. (2016). An alternative hybrid evolutionary technique focused on allocating machines and sequencing operations, International Journal of Industrial Engineering Computations, Vol. 7, No. 4, 585-596, doi:10.5267/j.ijiec.2016.4.002

[15] Xiang, W.; Yin, J. & Lim, G. (2015). An ant colony optimization approach for solving an operating room surgery scheduling problem, Computers & Industrial Engineering, Vol. 85, 335-345, doi:https://doi.org/10.1016/j.cie.2015.04.010

[16] Hao, X.; Gen, M.; Lin, L. & Suer, G. A. (2017). Effective multiobjective EDA for bi-criteria stochastic job-shop scheduling problem, Journal of Intelligent Manufacturing, Vol. 28, No. 3, 833-845, doi:10.1007/s10845-014-1026-0

[17] Ojstersek, R.; Tang, M. & Buchmeister, B. (2020). Due date optimization in multi-objective scheduling of flexible job shop production, Advances in Production Engineering & Management, Vol. 15, No. 4, 481-492, doi:10.14743/apem2020.4.380

[18] Huang, J. & Süer, G. A. (2015). A dispatching rule-based genetic algorithm for multi-objective job shop scheduling using fuzzy satisfaction levels, Computers & Industrial Engineering, Vol. 86, 29-42, doi:10.1016/j.cie.2014.12.001

[19] Wang, L.; Ng, A. H. C. & Deb, K. (2011). Multi-objective Evolutionary Optimisation for Product Design and Manufacturing, Springer, London, doi:10.1007/978-0-85729-652-8

[20] Zhang, Y.; Gong, D.-W. & Ding, Z. (2012). A bare-bones multi-objective particle swarm optimization algorithm for environmental/economic dispatch, Information sciences, Vol. 192, 213-227, doi:10.1016/j.ins.2011.06.004

[21] Yin, Y.; Wang, D.-J.; Wu, C.-C. & Cheng, T. C. E. (2016). CON/SLK due date assignment and scheduling on a single machine with two agents, Naval Research Logistics, Vol. 63, No. 5, 416-429, doi:10.1002/nav.21700

[22] Zhang, Y. Q. & Zhang, H. (2020). Dynamic scheduling of blocking flow-shop based on multi-population aco algorithm, International Journal of Simulation Modelling, Vol. 19, No. 3, 529-539, doi:10.2507/ijsimm19-3-co15

DOI: 10.2507/32nd.daaam.proceedings.034

INFLUENCE OF SELECTED PHOTOPOLYMERS ON THE RESULTING ACCURACY AND SURFACE ROUGHNESS OF THE COMPONENT IN DIGITAL LIGHT PROCESSING TECHNOLOGY

Jan Milde, Jozef Peterka, Frantisek Jurina, Patrik Dobrovszky,
Jakub Hrbal, Peter Pitek & Jozef Martinovic

This Publication has to be referred as: Milde, J[an]; Peterka, J[ozef]; Jurina, F[rantisek]; Dobrovszky, P[atrik]; Hrbal, J[akub]; Pitek, P[eter] & Martinovic, J[ozef] (2021). Influence of Selected Photopolymers on the Resulting Accuracy and Surface Roughness of the Component in Digital Light Processing Technology, Proceedings of the 32nd DAAAM International Symposium, pp.0235-0245, B. Katalinic (Ed.), Published by DAAAM International, ISBN 978-3-902734-33-4, ISSN 1726-9679, Vienna, Austria
DOI: 10.2507/32nd.daaam.proceedings.034

Abstract

The article focuses on the influence of the used photopolymers on the dimensional accuracy and surface roughness of the component in the production process using DLP technology. Thin square parts containing four holes and four roundings were used for the experiment. According to previous research in the field of additive technologies, the authors examined the diameters of holes and surface roughness dependences. A GOM ATOS II TripleScan optical 3D scanner with a measuring volume of MV170 was used for the 3D scanning of samples. The scans were evaluated in the GOM INSPECT software. The 3D printing method was Digital Light Processing (DLP) technology and a total of five types of photopolymers from Zortrax were used. Surface roughness was investigated by portable surface roughness tester Mitutoyo SJ-210. Five samples were printed from each material used (thirty copies in total). Prints were made on a Zortrax Inkspire 3D printer. Dimensional accuracy, ie diameters of all holes and surface roughness are shown in the article in the form of graphs.

Keywords: DLP; 3D scanning; Photopolymer; Surface Roughness

1. Introduction

Digital light processing (DLP) is the technology of additive manufacturing allowing users to create complex parts out of photo-curable resins. DLP is highly similar to Stereolithography (SLA). When compared, both technologies use the same resins and create parts layer by layer. The main difference is that while SLA uses a laser to cure the resin, DLP uses a projector instead. Using a projector instead of the laser allows the DLP printer to illuminate the whole layer at once and by that reduces printing time. The DLP technology uses a digital micromirror device (DMD) that reflects and focuses ultraviolet (UV) light on the surfaces of photoreactive materials that polymerize in a layer-by-layer fashion. While DLP offers faster printing time, SLA on the other hand can produce parts with higher detail, since the resolution in SLA is defined by minimum laser spot size as opposed to minimum pixel size in DLP [1] [2].

When producing parts with additive manufacturing technologies, the produced 3D parts are sliced into 2D layers, which are then layered on top of each other. This layering can produce so called stair stepping effect. When producing parts by layering them, the stair stepping effect must always be taken into consideration, as it is main contributor to the resulting surface finish. The stair stepping effect can be generally reduced by decreasing layer thickness, however this causes the building process to be significantly slower [3]. The quality of the newly formed additive surface is affected by the layer thickness, the additive head step, the pixel size or the size of the laser beam. Here you can see an analogy with the technology of material removal - machining with machining parameters: feed rate [4], [5], [6] and tool diameter size (tip radius) [7], [8], [9]. These machining parameters also affect the accuracy of the achieved dimensions [10]. The dimensional accuracy is mainly affected by shrinking and warping of the produced part. Warpage and shrinkage in resin based additive technologies is a frequent issue, caused by incomplete polymerization of resin during the curing process [11].

Several authors already studied the surface finish and dimension accuracy of DLP produced parts. Keßler et al. studied the influence of different additive manufacturing methods, resins, and sterilizations on the accuracy of surgical implants, when compared to traditionally milled implants. Their study found out that the accuracy of printed implants was almost comparable with milled implants and that used printing devices and materials affected the precision of parts [12].

Shim et al. studied the effect of various printing orientations on printing accuracy, mechanical properties, surface characteristics, and microbial adhesion of 3D-printed resins. Their findings were that specimens printed at 90-degree orientation showed the lowest error rates for length and specimens printed at 45 degrees showed the higher surface roughness when compared with other orientations. Therefore, printing orientation should be carefully decided to fabricate products with appropriate properties [13].

Sabbah et al. studied the impact of layer thickness and storage time on the properties of dental dies produced by SLA and found out that storage time affected dimensional stability of dies, with most significant changes in the first 3 weeks of storing [14].

The author Saed produced a scaffold from the hard fabric of L-lactic acid (PLLA) synthetic resin compatible with 3D printing for digital light processing (DLP). Porous models with a pore size of 600 microns and 70% nominal porosity were designed and manufactured using the DLP technique. The effects of two process parameters, light exposure time and dye concentration, and their effect on compressive strength and morphological properties of the printed sample were investigated on these samples. The author was concluded that the synthesized polymer and the used method of 3D printing are suitable for fabricating scaffolds with intricate structures [15].

3D printing or additive manufacturing allows the direct production of complex 3D structures according to a computer-aided design [16]. DLP technology can be used in the production of prototypes, electronics [17], and biomedical devices [18]. Technology DLP is beginning to find application in the medical industry as well. Krkobabić investigated the effect of poly (ethylene glycol) 400 (PEG 400), sodium chloride (NaCl), and mannitol as hydrophilic excipients to prevent very slow and incomplete drug release from tablets made by photopolymerization using digital light processing technology (DLP). The results of this paper suggest the possibility of implementing traditional excipients in various formulations for 3D printing based on photopolymerization [19].

The quality of printed components is limited by a stair-stepping phenomenon owing to the layer-by-layer production [3], [20]. Barone researched the technology of photopolymerization of a bath based on digital light processing (DLP) to produce polymeric orthodontic appliances. A special 3D printer DLP was developed using a standard digital projector to study the effect of printing parameters on surface roughness. Dental LT Clear Resin was used to make specimens and dental appliances. Roughness measurements were performed using a Hommel Etamic T8000 measuring system, with a feed length of 1.5 mm and a measurement speed of 0.15 mm/s. Roughness measurements were performed on 20 x 15 x 12 mm 3 (XYZ) printed by using two different building orientations. DLP and SLA technologies were used for experiments and sample production. The roughness of the samples produced by the developed DLP printer is very close and, in some cases, better than the roughness obtained for the samples printed by the SLA technology [21].

The Jacobs working curve is very important in obtaining highly accurate printing results in the digital light (DLP) process. The curve expresses the relationship between the absorbed UV energy and the cured thickness of the photocurable material during exposure to UV light for a particular photocurable material. The use of the Jacobson curve can significantly save time and reduce the waste of printing materials when developing new DLP printing materials. In DLP studies [22], [23] the Jacobs working curve was used to indicate the relationship between the absorbed energy of the photocurable material and the cured thickness.

Mostafa examined the roughness of the surface on the manufactured parts by PμSLA. The author presented a physics-based analytical model to predict the surface roughness of the parts produced by the projection micro-stereolithography additive manufacturing. The model was a function of the layer thickness and the exposure time [24].

Milde et al. investigated the influence of part orientation on the surface roughness in process of Fused Deposition Modeling (FDM). The surface roughness was measured on the simple cuboid part with the dimension 15 x 15 x 30 mm. During the production of the samples, five different angles were set to the print orientation in the X-axis. It was found that the print orientation angle has a significant effect on the surface roughness [25].

Kowsari et al. choose a more material-based approach, in their work they tried to produce an ideal formulation of photopolymers to achieve optimal resolution of final parts. Examination of in-house formulated photopolymers was performed. Results of this study showed that di(meth) acrylate-based resins should be used when production of optically clear smooth surfaces is desired, use of these resins also helped to minimize stair-stepping effect [20].

Mentioned authors did provide us with valuable information, which will be used in the analysis of the influence of photopolymers on surface quality and dimensional accuracy of produced parts. Results of this analysis can be considered in future research aimed at parts produced by DLP, specifically at parts produced with Zortrax branded resins. The aim of this paper is to analyze the influence of different resins on the surface roughness and dimensional accuracy of parts produced on DLP devices. Surface roughness was measured with surface roughness tester Mitutoyo SJ-210 and dimensional accuracy was analyzed with GOM ATOS II TripleScan.

2. Statement of the problem

The purpose of this research is to compare dimensional accuracy and surface roughness of thin parts with holes and roundings of different diameters printed from five different photopolymer materials by Zortrax Inkspire LCD / DLP 3D printer. These materials have the following designation: Basic - Grey, Bluecast - X10, PRO Black, Basic - Pigmentfree, Flexibile – Transparent Yellow. The goal of this research is to evaluate the limitations of the used photopolymers, to rule out factors that do not contribute significantly to print accuracy and to provide a practical, quantitative guide for accuracy measurement as an engineering tool. A methodology or a practical approach for calculating the accuracy and surface roughness of samples will be developed which can be used in the future to test accuracy and surface roughness. It will establish which material is more accurate and what is the reason behind it.

3. Photopolymers

Photopolymers are materials based on polymers / oligomers / monomers that can be selectively polymerized and / or bonded by exposure to light radiation of other wavelengths. Polymerization is very attractive today because it produces substances that can be produced by photopolymerization reactions of monomers / oligomers, which have environmental, economic and production examples. Photoinitiators absorb the irradiating light of 3D printers and initiate layer-by-layer photopolymerization reactions of monomers / oligomers [26]. Distribution of photopolymers according to the resulting properties after curing:

- Solid,
- Hard,
- Flexi,
- Castable,
- Tough,
- High strength.

Curing mechanism

The curing reaction of stereolithographic resins is an exothermic polymerization process characterized by chemical crosslinking reactions that form an infusible, insoluble, and highly crosslinked 3D network. This reaction is initiated by delivery appropriate form of energy, which depends on the stereolithographic strategy used. A general illustration of the solidification process is shown in the figure below (Fig. 1) [27]. During the curing reaction, two main events occur, independently of the stereolithographic strategy used to initiate the polymerisation process:

- Gelation - the transition from liquid to rubber,
- Vitrification - ie. transition of liquid or rubber to glass [27].

The gelation described by *Gilham* as a "chemical explosion" is an irreversible phenomenon and corresponds to the incipient formation of an infinite molecular network (molecular gelation), which is associated with a dramatic increase in viscosity (macroscopic gelation). At this point, the system will no longer flow and there will be two phases simultaneously: the gel phase and the salt phase. The gel phase is the gelatinized part, insoluble in non-degrading solvents, while the salt phase, which remains soluble, can be extracted with solvents. Molecular gelation is defined as the initial formation of branched molecules with infinity molecular weight and occurs at a specific fractional conversion for each system, which depends on the functionality, reactivity, and stoichiometry of the reactive substances [27], [28].

Macroscopic gel formation is associated with a significant increase in viscosity and a corresponding decrease in processability. The sudden transformation from a viscous liquid to an elastic gel or rubber corresponds to the point of the gel and defines the onset of gelatinization. As a reaction further, the amount of salt phase present in the system decreases, and the polymer becomes more and more crosslinked (Fig. 8). Molecular mobility also decreases due to limitations due to the crosslinked structure [28].

Vitrification is a gradual, thermally reversible process that corresponds to the formation of a glassy solid material due to an increase in both the crosslinking density and the molecular weight of the cured polymer. The vitrification will significantly reduce the reaction rate and the reaction will be very slow because it will be regulated by the diffusion of reactive substances. In this way, vitrification refers to the change from a reaction that is preferably kinetically controlled to a reaction that is controlled by diffusion. In this mode, the time required for reactive groups to be found is much longer [28].

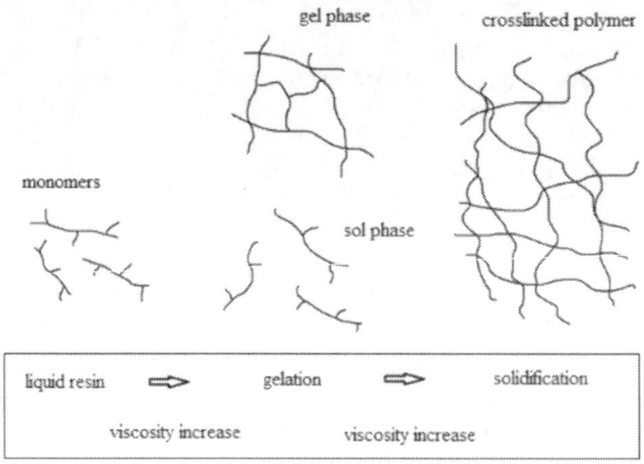

Fig. 1. Curing process

4. Digital Light Processing (DLP)

The main difference between SLA and DLP is in the light source used. While the SLA is a laser coordinated by galvanometers, the DLP technology is a specially designed digital light projector that cures the entire layer simultaneously. From this point of view, DLP is therefore considered to be much faster than SLA. In SLA, the laser must cure the resin using a point-to-point technique. On the other hand, the DLP projector screen flashes on the entire layer at the same time. Because DLP is a digital technology, a projected 2D image is made up of pixels. When translated into 3D, they become voxels. The DLP 3D printer's light source alone and LED screen would not work without the digital micromirror device (DMD) that is the basis of every chipset. DMD contains hundreds of thousands or even millions of small micromirrors that direct light and form a pattern of a particular layer at the bottom of the resin tank. The resolution of the printed part by DLP printing usually corresponds to the number of micromirrors inside the DMD device.

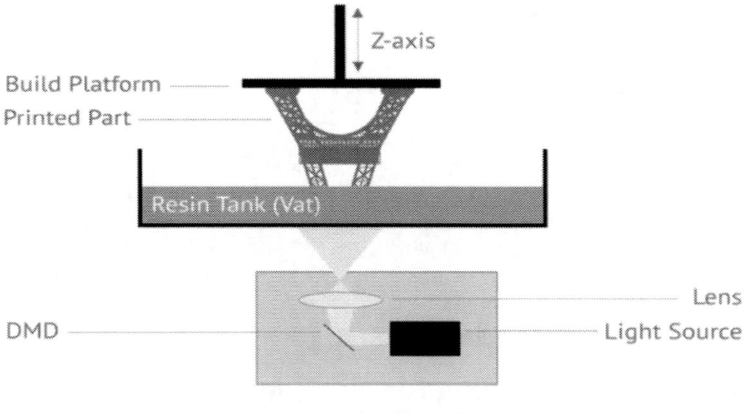

Fig. 2. Scheme of DLP technology [29]

5. Parameters settings of DLP technology

In this experiment was used LCD/DLP 3D printer Zortrax Inkspire. Manufacturing parameters used for all mentioned materials are shown in Tab. 1.

Manufacturing parameters	Basic - Grey	Bluecast X10	PRO Black	Basic - Pigmentfree	Flexibile - TransparentYellow
Layer thickness [mm]	0.05	0.05	0.05	0.05	0.05
Orientation in X/Y axis [°]	0	0	0	0	0
Infill [%]	100	100	100	100	100
Z lift distance [mm]	5	5	5	5	5
Layer exposure time [s]	10	10	10	10	10

Table 1. Parameters used for manufacturing of samples

6. Process of digitizing

The basis of the 3D scanning process is a scanning device, in this case, an optical 3D scanner GOM ATOS II Triple Scan and a physical component, in our case a scanned sample. 3D scanning obtains the coordinates of the scanned surface, which is referred to as a point cloud. In the next step, this point cloud is converted into a polygonal grid using special software, in other words, the points are connected into polygons (squares, triangles, or polygons). This process is called polygonization. The most used polygon format is STL.

Before scanning the sample placed uncoded reference points that are used to connect multiple waveform scanning. They also serve for better orientation of the drawn model in the software. When the reference points are in place, the sample is sprayed with anti-reflective (chalk) powder. In this way, a thin layer of coating is applied to the surface of the sample, which has the task of creating a matt or at least substantially duller surface than the scanned sample. The function of this powder is to eliminate the diffraction and refraction of incident rays, and thus to ensure a more accurate scan. The scheme of projecting the strips of structured light is shown in Fig 3.

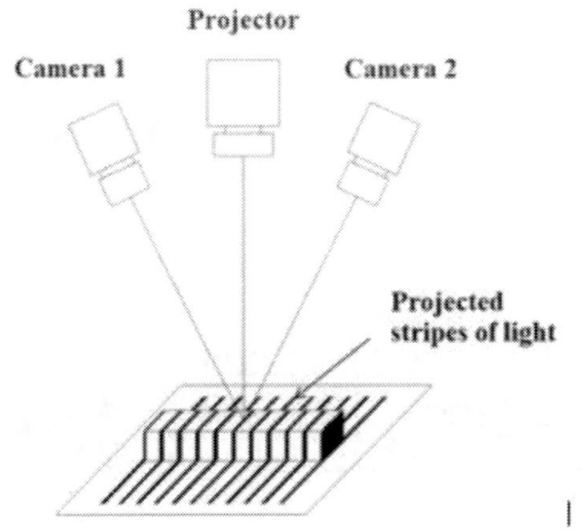

Fig. 3. Scheme of projecting stripes of structured light on the scanned object [30]

7. Analysis of results

Used materials in this experiment were photo-polymeric resins of Zortrax company: PRO black, basic gray, bluecast X10, transparent yellow, and pigment-free resins. Five specimens of each type of resin were printed (25 samples in total). When determining dimensional accuracy, flatness, and surface roughness of the component, we can measure it using inspection software for measuring. Examination of all of the examples was performed via GOM Inspect software. The analyzed sample is a thin part with holes and roundings of various diameters. These parameters were measured and evaluated for print accuracy. Otherwise, the diameter of the holes, roughness, and the overall flatness of the part was measured.

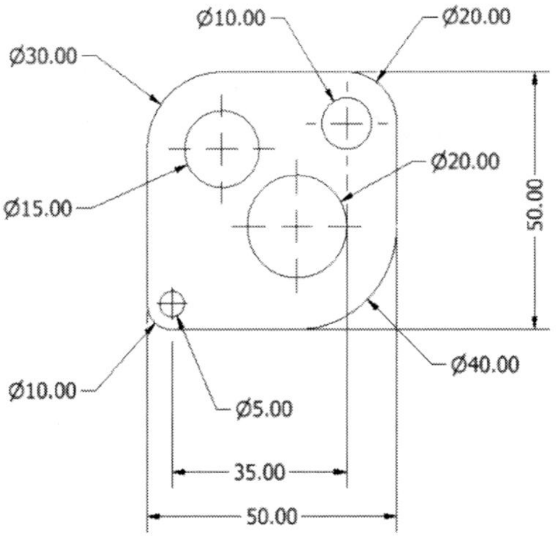

Fig. 4. Measured sample

All measurements of internal holes showed negative values. The largest negative deviation, in this case, was for the Bluecast original material with an average value of -0.236 mm, while the smallest average deviation was measured for the Basic grey material with a value of -0.059 mm. In terms of flatness, the worst came out measured sample of Transparent yellow material, where a value of 0.760 mm was measured, which is almost one millimeter. This fact can be justified by the flexibility and softness of this type of material.

Photopolymer type	**sample**	**hole Ø20 [mm]**	**hole Ø15 [mm]**	**hole Ø10 [mm]**	**hole Ø5 [mm]**	**Flatness [mm]**
Basic grey	1	19.888	14.865	9.864	4.918	1.190
	2	19.901	14.887	9.859	4.948	1.140
	3	19.883	14.954	9.933	4.958	1.290
	4	19.849	14.885	9.887	4.926	1.500
	5	19.883	14.881	9.857	4.956	0.840
	Σ	**19.881**	**14.894**	**9.880**	**4.941**	**1.192**
	(+/-)	**-0.119**	**-0.106**	**-0.120**	**-0.059**	**0.192**
PRO black	1	19.824	14.833	9.820	4.806	1.270
	2	19.852	14.846	9.859	4.820	1.260
	3	19.827	14.837	9.819	4.813	1.150
	4	19.812	14.843	9.842	4.855	1.140
	5	19.988	15.056	10.005	4.962	0.570
	Σ	**19.861**	**14.883**	**9.869**	**4.851**	**1.078**
	(+/-)	**-0.139**	**-0.117**	**-0.131**	**-0.149**	**0.078**
Bluecast X10	1	19.744	14.814	9.857	4.797	1.360
	2	19.727	14.793	9.809	4.843	1.330
	3	19.817	14.907	9.898	4.931	1.340
	4	19.765	14.861	9.850	4.852	1.470
	5	19.765	14.860	9.848	4.874	1.450
	Σ	**19.764**	**14.847**	**9.852**	**4.859**	**1.390**
	(+/-)	**-0.236**	**-0.153**	**-0.148**	**-0.141**	**0.390**
Pigmentfree	1	19.819	14.815	9.823	4.800	1.580
	2	19.829	14.849	9.833	4.881	1.490
	3	19.811	14.818	9.796	4.872	1.480
	4	19.840	14.874	9.867	4.917	1.710
	5	19.826	14.804	9.803	4.875	1.200
	Σ	**19.825**	**14.832**	**9.824**	**4.869**	**1.492**
	(+/-)	**-0.175**	**-0.168**	**-0.176**	**-0.131**	**0.492**
Flexibile - Transparent yellow	1	19.846	14.860	9.890	4.906	1.740
	2	19.866	14.906	9.855	4.956	1.730
	3	19.827	14.856	9.828	4.927	1.690
	4	19.839	14.822	9.820	4.852	1.590
	5	19.865	14.889	9.865	4.935	2.050
	Σ	**19.849**	**14.867**	**9.852**	**4.915**	**1.760**
	(+/-)	**-0.151**	**-0.133**	**-0.148**	**-0.085**	**0.760**

Table 2. Evaluation of holes of samples

From the types of materials mentioned and compared, it can be concluded that the accuracy of the printout is best achieved from a resin of a photo-polymeric material with higher hardness. Thus, in this case, the black and basic grey materials achieved the best results in terms of the accuracy of the measured deviations and properties. This phenomenon may be due to their material structure and crosslinking bonding during the curing process. It is also worth noting that harder types of materials have a better finish, and it is easier to remove excess parts or supports.

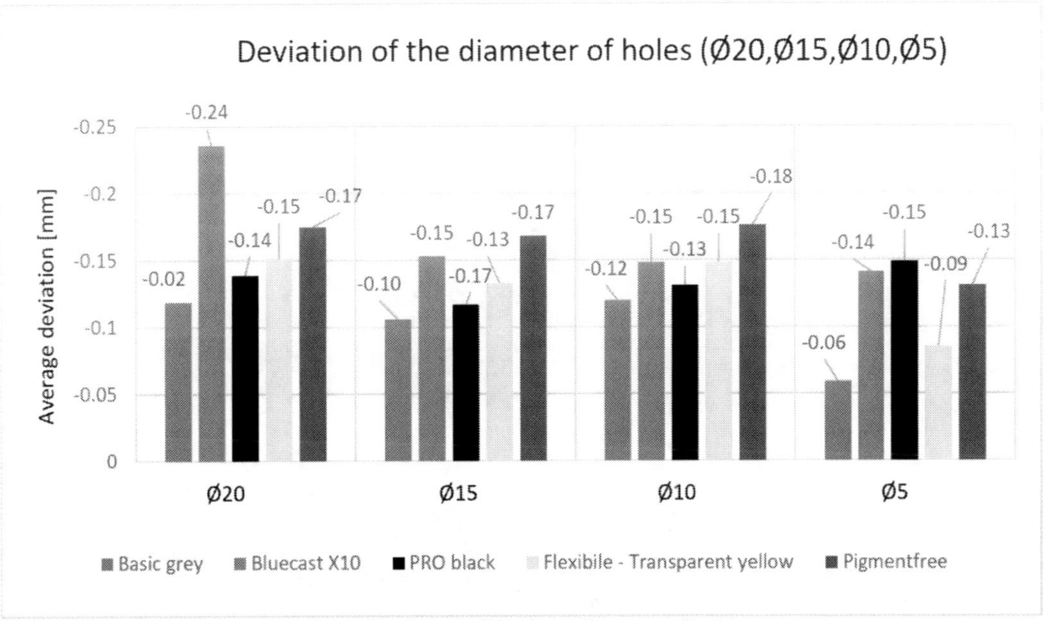

Fig. 5. Deviation of the inner diameter of holes

8. Roughness measurement

Geometrical model of roughness

For the theoretical calculation of the roughness of the DLP additive surface, we used analytical pixel analysis. We used the following theoretical assumptions for our analysis:

- we do not consider the inaccuracy of pixel dimensions,
- we do not consider deformation of the material during cooling,
- we do not consider the oscillations and vibrations of the 3D display in any direction (X, Y).

Our geometric model is based on the arrangement of pixels around the theoretical profile (shape). The squares represent voxels with a height *Lt* and with the dimensions of the base in the XY plane with the size of pixels of the same resolution. Gray squares represent the material after curing. White squares represent uncured material. Figure 5 shows a geometric model of the DLP profile with examples of the influence of the layer thickness *Lt* for different angles of the profile inclination α = 0°, 30°, 45°, 60°, 90°.

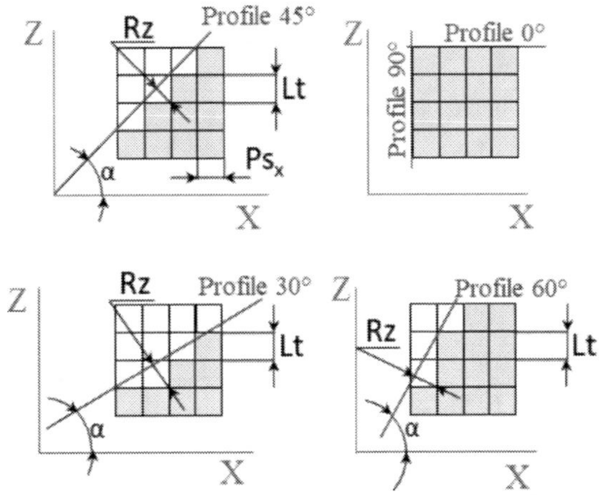

Fig. 6. Geometric model of LCD/DLP profile, influence of layer thickness for different angles of profile inclination α.

Where Rz is the theoretical maximum height of the irregularities of the newly formed additive DLP surface, *Lt* is layer thickness and PS_x is the size of the pixel in the X-axis direction. We can say that the surface roughness in this case is a function of the layer thickness Lt, the pixel size Psx and the angle of inclination of the profile:

$$R_Z = f(L_t, Ps_x, \alpha) \quad (1)$$

Fig.6 shows the geometric model of the DLP profile - holes.

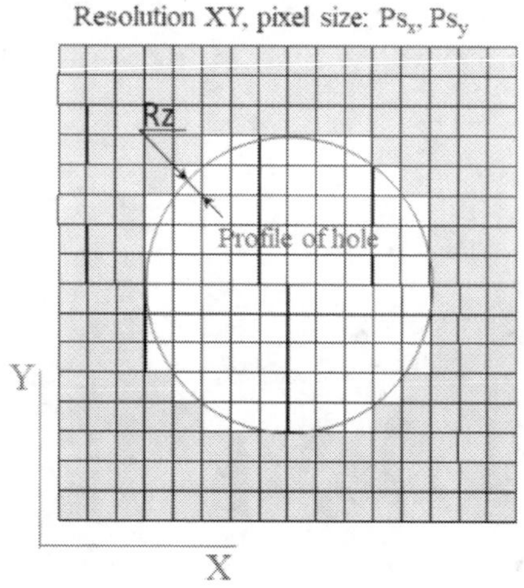

Fig. 7. Geometric model of DLP profile - hole, in XY plane

We can say that the surface roughness in this case is a function of the thickness of the pixel size Ps_x and Ps_y, ie from the pixel density resp. resolution.

$$R_Z = f\left(Ps_x, Ps_y\right) \quad (2)$$

Overall, it can be stated that for horizontal and perpendicular surfaces, from a geometric point of view, the roughness of the additive surfaces should be better than for surfaces with a different angle of inclination.

Experimental results

In this experiment, the surface roughness of the samples was measured with a Mitutoyo SJ 210 surface tester. Surface roughness was measured on every part on the three places to eliminate bias and random errors. It was measured at a distance of 2.4 mm. Overall, the two parameters measured roughness (Ra and Rz). From the data were evaluated by the average roughness values for the various parameters of roughness. Roughness was measured on the face of components made from a total of five different photopolymers. The surface roughness of all materials in the experiment showed the same dependencies except for Transparent Yellow. The arithmetical mean of surface roughness is shown in Table 3. – Table 7.

Basic grey						
Roughness parameter	Print 1 [μm]	Print 2 [μm]	Print 3 [μm]	Print 4 [μm]	Print 5 [μm]	Arithmetical mean [μm]
Ra	0.233	0.184	0.232	0.943	0.881	**0.494**
Rz	1.700	0.976	1.626	5.233	5.103	**2.927**

Table 3. Ra and Rz values for material Basic grey

Bluecast X10						
Roughness parameter	Print 1 [μm]	Print 2 [μm]	Print 3 [μm]	Print 4 [μm]	Print 5 [μm]	Arithmetical mean [μm]
Ra	0.787	0.282	0.334	0.287	0.554	**0.449**
Rz	4.781	1.388	1.886	1.531	2.551	**2.427**

Table 4. Ra and Rz values for material Bluecast X10

PRO Black						
Roughness parameter	Print 1 [µm]	Print 2 [µm]	Print 3 [µm]	Print 4 [µm]	Print 5 [µm]	Arithmetical mean [µm]
Ra	0.374	0.182	0.515	0.154	0.824	**0.410**
Rz	2.465	1.099	3.055	0.927	4.306	**2.370**

Table 5. Ra and Rz values for material PRO Black

Transparent Yellow						
Roughness parameter	Print 1 [µm]	Print 2 [µm]	Print 3 [µm]	Print 4 [µm]	Print 5 [µm]	Arithmetical mean [µm]
Ra	1.123	0.865	1.076	1.118	0.951	**1.027**
Rz	6.593	4.834	6.201	6.413	5.182	**5.844**

Table 6. Ra and Rz values for material Transparent Yellow

Pigment-free						
Roughness parameter	Print 1 [µm]	Print 2 [µm]	Print 3 [µm]	Print 4 [µm]	Print 5 [µm]	Arithmetical mean [µm]
Ra	1.131	0.236	0.820	0.230	0,194	**0.522**
Rz	6.413	1.161	5.306	1.474	1.360	**3.143**

Table 7. Ra and Rz values for material Pigment-free

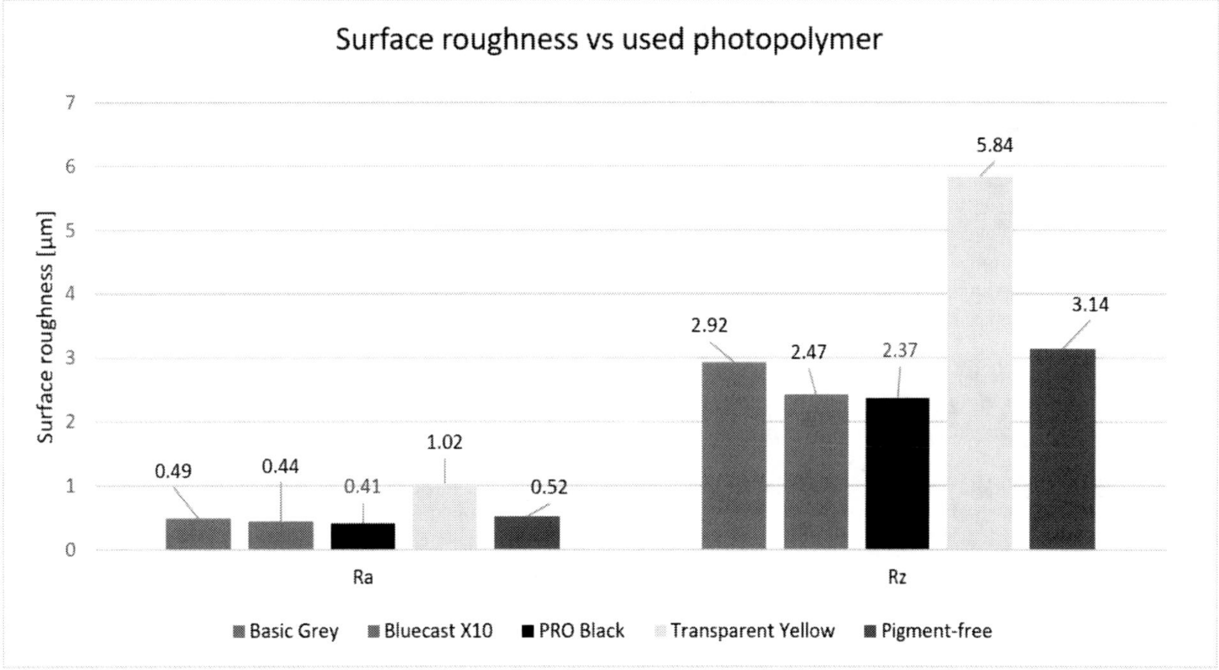

Fig. 8. Ra and Rz vs used photopolymer

9. Conclusion

These measured facts result from a slight bending due to shrinkage of the plastic material. Harder materials (Black, Basic grey) showed better properties in overall flatness. These materials have better bonding during the solidification process. Therefore, they can reach a better finish with a greater grade of accuracy. Materials with the harder structures are evaluated as better in terms of scratch resistance or overall resistance to surface damage. Softer materials as Transparent yellow or Pigmentfree showed similar deviation to harder materials in terms of accuracy of the holes.

On the other hand, these materials are flexible and softer, therefore their crosslinking process is with long chain segments, which makes these materials reminiscent of hard rubber structure. All in all, compared materials do not have an extreme difference in deviation except for flatness.

As can be seen in Fig 5, surface roughness values are similar in 4 out of 5 resins. In the case of Transparent Yellow resin, the measured values of surface roughness when compared with other resins are roughly twice as high. This fact can be contributed to the unique optical property of resin, since it is the only one of tried resins that were transparent. Therefore, projected UV light from DLP device while curing next layer in order might have reached already cured layers, resulting in overcuring, respectively partially increased layer exposure time in comparison to non-transparent resins. This applies to the possible process of post-processing in IPA and UV chambers as well. However, more research will be needed to address and properly analyze this outcome. It must also be noted, that while mean values of roughness are similar, the deviation in roughness values of individual prints is observable. Although the individual surface roughness values of Ra are erratic, they generally do not reach values higher than 1 μm, in most cases not reaching 0.5 μm.

10. Acknowledgments

The article was written with the support of the project VEGA 1/0019/20.

11. References

[1] J. Borrello, P. Nasser, J. C. Iatridis, and K. D. Costa, (2018). "3D printing a mechanically-tunable acrylate resin on a commercial DLP-SLA printer," Addit. Manuf., vol. 23, pp. 374–380, Oct., doi: 10.1016/j.addma.2018.08.019.

[2] J. Schmidt, H. Elsayed, E. Bernardo, and P. Colombo, (2018). "Digital light processing of wollastonite-diopside glass-ceramic complex structures," J. Eur. Ceram. Soc., vol. 38, no. 13, pp. 4580–4584, , doi: 10.1016/j.jeurceramsoc.2018.06.004.

[3] Y. Pan, X. Zhao, C. Zhou, and Y. Chen, (2012). "Smooth surface fabrication in mask projection based stereolithography," in Journal of Manufacturing Processes, vol. 14, no. 4, pp. 460–470, doi: 10.1016/j.jmapro.2012.09.003.

[4] M. Beno, M. Zvoncan, M. Kovac, et al., (2013). "Circular interpolation and positioning accuracy deviation measurement on five axis machine tools with different structures". Tehnicki vjesnik-technical gazette, Volume: 2, Issue: 3, Pages: 479-484,

[5] T. Vopat, J. Peterka, V. Simna et al., (2014). "The Influence of Different Types of Copy Milling on the Surface Roughness and Tool Life of End Mills". Conference: 25th DAAAM International Symposium on Intelligent Manufacturing and Automation Location: oldest technical univ, Vienna, AUSTRIA Date: NOV 26-29, Sponsor(s): Vienna Univ Technology, 25TH DAAAM International Symposium On Intelligent Manufacturing And Automation, 2014, Book Series: Procedia Engineering, Volume: 10, Pages: 868-876

[6] J. Peterka. (2004). "A new approach to calculating the arithmetical mean deviation of a profile during copy milling". Strojniski vestnik - journal of mechanical engineering. Volume: 50, Issue: 12, Pages: 594-597

[7] M. Polakovic, I. Buransky, j. Peterka, (2008). "Simulation concept for machined surface roughness and shape deviation prediction". Conf. 19th Intern. Sympos. of DAAAM, Trnava, Slovakia, Oct 22-25, Annals of DAAAM Proceedings of the 19th Int. DAAAM Symp. Book Series: Annals of DAAAM and Proceedings, pages: 1089-1090

[8] J. Peterka, P. Pokorny, (2013). " Influence of the Lead Angle from the Vertical Axis Milling on Effective Radius of the Cutter " Conference: 7th International Congress of Precision Machining Location: Miskolc, HUNGARY Date: OCT 03-05, 2013 Sponsor(s): Int Visegrad Fund; SANDVIK Cromant. Precision Machining VII, Book Series: Key Engineering Materials, Volume: 581, Pages: 44-49

[9] K. Sentyakov, J. Peterka, V. Smirnov et al. (2020). "Modeling of Boring Mandrel Working Process with Vibration Damper" Materials, Volume: 13, Issue: 8, rticle Number: 1931

[10] J. Peterka, P. Pokorny, S. Vaclav. (2008). "CAM strategies and surfaces accuracy". Conference: 19th International Symposium of the Danube-Adria-Association-for-Automation-and-Manufacturing Location: Trnava, SLOVAKIA Date: OCT 22-25, ANNALS OF DAAAM FOR 2008 & Proceedings Of The 19th International Daaam Symposium, Book Series: Annals of DAAAM and Proceedings, Pages: 1061-1062, Published: 2008

[11] N. Nestler, C. Wesemann, B. C. Spies, F. Beuer, and A. Bumann, (2021). "Dimensional accuracy of extrusion- and photopolymerization-based 3D printers: In vitro study comparing printed casts," J. Prosthet. Dent., vol. 125, no. 1, pp. 103–110, doi: 10.1016/j.prosdent.2019.11.011.

[12] A. Keßler, M. Dosch, M. Reymus, and M. Folwaczny, (2021). "Influence of 3D- printing method, resin material, and sterilization on the accuracy of virtually designed surgical implant guides," J. Prosthet. Dent., , doi: 10.1016/j.prosdent.2020.08.038.

[13] J. S. Shim, J. E. Kim, S. H. Jeong, Y. J. Choi, and J. J. Ryu,(2020). "Printing accuracy, mechanical properties, surface characteristics, and microbial adhesion of 3D-printed resins with various printing orientations," J. Prosthet. Dent., vol. 124, no. 4, pp. 468–475, doi: 10.1016/j.prosdent.2019.05.034.

[14] A. Sabbah, G. Romanos, and R. Delgado-Ruiz, (2012). "Impact of layer thickness and storage time on the properties of 3D-printed dental dies," Materials (Basel)., vol. 14, no. 3, pp. 1–10, doi: 10.3390/ma14030509.

[15] A. Bagheri Saed, A. H. Behravesh, S. Hasannia, S. A. Alavinasab Ardebili, B. Akhoundi, and M. Pourghayoumi, (2019). "Functionalized poly L-lactic acid synthesis and optimization of process parameters for 3D printing of porous scaffolds via digital light processing (DLP) method," J. Manuf. Process., vol. 56, no., pp. 550–561, 2020, doi: 10.1016/j.jmapro.2020.04.076.

[16] I. Buransky, L. Morovic, J. Peterka, (2013). "Application of Reverse Engineering for Redesigning and Manufacturing of a Printer Spare Part". Conference: 4th International Conference on Manufacturing Science and Engineering (ICMSE 2013) Location: Dalian, People R China, Date: MAR 30-31, 2013, Sponsor(s): NE Univ; Harbin Inst Technol; Jilin Univ., Material Design, Processing And Applications, Parts 1-4, Book Series: Advanced Materials Research, Volume: 690-693, Pages: 2708-2712,

[17] Q. Mu et al., (2017). "Digital light processing 3D printing of conductive complex structures," Addit. Manuf., vol. 18, pp. 74–83, doi: 10.1016/j.addma.2017.08.011.

[18] F. Chen et al., "Preparation and biological evaluation of ZrO2 all-ceramic teeth by DLP technology, (2020). " Ceram. Int., vol. 46, no. 8, pp. 11268–11274, doi: 10.1016/j.ceramint.2020.01.152.

[19] M. Krkobabić, D. Medarević, S. Cvijić, B. Grujić, and S. Ibrić, (2019). "Hydrophilic excipients in digital light processing (DLP) printing of sustained release tablets: Impact on internal structure and drug dissolution rate," Int. J. Pharm., vol. 572, no. July, p. 118790, doi: 10.1016/j.ijpharm.2019.118790.

[20] K. Kowsari et al., (2018). "Photopolymer formulation to minimize feature size, surface roughness, and stair-stepping in digital light processing-based three-dimensional printing," Addit. Manuf., vol. 24, pp. 627–638, doi: 10.1016/j.addma.2018.10.037.

[21] S. Barone, P. Neri, A. Paoli, A. V. Razionale, and F. Tamburrino, (2019). "Development of a DLP 3D printer for orthodontic applications," Procedia Manuf., vol. 38, no. 2019, pp. 1017–1025, doi: 10.1016/j.promfg.2020.01.187.

[22] Y. Li, Q. Mao, J. Yin, Y. Wang, J. Fu, and Y. Huang, (2021). "Theoretical prediction and experimental validation of the digital light processing (DLP) working curve for photocurable materials," Addit. Manuf., vol. 37, no. August 2020, p. 101716, doi: 10.1016/j.addma.2020.101716.

[23] C. Hofstetter, S. Orman, S. Baudis, and J. Stampfl, (2018). "Combining cure depth and cure degree, a new way to fully characterize novel photopolymers," Addit. Manuf., vol. 24, no. September, pp. 166–172, doi: 10.1016/j.addma.2018.09.025.

[24] P. Cirp et al., (2020). "Investigation of Light-Induced of Surface Roughness in Projection Micro- Stereolithography Additive Manufacturing (PμSLA) Stereolithography Additive Manufacturing A new method," Procedia CIRP, vol. 92, no. ii, pp. 187–193, doi: 10.1016/j.procir.2020.05.177.

[25] J. Milde, F. Jurina, J. Peterka, P. Dobrovszký, J. Hrbál and J. Martinovič, (2021). "Influence of part orientation on the surface roughness in the process of Fused Deposition Modeling." ICIEM 2021 RUDN University, moscow, Russia, April 28-30, 2021, in the press.

[26] Materialpro3d, (2019). materialpro3d. [Online] Available at: https://www.materialpro3d.sk/slovnik-materialov/fotopolymery

[27] Bártolo, P. J., (2011). Stereolitography: Materials, processes and applications. s.l.:s.n.

[28] Gillham, J. K., (1979). Award address information and properties of network polymeric materials. s.l.:s.n.

[29] Gregurić, L., (2018). all3dp. [Online] Available at: https://all3dp.com/2/what-is-a-dlp-3d-printer-3d-printing-simply-explained/

[30] I. Buransky, (2020). Ateliér počítačovej podpory III, Trnava: s.n

DOI: 10.2507/32nd.daaam.proceedings.035

DIFFERENTIAL PRESSURE CONTROLLER MODEL AND ITS APPLICATION IN DIFFERENT HEATING SYSTEMS

Armin Teskeredzic & Rejhana Blazevic

This Publication has to be referred as: Teskeredzic, A[rmin] & Blazevic, R[ejhana] (2021). Differential Pressure Controller Model and its Application in Different Heating Systems, Proceedings of the 32nd DAAAM International Symposium, pp.0246-0255, B. Katalinic (Ed.), Published by DAAAM International, ISBN 978-3-902734-33-4, ISSN 1726-9679, Vienna, Austria
DOI: 10.2507/32nd.daaam.proceedings.035

Abstract

Variable flow distribution systems are promoted in complex heating and cooling systems as energy efficient solutions for new systems and as an upgrade of the existing constant distribution flow systems. In the case of variable flow, the pumping energy costs decreases and the preferable return temperature can be achieved which is beneficial for operation of the condensing boilers or district heating connection. As a result of different demand in heating/cooling zones, the pressure variations in the system can lead to hydraulic problems of single zones/consumers, which at the end result in control stability problems and decrease of thermal comfort for end users. High differential pressures at zones connections can in case of mixing circuits lead to back flows in the bypass ports of the three-way valves or backflow at the injection circuits with two port valves are used, increasing the return temperature and preventing the work of control valves in the full stroke. In these cases, the differential pressure controllers are necessary in order to stabilize the strong variation of differential pressure by preserving the high value of the regulating valve authority which guaranties the stable control over the entire stroke of the control valves, very close to the nominal conditions for which they are designed. This paper presents the simple algebraic model of the differential pressure controller as part of the hydraulic system and demonstrates the use of the proposed model at four characteristic examples applied at three different systems: Fixed flow radiators system, variable flow for injection circuit consumer connection and variable flow connection of indirect heating substation. The results of calculations demonstrate the applicability of the developed model for calculation of more complex cases.

Keywords: variable flow; differential pressure controllers; bivalent temperature; seasonal coefficient of performance

1. Introduction

In many practical situations high differential pressures are developed in the distribution systems which in turn affects the optimal work of control valves, expensive modulating PI controllers starts to work in introduces the ON/OFF mode and the benefits of modulating control are lost. This usually results with an increased energy consumption and negatively influence the thermal comfort of consumers.

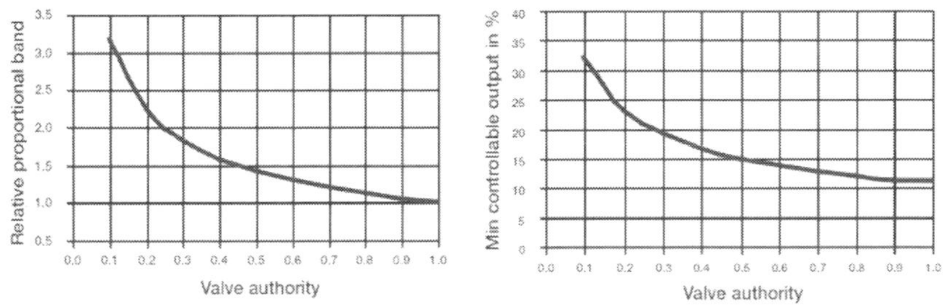

Fig. 1. Relative proportional band (left) and minimum controllable flow (right), both as a function of valve authority [1]

Figure 1 (left) shows the relation between the control valve authority and relative proportional band of the controller, while Figure 1(right) provides the dependency of minimum controllable output of terminal as a function of control valve authority. It can be seen that the minimum controllable relative output is 32% for the valve authority of 0,1. It means that for the loads below 32% the controller works in ON/OFF mode and is not modulating. The final result is the temperature swings at consumer taking into consideration that proportional band has to be multiplied by 3 for this authority (Figure 1 left).

It means that if the proportional band is 2°C for control valve authority of 1, it should be 6°C for an authority of 0,1! One can suppose that the I function of the controller will compensate this deviation by slowly resetting the proportional controller but an integral function can only reduce such a deviation by half in 30 minutes [1]. This is usually the time during which the consumer will change the set point at the room or home controller and this may result in control loop instability and hunting the set point and increased energy consumption by overheating of the room/home.

Even if the control valve has been chosen properly with relatively high authority for design conditions, due to high variations of differential pressures the authority decreases dramatically and the described situation can happen. These variations of differential pressures are present in variable flow distribution networks, such as district heating or at larger buildings with multi-zone consumers. According to current literature review and recommendations for energy efficient design of HVAC systems, the variable flow distribution system is preferred.

In district heating systems and indirect connections of consumers, the distribution network is hydraulically decoupled from the consumer network via heat exchanger and differential pressure controllers are must-to-have elements of the substation connections. In that light, the simple algebraic model of the differential pressure controller should support the design engineers to analyse and optimize control strategies in order to decrease energy losses. These interventions and optimization can be the variation of the primary supply temperature, installation and control of variable speed drive pumps and installation of differential pressure controllers or any combination of these. One of the outcomes of these analysis should be the work of control valves in their full stroke, the possibility to control low loads with modulation and not ON/OFF, while providing the thermal comfort at consumers. Numerous studies have been carried out on improving the efficiency of heating systems that are focused on various heating elements, but control strategy was rarely analysed. Some of the suggestions for improved control strategy in case of decentralized heat supply are given in [2]. Also, operation and testing of the Danfoss pressure controller with integrated anti-oscillation system was done on district heating substation and results are presented in [3]. Results shows that developed system successfully removes oscillations in hydraulic system and automatically adjusts the pressure according to the system load.

In heating systems with variable flows in the distribution network, the connection to consumer can be made by installing two port valves with the bypass which provides the constant flow in consumer circuit or the three way – usually mixing valves. In both these cases it is supposed that the consumer requires the constant flow in the secondary network. In the first case, an increased differential pressure at the connecting point can result in reverse flow in the bypass, while in case of connection with mixing valve the mixing function of the three-way valve can be cancelled and the valve should start working as diverting valve, providing the counter flow at the bypass port of the three-way valve. In both cases the return water temperature will increase and in the second case the control of the consumer thermal comfort can become instable. Increased water return temperature will decrease the efficiency of condensing boiler and will result in increased energy consumption. In the second case, if the consumer with constant water flow is controlled with weather compensation control the supply temperature would be hard to achieve, since the mixing function is excluded.

Optimising the hydronic distribution of an existing system and thus reducing energy consumption can be achieved by combining a series of measures that bring small improvements one by one. Some of these measures were considered in [4]. Investigation conducted in [5] was used to analyze possibility for achieving hydraulic balance in building distribution systems, and consequently save energy and reduce the emissions by using pre-set radiator valves combined with differential pressure (DP) controllers. For this purpose, authors developed an integrated model for the simulation of the heat performance and hydraulics and applied it for pressure and flow conditions analysis of a two-pipe multistorey building heating system in China. Also, two solutions which involve the use of automatic differential pressure controller or proportional relief valves in order to avoid strong variation of differential pressure with load change are presented in [6].

This paper is designed to support the design engineers to understand and use this simple model as part of their calculation tool of the whole network. Equations which will be outlined in the next section are valid irrespective of the nominal pressure values and the model can be applied for one zone differential pressure controller or can be used as the main differential pressure controller positioned at the primary side of the large capacity heating substation in the district heating.

2. Mathematical model

In this subsection a mathematical model for the problem is outlined as well as the producer's data on available differential controller properties.

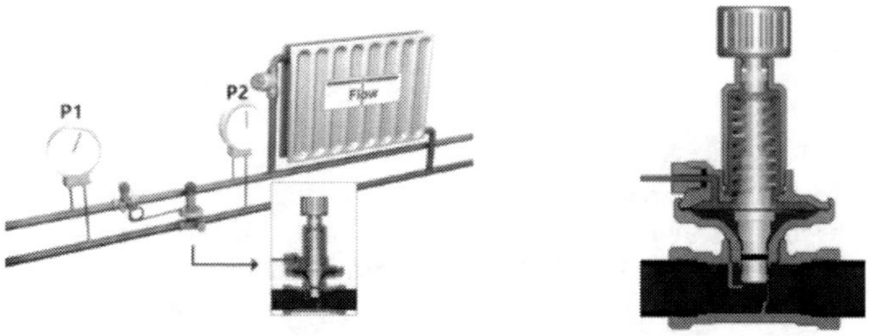

Fig. 2. Position of differential pressure controller in radiator heating system (left) and cross-section view at the valve body (right) [7]

Figure 1 (left) shows the differential pressure controller in the return pipe which upper part is connected to the supply pipe with capillary tube. Its main task is to keep constant differential pressure (P1/P2). At the right-hand side of this Figure the cross section of the valve body is presented. Valve seat position is determined by the resulting force coming from the spring and pressure difference at the membrane which separates the upper and lower parts of the valve at which corresponding pressure acts. In case that pressure P1 in the supply pipe increases, the resulting force, due to higher pressure at the upper side of the membrane, moves the valve seat downwards. As the valve closes, the pressure P2 increases and the valve seat find the new equilibrium position at which again the forces are in balance.

If pressure P1 drops, the force from the lower side of the membrane prevails and moves valve seat upwards, decreasing the pressure P2 and again establish new equilibrium position determined as new balance of the resulting force in the valve. The process is continuous, there is no additional energy provided to move the valve seat, everything is done by the resulting force coming from the spring and pressure difference between supply and return. Desired differential pressure value which needs to be maintained constant is determined by the choice of appropriate valve and the neutral position of the spring.

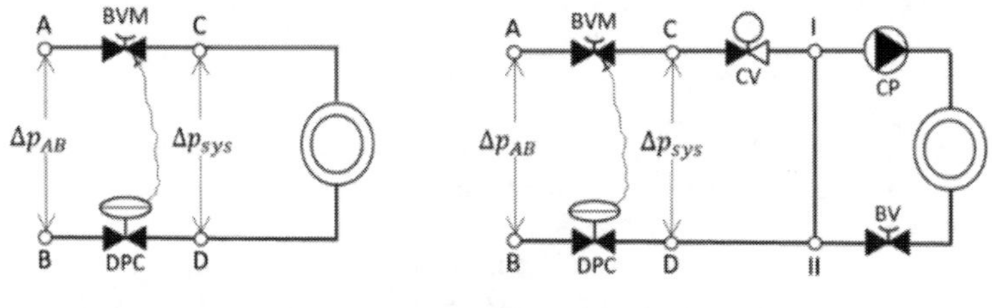

Fig. 3. System without the circulation pump (left) and typical connection with two/port valve and circulation pump (right) with positions of the differential pressure controllers for both cases

Figure 3 shows two characteristic connections of the consumers. Figure 3 (left) describes the connection without additional circulation pump in which the primary differential pressure is responsible for the water circulation towards consumer. Differential pressure controller (DPC) is positioned in the return pipe and is via measuring balancing valve (BVM) connected to primary side via capillary pipe. It is the DPC task to keep the constant differential pressure between points C and D or to keep the Δp_{sys} at the constant value. In Figure 3 (right), typical connection with two-port valve of consumer with constant flow is presented. In this case, DPC maintains the constant differential pressure and the control valve (CV) is working with practically constant authority.

Figure 4 presents the typical connection of consumer in district heating systems. This is indirect connection since the primary from secondary circuit is separated via heat exchanger. Control valve (CV) controls the primary flow and match the desired temperature in the secondary supply. It is in most cases weather compensated control [8]. Because of the strong pressure variations in the district heating network the differential pressure controller is positioned before control valve as shown in Figure 4.

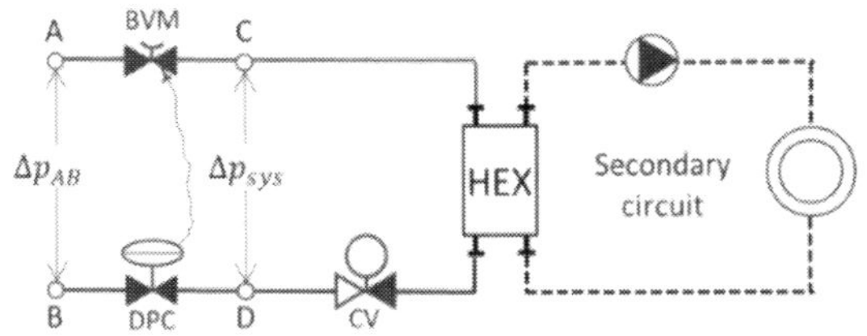

Fig. 4. Indirect connection of consumer in district heating where heat exchanger separates primary from secondary circuit

We will first focus on the system shown in Figure 3 (left) but before we proceed, we will adopt the following assumptions, valid for all cases:

- It is assumed that the local hydraulic resistances prevail between points C and D in Figure 3. It means that the assumption is to keep the constant value of the friction factor in the pipeline. With the change of flow the friction factor will be assumed constant,
- Differential pressure controller will be approximated as the proportional controller with linear relationship between the pressure differential and the relative position of the valve (Y),
- Differential pressure valve is treated as the linear valve.

All three given assumptions are reasonable from the engineering side and introduce simplifications in the mathematical model which will be presented. First the corresponding equations and the problem formulation for the system given in figure 3 (left) will be given. The available differential pressure Δp_{AB} should be equal to sum of hydraulic resistances put in series.

$$\Delta p_{AB} = \Delta p_{BVM} + \Delta p_{sys} + \Delta p_{DPC} = \left(\frac{\dot{V}}{\mathrm{K_v^{BVM}}}\right)^2 + \left(\frac{\dot{V}}{\mathrm{K_v^{SYS}}}\right)^2 + \left(\frac{\dot{V}}{\mathrm{Y\,K_{vs}^{DPC}}}\right)^2 \tag{1}$$

Where Δp_{AB} is the differential pressure between point A and B, Δp_{BVM} is the pressure drop of the measuring balancing valve, Δp_{DPC} is the pressure drop of the differential pressure valve. Note that on the right-hand side of equation (1) the pressure drops are expressed in the form dependent on the volume flow $\dot{V}$ and K_v values of the measuring balancing valve with index BVM, system SYS and differential pressure valve DPC, where Y denotes valve relative position, Y = 1 (valve fully open) and Y = 0 (valve fully closed). Volume flow in equation (1) is given in m^3/h, while the pressure drop is given in bar. It is simple to convert the equation in other units, for example if the pressure drop is to be expressed in kPa than every expression in the brackets on the right-hand side of equation should be multiplied by 100.

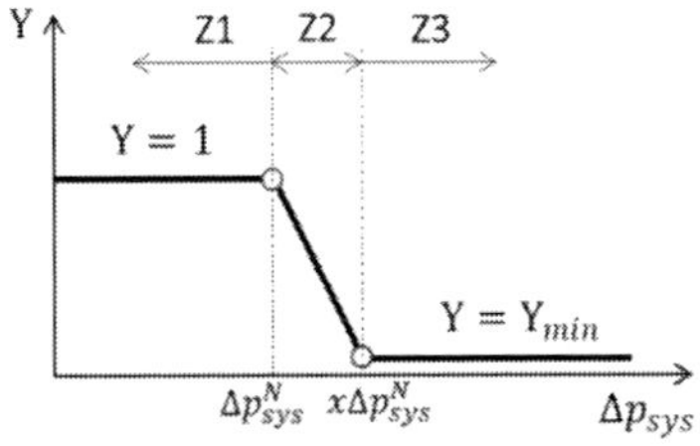

Fig. 5. Graphical dependency of the position of differential pressure valve on the system differential pressure Δp_{sys}

With the presented assumptions listed above, it is possible to introduce the $\mathrm{K_v^{SYS}}$ value of the system, although this is not the usual way how to present the pressure drop at the consumer side. However, the fact that we assumed that our consumer or the system behaves as the fixed hydraulic resistance, our approach is justified. Equation (1) cannot be solved since we have two unknown variables, the volume flow $\dot{V}$ and the position of the DPC valve Y. In order to close the system, we will introduce the relationship which links the position of the DPC valve with the system differential pressure Δp_{sys}. The assumed relationship is presented graphically as shown in Figure 5.

In Figure 5 a graphical representation of the position of the DPC valve on Δp_{sys} is given. One can note three characteristic zones. In the Zone 1 (Z1) the differential pressure Δp_{sys} is lower than the nominal value Δp_{sys}^{N} . As it can be seen in this zone the DPC valve is fully open (Y = 1) and it is expected that the nominal flow in the system cannot be achieved. In the Zone 3 (Z3) the DPC valve is in its minimal position defined as $Y = Y_{min}$ and it does not depend on increase of the system differential pressure. Zone 2 is the most interesting for us, since it is expected that the differential pressure controller will operate in this zone for the most of the time during the system operation.

It is obvious that the linear dependency of the system differential pressure is defined. In Figure 5 the width of the zone 2 is the proportional band of the controller, while the value $x\Delta p_{sys}^N$ is the maximum value of the system differential pressure for which the DPC valve comes into the minimum position (Y_{min}). Note that the value of this differential pressure is given in relative numbers while it is obvious that $x > 1$. Since the DPC valve is proportional controller, it cannot provide the fix value of the system differential pressure value. Usually, relative variation of controlled differential pressure varies within $\pm 25\%$ which is acceptable. Equation of $Y = f(\Delta p_{sys})$ within zone 2 is given as follows:

$$Y - 1 = \frac{(Y_{min} - 1)}{(x - 1)\Delta p_{sys}^N}\left(\Delta p_{sys} - \Delta p_{sys}^N\right) \tag{2}$$

Where we assumed that the fully open DPC corresponds to the nominal differential pressure Δp_{sys}^N.

We will for the simplicity of presentation express the following:

$$\alpha_{DPC} = \frac{(1 - Y_{min})}{(x - 1)} \tag{3}$$

Where α_{DPC} constant is introduced, which is having in mind our assumption always positive value. If this simplification is introduced into equation (2) together with expressing Δp_{sys} via volume flow $\dot{V}$ it becomes:

$$Y = (\alpha_{DPC} + 1) - \frac{\alpha_{DPC}}{\Delta p_{sys}^N}\left(\frac{\dot{V}}{K_V^{SYS}}\right)^2 \tag{4}$$

Which now defines relative position of the DPC valve only in terms of the volume flow.

Once the nominal conditions are known, balance valve is set to provide the nominal flow in the system with fully open and set differential pressure control valve, we can define final form of the equation which will be solved in the form of Δe_{err}, so that:

$$\Delta e_{err} = \Delta p_{AB} - \left[\left(\frac{\dot{V}}{K_V^{BVM}}\right)^2 + \left(\frac{\dot{V}}{K_V^{SYS}}\right)^2 + \left(\frac{\dot{V}}{\boxed{Y}\,K_{VS}^{DPC}}\right)^2\right] \to 0 \tag{5}$$

Where $\boxed{Y}$ denotes the value of relative position of the DPC valve which is defined by equation (3). It can be noted that the equation (5) cannot be solved explicitly and iterative solution is necessary.

Calculation methodology applicable for the system presented in Figure 3 (left) is as follows:

- Based on known nominal conditions calculate values K_V^{BVM}, K_V^{SYS} and take K_{VS}^{DPC} value from the supplier catalogue,
- Expose the system to differential pressure Δp_{AB} and solve equation (5) iteratively until the prescribed criterion is met – for example $\Delta e_{err} < 1.e^{-03}$,
- Calculate the relative position of DPC valve (Y) from equation (4) and calculate pressure drops of each component Δp_{BVM}, Δp_{sys} and Δp_{DPC}.

Please note that presented model does not take into account the dynamics of the DPC valve, it is rather model which can cover quasi-steady state calculations. It means that it is assumed that the under new imposed conditions (change of Δp_{AB}) system has enough time to reach new steady state condition. Proposed calculation methodology can be applied to practical situations where the designer knows all parameters in nominal conditions and is interested what happens to the system when it is exposed to change in differential pressure at its connection.

Let us now analyse second system which is more applicable for heating systems and connections of indirect heating substations in district heating. In this case it is assumed that the consumer water flow is constant and that weather compensated control is applied in which the control variable is supply temperature in the secondary circuit (see Figure 3 right). Desired supply temperature is dependent on the outside temperature and settings at the controller. Required supply temperature in the secondary circuit is achieved by action of the two-port valve, which determines the mixing ratio between the supply hot water and return cold water. The mixing point is in point I in Figure 3 (right) and the temperature sensor is placed after point I towards the consumer. As the outside temperature increases, less flow from the primary side is necessary to achieve desired supply temperature and control valve closes. With closing of the control valve, pressure at point C increases and resulting differential pressure measured between points C and D increases as well. At this stage, differential pressure valve, due to forces disbalance, acts and closes DPC valve trying to keep the differential pressure at the set value.

In case when the outside temperature decreases, control valve opens and pressure at point C decreases, the resulting force within the DPC valve changes the valve seat position and opens the valve until new equilibrium position is found. In terms of the definition of the problem, we here face a bit reverse problem.

So, even if the connection differential pressure is kept constant at the nominal value, as a result of change in the valve relative position, the change in differential pressure in the system is expected. The task of the differential pressure controller is to react on the control valve action and to provide the nearly constant differential pressure keeping the authority of the control valve close to its nominal value. Also, in terms of the definition of the engineering problem here, we have a different formulation. Namely, since the weather compensated control is in most cases applied as a control strategy in the secondary circuit, we know the volume flow both from primary supply line via two-port valve and from the circuit return. In order to keep the problem strictly on hydraulics here, we may confirm that the volume flow from the primary side is known.

Although, the system shown in Figure 3 (right) seems to be more complex because of the existence of the control valve which is modulating based on demand, it can be shown that the calculation procedure is even simpler in this case. Namely, since the pressure drop in the bypass between points I and II is negligible ($p_I \cong p_{II}$, or $\Delta p_{I-II} \cong 0$) we can confirm that the pressure drop at the control valve is equal to the Δp_{sys}. When this connection is designed, it is required that the nominal pressure drop of the control valve (fully open) compensates the available differential pressure at the connection subtracted by the pressure drop at the differential pressure controller and the pressure drop at the measuring balancing valve. If we put this in the form of equation, it is given as follows:

$$\Delta p_{AB} = \left(\frac{\dot{V}}{\mathrm{K_V^{BVM}}}\right)^2 + \boxed{\left(\frac{\dot{V}}{\mathrm{Y_{CV}K_{VS}^{CV}}}\right)^2} + \left(\frac{\dot{V}}{\mathrm{Y\,K_{VS}^{DPC}}}\right)^2 \tag{6}$$

Where the second term at the right-hand side of equation is practically the Δp_{sys} and contains flow coefficient of the control valve and its relative position as $\mathrm{Y_{CV}}$. Note that in the nominal condition volume flow becomes nominal flow and the relative position of the control valve is $\mathrm{Y_{CV}} = 1$.

Again, we have two unknown variables, relative position of the differential pressure control valve Y and relative position of the control valve $\mathrm{Y_{CV}}$. However, we propose decoupling of the differential pressure controller and control valve in a way that the term presented in box in equation (6) is defined as Δp_{sys} which, according to our assumptions and discussion above, can be expressed as:

$$\Delta p_{sys} = \Delta p_{AB} - \Delta p_{BVM} - \Delta p_{DPC} = \Delta p_{AB} - \left(\frac{\dot{V}}{\mathrm{K_V^{BVM}}}\right)^2 - \left(\frac{\dot{V}}{\mathrm{Y\,K_{VS}^{DPC}}}\right)^2 \tag{7}$$

Note that we again have two unknown values but we can use already defined relation for differential pressure controller in equations (2) and (4). By combining equations (2, 4 and 6) we can reformulate our expression for relative position of the differential pressure controller and write it in the form of error function such as:

$$\Delta e_{gr} = \mathrm{Y} - \left\{(\alpha_{DPC} + 1) - \frac{\alpha_{DPC}}{\Delta p_{sys}^N}\left[\Delta p_{AB} - 100\left(\frac{\dot{V}}{\mathrm{K_V^{BVM}}}\right)^2 - 100\left(\frac{\dot{V}}{\mathrm{Y\,K_{VS}^{DPC}}}\right)^2\right]\right\} \to 0 \tag{8}$$

Where we eliminated one unknown variable, wrote the equation in terms of the relative position of the differential pressure control valve Y (second term at the right-hand side of equation (6) and, since we again have equation in the form which cannot be explicitly expressed, we put it in the form of the error function. The solution is again to be sought iteratively until the prescribed tolerance is met. Equation (6) presents the key point for this system. At the end, we can calculate also the relative position of the control valve as:

$$\mathrm{Y_{CV}} = \frac{1}{\sqrt{\Delta p_{CV}}}\left(\frac{\dot{V}}{\mathrm{K_{VS}^{CV}}}\right) \tag{9}$$

Where volume flow is given in m³/h and pressure drop in bar and Δp_{CV} is the pressure drop of the control valve. Please note that it has been already said that the $\Delta p_{CV} = \Delta p_{sys}$ and this is valid for the system in Figure 3 (right).

If we take a look at the indirect heating substation and the problem formulation in this case, it is obvious that the model presented for system in Figure 3 (right) is applicable in this case as well. The only difference between these two systems is that system differential pressure Δp_{sys} in case of system given on Figure 4 can be calculated as:

$$\Delta p_{sys} = \Delta p_{HEX} + \Delta p_{CV} = \left(\frac{\dot{V}}{\mathrm{K_V^{HEX}}}\right)^2 + \left(\frac{\dot{V}}{\mathrm{Y_{CV}K_{VS}^{CV}}}\right)^2 \tag{10}$$

Where Δp_{HEX} is the pressure drop of the heat exchanger and since it is treated as the local resistance in which pressure drop depends on the square of the volume flow and can be written via flow coefficient $\mathrm{K_V^{HEX}}$.

Calculation methodology applicable for the system presented in Figure 3 (right) and system in Figure 4 is as follows:

- Based on known nominal conditions calculate value K_v^{BVM}, calculate Δp_{sys}^N (based on K_{vs}^{CV} and nominal flow) take K_{vs}^{DPC} value from the supplier catalogue, and in case of system in Figure 4 calculate K_v^{HEX}
- Expose the system to differential pressure Δp_{AB} and solve equation (8) iteratively by varying the Y until the prescribed criterion is met – for example $\Delta e_{err} < 1.e^{-06}$,
- Calculate the pressure drops of each component Δp_{BVM} (already known by the definition of the problem), Δp_{DPC} and Δp_{sys} as well as the Δp_{HEX} (for system in Figure 4),
- Calculate relative position of the control valve from equation (9) and for system in Figure 4 use equation (10) to get the Δp_{CV}.

Note that equation (8) poses no restriction in case of changing differential pressure Δp_{AB} and offers the possibility to take into account the simultaneous change in volume flow, as well as the change in differential pressure Δp_{AB}.

3. Results and discussion

For the validation of proposed approach four calculated cases are presented to demonstrate the applicability of the proposed methodology. First and second cases are related to system shown in Figure 3 (left). In the first case, the system is exposed to linear variation of the connection differential pressure and the volume flow and system differential pressure is calculated. Second case demonstrates the behaviour of the differential pressure controller when the system is exposed to the arbitrary variation of connection differential pressure. Third case is related to the system described in Figure 3 (right) where the periodical variation of the connection differential pressure and gradual change in flow are simultaneously applied.

The basis of the fourth case are measurements and results published in HERZ material [9] in which the indirect heating substation was taken (Figure 4). The connection differential pressure was kept constant while the three cycles of opening and closing the control valve are applied and the system differential pressure is measured. In this case, since we did not have data on exact measurements setup, an improvisation is used to check the applicability of our methodology. Even so, the hydraulic parameters can be clearly seen from the presented graph and our results correspond with the published data on measurements.

3.1. Linear increase of connection differential pressure

In this case the system in Figure 3 (left) is taken. The system is exposed to the linear increase of the connection differential pressure starting from 50 kPa up to 150 kPa, while the nominal connection differential pressure is 50 kPa. The following numbers were input for calculation:

$$\Delta p_{AB}^N = 50 \text{ kPa}, \Delta p_{BVM}^N = 5 \text{ kPa}, \Delta p_{sys}^N = 20 \text{ kPa}, \Delta p_{DPC}^N = 25 \text{ kPa}, \text{nominal volume flow rate is } \dot{V} = 1{,}786\ \frac{\text{m}^3}{\text{h}}.$$

For the illustration purposes this system would be equivalent to the consumer with 40 kW in nominal conditions. In this case we will assume that the parameter $\alpha_{DPC} = 3$. Now, it is very important to demonstrate the ratios of pressure drops for individual components of the system.

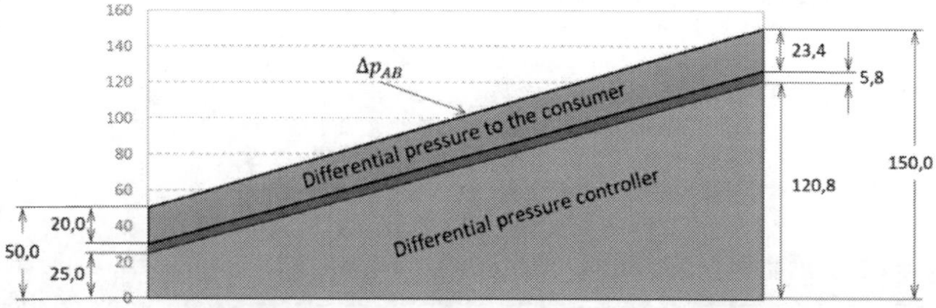

Fig. 6. Contribution of pressure drop for each component of the system in Figure 3 (left) when the system is exposed to linear increase of the connection differential pressure Δp_{AB} from 50 kPa to 150 kPa.

Figure 6 shows the contribution of the pressure drops of the DPC valve (area in orange colour), BMV valve (area in blue colour) and available Δp_{sys} (area in green colour). Note that the increase of the connection differential pressure is given in black line at the top of the diagram. It can be seen that DPC valve compensates the increase of the connection differential pressure, while preserving the system differential pressure almost at constant level.

At the left-hand side of the diagram given in Figure 6 the numerical nominal values of pressure drops are given while at the right-hand side the final values of all components are presented when the system is exposed to $\Delta p_{AB} = 150$ kPa.

Final remark is that the increase of the connection differential pressure of 100 kPa, over the nominal value resulted in increase of the system differential pressure in 3,4 kPa. In other words, due to the DPC valve action the disturbance has been reduced 30 times! This is exactly was is expected from the differential pressure controller.

3.2. Arbitrary variation of connection differential pressure

In this case, the same system with identical characteristic values is used and an arbitrary variation of the connection differential pressure is applied. The goal is to proof that the variations of the system differential pressures will remain within prescribed tolerances.

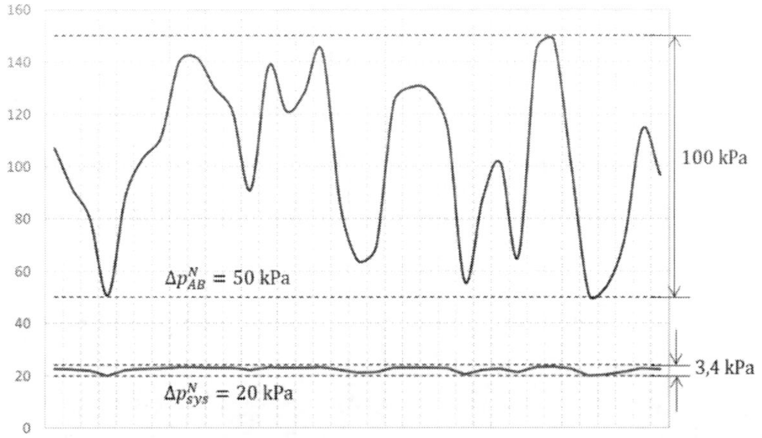

Fig. 7. Arbitrary variation of connection differential pressure Δp_{AB} (blue line) and corresponding system differential pressure Δp_{sys} (red line) in time

Figure 7 shows an arbitrary variation of connection differential pressure Δp_{AB} (blue line) and corresponding system differential pressure Δp_{sys} in time. Note that although the time is one of the axes, the dynamic behaviour of the differential pressure controller is not taken into account and the results are presented as the quasi-steady-state solutions. As it can be seen the variation of the connection differential pressure is 100 kPa while the connection differential pressure never falls below nominal connection differential pressure. In this case the maximum amplitude of the system differential pressure is 3,4 kPa as shown in diagram in Figure 7.

3.3. Variation of connection differential pressure and step change in flow

This case is applied to the system shown in Figure 3 (right). In order to test the proposed calculation methodology simultaneous change in connection differential pressure and change in volume flow is applied. Sinusoidal variation of the connection differential pressure is generated, while the step change in the relative volume flow is applied. The flow is changed from the starting 10% , with increment of 10% and ends with 100% value or the nominal flow. The following data are used for the calculations:

$\Delta p_{AB}^N = 50$ kPa, $\Delta p_{BVM}^N = 5$ kPa, $\Delta p_{sys}^N = 20$ kPa, $\Delta p_{DPC}^N = 25$ kPa, nominal volume flow rate is $\dot{V} = 1{,}786\ \frac{\mathrm{m}^3}{\mathrm{h}}$. Here we also supposed that minimal relative position of the differential pressure control valve is $Y_{min} = 0{,}045$, which corresponds to the maximal differential pressure $x\Delta p_{sys}^N = 1{,}32\ \Delta p_{sys}^N = 26{,}4$ kPa. So, the proportional band for DPC valve is in this case is 32% (see Figure 5). In this case, we said that pressure drop at the fully open control valve should be equal to the available nominal differential pressure $\Delta p_{sys}^N = \Delta p_{CV}^N = 20$ kPa.

If we now apply preliminary analysis, it is obvious that the authority of the control valve in nominal conditions is 1. With maximal differential pressure (26,4 kPa) the authority will change to 0,76, which is again very good. At contrary, if we assume that no differential pressure control is installed, the authority in nominal conditions would be 0,4 and when system is exposed to maximal connection differential pressure of 150 kPa, an authority would decrease to only 0,13 which brings the problems elaborated in the introductory part of this paper.

Figure 8 presents results of calculation for applied boundary conditions for the system given in Figure 3 (right). It is visible that despite the variation of the connection differential pressure and step change in the volume flow presented in red line and given at the right-hand side vertical axes, the system differential pressure remains almost constant and that the differential pressure control valve takes over almost all excessive pressure.

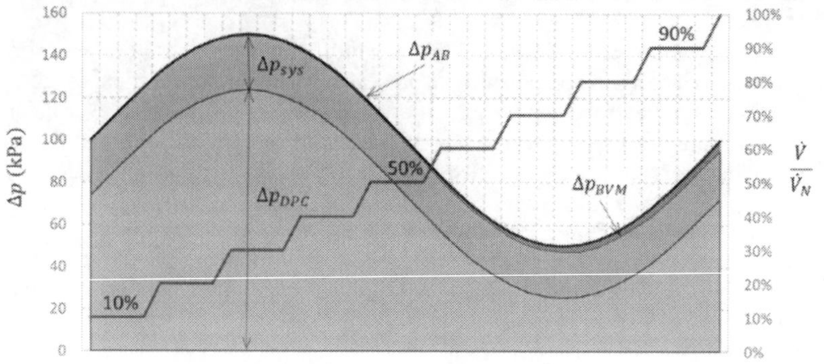

Fig. 8. Sinusoidal variation of connection differential pressure (Δp_{AB}), step change in relative flow (red line) and contributions of pressure drops for Δp_{DPC}, Δp_{sys} (or Δp_{CV}) and Δp_{BMV}

It should be said that the nominal conditions have never been met in this example. Namely, when the connection differential pressure drops to the nominal value, the relative flow through the primary side is at the level of 80% and also when the flow is nominal which corresponds to the last point in diagram in Figure 8, the connection differential pressure is at 100 kPa.

One can also note that the blue area which presents the pressure drop coming from the balance measuring valve is almost invisible until the relative flow exceeds 50% which is reasonable, since the balance valve is local resistance and its pressure drop is proportional to the square of the flow. In cases where flow is small the pressure drop at this valve is almost negligible.

3.4. Indirect heating substation with differential pressure controller

This case was chosen to show the applicability of the developed calculation methodology to treat the stabilisation of the differential pressure in indirect connection in district heating application. Results of measurements for comparison purposes are taken from the published material [9]. Although exact experimental setup was not available, based on the presented results the case was reconstructed and calculated.

This case deals with the system in Figure 4, where differential pressure controller stabilises the differential pressure for the control valve which is in series placed in the return pipe after the heat exchanger. In this case the connection differential pressure was kept constant during the measurements and the control valve controlled the flow through the heat exchanger in three full cycles from the fully open to the almost closed position. As a result of changes in system differential pressure, the differential pressure control valve reacted and was trying to keep the differential pressure at the almost constant value. The following data are used for the calculations:

$\Delta p_{AB}^N = 250$ kPa, $\Delta p_{BVM}^N = 35$ kPa, $\Delta p_{DPC}^N = 25$ kPa, $\Delta p_{sys}^N = 19$ kPa, $\Delta p_{CV}^N = 12{,}25$ kPa, $\Delta p_{HEX}^N = 6{,}75$ kPa, nominal volume flow rate is $\dot{V} = 3500\ \frac{\mathrm{l}}{\mathrm{h}} = 3{,}5\ \frac{\mathrm{m^3}}{\mathrm{h}}$.

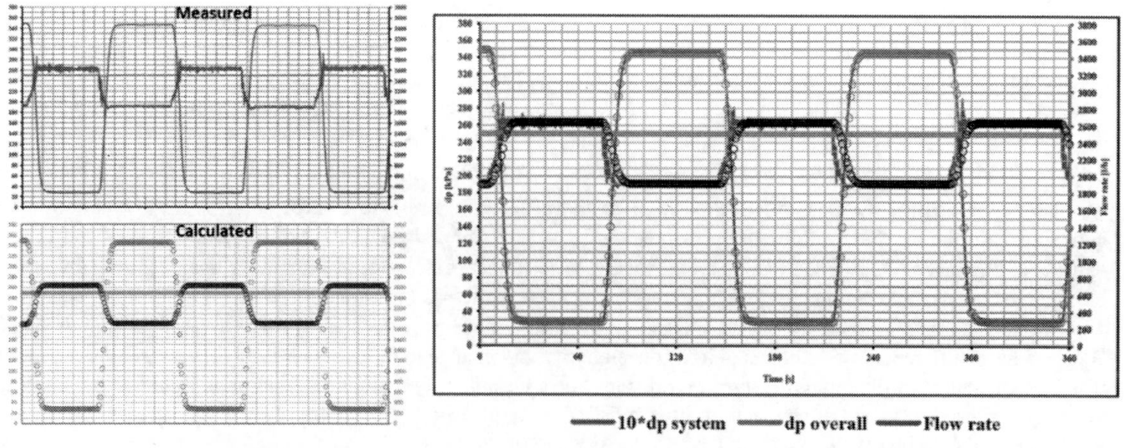

Fig. 9. Volume flow, connection differential pressure and system differential pressure / measured (top left) [9], calculated (bottom left) and comparison measured vs. calculated (right)

It is estimated that based on the assumed nominal temperature difference at the primary side of $\Delta t = 55°C, (110/55)$ the heat exchanger capacity was close to 230 kW, and the pressure drop for the heat exchanger together with connections was estimated by the available producer's data. Control valve and the DPC valves are also chosen according to the market available K_{vs} values. In our case, the volume flow was read from the diagram and this volume flow was used to solve the equations (8, 9 and 10) and to determine the system differential pressure for every given volume flow.

Also, based on our estimation maximum differential pressure for the almost closed DPC valve ($Y_{min} = 0{,}05$) is $x\Delta p_{sys}^{N} = 1{,}4\ \Delta p_{sys}^{N}$.

Figure 9 (left) shows the results of measurements (up) and results of calculations (down) and diagrams one over the other (right). Variation of the flow (blue line) is given in litres per hour at the right-hand side vertical axes, while the differential pressure (red line) is given in kPa at vertical left-hand side axes. Please note that the system differential pressure is multiplied by 10 in order to make measured variable clearly visible on the graph. It is obvious that our estimation was good and the results in terms of the system differential pressure are very good. Discrepancies are visible only at the time intervals where the flow is changing from minimum to the maximum position or vice versa. This is expected since our proposed model do not treat the dynamics of the differential pressure controller than rather treat steady state conditions which means that DPC valve reacts instantly on any disturbance.

4. Conclusions and further work

Simplified algebraic model of the differential pressure controller is presented in this paper, which can be of use for design engineers in preliminary and design phases. This model can be incorporated into the modelling tools which then treat interaction of different components in the heating/cooling systems or indirect heating substations in district heating. It has been demonstrated on four examples that the proposed calculation methodology is useful for calculation of different properties such as individual pressure drops of different components in the heating system and relative positions of DPC and control valve.

Future work is focused on implementation of this model to Modelica Buildings library, which then allows analysis of different design concepts and include, but is not limited to, the variable speed drive pumps, advanced models of plate heat exchangers, different control strategies etc. Preliminary work has been already done and the results are promising.

5. References

[1] Petitjean, R. (2002). Hydronic balancing with differential pressure controllers, Tour & Andersson AB, Ljung, Sweden

[2] Lennermo, G., Lauenburg, P. & Brand, L. (2014). Decentralised heat supply in district heating systems: Implications of varying differential pressure. The 14th International Symposium on District Heating and Cooling, 07 – 09 September 2014, Stockholm, Sweden.

[3] Kos, T., Petrovčič, J., Šaponia, Z., Musizza, B., Podržaj, G., Krančan, S. & Vramčić, D. (2019). Intelligent differential pressure drive with anti-oscillation system iSET, AIG'19, 09 – 10 April 2019, Maribor, Slovenia, ISBN 978-961-93818-6-1.

[4] Carette, J.C. & Bernadou, E. Optimising hydronic distributions for energy efficiency, Available from: file:///C:/Users/muham/AppData/Local/Temp/Optimising-hydronic-distributions-for-energy-efficiency-1.pdf Accessed: 2021-08-21

[5] Zhang, L., Xia, J., Thorsen, J.E., Gudmundsson, O., Li, H. & Svendsen, S. (2017). Method for achieving hydraulic balance in typical Chinese building heating system by managing differential pressure and flow, Building Simulation, Vol.10, No.1., pp. 51 – 63, DOI: https://doi.org/10.1007/s12273-016-0307-2

[6] Dinu, R.C. & Popescu, D. (2016). Solutions with differential pressure in heating with radiators, Journal of Sustainable Energy, Vol. 7, No. 4, 2015, pp. 137-142., ISSN: 2067-5534.

[7] https://cupdf.com/document/danfoss-training-module-3-v1-understanding-pressure-control-valves-compressedpps.html, (2014). Danfoss Training module-3, Understanding differential pressure control valves, Power Point Presentation, Accessed on: 2021-08-13

[8] Teskeredzic, A., Blazevic, R. & Dzino, S. (2020). Application of air-water heat pump in refurbished building with existing radiator system, Proceedings of the 31th DAAAM International Symposium, 21-24 October 2020 Mostar, ISBN 978-3-902734-29-7, ISSN 1726-9679, B. Katalinic (Ed.), pp. 0510 – 0519, Published by DAAAM International, Vienna, Austria, DOI: 10.2507/31st.daaam.proceedings.071

[9] file:///C:/Users/muham/AppData/Local/Temp/4007_differenzdruckregler-en.pdf. Differential pressure controller-dynamic control for hydraulic balancing, Herz manual, Accessed on: 2021-08-27

DOI: 10.2507/32nd.daaam.proceedings.036

SAFETY LINES DETECTION AS A MEANS OF NAVIGATION IN INDUSTRIAL FACILITIES

Karamanos Xenofon, Bechtsis Dimitrios,
Vlachos Dimitrios & Mastos Theofilos

This Publication has to be referred as: Karamanos, X[enofon]; Bechtsis, D[imitrios]; Vlachos, D[imitrios] & Mastos, T[heofilos] (2021). Safety Lines Detection as a Means of Navigation in Industrial Facilities, Proceedings of the 32nd DAAAM International Symposium, pp.0256-0261, B. Katalinic (Ed.), Published by DAAAM International, ISBN 978-3-902734-33-4, ISSN 1726-9679, Vienna, Austria
DOI: 10.2507/32nd.daaam.proceedings.036

Abstract

Autonomous Mobile Robots (AMRs) are becoming a crucial part of industrial environments. Path planning and navigation activities are still core components of AMR literature. AGV navigation methods traditionally utilize floor markings, while state of the art systems are using cameras to identify the vehicle's position and direction and map the facility layout. This paper focuses on how an AMR with an RGB camera could take advantage of floor markings and improve the vehicle's navigation and safety conditions, especially in manufacturing environments with large empty spaces where cameras have limited positioning capabilities. The system is built on top of the Robot Operating System (ROS). A safety line detector is trained to recognise floor markings and the AMR uses them as a guidance line in order to safely navigate at a specific distance from it. The detector is a MASK-RCNN network, trained on real-world images captured at the industrial facility. Each image was manually annotated in order to feed the network. The proposed system was extensively tested in simulation and in real world environment.

Keywords: ROS; MASK-RCNN; Navigation; Intralogistics; Lane Detection

1. Introduction

Autonomous Mobile Robots (AMRs) are becoming increasingly valuable assets to warehouses and industrial facilities, as part of the fourth industrial revolution (Industry 4.0). AMRs can be used for intra-logistics and material handling activities without being manipulated by human agents. Incorporating AMRs can increase productivity and flexibility [1], reduce costs and enable agile manufacturing. Moreover AMRs flexibility allows them to be easily customized for many tasks. In this context, AMRs must be able to work along with human agents and safely travel through complex layouts and industrial spaces. Of course, simulations tools verify the added value and help identify the operational needs of facilities[2] . Additionally, visual tools are proposed in order to facilitate the AMRs by non-expert personnel [3].

Most industrial facilities are required to clearly indicate goods, the transfer zones that are used as buffers and the cells were the workstations are installed. Colour labelling is one of the most efficient ways to divide the different segments of a workplace and is also considered as a part of 5S [4] and lean manufacturing.

Colour coding gives some additional benefits such as speeding up visual search, emphasizing information like aisle-way and exit paths and notates hazardous areas. A main aspect of this colour labelling is to separate forklifts' traffic flows from pedestrians' flows. Pedestrians' pathways are clearly marked and visible as the forklifts' pathways. This ensures a significant safety level and reduces the accidents by making the workspace safer, more pleasant, and well-organized [5]. The proposed research work, uses the already established floor markings in order to facilitate the vehicle's navigation with increased safety. Recognizing the floor markings and safety lines, improves safety and navigation of the AMR, and this can be achieved with a single RGB camera. The rest of this article is organized as follows. In Section 2, a brief literature review on AMR systems and lane detection systems is presented, Section 3 analyses the proposed system and Section 4 focuses on the experimental results. Finally, Section 5 presents the conclusions and future research directions.

2. Literature Review

Many methods, are actually used to detect a lane or a lane's lines in a road or an industrial transportation path. There are many image processing algorithms that can be used to detect lane lines such as feature-based detections, model-based and more modern ones with semantic segmentation networks.

A feature-based method operates most of the time using the characteristics of line shape and/or colour features. A simple algorithm is thresholding [6], global or dynamic. In global thresholding algorithms, grey level histogram of the image is used and the algorithm includes this information to detect lane markers. This is not very effective, though, due to that a similar grey scale histogram is required [7]. Other methods, include a pre-process step, like edge detectors, such as Canny, Gabor, Sobel and Laplacian operators. After the edge detection is completed, image enhancement is performed to improve the next step of lane line detection. The algorithms are then able to detect lines based on the vanishing point estimation [[8] and curve fitting [9].

Model-based methods are usually trying to fit the detection into some geometric or curve model. [10] utilizes the camera parameters to form a lane geometrical model. The initial stage is to estimate the vanishing point, followed by the lane boundary detection using Canny edge detection and Hough transforms. The final step is to match the lane's curvature from a set of candidates. In [11] the authors propose a lane detection based on spatiotemporal images. Points of lanes are detected by aligning scanlines along the time axis, followed by a Hough transform and the prediction of an offset for the scanline in the next frame. All the points detected are fitted to a cubic curve model.

Deep neural networks are not a recent breakthrough [12]although computational advances enabled the heavy usage of such networks. The semantic segmentation network is based on neural deep convolutional networks. Its main purpose is to segment the pixels in an image into distinguishable targets, like roads, persons, pavements and others. The authors in [13] based their method on Convolution Neural Networks (CNN). The lane line segmentation is split into two stages. The first stage is responsible to semantically segment the lane boundaries and the second to classify the lane. [14]performs lane detection using a modified encoder-decoder based model ENet [15] with two distinct branches for binary and instance segmentation. Binary segmentation is performed to detect all lines in the frame. Instance segmentation outputs an embedding for each lane pixel so that the distance between pixel embedding belonging to the same lane is small, where to different lanes is maximized.

3. Safety Line Detection System

Safety Line Detection System is a pretrained Mask-RCNN [16] model, trained further on images taken from industrial warehouses. First, we must collect images (frames) to train the system. Then, the network is trained, and finally its performance is validated in order to be deployed on the robot and perform in real-world environments. The whole pipeline of training the system is explained below.

3.1. Annotating Data

To train the system, we manually annotated and labelled 1145 frames. Due to that Mask-RCNN is already pretrained, there is no need for huge amount of data. The whole annotation process was conducted with a user-friendly open-source tool called VGG Image Annotator [17] . VGG is a web-based tool and offers various shapes to label specific regions within the images, namely rectangles, circles, polygons and offers multiple types of attributes (indicatively text, checkboxes, radio buttons).

After the annotation is completed as indicated in Figure 1, the tool allows us to save the project for future use and export annotations to various formats like CSV, JSON and COCO [18] annotation format. We export the annotations as COCO, since our next step, training, requires the COCO format. COCO format is a type of JSON file type with specific fields for different annotation types.

Fig. 1. VGG Image annotator GUI with two safety lines annotations (left and right lines)

3.2. Training MASK-RCNN

Having the annotations, the next step of the pipeline is to actually train it. This is possible with frameworks such as TensorFlow [19] and PyTorch [20]. The PyTorch framework was selected due to its clarity and being developer friendly. Also, it is written with keeping in mind the pythonic programming style native python and as a result, it is an object-oriented framework.

At the beginning, the annotation that was created in the previous step must be used to create the Dataset that the network will be trained on. PyTorch offers the ability to create custom datasets in an extremely easy and friendly way. It provides various classes that a user can derive from conforming to different annotation formats. Since the annotations was exported in COCO format, we can create a new class deriving from the *torchvision.datasets.CocoDetection* class. We then load our annotations with the package *pycocotools*. We also split the dataset to a training and a validation dataset to prevent overfitting the model.

The next step is to define our model, specifically Mask-RCNN model. Mask-RCNN is an extension to Faster-RCNN [21]. Both can use as backbone, different feature detectors. such as Resnet {50,101} [22], Yolov3 [23] and others. Faster-RCNN can detect regions of interest and place bounding boxes around them. Mask-RCNN adds an extra branch that can not only place that bounding box but also segment the instance and provide a mask of it. Moreover, all this is possible only with a small overhead to Faster-RCNN. Again, the simplicity and module reusability of PyTorch, allows the user to create the model very easily and modify specific parameters for any use case, such as number of classes to detect and number of hidden layers. We then define the optimizer and learning rate scheduler, where Stochastic Gradient Descent (SGD) and Step Learning Rate (StepLR) classes were used respectively.

Training the model was performed on a CUDA capable device, a GeForce GTX 1060 6GB/PCIe/SSE2 and total time for an epoch was around 2 min and 10 secs with batch size of 4. The system was trained for a total of 8 epoch with the final metrics of average precision shown in Table 1.

Safety Line Detection	**Backbone**	**AP**	**AP_{50}**	**AP_{75}**
Bounding Box	Resnet50-FPN	0.834	0.987	0.935
Segmentation	Resnet50-FPN	0.451	0.872	0.448

Table 1. Average Precision (AP) Metrics

3.3. Deploy on Robot

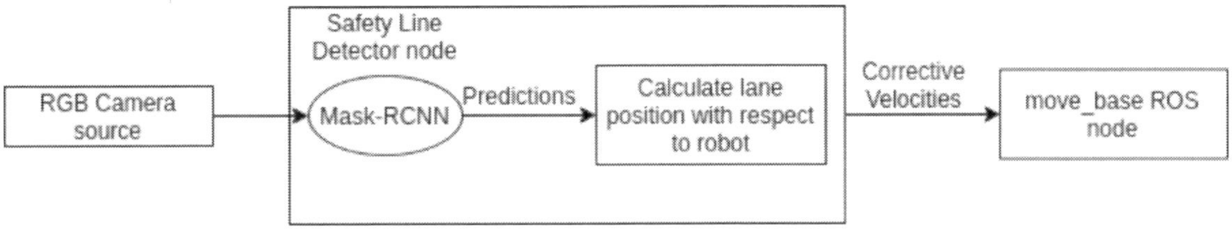

Fig. 2. ROS Node Architecture

As the robot uses the ROS operating system, a wrapper for the model must be introduced. A brief architecture of the node is displayed in Fig. 2. Firstly, we create a node that listens to robot's camera stream. That node also contains a member variable of our neural network model's class. The camera stream is also converted to a data-type acceptable by our model and then is fed to the model. The output of the model is a list of predictions containing the masks of the identified safety lines. Since, the robot should travel near the safety line, one can calculate how far the line is from the centre of the image. If it is furthest from a certain threshold, the velocities are corrected and sent to robot's drivers in order to retain the safety line as centred as possible in the current frame.

Another useful feature of the wrapper is its ability to turn the safety line detection on and off, in order to save computation resources as well as to expand the robot's battery life. This is applicable by using the ROS Framework and its *services* module.

4. Experimental results

The entire system is built on top of ROS [24] ecosystem. ROS is a middleware suite that provides various packages and software frameworks for easy robot software development. It also provides ready to use tools, that enable the robot to communicate, sense and act (indicatively tools for localization, mapping, visualization, debugging).

All simulations were conducted in a custom-made Gazebo simulation (Fig. 3) and the real-world experiments used the Turtlebot 2 with a kobuki base, equipped with RPLidar A3 sensor and Intel RealSense D435i camera sensor. The laptop that was used for controlling the robot has the following specifications: Intel Core i7-8750h CPU @ 2.2 GHZ, an Nvidia GeForce GTX 1060 and 8 GB RAM.

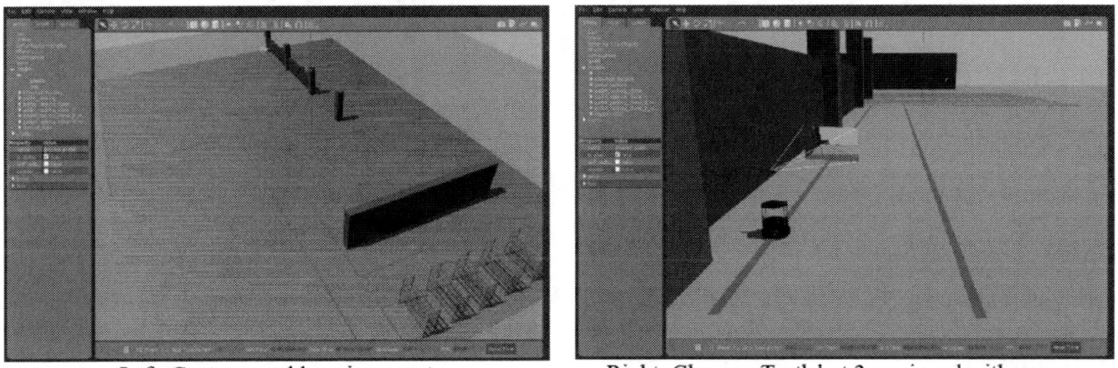

Left: Custom world environment Right: Close up Turtlebot 2, equipped with camera

Fig. 3. Gazebo Simulation Environment

Gazebo environment simulates an industrial environment and is based on a real-world one. The only difference is that our simulation world has no dynamic or static objects like in the real-world one. Nevertheless, it is a good starting point to test if our node, safety line detection and course correction is working. To better visualize the results, the RViz tool is used. As we can see in Fig. 4, our model detects the mask of safety lines and for presenting the results we overlap them with the original image. We then procced to give a move goal to the robot, generating a path from its current pose to the goal. As the robot is moving and follows the path produced by the global planner, the corrected velocities are provided from the developed node and enable the vehicle to stay near the safety line.

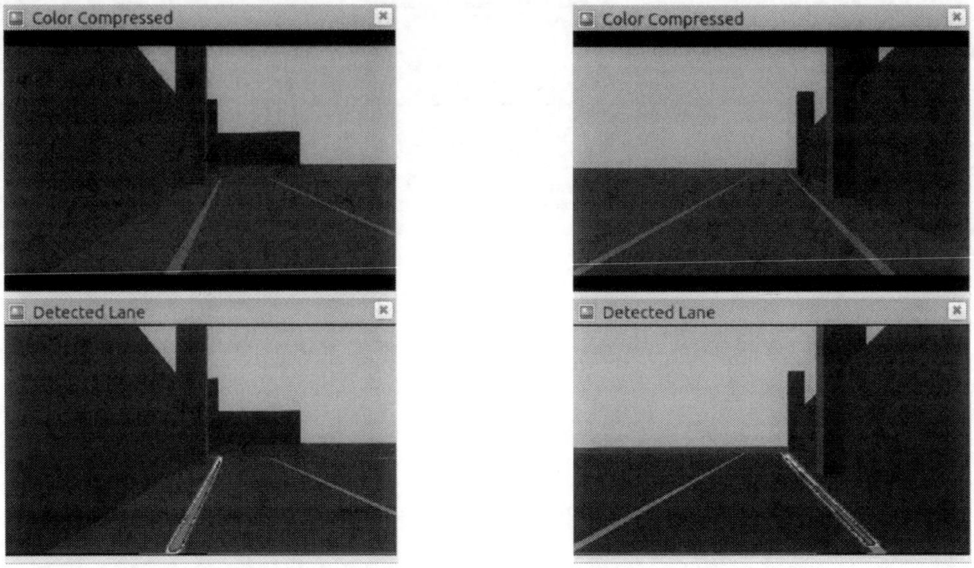

Fig. 4. Two snapshots of detected lines.

Likewise, when the node was used in the real-life scenario, the robot was able to detect the safety lines and keep close distance to them.

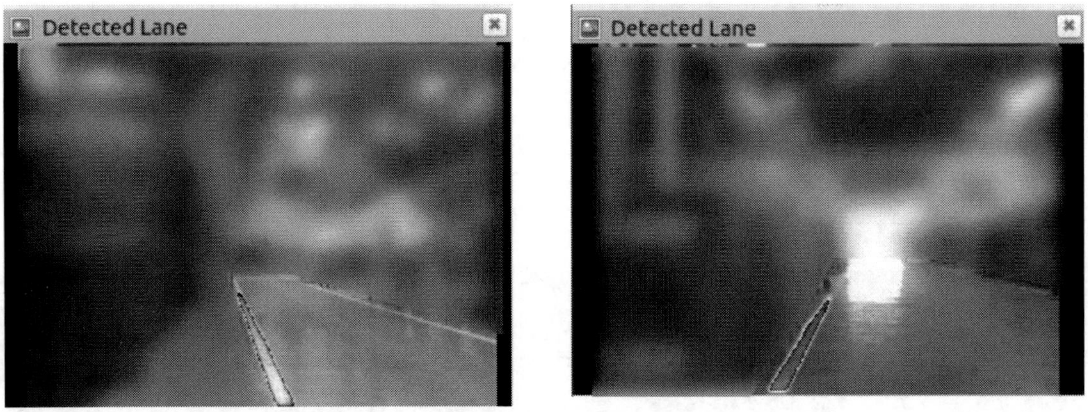

Fig. 5. Two snapshots of detected lines in real warehouse environment

The inference time of a single frame in the system described above is about 0.14 seconds making it almost suitable for real-time operation. Inferencing continuously though on the latest frame, resulted to some dropped frames and safety line following was unreliable. Instead, we operated our node at a fixed rate of 2 HZ in order to improve reliability.

5. Conclusion

In this article, the authors proposed a method for safe navigation in industrial facilities. A lane detection system is introduced based on state-of-the-art Mask-RCNN model for AMR systems. Lane detection is a realistically identified as feasible method since floor markings are found in most industrial facilities and warehouses. The system is trained on manually annotated frames containing floor markings. Following the lane corrective commands are provided to the AMR to closely follow the line. The whole system is implemented on top of ROS and PyTorch frameworks. Experiments has shown that the system, can achieve almost-real time performance and that the AMR is successfully following the lanes' line.

Future work could incorporate the lanes detection system into an industrial SLAM (Simultaneous Localization and Mapping) solution. Mapping the floor markings, would greatly improve the localization and semantic information produced by SLAM system. Moreover, the system can be further improved, by training on more data from industrial facilities using multiple line types.

6. Acknowledgments

This paper was partially supported by the Greek Secretariat of Research and Technology, Operational Programme Competitiveness, Entrepreneurship and Innovation 2014-2020, project "Dynamic Quality CONtrol on Production lines using intelligent AutonomouS vehicleS (Q-CONPASS)".

7. References

[1] Fragapane, G.; Ivanov, D.; Peron, M.; Sgarbossa, F.; Strandhagen, J. O. (2020). Increasing flexibility and productivity in Industry 4.0 production networks with autonomous mobile robots and smart intralogistics, Annals of Operations Research 2020, 1–19. doi:10.1007/S10479-020-03526-7

[2] Bechtsis, D.; Tsolakis, N.; Vlachos, D.; Srai, J. S. (2018). Intelligent Autonomous Vehicles in digital supply chains: A framework for integrating innovations towards sustainable value networks, Journal of Cleaner Production, Vol. 181, 60–71. doi:10.1016/J.JCLEPRO.2018.01.173

[3] Karamanos, X.; Mallioris, P.; Poulimenos, D.; Bechtsis, D.; Vlachos, D. (2019). A ros tool for optimal routing in intralogistics, Annals of DAAAM and Proceedings of the International DAAAM Symposium (Vol. 30). doi:10.2507/30th.daaam.proceedings.115

[4] Hirano, Hiroyuki. (1995). 5 pillars of the visual workplace : the sourcebook for 5S implementation, 353

[5] Joshi, A. A. (n.d.). A Review on Seven S (7S) as a tool of Workplace Organization

[6] Gonzalez, J. P.; Ozguner, U. (2000). Lane detection using histogram-based segmentation and decision trees, IEEE Conference on Intelligent Transportation Systems, Proceedings, ITSC, 346–351. doi:10.1109/ITSC.2000.881084

[7] Hu, J.; Xiong, S.; Sun, Y.; Zha, J.; Fu, C. (2020). Research on Lane Detection Based on Global Search of Dynamic Region of Interest (DROI), Applied Sciences 2020, Vol. 10, Page 2543, Vol. 10, No. 7, 2543. doi:10.3390/APP10072543

[8] Ozgunalp, U.; Fan, R.; Ai, X.; Dahnoun, N. (2017). Multiple Lane Detection Algorithm Based on Novel Dense Vanishing Point Estimation, IEEE Transactions on Intelligent Transportation Systems, Vol. 18, No. 3, 621–632. doi:10.1109/TITS.2016.2586187

[9] Niu, J.; Lu, J.; Xu, M.; Lv, P.; Zhao, X. (2016). Robust Lane Detection using Two-stage Feature Extraction with Curve Fitting, Pattern Recognition, Vol. 59, 225–233. doi:10.1016/J.PATCOG.2015.12.010

[10] Zhou, S.; Jiang, Y.; Xi, J.; Gong, J.; Xiong, G.; Chen, H. (2010). A novel lane detection based on geometrical model and Gabor filter, IEEE Intelligent Vehicles Symposium, Proceedings, 59–64. doi:10.1109/IVS.2010.5548087

[11] Jung, S.; Youn, J.; Sull, S. (2016). Efficient lane detection based on spatiotemporal images, IEEE Transactions on Intelligent Transportation Systems, Vol. 17, No. 1, 289–295. doi:10.1109/TITS.2015.2464253

[12] Wang, H.; Raj, B. (2017). On the Origin of Deep Learning

[13] Pizzati, F.; Allodi, M.; Barrera, A.; García, F. (2019). Lane Detection and Classification using Cascaded CNNs, Lecture Notes in Computer Science (Including Subseries Lecture Notes in Artificial Intelligence and Lecture Notes in Bioinformatics), Vol. 12014 LNCS, 95–103

[14] Neven, D.; de Brabandere, B.; Georgoulis, S.; Proesmans, M.; van Gool, L. (2018). Towards End-to-End Lane Detection: An Instance Segmentation Approach, IEEE Intelligent Vehicles Symposium, Proceedings, Vols. 2018-June, 286–291. doi:10.1109/IVS.2018.8500547

[15] Paszke, A.; Chaurasia, A.; Kim, S.; Culurciello, E. (2016). ENet: A Deep Neural Network Architecture for Real-Time Semantic Segmentation

[16] He, K.; Gkioxari, G.; Dollár, P.; Girshick, R. (2017). Mask R-CNN, IEEE Transactions on Pattern Analysis and Machine Intelligence, Vol. 42, No. 2, 386–397

[17] Dutta, A.; Zisserman, A. (n.d.). The VIA Annotation Software for Images, Audio and Video

[18] Lin, T.-Y.; Maire, M.; Belongie, S.; Bourdev, L.; Girshick, R.; Hays, J.; Perona, P.; Ramanan, D.; Zitnick, C. L.; Dolí, P. (n.d.). Microsoft COCO: Common Objects in Context

[19] Abadi, M.; Barham, P.; Chen, J.; Chen, Z.; Davis, A.; Dean, J.; Devin, M.; Ghemawat, S.; Irving, G.; Isard, M.; Kudlur, M.; Levenberg, J.; Monga, R.; Moore, S.; Murray, D. G.; Steiner, B.; Tucker, P.; Vasudevan, V.; Warden, P.; Wicke, M.; Yu, Y.; Zheng, X. (2016). TensorFlow: A system for large-scale machine learning, Proceedings of the 12th USENIX Symposium on Operating Systems Design and Implementation, OSDI 2016, 265–283

[20] Paszke, A.; Gross, S.; Massa, F.; Lerer, A.; Bradbury, J.; Chanan, G.; Killeen, T.; Lin, Z.; Gimelshein, N.; Antiga, L.; Desmaison, A.; Köpf, A.; Yang, E.; DeVito, Z.; Raison, M.; Tejani, A.; Chilamkurthy, S.; Steiner, B.; Fang, L.; Bai, J.; Chintala, S. (2019). PyTorch: An Imperative Style, High-Performance Deep Learning Library, Advances in Neural Information Processing Systems, Vol. 32

[21] Ren, S.; He, K.; Girshick, R.; Sun, J. (2015). Faster R-CNN: Towards Real-Time Object Detection with Region Proposal Networks, IEEE Transactions on Pattern Analysis and Machine Intelligence, Vol. 39, No. 6, 1137–1149

[22] He, K.; Zhang, X.; Ren, S.; Sun, J. (2015). Deep Residual Learning for Image Recognition, Proceedings of the IEEE Computer Society Conference on Computer Vision and Pattern Recognition, Vols. 2016-December, 770–778

[23] Redmon, J.; Farhadi, A. (n.d.). YOLOv3: An Incremental Improvement

[24] Quigley, M.; Gerkey, B.; Conley, K.; Faust, J.; Foote, T.; Leibs, J.; Berger, E.; Wheeler, R.; Ng, A. (n.d.). ROS: an open-source Robot Operating System

DOI: 10.2507/32nd.daaam.proceedings.037

FRUGAL INNOVATIONS IN INDUSTRY 4.0: THE IDENTIFICATION OF KEY INDICATORS

Sinisa Tesic, Bojana Bajic, Ilija Cosic & Aleksandar Rikalovic

This Publication has to be referred as: Tesic, S[inisa]; Bajic, B[ojana]; Cosic, I[lija] & Rikalovic, A[leksandar] (2021). Frugal innovations in Industry 4.0: The Identification of Key Indicatiors, Proceedings of the 32nd DAAAM International Symposium, pp.0262-0269, B. Katalinic (Ed.), Published by DAAAM International, ISBN 978-3-902734-33-4, ISSN 1726-9679, Vienna, Austria
DOI: 10.2507/32nd.daaam.proceedings.037

Abstract

Innovations refer to the actions required to create new ideas, processes or products. The implementation of new ideas, processes or products leads to positive effective change in a company. The positive effective change in a company is accomplished through savings. Thus, company makes savings in different types of resources via innovations, contribute to social goal or influence the development of technological solutions. In industry, savings via innovations are mostly reflected in redesigning products and processes to cut out unnecessary costs. That type of innovations are called frugal innovations. Frugal innovations represent innovations of existing solutions with accent of achieving resource savings. Nowadays, frugal innovations, which are used in industry, play a vital role in era of fourth industrial revolution, named Industry 4.0. From the innovations point of view, the core of Industry 4.0 is the development of advanced technologies with the goal of achieving savings in processes, products or services The value of frugal innovation analysis in Industry 4.0 represents the contribution of technology transfer in developing countries with the goal of strengthening the competitiveness of companies and the development of new ways of innovations through the impact on society as a form of savings that the company could achieve. Therefore, this paper deals with the analysis of frugal innovations in Industry 4.0 with the goal of the identification of key indicators that have a high influence on their development.

Keywords: Frugal innovation; Industry 4.0; Indicators; Cyber Physical Systems (CPS); Industrial Internet of Things (IIoT);

1. Introduction

Any positive change in the management, process, production or service in the company is an innovation. Innovation does not have to be a revolutionary invention, it could be a minor change that contributes to the ease of work or makes the process, product or service different from the existing one [1]. However, if the focus is put on savings in resources through innovations which has benefits to society, then we talk about frugal innovations. Frugal innovations represent a process, product or service that arises as savings in finance, human resources, technology or another type of savings where the end result is cheaper solutions than the competitions' which meets customer need [2]. Thus, frugal innovations can be described as innovations that achieve significant savings with minimal investment [3].

In the present, the development of frugal innovation is influenced by means of savings in new products, processes, service and business models due to impact of Industry 4.0 era. The fundaments of Industry 4.0 is based on innovations that directly affect the improvement of the company's business. Innovations in Industry 4.0 are related to technological and managerial innovations in the company. The Industry 4.0 concept is driven by Industry 4.0 technology categories, namely: cyber physical systems (CPS), Internet of Things (IoT), big data analytics (BDA), cloud computing, fog and edge computing, augmented and virtual reality (AR/VR), robotics, cyber security, semantic web technologies, and additive manufacturing (AM) [4], in order to improve products, processes, service and business models [5]. The technology categories have emerged as a result of technological innovation. Technological innovations via technology categories in Industry 4.0 have paved the way for prediction, control, maintainance and integration of production processes [6], [7]. However, the recent literature [8], [9], [10] recognized that the frugal innovations in Industry 4.0 are driven by several indicators, namely: the development of the economic environment, enterprise size, sustainability of furgal innovation, the degree of digital transformation in the environment in which the company operates, and so on [8], [9],[11].

Since the relevant literature lack of the observations of frugal innovations from the key indicators point of view in the Industry 4.0, this paper tries to fullfil this gap. That is done by indentifieing and analysing the indicators that influence the development of frugal innovation in Industry 4.0. The development of frugal innovation is based on the needs of the industry for continuous savings, but also on customer requirements for good and affordable products. To answer the question whether the frugal innovations have an impact on Industry 4.0, the paper analyzes the connection between Industry 4.0 and frugal innovation through key indicators. The research questions that this paper is answering are: How frugal innovation contribute to the better development of Industry 4.0 and which indicators influence the development of frugal innovations in Industry 4.0?

The rest of the article is organized as follows. Section 2 provides a theoretical background of innovations in Industry 4.0 and frugal innovations. Section 3 presents the identification of key indicatiors of frugal innovation in Industry 4.0. Section 4 discusses importance of key indicators of frugal innovation in Industry 4.0. Finally, Section 5 provides conclusion and future work.

2. Background

2.1. Frugal innovations

The history of frugal innovation is linked to Schumacher. He aimed to reduce class differences, not maximize profits. Therefore, its understanding and connection with new technologies have led to a reduction in class differences through the application of frugal innovations. According to Schumacher, the beginning of frugal innovations was associated with the transfer of technology from advanced countries to developing countries, and these innovations, in addition to savings, also contribute to the social goal [10]. However, the transfer of technology based on frugal innovations gained importance in the first decade of the 21st century [12].

Even though, the beginnings of research in the field of frugal innovation was related to developing countries, the recent researches of frugal innovation are recognized as a global potential for development in advanced countries as well. Since 2014, the European Union (EU) has acknowledged the potential of frugal innovations development described in their official document [13]. The EU recognizes the first generation of frugal innovation as an approach to global challenges for certain areas of the economy based on savings and contributing to society goals that rely on advanced technologies [8]. According to the EU [13] some of the features of frugal innovation are reduction of a unit cost, addition of a value of innovation and achievement of the acceptable performance of innovation. An important aspect of frugal innovation is to adjust the achieved savings of innovation and give value to society [14]. Frugal innovation does not necessarily represent the core activity of the company, but can relate to one area or department in the company. If we look at company management, frugal innovation can enhance a company's performance by managing a company based on savings [15] and cutting out business costs [16].

Within the second generation of definitions of frugal innovation, different authors suggest a different perspective and provide a process-oriented definition. In recent times, frugal innovations are seen as significant savings with minimal investments [17]. Finally, the third generation represents a breaking point for frugal innovation and identified three elements that characterize frugal innovation, both in emerging and developed markets: substantial cost reduction, concentration on core functionalities and optimized performance level [18]. Another important research identified three fundamental dimensions of constraint-based innovation: cost-effectiveness, ease-of-use, and prescriptive variables [19].

Frugal innovations can affect economic development by using and adapting developed IT technologies owned by advanced countries in underdeveloped countries. Based on the above, totally new innovative solutions can be developed in advanced countries, which can cause significant changes and lead to disruption of technology [20]. Thus, frugal innovations represent the potential for the development of innovative responsible business of companies. Responsible business of the company means the contribution that the company gives through solving a certain society problem. In addition, frugal innovations should offer innovative solutions based on savings in finance, human resources, or technology, which result in cheaper than the current solutions [2]. If we connect the frugal innovations with production then frugal innovations apply as a change in the way of thinking through the entire production process, i.e. product development. The impact of frugal innovation can be extended, so development of frugal innovation in production represents a change in the mindset of the entire business model [21].

The development of frugal innovations can be influenced by various factors in order to be successful. Successfulness of implemented innovation depends of its type. Types of innovation can be distinguished by the use of three indicators, namely sophistication, sustainability and market orientation [22]. Frugal innovation can contain several different types of innovations and most often frugal innovations can include economic innovations, social innovations and technological innovations or their combination [23]. However, the most important characteristic of all types of frugal innovations are savings. Thus, the definition of frugal innovations that is used in this paper focuses on saving, and reads: *frugal innovations are innovations that reduce the total cost of the product by eliminating an addition value of the product and opens different perspectives to industries regarding markets.*

2.2. Innovations in Industry 4.0

Industry 4.0 represents the concept of the fourth industrial revolution that is relevant to the manufacturing sector. Before Industry 4.0, the first three industrial revolutions took place. The first industrial revolution happened at the end of the 17th century, and was initiated by the appearance of steam engines, water power and mechanization. After that, the second industrial revolution was forced by assembly lines and the introduction of mass production by Henry Ford. The third industrial revolution was driven by the use of computers and automation in production processes. Finally, the fourth industrial revolution, known as Industry 4.0, is a concept coined and introduced by the German federal government as an innovative high-tech strategy in late 2011 [22].

The term 'revolution' in the Industry 4.0 concept suggests the existence of a differentiating patterns of product innovation, process innovation and business model innovation. This concept bases on an innovative approach and is driven by the development of digital technology. Digital technology in Industry 4.0 concept brings together the physical parts of a particular production process and the cyber parts through smart networking [24]. Physical parts of the production process consist of machinery, various devices and sensors, while smart networked cyber parts consist mainly of advanced software [24]. Therefore, the current technological benefits in companies are directly connected with Industry 4.0 concept [22].

The development of Industry 4.0 is conditioned by the difficulties of implementation of Industry 4.0 in practice. The challenges of practical implementation of Industry 4.0 represent the potential for the development of innovations in Industry 4.0 that can be divided into managerial and technological challenges [6]. With the development of Industry 4.0, a new form of innovation have developed. Before Industry 4.0, innovations in manufacturing sectors were focused on savings in automation and digitization of production processes. Also, innovations in organization have been developed, but Industry 4.0 has contributed to greater opportunities and a wider range of innovations in management. The possibility of developing innovations in Industry 4.0 depends on technological innovation capability, which includes not only engineerings' capability of utilizing new information technology and operation technology, but also innovation management capability of managing the process of linking market with technological development and strategy [5].

In order to identify all stakeholders in the analysis of frugal innovation and Industry 4.0, it is necessary to identify the framework for action and significant sources of frugal innovation in Industry 4.0. We can identify stakeholders who are interested in frugal innovation and we can divide them into institutions that develop the eco-system of frugal innovations, innovators involved in the process of development of frugal innovations and end users of frugal innovations. Stakeholders in the field of frugal innovation in the Industry 4.0 era could be grouped into:

- *Institutions and relations among ecosystem actors* – here we identify the role played by institutions, considering both the policy-making level, and the role of education and research institutions in the social context;
- *Innovators involved in the frugal innovation process* – where we distinguished two levels of innovators:
    - Firm-level innovators – innovation involving different types of organizations, such as small, micro, and medium-sized enterprises, as well as multinational enterprises;
    - Grassroots-level innovators – innovation involving individuals, informal enterprises, and local communities.
- *Consumers* – they play a key role as the final beneficiaries of frugal innovations [8].

Industry 4.0 is based on the digitization of industrial value creation, especially connecting vertical and horizontaapproaches in industry that are reflected through the networking of humans, machines, and products [25]. Vertical connectivity refers to the development and management of a company's activities, such as: organizational structure, human factor, departmental relations, technological and management level. Horizontal connectivity includes external relations, establishes the integration of supplier and customer networks, information and management systems[25]. Vertical and horizontal integration will allow industry to identifying gaps and find solution to implement Industry 4.0 concept [26].

The consequences of the Covid 19 pandemic during 2020 influenced the global usage of digital technology. Although, the usage of digital technology increased, the pandemic disabled the daily activities in production companies. This has had the effect of reducing the company's profits and thus the improvements of the company has been lagging behind. This lag slowed down the global development of the company and significantly affected the development of digital technology along with Industry 4.0 concept. Therefore, it is important to learn a lesson from this situation, and to improve the production system in the direction of digital technology development, where the influence of the human factor would be reduced, and the level of system automation would develop in the direction of "self-awareness", "self-predicting", "self-maintaining" and "self-learning" production [23].

3. Key indicators of frugal innovation in Industry 4.0

By their definition, the frugal innovations create savings and contribute to socio-economic development, while Industry 4.0 contributes to technological development. The analysis of literature shows that there is a connection between frugal innovations and Industry 4.0 [27], [28]. Their connection is reflected in the creation of savings. Savings are defined through key indicators for the development of frugal innovation in Industry 4.0. However, identifying indicators in the development of frugal innovation in Industry 4.0 is a challenge. The indicators of frugal innovation in Industry 4.0 represent a combination of advanced technologies and social/economics development. By searching the existing literature of the development of frugal innovations in Industry 4.0, we identified 7 indicators (Fig. 1), namely:

- artificial intelligence (AI),
- cyber physical system (CPS)
- industrial Internet of Things (IIoT),
- big data analytics (BDA),
- country development level,
- return of the investment (RoI), and
- a company size.

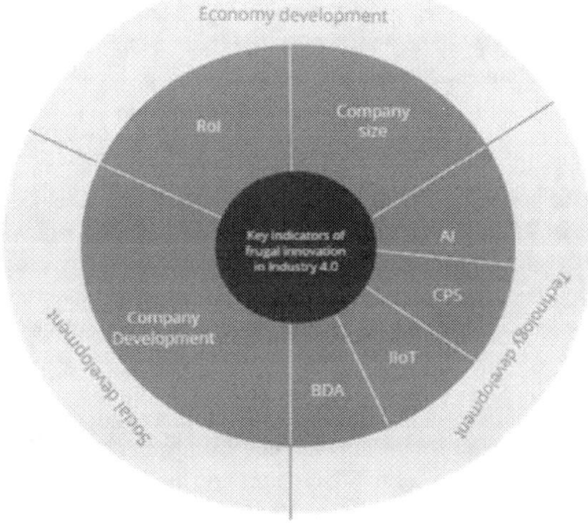

Fig. 1. Key indicators of frugal innovation in Industry 4.0 with their impact on development

Artificial intelligence – AI can be described as giving tasks to computers to solve them that would normally require human intelligence [29]. The member states of the EU are working together on AI to improve their competitive advantages, but AI can be used to advance underdeveloped countries as well. Artificial technologies can be applied as a tool for frugal innovation to solve the problems and challenges that companies are facing with [30]. Rapid development of AI within Industry 4.0 will affect the multiple innovations and creativity which results in health, robotics, computing technology and even automation of various sectors [31]. It can be concluded that AI is an indicator of the development of frugal innovation in Industry 4.0. The sensitivity of the development of frugal innovations within AI can lead to the development of disruptive technologies, i.e. disruptive innovation.

Cyber physical systems – CPS consist of both computational and physical components working together with the goal of implementation of a process monitoring and managing in real time. Thus, CPS is a very important indicator in the development of frugal innovation in Industry 4.0. If we analyze the potential for the development of frugal innovation through the development of CPS at the time of Covid 19, we can see that Industry 4.0 as a goal wants to achieve savings, which is also the goal of frugal innovation. The importance of the transformation of Industry 4.0 in influential innovations is related to the globalization of technologies and innovations, as well as to the interruptions that occur due to technological progress. In the rapid development of Industry 4.0, individualized demand and a small series of production are based on CPS and technology that lead to frugal innovations [32]. Technology and innovation management needs to find new ways related to business models based on savings, quasi-public goods, responsible earnings, digital ecosystems, platform solutions and relevance to competition [33].

Industrial Internet of Things – IIoT is an innovative technology consisting of applications that realize functions and services in a wide range of domains [34]. IIoT connects the physical world with the digital world [35]. IIoT is a collection of smart devices and objects that detect, collect, process and communicate in real time in industrial settings [36].

It leads to resource savings that have a connection with frugal innovation through digitalization and automation of industrial production, organization or process. Rapid development of IIoT, due to the specific requirements of the market and customers, is driven by savings that lead to the development of frugal innovation. However, in the development process there can be major changes, namely development of disruptive innovations.

Big data analytics (BDA) – big data can be defined as advanced software or mathematical model that has the function to analyze, process and extract information from complex and huge data sets. BDA is associated with three aspects of technical innovation and extremely large data sets: automated parallel computing, data management schemes, and data mining. [37]. The development of BDA has an impact on value creation and savings. The use of big data enables the company to develop efficient and effective innovations [38]. BDA can help company to collect and process market information to better understand consumers' preferences, which can play a critical role in innovation performance. Based on the link between innovation and the savings they make, BDA is an indicator for the development of frugal innovation in Industry 4.0. The ability to develop innovations in the company is one of the possibilities in the use of new sources, such as BDA that contribute to the improved performances of the company [39].

Country development level – indicator for the development of frugal innovation in Industry 4.0 is the level of the country development [40]. The development of frugal innovation depends whether the company is located in developed country or an underdeveloped country. The country's development directly affects the development of Industry 4.0 due to the rather limited development of advanced technologies and smart processes in underdeveloped countries. When a company's production and services move from developed country to underdeveloped country, it poses a threat to the competitiveness of cheap production in underdeveloped country based on cheap labor [5]. Certain research claims that the production revolution will affect the fact that less and less production in the future will be in underdeveloped countries [41]. The development of frugal innovations is influenced by various elements related to the level of country development in which frugal innovations are developed, such as socio-economic characteristics, institutional capacities and elements of the environment. The dependence of frugal innovations in advanced industry is directly related to the level of the country development. The specificity and development of a country that develops frugal innovation have a great influence on frugal innovations.[8]. The need for frugal innovation has been recognized in underdeveloped countries and in developed economies throw the innovation programs which have a focus on saving resource, availability, and sustainability [42].

Return of the investment – RoI represents an important indicator in Industry 4.0 which support sustainability of frugal innovation. High RoI is almost as likely to result from investments in innovation. The development of innovation in companies is often based on technological development. Evaluation of profitability of investments in technological development is realized through RoI analyzes [43]. Industry 4.0 is based on the technological development and sustainability of investments in development and innovation. According to [44] investments in technological development that contribute to innovation in 51% of cases have a great importance to companies. The investments in technological development that contribute to better productivity of the company are of great importance for the sustainability. Investments in technological development that contribute to innovation and productivity affect the development of frugal innovations in Industry 4.0.

Company size – One of the important indicator of the development of frugal innovation in advanced industries represents the company size that develops frugal innovations. Companies can be divided into micro, small, medium, multinational based on size.There are several factors that determine the company size and one of them is the number of employees in the company. Micro companies have up to 10 employees, small companies up to 50 employees, medium companies between 50 - 250 employees while large or multinational companies have over 250 employees. It is not the same if frugal innovation should be developed in a small company or it wants to be developed by a multinational company that has a special department for innovation development. Multinational companies often implement frugal innovations in developing countries. Based on the research we could identify three types of organizations that influence on frugal innovation: local/micro companies, small and medium enterprises, and multinational enterprises [9].

4. Discussion

Frugal innovations are much more discussed from the non-governmental organizations (NGOs) and government agencies perspectives rather than industry. The NGOs emphasize the importance of frugal innovations from the aspect of social savings. The social savings goals have been recognized by national institutions as the potential for practical improvements in society through frugal innovation. During 2020, the Austrian Development Agency piloted a program to support the development of frugal innovation and cooperation with underdeveloped countries (Program: Innovation for a Better World - Advanced Solutions for Sustainable Development). An important element of the Program for the development of frugal innovations is the obligation to include companies that must be the bearers of development.

However, the Creativity Era of business motivates drivers such as hyper-competition, technology disruption and new customer power. Therefore, companies are increasingly looking to achieve growth from new non-core areas, requiring more focus on innovation - especially breakthrough innovation - in order to survive and prosper [45]. Today, the industry bases its improvements on innovations that affect the development of all sectors. Such innovations include technological and non-technological innovations. The industies improvements based on concept of Industry 4.0 focuses on the development of technological innovation but the need for savings affects the development of non-technological innovation in some sectors, e.g. management in the company.

The existing literature does not provide enough information on the impact of frugal innovations in Industry 4.0. On the one hand, analysis of the existing literature and examples from the practice of frugal innovations disscusses that frugal innovations should enable more inclusive development processes and contribute to responsible business [46], [47]. On the other hand, the main goal of the implementation of the Industry 4.0 concept is improvement of the company.

However, improvements are achieved and realized through savings via innovations, where frugal innovations play a significant role in Industry 4.0. Therefore, viewed from this aspect frugal innovations represent a need for Industry 4.0 development. The significance of indicators regarding frugal innovations is their prediction abilities of future success of a company. Frugal innovations in Industry 4.0 are defined by indicators that contribute to technological, economic and social development that is presented in table 1. This paper recognizes the key indicators of technological development of frugal innovations in Industry 4.0 are: AI, CPS, IIoT, BDA while the social development indicator is country development level and economic development indicators are: RoI and a company size (Fig. 1). Defined key indicators of frugal innovation create value and lead to new trends in the development of Industry 4.0. The value of frugal innovations in Industry 4.0 is defined through the savings that innovation creates and new development trends arise from the synergy of technological, economic and social development that in some cases lead to disruption.

Key Indicators		*Frugal innovation in Industry 4.0*
Technological	**AI**	The impact of AI on frugal innovation in Industry 4.0 leads to disruptive innovation.
	CPS	The impact of CPS on frugal innovation in Industry 4.0 leads to development of new technology.
	IIoT	The impact IIoT driven by savings lead to disruptive innovation.
	BDA	BDA driven by savings leads to efficiency and effectiveness of the company.
Social	**Country Development**	The impact of Country development on frugal innovation in Industry 4.0 leads to new innovations programs.
Economic	**RoI**	The impact of RoI on frugal innovation in Industry 4.0 leads to greater investment in frugal innovation.
	Company size	The impact of company size in Industry 4.0 leads to different speeds of frugal innovation development.

Table 1. The impact of the key indicators of frugal innovation in Industry 4.0

5. Conclusion

The frugal innovations have the word innovation in their name, but they behave differently from traditional innovations. While traditional innovations represent improvements to existing solutions without additional impact, the frugal innovations must contribute to savings and represent an improvement over the existing solution. That can be achieved by thorough analysis of an existing solution with addition of an element of the positive cost-effective performances. Nowadays, positive cost-efective preformances are accomplished via advanced technologies of Industry 4.0 concept.

Companies' development of frugal innovations in business models based on Industry 4.0 concept, changes the way of thinking in addition to the innovation advantage. Frugal innovation contributes to the development and advancement of technology as it is often subject to transfer from developed countries to underdeveloped countries. The real profit of a company largely depends on the innovation in that company. In many ways, frugal innovations give the company the opportunity to make a profit in a responsible way based on resource and cost savings.

This paper is considering the impact of Industry 4.0 concept and frugal innovation as well as the importance of indicators that influence the development of frugal innovation in Industry 4.0. In this paper we argue that the Industry 4.0 concept needs frugal innovations in order for progress of savings development, improvement of existing processes and future industry development. The concept of Industry 4.0 in combination with frugal innovations simplifies operations, reducing operating costs, and making products and services available. That is achieved by identifying the key indicators of frugal innovation that enable Industry 4.0 concept. Thus, this paper identifies the key indicators of frugal innovation in Industry 4.0, namely: artificial intelligence (AI), cyber physical system (CPS), industrial Internet of Things (IIoT), big data analytics (BDA), level of the country's development, return of the investment, and company size. The indentified indicators are grouped into three groups depending of the acpect of their contribution of development, namely: technological, economic and social development. The indentified technological development inicators are AI, CPS, IIoT, BDA. The economic development indicators are RoI and a company size. And last, but not the least is social development indicator refering to country development level. Based on the identified key indicators and the aspetc of their contribution, we argue that frugal innovation could give a competitive advantage to the less developed companies over developed companies through savings. In order to develop frugal innovations in Industry 4.0, it is necessary to define an eco-system. An eco-system can be defined as a set of elements that have an impact on frugal innovation in Industry 4.0. Existing literature claims that frugal innovation ecosystems are including local firms, its suppliers, a network of customers and vendors, government agencies, business support services, logistics and delivery providers, alliance partners, and often even rival firms. Eventhough the frugal innovation eco-system is out of scope of this research, it represents a very important aspect of frugal innovation development. Therefore, in future work, an eco-system for the development of frugal innovations in Industry 4.0 will be developed. Frugal innovations in Industry 4.0 represent development potential and a multidimensional phenomenon that affects products, processes, service and business models. The phenomenon of frugal innovation in Industry 4.0 characterizes the development of savings in an innovative way.

Savings-based innovations represent a great potential for Industry 4.0, but an ecosystem is needed for their development. The ecosystem for the development of furgal innovations in Industry 4.0 in future works will be confirmed from a real-life case study.

6. References

[1] Dahlin, K. B. & Behrens, D. M. (2005). When is an invention really radical?: Defining and measuring technological radicalness, Research Policy, Vol. 34, No. 5, (June 2005), pp. 717-737, 00487333 [2] Simula, H.; Hossain, M. & Halme, M.(2015). Frugal and reverse innovations - Quo vadis?, Vol. 109., (September 2015) pp. 1567–1572, ISSN: 00113891

[3] Radjou, N. & Prabhu, J. (2015). Frugal Innovation: How to do more with less, Profile Books Limited, 1781253757, 9781781253755, 29 Cloth Fair London, United Kingdom

[4] Kamble, S. S.; Gunasekaran, A. & Gawankar, S. A. (2018). Sustainable Industry 4.0 framework: A systematic literature review identifying the current trends and future perspectives, Process Safety and Environmental Protection, Vol., 117., (July 2018) pp. 408-425, 09575820

[5] Lim, C.; Lee, J. H.; Sonthikorn, P. & Vongbuyong, S. (2020). Frugal innovation and leapfrogging innovation approach to the Industry 4.0 challenge for a developing country, Asian Journal of Technology Innovation, (2020) pp. 1-22, ISSN: 19761597

[6] Lu, Y. (2017). Industry 4.0: A survey on technologies, applications and open research issues, Journal of Industrial Information Integration, Vol., 6., (June 2017) pp. 1-10, 2452414X

[7] Bajic, B.; Rikalovic, A.; Suzic, N. & Piuri, V. (2020). Industry 4.0 Implementation Challenges and Opportunities: A Managerial Perspective, IEEE Systems Journal, ISSN 1932-8184, pp. 546-559, IEEE Systems Journal · March 2021, 10.1109/JSYST.2020.3023041

[8] Prabhu, J. & Jain, S. (2015). Innovation and entrepreneurship in India: understanding jugaad, Asia Pacific Journal of Management, (November 2015) pp. 843- 868, 02174561

[9] Neal, R. (2018). Funding Frugal Innovation in Global Health: Philanthropy, Aid, and Industry, In: Frugal Innovation in Bioengineering for the Detection of Infectious Diseases, Chavali, A.K. & Ramji, R. (Ed.), pp. 99-112, 978-3-319-66645-7, Springer International Publishing AG, Switzeland

[10] Rosca, E., Reedya, J. & Bendul, J. C. (2018). Does Frugal Innovation Enable Sustainable Development? A Systematic Literature, Review European Association of Development Research and Training Institutes, The European Journal of Development Research, Vol. 30, No. 1 , (Januar 2018) pp. 136-157, ISSN: 09578811

[11] Pisoni, A. B.; Michelini, L. & Martignoni, G. (2018). Frugal approach to innovation: State of the art and future perspectives, Journal of Cleaner Production, Vol. 171 , (January 2018) 107-126, ISSN: 09596526

[12] Kaplinsky, R (2011). Schumacher meets Schumpeter: Appropriate technology below the radar," Res. Policy, vol. 40, pp. 193–203, doi: 10.1016/j.respol.2010.10.003

[13] McGrath, C. H.; Horvath, V.; Baruch, B.; Gunashekar, S.; Lu, H.; Culbertson, S.; Pankowska, P. & Chataway, J. (2014). The international dimension of research and innovation cooperation addressing the grand challenges in the global context, European Commission, 2014. RR-725-EC

[14] Chaisung, L. & Takahiro, F. (2019). Frugal innovation and design changes expanding the cost-performance frontier: A Schumpeterian approach, Research Policy, Vol. 48, No. 4, (May 2019), pp. 1016-1029, 00487333.

[15] Bound, K. & Thornton, I. (2012). Our frugal future: lessons from India's innovation, Available from: https://media.nesta.org.uk/documents/our_frugal_future.pdf Accessed: 2012-07-01

[16] Tjahjana, D.; Manurung, A. H.; Setiadi, N. J. & Kosasih, W. (2020). Innovation, Digital Business and Frugal Innovation Type, European Journal of Molecular & Clinical Medicine Vol. 07., (July 2020) pp. 820-831, ISSN: 2515-8260

[17] Lozanoa, C. V. & Vijayana, K. K. (2020). Literature review on Cyber Physical Systems Design, Procedia Manufacturing, Vol. 45, (April 2020), pp. 295-300, ISSN: 2351-9789

[18] Sean, T. (2012). High-speed rail, the knowledge economy and the next growth wave, Journal of Transport Geography, Vol. 22 (May 2012), pp. 285-287, ISSN: 09666923

[19] Agarwal, N. & Brem, A. (2017). The Frugal Innovation Case of Solar-powered Automated Teller Machines (ATMs) of Vortex Engineering in India, Journal of Entrepreneurship and Innovation in Emerging Economies, Vol. 3, No. 2, (October 2017), pp. 115-126, ISSN: 2393-9575

[20] Jucevicius, G.; Sajeva, S. & Stankevice, I. (2012). Patterns of Strategic Innovation in the Latecomer Economies: Beyond the Frugality, International Conference on Business Management & Information Systems, No. 1, (January 2012), 2164-263X

[21] Soni, P. & Krishnan, R.T. (2014). Frugal innovation: aligning theory, practice, and public policy, Vol. 6, No. 1 , (Januar 2014) pp.29-47, ISSN: 1755-4195

[22] Brem, A. & Wolfram, P. (2014). Research and development from the bottom up - introduction of terminologies for new product development in emerging markets, Journal of Innovation and Entrepreneurship, Vol. 3, No. 1, (December 2014) pp. 3-9, ISSN: 21925372

[23] Ahuja, S & Chan, Y. (2016). Digital Innovation: A Frugal Ecosystem Perspective, Proseeding of Thirty Seventh International Conference on Information Systems 11-14. December 2016, Dublin Ireland, 978-099668313-5,

Agerfalk, P. J.; Levina, N. & Kien, S. S. (Ed.) pp. 1-22, Association for Information Systems, Dublin, https://aisel.aisnet.org/icis2016/

[24] Nikolic, B.; Ignjatic, J.; Suzic, N.; Stevanov, B. & Rikalovic, A. (2017). Predictive manufacturing systems in Industry 4.0: Trends, benefits and challenges, Annals of DAAAM and Proceedings of the International DAAAM Symposium, 17269679

[25] Müller, J. M. & Däschle, S. (2018). Business model innovation of Industry 4.0 solution providers towards customer process innovation, Processes, Vol. 6, No. 12, (December 2018), pp. 1-19, ISSN: 222797177

[26] Schumacher, A.; Erol, S. & Sihn, W. (2016). A Maturity Model for Assessing Industry 4.0 Readiness and Maturity of Manufacturing Enterprises," Procedia CIRP, vol. 52, pp. 161–166, 2016, doi: 10.1016/j.procir.2016.07.040.

[27] Himang, C. (2020). Defining stages of the Industry 4.0 adoption via indicator sets. Engineering Management in Production and Services, Vol. 12, No. 2 (June 2020) pp. 32–55, ISSN 2543-6597

[28] Švarcová, J.; Urbánek, T.; Povolná, L. & Sobotková, E. (2019) Implementation of R&D Results and Industry 4.0 Influenced by Selected Macroeconomic Indicators, Vol. 5 (May 2019) pp. 2-14, ISSN 2076-3417.

[29] Hofmann, E. & Rüsch, M. (2017). Industry 4.0 and the current status as well as future prospects on logistics, Computers in Industry, Vol. 89, (August 2017), pp. 23-34, ISSN: 01663615

[30] Lucci, S. & Kopec, D. (2016). Artificial intelligence in the 21st century : a living introduction, Mercury Learning and Information, ISBN: 9781523101320, Dulles, Virginia

[31] Masanja, N. M. & Mkumbo, H. (2020). The Application of Open Source Artificial Intelligence as an Approach to Frugal Innovation in Tanzania, International Journal of Research and Innovation in Applied Science (IJRIAS), Vol. 5 No. 3 (March 2020), pp.36-46, ISSN 2454-6194

[32] Kamble, R. & Shah, D. (2018). Applications of artificial intelligence in human life, International journal of research, Vol. 6 No. 6 (June 2018) pp. 178-188, ISSN: 2394- 3629

[33] Lua, C.; Changb, F.; Rongc, K.; Shi, Y. & Yu, X. (2020). Deprecated in policy, abundant in market? The frugal innovation of Chinese low-speed EV industry, International Journal of Production Economics. Vol. 225, (July 2020), pp. 1-14, ISSN107583

[34] Tiwari, R. & Buse, S. (2019). Managing Innovation in a Global and Digital World: Meeting Societal Challenges and Enhancing Competitiveness, SprignerGabler, 978-3-658-27241-8, Switzerland

[35] Perera, C., Zaslavsky, A., Christen, P., & Georgakopoulos, D. (2013). Context aware computing for the internet of things: A survey. IEEE communications surveys & tutorials, Vol. 16, No. 1,(May 2013), pp. 414-454, 1553-877X.

[36] Lampropoulos, G.; Siakas, K. & Anastasiadis, T. (2019). Internet of things in the context of Industry 4.0: an overview, International Journal of Entrepreneurial Knowledge, Vol. 19, No. 1, (January 2019), pp. 4-19, ISSN: 2336-2952

[37] Liao, X.; Faisal, M.; Chang, Q. Q. & Ali, A. (2020). Evaluating the Role of Big Data in IIOT-Industrial Internet of Things for Executing Ranks Using the Analytic Network Process Approach. Hindawai, Vol.2020, (October 2020), pp.1-7 , 8859454

[38] Khalandar, S. A.; Muzamil, S.; Vincent, D. R. & Rajput D. S. (2019). Chapter 11 - Challenges in Storing and Processing Big Data Using Hadoop and Spark in book: Deep Learning and Parallel Computing Environment for Bioengineering Systems, Sangaiah, A. K. (August 2019) pp. 179-187, Elsevier Inc. 978-0-12-816718-2, St. Louis, Missouri

[39] Ghasemaghaei, M. & Calic, G. (2020). Assessing the impact of big data on firm innovation performance: Big data is not always better data, Journal of Business Research, Vol. 108, (January 2020), pp. 147-162, ISSN: 0148-2963

[40] Venkatraman, N. V. ; El Sawy, O. A.; Pavlou, P. A. & Bharadwaj A. (2014). Theorizing Digital Business Innovation: Platforms and Capabilities in Ecosystems, Available at SSRN: https://ssrn.com/abstract=2510111 2014-09-01

[41] OECD (2017). The Next Production Revolution: Implications for Governments and Business, Avilable from: https://www.oecd-ilibrary.org/science-and-technology/the-next-production-revolution_9789264271036-en Accessed: 2017-05-10

[42] Prahalad, C.K. & Mashelkar, R.A. (2010). Innovation's Holy Grail, Harvard Business Review, Vol., 88., (July 2010). pp. 1-11, ISSN: 00178012.

[43] Roulstone, B. & Phillips, J. J. (2008). ROI for Technology Projects, Butterworth Heinemann, ISBN: 9780-0-7506-8588-7, London

[44] Rao, A.; Kaji, J.; Hurley, B.; Bhat, R. &Khan, A. (2018). The Industry 4.0 paradox Overcoming disconnects on the path to digital transformation, Avaliable from: www.deloitte.com/insights Accessed: 2021-01-18

[45] Lenarduzzi, M. & Taibi D. (2016). MVP Explained: A Systematic Mapping Study on the Definitions of Minimal Viable Product, Euromicro SEAA, (August 2016), pp.1-8, ISSN: 2376-9505

[46] Granqvist, K. (2017). Policy brief: Funding frugal innovation Lessons on design and implementation of public funding schemes stimulating frugal innovation, INDIGO Policy , Austria

[47] Baud, I. (2016). Moving Towards Inclusive Development? Recent Views on Inequalities, Frugal Innovations, Urban Geo-Technologies, Gender and Hybrid Governance, European Journal of Development Research, Vol. 28, No. 2, (April 2016), pp. 119-129, ISSN: 17439728

DOI: 10.2507/32nd.daaam.proceedings.038

INFLUENCE OF MOISTURE CONTENT ON THE STRENGTH OF FINGER JOINTS BONDED BEECH WOOD

Alen Ibrisevic, Murco Obucina, Seid Hajdarevic & Goran Mihulja

This Publication has to be referred as: Ibrisevic, A[len]; Obucina, M[urco]; Hajdarevic, S[eid] & Mihulja, G[oran] (2021). Influence of Moisture Content on the Strength of Finger Joints Bonded Beech Wood, Proceedings of the 32nd DAAAM International Symposium, pp.0270-0274, B. Katalinic (Ed.), Published by DAAAM International, ISBN 978-3-902734-33-4, ISSN 1726-9679, Vienna, Austria
DOI: 10.2507/32nd.daaam.proceedings.038

Abstract

Wood extension is a technological process that has enabled better use of solid wood, reaching longer element lengths and greater stability of solid wood shapes. The paper examines the influence of humidity and adhesive types on the bending strength and modulus of elasticity of the finger joint beech wood. Samples of beech wood of different moisture were lengthwise joined using polyvinyl acetate (PVAC) and polyurethane (PU) adhesive. The bending strength and modulus of elasticity test was performed on a testing machine, and the obtained results were statistically processed by analysis of variance method (ANOVA). The test results showed that when glued in the optimal humidity range of 10-12%, the best results of bending strength and modulus of elasticity are obtained, while with increasing humidity, the mechanical properties of the joint are reduced. Adhesive bonding with PVAC adhesive gives better results of bending strength and modulus of elasticity compared to bonding with PU adhesive.

Keywords: bending strength; modulus of elasticity; moisture; adhesive

1. Introduction

The lack of quality of technical wood as a raw material for the wood industry is evident and is present all over the world. Therefore, wood extension techniques to be more sophisticated, in order to increase the utilization of sawn timber as well as the dimensional stability and strength of extended wood. The most common way to extend wood is to use a finger joint.

Wood extension is a technological process that has enabled better use of solid wood, reaching longer element lengths and greater stability of solid wood shapes. Wood extension is applied in various constructions and products, such as furniture, construction joinery, various laminated beams and solid boards. Longitudinal joining of wood is most often used in the production of laminated beams [1]. In general, the joint profile and the adhesive used play an important role in the joint strength [2] [6] [7]. The joint profile has the greatest influence on the joint strength. The main geometric characteristics of the finger joint are step (p), length (L), angle (s) and width (T) [9] (Figure 1). The joint strength is a mechanical property that significantly affects the strength of the entire structure. Each structure is exposed to different stresses during operation. The most common stress that occurs in structures is the bending stress [8].

Therefore, the bending strength of the finger joint for example in structural finger jointed solid timbers is the most important mechanical property. The joint strength is affected by several factors, primarily the construction of the joint and the adhesive used. The joint strength depends on the cohesion of the adhesive and the internal stresses in the joint [10]. Polyvinyl acetate adhesive (PVAC) and polyurethane (PU) adhesive are commonly used to extend the wood used to make furniture.

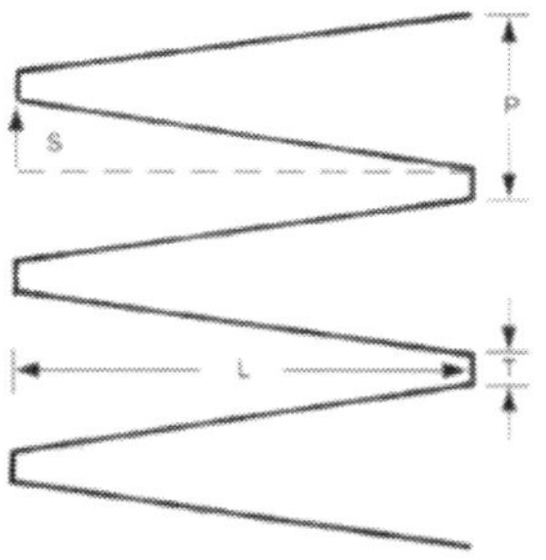

Fig, 1. Geometric characteristics of the finger joint

In addition, climatic factors, such as moisture content and temperature, have a significant impact on product performance [3] [4]. The moisture content in the wood is an important factor that affects the strength of the joint, because some adhesives are affected by the moisture content of the wood [3] [5] [6]. Humidity of wood in wood products intended for indoor use must be dried to moisture content in the range from 8-12%. If the wood is used for outdoor use, where average air humidity is higher, then the moisture content of the wood used for gluing should also be higher. It is necessary to avoid increased joint stresses due to dimensional changes of the wood.

The aim of this paper is to examine the influence of moisture and type of adhesive on the bending strength and modulus of elasticity of finger joints beech wood.

2. Materials and methods

The first part of the experiment involved is making samples. The elements were made of slighty steamed beech wood (Fagus sylvatica L), without visible defects. The dimensions of the elements were 265x42x31 mm, made in a combination of radial and tangential cut. The elements were planned on a four-sided planer "Weinig Profimat 26" with displacement speed of 15 m / min.

The finger joint elements for longitudinal jointing were made on a "Weinig Grecon Ultra 4" machine. The finger joint profile is shown in Figure 2. The joints were made in the way they are usually done in production of boards with length and width assembly. It is important to notice the tree construction parts/zones of that finger joint. There are three fingers placed in the middle of joint, while in the upper and lower part there is a zone of butt gluing. In this way, the finger joint is hidden, and only a glue line is visible on the product surface.The elements were then bonded with polyvinyl acetate (PVAC) adhesive (AkzoNobel 3311) or with polyurethane (PU) adhesive (Jowatherm-Reactant MR 607.90). Joining was done in such a way that the two individual lamellas were joined together so the final samples measured 530x42x31 mm (Figure 2).

Totally there was 48 bonded samples, 25 samples were bonded with PVAC, and 23 samples were bonded with PU. Samples were prepared in 5 humidity classes: 9, 10, 11, 12, 14%, and wrapped in self-adhesive foil to prevent receiving and releasing of moisture. Moisture content was determinate by the gravimetric method.

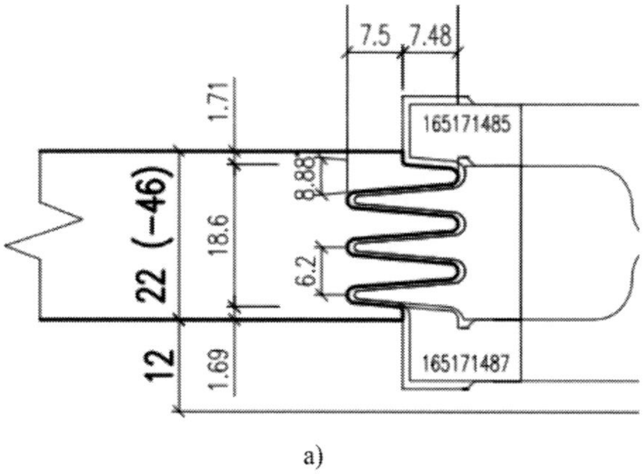

a)

b)

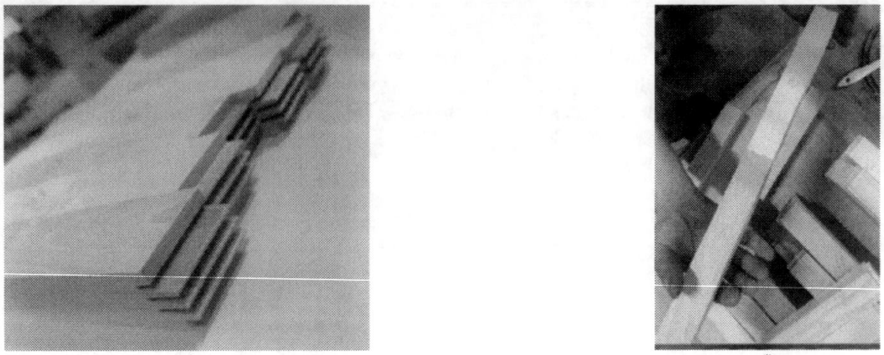

c) d)

Fig. 2. a) joint geometry, b) tool, c) joint layout, d) test

The second part of the experiment was done in the laboratory located at the Faculty of Mechanical Engineering in Sarajevo. In this phase, the bending strength and modulus of elasticity were tested on samples with different moisture content bonded with PU and PVAC adhesive. The test was performed according to the standard BAS EN 408. The testing was done on "ZWICK + CO.KG Einseingen Uber Ulm Type 1282" machine with connected force and displacement sensors run by CATMAN software (Figure 3). The distance between the supports was 465 mm, while the distance between the presses was 186 mm. The test was performed at a temperature of 20 °C and a relative humidity of 65%. The obtained results were processed by the analysis of variance method.

Fig. 3. Bending strength test

3. Results and discussion

Results of bending strength according to BAS EN 408 standard and moisture content obtained by gravimetric method are presented in Figure 4.

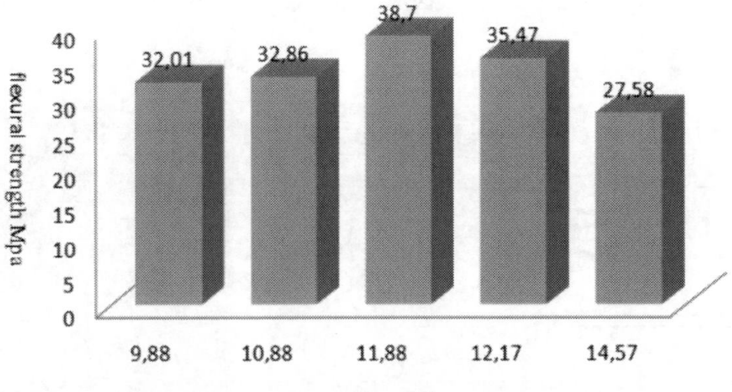

Fig. 4. Bending strength depending on wood moisture (PVAC bonds)

Results shows that the highest strength is obtained at moisture content of 11.13%, which is in the range for optimal gluing moisture content of 8-12% according to literature. With an increase in moisture content over 12%, the bending strength decreases. Results were additionally processed by the method of analysis of variance (ANOVA), which showed that there is a statistically significiant difference in bending strength with increasing moisture content (Table 1).

Source	DF	Adj SS	Adj MS	F-Value	P-Value
Sample	4	331,4	82,85	7,71	0,001
Error	20	214,9	10,75		
Total	24	546,3			

Table 1. Analysis of variance of bending strength of samples with PVAc adhesive

In the case of samples bonded with polyurethane adhesive, the results of bending strength are shown in Figure 5. Figure 5 shows that the highest bending strength is present in samples with a moisture content of 10.47%. It is also visible that best results are obtained in the limits of optimal moisture content of 8-12%, while with the increase of moisture content over 12%, there is a decrease in bending strength. The results were additionally processed by the method of analysis of variance (Table 2), which shows that there is a statistically significant difference between individual samples.

Source	DF	Adj SS	Adj MS	F-Value	P-Value
Sample	4	191,43	47,857	8,92	0,000
Error	18	96,57	5,365		
Total	22	288,00			

Table 2. Analysis of variance of bending strength for samples bonded with PU adhesive

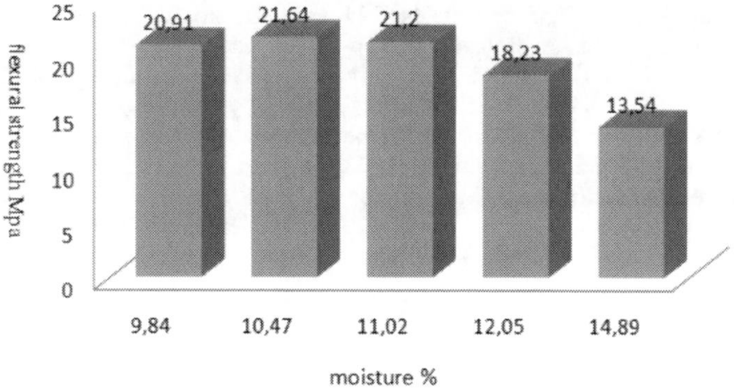

Fig. 5. Bending strength depending on wood moisture (PU bonds)

Based on the presented results, it is evidently that the samples bonded with PVAC adhesive have a higher bending strength compared to the samples bonded with PU adhesive. The maximum bending strength of PVAC samples are 44% higher than the maximum bending strength of PU samples. Based on the graph and data processing, it can be concluded that in the same moisture range, both PVAC and PU adhesives give the best bending strength results (moisture around 11%). Table 3 gives comparative analyses of bending strength and modulus of elasticity for samples bonded with PVAC and PU adhesives. It can be noticed that, the values of bending strength are higher in samples with PVAC adhesive compared to PU adhesive. It is also evident that the type of adhesive does not affect the change of the modulus of elasticity, because the values of the modulus of elasticity are approximately the same regardless of the type of adhesive. The result of the influence of the moisture content on modulus of elasticity largely coincides with the result of its influence on bending strengths.

Adhesive	Moisture (%)	9	10	11	12	14
PVAc	Bending strength (Mpa)	32,01	32,86	38,7	35,47	27,58
	Modulus of elasticity (GPa)	1,93	2,12	2,65	2,78	2,41
PU	Bending strength (Mpa)	20,91	21,64	21,2	18,23	13,54
	Modulus of elasticity (GPa)	2,18	2,36	2,6	2,77	2,59

Table 3. Comparative values of bending strength and modulus of elasticity

Experimental testing of bending strength and modulus of elasticity yielded small values general. The values obtained are probably the result of the joint profile. Practically, only three fingers were made in by the thickness value, and a large gluing surface was not achieved. On the other hand, the area of the butt adhesive bonding is quite large compared to total bonding surface, and consider it low bonding strength it expected that it is going to is reflected in the lower value of the bending strength and modulus of elasticity of complete joint.

4. Conclusion

Testing the bending strength on samples with PVAC and PU adhesive at different wood moisture content showed that the moisture in wood greatly affects the joint strength. Wood glued within the optimum humidity range of 10-12% showed the highest values of bending strength. When the moisture content increases over 12%, the strength of the joint decreases.

Finger joints bonded with PVAC adhesive, have a higher joint strength compared to elements bonded with PU adhesive with the difference in strength of 44%. If we take into account that PVAC adhesive has a much lower price compared to PU adhesive, then we come to the conclusion that finger joints with PVAC adhesive gives a more competitive product. The value of the modulus of elasticity is not affected by the type of adhesive, while moisture affects it in the same way as the bending strength. The highest values of the modulus of elasticity are obtained at a wood moisture content of 10-12%, while with an increase in humidity over 12%, the modulus of elasticity changes.

The results obtained small values of bending strength and modulus of elasticity than expected. These small values are for the most part influenced by geometry (joint profile).

5. References

[1] Frangi, A; Bertocchi, M ; Clauss, S; Niemz, P. (2012). Mechanical behaviour of finger joints at elevated temperatures, Wood Science and Technology, Vol. 46, No.5, 793-812, DOI: 10.1007/s00226-011-0444-9

[2] Kumar, V.S.K; Sharma, C.M; Gupta, S (2015). Compression and flexular properties of finger jointed mango wood sections, Maderas-Ciencia y Technologia, Vol.17, No.1, 151-160, DOI: 10.4067/S0718-221X2015005000015

[3] Liu, J; Yue, K; Wang, F; Wu, JH; Tang, ZQ;Chen, ZJ; Lu, WD; Liu, WQ (2020). Effects of Moisture Content on Lap-shear, Bending, and Tensile Strength of Lap-jointed and Finger-Jointed Southern Pine using Phenol Resorcinol Formaldehyde and Melamine Urea Formaldehyde, BioResources, Vol.15, No.2, 3534-3544, DOI: 10.15376/biores.15.2.3534-3544

[4] Liu, J; Yue, K; Xu, LQ; Wu, JH; Chen, ZJ; Wang, L; Liu, WQ; Lu, WD. (2020).Bonding performance of melamine-urea–formaldehyde and phenol-resorcinol– formaldehyde adhesive glulams at elevated temperatures, International Journal of Adhesion and Adhesives, Vol. 98. DOI10.1016/j.ijadhadh.2019.102500

[5] Bomba, J., Šedivka, P., Böhm, M., and Devera, M. (2014). Influence of moisture content on the bond strength and water resistance of bonded wood joints, BioResources Vol. 9, No.3, 5208-5218. DOI: 10.15376/biores.9.3.5208-5218

[6] Máchová, E., Langová, N., Réh, R., Joščák, P., Krišťák, Ľ., Holouš, Z., Igaz, R., and Hitka, M. (2019). Effect of moisture content on the load carrying capacity and stiffness of corner wood-based and plastic joints, BioResources, Vol.14, No.4, 8640-8655. DOI: 10.15376/biores.14.4.8640-8655

[7] Fortuna, B; Azinovic, B; Plos, M; Suligoj, T; Turk, G. (2020). Tension strength capacity of finger joined beech lamella, EUROPEAN JOURNAL OF WOOD AND WOOD PRODUCTS, Vol.78, No.5, 985-994, DOI 10.1007/s00107-020-01588-9

[8] Ibrisevic,A; Obucina,M; Hajdarevic,S.(2019). Effect of Particleboard Coating on Flexual Strength and Modulus of Elasticity, Proceedings of the 30th International DAAAM Symposium "Intelligent Manufacturing & Automation", Zadar, ISBN 978-3-902734-xx-x, ISSN 1726-9679, Katalinic, B. (Ed.), Published by DAAAM International, Vienna, Austria, DOI: 10.2507/30th.daaam.proceedings.124

[9] Barboutiss, I; Vassiliou, V.(2019). Strength of finger-jointed beech wood (Fagus sylvatica) constructed with small finger lengths, ProLigno, Vol.9, No.4, 359-364, ISSN 2069-7430

[10] Obucina, M; Gondzic, E; Smajic, S.(2014). The Influence of Amount of Layer on the Bending Strenght by Longitudinal Finger-Jointing Wood Elements, Procedia Engineering. Vol.69, 1094-1099

DOI: 10.2507/32nd.daaam.proceedings.039

DETERMINING THE GEOMETRICAL ELEMENTS IN THE PHOTOGRAMMETRIC METHOD FOR MEASURING LINEAR DIMENSIONS

Branko Sotirov, Danko Tonev,
Svetlin Parvanov & Miroslav Kokalarov

This Publication has to be referred as: Sotirov, B[ranko]; Tonev, D[anko]; Parvanov, S[vetlin] & Kokalarov, M[iroslav] (2021). Determining the Geometrical Elements in the Photogrammetric Method for Measuring Linear Dimensions, Proceedings of the 32nd DAAAM International Symposium, pp.0275-0280, B. Katalinic (Ed.), Published by DAAAM International, ISBN 978-3-902734-33-4, ISSN 1726-9679, Vienna, Austria
DOI: 10.2507/32nd.daaam.proceedings.039

Abstract

This paper presents the specific characteristics of the photogrammetric measurement method applied in industry. The focus of the study are linear and angular dimensions of mechanical parts from heavy industry. The definitions of sizes to be determined using the photogrammetric measurement method are based on the international standards ISO 14405-1:2016 and ISO 286-1:2010 dealing with geometrical product specifications. The main objective is to define the algorithms for processing the results obtained by the photogrammetric method, and to use the method without the need of a specialized software which is costly and limits the applicability of the method in small and medium-sized companies.

Keywords: photogrammetry; photogrammetric measurement method; the direct global associated size; direct global size.

1. Introduction

Photogrammetry is an applied science used widely in various areas of metrological practice – construction, architecture, geology, agriculture and forestry, mapping, mechanical engineering, historic research, documentaries, etc. It is known that some of the most critical and frequently controlled parameters in heavy manufacturing are the deviations of linear and angular features of size (dimensions), form, location and orientation of surfaces and axes.

Even though the photogrammetric measurement method (PMM) is relatively new in respect to its application in mechanical engineering, it widens its application in manufacturing technologies and reengineering, where it is used for measuring small and very small, as well as large (more than 500 mm) measured objects. It is known that there exist specialized methods, equipment and software for such measurements. They develop continuously but currently their high prices present a serious obstacle to their distribution in Bulgaria [5], [6]. It is established that when controlling small dimensions of large structures, the total measurement uncertainty shall not exceed 10 ÷ 20 μm, whereas for large dimensions it may be several times greater. At the same time, there is no sufficient information about the methodology for processing measurement information related to determining the necessary geometrical elements and parameters of the measured objects.

This methodology is directly related to the accuracy and reliability of the results obtained and immediately affects the measurement quality of the photogrammetric method, therefore, affects the quality of the assessed products.

2. Linear distance in the photogrammetric measurement methods

The photogrammetric measurement method (PMM) is classifies within the group of contactless methods that best illustrate (visually) terms such as 'integral feature', 'real feature', 'extracted integral feature', 'associated integral feature' [4]. In the case of PMM, these terms characterize the existence and the type of a specific repeating fragment in the image. The subjective factor plays a significant role in the interpretation of the limits/contours of measured objects made of different materials (metallic and non-metallic).

The presence of a texture that characterizes the existence and the type of a specific repeating fragment in the image depends on the sizes of identical objects that produce it and on the scale of the image, respectively. This allows to differentiate objects that cannot be defined otherwise, because practically they blend into the background. Fig. 1a presents a fragment of the contour of a metallic object, shot a close range, and Fig. 1b - presents a fragment of the contour of a non-metallic object, both images magnified 2400 times. The figure clearly demonstrates that even though under normal magnification a line theoretically is described, in the worst case, by 3 px, in reality, especially for non-metallic materials, this becomes more complicated. When the photographic image is greatly enlarged, the constituent pixels can be clearly seen.

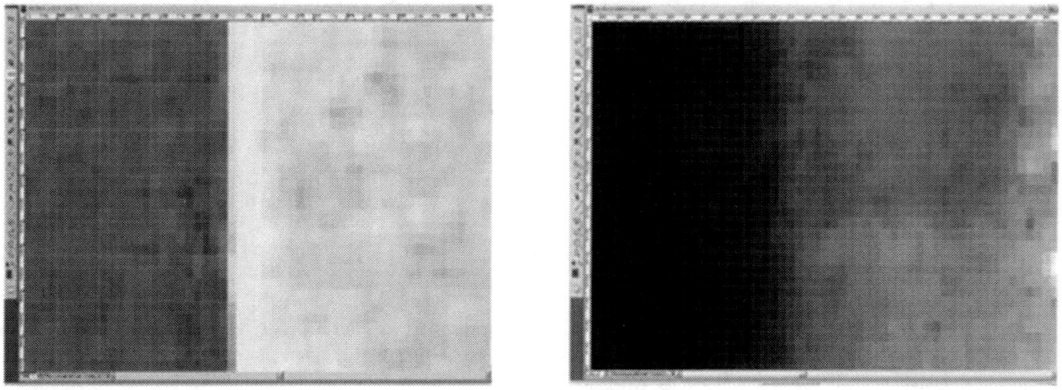

a) b)

Fig. 1. A fragment of the contour of: a) a metallic object, and b) a non-metallic object

The main reason for this fact is the circumstance that the dimension, determined using PMM, and respectively 'industrial photogrammetry', is the result of a quantitative piece of information, obtained from a multitude of pixels of a photographic image of the measured object. This finite number of pixels determines the extracted integral feature, obtained by taking a photo. It is known that the photogrammetric determining of geometrical dimensions is a combination of two activities – taking a photo and measuring. In this sense, the dimension determined using the PMM has to be defined as a direct global size, obtained **as associated in a direct measurement method** [1], [3], [5] (Fig. 2).

In certain isolated cases, the discussed dimension conditionally could also be defined as indirect global size, calculated from a homogeneous group of local dimensions. The main problem in this case is to justify the definition of local dimensions, obtained by using PMM. The **direct global associated size** (DGAS) is a geometrical feature of a nominal surface approximating (associated with) a real surface. Therefore, DGAS can be placed within the material of the measured object or outside of it, whereby the nominal surface is enveloping the real profile. It is a unique measurement and assessment result that eliminates the influences of the form, location and orientation of the object [2].

In the first case, the DGAS is a mean feature obtained from the extracted integral feature (profile) using the least-squares association criteria. It is known that this method provides stable and easily reproducible results [2]. In the second case, the DGAS can be obtained as the maximum inscribed size, the minimum circumscribed size or as a minimax size of the extracted integral feature (Fig. 2). It is known that the envelope requirement is the ideal form for maximum material sizes. The choice between the two approaches is determined mainly by the exploitation purpose of the measured surface, the requirements and/or the possibility for assembly, etc.

3. Specifics of determining sizes in the photogrammetric measurement method

The specifics of determining sizes are defined by the geometrical form of the measured object. In mechanical engineering, the generalized form of object in respect to their measurement can be cylindrical, cuboid or, relatively less frequently – spherical form, regardless of their size- small, average or large. Any of the generalized cases mentioned above has features that are strictly characteristic of it.

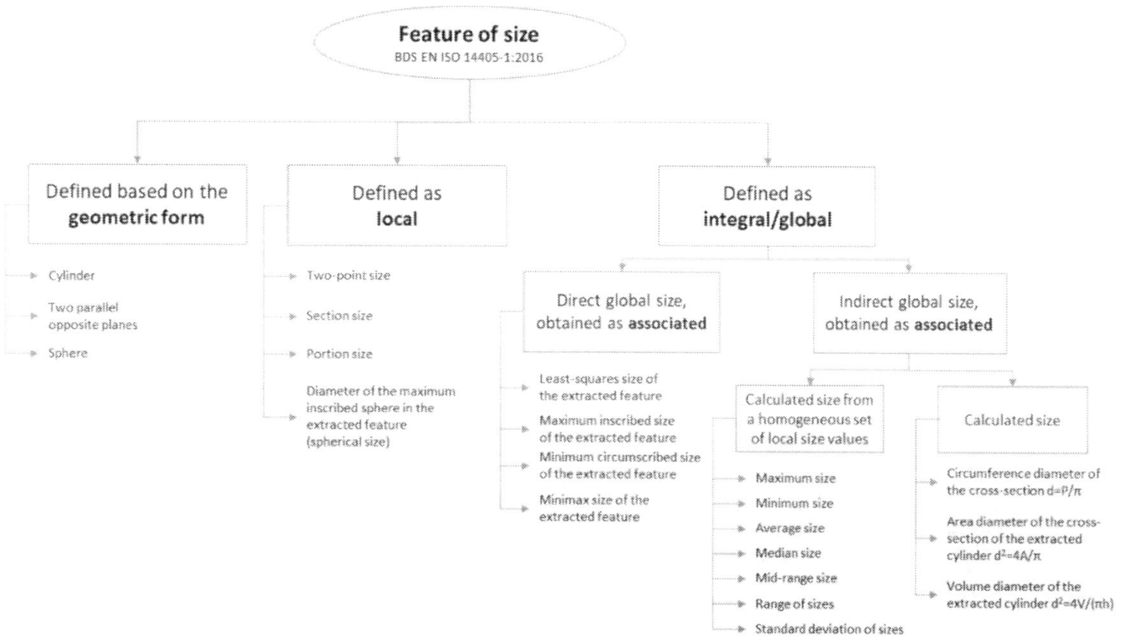

Fig. 2. Defining of features of size

3.1. Specifics of developing the contour description of the image in the photogrammetric measurement method

As mentioned in item 2, PMM is a contactless method that is applied on a photograph (image) of the object/part having a reading resolution of 1 px by using software. The developing of a contour description of the image can be obtained by segmentation of the image area into areas and by determining the contours in the image. It is based on the assumption that the individual areas are separated by a more or less pronounced boundary. On the two sides of this boundary can be seen a sharp change in the parameters of the image (for example, dark-light). This change may also manifest itself by a change in the average grey level or of the statistical parameters that describe in colour the texture areas. The contour is the line where the change in the parameters occurs. Depending on the character of this change one can see edges or a zone which contains the edge. The edge is a boundary which separates areas with different properties [8].

This change in the characteristics, observed in the normal direction to the boundary line, is approximated with a sharp transition – a step (Fig. 3a), ideally two adjacent pixels) or a smooth transition (Fig. 3b), in reality, a minimum of 3 pixels). The situation is even more complex because of the lack of sharp edges and the presence of relatively large rounding radii at the ends of the measured surface (Fig. 3). This change in the characteristics, observed in the normal direction to the boundary line, is approximated with a sharp transition – a step (Fig. 3a), ideally two adjacent pixels) or a smooth transition (Fig. 3b), in reality, a minimum of 3 pixels). The situation is even more complex because of the lack of sharp edges and the presence of relatively large rounding radii at the ends of the measured surface (Fig. 3).

Based on the experience gained to date it is known that the measurement uncertainty is significantly influenced by the parameters of the photographed object, the parameters of the matrix and the parameters of the optical system of the camera used. Under a certain unfavourable circumstances, their influence may be decisive for the application of PMM. The results of experimental studies demonstrate that the measurement uncertainty increases significantly in case of poor focusing combined with shading and large rounding radii, regardless of image processing and the software used [6], [8]. The main reason for this is the subjective interpretation / assessment of the boundary of the measured object. This boundary is the visualized (on the computer display) thickness of the marker line of the software used. In this context, it is influenced by the positioning (aiming) error of the start marker or the end marker in the process of defining the measurement distance between the pixels of the display [9]. This significant circumstance requires the development of a justified approach for determining the specific feature of size, and then its value from the image in PMM.

3.2. Determining linear features of size of cylindrical parts

In this case, the part is defined by two basic dimensions – the diameter of the cross-section, and the length of the generatrix of the cylinder. The diameter of the cross-section in the case of a 2D photogrammetric measurement can be determined only in the two end sections, and in the case of a 3D measurement – in a random number of cross-sections. Finally, the required diameter can be calculated as the average value of the measured values.

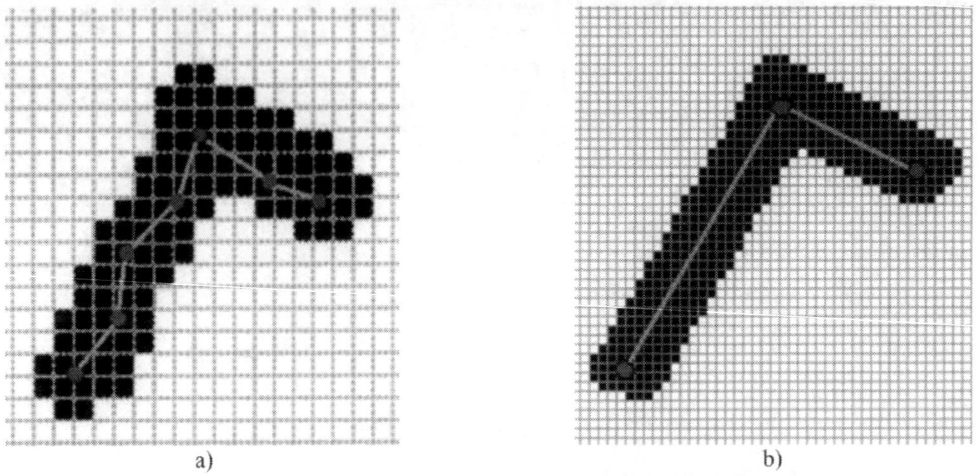

Fig. 3. Forming a contour image: a) real, b) ideal

The determining of the diameter of a cylindrical part in an end section as a DGAS is performed in the following sequence:

a) determining the pixels having the smallest and the largest radius (the outermost and the innermost point of the extracted feature) (Fig. 4). These pixels have the greatest potential for outliers (gross errors), thus they are excluded from the calculations. Naturally, in this case one can use a standard criterion for discovering outliers, which is applied to pixels located about the average of value of the coordinated of the obtained N pixels;

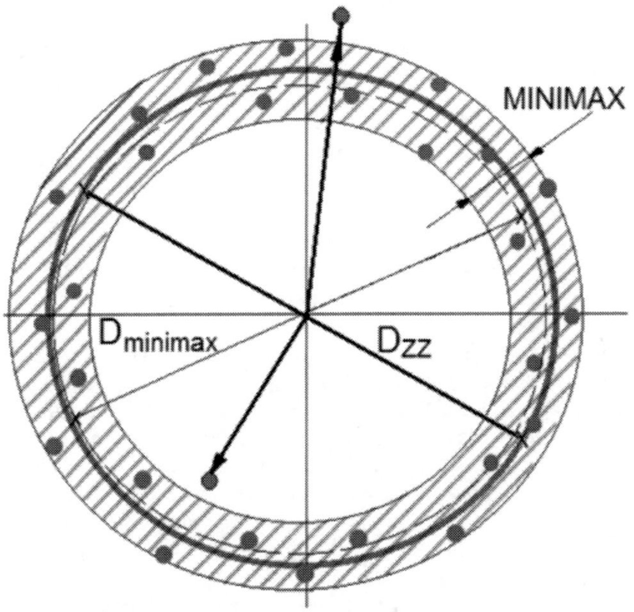

Fig. 4. Diameter of a cylindrical part

b) building a mean circle for the remaining N-2 pixels by applying the least-squares association criteria. The diameter of the circle D_{ZZ} is the required direct global associated size of the diameter of the measured cylindrical object;

c) if the dimension is determined graphically, the mean circle can be built based on the middle of the dimension of the minimal zone MINIMAX of the extracted integral feature, and the required diameter Dminimax is the distance between two opposing pixels, obtained by the intersection between the mean circle and a line passing through the centre of the extracted feature (Fig. 4). The application of this approach does not require neither the use of specialized photographic equipment, nor the use of specialized software.

d) if the dimension being determined is part of a fit, the required diameter is determined as the diameter of the largest inscribed circle D_{GX} of the extracted integral feature for holes (internal features of size) or the diameter of the smallest circumscribed circle D_{GN} of the extracted integral feature for shafts (external features of size).

The length of the generatrix of the cylindrical part can be determined by following the methodology, described in item 3.3 below.

3.3. Determining linear features of size of cuboid parts

The discussed measurement object most often are parts having the form of a rectangular parallelepiped (cuboid) and dimensions up to 500 (3150) mm or prismatic-housing parts with average or large dimensions (up to and more than 10000 mm), used in heavy machinery. In this case, the most frequently measured dimensions are linear distances between surfaces, which nominally are parallel to each other, and often are nominally parallel and/or perpendicular to other datum surfaces. The linear distance between the associated features of two nominally parallel surfaces is the DGAS. It is determined in the following sequence:

a) Build a mean straight line for the photographed (visualized) pixels of the extracted integral feature (surface) by applying the least-squares association criteria. Identify the two extreme pixels and exclude them from further calculations. Here, similar to item 3.2 a), one could also apply a standard criterion for discovering outliers. After eliminating the outliers, by applying the least-squares association criteria once again determine the final mean straight line of the extracted integral feature (Fig. 5). Objectively considered, the resulting associated straight line may not fulfil the requirements for deviation from form and/or location in relation to the datum surfaces.

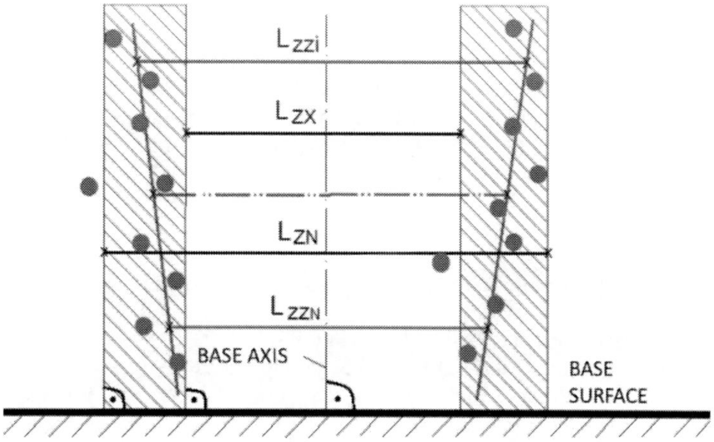

Fig. 5. Direct global size of a cuboid part

The latter are used for positioning the axis of symmetry (or another datum axis) between the parallel surfaces of the nominal feature;

b) The procedure for determining the mean straight line is repeated for the pixels of the extracted integral feature that describe the opposite, nominally parallel surface. The obtained second straight line probably will not be parallel to the first associated mean straight line, determining the first pixel of the required dimension. Consequently, it also may not fulfil the requirements for deviation from form and/or location in relation to the datum surfaces, indicated on the nominal features of size – the part drawing.

This circumstance allows to determine some deviations from form and/or location in relation to the datum surfaces.

c) Build a datum axis that meets the requirements for parallelism or perpendicularity of a specified nominal datum surface. It is appropriate that the datum surface is built as an enveloping surface of the extracted integral datum feature. If such a datum is not specified, then for the purposes of measurement it is necessary to establish a simulated datum feature [4]. A possible solution when measuring cuboid parts is to use a real, precise enough surface, that is perpendicular or parallel to the measured ones, and is associated to a straight line built according to the least-squares association criteria or as an enveloping surface;

d) If the measured dimensions belong to surfaces that are part of functional fits, instead of mean straight lines one has to use the maximum inscribed size L_{ZX} of the extracted integral feature for internal features of size and/or the minimum circumscribed size L_{ZN} of the extracted integral feature for external features of size. The resulting envelope associated features (lines) have to be parallel (or perpendicular) to the corresponding datum surfaces (or simulated datum features);

e) DGAS is determined as the distance L_{ZZ} between two opposing pixels, obtained by the intersection of the two associated mean lines or L_{ZX}/L_{ZN} for circumscribed associated straight lines and a straight line that is perpendicular to the datum axis. In order to improve the accuracy of the estimate for L_{ZZ}, the latter can be calculated as the average of $N \geq 3$ values L_{ZZi} (i=N) of features of size that are equally spaced on the image (Fig. 5);

f) If the dimension is determined graphically, one uses the mean straight line, built based on the mean dimension of the minimal MINIMAX zone of the extracted integral element, or uses one of the straight lines that limit the minimal zone. For external dimensional entities this is the outer straight line, and for internal features dimensional entities this is the inner straight line. The approach can be applied without the need of using specialized photographic equipment or specialized software.

3.4. Determining the diameter of spherical parts

The problems of measuring objects having spherical form are among the most interesting and yet the hardest ones to solve in dimensional metrology. It is known that their geometric accuracy is determined by the deviations of the diameter and/or the form deviation from the spherical surface whose total measurement is practically impossible [7]. Nevertheless, PMM can be used for partial and/or compromise solution to the problem. In this case the method needs to be applied consecutively $N \geq 3$ times, and the desired dimension will be determined as an **indirect global calculated size from a homogeneous set local sizes**. For this purpose the measurement is performed in the following sequence:

a) the diameter of the sphere $D_{ZZsphere}$ is determined as a direct global size from a 2D photo according to the methodology described in item 3.2. The procedure is repeated N times, where the value of N is determined by the necessary accuracy of the estimate of the measured diameter;
b) the obtained N values of D represent **a homogeneous set of local sizes** of N **realizations of a local dimension**. Using them one can calculate the final **indirect generalized calculated dimension** of the diameter of the sphere;
c) the diameter of the sphere is calculated as the average, median or mean of the range of the group or is assumed to be the largest one or the smallest one from the group of local dimensions.

4. Conclusion

The dimensions obtained when measuring using the photogrammetric measurement method shall be defined as generalized (integral) dimensions obtained as associated using a direct measurement method. Proposed are algorithms for processing the results obtained by the photogrammetric method for manufacturing parts having a general geometrical form of a cylinder, cuboid or sphere. Proposals are made how to process similar results using a graphical method, without the need to use specialized equipment and specialized software.

5. References

[1] BDS EN ISO 14405-1:2016. Geometrical Product Specification (GPS) – Dimensional tolerancing – Part 1: Linear sizes.

[2] Effenberger, G. Geometrical Product Specifications (GPS) – ISO 14405-1, the General GPS Standard for Dimensional Tolerancing of Linear Sizes. Q-DAS Web Portal. May 2013.

[3] ISO 286-1:2010. Geometrical Product Specification (GPS) – ISO code system for tolerances on linear sizes – Part 1: Basis of tolerances of deviations and fits.

[4] Metrology and measurement equipment. A reference book in three volumes. (2010). Vol. 2. H. Radev (Ed.), Published by Softrade, pp. 987, ISBN 978-954-334-093-4, Sofia, Bulgaria. (Метрология и измервателна техника. Книга справочник в три тома. Том 2, под общата редакция на проф. Христо Радев - София: Софтрейд, 2010, 987 стр.).

[5] Sakakushev, B.; Kokalarov, M.; Parvanov S. (2018). Perspectives for the application of the photogrammetric method for large-scale details and structures. 28th International scientific symposium "Metrology and metrology assurance 2018", No. 28, pp. 160 - 162, ISSN 1313-9126.

[6] Sakakushev, B.; Parvanov, S.; Gueorguiev, Tz. (2018) The Real Capabilities of the Photogrammetric and the Stereophotogrammetric Methods for Measurement in Mechanical and Manufacturing Engineering. Proceedings of SOCIOINT 2018, 5th International Conference on Education, Social Sciences and Humanities, Dubai, U.A.E. ISBN: 978-605-82433-3-0.

[7] Sotirov B.; Tonev D. (2020). Measurement of geometric features of spherical objects. International Conference Knowledge-based organization - Sibiu, Romania, Land Forces Academy "Nicolae Balcescu", Conference proceedings Vol. XXVI, No. 3, 2020, pp. 167-172, ISSN: 1843-682X.

[8] Kokalarov, M. (2019). Predetermination of the Uncertainty Budget and Error. Analysis of the Photogrammetric Method in Heavy Machinery. Proceedings of the 30th DAAAM International Symposium. Published by DAAAM International, 2019, Vienna, Austria. ISBN 978-3-902734.

[9] Georgiev G.; Sakakushev B.; Georgieva K. (2015). Study of the applicability of the photogrammetric mehod for measurement of very small dimensions. Proceedings of the University of Ruse, Vol. 54, Series 2. ISSN 1311-3321, Ruse, Bulgaria. (Георгиев, Г., Сакакушев, Б., Георгиева, К. Изследване на приложимостта на фотограметричния метод за измерване на много малки размери. Научни трудове на Русенския университет – 2015, том 54, серия 2.).

DOI: 10.2507/32nd.daaam.proceedings.040

EVALUATION OF CUTTING STABILITY AND ELECTRONIC ACCURACY OF CNC MILLING MACHINES

Mailis Zimmermann, Marten Madissoo & Tõnu Leemet

This Publication has to be referred as: Zimmermann, M[ailis]; Madissoo, M[arten] & Leemet, T[onu] (2021). Evaluation of Cutting Stability and Electronic Accuracy of CNC Milling Machines, Proceedings of the 32nd DAAAM International Symposium, pp.0281-0288, B. Katalinic (Ed.), Published by DAAAM International, ISBN 978-3-902734-33-4, ISSN 1726-9679, Vienna, Austria
DOI: 10.2507/32nd.daaam.proceedings.040

Abstract

Numerically controlled machine tools are a common phenomenon today and a norm in the manufacturing industry. However, these devices still have features and parts that need to be researched and improved. Therefore, this article focuses on the accuracy features of machine tools, but not so much on mechanical rigidity but rather on electronic rigidity and precision. It is important to know how a numerically controlled system converts input signals into physical movements, resulting in a real machine part. In this work, we focus on milling test specimens with circular shape for determining and analysing shape deviations. Each error has its description, possible causes, the effect that it has on the machined part and corrective action. The purpose of this work is to provide a validated methods and tests for assessing the accuracy of a machine tool. It is important to understand how the shape of the part can be affected by the material of the work piece and the cutting parameters according to the change in the cutting forces. An overview of methods and experiments for assessing the accuracy of a machine tool is given and helps to improve the understanding of how the input signals entered on the machine controller are interpreted into the actual movements of the milling machine.

Keywords: CNC milling; workpiece geometry; electronic rigidity; cutting forces; circularity

1. Introduction

The precision of machining process has always been crucial in mechanical engineering. With ever-increasing demands on machine elements and in the search for solutions to real-time changes in cutting process of machine tools, it is crucial to study this topic more in depth. In general, it can be said that each part produced by cutting and especially chip cutting has specific requirements to ensure the accuracy of these products, such as surface roughness, dimensions and shape deviations with high productivity in mind. Consequently, it is not only the mechanical properties and rigidity of the machine, but also the electronic accuracy, and to evaluate their ability to respond to cutting forces to ensure a stable and accurate cutter movement [1]. Precision of the machine drives assure the accuracy between the actual movements of the machine and the entered code. However, movement of the motors and their control by encoders may be related with certain errors.

The design of servo control systems that generate accurate coordinated multi-axis motion is therefore of great importance in high precision machining applications. Contour errors can often arise from unmatched dynamics, limited axial bandwidth and undesirable nonlinearities [2]. Numerically controlled machines primarily use both velocity and position control loops in monitoring and controlling the axes with the velocity feedback normally provided by a tacho generator and the position feedback by digital encoders. Several factors such as the dynamic response of the machine axes, stability of the control systems, time and frequency response of the servo system are involved in determining the final performance of the machine tool. The performance criteria for such servo control systems needs to be clearly established as well as closely monitored to get the best results and be aware of the capabilities of your processing equipment. 'Tracking error' and 'contour error' are two important aspects that need to be effectively handled by any servo control system [3]. Several systematic geometrical deviations are distinguished due to the electronic control of the machine tool. In order to explain the systematic errors that result from operating a milling machine, it is necessary to understand their nature [4]

Consequently, a circular test is recognized as one of the more convenient methods to test the motion and controller accuracy of CNC machines [1]. Although the circular test is used for checking motion accuracy of CNC machine tools or coordinate measuring machines, it has also capability to diagnose motion error sources. The motion error in the circular test shows possible servo system errors and geometric errors hidden in a CNC machine. These error sources can be identified from the motion error signal. The main errors patterns related to the various motion error sources in the circular test identify core motion error sources of CNC machines in circular test. Most of motion error patterns have been widely studied in the literature is possible to extract major motion error patterns from the circular test results. However, since the method requires an insight into the motion error patterns, it may be limited in use if many error sources simultaneously influence motion error signals in the circular test [5]. Some data-processing methods such as finite/fast Fourier transform (FFT) analysis and least-squares methods have been proposed, but did not provide sufficient flexibility to take into account the actual situation of the circular test [6]. A systematic approach is still desirable to decompose motion error sources from the circular test results.

The present paper proposes a strategy to identify motion error sources of CNC machines from the circular test results and the possible relationship between cutting forces in the circular error patterns. The possible machine geometric errors are discussed and a way to diagnose them is presented. Each error has its description, possible causes, the effect that it has on the machined part and corrective action. In this research a roundness test on two different milling machine has been made. This paper focuses on finding circular errors, which may include positive backlash in the Y axis, unequal backlash in the Y axis, reverse spikes in the X axis, plot rotation error, and cyclic error in the X axis, to name a few [7]. In order to find systematic errors, it is necessary to perform cutting tests for milling a cylindrical test piece and measure cutting forces and circular accuracy of the cut specimen.

2. Materials and Methods

There are many factors that affect the accuracy of a part: material, cutting parameters, coding and etc. making it difficult to assess which factor caused problems in the outcome of the part (Fig. 1.). The main complication is linking errors with altering factors and methodology.

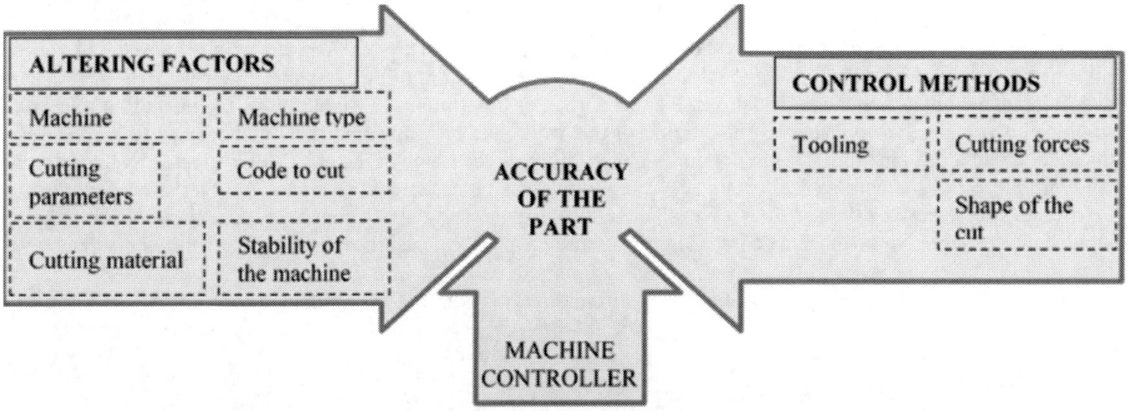

Fig. 1. Different factors affecting part accuracy and what are the methods of controls

Milling in essence is dynamic cutting, meaning during cutting process there are several different adjustments happening at same time like change in the position of the tool, force values and machine load. Total rigidity of a milling machine depends on geometrical characteristics of manufacturing system layout. However, errors in surface machining depend highly on tool deflection, electronic rigidity, coding and workpiece distortion [8].

In the evaluation of the machine accuracy analysing individual effect of altering factor will not show all possible causes of the errors in the cutting process, but investigation of various interactions between altering factors do. Therefore, it is necessary to analyse workpiece and tool loads together with milled workpiece geometry.

2.1. Machine accuracy

For the controlment of machine axes many machining applications involve generation of complex profiles to synchronize the motions of different axes. This in turn makes it possible for the machine to generate required machined surface and form the shape of the workpiece profile [9]. In a CNC machine tool, it is mostly done with help of a multi-axis servo control system that consists of an interpolator and several machine axis controllers. The interpolator generates the desired tool motion relative to the workpiece and then decomposes the desired motion into the reference position commands for the separate machine axes. The individual axis controllers do actual positioning of the axes. The function of an axis controller is to provide an appropriate drive signal to the electric motor, so that the actual position of the axis accurately tracks desired axis reference command to eliminate the position error along each driving axis [10]. Generally, two types of controllers are used for each machine axis in a CNC machine: open-loop or closed-loop controls. Open-loop controls use stepping motors to drive the machine axis while in closed-loop control systems, a position feedback element such as an encoder or a resolver, is used to provide information of actual position of the axis. There are two types of closed-loop position control systems [11]. Point-to-point is mainly for auxiliary movements while for the path of the machine tool axis from one point to the other and its velocity are not significant. This is because the tool is not machining or in contact with the workpiece during the motion. The accuracy of positioning axis to desired end is what only matters. On the other hand, in contouring systems the path of the cutting tool, relative to the workpiece, is of prior importance. This is because the machined part depends primarily on this path [12]. The most significant factor in performance of contouring systems is the contour error defined as difference in distance between the programmed path and the actual tool path.

Question of contour error arises in linear motion, circular motion, cutting of the corner radius and any desired contour. Reduce of contour error in linear motions is achieved by having high average gains and match of the individual axis. The difference in axis open-loop gain values represents the mismatching of the system, which creates the contour error. It is known that lower axis feed rates will reduce the contour error. However, to obtain high contouring accuracy at high speeds, the system should be critically damped and the system gains well matched [12]. High speed machining techniques mean faster feed motion between the tool and workpiece, in accordance with the increased spindle speed. Most common servo controllers used in industrial CNC machines, proportional (P), proportional-derivative (PD) or proportional-integral-derivative (PID), have limited achievable bandwidth due to tracking errors in each axis as the closed-loop control system is not able to follow rapidly varying position commands. Together with the friction in the guideways and cutting forces disturbances to the servo loop are formed opposite to the tracking accuracy. Furthermore, even if the servo controller is tuned to maintain accurate tracking at high speeds, the performance may degrade if the drive parameters change according to varying workpiece mass and lubrication conditions. The disturbance and parameter variation sensitivity are much more severe in direct drive systems compared to ball screw drives and a well-designed servo controller must be capable of handling both types without requiring any major changes [9].

In this experiment, two different machines were used to evaluate how machine itself affect quality of the part. First machine was a Haas Super Mini Mill (Haas) and second machine was a DMG Mori M1 (DMG). Haas has its own operating software and system opposite to DMG, which uses Siemens's control system. Another difference between two machines is that in Haas the worktable moves and tool moves only in z direction and in DMG Mori worktable is stationery and tool moves along axises. Machines are fixed with bolts onto the concrete floor. Haas has a direct speed, belt drive ball screw system, regards to DMG, which has an inline spindle and similar belt rive ball screw system.

2.2. Measuring circularity

Ideal roundness, according to ANSI standard B89.3.1, is the representation of 2D profile where all points of which are equidistant from a canter in the plane. Out-of-roundness is the deviation of the actual measured circular profile from ideal defined round profile and the out-of-roundness value is the difference between the largest radius and the smallest radius of a measured profile. There are three major methods for performing a circular test: one method is using a rotating spindle in connection with a one-dimensional probe; the second method is using a disk and a two- dimensional probe; and the third method is using a double ball bar system, which consists of two magnetic balls constrained by special sockets and a flexural bar measuring the deviation of the relative distance between balls. [5]. One of the best methods for measuring out-of-roundness is through circular geometry gages with precision spindles [13], [14].

Out-of-roundness is either symmetrical or asymmetrical involving regular geometrically arranged deviations and points on the part's circumference, where the fluttering is not regular. Most machining processes create symmetrical fluttering, producing either an even or an odd number of irregularities. Even number flutter is sometimes seen in precision boring operations, caused by a worn or out-of-balance spindle. Odd number flutter may be caused by a 3-jaw chuck, which can produce a 3-flutted workpiece or a centreless grinder creating 5-flutted part [15], [16].

Roundness can be measured by rotation of a pick-up measuring probe or the workpiece on a precisely generated axis. Variations in the radius of the workpiece are plotted on a polar chart on which it is also possible to represent a reference of a perfectly round circle with determined radial deviations. It is expressed in terms of the separation of two circles, drawn from the same centre, that just contain the undulations. There are two standardized comparisons methods; the minimum separation of given center and the least-squares centre. The error measurement is the maximum value of the peak's valleys on the measured circle.

For the analyse of shape deviations and accuracy of cutting circular contours circularity tests were carried out. All measurements were made with Mahr MMQ 100 form tester, which uses a rotating spindle in connection with a one-dimensional probe. Reference measurement was carried out to control machine accuracy before running the tests.

2.3. Measuring cutting forces

Cutting forces are measured to evaluate machine system dynamics and cutting process during milling. Any kind of system that measures forces can be used to assess fluctuations in force values. However, to see and thoroughly examine what takes place during cutting process and linking an error with possible cause, need of much accurate measuring machines are essential. Abnormalities in the value of cutting forces can be an indication of system error, tool or chip breakage. Whether or not the error is straightforward, important is to find an explanation why it occurred. In general, there are direct and indirect force measurement methods. Direct measuring method is usually more accurate rather than indirect method, which accuracy is highly affected by the change in environment [17], [18]. Therefore, the study approach was to use direct measuring method such as most known direct force measurement machine, table dynamometer.

The reliability of measuring process is highly affected by tool passing frequency (TPF) and sample frequency (SF) of measuring machine. The value of TPF, calculated with spindle speed, time and number of teeth on the cutting tool, is a base for estimation of right sample frequency, which should be at least two or even three times over the value of TPF. In analyse of machine accuracy the value of SF should be higher than two or three times over TPF value for collection of background data during milling. This will give enough measuring points to get reliable results and width of bandwith [19], [17]. With given data it is possible to see how the cutting forces excited the system and what is the shape of the peak. Another thing to look is how the data is presented, being periodical, episodic or serial. Also, it shows what method of cutting, climb or conventional milling, was used to cut and where there backlash at the very beginning of the cut. Received force values sign, negative or positive, give an indication where the cutting took place in the coordinate system. During milling, cutting forces were measured with table dynamometer Kistler 9257B, to see correlation between value of the cutting force and shape deviations in circular shapes. Signal processing was made with program Dynoware. All runs where filtered with low pass filter for 600 Hz.

2.4. Experimental setup, coding

The shape of the machined part is affected by the material and cutting parameters of the workpiece according to the machine electronic rigidity and coding [20].

It was important to assure repeatability of test regardless of CNC milling machine. Therefore, material used in this experiment was widely used S355JR and the size of the specimen was chosen accordance the size of the dynamometer used in this experiment and small enough to fit in smaller CNC machines work area. Dynamometer was fixed with polts directly onto the milling machine table. On top of that was specimen also fixed with polts. For this experiment 10 specimens where cut of which 5 for each milling machine. All specimens where pre-processed to shape and size before running the experiment (Fig. 2).

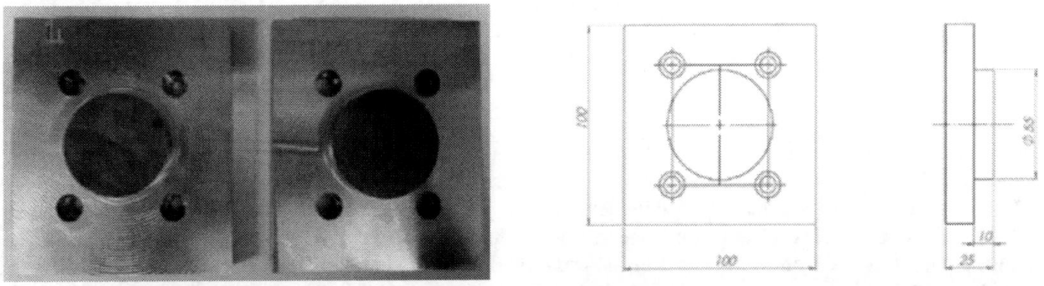

Fig. 2. a) Specimen after first run in Haas and DMG milling machines b) drawn shape and size of the specimen

To get reliable data different factors like tooling and fixing of the workpiece must be stable. With that in mind, a solid end mill with diameter 10 mm was chosen. Widely used milling cutter PLSC4L10045V had a maximum cut length of 22 mm and AlCrN coating. Recommended cutting parameters for stable cutting (Table 1.) were used to prevent increase in the risk of rapid failure of the cutting tool. Coolant was used to reduce heat process during cutting.

End mill Ø	Parameters			
	f_c (mm)	a_p (mm)	v_c (m/min)	a_e (%)
10	0.05	9	220	20

Table 1. Experimental values of cutting parameters

The NC-code for milling toolpaths was hand coded for the Haas. In the case of DMG machine tool the built-in ShopMill programmig tool was used to program the exact same movements of the cutting tool. In all of the experiments the instrument approached and exited the cut along a tangential arc. Experiments were divided into two parts where circular contour was first cut down to diameter value of 51 mm. After what circularity was measured and test was repeated cutting down to diameter value of 47 mm. Parameters f_c, a_p, v_c and a_e were constants and were not changed when performing tests on two separate machine tools or when diameter of the cut changed. During milling, cutting forces were measured to see correlation between value of the cutting force and shape deviations on the circular contour.

3. Results and discussions

A row of experiments were performed with the parameters described above and results are shown in Table 2. The table shows the diameter of the cut circle in mm and how large the deviations were in each run in µm. In specimen Exp_2 Ø47_H measurements for circularity were not carried out because of the incision made during milling.

Diameter of the shape before cutting (mm)	Experimental runs and results (µm)				
	Exp_1	Exp_2	Exp_3	Exp_4	Exp_5
Ø51_H	37	43	47	35	45
Ø47_H	46	-	42	49	27
Ø51_D	50	51	48	45	44
Ø47_D	27	32	28	27	30

Table 2. Mean deviation of circularity of given specimen roundness Ø51 and Ø47 in Haas Super Mini Mill and DMG Mori M1 milling machines

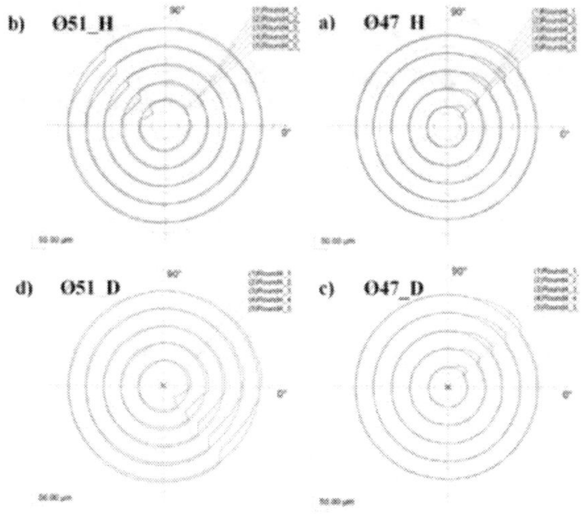

Fig. 3. Measured roundness and deviations direction of cut contours a) Ø51 in Haas, b) Ø47 in Haas, Ø51 in DMG, d) Ø47 in DMG

First part of the experiment foresaw cutting a circular shape workpiece to get specific diameter (51 mm and 47 mm) two times. It showed how machine interprets the code without making any changes in machine settings. The same experiment was repeated in both Haas and DMG milling machine. Between test runs roundness of the cut contour was measured with Mahr coordinate measuring machine. In both machines the same code was used. Results obtained show clearly the point of entry into the material in machine A (Fig. 3). Similar results were with machine B. However, the difference between machine A and B was evident in the accuracy of the circle. Accuracy of the test runs in machine A was between ~39-49 µm and in machine B it was ~45-56 µm. Also, this part of the experiment showed that the accuracy of the machine grows when the circular contour decreased. Another difference between runs was noticeable notch appeared when cut to Ø51 and bump appeared when cut to Ø47. The same error occurred in both milling machines.

After measuring the circumference and detecting the deviations, work with literature for possible causes of the occurred shape deviation was made. It was important to know whether the error was caused by the mechanical parts of the milling machine or by electronic means. Comparing the various standard mechanical and possible electronic shape deviations that can occur when cutting a circle, it was found that this phenomenon is not related to any of the known errors related to the machine tool mechanics described in the literature. Since both milling machines had the same error and both machines use the same technology to move the axes, it could be concluded that the error is due to the interpretation of the machine input signals is related rather to electronic accuracy [5]-[7].

Second part of the experiment was analysis of measured cutting forces during milling to see how forces describe forming of the circle. Fig. 4 presents filtered data of experimental run no. 4, cut in Haas, where graphs a) and c) show whole measured cutting forces cut to Ø51 and Ø47 mm. Graphs b), and d) show sections of measured data zoomed in to bring closer to what is called to be "region of interest". In Fig. 5 it is showed filtered data of experimental run no. 4, cut in DMG, where graphs a) and c) show whole measured cutting forces cut to Ø51 and Ø47 mm. Graphs b), and d) show section of measured data zoomed in to bring closer to "region of interest".

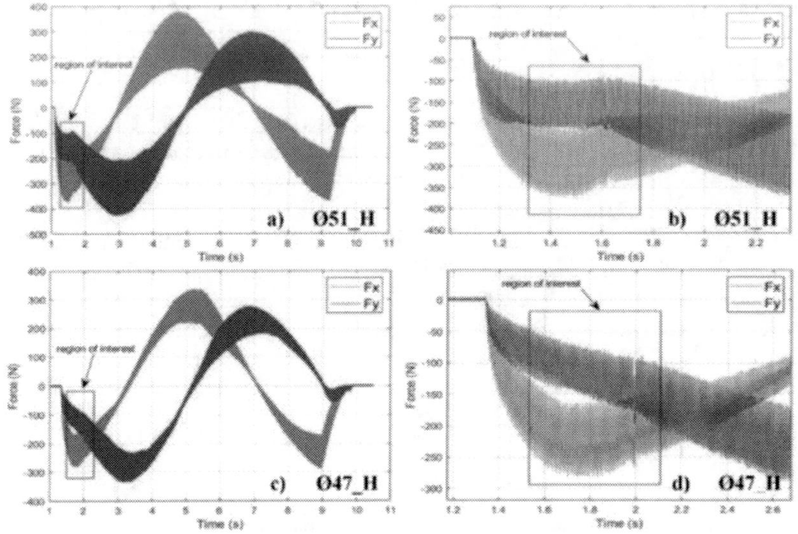

Fig. 4. a) Force components Fx and Fy filtered data of Exp_4, cut in Haas Super Mini Mill (Ø51), b) "region of interest" filtered data of Exp_4, in Haas Super Mini Mill (Ø51), c) Force components Fx and Fy filtered data of Exp_4, cut in Haas Super Mini Mill (Ø47), d) "region of interest" filtered data of Exp_4, in Haas Super Mini Mill (Ø47)

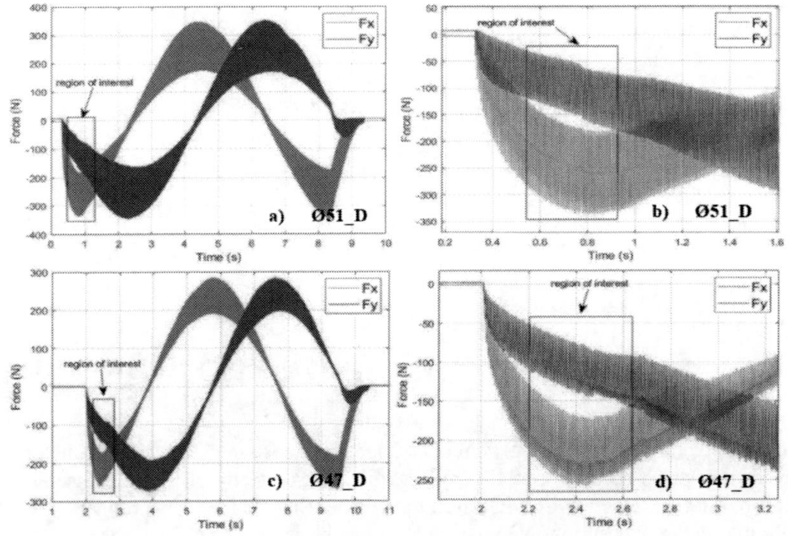

Fig. 5. a) Force components Fx and Fy filtered data cut in DMG Mori M1 (Ø51), b) "region of interest" of filtered data in DMG Mori M1 (Ø51), c) Force components Fx and Fy filtered data cut in DMG Mori M1 (Ø47), d) "region of interest" of filtered data in DMG Mori M1 (Ø47)

The region of interest validate that the results obtained in circularity measurements and measured cutting forces show similar region of error. The error is most likely the place where previously described notch and bump appear to be. There are several possible assumptions for this error. One is that the value of a_e increased up until a point where maximum contact area was reached and machine had difficulty to obtain desired tool path for approximately half a second to move forward. Another possible reason is tool deflection creating backlash in machine system. Further cutting demonstrates that shear force stabilized after initial disturbances confirmed by the results of the shear force analysis given in Fig 4. and 5. Together with the friction in the guideway, tool deflection and cutting force disturbances influenced the tracking accuracy at the servo loop control. In this work only one error pattern occurred, but in order to reach final clarity on the depth of inaccuracies in shape deviations, further work with different machine tools and tool paths are needed.

4. Conclusion

This work focused on the evaluation of milling accuracy and electronic accuracy. To this end, an overview was given of the interpretation of electronic signals and real movements of the machine tool. To evaluate the accuracy of the machine tool, a method was developed in which a round cylinder was milled and its shape deviation was measured. The tests were performed with two different machines Haas and DMG, which are similar in terms of moving their axes and feedback. In addition, the cutting forces were measured during the tests to establish a relationship with possible shape deviations and changes in forces on the cut workpiece. The patterns of shape deviations caused by milling a circular shape and possible numerically controlled machine tool errors were thoroughly studied in the literature to find out a definite difference between the deviations caused by the mechanical parts of the machine and electronics. The aim was to eliminate mechanical deviations and to focus on finding electronic accuracy or possible deviations. Based on the literature and experimental results, it can be stated that no shape deviation patterns caused by mechanical components were detected on these machines, but there was still an iterative error on the machined circular specimens.

The error was the same for both Haas and DMG machine tools. A systematic error occurred in all experiments, especially when the cutting instrument entered the workpiece material, leaving a notch when cutting the diameter of 51 mm and a bump when cutting the diameter of 47 mm. The instrument entered and exited the cut along a tangential arc, the cutting force increased smoothly until the maximum cutting depth was reached. Evaluation of the shape error and cutting forces, it became clear that at the time when the cutting tool reached the maximum a_e value, there were also changes in the cutting forces. Consequently, it can be assumed that there is a direct relationship between the formation of shape deviation and the change in shear force. In addition, the geometrical error occurred precisely because of this change in cutting force, causing such an error in the circle profile. As further work on this topic, it is planned to widen the range of machine tools used for cutting the test workpiece. Also, it is needed to study different material cutting methods and change the cutting parameters to examine the effect of high speed and feed rates to the occurred shape deviation profile. It is definitely necessary to do additional work to find out possible solutions to eliminate this error and to offer solutions for setting up the milling machines to avoid and eliminate the potential of reoccurrence of such errors.

5. References

[1] Tools, J.M. & Vol, M. (1997). “An efficient method for identification of motion”, International Journal of Machine Tools & Manufacture, Vol. 37, No. 3, 1997, pp. 327-340

[2] Xi, X. C., Poo, A. N. & Hong, G. S. (2009). “Tracking error-based static friction compensation for a bi-axial CNC machine”, Precision Engineering, Vol. 34, No. 3, 2010, pp. 480-488, DOI:10.1016/j.precisioneng.2009.12.003

[3] Ramesh, R., Mannan, M. A. & Poo, A. N. (2005), “Tracking and contour error control in CNC servo systems”, International Journal of Machine Tools & Manufacture, Vol. 45, No. 3, 2005 pp. 301-326, DOI:10.1016/j.ijmachtools.2004.08.008

[4] Chiu, H. W. & Lee, C. H. (2017). “Prediction of machining accuracy and surface quality for CNC machine tools using data driven approach”, Advances in Engineering Sodtware, Vol. 114, 2017, pp. 246-257, DOI:10.1016/j.advengsoft.2017.07.008

[5] Knapp, W. (1983). “Circular test for three-coordinate measuring machines and machine tools”, Precision Engineering journal of The International Societies for Precision Engineering and Nanotechnology, Vol. 5, No. 3, 1983, pp. 115-124, DOI: 10.1016/0141-6359(83)90007-7

[6] Usop, Z., Sarhan, A. A. D., Mardi, N. A. & Wahab, M. N. A. (2015). “Measuring of positioning, circularity and static errors of a CNC Vertical Machining Centre for validating the machining accuracy”, Measurement, Vol. 61, 2015, pp. 39-50, DOI: 10.1016/j.measurement.2014.10.025

[7] https://www.haascnc.com/, (2021). Haas Automation Inc., “Renishaw Ballbar Test - Plot Interpretation – Mills”, Accessed on: 2021-08-17

[8] Cao, H., Li, D. & Yue, Y. (2017) “Root cause identification of machining error based on statistical process control and fault diagnosis of machine tools”, Machines, Vol. 5, 2017, DOI:10.3390/machines5030020

[9] Erkorkmaz, K. & Altintas, Y. (2001). “High speed CNC system design. Part III: High speed tracking and contouring control of feed drives”, International Journal of Machine Tools and Manufacturing, Vol. 41, No. 11, 2001 pp. 1637-1658, DOI: 10.1016/S0890-6955(01)00004-9

[10] Lo, C. C. (2002). "A tool-path control scheme for five-axis machine tools", International Journal of Machine Tools and Manufacturing, Vol. 42, No. 1, 2002, pp. 79-88, DOI: 10.1016/S0890-6955(01)00092-X

[11] Borovic, B., Liu, A. Q., Popa, D., Cai, H. & Lewis, F. L. (2005). "Open-loop versus closed-loop control of MEMS devices: Choices and issues", Journal of Micromechanics and Microengineering, Vol. 15, No. 10, 2005, pp. 1917-1924, DOI: 10.1088/0960-1317/15/10/018

[12] Koren, Y. (1983). Computer control of manufacturing systems, McGraw-Hill Book Company, ISBN:0070353417, New York, USA

[13] Kumar, M. R. P., Kumaar, P. P. & Kameshwaranath, R. (2018). "Roundness error measurement using teaching learning based optimization algorithm and com- parison with particle swarm optimization algorithm", International Journal of Data and Network Science, Vol. 2, 2018, pp. 63-70, DOI:10.5267/j.ijdns.2018.8.003.

[14] Sui, W. & Zhang, D. (2012). Four Methods for Roundness Evaluation, International Conference on Applied Physics and Industrial Engineering, pp. 2159-2164, D. Yang (Ed.), Published by Physics Procedia, DOI: 10.1016/j.phpro.2012.02.317

[15] Herrera-Navarro, A. M., Hernández, H., Peregrina-Barreto, H., Manríquez-Guerrero, F. & Terol-Villalobos, I. R. (2013). "A new measure of circularity based on distribution of the radius", Computacion y Sistemas, Vol. 17, No. 4, 2013, pp. 515-526, DOI: 10.13053/CyS-17-4-2013-005

[16] Naughtie, J. (1988). "The shape of things to come", Nursing times, Vol. 84, No. 21, 1988, pp. 24-25

[17] Scippa, A., Sallese, L., Grossi, N. & Campatelli, G. (2015). "Improved dynamic compensation for accurate cutting force measurements in milling applications", Mechanical Systems and Signal Processing, Vol. 54, 2015, pp. 314-324, DOI: 10.1016/j.ymssp.2014.08.019

[18] Albrecht, A., Park, S. S., Altintas, Y. & Pritschow, G. (2005). "High frequency bandwidth cutting force measurement in milling using capacitance displacement sensors", International Journal of Machine Tools and Manufacturing, Vol. 45, No. 9, 2005, pp. 993-1008, DOI: 10.1016/j.ijmachtools.2004.11.028

[19] Altintas, Y. (2000). Manufacturing Automation, Metal cutting Mechanics, Machine Tool Vibrations, and CNC Design, Cambridge University Press, ISBN: 052165029-6, New York, USA

[20] Wolny, R (2018). Evaluation of Accuracy of Five-Axis CNC Milling Machine on the Basis of Test Piece Machining, Proceedings of the 29th DAAAM International Symposium, pp.0164 – 0168, B. Katalinic (Ed.), Published by DAAAM International, ISBN 978-3-902734-20-4, ISSN 1726-9679, Vienna, Austria DOI: 10.2507/29th.daaam.proceedings.023

DOI: 10.2507/32nd.daaam.proceedings.041

EVALUATION OF EFFECT OF ATTRIBUTIVE MEASUREMENTS IN A CLIME COMPARTMENT ON THE VALUE OF DIMENSIONAL MEASUREMENT UNCERTAINTY IN AN ACCRETED TEST LABORATORY

Nikola Skrivanova & Martin Melichar

This Publication has to be referred as: Skrivanova, N[ikola] & Melichar, M[artin] (2021). Evaluation of Effect of Attributive Measurements in a Clime Compartment on the Value of Dimensional Measurement Uncertainty in an Accreted Test Laboratory, Proceedings of the 32nd DAAAM International Symposium, pp.0289-0293, B. Katalinic (Ed.), Published by DAAAM International, ISBN 978-3-902734-33-4, ISSN 1726-9679, Vienna, Austria
DOI: 10.2507/32nd.daaam.proceedings.041

Abstract

In the field of precision measurement and testing by state-accredited laboratories, the area of measurement uncertainty has a clear and indispensable role. This article builds on a 2018 study (The Influence of Humidity on ABS Plastic Measurement Result). This study addressed the variability in results occurring with changes in otherwise constant humidity in an accredited testing laboratory. This paper builds on the results obtained and evaluates the effect on the components of the standard uncertainty of the Type A measurement and the uncertainty of the Type B measurement). In order to understand the milestones of the process and their limitations. An analysis of the measurement system according to the VDA5 methodology was carried out and a clear limit was established beyond which the resulting values are no longer reliable. The results will lead to an increase in the relevance and accuracy of the measurements and the awareness of the professional metrology community.

Keywords: Metrology; Uncertainty; Measurement; Attributive; Clime Compartment

1. Introduction

Within the framework of the measurement methodology on a coordinate measuring machine of a state-accredited laboratory (ISO/IEC 17025 standard), measurement uncertainties represent a mandatory area of discussion between the metrology service provider, i.e. the customer, and the metrology service provider. Without at least a basic knowledge, the results are misinterpreted or at least distorted by the customer. [1], [2]

This analysis builds on a previous experiment from 2018, which was published as "The Influence of Humidity on ABS Plastic Measurement Results“. Humidity environment parameter remains a specific requirement. The standard for the CMM according to standard ISO 10360-2, however, indicates the reference value of 50±10%, but this is the recommended value for 3D measuring machines and not for the products. [3] Our follow-up paper operates one level down, where the effect that is potentially introduced when simulating environmental conditions using a climate chamber is investigated.

2. Experimental analysis

Within the accredited metrology laboratory L-1718 of the Regional Institute of Technology at the Faculty of Mechanical Engineering, University of West Bohemia in Pilsen, a comprehensive analysis was carried out to investigate the influence of clime compartment device used to simulate required air pressure condition. All measurements were subject to DIN ISO 1101 environmental conditions, i.e. a temperature of 20 °C ± 1 °C in a vibration-neutral environment and humidity. [4]

Fig. 1. Clime Compartment

The experiment was conducted under controlled conditions at a constant temperature of 20 ± 2 ° C. During the experiment only 1 operator performed measurements to eliminate the reproducibility error. This operator repeatedly measured the selected parameters on samples of ABS plastic part. Selected parameters = Inner diameter, wall thickness, length describe main metrology tasks for ABS evaluation. The measurement itself was performed on the CMM Carl Zeiss Prismo 7 Navigator equipped with a fixed sensor head with an measurement uncertainty of 0.9μL / 350μm. The measurements were carried out gradually at 30%, 45%, 60% humidity. Total of 5 samples of the same test sample being performed at each value due to the relevance of subsequent statistical data processing. [5], [6]

Fig. 2. Etalon

3. Data processing

As it is clear from the output, the resulting measurement value changes at 30%, 45%, 60%. The specified mathematical formula may be a useful tool for industrial practice, especially for the situations where measurement does not take place in a controlled environment of a metrology laboratory. [7]

3.1 Parameter length_1_Y (chosen example)

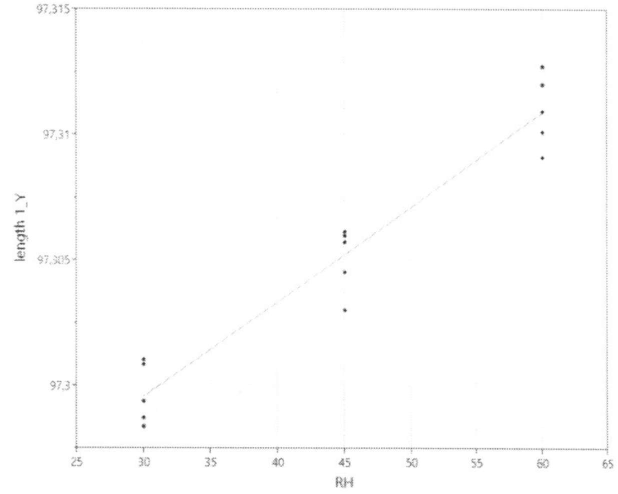

Fig. 3. Chart dependence for length_1Y

Parameter	Value
RSquare	0,937721
RSquareAdj	0,93293
Root Mean Square Error	0,001279
Mean of Response	97,30526
Observations (or Sum Wgts)	15

Table 1. Parameters of dependence for length_1Y

The predictive model of the length dependence of the change of the humidity value RH (1) describes 93.293% of the variability of the studied variable on the basis of the adjusted index of determination (RSquareAdj).[9] Therefore, the model predicts within the moisture interval 30 to 60% with an accuracy of 93.293%, with an average value of the measured quantity being 97.30526. [8]

Source	DF	Sum of Squares	Mean Square	F Ratio	Prob> F
Model	1	0,00032	0,00032	195,7368	<,0001
Error	13	2,13E-05	1,64E-06		
C. Total	14	0,000342			

Table 2. Prediction model for length_1Y

From the ANOVA analysis table, the model used is adequate based on the Fischer-Snedecor test criterion and therefore we can say that at the selected level of significance $\alpha = 5\%$, there is at least one factor (predictor) in the model that varies from zero.

Term	Estimate	Std Error	t Ratio	Prob>|t|
Intercept	97,28827	0,001258	77338	<,0001
RH	0,000377	0,000027	13,99	<,0001

Table 3. Parameters of dependence formula

The table shows that the greatest effect on the change in the value of length 1Y is the absolute intercept. It hides all the neglected factors we have not considered in the experiment. But, at the chosen level of significance $\alpha = 5\%$, the RH moisture value is statistically significant. The model itself can therefore be expressed in the form of:

$$lenght\_1\_Y = 97,28827 + 3,77.10^{-4}.RH$$

4. Results confirmation

Due the fact motivation for this paper is from automotive industry, the VDA5 methodology was chosen as the uncertainty effect evaluation methodology. Considering the lengthy calculations and the need for a considerable amount of data (MSA I + II analysis was also performed as part of the validation calculations), the supporting tool Yarvyn was used. [9]

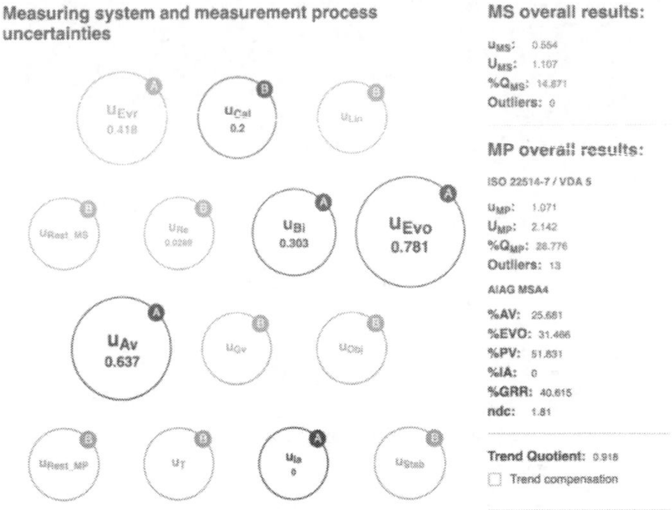

Fig. 4. Yarvyn output with uncertainties

In the case of the clime compartment: the influence and experimentally verified substantial increase of the A-type uncertainty components, namely: uEvo, uBias, uCal, uAv, has been unequivocally proven, thus leading to the degradation of the entire measuring system, including its accuracy, which drops by up to 3 orders of magnitude. [10]

5. Conclusion

The motivation for developing this paper was a consideration of the effect of the attributive clime compartment device on the parameters of measurement uncertainty. The subject chosen for testing was described in previous study "The Influence of Humidity on ABS Plastic Measurement Results". The measurement system analysis according VDA5 methodology and evaluation of single effects on the uncertainty of the uB type and thus on the overall uncertainty of the uc was carried out to understand the process and its limits. All inspection processes are characterized in a table of result values. This paper is part of a comprehensive study of the effects of human factors on the accuracy and reliability of results. The individual sub-experiments carried out in 2021 and planned for 2022 will be published in a comprehensive method manual on workshop metrology.

6. Acknowledgments

The article contribution has been prepared under project SGS-2019-008

7. References

[1] Svingerova, M. & Melichar, (2017). Evaluation of Process Risks in Industry 4.0 Environment, Proceedings of the 28th DAAAM International Symposium, pp.1021-1029, B. Katalinic (Ed.), Published by DAAAM International, ISBN 978-3-902734-11-2, ISSN 1726-9679, Vienna, Austria DOI: 10.2507/28th.daaam.proceedings.142

[2] ZEISS [online]. [cit. 2018-09-15]. [2] Jirouek, P.: Eligibility of measurement system in the gearboxes production of Škoda auto a.s (2012) [online]. [cit. 2021-01-10]

[3] ISO 14001 Certification, iso 14001 requirements, iso 17025 standard. ISO 14001 Certification, iso 14001 requirements, iso 17025 standard [online]. Copyright © 2017, ISO 17025 CONSULTANT. [cit. 2021-01-10]. Dostupné z: http://www.consultantiso17025.com/

[4] Melichar M., Kubátová D., Kutlwaše J.: (2016). CMM measuring cycle and human factor, Proceedings of the 27th DAAAM International Symposium, ISBN 978-3-902734-08-2

[5] Pernikař, J.: Assessment of the competence of control means [online]. [cit. 2021-01-10]. Available from: http://gps.fme.vutbr.cz/STAH_IN FO/31_Pernikar_VUTBR.pdf

[6] Płowucha W., Jakubiec W., Wojtyła M.: Possibilities of CMM Software to Support Proper Geometrical Product Verification, Procedia CIRP, Volume 43, 2016, Pages 303-308, ISSN 2212-8271, http://dx.doi.org/10.1016/j.procir.2016.02.124

[7] Barini M. E., Tosello G. Chiffre d L.: Uncertainty analysis of point-by-point sampling complex surfaces using touch probe CMMs: DOE for complex surfaces verification with CMM, Precision Engineering, Volume 34, Issue 1, January 2010, Pages 16-21, ISSN 0141-6359, http://dx.doi.org/10.1016/j.precisioneng.2009.06.009

[8] D. Kubátová, M. Melichar, J. Kutlwašer, Evaluation of Repeatability and reproducibility of CMM equipment, In Procedia Manufacturing, Volume 13, 2017, Pages 558-564, ISSN 2351-9789, https://doi.org/10.1016/j.promfg.2017.09.091

[9] Stojkic, Z[eljko]; Culjak, E[va] & Saravanja, L[uka] (2020). 3D Measurement - Comparison of CMM and 3D Scanner, Proceedings of the 31st DAAAM International Symposium, pp.0780-0787, B. Katalinic (Ed.), Published by DAAAM International, ISBN 978-3-902734-29-7, ISSN 1726-9679, Vienna, Austria DOI: 10.2507/31st.daaam.proceedings.108

[10] Kubatova, D; Melichar, M; (2019). Influence of the Number of Points on the Evaluated Element when Measuring on CMM, Proceedings of the 30th DAAAM International Symposium, pp.0476-0483, B. Katalinic (Ed.), Published by DAAAM International, ISBN 978-3-902734-22-8, ISSN 1726-9679, Vienna, Austria DOI: 10.2507/30th.daaam.proceedings.064

DOI: 10.2507/32nd.daaam.proceedings.042

ANALYSIS OF AIR VALVE ASSEMBLY TECHNOLOGY

Tomas Zatloukal, Vaclava Pokorna, Ondrej Marsalek & Simon Syrovatka

This Publication has to be referred as: Zatloukal, T[omas]; Pokorna, V[aclava]; Marsalek, O[ndrej] & Syrovatka, S[imon] (2021). Analysis of Air Valve Assembly Technology, Proceedings of the 32nd DAAAM International Symposium, pp.0294-0298, B. Katalinic (Ed.), Published by DAAAM International, ISBN 978-3-902734-33-4, ISSN 1726-9679, Vienna, Austria
DOI: 10.2507/32nd.daaam.proceedings.042

Abstract

This article focuses on assembly technology, which is an important part of the production process. When assembling a new product, emphasis is placed on the expected speed of product assembly and the associated development of the training curve. The training curve is an event in which the value of time decreases with the number of increases in repetitions of activity over a period. This is an important indicator for planning production and the number of assembled products over a period of time. The article contains his own design and implementation of an experiment in which an air valve from Bosch. The following is an evaluation of the experiment focused on the assembly speed of one valve, from which the training curve is created. Finally, there is a comparison with the theoretical training curve based on the literature.

Keywords: Assembly technology; training curve; productivity; air valve; Bosch

1. Introduction

Assembly is a set of operations of people, machines and equipment in the assembly system, the execution of which in a specified order creates a finished product from individual components and assembly units. Assembly is an important part of the production process, where individual technical, technological, organizational, but also economic factors from previous preparation and planning meet. The importance of assembly in mechanical engineering can be described by the share in the total laboriousness of engineering products. This share is around 40% for the textile and consumer industries. In the automotive industry, this share is smaller and is around 32%. When assembling electrical equipment, the assembly can reach up to 80% of the total production workload.

The importance of assembly can also be characterized by the percentage of assembly workers to the total number of employees in the company. In the automotive industry, this value is around 40%, in the textile industry it is 30% and the consumer industry has the smallest share of assembly workers in the total number of workers, namely 20%. The values depend on several factors, such as the type and organization of production, the degree of mechanization or the design of the product. For more complex products, the share of assembly workload decreases, which is mainly due to the mechanization of assembly operations due to higher demands on accuracy, which is also related to surface quality. Assembly includes not only connecting and assembling components, but also transport, handling and possible inspection of products. [1] [3] [5] [7]

The assembly is divided into several independent phases, depending on the type of product. These phases are manual handling, coupling, inspection, positioning and disassembly, and associated reassembly. With the increasing series of assembly, the share of manual handling and joining of components increases. On the other hand, the share of position control and adjustment is reduced. This growth and decline in values is mainly influenced by the increasing level and sophistication of the production process. [1] [2] [5] [7]

The assembly activity is composed of several stages. The first stage is preparatory, in which cleaning, shape adjustment, balancing, marking, etc. are performed. The next stage is handling. This stage includes inserting, removing, inserting, setting up and moving. The next stage is the connection. In this part, screwing is performed, which is the most commonly used way of joining components, it also includes pressing, riveting, welding, soldering or gluing. The last in the series is the control stage, during which the functionality of the assembled product is measured and tested. The decisive activities in piece and small series production are preparatory and adaptation work, and adjustment, inspection and disassembly work play a significant role in the actual assembly. In total, these activities cover approximately 80% of the assembly workload. In mass and series assembly, the share of typical assembly activities, such as joining and handling, will increase, as complex assembly activities will be eliminated due to the higher sophistication of production. [1] [5] [10]

The assembly technology is an important part of the production process. When assembling new products, the expected speed of product assembly and the associated development of the training curve is taken into account. Therefore, it is a very important indicator for production planning and the number of assembled products for a given period. The learning curve is an event in which the value of time decreases with the number of increases in repetitions of activity over a period. Previously, this dependency was called the "production function." The theoretical learning curve is mathematically designed as a smooth dependence. Equation (1) is a mathematical notation for a learning curve, where Y is the time at the xth assembly, K is the assembly time of the first piece, and X is the number of assemblies supplemented by the constant A for a specific situation. [6]

$$Y = K \cdot X^{-A} \quad (1)$$

The assembly worker is usually very insecure and nervous during training. For this reason, he uses visual inspection more than a worker who is already trained. This visual inspection decreases with increasing number of repetitions. This is one of the main reasons why the curve decreases with increasing repetition. The motivation of a person is to strive for the shortest possible use of sight and to consciously reduce position control. When observing experienced employees at work, it was found that they use sight to a minimum. An operator who can look around while still working is called skilled. This fact sometimes causes difficulties for a manager who feels that such an operator is not fully focused on the activity being performed. If the operator responds to these warnings, its productivity may even decrease. [8] [9] [10]

However, the actual course of training is not a smooth dependence, see Figure 1. There are small errors that affect the resulting assembly time and thus the smoothness of the curve. Only after interpolating the real curve can we get closer to the shape of the theoretical curve. The Y-axis shows the assembly time of the measured piece or unit and the X-axis shows the number of repetitions. For short periods of time, however, the learning curve is less significant [4] [6]

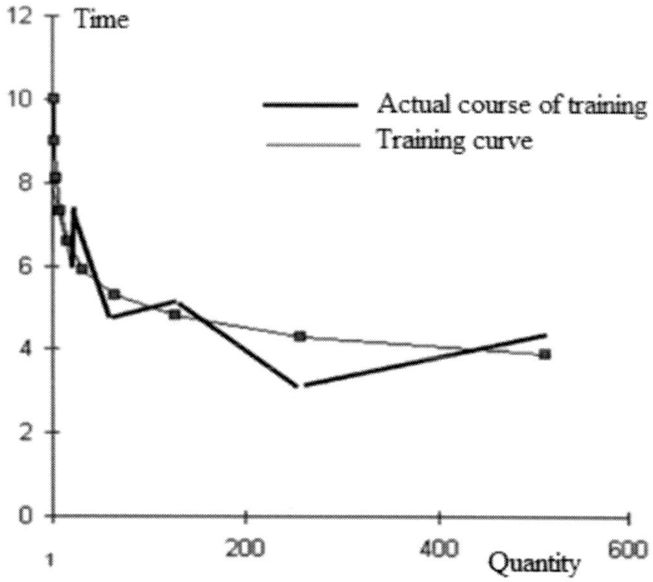

Fig. 1. – The actual process of the training curve [6]

2. Description and preparation of the experiment

The experiment was performed at the Department of Machining Technology at the University of West Bohemia in Pilsen. In the room there is a training workplace for the installation of an air valve from Bosch. An air valve is a device that regulates the air flow. The air valve consists of ten components and requires manual assembly to store them. The workplace consists of several boxes for components and a box with the main valve cover, see figure 2. There is also a hand lever on the table to help hold the valve during final assembly.

Fig. 2. – Workplace for air valve assembly

When assembling the valves, it is possible to use two assembly instructions. The first instructions are written only in writing in hand. These assembly instructions are very confusing and illegible, and in some cases more detailed descriptions would be needed to better understand the individual steps. The second assembly instructions are pictorial, supplemented by a description of the individual steps. According to the second instruction, the air valve was assembled very quickly and easily, so this procedure would be a more suitable option for employee training. Due to the fact that the labels of the containers were incorrectly marked, so without the pictures in the assembly instructions, it would be very difficult to fold the valve correctly. The workflow was no longer required to further assemble the valves, as the assembly was very easy to remember.

Aids for assembling the valve are a classic screwdriver, a mounting lever, which is already part of the work table. In addition, tools will be needed to measure and record the resulting times. An electric drill is available for faster disassembly of completed valves and a camera for photo documentation. The experiment will be performed on three consecutive days. The experiment will involve two people who will measure the final assembly time of one piece. On the first day, the operators assemble ten pieces, four times in a row with ten-minute breaks. The aim of the observation will be the time of assembling the individual pieces and observing the errors during training. On the second and third day, the operators assemble 35 pieces at once twice, with a thirty minute break. The aim of the observation will again be the individual assembly times, the observation of errors and certain shortcomings in the workplace.

3. Evaluation and result of the experiment

When assembling the first pieces, the operators often forgot about the components, looked confused and were not 100% sure if they had forgotten something during the assembly. However, this is a common practice during training. After training the operators and stabilizing their final valve assembly times, the following errors occurred. Especially on the first day of assembly, the operator forgot about the large seal that was to be put on the roller. This minor error is very fundamental in the function of the valve, which can be considered faulty. If an error is detected, the valve can of course be disassembled, the missing seal can be added and then the valve reassembled. This would be time consuming, which would have a negative effect on production efficiency and employee performance. In addition, the operator did not even notice this error in most cases.

Another error found, which often occurred in both operators, was the handling of small screws. Since half the assembly time of the valve was that the screws had to be grasped, inserted into the hole and screwed in. This offers the possibility of improvements to increase production efficiency. During handling, it sometimes happened that the screws fit into the valve when inserted into the holes. The countersunk screw then most often had to be removed with a screwdriver and re-inserted into the hole, which had a fundamental effect on the final assembly time. This problem can be easily eliminated by replacing the original screwdriver with a screwdriver with a magnetic end, which would make inserting screws much easier and faster. The last significant mistake, which occurred especially with the operator, was the incorrect placement of the face plate on the lever. When the face plate was placed incorrectly, the lever could not be pushed all the way to the table, because it did not fit the shape of the piston body. This error meant an increase of at least 15 seconds in total time.

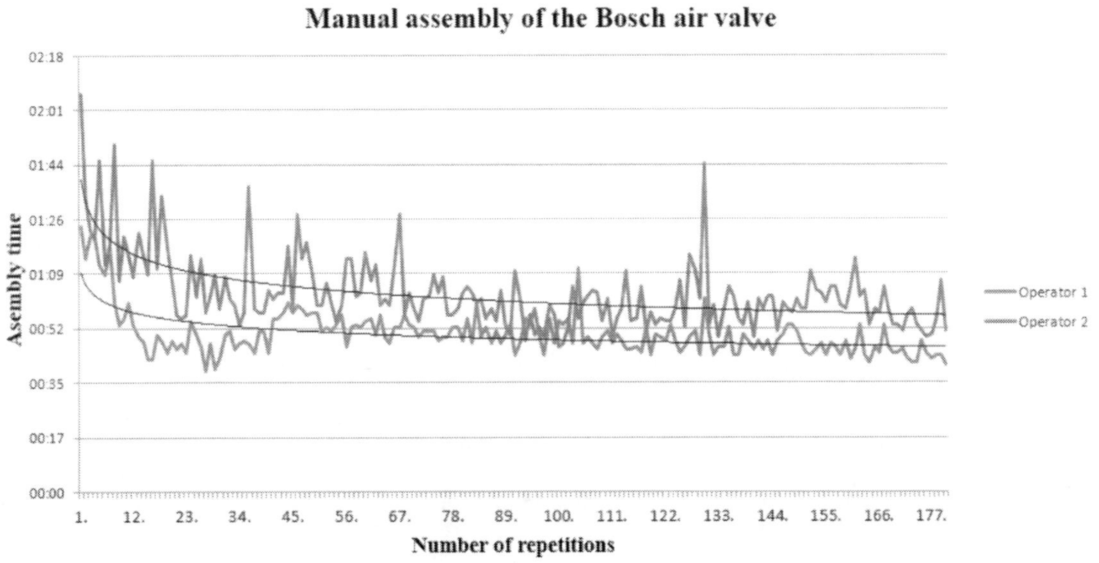

Fig. 3. Manual assembly of the Bosch ai valve

As already mentioned, the training curve is an event in which the time value decreases with the number of increases in the repetition of a certain activity per period. The graph above was compiled from the record sheet, which consists of measured time values for each assembled pieces for both assembly operators. After interleaving, this graph creates a training curve, which with its shape is very close to the theoretical curve, which is shown in Figure 1. The experiment thus confirmed the statement that with increasing number of assembled pieces, the time requirement decreases. Therefore, the training curve is a very important factor in planning the production and cycle time of assembly lines.

4. Conclusion

The paper was focused on assembly technology, which is an important part of the production process. The assembly can be final, at the end of which the product is already finished, but also continuous during the assembly of subassemblies. When assembling new products, the expected speed of product assembly and the associated development of the training curve is taken into account. A training curve is an event in which the value of time decreases with the number of increases in repetition of activity over a period of time. Therefore, it is a very important indicator for production planning and the number of assembled products for a given period.

The experiment consisted of assembling an air valve from Bosch. The workplace is located in one of the teaching laboratories of the Department of Machining Technology. The whole experiment lasted for three consecutive days. Two people took part in the assembly. The main monitored parameters were: the total time required to assemble one air valve, then the number of assembled pieces and the analysis of errors made by operators during assembly. These parameters were written into record sheets, from which graphs were subsequently made.

The aim of the evaluation of the experiment was to determine the errors during assembly, to describe the assembly workplace and to compare the resulting training curve with the theoretical training curve. The most common errors included prolonged handling of the bolts, forgetting about the component called the "big seal" and problems assembling the valve with the auxiliary lever. For operator 1, the assembly time of one valve averaged around 46 seconds after training. For operator 2, this time was higher and was around 56 seconds after training. In my opinion, the main reason for this time difference was the unsatisfactory height of the table for operator 2 with a height of 162 cm. A curve was created from the graph, which was created from the times of the individual assembled valves, which with its shape is very close to the shape of the theoretical training curve. The whole experiment proved the fact that it is important to devote the necessary time and effort when training a new employee.

5. Acknowledgments

This article was created under the project SGS-2019-008: Research and Development for Innovation in the Field of Manufacturing Technology - Machining Technology III.

6. References

[1] Vigner, M.; Kral, M. & Zelenka, A. (1984). Methodology of Designing Production Processes 1, Published by Prague 1984, SNTL, 1984. 588^S.

[2] Zelenka, A.; Preclik, V. & Haninger, M. (1999). Design Of Machining and Assembly Processes. Published by Prague ČVUT, 1999, 190 S. ISBN 80-010-2013-4.

[3] Mickal, K. & Kolar, P. (1989). Mechanical Assembly: The Field Of Mechanical Engineering with a Focus on Processing. Metals and Assembly of Machinery and Equipment, Published by Prague: SNTL, 1989. 205 S.

[4] Salvendy, G. (1992) Handbook of Industrial Engineering. 2nd Ed. Published by New York: Wiley, C1992. ISBN 0471502766.

[5] Malik, A. A. & Bilberg, A. (2014). Augmented Reality System for Virtual Training of Parts Assembly, Annals of DAAAM for 2014 & Proceedings of the 25th International DAAAM Symposium, Published by DAAAM International, Vienna, 2014, ISSN 1877-7058

[6] Matejka, J. (2016). Rationalization of Work. ZCU in Pilsen [Internal Documentation], Accessed: 2020-08-08

[7] Malik, A. A. & Bilberg, A. (2017). Framework to Implement Collaborative Robots in Manual Assembly: A Lean Automation Approach, Annals of DAAAM for 2017 & Proceedings of the 28th International DAAAM Symposium, Published by DAAAM International, Vienna, 2017, ISBN 978-3-902734-11-2

[8] https://eluc.kr-olomoucky.cz/verejne/lekce/1074 (2020). Abilities, Measurement errors and their causes, Accessed 2021-01-10

[9] Http://Www.Univerzita-Online.Cz/Mng/Psychologie-V-Ekonomicke-Praxi/Vykon-Vykonnost-Pracovnika-Pracovni-Rezim (2021). Worker Performance and Efficiency Accessed 2021-01-10

[10] Buransky, I.; Vaclav, S.; Pokorny, P. & Benovic, M. (2011). Fundamental Facts about Manual Assembly Systems, Annals of DAAAM for 2011 & Proceedings of the 22nd International DAAAM Symposium, Published by DAAAM International, Vienna, 2011, ISBN 978-3-901509-83-4

DOI: 10.2507/32nd.daaam.proceedings.043

CONSTRUCTION OF TOOL FOR MACHINING HARD MATERIALS

Michal Povolny, Tomas Zatloukal & Jan Rehor

This Publication has to be referred as: Povolny, M[ichal]; Zatloukal, T[omas] & Rehor, J[an] (2021). Construction of Tool for Machining Hard Materials, Proceedings of the 32nd DAAAM International Symposium, pp.0299-0305, B. Katalinic (Ed.), Published by DAAAM International, ISBN 978-3-902734-33-4, ISSN 1726-9679, Vienna, Austria
DOI: 10.2507/32nd.daaam.proceedings.043

Abstract

This paper is based on the knowledge gained in the field of development and construction of tools for machining very hard and difficult to machine materials based on thermal spraying (for example NiCrBSi, Fe13Cr, Stellite 6 and others) and printed materials such as Inconel 718. Due to the methods of coating production, using heat technologies and their specific composition due to increased hardness and wear resistance, their conventional machining is very difficult. A common way of machining the coatings thus formed is by grinding. However, grinding productivity is low and production costs are relatively high. Therefore, it is necessary to explore the possibilities of machinability of thermal sprays and increase the cutting ability of tools, especially through more productive approaches and technologies. When machining thermal sprays, a large heat is also generated in the cutting area, which further helps to wear the tool edge faster. In the conventional method of machining by turning and milling, cutting materials such as cubic boron nitride (CBN) or polycrystalline diamond (PKD) are often used, precisely because of their high hardness or heat resistance.

Keywords: tool design; construction of tool; machining; thermal spraying

1. Introduction

Certain problems arise when milling or milling difficult to machine materials, especially hot-dipped materials. One of the problems is plastic deformation affecting the base material due to the effect of the small depth of cut. This poses a risk of possible cracking or partial tearing of the hot-dip coating. Another issue is the high temperature in the cutting area, which causes in particular an increase in the wear intensity of the inserts. This leads to a deterioration of machinability and especially to an increase in the wear intensity of the insert. High hardness also increases the requirements for tool rigidity due to high cutting forces. [1] [2] [17]

Because of plastic deformation and the need for very small cutting depths of up to 0.5mm, tool tip machining is a problem. Therefore, it is advisable to use a so-called "linear cutting edge". This way the tool does not work through the tip, but the whole length of the cutting edge can be used. Due to the small cutting edge radius, very little wear can be achieved and it is also possible to use the entire length of the cutting edge and thus increase the life of the insert. [1] [2] [17]

On the basis of the experiments already carried out, high wear and low lifetime of the inserts during hot spray machining by conventional turning method, i.e. in the case where the tip of the insert enters the cutting point, was evaluated as one of the problems. Based on these findings, due to the high temperature of the cutting point and the low life of the insert, it is proposed to optimize the cutting tool to bring the coolant as far as possible into the cutting point and to use the full length of the inserts (linear cutting edge) to eliminate the aforementioned shortcomings. [1][2]

Thus, the main idea is to make the wafer cooling more efficient and to use the full length of the cutting edge of the wafer, thus making the machining of the hot stampings more efficient and increasing the life of the wafer. The next point is then to increase the rigidity of the tool and to develop a tool that meets the above requirements. [1][2]. Since the design of the ideal cooling and placement of the insert so that the entire length of the cutting edge can be utilized while maintaining or even increasing tool stiffness is a rather complex design task, the use of topological optimization using additive technologies appears to be the ideal design method. [1][2] [15]

2. Summary of the current state in the field of printed tools and cooling

When machining certain types of materials, it is necessary to use a cooling or lubricating medium during the cutting process. During cutting, heat is generally generated by the friction of the surfaces between the tool and the material being cut, as well as by plastic deformation at the point of chip formation. The dissipation and reduction of heat at the cutting point then has positive results on the durability of the cutting edge and the resulting quality of the machined surface. The cooling system then has positive effects on the actual cutting, lubrication and cleaning of the cutting area. The medium, usually a liquid, can be fed to the cutting point in a conventional external way. The second way (internal), which is increasingly used nowadays, is to feed it directly through the cutting tool, which increases the cooling efficiency and thus the machining performance, thanks to the possibility of more efficient and precise delivery of the fluid to the cutting point. The fluid can be supplied at standard or elevated pressure, referred to as high pressure cooling. Where this type of high pressure cooling allows better penetration of the fluid between the chip and the tool face/back, thus aiding chip separation and better heat dissipation from the cutting point. [3] [17]

Nowadays it is common to design tools with internal cooling, such as sintered carbide drills, where it is possible to produce cooling channels in the helix, or milling and turning knives with straight drilled channels. The disadvantage is the impossibility of bringing the channels to exactly the required locations to make cooling more efficient. It is this disadvantage that is eliminated by topological optimization using additive technology, where it is possible to achieve the ideal shape and course of the channel as well as its precise alignment to achieve the most efficient cooling and reduce turbulent fluid flow in the channel and fluid pressure losses. [3] [17]

The following text describes the state of the art in the field of process fluid feed solutions and control options, as well as tools constructed using topological optimization and additive technologies. The first tool that is already conventionally offered is the CoroMill® 390 lightweight cutter manufactured using additive technology by Sandvik Coromant. In designing the cutter, topological optimization was used to create an optimum cutter design to achieve minimum weight and compactness of the tool. The cutter is printed in titanium alloy, which further helps to reduce the resulting weight. The lower overall weight in this case helps to minimize vibrations and increase machining safety when machining with long tool unloading, thus also increasing machining productivity. The milling cutter is particularly used in milling operations with long tool unloading, where the adapter houses the damping mechanism. The low weight of the tool as well as the short distance between the damping mechanism and the tool cutting edge thus helps to reduce vibrations and thus increase the performance and improve the safety of the machining process. The milling cutter can be used for face and side milling of deep corners, cavities and grooves and is available in 40 or 50mm diameter, with a choice of three or four teeth with uneven tooth spacing and internal coolant supply. [4]

Fig. 1. Sandvik CoroMill 390 [4]

LMT's tool is based on a similar principle. This tool uses additive technology to create curved cooling channels. This delivers the fluid exactly where it is needed. According to tests conducted, this design change has resulted in a 50% increase in VBD life when milling titanium. The following question is a prototype of LMT's coolant channel tool. [5]

Fig. 2. A tool developed using 3D printing and optimization of cooling channels [5]

Makino also has a specific approach to lightweighting the tool. The tool pictured below is designed for machining gearboxes and motor housings. Given the size of the tool, the aim is to make it as light as possible. Particularly because of the ability to change cutting speeds quickly and also to move the tool quickly in the tool magazine. Makino uses additive technologies in conjunction with a carbon fibre composite tool body. This achieves very low weight while maintaining high strength and rigidity. [5]

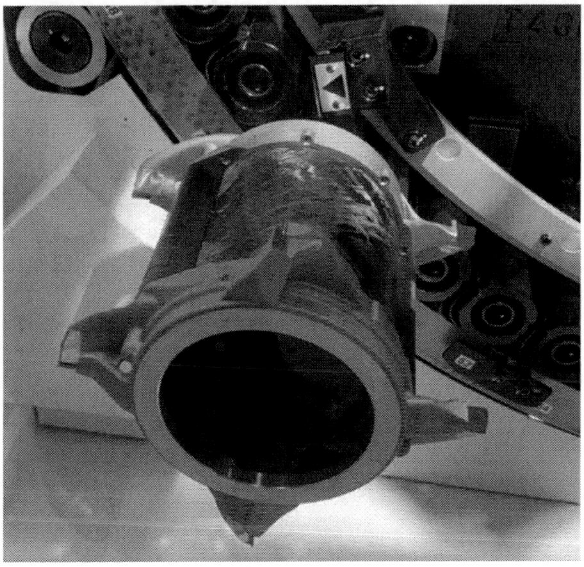

Fig. 3. Lightweight tool with composite body [5]

Another tool that is based on the principle of optimization and 3D printing is a cutter developed at the University of West Bohemia. This milling machine is based on the principle of a bar construction, where the individual insert carriers and the tool body are connected by spacers and struts are incorporated between them. This achieves sufficient rigidity of the tool while reducing its weight as much as possible. The body of the tool is also fitted with nozzles which are directed onto the face and back of the interchangeable inserts, serving to cool the tool more efficiently. The entire tool is printed in tool steel, due to the difficulty of machining and high cost in the case of titanium printing. The tool is designed primarily for roughing operations. The tool's bar structure is not manufacturable by conventional manufacturing methods, so additive manufacturing is used. Another advantage is the material savings, up to 60% less weight compared to standard tools, while maintaining the rigidity and strength of the tool. [6] [7] [14] [16]

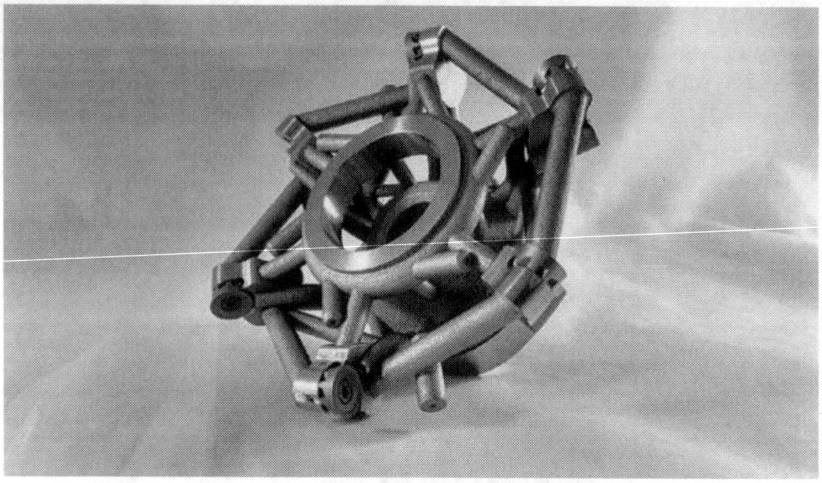

Fig. 4. KRAKEN milling head [7]

The following patents are based on the principle of controlling, opening and closing the cooling channels. The first patent described in US8696252 is a cutter with an adjustable cooling mechanism. Where, on the circumference of the tool, helically mounted inserts with cooling channels are fed to the inserts faces. The tool is designed as a conventional cutter without the use of optimization and 3D printing. A center pin is located in the center of the cutter and serves as a mechanism to open and close certain cooling channels. The pin is hollow and has grooves in which the outlet of the cooling channels is located. These can then be further connected to the channels in the body of the tool. The grooves on the body of the bolt can be provided with a sealing ring and thus the flow of coolant into the tool body and to the replaceable inserts can be sealed (interrupted). The stud is threaded and the stud must be unscrewed each time the cooling is adjusted, meaning adding or rearranging the seal. This solution is therefore quite tedious, especially in a situation where there is a frequent need to open and close certain channels according to the type of material being machined, the machining method and the cutting conditions. The following figure shows a prototype tool with a pin controlling the fluid supply. [8]

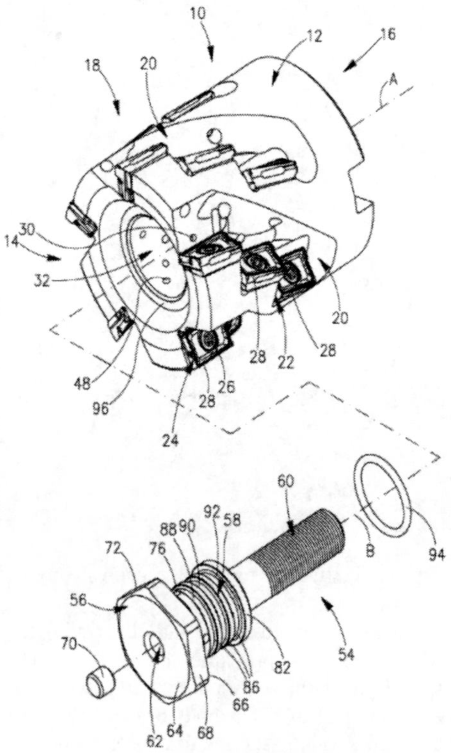

Fig. 5. Patent of tool with adjustable cooling [8]

The invention CZ 306957 is based on a similar principle. It is a cutter which is attached to a chuck by means of a hollow screw equipped with a system for supplying coolant to the cutting point. The cutter thus has an internal fluid supply through the tool body and through the hollow screw, thus achieving more efficient cooling at the face and directly at the cutting point. However, this design of the tool does not allow for control in the form of partial closure of the cooling channels compared to the previous design. The design of the tool with the hollow locking screw is shown in the following figure. [10] [13]

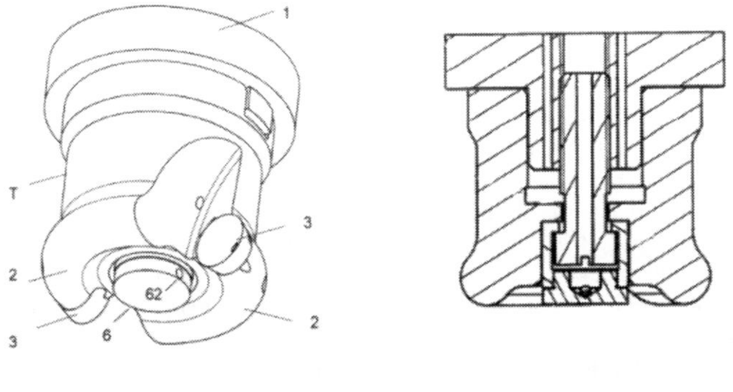

Fig. 6. Design of rotary tool with internal cooling [10]

A different view of coolant control is offered by patent US9434011.This patent addresses the "spray" nozzle that is located at the outlet of the coolant channel facing the face of the plate. The nozzle consists of an insert that is screwed into the body of the tool and then an adjustable nozzle itself with a slot-shaped outlet. The advantage of this nozzle design is both the two-axis adjustability and rotation and the slotted outlet shape. This makes it possible to direct the coolant directly to the cutting point and the entire length of the blade, thus achieving better cooling and chip breaking. [9]

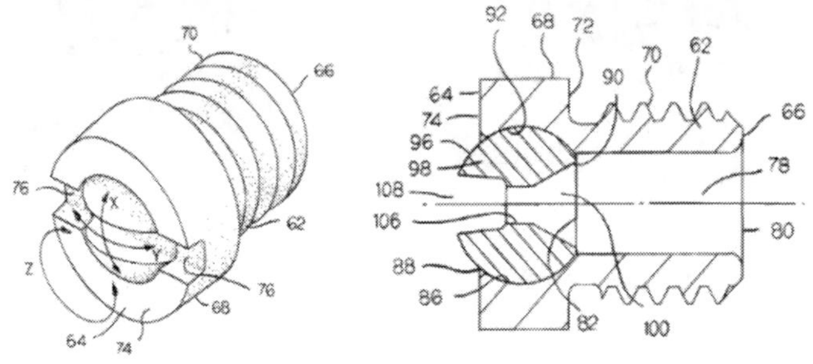

Fig. 7. Coolant spray nozzle [9]

Many other patents and inventions, such as US8573098B2 or US7125207B2, are based on the principle of feeding the cutting fluid through the central clamping screw. However, none of these solutions address the possibility of regulating the supply of cutting fluid to the cutting point, or the possibility of supplying fluid only to the face, only to the back, or both at the same time, depending on the need and method of machining. [11] [12]

3. Prototype of a machining tool

From the findings presented in the previous chapter, it is clear that additive technologies combined with topological optimization enables a new and completely different perspective on the design of machine tools. By using these technologies, it is possible to achieve improvements in tool properties, in particular in terms of addressing the cooling of the cutting edge of the tool, increasing its strength and stiffness, or improving its damping properties, for example. A major benefit of 3D printing is the ability to optimise the cooling channels, thus achieving the desired shapes and the supply of cutting fluid exactly to the cutting point with the desired channel pattern in the tool. Losses caused by, for example, perpendicular drilling of the channels or inaccurate direction of the cutting fluid are then eliminated. [15]

None of the tooling solutions analysed offers a combination of features and technical solutions that are suitable for use in hot runner machining. This means designing the tool to be able to use the full length of the cutting edge during machining, and to be able to mill and also possibly turn when fitted with the same inserts for both operations. For this reason, it is also necessary to regulate the cutting fluid supply, specifically the ability to close and open the cooling channels according to the required operation and other machining parameters. This is also related to the optimised coolant supply to the back and front of the insert along the entire length of its blade, precisely because of the use of the entire length of the blade.

The second, and no less important, part is the optimisation of the tool design to increase its strength and reduce its susceptibility to instability in the cutting process, for example due to vibrations from the cutting forces. This can be achieved precisely on the basis of topological optimisation, where the design is designed to eliminate as far as possible the influence of cutting forces on the instability of the tool and the associated vibrations. [15]

On the basis of the obtained knowledge and the shortcomings identified from it, a prototype of a special tool was designed, especially for machining of hot spray. I participated in the development and design of this tool myself and also carried out and pre-experiments and experiments of NiCrBSi and Fe13Cr spray machining. The tool will then be further optimized based on the evaluation of the experiments and machining experience of these injections. As already mentioned, the tool is designed for both milling and turning, for this reason it has a specific position of the interchangeable inserts in relation to the tool axis. This makes it possible to machine with a linear cutting edge and use the full length of the cutting edge. Thanks to additive technologies, it was possible to design the cooling channels with the smoothest possible flow so that there is no loss of fluid flow. Another advantage of using additive technology is the change in cross-section of the channels from an inlet circular to an outlet oval, which covers the entire length of the blade, a very important feature in linear cutting edge machining. The channels are also brought to the face and back of the insert for more efficient cooling. One of the main benefits is then the ability to regulate the coolant using the control elements on the side of the tool. With these members, it is possible to cool either the face/back or both at the same time, independently for each inserts separately according to the needs for the specific machining operation. Each control member therefore has four positions: all open, face only open, back only open and all closed. This makes it possible, for example, to open the cooling on only one insert in the case of turning, and thus achieve higher fluid pressure than when the cooling is also open on inserts that are not being machined. The following figures show the design of a prototype tool. [16] [17] [18]

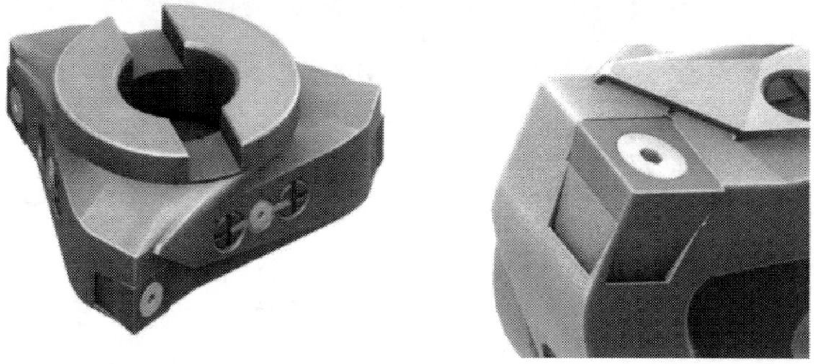

Fig. 8. prototype of tool

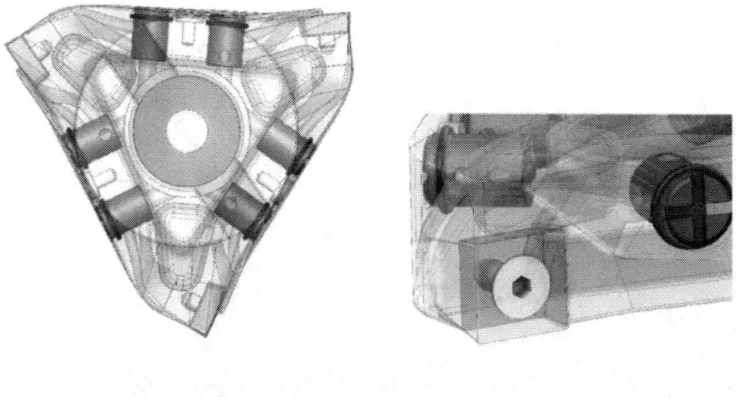

Fig. 9. Optimized channel guidance inside the tool

4. Conclusion

The main objective of the thesis was to elaborate the knowledge and introduction to the problem of design optimization. Due to the scope of the thesis, it does not contain a comprehensive view of the issue, but only a basic introduction to it. The review part is then devoted to the analysis of the current state and level in the field of development and design of tools using topological optimization and additive manufacturing, some part is also devoted to the solution of coolant supply to the cutting point and its possible regulation. On the basis of the research, the design of the tool design, evaluation of the knowledge and setting of objectives for the follow-up work, which will be focused on the development and design of the tool for machining not only for hot spraying, and on the experimental activities associated with it. The aim of the follow-up work will be to link the knowledge gained so far in the given issue and to deepen it. [17] [18]

5. Acknowledgments

This article was created under the project SGS-2019-008: Research and Development for Innovation in the Field of Manufacturing Technology - Machining Technology III.

6. References

[1] Zatloukal, T., Řehoř, J., Fulemová, J. & Povolný, M. (2020) Optimization of the machining process of NiCrBSi thermal spraying. In Annals of DAAAM and Proceedings of the International DAAAM Symposium. Vienna: DAAAM International Vienna, s. 632-637. ISBN: 978-3-902734-29-7, ISSN: 1726-9679

[2] Zatloukal, T., Gombár, M., Fulemová, J., Řehoř, J. & Povolný, M. (2019) Optimization of the thermal spraying process depending on the surface roughnes. In Proceedings of the 30th International DAAAM Symposium "Intelligent Manufacturing & Automation". Vienna: DAAAM International Vienna, s. 1026-1033. ISBN: 978-3-902734-22-8, ISSN: 1726-9679

[3] Humár,A., (2004) „Technologie obrábění - 2.část," Brno 2004. [Online]. Available from: http://ust.fme.vutbr.cz/obrabeni/opory-save/TI_TO-2cast.pdf.

[4] https://www.sandvik.coromant.com. Accessed on: 2020-08-26

[5] https://www.additivemanufacturing.media/blog/post/am-is-advancing-into-tools-for-machining-examples-from-emo, (2020) Accessed on: 2020-08-27.

[6] https://rti.zcu.cz/export/sites/rti/dokumenty/dokumenty_public/KRAKEN-Special-light-milling-head.pdf (2020). Accessed on: 2020-08-26.

[7] https://rti.zcu.cz/laboratore/obrabeni/index.html (2020). Accessed on: 2020-08-26.

[8] Bonnarang, F. & Christian, G. (2014) Rotary cutting tool having an adjustable cooling mechanism: US8696252. Available from: https://worldwide.espacenet.com/publicationDetails/biblio?DB=EPODOC&II=0&ND=3&adjacent= true&locale= enEP&FT=D&CC=US&NR=8696252B2&KC=B2 . Accessed on: 2020-08-26.

[9] Morrison, M., Verellen, J. & Long, T. (2016) Coolant spray nozzle and cutter with coolant spray nozzle: US9434011. Available from: https://worldwide.espacenet.com/publicationDetails/biblio?DB=EPODOC&II=0&ND=3&adjacent= true&locale=en_EP&FT=D&CC=US&NR=9434011B2&KC=B2 (2020). Accessed on: 2020-08-27.

[10] Zetek, M. & Schorník, V. (2016) Fréza s chlazením Available from: https://patentimages.storage.googleapis.com/78/72/15/79a7a0195505f4/CZ306910B6.pdf (2020). Accessed on: 2020-08-27.

[11] Craig, K., Long, T. & Frota De Souza, F. (2004) Tool holder with integral coolant channel and locking screw therefor: US7125207B2. Available from: https://worldwide.espacenet.com/patent/search/family/035757578/publication/ US7125207B2?q=pn%3DUS7125207B2&queryLang=en%3Ade%3Afr Accessed on: 2020-08-27.

[12] Hoffer, B., Long, T. & Frota De Souza, F. (2011) Cutting tool including a locking screw and adapter with coolant delivery: US8573098B2. Available from: https://worldwide.espacenet.com/patent/search/family/046705531/publication/US8573098B2?q=pn%3DUS8573098B2&queryLang=en%3Ade%3Afr. Accessed on: 2020-08-27.

[13] Hanzl, P. & Zetková, I. (2019) Benefits of a new approach to designing milling cutter using Metal Additive Manufacturing, Published by Manufacturing Technology, ISSN 1213-2489, Vol. 18, No. 2

[14] Hanzl, P., Zetková, I. & Daňa, M. (2018) Uniaxial Tensile Load of Lattice Structures Produced by Metal Additive Manufacturing. In ICKTI 2018 nové poznatky v technologiích a technologické informace '18 sborník příspěvků. Ústí nad Labem, ČR: Ústav technologií a materiálů, Fakulta strojního inženýrství, Univerzita J. E. Purkyně v Ústí nad Labem, Česká republika, s. 57-61. ISBN 978-80-7561-114-7.

[15] Hanzl, P., Zetek, M., Rulc, V., Purš, H. & Zetková, I. (2019) Finite Element Analysis of a Light-weight Milling Cutter for Metal Additive Manufacturing, Published by Manufacturing Technology, ISSN 1213-2489, October 2019, Vol. 19, No. 5, Pp. 753-758

[16] Zhang, W.; Jihong Z. & Tong G. (2020). Topology Optimalization in Engineering Structure Desing. iSTE Press . ISBN 9781-178548-224-3.

[17] Liborius, H., Grund, T., Nestler, A., Paczkowski, G., Schubert, A., & Lampke, T. (2021). Influence of the finish-machining by turning and diamond smoothing on the tribological properties of Fe17Cr2Ni0.2C thermally sprayed coatings. Surface and Coatings Technology, 405 doi:10.1016/j.surfcoat.2020.126731

[18] Kim, S. H., & Park, M. S. (2017). Improvement of STS316L milling characteristics according to coolant spray position. Transactions of the Korean Society of Mechanical Engineers, A, 41(5), 427-433. doi:10.3795/KSME-A.2017.41.5.427

DOI: 10.2507/32nd.daaam.proceedings.044

BENEFITS AND RISKS OF MULTIMODAL TRANSPORT WITH A MAIN FOCUS ON THE EU

Barbora Kotkova & Martin Hromada

This Publication has to be referred as: Kotkova, B[arbora] & Hromada, M[artin] (2021). Benefits and Risks of Multimodal Transport with a Main Focus on the EU, Proceedings of the 32nd DAAAM International Symposium, pp.0306-0310, B. Katalinic (Ed.), Published by DAAAM International, ISBN 978-3-902734-33-4, ISSN 1726-9679, Vienna, Austria
DOI: 10.2507/32nd.daaam.proceedings.044

Abstract

The article provides an overview of the main benefits and risks of multimodal transport, focusing not only on the European Union. The following is a list of possible measures that can be implemented shortly and those that the states of the Community of the European Union must reach together. Finally, the article presents a model case of the application of multimodal transport and common recommendations leading to the expansion of multimodal transport. The realization of these goals should be common to all states to stop the devastation of the natural environment and improve living conditions.

Keywords: environment; risk; safety; transport.

1. Introduction

International trade could not exist without a substantial part of it, namely transport. The amount of goods transported is constantly increasing and this trend will continue. It is, therefore, necessary to work now to increase the capacity of individual transport sectors. Part of this task is to improve transport services and speed up transport times. The volumes of passenger and freight transport are constantly increasing, not only within the Czech Republic but also in other European Union countries. The ecological consequences of this development have for a long time been noticeably negatively reflected in human life and the environment. One of the methods that reduce these adverse effects and that can also be applied immediately and effectively is multimodal transport.

Multimodal transport has been defined by the European Conference of Ministers of Transport (ECMT) as transport in which more than two different modes of transport are used to transport goods. The UN Convention on Multimodal Transport is also concluded, concluded in 1981. Multimodal transport is also characterized here as the transport of goods by at least two different modes of transport. For comparison with combined (intermodal) transport, in which the use of the same freight unit is assumed when changing the type of transport, in multimodal transport, there may also be a change in the freight unit, means, or distribution of cargo. In most European countries, road transport predominates in the transport of goods and materials. The total volume of road transport in the European Union decreased from 57,092 million thousand km in 2007 to 42,588 million thousand km in 2016.

However, although the overall indicator for the EU has decreased, this trend does not apply in all countries. In Germany, Spain, France, Portugal, and Poland, for example, there has been an increase. The main reasons for not using multimodal transport to a greater extent are low transport volumes and / or short transport distances. Within Europe, transport distances are indeed very short or there are insufficient and high-quality connections.

It is therefore clear that European countries must actively cooperate in solving the following problems, both in terms of passenger and freight transport:

- Air and road transport is permanently congested. This congestion will increase in the future and the volumes of freight and passenger transport will continue to grow.
- Transport is still 96% dependent on oil. It has long been known that there will be a clear shortage of oil soon, which is also found in fragile countries.
- Greenhouse gas emissions - global warming is taking place and this trend must be significantly and effectively curtailed
- Transport infrastructure varies significantly within the European Union. In some countries, there are also large differences and shortcomings within a territorial unit.
- Non-European markets are experiencing a boom in transport development and these European markets are ultimately difficult to compete with.

In the Member States of the European Union, road transport predominates in freight transport. The average value of the share of road freight transport in 2016 is 76.4% of the total volume of transport. Due to the larger share of rail transport, two countries are an exception, namely Latvia. Here, the share of road freight transport is only 23.4%. The second country is Lithuania, where the share of road freight transport in 2016 corresponded to 35.0%, compared to rail, which was 65.0%. For other modes of transport, the Netherlands uses inland waterway transport for freight transport, with a significant share of 44.6%, Romania 29.4%, and Bulgaria 27.2% in 2016. [1]

It is therefore highly desirable to carry out studies such as this article, collecting and comparing national and community objectives. These can then be used in individual international meetings to pass on the experience that helped to achieve the goal or not. Each database, which is complete and accurate, provides the basis for another professional researcher to eliminate dead ends or, conversely, to develop those that appear to be effective.

2. Characteristics of basic types of transport and their negative effects

Transport consists of the following elements: means of transport, transport infrastructure, transport facilities and legislation. [2], [3] The product of transport is transport, which is divided into several types:

- Road transport - it is mainly used for short and medium distances. The advantages include speed and availability, the disadvantage is the high negative burden on the environment and higher accidents.
- Rail transport - is used to transport large volumes, the range is rather medium to long distances and there is a relatively high level of safety. Compared to other modes of transport, electricians have lower energy consumption and emissions.
- Water transport - The high cost of building inland canals does not allow for the necessary expansion, so in the future, it will be more of an effort to maintain and expand the existing ones (applies to the EU).
- Air Transport - It is used primarily for the air transport of passengers, mail, and small parcels. The disadvantage is the high burden on the environment.
- Transport of bicycles and pedestrians - It is mainly used for short transfers, especially in combination with other modes of transport. Both are environmentally friendly, do not cause disturbing noise, and have a positive effect on human health.

Emissions individual transport modes	2019							
	CO2	CO	NOx	N2O	*compound	CH4	SO2	particles
Transport - total	21 118	75 974	59 921	705	14 209	949	215	4 400
Individual car transport	12 223	62 948	29 050	344	11 146	734	85	2 521
Public transport by road	601	768	2 749	12	93	82	3	103
Road freight transport	6 693	7 030	20 119	209	942	68	43	1 630
Motorbikes	46	2 197	59	1	891	40	0	10
Rail transport	270	1 657	2 848	103	394	15	2	128
Water transport	10	59	102	0,3	14	0,9	0,1	8
Air transport	1 275	1 314	4 994	36	728	9	82	0

*volatile organic compounds

Fig. 1. Overview of the production of emissions of individual modes of transport in 2019 [4]

The Czech Republic was selected as a reference element for comparing emissions from transport. The results published in the 2019 yearbook by the SYDOS traffic statistics system were used. Total emissions from individual modes of transport for 2019 are shown in Table 1. [4] The largest amount of emissions, taking into account the volume of transport, is produced by air transport. Given that water transport is minimally represented in the Czech Republic and its transport, it appears to be the most environmentally friendly mode of transport in terms of emissions.

3. Multimodal transport in the EU

The economic development of the European Union depends to a large extent on the development of transport and related services. The market and its development must be ensured by the existence and support of quality infrastructure across the continent. It is necessary to remove narrow and problematic places of the so-called transport network neck. These are due to the lack of harmonization and fragmentation of Member States' transport infrastructures. The European Union has set up the EU's Common Transport Policy project to eliminate these obstacles. The aim of European transport policy until 2030 is to transfer 30% of current road freight transport for distances longer than 300 km in the European Union to rail or waterborne transport. Other priorities of the European Union include reducing greenhouse gases produced by individual transport areas, according to the Commission's so-called White Paper by 60% by 2050 compared to 1990. Multimodal transport and its main benefit - reducing the negative impact of road transport on the environment are offered as one of the current possible solutions.

To sufficiently understand the process of implementing multimodal transport, the following table was created by SWOT analysis. It shows the strengths and weaknesses of this transport, the risks, and future opportunities, influencing the development in a positive direction. The assessment of risks related to multimodal transport will vary according to the specific dispositions of the evaluated EU countries. [5], [6].

Strenghts	Opportunities
• Current market and future market development • Existing number and suitable location of transshipment depots, logistics and distribution centers, available storage area, logistics operators, number of transport companies • An increasing number of multimodal operators • Sufficient number of trucks, trailers, and semi-trailers • Infrastructure availability • High level of education of employees and future ones • High demand for transport and logistics services • Market saturation level • Market dynamics and new investments • Financial complexity of entering the market • Medium level of company cooperation • • High competition in freight transport and logistics	• Economic growth • Promoting greener transport modes due to the high price of oil • Strategic transport position, new investors, state investments, and subsidies • Taxes and tolls - increase • Stability of EU policy - security, duty-free union • Intensification of cooperation between entities • Transit through international corridors • Taking environmental aspects into account in sustainable transport policy • Development of multimodal transport to reduce transport costs • Support for the development of multimodal transport from the EU • Development of transport infrastructure within individual transport modes • Development of shipping containers and their standardization • Development of new technologies
Weaknesses	**Threats**
• Low number of employees • Low wages for professionals • High employee turnover • Low quality of roads, waterways, and railways • Insufficient level of investment in the development of freight transport by the state • Support for financial institutions, government institutions • Emergency condition of some roads • High purchase price of ecological cars • Higher price of alternative fuels • Limited offer of means of transport • Lack of joint European projects • Small infrastructure condition check • Inconsistent transport rules in the EU • Different amounts of investment in different countries • Absence of high-speed rail	• Lack of employees • High labor costs • Financial risks associated with long-term projects • Lack of transparency of political influences • Obstruction of institutions in investments in transport • Roadblocks ecologists, landowners, strikes • Infrastructure problems such as poor quality, low capacity, construction delays, and infrastructure upgrades • Lack of financial projects in the national budget • Legislative delays, restrictions, and heavy bureaucracy in terms of time • Growth in passenger transport • Slow rail transport

Fig. 2. SWOT analysis of the Multimodal transport in EU - Author

The Czech Republic strives above all for the innovation of international transport in particular, which is only partially carried out here. The increased use would therefore mainly concern rail transport, to a lesser extent also water transport. This would reduce exhaust emissions, the largest production of which is road transport. If road transport cannot be replaced, either due to infrastructure or the type of goods transported, the use of other types of propulsion (hydrogen, electricity) is an alternative. [7] There are 13 currently used multimodal terminals in the Czech Republic. On four of them, it is possible to realize transshipment between three modes of transport, at other terminals, it is possible to tranship only between road and rail transport. The benefits of intermodal transportation development lie in the process of connecting all the operators and parties involved in the international trade between Southeast Asia and the EU countries. [8]

4. Multimodal transport chain and its organization

Transport is part of the logistics chain, which consists of activities of an administrative, legal and technological nature. In the case of the transport of products by multimodal transport, in which several modes of transport are interconnected, the organization and coordination of the whole transport process are considerably more complicated.

In normal transport, the chain works in such a way that the consignor, consignee, or trade organization enters into a transport contract with a specific carrier. The selected carrier will perform the transport using the means of transport for the given type of transport and the given type of product. In more complicated multimodal transports, another entity enters the process. This is a freight forwarder who concludes a contract with the sender. Its function is to mediate and coordinate all activities related to the transport - warehousing, customs procedures, concluding transport contracts with carriers, insurance. At the same time, he is responsible for the given transport to the sender according to the concluded contract. In the case of multimodal transport, there is a so-called multimodal transport operator (MTO), which ensures the implementation of the entire transport chain. They provide transport through their own or leased terminals and cooperate with foreign multimodal transport operators.

Model case

- The customer who is interested in the implementation of the transport will contact the forwarder (MTO) with a request for a price offer for transport. This is based on the address of loading and unloading, quantity, weight, and time conditions.
- After receiving the offers, a comparison and evaluation will take place. Confirms the selected order, concludes a transport contract. The forwarder hereby undertakes to transport the goods properly to the designated place for the ordering party, on time and with optimal conditions. The goods must be delivered undamaged and in unchanged quantity and quality.
- The carrier will provide a suitable transport unit, which must be manageable at both the consignee and the consignor and must be accessible to the consignor. Large ISO 1 deep-sea containers for deep-sea transport are available in larger multimodal transport terminals.
- The road carrier will take over this transport unit from the sender and transport it to the nearest agreed multimodal transport terminal. After loading, the combination goes to the nearest multimodal transport terminal, where the semi-trailer or swap body is transferred to the railway car and the vehicle (tractor) leaves the terminal. It is very important to follow the departures of combined transport trains from the terminal so that it is still possible to ensure their transfer to the railway car of the respective train from the multimodal transport terminal. In case of non-compliance with the cut-off time in the terminal, the containers will be placed on the storage area of the multimodal transport terminal, where the containers are left until the departure of the next train. This storage service is charged, especially for longer periods.
- At the destination terminal, the road hauler or the transport operator contracts another road hauler to transport the goods to the consignee.

In the transport process, continuous information about any extraordinary event or change on the part of the sender, carrier, or final consignee is required. In the event of an emergency, it is necessary to immediately activate the relevant and pre-arranged emergency scenario. The final evaluation of the transport should always be performed. This can also be due to unplanned logistics costs, non-compliance with agreed parameters, or quality Databases with environmental performance data for transport chains should be developed. [9]

5. Conclusion

Interest in multimodal transport in the European Union is still growing. The use of multimodal transport will divide the transport into more transport modes. This situation then makes it possible to take advantage of their benefits and at the same time avoid some of the problems and risks associated with specific modes of transport. The shift of costs from the most frequently used road transport to multimodal transport is supported by the entire European Union. It committed itself to this in the document Concept of Freight Transport for the period 2017 to 2023 with a view to 2030. The expansion of multimodal transport is complicated by a number of obstacles that hinder the necessary and rapid development.

The main reason is the critical emergency condition of the infrastructure in some regions, the existence of equipped transhipments, especially railways. These include, for example, narrow capacity bottlenecks on important main lines. This reduces the reliability and timeliness of train journeys to destination terminals. The solution is, for example, the construction of corporate terminals, which will enable the use of new transport solutions with universal cars and interchangeable superstructures. It is possible to use the financial support of the European Union for their construction. The development of transport in the EU is identified in the White Paper on Transport and in the strategic documents of individual member states. Transport is the basis of the European economy and society, and the mobility of goods and people is very important. It is important to increase traffic, promote mobility and at the same time aim to reduce emissions by up to 60%. To this end, it is necessary in the future to create a new common pattern of transport, according to which transport will be realized by means of the most efficient fastening means or their interconnection.

The European Union for Transport Development must meet the following main requirements:

- increasing the energy efficiency of vehicles,
- optimization of multimodal logistics chains,
- greater involvement of road traffic management systems and information,
- financial support from public funds of the state and its investments in the development of road transport and infrastructure
- construction and modernization of multimodal terminals and logistics centers, as well as improving access to the use of EU funds in this area
- the creation of common European funds for multimodal transport to support the development of this transport system,
- favoring multimodal chain carriers
- support for multimodal concessions

Another EU transport target for 2050 is to shift 30% of road transport to 2030 and to 50% by 2050. Rail or water transport will help to do this, which is an alternative in terms of cost and less negative effects on the environment. In particular, road traffic will be reduced and thus carbon monoxide emissions will be reduced. The issue of multimodal transport will continue to require increased attention and effective solutions in the future. Therefore, individual progress will be further monitored, and comparisons will be made between individual flights and countries. These will be further published as a continuation of the article assessing the fulfillment of the objectives, how the individual states have committed themselves in the future, and recommendations on which of the successes other states and communities should adopt and which, on the contrary, did not lead to the expected results.

6. Acknowledgments

This research was based on the support of the Internal Grant Agency of Tomas Bata University in Zlín, the IGA / FAI / 2021/002 project and the Institute of Safety Engineering, Faculty of Applied Informatics.

7. References

[1] https://ec.europa.eu/eurostat/data/database (2021) Accessed on: 2021-02-15

[2] Hanus, J. (2014). Comparison of individual modes of transport with regard to the impact on the environment. Universita Jan Perner. Pardubice

[3] Besta, P. (2021) Comparison of individual modes of transport. College logistics. ISBN 2021-03-28. Prerov, Czech republic

[4] https://www.sydos.cz/cs/rocenka-2020/rocenka/htm_cz/index.html (2021) Accessed on: 2009-08-13

[5] Tichy, M. (2006). Risk management: Analysis and management. ISBN 80-7179-415 Prague

[6] Management Mania, Risks, (2021). Accessed on: 2021-03-28.

[7] Ministry of transport of the Czech Republic. (2016) Freight concept for the period 2017–2023 with wiew to 2030. Accessed on: 2021-03-28.

[8] Bendekovic, J., Rogozar, Z., Naletina, D. (2014) Intermodal transport is brand of Croatia transport offer, Proceeding of DAAAM International Symposium, ISBN 978-3-90150998-8, Editor B. Katalinic, Published by DAAAM International, Vienna, Austria.

[9] Nicolae, F., Popa C., Beizaheda, H. & Nistor, F. (2010) Enviromental performance assesment of multimodal transport system. Proceeding of DAAAM International Symposium, ISBN 978-3-901509-73-5. Editor B. Katalinic, Published by DAAAM International, Vienna, Austria.

DOI: 10.2507/32nd.daaam.proceedings.045

INTEGRATION SOFTWARE FOR A DISTRIBUTED GROUP OF MOBILE SERVICE ROBOTS

Pryanitchnikov Valentin, Yuri Kolesov,
Radomir Tarasov & Plotnikov Aleksey

This Publication has to be referred as: Pryanichnikov, V[alentin]; Kolesov, Y[uriy]; Tarasov, R[adomir] & Plotnikov, A[leksey] (2021). Integration Software for a Distributed Group of Mobile Service Robots, Proceedings of the 32nd DAAAM International Symposium, pp.0311-0315, B. Katalinic (Ed.), Published by DAAAM International, ISBN 978-3-902734-33-4, ISSN 1726-9679, Vienna, Austria
DOI: 10.2507/32nd.daaam.proceedings.045

Abstract

This article describes the study of the problems of testing and integration of the group of mobile service robots (Amur-307 type) hardware and software complex using the virtualization support library. The architecture of the software solution and its features, that simplify integration testing and speed up debugging of compilation and reliability errors are described. An approach to emulation of hardware, sensors and electric driver of robots with manipulators specific to this software and hardware complex is also outlined, which allows creating a complete virtual analogue of it and speeding up software development (especially in terms of routing a robotic complex) using unit tests.

Keywords: integration software; service mobile robots; emulation and testing of hardware and software; based on microprocessors and on-board computers.

1. Introduction

Currently, one of the most successful approaches to software development is continuous integration. The software must be ready at any time for launch and execution, so that execution errors do not occur at any arbitrary time [15]. This approach also justifies itself in the development of client software for the Amur-307 robotic complex[1, 2]. In particular, the developed software can be used for the entire spectrum of robotic systems, including underwater ones [17-22], providing better results compared, for example, with work [23].

When developing software in a team, it is necessary to conduct testing at various levels of abstraction: unit testing, integration testing and user interface testing. However, when the program code interacts with various robot equipment, there are moments, that add the need to conduct additional testing using physical equipment [16]. Hardware virtualization is often used to speed up full-featured testing and its automation, which allows not using physical hardware [6]. It is this method, that we have relied on in this work.

2. Applied software for development and integration

When developing client software for the Amur-307 robot, the following structure of applied technologies was formed (Fig.1.):

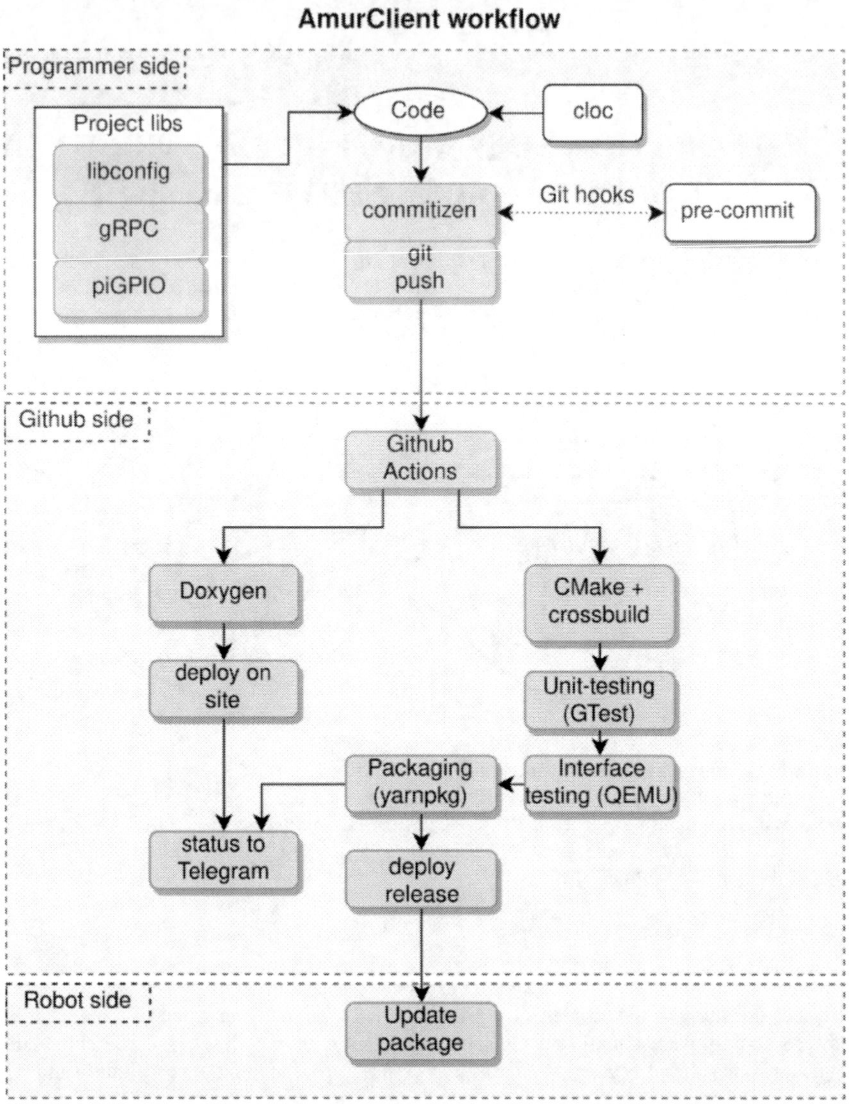

Fig. 1. The structure of applied technologies in the life cycle of client software.

The proposed architecture for testing and debugging client software allows, after making edits to the source codes, to automatically assemble and test software for the Rapsberry Pi platform without resorting to manual testing on a real microcomputer. Also, due to the need for fast and convenient work with the program code, technologies of continuous integration and continuous testing in the repository were used.

For the functioning of the client software, the need for the use of specialized libraries was identified, which would allow to root the development, simplify the testing of system modules, as well as reduce the load on the processor and network [5]. After comparative tests , the following were selected:

1. Network framework gRPC + Protobuf serialize library. This framework was needed for convenient and fast serialization of data into binary format, to reduce the transmitted data and increase the speed of exchange in comparison with the REST API. Also, this framework made it possible to maintain backward compatibility with older versions of microservices, which in turn made it possible to simply change the structures of the transmitted data.
2. piGPIO library. This library provides software for working with peripheral equipment of the Raspberry Pi microcomputer. For this microcomputer, there are several programs for working with GPIO ports, the most popular are WiringPi and piGPIO. WiringPi - at the moment the library is not supported and does not have the necessary DMA modulation to increase the number of PWM signals. At the moment, piGPIO is most often used, due to its functionality and speed of operation. For these reasons, this particular library was chosen.
3. Library libconfig. This library is used to work with configuration files. Its main advantage is cross-platform and entry into the POSIX standard. The configuration file format is readable, and the markup language itself has good functionality.

3. Applied technologies of continuous integration and deployment of the project.

For representativeness, versioning and unification of commits, a separate *commitizen* program is used. This program significantly saves time when accepting a request to add code to the repository, when viewing the commit history, and also allows you to do some work with the project in semi-automatic mode, such as assigning the current software version and generating a change file. Also, the cloc software package is used to estimate the amount of code, comments and obtain some statistics on the amount of program code. The Git system has a built-in git hooks mechanism, that allows you to perform an action on a specific event. This mechanism is used for additional verification of the commit before it is accepted and access is restricted.

Continuous Integration and Continuous Deployment technologies based on the *Github Actions* service are used to accelerate development and testing. These technologies allow you to find out the test result, generate documentation and assemble the application package on a third-party server, which in turn greatly reduced the time spent to obtain results. When a new commit is received, two YAML scripts are run on the server, one of which describes the assembly of documentation, and the second describes the assembly and testing of the project itself. The documentation is assembled by *Doxygen* [13]. Documentation and diagrams are generated based on the client software program code. After the documentation is assembled, build status alerts are sent to the Telegram work chat and if the build was successful, the documentation is updated on the project website.

To compile a project consisting of a large number of code files, it is necessary to use an assembly system that, according to a certain algorithm, will assemble all the files in the correct sequence, and also allow you to run additional precompilation steps. There are several build systems for the C++ language: *qmake, make, cmake, autotools*. Since there is no pre-installed *Qt library* in the Raspberry Pi distribution, the qmake build system is not suitable due to the large size of the library, the make system is not cross-platform like autotools. Therefore, the *CMake* build system is used to build the program, which provides a more user-friendly interface than the Makefile and is cross-platform. Since the final program must be run on an ARM processor, cross-compilation is used, which allows you to compile the project on the server and not load a low-power Raspberry Pi processor. After that, unit testing is required to evaluate the performance of individual modules using the *GTest library*. After successfully passing the unit tests, it is necessary to run interface testing to make sure that the system as a whole is working properly. For this, a QEMU virtual machine with emulation of Raspberry Pi hardware, including its I/O ports, is used. With the help of a set of programs, an image is created with the recorded Amur-307 [17].

4. Implementation of a virtual environment for integration testing.

For virtualization, the following technologies are used to provide interfaces for interaction with virtual hardware: tun/tap, iproute2, iptables, qemu, socat. An example of hardware virtualization can be found in the article [7]. Virtualization in this case uses the full implementation of the TCP/IP stack (this is necessary, among other things, for debugging the server part of the software, which remains outside the scope of this article). When creating a test virtual machine, the topology of the test network is described, since virtualization does not provide automatic access to the local computer network, so it became necessary to add support for the network stack by automatically assigning virtual network interfaces. For this purpose, the TUN/TAP virtual network interface mechanism was chosen [8] since it is already implemented in the Linux OC kernel. To solve the problem of creating, deleting and assigning an ip address, routing, and managing network bridges (the implementation of a virtual network router, also implemented in the OS kernel), the iproute2 package is used [12]. In this case, this interface was chosen for the speed of development and maximum portability of the code, due to the fact, that ifconfig and route tools similar in purpose are no longer supported in new versions of Linux-based distributions, which in turn will complicate project support.

To connect a virtual machine to a local computer network, the NAT (network address translation) mechanism is used using the iptables package[14]. In this case, the choice of the package is unambiguous: ipfwadm and ipchains are outdated and no longer supported, while the choice of iptables is due to the need to isolate the subnet of the test virtual machine for security and the implementation of an arbitrary network topology.

With the help of a *qemu-based* virtual machine, it is also launched via the socat interface [9, 10] (which emulates the serial port) we get the result of integration testing, which shows the absence or presence of execution errors when running the Amur-307 software on a virtual machine [3, 4]. The result of this launch falls into the pipeline of the sequential integration environment, in our case it is *Github Actions*.

5. Mechanism for checking the correctness of source codes and generating executable code.

Next, debugging of execution errors and new functionality is performed using the integration software described above, if any. The *qemu* virtual machine allows the use of *gpio* virtual sensors, which allows us to debug the functionality of the components of the Amur-307 client software on a virtual machine, for example, basic tests of the functionality of various sensors. In addition, it is possible to emulate the work of the robotic complex as a whole, namely, movement along the route, interaction with manipulators and much more, which significantly reduces the debugging time directly on the robot's equipment and allows you to identify machine-independent errors.

After successful testing, the installation *apk* package for *Raspbian OS* is automatically assembled, which is published in the repository releases and also published on the project website.

6. Conclusion

As a result of the conducted research, an environment for continuous development with the possibility of testing software on virtual hardware was identified, created and configured. This system allows you to save time developing software for the AMUR-307 robot by eliminating constant testing on real robot hardware, as well as counting the automatic launch of tests with each new commit. As a result of the work of this software package, after the assembly and publication of the release, the package is installed by the user on the robot or updated from the project repository. Thus, the mechanism of continuous development, continuous integration and continuous deployment was implemented.

7. Acknowledgments

The work was supported by International Institute of New Education Technologies RSUH and Keldysh IAM Russian academy of sciences in accordance with the common project "Intelligent robotronics"/

8. References

[1] Aryskin, A[lexander]; Grigorev, A[leksei]; Khelemendik, R[oman]; Petrakov, M[axim]; Pryanichnikov, V[alentin] & Tarasov, R[adomir] (2020). System for Monitoring and Control in Industrial Technological Processes, Proceedings of the 31st DAAAM International Symposium, pp.0644-0649, B. Katalinic (Ed.), Published by DAAAM International, ISBN 978-3-902734-29-7, ISSN 1726-9679, Vienna, AustriaDOI: 10.2507/31st.daaam.proceedings.089

[2] Aryskin, A[lexander]; Davydov, O[leg]; Eprikov S[tanislav]; Ksenzenko A[lexander]; Pryanichnikov, V[alentin]; Punenkov O[leg]; Shipovalov E[gor]; Solovyova M[ariia]; Stepanova D[aria]; Tarasov R[adomir] & Tikhomirov A[lexey](2020). Robotariums implementations witn the Elements of Artificial Intelligence Control, Proceedings of the 31st DAAAM International Symposium, pp.0973-0985, B. Katalinic (Ed.), Published by DAAAM International, ISBN 978-3-902734-29-7, ISSN 1726-9679, Vienna, AustriaDOI: 10.2507/31st.daaam.proceedings.1351 (P. 3)

[3] Aryskin, A[lexander]; Bogdanovich, A[lena]; Davydov, O[leg]; Grigoriev, A[leksey]; Khelemendik, R[oman]; Kharin, K[onstantin]; Kuvshinov, S[ergey]; Petrakov, M[aksim]; Plotnikov A[leksey], Pryanichnikov, V[alentin]; & Tarasov R[adomir] (2019). Control algorithms for service and industrial transport robots, Proceedings of the 30th DAAAM International Symposium, pp.1166-1173, B. Katalinic (Ed.), Published by DAAAM International, ISBN 978-3-902734-22-8, ISSN 1726-9679, Vienna, Austria. DOI: 10.2507/30th.daaam.proceedings.164

[4] Aryskin, A[leksander] A[.]; Bogdanovich, A[lena]; Davydov, O[leg]; Khelemendik, R[oman]; Petrakov,M[axim] & Pryanichnikov, V[alentin] (2019). Control and Interaction Algorithms for Industrial and Service Robots, Chapter 33 in DAAAM International Scientific Book 2019, pp.377-384, B. Katalinic (Ed.), Published by DAAAM International, ISBN 978-3-902734-24-2, ISSN 1726-9687, Vienna, AustriaDOI: 10.2507/daaam.scibook.2019.33

[5] R.B. Tarasov, O.I. Davydov, V.E. Pryanichnikov, M.D. Solovyova, A.N. Tikhomirov, E.A. Shapovalov, S.R. Epikov // Software-Control Shell for Service Robots With Parallel Computing. Extreme robotics. // Collection of abstracts of the 31st International Scientific and Technical Conference "Extreme Robotics". - St. Petersburg: CPI LLC "Polytechnic-print", 2020. - 86 p., pp.41-42. ISBN 978-5-85875-602-6.

[6] V.E. Pryanichnikov, E.A. Shapovalov, R.B. Tarasov // Intellectual Support of Logistics Missions in Medical Institutions With Mobile Service Robots AMUR-307. Extreme robotics. // Collection of abstracts of the 31st International Scientific and Technical Conference "Extreme Robotics". - Saint Petersburg: CPI LLC "Polytechnic-print", 2020. - 86 p., p.73. ISBN 978-5-85875-602-6.

[7] Simple, Painless Application Testing on Virtualized Hardware – Star Lab Software. Available at: https://www.starlab.io/blog/simple-painless-application-testing-on-virtualized-hardwarenbsp (Accessed 20 October 2011).

[8] Documentation/Networking - QEMU. Available at: https://wiki.qemu.org/Documentation/Networking#Tap (Accessed 20 October 2011).

[9] Features/ChardevFlowControl – QEMU. Available at: https://wiki.qemu.org/Features/ChardevFlowControl (Accessed 20 October 2011).

[10] Getting started with socat, a multipurpose relay tool for Linux. Available at: https://www.redhat.com/sysadmin/getting-started-socat (Accessed 20 October 2011).

[11] Ushakov Yu.A., Polezhaev P.N., Shukhman A.E. Creation of a multiservice multipoint VPN network with dynamic auto-tuning. // Bulletin of Orenburg State University. - 2015. - No. 9. - pp. 170-178.

[12] M. Shahin, M. Ali Babar, and L. Zhu, —Continuous Integration, Delivery and Deployment: A Systematic Review on Approaches, Tools, Challenges and Practices", IEEE Access, 2017. https://arxiv.org/pdf/1703.07019.pdf

[13] Doxygen, a documentation generator mainly for C++. Available at: https://starkovden.github.io/Doxygen.html (Accessed 20 October 2011).

[14] Mathematical Models and Software Protection Tools for Distributed Computer Systems: Galatenko A.V. abstract of the dissertation for the degree of candidate of Physical and Mathematical Sciences / Institute of System Programming. Moscow, 2004

[15] Davydov D.V., Eprikov S.R., Kirsanov K.B., Pryanichnikov V.E. (2017). Service Robots Integrating Software and Remote Reprogramming, Proceedings of the 28th DAAAM International Symposium, pp.1234-1240, B. Katalinic (Ed.), Published by DAAAM International, ISBN 978-3-902734-11-2, ISSN 1726-9679, Vienna, Austria, DOI: 10.2507/28th.daaam.proceedings.172.

[16] Pryanichnikov V. E. Artificial intelligence and software and hardware robotic systems. Information-measuring and control systems. - 2018. — Vol. 16, No. 12. - Pp. 3-11, ISSN 2070–0814.

[17] Aryskin, A[lexander]; Davydov, O[leg]; Eprikov S[tanislav]; Ksenzenko A[lexander]; Pryanichnikov, V[alentin]; Punenkov O[leg]; Shipovalov E[gor]; Solovyova M[ariia]; Stepanova D[aria]; Tarasov R[adomir] & Tikhomirov A[lexey] (2020). Robotariums implementations witn the Elements of Artificial Intelligence Control, Proceedings of the 31st DAAAM International Symposium, pp.0973-0985, B. Katalinic (Ed.), Published by DAAAM International, ISBN 978-3-902734-29-7, ISSN 1726-9679, Vienna, Austria DOI: 10.2507/31st.daaam.proceedings.135

[18] V.V. Arykantsev, A. A. Ariskin, O.O. Belyaev, A.Ya. Ksenzenko, E.A.Prysev, V.E. Pryanichnikov, V.V. Chernyshev, S.R.Eprikov. Supervisory control of underwater walking apparatus, Extreme robotics and conversion trends. Collection of abstracts of the international scientific and technical conference, 2018, St.Petersburg, p. 203.

[19] A.Y. Ksenzenko, Yu.S. Marzanov, E.A. Prysev, V.E. Pryanichnikov, V.V. Chernyshev, Prototyping of contactless data exchange and power supply of underwater satellite robots with a base station walking on the bottom, Extreme robotics. Collection of abstracts of the international scientific and technical conference, 2017, St. Petersburg, p. 268.

[20] Bielich T., Ksenzenko A. Ya., Kirsanov K. B., Pryanichnikov V. E. Increasing the time of continuous operation of the underwater inspection robotic complex. Information-measuring and control systems. 2015, No. 7, pp. 51-56, ISSN 2070–0814.

[21] Ksenzenko A.Y, Prysev E.A., Pryanichnikov V.E., Chernyshev V.V., (2017). Intelligent Robotronics: Design the Contactlees Charger and Contactless Data transfer between underwater Robot-Satellits and Underwater 6-legged Vehicle. Proceedings of the 28 th DAAAM International Symposium on Intelligent Manufacturing and Automation, pp. 1197-1201, B. Katalinic (Ed), Published by DAAAD International, ISBN 978-3-902734-08-2, ISSN 1726-9679, Vienna, Austria, DOI: 10.2507/28th.daaam.proceedings.166.

[22] V.V. Chernyshev, V.E. Pryanichnikov, V.V. Arykantsev, I.P. Vershinina, Ya.V. Kalinin, “Research of the walking type of movement in underwater conditions”, OCEANS 2019 – Marseille, IEEE, Marseille, France, 2019, pp. 1–6, https://doi.org/10.1109/OCEANSE.2019.8867233.

[23] S.Y. Yoo, B.H. Jun, H. Shim, “Design of Static Gait Algorithm for Hexapod Subsea Walking Robot: Crabster”, Transactions of the Korean Society of Mechanical Engineers A. Vol.38, 2014, pp. 989-997, 10.3795/KSME-A.2014.38.9.989.

DOI: 10.2507/32nd.daaam.proceedings.046

ONE APPROACH FOR DETERMINING AVAILABILITY OF PARKING LOTS IN SHARED PARKING FOR DISABLED PERSON

Stevan Stankovski*, Srdjan Tegeltija*, Gordana Ostojic*, Igor Baranovski*, Milos Stanojevic* & Mladen Babic*

*University of Novi Sad, Faculty of Technical Sciences, Centar za identifikacione tehnologije
Trg Dositeja Obradovica 6, 21000 Novi Sad, Republic of Serbia

This Publication has to be referred as: Stankovski, S[tevan]; Tegeltija, S[rdjan]; Ostojic, G[ordana]; Baranovski, I[gor]; Stanojevic, M[ilos] & Babic, M[laden] (2021). One Approach for Determining Availability of Parking Lots in Shared Parking For Disabled Person, Proceedings of the 32nd DAAAM International Symposium, pp.0316-0320, B. Katalinic (Ed.), Published by DAAAM International, ISBN 978-3-902734-33-4, ISSN 1726-9679, Vienna, Austria
DOI: 10.2507/32nd.daaam.proceedings.046

Abstract

Due to the growing number of vehicles in cities, many human activities have to accommodate this fact. One of these activities is finding a free parking lot. This activity can be very stressful, especially for disabled persons, because the number of parking lots reserved for these persons is usually limited and cannot be found at every street, garage, or other parking space. There are parking lots intended and reserved for people with disabilities to make it easier for people with disabilities to perform various activities. However, in some cases, these reserved parking lots are already occupied by people with disabilities, or unscrupulous drivers occupy these parking lots. This paper presents one approach for determining the availability of parking lots in shared parking for disabled people, based on different kinds of recognition of vehicles in which owners are disabled persons or recognition of tags in which owners are disabled persons.

Keywords: parking lot; parking sensor; RFID tag; Bluetooth tag license plate recognition

1. Introduction

Finding free parking lots is a daily problem for drivers [1]–[3]. Finding a free parking lot can be very stressful, especially for disabled persons. The number of parking lots for these persons is usually limited and cannot be found at every street, garage, or other parking space. Limited and insufficient parking lots in cities result in the conscious or unconscious occupation of reserved parking lots reserved for people with disabilities by other drivers. In research published in papers [4]-[6], some potential solutions to the problem of detecting the occupancy of parking spaces reserved for people with disabilities have been proposed. However, monitoring and detecting illegal occupancy of parking spaces reserved for people with disabilities is only a tiny part of the problems these people face. The issues faced by people with disabilities are much broader.

They consist of the registration of vehicles owned by people with disabilities, the registration of people with disabilities to obtain dedicated permits for the use of parking spaces reserved for people with disabilities, the search for free parking spaces reserved for persons with disabilities, removal of illegally parked vehicles from places reserved for persons with disabilities, etc.

This paper presents one approach for determining the availability of parking spaces for people with disabilities by defining a model that will connect people with disabilities, organizations for people with disabilities located in different cities, all public companies that provide parking services, all organizations that define parking policies in cities and systems for collecting data on the state of occupancy of parking spaces reserved for people with disabilities based on Internet of Things (IoT) technologies into one whole. The Internet of Things (IoT) describes the network of physical objects (things) that are embedded with sensors, software, and other technologies to connect and exchange data with other devices and systems over the Internet [7]-[11]. The proposed model would enable faster and easier obtaining of dedicated parking permits for persons with disabilities and a quick insight into the current state of parking spaces intended for persons with disabilities. Furthermore, integrating the Lean concept will further improve the proposed model [12], [13].

2. Model for SDPS (Shared Disabled Parking System)

The current situation in the city of Novi Sad was taken as the basis for the development of the SDPS model for parking space sharing. In Novi Sad, people with disabilities are organized in a local organization. This local organization has its database of people with disabilities, and for those who have vehicles, appropriate parking permits are issued in the form of stickers. Each sticker is designated for a valid vehicle license plate and is affixed to the vehicle's windscreen. The validity period of the sticker is determined by the validity period of the license plate. At the same time, there is information on the vehicle's license plate in the database. From the public company "Parking Service"(provider of parking services in Novi Sad), data on the number and location of parking spaces for people with disabilities are obtained. This company also takes care of the maintenance and marking of these parking spaces. City policies define the manner of using the parking spaces for people with disabilities. These policies determine in what time interval parking spaces for the people with disabilities can be used. The situation is similar in other cities in Serbia. The idea of SDPS is that all organizations for people with disabilities located in different cities, all public companies that provide parking services, and all organizations that define parking policies in cities are connected into one whole. A schematic representation of the model for SDPS is shown in figure 1.

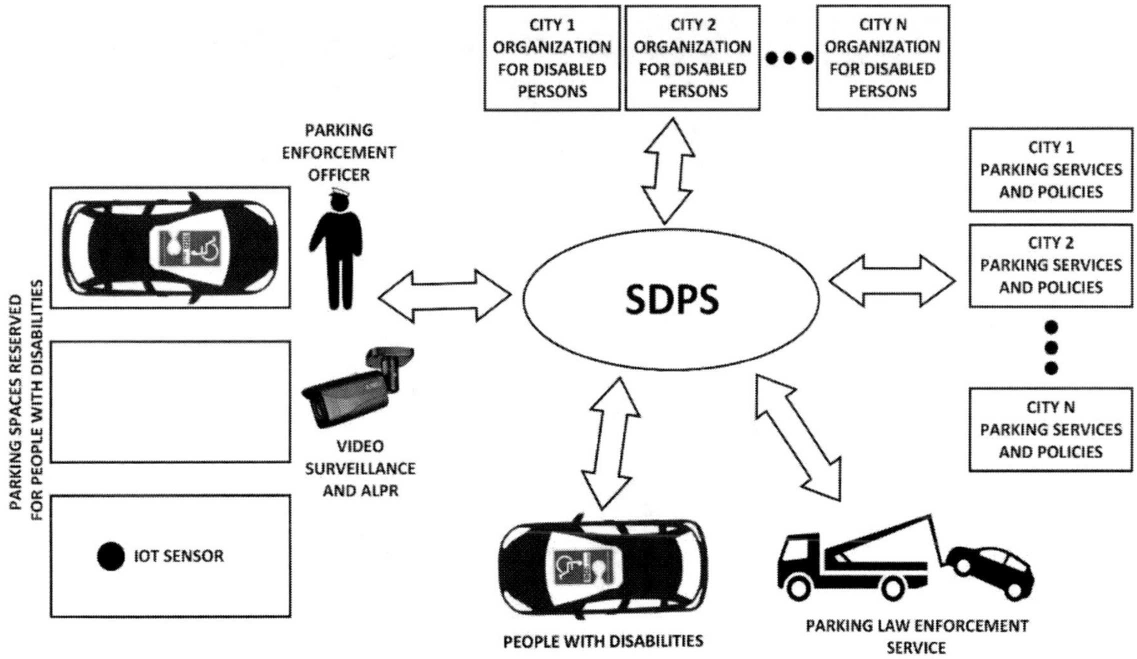

Fig. 1. Schematic representation of the model for SDPS

There are several advantages that the proposed model for SDPS provides. Granting special parking permits to people with disabilities would be simplified and accelerated. City policies and additional services that companies offer to users with disabilities would be harmonized in all cities. Such a centralized system would also make it possible to control the occupancy of parking spaces for people with disabilities. People with disabilities would have an insight into the condition of the parking space in real-time, and, based on that, they could plan to perform various activities more efficiently.

Depending on the possibility of implementation, parking occupancy data can be obtained in three ways:
1- By reading the number of the disability sticker/vehicle license plate number by the parking enforcement officer
2- By reading the car registration number using a video surveillance camera
3- By reading RFID/Bluetooth tags and IoT sensors.

In the event of illegal occupation of a parking space, the parking law enforcement officers would be automatically notified. Then, with an additional check, the parking law enforcement officers would apply a defined penalty policy, after which unscrupulous drivers would be fined, and their vehicle would be removed.

3. Experimental setup

An experimental environment was formed to test the proposed SDPS model. The experimental environment consists of a server application, two parking spaces with IoT parking sensors, and a video surveillance camera to recognize vehicle license plates. The server application with a database is in charge of storing data on test users of the system and for storing and processing data received from IoT parking sensors and video surveillance cameras. The server application was implemented in the MS Visual Studio 2019 programming environment using the C# programming language. Two types of parking sensors are installed in each parking space. Figure 2. shows the positions of IoT parking sensors in one parking space. IoT sensor type 1 is Sensit by Nedap [14]. These sensors detect the presence of vehicles based on dual detection technology (infrared and magnetic). The sensors are interconnected to form a local wireless sensor network to send parking spot occupancy data. A dedicated device, a data collector, must be present to collect data from the sensors and further forwards it over the Internet to the server application. IoT parking sensor type 2 is an IoT parking sensor by Fleximodo [15]. This sensor also detects the presence of a vehicle based on magnetic detection technology. These sensors use LoraWAN technology for communication [16]. The sensors are connected via the Lora infrastructure of the public company Telekom Srbija and the installation of additional devices for communication over the Internet is not required. Both types of IoT sensors have a battery power supply, making it easier to install because it is unnecessary to implement the power supply installations when installing the sensors. An additional advantage of the IoT parking sensor type 2 is that it can detect the IoT Permit user tags, which allows additional verification of whether a person with disabilities occupies the parking space with a disability (figure 3.) [17].

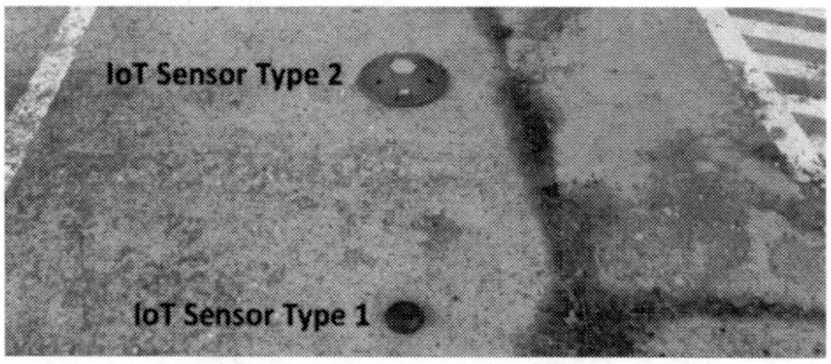

Fig. 2. Position of two types of IoT parking sensors in one parking space

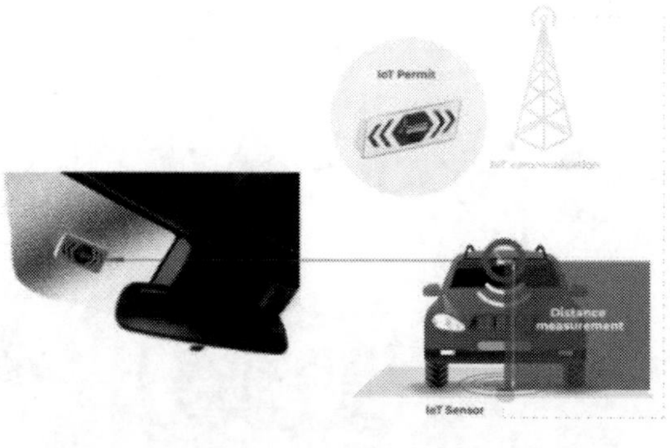

Fig. 3. IoT Permit user tag

An ALPR (Automatic Licence Plate Recognition) camera DS-2CD4A26FWD-IZS/P by Hikvision [18] is used for video surveillance and license plate recognition. This camera can capture vehicles and recognize license plate numbers of various countries and regions. This camera offers high resolution, 2 MegaPixel, 120 dB WDR (Wide Dynamic Range), Auto-iris, PoE (Power over Ethernet), EIS (Electronic Image Stabilization), and 3D DNR (Digital Noise Reduction). Moreover, the camera can recognize a vehicle's plate number in four different lanes automatically. Figure 4. shows a video surveillance camera that covers both parking spaces.

Fig. 4. Position of the camera for video surveillance

For the video surveillance camera and type 1 IoT parking sensors to be able to send data to the server application, the camera and the data collector for type 1 IoT parking sensors are connected to the local network, and the data is sent via the HUAWEI B315S-22 modem configured to work in Telenor mobile network [19]. By setting up such an experiment, we have the opportunity to check all three ways of obtaining data on vehicles in the parking lot, as provided in the SDPS model.

4. Discussion

The crucial part of the successful work of SDPS is to obtain timely data on the occupancy of parking spaces reserved for people with disabilities. The worst-case scenario is that there is only reserved parking space for people with disabilities, without IoT parking sensors or video surveillance cameras for vehicle license plates recognition. In that case, the parking enforcement officers must send the data, which can only be expected in paid parking areas. In addition, this information does not have to correspond to the actual situation since the parking enforcement officers stay around one parking space for only ten seconds. There is a bigger problem at the parking areas with free parking because the parking enforcement officers do not control those areas. If the parking spaces are equipped with a camera for recognizing vehicle license plates, then the success of reading license plates in daylight is at best 98%. At the same time, the worst case, which was experimentally determined, is 50% when night falls and street lighting is not suitable. The best accuracy of parking space occupancy detection is when the parking space is equipped with an IoT parking sensor. In that case, the occupancy detection percentage is over 99% accurate (only in exceptional circumstances when the battery of the IoT parking sensor is running low and a vehicle with high ground clearance is present, the readings are not reliable). Such a high reading accuracy is present in both types of IoT parking sensors. It took 5 to 10 seconds from changing the state of the parking space to updating the information on the server application when the IoT parking sensor type 1 was used. With a type 2 IoT parking sensor, this time was longer and was 15 to 20 seconds. Although the type 2 IoT parking sensor has a slightly longer response time, its great advantage is that it allows simultaneous checking whether the vehicle belonging to a person with disabilities is parked in a place reserved for them or whether it is illegal parking.

The fact is the number of reserved parking spaces for people with disabilities, especially in paid parking areas, is usually not enough. One way to alleviate this problem is to implement a system (using edge computing and artificial intelligence, including the possibility of different kinds of payments [20]-[24] that allows parking spaces to be shared over time depending on the interest and needs of users. For that purpose, the proposed SDPS model enables the distribution of the time of use of parking spaces reserved for persons with disabilities. The essential condition is that people with disabilities show interest in a specific parking place through the mobile application for SDPS. Then, based on the established priority and the rules determined for using a particular parking space, the SDPS mobile application sends information to users whether the parking space is free or a notification that the period of use of the parking space expires. At the same time, in case of unauthorized use of the parking space, the SDPS system sends a notification to the parking service provider to execute the defined penalty policy (issuance of a fine, removal of the vehicle, etc.).

5. Conclusion

People with disabilities often have trouble finding a parking space, especially in paid parking lots. An SDPS model has been proposed to solve this problem. The proposed SDPS model allows time-sharing of parking space usage. The model is made according to the conditions in cities in Serbia, but it can be applied in all cities with minor adjustments. For future work, the realization of a system based on the SDPS model in Novi Sad is planned to perform testing in real conditions. For the proposed SDPS model to function well, two primary conditions must be met. The first condition is that data on the occupancy of parking spaces intended for persons with disabilities are collected in real-time. Another condition is that people with disabilities want to use the applications proposed by the SDPS model. Additional activities in future work will be aimed at exploring new ways of detecting the occupancy of parking spaces reserved for people with disabilities and promoting the proposed solution so that as many people with disabilities as possible use the implemented systems.

6. Acknowledgments

This research has been supported by the Ministry of Education, Science and Technological Development, Government of the Republic of Serbia, through the project: "Innovative scientific and artistic research from the FTN activity domain."

7. References

[1] Babic, M.; Vekic, A.; Stanojevic, M.; Ostojic, G.; Borocki, J. & Stankovski, S. (2019). Modern Parking Solutions for Smart Cities, Proceedings of the 30th DAAAM International Symposium, pp.1075-1083, B. Katalinic (Ed.), Published by DAAAM International, ISBN 978-3-902734-22-8, ISSN 1726-9679, Vienna, Austria DOI: 10.2507/30th.daaam.proceedings.150

[2] I. Baranovski, S. Stankovski, G. Ostojić, and S. Horvat (2020). "Support for Self-service Automated Parking Systems," in 2020 19th International Symposium INFOTEH-JAHORINA (INFOTEH), pp. 1–5.

[3] D. C. Shoup, (2006). "Cruising for parking," Transp. Policy, vol. 13, no. 6, pp. 479–486

[4] Katanalp, B. Y. , Nennioğlu, A. K. , Özinal, Y. , Eren, E. & Yıldırım, Z. B. (2019). "A design for prevention of violation of disabled parking spaces: DDPS (Deterrent Disabled Parking System)," Eskişehir Osmangazi Üniversitesi Mühendislik ve Mimarlık Fakültesi Dergisi, 27(3), pp. 242-251. DOI: 10.31796/ogummf.594213

[5] Tegeltija, S.; Radovanovic, M.; Babic, M.; Stanojevic, M.; Ostojic, G. & Stankovski, S. (2020). One Approach to the Detection of Illegal Occupation of Parking Spaces Reserved for Persons with Disabilities, Proceedings of the 31st DAAAM International Symposium, pp.0650- 0655, B. Katalinic (Ed.), Published by DAAAM International, ISBN 978-3-902734-29-7, ISSN 1726-9679, Vienna, Austria DOI: 10.2507/31st.daaam.proceedings.090

[6] S. Tegeltija, M. Babić, L. Tarjan, I. Baranovski and G. Stojanović (2021). "One Solution for Validation of Legal Usage Of Reserved Parking Spaces For People With Disabilities," 2021 20th International Symposium Infoteh-Jahorina (INFOTEH), pp. 1-5, doi: 10.1109/INFOTEH51037.2021.9400689.

[7] https://www.oracle.com/internet-of-things/what-is-iot/. What is IoT? Accessed on: 2021-06-01

[8] Stankovski, S.; Ostojić, G.; Šaponjić, M.; Stanojević, M. & Babić, M. (2020). "Using micro/mini PLC/PAC in the Edge Computing Architecture," in 2020 19th International Symposium Infoteh-Jahorina (Infoteh), pp. 1–4.

[9] Stankovski, S.; Ostojić, G. & Zhang, X. (2016). "Influence of Industrial Internet of Things on Mechatronics," Journal of Mechatronics, Automation and Identification Technolgy, vol. 1, no. 1, pp. 1–6.

[10] Stankovski, S.; Ostojić, G.; Baranovski, I.; Babić, M. & Stanojević, M. (2020). "The Impact of Edge Computing on Industrial Automation," in 2020 19th International Symposium Infoteh-Jahorina (Infoteh), pp. 1–4.

[11] Stankovski, S.; Ostojić, G.; Zhang, X.; Baranovski, I.; Tegeltija, S. & Horvat, S. (2019). "Mechatronics, Identification Technology, Industry 4.0 and Education," in 2019 18th International Symposium Infoteh-Jahorina (Infoteh) 2019, pp. 1–4.

[12] Sremcev, N., Stevanov, B., Lazarevic, M., Mandic, J., Tesic, Z., Kuzmanovic, B. (2019). "Improving Process of Quotation Creation Through Value Stream Mapping and Simulation, International Journal of Simulation Modelling", Vol. 18, No. 4, pp. 563-573, ISSN 1726-4529

[13] Lazarevic, M., Mandic, J., Sremcev, N., Vukelic, D., Debevec, M. (2019). "A systematic literature review of Poka-Yoke and novel approach to theoretical aspects", Strojniski vestnik - Journal of Mechanical Engineering, Vol. 65, No. 7-8, pp. 454-467, ISSN 0039-2480

[14] https://portal.nedapidentification.com/download/SENSIT/Datasheet/English/SENSIT%20IR%20Flush%20Mount%20datasheet/ . SENSIT IR Flush Mount Accessed on: 2021-06-01

[15] https://www.fleximodo.com/. Modular parking solution Accessed on: 2021-06-01

[16] https://lora-alliance.org/about-lorawan/. LoRaWAN Specification Accessed on: 2021-06-01

[17] https://www.fleximodo.com/wp-content/uploads/2021/02/ipermit_datasheet.pdf. IoT Permit card Accessed on: 2021-06-1

[18] https://www.hikvision.com/my/products/IP-Products/Network-Cameras/Ultra-Series-SmartIP-/ds-2cd4a26fwd-lzs-p/. Hikvision DS-2CD4A26FWD-IZS/P Accessed on: 2021-06-01

[19] https://www.telenor.rs/webshop/en/Business/Modems/Huawei/B315s-22?purchase=subsidy&tariff=410. HUAWEI B315S-22 Accessed on: 2021-06-01

[20] S. Nemet, G. Ostojić, D. Kukolj, S. Stankovski, and D. Jovanovic (2019). "Feature Selection Using Combined Particle Swarm Optimization and Artificial Neural Network Approach," Journal of Mechatronics, Automation and Identification Technolgy, vol. 4, no. 1, pp. 7–11

[21] S. Stankovski, G. Ostojić, M. Šaponjić, M. Stanojević, and M. Babić (2020). "Using micro/mini PLC/PAC in the Edge Computing Architecture," in 2020 19th International Symposium Infoteh-Jahorina (INFOTEH), pp. 1–4.

[22] S. Stankovski, G. Ostojić, L. Tarjan, M. Stanojević, and M. Babić (2019). "Challenges of IoT payments in smart services," Ann. DAAAM Proc., vol. 30, 2019.

[23] Baranovski, S. Stankovski, G. Ostojić, S. Horvat, and T. Srdjan (2020). "Augmented reality support for self-service automated systems," J. Graph. Eng. Des., vol. 11, no. 1, p. 63

[24] B. Tejić, S. Tegeltija, S. Horvat, M. Nićin, M. Stanojevič, and M. Babić (2019). "Payment Methods in Vending Machines," Journal of Mechatronics, Automation and Identification Technolgy, vol. 4, no. 3, pp. 20–25

DOI: 10.2507/32nd.daaam.proceedings.047

ANALYSIS OF THE INDOOR AIR QUALITY INSIDE MOTOR VEHICLES

Laura Dzunova, Ruzena Kralikova & Veronika Gumanova

This Publication has to be referred as: Dzunova, L[aura]; Kralikova, R[uzena] & Gumanova, V[eronika] (2021). Analysis of the Indoor Air Quality Inside Motor Vehicles, Proceedings of the 32nd DAAAM International Symposium, pp.0321-0326, B. Katalinic (Ed.), Published by DAAAM International, ISBN 978-3-902734-33-4, ISSN 1726-9679, Vienna, Austria

DOI: 10.2507/32nd.daaam.proceedings.047

Abstract

Air quality inside motor vehicles is becoming an important issue worldwide because of the ever-increasing number of motor vehicles on the road. These vehicles are a source of exhaust emissions such as carbon dioxide (CO_2), carbon monoxide (CO), nitrogen dioxide (NO_2), volatile organic compounds (VOCs) and others. Motor vehicles are also source of a non-exhaust emissions such as particulate matter (PM). This article deals with the issue of air pollution in car interiors caused by the pollutants from both external and internal sources. Article focuses on the issue of air quality in motor vehicles interiors and the impact of this indoor environment on the health and of passengers. The paper also provides an overview of recently published studies on the issue and their results, as well as the information about the impact of these pollutants on human health. The research will serve as a theoretical basis for future research in this area.

Keywords: Indoor Air Quality; Motor Vehicle; VOCs; CO_2; Particulate Matter

1. Introduction

Modern lifestyle is related to indoor environment, whether it is time spent at home, at work, in a shopping centre, or in a vehicle, many activities take place indoors. Such way of life with a lack of fresh air and natural daylight contributes to the origin and development of many diseases.

In developed countries, many times one family has more than one car. Overall, car transport is one of the most used, mainly due to the many benefits it offers, such as convenience, time savings, privacy and the possibility of direct transport to a specific location, almost anywhere in the world. We can say that the benefits that this transport offers to people are so significant that it is worth the extra expenses. However, the increasing number of cars on the road lead to burden on the environment and reduction in the effectiveness of the above-mentioned benefits. Travel time is increasing due to traffic congestion, lack of parking spaces, drivers are becoming more and more aggressive, impatient and ruthless, and also for these reasons, the number of accidents is rising. This which contributes to the fact that motorcycle and car travel is ranked as most dangerous. From an environmental point of view, road transport makes a significant contribution to the deterioration of the quality of the environment. With the growing number of cars, the demand for raw materials to produce automotive components and fuels.

Mining and industrial production generate large amounts of greenhouse gases and waste as a by-product. The operation and maintenance of motor vehicles does not lag behind. The air quality around busy roads, especially in big cities, is poor. Motor vehicles pollute the air with toxic particles and vapours. Drops of oil, fuel and other toxic substances, together with tire fragments, are trapped on the roads and with the rain they get to pollute the groundwater and soil around the roads. The modern way of life is taking its toll and the quality of life in the future will reflect the decisions of individuals, large corporations and whole society.

With ever-increasing information about the consequences of air pollution and its leading position in the ranking of the most common causes of premature death, the issue of the quality of the air we breathe in our motor vehicle is gaining in importance. The air quality in the car's interior is influenced by various factors, such as the way and intensity of use of the ventilation system, air conditioning and heating, the ambient air quality and the individual characteristics of the car (make, age, model ...).

2. Air pollutants inside the motor vehicles

The most common external air pollutants are those emitted by motor vehicles include particulate matter (PM), nitrogen dioxide (NO_2), carbon dioxide (CO_2), carbon monoxide (CO) and volatile organic compounds (VOCs) and they are most concentrated in the vicinity of frequently used roads and in the environment and can penetrate car interior. Internal pollutants include cigarette smoke, CO_2 accumulation as a result of passenger breathing, volatile organic compounds from plastic components and the use of various chemicals to clean the interior. Various bacteria and fungi may also be present in the interior. Exposure to these pollutants is associated with short-term or long-term adverse effects on human health and in the worst case can lead to premature death.

2.1. Volatile organic compounds (VOCs)

Volatile organic compounds (VOCs) are an important group of pollutants. VOCs include benzene, formaldehyde, acetone, toluene, and others. They also participate in the formation of photochemical smog, of which the most harmful component is ground-level ozone. Sources of VOC emissions caused by human activity include oil and gas production, fuel combustion, chemical and industrial processes (production and use of paints and organic solvents, processing of synthetic rubber ...), use of pesticides and cigarette smoke. VOCs can cause a variety of health problems, such as headaches, breathing problems, skin and eye irritation, cardiovascular disease, and more. Some volatile organic compounds are carcinogenic, such as formaldehyde and benzene [1].

Sources of VOCs in the car interior may include various plastic components, textiles, the use of chemicals to clean the interior and leather seats, various car fragrances, and more. Recommended exposure values for individual VOCs vary depending on the toxicity of the volatile substance, the exposure values of tVOC (total VOC concentration) in the indoor environment are given in the following Table 1. None of the individual VOCs should exceed 10% of the concentration in the total VOC concentration [1], [2].

tVOC [ppb]	**Exposure**	**Recommended exposure time**
0-65	recommended range	No limit
65-220	low danger, recommendation for increased ventilation	No limit
220-660	medium danger, recommendation for increased ventilation and removal of the source	< 12 months
660-2200	high danger, need for ventilation and removal of resources	< 1 months
>2200	unacceptable level, need for intensive ventilation and immediate removal of the source	Several hours

Table 1. tVOC Exposure levels

2.2. Particulate matter (PM)

Airborne dust is the sum of solid particles of various sizes that are freely dispersed in the air. They enter the air in a natural way, but also as a result of human activity (combustion of solids, tire wear, exhaust gases ...). Their impact on human health depends on the toxicity, number and size of these particles. While larger particles above 10 µm (PM10) are trapped in the upper respiratory tract and can cause irritation of the upper respiratory tract and eyes, particles smaller than 10 µm reach the lower respiratory tract, where they can settle and cause health problems. The most dangerous are particles smaller than 2.5 µm (PM2.5), which can penetrate into the lung vesicles and settle there or continue into the bloodstream. It is proven that air pollution with dust particles causes increased mortality of the population from diseases of the respiratory and cardiovascular system [3].

Air pollution by solid particles around roads is caused mainly by the wear of tires, followed by exhaust gases. The number of particles released by tire wear depends on the size and weight of the vehicle as well as the driving style. For example, SUVs and electric cars experience increased tire wear. In addition to causing air pollution, rain also causes contamination of the surrounding soil, water sources and groundwater with microplastics [4]. The WHO (World Health Organization) guideline stipulates that the concentration of PM2.5 must not exceed an annual average of 10 $\mu g/m^3$ or a daily average (24 hours) of 25 $\mu g/m^3$. And the concentration of PM10 must not exceed an annual average of 20 $\mu g/m^3$ or an average of 50 $\mu g/m^3$ [5].

2.3. Carbon Dioxide (CO_2)

Carbon dioxide (CO_2) is a colourless, tasteless, odourless gas that is heavier than air and can accumulate in lower spaces and cause a lack of oxygen. It is naturally present in the Earth's atmosphere as a trace gas and a product of cellular respiration or the burning of fossil fuels. Its unstable solid form is referred to as dry ice.

Typical outdoor CO_2 concentrations are approximately 380 ppm (0.038%), with air concentrations depending on local conditions, altitude and relative humidity. As a result of human activities such as burning fossil fuels, cement production, deforestation, the concentration of CO_2 in the atmosphere is increasing. The main source of CO_2 in interiors is human respiration. It has a negligible toxic effect at low concentrations. At higher concentrations, it leads to an increase in respiratory rate, arrhythmias, nausea and loss of consciousness. The severity of symptoms depends on the concentration of CO_2 and the time of exposure. At high concentrations (> 4%) it causes immediate danger to life and health [6], [7].

CO_2 [ppm]	**Exposure**
250 - 400	concentration similar to ambient air
401 - 1000	normal concentration for indoor living areas with good air exchange
1001 - 2000	higher concentrations, on prolonged exposure causing drowsiness
2001 - 5000	high concentration with prolonged exposure causing headache, drowsiness, loss of attention, nausea
> 5000	represents the occupational exposure limit value for 8 hours. change
> 40000	threat to life and health

Table 2. CO_2 Exposure levels

2.4. Carbon Monoxide (CO)

Carbon monoxide (CO) is a colourless, tasteless and odourless gas produced by the incomplete combustion of carbon-containing materials. It is lighter than air, is present in the atmosphere in a small amount and is formed by photolysis of carbon dioxide by ultraviolet radiation. Natural resources also include volcanic activity and fires. Carbon monoxide is the gas that has caused the most poisoning in human history. It is present in the exhaust gases of cars, it is also formed during heating by gas, diesel and petrol heaters and accumulates especially in small and poorly ventilated rooms. Other important sources include smoking and wooden grills.

Carbon monoxide (CO) is released as a by-product of both diesel and gasoline internal combustion engines and is toxic to humans and animals in concentrations above ~ 35 ppm. A properly functioning exhaust system must prevent CO from entering the cab. Broken system exhaust seals, cracks or openings along the exhaust pipe, open trunk, or improperly tightened connections can result in toxic gases entering the cab. Pregnant women, young children, people with cardiovascular diseases and the elderly are the most sensitive to carbon monoxide. To protect health, it is important to avoid cigarettes, long-term stays in smoking areas and places with a high concentration of road traffic emissions [8].

At high concentrations of carbon monoxide, intoxication occurs, which seriously impairs the ability to drive. People with CO intoxication think slowly and irrationally, are confused and unable to drive a motor vehicle safely. This condition is especially dangerous because one does not realize it.

CO [ppm]	Exposure
9	maximum long-term exposure (ASHRAE standard)
35	maximum exposure during 8 hours shift (OSHA)
>800	death within hours
>12 800	death within 1-2 minutes

Table 3. CO Exposure levels

2.5. Nitrogen Dioxide (NO_2)

Nitrogen dioxide (NO_2) is a yellow-brown, aggressive, toxic gas. It causes acid rain and smog in the air. It irritates the mucous membranes of the respiratory tract and reduces their immunity against infections.

Up to 50% of nitrogen dioxide comes from road transport, another important source is the combustion of natural gas. The transition from petrol to diesel vehicles exacerbates the danger of NO_x. NO_x is a mixture of nitric oxide (NO), which is harmless, and nitrogen dioxide (NO_2). Combustion of diesel in the engine instead of petrol produces more NO_x overall, of which up to 70% is NO_2, compared to only 10 to 15 percent of NO_2 when burning petrol. From certain concentrations (> 20 ppm for a few minutes), inhalation of NO_2 causes poisoning associated with toxic lung edema, and immediate life-threatening [9], [10].

NO_2 [ppm]	Exposure
< 0,1	minimal impact on health
0,1 - 5	adverse health effects, especially in vulnerable populations (asthmatics, the elderly, people with respiratory problems…)
>5	threat to life and health

Table 4. NO_2 Exposure levels

3. Research review

The issue of the impact of road transport on human health and the environment is extensive and current. It is the subject of several international studies and research. "Bad air" in the interior is often a cause of discomfort, but it can also lead to serious health problems. By breathing, harmful substances from get into the lungs and blood circulation. Air quality therefore plays an important role in protecting human health as well as environment. This part of the paper presents review of studies dealing with issue of air quality inside motor vehicles.

Recently published studies in this area include the study by Bakhtiari et al. [11], which focused on the evaluation of volatile organic compound (VOC) concentrations in taxi interiors depending on the age of the vehicle and the type of fuel. The study was performed on four models of taxis in three age groups and powered by three different types of fuel (petrol, CNG - compressed natural gas and LPG - liquefied petroleum gas). The amount of volatile organic compounds in the interiors of taxis was measured before and after refuelling. The results of the study show that refuelling increased the concentrations of pollutants in the vehicle, especially for CNG and LPG fuels, the concentrations of volatile organic compounds were most pronounced in petrol models and the age of the taxi indirectly affected the concentration of formaldehyde and acetaldehyde in the interior.

Barnes et al. [12], in their study tested fifty-one vehicles for the presence of particulate matter (PM0.3 and PM2.5), total volatile organic compounds (tVOC), carbon monoxide (CO), carbon dioxide (CO_2), airborne bacteria and fungi. Of the total number of tested vehicles, up to 24% of them exceeded the recommended level of volatile organic compounds and this concentration correlated positively with the age of the vehicle. Also, the presence of tVOC was significantly higher when the engine was idling than while driving. The level of carbon monoxide (CO) in all vehicles did not exceed the recommended values, while 96% of vehicles exceeded the recommended level of CO_2 1000 ppm and 16% of vehicles even the value of 5000 ppm. The presence of microbes, bacteria and fungi in the interior was relatively low. Further in the study by Barnes et al. up to 96% of the tested cars exceeded the recommended CO_2 values (1000 ppm) while driving, the values ranged on average around 3 413 ppm, 16% of vehicles exceeded the value of 5,000 ppm. A total of 90% of the cars tested exceeded the recommended CO_2 values in the cab during engine idling.

According to Chen et al. [13] VOC concentrations increase with temperature and relative humidity and decrease with the age of the car and the total mileage. They are also higher for cars with small cabs and depend on the car model and the sampling location.

Particulate matter (PM) pollution from transport emissions has a significant impact on human health. Dröge et al. [14] examined PM levels in the car cabin with a mobile aerosol spectrometer under various conditions. The particle concentrations depended on the particle size and the concentration of PM in the ambient air. The concentration of smaller particles showed small fluctuations, but the concentration of coarse particles showed high fluctuations with maximum values on busy roads. Closing the vehicle window reduced PM exposure, especially to coarse particles.

Particulate matter (PM) concentrations inside vehicle cabs are often extremely high. Another study from 2019 [15] looked at the relationship between the emissions of the lead vehicle and the PM concentration levels in the cab of the immediately following vehicle. The study was performed on a 26 km long fixed route, where 10 repeated tests were performed during a 60-minute drive. Monitoring of particle concentrations of different sizes (PM0.3-1 and PM1-2.5) was performed inside the cab of the vehicle with a fixed ventilation setting (i.e. windows closed, air conditioning off and recirculation off). The results showed that the PM levels in the cab were significantly affected by the emissions of the leading vehicle.

A study by Hudda, N. & Fruin, S.A. of 2018 [16], examined the effect of driving and ventilation on cab CO_2 accumulation. The results indicate that recirculation of the interior air and closing of the windows reduces the levels of fine particles in the vehicle but increases the CO_2 concentration as a result of the passengers' breathing. The CO_2 concentration is greatly affected by circumstances such as the number of passengers, the age of the vehicle, the ventilation setting, the volume of the cab, the speed and the driving time. By adjusting the internal air circuit, the CO_2 level can exceed a threshold value (2,500 ppm), which has negative cognitive or physiological consequences such as difficulty concentrating and fatigue). For longer journeys or number of passengers, the use of air recirculation should be regularly interrupted or partially mixed with the outside air, ideally so that the CO_2 concentration is below 1000 ppm.

Nitrogen dioxide (NO_2) is also a pollutant and harmful substance in the exhaust gases of motor vehicles. A Polish study [17] examined the exposure of drivers, passengers, and pedestrians to NO_2 during various excursion scenarios. NO_2 samples were collected in the interiors of cars, trucks and buses, as well as in the ambient air using passive samplers, and other NO_2 concentrations were analysed using spectrophotometric techniques. The results of the research suggest that the highest concentrations of NO_2 were associated with travel on urban roads, while lower concentrations were associated with movement in suburban traffic, on roads with low traffic intensity.

A Korean study [18] also looked at the personal NO_2 exposure of 31 taxi drivers in two Korean cities. Exposure was estimated based on measurements in the outdoor environment and the indoor environment of the home and taxi. Based on the results, we can consider driving a car at work as the main source of NO_2 exposure for taxi drivers. Research also shows that NO_2 exposures in LPG taxis are significantly lower compared to diesel vehicles.

4. Conclusion

Although we often do not even realize it, air quality has a significant impact on the environment and our health. For several years now, road transport has been at the top of the list of air pollutants. Every year we record an increase in road mobility and transport, which leads to a deterioration in air quality. This can be felt especially by people living in big cities and near busy roads. Modern lifestyle is associated with motor vehicles and long time spent on the road, so maintaining a suitable indoor air quality is gaining in importance. Regular ventilation, proper use of air conditioning systems, inspection and replacement of filters can be useful tools to improve the quality and hygiene inside the vehicles.

The issue of air quality in the car interior is very broad and, despite the studies carried out so far, provides space for further experiments, such as obtaining information on the influence of individual characteristics of the vehicle, such as age, model, manufacturer etc. and performing tests under various conditions.

The results of the research should lead to an increase in awareness of the possible risks associated with poor air quality in motor vehicle interiors. Reducing the adverse effects of transport is also an important goal of EU policy. EU strategy papers focus on zero greenhouse gas emissions, emphasize the importance of moving to low-carbon regimes and zero-emission vehicles, emphasize the central role of electrification and renewable energy sources, and prioritize operational efficiency improvements. Better spatial planning and full use of public transport are also required. All of this can help reduce the health risks of poor air quality. An accelerated shift to the use of hydrogen-powered vehicles can also play an important role in reducing the negative impact of road traffic on overall air quality.

5. Acknowledgments

This paper was written in the framework of the projects KEGA 011TUKE-4/2021 and KEGA 009TUKE-4/2021, solved at the Faculty of Mechanical Engineering - Technical University of Košice.

6. References

[1] Montero-Montoya, R. et al. (2018). Volatile Organic Compounds in Air: Sources, Distribution, Exposure and Associated Illnesses in Children. Ann Glob Health. 2018. Vol. 84, No. 2, pp. 225-238. doi:10.29024/aogh.910

[2] Indoor air quality and volatile organic compounds.Sensirion. (2017) Online: https://www.repcomsrl.com/wp-content/uploads/2017/06/Environmental_Sensing_VOC_Product_Brochure_EN.pdf
[3] Lozhkin, V. et al. (2018). A study of air pollution by exhaust gases from cars in well courtyards of Saint Petersburg. Transportation Research Procedia. 2018. Vol. 36, pp. 453-458. ISSN 2352-1465. doi: 10.1016/j.trpro.2018.12.124.
[4] Tyres not tailpipe. Emission analytics. (2020). Online: https://www.emissionsanalytics.com/news/2020/1/28/tyres-not-tailpipe
[5] World Health Organization, Occupational and Environmental Health (2005). "WHO Air quality guidelines for particulate matter, ozone, nitrogen dioxide and sulfur dioxide: Global update 2005: Summary of Risk Assessment". Online: https://apps.who.int/iris/bitstream/handle/10665/69477/WHO_SDE_PHE_OEH_06.02_eng.pdf;jsessionid=3252302599942E092D270818EDBB969F?sequence=1
[6] Azuma, K. et al. (2018). Effects of low-level inhalation exposure to carbon dioxide in indoor environments: A short review on human health and psychomotor performance, Environment International. 2018. Vol. 121, No.1, pp. 51-56. ISSN 0160-4120. doi:10.1016/j.envint.2018.08.059.
[7] What are safe levels of CO and CO2 in rooms? KANE. Online: https://www.kane.co.uk/knowledge-centre/what-are-safe-levels-of-co-and-co2-in-rooms
[8] Dirks, K. et al. (2018). In-Cabin Vehicle Carbon Monoxide Concentrations under Different Ventilation Settings. Atmosphere. 2018. Vol. 9. doi: 338. 10.3390/atmos9090338.
[9] Carslaw, D. et al. (2013). Remote sensing of NO2 exhaust emissions from road vehicles. A report to the City of London Corporation and London Borough of Ealing. Online: https://uk-air.defra.gov.uk/assets/documents/reports/cat05/1307161149_130715_DefraRemoteSensingReport_Final.pdf
[10] Adverse Health Effects of Elevated Nitrogen Dioxide (NO2) Levels in Homes.E Instruments. Online: https://www.e-inst.com/applications/adverse-effects-of-elevated-no2/
[11] Bakhtiari, R. et al. (2018). Investigation of in-cabin volatile organic compounds (VOCs) in taxis; influence of vehicle's age, model, fuel, and refueling. Environ Pollut. 2018. Vol. 237, pp. 348-355. doi: 10.1016/j.envpol.2018.02.063.
[12] Barnes, N.M. et al. (2018). In-Cabin Air Quality during Driving and Engine Idling in Air-Conditioned Private Vehicles in Hong Kong. Int J Environ Res Public Health. 2018. Vol.15, No. 4, p. 13. doi: 10.3390/ijerph15040611.
[13] Chen X. et al. (2014). Analyses on influencing factors of airborne VOCS pollution in taxi cabins. Environ Sci Pollut Res Int. 2014. Vol. 22. doi: 10.1007/s11356-014-3223-y.
[14] Dröge, J. et al.(2018). Mobile Measurements of Particulate Matter in a Car Cabin: Local Variations, Contrasting Data from Mobile versus Stationary Measurements and the Effect of an Opened versus a Closed Window. Int J Environ Res Public Health. 2018. Vol. 15, No. 12. doi: 10.3390/ijerph15122642.
[15] Campagnolo, D. et al. (2019). In-vehicle airborne fine and ultra-fine particulate matter exposure: The impact of leading vehicle emissions. Environ Int. 2019 Vol. 123, pp. 407-416. doi: 10.1016/j.envint.2018.12.020.
[16] Hudda, N. & Fruin, S.A. 2018. Carbon dioxide accumulation inside vehicles: The effect of ventilation and driving conditions. Sci Total Environ. 2018. doi: 10.1016/j.scitotenv.2017.08.105.
[17] Zak, M. et al. (2016). The exposure of pedestrians, drivers and road transport passengers to nitrogen dioxide. Atmospheric Pollution Research. 2016. Vol. 8, pp.781–790. doi: 10.1016/j.apr.2016.10.011
[18] Son, B. et al. (2004). Estimation of occupational and nonoccupational nitrogen dioxide exposure for Korean taxi drivers using a microenvironmental model. Environ Res. 2004.Vol. 94, No. 3, pp. 291-6. doi: 10.1016/j.envres.2003.08.004.

DOI: 10.2507/32nd.daaam.proceedings.048

DETERMINATION OF THE VISCOSITY CLASS OF MOTOR OILS USING COLOR CHARACTERISTICS AND STATISTICAL ANALYSIS

Tsvetelina Georgieva, Albena Ivanova-Vasileva & Plamen Daskalov

This Publication has to be referred as: Georgieva, T[svetelina]; Ivanova-Vasileva, A[lbena] & Daskalov, P[lamen] (2021). Determination of the Viscosity Class of Motor Oils using Color Characteristics and Statistical Analysis, Proceedings of the 32nd DAAAM International Symposium, pp.0327-0332, B. Katalinic (Ed.), Published by DAAAM International, ISBN 978-3-902734-33-4, ISSN 1726-9679, Vienna, Austria
DOI: 10.2507/32nd.daaam.proceedings.048

Abstract

The viscosity class of motor oils is one of the most important indicators for their quality and identification. This paper describes a fast, inexpensive, non-destructive, image-based approach to analyzing and determining the viscosity grade. Two groups of motor oils of different viscosity class SAE XW-30 & SAE XW-40 were analyzied. For each sample of its digital image, the values of the three RGB components of the basic color model are obtained. The obtained images were converted to HSV and Lab color models in MATLAB. The digital images of the samples were obtained using digital camera, document camera and colorimeter. The obtained values of the color components were analyzed statistically. The correlation coefficient was used to assess the informativeness of the characteristics in relation to the viscosity of the oil. The selection of the camera and algorithm for obtaining the color digital images of the oils a comparative analysis was made.

Keywords: motor oils; image analysis; Matlab; statistical analysis

1. Introduction

The viscosity grade of motor oils is one of the most important parameters for their quality, classification and identification. Method based on spectroscopy is used for assessment of oil characteristics [11]. Other method is based on color assessment of oil samples. This paper describes a fast, inexpensive and non-destructive approach based on image analysis and image recognition techniques to determine the viscosity class according to their respective color histograms. For the purposes of the study, color histograms were used in different color spaces, such as red, green, blue (RGB), intensity of nuanced violence (HSL) and others. Two groups from different viscosity grades motor oils (SAE XW-30 & SAE XW-40) were studied. For each sample, we applied a color histogram in different color spaces, which can help with the classification procedure. These color histograms and their combinations were used as input for model development. The resulting images were converted to HSV and Lab color models in the middle of MATLAB.

2. Brief description of the problem

Motor oils are used to lubricate internal combustion engines in cars and vans, trucks, motorcycles, mowers, engine generators and many other machines.

The most common motor oils are produced by mixing base oil components, petroleum derivatives and additives. The base oil components used to produce the different viscosity classes of motor oils may be of groups I, II and III according to the API classification. In the production of oils of the highest class, synthetic base components are also used, which can be polyalphaolefins and / or esters. The main function of oils is to reduce wear on moving engine parts. In addition to this function, the oils have detergent properties, protect against rust and corrosion, improve sealing and cool the engine by removing heat from moving parts. The classification of motor oils is important for quality control, identification of oils, and to a large extent for determining their performance.

The viscosity classification for motor oils was developed in 1995 by the Association of Automotive Engineers SAE and is known to this day as SAE J-300. The current viscosity classification of motor oils is from 2015 and retains the two main classes originally introduced: single-season and all-season oils. The separation or grouping of oils into certain groups, which are called viscosity classes, is done through the indicators Kinematic and dynamic viscosity. Kinematic viscosity is a measure of the resistance to leakage of a liquid under the action of gravitational force. It is measured at 100 ° C and 40 ° C. Dynamic viscosity is the ratio between the applied tangential stress and the velocity gradient. The determination of the indicators kinematic viscosity and dynamic viscosity is performed in laboratory conditions and takes time. Therefore, it would be a major step forward to find a non-destructive method to determine the viscosity class of oils in a faster and easier way.

Various analytical instruments such as infrared spectroscopy, laser-induced fluorescence and phosphorescence spectroscopy are widely used to analyze and control the quality of petroleum products. All these testing techniques require special equipment, sample preparation and, of course, serious knowledge of the chemistry of the analytes. In today's world, where rapid diagnostics and results are sought, it is a challenge for science to develop a more accessible and rapid method for classifying oils. This can be done after a good knowledge of the characteristics of the oils, their significance and the currently used test methods. The only applied, fast method that uses the color of the oils in the visible spectrum is the color indicator. The procedure for visually determining the color of different lubricating oils is performed by observing the test sample using a daylight source and comparing the color with a scale of reference-colored glasses (Fig. 1). The test is performed with a colorimeter, which allows the sample and the reference samples to be illuminated and monitored simultaneously.

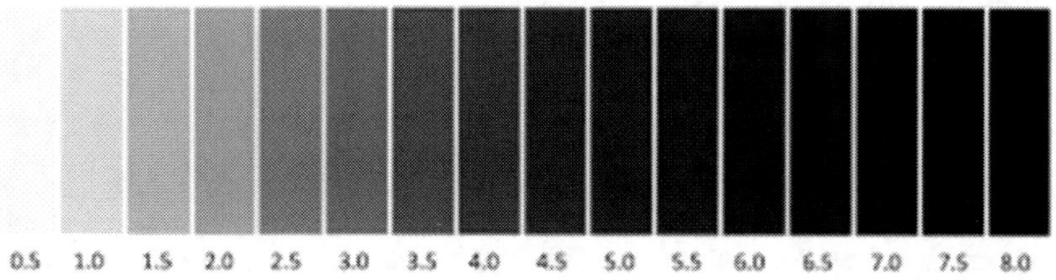

Fig. 1. ASTM scale

This indicator applies to the testing of both base oils and finished, additive oils. Depending on the type of test sample, information can be obtained on the degree of purification of base oils, their degree of deep processing in the refining process. For example, hydrotreated base oils are water white or very pale yellow (Fig. 2, sample <0.5 and 1). The oils obtained by conventional selective purification are yellow (Fig. 2, sample 3) to dark brown (Fig. 2 samples 5 and 7).

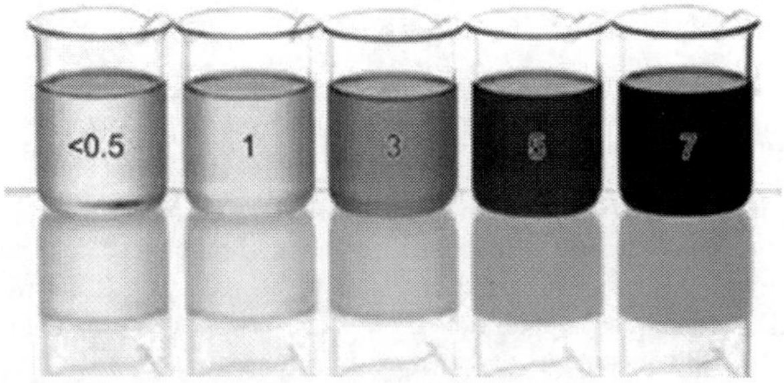

Fig. 2. Oils with different ASTM colors

However, when it comes to inhibited oils, the color indicator cannot be used unambiguously to determine the quality of the oil and its affiliation. The reason is that the additives used in the inhibited oils are dark in color and usually color the oils significantly. Thus, regardless of whether we have motor, transmission or hydraulic oils, depending on the percentage of application of the additive and their properties, they can have the same color determined by the ASMT scale.

Nowadays, digital image processing is becoming increasingly important due to its ability to perform fast and non-invasive, inexpensive analysis of products and materials. Digital image processing has been used successfully to classify honey [1], coffee [2], bacteria [3], tea [4], vegetable oil [5] and biodiesel [6]. In this sense, it would be interesting to develop a simple methodology based on digital images and pattern recognition techniques for the classification of motor oils, which is fast, inexpensive, and non-destructive.

Here, as in the case where oils with different applications can be of similar or the same color, there is a tendency for oils of different viscosity classes to have the same color. This means that it is necessary to look for such a characteristic of light that allows to distinguish between different viscosity classes of oils. Studies have so far been done by analyzing histograms of oil images [7]. The histogram shows the distribution of color values in the image. It is often used in image processing. Another option is to use models such as RGB, HSL and others.

The experiments are based on a report that examines the problem of classifying motor oils based on image processing techniques [8]. The study examines the most used motor oil products in Iran (Kian, CNG, Total 5000, Total 7000, Seman, Yaghout, Alp, Speedy and Behran). To this end, the authors analyzed a set of image functions based on three color spaces, [9], red, green, blue (RGB), [10] hue, saturation, intensity (HSI), and [7], gray scale.

The RGB system is based on the mechanism of color formation in the human eye, in which combinations of different levels of light emission at red, green and blue wavelengths provide a complete system for identifying and recognizing colors in the eyes. This model is an additive color model and has been used in computer environments and in color television imaging. In it, the colors red, green and blue are combined in different ways to obtain other colors.

RGB values, which are encrypted in 24 bits per pixel, are determined using numbers from 0 to 255. This value indicates the saturation of the color element (red, green or blue). For example, different values of the RGB color system and their corresponding colors are as follows:

- (0, 0, 0) black;
- (255, 255, 255) white;
- (255, 0, 0) red;
- (0, 255, 0) green;
- (0, 0, 255) blue;
- (255, 255, 0) yellow;
- (0, 255, 255) cian;
- (255, 0, 255) magenta;

In HSL, the color pattern (hue, saturation and luminance) hue determines the main color hue (for example, pastel pink has a red base tone); saturation determines the purity of the color compared to a neutral gray tone (the less saturated the colors of an image, the more it resembles a monochrome gray halftone image) and luminance (value, intensity) determines the brightness of the color compared to black color and can be considered as a quantitative participation of the black ingredient in the color in question. (14).

3. Experiments

56 samples (Fig. 3) from two viscosity grades of motor oils XW-30 and XW-40 were analyzed. The samples were selected from different batches of motor oils with different applications and ACEA performance levels. 5.0 ml of each sample was used to obtain an image. The device used for capturing images is constructed using a styrofoam box measuring 45 × 40 × 35 cm, which is covered by a black foil closure to avoid the unfolding of the interference light, ensuring the uniformity of the captured image.

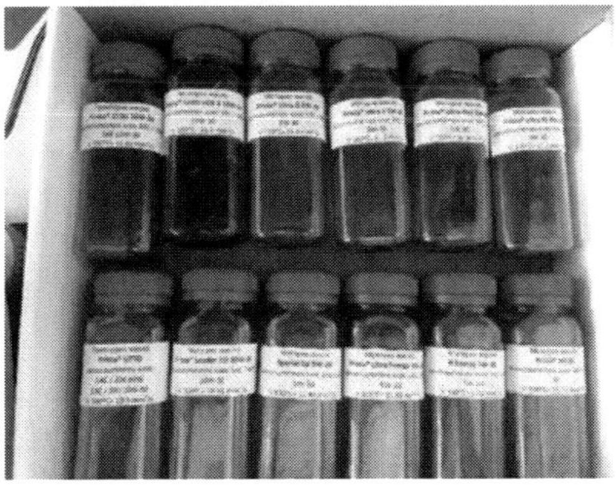

Fig. 3. Oil Samples

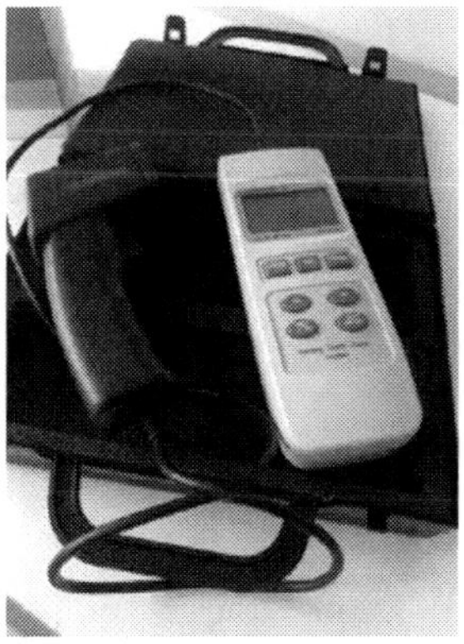

Fig. 4. Colorimeter

The Triumph Board A 405 document camera, a 25-megapixel Nikon camera (D5300, DLSR) with a sensor size of 23.5 × 15.6 mm and a PCE-RGB-2 colorimeter (Fig. 4) were used to capture images. The cameras were placed vertically in the center of the box and above the sample state. The distance between the chamber and the sample holder was 20 cm. A lamp built into the document camera is used for shooting. A total of 112 images were obtained (two for each engine oil sample). Then, in the center of each image, a region of interest (ROI) with × pixels was selected (Fig. 5). Because color histograms describe the statistical determination of colors in pixels, the choice of ROI provides sufficient information to obtain an appropriate classification model.

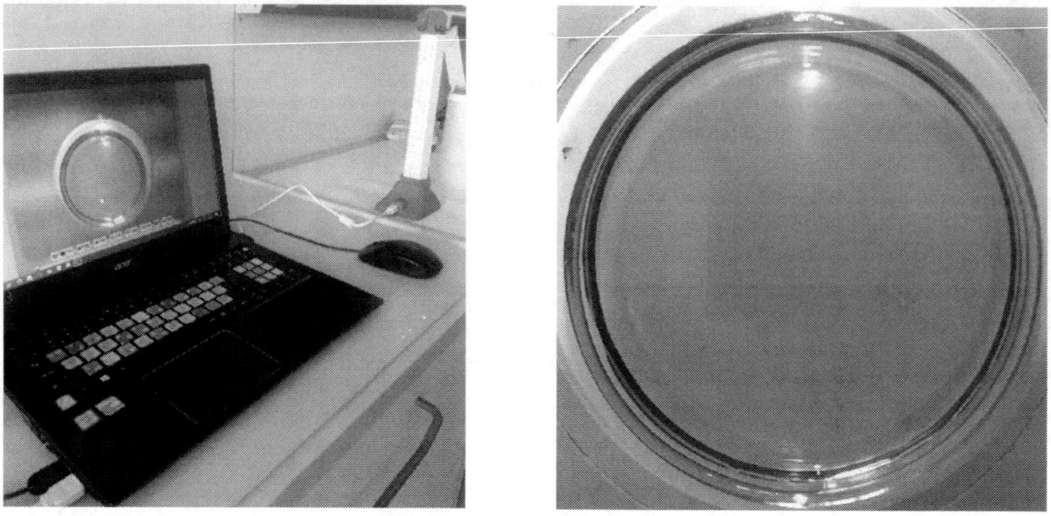

Fig. 5. Obtained Image

4. Results and discussions

The obtained results of the color components of all images are saved. Part of them (using Nikon camera) are shown in Table 1. The obtained values of the color components were analyzed statistically. The obtained values of the digital images using digital camera, document camera and colorimeter are compared between two viscosity grades of motor oils XW-30 and XW-40 and relations are shown on Fig. 6 (a-using Nikon camera; b-using document camera; XW-40 –blue color, XW-30 – red color). Obtained experimental results show that the viscosity classes are similar but definitly could be separate. But color features have values in the similar range and couldn't be used as a features for oil viscosity grading using oil color. Also correlation coefficient was analaized to define informative features for color grading of oil viscosity. The results for XW-30 using three types of device for image acquisition are shown on Table 2. Device 1 is document camera with light, 2 - document camera without light and 3 – Nikon. The results show that the maximum value of the correlation coefficient is 0,396014. It is not appropriate to use these color components for separating motor oils in viscosity classes by color. But another option is to apply this experiment for separating the motor oils in one viscosity class and to recognize and qualify them by viscosity.

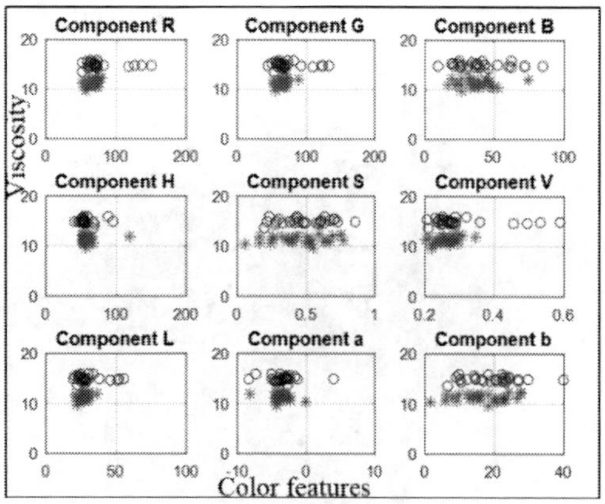

a) Obtained using Nikon camera

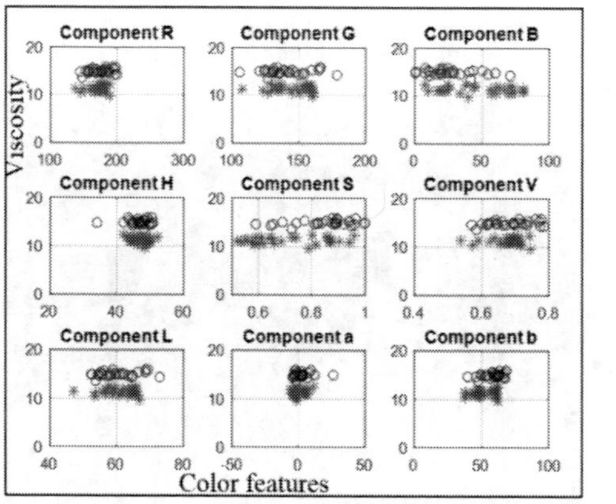

b) obtained using document camera

Fig. 6. Relation between color characteristics and viscosity of motor oil

N	R	G	B	H	S	V	L	a	b	KV at 100, cSt
1	116,5462	108,1743	61,7438	50,8889	0,4729	0,4570	45,4122	-3,8054	27,0690	14,6
2	135,8006	122,5900	71,3114	47,7071	0,4754	0,5326	51,5391	-2,4090	29,7911	14,77
3	150,1308	130,5117	60,3200	46,8772	0,5981	0,5887	54,9892	-1,4306	39,8921	14,93
4	125,9721	121,2541	84,7526	53,1467	0,3277	0,4940	50,4726	-4,2875	20,7075	14,73
5	68,8650	66,0941	42,8895	53,6878	0,3777	0,2701	27,6802	-3,0628	14,5736	14,96
6	66,4192	62,4750	37,0088	52,1350	0,4431	0,2605	26,1620	-2,7894	16,2483	14,99
7	62,9637	60,0970	37,6084	53,3286	0,4045	0,2469	25,0516	-2,8793	14,3712	15,2
8	52,0753	53,3407	43,1141	67,0074	0,1931	0,2092	21,8801	-2,7623	6,2774	13,62
9	52,1517	50,0519	36,4757	52,1043	0,3026	0,2046	20,6263	-1,7966	9,0058	15,4
10	58,5986	55,3883	33,8594	52,3421	0,4233	0,2298	22,9998	-2,5407	14,0050	15,6
11	80,5437	91,5948	72,0585	93,9227	0,2136	0,3592	37,4449	-8,3912	9,8459	14,78
12	70,6232	72,9044	52,3798	66,5864	0,2823	0,2859	30,2230	-4,9818	12,0177	14,96
13	72,6152	71,9573	41,6034	58,8023	0,4307	0,2858	29,8900	-5,1005	18,3188	15,78
14	71,3286	66,7925	30,5654	53,3329	0,5730	0,2797	27,9407	-3,8551	22,4739	14,74
15	66,7418	62,6721	26,4888	53,9425	0,6054	0,2617	26,0885	-4,0016	22,5410	15
16	68,2593	67,4590	39,9731	58,3191	0,4160	0,2677	27,9990	-4,6259	16,8111	14,64
17	72,7363	68,6167	30,2004	54,2080	0,5870	0,2852	28,6594	-4,3199	23,5697	14,26
18	60,5597	56,5324	24,6857	53,2878	0,5949	0,2375	23,4198	-3,4457	20,2958	14,79
19	60,8601	63,3054	45,8533	68,3352	0,2778	0,2483	26,0591	-4,5872	10,3915	14,43
20	72,2990	80,7486	62,3737	87,6350	0,2279	0,3167	33,1060	-7,2317	9,7154	15,78
21	64,2558	60,2280	24,4639	53,9416	0,6221	0,2520	25,0103	-3,9272	22,3879	14,57
22	61,4250	55,8688	19,2418	52,1185	0,6904	0,2409	23,2225	-3,1015	23,2124	15,35
23	68,1379	62,5046	23,1152	52,5206	0,6621	0,2672	26,1437	-3,5080	24,5468	15,21
24	63,9197	59,6464	23,6323	53,6606	0,6319	0,2507	24,7801	-3,8148	22,5882	15,91
25	63,5665	57,4385	19,1070	51,7735	0,7029	0,2493	23,9638	-2,9987	24,1787	15,31
26	70,7015	66,8725	36,3568	53,3664	0,4873	0,2773	27,9852	-3,5524	19,0671	14,62
27	71,5323	63,9305	20,3657	51,0976	0,7179	0,2773	26,9274	-2,8834	26,9735	14,8
28	60,7008	43,6620	9,4775	40,0452	0,8469	0,2380	19,1335	3,8002	23,4720	14,89

Table 1. The obtained results of the color components using Nikon camera

Device	R	G	B	H	S	V	L	a	b
1	0,087574	-0,102633	-0,151744	-0,239350	0,165683	0,087681	-0,059532	0,248703	0,143128
2	0,079867	-0,102006	-0,147627	-0,222355	0,159583	0,079801	-0,061005	0,237036	0,137792
3	0,372072	0,350589	0,002372	0,149842	0,159709	0,396014	0,355204	-0,253879	0,207868

Table 2. The calculated results for correlation coefficient

Plans for the future research include additional experiments for assessment of other color features for oil quality grading and development of classification procedures based on neural networks. The research described in the paper will lead to find informative color features for oil quality assessment. Future work will be development of tool for fast quality assessment of oil viscosity class based on these color features.

5. Conclusion

The problem is how to assess oil viscosity with fast method and tool without expert. The paper describes a fast, inexpensive, non-destructive, image-based approach to analyzing and determining the viscosity grade. Two groups of motor oils of different viscosity class SAE XW-30 & SAE XW-40 were analyzed. For each sample of its digital image, the values of the three RGB components of the basic color model are obtained. The obtained images were converted to HSV and Lab color models in MATLAB. The digital images of the samples were obtained using digital camera, document camera and colorimeter. The obtained values of the color components were analyzed statistically. Obtained experimental results show that the viscosity classes are similar but definitely could be separate.

But color features have values in the similar range and couldn't be used as a features for oil viscosity grading using oil color. The maximum value of correlation coefficient used to assess the informativeness of the characteristics in relation to the viscosity of the oil is 0,396014. It is not appropriate to use these color components for separating motor oils in viscosity classes by color. But another option is to apply this experiment for separating the motor oils in one viscosity class and to recognize and qualify them by viscosity. Future research will include additional experiments for assessment of other color features for oil quality grading and development of classification procedures based on neural networks.

6. Acknowledgments

The research is supported by a contract with Ruse University "Angel Kanchev" under NSF – 2021 - EEA – 05.

7. References

[1] Domínguez, M.A., Diniz, P.H.G.D, DiNezio, M.S., Araújo, M.C.U, & Centurión, M.E. (2014) Geographical origin classification of Argentinean honeys using a digital image-based flow-batch system, Microchem. J. 112, 104–108

[2] Souto, U.T.C.P., Barbosa, M.F., Dantas, H.V., Pontes, A.S., Lyra, W.S., Diniz, P.H.G.D., Araújo, M.C.U., & Silva, E.C. (2015) Screening for Coffee Adulteration Using Digital Images and SPA-LDA, Food Anal. Method 8, 1515–1521

[3] Almeida, V.E., Costa, G.B., Fernandes, D.D.S., Diniz, P.H.G.D., Brandão, D., Medeiros, A.C.D., & Véras, G. (2014) Using color histograms and SPA-LDA to classify bacteria, Anal. Bioanal. Chem. 406, 5989–5995

[4] Diniz, P.H.G.D., Dantas, H.V., Melo, K.D.T., Barbosa, M.F., Harding, D.P., Nascimento, E.C.L., Pistonesi, M.F., Band, B.S.F., & Araújo, M.C.U. (2012) Using a simple digital camera and SPA-LDA modeling to screen teas, Anal. Methods 4, 2648–2652

[5] Milanez, K.D.T.M., & Pontes, M.J.C. (2014) Classification of edible vegetable oil using digital image and pattern recognition techniques, Microchem. J. 113, 10–16

[6] Costa, G.B., Fernandes, D.D.S., Almeida, V.E., Araújo, T.S.P., Melo, J.P. (2015) Digital image-based classification of biodiesel, Talanta, 139, 50-55

[7] Balabin, R.M, & Safieva, R.Z. (2008) Motor oil classification by base stock and viscosity based on near infrared (NIR) spectroscopy data, Fuel 87, 2745–2752

[8] Ahmadi, S., Ahmad, M.V., Habibi,B. (2018) Motor Oil Classification Using Color Histograms and Pattern Recognition Techniques, Journal of AOAC International 101(6)

[9] Brown, S.F. (2015) Base oil groups: manufacture, properties and performance, Tribol. Lubr. Technol. 4, 1–32

[10] Bassbasi, M., Hafid, A., Platikanov, S., Tauler, R., & Oussama, A. (2013) Research articleFull text access. Study of motor oil adulteration by infrared spectroscopy and chemometrics methods, Fuel 104, 798–804

[11] Prosr, P.; Polansky, R. & Mentlik, V. (2011) Infrared Spectroscopy of the Regenerated Mineral Insulating Oil, Proceedings of the 22nd International DAAAM Symposium "Intelligent Manufacturing & Automation", Volume 22, No.1

DOI: 10.2507/32nd.daaam.proceedings.049

REDESIGN OF IMPRESSION TRAYS USING REVERSE ENGINEERING AND FINITE ELEMENT ANALYSIS

Vesna Raspudic & Marijana Marusic

This Publication has to be referred as: Raspudic, V[esna] & Marusic, M[arijana] (2021). Redesign of Impression Trays using Reverse Engineering and Finite Element Analysis, Proceedings of the 32nd DAAAM International Symposium, pp.0333-0338, B. Katalinic (Ed.), Published by DAAAM International, ISBN 978-3-902734-33-4, ISSN 1726-9679, Vienna, Austria
DOI: 10.2507/32nd.daaam.proceedings.049

Abstract

The reverse engineering technologies facilitate development of new or redesign of existing products, which has been presented in this paper on an example of dental impression trays redesign. There are different types of trays available today, in order to ensure proper retention, stability and support in the impression taking procedure. Disposable lower and upper trays have been used as the initial samples for 3D scanning, which was performed by the GOM Atos Compact Scan 8M optical scanner. The CAD models were build using Geomagic and SolidWorks, while the redesign consisted of introducing new perforations in order to allow the impression material to expand through the holes and so be held securely in place. Finite element analysis has been used to investigate the effect of introducing these design changes to structural rigidity. Due to the need to consult a prosthetist about the acceptability of the redesign, the models have been printed by the 3D printer Zortrax M200 Plus.

Keywords: Dental impression tray; Reverse Engineering; FEA; Rapid Prototyping

1. Introduction

The development of CAD/CAM/CAE has greatly automated and facilitated a range of activities related to design, simulation and optimization processes in a number of areas of industrial application. There are two different approaches to the engineering design process, and these are traditional and reverse engineering. In traditional engineering, product design is generally done by forward design, which proposes conceptual ideas according to the function and use of the product, models it and finally processes and shapes it by technical means. The design model is mostly done by the designer on CAD software, and this method often has a long cycle time and low efficiency. In contrast to this approach, reverse engineering (RE) is the process of creating a CAD model from a physical part or prototype. RE techniques convert point cloud data, obtained by digitization by contact or noncontact scanning, into CAD models either in NURBS or STL format. These CAD models can subsequently be used to fabricate using either conventional material removal methods or material incremental methods, i.e. rapid prototyping (RP).

Engineering modelling often combines advances of both of these methods, which is a hybrid process of reverse engineering and forward modelling, to shorten the product design cycle, improve efficiency and meet the fast-paced changes of the modern market.

The technology advances of CAD/CAM/CEA have also made a significant impact in medical fields, providing especially many benefits for restorative dentistry [1], [2]. In prosthodontics, the digital method of obtaining dental impressions has proven to be more efficient than conventional one when assessing the total treatment time and patient comfort [3], [4]. Nevertheless, there are studies showing that digital scanning systems are not superior to conventional methods in comparing fidelity, accuracy and detail reproduction [5]. Also, the high cost of digital systems often makes their use difficult in clinical practice, so the conventional impression method is still a standard procedure in many dental practices.

The purpose of the dental impression is to copy the patient's intraoral situation, transforming it into a model. Obtaining a model of good quality and true to the original is extremely important for the success of the restoration. In conventional impression methods, different types of materials and impression techniques have been used over the years to achieve the desired accuracy. Independent studies of the quality and accuracy of impressions for crown and bridge works have shown that about 40% to 50% of impressions were not good enough to reliably obtain a satisfactory, well-fitting restoration for the first time [6].

Looking at the reasons for the impression defects, the researchers found that one of the most interesting factors is the use of an appropriate impression tray. The tray is required to give rigid support to the impression material and to facilitate its introduction into the mouth. They can be categorized as custom trays made specifically for a patient or stock trays that are available in a variety of sizes from the manufacturer. Custom trays are used to make a final impression, while stock trays are used for an initial or preliminary impression.

In this paper a procedure based on reverse engineering is implemented, where 3D scanning has been used for creating CAD models of disposable perforated lower and upper impression trays. Based on them, new models with additional perforations were designed using Geomagic and SolidWorks software. Static analysis in SolidWorks Simulation has been used to predict the physical behaviour of products by virtually testing models.

2. Rigidity of disposable stock trays

Stock trays can be subcategorized into two general types: metal stock trays and disposable plastic stock trays. It is important that the impression material remains attached to the tray during the impression. Partial detachment can cause gross distortions of the impression that may remain undetected and will almost certainly lead to ill-fitting appliances or restorations. Manufacturers of impression materials often supply tray adhesives that are used to enhance bonding. Additional retention is achieved by using perforated trays. The impression material flows through the perforations allowing a more even distribution of the impression material and giving a mechanical lock to the tray.

There is a strong preference between metal and plastic trays for disposable plastic ones, as they do not need to be returned from the laboratory and do not have to be cleaned and disinfected after use. An important question, however, is whether there is a significant difference in the precision of impressions made with metal and plastic stock trays, as they should be rigid enough to withstand the forces generated during impression procedures without permanent deformation. Studies have shown that most disposable plastic trays are too flexible to ensure accuracy, especially when the two-stage putty/wash silicone impression material is used [6]. The procedure requires light-bodied material to be added to the putty silicone, and when this occurs stresses can be induced in the putty. If a stock tray made of flexible plastic is used, these stresses can deform the tray. The critical area for tray stiffness is bucco-lingually. During seating, the highly viscous material will expand disposable tray (Fig. 1a). If the tray deforms, it will spring back to a smaller bucco-lingual width when removed from the mouth (Fig. 1b), resulting in castings which are then too narrow along this dimension.

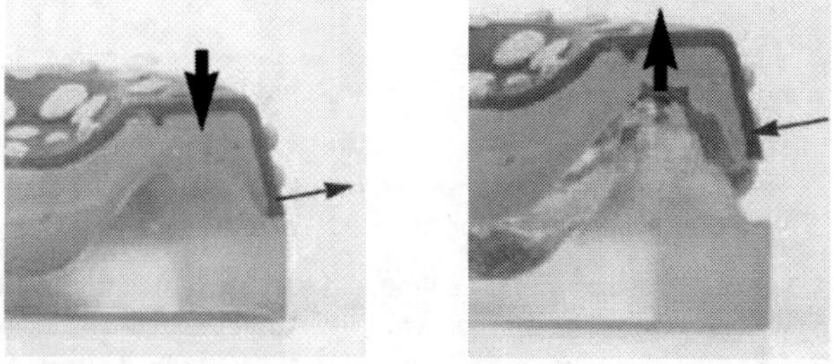

a) b)

Fig. 1. Bucco-lingually deformation of disposable tray [6]

Dimensional distortion of the impression is also more likely when a flexible tray is used, if excessive pressure is applied during impression making. An adequate impression tray should be rigid, i.e. flexion and torsion resistant, in order to resist the forces generated during impression procedures without permanent deformation. In [7], the authors studied the cross-sectional and cross-arch changes of six disposable plastic trays and compared them with a metal stock tray. The tested metal tray did not produce any significant changes, while all related disposable plastic trays evaluated produced significant dimensional changes when measured without impression material, on seating and polymerization of the impression material, and on removal of the impression material.

3. Generating CAD models based on RE

The 3D scanning system includes the GOM Atos Compact Scan 8M optical scanner, a movable stand to which the 3D scanner is attached, a laptop, a rotary table, an optical cable that connects the 3D scanner to the computer and a power adapter (Fig. 2). The software used for 3D scanning was ATOS Professional 2020.

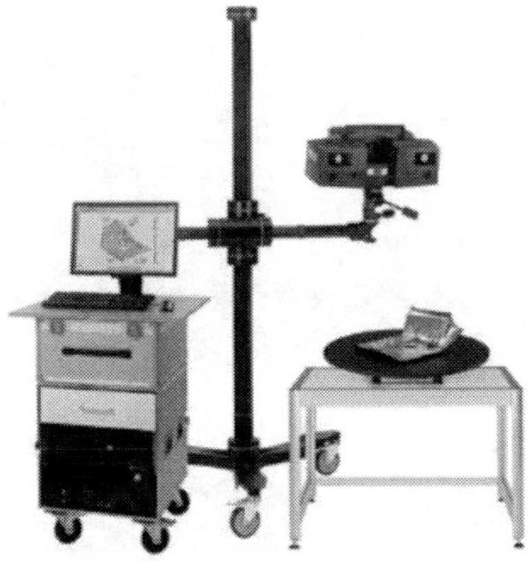

Fig. 2. The 3D scanning system [8]

GOM Atos Compact Scan 8M is a mobile 3D scanner that provides adaptability to different applications, part sizes and environments [8]. It is a type of coordinate measuring machine that uses a contactless blue light projector and two cameras that record line projections. On the scanner, there is a possibility of replacing the lenses on the cameras, which changes the measuring volume depending on the size of the object being scanned. In this case, the measuring volume 350 has been used. Before starting the scanning process, it is necessary to prepare the object. The first step begins with its cleaning, i.e. degreasing in case there are certain physical impurities. Reference point markers are then placed on its surface, which are of a specific shape (white dot on a black background) and diameter, depending on the measured volume. They must not be set near the edges of the scanning object. The 3D scanner automatically recognizes and encodes these markers, using them as reference points to properly align subsequent scans. The prepared object needs to be placed on a rotary table on which measuring points have been previously prepared and glued, which will later serve as a reference for separating the scanning object from the environment. The scanning object can be attached to a table to prevent it from moving and thus prevent the image quality from deteriorating.

It is also necessary to check that the distance between the sensor and the object is satisfactory, followed by setting the image saturation parameters. The angle of rotation of the rotary table, and thus the object itself, was set with a step of 72° between two consecutive images, and if necessary, two more images could be added to further fill the gaps created in the scan. After scanning one side of the object, the same procedure is performed for the other side.

The obtained scan results are further processed in the GOM software, which includes the Transform Measurement Series option where the software can recognize points to be connected or they can be manually selected. When merging multiple scanned sides of an object, overlapping scanned parts and points that potentially deviate from the geometry are automatically removed. The connection is made by polygonization of points, which gives a mesh of polygonized triangles.

Before solid modelling, the mesh is further processed. This primarily refers to the alignment of the workpiece and the filtering of defects that may occur on the surface (roughness, glare, etc.) or may be systemic, which are manifested in the appearance of spikes due to non-smoothing of the surface. The final step is to export the mesh in STL format and import it into Geomagic, which is a reverse engineering software toolset that allows importing scan data directly in SolidWorks. Using Geomagic ability to extract reference geometries and cross-sections directly from point cloud data, multiple cross-sectional curves associated with the original models were created. They were used as a basis for defining profile and guide curves for the standard set of SolidWorks tools for creating 3D feature based, parametric models. Afterwards, deviation analysis was performed. Based on these CAD models, the redesigned impression trays were constructed on which posterior area became less rounded and new perforations were generated. They should allow the impression material to escape through them, which contributes to a more even material distribution and better retention.

The redesigned models have been printed by the 3D printer Zortrax M200 Plus, in order to consult a prosthetist on the acceptability of the redesign (Fig. 3).

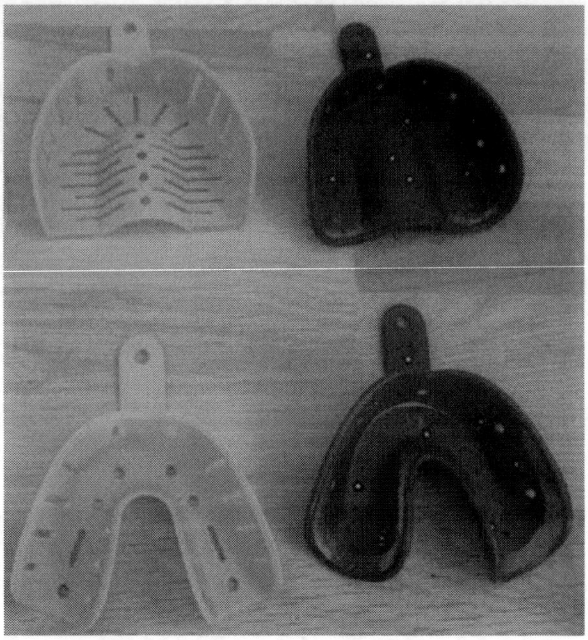

Fig. 3. The printed redesigned and original impression trays

4. Finite element analysis and discussion of the results

The CAD models obtained by the scanning process, Geomagic and SolidWorks software were used as a basis for creating FEM models in SolidWorks Simulation static study. The material used in this study was Polystyrene PS Medium/High Flow from SolidWorks material database. It was assumed to be homogenous, isotropic and to have linear elasticity, with elastic modulus E=2280 N/mm2 and Poisson's Ratio ν=0.387.

The mesh was created using the second order tetrahedral elements, with mesh controls defined on geometry elements where a high stress concentration was expected.

The forces required to insert and remove an impression from mouth have been considered by some investigators [9]. It has been shown that they depend on the properties of the impression and tray materials, the tray design, the method used to seat and remove the completed impression, and the shape and consistency of the soft tissues. In [9], Sotiriou and Hobkirk examined forces exerted during in vivo recording of impressions of dentate maxillary arches. They used four different impression materials in metal and disposable plastic trays. The load rate and the maximum insertion and removal forces were measured with strain gauges mounted on the tray handles. When the impression tray was inserted and removed from the mouth, the tray was held with the thumb and the index finger of the right hand of the operator, and the necessary forces were applied only on the handle of the tray. In this study, a boundary condition was applied that prevents the nodes from moving on the surfaces of the handle on which it is held.

There are a number of studies that have investigated the pressure applied to the tissue during final impression making [10], [11]. These studies were mainly conducted in vitro and with custom impression trays. The results show that when impression material is placed in a tray and compressed, the highest pressure is located in the central region, and pressure gradually decreases in the direction of the periphery. This is due to the rheological properties of the impression material, as the internal pressure of the material decreases in the direction of the peripheral area where the pressure escapes [12].

Based on the results of the aforementioned studies, we made an assumption that the pressure acting on the tray during initial impression was distributed by the equation:

$$p(x,y)=\frac{p_0}{L}x\left(1-\frac{y}{H}\right) \tag{1}$$

where p_0 is the pressure acting in the most anterior region of the tray body; L and H are the length and height of the tray; x and y are coordinates relative to the local coordinate system with the origin placed at the most posterior point of the mid-sagittal line on the plane coinciding with the tray floor, as shown for the upper trays in Fig. 4. The equation (1) assumes linear increase in pressure from the posterior towards anterior part, as well as linear pressure decrease from the tray floor towards the tray flanges borders. The maximum value of pressure p_0 is estimated based on the resulting force on the tray handle, which corresponds to the average force records for plastic trays in [9]. The value used in this study was p_0= 5 kPa.

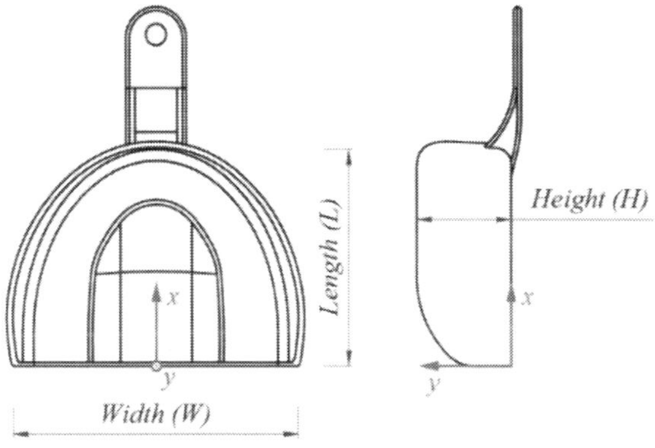

Fig. 4. Local coordinate system for the upper trays

In this study, a qualitative analysis of stress distribution using Von Mises criteria and principle stresses to measure the overall stresses has been performed. The stress distribution for all tray designs showed that for the assumed load model the highest stresses occurred on the tray handle, with values within the safe limits of the material used, which ensured the safe use of all designs.

Deformation plots are shown in Fig. 5-7. They are presented with undeformed models (in grey colour) superimposed on the deformed shapes, which helps to capture the deformed shape relative to the original shape. This type of display can also facilitate insight into which areas of the trays most deviate from the initial undeformed shape, and accordingly in which areas of the cast model deviations from the anatomical shape can be expected. The maximum deformations for the upper trays are in the vertical direction at the most posterior point of the mid-sagittal line (Fig. 5-6). The difference between these values for the original and redesigned tray is about 10 µm, and the same difference in deformation values is also present in their bucco-palatal direction. For the lower trays, maximal deformations are present at the posterior areas of the lingual flanges, with a value difference of about 20 µm between the two tray configurations (Fig. 7).

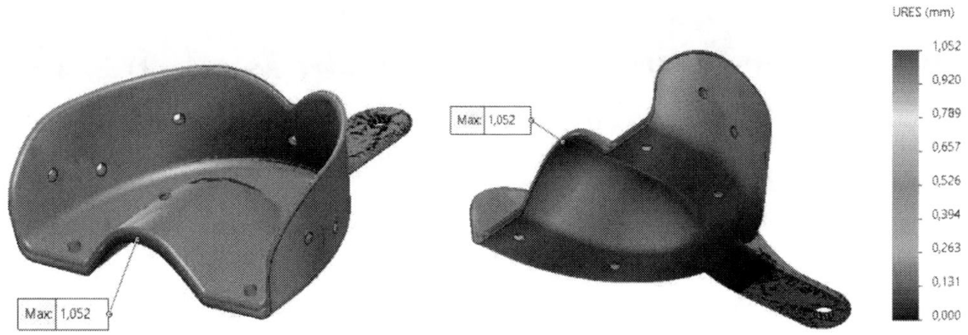

Fig. 5. Total deformation of the upper original tray

Fig. 6. Total deformation of the upper redesigned tray

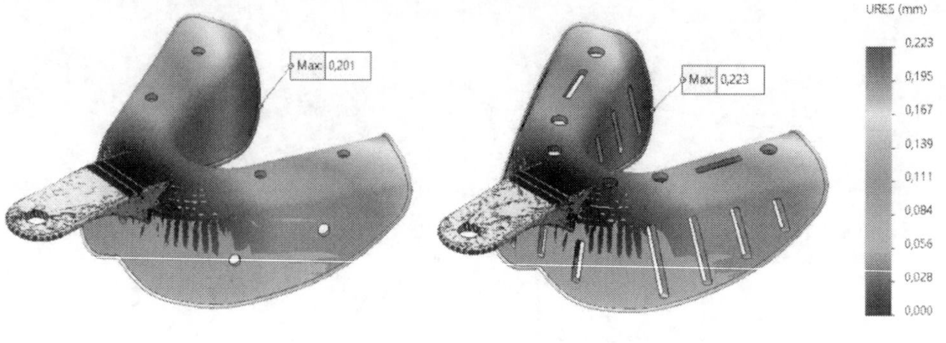

Fig. 7. Total deformation of the lower original and redesigned tray

5. Conclusion

RE can imitate and redesign products based on existing physical parts or prototypes, while RP can produce parts of a very complex geometric shape in a short time compared to traditional methods of producing spatial objects. In this paper, RE and RP have been used in order to redesign disposable upper and lower impression trays. Tray rigidity is an important parameter when taking impressions because even slightly flexing of the tray causes distortion of the impression. The FEA has been used to test the structural rigidity for the proposed load model to which the trays are exposed when taking impressions. Based on the results of the numerical studies, both tray configurations, the original and redesigned ones, have similar deformation pattern, with the displacement values of the redesigned models being higher because their rigidity decreases due to the additional perforations.

Further research into distortion of impressions done with the trays would provide data to define more suitable numerical models for obtaining values that may be closely correlated with values measured intraorally or with in vitro laboratory tests. Thus, the application of FEA creates an opportunity to recreate mechanical forces corresponding to the real clinical condition and to obtain stresses and strains whose direct measurement would be difficult if an adequate computational model was not available. This would enable the design to be optimized in terms of a more suitable perforation arrangement to achieve proper retention, stability and support in the impression taking procedure while maintaining maximum tray rigidity.

6. References

[1] Abduo, J.; Lyons, K. & Bennamoun, M. (2014). Trends in computer-aided manufacturing in prosthodontics: A review of the available streams, International Journal of Dentistry, Volume 2014, Article ID 783948, http://dx.doi.org/10.1155/2014/783948

[2] Beuer, F.; Edelhoff, D. & Schweiger, J. (2008). Digital dentistry: an overview of recent developments for CAD/CAM generated restorations, British Dental Journal, Vol. 204(9), pp. 505-511, DOI: 10.1038/sj.bdj.2008.350

[3] Lima, J., et al. (2019). Evaluation of the accuracy of conventional and digital methods of obtaining dental impressions, International Journal of Odontostomatology, Vol. 12(4), pp. 368-375, DOI: 10.4067/S0718-381X2018000400368

[4] Park, H.S. & Shah, C. (2014). Development of high speed and high accuracy 3D dental intra oral scanner, Proceedings of the 25th International DAAAM Symposium "Intelligent Manufacturing & Automation", 26-29th November 2014, Vienna, Austria, Procedia Engineering 100 (2015) 1174 – 1181

[5] Chandran, S., et al. (2019). Digital versus conventional impressions in dentistry: A systematic review, Journal of Clinical and Diagnostic Research, Vol. 13(4): ZE01-ZE06, DOI: 10.7860/JCDR/2019/38494.12756

[6] https://europe.gc.dental/sites/europe.gc.dental/files/products/downloads/coeimpressiontrays/manual/, The foundation of a good restoration. Impression Trays. Guide to correct tray selection, Accessed on: 2021-06-20

[7] Cho, G. & Chee, W. (2004). Distortion of disposable plastic stock trays when used with putty vinyl polysiloxane impression materials, The Journal of Prosthetic Dentistry, Vol. 92(4), pp. 354-358, DOI: 10.1016/j.prosdent.2004.07.020

[8] https://www.gom.com/en/products/3d-scanning/, Industrial 3D scanning technology, Accessed on: 2021-06-15

[9] Sotiriou, M. & Hobkirk, J. (1995). An in vivo investigation of seating and removal forces associated with recording impressions in dentate patients, The Journal of Prosthetic Dentistry, Vol. 74(5), pp. 455-462

[10] Chang, Y., et al. (2019). Influence of edentulous tray modification on the pressure distribution of the impression: An in vitro study, The International Journal of Prosthodontics, Vol. 32(3), pp. 278-280

[11] Iwasaki, M., et al. (2016). Pressure dynamics in the trays caused by differences of the various impression materials and thickness of the relief in the maxillary edentulous model, Journal of Prosthodontic Research

[12] Komiyama, O., et al. (2004). Effects of relief space and escape holes on pressure characteristics of maxillary edentulous impressions, The Journal of Prosthetic Dentistry, Vol. 91(6), pp. 570-576

DOI: 10.2507/32nd.daaam.proceedings.050

BIG DATA AND BUSINESS INTELLIGENCE: RESEARCH AND CHALLENGES IN TELECOM INDUSTRY

Ivan Krasic, Stipe Celar & Zeljko Seremet

This Publication has to be referred as: Krasic, I[van]; Celar, S[tipe] & Seremet, Z[eljko] (2021). Big Data and Business Intelligence: Research and Challenges in Telecom Industry, Proceedings of the 32nd DAAAM International Symposium, pp.0339-0348, B. Katalinic (Ed.), Published by DAAAM International, ISBN 978-3-902734-33-4, ISSN 1726-9679, Vienna, Austria
DOI: 10.2507/32nd.daaam.proceedings.050

Abstract

This paper focuses on the Big Data (BD) and Business Intelligence (BI) application in the telecom domain and the benefits it offers. Telecom companies themselves are the producers of big data on daily basis – structured and unstructured data formats from many sources. Already some level of Business Intelligences is implemented and in-use for supporting telecom daily business. We provide literature review of BD and BI state-of-art in telecom domain, example of real world use cases of BD and BI analytic, platforms, frameworks, practices and the related challenges of the further development and research trends.

Keywords: big data; business intelligence; telecom; churn; literature review.

1. Introduction

The rapid development of telecommunications and other large systems over the last decades has resulted in increased system complexity and data itself. Telecom operators striving to analyse large volumes of data in order to support their core business and operations. Solution was to deploy sustainable storage data systems - which differs from operating/transactional systems of historical, unaltered and multidimensional organized data that are intuitive and adapted to thinking and human inference and analysis. Today, in highly competitive market, rising era of *big data* (BD) is characterization of all operatory whole around world. In addition to telecom changes, proliferation of other systems directly related to telecom like Internet, Internet of Things, and Cloud Computing have led to the explosive growth of data in almost every industry and business area, affecting telecom operators too.

This paper provides a short overview on how telecom operators have been striving to analyse large volumes of data in past. Thus, we explain big data paradigm with relation to business intelligence (BI) process. We then review state-of-art in BD and BI in telecom domain. Finally, we provide example of real world use cases of BD and BI analytic, platforms, frameworks, practices and the related challenges of the further development and research trends.

2. Business intelligence

Business intelligence, even used in more decades, is not a well-defined term, depends on who is defining. Some authors as [1] define BI as "data reporting and visualization, some consider as business performance management. Database vendors emphasize data extraction, transformation and integration features. Analytics vendors highlight statistical analysis and data mining". Also it can be defined as system that can be used for making decision from information based on available data – process which covers: *integration* and *collection* of *data sources*, *storage* and *information database*, *management* of *knowledge* with the *use* of *different systematic* and *analytical tools* in *presenting information* to the organization's decision makers and stakeholders. They should take an action (decision) based on BI environment back to origin systems (OSS, Network etc.) or send it as a input of toll based on artificial intelligence, releasing action too. Typical BI architecture consist of following components, connected by unidirectional communication flow between different components (Figure 2):

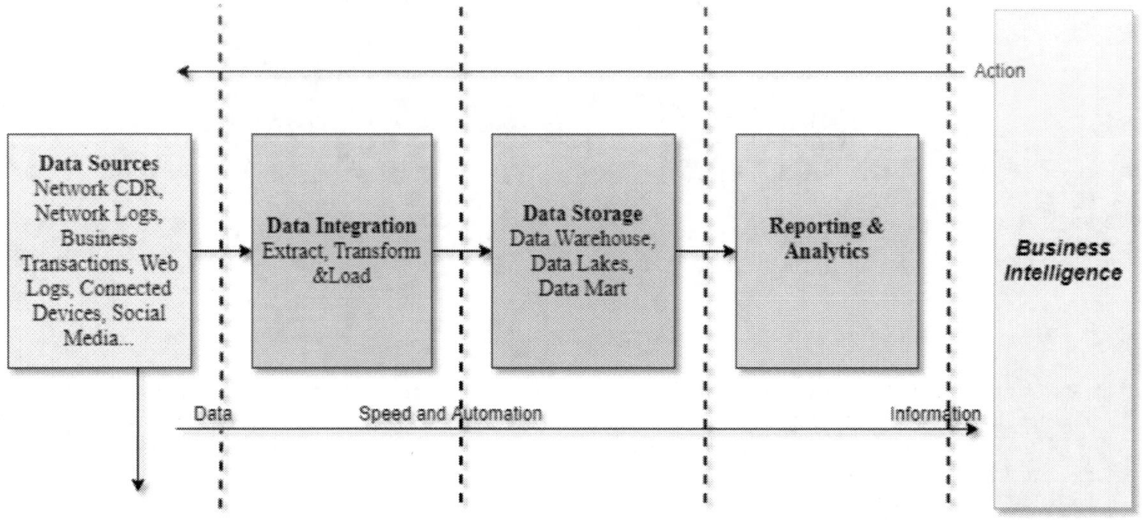

Fig. 1. Typical BI architecture

Typical BI architecture (presented in Figure 1) consist of the following layers:

- Data Sources – include all data from organization's systems, external sources too, structured on unstructured, many different formats, with different nature in sense of characteristics (volume, velocity,…).
- Data Integration – provide extraction, transformation and integration of data via appropriate tools, from many sources, automated tasks with remark of speed and automation level.
- Data Storage – layer responsible for data storage, on premise or cloud, organized in classic way (relational database management system) or specialized databases that provide control over selected activities and selected business process (data warehouse, data mart, data layers etc.).
- Reporting & Analytics – subset of different software tools exploiting stored data, in order to make different reports, analyse specific or ordinary business needs, or even make proactive tasks. In addition, this layer provide specialized different W analyses (Why, When, Who, What, What If…).

This simplified architecture present classic and typical BI framework deployed in many telecommunication companies today. Typical BI deployment and framework are widely accepted and are de-facto standard for business and corporate analytics and decision support systems (DSS). On the other side this deployment scenario is not appropriate for upcoming challenges in rapid growth of data volume, velocity and variety.

3. Big data

The term "*big data*" has long history and there is no single standard definition. Some authors finding origins in academic and non-academic community from 1990-th, even earlier, but truly come of age in 2008. – report "*Big Data Computing: Creating revolutionary breakthroughs in commerce, science, and society*" from *Computing Community Consortium* and in age of 2013 – when Oxford English Dictionary introduced the term "*Big Data*" for the first time in its dictionary [2]. In the last decade big data research has significantly increased (Figure 1).

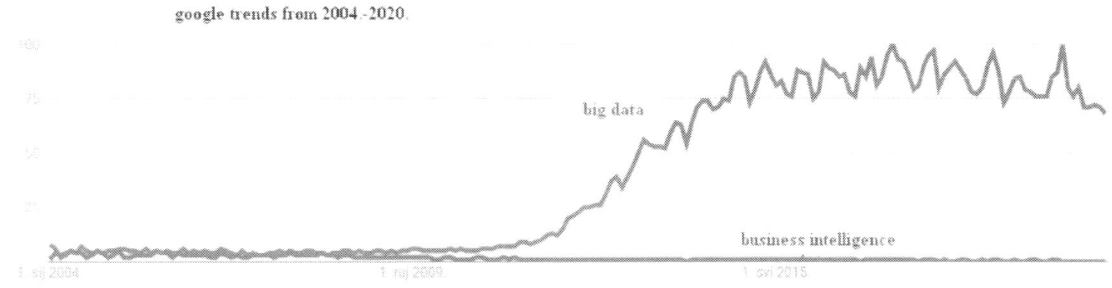

Fig. 2. Google trends for "big data" and "business intelligence" (2004-2020)

Big data trends have been enabled by advances in computing technologies, which facilitated the sudden explosion of data from various sources such: Internet and Web, social medias, sensors etc. The large amount of data emergence application of a data-driven paradigm to take advantage of the newly available computing technologies. Big data technologies materialized the data-driven paradigm, making it increasingly sophisticated and useful [3].

Today, Big Data (big data or BD) is a comprehensive term that primarily refers to the sources of large amounts of data and the infrastructure that supports their storage. Data access requires new tools and technologies (framework) for optimized access. Numerous languages have been developed for the analysis of new sources, real time solutions, security concepts, etc. Big data includes technology that serves to collect process and complex analyse a large amount of data. Data are diverse, structured and unstructured, they are generated and arrive at high speed at different intervals (asynchronous or real time).

Big data is often characterized by different V's, according to leading letter "V" in characterize - originally 3V, then some authors proposed 4V, 5V, 6V, 7V...:

Vs	**Characheristic**	**Description**
3V	**Volume**	*Scaling* – data volume is increasing exponentially, quantity measure
	Variety	*Complexity* – various formats, types, and structures,
	Velocity	*Speed* – data is generated fast and need to be processed fast (real time application, etc…)
4V	Veracity	*Confidence* or *Completeness* or trust in the data. This is similar to, but not the same as, validity or volatility.
5V	Value	Need to derive *business value* from the data – like customers data, targeting them, optimizing processes or improving machine or business performance.
6V	Validity	Refers to how *accurate* and *correct* the data is for its further use.
7V	Volatility	Refers to *how long is data valid* and *how long should it be stored.*
10V …	Variability	Refers to *number* of *inconsistencies* in the data and *inconsistent speed* of data loading into database or any storage layer.
	Vulnerability	Big data requires *new security concerns.*
	Visualization	Characteristic of big data is *how* solve new *visualization chellenges.*

Table 1. List of V's of Big Data

Concerning that big data access is primary data-driven based, similar as in past, where storing data is integrated from one or more sources (operating systems etc.), building any data storage is a complex and thankless job. It follows the steps like other information systems but with one difference that the impact of the project warehouse is important and implies specific knowledge about certain problems and other skills. The projector warehouse was responsible for the model warehouse and later (un)/successfully used it. The choice of measures / measurement data to be analysed and contained in the multidimensional model is the role of the designer.

Rising of big data technologies other roles should be introduced and developed, leaving the past known human roles and applied technologies. Comprehensive review study about BI and BD research evolution and interaction and can be found in [4].

4. Literature review search and analysis

In this paper we concentrates on searching in scientific databases – rather than in specific books or technical reports.

4.1. Literature review search process

This paper has selected next databases to perform the search process as follows:
1. Science Direct (www.sciencedirect.com)
2. Scopus (https://www.scopus.com/home.uri)
3. Google Scholar (www.scholar.google.com)
4. IEEE Explore (https://ieeexplore.ieee.org/)
5. ACM Digital Library (https://dl.acm.org/)
6. Springer (https://www.springer.com/gp)

The databases were chosen as they offer the most important and highest impact full-text journals and conference proceedings, covering the requested fields (big data, telecom/(munication)). For this work paper we found and include scientific papers from Science Direct, Scopus and other listed above sources. In order to define a search string, we experimented with a variety of combinations of keywords to test synonyms used in literature (telecom or telco or telecommunication(s)). After few iterations we finally ended up with the following search string:

"Big data" AND ("telecom" OR "telecommunication")

Search is done by simple keywords in article title, abstract or content, regardless of how specific search engines works at by itself. We started our search all well-known digital libraries, namely Science Direct, Scopus, IEEE Explorer, ACM, Springer, and Google Scholar. This whole process, including initial and result set, consist of following steps (Figure 3).

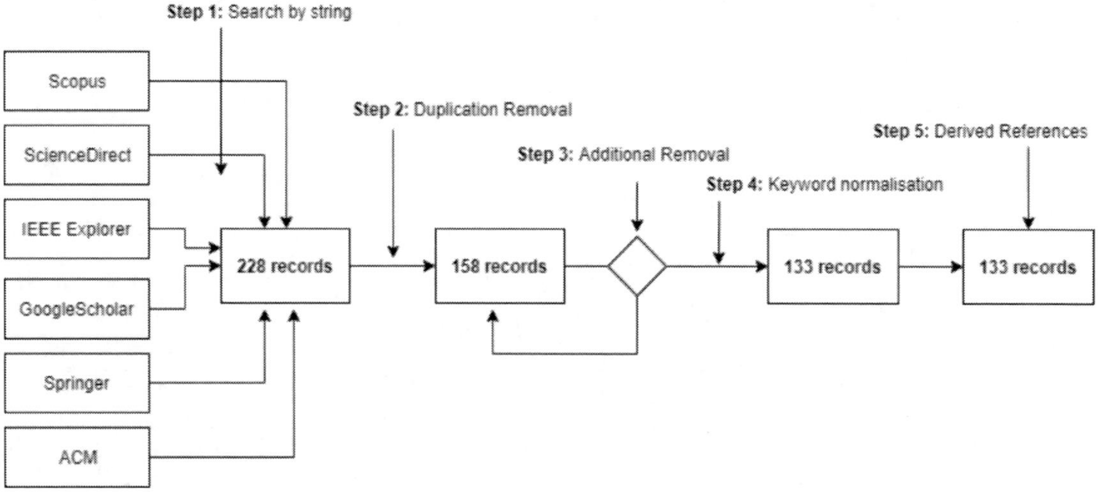

Fig. 3. Literature search process

We followed these procedure steps:

Step 1 **Search digital libraries**: search by keywords;

Step 2 **Duplication Removal**: After consideration of the 228 publications we found that 30,7% papers were duplicate. After discarding duplicates, we were left with 158 articles;

Step 3 **Additional Removal**: this step is done by following simple inclusion/exclusion criteria: "*The language of the literature is English*", "*Author keywords must be defined*", "*Paper has to be either journal article, conference paper of book chapter and publisher*".

Step 4 **Keywords normalisation:** update keywords, from plural to singular ({telecommunications}, {telecommunication}...), abbreviations ({*call detail record(s)*, {*CDR*}, {*cdr*}...}

Step 5 **Derived references**: final reference list is available for further processing in different formats (*.ris, *.bib...)

The most of them are journal articles, with one or more authors, order by autonomous search engine ranking (best match etc...). Furthermore, next step in analysis is using Mendelay tool, reference manager desktop software, for constructing library of selected articles. Exported reference list is available in BibTeX (*.bib) or Research Information System (*.ris) document format.

4.2. Literature review analysis

In this review, we identify and analyse the selected papers in order to a knowledge base that aims to summarize more clearly the key aspects of big data and analytics application in the telecom industry. This should bring values to both academia and practice, by providing both insights into what has already been studied with respect to the telecom development trends and a motivation for future research. Figure 4 indicates the distribution of selected papers by type – notice that journal articles has the higher selected papers per type among conference proceeding and book chapter.

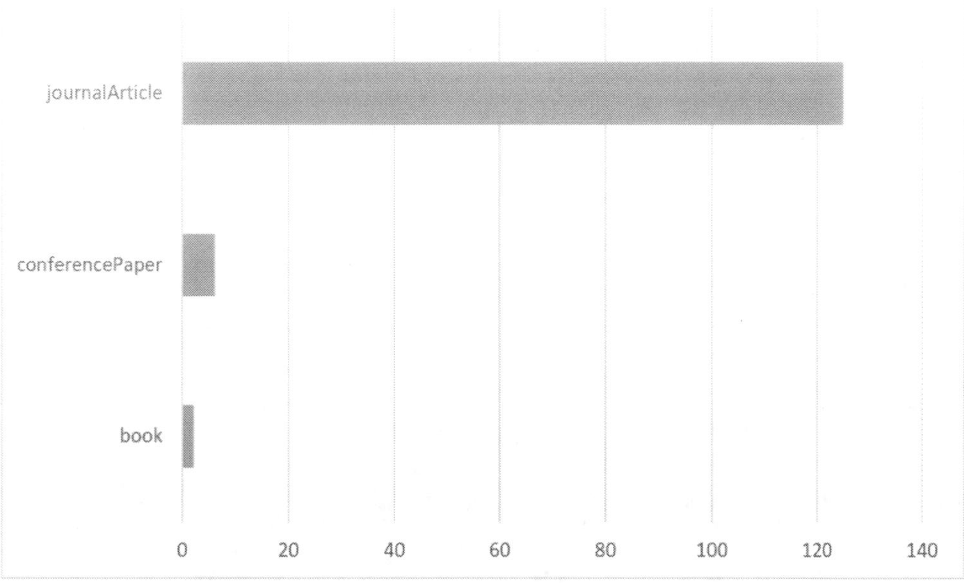

Fig. 4. Distribution of papers by type

Figure 5 shows distribution of papers by year in study. Regardless of the common perception that the "*big data*" started in the late 2000s when a dozen number of most prominent technology companies developed architectures to handle large-scale data (e.g., MapReduce by Google in 2004) "big data" is still a new term to the public and to academia. Joining the term "Telecom" to search we can detect increase in the number of scientific articles after 2015.

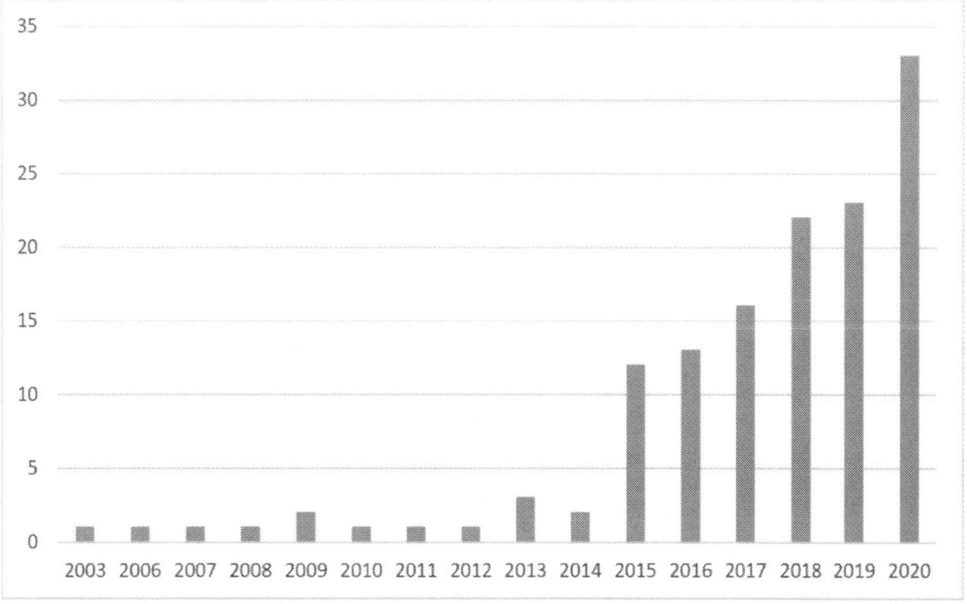

Fig. 5. Distribution of selected papers by year of publishing

All of selected articles are published from year 2003, showing relatively weak interest up to 2015, excluding years 2009 and 2013. Thus, the number of publication is extremely increased in year 2015 and growing up to 2020, showing more research interest. Therefore, we are expected that years 2020- 2021 can reach higher than past few years, since this article list and review is based on export in May 2021.

5. Results analysis

In this section, we quantitatively and qualitatively analyse selected papers. The results of our work include the discovery of interactions in application of big data in telecom domain, in order to provide recommendations for all stakeholders (e.g., academic researchers, policy-makers, entrepreneurs…).

For this analysis we use tool VOSViewer for constructing and visualizing bibliometric networks. Network visualisation of keywords (by authors) is illustrated in Figure 3. The size of each circle corresponds to the proportion of keyword occurrences from picked papers.

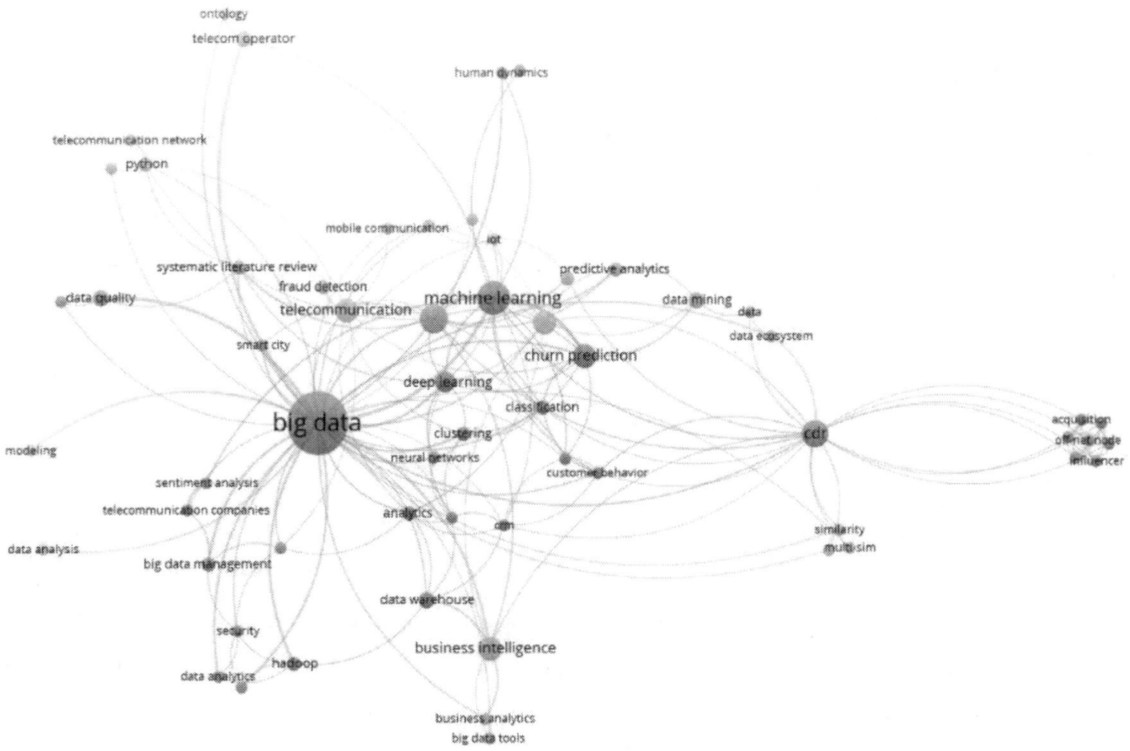

Fig. 6. Network visualization of keywords – word cloud of associated key words

For graph visualisation following values are exported:

- Total of different keywords (472)
- Minimum number of keyword occurrence (2) – exclude keywords with only one occurrence)
- Remaining keywords (62)

Decreasing minimum number of keyword occurrences can fit more specific use cases, maybe more interesting for telecom too but in this paper we start with default values. We can note relatively diversity of exported keywords,

Furthermore, the related keywords are grouped into different clusters categories and coloured with same colour (total number of clusters is 13). Following clusters are clearly detected in Figure 6, i.e., cloud computing (with MapReduce, Hadoop, and related techniques), machine learning (with a broad range of artificial intelligence techniques and applications), CDR (Call Detail Record), business intelligence and churn prediction term. Generally, there were no unexpected terms, all terms and their synonyms can be found on the other lists and are well known research topics. Looking into clusters distribution defined by VOS (13 in total) we can observe larger 5 clusters. Figure 5 displays the first cluster.

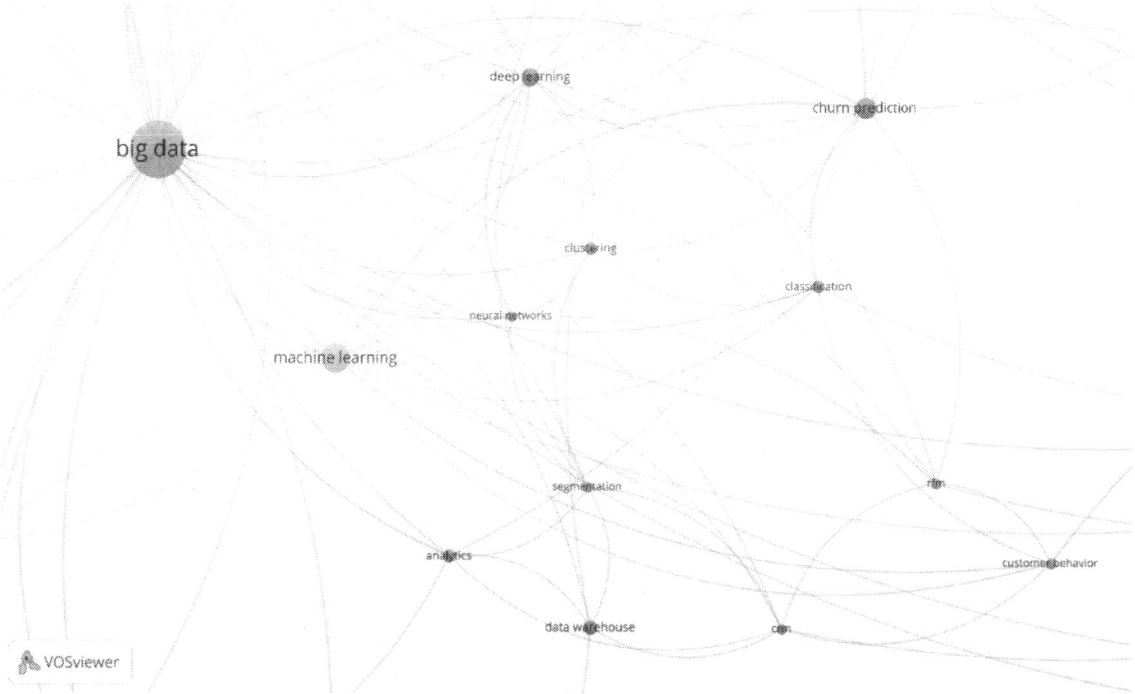

Fig. 7. Cluster_1 with prominent keyword categories big data and customer analysis

In above cluster we can see that big data is in strong connection with machine learning methods (clustering, deep learning, classification,…) in order to provide better way to customer related issues (churn prediction, customer related management,…).

The following cluster is shown in Figure 8.

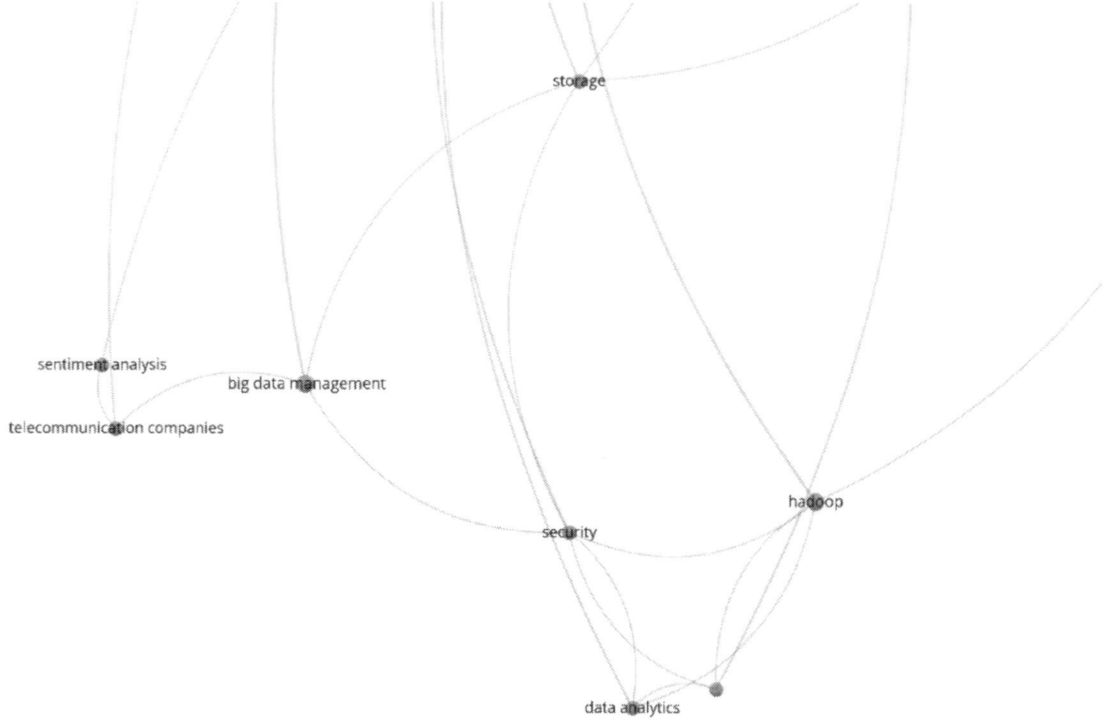

Fig. 8. Cluster_2 with prominent keyword categories big data and technology issues

Cluster_2 is related to technology issues. For instance, it includes issues related to data storage, collection, analytics, processing and integration infrastructure (Hadoop). For example, sentimental analysis – collecting and analysing pure textual data in order to defining positive or negative reactions are also key research issues.

Fig. 9. Cluster_3 – CDR related analysis and research

In telecom industry, all data are stored in CDR files (Call Detail Records) and can be classified into three main groups: *call detail data* (related to communication between subscribers or to provider, calls, messages, data session), *network data* (all data from network systems and networks) and *customer data* (includes information about the personal data of customer such as name, address details, subscription type, payment history and other customer attributes).

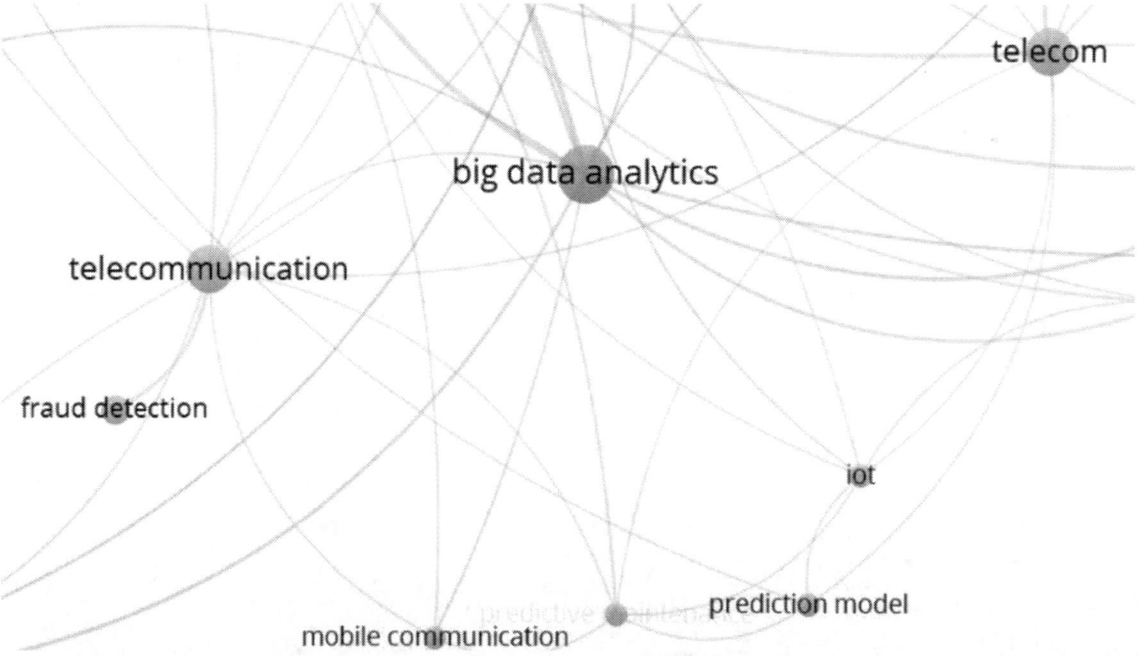

Fig. 10. Cluster_4 – Cluster related to specific telecom application

This cluster (Figure 10) is related to papers and research related to three specific areas: (1) specific telecom application like fraud detection and mobile communication, (2) newer trends like 5G and IoT (Internet of Things), and (3) network operation like predictive maintenance and prediction model (traffic), mainly in OSS domain (Operations Support System)

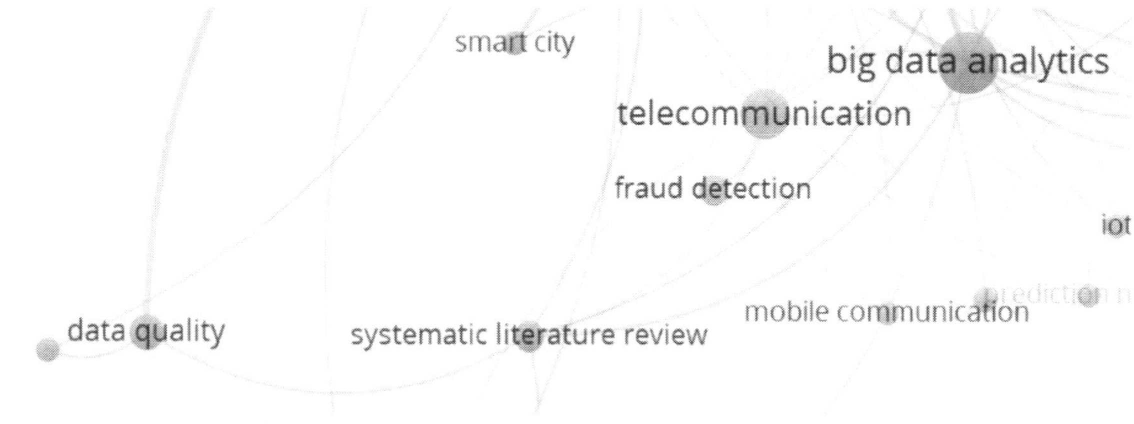

Fig. 11. Cluster_5 – Cluster related to big data analytics

This cluster (Figure 11, purple lines) is related to big data analytics in general, highlighting the topics related to data quality issues. In telecom operators multiple and variety data sources leads to data quality problems. Data quality in telecommunication systems has been studied from various perspectives, related to customer and network data.

Other clusters have smaller number of items (<5) with larger diversity, and can observe them like more specific research interest, out of this paper.

6. Big data challenges and research motivation

According to presented correlation between research interests we can propose following challenges in telecom domain:

1. Challenges related to network services and management
2. Challenges related to customer acquisition and retention based
3. Challenges related to security issues
4. Challenges related to big data technology and management

6.1 Challenges related to network services and management

We can note several interesting research topics related to big data and application of machine learning methods to propose novel predictive maintenance frameworks modelling and implementation, reducing human actions and better error prevention. In case of service management, big data analytics can improve quality of service indicators (QoS KPI), provide more independent end to end service provision and management, also like shortening time to market in case of introducing new services and related products.

One more interesting research area is Robotic Process Automation (RPA) adopting in big data for automating telecom network management. Legacy telecom OSS and BSS are inadequate for the needs of the modern service providers because they were primarily designed to assist the humans in executing the processes that ran the business. To support the modern telecom needs, these systems need stronger automation and orchestration to operate the business quickly and autonomously under the control and oversight of people. This leads to roadmap for ultimately automation of all processes that run the telco business and management of the network.

6.2. Challenges related to customer churn, acquisition and retention

This group of challenges are well known topics like customer churn (unexpected leaving of operator), improving customer retention rates (customers who are retained over a specific time period) and better customer experience. This research topics are well none in past decades, data mining and business intelligence techniques are mainly applied to this issues. Introducing big data concepts and stronger machine learning methods can produce significant research contribution in future. According to [5], big data approach is applied for telecom churn prediction instead of use of data warehouse and data mining techniques, due to challenges of data volume, processing power and capacity management. Same authors quote that "acquiring a new customer costs six times higher than the cost of retaining the customer likely to churn". This indicates to direct conclusion that cost reduction and profit growth are well reasons to stronger research and real use case applications in this field.

6.3. Challenges related to security issues

We can note several terms as Fraud detection (detection of fraud activities), Fraud Prevention (active and real time system for prevention of fraud, proactive approach), and other security issues that are related to data protection and privacy, as standard big data research problem. Telecom providers care many private and protected customer information like patterns of social behaviour, communication needs, location based data etc. From one hand this data can be very useful to take into account for providing future better customer experience, personalized service and products, but on other hand this requires very high level of security and privacy protection. Concerning security and privacy issues in big data environments for further research systematic literate review can be fine starting point [6]. Telecom domain fit all conditions to standard security and privacy problems research.

6.4. Challenges related to big data technology and management

This topics are already in research's focus for years. The technology dimension of big data includes issues related to data collection, storage, analytics, integration infrastructure and architecture. Finding best and appropriate infrastructure for telecom domain reflects the huge research interests, for example the cost-benefit assessment. The complexity of big data processing is one of the fastest growing technology challenges. Many tools and platforms are available for choice: open source or commercial, batch or stream or hybrid processing tools. We recommend further reading in [7] and [8].

6.5 Other research challenges

In this group web can place some research topics who are applicable for any other research group but also very interesting by itself. First of them is Data Quality (DQ), an active and attractive research area for many years. Owning a huge amount of data could increase the risk of quality degradation and require efficient mechanisms to check data quality and manage quality throw whole process of data lifecycle, from generation to information knowledge extraction. Basics and further reading can be found in [9].

One more interesting area could be Big Data Analytics (BDA), as new way to business analytics then traditional Business Intelligence does. BDA We recommend further reading in [4] and [10].

7. Conclusion

The goal of this paper is to conduct a current state-of-art on big data application in telecom domain. Our purpose is research and identify the features of big data technology which can be help for telecom companies in their daily business. Our results indicate that there are at least four areas which need to be considered in order to applicate big data approach for improvement of analytics and business at all. Further research should also consider technical reports and real world use cases in order to better understand telecom needs and new solutions. This review shows that, despite the academic attention that the big data applied in telecoms has attracted to date, there is considerable room for further research on this area.

8. References

[1] Azvine, B.; Cui, Z.; Nauck, D.; Majeed, B. (2006). Real Time Business Intelligence for the Adaptive Enterprise. CEC/EEE 2006 Joint Conferences. 2006. 29-29. 10.1109/CEC-EEE.2006.73.

[2] Kocijan, K. (2014). Big Data: how we got the big data and where they lead us, Department of Information Studies, pp. 37-62, Zagreb (in Croatian)

[3] Veda C. S.; Il-Yeol Song (2017). Big data technologies and Management: What conceptual modeling can do, Data & Knowledge Engineering, Volume 108 (2017), pp. 50–67.

[4] Ting-Penn. L.; Yu-His, L. (2018). Research Landscape of Business Intelligence and Big Data analytics: A bibliometrics study, Expert Systems with Applications, Volume 111(2018), pp. 2-10, ISSN 0957-4174.

[5] Ahmad, A.K., Jafar, A. & Aljoumaa, K. Customer churn prediction in telecom using machine learning in big data platform. Journal of Big Data 6, article 28 (2019).

[6] Boel, N.; Tomas, O. (2016). Security and Privacy for Big Data: A Systematic Literature Review, 2016 IEEE, International Conference on Big Data (Big Data) 978-1-4673-9005-7/16, IEEE 369

[7] Suleykin, A[lexander] & Panfilov, P[eter] (2019). Implementing Big Data Processing Workflows Using Open Source Technologies, Proceedings of the 30th DAAAM International Symposium, pp.0394-0404, B. Katalinic (Ed.), Published by DAAAM International, ISBN 978-3-902734-22-8, ISSN 1726-9679, Vienna, Austria

[8] Mohamed, A., Najafabadi, M.K., Wah, Y.B. et al. The state of the art and taxonomy of big data analytics: view from new big data framework. Artif Intell Rev 53, 989–1037 (2020). https://doi.org/10.1007/s10462-019-09685-9

[9] Taleb, I.; Serhani, M.A.; Bouhaddioui, C. et al. (2021). Big data quality framework: a holistic approach to continuous quality management. J Big Data 8, 76 (2021).

[10] Ajah, I.A.; Nweke, H.F. (2019). Big Data and Business Analytics: Trends, Platforms, Success Factors and Applications. Big Data Cogn. Comput. (2019), 3, 32.

DOI: 10.2507/32nd.daaam.proceedings.051

ANALYSIS OF THE INTERACTION BETWEEN SAFETY AND SECURITY DEMONSTRATED ON A MOBILE ROBOT AND A PRODUCTION NETWORK

Isabella Reithner*[1], Maximilian Papa[1,2], Mohamed Aburaia[1], Wilfried Wöber[1,3] & Clemens Ambros[1]

This Publication has to be referred as: Reithner, I[sabella]; Papa, M[aximilian]; Aburaia, M[ohamed]; Woeber, W[ilfried] & Ambros, C[lemens] (2021). Analysis of the Interaction Between Safety and Security Demonstrated on a Mobile Robot and a Production Network, Proceedings of the 32nd DAAAM International Symposium, pp.0349-0355, B. Katalinic (Ed.), Published by DAAAM International, ISBN 978-3-902734-33-4, ISSN 1726-9679, Vienna, Austria
DOI: 10.2507/32nd.daaam.proceedings.051

Abstract

The 4th Industrial Revolution increases connectivity within the production environment by transforming traditional automation systems to cyber-physical production systems. This conversion impacts intralogistics of a factory: Intelligent and interconnected transport systems, in the form of autonomous mobile robots, are increasingly integrated into the production process. This enhanced connectivity expands the attack surface of production facilities for cyber-attacks. In addition, the interconnection of safety and security poses a risk for people working on machines or around mobile robots in case of an attack. In order to highlight the interaction between safety and security, this study discusses the requirements of a mobile robot from both perspectives. Verification in practice is achieved by performing a penetration test on a test environment of a production network with an integrated mobile robot. The discussion of the detected vulnerabilities outlines the hazards of operating a mobile robot in an interconnected production network, causing a loss of *Confidentiality*, *Integrity* and *Availability*. Finally, these results are providing a basis for trustworthy mobile robotics implementations in production plants.

Keywords: Mobile Robotics; Safety & Security; Penetration Test; Production Network

1. Introduction

The digitalization within production as part of the 4th Industrial Revolution leads to a high degree of connectivity in industry [1]. Existing devices are replaced by Cyber-Physical Production Systems (CPPS) [1] and the material flow within a plant is realized with automated guided vehicles or autonomous mobile robots [2].

[1] Department of Industrial Engineering, University of Applied Science Technikum Wien, Vienna, Austria
[2] Institute of Management Science, TU Wien, Vienna, Austria
[3] Institute for Integrative Nature Conservation Research, University of Natural Resources and Life Sciences, Vienna, Austria

This new approach requires a revision of outdated hierarchical communication models such as the *Automation Pyramid* and the isolation of Information Technology (IT) and Operational Technology (OT), in order to cover all aspects. However, with the increasing connectivity of industrial systems, the attack surface for cyber-attacks affecting CPPS, e.g., mobile robots, is increasing [3]. In 2019, a survey pointed out that 59% of production companies expected an increase in security incidents [4]. Furthermore, the priorities of IT and OT systems are different, as the protection goals of *Confidentiality*, *Integrity* and *Availability*, also known as the CIA triad, indicate. *Confidentiality* has the highest priority within an IT system, followed by *Integrity* of information. *Availability* is only in third place. However, this ranking is different for an Industrial Automation and Control System (IACS). Here, *Availability* represents the highest objective, followed by *Integrity* and *Confidentiality* [5].

Despite its relevance today, security is not the only issue in robotics. Technical safety measures are intended to reduce threats to functional safety and therefore safety of persons is enhanced. If these technical safety measures are threatened by security vulnerabilities, they can be compromised in case of an attack [5]. Therefore, the two-way interaction between safety and security has recently been recognized as an issue by companies [6][7] and is used as motivation for the following research.

In order to solve the problems mentioned above, the requirements for an application involving mobile robotics in Austria are first analysed followed by a focus on cyber-security. The knowledge acquired is applied to a dedicated demonstrator with a mobile robot by conducting a risk assessment. Subsequently, a vulnerability analysis and selected penetration test procedures are executed on the demonstrator and the mobile robot. In this concrete scenario, the connection between safety and security as well as the impact of security on safety in mobile robotics is to be analysed.

2. Current Requirements for an Interconnected Production Environment Including Mobile Robotics

In order to meet the requirements of the 4th Industrial Revolution, it is necessary to shift away from traditional automation systems and focus on a higher degree of flexibility. Custom requirements can be fulfilled using mobile robots, offering a high degree of flexibility in navigation, task performance and organization [8][9]. With this increase in flexibility, a growth phase in the market of mobile robotics is currently evident [10]. However, the implementation of flexible transport systems in interconnected production environments poses a new challenge in the context of security as the focus in robotics has long been on safety, development costs, speed to market and customer characteristics [11]. Requirements are therefore not only placed on the performance but also safety of the implemented vehicles which can be divided into mandatory laws and optional requirements based on the respective application and process.

Figure 1 outlines legal framework conditions with relevance for the development and implementation of mobile robotics in Austria [12]. National legislation defines the basic regulations for this purpose. In Austria, these are the *Maschinen-Sicherheitsverordnung 2010* (BGBl. II Nr. 282/2008)[4], *ArbeitnehmerInnenschutzgesetz 1994 (BGBl. Nr. 450/1994)*[5] and *Netz- und Informationssystemsicherheitsgesetz* 2018 (BGBl. I Nr. 111/2018)[6].

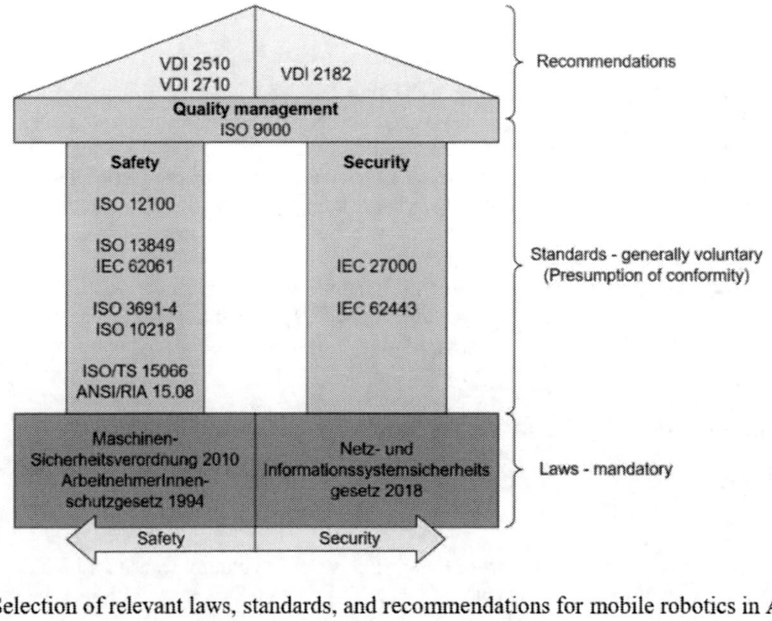

Fig. 1. Selection of relevant laws, standards, and recommendations for mobile robotics in Austria

[4] Equivalent EU regulation: Directive 2006/42/EC

[5] Worker Protection Act 1994

[6] Equivalent EU regulation: Directive (EU) 2016/1148

Apart from national laws, international and national standards concerning the current state-of-the-art in safety and security are providing a framework for the operation and marketing of mobile robots. Despite the voluntary nature of standards, compliance may be mandatory by law (e.g., for the presumption of conformity) [13]. The implementation of a quality management mechanism is a recommended way to combine safety and security as the essential aspects for fully automated vehicles are: Quality, safety and security [14]. Furthermore, Figure 1 shows the guidelines of the Association of German Engineers (Verein Deutscher Ingenieure - VDI). These guidelines similarly provide state-of-the-art information elaborated by experts [15].

In addition to legal requirements, key requirements for a mobile robot, identified by a literature study, are *Hardware*, *Navigation*, and the *Control System*. The requirements for the type of the mobile robot [16][17], its geometric dimensions [18], load size [17] as well as the sensor system [19][20] and the power supply [19][18] are grouped in the category *Hardware*. *Navigation* includes the use of a guide path [17] and the integrated path planning functionalities [21]. The *Control System* is closely related to this category. Here, a distinction can be made between a central high-level control system for integrating several mobile robots into a production plant and a low-level control system directly on the vehicle [21]. Within the production process, indicators like idle time [17], average operation time [21], delivery reliability [17] and resource utilization [17][21] are identified as key requirements.

Besides safety, which is covered by laws and standards, security must also be taken into account. Unprotected, highly interconnected CPPS, as mobile robots are, provide increased attack surfaces for cyber-attacks. In addition, a large number of different security threats can amplify vulnerabilities of insufficiently protected CPPS during an attack. Typical security breaches lead to the loss of one of the three security objectives: *Confidentiality*, *Integrity* and *Availability*. But for mobile robotics, there is an additional component: Physical interventions in the environment may result in safety hazards.

The Robot Operating System (ROS) [3], a framework for mobile robots, has not only become established in the research sector, but is starting to develop a stronger position in the industrial sector as well [22]. The advantage of using ROS is the possibility of controlling mobile platforms combined with manipulators (mobile manipulators) in a single software framework [12]. However, there are several vulnerabilities in the implementation ROS such as missing authentication, authorization and encryption procedures [3], making it possible to bypass or deactivate the safety system in the event of an attack [23].

In order to counteract these challenges, a holistic security approach is necessary when designing a system from the ground up (*Security by Design*). The IEC 62443 series provides seven basic requirements in addition to the security principle *Defense in Depth* (also called the *Zones and Conduits* model) which are intended to extend the CIA triad [5]:

1. *Access Control*: Passwords and user authentication
2. *Use Control*: User roles and authorization enforcement
3. *Data Integrity*: Session handling, cryptography
4. *Data Confidentiality*: Encryption
5. *Restrict Data Flow*: Network segmentation
6. *Timely Response to Event*: Logs
7. *Resource Availability*: System backup and recovery

3. Materials and Methods: Analysis of Safety and Security for Mobile Robotics

In order to practically analyse the cyber-security risks of mobile robotics in terms of safety, a demonstrator is developed. It is a test environment for a production network with an OPC UA server[7] (*Process Control Server*), a Human Machine Interface (HMI) and a MiR100[8] mobile robot. The demonstrator is an enhancement of the system presented in Reithner et al. [24] and the network topology used for the analysis performed is shown in Figure 2.

Compared to the previous version of Reithner et al. [24], the desktop PCs have been replaced by industry-compatible network devices from Siemens AG. These include a Siemens Scalance S615 router and a Siemens Scalance W774-1 access point. The Process Control Server has been implemented with a Revolution Pi Core 3+ 32GB by Kunbus GmbH. Additionally, the OPC UA communication with the robot has been revised. In order to trigger the individual missions already predefined on the MiR100, it is necessary to send a POST request to the REST interface of the MiR100. Since this is not possible directly from the OPC UA server, a Raspberry Pi (*MiRWrapper*) is used as a gateway.

The demonstrator as described above provides the basis for the following practical analysis of the interaction between safety and security. A risk assessment is performed first, followed by a penetration test and vulnerability analysis.

[7] https://github.com/open62541/open62541
[8] https://www.mobile-industrial-robots.com/de/solutions/robots/mir100/

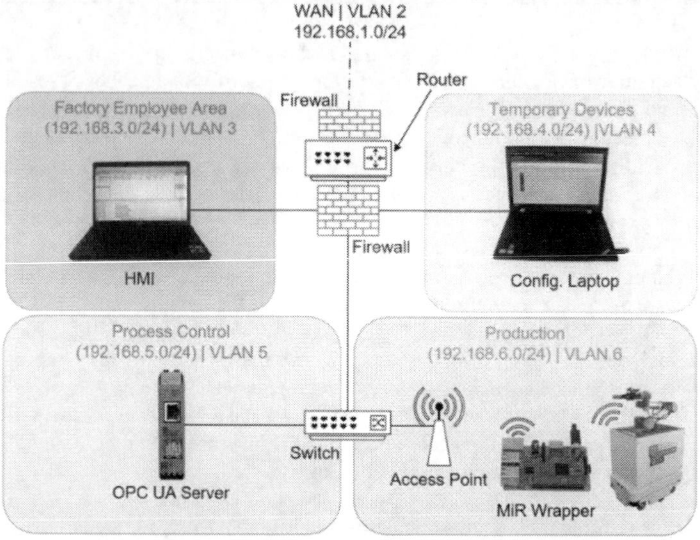

Fig. 2. Zones configured as VLANs at the demonstrator

3.1. Risk Assessment of the Demonstrator

The risk assessment follows the process provided by the OVE EN 62443-3-2:2018 [25]. The first step is to identify the assets of the demonstrator, which are listed Table 1. Based on the identified assets and access points, the risks are identified and assessed within a multistage process. The most significant threats to the demonstrator, which may lead to limitations or failures, are identified during a *High-Level Risk Assessment*. After revising the *Zones and Conduits* model used for the demonstrator, all threats previously classified as *High* or *Extreme* are re-examined in the *Detailed Cyber-Security Risk Assessment*. In order to assess the security risk of the respective vulnerability, the risk classes identified during the risk assessment are used for the concrete evidence at the demonstrator.

1. *Industrial Automation and Control System (IACS) components:*

- Router [Siemens Scalance S615]
- Switch [CISCO Catalyst 2950]
- Access point [Siemens Scalance W774-1]
- Revolution Pi "Process Control Server" [Revolution Pi Core 3+ 32GB]
- Mobile robot [MiR100]
- Raspberry Pi "MiRWrapper" [Raspberry Pi 3B V1.2]
- Laptop Win10 "HMI" [Sony VAIO PCG-71211M]
- Laptop Win10 "Configuration Laptop" [Lenovo ThinkPad T460p]

2. *Communication infrastructure:*

2.1. Internal dataflow - Protocol:
- OPC UA (between the Process Control Server, HMI and *MiRWrapper*)
- Hypertext Transfer Protocol (HTTP) (between the *MiRWrapper* and mobile robot)

2.2. Internal dataflow - Medium:
- LAN (connection of the server, switch, access point, HMI and configuration laptop)
- WLAN (access point, *MiRWrapper* and MiR100)

2.3. External data flow:
- Access to the internet (only during configuration)

3. *Legal position:*

- Assurance of functional safety

Table 1. Identified assets concerning the demonstrator

3.2. Penetration Test and Vulnerability Analysis executed on the Demonstrator

The penetration test is executed according to the methodology presented in Shebli & Beheshti [26] combined with the Open-Source Security Testing Methodology Manual (OSSTMM) [27]. The first step, the preparation, includes the previously mentioned risk assessment, the choice of tools involved and the calculation of the robot's speed. Afterwards, information about the system is gathered and an automatic vulnerability analysis is carried out by OpenVAS[9]. The vulnerabilities found are then exploited to disrupt the operation of the demonstrator or gain access to sensitive information. Tools deployed for this purpose are: hping(3), hydra, kisment(1), netcat(1), nmap(1), tcpdump(8) and Wireshark.

The following limitations became apparent when carrying out the penetration test: The MiR100's safety system was examined during the risk assessment but not considered as a target of the penetration test. The robot poses a potential danger to humans in the case of a successful attack or unsuccessful reversal of the test. Likewise, the MiR100's internal router and computer were not considered a target.

4. Discussion of the Results

The results of the risk assessment based on the OVE EN 62443-3-2:2018 [25] indicate a high to extreme cyber-security risk. The classification of 63% of the risks identified in the *Detailed Risk Assessment* is *High*. Another 20% are considered extreme, as human injury may result. At the same time, none of the risks are classified as *Low*, as they represent a more detailed analysis of the High and Extreme risks of the *High-Level Risk Assessment*.

Additionally, the penetration test revealed that the demonstrator discloses a lot of information (see Table 1). The services of the individual ports reveal which (insecure) protocols are used. Not only that the MiR100 is controlled via a HTTP connection, but also that expired certificates and outdated software or protocol standards are used. The Telnet port is not blocked at the access point making a cleartext login possible. Furthermore, the *Process Control Server*, router and access point are using outdated TLS versions (TLSv1.0 and TLS1.1). Another vulnerability affecting SSL/TLS is the expired certificate on the *Process Control Server* and the used signature algorithm of the certificate. SHA1 with RSA encryption is now considered weak and should therefore no longer be used. However, these are security breaches in the area of *Confidentiality*. Using the information gathered, further attacks could be initiated. These resulted in a loss of *Integrity*, e.g., by sending previously sniffed REST requests to the MiR100. A loss of *Availability* was triggered either by a Denial of Service (DoS) attack on web interfaces or directly on the *Process Control Server*. Additionally, a loss of *Availability* was caused by the execution of different scans (e.g., OpenVAS, nmap (1)) or the manipulation of the MiR100 using the ROSPenTo, a penetration testing tool for ROS [28].

Vulnerability / Threat	Affected Device	Risk Classification
Encryption of the OPC UA communication	*MiRWrapper*, *Process Control Server* and HMI	Low
Expired SSL/TLS certificate	*Process Control Server*	Medium
Disclosure of the use of insecure services	*Process Control Server*, access point and MiR100	Medium
Telnet Unencrypted Login	Access point	Medium
Service failure triggered by an intrusion	*MiRWrapper* and *Process Control Server*	High
ROSPentTo Stealth Publisher Attack [28]	*MiRWrapper* and MiR100	High
Choice of insecure passwords for the devices	*MiRWrapper* and MiR100	Extreme

Table 2. Identified and most relevant vulnerabilities of the demonstrator, affected devices and respective risk classification

None of the actively conducted attacks resulted in a movement of the MiR100 or a failure of the safety system, primarily because it was not considered to be a potential target. In conclusion, the MiR100 within the demonstrator does not pose a danger related to machine safety per se. Nevertheless, this combination poses a high risk for cyber-attacks.

5. Conclusion

By integrating CPS, such as mobile robots, as part of modern production systems, new opportunities for attacks and potential hazards arise. A high level of expertise is required in order to identify and mitigate these with suitable measures. Laws, standards and guidelines, especially in the area of safety, already form a well-developed basis. However, there is also an interaction with IT security. The lack of awareness concerning security in production facilities is still a challenge. Therefore, this study practically analyses the cyber-security threats associated with mobile robotics and the impact of selected attacks on safety. Based on a risk assessment, a penetration test was performed on the demonstrator and the MiR100. It revealed vulnerabilities already identified in the previous step by practical application.

[9] https://www.openvas.org/

In particular, the security aspect is endangered by the high vulnerability of *Confidentiality*. Breaches in the security goals of *Integrity* and *Availability* were also revealed which had no impact directly on the safety of the mobile robot, as the critical parts were excluded from the penetration test.

Concluding, it can be said that the MiR100 within the demonstrator does not pose a danger related to machine safety per se. However, this does not apply to other scenarios where hazards for persons in the area around the robot triggered by security breaches may certainly occur. In order to accomplish an integration of mobile robotics in a production environment, a mitigation of the identified risks and a subsequent re-evaluation is necessary. This applies in particular to cyber-security risks, as these are not yet covered by legal regulations. Therefore, it is recommended to apply the IEC 62443 series as a basis for securing industrial communication networks.

Further research can be conducted by the extension with a real production process, consisting of a machine and transported goods. Other and more extensive scenarios can be realised, which may also pose a greater threat to safety. In addition, removing the restrictions for the penetration test (safety PLC, MiR's computer and MiR's internal router) offers three entirely new attack targets, providing a definite greater impact on safety. In addition, the attack surface of Industry 4.0 plants is increasing due to the implementation of additional Cyber-Physical Systems. Therefore, the connection between safety and security cannot be neglected in industry. A general awareness of the connection between safety and security has to be created.

6. Acknowledgments

This work was partly supported by the MA23 of the city of Vienna as a part of the project 19-05 "Sicherheit in intelligenten Produktionsumgebungen (SIP4.0)". We would like to express our special thanks to our sponsor MA23 Vienna, Austria.

7. References

[1] Martins, G.; Bhatia, S.; Koutsoukos, X.; Stouffer, K.; Tang, C. & Candell, R. (2015). Towards a systematic threat modeling approach for cyber-physical systems, Proceedings of Resilience Week (RWS), 18-20 Aug. 2015, Philadelphia, PA, USA, ISBN 978-1-4799-8594-4, pp. 1-6, IEEE, DOI: 10.1109/RWEEK.2015.7287428

[2] Papa, M; Kaselautzke, D; Stuja, K & Wolfel, W (2018). Different Safety Certifiable Concepts for Mobile Robots in Industrial Environments, Proceedings of the 29th DAAAM International Symposium, pp.0791-0800, B. Katalinic (Ed.), Published by DAAAM International, ISBN 978-3-902734-20-4, ISSN 1726-9679, Vienna, Austria DOI: 10.2507/29th.daaam.proceedings.115

[3] Dieber, B. & Breitling, B. (2019). Security considerations in modular mobile manipulation, Proceedings of the Third IEEE International Conference on Robotic Computing (IRC), 25-27 Feb. 2019, Naples, Italy, ISBN 978-1-5386-9245-5, pp. 70-77, IEEE, DOI: 10.1109/IRC.2019.00019

[4] Zimmermann, S.; Reimann, G.; Bock, H.-P.; Ditting, S.; Gernhard, H.-U.; Mehrfeld, J.; Stadler, W.; Straßer, F.; Wittwer, B.-U. & Wirth, H. (2019). Industrial security in machinery and plant engineering - results of the VDMA study and recommendations for action (Industrial Security im Maschinen- und Anlagenbau Ergebnisse der VDMA-Studie und Handlungsempfehlungen), VDMA Competence Center Industrial Security, Frankfurt

[5] International Electrotechnical Commission (2009). IEC/TS 62443-1-1 Ed. 1.0:2009 Industrial communication networks - Network and system security - Part 1-1: Terminology, concepts and models

[6] https://www.tuv.at/next-generation/white-paper/, (2018). White Paper III - safety and security in human-robot collaboration - Influence of IT-security, TÜV AUSTRIA HOLDING AG & FRAUNHOFER AUSTRIA RESEARCH GMBH, Accessed on: 2021-09-20

[7] https://www.trendmicro.com/vinfo/us/security/news/internet-of-things/rogue-robots-testing-industrial-robot-security, (2017). Rogue Robots: Testing the limits of an industrial robot's security, Trend Micro Incorporated., Accessed on: 2021-09-20

[8] Lottenmoser, A.; Berger, C.; Braunreuther, S. & Reinhart, G. (2017). Method of usability for mobile robotics in a manufacturing environment. Procedia CIRP, Vol. 62 (July 2017) pp. 594-599, ISSN 2212-8271

[9] Siegwart, R. & Nourbakhsh, N. (2004). Introduction to autonomous mobile robots, MIT Press, ISBN 0-262-19502-X, Cambridge, USA

[10] https://www.mobile-industrial-robots.com/en/insights/amr-safety/your-guide-to-a-safe-mobile-robot-installation/, (no date) Your guide to a safe mobile robot installation, Mobile Industrial Robots A/S, Accessed on: 2021-09-20

[11] Clark, G. W.; Doran, M.V. & Andel, T. R. (2017). Cybersecurity issues in robotics, Proceedings of the IEEE Conference on Cognitive and Computational Aspects of Situation Management (CogSIMA), 27-31 March 2017, Savannah, GA, USA, ISSN 2379-1675, ISBN 978-1-5090-6380-2, pp. 1-5, IEEE, DOI: 10.1109/COGSIMA.2017.7929597

[12] Markis, A.; Papa, M.; Kaselautzke, D.; Rathmair, M.; Sattinger, V. & Brandstötter, M. (2019). Safety of Mobile Robot Systems in Industrial Applications, Proceedings of the Joint ARW & OAGM Workshop 2019, 9-10 May 2019, Steyr, Austria, ISBN 978-3-85125-663-5, pp. 26-31, Verlag der Technischen Universität Graz, DOI: 10.3217/978-3-85125-663-5-04

[13] https://europa.eu/youreurope/business/product-requirements/standards/standards-in-europe/index_en.htm, (2021). Standards in Europe, European Union, Accessed on: 2021-09-20
[14] Gifei, S. & Salceanu, A. (2017). Integrated management system for quality, safety and security in developing autonomous vehicles, Proceedings of the 10th International Symposium on Advanced Topics in Electrical Engineering (ATEE), 23-25 March 2017, Bucharest, Romania, ISBN 978-1-5090-5160-1, pp. 673-676, IEEE, DOI: 10.1109/ATEE.2017.7905041
[15] https://www.vdi.de/richtlinien, (2019). VDI guidelines: Setting standards - at the current state of the art (VDI-Richtlinien: Standards setzen – auf dem aktuellen Stand der Technik), VDI Verein Deutscher Ingenieure e.V., Accessed on: 2021-09-20
[16] Bostelman, R. V. & Messina, E. R. (2016). Autonomous industrial vehicles: From the laboratory to the factory floor, National Institute of Standards and Technology (NIST), West Conshohocken, Pennsylvania, USA
[17] Reith, K.; Boden, P.; Däumler, M.; Rank, S.; Schmidt, T. & Hupfer, R. (2019). Evaluating automated guided vehicle system characteristics in semiconductor fab automated material handling systems, Proceeding of the 30th Annual SEMI Advanced Semiconductor Manufacturing Conference (ASMC), 6-9 May 2019, Saratoga Springs, NY, USA, ISSN 2376-6697, ISBN 978-1-5386-7601-1, pp. 1-6, IEEE, DOI: 10.1109/ASMC.2019.8791758
[18] Capak, T.; Kohar, R.; Škrabala, J. & Galík. J. (2019). Optimization of traction unit for lowcost automated guided vehicle. IOP Conference Series: Materials Science and Engineering, Vol. 659, No. 1, pp. 012064, ISSN 17578981
[19] Lee, C. K. M.; Keung, K. L.; Ng, K. K. H. & Lai, D. C. P. (2018). Simulation-based multiple automated guided vehicles considering charging and collision-free requirements in automatic warehouse, Proceedings of the IEEE International Conference on Industrial Engineering and Engineering Management (IEEM), 16-19 Dec. 2018, Bangkok, Thailand, ISSN 2157-362X, ISBN 978-1-5386-6786-6, pp. 1367-1380, IEEE, DOI: 10.1109/IEEM.2018.8607396
[20] Fedorko, G.; Honus, S. & Salai. R. (2017). Comparison of the traditional and autonomous AGV systems. MATEC Web of Conferences, Vol. 134, pp. 00013, ISSN 2261-236X
[21] Watfa, M. K. & Karmadi. K. A. (2019). Connected automated guided vehicles in a smart container terminal. OP Conference Series: Materials Science and Engineering, Vol. 659, pp. 012064, ISSN 1757-899X
[22] Breiling, B.; Dieber, B. & Schartner, P. (2017). Secure communication for the robot operating system, Proceedings of the Annual IEEE International Systems Conference (SysCon), 24-27 April 2017, Montreal, QC, Canada, ISSN 2472-9647, ISBN 978-1-5090-4623-2, pp. 1-6, IEEE, DOI 10.1109/SYSCON.2017.7934755
[23] Taurer, S.; Breiling, B.; Svrta, S. & Dieber, B. (2019). Case study: Remote attack to disable MiR100 safety, Proceedings of the first Cybersecurity for Robotics 2019 Conference (CSfR2019), Available from: https://www.joanneum.at/robotics/publikationen/detail/case-study-remote-attack-to-disable-mir100-safety Accessed on: 2021-09-20
[24] Reithner, I.; Papa, M.; Lueger, B.; Ćato, M.; Hollerer, S. & Seemann, R. (2020). Development and implementation of a secure production network, Proceedings of the 31st DAAAM International Symposium, pp. 0736-0745, B. Katalinic (Ed.), Published by DAAAM International, ISBN 978-3-902734-29-7, ISSN 1726-9679, Vienna, Austria DOI: 10.2507/31st.daaam.proceedings.102
[25] Austrian Association for Electrical Engineering (2018). OVE EN 62443-3-2:2018 Security for industrial automation and control systems – Part 3-2: Security risk assessment and system design
[26] Shebli, H. M. Z. A. & Beheshti, B. D. (2018). A study on penetration testing process and tools, Proceedings of the IEEE Long Island Systems, Applications and Technology Conference (LISAT), 4 May 2018, Farmingdale, NY, USA, ISBN 978-1-5386-5029-5, pp. 1-7, IEEE, DOI: 10.1109/LISAT.2018.8378035
[27] Barceló, M.; Herzog, P.;Chiesa, R.; Ivaldi, M.; Guasconi, F.; Sensibile, F.; Rudolph, H.; Brown, A.; Mitchell, R.; Feist, R.; Mayencourt, N.; Weber, U.; Gschwend, A.; Bader, T.; Lee, R.-E.; Puente, F.; Truett, K.; Abella, J.; Rosielle, C.; Baumgartner, C. & Louis, J.-C. (2010) OSSTMM 3 – The Open Source Security Testing Methodology Manual, Institute for Security and Open Methodologies (ISECOM), Cardedeu, Spain
[28] Dieber, B.; White, R.; Taurer, S.; Breiling, B.; Caiazza, G.; Christensen, H. & Cortesi, A. (2020). Penetration testing ROS, In: Robot Operating System (ROS): The Complete Reference (Volume 4), Springer International Publishing, ISBN 978-3-030-20190-6, Cham, Germany

DOI: 10.2507/32nd.daaam.proceedings.052

PREDICTIVE MAINTENANCE SUPPORTED BY IIOT TECHNOLOGIES

Zeljko Stojkic, Igor Bosnjak, Luka Saravanja & Eva Culjak

This Publication has to be referred as: Stojkic, Z[eljko]; Bosnjak, I[gor]; Saravanja, L[uka] & Culjak, E[va] (2021). Predictive Maintenance Supported by IIoT Technologies, Proceedings of the 32nd DAAAM International Symposium, pp.0356-0362, B. Katalinic (Ed.), Published by DAAAM International, ISBN 978-3-902734-33-4, ISSN 1726-9679, Vienna, Austria

DOI: 10.2507/32nd.daaam.proceedings.052

Abstract

This paper describes new predictive maintenance trends and techniques. By applying new technologies within Industry 4.0, such as the Internet of Things, i.e., by processing data from many sensors, it is possible to identify potential problems in production facilities much before a failure occurs. Therefore, the paper describes a model of predictive maintenance using Industrial Internet of Things. It shows the architecture of predictive maintenance and its implementation on the 3D printer at the Learning Factory, University of Mostar. In addition to these technologies, for the full functioning of predictive maintenance, tools for processing and analysis of collected data such as machine learning methods are needed. Therefore, the paper presents ready-made machine learning solution. The paper shows that with modern technologies, such as the Internet of Things and Big Data Analysis, it is possible to integrate automated systems with IT systems, to better control and manage them, especially in the field of maintenance. Also, the goal of this paper is to contribute to the development of modern solutions in the field of predictive maintenance, the implementation of which in companies is one of the basic foundations for sustainable progress on the path of Industry 4.0.

Keywords: Predictive maintenance; Internet of Things; Industrial Internet of Things; Machine Learning

1. Introduction

Today's requirements in production, such as shorter production times, smaller series of products and higher quality, while keeping costs as low as possible, represent a great challenge for managing and organizing production. In order to be able to meet these trends, it is necessary to provide reliable production capacities, i.e., production machines and equipment that will be able to meet the planned production at the required time. Therefore, maintenance plays a key role in achieving these goals. However, today maintenance is also undergoing a transformation, previous strategies such as corrective and preventive maintenance do not meet the needs and requirements of production on the way to Industry 4.0. Industry 4.0 proposes the predictive manufacturing in the future industry. The machines are connected as a collaborative community.

Such evolution requires the utilization of advance prediction tools, so that data can be systematically processed into information that can explain the uncertainties and thereby make more "informed" decisions [1]. The application of Industry 4.0 technologies such as the Internet of Things (IoT) and Big Data Analysis and the development of machine learning methods have enabled the development of predictive maintenance. Collecting data from various production equipment using the Industrial Internet of Things (IIoT) platform is a major step in building a predictive maintenance system, simplifying the maintenance process at industrial plants. Predictive maintenance is the direct monitoring of the condition and performance of equipment during normal operation to reduce the likelihood of failures and thus reduce costs by reducing the frequency of maintenance activity, reducing unplanned failures, and eliminating unnecessary preventive maintenance. Specific types of predictive maintenance include:

- Vibration analysis: Vibration sensors can be used to detect deterioration in the performance of equipment such as pumps and motors,
- Infrared analysis: Infrared cameras are often used to detect unusually high temperature conditions,
- Acoustic analysis: Acoustic analysis is performed by sound or ultrasound tests, to find gas or liquid leaks,
- Oil analysis: Oil analysis determines equipment wear by measuring the number and size of particles.

Predictive maintenance is an important enabler of Industry 4.0 concept. It is based on Computerized Maintenance Management System (CMMS) concept that takes advantage of state-of-the-art technological innovations. CMMS coordinates all activities related to the availability, productivity, and maintainability of Cyber-Physical Systems (CPSs). CMMS functioning is a data-driven process based on condition monitoring. High degree of automation is another crucial characteristic of such systems. All steps of a computerized maintenance process happen without or with minimal human involvement. Such approach prevents or minimises effects of human mistakes. To implement it, complex CMMS are required, which work fully autonomously, and are able to collect, store and analyse data continuously [2]. Moreover, when processed and analysed, these data can bring out valuable information and knowledge from manufacturing process and system dynamics. By applying analytic approaches based on data, it is possible to find interpretive results for strategic decision-making, providing advantages such as, maintenance cost reduction, machine fault reduction, repair stop reduction, spare parts inventory reduction, spare part life increasing, increased production, improvement in operator safety, repair verification, overall profit among others [3].

The same data can also be used to make energy consumption more transparent, allowing for further potential cost savings. Together, these forecasting mechanisms provide benefits to industry in the form of increased productivity and efficiency. Data generated from large sensor installations and other IIoT endpoints can be the basis for analytics to help accurately plan maintenance, based on changing equipment conditions and related processes. This approach can be extremely valuable in modernizing maintenance, reducing downtime, and optimizing the use of critical capital equipment. It is also important to emphasize the Prognostic and Health Management concept (PHM) that is closely related to maintenance approaches such as predictive maintenance. PHM is an engineering process where algorithms are used to detect anomalies, diagnose faults, and predict Remaining Useful Lifetime (RUL). Although the main goal of PHM is to provide the health state and estimate the RUL of the components or equipment, also financial benefits such as operational and maintenance cost reductions and extended lifetime are achieved [4].

2. IoT architecture for predictive maintenance

The paper proposes an architecture for predictive maintenance (PdM) based on IIoT, with all included modules that make it functional and as such can be implemented in intelligent and predictive maintenance. The scheme of the developed concept is shown in Figure 1. IoT sensors are used to monitor and collect data in real time. The collected data are then modelled into a PdM model using statistical tools and software. The components of PdM will be briefly described below.

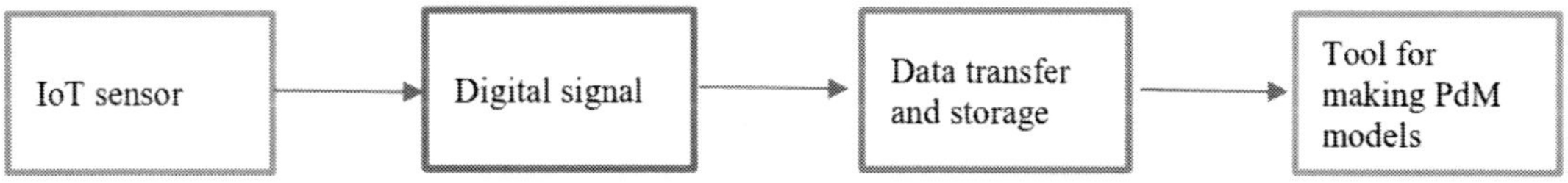

Fig. 1. IoT architecture for predictive maintenance

Sensor devices constitute at the same time the lowest level and the most important part of the system. In fact, they directly interact with the environment by measuring variables of interest and by processing acquired data to perform on board analysis tasks [5]. Few Popular Industrial IoT enabled sensors measure temperature, humidity, pressure, current, vibration, air quality, footfall, gas, and weight. Another commonly used sensor is the photoionization sensor, which is used to measure levels of volatile inorganic compounds and gases. The ultrasonic sound sensor is used to translate high frequency sounds due to leaks [6]. Data from sensors collected in the form of analogue signals are converted into a digital signal using an analogue digital converter. The digital signal is machine-readable and is used for further analysis.

Digital data stream generated is stored on organizations' local intranet or a secure centralized or remote cloud location in the form of a database file. IoT sensors which are installed in remote locations without internet connectivity have inbuilt Random-Access Memory (RAM) and Storage space where data is stored. Data from IoT sensors is transmitted to a centralized cloud server or decentralized fog nodes via Wi-Fi, or in remote locations through Cellular networks or Bluetooth Low Energy (BLE) technology [6].

Machine Learning (ML), within artificial intelligence, has emerged as a powerful tool for developing intelligent predictive algorithms in many applications. ML approaches have the ability to handle high dimensional and multivariate data, and to extract hidden relationships within data in complex and dynamic environments (such as industrial environments) [3]. Based on the data collected and stored, it is possible to develop models and algorithms using machine learning that can help us predict future events based on past data. Machine learning means acquiring knowledge based on experience. It is not about remembering, but about finding rules in learning data. ML algorithms are categorized into three different types; supervised, unsupervised, and reinforcement learning (RL). The aim is to show how complex the structure can be and the commonly used available learning techniques. Moreover, different algorithms can be combined in order to maximize the classification power. To add on, some among the ML algorithms are both applicable to unsupervised and supervised learning [7].

2.1. Selection of equipment to be applied for the case study at the Learning Factory

For overview and presentation, the predictive maintenance architecture will be implemented on a 3D printer, inside the Learning Factory, from which data will be collected, i.e., temperature and vibrations. Based on the collected data, predictive analytics will be performed. By analysing the literature, possible solutions for the equipment needed for the realization of the definition of architecture were collected and systematized.

<table>
<tr><th>Architecture level</th><th colspan="7">Possible examples of a hardware solution</th></tr>
<tr><td>Integration into other software</td><td>Grafana</td><td>Tableau</td><td rowspan="4">Siemens Mindsphere</td><td>Integration in MES</td><td rowspan="3">Splunk</td><td>Production applications</td><td>PTC ThingWorx (Production applications)</td></tr>
<tr><td>Data analysis</td><td rowspan="2">Rapidminer</td><td rowspan="2">Python, R</td><td>Matlab Statistics</td><td rowspan="3">Microsoft Azure</td><td rowspan="2">PTC ThingWorx (IIoT platform)</td></tr>
<tr><td>Data aggregation</td><td>NODE-RED</td></tr>
<tr><td>Data storing</td><td>Relational database</td><td>Hadoop Cluster</td><td>Bosch IoT Cloud</td><td>Edge server</td><td>PTC ThingWorx (KEPServerEX)</td></tr>
<tr><td>Protocol (data)</td><td>MQTT</td><td>AMQP</td><td>OPC-UA</td><td>TCP/IP</td><td>LoRaWAN</td><td>HTTP</td><td>Modbus</td></tr>
<tr><td>Network (Data)</td><td>LAN</td><td>LAN</td><td>BLE (Bluetooth)</td><td>5G</td><td>NFC</td><td>6LOWPAN</td><td>ZigBEE</td></tr>
<tr><td>Preprocessor software</td><td>Node-RED</td><td>RapidMiner</td><td>R i Python</td><td>Siemens Simatic</td><td rowspan="4">Bosch XDK, Bosch Connected Industrial Sensor (CISS)</td><td>Microsoft Azura IoT-Edge</td><td>PTC ThingWorx</td></tr>
<tr><td>Preprocessor hardware</td><td>S7 1200</td><td>Harting Mica</td><td>Balluf IO Link</td><td>MindCOnnect 2040</td><td rowspan="2">PLC</td><td>...</td></tr>
<tr><td>Interface for embedded systems</td><td>S7 1200</td><td>PC</td><td>CAN-Bus</td><td>SPI-Bus</td><td>RS-232</td></tr>
<tr><td>Sensors</td><td>Siemens sensors</td><td>External sensors</td><td>Internal sensors</td><td>Bosch Phantom</td><td>...</td><td>...</td></tr>
</table>

Table 1. Possible examples of hardware solution by levels of architecture [4], [5], [6], [7]

Control unit

- Power supply for PLC, display and sensor, PM 190W
- Siemens S7-1200 PLC

Sensors

- Temperature transmitter AQZ-025, input Pt100, output 4-20mA, 24VDC
- Probe type Pt100, Ø3 x 50 mm, I = 1 m, 0-200°C
- Humidity sensor with measuring transducer B + B Thermo-Technik
- Vibration monitoring module
- Vibration measuring sensor

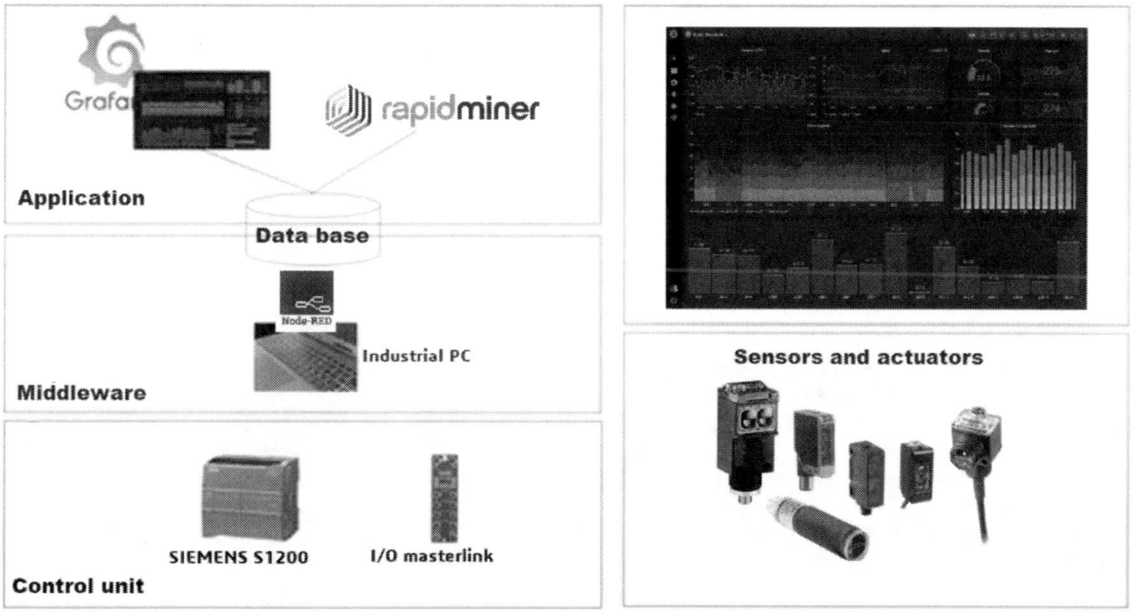

Fig. 2. Selected equipment architecture

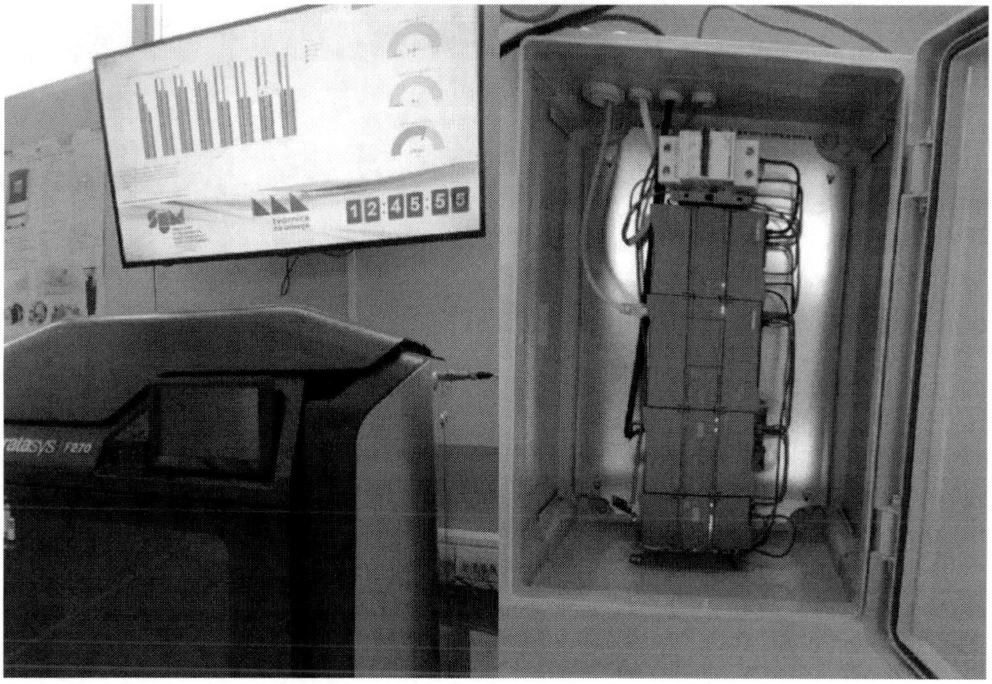

Fig. 3. a) The screen presents the data collected from the 3D printer, b) a cabinet with the entire control unit

2.2. Middleware - connection between Siemens PLC and database

An industrial computer is a standard computer that is used to install the necessary software for Node-RED and connect data from sensors via Siemens PLC via Node-RED and MQTT protocol to the database.

Node-RED is a flow-based development tool for visual programming developed originally by IBM for wiring together hardware devices, APIs, and online services as part of the Internet of Things [8]. Node-RED offers a web-based flow editor that can be used to create JavaScript functions. Application elements can be saved or shared for reuse. Runtime is built on Node.js. Streams created in Node-RED are stored using JSON. Node-Red in our architecture is used to connect between control hardware and the database.

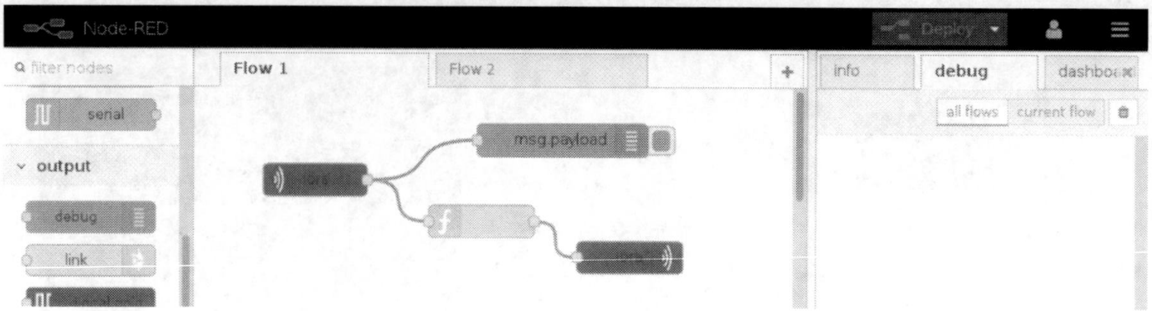

Fig. 4. Display the Node-RED tool interface

The MQTT protocol is an open, simple and lightweight protocol that allows distributed messaging between end nodes on a publish / subscribe basis. These features make it ideal for use in limited environments and low-bandwidth networks with limited processing capabilities, low memory capacity, and high latency. Messages are published on a specific channel through which the message is forwarded to all subscribers of that channel. The MQTT broker (MQTT broker) is in charge of comparing posts and subscriptions, which forwards published messages and remembers client subscriptions.

A node for Siemens PLC is installed within the Node-RED interface to allow access to data from the PLC. Each connection to the PLC is represented by the endpoint configuration node S7. We can configure the PLC address, the available variables and their addresses, and the variable read cycle time. Node-RED allows us to save data from nodes to the desired database, thus ensuring the periodic storage of data read from the sensor to the database.

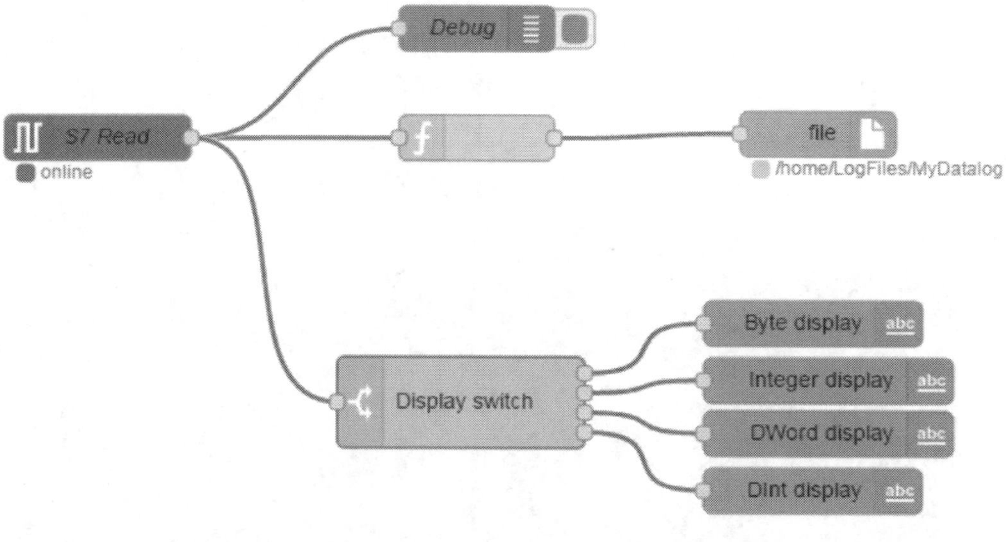

Fig. 5. Integration of Siemens PLC into Node-RED

2.3. *Tools for processing the collected data and presentation of predictive maintenance models*

It is planned to use the RapidMiner software tool, which integrates different machine learning methods, to create a predictive maintenance model. In the time of writing, we still did not develop a predictive maintenance model, which should process and analyse the data collected in the RapidMiner. But in the coming period, we plan to complete the model and describe it in future papers. Also, it is planned to use the Grafana software tool, to graphically present the obtained data and results. Below are briefly described software packages.

2.3.1. *RapidMiner - a software package for machine learning*

RapidMiner (RM) is a software platform for scientific data processing developed by the company of the same name. Provides an integrated environment for data preparation, machine learning, deep learning, text mining and predictive analytics.

It is used for business and commercial applications, as well as for research, education, training, rapid prototyping and application development, and supports all steps of the machine learning process, including data preparation, visualization of results, validation and model optimization [9]. RapidMiner is an environment for machine learning and data mining processes. It is open-source, free project implemented in Java. It represents a new approach to design even very complicated problems - a modular operator concept which allows the design of complex nested operator chains for a huge number of learning problems. RM uses XML to describe the operator trees modelling knowledge discovery (KD) processes. RM has flexible operators for data input and output in different file formats. It contains more than 100 learning schemes for regression, classification, and clustering tasks [10].

The work in the software itself requires experience and prior preparation of the collected data, which must be adjusted before the analysis, i.e., perform the necessary data operations, which include data cleaning, data collection, data sorting and data formatting. After preparing the data, machine learning operations can be performed, in order to obtain a predictive model These operations are performed to draw out meaning information from data in order to make better decisions towards business. Analytics can be performed with various machine learning or custom algorithmic concepts like regression, classification, clustering, model-based recommendation etc. These models are further evaluated and improved by various stages of machine learning concepts. Improved or optimized algorithms can provide better results [11].

After creating the model, it needs to be visualized. RM provides a high level of visualization support. It is possible within the software to create detailed results of data analyses. In addition, the visualization of nodes and other information can be highly colourful and appealing to the eye. In this way, RapidMiner can allow users with higher level programming and coding skills to have increased output from their efforts in terms of being able to visualize the data and results. However, without at least some basic programming and coding skills, it seems unlikely that the full performance of the visualization abilities of the software could be achieved. The reason for this is that the visualization support within the software is connected to the functions that are performed [12].

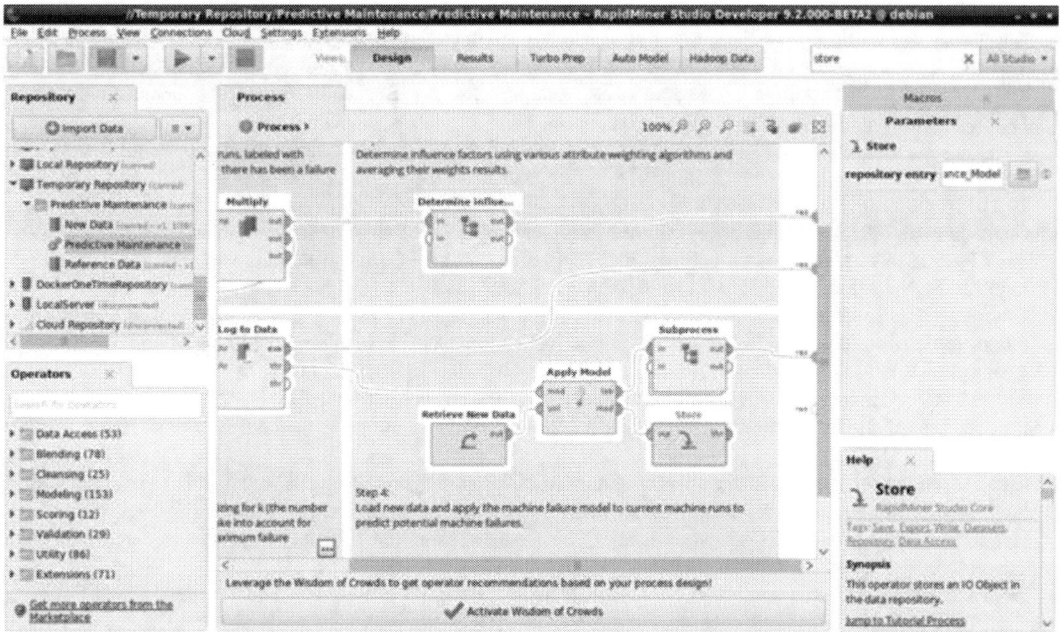

Fig. 6. RapidMiner software tool interface

2.3.2. Grafana - software package for visualization and display of data

Grafana is a multiplatform open-source analytical and interactive visualization web tool. Allows you to create charts, graphs, and alerts for the web when connected to supported data sources. It is extensible via a plugin system. End users can create complex dashboards using interactive query tools.

3. Conclusion

Finally, we can conclude that in today's conditions and requirements in production, predictive maintenance of production plants and machines plays an important role in the sustainability and profitability of the company's business. The development and application of new technologies in industry, i.e., the application of IIoT, has enabled the development and implementation of predictive systems in maintenance.

With such systems, cost savings can be expected, as there should be no sudden breakdowns and unnecessary preventive maintenance activities. One of the problems the companies encounter in migrating to IIoT is the fact that different devices have used different protocols to send and receive data to each other in the past. Although several different communication protocols are currently used, due to their low cost and security in communication, the MQTT protocol is the standard for IIoT. In order to remain competitive, companies are committed to modernizing systems and equipment to meet new regulations and keep pace with growing market demands and stability. Companies that have embraced IIoT have made significant improvements in safety, efficiency, and profitability, and this trend is expected to continue with the spread and adoption of new IIoT technologies. Also, in order to complete the architecture of the predictive model, customized solutions in the field of artificial intelligence such as machine learning methods are needed, which will be able to accept and process the collected data and make decisions in the maintenance and production function. One of the limitations of such a model is the need for machine learning experts, who will be able to develop and optimize prediction models in a ready-made tool such as RapidMiner. To date, a large number of algorithms have been developed in the field of machine learning, but with the constant development of new digital technologies, algorithms often need to be adapted and optimized. Therefore, future work will be based on further development of the architecture of the predictive maintenance model, which will include the use of new solutions in IIoT and the development and optimization of algorithms for predicting and forecasting the condition of individual machines on the basis of which faster and more accurate decisions will be made regarding the reliability and condition of individual plants and machines. Also, future works will monitor the implementation of such systems in production facilities.

4. References

[1] Deac, C. N.; Deac, C. D.; Popa, C. L.; Ghinea, M & Cotet, E. C. (2017). Using augmented reality in smart manufacturing, Proceedings of the 28th DAAAM International Symposiums, pp. 0727-0732, B. Katalinic (Ed.), Published by DAAAM International, ISBN 978-3-902734-11-2, ISSN 1726-9679, Vienna, Austria DOI: 10.2507/28th.daaam.proceedings.102

[2] Katona, A & Panfilov, P (2018). Building predictive maintenance framework for smart environment application systems, Proceedings of the 29th DAAAM International Symposium, pp.0460-0470, B. Katalinic (Ed.), Published by DAAAM International, ISBN 978-3-902734-20-4, ISSN 1726-9679, Vienna, Austria DOI: 10.2507/29th.daaam.proceedings.068

[3] Carvalho, T. P. et al. (2019). A systematic literature review of machine learning methods applied to predictive maintenance. Computers & Industrial Engineering Vol. 137, November 2019, 106024, DOI: 10.1016/j.cie.2019.106024

[4] Cachada et al. (2018). Maintenance 4.0: intelligent and predictive maintenance system architecture, Proceedings of 23rd International Conference on Emerging Technologies and Factory Automation (ETFA), ISBN 978-1-5386-7109-2, ISSN 1946-0740, DOI: 10.1109/ETFA.2018.8502489

[5] Civerchia, F.; Bocchino, S.; Salvadori, C.; Rossi, E.; Maggiani, L. & Petracca, M. (2017). Industrial internet of things monitoring solution for advanced predictive maintenance applications. Journal of Industrial Information Integration Vol. 7, September 2017, Pages 4-12 DOI: 10.1016/j.jii.2017.02.003

[6] Nangia, S.; Makkar, S. & Hassan R. (2020). IoT based predictive maintenance in manufacturing sector, Proceedings of the International Conference on Innovative Computing & Communications (ICICC) 2020, DOI: 10.2139/ssrn.3563559

[7] Cinar, Z. M. et al. (2020). Machine learning in predictive maintenance towards sustainable smart manufacturing in industry 4.0. Sustainability Vol. 12, October 2020, DOI: 10.3390/su12198211

[8] Heath, N. (2014). "How IBM's Node-RED is hacking together the internet of things". techrepublic.com. CBS Interactive. Retrieved January 16, 2017

[9] Hofmann, M. & Klinkenberg, R. (2013). RapidMiner: data mining use cases and business analytics application, CRC Press, ISBN 9781482205497

[10] Graczyk, M.; Lasota, T. & Trawinski, B. (2009) Comparative analysis of premises valuation models using KEEL, RapidMiner and WEKA, Proceedings of the International Conference on Computational Collective Intelligence 2009, Wroclaw Poland, DOI: 10.1007/978-3-642-04441-0_70

[11] Dwivedi, S.; kasliwal, P. & Soni, S. (2016). Comprehensive study of data analytics tool (RapidMIner, Weka, R tool, Knime), Symposium on Colossal Data Analysis and Networking (CDAN) Indore India, ISBN 978-1-5090-0670-0 DOI: 10.1109/CDAN.2016.7570894

[12] Hirudkar, A. M. & Sherekar, S. S. (2013). Comparative analysis of data mining tools and techniques for evaluating of database systems, International Journal Of Computer Science And Applications Vol. 6, No.2, April 2013, ISSN: 0974-1011

DOI: 10.2507/32nd.daaam.proceedings.053

DEVELOPMENT OF LOW COST SMALL 3D SCANNER FOR ROBOTIC APPLICATIONS

Aleksandr Levin, Andrei Vorotnikov & Yuri Poduraev

This Publication has to be referred as: Levin, A[leksandr]; Vorotnikov, A[ndrei] & Poduraev, Y[uri] (2021). Development of Low Cost Small 3D Scanner for Robotic Applications, Proceedings of the 32nd DAAAM International Symposium, pp.0363-0368, B. Katalinic (Ed.), Published by DAAAM International, ISBN 978-3-902734-33-4, ISSN 1726-9679, Vienna, Austria

DOI: 10.2507/32nd.daaam.proceedings.053

Abstract

The integration of digital technologies, such as three-dimensional scanning, in production and everyday life can dramatically improve the quality of life. The low cost and ease of use of a 3D scanner can accelerate the implementation of this technology. In this paper, the existing solutions in the field of inexpensive 3d scanners are considered. The development of a prototype 3D scanner based on the principle of laser triangulation and having 2 cameras is presented. A calibration method is described that allows reducing the number of images required for calibration. The algorithm for detecting the contour of the laser line in the image is shown. The accuracy of the developed 3D scanner is checked and conclusions are drawn about the work done.

Keywords: 3D scanner; stereo vision; 3D measurement; robot application.

1. Introduction

The ability to work with a digital copy of a real object is one of the most important needs of production and automation today. The reconstruction of a real object in the form of a digital copy in modern production is carried out using 3D scanning technologies. Unfortunately, 3D scanners have a high cost ($1000 and above). There is also a segment of scanners that differ in relatively low cost (about $200). Usually, such scanners have a rotating platform that limits the size of the possible surface for scanning. For example, they do not allow to scan a part of a wall, or a person.

For robotic scanning tasks, manual scanners are inconvenient to use, since they take up a large amount of space, and stationary ones, as mentioned above, have limited scanning capabilities in terms of the size of the object being scanned. Robotic scanning is used in bioprinting [1], in maxillofacial surgery [2],[3],[4],[5],[6]. Scanning is also used in a wide variety of areas: wound scans [7], tire scans [8]. Virtual models of 3D scanners are also being created [9]. Studies have been conducted on the quality of such scanners [10]. But these scanners are used either permanently, or a large scanner is used on a robotic platform, which will not allow installing another unit of technological equipment on this robot. Therefore, it was decided to create a new device. To solve this problem, a miniature 3d scanner was developed. The development included two main stages: prototype development and software development. The last stage, in turn, was divided into several more sub-stages: calibration of the camera, determining the position of the laser surface and obtaining the laser contour from the image. Then the developed 3D scanner was tested on trial samples to confirm its operability.

2. Prototyping

There are several methods of 3D scanning. The most widespread are photogrammetry, stereometry, structured light [11] and laser triangulation. In this work, the structured light method was not used, since it requires a projector, which takes up a lot of space, and also has a relatively high cost. And for acceptable accuracy using stereometry and photogrammetry, expensive cameras are required. Therefore, the method of laser triangulation was chosen, but using two cameras, to increase the accuracy of scanning and increase reliability.

In this prototype, the Espada ENDSC2M endoscope was used as cameras (Fig. 1). Its characteristics are shown in Table 1. This camera has a small size and good quality. At the same time, the focal length of this camera does not change, which is necessary for the correct calibration of the internal parameters of the camera, and the laser module with the red laser line. The housing was made of plastic on a 3d printer.

The cameras are installed at a distance B (set within 40 ... 100 mm) and at an angle α relative to each other. The laser module is installed at an angle β between the plane formed by the main axes of the cameras and the main axis of the laser module. The angles α and β are adjusted so that the laser line completely fills the image width of each camera. The price of one camera is \$5, the price of a laser module is \$2, so the price of a scanner is \$15, which is an order of magnitude cheaper than any other similar scanning system. It is assumed that the user has a personal computer.

Name of the parameter	Value
Cable diameter	4 mm
Cable length	1 m
Lens diameter	7.8 mm
Size	8 x 46 mm
Maximum resolution	1280 x 960 pix.
Weight	120 g

Table 1. Camera characteristics

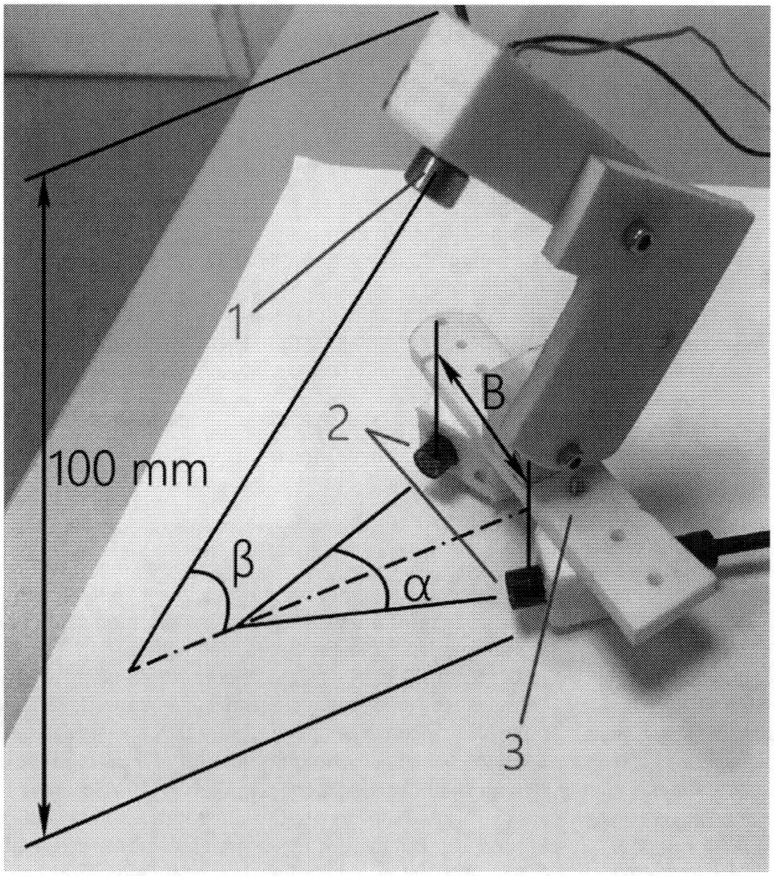

Fig. 1. Photo of the prototype (1 - laser module, 2 - cameras, 3 - housing, α - angle between the cameras, β - angle between the laser scanning plane and the cameras)

3. Camera position

To perform automatic scanning, the developed 3d scanner is installed on a controlled platform (robot, machine, 3d printer). To control such a device, software was developed in the C# language. Also, this program performs all the calculations for the camera, allows to view the scanned model. To work with three-dimensional graphics, the OpenGL framework for C# was used. The interaction of the user and the scanning system is carried out through the graphical interface (Fig. 2).

When scanning, the user sets the coordinates of the start and end points of the scan and the number of shooting points. The higher the number of shooting points, the higher the resolution of the model along the scan line will be. In order to increase the accuracy of scanning, the robot makes a stop before each image, and transmits the values about the coordinates to the computer at the time of shooting. At the same time, before taking a picture, the robot waits for a complete stop and only then sends a signal that it needs to be saved. The robot then proceeds to the next shooting point. Thus, the shooting speed depends on how far away the neighboring scanning points are. For example, if the distance between the points is 500 microns and 80 scanning points, the total scanning time will be about 100 seconds.

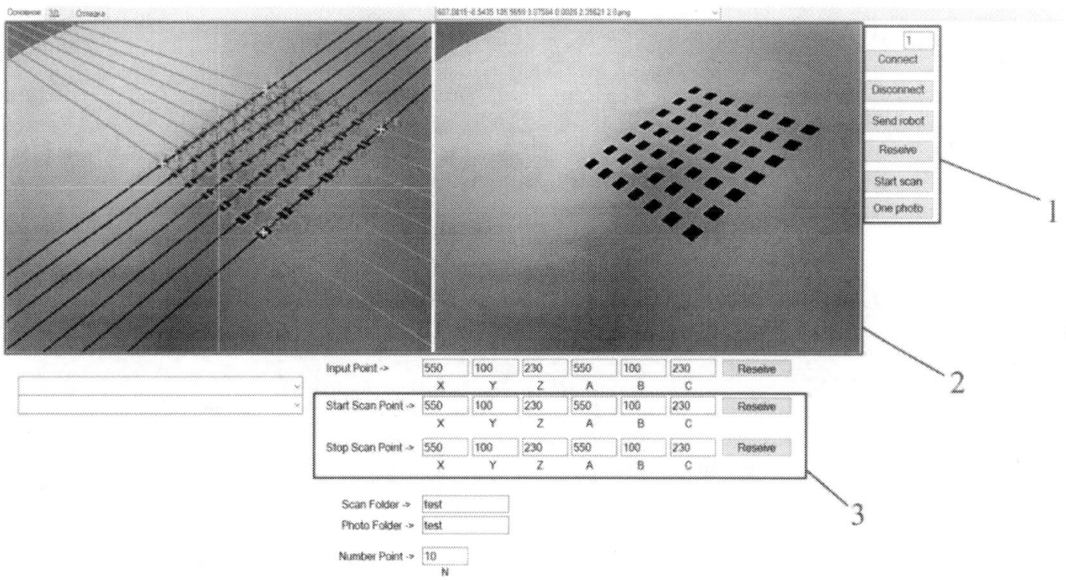

Fig. 2. The image of the user interface of the program (1 – The main control panel of the scanning system, 2-The window displaying the image from the cameras, 3-The panel for setting the coordinates of the start and end points of the scan)

In most cases, the calibration algorithm [12] for the 3D scanner camera of this price segment is used, implemented in the OpenCV library in a variety of languages: C++, Python, C#. Calibration using this algorithm involves the use of about 10-15 images taken manually to determine the internal parameters of the camera. At the same time, there is a possibility that the program does not correctly calibrate the camera. A calibration algorithm has been developed that allows to calculate the camera position from four images of the calibration mark.

This result is achieved due to several simplifications: the distortion of these cameras is very small (they are calibrated in advance at the manufacturer's enterprise) and it was set in advance in which octant the camera is located relative to the coordinate system of the mark. An algorithm was also developed to automatically detect the mark on the image. To test the position detection algorithm, a virtual camera was created using OpenGL tools. A virtual experiment was conducted to determine the position of the calibration mark and compare it with the value of the camera position. The results of the experiment are shown in Table 2.

N	Vx	Cx	Vy	Cy	Vz	Cz	δVC
1	0	-0.4289	0	0.0557	113.475	113.3854	0.4417
2	0	-0.2552	0	0.0335	67.15	66.9896	0.3032
3	0	-0.3287	0	0.0433	87.295	87.1577	0.3588
4	17.4133	17.7244	-74.9847	-74.7448	42.3033	42.5890	0.4857
5	90.3101	90.2515	-60.9709	-61.2016	78.5054	78.5675	0.2459

Table 2. Checking on a virtual model (Vx,Vy,Vz – the value of the camera coordinate, Cx,Cy, Cz-the calculated value of the camera coordinates, δVC - the distance between points)

For the experiment, 5 different positions of the calibration mark were selected. Next, we determined how much the calculated coordinates differ from those set virtually:

$$\delta VC = \sqrt{(V_x - C_x)^2 + (V_y - C_y)^2 + (V_z - C_z)^2} \qquad (1)$$

Then it was determined in the average distance between these points:

$$\overline{\delta VC} = \sum_{i=1}^{5} \delta VC_i = 0.36706 \qquad (2)$$

This error is caused by the discreteness of the pixel image.

4. Detection laser line

The laser line is represented in the image as a light strip. We are interested in the contour of the object, that is, a certain clear line. The lower border of the light strip was chosen as the contour in the image. The image of the line, depending on the object, may have breaks and noises. To build a 3D model that is closest to reality, it is necessary to eliminate these problems. First of all, the image was binarized according to a certain threshold value set by the user. It depends on the ambient lighting, camera settings, scanner configuration (distance B, angles α, β).

To obtain the lower contour of the light strip in each column of the image, only the lowest pixel was left. Figure 3 shows areas where there is not a single pixel, each such area was filled with a line connecting its extreme points. Then the entire contour is smoothed so that the model surface is smooth. The coordinate of the *i*-th pixel during smoothing is calculated as follows:

$$p_i = \sum_{i=-k/2}^{k/2} p_{i+k} \qquad (3)$$

In this expression, *k* is the smoothing width, the higher the value of this parameter, the smoother the model will turn out. But at the same time, sharp corners will also be rounded off. Figure 3 shows the step-by-step operation of the algorithm for determining the contour from the image.

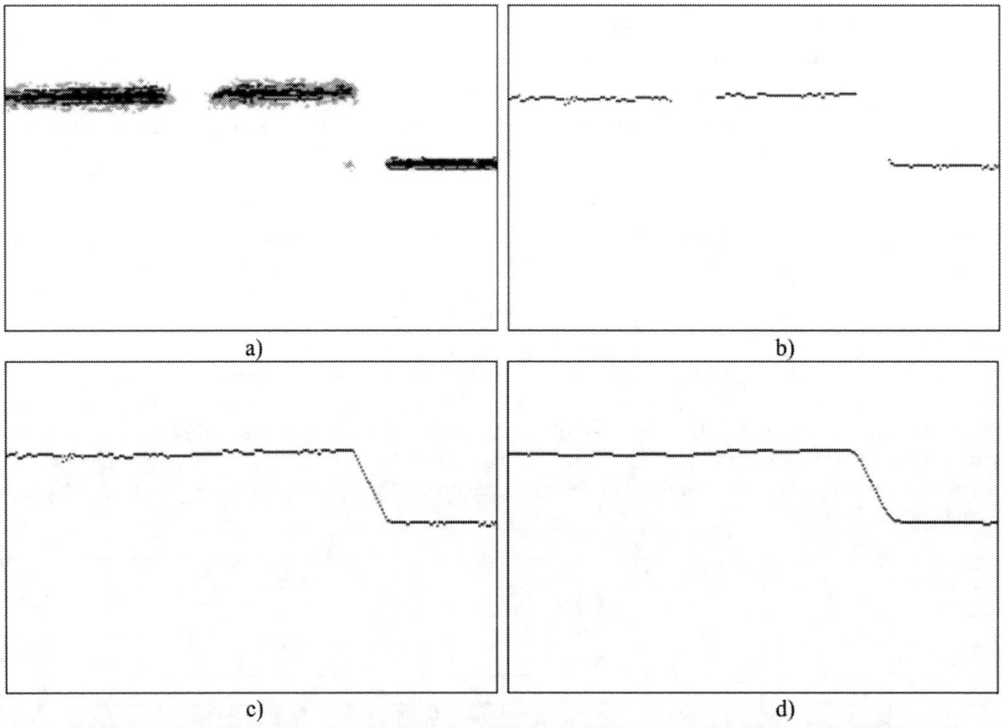

Fig. 3. a) The image of the laser line after binarization, b) The definition of only one pixel in each column, c) The contour after filling in the missing sections, d) the contour after applying smoothing

5. Measuring accuracy

The developed scanner was tested by scanning two test samples. For testing, 2 objects were selected as an example (Fig.4) made of different materials – metal and plastic. The measurement results are shown in Table 3. An example of the scanned parts is shown in Figure 4. The measurement was performed for 3 different distances between the cameras.

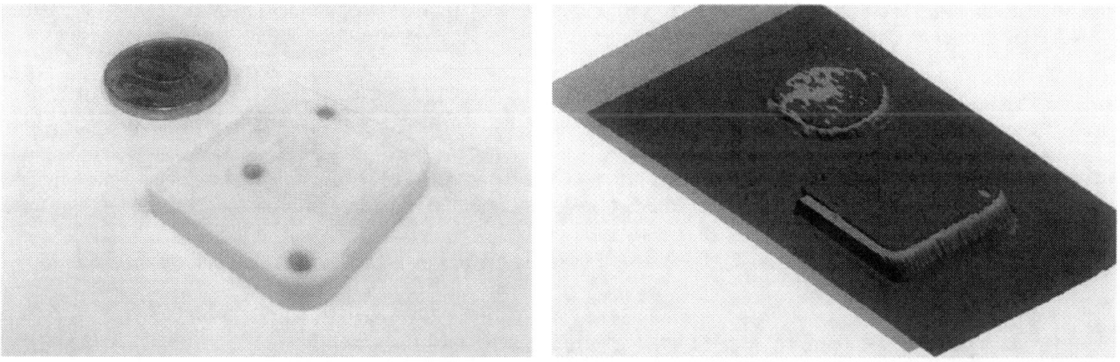

a) b)

Fig. 4. a) the scanned object b) the recreated 3d model (green is the first camera, blue is the second camera)

	Measured values (Caliper, 0.1 mm error)	3D model values (B = 42 mm)	3D model values (B = 68 mm)	3D model values (B = 102 mm)
Length, mm	31.9…32.0	31.5…32.5	31.05…32.8	30.4…33.4
Width, mm	27.9…28.0	27.5…28.5	27.1…28.8	26.6…29.3
Height, mm	5.4…5.5	5.2…5.8	5.1…5.8	4.9…6.1

Table 3. Results of scanning the test part (B – the distance between the cameras)

It can be seen from the scanning results that the scanner is able to obtain 3D models with an uncertainty [13] of about 1 mm, which is acceptable for a $15 scanner.

6. Conclusion

The task of this work was to develop a 3d scanner with a low cost and acceptable resolution. The existing solutions were reviewed, and it was concluded that it is necessary to develop own more inexpensive solution. We managed to solve this problem by choosing a laser triangulation method for this scanner, as well as two very low-cost cameras. Software has also been developed that allows to control both the movement of the platform and the shooting of the camera. The introduction of such a device will accelerate the development of computer technology. Due to the low cost, the number of consumers of the technology will expand, as well as the number of people who will want to develop it further. It is also possible that this will lead to a slight drop in the prices of professional scanners.

In the future, it is planned to improve one of the aspects of the software, namely the software post-processing of the scanned model. For example, face recognition, increasing the accuracy of the model by improving the algorithms for calculating the contour.

7. Acknowledgments

Scientific Research by Educational Organizations in 2020–2022 Project under Grant NoFSFS-2020-0031.

8. References

[1] Levin, A.; Vorotnikov, A. & Poduraev, Y. (2020). A Robotic System for Hydrogel 3D Printing on Complex Surfaces, pp. 0793–0799. DOI:10.2507/31st.daaam.proceedings.110

[2] Vorotnikov, A. A.; Klimov, D. D. K. & Bazykyan, E. A. (2018). Criteria for Comparison of Robot Movement Trajectories and Manual Movements of a Doctor for Performing Maxillofacial Surgeries, International Journal of Mechanical Engineering and Robotics Research, Vol. 7, No. 4, pp. 361–366. DOI:10.18178/ijmerr.7.4.361-366

[3] Vorotnikov, A. A.; Klimov, D. D. & Chunihin, A. A. (2017). Cutting velocity accuracy as a criterion for comparing robot trajectories and manual movements for medical industry, Mechanics & Industry, Vol. 18, No. 7, pp. 712. DOI:10.1051/meca/2017047

[4] Chunikhin, A. A.; Bazikyan, E. A. & A. A.; Klimov, D. D. (2019). Comparative experimental assessment of the accuracy of nanosecond laser surgery of the oral cavity when the instrument is moved by a robotic complex and a surgeon, Russian Open Medical Journal, Vol. 8, No. 3, pp. e0307. DOI:10.15275/rusomj.2019.0307

[5] Vorotnikov, A. A.; Buinov, M. A. & Chunihin, A. A. (2016). Standard Deviation from the Average Cutting Velocity as a Criterion for Comparing Robot Trajectories and Manual Movements of a Doctor for Performing Surgical Operations in Maxillofacial Surgery, International Journal of Mechanical Engineering and Robotics Research, Vol. 7, No. 2, pp. 319–323. DOI:10.18178/ijmerr.7.3.319-323

[6] Chunikhin, A. A.; Poduraev, Y. V. & Bazikyan, E. A. (2017). Efficiency Assessment of Nanosecond Laser Robotic Maxillofacial Area Surgery in Experiment, Sovremennye Tehnologii v Medicine, Vol. 9, No. 4, pp. 123. DOI:10.17691/stm2017.9.4.15

[7] Zhao, Y. M.; Currie, E. H.; Kavoussi, L.; Rabbany, S. Y. (2021). Laser scanner for 3D reconstruction of a wound's edge and topology, International Journal of Computer Assisted Radiology and Surgery. DOI:10.1007/s11548-021-02459-1

[8] Frosio, I.; Borghese, N. A.; Tirelli, P.; Venturino, G.; Rotondo, G. (2011). Flexible and low cost laser scanner for automatic tyre inspection, 2011 IEEE International Instrumentation and Measurement Technology Conference, IEEE, pp. 1–5. DOI:10.1109/IMTC.2011.5944190

[9] Lima, J.; Gonçalves, J.; Costa, P. J. (2015). Modeling of a Low Cost Laser Scanner Sensor, pp. 697–705. doi:10.1007/978-3-319-10380-8_67

[10] Polo, M.-E.; Felicísimo, Á. M. (2012). Analysis of Uncertainty and Repeatability of a Low-Cost 3D Laser Scanner, Sensors, Vol. 12, No. 7, pp. 9046–9054. DOI:10.3390/s120709046

[11] Straub, J.; Kerlin, S. (2014). Development of a Large, Low-Cost, Instant 3D Scanner, Technologies, Vol. 2, No. 2, pp. 76–95. DOI:10.3390/technologies2020076

[12] Zhang, Z. (2000). A flexible new technique for camera calibration, IEEE Transactions on Pattern Analysis and Machine Intelligence, Vol. 22, No. 11, pp. 1330–1334. DOI:10.1109/34.888718

[13] Vorotnikov, A.; Romash, E. & Poduraev, Y. (2016). Uncertainty Estimation of Axes Direction Determination of Industrial Robot Using an Ellipsoid Concentration Model, pp.0480–0486. DOI:10.2507/27th.daaam.proceedings.072

DOI: 10.2507/32nd.daaam.proceedings.054

TESTING OF A WOODEN HIGH-WALL GIRDER LOADED OVER A STEEL PLATE ON THE TOP

Stanko Colak & Damir Varevac

This Publication has to be referred as: Colak, S[tanko] & Varevac, D[amir] (2021). Testing of a Wooden High-wall Girder Loaded over a Steel Plate on the Top, Proceedings of the 32nd DAAAM International Symposium, pp.0369-0376, B. Katalinic (Ed.), Published by DAAAM International, ISBN 978-3-902734-33-4, ISSN 1726-9679, Vienna, Austria DOI: 10.2507/32nd.daaam.proceedings.054

Abstract

In this paper, a test of wooden high-wall girders 5.0 m high, 0.18 m thick and of different widths compressed by a steel plate made of structural steel 20 cm wide and 2 cm thick was performed in the program Ansys. We have a total of 12 cases of loading the high-rise girder over the steel plate. The results of 3 high-wall girders were compared, the first 1.5 m wide, the second 1.15 m and the third 0.8 m loaded over a steel plate first with a force of 250 kN then with a force of 500 kN, with and without friction. The results show that the friction between the steel plate and the wooden high-wall girder has a small effect on the increase of the longitudinal stress of the high-wall girder. Also the width of the high-wall girder affects the normal stress in the length direction.

Keywords: wooden high-wall girder; steel plate; normal stress; Ansys.

1. Introduction

The task was to examine the normal stress of wooden high-wall girders in the direction of the length of the high-wall girders when they are loaded over the steel plate. The aim and purpose of the test is to describe the test results of high-wall girders loaded over a steel plate and to compare the case where there is friction at the contact between the steel plate and the high-wall girder with the case when there is no friction at the contact between the high-wall girder and the plate. The problem to be solved by this work is to find out the place on the high-wall girder where the highest stress occurs and whether there are too many differences between the contacts between the high-wall girder and the plate without friction and when there is a friction coefficient of 0.4.

The stress calculation of high-wall girders was performed in the program Ansys [1]. Further tests in this area will be the loading of wooden high-wall girders via a steel plate connected to the high-wall girder by means of screws. After the introduction, the dimensions and mechanical properties of the wooden high-wall girder are described in the second part. The procedure for testing high-wall girder samples in Ansys was then described. The test results are presented and described. A conclusion was written at the end of the paper.

2. Dimensions and mechanical properties of wooden high-wall girder

Figure 1 shows a model of a wooden high-wall girder with a steel plate on top from the program Ansys [1]. The high-wall girder is 5.0 m high, 0.18 m thick and of various widths loaded over a steel plate made of structural steel 20 cm wide and 2 cm thick. We have three different high-wall girders according to the width dimensions: high-wall girders with a width of 1,50, 1,15 and 0,8 m.

Fig. 1. High-wall girder from program Ansys [1]

Figure 2 shows the mechanical properties of a wooden high-wall girder. The density of the wood from which the high-wall girder is made is 410 kg / m2, and the Poisson's ratio is 0.3. The Young's modulus of elasticity of wood in the fiber direction is 12600 MPa, and the Young's modulus of elasticity of wood perpendicular to the fiber direction is 420 MPa. The shear modulus in all three directions is equal to 420 MPa [2].

1	Property	Value	Unit		
2	Material Field Variables	Table			
3	Density	410	kg m^-3		
4	Orthotropic Elasticity				
5	Young's Modulus X direction	12600	MPa		
6	Young's Modulus Y direction	420	MPa		
7	Young's Modulus Z direction	420	MPa		
8	Poisson's Ratio XY	0,3			
9	Poisson's Ratio YZ	0,3			
10	Poisson's Ratio XZ	0,3			
11	Shear Modulus XY	420	MPa		
12	Shear Modulus YZ	420	MPa		
13	Shear Modulus XZ	420	MPa		

Fig. 2. Mechanical properties of the wood from which the high-wall girder is made [1]

3. Test procedure and results

The test procedure was performed in such a way that the high-wall girder support was clamped on the underside and compressed on the upper side via a 2 cm thick steel plate. We have a total of 12 samples of high-wall girder:

1. high-wall girder 5 m high and 1,5 m long loaded over the steel plate with the force of 250 kN without friction at the contact between the high-wall girder and the steel plate,
2. high-wall girder 5 m high and 1,5 m long loaded over the steel plate with the force of 500 kN without friction at the contact between the high-wall girder and the steel plate,
3. high-wall girder 5 m high and 1,5 m long loaded over the steel plate with the force of 250 kN with friction at the contact between the high-wall girder and the steel plate (coefficient of friction on the contact surface is equal 0,4),
4. high-wall girder 5 m high and 1,5 m long loaded over the steel plate with the force of 500 kN with friction at the contact between the high-wall girder and the steel plate (coefficient of friction on the contact surface is equal 0,4),
5. high-wall girder 5 m high and 1,15 m long loaded over the steel plate with the force of 250 kN without friction at the contact between the high-wall girder and the steel plate,
6. high-wall girder 5 m high and 1,15 m long loaded over the steel plate with the force of 500 kN without friction at the contact between the high-wall girder and the steel plate,
7. high-wall girder 5 m high and 1,15 m long loaded over the steel plate with the force of 250 kN with friction at the contact between the high-wall girder and the steel plate (coefficient of friction on the contact surface is equal 0,4),
8. high-wall girder 5 m high and 1,15 m long loaded over the steel plate with the force of 500 kN with friction at the contact between the high-wall girder and the steel plate (coefficient of friction on the contact surface is equal 0,4),
9. high-wall girder 5 m high and 0,8 m long loaded over the steel plate with the force of 250 kN without friction at the contact between the high-wall girder and the steel plate,
10. high-wall girder 5 m high and 0,8 m long loaded over the steel plate with the force of 500 kN without friction at the contact between the high-wall girder and the steel plate,
11. high-wall girder 5 m high and 0,8 m long loaded over the steel plate with the force of 250 kN with friction at the contact between the high-wall girder and the steel plate (coefficient of friction on the contact surface is equal 0,4),
12. high-wall girder 5 m high and 1,5 m long loaded over the steel plate with the force of 500 kN with friction at the contact between the high-wall girder and the steel plate (coefficient of friction on the contact surface is equal 0,4).

The calculation was performed in the program Ansys and the results of the normal stress of the high-rise girder in the direction of its width are presented.

3.1. High-wall girder 1,5 m wide and loaded with the force of 250 kN without friction

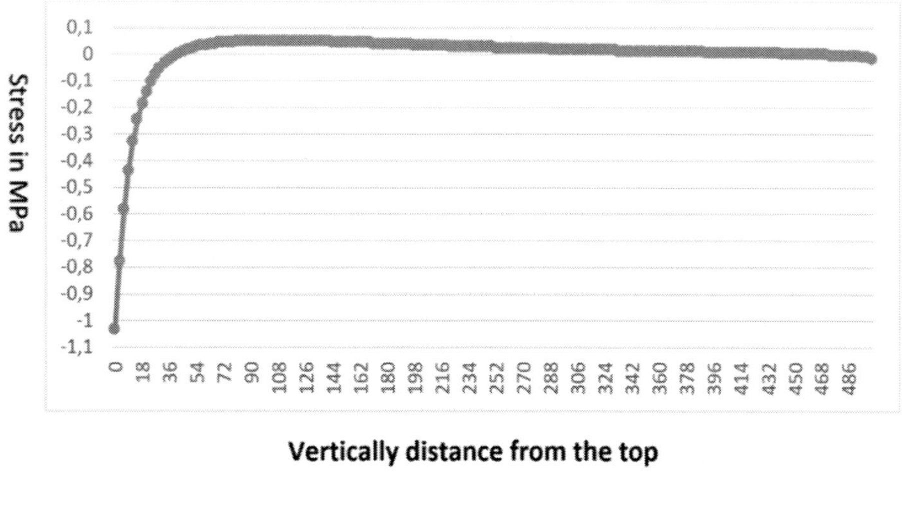

Fig. 3. Stress along the height of the high-wall girder

The maximum normal stress is at a distance of 105 cm from the top of the high-wall girder and is 0,050912 MPa.

3.2. High-wall girder 1,5 m wide and loaded with the force of 500 kN without friction

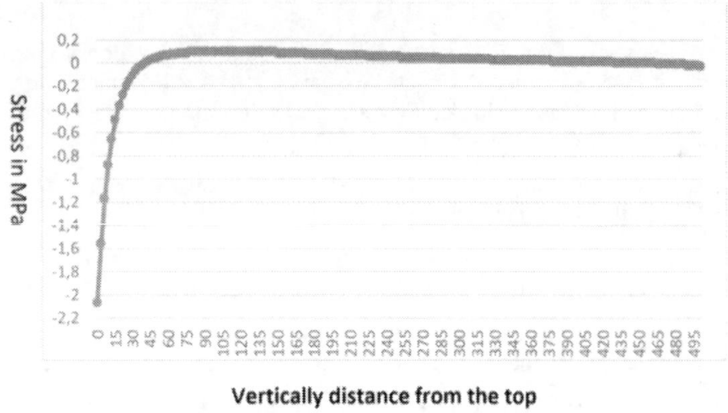

Fig. 4. Stress along the height of the high-wall girder

The maximum normal stress is at a distance of 105 cm from the top of the high-wall girder and is 0,10182 MPa.

3.3. High-wall girder 1,5 m wide and loaded with the force of 250 kN with friction

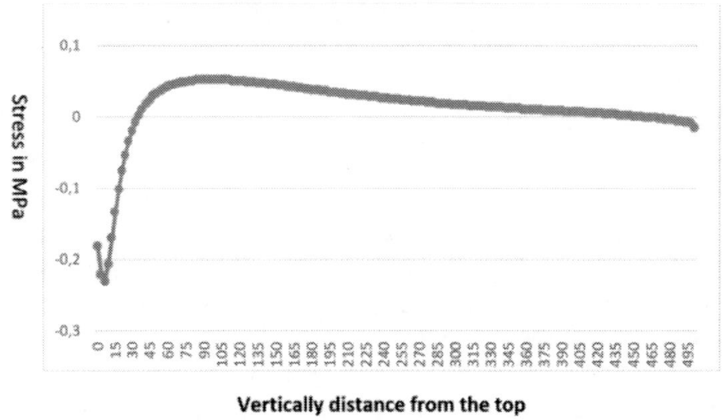

Fig. 5. Stress along the height of the high-wall girder

The maximum normal stress is at a distance of 96 cm from the top of the high-wall girder and is 0,051692 MPa.

3.4. High-wall girder 1,5 m wide and loaded with the force of 500 kN with friction

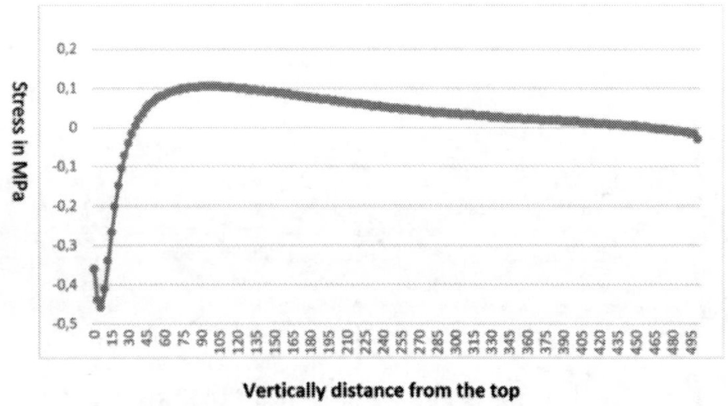

Fig. 6. Stress along the height of the high-wall girder

The maximum normal stress is at a distance of 96 cm from the top of the high-wall girder and is 0,10338 MPa.

3.5. High-wall girder 1,15 m wide and loaded with the force of 250 kN without friction

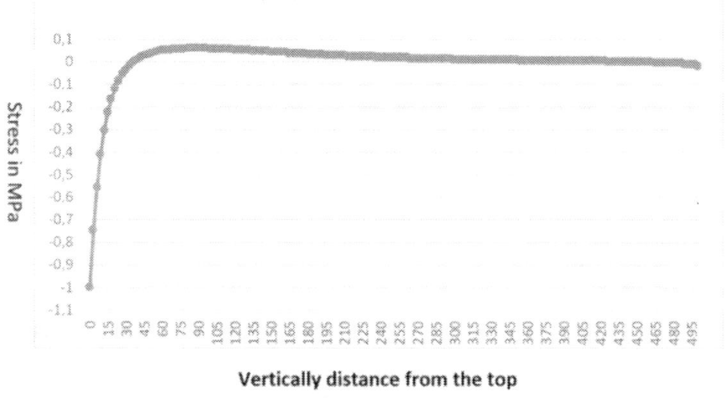

Fig. 7. Stress along the height of the high-wall girder

The maximum normal stress is at a distance of 87 cm from the top of the high-wall girder and is 0,06066 MPa.

3.6. High-wall girder 1,15 m wide and loaded with the force of 500 kN without friction

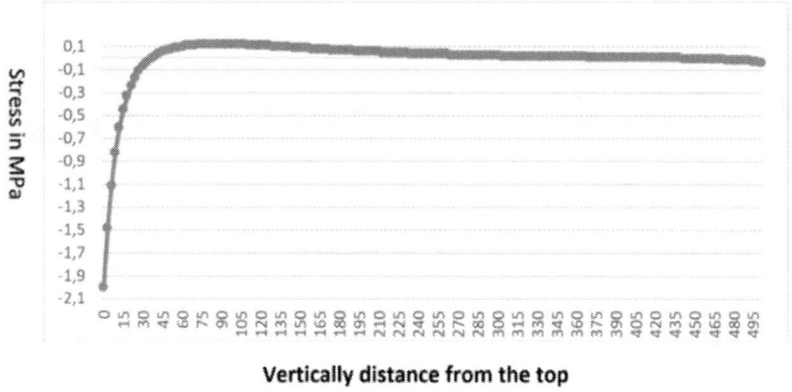

Fig. 8. Stress along the height of the high-wall girder

The maximum normal stress is at a distance of 87 cm from the top of the high-wall girder and is 0,12132 MPa.

3.7. High-wall girder 1,15 m wide and loaded with the force of 250 kN with friction

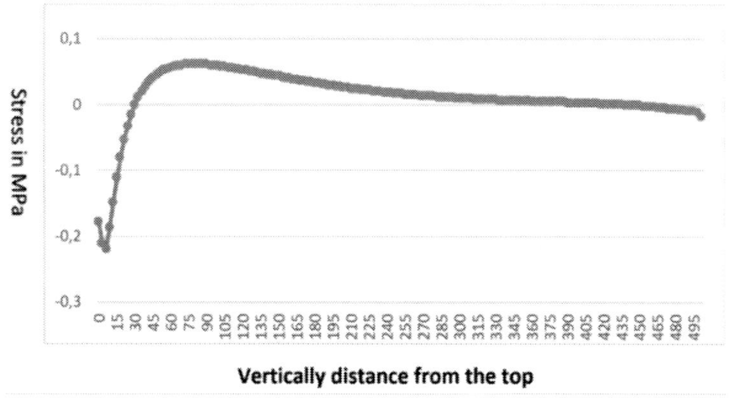

Fig. 9. Stress along the height of the high-wall girder

The maximum normal stress is at a distance of 81 cm from the top of the high-wall girder and is 0,061526 MPa.

3.8. High-wall girder 1,15 m wide and loaded with the force of 500 kN with friction

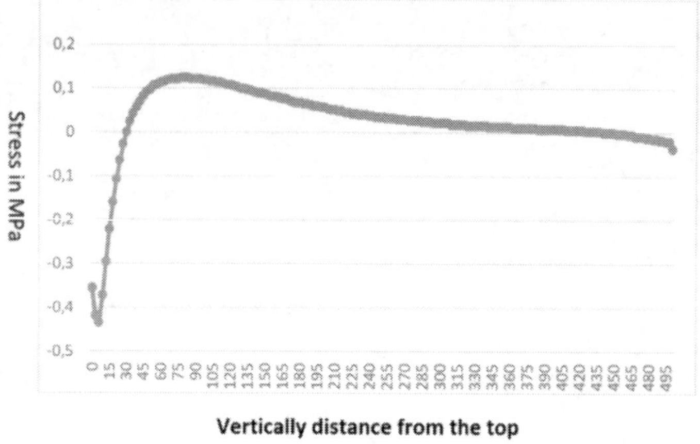

Fig. 10. Stress along the height of the high-wall girder

The maximum normal stress is at a distance of 81 cm from the top of the high-wall girder and is 0,12307 MPa.

3.9. High-wall girder 0,8 m wide and loaded with the force of 250 kN without friction

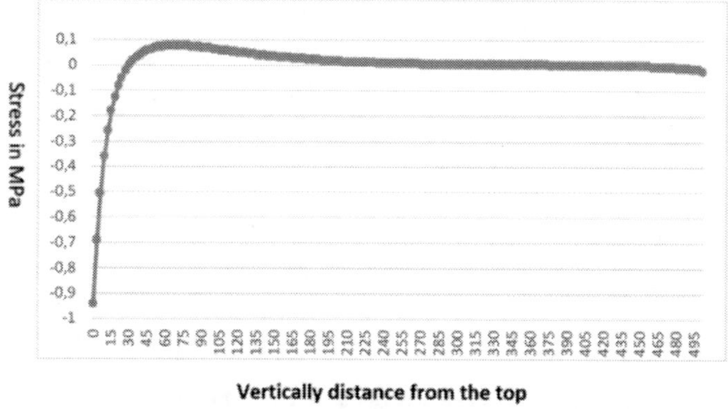

Fig. 11. Stress along the height of the high-wall girder

The maximum normal stress is at a distance of 69 cm from the top of the high-wall girder and is 0.075191 MPa.

3.10. High-wall girder 0,8 m wide and loaded with the force of 500 kN without friction

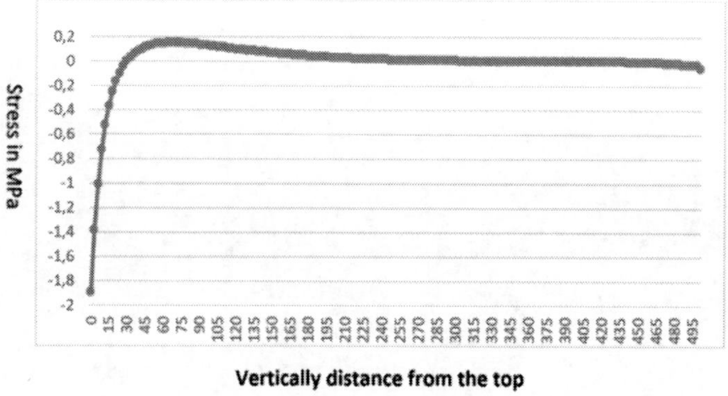

Fig. 12. Stress along the height of the high-wall girder

The maximum normal stress is at a distance of 69 cm from the top of the high-wall girder and is 0,15038 MPa.

3.11. High-wall girder 0,8 m wide and loaded with the force of 250 kN with friction

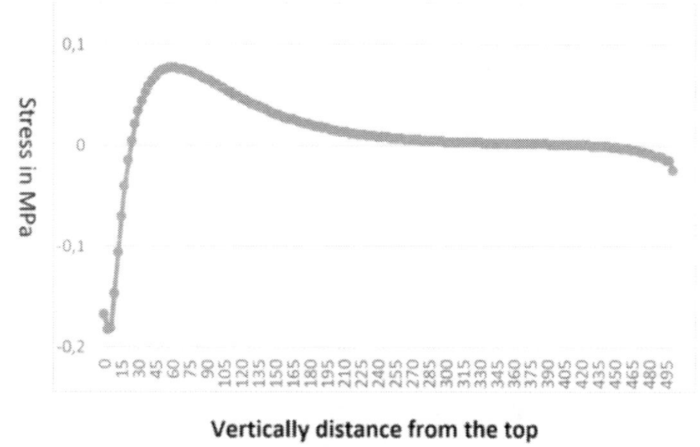

Fig. 13. Stress along the height of the high-wall girder

The maximum normal stress is at a distance of 60 cm from the top of the high-wall girder and is 0,076125 MPa.

3.12. High-wall girder 0,8 m wide and loaded with the force of 500 kN with friction

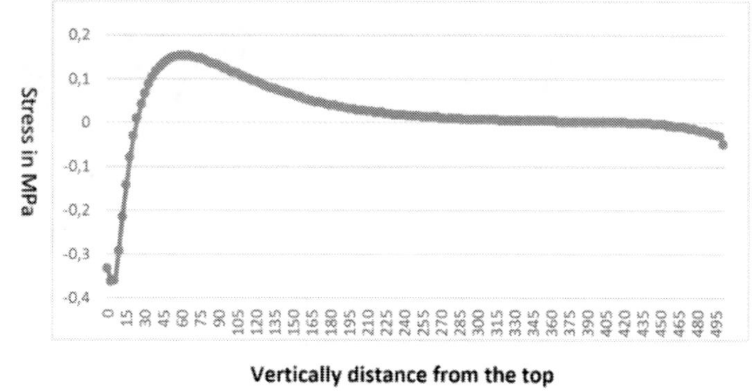

Fig. 14. Stress along the height of the high-wall girder

The maximum normal stress is at a distance of 60 cm from the top of the high-wall girder and is 0,15227 MPa.

4. Interpretation of results

It can be seen from the results that in the case of loading a wooden high-wall girder over a steel plate, the longitudinal stress along the length of the high-wall girder is higher in the case of friction between the steel plate and the high-wall girder than when the connection is frictionless. It has been shown that the maximum normal stress is at a greater distance from the top of the high-rise girder in the case of load if the connection between the steel plate and the high-wall girder is frictionless than if the connection is with friction. Also from the results it can be seen that the greater the width of the high-wall girder the maximum stress is at a greater distance from the top.

5. Conclusion

In the case of loading a wooden high-wall girder over a steel plate, it is shown that there is not much difference if there is friction at the contact between the high-wall girder and the steel plate and in the absence of friction, although the stress is slightly higher if there is friction. It can be seen from the results that in all cases the maximum normal stress is at a distance of approximately one tenth to one fifth of the total height of the high-wall girder seen from the top. The results showed that there would be no breakage or dangerous deformation of the wooden high-wall girder. The stresses of the wooden high-wall girders are in line with expectations. The research in this paper will lead to further tests in which the load of the wooden high-wall girder will be observed via a steel plate that will be bolted to the wooden high-wall girder, the case is higher than the maximum normal stress.

6. References

[1] Ansys 2019 R3. (2019). South of Pittsburgh in Canonsburg, Pittsburgh, Pennsylvania, United States of America

[2] Ćubela, D. (2020). Wooden high-wall girders, Lecture from Basics of wooden structures, Faculty of civil engineering, University of Mostar, Mostar, Bosnia and Herzegovina

[3] Takač, S., Varevac, D., Markulak, D., Analysis of the Stress Distribution in the Semispace of the Steel Pinned Joint, The University of J. J. Strossmayer in Osijek, Civil Engineering Faculty, Osijek, Croatia

[4] https://www.grad.unizg.hr/_download/repository/Svojstva_drva_2014%5B2%5D.pdf, 2014, Faculty of civil engineering, University of Zagreb, Accessed on: 2009-09-27

[5] https://drvene-konstrukcije.hr/wp-content/uploads/2010/02/14.pdf, 2010, Accessed on: 2009-09-28

[6] Newaz, G., Mayeed, M., Rasul, A. (2016). Characterization of balsa wood mechanical properties required for continuum damage mechanics analysis, Journal of Materials: Design and Applications, Vol. 230, No. 1, 2016, pp. 206-218, DOI: 10.1177/1464420714564711

[7] Brauns, J., Rocens, K., Lawal, S. A. (2018). Modification of Wood: Mechanical Properties and Application, Reference Module in Materials Science and Materials Engineering, DOI: 10.1016/B978-0-12-803581-8.11226-3

[8] https://en.wikipedia.org/wiki/Wood, Wood, Wikipedia, the free encyclopedia, Accessed on: 2009-09-28

DOI: 10.2507/32nd.daaam.proceedings.055

COMPARISON OF THE LIGHTING CONDITION OF THE INTERIOR TO CREATE A 3D BACKGROUND IN VIRTUAL REALITY

Irena Drofova, Milan Adamek, Lucie Sousedikova,
Adam Malatinsky & Pavel Valasek

This Publication has to be referred as: Drofova, I[rena]; Adamek, M[ilan]; Sousedikova, L[ucie]; Malatinsky, A[dam] & Valasek, P[avel] (2021). Comparison of the Lighting Condition of the Interior to Create a 3D Background in Virtual Reality, Proceedings of the 32nd DAAAM International Symposium, pp.0377-0383, B. Katalinic (Ed.), Published by DAAAM International, ISBN 978-3-902734-33-4, ISSN 1726-9679, Vienna, Austria
DOI: 10.2507/32nd.daaam.proceedings.055

Abstract

This article deals with capturing the image in the gallery's interior with a 3D camera in different lighting conditions. Currently, 3D backgrounds are a standard part of modelling and virtual reality. There are several ways to create a background, and there are various high-resolution 3D scanning devices. This study deals with capturing an image in the interior of a gallery with a 3D camera in different lighting conditions. The text lists the fundamental properties of light and lighting. The parameters and possibilities of capturing the image with a 3D camera are also described. The article further describes the indoor lighting conditions in the interior when using a 3D camera and compares the lighting conditions and their effect on the quality of the captured image and the goal of obtaining the best possible quality of the captured image.

Keywords: 3D camera; digital image; lighting condition; virtual reality; art gallery

1. Introduction

This article focuses on the preparation and acquisition of 3D image data for further processing in virtual reality. [1] Currently, 3D image data is used in many fields. There are many devices and techniques for capturing 3D images. Subsequent use of 3D image data then determines their further processing. Created 3D data and objects are used in many fields. They are used for technical documentation, simulations in industry, construction, architecture, design, and art. 3D models and 3D scanned images are also used in virtual reality. Image analysis and high-quality image data are essential, for example, in the field of medicine. The created 3D background as an interior simulation is essential for the degree of nesting in the virtual environment. Especially when simulating events or for training simulators. These are used in the army, rescue system units, or the police. 3D-created backgrounds based on actual data can thus significantly contribute to all the mentioned areas.

In this study, attention is focused on the field of art and works of art. In this field, great importance is placed on the image digital reproduction of materials and colours. Assessing the quality of the 3D image with respect to the lighting type is an essential step in image analysis, especially for the details captured in the picture. [2]

The high image quality will also enable accurate image reproduction in 3D modelling and virtual reality. Digitization and its progressive development have also manifested themselves in this field. Online exhibitions and auctions are now a standard part of the art world and the art trade. High demands are placed on the reproduction of art and the quality of its depiction. Methods for identifying counterfeits are also changing in this area. There are many methods for distinguishing the original from the counterfeit. One of these methods is visual assessment. This method focuses on the artist's painting style, colours, scene, light. [3]

Modern computer vision techniques are currently used to obtain physical evidence of possible manipulations of a painting or the resolution of a forgery. [4] Therefore, it is essential to reproduce a work of art in a digital environment so that the reproduction of the work corresponds highly to the physical original. Significant differences in the observer's perception of a work of art can be caused by the type of lighting in the room and the illumination of the work of art. In this study, attention is focused on working with a 3D camera in the interior of the gallery. The image in the room was taken mainly in daylight.

1.1. The light in the visible radiation

The colour of the human eye perceives depends on the external environment and is variable in connection with its change. These are mainly lighting conditions and the environment, which significantly affect the resulting colour perception. In this chapter, attention is focused on the issue of *Photometry*, respectively methods that deal with visible radiation and in which the properties of the eye play an important role. Visible radiation is only a tiny part of the range of *electromagnetic radiation* and is a range of wavelengths of 380 to 760 nm that act on the retina of the human eye. Below are some photometric quantities, of which it is considered to be the central *luminosity* (I) in *Photometry*. [5]

The luminosity (I) is, therefore, the primary photometric quantity. The unit of *luminosity* is a *candela (cd). Luminosity expresses* the ability of point sources to radiate in a given direction and to evoke visual perception.

Brightness / Luminance (L) is a specific measure of the *luminosity* of a planar source. *Brightness* is the required ratio of the *luminosity* of the elementary surfaces of the *sources (dI)* in the selected direction and the perpendicular projection of the surface in this direction (dS cosα):

$$L = \frac{dI}{dS \cos \alpha} \tag{1}$$

Luminous flux (Φ) is a photometric quantity that expresses the ability of a given radiant flux to evoke visual perception. The unit is *lumen (lm)* and is a derived unit. From approximately point sources with *the omnidirectional luminosity* of one *candela (I)*, the transmitter has a *luminous flux (Φ)* of the size of one *lumen (lm)* up to the spatial angle of one steradian:

$$\Phi = I\Omega \tag{2}$$

Illumination (E) is a specific measure of the *luminous flux (Φ)* on an *illuminated surface*. The unit of *illumination* is *lux (lx)*:

$$E = \frac{d\Phi}{dA} \tag{3}$$

Image capture and its subsequent quality are affected by light. The image captured outdoors will be different from the image captured indoors. The resulting image will differ when shooting day and night. Other techniques used in photography also affect image quality. An object illuminated by artificial light or shadows suppressed by photo plates when shooting will be of a different quality. Complementary techniques also affect the shadows in the overall image. All these aspects and many more affect the final quality. All attributes affecting the resulting scene are adjustable during subsequent 3D modelling and light settings when creating a virtual environment.

1.2. Type of the light sources

Light sources are devices that emit visible radiation. These sources allow the creation of a particular lighting environment. This environment must comply with hygienic, environmental, and colorimetric standards. Standardized observation conditions are thus ensured. We divide electric light sources: [5]

- Heat sources: the electric current flowing through these sources heats the metal fiber. That excites the metal atoms. The spontaneous emission of photons produces thermal radiation with a continuous spectrum.
- Discharge sources: c electric discharge causes electrons and ions to collide with gas atoms or metal vapors. The energy of particles changes into optical radiation. Discharge lamps are often equipped with phosphors on the inner wall. These emit visible light when excited by UV radiation. The spectrum is linear or continuous with the spectral lines.

- Laser radiation: monochromatic with high temporal and spatial coherence. This radiation is created by stimulated emission.
- Daylight: The source of daylight is the Sun. This light is significantly different from artificial light. The difference is mainly in the spectral composition and variability of radiation intensity. The spectrum is continuous with maximum intensity in the visible region.

2. Type of 3D camera for capture image

There are currently many 3D scanning devices on the world market. [6] For example, laser 3D scanners are efficient and straightforward devices for instant 3D imaging with an extensive range in scanning angle, distance, and scanning speed. [7] Another option for capturing a 3D image is 3D cameras. There are also a large number of these devices with different applications. This text focuses on capturing the image in the gallery's interior using the 3D camera Insta360 Pro. [8]

Fig. 1. The body of 3D camera Insta360 Pro [8]

This 3D and panoramic camera have 6 "fish eyes" with 200° F2 lenses around the circumference of the aluminum spherical body. That allows capturing high-quality 360° images. The four microphones also provide good sound when recording video. In normal mode, the camera takes six still photos at a resolution of 4000 x 3000 px. These images can then be used to staple and subsequently process 8K monoscopic and stereoscopic spherical images. The Insta360 stapling application is used for subsequent processing and stapling of the image into one unit. Insta360 Player is used to playing back stitched images. Insta360 Pro records data captured using an SD card (speed class V30 + necessary for video recording) and a USB SSD (USB3.0 interface). Insta360 Pro have static image modes:

- Normal mode captures six shots at 4000 x 3000 px, which are stapled into 8K monoscopic or stereoscopic images in post-processing.
- RAW mode is the primary mode for post-image processing for color calibration and other adjustments depending on image purpose. The file format is (.dng). Take six shots to staple 8K monoscopic or stereoscopic spherical images sequentially.
- HDR mode is the mode is used to eliminate the difference in brightness. It takes three series of 6 shots at a resolution of 4000 x 3000 px, which are then used to staple an 8K highly dynamic spherical image. This mode does not allow automatic stapling directly in the camera.
- Burst mode is for a series of 10 with six images in resolution 4000 x 3000 px in one series. This mode does not allow automatic stapling of the captured image directly in the camera. Subsequent sewing again gives an 8K image in monochromatic or stereoscopic.
- Static images can be captured by all six cameras manually by selecting one of the modes or remotely using the mobile app, ensuring convenient shooting in rugged terrain or confined conditions where there may be restrictions on manual handling and camera settings. Selected modes also enable real-time stapling. That will directly display the resulting panoramic image and can also display the image in virtual reality. The app needs to be reset every time use again. Likewise, when pairing mobile devices with wifi cameras.

The application for controlling and setting the 3D camera allows displaying the captured image directly in virtual reality. In this way, the camera also allows previewing the image before capturing it. That not only allows to adjust the lighting and scene before shooting. In the application and the display in virtual reality, it is possible to see the image's quality and the fidelity of the reproduction of works of art. These can be seen in digital form and partly affect the quality of their digitization. Colour vision and colorimetry are essential for subsequent image processing in digital form. [9] The colours transferred between colour spaces [10] and the definition of colours in the digital environment will allow a high-quality reproduction of the original work of art. The following chapter describes the process of capturing the gallery interior in different lighting.

3. 3D scanning by 3D camera in the interior

The Insta360 Pro 3D camera has 6 lenses around the perimeter of the body. These are 200° F2 "fish eye" lenses. These types of objects are not fully suitable for 3D modelling by photogrammetry. For example, they are not recommended for processing in Agisoft Metashape Professional software. [11] Editing a 3D model would be very challenging. The model developed in this software is shown in Figure 2.

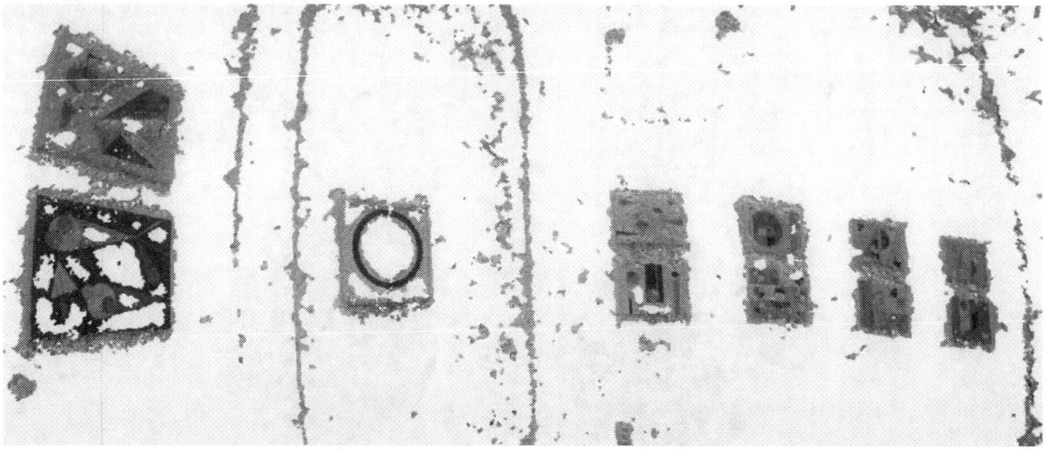

Fig. 2. The 3D model gallery's interior by software Agisoft Metashape Professional

3.1. Capturing an images and interior light

Indoor image capture has its specifics. In particular, lighting and scanning devices have a significant influence on the quality of the acquired image. The lighting in the room can be daylight or artificial. Both types of lighting may also be required. In art galleries, it is also possible to illuminate individual exhibits separately. This lighting is chosen to underline the artwork. As a rule, however, works of art are very well observed in daylight. The lighting of works of art usually chooses to be as close as possible to daylight. For the colour vision of the human eye, the colours of the works of art are better seen in interior lighting close to daylight. In this study, the interior of the gallery in a historic building was chosen. Daylight is natural in this interior. Artificial lighting is used to dim the room. That is a classic interior lamp with discharge sources. In the subsequent text, both types of lighting are compared to the display of works of art in the gallery's interior.

Fig. 3. Picture captured in the gallery's interior: daylight / discharge sources lamp / without light

Figure 3 shows a comparison of the image captured during three different lighting conditions. The first image shows the quality of the image captured in daylight, the second figures shows quality of the image in discharge sources lighting and the third figure shows image in the dark interior. The room has one large door towards the terrace. It was sunny when the picture was taken indoors. However, sunlight did not fall into the room. It can be seen that the captured image is directed in blue tones in the nature day illumination. This image will need to be subsequently adjusted to minimize the blue tone and the noise created in the shadows of the corner in the room. That image captured in discharge lighting lamp in the interior have more high quality. There is no blue tone in the image. At first glance, however, it is clear that the image is tinted slightly to a red hue. Also, this image and its quality can be improved by further editing in the graphics software. In the case of an image captured without lighting, i.e. in the dark, it is clear that the image quality is very low. Works of art can only be minimally recognized in the image. The image captured in this way is not suitable for further processing.

3.2. Types and possibilities of capturing the image with a 3D camera

The previous chapter lists the formats that can capture an image with the Insta360 Pro 3D camera. This scanning device has many uses. The camera can take static photos with this 3D camera. It will create a 3D image with these photos and their stitching. The camera can also take panoramic photos. For 3D modelling, it is appropriate to use the 3D format. These are three ways to create an environment or background in 3D models and a virtual environment. Last but not least, the camera is also suitable for recording high-quality video and for a live broadcast or stream. The following figures show some of the above types of captured images. The images were obtained in the gallery's interior with a discharge light source.

Fig. 4. The type of six static photos (200° F2 "fish eye" lenses) and their stitching

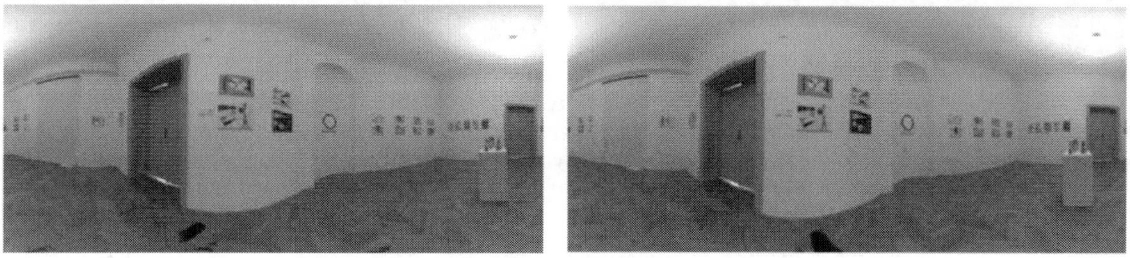

Fig. 5. The type of the panoramatic photo: the left and the right panoramatic photo

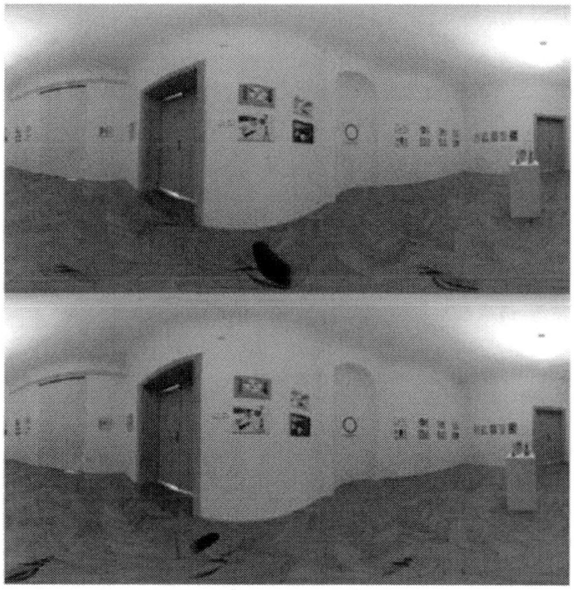

Fig. 6. The type of 3D photo

In this particular case, the Insta360 Pro's 3D camera captured the image using a natural interior environment with one discharge lamp in the middle of the room. The room also has a large wing glass door facing west with access to the terrace. The paintings were taken in August 2019 at the time of year with the highest intensity of daylight. The aim was to determine what image quality would be captured by this particular device and to what extent it would be necessary to modify the captured image in graphics software. Therefore, no photographic accessories have been used to improve image quality and suppress side effects. These accessories and tools would increase the quality of the scanned image.

4. Conclusion

Many methods are currently used in modern computer graphics to create backgrounds, material structures, and virtual environments. In many fields, it is desirable to create a real digital image. In this study, attention was focused on the field of art. In the modern digital world, there is a need for the high-quality digital reproduction of works of art for online exhibitions, online art auctions, or virtual reality art.

This work aimed to find out to what extent a quality image of the interior can be obtained so that this image does not have to be complicatedly modified. Emphasis was placed on the possibility of capturing an image in different lighting of a room with works of art. A high-quality 3D camera captured this interior with several modes for further use. This article focuses on the possibilities of creating a 3D image for further use in the field of 3D graphics and virtual reality. The image was taken on a sunny day, and the natural daylight sufficiently illuminated the entire interior with the images. Another capture of the room was under artificial lighting. A discharge light source illuminated the space. In addition, an image of the interior with minimal lighting was taken for further assessment.

The assumption of the high quality of the captured image in daylight has not been confirmed. The image thus obtained showed a high degree of blue tone and noise in the shadows. In an environment with artificial lighting, a worse red tone appeared but minimal. These phenomena can be easily eliminated in computer graphics and thus obtain a higher quality digital image, especially for quality reproduction of colours used to create works of art in the interior. For these needs, it is still necessary to define the colours in the colour spaces so that the colours tones are not fundamentally changed during further adjustments. Both types of scanned images could be further detected after adjustments. They may be appropriate for an initial view in virtual reality, where you can continue working with light and its properties. In the case of creating a real virtual gallery, this would have to be necessary. On the contrary, a captured image in a dark environment cannot sufficiently perform works of art. In this case, it would be appropriate to use external lighting and other photographic accessories, especially for capturing the image at night.

All three image capture methods in Chapter 3 could be used to create backgrounds in 3D modelling and virtual reality. Definition of colours and quality processing of works of art would precede further processing. In many modelling software and virtual reality programs, working with light is very important. That would also make it possible to suppress some of the undesirable phenomena and, on the contrary, to add others, which would make the space more pleasant for observers.

The 3D camera used in this study is suitable for both interior and exterior. No photographic accessories or image enhancement devices were used to capture the image indoors. The aim was to find out what it would look like in the live stream in an actual image into virtual reality. Subtle tonal variations could partially limit the calibration of colours on the virtual reality headset, however, not in the case of a captured image without lighting, which does not show the exhibited works of art insufficient quality and are displayed only minimally. An image captured in daylight and artificial lighting would be suitable for the live broadcast from the opening of exhibitions by the artist. On this occasion, the artist presents his works and adds commentary to them. This way of transmitting the captured image will be further research into the opening of another art exhibition.

The pictures were taken in the daytime, and artificial lighting shows incomparable colour tones. This reduced image quality can increase image editing and correction in graphics software. Use the process to build seamlessly on this study. The digital image will increasingly be used as a background in the created model of the virtual gallery with an exhibition of the authors of the created more miniature sculptures in the same artistic style.

5. Acknowledgments

This research was based on the support of the author of the paintings, Mr. Michal Pasma, Vaclav Chad Gallery in Zlin, and of the International Grand Agency of Thomas Bata University in Zlin, IGA/CebiaTech/2021/004, and the Department of Security Engineering, Faculty of Applied Informatics.

6. References

[1] S. Martirosov & P. Kopecek, (2017). Virtual Reality and its Influence on Training and Education - Literature Review, Proceedings of the 28th DAAAM International Symposium, pp.0708- 0717, B. Katalinic (Ed.), Published by DAAAM International, ISBN 978-3-902734-11-2, ISSN 1726-9679, Vienna, Austria DOI: 10.2507/28th.daaam.proceedings.100

[2] V. Huskova, V. Adamova, and K. Boc, (2021) Use of 3D Screening Technology Traks in the Process Criminal Identification (SK), Young Science Conference, Tomas Bata University in Zlin, Facilty of Applied Informatics, Trilobit, [online], http://trilobit.fai.utb.cz/Data/Articles/PDF/8dcada30-4b45-403b-a048-2c5d1026855a.pdf

[3] E. Gultebpe, E. Thomas, and M. M. Conturo, (2018) Predicting and grouping digitized paintings by style using unsupervised feature learning, Journal of Cultural Heritage., vol. 31, 13-23., ISSN 1296-2074., DOI:10.1016/j.culher.2017.11.008
[4] A. Pelagotti, A. Piva, F. Uccheddu, D. Shullani, M. F. Alberghina, S. Schiavone, E. Massa and C. M. Menchetti, (2020) Forensic Imaging for Art Diagnostics. What Evidence Should. We Trust?, IOP Publishing, IOP Conf. Series: Materials Science and Engineerin, 012076, 949 doi:10.1088/1757-899X/949/1/012076
[5] M. Dohnal, (2007) Physical Bases of Image Reproduction, University of Pardubice , ISBN: 978-80-7194-945-9
[6] Z. Stojkic, E. Culjak & L. Saravanja, (2020), 3D Measurement - Comparison of CMM and 3D Scanner, Proceedings of the 31st DAAAM International Symposium, pp.0780-0787, B. Katalinic (Ed.), Published by DAAAM International, ISBN 978-3-902734-29-7, ISSN 1726-9679, Vienna, Austria. DOI: 10.2507/31st.daaam.proceedings.108
[7] Gefos ČR Leica BLK360 (2021). [online], https://www.gefos-leica.cz/o-produktech/3d-laserove-skenovani/3d-laserove-skenery/blk360
[8] Insta 360 Pro, (2021), [online], https://www.insta360.com/product/insta360-pro/
[9] M. Dohnal, (2019) Color vision. Colorimetry, University of Pardubice, ISBN: 978-80-7560-246-6
[10] P. Ehkan, S. V. Siew, F. F. Zakaria, M. N. M. Warip and M. Z. Ilyas, (2020) Comparative Study of Parallelismand Pipelining of RGB to HSL Colour Space Conversion Architectureon FPGA, IOP Conf. Series: Materials Science and Engineering., DOI:10.1088/1757-899X/767/1/012054
[11] Agisoft (2021). [online], https://www.agisoft.com/features/professional-edition/

DOI: 10.2507/32nd.daaam.proceedings.056

THE ROLE OF LIE DETECTION BASED SYSTEM IN CONTROLLING BORDERS

Lucie Sousedikova, Adam Malatinsky, Irena Drofova & Milan Adamek

This Publication has to be referred as: Sousedikova, L[ucie]; Malatinsky, A[dam]; Drofova, I[rena] & Adamek, M[ilan] (2021). The Role of Lie Detection Based System in Controlling Borders, Proceedings of the 32nd DAAAM International Symposium, pp.0384-0388, B. Katalinic (Ed.), Published by DAAAM International, ISBN 978-3-902734-33-4, ISSN 1726-9679, Vienna, Austria

DOI: 10.2507/32nd.daaam.proceedings.056

Abstract

Security engineering refers to the latest techniques and methods used to protect large numbers of people concentrated in public crowded places every day. Such a new modern method represents the Intelligent Portable Border Control System (iBorderCtrl) based on lie detection. This paper describes and analyses this automated deception detection border security system developed mainly for airport security but also for the control of the land border crossing points as roads, walkways, or train stations. The system is built using advanced and high-performance deep learning models with the aim of effectively prevent any threats or potentially dangerous situations from arising or entering the country.

Keywords: Border Control; Deception Detection; Lie Detection; Analysis.

.

1. Introduction

Nowadays, border agencies come under considerable pressure with continuous traffic growth, combined with the increased threat of terrorist attacks and refugee crisis. That is why the areas of borders and migration management have become more and more prominent as sites of experimentation and investment. The European Commission has set aside a proposed €34.9 billion for border control and migration management between 2021 and 2027 (Gallagher & Jona, 2019). Recently, it was developed and tested new border control system called Intelligent Portable Border Control System (iBorderCtrl) which was funded by €4.5 million under the EU Horizon 2020 program [2].

The iBorderCtrl is a comprehensive system that consists of a two-stage procedure, designed to reduce cost and time spent per traveller at the border crossing station. It is an innovative automated border security system that detects deception based on lie detection, facial recognition technology, and the measurement of human micro gestures and non-verbal behaviour due to an artificial intelligence system known as an avatar. Although various works that used artificial neural networks have been done [3], [4] they have been too limited. Currently, there is a lot of publications describing the iBorderCtrl project [5]. This paper examines a recently drafted border control system by using critical analysis and highlights its main shortcomings.

2. Critical Analysis Framework

In this research, the iBorderCtrl system is analysed by using critical analysis which depicts strengths but above all limits of the system. Figure 1 introduces the method of critical analysis and shows its architecture for understanding. As shown, the outcome of the critical analysis is the entire evaluation of the system.

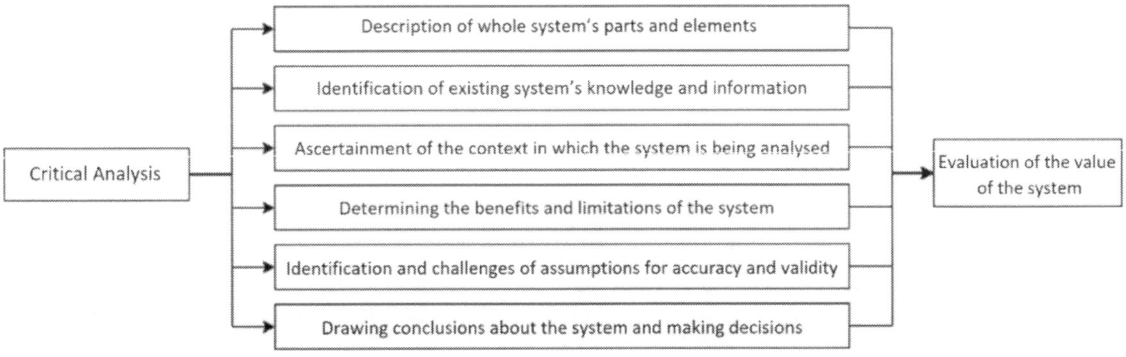

Fig. 1. Critical Analysis Framework

3. Technical Framework

The iBorderCtrl specific objective is to significantly increase traveller throughput at the borders in reduced time and at the same time enhance the security by bringing together many states of the art hardware and software technologies. This section provides a general overview of the iBorderCtrl's interfaces and subsystems with their technical description.

Fig. 2. Intelligent Portable Border Control System [6]

The system has three different interfaces, one for each type of user, and different modules for processing the gathered information:

- Traveller User Application for the traveller
- Border Guard User Application for the border guard agents
- Border Manager User Application for the border guard managers [7].

The modules that complete the system are in charge of processing the information, analyse the possible risks associated and provide useful information to the border guards to manage more effectively the Border Crossing process. The main modules composed the overall system solution are the following:

- **The Automatic Deception Detection System (ADDS).** During the pre-travel stage, ADDS requires each traveller to undertake an interview with an avatar, a sophisticated artificial intelligence system utilizing non-verbal micro-gestures to detect deception. Figure 1 presents the ADDS architecture. Each traveller has a QR code that uses throughout the interview. Control is passed from the Traveller User Application to ADDS using an iframe, at which point ADDS requests sufficient information to be able to ask 16 questions by the actual border guard in real-life crossing scenario. ADDS interviews with an avatar (male or female) that utilizes three attitudes (puzzled, neutral, positive) depend upon the deception score of the last question answered. ADDS requests to the avatar database to receive the requested corresponding Avatar video and updates the local ADDS database respectively. At the end of the interview the video frames, questions, and interview scores are uploaded to the iBorderCtrl databases to be used by the RBAT, FMT, and BCAT modules [6], [8].

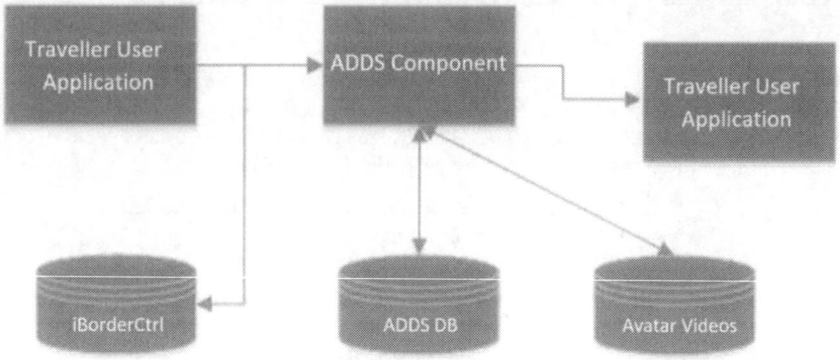

Fig. 3. ADDS Modul [6]

- **The Risk Based Assessment Tool (RBAT)** calculates and manages-traveller related risks in the border check procedure. It is a robust and flexible tool to support the decision-making process of the border authorities.
- **The Integrated Border Control Analytics Tool (BCAT)** includes advanced analytics methodologies for border managers. BCAT analysis detects illicit activities or patterns of behaviour associated with illegal activities. It also evaluates the effectiveness of each iBorderCtrl system.
- **The Biometric Module** consists of the fingerprint and palm vein modules validating the identification of the traveller. The aim of both modules is to compare the fingerprints and palm vein images to relevant information stored in European databases. Then the risk scores according to the match percentage through the central server software for matching are provided and forwarded to the iBorderCtrl database for the RBAT overall risk score calculation and assessment.
- **The Face Matching Tool (FMT)** is the facial recognition system for iBorderCtrl which is used both at the pre-registration and the border crossing phase.
- **The Hidden Human Detection Tool (HHD)** detects any hidden people inside various vehicles. For detection are used a radar/electromagnetic or an acoustic sensor depending on detecting space. The radar sensor is suited for the detection of all non-metallic spaces and compartments. In contrast, the acoustic sensor is used mainly for metallic cargo containers. Both types of sensors are small, and light weighted and support the Border Guard.
- **The Document Authenticity Tool (DAAT)** is used during the pre-registration as well as the border crossing stage. At the pre-registration stage, DAAT deals with the information (passport, visa, ID, residence permit) provided by the traveller using the Traveller User Application. At the border crossing point, the border guard uses the document scanner integrated into the Portable Unit to scan travel documents. The DAAT system checks the security features of documents and assesses the validation of documents and detects its counterfeiting.
- **The External Legacy and Social interfaces systems (ELSI)** checks the traveller's information from legacy systems (SIS II etc.) [6], [9].

4. Border Control Procedure

The iBorderCtrl changes the traditional approach followed in border management with a new one based on the integration of advanced technologies for data analysis for Border Guards and the two-stage procedure for the travellers:

- **Pre-registration stage** when all travellers are linked to any pre-existing authority data and through the Traveller User Application (web, mobile application), that verifies the authenticity are able to update their personal and travel information, upload travel-related documents, see expected traffic statistics regarding the selected country and border control point, pass an avatar interview conducted by ADDS. During the interview is filmed by his/her video camera and computer software obtains his/her micro-gestures to detect deceptive behaviour. The FMT is used and the traveller´s passport photo is compared to a short video sequence of the traveller.
- **The border crossing stage** when all registered documents and information are re-evaluated and the security controls are performed by a portable iBorderCtrl unit which consists of the devices needed to capture information (camera, fingerprint sensor, document reader, etc.) and provides the Border Guard Application for border guards. The Traveller User Application verifies the authenticity and truth of information forwarded. At this stage, the travellers present themselves by QR code to the border guard could obtain his/her personal information and the risk assessment score through the Border Guard User Application bringing all analytic results from each technology together to identify risks to the border guard. It enables overall evaluation of the traveller as well as highlighting specific potential issues the agent should focus on. In the end, like in the pre-registration phase, the risk is calculated. Furthermore, biometric checks (face matching, fingerprints, palm vein) and hidden human detection (crossing with the vehicle) take place. The face-matching tool performs the validation of the identity of the traveler to assign the risk of deception. It includes three different checks:

- Matching the images taken by the portable unit's camera and the high-definition image obtained from the passport or any external system such as VIS.
- Measuring the risk that the person appearing in the documents is the same one that undertook the avatar interview.
- Measuring the risk that the person at the border crossing point (images taken by the camera of the portable unit) is the same one that performed the interview during the pre-travel stage. [6], [9], [10].

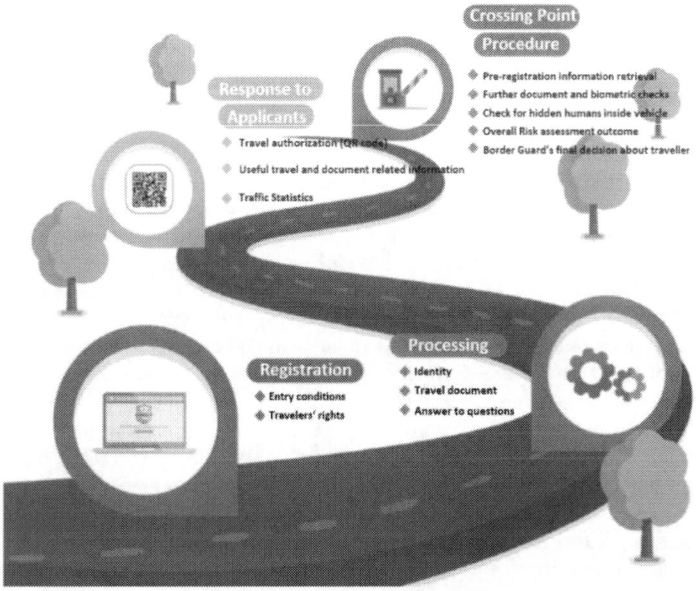

Fig. 4. Border Control Procedure [7]

5. Critical Analysis

The iBorderCtrl system solution changes the traditional approach followed in border management with a new one based on the two-stage procedure for travellers and the integration of modern technologies for data analysis. Since the end of the project testing phase, the iBorderCtrl is still considered experimental. This section provides a critical analysis of this system and outlines its benefits and limitations.

The iBorderCtrl system provides a wide range of capabilities and advantages such as:

- Document validation
- Face recognition
- Hidden human detection
- Data Analytics and statistics
- Provide early-warning capabilities due to risk assessment
- Three different interfaces
- Portable unit with Commercial-Off-The-Shelf technologies
- Easily expansion with new technologies or functionalities
- The possibility of integrating the similar functionalities of the new ETIAS system
- New two-stage procedure for border crossing
- Extend the capabilities of taking the border crossing procedure to where the travellers are (bus, car, train, etc.) instead of moving the travellers to where the checking booth is [7].

Since the end of the project testing phase, the iBorderCtrl is still considered experimental. Although it has its benefits, there is still a lot of shortcomings of this system:

- **Imperfect deception detection system** based on facial recognition which is still in progress in development. Currently, the system is not suitable - it is biased towards minorities and people with a range of nervous system abnormalities and correlated differences in social behaviour (attention-deficit disorder, post-traumatic stress disorder, people with depression state, or chronic pain conditions, etc.). Because researchers trained their artificial intelligence on a small sample of mostly European men, the system can discriminate against people based on their ethnic origin, and it has had a higher accuracy rate for the mentioned group.

- **Success rate of 75%** although it was assumed the success rate of 85%. On the contrary, the accuracy of the polygraph is up to over 90% according to the American Polygraph Association. But it is necessary the trained polygraph examiners who conduct time-consuming interviews that depend on the subject and the object of the investigation. On the other hand, this artificial lie detector is automated and has a great potential for easy use on many people, there is still a wide scope for errors. Many reporters and journalists have already tried to test the system and however, they answered honestly to all questions, they were evaluated as liars.
- **Need to establish an extensive biometric database** called Common Identity Repository if the EU wants to implement it continent-wide.
- **Threat to freedom of thought** is in danger with this technology. That is a big question if the system can look inside people's minds, it is legal to set it as mandatory. Together with new neurological applications, this requires a new legal framework in the future to protect the old human rights as the right to mental privacy, the right to mental integrity, the right to psychological continuity cognitive liberty.

6. Conclusion

The iBorderCtrl system was designed and developed with the specific objective to significantly increase traveller throughput at the borders in reduced time as well as security in terms of significantly fewer successful crimes. The system consists of three user-friendly applications and several modules that are in charge of processing the information, analyse the possible risks associated, and provide useful information to the border guards to manage more effectively the border crossing process. Despite being considered as a new modern solution - automated, unobtrusive, and easily used, the system is still limited. The main purpose of this research was to analyse it and highlight its limitations. After the completion of the pilot phase of the iBorderCtrl in Greece, Hungary, and Latvia, the system is still considered experimental and does not prevent anyone from entering the European Union for now. However, since august 2019 everyone who crossed the borders of a member state for the first time has been already provided his/her personal data including his/her privacy data on social media. If the shortcomings of the system presented in this article were corrected, the system would represent an automated, simple, and easily used solution for faster border crossing processes of EU for third-country nationals. Every passenger could be checked quickly but carefully whilst keeping the strict regulations. In this case, the iBorderCtrl could complement existing border control technology. Also, it would facilitate the border guards' working in spotting illegal immigrants, and it would prevent crime and terrorism. But on the other hand, there is still a lot of limitations and disadvantages. For the future, research would need to use more diverse study samples, or its accuracy rates will drop for racial reasons. Also, the people´s right to privacy should be solved.

7. Acknowledgement

This work was supported by the Internal Grant Agency of Tomas Bata University in Zlin, the Department of Security Engineering, Faculty of Applied Informatics, under the project No. IGA/CebiaTech/2021/004.

8. References

[1] Gallagher, R.. & Jona, L. (2019). We Tested Europe's New Lie Detector for Travelers—And Immediately Triggered a False Positive. The Intercept. Available from: https://theintercept.com/2019/07/26/europe-border-control-ai-lie-detector/ Accessed: 2021/09/20

[2] Sánchez-Monedero, J. & Dencik, L. (2020): The politics of deceptive borders: 'biomarkers of deceit' and the case of iBorderCtrl, Information, Communication & Society, DOI 10.1080/1369118X.2020.1792530

[3] Holmes, M.; Latham, A.; Crockett, K. & O'Shea, J. D. (2017). "Near Real-Time Comprehension Classification with Artificial Neural Networks: Decoding e-Learner Non-Verbal Behavior," IEEE Transactions on Learning Technologies, Vol. 11, No. 1, pp. 5-12, 2018, DOI 10.1109/TLT.2017.2754497.

[4] Rothwell, J.; Bandar, Z.; O'Shea, J. & McLean, D. (2006). Silent talker: A new computer-based system for the analysis of facial cues to deception. Applied Cognitive Psychology. 20. 757 - 777. 10.1002/acp.1204.

[5] The iBorderCtrl (2020). Publications, Available from: Publications | iBorderCtrl Accessed: 2021/09/20

[6] The iBorderCtrl Consortium (2015). Intelligent Portable Control System, Available from: https://www.asktheeu.org/es/request/6087/response/19711/attach/4/8%20D3%202%20First%20version%20of%20tech%20tools%20and%20subsystems%20redacted.pdf?cookie_passthrough=1 Accessed: 2021-09-20

[7] The iBorderCtrl (2020). Outcomes, Available from: Publications | iBorderCtrl Accessed: 2021/09/20

[8] OrShea, J.; Crockett, K.; Khan, W.; Kindynis, P.; Antoniades, A. & Boultadakis, G. (2018). Intelligent Deception Detection through Machine Based Interviewing, Proceedings of International Joint Conference on Neural Networks, Rio de Janeiro, Brazil, ISSN 2161-4393, ISBN 978-1-5090-6014-6, pp 1-8, DOI 10.1109/IJCNN.2018.8489392

[9] The iBorderCtrl (2020). Technical Framework, Available from: Publications | iBorderCtrl Accessed: 2021/09/20

[10] Rodríguez Carlos-Roca, L.; Hupont Torres, I. & Fernandez, C.(2018). Facial recognition application for border control, Proceedings of International Joint Conference on Neural Networks, Rio de Janeiro, Brazil, ISSN 2161-4393, ISBN 978-1-5090-6014-6, pp 1-7, DOI 10.1109/IJCNN.2018.8489113

DOI: 10.2507/32nd.daaam.proceedings.057

ADDITIVE MANUFACTURING OF INDEXABLE INSERTS: NEW POSSIBILITIES FOR MORE EFFECTIVE COOLING

Jindrich Sykora & Lubos Kroft

This Publication has to be referred as: Sykora, J[indrich] & Kroft, L[ubos] (2021). Additive Manufacturing of Indexable Inserts: New Possibilities for more Effective Cooling, Proceedings of the 32nd DAAAM International Symposium, pp.0389-0394, B. Katalinic (Ed.), Published by DAAAM International, ISBN 978-3-902734-33-4, ISSN 1726-9679, Vienna, Austria
DOI: 10.2507/32nd.daaam.proceedings.057

Abstract

Modern additive manufacturing (AM) processes allow us to create parts of fairly high precision and various shapes with diverse internal features. These possibilities are, for example, often used when designing the conformally cooled parts of injection moulds. However, this research is focused on applying an additive approach to the design and manufacturing of a standard CCMT indexable insert to improve tool life when turning Inconel 718. Machining materials such as Inconel or titanium alloys requires intensive cooling of the cutting edge to maximize the tool life. The system proposed in this paper delivers a precisely directed coolant through internal channels in the indexable insert and its holder to the cutting edge from the rake face as well as the flank face of the insert. This has a significant positive influence on the life of the tool when machining Inconel 718 and other superalloys. This paper provides an extensive overview of the problem and then proposes three possible approaches to manufacture said indexable insert. They are additive manufacturing out of tungsten carbide, AM out of steel and brazing cutting edge out of tungsten carbide or CBN and AM out of tool steel and thermally spraying a wear resistant layer.

Keywords: indexable insert; cutting tool; additive manufacturing; Inconel 718

1. Introduction

Modern superalloys such as Inconel, Hastelloy or titanium alloys provide excellent properties for high demand applications in energy turbines and the aerospace industry. Using these advanced alloys allows for lighter and more efficient designs and, therefore, the environmental impact of these industries is decreased. Most of the parts for these applications have to be machined and the requirements for precision and surface quality are very strict. [1], [2] The innate properties of mentioned alloys like low heat conductivity or high strength at elevated temperature drastically decrease the tool life. One of the ways to increase wear resistance of a cutting tool is to increase the efficiency of cooling. This could, for example, be done by introducing high pressure coolant which provides better penetration of the coolant jet through the vapour barrier that is surrounding the area of the cut. [3] This also has a secondary positive effect for turning operations of breaking the chips into smaller and more manageable pieces.

Modern cutting tool holders are equipped with precision coolant nozzles that are aimed at the cutting zone and feature a reduction in diameter to maximize the velocity of the coolant jet while reducing the energy required in comparison with more usual flood coolant systems. [4], [5] Moreover, tool life when machining superalloys such as Inconel 718 is known to be improved by spraying coolant not only from the rake face but also from the flank face of the tool. This combination reduces the flank wear of the tool by 20% when an 8 MPa flank face coolant jet is added to the conventional flood cooling. [3]

Another option to combat the high heat in the cutting zone when machining superalloys is to use cutting materials that show a high abrasion resistance and strength at elevated temperatures such as ceramics and CBN. These materials can be generally used at higher cutting speeds and show a longer tool life, however, they are not well suitable for interrupted cut and their price is much higher in comparison to the more traditional tungsten carbide tools.

Additive manufacturing (AM) provides a solution for creating parts of complex shapes with intricate internal features. The DMLS method of AM can work with a large variety of materials, ranging from aluminium alloys to tool steel. [6] For example, it is nowadays used for conformal cooling of certain parts of commercial injection moulds. [7] A similar idea can be applied to regular machining tools such as drills, where the body of the drill is additively manufactured and equipped with channels for coolant that can have unconventional shapes to maximize the cooling effect and minimize the loss of rigidity. [8]

This paper proposes a combination of the methods mentioned above and their consolidation into a complex solution for increasing tool life and productivity when turning parts from high temperature superalloys and investigates three possible solutions. An increase in tool life reduces the number of insert changes in the manufacturing process and therefore minimizes setup uptime and the need for unnecessary manual labour, which is one of the attributes of the Industry 4.0 concept.

2. Proposed approach to additive manufacturing of indexable inserts

As clearly stated by the current research the coolant at the rake and flank face of the insert has a large effect on the wear rate of the cutting edge when turning Inconel 718. When the coolant exits a coolant channel the jet naturally expands and loses its kinetic energy. For this reason, its penetration into the cutting zone is decreased and thus the efficiency of the cooling. Therefore the distance between the coolant nozzle and the area of the tool that has to be cooled is critically important, the further away the nozzle is located the less efficient the coolant jet becomes. For this reason, the nozzle should be placed as close to the cutting edge as possible.

The novel idea is to use AM to create an indexable insert that has internal channels for coolant and nozzles at the rake and flak face extremely close to the cutting edge. This should not only improve the tool life but also reduce the need for high coolant pressure. Meaning that the solution could be successfully implemented even to older machines that don't have high pressure coolant option, moreover, additional costs and maintenance issues associated with high pressure coolant systems could be spared. Additive manufacturing provides a practical solution to this problem. As mentioned in the introduction, intricate channels can be created when additive manufacturing is used. The new freedom of design allows to create passageways for coolant of hydrodynamic shape and therefore minimize any energy loss when coolant medium flows through the channels. Directing the coolant nozzles at the cutting edge requires several bends in the trajectory of the channel. As shown in Figure 2, when manufacturing additively the bends can be a smooth transition and the energy loss should be minimal in comparison to the more conventional subtractive manufacturing method.

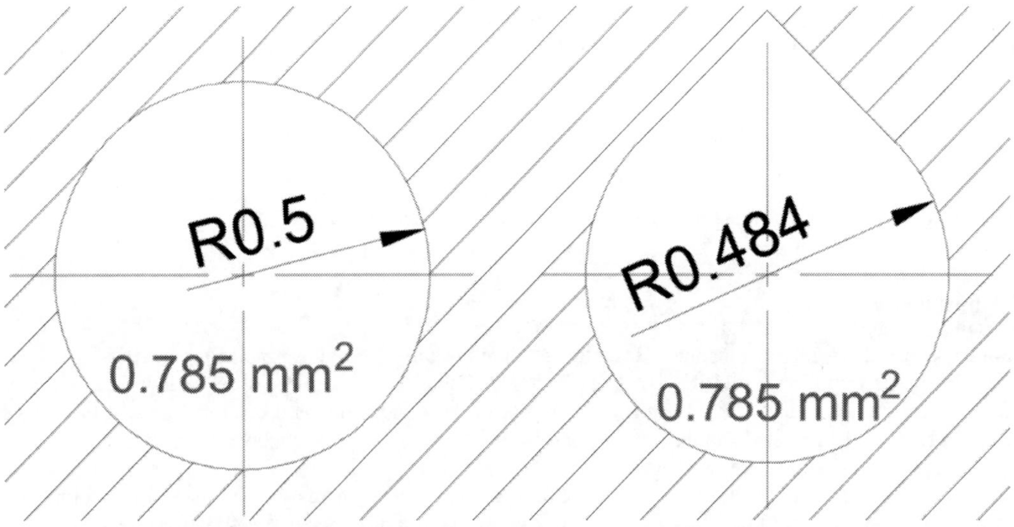

Fig. 1. Comparison of regular circular channel cross section and cross section optimized for AM

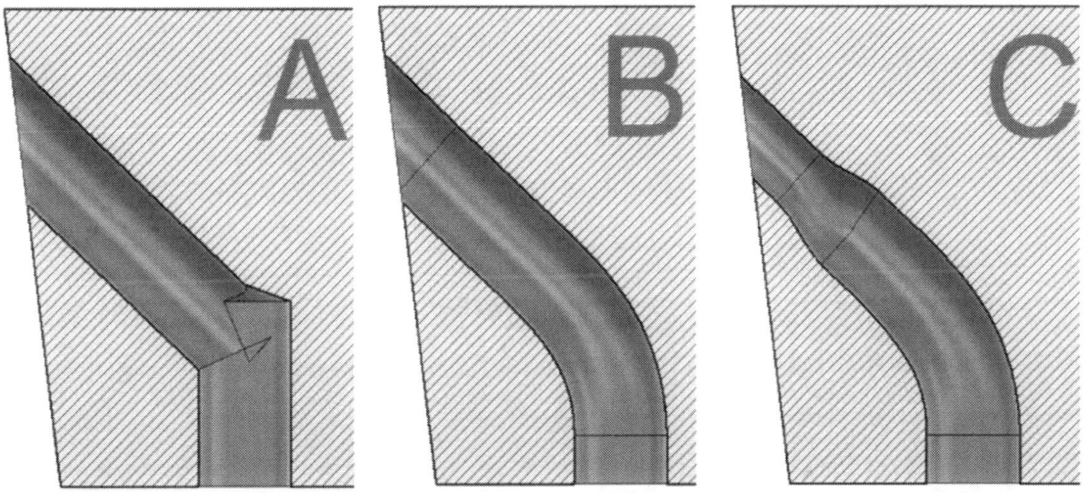

Fig. 2. Comparison of coolant channels (A) Conventionally manufactured; (B) Additively manufactured; (C) Optimized for AM

The design of the coolant channels has several basic boundary conditions. There has to be an inlet through which the coolant will flow into the insert, next there has to be a junction that will divide the single inlet channel into flank and rake channels. The combined surface area of the cross section of the channels has to remain the same or decrease in the direction of the coolant flow, this should ensure a favourable increase in velocity of the fluid. Lastly, exit nozzles should be aimed at the cutting as well as possible. Furthermore, when designing the internal channels, known restrictions of AM have to be taken into account. In this case, the minimal diameter of a hole has to be considered and the literature shows that the diameter should be greater than 0.7 mm when manufacturing out of tool steel. Although, this limitation depends on the process parameters and the orientation of the channel in space. Another aspect to consider is the surface roughness on the top side of a round hole when printed in the horizontal direction (see Fig. 3.). [9], [10] This would cause unwanted turbulence in the coolant channel, cause an energy loss and in fact decrease the throughput of the channel.

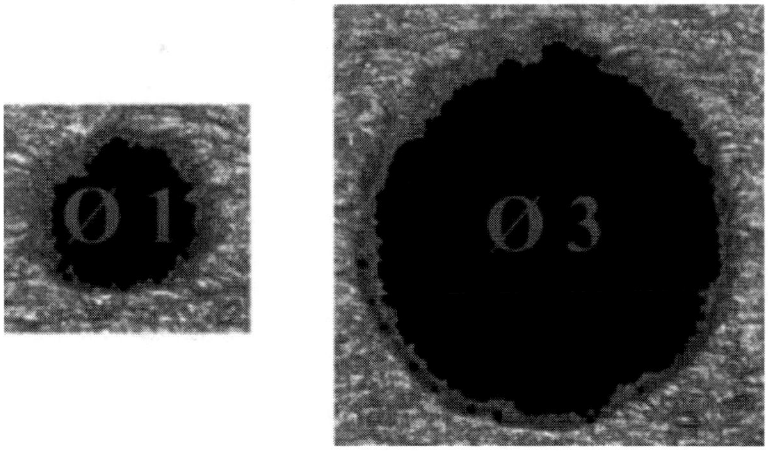

Fig. 3. Surface roughness of holes additively manufactured in the horizontal direction [10]

This can however be easily mitigated by using a different shape of the cross section of the channel than a circle. Shapes like triangles or drops that are closing at a steeper angle than a circle usually provide a significantly better surface inside the channel. The quality surface of the channels could be also significantly improved in the postprocessing stage by running an abrasive medium through them, but this step will increase the overall cost of manufacturing the insert and so this step would only be taken if necessary.

Finally, the literature shows that reduction of the cross section area of the nozzle is very favourable as it increases the velocity of the coolant and provides a precise laminar jet (see Fig. 4.), AM allows to easily create a nozzle with a seamless and gradual area reduction which reduces energy loss.

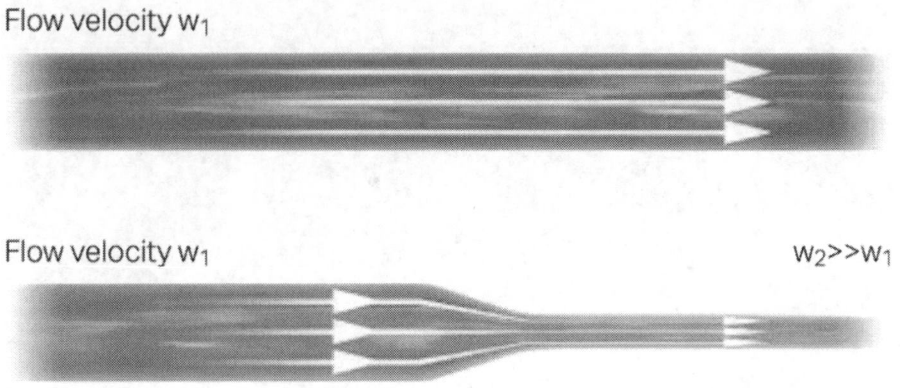

Fig. 4. The effect of nozzle diametre reduction on the jet velocity [4]

This is the general idea of the proposed solution for improving the process of turning components made out of Inconel 718, but there are several practical approaches one can take. Each concept will be discussed in detail in the following chapters.

2.1. Concept 1 – AM steel inserts with carbide or CBN cutting tip

The first option is to manufacture the inserts out of tool steel using a DMLS method. This process is well understood, reliable and creates parts of mechanical properties similar to the conventional material. There are several options to choose from and 1.2709 and 1.2379 tool steels seem to be the best options. The properties of these tool steels are suitable for form tooling and injection mould tooling, however, they are not sufficient for cutting tools. The process of machining generates heat and the tool has to be able to withstand it together with resisting the abrasion from various hard particles that are usually present in the workpiece. [6], [11] This is even more pronounced when machining Inconel 718 which contains various hard carbide particles and when machined generates a large amount of heat compared to, for example, free machining steel. For these reasons, the first concept relies on attaching a cutting tip to the insert made out of a different material. There will be a printed or machined pocket on the insert that accepts a cutting tip. This can be just a regular micro grain tungsten carbide that is suitable for machining Inconel 718. The best solution for securing the tip in place is probably silver brazing, this process is well understood, creates a very strong bond and can be easily automated. As a final operation, the insert should be equipped with a coating appropriate for this application. For example, carbide grade IC806 from Iscar with PVD TiAlN coating would be a good choice for this application. [12] When high material removal rate is required the cutting tip can be made out of CBN, which can tolerate much higher temperature and thus cutting speed, but the price of this tip will also be significantly higher. [13]

2.2. Concept 2 – AM steel insert with Stellite 6 coating

The second concept is the same in its basic structure, the body of the insert would be made out of tool steel and equipped with coolant channels. However, there will be no pocket for cutting tip present, but the outside faces of the insert will be offset by a negative 0.2 mm. The next step after AM and postprocessing will be thermal spraying of a layer of Stellite 6. This approach is usually used to create a wear resistant bearing layer on large shafts for steam turbines, nonetheless, it had been demonstrated that this material can be used in machining applications as well. [14]

A layer of approximately 0.4 - 0.5 mm of Stellite 6 would be applied on the outside faces of the 3D printed blank. The coolant channels will be protected with a non-stick compound. In the next step, the excess 0.2 – 0.3 mm would be removed by precision grinding of the insert to its final shape and dimension. The final result should be an indexable insert with an extremely hard and wear resistant case and a comparatively very tough core. This combination should make the insert very impact resistant and therefore ideal for interrupted cuts. This might be a great choice for roughing castings and forgings.

2.3. Concept 3 – AM Carbide insert

The last concept that will be discussed in this paper is additively manufacturing the indexable insert directly out of suitable cutting material. This should significantly simplify the whole process of manufacturing the insert and cut the manufacturing time. Tungsten carbide currently seems to be the best option for this application, since there already are many manufactures of suitable WC-Co raw powder and there was some research done in the field. Therefore, there are basic process parameters for AM known, however, they need to be further developed. The main problems with additively manufactured WC-Co are porosity, cracking and formation of foreign brittle phases (e.g. Eta-phase).

However, the research shows that these problems can be to a certain degree mitigated by selecting suitable process parameters for given raw powder composition. For example, increasing the build platform temperature minimizes the amount of cracking in the samples. Furthermore, additional processes like hot isostatic pressing can be used to further improve the properties of AM WC-Co parts. No sufficient research focused on manufacturing internal channels and cavities is present and this will have to be further investigated to find a suitable channel design for optimal coolant flow. [15], [16]

3. Conclusion

Machining heat resistant super alloys like Inconel 718 is certainly challenging for several reasons, and short tool life is one of the most significant ones. The combination of findings gained when machining these materials and cutting edge additive manufacturing approaches and strategies enable the development of innovative cutting tools that meaningfully improve the machining process. When tool life is improved, valuable time can be saved eider by increasing the cutting speed or by minimizing the time lost when a tool needs to be changed. Furthermore, an important benefit of this system is the improvement of the cooling process without the need for high pressure coolant. Accordingly, the energy expenditure and coolant loss due to evaporation is reduced, which has a positive effect on the environmental aspects of the machining process.

This work presents three possible approaches to solving this problem and discusses the main considerations for each of them. This provides the necessary insight into the matter and enables one to create a practical solution. Therefore, the next step in this research will be to choose one of the concepts and create a physical product. Make a design suitable for AM, manufacture all of the necessary components and at last test the new cutting tool and measure the benefit of the proposed innovative cooling system in comparison with the industry standard solutions.

To conclude, the fast development of additive manufacturing provides new opportunities for significant innovation in many areas and cutting tools for machining heat resistant super alloys can benefit. The presented approach to manufacturing can introduce a new generation of more productive tools that will widen one of the bottlenecks of the aerospace and energy industry.

4. Acknowledgements

This paper was created under the project SGS-2019-008: Research and Development for Innovation in the Field of Manufacturing Technology - Machining Technology III.

5. References

[1] Farinia Group (2014). 'Why Choosing Inconel 718 for Aerospace Additive Manufacturing?', Farinia Group, [Online]. Available: https://www.farinia.com/blog/why-choosing-inconel-718-aerospace-additive-manufacturing. [Accessed: 17-Sep-2021]

[2] R. Schafrik, D. D. Ward, and J. Groh (2001). 'Application of Alloy 718 in GE Aircraft Engines: Past, Present and Next Five Years', doi: 10.7449/2001/SUPERALLOYS_2001_1_11.

[3] N. Tamil Alagan, P. Hoier, P. Zeman, U. Klement, T. Beno, and A. Wretland (2019). 'Effects of high-pressure cooling in the flank and rake faces of WC tool on the tool wear mechanism and process conditions in turning of alloy 718', Wear, vol. 434–435, p. 102922, doi: 10.1016/j.wear.2019.05.037.

[4] Sandvik Coromant, 'Coolant aspects', Sandvik Coromant. [Online]. Available: https://www.sandvik.coromant.com/en-us/knowledge/machine-tooling-solutions/tooling-considerations/pages/coolant.aspx. [Accessed: 17-Sep-2021]

[5] Sandvik Coromant, 'Innovations have changed the use of coolants', Sandvik Coromant. [Online]. Available: https://www.sandvik.coromant.com/hk/news/press_releases/pages/innovations-have-changed-the-use-of-coolants-en.aspx. [Accessed: 17-Sep-2021]

[6] EOS GmbH, 'EOS Materials Metal Portfolio Overview'. [Online]. Available: https://www.eos.info/03_system-related-assets/material-related-contents/material_pdf/eos_materials_overview_metal_en.pdf. [Accessed: 17-Sep-2021]

[7] J. Navratil, M. Stanek, M. Manas, D. Manas, M. Bednarik, and A. Mizera (2011). 'Utilization of DMLS in Injection Mold Design', presented at the 26th DAAAM International Symposium On Intelligent Manufacturing And Automation, pp. 1507–1508, doi: 10.2507/22nd.daaam.proceedings.738.

[8] S. HENDRIXSON, 'Additive Manufacturing Makes Subtractive Cutting Tools', additivemanufacturing.media. [Online]. Available: https://www.additivemanufacturing.media/articles/additive-manufacturing-makes-subtractive-cutting-tools. [Accessed: 02-Jun-2020]

[9] Redwood, Schoffer, and Garret (2018). The 3D printing handbook. 3D hubs B.V.

[10] M. Dana, I. Zetková, P. Hanzl, and O. Hronek (2017). 'Accuracy of Holes Created by 3D Printing (DMLS)', presented at the 28TH DAAAM International Symposium On Intelligent Manufacturing And Automation, pp. 0467–0473, doi: 10.2507/28th.daaam.proceedings.065.

[11] L. Kučerová and I. Zetková (2016). 'Metallography of 3D Printed 1.2709 Tool Steel', Manufacturing Technology, vol. 16, pp. 140–144, doi: 10.21062/ujep/x.2016/a/1213-2489/MT/16/1/140.

[12] ISCAR ČR s.r.o., 'IC806 SUMO TEC PVD for Inconel 718', www.iscar.com. [Online]. Available: https://www.iscar.com/Products.aspx/countryid/6/ProductId/5392. [Accessed: 07-Jun-2020]

[13] J. Zhou, V. Bushlya, P. Avdovic, and J.-E. Ståhl (2011). 'Study of surface quality in high speed turning of Inconel 718 with uncoated and coated CBN tools', International Journal of Advanced Manufacturing Technology, vol. 58, pp. 141–151, doi: 10.1007/s00170-011-3374-7.

[14] K. Traxel and A. Bandyopadhyay (2018). 'First Demonstration of Additive Manufacturing of Cutting Tools using Directed Energy Deposition System: StelliteTM-Based Cutting Tools', Additive Manufacturing, vol. 25, doi: 10.1016/j.addma.2018.11.019.

[15] D. Bricín and A. Kriz (2019). 'Processability of wc-co powder mixtures using slm additive technology', MM Science Journal, vol. 2019, pp. 2939–2945, doi: 10.17973/MMSJ.2019_06_2018115.

[16] T. Josef, S. Tim, B. Timo, M. Markus, S. Gerhard, and S. Dirk (2018). 'Laser sintering of tungsten carbide cutter shafts with integrated cooling channels', presented at the Pro-AM 2018, doi: 10.25341/D48G6C [Online]. Available: https://dr.ntu.edu.sg//handle/10356/88554. [Accessed: 09-Nov-2020]

DOI: 10.2507/32nd.daaam.proceedings.058

STUDY OF THE INFLUENCE OF CUTTING TOOL GEOMETRY DURING TURNING OF SUBMILLIMETRE NICKEL PARTS

Jindrich Sykora, Michal Povolny, Tomas Zatloukal,
Lubos Kroft & Simon Syrovatka

This Publication has to be referred as: Sykora, J[indrich]; Povolny, M[ichal]; Zatloukal, T[omas]; Kroft, L[ubos] & Syrovatka, S[imon] (2021). Study of The Influence of Cutting Tool Geometry During Turning of Submillimetre Nickel Parts, Proceedings of the 32nd DAAAM International Symposium, pp.0395-0401, B. Katalinic (Ed.), Published by DAAAM International, ISBN 978-3-902734-33-4, ISSN 1726-9679, Vienna, Austria
DOI: 10.2507/32nd.daaam.proceedings.058

Abstract

The subject of this experiment was to produce parts with a specific shape and submillimetre elements. The parts represented components of commercial sparkplugs and the machined section was made of almost pure nickel. Relatively narrow required tolerance of parts (hundreds of micrometres), in combination with a large addition of material to the semi-finished product (several millimetres) and the required number of parts manufactured by one cutting insert for a very small time cycle led to the implementation of several tests to find ideal cutting inserts and conditions for production. All of the machining was done under dry conditions and the lowest possible machining time was also taken into account. Seven inserts with potentially suitable cutting-edge geometries were selected for testing. They were positive ISO C and ISO D inserts. After initial testing and determining the most suitable conditions for the most inserts, cutting forces during machining under constant conditions were measured. After the machining, each part was closely inspected by the microscope and all the critical dimensions, general shape and state of the surface were evaluated. In the course of the experiment, a significant effect of the geometry of the cutting insert was demonstrated. The best performing of used cutting inserts for these conditions was ISO D shape with a nose radius of 0.2 mm and a highly positive rake angle. In terms of the thin layer used, a TiAlN PVD coated insert seems to be the most ideal as it should significantly increase tool life. To verify this assumption extensive long-term testing would have to be done. These results confirm the importance of selecting appropriately positive geometry for turning small features in soft alloys along with the influence of nose radius with regard to depth of cut.

Keywords: Cutting forces; nickel; turning; cutting tool geometry

1. Introduction

The automotive industry is one of the main drivers of the economy today. The production of parts for this industry is now widespread and the automotive industry extends into a large range of other sectors. Here too, there is a considerable amount of innovation and new, unconventional solutions. The machining of such components must also be adapted to these new developments, both in terms of the machine and the tooling and machining strategy.

The combination of non-traditional machining materials and workpiece shapes is now quite common. The problem arises when the component is really small in diameter but large in overall length. In engineering practice, we call such parts with a large slenderness ratio and machining them is always problematic. These are usually turned parts produced in large quantities. If this part needs to be machined in a material with high plasticity in a very short time with several other constraints, the problem becomes very complicated.

The machining of slender components brings with it many problematic factors to consider when designing a tool and machining strategy. Components with high slenderness ratios are prone to vibration, residual stress deformation or thermal deformation. All these phenomena are usually associated with dimensional instability or damage to the workpiece or tool. Another problem can be deformation caused by improper tool geometry. The choice of geometry of a given cutting tool has a dominant influence on the overall result and should be given considerable importance. [1], [2]

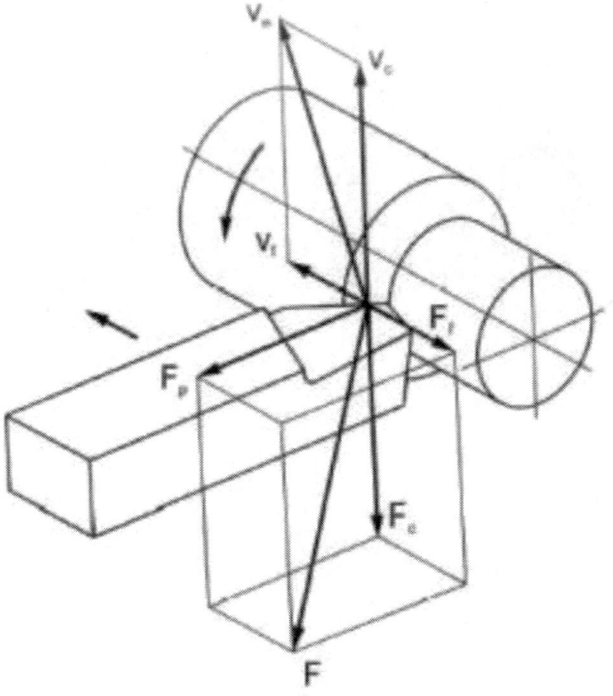

Fig. 1. Scheme of distribution of cutting forces during turning [6]

The geometry of the cutting edge, together with the cutting conditions and the material being machined, or its mechanical properties, gives us the initial force load. The total cutting force F required to cut the material must be greater than the resistance R given by the material properties and is influenced by a number of factors. However, the total cutting force is not very revealing for a deeper investigation of the machining process, so in most cases, it is necessary to measure the individual components as shown in Figure 1. These cutting force components are oriented according to an orthogonal coordinate system in a plane labelled by face, back and shear.

The total force can be calculated based on relation 1.

$$F = \sqrt{F_c^{\,2} + F_p^{\,2} + F_f^{\,2}}\ [N] \quad (1)$$

Each of these ingredients has a slightly different effect. From the point of view of part precision, the Fp component is essential - the feed force or radial force. Its value significantly influences the part's push-off from the cutting edge and thus especially the geometric accuracy of the part. Its value can be influenced by the appropriate choice of tool geometry, in particular the angle of the face and back and the angle of the main blade setting.

All the influences are described in Table 1 and illustrated in Figure 2. During the experiments described below, all components of the cutting force were measured and subsequently evaluated with respect to the qualitative results obtained. [3], [4], [5]

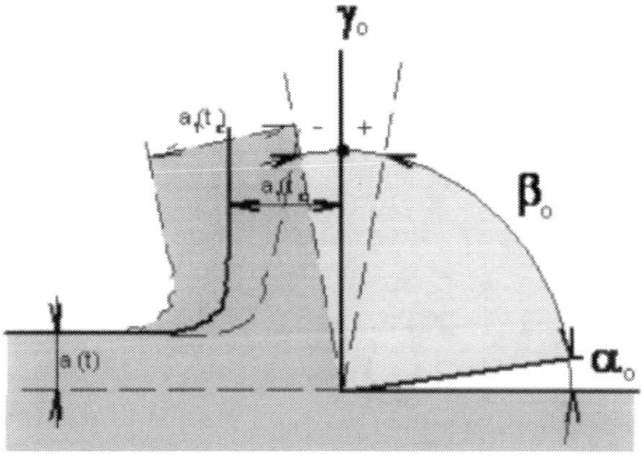

Fig. 2. Scheme of cutting angles in the cutting process [6]

Individual cutting angles	Influence on the machining process
α_0	effect on the strength of the cutting edge
	effect on the friction between the back and the workpiece
β_0	effect on the strength of the cutting edge
	effect on thermal conductive properties
γ_0	effect on plastic deformation in the area of chip formation
	effect on friction between the chip and the cutting edge face
	effect on the strength of the cutting edge

Table 1. Table the effect of cutting angles on the cutting process [6]

1.1. Large-scale production of parts

As already mentioned, the entire research was focused on turning very small parts for automotive applications. This meant in this case that the production of a large number of parts was expected. For this type of production, machines referred to as long-running machines (swiss type) are normally used. This type of machine is very suitable for the production of thin parts, as the part is continuously ejected from the spindle, thus eliminating most of the problematic phenomena occurring when machining parts with a large slenderness ratio. Unfortunately, this type of machine cannot be used in this experiment because the parts to be machined based on the results below are already in piece form after previous manufacturing operations and do not have a full-length rotational shape. [7]

Therefore, for the mass production of this type of workpiece, a machine with a vertical spindle was designed so that the part faces downwards after automatic foundation. Here, the parts are inserted and clamped by air and the whole machining cycle takes 8 seconds. This very short cycle time results in a clamping time of approximately 3 seconds after deducting the clamping time in which the part must be machined. With this in mind, an experiment was designed so that the entire part is machined to completion with only one pass of the tool.

The last limiting factor is that the parts will be machined without process fluid. The only possible cooling method is pressurized air. Due to the material to be machined, this is a rather complex issue, the solution to which is outlined in the next section of this article.

2. Materials and methods

2.1. Workpiece and setup

The workpiece used for this experiment was the centre electrode of a combustion engine spark plug. The goal is to create a shape on the end of the electrode that can be seen on Fig. 1 in less than 3 seconds under dry conditions as this operation will be a part of high volume automated production. The exact dimensional tolerances can not be revealed, but they are in the 0.05 – 0.1 mm range. This part has several specific characteristics that can make machining a complex problem. Firstly, the mechanical properties of the material are very close to a pure nickel (the exact alloy is classified), which means a high ductility, low tensile strength, and austenitic structure. Secondly, the machined features should not have any burrs, which is always a challenge when working with austenitic materials.

Lastly, due to the shape of the blank part, the overhang from the work holding must be at least 6 mm, this combined with the diameter of the part means a 3 times diameter extension. Therefore, the lever is significant, and the cutting force and their distribution will have a large influence on the final shape of the part.

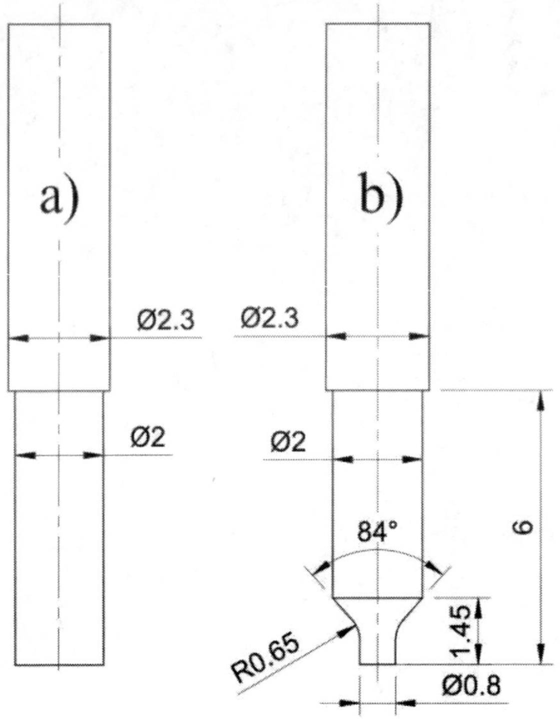

Fig. 3. Drawing of the blank workpiece (a) and the final part (b).

The parts were machined on an Emco MAXXTURN 25 lathe and held in an appropriate collet. Both of the tool holders SDJCL 1212F-11 and SCLCL 1212F-09 were held in a Kistler 9119AA2 dynamometer to measure the cutting forces. Suitable cutting conditions were determined as a result of pre-experiment, the feed rate per revolution was set to 0.07 mm/rev, the speed to 5000 rpm and the maximum depth of cut was 0.6 mm. After machining the parts were inspected on the Keyence VHX-6000 microscope. Before starting a cutting force measurement for each insert a test part was machined, measured and the inaccuracy was compensated in order to keep cutting conditions the same for each insert.

2.2. Cutting tools

The performance of inserts ISO C held in SCLCL 1212F-09 and ISO D held SDJCL 1212F-11 holder was tested. Inserts of varying geometries and coatings were sourced from one of the world's largest manufactures of cutting tools. The geometry of the cutting edge greatly influences the shearing action when material is removed and therefore, the resulting cutting forces and deformation of the workpiece. For this reason, the microgeometry of each insert was measured on Alicona ifm g4 (see Table 2).

Insert	Grade	Edge radius [μm]	Rake angle [°]	Clearance angle [°]	Nose radius [mm]
DCGT - AS	IC 20	4	16	7	0.2
DCMT - PF	IC 807	43	5	7	0.2
DCGT – AS	IC 907	6	20	7	0.2
DCGT – AF	IC 20	5	13	7	0.4
CCGT - AS	IC 320	8	18	7	0.2
CCGT - AS	IC 520	11	20	7	0.2
CCMT - F3M	IC 807	39	12	7	0.2

Table 2. The microgeometry of each tested insert was measured on Alicona ifm g4.

3. Experiments

In initial tests, it was found that all ISO C inserts tend to undercut the initial few tenths of a millimetre of the cylindrical part, creating a significant radius at the edge. The part with the undesirable geometry is shown in figure 4. ISO D-shaped inserts with a significantly positive face geometry (AS designation) and a tip radius of 0.2 mm proved to be more suitable. These VBDs were subsequently tested at several variations of cutting speeds and feed rates. This comparison was made with the VBD - DCGT 11T302-AS IC907, which appears to be the most suitable for production. The IC907 grade is coated with a thin layer of TiAlN + TiN, which should provide longer cutting edge life.

Fig. 4. Part machined with ISO C insert has unwanted curvature on the face

3.1. Measurement of cutting forces

The next step of the experiment was the measurement of cutting forces and their subsequent evaluation. Four measurements were recorded for all inserts, from which the resulting average and maximum machining forces were evaluated in the form of graphs (see figures below). The evaluation was performed based on the maximum cutting forces for Fp, Fc and Ff. The first graph describes the maximum cutting force components for each insert. The following graphs then show the cutting force components as a function of the edge radius on the insert. [8]

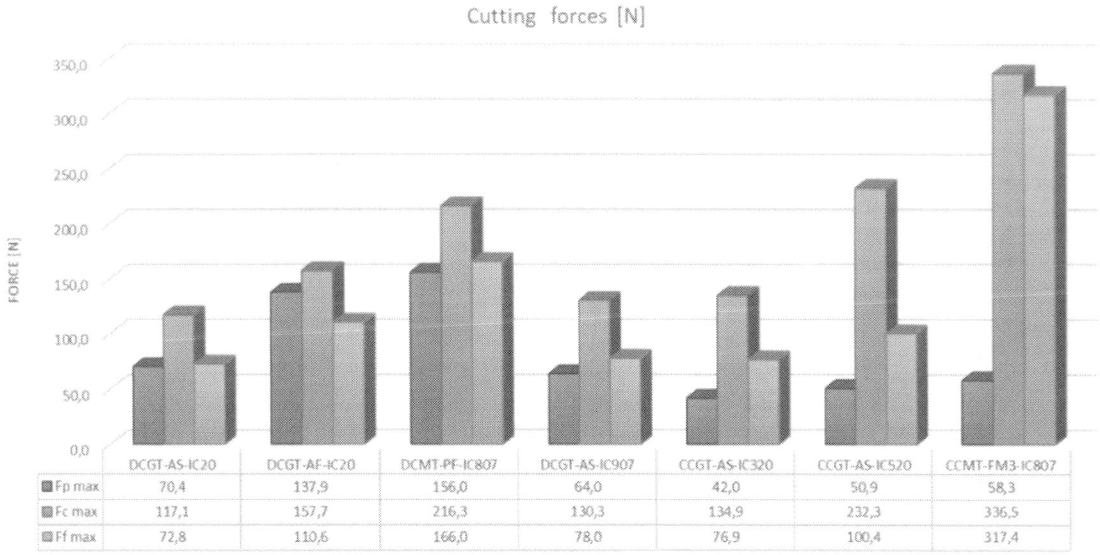

Fig. 5. Maximum values of measured forces

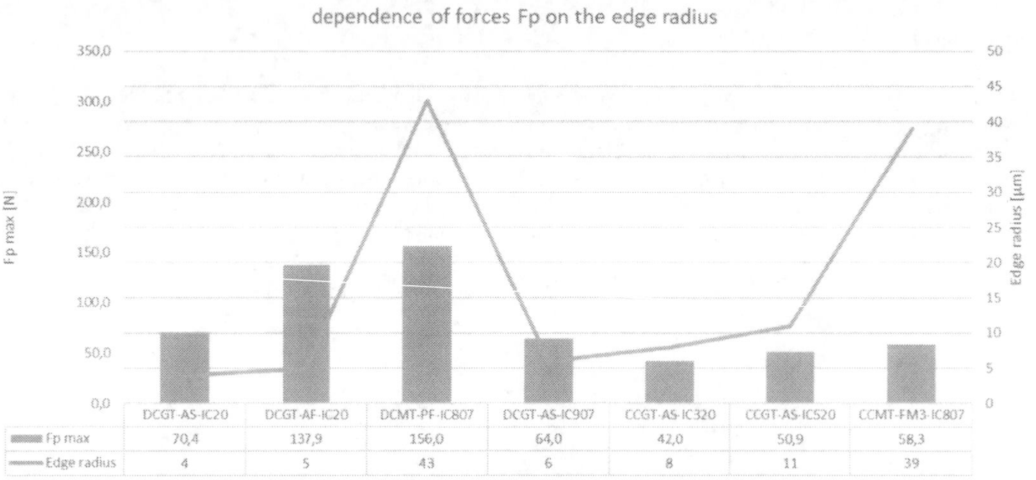

Fig. 6. Influence of cutting edge radius on Fp force

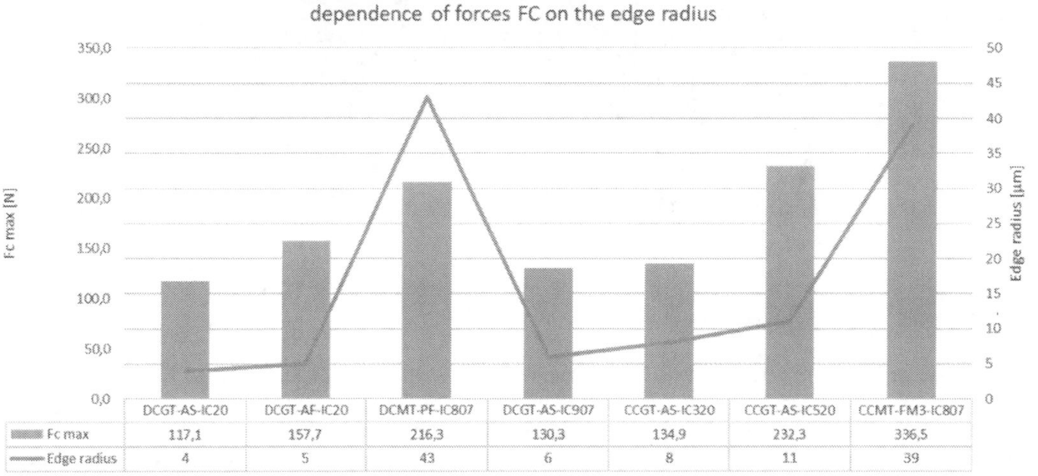

Fig. 7. Influence of cutting edge radius on Fc force

It is clear from the graphs that there is an increase in cutting forces when using a larger nose radius for the same DCGT-AS-IC20 (nose radius 0.2 mm) vs DCGT-AF-IC20 (nose radius 0.4 mm) inserts. The reason being that the DCGT-AF-IC20 has a larger nose radius which effectively changes the side cutting edge angle when using small depths of cut. The large increase in forces is then evident when using inserts with a large edge radius, especially the DCMT-PF-IC807 and CCMT-FM3-IC807 types. This large difference in cutting forces only confirms the original finding that inserts with a larger nose radius but also with a larger edge radius tend to undercut. This results in large cutting forces and undesirable rounding at the face of the manufactured part. The graphs continue to show that the DCGT-AS-IC20 and DCGT-AS-IC320 with a nose radius of R 0.2 and an edge radius of 4 and 6 µm achieve the lowest cutting forces. This only confirms the results obtained during the measurement and inspection of the manufactured parts. When parts produced with these inserts achieved the most accurate dimensions and the best surface roughness parameters.

4. Conclusion

Research on machining strategy for machining the tip of a combustion spark plug was presented in this paper. This is a section with a large slenderness ratio produced in large batches from nearly pure nickel. This strategy will be implemented on a special single-purpose machine with no process fluid supply, with a single tool pass in a very short time of about 3 seconds. The problem was addressed mainly in terms of tooling and also the choice of suitable cutting conditions. Several conclusions can then be drawn from the experiments concerning the choice of tool geometry and the choice of cutting conditions. The choice of the edge radius is of fundamental importance. The larger the radius, the greater the cutting force and thus greater workpiece deformation and consequently increase in geometric inaccuracies.

Furthermore, the size of the edge radius is very crucial from the micro-geometry point of view. The smaller the value of the edge radius the better the resulting surface roughness. The results of the experiments give us the most suitable combination of cutting tool, cutting conditions used and machining strategy is chosen. These results will then have to be verified during real production and other important parameters such as tool life will have to be measured. The result of the whole development is of considerable benefit not only to the customer but also to the research department and the results will be used as starting point in further research activities.

5. Acknowledgements

This article was created under the project SGS-2019-008: Research and Development for Innovation in the Field of Manufacturing Technology - Machining Technology III.

6. References

[1] Kaufnerova, A. & Kutlwaser, J. (2019). The Course 'Practice of Experimental Methods in Machining', Annals of DAAAM for 2019 & Proceedings of the 30th International DAAAM Symposium, Published by DAAAM International, Vienna, 2019, ISBN 978-3-902734-22-8

[2] Hronek, O. & Zetek, M. (2017). The Influences of Cutting Edge Radius on Surface Roughness when Milling Nickel Alloy, Annals of DAAAM for 2017 & Proceedings of the 28th International DAAAM Symposium, Published by DAAAM International, Vienna, 2017, ISBN 978-3-902734-11-2

[3] https://www.sandvik.coromant.com/cs-cz/knowledge/materials/pages/workpiece-materials.aspx, (2018). Machined materials, Accessed: 2021-10-10

[4] Cutting Conditions. (2017). https://eluc.kr-olomoucky.cz/verejne/lekce/1202, Accessed: 2021-10-10

[5] Brychta, J.; Čep, R.; Nováková, J. & Petřkovská, L. (2007). Technology II - Part 1, Published by Ostrava VŠB-TU, 2007, ISBN 978-80-248-1641-8.

[6] VYŠATA, Jiří. Materials for the subject "engineering technology of machining" [online]. Pilsen, 2015 [cit. 2021-10-17]. University of West Bohemia.

[7] Elkins D. A., Huang N., and Alden J. M. (2004). 'Agile manufacturing systems in the automotive industry', International Journal of Production Economics, vol. 91, no. 3, pp. 201–214, doi: 10.1016/j.ijpe.2003.07.006.

[8] Hnatik, J; Kutlwaser, J & Sklenicka, J (2016). iMachining Cutting Force Measurement, Proceedings of the 26th DAAAM International Symposium, pp.0955-0959, B. Katalinic (Ed.),Published by DAAAM International, ISBN 978-3-902734-07-5, ISSN 1726-9679, Vienna, Austria DOI: 10.2507/26th.daaam.proceedings.134

DOI: 10.2507/32nd.daaam.proceedings.059

THE NEW ROLE OF TEACHERS AFTER THE COVID19 PANDEMIC AND THE APPLICATION OF THE LEAN CONCEPT IN EDUCATION

Jelena Milicevic, Nemanja Sremcev, Ilija Cosic, Milovan Lazarevic & Nemanja Tasic

This Publication has to be referred as: Milicevic, J[elena]; Sremcev, N[emanja]; Cosic, I[lija]; Lazarevic, M[ilovan] & Tasic, N[emanja] (2021). The New Role of Teachers After the Covid19 Pandemic and the Application of the Lean Concept in Education, Proceedings of the 32nd DAAAM International Symposium, pp.0402-0407, B. Katalinic (Ed.), Published by DAAAM International, ISBN 978-3-902734-33-4, ISSN 1726-9679, Vienna, Austria
DOI: 10.2507/32nd.daaam.proceedings.059

Abstract

In today's increasingly turbulent and unpredictable environment, it has been shown that at the very top of priorities, in terms of technological, social and economic development, as well as in the national global strategy, the state of the production and service sector, human resource structure and education of the country are at the very top. Thanks to the advancement of engineering and technology and the advantages offered by the Internet, it has been much easier to work in many professions, especially for professors during the Covid19 pandemic. Professors are also able to improve their knowledge, better design their lectures and compare their methodology with the work of their colleagues in any part of the world. By using technology, work operations are shortened, complex organizations are simplified, and time savings are certainly evident. One of the important, positive, impacts of the Internet is precisely that, reducing costs, such as the cost of telephone, fax, and photocopy through deliberate use of the Internet. The same can be applied to higher education, such as in terms of the time of preparation of lectures and "dissemination" of educational material, printing content and similar things. In this sense, we can notice a new role of professors in the new work environment as well as the importance, which has recently focused on two "phenomena", namely "smart" or "green" university, and which will be presented in the paper through a lean concept that through its philosophy can improve the university so that it becomes at least one step closer to the goal that its business and regulated processes can bear that exact name.

Keywords: Education technologies; Covid19 pandemic; costs reduction; "smart" and "green" university; lean concept

1. Introduction

1.1. Needed

One of the great advantages of the Internet is its "connection" with the whole world, so professors, students, as well as the administrative staff of schools, colleges, can constantly monitor new developments in the field of education. Through constant contacts with their colleagues around the world, professors get to know new teaching techniques, learn news from the field in which they work, i.e. which they teach and, in that way, provide their students with the latest information, which is certainly a big plus in education. If there is one domain of activity that benefit, especially from the Internet, it is the domain of research and development of knowledge, because the Internet was born in scientific environments in order to stimulate cooperation in that domain [1], [2]. The use of the Internet makes it easier for professors around the world to compare their work methodology with the work of their foreign colleagues, to exchange personal experiences and thus improve their work. The innovative process is a basic research and development activity. There are no secrets to the way successful distance education programs are made. They are not created spontaneously; they are created by hard work and conscious effort of many individuals and organizations [2].

1.2. Issue

A large number of teachers still resisted the use of an online pedagogical approach and believed that online learning could decrease the engagement of students, thus leading to less favourable academic success than traditional face-to-face instruction [3]. Very few online teaching and learning activities could follow a rigid schedule and design [3]. Although online or distance education is not something completely new, both in the world and in our country, several challenges arose during the Covid19 pandemic regarding this type of education, such as: a) Lack of preparation time: Teachers have not prepared their learning content to adapt to online learning, and preparing such content will take time. Teacher/student isolation: In this first-ever application of pure long-term online learning (without face-to-face learning or blended learning), both teachers and students should not feel that they are left alone during the teaching and learning processes. b) Need for effective pedagogical approaches: Need for effective pedagogical approaches: New effective pedagogical approaches are needed to keep students motivated and engaged during this long period of online learning, especially that drop-out rates of distance learning is generally higher than in-campus based learning [4]. Distance learning professors face special challenges such as that the teacher must understand the characteristics and needs of distant learning students with little first-hand experience and limited, or no, face-to-face contact, and adjust the teaching style where necessary. The needs and expectations of multiple, often different, audiences are also to be taken into account. A prerequisite is to develop a working understanding of the technology he/she needs, but to stay focused on his/her teaching role which is a priority and to function successfully as a skilled assistant and lecturer.

1.3. Distance learning and the new role of the professor

At the suggestion of the European Union, regarding the new process of education, and how it could stay in the race with the revolution in information and communication technologies and be able to follow the progress of technology in the near future, is the reform of educational systems. In order for teachers to become active actors using the full potential of information and communication technology, these new education systems force us to redefine the role of teachers and it is necessary to understand that the two teaching methodologies (face to face and online) are not in conflict but are, in fact, complementary. All of this conditions a completely new role of teachers in the 21st century, especially in the time after the pandemic, which has led to a simultaneous reduction in costs in terms of time, but also the cost of tuition, materials used, etc. Given that we live in a time that is rich in information and resources for acquiring knowledge, it is important to develop the ability to acquire knowledge faster and better [5]. In accordance with that, it is necessary to change the contents of education, acquaint teachers with efficient methods of work and shaping students' personalities, develop and expand systems for measuring and efficiency of teaching, adequately reward teachers, single out and encourage those who achieve the best results.

The new role of the teacher, in line with online teaching, allows the teacher a diverse and perhaps inspiring role, however, it is difficult to maintain a stimulating discussion between teacher and student, when spontaneity is conditioned by technical requirements and distance. For that reason, in recent times, the teacher is both a lecturer and a designer, producer, media specialist, provider. In order to achieve the best possible quality, it is necessary to use modern educational technology. The integrative role of teachers is reflected in the ability to connect the school environment, curriculum, student behaviour and their way of working into a single action unit [5].

The entire life and work in educational institutions depends a lot on the personality and activities of teachers: the organization and performance of educational work, the quality of acquired facts as well as their functional connection with objective reality. Teachers should have an active role in building a trustworthy learning environment by continuously encouraging their students to share their opinions and answers [4]. Successful realization of these roles requires that the teacher should be oriented to the new pedagogical laws of educational activities based on scientific principles, and that the role of teachers in modern teaching is contained in: inclusion in the informatization of education; the transition of the work of teachers from the stage of realization to the stage of preparation of teaching; increased engagement in planning and realization of goals and tasks of upbringing and education; optimal use of teaching aids, teaching films, computers; greater focus on student motivation; the need to make more use of educational technology in order to better perform teaching, especially in the online environment.

Given that modern man, ie the student of the 21st century, is flooded with a variety of information, we can distinguish short-term knowledge that students forget immediately after the course and which do not really serve anything. In this sense, such knowledge and information should be replaced by activities that both teachers and students need to work on and from which both can benefit. Therefore, everything that students provide as learning material as a result of their work (eg papers, presentations, videos), under the supervision of teachers, can be collected and transmitted online so that other students and teachers (future generations) can always have as an example of how you should (or should not) work and certainly benefit from them. [6].

2. Research review

2.1 Methods

In researching the impact of the impact of Digitization of education in the Republic of Serbia and the world as well as the problems school systems face worldwide, this paper used information from various websites and institutions, as well as articles and journals. Certainly research through Google academic helped to get information about how are Since 2012, various activities have been undertaken in the Republic of Serbia to improve the level of application of information and communication technologies (ICT) in teaching, which was especially influenced by the adoption of the Guidelines for Improving the Role of Information and Communication Technologies in 2013, as well as the Strategy of the development of education in Serbia until 2020. In accordance with this Strategy and guidelines, the "Digital Competence Framework - Teacher for the Digital Age" was adopted. However, specific activities of the competent ministry were undertaken only in the 2018/2019 school year, when a pilot project "2000 digital classrooms" was organized in primary schools, as part of which primary school teachers went through a two-day training for the use of electronic textbooks and digital educational content [7]. UNESCO (2020) pointed out that more than 1.5 billion students worldwide could not attend school or university due to the outbreak of Covid19 on April 4, 2020 [4]. In this context, the Chinese Ministry of Education (Moe) and university experts began to discuss the use of information and communication technologies (ICT) to reform the entire education system in the midst of this pandemic and to enable online and distance learning, even during sporadic classes [4]. The Covid19 virus pandemic has forced a rapid transition to online teaching, regardless of the readiness of students and teachers, and the digital technology available in schools. Therefore, the readiness of teachers and students for online teaching in the Republic of Serbia is an important problem, because it is necessary to identify obstacles to the effective implementation of online teaching and design training programs to overcome them. In addition, the benefits and advantages of online teaching can be permanently integrated into traditional teaching after the end of the Covid19 virus pandemic.

2.2 Research

The research on the attitudes of high school students about online education in Serbia arose from the need to look at the technical readiness of both students and teachers from the point of view of students and their expectations, and to raise readiness in that sense to a higher level so that they can learn and progress [7]. The research instrument is a questionnaire that the respondents filled out, which contains questions related to satisfaction related to the online teaching conducted in the high schools they attend. The students answered the closed-ended questions by choosing the answer that best suited their attitude. The research procedure was initially invited to participate in the research towards 300 schools from the territory of the Republic of Serbia. The invitation to participate in the research contained a link to a questionnaire that was directly forwarded to the students by the teachers [5].

Online classes:	**Percentage:**
Lack of interaction with teachers	26,5 %
Lack of interaction with other students	31%
Lack of various content (movies, simulations, animations, interactive exercises, quizzes, games, etc.)	19,6%
Monotonous teaching	18,6%
Insufficient knowledge of ICT by teachers	4,5%

Table 1. What bothers high school students in Serbia in conducting online classes in the schools they attend?

What is obviously the most urgent segment for improvement from Table 1 is the interaction in online teaching, both the interaction of students with teachers and the interaction between students [5]. Also, it is necessary to improve the teaching content, make them more diverse and interesting to students.

Teacher access to online teaching	**Percentage**
I am not satisfied at all	5,7%
I am not satisfied	5,7%
I am very satisfied	11%
I am neither satisfied nor dissatisfied	32,5%
I am satisfied	45,1%

Table 2. Student satisfaction with teachers' access to online teaching

In terms of student satisfaction with teachers' access to online teaching, we see that 62.6% of students are satisfied with the way teachers approach online teaching, which still shows that students recognize and valorise teachers' efforts to cope with new circumstances, while 11,4% of students are not satisfied with teachers' approach to online teaching, and 32.5% of them are undecided (neither satisfied nor dissatisfied).

Measures to improve the quality of online teaching	Percentage
To use other teaching platforms / applications in teaching	20,8%
That schools organize additional trainings for teachers	20,3%
For teachers to say which platforms / applications for online teaching they know, so that they can use them	31,7%
That schools hire administrators who would facilitate the work with teachers	27,3%

Table 3. What should schools and teachers do to improve the quality of online teaching?

According to the presented data, it is noticed that students believe that online teaching can be improved through several segments. The largest share of students (31.7%) believe that teachers should say which platforms they know so that these platforms / applications are used in online teaching, which is observed with the fact that 20.3% of respondents believe that schools should organize additional training for teachers, shows that more than half of the respondents believe that teachers need to improve their digital competencies and knowledge of platforms / applications used in online teaching, which would contribute to the quality of online teaching [5]. Certainly, this can be added to 27.3% of respondents who believe that schools should hire an administrator to facilitate work with teachers, and this attitude of almost a third of respondents indicates that students recognize and valorise the work and commitment of teachers, and have an understanding of the circumstances. in which online teaching took place at the time of the pandemic. Additionally, the Covid19 pandemic, and the sudden, widespread transition to online instruction, has created an even stronger need for faculty support, and in particular, support in a form that is compatible with remote working and teaching [8]. More broadly, there is a need for more and better models for supporting faculty's pedagogical growth and development. This need is acute in situations where faculty are attempting to use approaches that are substantially different from how they themselves were taught [8].

3. Discussion

3.1 Some of the solutions in conducting online classes as an introduction to the application of the lean concept as a saving of work time and learning materials

To overcome the problem of time to prepare online content, teachers could now use the thousands of open educational resources (OERs) published by the Ministry of Defence in China, for example, which are available in other national and international online "repositories", as well as public networking tools, platforms and available technologies. The term open educational resources was first used at the UNESCO Forum in 2002, and was defined in a recent UNESCO Recommendation on OER as "learning, teaching and research materials in any format and medium placed in the public domain or under copyright which is published under an open license that provides access at no cost" [4]. In order to solve problems related to teacher / student isolation, as well as the need for effective pedagogical approaches to activate and engage students, teachers should build their OER courses and require their students to find problem-solving content, written reports, or research on a given topic.

Specifically, Open educational practices (OEP) – including open pedagogy, open collaboration, and open assessment – should be implemented to keep the students motivated and engaged during this long period of online learning. To facilitate acquiring OEP, teachers should choose already known learning tools and technologies that students are already familiar with. They should also avoid overloading students by asking them to use too many tools, which in practice in most cases has resulted in a less desirable learning practice for them.

Learning is facilitated not only by teachers but also by peers. Therefore, to make the teaching process more interactive, teachers can build open learning communities where the students can openly exchange ideas, create discussions, and collaborate on different tasks [9]. To ensure interactive and open learning communities, teachers should use social networks during the learning process, such as Wechat, QQ, and Sina Weibo (Facebook or Twitter, for international readers).By using these social networks, teachers can share questions related to specific course materials, and students can discuss them to determine specific answers.

Consequently, students learn by exchanging ideas and opinions. Furthermore, the jigsaw classroom pedagogy can be applied online by dividing the assignment into several tasks and making each team work on a specific task. This will foster both individual accountability and the achievement of team goals. Additionally, the open learning within social networks can be gamified using Emojis to make the learning process more engaging and interactive. For instance, Saif used, during the learning process on Facebook, the number of given "likes" on a particular learner's answer as the score he/she gets for that answer [10].

3.2 The advantage of online learning and the spontaneous introduction of the lean concept in education

Online learning advantages included remote learning, comfort, and accessibility but online learning was limited to inefficiency and difficulty in supervising students. When conducting online teaching, teachers could try to monitor students and improve their learning efficiency [11]. It is the internet and online education that has conditioned the urgent need to modernize governance within higher education, which is extremely important for both efficient and effective business.

All this somehow spontaneously conditions the application of the lean concept in higher education, which will enable easier (re) definition of existing as well as future organizational and managerial structures. By modernizing the management and improvement of university / faculty processes, along with systematic and continuous research for the improvement of university / faculty processes, it directly affects the possibility for university faculties to receive the epithet "smart or green" university [12]. Therefore, to assess learners in open learning environments, teachers should use project-based assessments based on the OER delivered by learners. In this context, teachers can invite their students to open presentations (where parents and other teachers can attend) of their delivered projects for assessment and grading.

Modernisation of the learning process, especially in higher education institutions, can be achieved through several platforms that support live video communication, such as Skype and Zoom. If we talk about mobile technologies that facilitate access to the platform for online learning, these are all applications that are supported by smart phones as well as computers [13]. Certainly, what makes online learning much simpler, more efficient, easier are all the aids related to information technologies such as laptops, tablets, iPads, which are already widely used and well accepted in higher education institutions [14]. Online learning could bring many benefits to students such as focusing on really important knowledge as well as enabling them to engage in mutual cooperation and group exchange of opinions online when it comes to learning [15]. All this together saves time, space, and in a way automatically reduces costs.

3.3 Modernization of management within higher education and application of the lean concept

In every social community in the world, we can show the level of social development that it has reached by the educational level of the population. Higher education institutions will play a central role in solving the challenges of the 21st century, because education is of key importance for the development of a knowledge-based economy and the improvement of society as a whole [12]. The role of higher education institutions through leadership training, education of the growing population in all social skills needed to achieve a complex, modern society are vital in an increasingly dynamic climate of competitiveness and globalization, higher education is a significant link between national and international environment, especially in Covid19 we achieve primarily through the internet and online education [1].

It was the Covid19 pandemic that, like the internet [16] and online education, conditioned the urgent need to modernize governance within higher education, which is extremely important for both efficient and effective business. All this somehow conditions the application of the lean concept in higher education, which will enable easier (re) definition of existing as well as future organizational and managerial structures. By modernizing the management and improvement of university / faculty processes, along with systematic and continuous research for the improvement of university / faculty processes, it directly affects the possibility for university faculties to receive the epithet "smart or green" university.

3.4 Higher education and lean concept

Lean means working more with less human effort, equipment, time and space, while providing products that will meet the needs of consumers, which is actually the ideal solution in the newly created working conditions during a pandemic, when it comes to higher education. In order to implement the lean concept within higher education, it is necessary to fulfil several steps that can be presented as: loss detection and process mapping, implementation of various lean methods and tools for certain processes and continuous improvement of improved process processes [12]. In order to better apply the LEAN system in education, the initial step we must take is to identify the losses in education (WASTE) [17] and based on the identified losses it is necessary to define the strategy of LEAN system, ie the application of appropriate LEAN tools that can be used in the education process [18]. The final value of the product, in this case, is the knowledge and success of the students, which would be at the highest possible level, and thus the quality of school work would be higher and better. If we look at the education system, several possible losses can be noticed here: 1. lack of motivation of students and teachers; 2. insufficient teaching aids; 3. poor material conditions; 4. plan and program; 5. a good textbook; 6. low interest in professional development of teachers; 7. weak control over the work of teachers; 8. poor schedule; 9. information system exists but does not function; 10. poor cooperation on the teacher-parent line.

4. Conclusion

The Covid19 virus pandemic imposed a rapid transition to online teaching, regardless of the readiness of students and teachers and available digital technology in schools, which led to greater use of the Internet, which in these conditions was of great importance for quality teaching. Internet and online education have conditioned the urgent need for technological and technical modernization, especially in the management within higher education, which is extremely important for both efficient and effective business. All this spontaneously introduces the application of the Lean concept in higher education, which gives the professor a completely more complex role in education. The application of LEAN tools in education would provide better motivation of students, better grades of students, better average of all students in general, great satisfaction and motivation of teachers for better results. On the other hand, by using modern technologies in education, the professor becomes both a lecturer and a designer, producer, media specialist, provider.

In order to achieve the best quality in education in the 21st century, especially during the Covid19 pandemic and after it, it is necessary to invest in new methods of education which include the implementation of Lean methodology within which it is necessary to expand measurement systems and teaching efficiency, teachers, and especially to single out and encourage those who achieve the best results.

Currently, the work is limited by the small number of papers that exist on the application of the lean philosophy in higher education, especially related to online learning, and further research will be focused on the implementation of several different Lean tools in higher education institutions in Serbia in order to improve the current way of working through online platforms during the Covid19 pandemic.

5. References

[1] Radosevic, M.; Baosic, M.; Radicev, S.; Ikonov, D. & Matic, B. (2011). Advancing higher education with the "Lean" philosophy, XVII Conference development trends, Kopaonik, ISBN 978-86-7892-323-4, Publisher Fuculty of Technical Sciences Novi Sad

[2] Desnica E. (2010). Distance learning model as a factor in the efficiency of teaching graphic communications in higher education of technical professions, Ph.D. Dissertation, Technical faculty „Mihajlo Pupin", Zrenjanin, University Of Novi Sad, Novi Sad, Republic of Serbia

[3] Zhonggen, Yu. (2021). The effects of gender, educational level, and personality on online learning outcomes during the COVID-19 pandemic, International Journal of Educational Technology in Higher Education, 18, 14 (2021). https://doi.org/10.1186/s41239-021-00252-3

[4] Huang, R.; Tlili, A.; Chang, TW.; Zhang, X.; Nascimbeni, F. & Burgos, D. (2020) Disrupted classes, undisrupted learning during COVID-19 outbreak in China: application of open educational practices and resources. Smart Learn. Environ. 7, 19 https://doi.org/10.1186/s40561-020-00125-8

[5] http://zadecu.org/wp-content/uploads/2021/02/Rezultati-istrazivanje-o-stavovima-mladih-o-onlajn-obrazovanju-u-Srbiji-1.pdf, (2021). Friedrich Naumann Foundation for Freedom - Office for the Western Balkans, Research results Attitudes of young people about online teaching in Serbia, Accessed on: 2021-08-21

[6] Wiley, D., & Hilton III, J. L. (2018). Defining OER-enabled pedagogy, The International Review of Research in Open and Distance Learning, 19(4). DOI: https://doi.org/10.19173/irrodl.v19i4.3601

[7] https://digitalnaucionica.edu.rs, (2021). Digital classroom, Accessed on: 2021-09-05

[8] Price, E.; Lau, A.C.; Goldberg, F.; Turpen, Ch.; Smith, P.S.; Dancy, M., & Robinson, S. (2021). Analyzing a faculty online learning community as a mechanism for supporting faculty implementation of a guided-inquiry curriculum, IJ STEM Ed 8, 17 https://doi.org/10.1186/s40594-020-00268-7

[9] Hegarty, B. (2015). Attributes of open pedagogy: A model for using open educational resources, Educational Technology, pp. 3–13.7

[10] Saif, M. A. F.; Tlili, A.; Essalmi, F. & Jemni, M. (2019). Impact of Facebook as a learning tool on learning outcomes, technology acceptance, and attitude, Journal of Cases on Information Technology (JCIT), 21(4), pp. 46–61.

[11] Stankovski, S., Ostojić, G., Zhang, X., Baranovski, I., Tegeltija, S., Horvat, S. (2019). Mechatronics, Identification Tehnology, Industry 4.0 and Education, 18th International Symposium INFOTEH-JAHORINA, INFOTEH 2019 - Proceedings, art. no. 8717775

[12] Radošević M. (2014). LEAN concept to "SMART I GREEN" University, XX Conference development trends, Kopaonik, Publisher Fuculty of Technical Sciences Novi Sad, Available from: http://www.trend.uns.ac.rs/stskup/trend_2014/radovi/T6.2/T6.2-1.pdf Accessed: 2021-09-07

[13] Panigrahi, R.; Ranjan, P. & Sharma, D. (2018). Online learning: Adoption, continuance, and learning outcome - A review of literature, International Journal of Information Management, Vol. 43, pp. 1 - 14. https://doi.org/10.1016/j.ijinfomgt.2018.05.005

[14] Starr-Glass, D. (2013). From connectivity to connected learners: Transactional distance and social presence. Increasing student engagement and retention in e-learning environments: Web 2.0 and blended learning technologies, Bingley: Emerald Publishing. https://doi.org/10.1108/S2044-9968(2013)000006G007

[15] Alwi, N.; Adams, R. & Newton, J. (2012). Writing to learn via text chat: Task implementation and focus on form, Journal of Second Language Writing, Vol. 21, pp. 23–39.

[16] Gabriel, Dinu, and Dinu Loredana. "Using Internet as a Solution for Sales in Covid-19 Pandemic: E-Commerce." Annals of DAAAM & Proceedings, annual 2020, pp. 0104+. Gale Academic OneFile, link.gale.com/apps/doc/A653725459/AONE?u=anon~50641215&sid=googleScholar&xid=64853d91. Accessed 11 Oct. 2021.

[17] Tegeltija, S.; Lazarevic, M.; Stankovski, S.; Ćosic, I., Todorović, V. & Ostojić, G. (2016). Heating circulation pump disassembly process improved with augmented reality, Thermal Science, Vol. 20, pp. S611-S622.

[18] Manojlovic, G.; Macuzic, I.; Jeremic, B. & Arsovski, S. (2012). Lean koncept as a tool for education system improvement; Quality Festival 2012, ISBN 978-86-86663-83-2, Publisher Association for Quality and Standardization of Serbia

DOI: 10.2507/32nd.daaam.proceedings.060

USE OF AUGMENTED REALITY TECHNOLOGY IN POPULATION PROTECTION AND CRISIS MANAGEMENT

Martin Dzermansky, Lukas Snopek, Katerina Vichova, Martin Ficek & Jakub Rak

This Publication has to be referred as: Dzermansky, M[artin]; Snopek, L[ukas]; Vichova, K[aterina]; Ficek, M[artin] & Rak, J[akub] (2021). Use of Augmented Reality Technology in Population Protection and Crisis Management, Proceedings of the 32nd DAAAM International Symposium, pp.0408-0414, B. Katalinic (Ed.), Published by DAAAM International, ISBN 978-3-902734-33-4, ISSN 1726-9679, Vienna, Austria
DOI: 10.2507/32nd.daaam.proceedings.060

Abstract

Population protection and crisis management are some of the most-watched areas today. With the growing emergencies and crises, there is also a need to inform the population more about the possible effects, prevent them, or deal with them. For this purpose, the use of software applications is offering. In the last years, there is a growing trend in augmented reality technology, which can also be applied to mobile applications and can be slowly ranking among the most effective future technologies. This article describes the possible application of augmented reality technology in population protection and crisis management. The explanation, experimental, and modelling methods were used to make the article, by which the augmented reality model was applied to the Hololens 2 glasses. Blender and Unreal Engine software were used to create modelling and augmented reality.

Keywords: augmented reality; crisis management; Hololens; population protection; technology.

1. Introduction

Population protection represents the fulfilment of tasks in planning, organization, and activities to prevent the occurrence and ensure preparedness for an emergency and crisis management, which fulfils the role of security risk analysis. The preparation and the ability of the population to respond to emergencies is a significant problem today. People lack information on how to behave during individual emergencies and how to respond to them. Most of the data can inhabitants found on the websites of cities and municipalities that publish this information. Still, only a tiny part of the population finds this information and gets acquainted with it. Another aspect of the population does not know that they will find this information on their town or village websites.

One way to protect the population is to use SW support for population protection and crisis management. Today, it offers the possibility of using augmented reality technology. This technology is possible to apply to smart devices too. Therefore, it represents the easiest way to train the population or inform and respond to an emergency. This article presents the possible application of augmented reality technology in population protection and crisis management. The work aims to create a functional model based on accurate map data and shows critical infrastructure objects in the selected area. Hololens 2 glasses are used to display created models and applied in augmented reality.

Augmented reality

Augmented reality (AR) is one of the variations of the virtual environment (VE) or virtual reality (VR). The difference between augmented reality and the more commonly used virtual reality is that virtual reality aims to immerse the user in an artificial world. In contrast, augmented reality receives digital or computer-generated information. This information can take the form of images, audio, video, or tactile or haptic perceptions that overlap in the real world. Augmented reality, therefore, offers the user to perceive the natural world and project virtually created objects into it. Specific components are required for the proper application of augmented reality. The parts needed for augmented reality: [3]

- Computer or mobile device.
- Monitor or another display medium.
- Camera.
- Tracking and sensing systems (GPS, compass, accelerometer) - are included in almost all smart devices.
- Network infrastructure.
- Tags – these are physical objects or places where real and virtual environments are connected. Here, the computer identifies the area where the digital information is to be presented.
- An application or program that runs locally.
- Web services.
- Content Server.

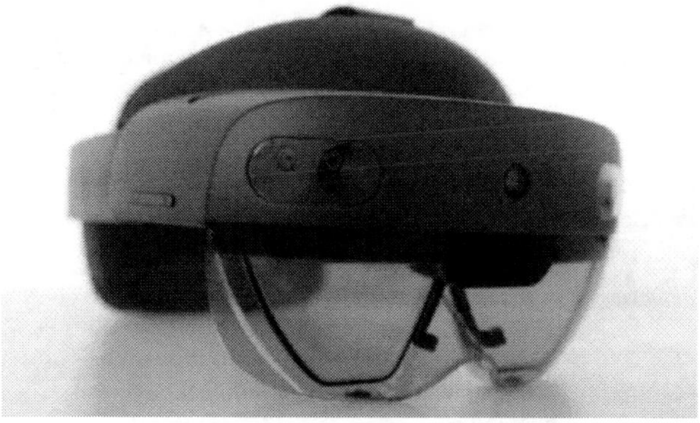

Fig. 1. Hololens 2

The beginnings of augmented reality can be dated back to the 60s and 70s of the last centuries. In 1968, the first head-worn display was created. This system was called "Damocles' Sword" and was designed by computer scientist and Harvard professor Ivan Sutherland. In 1974, the first artificial reality laboratory was built. The author was the artist and researcher Myron Kruger, and he called this laboratory "Videoplace." In this lab, projection and camera technology were used to emit silhouettes on the screen that surrounded the user for an interactive experience. Using this technology, for example, he was able to write by pointing his finger.

In the 1980s and 1990s, augmented reality flourished. In 1990, the term "augmented reality" was coined by Tom Caudell, and in 1992, one of the first fully functional augmented reality systems, the so-called "Virtual Fixtures," was created. Virtual Fixtures were developed for military personnel and allowed them to control and guide machinery to perform tasks such as training Air Force pilots to fly safer. The year 1998 brought the first application of augmented reality to a sports match. The 1st & ten virtual graphics system was used during the NFL game. The technology displayed a yellow line overlaid at the top of the canal and is still used today.

With the advent of the new millennium, new technologies also came. In 2000, an open-source software library called ARToolKit was developed by Hirokazu Kato. It is a package designed to help developers create augmented reality software programs. In 2009, augmented reality was used in print media for the first time, and ARToolKit brought augmented reality to web browsers. In 2013, the MARTA (Mobile Augmented Reality Technical Assistance) application was introduced, which provided Volkswagen with detailed repair instructions in the service manual. In 2014, Google continued to present the first augmented reality glasses, Google Glass. In 2016, competing glasses from Microsoft were introduced, specifically HoloLens glasses, and the year 2017 affected all retailers in the augmented reality sector forever. IKEA has launched an application that allows customers to see home decor options before making a purchase virtually. [4], [5]

An article by Alan Seal on the website vxchnge.com describes the importance of augmented reality technology and the bonuses it brings, describing its importance and the top seven statistics. Anyone could think about why augmented reality technology is important and come up with many things. In this article, Alan Seal mainly presents the contactless interface he has at his disposal compared to other technologies. Competitive virtual reality technology is among the most popular, but it requires a touch interface and other components from which augmented reality has freed itself.

Among the top seven statistics were:

- Achieve $ 70-75 billion in revenue by 2023.
- The most popular augmented reality game Pokemon GO has been downloaded more than 1 billion times.
- High awareness of augmented reality among people, 70-75% of people aged 16-44. knows the technology of augmented reality.
- 88% of midsize companies use augmented reality technology.
- Google Play and the App Store already offer thousands of mobile applications and games using this technology.
- Microsoft's first-generation augmented reality glasses, HoloLens, have sold more than 50,000.
- The Angel List, which gathers information on startups, lists more than 2,000 augmented reality startups. [6]

The future of augmented reality in population protection and crisis management is very open. This work aims to outline the possible use of augmented reality in these areas.

2. Possibilities of applying augmented reality technology in the field of crisis management and population protection

Augmented reality presents endless possibilities for expanding the real world. This technology can be found in the industry, such as a teaching background or procedure for employees, games for smart devices, or school teaching materials. Population protection and crisis management are areas that are an integral part of human lives. When considering the connection between augmented reality and these areas, developing much knowledge on working in these areas could be simplified and more efficient.

Population protection and crisis management are not only rules and regulations but also the bodies that represent them. These include the fire brigade, the police of the Czech Republic, and the medical rescue service, which form an integrated rescue system. Augmented reality could play a crucial role in the performance of their duties or exercises in the future. The document "Improving Our View of the World: Police and Augmented Reality Technology" by Thomas J. Cowper and Michael E. Buerger describes possible augmented reality applications to New York City police.

Potential applications of this technology include, for example:

- Intelligent implementation of real-time crime reporting.
- Face data, voiceprints, and other biometric data about known criminals.
- Scalable, three-dimensional maps, complete floor plans of buildings and sewers.
- Better coherence between team members and better coordination with the intervention commander.
- Modulation of sound effects of shots to increase concentration.
- Optical zoom, thermal and infrared imaging.
- AR video, audio, and sensing devices are used to visualize blood samples, bloodstains, and other sensor-detectable forensic data available at crime scenes.
- Coordinated use of robots, unmanned aerial vehicles (UAVs), and police.
- View location, activity, and status information projected into 3D.
- Coordination of widely dispersed units. [1]

One of the possible uses in population protection is the sector of training and education. An article appeared in The Science for Population Protection magazine, published by the Institute for the Population Protection Institute in Lázně Bohdaneč, which deals with augmented reality technology in this sector. This article presents the benefits of technology where it is possible to simulate floods, fires, plan contents, crisis management or disaster management systems and more. No one would intentionally create floods or fires for upbringing or education, so this technology is ideal for projecting these simulations to students.

The technology itself can revive teaching in this area. The authors describe the possibilities of applying this technology in education in two ways. The first is a purely educational journey, in which case it would be printed materials and textbooks that would be enriched with augmented reality elements. The second way is practical, therefore use in the field. [7]

This article aims to point out the possible use of augmented reality technology in population protection and crisis management. In these areas, this technology will be used by linking geographic information systems with augmented reality technology. Critical infrastructure objects will be displayed using models created by Blender and Unreal Engine software. Geographic information systems consist of hardware, software, data, applications, users, and processors. In addition to these elements, geographic information systems are created from raster and vector graphics. The most common use of these systems is in creating maps, which are used, for example, in navigation systems. [2] For this article, Blender software was used, which offers the creation of 3D graphics. This software provides the Blender GIS function, which was used to model the base map, in which building models were uploaded and then uploaded to the Unreal Engine software.

The basis of the model is data from open street maps. For better orientation, Uherske Hradiste was chosen, on which the objects of critical infrastructure are subsequently marked. To make the model as realistic as possible, height quotas are also created, as seen in the image below. The base of the map is wavy due to different altitude quotas.

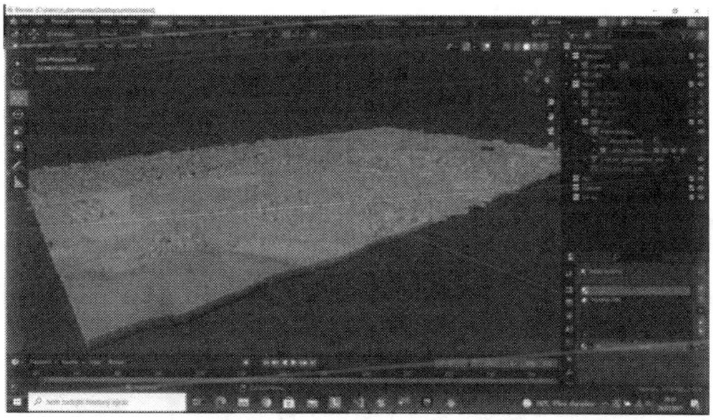

Fig. 2. Blender-GIS

After modification and adjustment of the model, the software Unreal Engine version 4.26.1 was implemented. This software offers work with augmented reality and virtual reality and is known primarily for creating games. After uploading the model to Unreal Engine, it was further modified. The initial settings for working with augmented reality have been altered, and the necessary plugins such as the Hololens plugin, Microsoft Windows Mixed Reality, and others have been downloaded. An important step was the creation of map components or so-called levels. In this case, a total of 5 levels were created.

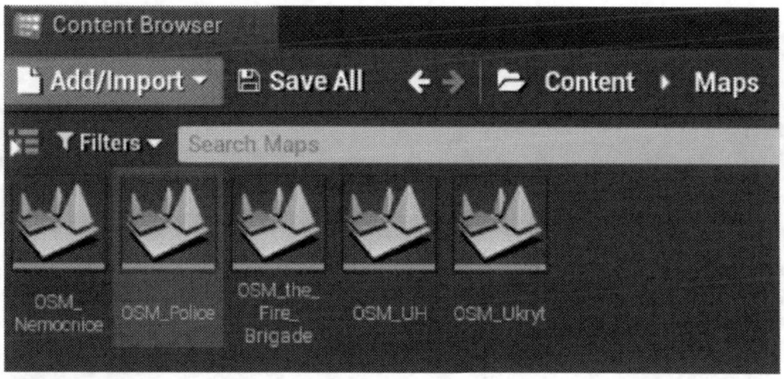

Fig. 3. Levels

Each of these levels contains the same background map but a different marked critical infrastructure object. These objects are:

- hospital,
- police,
- the fire brigade,
- permanent pressure shelter for the civilian population.

The fifth respective first level is the basic background map of Uherske Hradiste, in which the application begins.

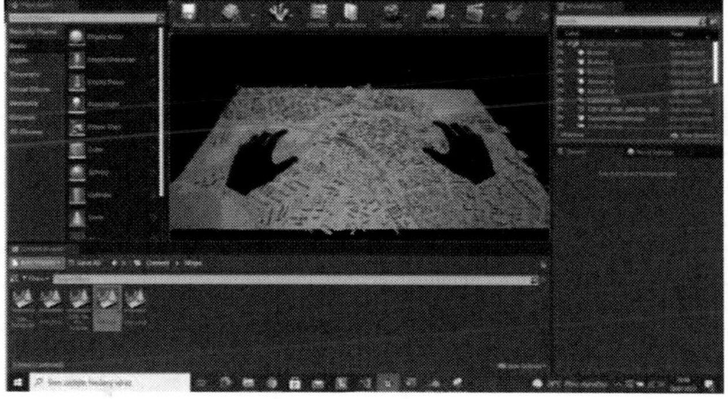

Fig. 4. Basic level

To switch between individual levels in the application and thus display critical infrastructure buildings, it was necessary to create a functional menu. This menu is created using blueprints and the "UxtPressableButton" function, allowing users to insert clickable buttons.

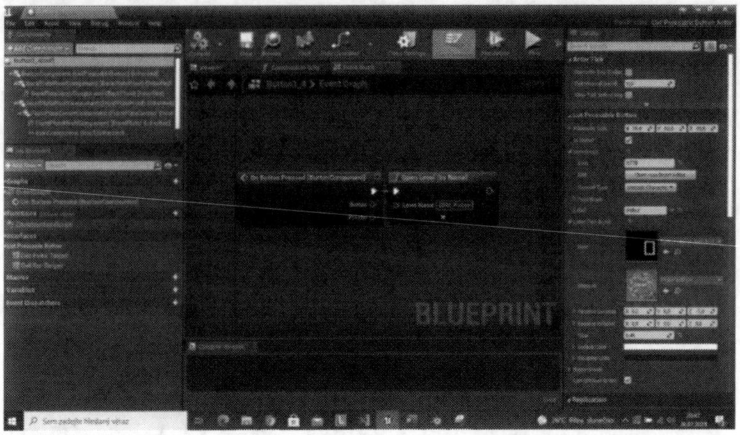

Fig. 5. Example of blueprint

To switch between the created levels (maps), the open level (by name) function was created using blueprints, thanks to which each of the buttons is oriented to a specific level.

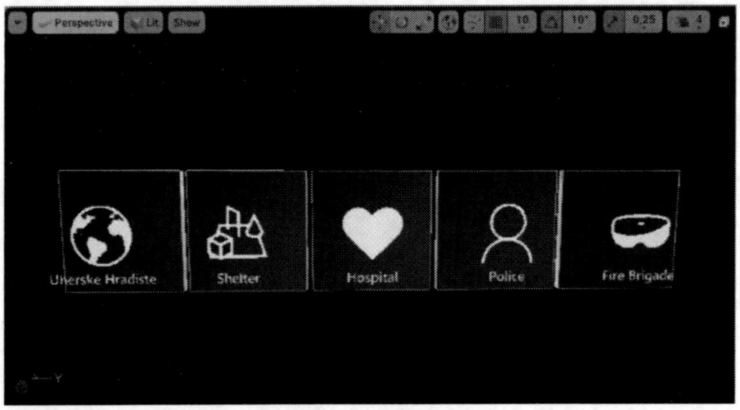

Fig. 6. Main menu

The menu is fully clickable, and the user can switch between individual objects. In the application, it then looks like this. The user launches the application and starts at the Uherske Hradiste level, i.e., on the essential map base.

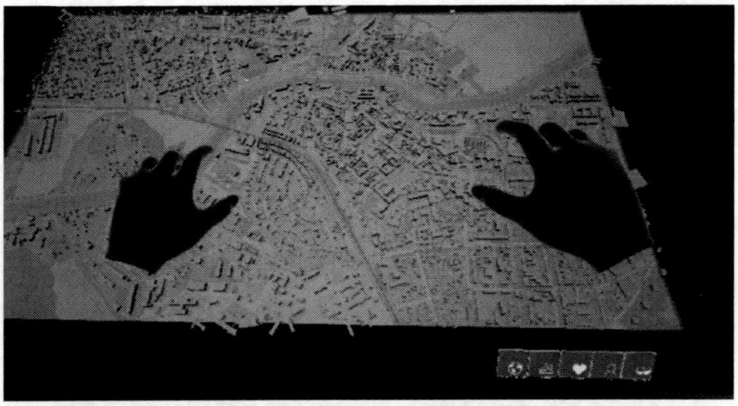

Fig. 7. Start of application in Unreal Engine

The application starts on the basic map of the Uherske Hradiste. Figure number 7 shows a view from the unreal engine environment. At this point, the user can move or zoom in and out as he wants. Suppose the user decides to view from one critical infrastructure element, which is listed in the menu attached to the map. In that case, all he has to do is click on the component, and the element will be color-coded on the map to be immediately apparent across the scale.

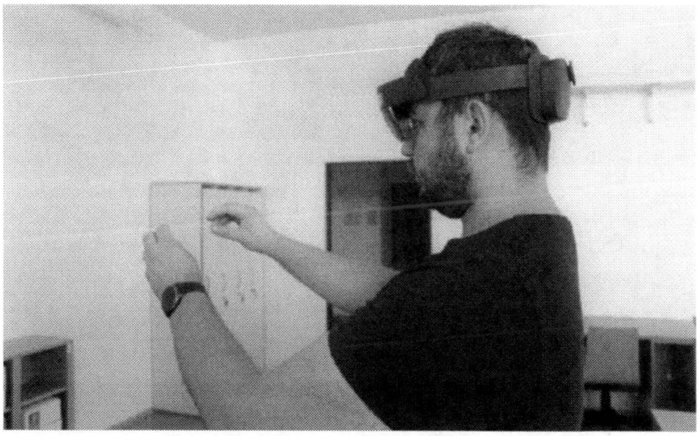

Fig. 8. Start of application in real world

Suppose it is decided to display one of the created elements. In that case, the user presses the button in the menu, and a different colored object will appear in the map base, which indicates the critical infrastructure, in this case, the hospital. The user can use the Hololens 2 glasses to press a button, as in ordinary life, or, if the object is farther away, to aim at it and press the thumb and forefinger to activate the button.

Fig. 9. Pressable buttons in the menu

As with other levels, it is possible to transform and move the object freely. The individual layers can be separated from each other.

Fig. 10. Changed level with the object of critical infrastructure

This application offers a more detailed view of the city of Uherske Hradiste in an augmented reality environment, which in this case, is created for the Hololens 2 glasses. These glasses offer fully interactive work with augmented reality. Compared to applications in smart devices, the use of Hololens is much more realistic. Applications created in this way can be used in the future, such as in crisis staff dealing with emergencies, planning logistics nodes for an integrated rescue system, or simulations of events that may affect the population. Thanks to the buildings created in geographical information systems, it is easier to imagine possible damages or better plan rescue and liquidation work.

3. Methods

Several scientific methods have been used in this article. Firstly, the explanation method was used to explain population protection, crisis management, and augmented reality technology. Secondly, a modelling method was used to create a model of the city of the Uherske Hradiste and buildings in which the objects of critical infrastructure were colour-coded. Modelling was used in Unreal Engine software for modifying the models too.

The models were created in Blender software, where was used the Blender GIS plugin, which enabled the connection of data from open street maps. The resulting models were exported to the Unreal Engine, in which the models were refined and then converted into augmented reality. Finally, the experimental method was used when the models were recorded in Hololens 2 glasses, and it was possible to project the model into the real world. [8]

4. Conclusion

The article aimed to use augmented reality technology in population protection and crisis management by linking geographic information systems and augmented reality technology using Hololens 2 glasses. Firstly, a literature review was conducted, where the authors in their professional publications emphasized the use of augmented reality and its possible implementation in population protection and crisis management. Secondly, for these purposes, the modelling method was used when a model of the city of Uherske Hradiste was created in the Blender software, in which buildings were created using the Blender GIS function. The central part of the article was to present the model that was uploaded to Unreal Engine software version 4.26.1, where levels were created. This model was modified, and critical infrastructure areas were created. Specifically, these are the hospital's buildings, the police, the fire brigade, and the permanent pressure shelter for the civilian population.

The limitations in this article were mainly in the display field options in the Hololens 2 glasses, which are limiting and thus fail to display models to a greater extent. For this reason, the models have been scaled down. The implications of a functional application can be used for emergency simulations, which could be used in practice, for example, in security or crisis staffs, which either analyze it or directly address the event. However, the purchase of glasses is limiting, when glasses would have to be provided for all staff members. Further research aims to create realistic textures and animations that can create an optimal model that could be used in practice and thus help, for example, the integrated rescue system and crisis staff components.

5. Acknowledgments

This article was supported by an internal grant project of Tomas Bata University under the number IGA/FAI/2021/003.

6. References

[1] Cowper, T. & Buerger, M. (2003). Improving Our View of the World: Police and Augmented Reality Technology [online]. In: . New York: FBI. Available from: https://www.fbi.gov/file-repository/stats-services-publications-police-augmented-reality-technology-pdf/view

[2] Dempsey, C. (2021). What is GIS. https://www.gislounge.com/what-is-gis/ [online]. California: www.gislounge.com. Available from: https://www.gislounge.com/what-is-gis/

[3] Kipper, G. & Rampolla, J. (c2013). Augmented reality: an emerging technologies guide to AR. Waltham: Syngress. ISBN 9781597497336.

[4] Poetker, B. (2019). History of Augmented Reality. Learning Hub [online]. Chicago: Learning Hub. Available from: https://learn.g2.com/history-of-augmented-reality

[5] Schmalstieg, D. & Höllerer, T. (2016). Augmented reality: principles and practice. Addison Wesley: Boston. ISBN 9780321883575.

[6] Seal, A. (2020). Top 7 Augmented Reality Statistics for 2020. Www.vxchnge.com [online]. Cleveland: www.vxchnge.com. Available from: https://www.vxchnge.com/blog/augmented-reality-statistics

[7] Tomanová, K.; Smetana, M.; Kavan, Š.; Rathauský, Z. & Mudrochová, S. (2020). Využití rozšířené reality pro přípravu a vzdělávání obyvatelstva. The Science for Population Protection. Available from http://www.population-protection.eu/prilohy/casopis/42/366.pdf

[8] Závěrečná práce - metodika, c2007-2013. Lorenc.info [online]. Praha: VŠE. From http://lorenc.info/zaverecne-prace/metodika.htm

DOI: 10.2507/32nd.daaam.proceedings.061

DESIGN OF A DESIGNATION FOR EMERGENCY MEDICAL SERVICES AT DEPARTMENT STORES

Martin Dzermansky, Ondrej Kostka & Marketa Habrova

This Publication has to be referred as: Dzermansky, M[artin]; Kostka, O[ndrej] & Habrova, M[arketa] (2021). Design of a Designation for Emergency Medical Services at Department Stores, Proceedings of the 32nd DAAAM International Symposium, pp.0415-0422, B. Katalinic (Ed.), Published by DAAAM International, ISBN 978-3-902734-33-4, ISSN 1726-9679, Vienna, Austria
DOI: 10.2507/32nd.daaam.proceedings.061

Abstract

With the advancement of technology and the architecture of buildings, more and more shopping centres are emerging, which attract hundreds to thousands of visitors every day. It is never clear in advance when any of these visitors will be in health danger or encounter any extraordinary events. If a person's health or life is endangered, urgent pre-hospital care is provided by the emergency medical service in the Czech Republic. For a smooth course and the fastest possible arrival and performance of the service, it is also necessary to have sufficiently planned logistics and boarding points at these department stores. This article aims to design signs and routes for the emergency medical service at the department store, aiming to streamline and speed up the trip and provide urgent pre-hospital care. Experimental and modelling methods are used in this article.

Keywords: emergencies; emergency medical service; model; signs.

1. Introduction

Today, department stores appear in every major city, but gradually even in smaller ones. Hundreds to thousands of visitors of all ages pass through these houses every day. Each of these visitors may encounter an emergency, such as nausea, limb injury, or even pledge, which would require an emergency medical service to provide the patient with urgent pre-hospital care and transportation to a hospital facility.

The arrival and placement of an ambulance at department stores is the frequent accumulation of visitors in front of department stores. Often, these people interfere with the smooth or complete passage of the ambulance with paramedics to the department store entrance. It would be possible to treat the patient and transport him to the hospital facility quickly. It is not always possible to manipulate the patient, and, for this reason, the emergency medical service must be brought as close as possible to the affected patient.

This article aims to design signs that would guide and clearly define the boarding point for the emergency medical service. This signs would have the effect of speeding up the travel time and informing the visitors of the department store about the possible arrival of the emergency medical service. Similar signs are already applied, for example, during the evacuation, which is one of the essential tasks of population protection.

Thanks to signs on the ground and warning elements, people can evacuate much faster as they follow a pre-set route. Thanks to the path marked in this way, there is a higher success rate of evacuation and thus more saved lives. The article "A virtual reality-based study of indoor fire evacuation after active or passive spatial exploration" by Lijun Cao, Jing Lin & Nan Li describes two evacuation methods. Passive and active, where a faster evacuation was achieved during passive with the help of created tags. The Šantovka department store was chosen for the experimental model. Signs will be created, and paths to individual entrances will be designed to make rescue work more efficient. [1], [3], [4]

2. Design of signage of places for emergency medical service

To design places for the emergency medical service in front of the department store, the Šantovka department store in Olomouc was chosen as an experimental model. It is a modern department store with one underground, one ground floor and two above-ground floors. Each floor has at least one entrance to the building, of which there are nine in total. For more detailed research, entrances from the underground floor and above-ground floors had to be excluded. The reason for excluding these entrances is to limit the precise height of vehicles to 2 meters at the entrances to underground car parks and 2.2 meters at the entrance to the roof car park.

Entrance on the first floor was excluded due to the need to leave the vehicle on the adjacent road at a distance of 200 meters from the entrance to the building. Furthermore, the emergency medical service would have to overcome the height difference of one floor over this bridge for pedestrians only. Four entrances on the ground floor were chosen as suitable places for the location of the proposed signage for the needs of the emergency medical service. For clarity, the individual entrances are numbered and marked with yellow arrows pointing to the building in Figure 1.

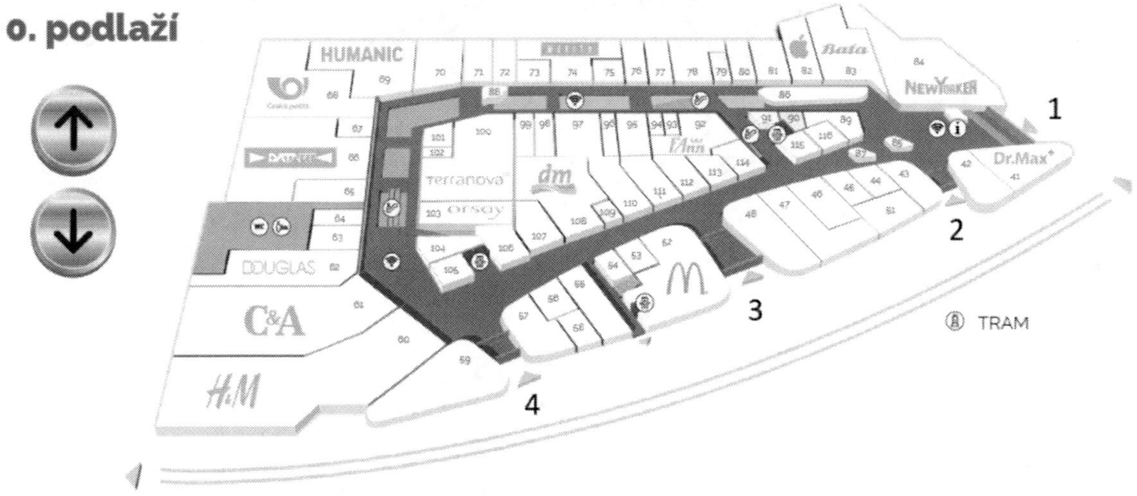

Fig. 1. Plan of the Šantovka department store [8]

A document issued by the Olomouc Region entitled "Plan for the coverage of the territory of the Olomouc Region by exit bases of the emergency medical service" divides the territory of the Olomouc Region into five territorial departments. These are the territorial departments of Jeseník, Šumperk, Přerov, Prostějov and Olomouc. There are six exit bases of the emergency medical service in the Olomouc Territorial Department, and three of these six bases are located in the city of Olomouc itself. Two are located on Hněvotínská Street, the base of the air rescue service and ambulance and rendezvous cars. There is an emergency medical service and ambulance base on Aksamitova Street.

The dislocation of the mentioned exit bases is shown in Figure 2. Geographic information systems were used to create map data in which the access routes and the distribution of the emergency medical service are shown. These systems are made up of hardware, software, people and data. A number of software are offered for the creation of geographic information systems, either paid or free.

QGIS software was used to create maps, which allows the display of geographical data in layers that overlap each other. The map also shows all access routes from both exit bases of the emergency services of the Olomouc region to the place of intervention at the Šantovka department store. Other maps contain a more detailed mapping of the emergency service's access routes to the department store's entrances. [2], [5], [7]

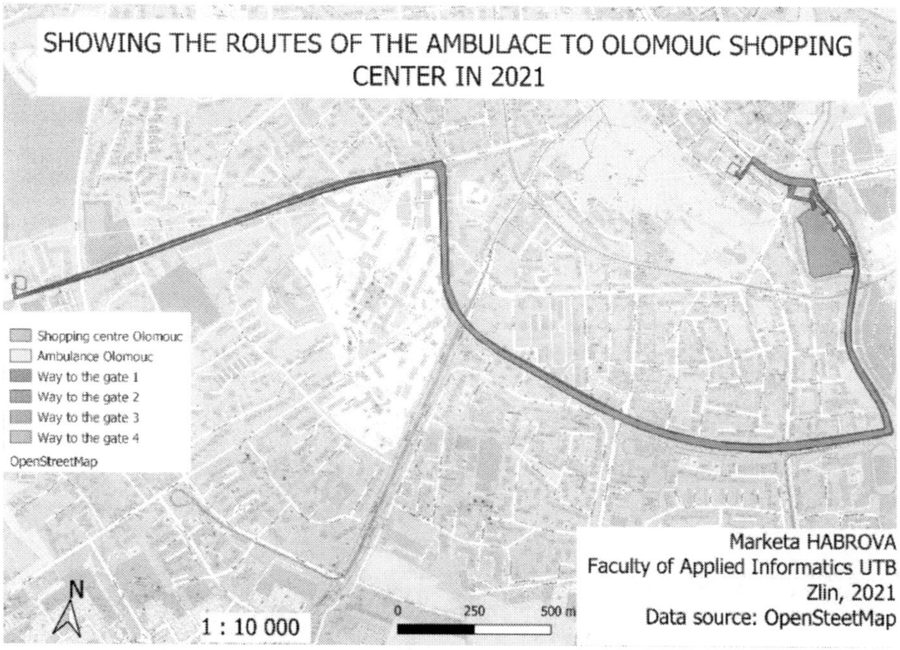

Fig. 2. Driveways of the emergency medical service to the department store Šantovka [6]

A yellow and black frame was chosen as a design for the emergency medical service at the Šantovka department store, which will alert passers-by to the possibility of an ambulance arriving. The text Place for emergency medical service is also given in this frame, and to preserve the visual of the business name, the logo of the Šantovka department store is also placed here.

Fig. 3. Design of signs for the place of emergency medical services

Entry No. 1

Entrance number one to the Šantovka department store is located on the northern facade of the building. Signs indicating the place for the emergency medical vehicle is proposed on the left side 4 meters from the entrance with a front view of the building. This location was chosen because it is located near the electronic sliding door. The second part of the entrance consists of a revolving door which is not suitable for the entry of paramedics with their emergency equipment. Furthermore, there are bicycle stands on the right side, which makes the space unsuitable for placing signs.

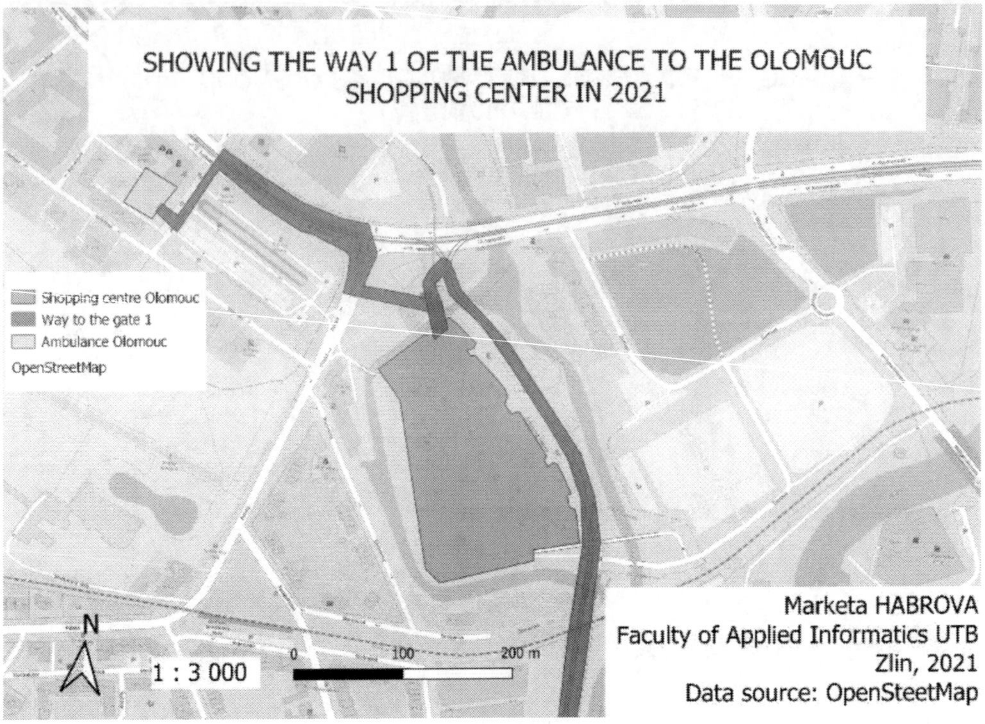

Fig. 4. The driveway of the emergency medical service to the first entry [6]

Figure 5 shows how an emergency medical service would be positioned if arrival at the first entrance were required. The car would have enough space for the performance, and a pre-marked place for their parking would alert visitors to a possible arrival.

Fig. 5. Design of a stand for the emergency medical service at the first entrance

Entry No. 2

Entrance number two is oriented to the east, as well as entrances 3 and 4. It is equipped with one electronic sliding door. We propose to place the emergency medical sign on the left side 2 meters from the entrance from the front view of the building. In the vicinity of our proposed space, there is a basket with an ashtray, which accumulates standing people. We, therefore, recommend moving the basket to the right at least 5 meters from the front door, which would reduce the number of people standing outside the proposed area for the emergency medical service.

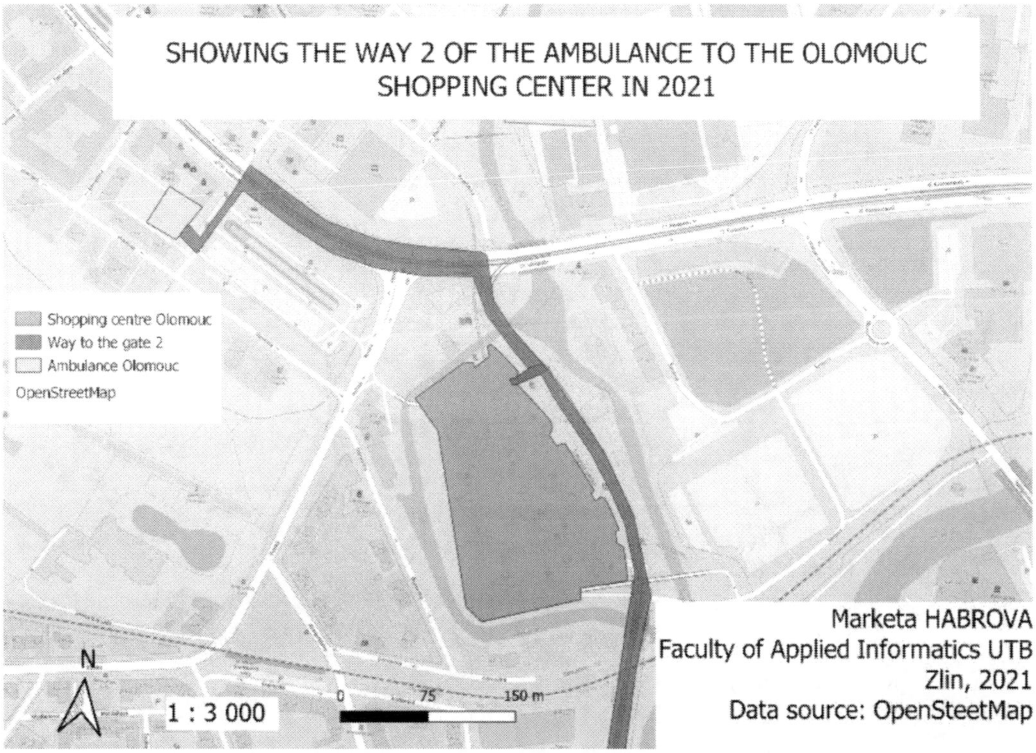

Fig. 6. The driveway of the emergency medical service to the second entry [6]

Figure 6 shows how an emergency medical service would be positioned if arrival at the second entrance were required. The car would have enough space for the performance, and a pre-marked place for their parking would alert visitors to a possible arrival.

Fig. 7. Design of a stand for the emergency medical service at the second entrance

Entry No. 3

It is equipped with two electronic sliding doors. On the right side from the front view is a summer sitting area. Therefore, we suggest placing the signage to the left side of the entrance at 2-3 meters. It would be appropriate to consider relocating the basket again due to the accumulation of people.

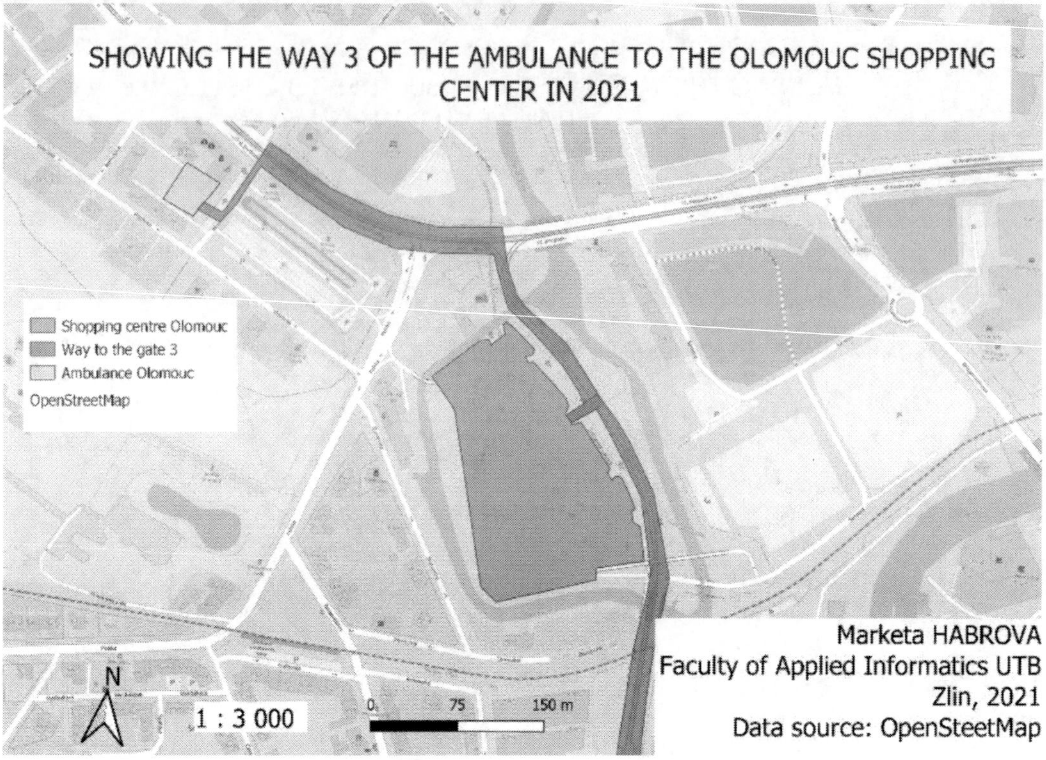

Fig. 8. The driveway of the emergency medical service to the third entry [6]

Figure 5 shows how an emergency medical service would be positioned if arrival at the third entrance were required. The car would have enough space for the performance, and a pre-marked place for their parking would alert visitors to a possible arrival.

Fig. 9. Design of a stand for the emergency medical service at the third entrance

Entry No. 4

Entrance 4 is equipped with one sliding door. On the front view side, to the right of the door, there is a waste bin. We suggest placing the signs and the left side of the entrance about 5 meters from the gate. It is worth mentioning the bicycle stand, which could cause complications when entering and exiting the vehicle.

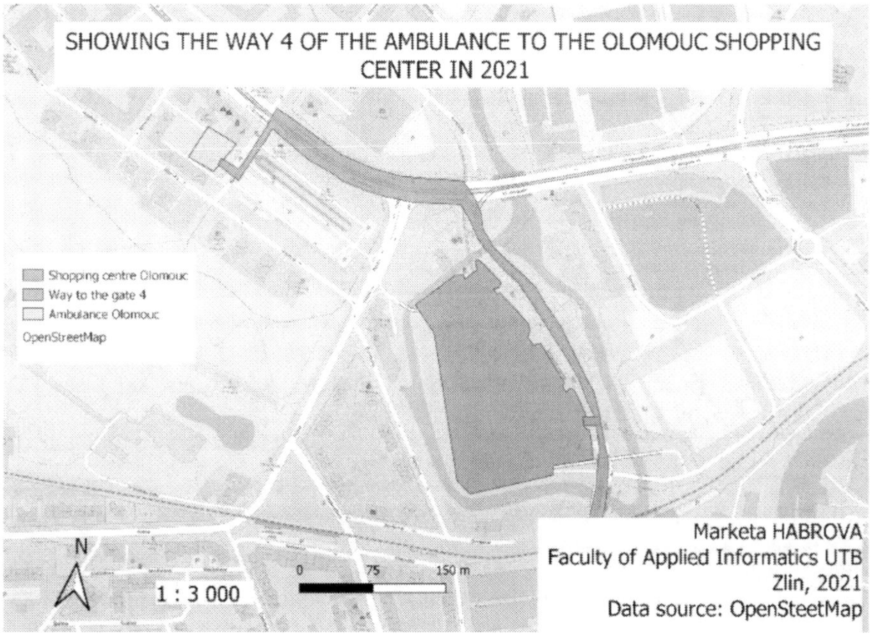

Fig. 10. The driveway of the emergency medical service to the fourth entry [6]

Figure 5 shows how an emergency medical service would be positioned if arrival at the fourth entrance were required. The car would have enough space for the performance, and a pre-marked place for their parking would alert visitors to a possible arrival.

Fig. 11. Design of a stand for the emergency medical service at the fourth entrance

When deciding which input to use, it is necessary to consider which part of the department store the intervention of the emergency medical service will be performed. Decisive factors may be the distance from the entrance, the need to use an elevator, or people's cumulation at individual entries. Not all proposed measures need to be implemented. The financial and aesthetic aspects of the adjustments must be considered when making decisions.

However, one of the proposals that would allow a smoother passage of the emergency medical service to the department store is the relocation of bicycle stands. As can be seen in the department store pictures, bicycle stands are placed on both sides of almost all access roads. These bicycles can make it more difficult for an ambulance to pass, so it would be better to move these stands, for example, between benches along a department store. Another proposal that would reduce the cumulation of people in the proposed places for the emergency medical service is the removal of ashtrays.

3. Methods

In the article, three scientific methods were used. The first was the explanation method, which explains the basic knowledge and objectives of the work. The second method is the modelling method, which was used to create map data in QGIS software. These map documents represent the routes of the emergency medical service to the individual entrances of the Šantovka department store and the general location of the emergency medical service in the city of Olomouc. The third method is the experiment method when the Šantovka department store was selected when the placement of signs for the emergency medical service was graphically designed. [9]

4. Conclusion

The article aimed to create a universal sign that would be placed at the main entrances to the department store because one of the problems of department stores is the boarding places for the emergency medical service or the generally integrated rescue system, where these places are not defined. The Šantovka department store in Olomouc was used as an experimental model. These signs are intended to indicate the possibility of the arrival of the emergency medical service. Therefore, people should not accumulate unnecessarily in these places and thus create more extensive groups of people. When an emergency medical service arrives, seconds can determine the patient's fate, and even these groups of people can delay paramedics during the procedure.

The article primarily describes the scientific problem and sets the article's goal, i.e., creating a proposal for signs places for emergency medical services at the department store. In this case, the Šantovka department store in Olomouc was used as an experimental pier. The next phase of the article was creating a signs design and creating map data in the QGIS software, which shows the emergency medical service routes to the department store's entrances. The last phase is the acknowledgments and references that were used to create this article. In further research, the authors aim to apply signs in a department store and perform a natural experiment, which would monitor whether visitors respect these signs and whether there is a decrease in the accumulation of people in these places.

5. Acknowledgments

This article was supported by an internal grant project of Tomas Bata University under the number IGA/FAI/2021/003 and an internal grant project of Tomas Bata University under the number IGA/FaME/2021/002.

6. References

[1] Coppola, Damon P. Introduction to international disaster management. Fourth edition. Amsterdam: Elsevier, [2021]. ISBN 978-0-12-817368-8.

[2] GIS Today. ESRI [online]. [cit. 2020-8-4]. Dostupné z: https://www.esri.com/en-us/what-is-gis/history-of-gis

[3] Haddow, George D., Jane A. Bullock a Damon P. Coppola. Introduction to emergency management. Amsterdam: Elsevier, [2021]. ISBN 978-0-12-817139-4.

[4] Lijun Cao, Jing Lin, Nan Li, A virtual reality based study of indoor fire evacuation after active or passive spatial exploration, Computers in Human Behavior, Volume 90, 2019, Pages 37-45, ISSN 0747-5632, https://doi.org/10.1016/j.chb.2018.08.041.

[5] Miklín, J., Dušek, R., Krtička, K., & Kaláb, O. (2018). Tvorba map. Ostrava: OSU

[6] Open Street Map [online]. Available from: https://www.openstreetmap.org/#map=15/49.3303/17.5290&layers=H

[7] Plán pokrytí území Olomouckého kraje výjezdovými základnami zdravotnické záchranné služby. Olomouc, 2021. Dostupné také z: https://www.zzsol.cz/download/ZZS-OK_mapa-pokryti-olomouckeho-kraje_1.7.2021.pdf

[8] Šantovka [online]. Olomouc: Galerie Šantovka, 2019. Dostupné z: https://www.galeriesantovka.cz/obchody/mapa-centra/

[9] Závěrečná práce - metodika, c2007-2013. Lorenc.info [online]. Praha: VŠE. From http://lorenc.info/zaverecne-prace/metodika.htm

DOI: 10.2507/32nd.daaam.proceedings.062

DRIVERS FOR SELECTING THE TYPE OF TRANSPORT IN THE MODAL SHIFT: COST, TRANSIT TIME AND ENVIRONMENTAL IMPACT ANALYSIS

Antonio Maria Coruzzolo[1], Francesco Lolli[1,2], Fabio Gennari[1], Samuele Marinello[2*] & Rita Gamberini[1,2]

[1] Department of Sciences and Methods for Engineering, University of Modena and Reggio Emilia, Via Amendola 2, Padiglione Morselli, 42122 Reggio Emilia, Italy

[2] En&Tech Interdepartmental Center of the University of Modena and Reggio Emilia, Piazzale Europa 1, 42124 Reggio Emilia, Italy, * Corresponding author

This Publication has to be referred as: Coruzzolo, A[ntonio] M[aria]; Lolli, F[rancesco]; Gennari, F[abio]; Marinello, S[amuele] & Gamberini, R[ita] (2021). Drivers for Selecting the Type of Transport in the Modal Shift: Cost, Transit Time and Environmental Impact Analysis, Proceedings of the 32nd DAAAM International Symposium, pp.0423-0432, B. Katalinic (Ed.), Published by DAAAM International, ISBN 978-3-902734-33-4, ISSN 1726-9679, Vienna, Austria
DOI: 10.2507/32nd.daaam.proceedings.062

Abstract

An increasing freight demand is putting pressure on the freight transport network, forced to pay high costs to ensure compliance with the conditions and delivery times, as well as to grow towards environmental sustainability. The purpose of this paper is to analyse the main drivers for choosing the type of transport in the planning of distribution logistics through the application of the Multinomial Logit Model (MLN) under the objective of CO_2 emission minimisation in order to identify the most sustainable transport solution as a function of distance along the same lines as Hoover's diagrams. The approach has been applied to a real case study (the logistics department of an Italian manufacturing company) in order to define the mode of transport that minimises the quantity of carbon dioxide emitted and therefore to analyse the effect that this modal shift strategy has on the other selection drivers. The results were encouraging in guiding the application of a modal shift strategy and the change to more sustainable transport modes: 62.5% of the company's transport shipments should be transferred from road to a different mode, with an overall reduction of 41% of CO_2 emissions, without any impact on costs and delivery time.

Keywords: Distribution logistics; Logit model; Multi-criteria; Optimisation; Sustainability

1. Introduction and state of the art

Nowadays an increasing freight demand is putting pressure on the freight transport network. Since most of this transport is based on fossil fuels, the change towards an efficient and sustainable freight transport system can no longer be postponed. In 2017 global primary energy consumption reached 9718 Mtoe (million tons of oil equivalent), among which the transportation sector accounted for 27.9% [1]. In addition to this high energy consumption, transportation contributes about 66% of the logistics cost [2].

Usually, the transportation of goods is carried out with different transportation modes: road, rail, air, sea. The combination of two or more modes of transportation with the same mobile storage unit is called intermodal [3]. It has been demonstrated that intermodal transport can reduce emissions. For example, intermodal transport comprising rail and road is 77.4% less polluting than full road transport [4]. For this reason, the development of a model that accounts also for the environmental impact as a selection driver for the modal shift is needed. In the literature, the environmental impacts are widely neglected by studies unless the study has a specific focus on the environment, as demonstrated in an empirical literature review on transportation buyers' choices [5], which highlighted clearly that the main driver considered is undoubtedly cost. This is the gap we want to fill, by providing an analysis of the main drivers for mode choices that also involve the environmental impact that can be useful for practitioners.

The problem of choosing different transportation modes, called the mode choice problem, is usually solved using the split model [6] in its disaggregate form. It relies on the utility theory [7] and presupposes that the option chosen by an individual is based on the characteristics of that individual and on the attributes of the option. The first application of this theory in the transportation sector was in 1985, by Ben-Akiva and Lerman [8] on transport demand. Since then, various extensions and applications can be found in the literature, and it is still in progress in the related literature. Della Chiara et al. [9] exploited a logit model as a split model, one of the most diffuse, to assess the influence of a ro-ro service between Italy and France, including the service frequency and total shipment time as drivers. Sivakumar et al. [10] used a fractional split model to estimate the fraction of the commodity consumed at each destination zone. Ferrari [11] extended the basic model in a dynamic way that supposes that the evolution over time of transport demand is coupled with a corresponding evolution of transport modes, and investigated the equilibrium of the resulting system and its characteristics (e.g., influences of system characteristics on the transition phase length). Hong et al. [12] evaluated the impact on emissions of five policies in South Korea. Zis and Psarafis [13] presented a modal split model that estimates modal shifts vis-a-vis competing maritime and land-based modes. Meers et al. [14] developed a choice-based conjoint experiment to increase the insight on the preferences of modal choice decision-makers in Belgium. Dong et al. [15] investigated a modal split transport model of two modes that integrates inventory controls to minimise the long-term expected total costs of transport, inventory holding, and backlogging. Wu et al. [6] applied the modal split model to analyse the attractiveness of different means of transportation in the trade transportation system between China and the Koreas.

As demonstrated in our literature review, there is an increasing interest in the literature in the split model, but to the best of our knowledge, there are no available works that applied a consignment-type multinomial logit model that comprises the cost, time, and CO_2 emissions. This is the gap we want to fill: the aim of the paper is to model transport mode choice to replicate the behaviour of decision-makers and to compare it with the minimisation of CO_2 emissions; that is, an analysis of the modal shift strategy between rail, road and sea with the MNL model. The other main contribution of the paper is to create, with the retrieved data similar to Hoover's cost and time diagrams [16], a model, "CO_2Choices", that makes it possible to choose the best transportation mode under the objective of CO_2 minimisation. The following graphs could be an important tool that can be used by practitioners to quickly identify the least polluting transport modes as a function of the linear distance between the origin and destination.

2. Methodology

2.1 Multinomial Logit Model

As already stated, the MNL model is one of the most important discrete models used in travel mode choice problems thanks to its simplicity and flexibility [17], and it supposes that the choice of the individual is based on his characteristics and the attributes options while exploiting utility theory [7]. The model is based on the hypothesis that each user q (q = 1...n) assigns to each alternative i (i = 1...m) from his set of choices *Iq* a certain utility and chooses the one with the highest utility value:

$$U_M = max_i \{U_i\} \tag{1}$$

Nevertheless, the decision-maker does not know all the factors that influence the final decision, so the utility is composed of both a systematic (V_i^q) and a random part (ε_i^q):

$$U_i^q = V_i^q + \varepsilon_i^q \tag{2}$$

The systematic utility V_i^q is a linear-in-parameters function that considers a series of determinants X_i^q (such as cost, travel time...) that affect the final rational choice based on the weights attributed to them by the decision maker (β_k):

$$V_i = \sum_k \beta_k X_{ki}^q \tag{3}$$

The error term ε_i^q represents the deviation from the average value of perceived utility caused by details that cannot be observed. The assumption about the distribution of the error term determines the model [18]: if the random component ε_i^q is identically and independently distributed (IID) according to the Weibull–Gumbel distribution, the discrete model is converted into a MNL model [7].

The IID principle involves two implications: the mean of the unpredicted terms ε_i is equal to zero, and the independence of irrelevant alternatives (IIA). The model with IID has a choice probability that depends on the utility function of each transport mode and is given by:

$$p[i] = \frac{e^{V_i/\theta}}{\sum_i e^{V_j/\theta}} \tag{4}$$

Accordingly, the probability of simulating the correct choices of all decision makers is measured by multiplying each probability p^q with the binary variable $i(q)$, which is equal to 1 when the alternative i is chosen by q and zero otherwise.

$$\prod_{a=1}^{n} p^q[i(q)]\,(X^q, \beta_k, \theta) \tag{5}$$

Exploiting the logarithms' properties and the IIA condition, the same result is obtained with the Maximum Log-Likelihood equation (logL):

$$logL(\beta_k, \theta) = \sum_q^n log\ p^q\ [i(q)](X^q, \beta_k, \theta) \tag{6}$$

To maximise the probability, so the function (6), the vector of parameters β_k/q has to be optimised by non-linear programming algorithms. Discrete mode choice models use various types of factors: generic variables are applied in the utility of all the alternatives, whereas specific attributes are evaluated in different way for each alternative. These kinds of attributes are called "Alternative-Specific Attributes" (ASA) and behave as binary variables and consider some qualitative features that cannot be quantified (such as the comfort of the mode of transport). To quantify all these factors, data can be collected through a Stated-Preference (SP) or a Revealed-Preference (RP) survey: the first technique consists in analysing some different scenarios, the second one collects real data on the effective behaviour of users.

In these types of applications, variables are fixed also for all decision-makers: this means that preferences are homogeneous across individuals. To solve this restriction, the use of a Mixed Logit (ML) model is preferred to MNL. To get a correct application of the MNL model, it is necessary to assign transports to third-party providers (3PL) and respect the IIA condition.

The rho-square value represents the level of accuracy of the model; it varies from zero to one (maximum accuracy) and it is measured as follows:

$$\rho^2 = 1 - \frac{logL(\beta^*)}{logL(0)} \tag{7}$$

where $logL(\beta^*)$ is the optimised Log-Likelihood (logL) equation (with the vector of optimised parameters β_k^*/θ), and $logL(0)$ is the Log-Likelihood equation with coefficients β_k set equal to zero, which has, by definition, no significance.

2.2 Hoover's diagram and extension

The objective of this paper is not only to use the MNL model to analyse the benefits obtained by the application of a modal shift strategy in a specific case study, but also to give some innovative indications to transportation research. The state of the art of transport modal choice does not provide any guidelines regarding transport modal assignment based on carbon emissions. It is important to introduce some new considerations about the topic of green transport modal choice in industry's transportation in light of the EU policy. Hoover defined diagrams regarding transport mode choice based on cost and time factors [16]. One of these is particularly useful because it shows in an approximate way the break-even points between the cost of transport by road, rail, air and sea when varying the distance (Figure 1 left). It also shows that the optimisation of the overall cost for long distances derives from the sequential combination of truck, rail, and ship modes. Sea transport dominates the transport of goods over medium and long distances [19], whereas land transport prevails over medium–short distances due to its flexibility and capillarity [20]. Another Hoover diagram in Figure 1 right compares various types of transport (rail, road and air) based on the transit time and distance.

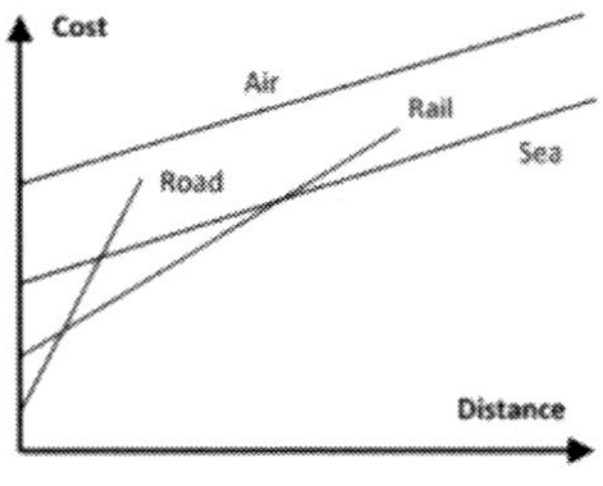

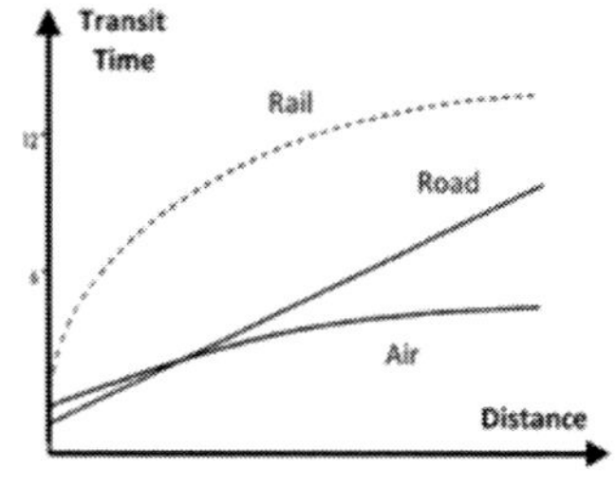

Fig. 1. Hoover's cost and distance diagram (left) and time and distance diagram (right)

$$Tot.Emissions[gCO_2] = Weight\ [tonne] \times Effective\ distance\ [km]\ Carbon\ emission\ factor\ [gCO_2/tonnekm] \quad (8)$$

To reach this goal, all transport modes are compared according to their environmental impact per kilometre. The total amount of emissions is calculated through Eq. (8). For the carbon emission factors, the reader can refer to those listed in Table 1, which are the ones used in our case study. In these calculations, weights represent a useless variable because they are identical for each transport alternative and directly proportional to CO_2 emissions, so they are assumed for simplicity as equal to one tonne. Our extension of the Hoover diagram is reported in the case study section, since it relies on the data presented there.

3. Model development and results discussion

The implementation of these kinds of models can be articulated through three basic steps:

- Specification: definition of the parameters, set of alternatives, and initial hypotheses.
- Calibration: optimisation of the coefficients (β_k, θ) to maximise the likelihood.
- Validation: verification of the goodness of the model's outputs.

The data included in the application refer to the transport schedules of one of the most important companies operating in the chemical sector in Europe, which has headquarters near Padova, Italy. Italian customers represent the leading destinations by sales volume and turnover: the database comprises almost all the national transport shipments made in 2018. In reality, the company assigns all its transport to third-party providers (3PL), but tankers, intercompany and multi-delivery shipments and express deliveries are excluded from the database for simplicity. The final sample totals 1361 despatches, divided between 282 Italian customers. The data have been collected through an RP survey since the model is based on despatches that have already happened. In addition to road transport, the only mode used by the company for this transport, the model will consider rail, sea and air modes. To calculate every single route of each mode, some Italian logistics terminals have been identified: Italy's 30 leading railway centres, 30 ports and 30 airports are introduced into the database. Since Padova is the location of departure, the port and the airport of Venice and the terminal of Padova represent the nearest logistic points for ocean, air and railway modes, respectively. According to the mode chosen, which is indicated by the binary variable $i(q)$, the model will calculate the correct weight (β_k/ θ) to assign to the determinants X_i^q to simulate the real mode choice at best. In the following case study, the freight service attributes that will influence the choice are cost, transit time, CO_2 emissions and availability (which defines if an alternative is available or not).

For the development, it was necessary to collect a lot of information: the value of each attribute for each option and destination. In doing this, some approximations have been made. The costs and transit times of the road transport consignments have been obtained from the rates of real carriers. Toll rates are expressed in terms of €/km and the rates are applied for weight ranges: 2500 < kg < 7500; 7500 < kg < 15000 and 15000 < kg < 30000. For rail and maritime modes, some quotations were collected: the information concerns both groupage and full container loads (20' and 40'). The data collected include the cost and time of the road transport necessary for the initial and final routes. Cost and transit time were generalised with Excel through trend lines as functions of the distance travelled to quantify every attribute for each destination. Although the price lists of the two leading companies in the sector were considered for air consignments, this alternative is not very useful due to the huge weights shipped and short distances that characterise the national territory. Moreover, their CO_2 emissions are very high, and run counter to the minimisation of the environmental impact. A parameter was introduced for availability: it is a binary variable that assumes a value equal to one if the alternative cannot be used, zero otherwise. Hence, the air mode is always set as unavailable a priori, while the availability of other types of transport depends on the province of destination. Obviously, the road mode, thanks to its flexibility and extensive coverage, always represents a feasible solution. The objective of the application concerns the minimisation of CO_2 emissions, and with this aim a parameter is introduced into the model: each type of transport emits a specific amount of CO_2 emissions and there are several methods in the literature to quantify these values. Here the quantification is based on the Activity-Based approach [21]. In Table 1 the recommended carbon emission factors that will be used in the subsequent application are shown. These coefficients are taken directly from [22].

Transport mode	gCO_2 / tonne km
Road	62
Rail	22
Sea	8
Air	602

Table 1. Recommended emission factors

3.1 Relation between linear and real distance

The calculation of real distances represents crucial and indispensable information for all types of transport and attributes. The huge potential of Google Maps Platform has been exploited using the Google Maps Distance Matrix

Application Programming Interface (API) [23] on Excel to compute almost all the road routes and some rail distances. Nevertheless, this technique cannot be exploited for sea and air. Thus, each address was converted into geographic coordinates (latitude and longitude) and the linear distance was measured with the Google Maps Geocoding API. The linear distance between two locations (A,B with latitudes and longitudes defined as a_i and b_i, respectively) is calculated with Equation (9), where 6.378 is the length of the radius of the Earth using the Spherical Law of Cosines [24].

$$d(A,B) = \cos^{-1}(\cos(a_1 - a_2) \times \cos(b_1) \times \cos(b_2) + \sin(b_1) \times \sin(b_2)) * 6.378 \qquad (9)$$

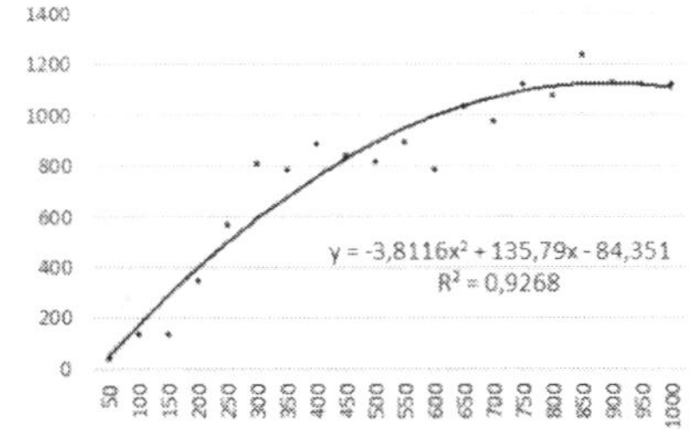

Fig. 2. Best trend between linear and real distances for sea mode

Linear distances between airports were assumed as the routes of the air transport mode. For the unavailable sea routes, we compared some real maritime routes with linear routes to get trend lines to predict sea routes as a function of the linear distances between ports. More than 400 real sea routes were collected in the initial sample, which is significant [25], and the linear distances were categorised into some numbered sections. Then the best trend (exponential, linear, logarithmic, polynomial or potential trend), with the best R^2, was chosen. Figure 2 shows the result for sea transports: the X-axis contains the linear distance categories (in this case, they were divided into classes of 50 km) and the Y-axis reports the averages of the real distances for each category. The same technique is extended also to the other alternatives and samples of almost 500 routes were gathered using Google Maps and categories of 25 km were set (Figure 3).

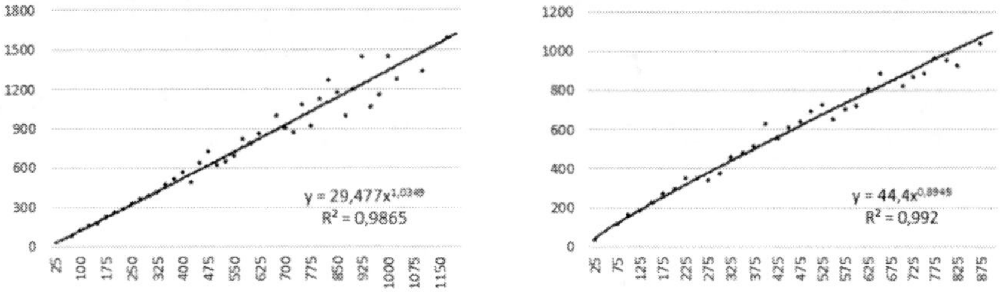

Fig. 3. Best trend between linear and real distances for road mode (left) and rail mode (right)

Table 2 reports for each transport mode the dimension of the categories, the best trend line and the corresponding correlation coefficient (R^2). Thus, the trend line makes it possible to simulate real distances y starting from the linear distance classes, because x represents the number of each category. For example, in the first category of 50 km x is equal to one, in the second category (100 km) x is equal to two and so on. Despite the approximation, R^2 shows very high values that make the use of these distances very reliable in CO_2 calculation.

Transport mode	Class dimensions	Best trend line	Correlation coefficient (R^2)
Road	25 km	$y = 29.477x^{1.03}$	$R^2 = 0.98$
Rail	25 km	$y = 44.4x^{0.89}$	$R^2 = 0.99$
Sea	50 km	$y = -3.81x^2 + 135.79x - 84.35$	$R^2 = 0.97$

Table 2. Best trends for national transport modes

Hence, every distance between two transhipment centres or two ports will be approximated starting from the linear distance, and the connections with the origin and the destination will be rightly measured as road routes and multiplied by the specific carbon emission factor (62 gCO_2/tkm).

3.2 Results

As a first goal, the MNL model to simulate the company's decision about the model transport mode choice has been applied. The MNL assigns an optimised weight to the model's attributes in order to replicate the decision-maker's choices. Alternatives included in the model are road, rail, sea and air, whereas attributes are cost [€], transit time availability [days] and CO_2 emissions [$kgCO_2$/tonnekm]. In 2018 the only means of transport used by the analysed company was trucks, so the model will have to assess this choice in all of the 1361 instances. Table 3 shows the most interesting values to analyse: β_k/q coefficients of each attribute, LogL (0), LogL (β*), the value of the ρ^2 test (which is the best indicator of the effectiveness of MNL models) and the number of errors committed by the specific application. Although the values of the attributes' coefficients depend on the unit of measures with which the parameters are quantified, the results reveal that the availability correctly represents the first determinant of choice, whereas time and cost are used secondly by the model to distinguish the alternatives. Emissions are not much considered, so the sign of its coefficient is wrongly positive: this means that the model does not replicate correctly the real choices. Although the value of the ρ^2 test is almost perfect, the model chooses the wrong alternative 11 times (0.8%). To reduce errors, the air mode is eliminated from all the calculations and the results are shown in Table 3. In this last case, all the signs are correct, there are no errors in the simulation, and the result of the test is even better than before. This means that the model works perfectly without considering the air mode, which is a useless alternative also in real national transport consignments.

In order to minimise CO_2 emissions, it is necessary to set a priori $i(q)$=1 in correspondence with the most environmentally friendly alternative so that the approach returns for each shipment the choice probability of each mode if the company in 2018 had decided to minimise the carbon dioxide emissions of its national transport schedule. This application represents a novelty compared to the state of the art because nobody has ever used the MNL model with this particular objective. Emissions' calculation refers to Equation (8) and depends on the recommended carbon emission factors ($kgCO_2$/tonnekm), transport weights and effective distances travelled. Table 4 shows the results derived from CO_2 optimisation: it is evident that the exclusion of the air mode led to better values also in this case. All the coefficients show signs and values that fit with the initial hypothesis: the model recognises availability as the most important attribute, then the choice probability depends on the emissions factor, as expected; cost and time have a rather low incidence on the selection and all the negative signs show that the MNL model works correctly. However, the ρ^2 test assumes a worse, but almost acceptable, value (~0.95). The instances that present a wrong choice are only three, and they are caused by the very small difference in carbon production: in fact, trucks prevail over the railway even though the latter emits less $kgCO_2$ than the former. The ratio of β coefficients represents another good indicator of the model's accuracy: if we divide β_{emiss} and β_{cost} we will get a sort of "Value of Emissions" (VoE) that indicates the importance given to the emissions factor. Excluding the air mode from the calculation, the VoE grows from 1079 €/$kgCO_2$ to 1530 €/$kgCO_2$ and this means that the carbon emissions assume a greater importance in the last application. These values are distantly comparable with those of the first simulation, where the VoE (without air mode) is equal to 1.84 €/$kgCO_2$.

With air mode		**Without air mode**	
Values of β	**Values of accuracy**	**Values of β**	**Values of accuracy**
β_{cost} = - 0.52	*LogL (0)* = - 819.40	β_{cost} = - 0.0057	*LogL (0)* = - 649.36
β_{time} = - 52.16	*LogL (β*)* = - 4.83	β_{time} = - 23.97	*LogL (β*)* = - 0.0034
β_{avail} = - 279.27	ρ^2 = 0.99	β_{avail} = - 101.91	ρ^2 = 0.99
β_{emiss} = + 0.73	N° errors = 6 (0.4%)	β_{emiss} = - 0.01	N° errors = 0

Table 3. Company's choices with and without air mode

With air mode		**Without air mode**	
Values of β	**Values of accuracy**	**Values of β**	**Values of accuracy**
β_{cost} = - 0.0001	*LogL (0)* = - 819.40	β_{cost} = - 0.0003	*LogL (0)* = - 649.36
β_{time} = - 0.0753	*LogL (β*)* = - 70.72	β_{time} = - 0.1596	*LogL (β*)* = - 33.26
β_{avail} = - 110.23	ρ^2 = 0.91	β_{avail} = - 135.6015	ρ^2 = 0.95
β_{emiss} = - 0.1897	N° errors = 11 (0.8%)	β_{emiss} = - 0.5699	N° errors = 3 (0.2%)

Table 4. CO_2 minimisation with and without air mode

The model is able to express both cases with great significance. The information that can be extrapolated from this latter numerical case is numerous and very interesting because we can analyse the effects of a shift in modal strategy. In 2018 the company chose only road transport, and the model makes it possible to study the consequences that a different choice would have entailed. Excluding the three errors committed by the model, the CO_2 minimisation led to the following optimisations:

- 62.5% of transport consignments should be transferred from road to a different mode, or rather:
  - 642 deliveries (47.3%) should be transferred to rail;
  - 206 deliveries (15.2%) should be made by sea mode.
- The emissions saved in a year would be at most 138.74 $tonCO_2$, which means an overall reduction of 41% (if all the national transport consignments of 2016 took place on the road, as really happened, the impact would be 336.11 $tonCO_2$), in particular:
  - 60.88 $tonCO_2$ (43,9%) thanks to the railway;
  - 77.86 $tonCO_2$ (56,1%) thanks to the sea transport;

- On average, the saving for each of the 1358 transport consignments would be equal to 102 $kgCO_2$, made up of:
  - 95 $kgCO_2$ for each of the 642 rail deliveries;
  - 378 $kgCO_2$ for each of the 206 sea deliveries.

The MNL model confirms that the least polluting modes, from which the highest emission savings are obtained, are rail and sea mode. Nevertheless, these alternative solutions lead to an increase in the transit time and cost of transport: in order to minimise carbon emissions, an increase in transit time of more than one day for each shipment and an increase of the costs equal to 50% (principally caused by the transfer to sea mode) are necessary.

In the following graphs, the x-axis considers the linear distance [km] between two generic points, whereas the y-axis reports the total amount of carbon emissions [$kgCO_2$/tonne]. The assessment was conducted with respect to the Italian territory, divided into 3 main areas: northern, central and southern Italy. Table 5 indicates that the average distances assume different values for northern, central and southern Italy, in particular for sea and rail modes (obviously, road does not provide any change of transport mode). It is derived that the higher the total road distance is, the greater the initial carbon footprint and, consequently, the value of the intercept with the y-axis. So, the CO_2 emissions caused by each mode are measured according to Equation (8) and the best approximation lines are chosen in order to maximise the correlation coefficient (R^2). Figure 4 shows the emissions–distance diagram for northern, central and southern Italy. Indeed, it is possible to outline some general principles that can help the user to choose the type of transport and apply a modal shift strategy: in northern Italy, as in the Hoover cost diagram, the optimal sequence is defined by the succession of road, rail and sea modes. Air mode can be excluded from this study due to its high pollution. In general, we can say that for short distance routes it is advisable to use road vehicles, while for long distances sea transports. Furthermore, the emissions balancing points between rail–road and rail–sea are located around 150 km and 450 km, respectively. For central and southern Italy, the situation is slightly different due to the presence of a single point of balance between truck and sea, which stands around 150 km. In fact, the lower distances of ports and higher distances with rail terminals give the sea mode an advantage such that it assumes a predominant role in the modal shift strategy. It is important to understand that these results depend strictly on the morphology and the specific infrastructure network of the territory: this analysis could be considered valid only within the Italian borders.

Zone	Air	Sea
Northern Italy	100	200
Central Italy	125	135
Southern Italy	100	135
Northern Italy	100	200

Table 5. Average distances for air, sea and rail modes

These considerations are not necessarily always true, and it would be interesting to compare them with a specific case or rather the MNL application. In the previous study, the model allowed us to assign to each destination the least polluting alternative, so it is possible to gather all the distances in different ranges and match the results with the emission–distance diagram of northern Italy, considering that the company is located in Padova (Veneto region). In Table 6 all 1358 transport consignments are categorised according to the road distance travelled (which differs slightly from linear distances but is still quite comparable) in eleven categories. Hence, for each zone the predominant alternative is selected according to the sum of all choices. We can see that the final results reflect the general guidelines quite well because the balancing points are very similar to those in the emissions–distance charts: we see that the values of these points are more or less 150 km and 450 km.

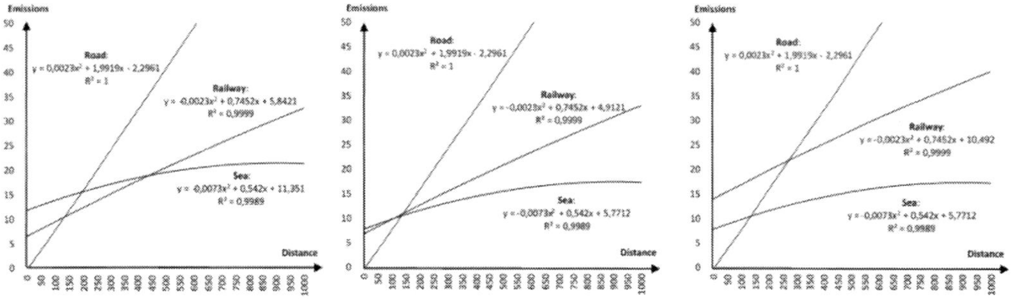

Fig. 4. Emissions–distance diagram for northern (left), central (centre) and southern (right) Italy origins

In this study the transition from road to rail is attested at around 100/200 km and the sea prevails over the railway in the 400<=km<500 category. However, between 200 and 300 km the gap between truck and rail assignments is almost null and there is a sort of equivalence. Despite this, it is evident that the element of greatest incongruity is represented by the choice of destinations that are about 600/700 km: the diagram indicates sea transport as the best alternative, while the

MNL model finds that rail is best. This incongruence could be caused by the Italian morphology: at this distance lie destinations in the Lazio region and the choice of sea would imply the circumnavigation of the whole eastern and southern part of the peninsula, which means an excessively long path compared to the direct link offered by the Italian railway network. Indeed, we can prove this affirmation by analysing the partition of modal choice according to each Italian region (Table 7 and Figure 5): the best solution for transport to the Lazio region is definitely the railway due to its choice probability of 98% (51 times out of 52 shipments). Although the Marche and Umbria regions are closer to the origin, it is better to use the sea mode in order to minimise CO_2 emissions thanks to the opportunity of landing in the port of Ancona.

Distance	Total	Road	Rail	Sea	Best
< 50km	167	167	-	-	Road
50<=km<100	74	62	12	-	Road
100<=km<200	349	119	228	2	Rail
200<=km<300	348	162	164	22	Rail
300<=km<400	188	-	146	42	Rail
400<=km<500	19	-	3	16	Sea
500<=km<600	62	-	28	34	Sea
600<=km<700	51	-	44	7	Rail
700<=km<800	96	-	17	79	Sea
800<=km<900	3	-	-	3	Sea
> 1000km	1	-	-	1	Sea
Total	**1358**	**510**	**642**	**206**	

Table 6. The partition of modal choice in the case study based on effective road distance

Italian Region	Total	Road	Rail	Sea	Best
Veneto	256	242	14	-	Road
Friuli Venezia Giulia	29	-	25	4	Rail
Trentino Alto Adige	22	-	22	-	Rail
Lombardia	485	234	251	-	Rail
Emilia Romagna	93	34	39	20	Rail
Piemonte	145	-	145	-	Rail
Toscana	66	-	57	9	Rail
Lazio	52	-	51	1	Rail
Marche	39	-	-	39	Sea
Umbria	10	-	-	10	Sea
Abruzzo	54	-	21	33	Sea
Molise	2	-	-	2	Sea
Campania	99	-	17	82	Sea
Basilicata	5	-	-	5	Sea
Calabria	1	-	-	1	Sea
Total	**1358**	**510**	**642**	**206**	

Table 7. The partition of modal choice in the case study based on Italian regions

Fig. 5. The map of the partition of modal choice in the case study based on Italian regions

The most important emissions reductions derive from the shift to sea mode for shipments in Campania (41.5%, or rather 57.56 tonCO_2) and to the use of rail mode for shipments directed to Lazio and Piemonte (17% and 16% reduction, respectively), whereas the shift from road to rail in the Veneto region for 14 shipments leads "only" to a reduction of 0.04% (5.5 tonCO_2). Except for Veneto, Lombardia and Emilia Romagna, which continue to need some shipments by truck, the other regions shift all their transport to more sustainable alternatives. Finally, other interesting information that can be obtained from this application is that the emissions caused by shipments at the current state could already be optimised by a modal shift strategy with the same costs: the result would be a reduction of 82.8 tonCO_2, so 60% of the maximum possible optimisation (138.74 tonCO_2) and 24.6% of what it would be if all the goods travelled by road (336.11 tonCO_2). The total emission reduction corresponds to 510 (37.5%) deliveries by truck, 642 (47.3%) and 206 (15.2%) deliveries by rail and sea, respectively. However, at the same costs it could be possible to have a reduction of 24.6%: in this case, the partition of road, rail and sea would become roughly 72%, 22% and 6%, respectively.

4. Conclusion

This paper offers a better understanding of freight transport modal choice and new contributions to the literature through the application of the MNL model in transport modal choice under the objective of CO_2 emission minimisation and the definition of new guidelines in order to identify the most sustainable transport solution as a function of distance along the same lines as Hoover's diagrams. The case study refers to part of the national outbound transport consignments of a company based near Padova (Italy), in 2018. The results obtained make it possible to highlight some significant aspects that can support the planning of shipments of goods and the choice of the mode of transport in relation to parameters that are already usually considered (particularly costs and distances), but also considering the aspect of environmental sustainability as a decision-making factor (represented specifically by CO_2 emissions).

The application of the MNL model shows a satisfactory overall fit ($\rho^2 \sim 0.95$). Except for three wrong instances (0.2%), the model manages to simulate the application of a modal shift strategy and the change to more sustainable transport modes. The information that could be extracted is very interesting; for example 62.5% of shipments should be transferred from truck to a different mode and the total emissions saved in a year would be at most 138.74 tonCO_2, or rather a reduction of 41%. The same results allow us to make a different and detailed analysis, and in fact it was possible to show another important contribution to the literature, along the same lines as Hoover's diagrams, which identify the cheapest and the fastest transport modes according to the distance travelled. In other words, Hoover offers some guidelines based on distances in order to choose the best transport mode that minimises the cost and the transit time, whereas this paper introduces some new general principles that can help users to choose the transport mode that minimises CO_2 emissions according to the linear distance between any origin and destination within Italy. In particular, for locations in northern Italy, the emissions balancing points between rail–road and rail–sea are located around 150 km and 450 km, respectively. Despite this, some questions remain outstanding, and there are some points that need to be further explored. A lot of the initial data are obtained from approximations: the model's attributes assume values that are collected through requests for quotations that have been generalised to complete the data set. In fact, for each transport mode (road, rail, sea, air) it is necessary to measure the value of each variable (cost, time, CO_2 emissions, availability): from the number of instances equal to 1361 consignments, it is evident that the amount of data needed is quite substantial. For simplicity, the weight categories that are considered in the model are $2500 < kg < 7500$, $7500 < kg < 15000$, and $15000 < kg < 30000$ and all road shipments are assumed as to be directly transported by lorry from the origin to the customer. Also, the costs and transit times should be measured specifically for each case and not approximated by trend lines.

5. References

[1] International Energy Agency (2017). World Energy Outlook 2017 (2017). ISBN: 9789264282308. https://doi.org/10.1787/WEO-2017-EN.

[2] Farahani, R.; Rezapour, S.; Kardar, L. (2011). Logistics Operations and Management, Elsevier 2011, ISBN: 9780323165204. https://doi.org/10.1016/C2010-0-67008-8.

[3] Steadieseifi, M.; Dellaert, N.P.; Nuijten, W.; Van Woensel, T.; Raoufi, R. (2014). Multimodal freight transportation planning: A literature review. Eur. J. Oper. Res., 233, 1–15. https://doi.org/10.1016/j.ejor.2013.06.055.

[4] de M. Pinto, J.T.; Mistage, O.; Bilotta, P.; Helmers, E. (2018). Road-rail intermodal freight transport as a strategy for climate change mitigation. Environ. Dev., 25, 100–110. https://doi.org/10.1016/j.envdev.2017.07.005.

[5] Flodén, L.; Bärthel, F.; Sorkina, E. (2017). Transport buyers choice of transport service – A literature review of empirical results. Res. Transp. Bus. Manag., 23, 35–45. https://doi.org/10.1016/j.rtbm.2017.02.001.

[6] Wu, S.; Sun, Y.; Yang, Z. (2020). Analyzing the trade transportation and its demand on multimodal transportation system between China and Koreas. Marit. Bus. Rev., 5, 84–98. https://doi.org/10.1108/MABR-08-2019-0033.

[7] McFadden, D. (1981) Econometric Models of Probabilistic Choice. In: Manski, C. and McFadden, D., Eds., Structural Analysis of Discrete Data with Econometric Applications, MIT Press, Cambridge, 198-272.

[8] Ben-Akiva, M.; Bierlaire, M. (1999). Discrete Choice Methods and their Applications to Short Term Travel Decisions. Handbook of Transportation Science, Kluwer, 5–33. https://doi.org/10.1007/978-1-4615-5203-1_2.

[9] Dalla Chiara, B.; Deflorio, F.P.; Spione, D. (2008). The rolling road between the Italian and French Alps: modeling the modal split. Transp. Res. Part E Logist. Transp. Rev., 44, 1162–1174.

[10] Sivakumar, A.; Bhat, C. (2002). Fractional Split-Distribution Model for Statewide Commodity-Flow Analysis. Transportation Research Record: Journal of the Transportation Research Board, 1790, 1, 80-88. https://doi.org/10.3141/1790-10.

[11] Ferrari, P. (2014). The dynamics of modal split for freight transport. Transp. Res. Part E Logist. Transp. Rev., 70, 163–176. https://doi.org/10.1016/J.TRE.2014.07.003.

[12] Hong, S.; Chung, Y.; Kim, J.; Chun, D. (2016). Analysis on the level of contribution to the national greenhouse gas reduction target in Korean transportation sector using LEAP model. Renew. Sustain. Energy Rev., 60, 549–559. https://doi.org/10.1016/J.RSER.2015.12.164.

[13] Zis, T.; Psaraftis, H.N. (2017). The implications of the new sulphur limits on the European Ro-Ro sector. Transp. Res. Part D Transp. Environ., 52, 185–201. https://doi.org/10.1016/J.TRD.2017.03.010.

[14] Meers, D.; Macharis, C.; Vermeiren, T.; van Lier, T. (2017). Modal choice preferences in short-distance hinterland container transport. Res. Transp. Bus. Manag., 23, 46–53. https://doi.org/10.1016/j.rtbm.2017.02.011.

[15] Dong, C.; Transchel, S.; Hoberg, K. (2018). An inventory control model for modal split transport: A tailored base-surge approach. Eur. J. Oper. Res., 264, 89–105. https://doi.org/10.1016/j.ejor.2017.05.010.

[16] Florence, P.S.; Hoover, E.M. (1949). The Location of Economic Activity. Econ. J., 59, 221. https://doi.org/10.2307/2226687.

[17] Arbués, P.; Baños, J.F.; Mayor, M.; Suárez, P. (2016). Determinants of ground transport modal choice in long-distance trips in Spain. Transp. Res. Part A Policy Pract., 84, 131–143. https://doi.org/10.1016/j.tra.2015.06.010.

[18] Patterson, Z.; Ewing, G.O.; Haider, M. (2010). How different is carrier choice for third party logistics companies?. Transp. Res. Part E Logist. Transp. Rev., 46, 764–774. https://doi.org/10.1016/j.tre.2010.01.005.

[19] Orlic Protega, A.; Rogic, K.; Vrdoljak, J. (2011). Logistic approaches to port management system. Ann. DAAAM Proc. Int. DAAAM Symp., 22, 1053–1054.

[20] Mlinaric, T.J.; Rados, B.; Vajdic, M. (2017). Proposal for the implementation of the european rail traffic management system (ertms) to the railway network in the republic of Croatia. Ann. DAAAM Proc. Int. DAAAM Symp., 109–117. https://doi.org/10.2507/28th.daaam.proceedings.014.

[21] European Commission (2017). COMMISSION REGULATION (EU) 2017/2400 of 12 December 2017 implementing Regulation (EC) No 595/2009 of the European Parliament and of the Council as regards the determination of the CO2 emissions and fuel consumption of heavy-duty vehicles and amending Directive 2007/46/EC of the European Parliament and of the Council and Commission Regulation (EU) No 582/2011.

[22] Cichosz, M.; Nowicka, K.; Pluta-Zaremba, A.; Moritz, N. (2018). Toolbox element: CO2 calculator deliverable 1.2.9 version 2 02/2018.

[23] Google Developers (2021). Distance Matrix API. https://developers.google.com/maps/documentation/distance-matrix/overview (accessed July 21, 2021).

[24] Gellert, W. (1989). The VNR concise encyclopedia of mathematics, 776.

[25] Sallauka, A.; Likaj, R. (2019). Performance of Dijkstra, Floyd and Astar Algorithms for Urban Transport Lines. Ann. DAAAM Proc. Int. DAAAM Symp., 21, 265-270.

DOI: 10.2507/32nd.daaam.proceedings.063

MACHINABILITY OF SPRING STEEL 42SICR

Ondrej Marsalek, Jaroslava Fulemova,
Tomas Zatloukal & Simon Syrovatka

This Publication has to be referred as: Marsalek, O[ndrej]; Fulemova, J[aroslava]; Zatloukal, T[omas] & Syrovátka, S[imon] (2021). Machinability of Spring Steel 42SiCr, Proceedings of the 32nd DAAAM International Symposium, pp.0433-0438, B. Katalinic (Ed.), Published by DAAAM International, ISBN 978-3-902734-33-4, ISSN 1726-9679, Vienna, Austria
DOI: 10.2507/32nd.daaam.proceedings.063

Abstract

Spring steels represent an important group of stainless structural steels. Thanks to their mechanical properties and especially their resistance to alternating, ie static and dynamic stresses are used for springs, mainly in the automotive industry. With constantly evolving technologies and tools, new materials and processes of heat treatment are also being developed. The aim of this work is to determine the machinability and its classification of spring steel 42SiCr, where one set of samples was heat-treated by conventional hardening and tempering and the other set of samples according to the Q-P process (Quenching and Partitioning Process) with extended residence time at redistribution temperature to obtain high strength while maintaining ductility. An experiment was performed in which the machinability of the tested spring steel samples was determined. The result of the experiment are graphs determining wears versus times measurement, graphs determining Taylor equation and a table with the determined machinability classes.

Keywords: steel 42SiCr; machinability; tool-life; flank wear

1. Introduction

With constantly evolving technologies and tools, new materials and processes for heat treatment are also being developed. Such materials, which by their chemical composition, physical and mechanical properties best suit certain machine components, thanks to which their functionality, efficiency, safety and other aspects will be better fulfilled. Due to the appropriate choice of heat treatment, it is possible to further improve the properties of the material. Heat treatments usually serve to enhance one particular property of a material. For example, hardening to increase hardness for parts requiring good abrasion resistance, or annealing to increase toughness for dynamically stressed parts.

Because it would be best to have materials with high strength and hardness and sufficient toughness, heat treatment techniques are therefore being developed to meet these properties. These properties are suitable, for example, for spring steels, which are used for these dynamically stressed components. However, high strength and ductility affect machinability. Machinability is important for a complete description of new material. Thanks to it, it is possible to determine suitable cutting conditions, which will increase the productivity of the cutting process and improve the quality of the surface of the part.

Machinability of a material is a complex technological property expressing the efficiency with which a material can be processed by machining technology at low costs under given technological conditions. This is one of the most important features in terms of machining technology. The machinability of a given material is difficult to define by the absolute value of a quantity. This is the overall influence of mechanical and physical properties (strength, hardness, thermal conductivity of the material), chemical composition (carbon content, alloying elements), the microstructure of the material, the heat treatment used and the method of manufacturing the semi-finished product. However, it cannot only be related to the material being machined, as it also depends on the cutting process and the properties of the machining tool. That is, on the method of machining, cutting environment, cutting conditions, tool material and its cutting ability, as well as tool geometry and other characteristics of the cutting proces [1].

The work deals with the determination of machinability and classification of spring steel 42SiCr during various heat treatments. These heat treatments are conventional quenching with subsequent tempering and a Q – P process with an extended residence time at the redistribution temperature, which ensures a suitable strength-to-ductility ratio [2]. The determination of machinability will be used to determine suitable cutting conditions, which will make it possible to machine this steel more efficiently.

2. Test material 42SiCr

Within the Department of Materials and Engineering Metallurgy, steel with the designation 42SiCr was developed in cooperation with other workplaces. It is an experimental high-strength low-alloy steel. The main alloying elements of this steel are carbon, silicon, manganese and chromium. More detailed chemical composition is contained in Table 1. By default, the 42SiCr material has a tensile strength Rm of 980 MPa, a hardness of 281 HV 10, which is a Vickers hardness with a load of 10 kiloponds, and a ductility A of about 15.4% [3]. By suitable heat treatment, it is possible to increase the strength and hardness limit, but at the expense of reducing the ductility.

C [%]	**Si** [%]	**Mn** [%]	**Cr** [%]	**S** [%]	**P** [%]	**Mo** [%]	**Nb** [%]
0.43	2.03	0.59	1.33	0.004	0.009	0.03	0.03

Table 1. Chemical composition of 42SiCr steel [3]

A total of two different samples were provided to the test material, which differed from each other in their chemical properties, which are the tensile strength Rm, hardness and ductility A. The chemical composition of all samples was the same. Different chemical properties of the samples were achieved using different types of heat treatment or different conditions during certain heat treatment. The semi-finished products of the samples were round bars with a diameter of 20 mm.

The types of heat treatment were:
- Hardening and tempering
- Q – P process with extended residence time at redistribution temperature

During quenching and tempering, the samples were heated in an oven to 950 °C for 21 minutes, followed by quenching in water to room temperature, then further heating in an oven to 250 °C for 10 minutes, and finally free cooling to room temperature. During the Q – P process with an extended residence time at the redistribution temperature, the samples were heated in an oven to 950 °C for 21 minutes, followed by quenching in water for 5 seconds. Next, immersion the samples in a salt bath at a temperature of 200 °C.

Then heat in an oven to 250 °C for 20 minutes and then cool to room temperature. The Q – P process is a process whose main goal is to stabilize untransformed austenite with carbon precipitated from supersaturated martensite and to suppress the formation of carbides by suitable alloying [2]. The following table shows the values of mechanical properties for the material 42SiCr in the basic state, in the state after hardening and tempering and in the state after the Q – P process.

State	**Tensile strength R_m** [MPa]	**Hardness HV 10**	**Ductility** [%]
Basic	980	281	15.4
Hardening and Tempering	2095	629	11.3
Q–P process	1960	575	15.7

Table 2. Mechanical properties of 42SiCr steel and tested samples

3. Method of determining machinability

A long-term machinability test was chosen to determine machinability. Its criterion is the value of the cutting speed and it can be done by turning or milling. The long-term test is performed under the same cutting conditions as during actual machining. Due to this fact, the results it shows are credible and corresponding to the facts. It is considered to be a basic test of machinability, according to which the objectivity of other tests is assessed. The principle is that the tool durability during machining of the tested material is determined using an experiment, which is then compared with the tool durability obtained during machining of the reference materiál [5].

It is desirable to perform a long-term machinability test under normative conditions. These mainly include the machining tool and the cutting conditions. This is to increase the reliability and comparability of test results, regardless of where and by whom the test was performed. Therefore, the experiment was designed according to the international standard ISO 3685. This international standard contains recommendations aimed at unifying procedures to increase the reliability and comparability of test results when comparing cutting tools, workpiece materials, cutting conditions or cutting fluids. To achieve these objectives as closely as possible, the standard includes and specifies in detail the recommended reference materials and the working conditions to be observed [6].

4. Experiment

The experiment was performed on a CNC lathe EMCO Maxxturn 25, which best met the defined conditions of the standard. The intensity of the wear was measured on a KEYENCE VHX-6000 digital microscope with a magnification of up to 200×. According to the specifications of the ISO 3685 standard, a tool with the designation SSRCL 1212F09 with a cutting insert SCMT 09T304-14 was chosen. The following figure (Fig. 1) shows a view of the working space of the machine. The picture shows a test sample that is clamped in a collet and support tip and a cutting tool.

Fig. 1. View into the working space of the machine during the test of machinability of 42SiCr steel

Before the start of each measurement, and every time the cutting edge and the cutting speed changed, a picture of a new blade was taken on a digital microscope, which can be seen in the following figure (Fig. 2). After a picture of the new cutting edge was taken, the insert was clamped in the tool holder and the cutting conditions were defined in the machine program. These conditions were selected from the ISO standard and verified by pre-experiment. The following were selected as the default cutting conditions:

- Cutting speed v_c = 70 m/min
- Feed f = 0.1 mm/rev
- Depth of cut a_p = 0.6 mm
- Flank wear criterium VB_{crit} = 0.3 mm

Fig. 2. New cutting edge

After machining the first layer, the insert was removed from the holder, cleaned and then measured on a microscope. The VBB max value was measured on the flank [4]. The measurement on the microscope was performed by plotting the measuring dimension from the auxiliary line defining the original cutting edge, which was used to measure the VBB max wear. The following figure shows the development of the flank wear under the conditions given above for samples heat-treated by the Q-P process. The figure shows the cutting edge from left to right after the first, third and ninth crossings for which the VBcrit value has been reached.

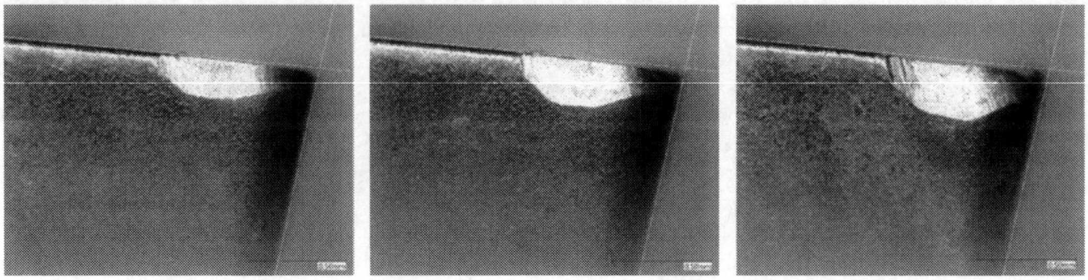

Fig. 3. Development of flank wear at v_c 70 m/min for sample after Q-P proces

This procedure was the same for all selected cutting speeds. According to the standard, it was necessary to select at least three values of the cutting speed. Cutting speeds were increased according to a standard defined index. All measured values were recorded in a table. The values of the diameter D and the cutting length L are needed to calculate the time in cut t and from that of the total time in cut Δt which is necessary to compile a graph of determining wears versus times measurement.

Cut No.	VB_B max. [μm]	D [mm]	L [mm]	Δt [min]
1.	140	18.8	110	0.928
2.	180	17.6	109.7	1.795
3.	200	16.4	109.7	2.602
4.	230	15.2	109.4	3.348
5.	240	14	109.4	4.036
6.	260	12.8	109.1	4.662
7.	290	11.6	109.1	5.230
8.	300	10.4	108.8	5.738

Table 3. Measured and calculated values VB_B max. and Δt [8]

5. Evaluation of the experiment

From the values of flank wear VBB max. and total cutting time Δt, it is possible to construct a graph showing the dependence of VBB flank wear on the cutting time t. After plotting all points, a curve can be constructed showing the course of the flank wear for the evaluated cutting speed. This was done for all selected cutting speeds, which were v_{c1} = 70 m/min, v_{c2} = 78 m/min, v_{c3} = 87 m/min and v_{c4} = 97 m/min. At the highest cutting speed, the VB_{crit} value was reached after the first cut. The resulting overall graph of VBB max versus t is shown below in the figure (Fig. 4.).

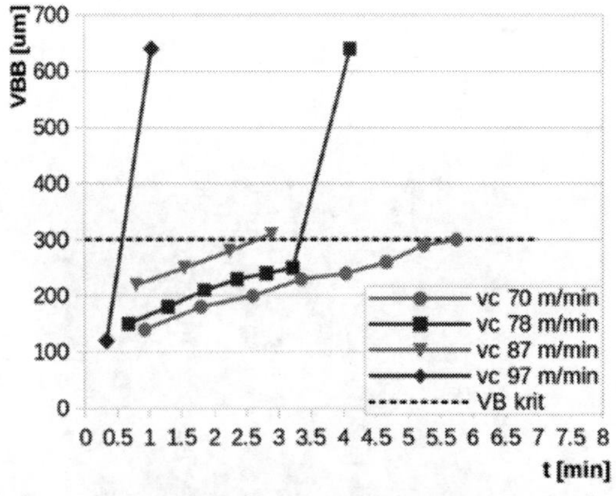

Fig. 4. Graph determining of flank wear versus time

From this graph, it was possible to determine tool life for certain cutting speeds. This was followed by the construction of a graph expressing a graphical representation of Taylor's equation. This graph can be seen in figure (Fig. 5.). From the graph in figure (Fig. 5.), the Taylor equation was determined and from it the machinability index, which is needed to determine the machinability class. The cutting speed vc1 was not considered in the graph because it achieved a tool-life of less than 1 min. Three cutting speeds values can be used to properly evaluate the experiment. The experiment also had to be performed on a reference material defined by the standard, which is carbon steel with the designation C45 [6].

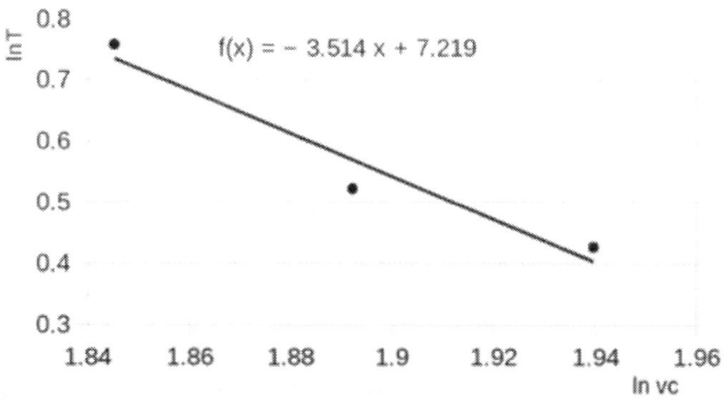

Fig. 5. Graph of dependence of durability on cutting speed

The following table shows the resulting machinability classes for the 42SiCr spring steel test samples heat treated by hardening and tempering and the Q-P process. The reference material has a machinability class of 14b, so both test specimens have been found to have poorer machinability because they have a lower machinability class [7].

Test sample	Machinability class
Hardened and tempered	12b
Q–P process	12b

Table 4. Specified machinability classes of test samples

6. Conclusion

This work aimed to determine the machinability and classification of various heat-treated test samples of 42SiCr spring steel. The machinability of the test samples was determined using a long-term machinability test, which was selected for reasons of the credibility of the measured results. The long-term machinability test was performed according to the ISO 3685 standard to increase the reliability and comparability of the results. The standard deals with tool-life testing with single-point turning tools. Experimental designs were performed from the knowledge gained from the standard. Furthermore, the types of cutting tools and cutting conditions were determined according to the characteristics specified by the standard and, last but not least, the methods of clamping the semi-finished products and the test procedures. The designs were followed by the actual implementation of experiments to determine machinability. To completely determine the machinability of the tested materials, it was necessary to perform a long-term machinability test for the reference steel marked C45. The machinability class 12b was issued for the tested samples, which indicates slightly worse machinability than that of the reference material.

This finding was very surprising because the chemical composition of the 42SiCr spring steel and the mechanical properties of the test specimens obtained by the various heat treatments indicated that the machinability would be significantly worse. In the future, this experiment needs to be repeated with a higher number of test specimens. It is assumed that these results were obtained due to the small number of test samples. With a larger amount of test material, a more thorough long-term machinability test will be performed, which will probably show that the machinability of the heat-treated 42SiCr steel samples is indeed much worse.

7. Acknowledgments

This article was created under the project SGS-2019-008: Research and Development for Innovation in the Field of Manufacturing Technology – Machining Technology III.

8. References

[1] Qehaja, N. E.; Salihu, A. H.; Zeqiri, H. M.; Osmani. H. & Zeqiri, F. (2012). Machinability of metals, methods and practical application. Proceedings of the 23rd DAAAM International Symposium, Volume 23, No.1, ISSN 2304-

1382 ISBN 978-3-901509-91-9, CDROM version, Ed. B. Katalinic, Published by DAAAM International, Vienna, Austria, EU, 2012

[2] Bublikova, D.; Jenicek, S.; Jirkova, H.; & Stehlik, A. (2018). Heat Treatment of Closed-Die Forgings of Low-Alloy AHSS Steels, Proceedings of the 29th DAAAM International Symposium, pp.0714-0718, B. Katalinic (Ed.), Published by DAAAM International, ISBN 978-3-902734- 20-4, ISSN 1726-9679, Vienna, Austria DOI: 10.2507/29th.daaam.proceedings.103

[3] Janda, T.; Jirkova, H. & Jenicek, S. (2018). Comparing Properties of the 42SiCr Steel After Conventional Heat Treatment and QP Processing, Proceedings of the 29th DAAAM International Symposium, pp.1164-1167, B. Katalinic (Ed.), Published by DAAAM International, ISBN 978- 3-902734-20-4, ISSN 1726-9679, Vienna, Austria DOI: 10.2507/29th.daaam.proceedings.167

[4] Zeqiri, H. M.; Qehaja, N. E.; Zeqiri, F.; Salihu, A. H. & Osmani, H. (2012). Research of flank wear in turning of steel CK60. . Proceedings of the 23rd DAAAM International Symposium, Volume 23, No.1, ISSN 2304-1382 ISBN 978-3-901509-91-9, CDROM version, Ed. B. Katalinic, Published by DAAAM International, Vienna, Austria, EU, 2012

[5] Rejshekhar, L.; Prasad, K. & Mohankumar, G. C. (2014). An Experiment Investigation on Machinability Studies of Steels by Face Turning. 3rd Internacional Conference on Materials Processing and Characterisation (ICMPC 2014). [Online]. Available: https://core.ac.uk/download/pdf/82785849.pdf

[6] International Standard ISO 3685. Tool-life testing with single-point turning tools. 2nd edit. Switzerland: International Organization for Standardization, 1993.

[7] Švarc, V. (2011). Obrobitelnost materiálů skupiny b – Machinability of group materials b. Master thesis. Department of Machining technology. University of West Bohemia. Pilsen. Czech Republic.

[8] Maršálek, O. (2020). Obrobitelnost pružinové oceli 42SiCr – Machinability of spring steel 42SiCr. [Online]. Available: https://dspace5.zcu.cz/bitstream/11025/41588/1/DP_Marsalek_Ondrej.pdf

DOI: 10.2507/32nd.daaam.proceedings.064

E-INTEGRATED MARKETING COMMUNICATION OF MICRO ENTERPRISES

Bruno Busic, Katarina Stavlic & Kresimir Lackovic

This Publication has to be referred as: Busic, B[runo]; Stavlic, K[atarina] & Lackovic, K[resimir] (2021). E-Integrated Marketing Communication of Micro Enterprises, Proceedings of the 32nd DAAAM International Symposium, pp.0439-0445, B. Katalinic (Ed.), Published by DAAAM International, ISBN 978-3-902734-33-4, ISSN 1726-9679, Vienna, Austria

DOI: 10.2507/32nd.daaam.proceedings.064

Abstract

The subject of research in this paper are micro-enterprises as important factors of any national economy. The importance of small and micro-enterprises was determined, and the process of e-integrated marketing communication was defined using methods of analysis and synthesis and empirical and modelling methods. The research aims to determine the characteristics of micro-enterprises in the national economy and define the marketing process using information technologies. In micro-enterprises, the owner is the manager and thus the bearer of all creative and managerial business responsibilities, even in marketing. Accordingly, the organization of integrated marketing is defined first, and then, based on the researched possibilities of Internet technologies, integrated marketing communication as an organizational model for the efficient conduct of the process. In this way, it is possible to conduct marketing activities with minimal costs in a micro company.

Keywords: Technology; Marketing communication; Information and communication process; Micro-enterprise

1. Introduction

In today's globalized and very dynamic market, various forms of entrepreneurial organization are emerging. In this sense, small businesses increasingly dominated the formation of economic entities of different sizes with several employees. In such micro-enterprises, all creative and executive responsibility is borne by one person as the owner and the manager. Therefore, the research object is a micro-enterprise and the methods of analysis, synthesis, and empirical were applied to the model. The research aims first to determine the characteristics and value of small enterprises in the national economy. Then explore the possibilities provided by internet technologies and define the marketing process in online terms. It is hypothesized that Internet technologies may be an important success factor in integrated micro-enterprise marketing. After defining and researching the strength of micro-enterprises, a marketing strategy and integrated marketing communication were defined. Furthermore, researching the technical and technological possibilities arising from the development of information Internet technologies. This process clearly shows that marketing activities can be carried out efficiently in a micro-enterprise with minimal costs. Thus, in micro-enterprise, internal marketing activity is focused on the availability of resources, institutional support and ease of use of Internet technologies in the market [1]. Limitations on the development of integrated information and communication processes in micro-enterprises may stem from the insufficient education and preparation of new micro-entrepreneurs. That is certainly a subject for future research on this topic.

2. Peculiarities of micro-enterprises and entrepreneurs

The object of this topic is an economic entity defined as a micro-enterprise. In the territory of the Republic of Croatia, this is regulated by the Accounting Act and based on the said Act [2]: Micro-entrepreneurs are those who do not exceed two of the following three conditions:

- Total assets HRK 2,600,000.00
- Income HRK 5,200,000.00
- Average number of employees during the business year - 10 employees.

In addition, the Republic of Croatia passed the Act on Encouraging the Development of Small Business with regulations according to which a micro-entrepreneur must meet the following criteria [3]:

- Number of employees less than ten employees.
- Annual income was less than 2,000,000.00 EUR.
- Assets of fixed assets less than EUR 2,000,000.00.

The data indicate the relative harmonization of the Micro Entrepreneurship Regulation Act between the Republic of Croatia and the European Union. In addition to the above data, there are research and results on other relevant indicators from which it can be concluded how much is the presence and importance of micro-enterprises in the Croatian national economy. In this sense, Figure 1 shows a survey on the number of employees in micro-enterprises in the Republic of Croatia. The graph in Figure 1 shows that, for example, in 22.60% of micro-enterprises, only one person is employed, and in 69.40%, up to five persons are employed. In any case, one person is responsible for creating and executing all management decisions, usually the business owner.

Fig. 1. Number of employees in surveyed micro-enterprises in the Republic of Croatia
Source: Author's empirical research in 2017 [4]

Description	In total	Micro	Small	Medium- sized	Big
Number of entrepreneurs	136.260	122.403	11.962	1.525	370
Number of entrepreneurs (part)	100,00%	89,84%	8,77%	1,11%	0,28%
Number of employees	969.776	286.603	249.826	183.189	250.158
Number of employees (part)	100%	29,55%	25,76%	18,89%	25,80%
Consolidated financial result - profit or loss for the period (in HRK million)	31.281	-152	8.596	7.127	15.711
Average monthly net salaries in HRK	5.815	4.217	5.668	6.393	7.369
Index 2018 = 100.0					
Number of employees	106,4	118,6	103	102,6	100,5
Consolidated financial result - profit or loss for the period (in HRK million)	100,7	-	103	97,4	110,5
Average monthly net salaries in HRK	104,2	109,8	105,5	103,5	104,5

Table 1. Business operations of entrepreneurs in the Republic of Croatia in 2019 by the size of the entrepreneur
Source: author's calculation based on FINA data, 2020

The social value of micro-enterprises in the Croatian economy is highlighted in the tables in Table 1, which shows the structure of the number of entrepreneurs and the number of employees in micro, small, medium and large enterprises in the Republic of Croatia in 2019. In addition to the previous exact indicators showing the value of small enterprises in the national economy, small enterprises are very successful and flexible, as can be seen from the following hypotheses [5]:

1. In many industries, small businesses can respond more quickly and at a lower cost to ever-faster changes in products and services, processes and markets.
2. A small business is attractive to gifted individuals.

To these features should be added the fact that the most significant success factor is human resources. In this sense, attention is focused on the quality and organization of employees and especially on the motivation and preparedness of the manager, who in most cases is also the owner of a micro company. The manager as an individual is in a situation to be responsible for creating, making and implementing all decisions relevant to the business. That is undoubtedly a significant advantage in terms of efficiency and the risk borne by the owner-manager. The success of a manager in a micro company and the manager's ability and motivation will depend on additional special education and the readiness to apply all new organizational and technical innovations. In any case, the manager must have a well-designed business strategy that will define the vision, elaborate goals and plan in line with the annual operational plans. This is followed by implementing and evaluating execution and analysis at the end of the strategic period as a basis for a new strategy. On this occasion, the starting point of consideration will be marketing because it is a window into the market world. Special attention is paid to marketing communication in the conditions of developed information Internet technologies. The starting position is the organization of marketing and the application of Internet technologies so that the manager can function effectively. The main goal in marketing is effective communication that includes the coordination of all marketing activities with a combination of all elements to achieve synergy effects in the creation, execution and control of marketing goals and objectives [6]. In this sense, the term integrated marketing communication has emerged.

3. Integrated marketing communication

In the early 1980s, there was a realization of the need for strategic integration of all marketing communication elements. The disintegration of mass markets into smaller ones, the development of new media, and the rise of consumer sophistication have resulted in a new concept in communication science known as integrated marketing communication. Until 1990, forms of marketing communication were studied and applied individually. However, changes in consumer behaviour caused by general social changes and technological advances resulted in an increasing need to integrate all communication activities and modified and fully customized communication messages. [7]. The essential facts of this term derive from its definition: "Integrated marketing communication is the process of developing and applying various forms of argumentative communication with consumers and potential customers at a given time. The goal is to influence or directly direct the behaviour of selected partners who express interest and use all acceptable forms of communication. The process starts with the potential customer and is directed backwards to develop a permanent process [8]." The ultimate goal is to build a continuous long-term communication process with customers, and this can be done efficiently with the logistical support of Internet technologies.

Fig. 2. The structure of the marketing process in a micro-enterprise

Given that the micro-enterprise is complete management related to one person, it is necessary to set up such an organization of the company, especially marketing, that information technology, with the entrepreneur's readiness and motivation, be effective logistics. That means that complete marketing communication takes place at a minimal cost. In this sense, the first step is to define the marketing process.

Figure 2 presents the organizational structure of a marketing process that is appropriate for a micro-enterprise. Every micro-entrepreneur, especially when founding a company, should define a business and thus a marketing strategy. The business strategy is to determine the primary long-term goals of the company, to adjust the directions of business activities, i.e. to determine the concepts and the choice of resources needed for the realization of the set goals [9]. In this sense, everything starts from a vision that expresses the desired opportunities in the future, widely enough understood and described to adapt to specific situations [10]. After a precisely determined desire, it is elaborated in the form of goals which explain the vision in a broader form. The final design of the marketing strategy is the definition of a plan where everything should be quantified, evaluated and set deadlines for implementation. Particular attention should be paid to the very concept of the previous process. Namely, an entrepreneur can have such a product defined that already has a market due to the established qualities. Another approach is sales, in which it is necessary to organize a certain level of logistics. Given the current conditions of globalization and significant market dynamics, most micro-entrepreneurs, in particular, must immediately approach the market to maximize information and communication to establish a synergistic long-term relationship, which is integrated marketing. Therefore, its strategic conception should be positioning, and in this case, it is directing the perception of customers to the product's position concerning the competition.

According to Figure 2, after the set marketing plan, market research activities can be approached. This activity helps the company gain information about consumers and non-consumers, competition and distribution channels, which serves as a basis for identifying needs and opportunities to sell in the market, business problems, and control [11]. A prerequisite for market research is to determine the marketing mix elements for a particular micro company. That means that the product or service must be precisely defined. After that, the price system and methods of distribution of products and services are determined. The basic instrument and tool of the marketing mix with most market research activities is promotion. The first phase of research is market segmentation as dividing the market into different groups of customers who might require special products or a marketing mix. The second phase is selecting the target market as a process of assessing and selecting one or more market segments to enter. That is followed by product positioning as formulating competitive positioning and detailed marketing mix [12]. It is a plan for possible placement in the market segment. The next phase of implementation is to maintain business relationships which is the ultimate goal of the business strategy. That can be made possible by information internet technologies in a micro-enterprise. The micro-entrepreneur must certainly approach the evaluation of the implementation of the strategic plan. Evaluation should specifically identify the relationships between the plan and implementation and the causes and consequences of certain differences. In addition to these differences, the evaluation results enable a more realistic setting of the future marketing strategy and its components and instruments. In the complete strategic process, great logistical assistance is provided by information internet technologies. In addition, as one of the most significant technological phenomena of our time, the Internet provides customers with entirely new opportunities, the most important of which is to provide immediate and always open access [13].

4. Internet logistics

The development of information technologies at the end of the twentieth and the beginning of the twenty-first century has greatly supported entrepreneurs in the creation and implementation or integration of marketing activities. On this occasion, some internet technological possibilities have been singled out to support integrated marketing communications at a low cost.

Every micro-entrepreneur should open a web portal when founding a company because in that way, they will inform future business partners about all necessary information at no cost. The sequel follows the personalization of the web, especially in market research. Personalization aims to deliver information that is important to an individual or group of individuals among customers and suppliers in a specific form, schedule, and time intervals [14]. Opening e-mail enables direct communication with the business public appropriately and mastering Internet technologies, especially after market segmentation, should open a blog, Twitter and Instagram. The blog's role is that it is actually an online diary, the author of which almost every day publishes various current comments and news, which describe the events about the company's business with photos and links. It should be emphasized that these are thematic blogs of particular interest to customers and suppliers, and the rest of the public [15]. For a shorter form of communication, especially when defining the market segment, Twitter or microblogging service is used, based on exchanging short messages and socializing or following other users [16]. After positioning a company or product, the entrepreneur can address customers directly. This type of discussion is known as a forum which is the technological evolution of bulletin boards. From a technological point of view, forums or boards are web applications that manage user-generated content [17]. Entrepreneur's education in the positioning phase enables synchronous communication through video conferencing, which is organized by agreement from time to time with well-known customers from the market environment. Videoconferencing is the connection of individuals and groups through telecommunications networks and video technology so that people have the impression of simultaneously attending meetings in the form of a live video link on a computer screen [18].

That completely establishes the connection of entrepreneurs with target customers from the market segment, so the maintenance of business relationships follows. One very high-quality internet approach enables Customer-Relationship-Management (CRM). It is an approach to e-business based on understanding customer behaviour through meaningful communication and all to attract and retain customers, develop their loyalty, and increase the company's profitability [19]. The primary goal of CRM is to build a long-term and profitable relationship with a selected market segment [20].

The basic components of CRM are building and managing customer relationships through marketing, observing interpersonal relationships by stages, then managing those relationships at each stage, and controlling the distribution of information. When establishing CRM through marketing, companies can use a variety of tools to help with organizational design. It should be noted in particular that interactions in multiple customer relationships can be linked to transactions. All this leads to cost reductions and thus higher profitability. Therefore, CRM enables the interaction of companies with a large number of Internet channels, portals, and platforms and thus achieves a very strong connection with current and former customers and makes it easier to establish new connections.

After completing the implementation of the marketing strategy and as part of the integrated marketing communication, the maintenance of customer relations can be completed through social networks with a special warning on Facebook and Linkedin. The concept of Facebook was designed to connect people who already know each other in real life via the Internet. The main feature of Facebook is numerous applications because, for example, it is possible to have virtual pets, receive virtual gifts and greeting cards, solve quizzes and play video games. Facebook offers the ability to control privacy, so users can hide their profile and photos and protect themselves from unknown visitors [21]. Linkedin is a business-oriented social networking service mainly used for professional networking, especially in recruitment and sales [22]. All of the above social networks provide an opportunity to personalize each other's information after discussions at video conferences. In this way, the entrepreneur creates even closer relationships with customers and suppliers and with the rest of the public. Such integrated marketing communication processes achieve a long-term connection with minimal costs and loss of time.

5. Integrated internet marketing information and communication process

Regardless of the company's size, every micro-entrepreneur should organize the necessary information and communication marketing process that will, with the help of Internet technologies, integrate the entire business and thus marketing activities. In this sense, Figure 4 shows a systematic information and communication process in which all marketing activities are fully integrated following the set strategy. The picture shows that the framework marketing plan can be distributed through a personalized web portal and e-mail, which goes to already recognized customers. Market research by collecting data using blogs and Twitter reduces the number of customers, and the forum already determines the market segment. After a video meeting with selected customers from the segment, fascinating customers are selected as the target group. After that, the market capacity of the segment is assessed, and the position or part of the market capacity is determined. Evaluating and maintaining customer relationships is most effective through CRM. Increasing the quality of relationships can take place through social networks. With the proposed process, a micro-entrepreneur can perform all market research activities, i.e. the implementation of a marketing plan, alone with minimal costs.

Fig. 3. Integrated information and communication process of marketing activities in a micro-enterprise

Figure 3 shows that the manager of a micro-enterprise can perform all marketing activities alone using applications and other capabilities provided by Internet information technologies. Each phase of marketing activities supports the appropriate portal or platform, and the complete marketing process takes place with minimal participation of managers. Therefore, it is clear from the course of activities and IT support that this is a very effective way of managing marketing in a micro-enterprise. The proposed model and the obtained results indicate the presence of continuous investment in new technologies and IT in micro-enterprises, which is evident from previous research [23], especially the presence of modern technologies in the business of micro-enterprises through investments in the acquisition of new technologies, application of IT and computer skills. The proposed model can be implemented in micro enterprises, provided that the entrepreneur or manager has a basic knowledge of information and marketing technologies, and that they understand the importance of the impact of the application of this model on their business. By applying this model, it is possible to improve existing marketing and communication processes, and positively influence the improvement of business. Disadvantages and possible negative outputs can be seen through the lack of interest of entrepreneurs in such activities and impatience in anticipation of visible results of the application of this model. Possible limitations of this research are visible in the lack of feedback from micro-entrepreneurs on the proposal of this model.

6. Conclusion

Based on the previous considerations, more can be concluded. First of all, micro-enterprises are a crucial factor in the national economy. That is why they deserve research to increase the efficiency of all business functions, especially marketing. In most cases, the owner is also responsible for all creative and executive decisions, even in marketing. One of the basic problems faced by entrepreneurs / owners is insufficiently developed marketing communication due to lack of their time and professional knowledge. In this sense, an integrated information and communication process is set up as an organizational form that allows the owner to communicate effectively with customers. Based on the research of the possibilities of Internet technologies, integrated marketing communication was defined as an instrument for the effective conduct of all marketing activities adapted to the micro-enterprise. That allows the micro-enterprise manager to perform all creative and management activities at a minimal cost, and the observed marketing communication problems with the market were solved. The development of this process can enable the education of entrepreneurs in business and especially IT. That is a potential area for further study of this topic, with an emphasis on exploring the impact of new or improved e-marketing communication on the business performance of micro-enterprises.

7. References

[1] Kurian, S.; Ramanathan, H. & Andrlic, B. (2019). The Role Internal Marketing in The Promotion of Technology Usage Among Teachers: An Empirical Investigation of the Mediating Role of Extrinsic Factors Leading to Improved Technology Use, Proceedings of the 30th DAAAM International Symposium, pp.0030-0037, B. Katalinic (Ed.), Published by DAAAM International, ISBN 978-3-902734-22-8, ISSN 1726-9679, Vienna, Austria. DOI: 10.2507/30th.daaam.proceedings.004

[2] Accounting law, National newspapers, Narodne novine no. 78/15,134/15,120/16, 116/18, 42/20, 47/20, Art. 5, Available from: https://www.zakon.hr/z/118/Zakon-o-ra%C4%8Dunovodstvu, Accessed on: 2021-06-15

[3] Small Business Development Encouragement Act (2016). National newspapers, Narodne novine no. 29/02, 63/07, 53/12, 56/13, 121/16, Art. 3, Available from: https: //www.zakon.hr/z/527/Zakon-o-poticanju-razvoja-maloggospodarstva, Accessed on: 2021-06-15

[4] Štavlić, K. (2018). Micro entrepreneurship – resources, potentials and success, Veleučilište u Požegi, pp. 64. ISBN 978-953-7744-32-8., Požega

[5] Siropolis, N. C. (1995). Menadžment malog poduzeća, prijevod, Mate, ISBN 953-6070-14-6, Zagreb

[6] Kesić, T. (2003). Integrirana marketinška komunikacija, Opinio, pp. 28. ISBN 953-98250-0-8, Zagreb

[7] Shimp, T. A. (2000). 'Advertising Promotion Supplemental Aspects of Integrated Marketing Communications (fifth edition). The Dryden Press: pp. 22-23.

[8] Schultz, D. E. (1993). Integrated Marketing Communication: Maybe definition is in the Point of View, Marketing News, January 18, 1963., pp 17

[9] Chandler, A. D. (1962) Strategy and Structure, pp 16, The MIT Presse, Cambridge

[10] Duličić, Ž. & Škaro, G. (2001). Redefiniranje vizije i misije hrvatskih brodarskih poduzeća, Ekonomska misao i praksa, Vol 10, no 2, pp 161. ISSN 1330-1039

[11] Marušić, M. & Vranešević, T. (2001). Istraživanje tržišta, ADECO, pp.6. ISBN 953-97228-3-7, Zagreb

[12] Kotler, P. (1988). Upravljanje marketingom 1, prijevod, Informator, pp. 258. ISBN 86-301-0142-3, Zagreb

[13] Marinović, M. & Andrlić, B. (2012). Social Geolocation in Tourism - Interrelationship of Web 2.0 and Geography. Proceedings of the 23rd International DAAAM Symposium, Volume 23, Br.1, pp. 0337, ISSN 2304-1382, ISBN 978-3-901509-91-9

[14] Kim, W. (2002). Personalization: Definition, Status and Challenges ahead, Journal of Object Technology , No.1, Available from: Http://www.jot.fm/issue_2002_05/colimn3/ Accessed on: 2021-04-21

[15] https://blog.dnevnik.hr/plazma/2006/05/1621035764/definicija-i-podjela blogova.html?page=blog&id=1621035764&subpage=0&subdomain=plazma, Accessed on: 2021-04-21

[16] Ružić, D.; Biloš, A. & Turkalj, D. (2009). e-Marketing, Faculty of Economics, pp. 26. CIP 120902038, Osijek
[17] "Bulletin Community Forum - FAQ: What is a bulletin board?". Available from: https://forum.vbulletin.com/articles, Accessed on: 2021-06-14
[18] Leksikografski zavod Miroslav Krleža, Available from: https://proleksis.lzmk.hr/50177/ , Accessed on: 2021-08-10
[19] Panian, Ž. (2003). Odnosi s klijentima (U e-poslovanju), Sinergija, ISBN 953-6895-11-0, Zagreb
[20] Andrlić, B.; Šostar, M. & Banožić, M. (2020). Application of E-Processes in Tourism Marketing, Proceedings of the 31st DAAAM International Symposium, pp.0004-0008, B. Katalinic (Ed.), Published by DAAAM International, ISBN 978-3-902734-29-7, ISSN 1726-9679, Vienna, Austria DOI: 10.2507/31st.daaam.proceedings.001
[21] Grbavac, J. & Grbavac, V. (2014). Pojava društvenih mreža kao globalnog komunikacijskog fenomena, Media, culture and public relations, No. 5, pp. 212. ISSN 1333-8371 (Print), ISSN 1848-8374 (Online)
[22] Nicholas Lemann, N. (2015). "Reid Hoffman's Big Dreams for LinkedIn - The New Yorker". The New Yorker
[23] Stavlic, K. (2020). Comparative Analysis of the Presence of the New and Modern Technologies in Micro Enterprises, Proceedings of the 31st DAAAM International Symposium, pp.0936-0942, B. Katalinic (Ed.), Published by DAAAM International, ISBN 978-3-902734-29-7, ISSN 1726-9679, Vienna, Austria DOI: 10.2507/31st.daaam.proceedings.130

DOI: 10.2507/32nd.daaam.proceedings.065

KINETOSTATIC ANALYSIS OF THE SERVICE MOBILE ROBOT WITH MANIPULATOR

Plotnikov Aleksey & Pryanichnikov Valentin

This Publication has to be referred as: Plotnikov, A[leksey] & Pryanichnikov, V[alentin] (2021). Kinetostatic Analysis of the Service Mobile Robot with Manipulator, Proceedings of the 32nd DAAAM International Symposium, pp.0446-0453, B. Katalinic (Ed.), Published by DAAAM International, ISBN 978-3-902734-33-4, ISSN 1726-9679, Vienna, Austria
DOI: 10.2507/32nd.daaam.proceedings.065

Abstract

When creating service mobile robots with manipulators for operation in cramped rooms (home/office/hospitals), there is a problem of capturing and transporting both small-sized objects and those, that make up a significant part of the robot's weight. When constructing the control of the lifting process of such objects, the task of dynamic capture arises while maintaining the stability of the robot. The article considers the solution of the dynamics problem for the AMUR-307 mobile robot with a two-armed manipulator circuit. The basis for the construction of such control is proposed to take the solution of the dynamics problem using the kinetostatics method. The kinematics of the robot is such, that to simplify calculations, the robot body system can be considered as two-dimensional. In the process of analyzing the dynamics of the system, three main parts are distinguished: the upper links of the manipulator with the object of manipulation, the lower links of the manipulator with the drive elements of the upper links, the transport platform of the robot. The distribution of forces in different parts of a robot with a manipulator is considered. The dependence of the forces and moments of inertia on the motion of the system are found. The resulting equations linking the generalized coordinates of the robot with the efforts of the drives are a reliable basis for generating synergies for capturing, loading and moving heavy objects by the robot.

Keywords: Robotics; Mobile Service robot; Movement synergy algorithms for heavy object manipulation; Kinetostatic dynamics analysis.

1. Introduction

A dictionary of motion synergies is currently being developed to control the manipulator of the Amur-307 mobile robot. The synergies of movements are coordinating changes in articular angles (kinematic synergy), articular moments (dynamic synergy) [1], as well as propulsion drives. In the process of creating the system, a dictionary of robot actions is formed, which includes a set of standard robot movements with a manipulator, depending on the task being performed and the situation. These elementary sequences of actions are formed by showing the operator executed movements of the manipulator and the mobile platform of the robot using the inverse kinematics solution method [3] and dynamically justified movements [2], using virtual models of the robot and/or a real service robot.

Let's take a closer look at one of the synergies - an algorithm for capturing an object by a robot, that has a mass comparable to the mass of the robot itself. This action is dangerous because in the process of its implementation, the robot may overturn due to a sharp forward shift of the center of mass of the robot + object system. The proposed object capture strategy is as follows (Fig. 1):

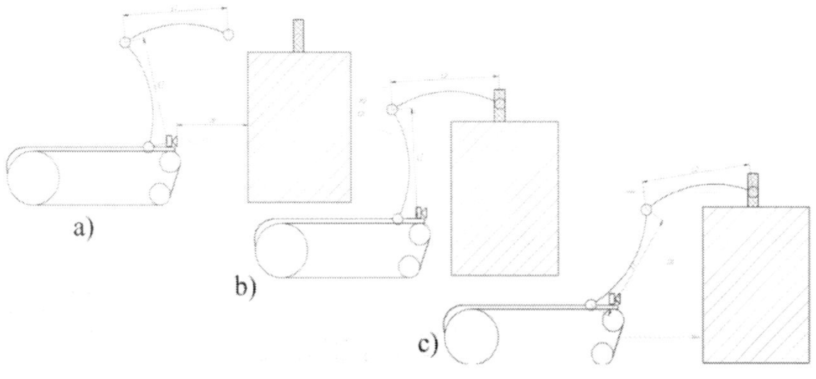

Fig. 1. The proposed object capture strategy

a) navigation system data is used to determine the safe area of movement during the capture;
b) the mass of the manipulating object is determined, when it is captured, using a special trial movement;
c) the object of manipulation is transferred inside the service area to the transport table, while the robot is moved forward in a coordinated manner to maintain balance.

To carry out the operations of estimating the mass of the object of manipulation and the implementation of a complex capture movement, it is necessary to solve the problem of system dynamics. In the literature, the solution of problems of robot dynamics by the kinetostatics method is widely described on the example of industrial robots [4], [5]. These techniques find their application in the work on the controling of manipulative robots [6]. For mobile robots, dynamics analysis is often performed using third-party systems for solving the dynamics problem [7] or using the Lagrange method [8], [9]. However, these techniques are poorly suited for solving dynamics problems on a mobile platform. The analytical solutions proposed in this work are obtained as a result of calculation by the kinetostatics method, which has a significant advantage in the speed of work and complexity of calculations. This is achieved by the absence of additional software (for example, Simulink) required for calculation by third-party software packages, and integration operations in the case of a Lagrange solution.

2. System description

The Amur 307 service robot consists of a tracked platform with a three-stage manipulator (Fig. 2).

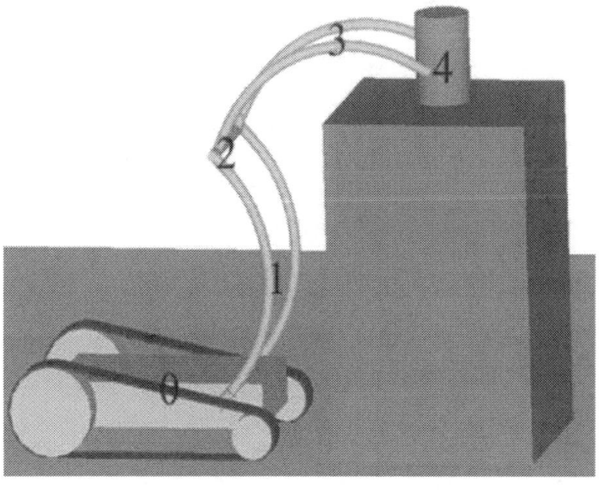

Fig. 2. Elements of the robot model

Due to the symmetry of the robot's design, as well as the fact that the movement of lifting the load occurs without turning the robot, analysis of the dynamics of the described movement can be reduced to a two-dimensional problem. When conducting a kinetostatic [4] analysis of a robot with a load, the system should be divided into 3 parts: the transport platform (Fig.2, pos. 0), the lower links of the manipulator (Fig.2, pos. 1-2), the upper links of the manipulator (Fig.2, pos. 3-4).

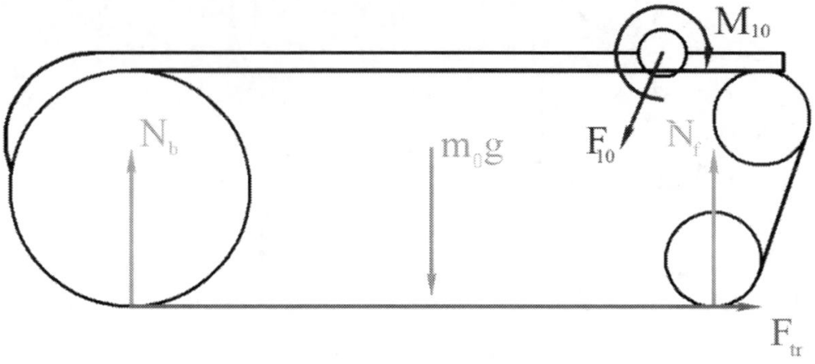

Fig. 3. Distribution of forces in the robot's transport platform

The following forces act on the transport platform (Fig.3.):
m_0g – gravity force of the transport platform;
N_f, N_b – the reaction forces of the support acting on the front and rear rollers, respectively. In view of the symmetry of the problem and the transition to two- dimensional consideration are taken as the sum of the corresponding left and right reactions;
F_{tr} – the traction force of the robot's tracked drive, equal to the sum of the traction forces of the left and right tracks;
F_{10}, M_{10} – force and moment of reaction of the connection from the lower links of the manipulator.

As shown (Fig. 2), the lower links of the robot manipulator (Fig. 2, pos. 1) connecting the transport platform with structural elements (Fig. 2, pos. 2). Structural elements (Fig. 2, pos. 2) provide a connection between the upper and lower links of the manipulator. Due to the compactness of the element (Fig. 2, pos. 2), it is considered without taking into account its dimensions together with the lower links of the manipulator. Based on the above, the following forces act on the lower link (Fig. 4.):

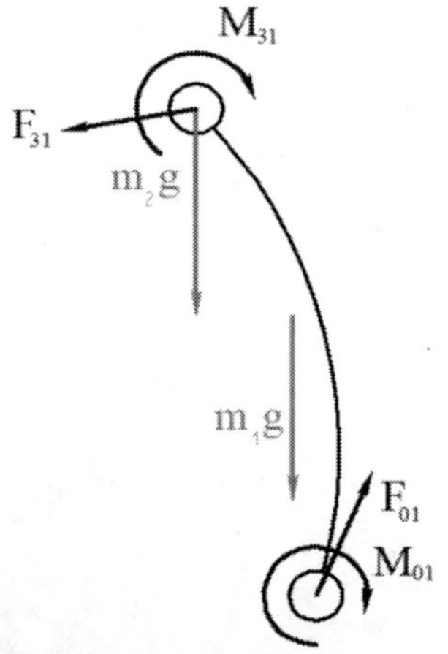

Fig. 4. The distribution of forces in the lower link of the robot manipulator.

m_1g – the gravity force of the lower link of the manipulator, due to the symmetry of the design and the transition to a two-dimensional consideration of the problem, the mass of the structural element in question is assumed to be equal to the sum of the masses of the lower links of the manipulator;
m_2g – gravity of structural elements;
F_{01} – the gravity force of the structural elements;
M_{01} – is the moment of reaction from the transport platform, due to the drive force of the robot manipulator.
F_{31}, M_{31} – force and moment of reaction of the connection from the upper links of the manipulator.

The upper links of the robot manipulator (Fig. 2, pos. 3), is connected to the lower links by means of structural elements (Fig. 2, pos. 2). At the ends of the upper links, there is a gripping device with the object of manipulation. The object of manipulation and gripping devices are considered without taking into account their size together with the upper links of the manipulator. Based on the above, the following forces act on the upper link (Fig. 6.):

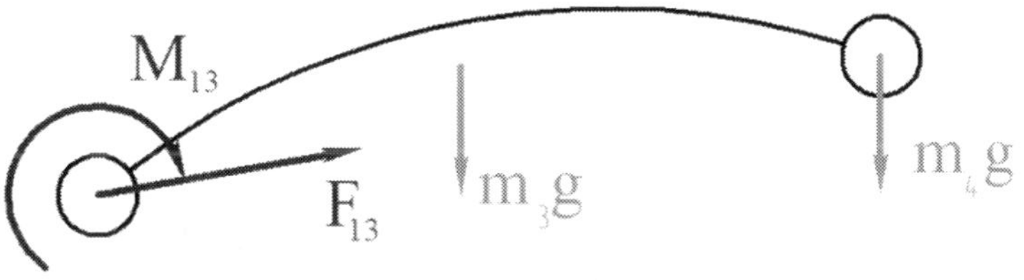

Fig. 5. Distribution of forces in the upper link of the robot manipulator

m_3g –the gravity force of the upper link of the manipulator, due to the symmetry of the design and the transition to a two-dimensional consideration of the problem, the mass of the considered structural element is assumed to be equal to the sum of the masses of the upper links of the manipulator;
m_4g – the total gravity force of the object of manipulation and gripping devices;
F_{13} – is the moment of reaction from the transport platform, due to the drive force of the robot manipulator;
M_{13} – the moment of reaction from the transport platform, due to the drive force of the robot manipulator.

It is necessary to find the relationship between the efforts of the manipulator drives $M_{13}(t)$, $M_{01}(t)$, the efforts of the caterpillar drive $F_{tr}(t)$ from the generalized coordinates $\varphi_1(t)$, $\varphi_2(t)$, $s_0(t)$. Where $\varphi_1(t)$ is the rotational coordinate of the lower link of the manipulator, $\varphi_2(t)$ is the rotational coordinate of the upper link of the manipulator, $s_0(t)$ is the translational movement of the transport platform during capture.

3. Objects acceleration and forces of inertia

To carry out kinetostatic analysis, in accordance with the Dalembert - Lagrange principle, the main vectors and moments of inertia forces applied at their centers of mass should be added to the forces acting on the manipulator links. To find the inertia forces, it is necessary to determine the acceleration of the masses of the system. The coordinate systems used in the calculations are shown in (Fig.7).

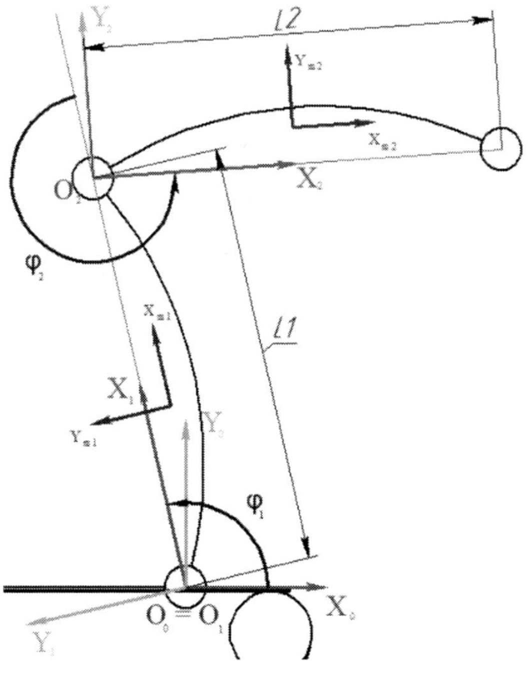

Fig. 6. Coordinate systems of the robot manipulator.

Acceleration of the robot's transport platform during linear movement $s_0(t)$ "(1)".

$$\vec{a}_{m0} = \begin{pmatrix} \frac{d^2 s_0}{dt^2} \\ 0 \\ 0 \end{pmatrix} \tag{1}$$

Equation of accelerations of the lower links of the manipulator "(2)".

$$\overrightarrow{a_{m1}} = \overrightarrow{a_{m1}^{\varphi_1}} + \overrightarrow{a_{m1}^{s_0}} = \overrightarrow{a_{m1}^{n}} + \overrightarrow{a_{m1}^{t}} + \overrightarrow{a_{m1}^{s_0}} \tag{2}$$

The position of the center of mass of the lower links of the manipulator is represented as the vector $\overrightarrow{R_{m1}}$ in the coordinate system of the lower links X_1Y_1. Taking into account the position of the center of mass of the links with the angle of location "(3)", the acceleration equations in the coordinate system associated with the lower link of the robot arm will take form "(4)".

$$\tau_{m1} = tan^{-1} R_{m1Y}/R_{m1X}, \tag{3}$$

$$\vec{a}_{m1} = \begin{pmatrix} cos\tau_{m1} & -sin\tau_{m1} & 0 \\ sin\tau_{m1} & cos\tau_{m1} & 0 \\ 0 & 0 & 1 \end{pmatrix} \begin{pmatrix} -\left(\frac{d\varphi_1}{dt}\right)^2 \overrightarrow{|R_{m1}|} \\ \frac{d^2\varphi_1}{dt^2} \overrightarrow{|R_{m1}|} \\ 0 \end{pmatrix} + \begin{pmatrix} cos\varphi_1 & sin\varphi_1 & 0 \\ -sin\varphi_1 & cos\varphi_1 & 0 \\ 0 & 0 & 1 \end{pmatrix} \vec{a}_{m0} \tag{4}$$

Acceleration equation for structural elements of the manipulator link "(5)".

$$\vec{a}_{m2} = \overrightarrow{a_{m2}^{\varphi_1}} + \overrightarrow{a_{m2}^{s_0}} = \overrightarrow{a_{m2}^{n}} + \overrightarrow{a_{m2}^{t}} + \overrightarrow{a_{m0}^{s_0}} \tag{5}$$

In the coordinate system associated with the lower link of the manipulator "(5)" takes the form "(6)".

$$\vec{a}_{m2} = \begin{pmatrix} -\left(\frac{d\varphi_1}{dt}\right)^2 L_1 \\ \frac{d^2\varphi_1}{dt^2} L_1 \\ 0 \end{pmatrix} + \begin{pmatrix} cos\varphi_1 & sin\varphi_1 & 0 \\ -sin\varphi_1 & cos\varphi_1 & 0 \\ 0 & 0 & 1 \end{pmatrix} \vec{a}_{m0} \tag{6}$$

Equation for accelerations of the upper links of the manipulator "(7)".

$$\overrightarrow{a_{m3}} = \overrightarrow{a_{m3}^{\varphi_2}} + \overrightarrow{a_{m2}^{\varphi_1}} + \overrightarrow{a_{m0}^{s_0}} \tag{7}$$

Taking into account the kinematic scheme of the manipulator, the equation " (7)" takes the form "(8)":

$$\begin{aligned} \overrightarrow{a_{m3}} = &\begin{pmatrix} cos\tau_{m3} & -sin\tau_{m3} & 0 \\ sin\tau_{m3} & cos\tau_{m3} & 0 \\ 0 & 0 & 1 \end{pmatrix} \begin{pmatrix} -\left(\frac{d(\varphi_1+\varphi_2)}{dt}\right)^2 \overrightarrow{|R_{m3}|} \\ \frac{d^2(\varphi_1+\varphi_2)}{dt^2} \overrightarrow{|R_{m3}|} \\ 0 \end{pmatrix} + \\ &+ \begin{pmatrix} cos\varphi_2 & sin\varphi_2 & 0 \\ -sin\varphi_2 & cos\varphi_2 & 0 \\ 0 & 0 & 1 \end{pmatrix} \begin{pmatrix} -\left(\frac{d\varphi_1}{dt}\right)^2 L_1 \\ \frac{d^2\varphi_1}{dt^2} L_1 \\ 0 \end{pmatrix} + \begin{pmatrix} cos(\varphi_1+\varphi_2) & sin(\varphi_1+\varphi_2) & 0 \\ -sin(\varphi_1+\varphi_2) & cos(\varphi_1+\varphi_2) & 0 \\ 0 & 0 & 1 \end{pmatrix} \vec{a}_{m0} \end{aligned} \tag{8}$$

Equation of accelerations of the manipulated object "(9)", "(10)".

$$\overrightarrow{a_{m4}} = \overrightarrow{a_{m4}^{\varphi 2}} + \overrightarrow{a_{m4}^{\varphi 1}} + \overrightarrow{a_{m4}^{s0}} \tag{9}$$

$$\overrightarrow{a_{m4}} = \begin{pmatrix} -\left(\frac{d\varphi_1}{dt}+\frac{d\varphi_2}{dt}\right)^2 L_2 \\ \left(\frac{d^2\varphi_1}{dt^2}+\frac{d^2\varphi_2}{dt^2}\right) L_2 \\ 0 \end{pmatrix} + \begin{pmatrix} cos\varphi_2 & sin\varphi_2 & 0 \\ -sin\varphi_2 & cos\varphi_2 & 0 \\ 0 & 0 & 1 \end{pmatrix} \begin{pmatrix} -\left(\frac{d\varphi_1}{dt}\right)^2 L_1 \\ \frac{d^2\varphi_1}{dt^2} L_1 \\ 0 \end{pmatrix} + + \begin{pmatrix} cos(\varphi_1+\varphi_2) & sin(\varphi_1+\varphi_2) & 0 \\ -sin(\varphi_1+\varphi_2) & cos(\varphi_1+\varphi_2) & 0 \\ 0 & 0 & 1 \end{pmatrix} \vec{a}_{m0} \tag{10}$$

The main moments of the inertia forces of the upper "(12)" and the lower links of the manipulator "(11)" relative to their centers of mass.

$$\overrightarrow{M_1^I} = -\begin{pmatrix} 0 \\ 0 \\ J_{1z} \end{pmatrix} \frac{d^2\varphi_1}{dt^2} \tag{11}$$

$$\overrightarrow{M_3^I} = -\begin{pmatrix} 0 \\ 0 \\ J_{3z} \end{pmatrix} \left(\frac{d^2\varphi_1}{dt^2}+\frac{d^2\varphi_2}{dt^2}\right) \tag{12}$$

The main vectors of inertia forces of the transport platform "(13)", lower "(14)" and upper "(16)" links, structural elements of the robot manipulator "(14)", object of manipulation "(17)":

$$\overrightarrow{F_0^I} = -m_0 \begin{pmatrix} \frac{d^2 s_0}{dt^2} \\ 0 \\ 0 \end{pmatrix} \tag{13}$$

$$\overrightarrow{F_1^I} = -m_1 \overrightarrow{a_1} \tag{14}$$

$$\overrightarrow{F_2^I} = -m_2 \overrightarrow{a_2} \tag{15}$$

$$\overrightarrow{F_3^I} = -m_3 \overrightarrow{a_3} \tag{16}$$

$$\overrightarrow{F_4^I} = -m_4 \overrightarrow{a_4} \tag{17}$$

4. Forces and moments equations

To find the values of reaction forces, in accordance with the principle of kinetostatics, the equations of forces and moments acting on each body of the system are written.
For the upper links of the manipulator, the "(18)", "(19)", "(20)" are valid.

$$\vec{F}_{13} + m_3\vec{g} + m_4\vec{g} - m_3\vec{a}_{m3} - m_4\vec{a}_{m4} = 0 \tag{18}$$

$$\vec{F}_{13} = m_3\vec{a}_{m3} + m_4\vec{a}_{m4} - m_3\vec{g} - m_4\vec{g} \tag{19}$$

$$\vec{M}_{13} = -1\left(\overrightarrow{R_{m3}} \times m_3\vec{g} + L_2\overrightarrow{X_2} \times m_4\vec{g} - \overrightarrow{R_{m3}} \times m_3\vec{a}_{m3} - L_2\overrightarrow{X_2} \times m_4\vec{a}_{m4} - - \begin{pmatrix} 0 \\ 0 \\ J_{3z} + m_3\left|\overrightarrow{R_{m3}}\right|^2 \end{pmatrix} \left(\frac{d^2\varphi_1}{dt^2}+\frac{d^2\varphi_2}{dt^2}\right)\right) \tag{20}$$

For the lower links of the manipulator, the "(21)", "(22)", "(23)", "(24)", "(25)" are valid.

$$\vec{F}_{31} = -\vec{F}_{13} \tag{21}$$

$$\vec{M}_{31} = -\vec{M}_{13} \tag{22}$$

$$\vec{F}_{01} + \vec{F}_{31} + m_1\vec{g} + m_2\vec{g} - m_1\vec{a}_{m1} - m_2\vec{a}_{m2} = 0 \quad (23)$$

$$\vec{F}_{01} = m_1\vec{a}_{m1} + m_2\vec{a}_{m2} - m_1\vec{g} - m_2\vec{g} + \vec{F}_{13} \quad (24)$$

$$\vec{M}_{01} = -1\Big(\overrightarrow{R_{m1}} \times m_1\vec{g} + L_1\overrightarrow{X_1} \times m_2\vec{g} - \overrightarrow{R_{m1}} \times m_1\vec{a}_{m1} - L_1\overrightarrow{X_1} \times m_2\vec{a}_{m2} - \\ - \begin{pmatrix} 0 \\ 0 \\ J_{1z} + m_1\left|\overrightarrow{R_{m1}}\right|^2 \end{pmatrix} \frac{d^2\varphi_1}{dt^2} + \vec{M}_{31}\Big) \quad (25)$$

Equations of moments of forces and forces of inertia, relative to its center of mass, are compiled for the transport platform.

$$\vec{M}_{10} = -\vec{M}_{01} \quad (26)$$

$$\vec{F}_{10} = -\vec{F}_{01} \quad (27)$$

$$\vec{M}_{10} + \vec{h}_{m0} \times \vec{F}_{tr} + \vec{r}_{Fm0} \times \vec{F}_{10} + \vec{r}_{Nbm0} \times \vec{N}_b + \vec{r}_{Nfm0} \times \vec{N}_f = 0 \quad (28)$$

$$\vec{h}_{m0} \times \vec{F}_{tr} = -1\left(\vec{M}_{10} + \vec{r}_{Fm0} \times \vec{F}_{10} + \vec{r}_{Nbm0} \times \vec{N}_b + \vec{r}_{Nfm0} \times \vec{N}_f\right) \quad (29)$$

The reaction forces of the supports are found from the equations of moments relative to the front "(31)" and rear "(30)" support points:

$$\vec{r}_{NfNb} \times \vec{N}_f = -1\left(\vec{M}_{10} + \vec{r}_{FNb} \times \vec{F}_{10} + \vec{r}_{mNb} \times (m_0\vec{g}) - \vec{r}_{m0Nb} \times m_0\vec{a}_{m0}\right) \quad (30)$$

$$\vec{r}_{NbNf} \times \vec{N}_b = -1\left(\vec{M}_{10} + \vec{r}_{FNf} \times \vec{F}_{10} + \vec{r}_{mNf} \times (m_0\vec{g}) - \vec{r}_{m0Nf} \times m_0\vec{a}_{m0}\right) \quad (31)$$

5. Conclusion

As a result of solving a system of equations "(20)", "(26)", "(30)" the dependencies connecting the generalized coordinates of the robot $s_0(t)$, $\varphi_1(t)$, $\varphi_2(t)$ with the moment of forces of the manipulator drives $M_1(t)$, $M_2(t)$, and the force of the tracks $F_{tr}(t)$ are found. Based on the obtained dependencies, it is possible to create algorithms for controlling the movements of the robot manipulator in the form of synergies. Equations "(18)", "(19)", "(20)" can also be used to determine the mass of the m_4 manipulation object. The solution of the problem of determining the mass of the manipulating object can be carried out both by direct measurements using force-moment sensors, and analytically using current feedbacks and drive positions.

In the AMUR-307 service robot, it is more expedient to use an analytical method for determining the mass of the manipulated object, based on a sequential increase in the lifting force of the object by the manipulator and comparing this with the measurement of the coordinates of the robot manipulator. The capture of a relatively heavy object is accompanied by the simultaneous movement of the robot with acceleration towards the captured object. Such a complex movement, which includes the simultaneous movement of the manipulator and the robot mover, allows you to create a moment of inertia forces that ensure the equilibrium of the system. Based on the values of reaction forces in the supports from "(31)", "(32)" and by solving "(30)" it is possible to determine the direction and desired acceleration of the balancing movement of the robot.

The work demonstrates the possibility of using kinetostatics to implement finite analytical formulas, that allow us to build synergies for manipulator control with minimal computational/time costs.

6. References

[1] Methods of biomechatronics of the human hand simulator / A.K. Platonov, A.A. Frolov, E.V. Biryukova, V.E. Pryanichnikov, S.N. Yemelyanov // Preprints of M.V.Keldysh IPM RAS 2012. No. 82. 40 p. URL: http://library.keldysh.ru/preprint.asp?id=2012-82
Методы биомехатроники тренажёра руки человека / А.К. Платонов, А.А. Фролов, Е.В. Бирюкова, В.Е. Пряничников, С.Н. Емельянов // Препринты ИПМ им. М.В.Келдыша РАН 2012. № 82. 40 с. URL: http://library.keldysh.ru/preprint.asp?id=2012-82

[2] Biologically adequate principles of human hand exoskeleton control / A.A. Frolov, E.V. Biryukova, P.D. Bobrov, A.K. Platonov, V.E. Pryanichnikov. // Information-measuring and control systems (vol. Intelligent adaptive Robots, vol.8, No. 1-2), Moscow: Radio Engineering, 2013, vol.11, No. 4. -pp.7-18, ISSN 2070-0814

[3] Биологически адекватные принципы управления экзоскелетом руки человека / А.А. Фролов, Е.В. Бирюова, П.Д. Бобров, А.К. Платонов, В.Е. Пряничников. // Информационно-измерительные и управляющие системы

(вып. Интеллектуальные адаптивные роботы, т.8, № 1-2), М.: Радиотехника, 2013, т.11, № 4. -С.7-18, ISSN 2070-0814

[4] Andreas Aristidou, Joan Lasenby, FABRIK: A fast, iterative solver for the Inverse Kinematics problem, Graphical Models, Volume 73, Issue 5, 2011, Pages 243-260, ISSN 1524-0703, https://doi.org/10.1016/j.gmod.2011.05.003.

[5] Mechanics of industrial robots: a textbook for higher education institutions: in 3 books. / Edited by K.V. Frolov and. I.Vorobyova. Book 1: Kinematics and dynamics / E.I. Vorobyov, S.A. Popov, G.I.Sheveleva. - M.: Higher School, 1988. - 304 p.

[6] Механика промышленных роботов: учебное пособие для втузов: в 3 кн. / Под ред. К.В. Фролова и. И. Воробьева. Кн. 1: Кинематика и динамика / Е.И. Воробьев, С.А. Попов, Г. И. Шевелева. – М.: Высш. шк., 1988. – 304 с.

[7] Kolovsky, M.Z. Fundamentals of the dynamics of industrial robots / M.Z. Kolovsky, A.V. Sloushch– - M.: Nauka. Gl. ed. phys. - mat. lit., 1988. - 240 p.

[8] Коловский, М.З. Основы динамики промышленных роботов / М.З. Коловский, А.В. Слоущ. – М.: Наука. Гл. ред. физ. - мат. лит., 1988. – 240 с.

[9] Anciporovich, P. P. Kinetostatic force analysis of a manipulator with 4 degrees of freedom / P. P. Anciporovich, V. K. Akulich, E. M. Dubovskaya // Theoretical and applied mechanics. Issue 28: International Scientific and Technical collection / edited by A.V. Chigarev; BNTU. - Minsk, 2013. - pp. 224-227

[10] Анципорович, П. П. Кинетостатический силовой анализ манипулятора с 4 степенями свободы / П. П. Анципорович, В. К. Акулич, Е. М. Дубовская // Теоретическая и прикладная механика. Выпуск 28: международный научно-технический сборник / под ред. А. В. Чигарева; БНТУ. – Минск, 2013. – С. 224-227

[11] Damic, V[jekoslav]; Cohodar, M[aida] & Muratovic, M[unir] (2017). Dynamic Modelling of Mobile Robots Based on its 3D CAD Model, Proceedings of the 28th DAAAM International Symposium, pp.0144-0149, B. Katalinic (Ed.), Published by DAAAM International, ISBN 978-3-902734-11-2, ISSN 1726-9679, Vienna, AustriaDOI: 10.2507/28th.daaam.proceedings.019

[12] Du, Y. (2016). Kinematics and Dynamic Modeling and Simulation Analysis of Three-wheeled Mobile Robot, International Conference on Mechanics Design, Manufacturing and Automation (MDM 2016) ISBN: 978-1-60595-354-0

[13] A. A. Hatab and R. Dhaouadi, "Dynamic modelling of differential-drive mobile robots using Lagrange and Newton-Euler methodologies: A unified framework," Adv. Robot. Autom., vol. 02, no. 02, 2013.

DOI: 10.2507/32nd.daaam.proceedings.066

INVESTIGATION OF THE OPERATION OF THE DEVELOPED PISTON MECHANISM OF THE AXIAL-PISTON HYDRAULIC SWASH PLATE MACHINE IN THE STRAINING AND CREEPING SPEEDS MODES

Victor Elchinsky, Branko Katalinic, Anton Kuzmin & Sergey Stazhkov

This Publication has to be referred as: Elchinsky, V[iktor]; Katalinic, B[ranko]; Kuzmin, A[nton] & Stazhkov, S[ergey] (2021). Investigation of the Operation of the Developed Piston Mechanism of the Axial-Piston Hydraulic Swash Plate Machine in the Straining and Creeping Speeds Modes, Proceedings of the 32nd DAAAM International Symposium, pp.0454-0461, B. Katalinic (Ed.), Published by DAAAM International, ISBN 978-3-902734-33-4, ISSN 1726-9679, Vienna, Austria

DOI: 10.2507/32nd.daaam.proceedings.066

Abstract

This article is devoted to the research and improvement of axial-piston hydraulic swash plate machines, widely used in power drives for various purposes. The paper presents a refined force analysis of the piston mechanism. Special attention is paid to the research of the friction forces in the "slipper - swash plate" pair. In the straining and "creeping speeds" modes these friction forces, as well as friction forces in the piston pair, can have a significant impact on the dead band of the hydraulic machine. The results of experimental researches of the piston mechanism are presented, which suggest a decrease in friction forces during straining and "creeping speeds" modes in the "slipper - swash plate" pair for the developed piston compared to a standard piston.

Keywords: axial-piston hydraulic swash plate machine; dead band; force analysis; friction force in the "slipper - swash plate" pair; drive.

1. Analysis of friction forces in a pair of "slipper-swash plate" in the straining and "creeping speeds" modes

The axial-piston hydraulic swash plate machine is widely used in energy-intensive drives for various purposes. It has high weight and size characteristics, simplicity of design and, as a result, higher reliability and service life, as well as low cost. However, these hydraulic machines are inferior to axial piston bent axis hydraulic machines in terms of such a parameter as the control range. Such difference is caused by the relatively large dead band when starting and reversing the operation of the hydraulic machine. In this regard, it is relevant to conduct research on axial-piston hydraulic swash plate machines and develop constructive and technological recommendations for the improvement and use of this type of hydraulic machines in high-precision drives. An indirect confirmation of the relevance of these researches is the recent increase in number of publications devoted to this topic.

Many researches are devoted to the power analysis of a hydraulic machine. In particular, Marek Woźniak, Tomasz Szydłowski and Krzysztof Siczek in their work [1] researched the nature of the movement of the slipper. They found that with a reciprocating sliding motion along the axis of the piston, the thickness of the lubricant between the slipper and the swash plate quickly reached a minimum and stabilized.

That is, hydrostatic unloading is realized. Xiangxu Meng, Chang Ge and Hongxi Liang in their research [2] show that the position of the slipper has a significant influence on the pressure distribution, and also propose a grooved sealing pad design. Heon-Sul Jeong and Hyoung-Eui Kim in their article [3] obtained two formulas for estimating the instantaneous frictional force of the piston and the average loss of the friction moment of the piston, taking into account the effects of mixed / boundary lubrication. As a result of further work [4], they presented a model of a hydraulic machine, which makes it possible to evaluate its performance based on the size of the hydraulic machine.

However, in the publications presented above, in the course of power analysis, as a rule, the modes of starting and "creeping velocities" are not considered. In this paper, experimental researches of the friction forces in the piston assembly axial-piston hydraulic swash plate machine in the straining and "creeping speed" modes were conducted. It was assumed, that higher friction forces in the "piston and cylinder" and "piston-slipper" interface lead to the tilt of the slipper. Therefore, the direct contact of its edge and the surface of the swash plate leads to a decrease in the lifting hydrostatic force, an increase in contact stresses and, as a consequence, a significant increase in the friction force in a given kinematic pair.

2. Experimental researches

To research the piston mechanism in the conditions of straining during the transition from pumping to motor mode and from motor to pumping mode, an experimental stand was used [5], shown in Figure 1. The experimental stand is based on the principle of inversion of movement of swash plate machines. The piston groups are assembled in a stationary cylinder block, and the swash plate is driven by an external drive.

Fig. 1. Experimental stand for the piston mechanisms of axial-piston hydraulic swash plate machines research

1 - pumping station;
2 - block of the rotating swash plate;
3 – stationary cylinder block;
4 – lever;
5 - piston groups;
6 - pressure sensor;
7 - weight sensor;
8 - analog-to-digital converter;
9 – computer.

The scheme of the stand is shown in figure 2.

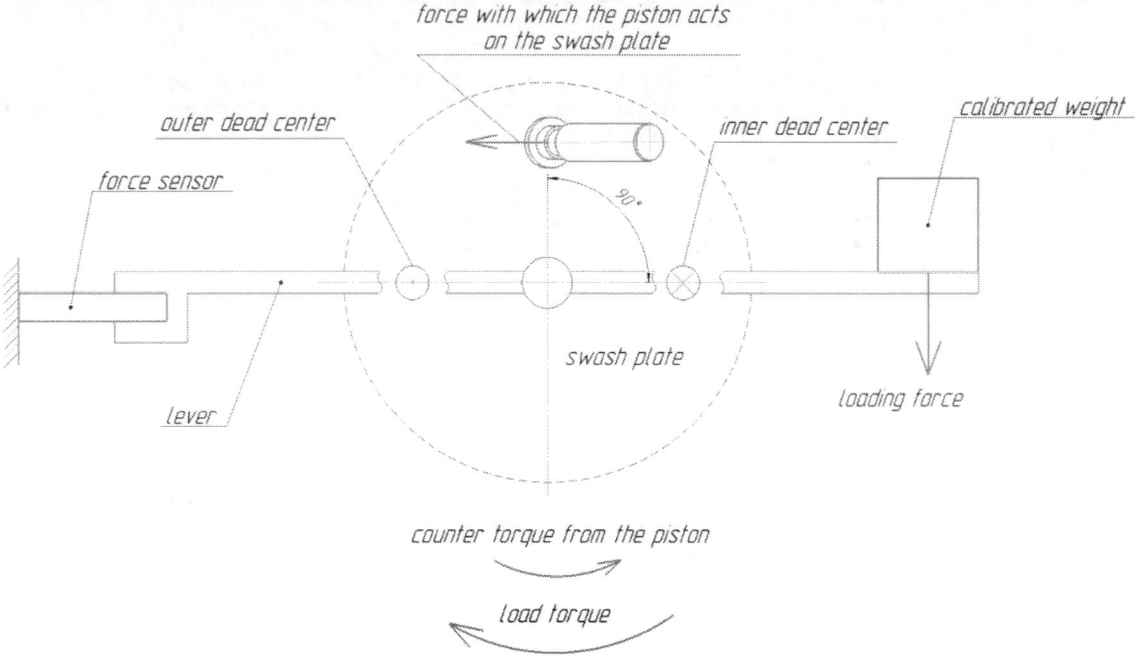

Fig. 2. Scheme of the experiment

A load moment, constant in direction and magnitude, was applied to the lever by a calibrated weight. Hydrostatic pressure acts on the piston, located in the middle position between the dead centers, creating a moment directed against the load moment. During the experiment, an increase in the hydrostatic pressure was carried out, and, consequently, an increase in the torque of the piston mechanism. When the value of the load torque was exceeded, the direction of the resulting torque on the lever changed (transition from pumping to motor mode). With a decrease in torque, a change took place in the opposite direction (transition from a motor to a pumping mod). As a result, the value of the dead band was measured during the transition from one mode to another. Figure 3 shows the researched samples of pistons of standard and developed design [6].

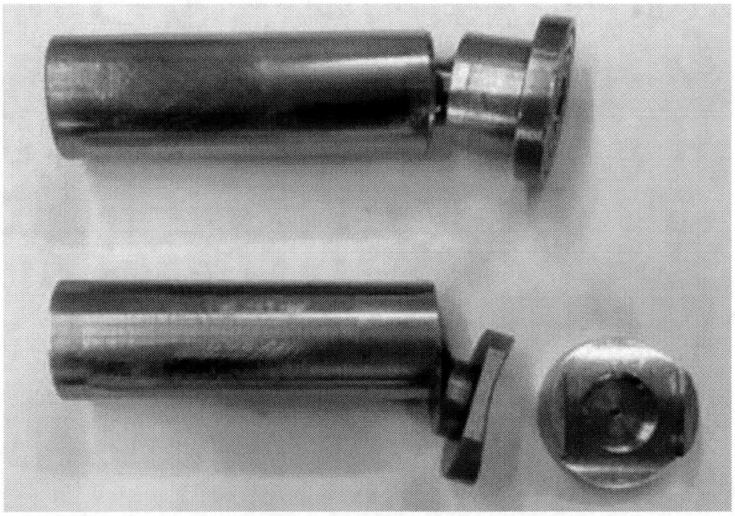

Fig. 3. Standard and developed piston mechanism

Figure 4 shows a graph of the transition from pumping to motor mode (resulting counterclockwise torque). The resulting torque is recorded by a weight sensor and, with the help of secondary equipment, is displayed on a computer. The part of the graph that lies below the abscissa axis corresponds to the pumping mode, above the abscissa axis - to the motor mode. A pronounced dead band is obvious from these graphs. In this case, the dead band of the modified piston mechanism is significantly less than for the standard one.

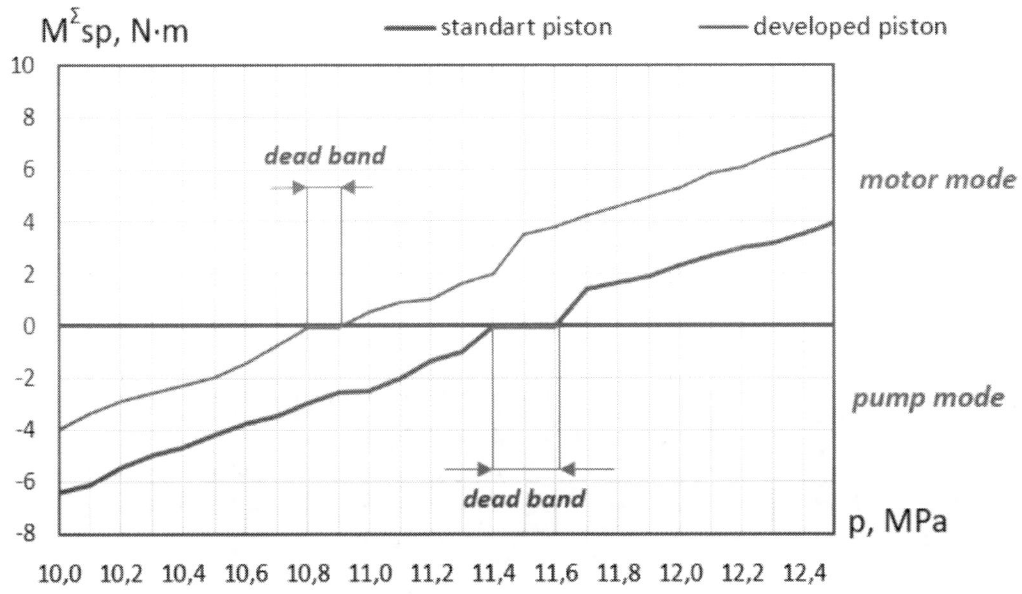

Fig. 4. The overall moment of force on the swash plate during the transition to the motor mode of operation

Figure 5 shows a graph of the transition from motor to pump mode. When the pressure in the sub-piston cavity decreases, the moment created by the piston mechanism decreases, while the moment from the load on the lever remains constant. Similar to the previous series of experiments, the curve values reflect the same operating modes of the hydraulic machine. At the same time, an increase in the value of the dead band was revealed for both the standard and the modified piston mechanism.

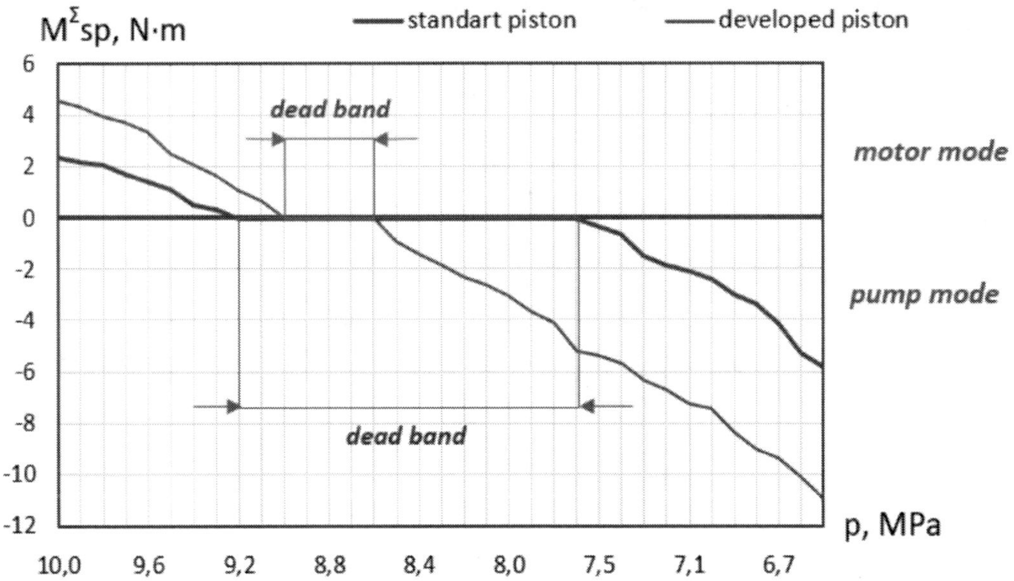

Fig. 5. The overall moment of force on the swash plate during the transition to the pumping mode of operation

A significant difference in magnitude between the dead bands during the transition from pumping to motor mode and from motor to pumping mode can be caused by the action of major and multidirectional friction forces in a pair of "slipper - swash plate". From a comparison of the graphs, it follows that the dead band for the developed piston mechanism changes less significantly than for the standard one. The difference in axial frictional forces between the pumping and motor modes of operation in the «piston-cylinder» pair for the developed piston mechanism in the straining and «creeping speeds» modes at a pressure of 10 MPa is clearly shown in [7] and is no more than 22%. Therefore, the bigger part of the dead band difference is due to friction in the "slipper - swash plate" pair.

3. Refined diagrams of forces acting on the piston mechanism

The experiments have shown that when straining and low speeds, friction between the slipper and the swash plate should be taken into account. The refined diagram of forces for a standard piston mechanism, taking into account the friction force of the slipper, is shown in figure 6 for motor and pumping modes. The point of application of the frictional force between the slipper and the swash plate depends on the degree of slipper skew. Therefore, the diagram shows a conventionally reduced friction force Fs located in the center of the slipper.

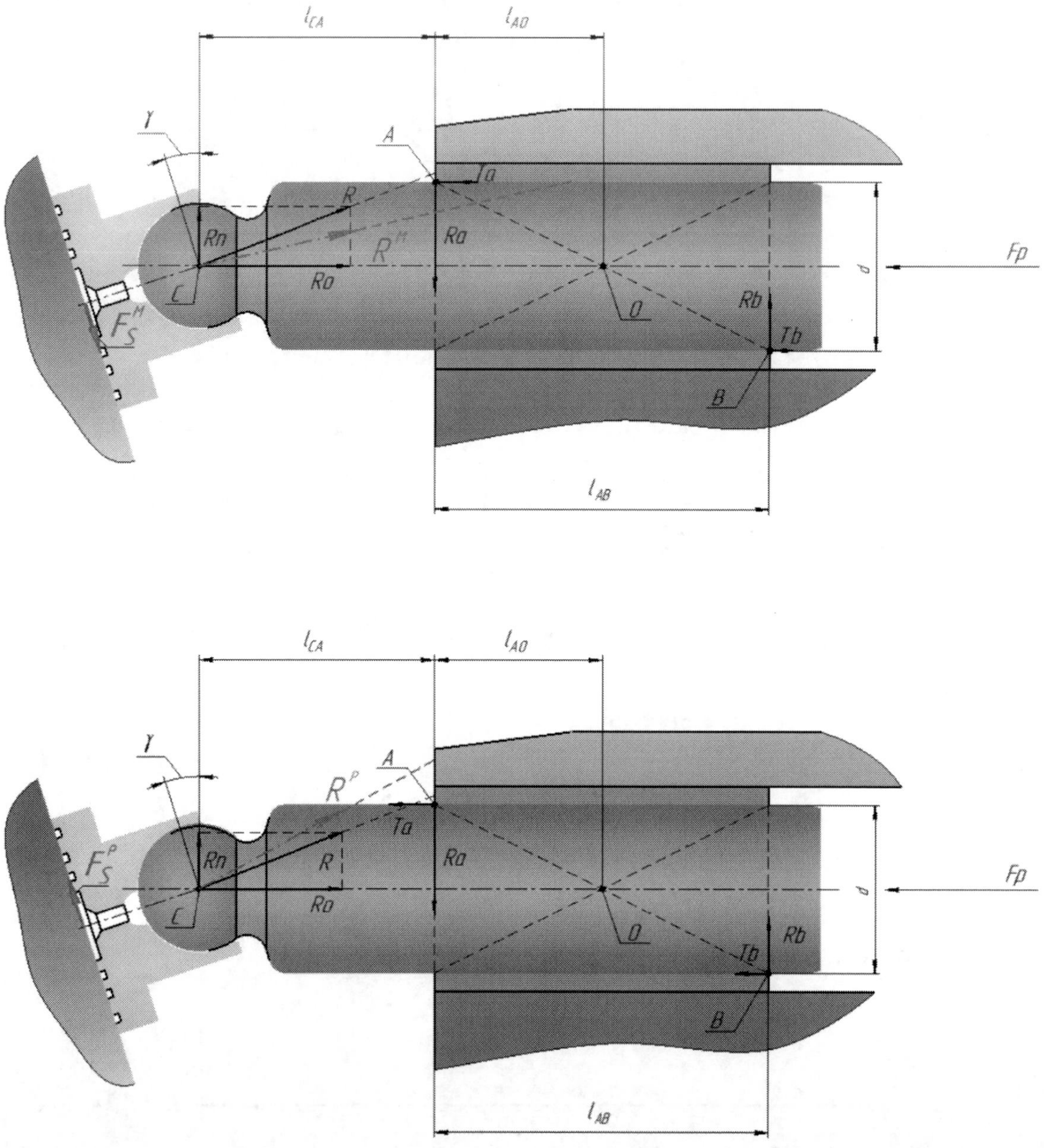

Fig. 6. Scheme of forces in a standard piston mechanism taking into account the friction forces between the slipper and the swash plate F_S^M for motor mode (upper figure) and F_S^P for pumping mode (lower figure)

Figure 7 shows the refined force diagram for a modified piston mechanism.

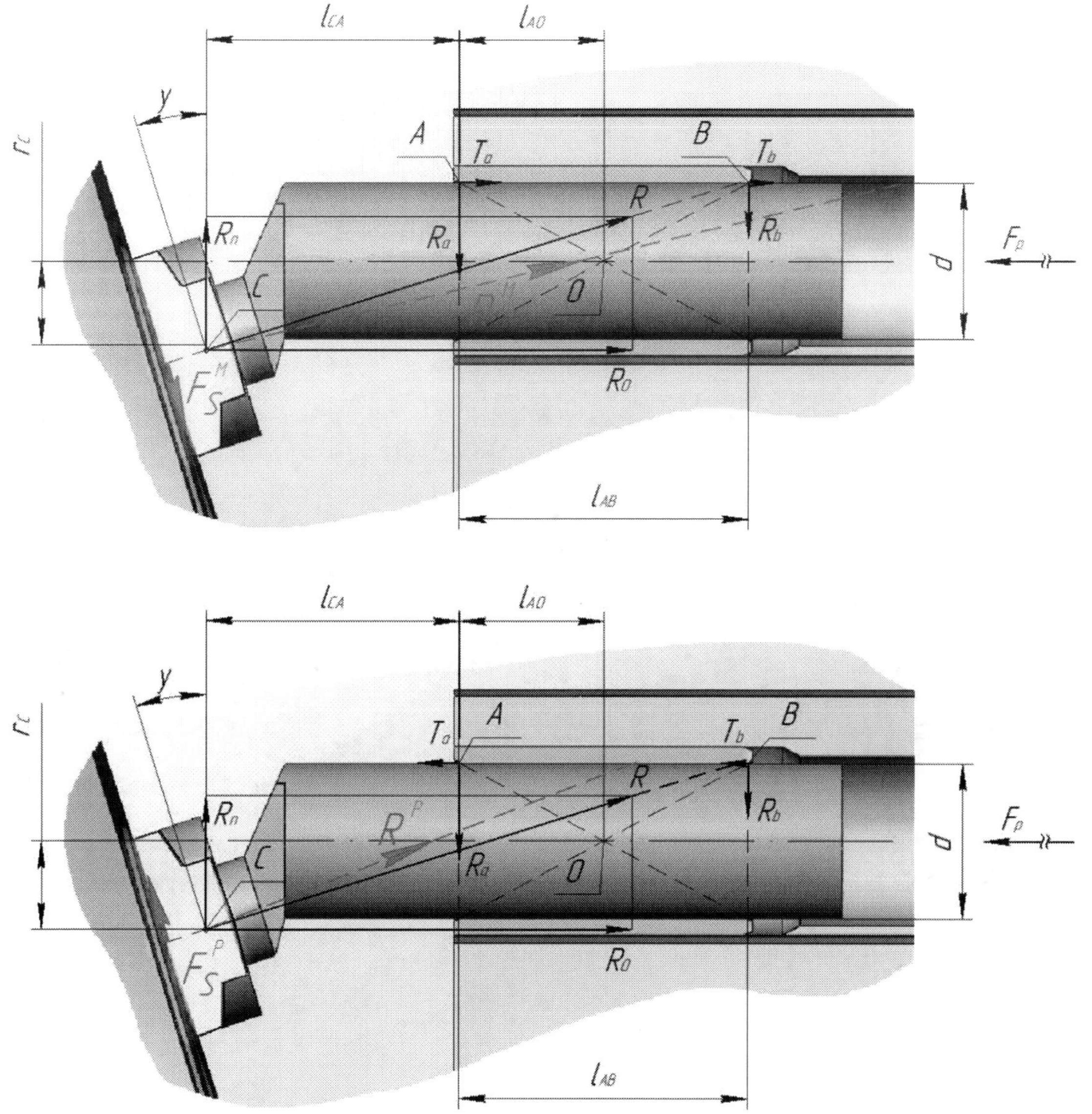

Fig. 7. Diagram of forces in the developed piston mechanism taking into account the friction forces between the slipper and the swash plate F_S^M for motor mode (upper figure) and F_S^P for pumping mode (lower figure)

Fp - force of pressure of the working fluid on the end surface of the piston
R - reaction force of the swash plate
Ro - axial component of the reaction force of the swash plate
Rn – transverse component of the reaction force of the swash plate
Ra – reaction force of the outer part of the guide sleeve
Rb – reaction force of the internal component of the guide sleeve
Ta - friction force of the outer part of the guide sleeve
Tb - reaction force of the inner part of the guide sleeve
γ - angle of inclination of the swash plate
C - center of the hinge
d - piston diameter
r - piston radius
rc - axial displacement of the piston hinge
O - geometric center of the guide sleeve
Fs - friction force between the slipper and the swash plate
R^M - reaction force taking into account F_S^M in motor operation
R^P - reaction force taking into account F_S^P in pumping operation mode

To determine the friction forces in the slipper and the coefficient of friction between the slipper and the swash plate, it should be assumed that the friction forces in the motor and pumping modes are equal. This assumption can be valid only conditionally, since the angle of inclination of the slipper relative to the plane of the swash plate in pumping and motor modes is unknown. The equation of moments for the piston pair from [8] for the motor mode and the standard piston design:

$$(F_P - T) \cdot tg\gamma \cdot (l_{CA} + l_{AO}) - R_B \cdot f \cdot r + R_A \cdot f \cdot r - R_A \cdot l_{AO} - R_B \cdot l_{AO} - F_S \cdot \cos\gamma \cdot (l_{CA} + l_{AO}) = 0 \quad (1)$$

$$2R_A \cdot l_{AO} = (F_P - T) \cdot tg\gamma \cdot f \cdot r + (F_P - T) \cdot tg\gamma \cdot (l_{CA} + l_{AB}) - F_S \cdot \cos\gamma \cdot (l_{CA} + l_{AO}) \quad (2)$$

$$R_A = \frac{(F_P - T) \cdot tg\gamma \cdot \big((l_{CA} + l_{AB}) + f \cdot r\big) - F_S \cdot \cos\gamma \cdot (l_{CA} + l_{AO})}{2l_{AO}} \quad (3)$$

$$R_B = \frac{(F_P - T) \cdot tg\gamma \cdot (l_{CA} + f \cdot r) - F_S \cdot \cos\gamma \cdot (l_{CA} + l_{AO})}{2l_{AO}} \quad (4)$$

$$T^0 \cdot F_p \cdot \frac{l_{AB}}{f} = \left(F_p - T^0 \cdot F_p\right) \cdot (tg\gamma \cdot ((2l_{CA} + l_{AB}) + 2f \cdot r)) - F_S \cdot \cos\gamma \cdot (l_{CA} + l_{AO}) \quad (5)$$

$$T_M^0 = \frac{(tg\gamma \cdot ((2l_{CA} + l_{AB}) + 2f \cdot r)) - \frac{F_S}{F_p} \cdot \cos\gamma \cdot (l_{CA} + l_{AO})}{\frac{l_{AB}}{f} + (tg\gamma \cdot ((2l_{CA} + l_{AB}) + 2f \cdot r))} \quad (6)$$

For pumping mode:

$$T_H^0 = \frac{(tg\gamma \cdot ((2l_{CA} + l_{AB}) - 2f \cdot r)) + \frac{F_S}{F_p} \cdot \cos\gamma \cdot (l_{CA} + l_{AO})}{\frac{l_{AB}}{f} - (tg\gamma \cdot ((2l_{CA} + l_{AB}) - 2f \cdot r))} \quad (7)$$

Equations 6 and 7 represent the effect of the friction force in the «slipper-swash plate» pair on the friction force in the «piston-guide sleeve» pair.

$$\begin{aligned}
&({p_{ЗН}}^{М} - {p_{ЗН}}^{Н}) \cdot S_П = \\
&{F_P}^{Н} \cdot \frac{(tg\gamma \cdot ((2l_{CA} + l_{AB}) - 2f \cdot r))}{\frac{l_{AB}}{f} - (tg\gamma \cdot ((2l_{CA} + l_{AB}) - 2f \cdot r))} - \\
&{F_P}^{М} \cdot \frac{(tg\gamma \cdot ((2l_{CA} + l_{AB}) + 2f \cdot r))}{\frac{l_{AB}}{f} + (tg\gamma \cdot ((2l_{CA} + l_{AB}) + 2f \cdot r))} + \\
&k_S \cdot \left(\begin{aligned} &{F_P}^{Н} \cdot \frac{\cos\gamma \cdot (l_{CA} + l_{AO})}{\frac{l_{AB}}{f} - (tg\gamma \cdot ((2l_{CA} + l_{AB}) - 2f \cdot r))} + \\ &+{F_P}^{М} \cdot \frac{\cos\gamma \cdot (l_{CA} + l_{AO})}{\frac{l_{AB}}{f} + (tg\gamma \cdot ((2l_{CA} + l_{AB}) + 2f \cdot r))} + \cos\gamma \cdot \left({F_P}^{М} + {F_P}^{Н}\right) \end{aligned} \right)
\end{aligned} \quad (8)$$

${p_{ЗН}}^{Н}$ - pressure of transition to pumping mode;
${p_{ЗН}}^{М}$ - pressure of transition to motor mode;
k_S - coefficient of friction in the «slipper-swash plate» pare.

From these dependencies it follows, that for a standard piston mechanism the pressure of the transition to the pumping mode is 7.34 MPa, the pressure of the transition to the motor mode is 11.72 MPa. Then the coefficient of friction between the slipper and the swash plate will be **0.21**, which corresponds to a pronounced boundary mode of friction with direct contact between the edge of the slipper and the swash plate.

For the developed piston mechanism, the transition pressure to the motor mode was 11.01 MPa, and to the pumping one - 8.46 MPa. Accordingly, the coefficient of friction in the «slipper–swash plate» pair was **0.12.**

4. Conclusions

In the power analysis of a hydraulic machine, as a rule, friction forces in the "slipper-swash plate" pare are not taken into account, implying the implementation of a liquid friction regime due to the presence of partial or complete hydrostatic unloading. In the course of theoretical and experimental researches, the assumption was confirmed about the need to take into account the friction forces in the "slipper-swash plate" pair in the straining and "creeping speeds" modes. In the course of experimental researches, a decrease in the value of the coefficient of friction in the developed piston mechanism in comparison with the standard one by 1.9 times was obtained. Such an improvement in piston characteristics makes it possible to predict a significant decrease in the value of the dead band of the hydraulic machine and, as a consequence, the expansion of its control range. In the future, our team plans to conduct a series of experiments during which the swash plate will work out the input sinusoidal signal. This will allow simulating the tracking mode of the hydraulic drive.

5. References

[1] Marek Woźniak, Tomasz Szydłowski, and Krzysztof Siczek (2020). The Effect of Dynamics of the Swash Plate–Slippers–Piston Assembly on Friction, Mechanics and Mechanical Engineering 2020, Vol. 24, pp. 42-55

[2] Xiangxu Meng, Chang Ge, Hongxi Liang (2020). Lubrication characteristics of the slipper–swash-plate interface in a swash-plate-type axial piston pump, Proceedings of the Institution of Mechanical Engineers, Part C: Journal of Mechanical Engineering Science 235(4)

[3] Heon-Sul Jeong, Hyoung-Eui Kim (2004). On the Instantaneous and Average Piston Friction of Swash Plate Type Hydraulic Axial Piston Machines , KSME International Journal, Vol. 18 No. 10, pp. 1700-1711

[4] Heon-Sul Jeong, Hyoung-Eui Kim (2007). A Novel Performance Model Given by the Physical Dimensions of Hydraulic Axial Piston Motors: Experimental Analysis, Journal of Mechanical Science and Technology, Vol. 21, pp. 630-641

[5] Kuzmin, A[nton]; Popov, V[alery] & Stazhkov, S[ergey] (2016). Advanced Axial Piston Swash Plate Pump Parameters Recommendations, Proceedings of the 27th DAAAM International Symposium, pp.0556-0561, B. Katalinic (Ed.), Published by DAAAM International, ISBN 978-3-902734-08-2, ISSN 1726-9679, Vienna, Austria

[6] Stazhkov, S[ergey]; Korobova, I[raida]; Korolev, V[ladimir] & Kuzmin, A[nton] (2019). Hydrodynamic Processes Research of the Positive Displacement Hydraulic Machines Piston Interface, Proceedings of the 30th DAAAM International Symposium, pp.0580-0585, B. Katalinic (Ed.), Published by DAAAM International, ISBN 978-3-902734-22-8, ISSN 1726-9679, Vienna, Austria

[7] Stazhkov, S[ergey]; Kuzmin, A[nton]; Elchinsky, V[iktor]; & Korolev, V[ladimir] (2020). Investigation of the Effect of the Gap on the Friction Forces in the Piston Pair of the Axial Piston Hydraulic Machine, Proceedings of the 31st DAAAM International Symposium, pp.0384-0390, B. Katalinic (Ed.), Published by DAAAM International, ISBN 978-3-902734-29-7, ISSN 1726-9679, Vienna, Austria

[8] Elchinsky, V[iktor]; Kuzmin, A[nton]; Popov, V[alery] & Stazhkov, S[ergey] (2020). Influence of the Design Parameters of the Piston Mechanism on the Dead Band of the Axial-Piston Hydraulic Machine, 16th International Conference on Industrial Manufacturing and Metallurgy, ICIMM 2021, IOP Conf. Series: Materials Science and Engineering, Ekaterinburg, Russia

DOI: 10.2507/32nd.daaam.proceedings.067

UNDERSTANDING THE DEMAND DRIVEN MATERIAL REQUIREMENTS PLANNING SCOPE OF APPLICATION: A CRITICAL LITERATURE REVIEW

Maria Angela Butturi*[,1], Giuseppina De Rosa[2], Elia Balugani[1] & Rita Gamberini[1,2]

1 En&Tech Interdipartimental Center of the University of Modena and Reggio Emilia, Via Amendola 2, Padiglione Morselli, 42122 Reggio Emilia, Italy

2 Department of Sciences and Methods for Engineering, University of Modena and Reggio Emilia, Via Amendola 2, Padiglione Morselli, 42122 Reggio Emilia, Italy

This Publication has to be referred as: Butturi, M[aria] A[ngela]; De Rosa, G[iuseppina]; Balugani, E[lia] & Gamberini, R[ita] (2021). Understanding the Demand Driven Material Requirements Planning Scope of Application: a Critical Literature Review, Proceedings of the 32nd DAAAM International Symposium, pp.0462-0471, B. Katalinic (Ed.), Published by DAAAM International, ISBN 978-3-902734-33-4, ISSN 1726-9679, Vienna, Austria
DOI: 10.2507/32nd.daaam.proceedings.067

Abstract

The supply chains complexity generated by the dynamic market demand imposes the improvement of the classical production control systems. A recently introduced method, the Demand Driven Material Requirements Planning (DDMRP), is proposed as an upgrade of the Material Requirements Planning (MRP), widely used in industry, capable of overcoming the nervousness of MRP environment and the bullwhip effect affecting supply chains under uncertainties. The DDMRP approach, however, is still not well established since the conditions for its application are still little investigated. Thus, this study aims at reviewing the existing scientific literature concerning DDMRP method in order to critically analyse its main scope of application as well as its real practical performance. From the reviewed literature three main research lines emerged: DDMRP basic principles, comparison with other methodologies, and case studies. The analysis of both research oriented papers and case studies points out some critical issues that are limiting the diffusion of the DDMRP method, including the additional costs necessary to adapt the in use control and planning software. The main criticality of the method is recognized to be the high subjectivity affecting the positioning of the buffers, and the need for classifying the suitable sectors of application.

Keywords: Production control systems; Demand and lead time uncertainties; Demand Driven Material Requirements Planning; Inventory level; Supply Chain complexity.

1. Introduction

The current evolution of the market, guided by the increasingly volatile needs of the customers, put companies in front of new challenges: the supply chain has become more complex, moving from mainly linear and vertically integrated structures to highly branched ones; greater volumes of material are requested and handled; the heterogeneity and complexity of the available products have significantly increased; technological advancement has contributed to rapidly changing consumer preferences, reducing the life cycles of products and making them more and more customizable. In the context of production planning, these issues translate into the need for improving the "classical" production control systems (PCS) to adjust them to the new market requirements. Many PCS's emerged in the last couple of decades to respond to the market evolution as well as to specific environments of application [1].

One of the most widely used production control and planning systems in the industry is MRP (Material Requirements Planning), and this system is often embodied in the enterprise resource planning (ERP) software [2]. MRP techniques use the bill of materials (BOM), the master production schedule (MPS), and the inventory records to schedule the required materials replenishment for the manufacturing of the final product for a given time horizon. It is based on a control system for the planning of stocks and production of the push-type, according to which organizations use forecasts to determine the market demand for their products. The MRP approach works well in a deterministic context, while it shows some weaknesses under uncertainties concerning the demand and lead time [3]. The continual adjustments in the scheduling due to even small changes of the codes of the higher levels of the BOM make the MRP unstable, causing the so-called MRP nervousness [4]. Moreover, the dependence between the nodes of the supply chain produces the bullwhip effect, transmitting and amplifying the variability that affects all the levels of the BOM to all the nodes in the supply chain [5].

Among the new approaches proposed to overcome MRP criticalities [6], [7], [8], the Demand Driven MRP (DDMRP) was firstly introduced in 2011 by Ptak and Smith [9], as a new PCS. According to the authors [10], the DDMRP is a hybrid push-pull method that combines some of the principles and practices of MRP, Lean, Six Sigma, and the Theory of Constraints, resolving possible conflicts among the different methods through innovative solutions. It uses Lean Manufacturing principles to allow a pull logic with approximation to real demand and the concept of signalling mechanisms to have visibility of stock levels; like TOC it focuses on products that limit the system performance and uses buffers to protect the process; from Six Sigma it considers the strong sensitivity to variations, trying to control them. DDMRP considers a decoupled lead time, keeping some pre-defined BOM components in stock, and uses the Available Stock Equation (ASE) to calculate, daily, the future stock based on actual demand and orders in production. The methodology is implemented through five steps: the strategic inventory positioning, where the positioning of the buffers responds to the need of decoupling the planning and execution horizons of the logistics chain reducing the bullwhip effect; the definition of the buffers profiles and levels, for sizing the three buffer layers (red, yellow and green) calculated by three factors: item type, lead time and variability; the dynamic adjustments of the buffers' layers with changes in demand; the demand-driven planning, for creating the production and purchase orders; the execution phase, to manage the orders once they are generated. For a more detailed analysis of the method see for example [11], [12].

In the last years the DDMRP method, initially developed by practitioners, gained new interest within the academic research that produced both theoretical and empirical studies [1]. However, though the method presents some innovative aspects, mainly regarding its ability in overcoming the bullwhip effect typical of supply chains and the MRP nervousness, the conditions for its application and its real practical performance are still little investigated. Only a few literature reviews tried to organise the available studies and analyse some specific issues related to DDMRP implementation. Thus, this study aims at reviewing the existing scientific literature concerning the DDMRP method, in order to investigate the more recent results on the methodology applicability and effectiveness in the industrial environment. It is organised as a critical literature review that updates the few available reviews, providing additional insights. Following the three main research lines that emerged from the literature, the paper discusses the advantages and limits of the methodology in the view of the various authors, analyses the critical issues that can limit its application, and outlines its possible future developments.

2. Literature review methodology

2.1. Scope of the review

Despite the recent attention to the study and application of the DDMRP and the number of articles available on the website of the Demand Driven Institute (DDI), the consulting company founded by Ptak and Smith operating in the DDMRP field, the method is still under-investigated by the scholars. Therefore, this literature review aims at exploring the academic perspective on the methodology and is driven by three main research questions:

1. Are the basic principles of DDMRP improved by academic research?
2. What are its required future evolutions according to the scientific community view?
3. Considering the benefits and difficulties encountered by the manufacturing industry applying the DDMRP, in what productive environments did this system prove useful?

Our study aims at updating and extending the three literature reviews published on the DDMRP methodology, which reveal different focuses. Marzougui et al. identify papers containing simulated or real case studies with the aim of comparing the academic results with the improvements achieved by companies implementing the DDMRP method as claimed by the DDI; master's theses and non-English paper are excluded from the review, which is updated to 2019 [11]. The systematic literature review of Orue et al. analyses the studies investigating the standardization of the procedure to implement the DDMRP model, to build up a standard application model enabling full exploitation of the methodology potential benefits, through reducing the implementation time and costs, the possible errors, and improving the quality of the process; also this review is updated to 2019 [13].

The more recent systematic review, presented by Azzamouri et al. and updated to April 2020, explores the literature contributions to three main issues providing knowledge on the method: the method validation, the improvement of the original techniques, and the proposal of new tools for its implementation [14].

2.2. Publications search and selection

The exploration of the literature was conducted using the online databases Science Direct, Web of Science and Google Scholar and completed on May 14, 2021. An initial set of papers was identified employing the keywords "Demand Driven Material Requirements Planning", "DDMRP", and "Demand Driven"; the saturation point occurred when the same articles kept appearing in search engines. Due to the relatively small number of collected papers and to ensure that our research was as comprehensive as possible, we further conducted a snowball search. At the end of the search process, we collected a total of 140 papers.

Then, we proceeded with the selection of the relevant papers according to the following exclusion criteria:

- Duplicates between search engines have been removed, i.e. 48 papers.
- Both qualitative and quantitative articles were selected, considering publications in academic journals or conferences, and Master and PhD degree theses; in this review, we included the theses because they made available the knowledge on the DDMRP methodology to the wide practitioners' community in their different national languages. Some magazines and newspapers articles written by consultants specialized in Supply Chain and logistics, mainly available on the DDI website, have been excluded since their main purpose could be marketing their product to companies and this could change the way of presenting the method, i.e. 33 papers.
- The publications without the specified keywords within the article were excluded, finding that they dealt with topics not directly related to the DDMRP (e.g. Decoupled lead time), i.e. 4 papers.
- The publications without an exclusive focus on the DDMRP methodology applied to the planning of material requirements in the supply chain (e.g. articles dealing with the broader Demand Driven Enterprise Model) were excluded, i.e. 7 papers.
- Since the scope of this review is to analyse the DDMRP related concepts that emerged in the literature in recent years, we defined as review temporal boundaries the period 2016-2021; in fact, starting from 2016 there is an increase in the number of scientific publications (Figure 1), following a few years the first introduction of the DDMRP methodology in 2011 [9]. Other 9 papers were thus excluded.

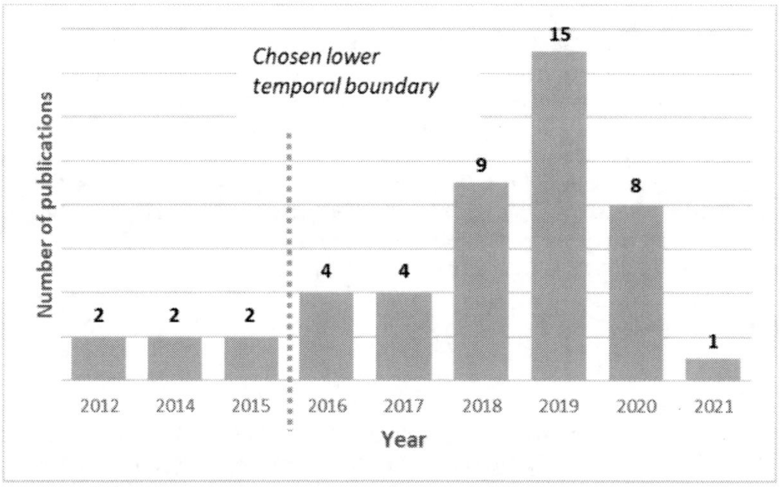

Fig. 1. Temporal distribution of the selected DDMRP publications.

Finally, a total of 39 relevant publications were selected for further analyses: 12 scientific articles, 6 conference papers, and 21 theses.

3. Results and discussion

3.1. Papers classification

To answer the research questions, the selected publications were classified analysing their content according to three main categories revealed by the review: DDMRP basic principles, comparison with other methodologies, and case studies (Table 1).

Content category	REF.	N. of studies
DDMRP basic principles	[15] [16] [17] [18] [19] [20] [21] [22] [23] [24] [25] [26] [27]	13
Comparison with other methodologies	[28] [29] [30] [12] [31] [32] [33] [34]	8
Case studies	[35] [36] [37] [38] [39] [40] [41] [42] [43] [44] [45] [30] [46] [47] [48] [49] [50]	17

Table 1. Classification of the selected studies according to the identified categories.

Table 1 shows that the research papers (including both the DDMRP basic principles and comparison with other methodologies categories) are more than the case studies, confirming that although some companies are now implementing DDMRP, still very few of them share their data and collaborate to scientifically study the advantages or critical issues related to the method within their specific environment [14]. The content of the papers is then analysed and discussed in relation to the three identified categories.

3.2. Group 1: DDMRP basic principles

The first group of publications includes 13 items, of which 7 articles were published in scientific journals or conferences and 6 Master degree theses. These publications mainly aim at examining the principles and basic concepts of the DDMRP methodology, showing future developments necessary for the optimization of the methodology.

An increasing interest in the DDMRP field by the Academy is observed, resulting in simulations, studies and methodology analyses. The Polytechnic University of Valencia uses DDMRP for students' group work according to the *problem-based learning* method [17] and many theses in different countries analyse how the DDMRP works to improve the theoretical knowledge on DDMRP methodology and introduce it to local companies.

Some studies address the main advantages and limitations of the method. According to Cuadra [16], the DDMRP allows to increase sales by a high percentage, to achieve a service rate of 99%, to limit the stock and to reduce the delivery time. Meinzel [21] uses a simulation to address the five steps of implementation of the method, examining the obtained benefits that result in + 13% of average service level, - 22% in average lead time and -31% average cost of inventory. De Biasi [25] shows that the DDMRP method allows to improve the information flows within the supply chain and the performance of the warehouses.

Many scholars investigate the reasons leading companies to choose the DDMRP and implement it in the manufacturing environment. According to Perälä [23], the key data needed for the DDMRP implementation within the company must be organized following the next points: to identify the product or item that can benefit from DDMRP application; to place the decoupling points in the BOM of the product; to calculate the zones and buffer levels. Marin [18] considers the novelties introduced by DDMRP the reasons why it finds opposition from some companies' staff members, who do not accept to change their habits. On the other hand, various authors consider essential the development of DDMRP-based software to be integrated into an ERP, for incentivizing organizations to choose the DDMPR method and allowing easier management of the tool by local companies [16], [18], [20].

According to the DDMRP methodology [10], the choice of some variables (namely the positioning of the buffers, the lead time and variability factors, as well as the determination of the safety stock) is based on the experience of the staff who must select the values of the parameters within a certain range. This procedure introduces a subjectivity in the method implementation that should be overcome by analyzing a larger sample of companies using the DDMRP to identify the most suitable sectors of application and provide a set of best practices and mathematical tools supporting the decision-making process [20], [18]. Lee et al. [19] propose a new safety stock formula for the calculation of the parameter, enhancing existing guidelines for the DDMRP implementation. Velasco Acosta et al. [27] developed a DDMRP model simulated through a discrete event software to evaluate the applicability of the method in a complex manufacturing environment, considering a number of BOM levels up to 4, finding that the customer satisfaction level can be improved by 41% and the average on-hand stock can be reduced by 18%.

Few papers consider the relation of the DDMRP to Supply Chain Management (SCM), though the level of service and the reduction of delivery times can directly affect the customer's relationship with the company. Achergui [26] use the DDMPR buffer positioning approach applied to a hybrid MTO/MTS manufacturing system, demonstrating a viable reduction of the inventory costs for small supply chain networks. Santos [24] evaluates the impact that DDMRP has on the SCM processes, through the AHP (Analytic Hierarchy Process) multi-criteria decision support technique; the author finds that the three most impacted processes are production flow management, demand management and order fulfilment. Pekarcíková et al. [22] analyse the use of multiple product-related BOMs to indicate different levels in the supply chain considering the demand behaviour, with the aim of extending the DDMRP knowledge base in the context of Industry 4.0.

A list of the required future research developments apt to foster the adoption of the DDMRP method, according to the reviewed publications, is sketched in the figure 2. In the circles, the number of papers referring to the required insight.

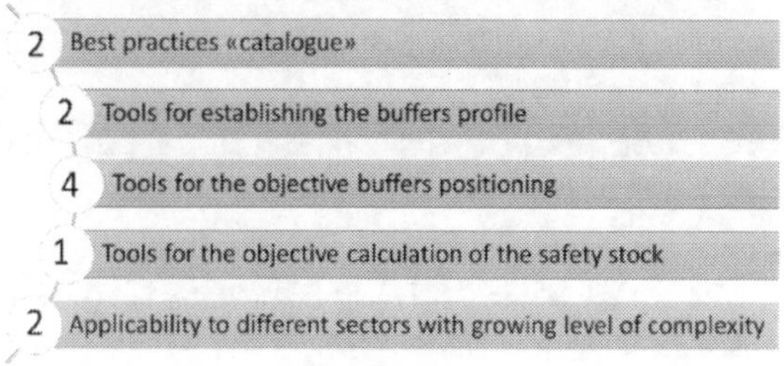

Fig. 2. Future developments of the studies on the DDMRP method, as expected by the papers included in the review.

3.3. Group 2: Comparison with other methodologies

The second group contains 8 publications, 6 papers and 2 theses, where DDMRP is compared with other methodologies, namely MRP and MRPII (Manufacturing Resources Planning), that it aims to overcome, the Economic Order Quantity (EOQ) tool and Kanban method.

Miclo et al. [31], [32], [33] use the discrete events simulation and the Kanban game[1] data as input data to compare the DDMRP with MRPII, using as main indicators the On-Time Delivery (OTD) and Working Capital (WC). The results obtained show that DDMRP achieves better output indicators than MRPII with less WC and WIP, but with a less satisfactory OTD. Reference orders are much smaller in the case of the DDMRP, but more frequent than in the MRPII; however, the increase in the number of orders does not generate a correspondent increase in the number of setups for the machines, which was set by the execution priorities of the DDMRP. Moreover, the DDMRP method seems to be more stable when facing the variability of real demand, although during the simulation some products went out of stock. As expected, the DDMRP efficiently anticipates the peaks in demand orienting the management to the pull logic for the normal demand and the push logic for the peaks. Still using a discrete events simulation, Shofa et al. [29] compare the DDMRP with the MRP focusing on the inventory level under long lead time and uncertain demand scenarios, finding that using the DDMRP method the inventory level is more stable and the stock level is reduced by 11%. Simulating the behaviour of four production planning and control systems (DDMRP, MRP, Kanban and Optimised Production Technology) applied to multi-stage assembly systems under different levels of bottleneck severity and delivery date tightness, Thurer et al. [34] found that DDMRP maintains a high level of performance also for severe bottleneck, with the lower inventory levels. Miclo et al. [12] simultaneously compare DDMRP with the MRPII and the Kanban systems at different levels of demand variability; they found that Kanban has a slightly lower WIP than DDMRP on average, while MRPII provides the worst results. According to the authors, the innovation of DDMRP lies in the fact that this methodology makes use of the best practices of the two previous approaches. The dynamic buffers have the function of absorbing the variability and allow the effective scheduling of the production, and a constant level of protection from the effect of nervousness. Finally, unlike the Kanban system, DDMRP uses buffers selectively, applying them only where they are needed. The comparison between DDMRP and EOQ, performed through a discrete event simulation in an ideal environment by Tounsi [28], shows how the DDMRP method uses less capital in stocks and WIP, with a higher percentage of orders fulfilled than the EOQ. Table 2 summarises the main advantages of the different methods as highlighted by the reviewed authors.

Method	Advantages	Respect to	REF
MRP	OTD slightly better	DDMRP	[31] [32] [33]
	Fewer stock outages even if of larger quantities	DDMRP	[29]
	Less responsive to small lead time changes	DDMRP	[29]
DDMRP	Lower working capital	MRP, MRPII and EOQ	[31] [32] [33]
	Lower WIP	MRPII	[31] [32] [33]
	Stable responding to the variability of real demand	MRPII	[31] [32] [33]
	Inventory trend centred	MRPII	[31] [32] [33]
	Lower inventory level	MRPII	[31] [32] [33]
	Smaller and more frequent orders	MRPII	[31] [32] [33]
	A higher percentage of orders fulfilled	EOQ	[28]
	Unfulfilled orders are fulfilled in a shorter time	EOQ	[28]
	Lower inventory level	MRP (KANBAN)	[29] [34]
	Shorter delivery times	EOQ	[28]
	Selective use of buffers	KANBAN	[12]
KANBAN	Lower WIP	MRP and DDMRP	[12]

Table 2. Comparison of the different methodologies for managing inventory planning.

[1] http://www.cipe.fr/jeux-et-seminaires/jeu-du-kanban

3.4. Group 3: Case studies

The third group collects 17 case studies retrieved from the literature of which 13 theses and 5 research articles, which describe 15 simulations on real data and only 2 cases reporting the method implementation in companies; the case studies explore the benefits and difficulties faced by the manufacturing industry. Table 3 lists the case studies and the company context and motivations for adopting or simulating the DDMRP.

Case study: company name / *sector*	Company context / needs / issues to overcome	REF.
COASA / *oils and lubricants for the aviation industry*	Obsolescence of the products (90%); Excessive stock; Unnecessary products purchase due to unplanned events	[35]
GSA / *kitchen tools and small appliances*	Need to improve the service level and reduce the costs to recover market segments	[36]
Consumer-Packaged Goods (Food sector)	Perishable goods and high inventory turns (1.6 to 2.5 weeks); simulating planning under uncertainty at finite capacity	[37]
White goods	Need to improve the service level and reduce the costs to recover market segments	[38]
Wa-Armonia / *cosmetics*	Need to improve the service level and reduce the inventory costs	[39]
Automotive Lighting / *Outdoor lighting for automotive*	Variable and unpredictable demand, due to the change in consumers' behaviour; growing structural complexity of the supply chain; architectural complexity of the products.	[40]
ISEO locks / *locking systems*	Need for higher flexibility and reduced delivery time, confirming the service level	[41]
Exquise1928 / *soft drinks and syrups*	Interest in improving the company's PCS	[42]
Dayco Ivrea / *components for the automotive sector*	Management of raw material stock not effective, causing failures in production planning; suppliers lead time and demand variability.	[43]
Science4you / *scientific and educational games/tools*	Market seasonality; high variability of the demand. Failures in raw materials stock management; long lead times.	[44]
Food	Excessive stock; obsolescence of the products.	[45]
Automotive	Need to improve inventory level management.	[30]
Food - sauces	High fluctuations in the product demand; unstable inventory level.	[46]
AeroCo / *Aircraft components*	Interest in analysing the applicability of the DDMRP.	[47]
Companies CS1 (*padlocks*), CS2 (*components for household appliances*), CS3 (*special parts for the automotive industry*)	CS1: considerable purchase lead times; only one experienced responsible for purchase orders. Need to adjust the inventory level to the market requirements, whilst maintaining the service level. CS2: complex supply chain; high levels of raw material at a significant cost; final products stored in the company warehouse before delivery to customers; demand and supply variability and lack of visibility; stock outs and rush orders. CS3: three production plants; complex supply chain, as well as purchasing and production planning.	[48]
Company CS1	(CS1 as above)	[49] [50]

Table 3. Case studies included in the review.

As can be seen in the table 3, the application of the DDMRP methodology to the industrial environment is mainly studied to demonstrate its efficacy in responding to demand and lead time variability, improving inventory management, with a resulting economic advantage, and service level. Some criticalities are also revealed.

The simulation performed by Castro J.E.N. [35], starting from an ABC classification of Coasa's products (oils and lubricants for the aviation industry), finds that excess stocks and expired products have been eliminated, reducing the stock by 25% and estimating an economic return of about $ 10,000 per year. A reduction in available stocks, with a simultaneous increase in the ADU and the reduction in stock coverage, is observed also by Kortabarria [48] in the companies where the DDMRP method was implemented. Salazar & Arenas [38] highlight a decrease in total inventories up to 60% with an increase in sales up to 20%, reducing the reactivity time to the market demand up to 80%, in the white goods company they analysed. Brahimi & Saadi [42] compare the planning system of a company bottling and selling soft drinks with and without DDMRP; they note that the DDMRP model improves the stock coverage and makes the inventory level more stable, while it lowers the inventory turnover rate. Leguizamo & Totaitive [32] in simulating a single product, talc, find that the conversion to the DDMRP methodology guarantees a reduction in costs. Dimas et al. [46] show how the DDMRP lower the inventory levels of a company producing food sauce allowing a savings of 53.5%.

According to Ptak and Smith [10], the reduction of stocks corresponds to an unchanged or even better level of service. Salazar & Arenas [38] find a level of service above 99%, while Provenzano [40], in his simulation on a group of finished products (Automotive Lighting), finds that using the DDMRP the customer demand is satisfied without corrective actions. Kortabarria et al. [49], [50] also stress that the reduction in stocks is not followed by a worsening of the level of service.

A reduction of the delivery times is observed by Abith Zachariah [41], together with a higher level of service and satisfaction for the customers in a company producing locking systems. Shofa et al. [30], through a simulation using the data of an automotive company, claim that DDMRP allows a reduction in delivery times and lead time (- 94%), improving the service level. Ducrot & Ahmed [37] simulate a European perishable food production company and compare conventional APS software and DDMRP planning, observing the service level and the inventory turns; according to their calculations, DDMRP performs better than the heuristics-based planning in three scenarios out of four and it provides similar results as an advanced mathematical solver. A service level approaching 98% (10% increase over the current service level) is obtained for one production line of the GSA company by Builes Henao through a discrete event simulation [36]. Perez Castro [45] argues that thanks to the implementation of DDMRP in a food company, the ROA improves by 2 to 3% and the products remain less time in stock.

Applying the DDMRP methodology to a company selling components for the automotive sector Hincapiè [43] shows that it makes planning management easier since managers can observe the status of stocks on a daily basis, prioritize critical orders and keep the stock in the desired range. According to Kortabarria [48], the implementation of DDMRP entails a fundamental change in the material planning process in each of the three companies (CS1, CS2, and CS3) he analysed after the migration from MRP to DDMRP. In all these companies, day-to-day planning was previously considered not practicable, as the information gathering process was time-consuming and not easy. After the implementation of the DDMRP method, the flow of information along the entire supply chain has significantly improved, and the three companies acquired access to the inventory daily, thus being able to plan in a simple and agile way the needs of materials and to manage the status of supplies, obtaining greater flexibility, saturation and synchronization of the production lines as well as the absence of urgent orders. Both the production and procurement planning should be adapted to the DDMRP implementation: for example, within the ISEO group, the ADU must be evaluated on a weekly instead of monthly basis as carried out previously; the sizing parameters must be continuously updated to dynamically adjust and optimize stock levels; when a new supplier is involved the related information must be promptly updated due to the dynamic nature of the size of the buffers [41].

The alerts dictated by the method, and the priority they represent for the decision-making process, allow to prevent critical situations of shortages or excess stocks, as pointed out by Tavares [44] in her simulation of a toys company; however, the author observes that the DDMRP methodology it is not intuitive to apply and requires preparatory steps as well as many input data. Moreover, the definition of the associated decision-making criteria depends on the practitioners' experience and the level of accuracy in defining the DDMPR parameters can impact the method applicability. So, the implementation of the DDMRP requires staff to be trained on the principles and operation of the method, an aspect that is not trivial since generally in the companies the orders and production planning is entrusted to staff members with extensive professional experience in this regard [40]. Abith Zachariah analyses the costs for the implementation of the new methodology within the ISEO group, quantifying the costs for the theoretical and practical training of the employees, as well as the costs for upgrading the used management software to support the planning and operational execution of the new method, finding that the overall expenditure is around 40-50 k€ [41].

Other criticalities of the DDMRP are emphasized in the reviewed case studies: the efficiency of DDMRP seems to be affected by the presence of long lead times in the intermediate levels of the supply chain [40], so it cannot be applied to all buffers [41]; moreover, the method is sensitive to the value of the decoupled delivery time and the orders' frequency and variability, which should not be empirically assigned [36]. Moreover, it is observed that the DMRP cannot be applied to all products or activities [22]: the purchase of raw materials become critical when delivery times are too long and cannot be related to the effective demand for long time ranges.

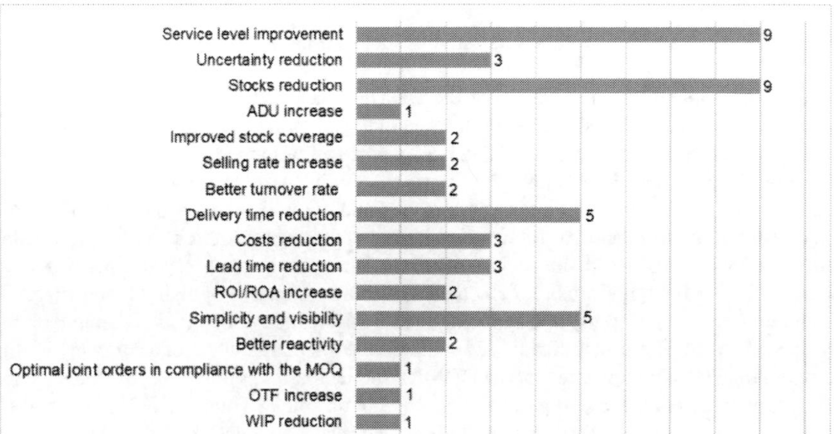

Fig. 3. DDMRP advantages as retrieved from the reviewed case studies.

The DDMRP supply logic does not apply if the quantity and not the price is negotiated, since if the price is low the company will buy more material than the actual requirement because the cost of keeping it in stock is lower than the higher prices the company would pay in the subsequent purchase period [41]. Figures 3 and 4 summarize the advantages and disadvantages of the DDMRP method as retrieved from the reviewed case studies. The bars represent the number of papers citing the considered advantage/disadvantage.

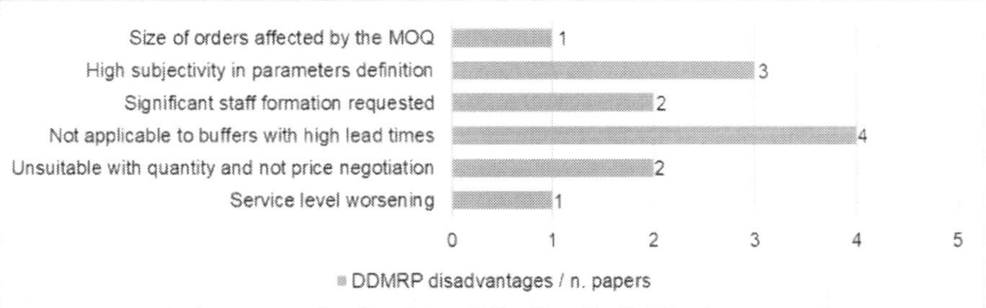

Fig. 4. DDMRP disadvantages as retrieved from the reviewed case studies.

4. Conclusion

The presented review of the available academic studies on the DDMRP methodology highlights the current academic research trends. The main efforts of the scholars are directed to the study of the basic principles of the method, often comparing it to the more established methods. Simulations are widely used to validate the original method and confirm some of the claimed advantages of the DDMRP, while since most of the organizations present complex scenarios the greatest challenge of the simulations is the setting of the parameters for the implementation of the DDMRP, still empirically based and poorly documented. Among these, the enormous subjectivity that dominates the positioning of the buffers is identified as particularly critical and this is still one of the gaps affecting the research papers on the application of this methodology and the implementation in the organizations as well, where a qualified and properly trained specialist should be dedicated to the activities of regularly monitor the demand and consequently adapt the production buffers.

A growing interest in the implementation of the DDMRP in the real industrial environment is proved by the number of case studies, mainly presented in Master or PhD theses after the candidates' internship in local companies; however, most of the retrieved case studies are simulations based on real data and only a few empirical ones. The analyses show that the method allows improving the organisations' production control system overcoming some of the identified weak points, while the costs of its implementation should be taken into account. However, the application of the method to the industrial context is limited to the analysis of the problems of a single company, often considering the possible transition from an MRP system to DDMRP. Moreover, the exiguity of studies reporting the real implementation of the method in the industrial context limits the possibility of compiling a catalogue of best practices.

From this study emerged that two main future research lines can improve the knowledge on the DDMRP applicability and effectiveness into the industrial environment: the development of tools and objective methodologies for the parameters set and the positioning of the buffers and the systematic analysis of the applicability of the method in relation to the sector and the complexity of the organization and its supply chain.

5. References

[1] Bagni, G.; Godinho Filho, M.; Thürer, M. & Stevenson, M. (2021). Systematic review and discussion of production control systems that emerged between 1999 and 2018. Prod. Plan. Control, Vol. 32, pp 511–525, ISSN: 0953-7287

[2] Katuu, S. (2020). Enterprise Resource Planning: Past, Present, and Future. New Rev. Inf. Netw., Vol. 25(1), pp 37–46, ISSN: 17407869.

[3] Dolgui, A. & Prodhon, C. (2007). Supply planning under uncertainties in MRP environments: A state of the art. Annu. Rev. Control., Vol. 31 (2), pp 269–279, ISSN: 13675788.

[4] Li, Q. & Disney, S.M. (2017). Revisiting rescheduling: MRP nervousness and the bullwhip effect. Int. J. Prod. Res., Vol. 55(7), pp 1992–2012, ISSN: 1366588X.

[5] Wang, X. & Disney, S.M. (2016). The bullwhip effect: Progress, trends and directions. Eur. J. Oper. Res., Vol. 250 (3), pp 691–701. ISSN: 03772217.

[6] Bregni, A.; D'Avino, M. & Schiraldi, M.M. (2011). A new approach to lower MRP nervousness. Ann. DAAAM Proc. Int. DAAAM Symp., 2011, Vienna (Austria), ISSN: 17269679, ISBN: 9783901509834, Katalinic B. (Ed.): pp. 1513–1514, DAAAM International, Vienna.

[7] Rossi, T.; Pozzi, R.; Pero, M. & Cigolini, R. (2017). Improving production planning through finite-capacity MRP. Int. J. Prod. Res., Vol.55 (2), pp 377–391, ISSN: 1366588X.

[8] Thevenin, S.; Adulyasak, Y. & Cordeau, J.F. (2021). Material Requirements Planning Under Demand Uncertainty Using Stochastic Optimization. Prod. Oper. Manag., Vol. 30 (2), pp 475–493. ISSN: 19375956.

[9] Ptak, C. & Smith, C. (2011). Orlicky's Material Requirements Planning (third ed.), McGraw Hill Professional, ISBN: 0071755632, 9780071755634.

[10] Ptak, C. & Smith, C. (2016). Demand Driven Material Requirements Planning (DDMRP). Industrial Press Incorporated, ISBN: 9780831193850, 0831193859.

[11] El Marzougui, M.; Messaoudi, N.; Dachry, W.; Sarir, H. & Bensassi, B. (2020). Demand Driven MRP: Literature review and research issues, Proceedings of the 13ème Conf. Int. Model. Optim. Simul., (MOSIM2020), 2020, AGADIR (virtual), Morocco. hal-03193163.

[12] Miclo, R.; Lauras, M.; Fontanili, F.; Lamothe, J. & Melnyk, S.A. (2019). Demand Driven MRP: assessment of a new approach to materials management. Int. J. Prod. Res., Vol. 57 (1), pp 166–181. ISSN: 1366588X.

[13] Orue, A.; Lizarralde, A. & Kortabarria, A. (2020). Demand driven MRP – the need to standardise an implementation process. Int. J. Prod. Manag. Eng., Vol. 8 (2), pp 65–73. ISSN: 23404876.

[14] Azzamouri, A.; Baptiste, P.; Dessevre, G. & Pellerin, R. (2021). Demand Driven Material Requirements Planning (DDMRP): A systematic review and classification. J. Ind. Eng. Manag., Vol 14 (3), pp 439-456. ISSN: 2013-0953.

[15] Miclo, R.; Fontanili, F.; Lauras, M.; Lamothe, J. & Milian, B. (2016). An empirical study of Demand-Driven MRP, Proceedings of the 6th International Conference on Information Systems Logistics and Supply Chain - ILS 2016, June 1-4, 2016, Bordeaux (France), pp 1–7.

[16] Cuadra, R.R. (2017) Estudio Del Ddmrp (Demand Driven Materials Requirement Planning), Master Thesis, Escuela de Ingenieria Industriales, University of Valladolid, Valladolid, Spain.

[17] Kortabarria, A. & Elizburu, A. (2018). Implementing Management Systems and Demand Driven MRP concepts: A Project Based Learning experience in Industrial Organization Engineering, Proceedings of the 4th International Conference on Higher Education Advances (HEAd'18), 2018, Universitat Politecnica de Valencia, Valencia (Spain), pp 543–551. doi:10.4995/head18.2018.8033.

[18] Marin, A. (2018). DDMRP: un approccio ibrido tra metodi push (MRP) e metodi pull (Lean), Mater Thesis, Corso di Laurea in Economia e Gestione delle Aziende, University Ca' Foscari, Venezia, Italy.

[19] Lee, C.J. & Rim, S.C. (2019) A Mathematical Safety Stock Model for DDMRP Inventory Replenishment, Math. Probl. Eng., Vol. 2019, Article ID 6496309. ISSN: 15635147.

[20] Bahu, B., Bironneau, L. & Hovelaque, V. (2019). Compréhension du DDMRP et de son adoption: premiers éléments empiriques. Logistique Manag., Vol. 27 (1), pp 20–32. ISSN: 1250-7970.

[21] Meinzel L. (2019). DDMRP: presentation of a new solution of stock management and master production scheduling, Master Thesis, Escola Tècnica Superior d'Enginyeria Industrial de Barcelona, Universitat Politecnica de Catalunya, Barcelona, Spain.

[22] Pekarcíková, M.; Trebuna, P.; Kliment, M. & Trojan, J. (2019). Demand driven material requirements planning. some methodical and practical comments. Manag. Prod. Eng. Rev., Vol. 10 (2), pp 50–59. ISSN: 20821344.

[23] Perälä, J. (2020). Demand Driven Material Requirement Planning Suitable components and data considerations, Master Thesis, School of Technology, University of Applied Sciences, Vaasa, Finland.

[24] Alves da Silva Santos, L. (2020). Demand-Driven Material Requirements Planning (DDMRP) Implementation Impacts on Supply Chain Management Strategic Processes: an AHP Approach, Master Thesis, Sao Carlos School of Engineering, University of Sao Paulo, Brazil.

[25] Gavaz De Biasi, F. (2020). Pianificazione della Produzione Secondo l'Approccio DDMRP. Un'Analisi Teorica. Master Thesis, Dipartimento di Tecnica e Gestione dei Sistemi Industriali, University of Padova, Padova, Italy.

[26] Achergui, A.; Allaoui, H. & Hsu, T. (2020). Strategic DDMRP's Buffer Positioning for hybrid MTO/MTS manufacturing. Proceedings of the 2020 IEEE Int. Conf. Technol. Manag. Oper. Decis. (ICTMOD), IEEE, pp. 1–8. doi:10.1109/ICTMOD49425.2020.9380588.

[27] Velasco Acosta, A.P.; Mascle, C. & Baptiste, P. (2020). Applicability of Demand-Driven MRP in a complex manufacturing environment. Intern. Journal of Prod. Research, Vol. 58 (14), pp 4233-4245. DOI: 10.1080/00207543.2019.1650978.

[28] Tounsi, W. (2018). Comparaison des approches DDMRP et EOQ: modélisation et simulation d'un cas d'étude, Master Thesis, Ecole Polytechnique de Montreal, Université de Montreal, Montreal, Canada.

[29] Shofa, M.J.; Moeis, A.O. & Restiana, N. (2018). Effective production planning for purchased part under long lead time and uncertain demand: MRP Vs demand-driven MRP. IOP Conf. Ser. Mater. Sci. Eng., Vol. 337, 012055. ISSN: 1757899X.

[30] Shofa, M.J. & Widyarto, W.O. (2017). Effective production control in an automotive industry: MRP vs. demand-driven MRP. AIP Conf. Proc. 1855, 020004. ISSN: 15517616.

[31] Miclo, R. (2016) Challenging the" Demand Driven MRP" Promises: a Discrete Event Simulation Approach, PhD Dissertation, Centre Génie Industriel, Ecole des Mines d'Albi-Carmaux, Albi, France.

[32] Miclo, R.; Fontanili, F.; Lauras, M.; Lamothe, J. & Milian, B. (2016). MRP vs. Demand-driven MRP: Towards an objective comparison. Proceedings of the 2015 Int. Conf. Ind. Eng. Syst. Manag. IEEE IESM 2015, pp: 1072–1080. doi:10.1109/IESM.2015.7380288.

[33] Miclo, R.; Fontanili, F.; Lauras, M.; Lamothe, J. & Milian, B. (2016). An empirical comparison of MRPII and Demand-Driven MRP, IFAC-PapersOnLine, Vol. 49 (12), pp 1725–1730. ISSN: 24058963.

[34] Thürer, M.; Fernandes, N.O. & Stevenson, M. (2020). Production planning and control in multi-stage assembly systems: an assessment of Kanban, MRP, OPT (DBR) and DDMRP by simulation, International Journal of Production Research, 1-15. DOI: 10.1080/00207543.2020.1849847.
[35] Castro, J.E.N. (2019). Diseño De Un Modelo De Gestión Demand Driven – Mrp, Para El Aprovisionamiento De Productos Aceites Y Lubricantes Aéreos Para Una Flota De Aviones Supersónicos, Master Thesis, Facultad De Ciencias Naturales Y Matemáticas Departamento De Matemáticas, Escuela Superior Politécnica del Litoral, Guayaquil, Ecuador.
[36] Builes Henao, J.L. (2019). Marco De Referencia Para La Implementacion De Ddmrp En La Planta De Non Stick Gsa, Master Thesis, Escuela de Ingenieria, Universidad EAFIT, Medellin, Colombia.
[37] Ducrot, L. & Ahmed, E. (2019). Investigation of potential added value of DDMRP in planning under uncertainty at finite capacity, Thesis, Master of Applied Science in Supply Chain Management, Massachusetts Institute of Technology, Cambridge (MA), USA.
[38] Salazar, M. & Arenas J. (2019). Disminución de niveles de inventario de repuestos de productos de línea blanca, basados en la metodología DDMRP, Thesis, Faculdad de Ingenieria, Universidad de Antioquia, Medellin, Colombia.
[39] Linares Leguizamo, M.D. & Totaitive M. (2017). Diseño del modelo DDMRP para la planeación de la Producción en la empresa Wa-Armonia, Thesis, Faculdad de Ingenieria, Pontificia Universidad Javeriana, Bogotà, Colombia.
[40] Provenzano, F. (2020). Modelli innovativi di gestione delle scorte nella filiera logistica: applicazione del metodo DDMRP nel settore automotive, Master Thesis, Collegio di Ingegneria Gestionale, Politecnico di Torino, Italy.
[41] Abith Zachariah, G. (2018). Demand Driven Materials Requirement Planning (DDMRP) A new method for production and planning management, Master Thesis, Management, Economics and Industrial Engineering, Politecnico di Milano, Milano, Italy.
[42] Brahimi, B. & Saadi, M.A. (2019). Amélioration de la planification de la production de la chaine logistique par l'application de la méthode DDMRP, Master Thesis, Faculté de TECHNOLOGIE, Université Aboubakr Belkaïd, Tlemcen, Algeria.
[43] Hincapiè, V.V. (2018). Application of Demand Driven Material Requirement Planning (DDMRP) methodology for components inventory management in a company of automotive industry, Master Thesis, Engineering and Management, Politecnico di Torino, Torino, Italy.
[44] Tavares, A.C.M. (2019). Demand Driven Material Requirements Planning (DDMRP): Estudo Exploratório, Master Thesis, Faculdade de Ciencias e Tecnologia, Universidade Nova de Lisboa, Lisboa, Portugal.
[45] Perez Castro, M.G. (2018). Diseño De Un Modelo De Gestión Demand Driven – MRP, Para el Proceso de Compras de Materias Primas de una Empresa de Insumos Alimenticios, Master Thesis, Facultad De Ciencias Naturales Y Matemáticas Departamento De Matemáticas, Escuela Superior Politécnica del Litoral, Guayaquil, Ecuador.
[46] Dimas Mukhlis, H.F.; Indraefrialdi, J. & Rimawan, E. (2019). Inventory Management using Demand Driven Material Requirement Planning for Analysis Food Industry. Int. J. Innov. Sci. Res. Technol., Vol. 4 (7), pp 495-499. ISSN:-2456-2165.
[47] Johansson, R. & Sodermark, G. (2019). Understanding the Requirements of Forecast in Demand Driven Material Requirements Planning, Master Thesis, Department of Technology Management and Economics Division of Supply and Operations Management, Chalmers University of Technology, Gothenburg, Sweden.
[48] Kortabarria Igartua, A. (2018). Analysis of the performance of Demand Driven Material Requirements Planning on supply chain management: A multiple case study, PhD Dissertation, Mechanical and Manufacturing Department, Mondragon Unibertsitatea, Bilbao, Spain.
[49] Kortabarria, A.; Apaolaza, U.; Lizarralde, A. & Amorrortu, I. (2018). Material management without forecasting: From MRP to demand driven MRP, J. Ind. Eng. Manag., Vol. 11 (4), pp 632–650. ISSN: 20130953.
[50] Kortabarria, A.; Apaolaza, U. & Lizarralde, A. (2019). Demand Driven MRP - Nuevo método para la gestión de la Cadena de Suministro: Un estudio de caso, Dir. y Organ., Vol. 67, pp 22–29. ISSN: 21716323.

DOI: 10.2507/32nd.daaam.proceedings.068

RESEARCH OF THE EFFECT OF MICROWAVE RADIATION ON THE SPEED OF THE WATER-ICE PHASE TRANSITION

Andrey Novikov, Nikolay Grebenshchikov, Denis Shilin, Dmitry Shestov, Pavel Ganin & Valery Moskvin

This Publication has to be referred as: Novikov, A[ndrey]; Grebenshchikov, N[ikolay]; Shilin, D[enis]; Shestov, D[mitry]; Ganin, P[avel] & Moskvin, V[alery] (2021). Research of the Effect of Microwave Radiation on the Speed of the Water-Ice Phase Transition, Proceedings of the 32nd DAAAM International Symposium, pp.0472-0476, B. Katalinic (Ed.), Published by DAAAM International, ISBN 978-3-902734-33-4, ISSN 1726-9679, Vienna, Austria
DOI: 10.2507/32nd.daaam.proceedings.068

Abstract

This article presents a study of the effect of microwave radiation on the speed of the water-ice phase transition using a laboratory heat pump installation. An experimental stand for research has been developed. Experimental techniques are described. The effect of microwave radiation on the speed of the water-ice phase transition is experimentally analyzed to increase the efficiency of the heat pump by increasing the rate of water crystallization. As a result of the experiments, it can be concluded that water exposed to uniform microwave radiation cools down to 0 °C 23 minutes earlier than water that has passed only filtration, and 74 minutes faster phase transitions to a solid-state. An experiment was carried out to obtain a potential difference to determine the phase state of an aqueous solution. There are 3 stages of signal change: cooling, phase transition, and overcooling.

Keywords: Water-Ice Phase Transition; Heat Pump; Experimental Stand; Thermal Energy.

1. Introduction

The use of heat pumps (HP) is becoming more and more relevant every year. In the global energy sector, there is a constant rise in energy prices and climate change caused by this greenhouse effect, countries such as Sweden, Germany, the United States, and Japan are switching to the use of heat pumps to reduce greenhouse gas emissions and save renewable energy resources [1], [2].

Heat pump installations can be widely used in agriculture both for heating the barn and for cooling milk [3]. Heat pumps are essential in a brewery for brewing beer is a complex and time-consuming process involving heating and cooling the product. In the country house, it is necessary to heat the room in winter and cooling in summer. The range of HP applications is very wide, every year there are many new technological processes in many areas of science and life, where thermal energy is needed [4], [5]. One of the problems of existing HP is the long crystallization time of aqueous solutions; in this article, we will consider the possibility of accelerating the freezing speed of water using microwave radiation. Controlling the speed of the phase transition of the coolant will increase the efficiency of the entire heat carrier installation, as well as reduce energy consumption by reducing the freezing and thawing time of the liquid.

The research purpose is to develop an experimental stand, to study the effect of microwave radiation on the crystallization rate of water, and also to conduct many experiments to obtain a potential difference across a pair of electrodes, depending on the temperature of the liquid, to determine the phase state of an aqueous solution.

One of the promising substances as a heat accumulator is water. It has a high specific heat capacity of 4200 kJ/(kg * K), which significantly exceeds the value of the specific heat of melting-solidification of most metals and organic substances [6]. This property of water makes it possible to widely use it as a cold and heat accumulator for subsequent practical use. There are many options for increasing the efficiency of the HP: regulation of the pump and compressor power, changing the parameters of the heat exchanger, affecting the coolant [7], [8], [9]. The article presents the results of experiments based on placing a container with bottled water and water subjected to microwave radiation in a refrigeration unit to determine the freezing rate of a liquid to obtain thermal energy, as well as experiments to obtain a potential difference on a pair of electrodes, depending on the temperature of the liquid, to determine the phase state of an aqueous solution.

2. Development of an experimental stand

The main criterion for conducting the experiment was the speed of the water-ice phase transition of bottled water and water passed through microwave radiation. For carrying out, we took a 90-liter freezer, including a Danfoss TLES4F compressor with a cooling capacity of 91 W, a programmable Arduino controller with four connected sealed DS18B20 temperature sensors that measure temperatures in the range from -55 °C to +125 °C [10], [11]. The liquid container is made of plastic. The outdoor air temperature was in the range from +18 ° C to +21 ° C. The scheme of the experimental stand is shown in Figure 1.

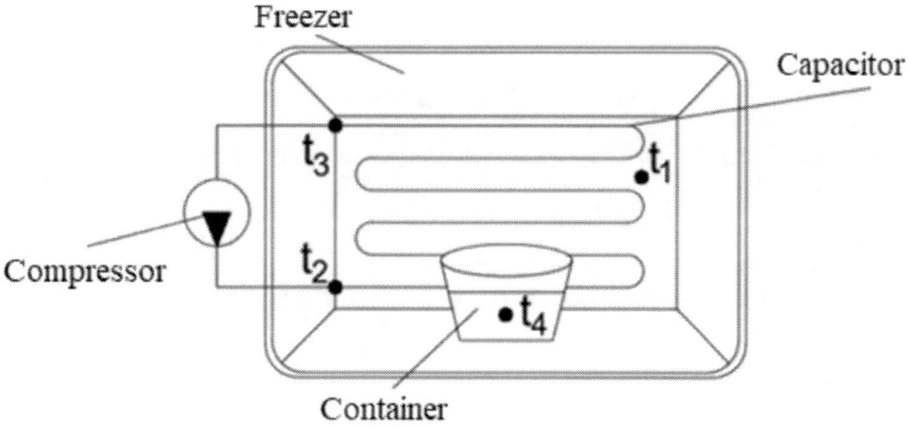

Fig. 1. The scheme of the experimental stand

The temperature sensor t1 is located in the freezer to determine the temperature inside the refrigeration unit. The temperature sensor t2 determines the temperature at the input to the capacitor grid, and the sensor t3 at the output of the capacitor grid. The t4 sensor is installed in the center of a container filled with water [12].

3. Carrying out experiments

To ensure the constancy of the physicochemical properties of process water, several stages of organic and chemical purification were carried out for each experiment. In each experiment, the water froze for the first time, since the properties of the water change with repeated freezing, and its reuse is impossible. Water after microwave radiation is identical in physical and chemical properties to water frozen without disturbing influences. For uniform treatment, each liter of water was subjected to microwave radiation sequentially, the thickness of the water layer was 4-5 mm, the processing time was 12 s, a 750 W magnetron was used [13]. The water treated with microwave radiation was placed in a container inside the freezer, the compressor was turned on only when the container was filled. The water temperature after treatment with microwave radiation was 27 ° C.

The experiments were carried out indoors at a room temperature of +23 ° C. The temperature of the refrigerating chamber before the start of the experiment is equal to the ambient temperature. The temperature limit for switching off the compressor was set in the range of –24 ° C ... –25 ° C.

Figure 2 shows a graph of the temperature of the sensor installed in the center of the tank. A thick line indicates the temperature of the water, 10 experiments were conducted with bottled water and with water after filtration and exposure to microwave radiation. The water used in each experiment has not previously been subjected to a freeze-thaw cycle. That has not been subjected to microwave radiation, a thin dotted line indicates water treated with microwave radiation.

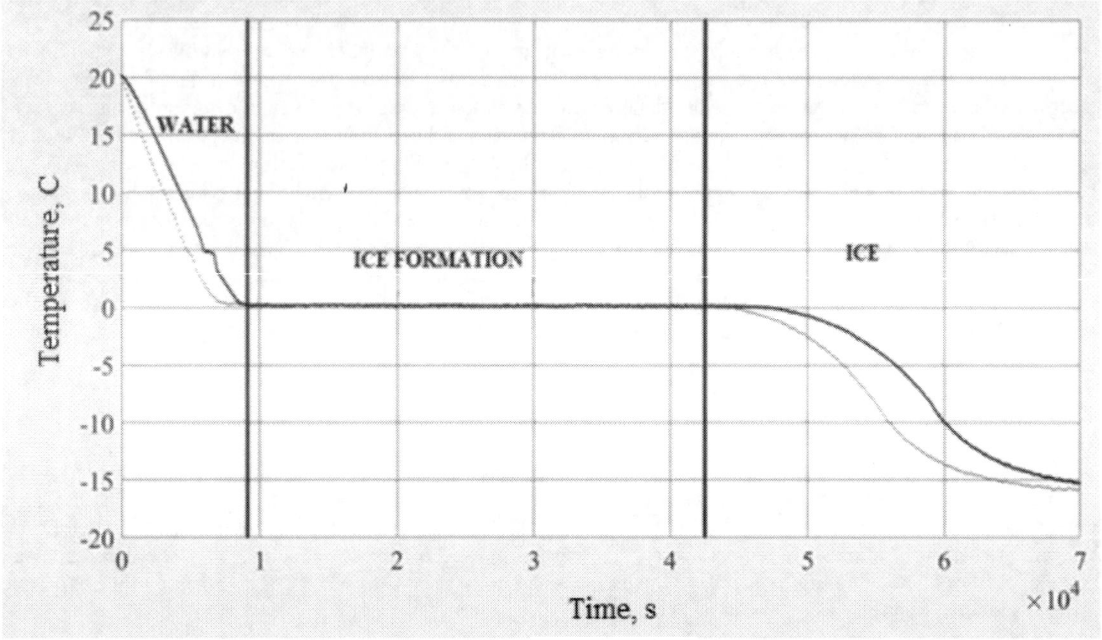

Fig. 2. Graph of the sensor temperature inside the water tank

After treatment with microwave radiation, the water was heated to 27 °C. It is known that the higher the initial temperature of the water, the faster it will freeze, so before the experiment, the water was brought to room temperature, identical to the temperature of the water that was not subjected to microwave radiation [14], [15]. As can be seen from the graph, the water exposed to microwave radiation cooled down to 0 °C 23 minutes faster. The crystallization process was faster than that of water that was not subjected to microwave radiation treatment. The water exposed to microwave radiation completely froze 74 minutes earlier than the usual filtered water. Due to the observed effect, it becomes possible to obtain the thermal energy of the water-ice phase transition faster. The temperature in the solid-state changed faster in the water exposed to microwave radiation than in ordinary filtered water.

The experiment involves obtaining the dependence of the change in the potential difference on a pair of electrodes on the temperature of the liquid in which the electrodes are placed. The location of the electrodes is shown in Figure 3. Several pairs of electrodes are placed in the container at an equal distance from each other, and the container is filled with a liquid medium.

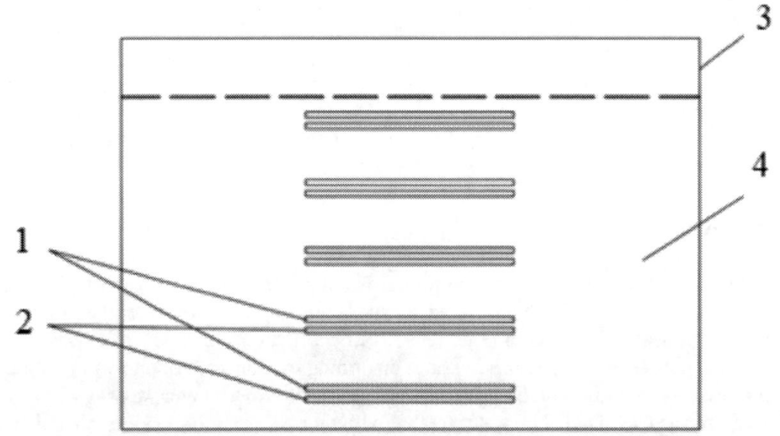

Fig. 3. The layout of the electrodes in the container. 1,2 – pairs of electrodes, 3-capacity, 4-liquid medium

A potential difference occurs on the electrodes placed in the liquid. The potential difference varies depending on the temperature of the liquid and its phase state. The study of this dependence will allow us to control the temperature and phase state of the liquid by the electric generating properties of the liquid. The experimental stand can be supplemented with radio components such as a capacitor, a resistance, or a high-frequency rectifier. This component base is designed to obtain a clearer characteristic of the change in the potential difference.

It also implies conducting experiments to determine the change in the potential difference depending on the distance between the electrodes, their shape, and material, as well as additional effects, such as microwave radiation or a magnetic field. An experiment was conducted to obtain a potential difference under the following conditions. The initial temperature of the liquid is +20 ° C, the final temperature of the liquid is -20 °C, the distance between the electrodes is 5 mm, the electrodes are made in the form of plates made of aluminum and copper. The obtained characteristic is shown in Figure 4.

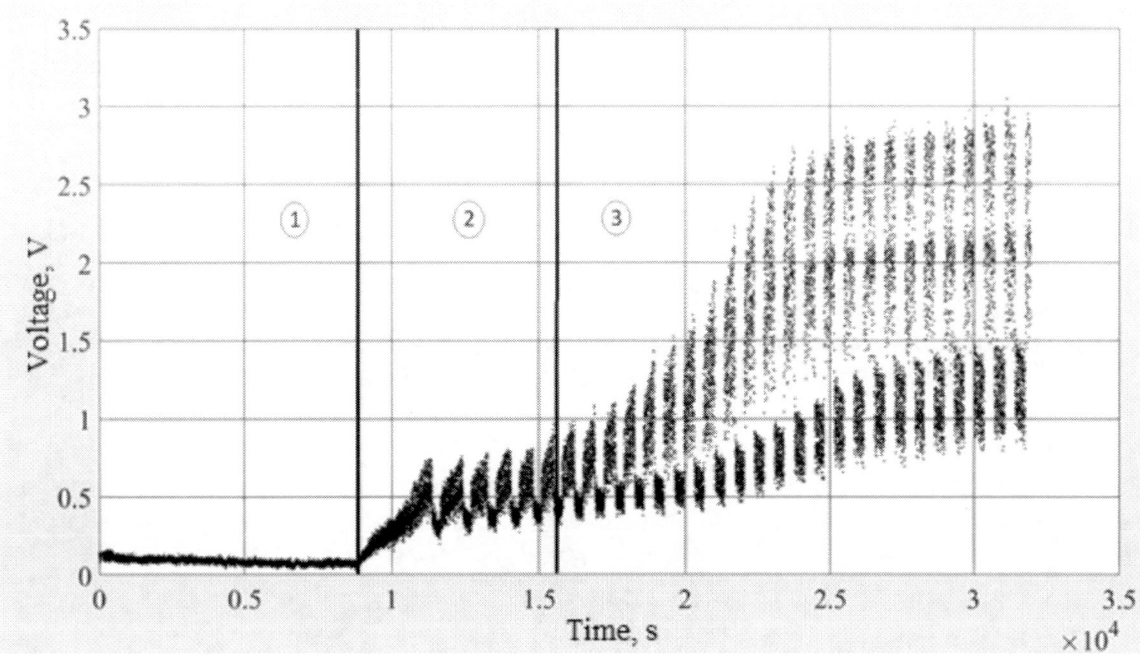

Fig. 4. Potential difference characteristics

There are 3 main stages of signal change. Stage 1-cooling, the average voltage value in this zone is 0.13 V, the oscillation of the character does not exceed 0.1 V, the medium in this area is in a liquid state. Stage 2 is a phase transition, the voltage at the electrode gradually increases to 1 V, the fluctuations of the characteristic increase and reach 0.6 V, the medium in this fouling passes from a liquid state to a solid-state. Stage 3-supercooling, as a result of icing of one of the plates, the movement of electrons stops, since the ice is a dielectric and zero or an error generated by the magnetic field is recorded on the measuring channel.

4. Conclusion

As a result of the experiments, it can be concluded that water exposed to uniform microwave radiation cools down to 0 °C 23 minutes earlier than water that has passed only filtration, and 74 minutes faster phase transitions to a solid-state. Microwave radiation can be used to increase the efficiency of a heat pump by using the energy of the water-ice phase transition, by accelerating the production of thermal energy from the heat carrier in the heat supply system. In the future, it is planned to conduct experiments with other disturbing influences to determine the parameters to obtain the thermal energy of a heat pump installation. As a result of the experiments, it was possible to obtain a potential difference and distinguish 3 stages of signal change: cooling, phase transition, and overcooling. In the future, it is planned to conduct experiments with other electrophysical effects.

5. References

[1] Volkov V. N., Kozina L. N., Dzyuban A.M. (2015) et al. Actual application of heat pumps for heating buildings and structures / / Vestnik NGIEI. N6(49). pp. 39-43.

[2] Luneva S. K. (2014) Application of heat pumps in energy-efficient heat supply systems. Technical and technological problems of the service. N3(29). pp. 74-81.

[3] Gornov V. F., Efremov M. N., Leskov V. A. (2019) et al. Heat pump systems for cottages // Energy saving. N1. pp. 30-37.

[4] Novikov A. A. (2020) Analysis of heat supply systems with a heat pump using water as a heat accumulator / / Innovations in agriculture. N2(35). pp. 28-43.

[5] Sahic, E[smina]; Husika, A[zrudin]; Zecevic, N[urin] & Dzaferovic, E[jub] (2020). Techno-Economic Analysis Solar Energy Use for Sanitary Water Heating in Institute ""Drin Fojnica"", Proceedings of the 31st DAAAM International Symposium, pp.0542-0549, B. Katalinic (Ed.), Published by DAAAM International, ISBN 978-3-902734-29-7, ISSN 1726-9679, Vienna, Austria DOI: 10.2507/31st.daaam.proceedings.075

[6] Golovin Yu. I. (2000) Water and ice – do we know enough about them / / Soros Educational Journal. Vol. 6. N9. pp. 66-72.

[7] Shitov V. V., Chernopyatova S. A., Malakhov A. N. (2012) Improving the efficiency of the heat pump // Bulletin of the Voronezh State University of Engineering Technologies. N2. pp. 37-39.

[8] Saken K. Sh., Ptashkina-Girina O. S., Nizamutdinova N. S. (2019) Economic indicators of renewable energy // Bulletin of the NGIEI. N2(93). pp. 48-58.

[9] Gasho E. G., Kozlov S. A., Puzakov V. S. (2017) et al. Heat pumps in modern industry and municipal infrastructure. M.: Publishing House "Pero". p. 204.

[10] Kazantsev S. A. (2015) Improving the accuracy of digital temperature sensors DS18B20 // Interexpo Geo-Siberia. Vol. 2. N2. pp. 79-81.

[11] Petin V. A. (2014) Projects using the Arduino controller. St. Petersburg: BHV-Petersburg. p. 400.

[12] Novikov A. A., Ershov I. G. (2020) Design of the experimental stand for the phase transition water-ice // Electrotechnology and electrical equipment in agriculture. T. 67. N2(39). S. 21-26. DOI 10.22314/2658-4859-2020-67-2-21-26

[13] Molodtsov, M. A., Sevastyanov V. (2017) the Possibilities and prospects of using microwave radiation in the industry (review) // the Forest magazine. N2(356). pp. 173-187.

[14] Mpemba E.B., Osborne D.G. (1969) Cool? Physics Education. Vol. 4. N3. pp. 172-175.

[15] Kell G.S. (1969) The Freezing of Hot and Cold Water. American Journal of Physics. Vol. 37. Iss. 5. pp. 564-565.

DOI: 10.2507/32nd.daaam.proceedings.069

EVALUATION OF STRATEGIES OVER STATIC CODE ANALYSIS TOOLS

Danilo Nikolic, Sara Havzi, Dusanka Dakic, Teodora Lolic & Darko Stefanovic

This Publication has to be referred as: Nikolic, D[anilo]; Havzi, S[ara]; Dakic, D[usanka]; Lolic, T[eodora] & Stefanovic, D[arko] (2021). Evaluation of Strategies over Static Code Analysis Tools, Proceedings of the 32nd DAAAM International Symposium, pp.0477-0485, B. Katalinic (Ed.), Published by DAAAM International, ISBN 978-3-902734-33-4, ISSN 1726-9679, Vienna, Austria
DOI: 10.2507/32nd.daaam.proceedings.069

Abstract

Source code quality is a key factor in any software product and requires constant inspection and supervision. Static code analysis tools are being increasingly used to improve code quality. Static code analysis is a valid way to infer the behavior of a program without executing it. Many tools allow static analysis in different programming languages, different frameworks, and detecting different defects in the source code. In this paper, based on the identified strategies over static code analysis tools, an experiment was conducted to evaluate the quality of the identified strategies based on established metrics. The aim of this paper is to present, analyze and evaluate the application of different strategies over different tools for static code analysis.

Keywords: static code analysis; tools; strategies; evaluation;

1. Introduction

Static analysis is used to maintain and improve the quality of the source code. Static analysis of program code has been used since the early 1960s to optimize the operation of compilers [1]. Later, it proved useful for debugging tools, as well as for software development frameworks. Various forms of application of information technologies are studied at the Faculty of Technical Sciences, such as [2], [3], [4], [5], [6], information technologies find application in areas such as [7], [8] too, in this paper, the application of information technologies and tools to static code analysis will be described.

Tools for static analysis have rapidly matured in the last decade, they have evolved from simple lexical analysis to employ much more complex techniques. Static analysis tools are used to produce reports and highlight specific deviations from the code quality requirements. These tools, on the other hand, do not allow for automated source code modifications. It is up to the software developers to decide whether or not to alter the composition of previously written code. A static code review tool assists software developers by creating a report that details the cause of a specific flaw as well as how to fix it. A systematic literature review in the field of static code analysis was performed as the first step in this research [9], and various static code analysis tools, programming languages for which they have support, and the types of defects they identify were presented and analyzed. The second step in the research was to identify different strategies for using static code analysis tools. Strategies are identified and described in [10].

The aim of this paper is to conduct an experiment over the identified strategies from [10] to analyze the use of different strategies over different tools for static code analysis and to evaluate strategies based on accuracy metrics and time required for implementation. The rest of the paper is organized as follows. Section 2 introduces fundamental principles of static code analysis tools; Section 3 describes related work based on a conducted systematic literature review; The methodology used to conduct the research is described in Section 4; Section 5 presents an experiment of the use of different strategies over static code analysis tools and Section 6 presents results and discussion of the conducted experiment and section 7 concludes the paper and suggests future research.

2. Static Code Analysis

This section will present how static code analysis is carried out, which tools are useful in the process, as well as what are the false positive and false negative results

2.1. Review of the source code

Code review is used to improve the software product maintenance process, increasing its reliability and security. Code review activity can reduce the number of required updates in source code and can be easily integrated into the software development lifecycle, so more experienced software developers often grow this activity in their work. However, manual review of source code takes a very long time [11]. In order to perform an effective manual code review, reviewers must know all possible flaws and inaccuracies before carefully checking all source code. As a substitute for manual source code review, the use of a static code analysis tool is recommended.

2.2. Static code analysis tools

Static code analysis is useful not just for optimizing the compiler's operation (which is what it was designed for), but also for detecting defects and potential flaws. It is possible to build software that will assist developers in understanding the behavior of a program and identifying numerous flaws in the program without having to run it [1]. Static analysis tools can often point to the root cause of a security problem, not just one of its symptoms, by examining the code itself. This is especially important if you want to make sure that any defects are fixed.

Static code analysis tools scan source code for a specific set of patterns or rules, similar to how antivirus software detects viruses. Practitioners can use more advanced tools to add new rules to a set of predefined ones. If patterns or rules do not specify such behavior, these tools will be unable to detect an error. The ability to identify common programming defects is the most valuable feature of a static code analysis tool.

Static analysis is suitable for identifying problems for several reasons [11]:

- Static analysis tools apply checks thoroughly and consistently, without any prejudice that the programmer may have about which parts of the code are "interesting" from a security perspective or which parts of the code can be easily performed by dynamic testing. It is very important that the code review is done without the subjective feeling of the programmer. Therefore, unbiased analysis can be valuable, which static code analysis tools provide.
- By examining the code itself, static analysis tools can often point to the root cause of a security problem, not just one of its symptoms. This is especially important to ensure that defects are corrected.
- Static analysis can find errors in early development, even before the program is run for the first time. Finding an error early not only reduces the cost of fixing the error, but a cycle of quick feedback can help guide the work of the programmer: The programmer has the opportunity to correct errors that he was not previously aware could even happen. Attack scenarios and code construct information used by the static analysis tool act as the means of transmitting knowledge.
- When analysis detects a new type of attack, static analysis tools make it easy to recheck a large portion of the code to detect where a new attack might occur. Some security bugs have been in the software for years before they were discovered, assessing outdated code of newly discovered types of defects is invaluable.

2.3. False positives and False negatives

Software developers point out that these tools produce too many bad results, these results fall into two categories [1]:

1. False Positives.
2. False Negatives.

False positive results occur when static code analysis tools report a defect in a program's source code that does not actually exist. A large number of false positive results can cause difficulties in operation. Not only is searching for a list of false positives tedious, but a software developer who needs to review a long list of false positives can easily overlook the important positive results found among them [12]. False positive results are certainly undesirable, but from a security perspective, false negative results are a much bigger problem. A false negative result implies that there is a defect in the source code of the program that the tool does not report.

When it comes to false positive results, they only expect software developers to review the list of detected defects for a long time. Whereas, when it comes to false-positive results, they imply that the source code still contains defects, even after the static analysis procedure has been completed. This leads to a false sense of security after generating a report that there are no defects in the source code and that the quality of the code is good, even though there are false negative results [12].

Paper [13] investigated the impact of false positive alarms generated by a static code analysis tool. The researchers discovered that, while the developers were effective at recognizing true positives, their false positive detection rates were no better than chance, and their experience had no bearing on false positive classification accuracy. The same research team utilized a single static code analysis method to assess four commercial Ericsson products in a more recent study [14]. The findings revealed that the majority of the tool's output was unrelated to security and that false positives were a problem. Surprisingly, just 37.5 percent of false positives were correctly classified as such by the developers, according to the study. Furthermore, it showed that a high number of false positives harmed the development process not only because of the time spent classifying them but also because of the risk of "correcting" incorrectly classified false positives [14]. Most static code analysis tools display both false positives and false negatives. However, during the period of development of these tools, the number of errors was getting significantly smaller and these tools were able to generate more reliable results.

3. Related Work

As the first step in the research, a systematic literature review in the field of static analysis of code and its tools was conducted. A paper describing the process of the systematic literature review was published in [9]. The paper review was conducted using Barbara Kitchenham's guidelines for a systematic review of software engineering literature [15], with the aim of presenting the most widely used resources, through programming languages for which they have support, and defects they detect.

Based on the set inclusion criteria, by reviewing abstracts and titles, 22 primary studies were included in the literature review. Some of the representative examples of the included studies are [12], [16], [17], [18], [19]. The tools for static code analysis that are most commonly used in research on different programming languages and detecting various defects were also examined as part of this study. The majority of static code analysis tools support the most widely used programming languages in the industry. This systematic review examines the most widely used static code analysis tools in science, focusing on the types of defects they detect and the programming languages they support. Six tools most commonly used in research by the review are presented in Fig. 1.

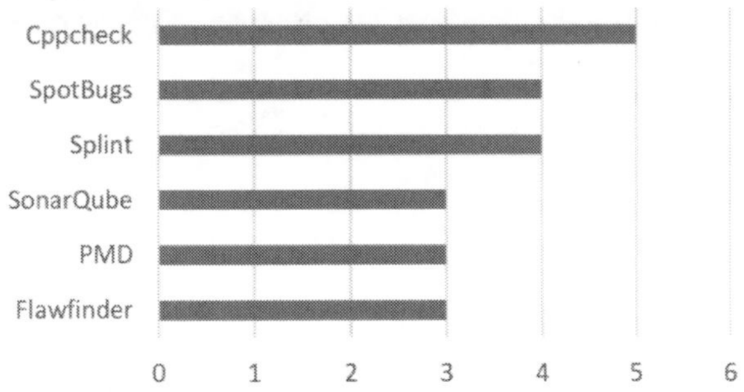

Fig. 1. The most common static code analysis tools presented in primary studies [9]

Based on the conducted literature review, the tools that will be used to conduct an experiment in this research have been selected. In addition to the literature review, support for the Java programming language and the possibility of including tools in the framework are provided as a criterion for selecting tools. Also, another research was conducted on the selected tools, in which these tools were evaluated by the DESMET method for evaluation of software tools [20]. The process of evaluation by the DESMET method and feature analysis of selected tools are described in [21].

4. Methodology

Based on the systematic review of the literature described in section 3 of this paper, four static code analysis tools were selected and used in the experiment.

The static code analysis tools used in the experiment are:

1. SonarQube,
2. SpotBugs,
3. PMD,
4. Checkstyle.

Two strategies and a total of four sub-strategies for the application of static analysis tools over program source code have been identified. Those are [10]:

1. Running a static analysis tool over a completed software project – **S1**.
   1.1. Starting the static analysis tool after correcting all defects – **S1.1.**
   1.2. Running the static analysis tool after correcting any defect – **S1.2.**
2. Running a static analysis tool during software project development – **S2**.
   2.1. Starting static analysis tools by external tool calling during software project development – **S2.1.**
   2.2. Running static analysis tools by including tools in the development environment in which the software project is developed – **S2.2.**

The research [10] described in detail how the proposed strategies were identified and elaborated on the defined steps for the implementation of these strategies in the static code analysis review cycle.

The research questions, based on which strategies will be analyzed, are:
RQ1: Is there a strategy for implementing the static code analysis process that gives the best results?
RQ2: Are the identified strategies applicable to all tools used in the experiment?
RQ3: Are some strategies faster to implement than others?
RQ4: Do some strategies give more accurate results than others?

Based on the research questions, an experiment of the application of all strategies over the selected tools for static code analysis was conducted. The metrics observed during the experiment were set based on defined research questions so that the analysis of the obtained results within the metrics could lead to conclusions and answers to the research questions. The source code of the project [22] written in the Java programming language (version 1.8), within the development environment Eclipse, was used to conduct the experiment, which is a condition for the selection of tools within this experiment.

5. Experiment

The experiment, which was performed on selected static code analysis tools, aims to compare the results of different strategies when applied to different static code analysis tools.

To compare the application of strategies over these tools, the following metrics were used:

- Time required to eliminate all defects detected by the static analysis tool - metrics measured in minutes required to eliminate all defects in the project. This metric was set up with the aim of answering research questions RQ3 and RQ1. Unit of time used - minute.
- A total number of detected defects - a metric that implies the total number of detected defects in the source code regardless of the moment of detection of defects during the process of static code analysis. This metric is set as a reference for other metrics.
- A total number of false positive results - defects detected by tools that do not need or cannot be eliminated in the project. This metric is set with the aim of answering research questions RQ4 and RQ1.
- A total number of false negative results - defects that are of great importance for the project and were not detected by the tool. This metric was set up with the aim of answering RQ4 and RQ1 research questions.

Identified strategies and sub-strategies were applied to each tool according to the defined steps described in [10] and during the implementation of this experiment, the obtained results were recorded in relation to the set metrics for each tool and for each strategy.The results of applying strategies to SonarQube are shown in Table 1.

Tool	**SonarQube**			
Metrics/Strategies	***S1.1***	***S1.2***	***S2.1***	***S2.2***
Time (min)	51	78	43	33
Defects	42	40	41	51
False positives	2	0	3	10
False negatives	0	0	2	1

Table 1. Implement strategies over SonarQube

The results of applying strategies to SpotBugs are shown in Table 2.

Tool	SpotBugs			
Metrics/Strategies	*S1.1*	*S1.2*	*S2.1*	*S2.2*
Time (min)	28	53	25	20
Defects	22	19	24	30
False positives	3	0	6	12
False negatives	0	0	1	1

Table 2. Implement strategies over SpotBugs

The results of applying strategies to PMD are shown in Table 3.

Tool	PMD			
Metrics/Strategies	*S1.1*	*S1.2*	*S2.1*	*S2.2*
Time (min)	140	161	98	95
Defects	127	122	131	135
False positives	14	10	20	26
False negatives	1	2	3	3

Table 3. Implement strategies over PMD

The results of applying strategies to Checkstyle are shown in Table 4.

Tool	Checkstyle			
Metrics/Strategies	*S1.1*	*S1.2*	*S2.1*	*S2.2*
Time (min)	122	128	96	88
Defects	224	224	238	242
False positives	18	20	35	39
False negatives	0	2	3	3

Table 4. Implement strategies over Checkstyle

In order to evaluate the results achieved by the selected tools by applying the strategies, all the tools will be presented together and the strategies identified with the modified metrics.

Now, the metrics will be labeled and interpreted as follows:

1. Average time required to eliminate one defect, expressed in seconds - **ATR**.
2. A total number of identified defects in the project - **ND**.
3. Share of false positive results in the total number of defects - **FPR**.
4. Share of false negative results in the total number of defects - **FNR**.

Aggregate results on selected tools, strategies, and defined metrics are shown in Table 5.

Tool	Strategies	Metrics			
		ATR	*ND*	*FPR*	*FNR*
SonarQube	S1.1	73	**42**	4.8	0
	S1.2	117	**40**	0	0
	S2.1	63	**41**	7.3	4.9
	S2.2	39	**51**	19.6	2
SpotBugs	S1.1	76	**22**	13.6	0
	S1.2	167	**19**	0	0
	S2.1	62	**24**	25	4.2
	S2.2	40	**30**	40	3.3
PMD	S1.1	66	**127**	11	0.8
	S1.2	79	**122**	8.2	1.6
	S2.1	45	**131**	15.2	2.3
	S2.2	42	**135**	19.3	2.2
Checkstyle	S1.1	30	**224**	8	0
	S1.2	34	**224**	9	0.9
	S2.1	24	**238**	14.7	1.3
	S2.2	21	**242**	16.1	1.2

Table 5. Implement strategies over static code analysis tools

6. Results and discussion

An experiment within this research, which implies the application of the identified strategies, was successfully conducted on all selected tools (RQ2). In order to answer the research question RQ3, the combined results related to the metric of the average time required to eliminate one defect will be observed. The aim of this paper is not to compare tools, but to compare strategies, and the results of the ATR metrics presented in Table 5 of this research will be calculated as the average time overall the tools used for a particular strategy. The obtained result is presented in Fig. 2.

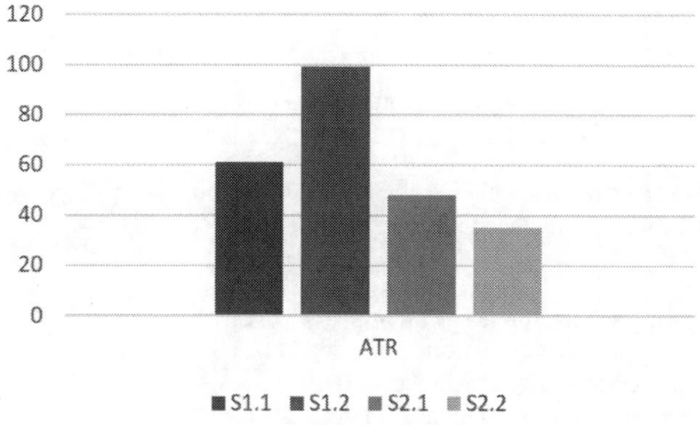

Fig. 2. Average time required to eliminate one defect overall tools according to the identified strategies

Based on the presented results, it can be concluded that the S2.2 strategy is the fastest to implement. For the RQ4 the same analysis of the results should be made as for the previous question, only now the metrics of the share of false positive defects and false negative defects in the total result of the number of defects are observed. These two metrics will be combined to analyze this result because both affect the accuracy of strategy implementation. The obtained results are presented in Fig. 3.

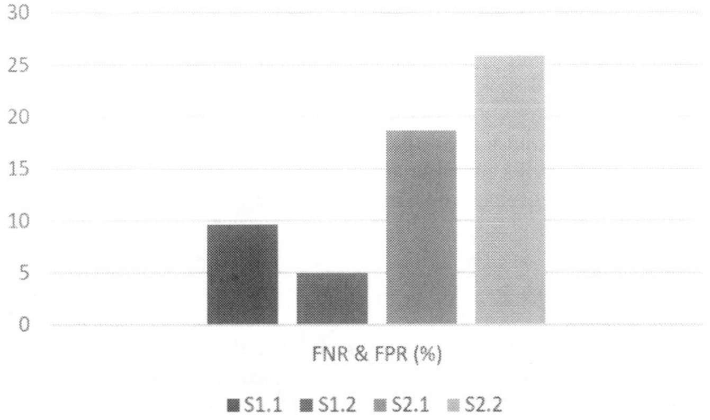

Fig. 3. Percentage of false positive and false negative results in the total number of defects overall tools according to the identified strategies

Based on the presented results, it can be concluded that strategy S2.2 has the highest share of false positive and false negative results in the total number of defects detected by the tools. On the other hand, strategy S1.2 generates the smallest number of such results (RQ4). The research question RQ2 can be answered in the affirmative, that is, the identified strategies have been successfully applied to all tools. This issue also has a clearly defined limitation, as the implementation of the strategy is checked only by the tools used in this research.

In order to answer the first research question RQ1, it is necessary to define what is the best result of the implementation of the strategy. Based on the defined metrics in the paper, the best strategy is the one that gives the most accurate results, in terms of the percentage of false positives and false negatives in the total number of detected defects, and the one with the fastest implementation, that is the one with the lowest elimination time. These two metrics were set up and compared in Fig. 4.

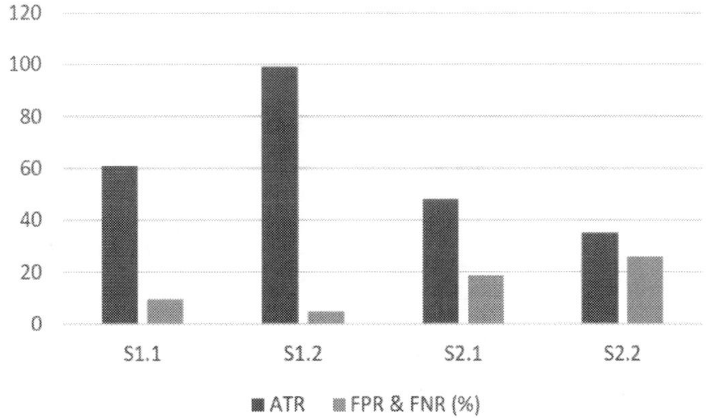

Fig. 4. Comparison of strategy implementation speed results and percentage of false positive and false negative results in the total number of identified defects

Based on the presented results, it can be concluded that strategies that are faster to implement over the tools used to generate a larger number of defects identified as FPR and FNR and strategies that are slower to implement generate a smaller number of erroneous results. The answer to research question RQ1 is that if both criteria are considered together, it is not possible to label one of the strategies as the strategy that gives the best results. It is concluded that the choice of strategy depends on the purpose of using the static code analysis tool.

7. Conclusion and further research

Static code analysis tools are increasingly finding their application and are increasingly used in the daily work of software developers in various positions. Also, these tools find their application in educating future software developers about potential defects that may occur in the code, as well as their elimination by reengineering the source code.

New and improved tools are being developed every day, offering their customers an ever-improving service by building trust in the results of static analysis and enabling them to avoid certain code defects in a quality way and make their software solution better. This is especially important since the lower use of static code analysis tools is caused precisely by the user's distrust of these tools and the time required to conduct such analyses over the source code of the program. When it comes to distrust in the results of static analysis, they primarily refer to the generation of false positive and false negative results. By developing and improving these tools, the percentage of errors is reduced and these tools give better results and therefore are used more often.

Based on the conducted experiment on these strategies, certain advantages and disadvantages were noticed on the basis of defined metrics. By analyzing and comparing strategies, it is defined which strategies are faster to implement (RQ3), and which strategies give more accurate results (RQ4). In further work, it is necessary to analyze which of the identified strategies are most often used by software developers, students, and other users within different positions by conducting a questionnaire on users of static code analysis tools on the strategies they use.

The authors limit themselves to the possibility of the existence of other strategies that are not identified in this paper and are not the subject of this research. It is also necessary to conduct additional research (on other tools, different programming languages, different projects) in order to determine the existence of other strategies and sub-strategies. It is also necessary to confirm the possibility of implementing the identified strategies on other tools and software projects. In conclusion, the use of static code analysis tools and continuous work on improving and maintaining the quality of source code provides better quality software solutions and represents a step forward in software development.

8. References

[1] A. Moller & I. Schwartzbach (2019). "Static program analysis"

[2] M. Arsenovic, S. Sladojevic, A. Anderla & D. Stefanovic (2018). “Deep neural network ensemble architecture for eye movements classification,” in 2018 17th International Symposium on INFOTEH-JAHORINA, INFOTEH 2018 - Proceedings, Jahorina

[3] D. Beric, S. Havzi, T. Lolic, N. Simeunovic & D. Stefanovic (2020). “Development of the MES software and Integration with an existing ERP Software in Industrial Enterprise,” in 2020 19th International Symposium INFOTEH-JAHORINA, INFOTEH 2020 - Proceedings, Jahorina

[4] D. Beric, D. Stefanovic, I. Cosic & B. Lalic (2018). “The implementation of ERP and MES Systems as a support to industrial management systems,” International Journal of Industrial Engineering and Management, vol. 9, no. 2, pp. 77-86

[5] D. Dakic, D. Stefanovic, I. Cosic, T. Lolic & M. Medojevic (2018). “ Business Process Mining Application: A Literature Review,” in Proceedings of the 29th DAAAM International Symposium, Vienna

[6] M. Jokic, D. Gracanin & B. Lalic (2019). “Sustainable Low Cost and High-Quality Supply Chain Assurance-A Systematic Literature Review,” in Proceedings of the 30th DAAAM International Symposium, Vienna, Austria

[7] M. Blahova, V. Mach, L. Pavlik, M. Hromada & M. Ficek (2019). “ The Information Security to Software of Crisis Management,” in Proceedings of the 30th DAAAM, Vienna

[8] A. Suleykin & P. Panfilov (2019). “Implementing Big Data Processing Workflows Using Open Source Technologies,” in Proceedings of the 30th DAAAM International Symposium, Vienna

[9] D. Stefanović, D. Nikolić, D. Dakić, I. Spasojević & S. Ristić (2020). “Static code analysis tools: A systematic Literature Review,” in Proceeding of of 31st DAAAM International Symposium, Vienna

[10] D. Stefanović, D. Nikolić, S. Havzi, T. Lolić & D. Dakić (2020). “Identification of strategies over tools for static code analysis,” in Proceedings of 9th International Conference On Engineering and Technology , Krabi

[11] J. West & B. Chess (2007). “Secure programming with static code analysis,” Pearson education

[12] K. Goseva-Popstojanova & A. Perhinschi (2015). “On the capability of static code analysis to detect security vulnerabilities,” Information and Software Technology, vol. 68, pp. 18-33

[13] K. P. B. C. L. L. D. Baca (2009). “Static code analysis to detect software security vulnerabilities,” in International Conference on Availability, Reliability and Security

[14] B. C. K. P. L. L. D. Baca (2013). “Improving software security with static automated code analysis in an industry setting,” in Software - Pract

[15] B. Kitchenham (2004). “Procedure for Undertaking Systematic Reviews,” Keele University, Computer Science Department

[16] A. Kaur & R. Nayyar (2020). "A Comparative Study of Static Code Analysis tools for Vulnerability Detection in C/C++ and JAVA Source Code," Procedia Computer Science, vol. 171, no. 2019, pp. 2023-2029

[17] D. Marcilio, C. Furia, R. Bonifacio & P. Gustavo (2020). "SpongeBugs: Automatically generating fix suggestions in response to static code analysis warnings," The Journal of Systems & Software, vol. 168,

[18] P. Nunes, I. Medeiros, J. Fonseca, N. Neves, M. Correia & M. Vieira (2019). "An empirical study on combining diverse static analysis tools for web security vulnerabilities based on development scenarios," Computing, vol. 101, no. 2, pp. 161-185

[19] N. Meghanathan (2013). "Identification and removal of software security vulnerabilities using source code analysis: a case study on a java file writer program with password validation features," Journal of Software, vol. 8, no. 10

[20] B. Kitchenham & S. Kubjnab (2000). "DESMET: A method for evaluating Software Engineering methods and tools"

[21] D. Nikolić, D. Stefanović, D. Dakić, S. Sladojević & S. Ristić (2021). "Analysis of the Tools for Static Code Analysis," in Proceedings of 20th International Symposium Infoteh-Jahorina, Jahorina

[22] "Github project from Author's repository" (2021). [Online]. Available from: https://github.com/nikolicdanilo/Java-Backend-Static-Code-Analysis. Accessed: 20 July 2021

DOI: 10.2507/32nd.daaam.proceedings.070

MEASURING INTERRELATIONS OF PROSPERITY PILLARS LEADING TOWARDS SUSTAINABLE SUCCESS

Mirjana Radman-Funaric, Barbara Pisker & Katarina Potnik Galic

This Publication has to be referred as: Radman-Funaric, M[irjana]; Pisker, B[arbara] & Potnik Galic, K[atarina] (2021). Measuring Interrelations of Prosperity Pillars Leading Towards Sustainable Success, Proceedings of the 32nd DAAAM International Symposium, pp.0486-0492, B. Katalinic (Ed.), Published by DAAAM International, ISBN 978-3-902734-33-4, ISSN 1726-9679, Vienna, Austria
DOI: 10.2507/32nd.daaam.proceedings.070

Abstract

This paper primary goal is set to measure the interrelations between different Legatum Prosperity Index (LPI) pillars in the global frame. Enterprise Conditions pillar is set as a dependent variable, a measure of sustainable success. The paper observes the relationship towards all other pillars of LPI namely: Social Capital, Safety & Security, Personal Freedoms, Governance, Investment Environment, Market Access and Infrastructure, Economic Quality, Living Conditions, Health, Education and Natural Environment. Pillars themselves are categorised into three main domains: Inclusive Societies, Open Economies and Empowered People aiming to ensure a holistic approach to the topic investigated. The panel sample included 167 world countries analysis for a period between 2007 and 2020. The overall number of observations is 2338 and the analysis included two methods: bivariate correlation and hierarchical multiple linear regression (through three models). The results gained show that Investment Environment, Market Access and Infrastructure strong and positively affect Enterprise Conditions. Health, Natural Environment, Governance and Social Capital pillars also show positive, but slightly weaker relation, while Education and Living Conditions as well as Safety and Security show negative relation towards entrepreneurship. Statistically significant but weak negative link between Personal Freedom towards Enterprise Conditions is also shown.

Keywords: Prosperity index; Enterprise conditions; Social capital; Sustainability; Wellbeing

1. Introduction

As defined by the Legatum Prosperity Index (LPI) „Prosperity is far more than wealth; it is when all people have the opportunity and freedom to thrive. Prosperity is underpinned by an inclusive society, with a strong contract that protects the fundamental liberties and security of every individual. It is driven by an open economy that harnesses ideas and talent to create sustainable pathways out of poverty. And it is built by empowered people, who contribute and play their part in creating a society that promotes wellbeing" [1]. Prosperous society is an ideal to be reached by upgrading individual, corporation, national and global prosperity. It is not an ultimate goal by itself, but a process of social transformation, a demanding venture measured by LPI from 2007.

The building blocks of prosperity constructed by the LPI are spread through three main domains: Inclusive societies, Open economies and Empowered people. Prosperity Index Domain Inclusive societies encompasses four key pillars: Safety and Security, Personal Freedom, Governance and Social Capital. Open economies domain includes Investment Environment, Enterprise Conditions, Market Access & Infrastructure and Economic Quality. The dimension of Empowered people includes Living Conditions, Health, Education and Natural Environment. Each of the four pillars developed under a specific domain is subdivided furthermore into specific elements and indicators. The value of the individual element and indicator contained in the overall prosperity is the result of the data specified in The Legatum Institute [2].

The purpose of this study is to measure the interrelations between different LPI pillars in the global frame enabling their practical implications in future creation of sustainably successful socio-economic environments. Enterprise Conditions pillar is considered a dependent variable in this paper, chosen as a prosperity determinant and a measure of sustainable success due to the advantages at the disposal of the nation and the local community as noted by Porter [3] and Brakman et al. [4]. Other pillars with all their belonging elements of the prosperity index (PI) are considered independent variables. The aim of the paper is to examine different PI relationship towards Enterprise Conditions and reveal answers the research question on whether there are variables that are more closely related to Enterprise Conditions, or whether some of the pillars of PI indicate that they are a guarantee of a greater entrepreneurial activity as a foundation of the wider societal community sustainable welfare (sustainable success, wellbeing, prosperity).

If the value of socio-economic performance can be directly related to the measured entrepreneurial environment and level of entrepreneurship, numerous studies are showing the growth effect of prosperity index variables towards Enterprise Conditions. The results of Mervar study [5] showed that economic growth in the early 1960s to the late 1980s was stronger in countries with higher investment and lower public spending, with fewer price distortions, less premium on the "black" foreign exchange market, lower inflation rate, stronger involvement in international flows, higher human capital, stable socio-political conditions. According to Mervar [5], the same is confirmed by the analyses of the International Monetary Fund conducted in 1993 and 1994 in developing countries. The subject of discussion is always which of the combinations of these instruments is the best, but in general, the above relations have not changed to date. Mervar [5] cites the distinction between the “tangible gap” of countries with a lack of capital or natural resources and the “gap of ideas” of countries suffering from a lack of knowledge needed to create value in a modern economy, as Romer wrote [6].

2. Research Methodology

2.1. Data and variables

The panel sample included 167 world countries in the last 14 years for a period between 2007 and 2020. All data have been taken from The Legatum Prosperity Index [7]. The value of the individual element and indicator contained in the overall prosperity is the result of the data specified in The Legatum Institute [2]. In 2019 the Business Environment pillar, measuring country’s entrepreneurial environment set, its business infrastructure, barriers to innovation, and labour market flexibility, was replaced by four new pillars introduced:

- The Enterprise Conditions pillar measures the degree to which regulations enable businesses to start, compete, and expand,
- The Investment Environment pillar measures the extent to which investments are adequately protected and are readily accessible,
- The Market Access and Infrastructure pillar measures the quality of the infrastructure that enables trade, and distortions in the market for goods and services,
- Living Conditions pillar encompasses a set of conditions and circumstances that are necessary for all individuals to attain wellbeing.

In the frame of dependent variable Enterprise Conditions pillar, the following elements have been researched: Domestic Market Contestability, Environment for Business Creation, Burden of Regulation, Labour Market Flexibility and Price Distortions.

LPI Enterprise Conditions elements	LPI Enterprise Conditions indicators
Domestic Market Contestability	Market-based competition
	Anti-monopoly policy
	Extent of market dominance
Environment for Business Creation	Private companies are protected and permitted
	Ease of starting a business
	State of cluster development

	Labour skill a business constraint
	Availability of skilled workers
	Burden of government regulation
	Time spent complying with regulations
	Number of tax payments
	Time spent filing taxes
	Burden of obtaining a building permit
Burden of Regulation	Building quality control index
	Cooperation in labour-employer relations
	Flexibility of hiring practices
	Redundancy costs
	Flexibility of employment contracts
Labour Market Flexibility	Flexibility of wage determination
	Distortive effect of taxes and subsidies
Price Distortions	Energy subsidies

Table 1. PI Enterprise Conditions pillar elements and indicators

2.2. Methods

The primary goal of bivariate correlation analysis is to check how the set of independent variables correlates with the Enterprise Conditions (ente). The aim is also to determine whether there is an interrelation between Enterprise Conditions and any other pillar or variable is more and/or significantly more interrelated to Enterprise Conditions pillar variable. The relationship between independent and Enterprise Conditions (ente) are examined using multiple regression. The formulas for the multiple regressions can be expressed as:

$$ente = a+\beta_1 soci+\beta_2 safe+\beta_3 pers+\beta_4 gove+\beta_5 inve+\beta_6 mark+\beta_7 econ+\beta_8 livi+\beta_9 heal+\beta_{10} educ+\beta_{11} envi+e \quad (1)$$

where Enterprise Conditions is dependent variable, and others are independent variables, e is the error term involved in using the linear model to predict the value of Y, a is the intercept of the slope, and β is the coefficient of the independent variable [8].

The first multiple regression model in Table 3 includes three independent variables: Economic Quality, Market Access and Infrastructure and Investment Environment, variables most commonly considered directly related to entrepreneurship in the frame of PI Open Economy domain. Sources and weights of each indicator are contained within The Legatum Institute [2]. Model 2 in the Table 3 adds four new variables: Living Conditions, Health, Education and Natural Environment, compounding the PI Empowered People domain. Sources and weights of each indicator of the Empowered people domain are contained within The Legatum Institute [2]. Model 3 adds additional four new variables: Safety and Security, Personal Freedom, Governance and Social Capital, seen as the societal, political and cultural surrounding. Sources and weights of each indicator of the Empowered people domain are contained within The Legatum Institute [2].

3. Results and discussion

Table 2 presents descriptive statistics for all variables included in the study. The number of observations is 2338. No standard deviation is greater than the mean, although descriptive statistics suggest that the sample is diverse. The greatest deviation from the average among the countries of the world was noted for the variables Market Access and Infrastructure, 37%, Governance 36,5%, Personal Freedom, 36% and Education 35%, and the least for the variables Natural Environment, 15,6%, Health 18,8%, Social Capital 18,9%. Enterprise Conditions variable deviates from the average 27%, where the highest value of the index is 88 and the lowest 19.7. The deviation from the mean value of the total PI is 23.8% while 50% of the countries in the world have PI index values between 45.8 and 65.6 and the Interquartile range of index values is 19.8.

The correlation matrix reveals that all the variables are significant at a 1% level. The correlation coefficient between some variables is greater than 0.8, which might be a sign of multi-collinearity between independent variables. Multi-collinearity is expected in the relation between overall PI and pillars due to individual PI results and is eliminated by investigating the relationship between the independent variables and the dependent variable Enterprise Conditions using multiple regressions.

Descriptive Statistic	N	Mean	St. Dev.	Min	Pctl (25)	Pctl (75)	Max
PI	2,338	56.2	13.4	27.8	45.8	65.6	84.5
Safety and Security	2,338	67.2	15.9	14.5	59.1	78.4	96.1
Personal Freedom	2,338	53.5	19.3	11.0	38.1	66.9	93.9
Governance	2,338	50.1	18.3	12.6	37.8	60.2	91.0
Social Capital	2,338	51.4	9.7	19.8	45.1	56.8	81.6
Investment Environment	2,338	52.1	15.2	22.3	39.7	62.8	87.0
Enterprise Conditions	2,338	53.3	14.4	19.7	43.2	62.4	88.0
Market Access and Infrastructure	2,338	49.3	18.3	14.9	33.2	65.2	88.2
Economic Quality	2,338	50.6	13.2	21.9	40.4	60.5	83.0
Living Conditions	2,338	67.3	21.0	17.2	48.8	84.9	97.1
Health	2,338	67.2	12.6	29.9	57.9	76.8	86.8
Education	2,338	57.0	20.1	10.4	41.6	73.6	91.5
Natural Environment	2,338	55.2	8.6	29.5	49.5	60.2	78.9

Table 2. Descriptive Statistics

The highest correlation of the PI index is with the Investment Environment (r = 0.95), Market Access and Infrastructure (r = 0.95), Governance (r = 0.94), and the lowest with Social Capital (r = 0.66) and Natural Environment (r = 0.67). Enterprise Conditions are very strongly related to PI index (r = 0.91), Investment Environment (r = 0.93) and Governance (r = 0.92). Enterprise Conditions correlate least with the Natural Environment (r = 0.55), and the variables in the Inclusive societies group; Social Capital (r = 0.64), Safety and Security (r = 0.66), Personal Freedom (r = 0.67).

Table 3. presents the result from the multiple regression on the three models outlined above. The overall fit of the models measured by the Adjusted R2 is very high. It ranges between 0.87 and 0.93. In other words, it seems that the models explain much of the variance in the dependent variable.

	Dependent variable: `**Enterprise Conditions**`		
	(1)	**(2)**	**(3)**
Investment Environment	**0.737*****	**0.688*****	**0.269*****
	(0.021)	(0.019)	(0.020)
Market Access and Infrastructure	**0.120*****	**0.386*****	**0.395*****
	(0.017)	(0.020)	(0.016)
Economic Quality	0.010	**0.066*****	**-0.045*****
	(0.019)	(0.018)	(0.015)
Living Conditions		**-0.213*****	**-0.200*****
		(0.017)	(0.014)
Health		**0.038****	**0.029***
		(0.019)	(0.016)
Education		**-0.073*****	**-0.056*****
		(0.015)	(0.013)
Natural Environment		**0.024***	**-0.038*****
		(0.014)	(0.014)
Safety and Security			**-0.041*****
			(0.008)
Personal Freedom			**-0.142*****
			(0.010)

Governance			**0.535*****
			(0.019)
Social Capital			**0.125*****
			(0.011)
Constant	8.515***	9.669***	16.039***
	(0.485)	(1.022)	(0.877)
Observations	2,338	2,338	2,338
R^2	0.868	0.891	0.930
Adjusted R^2	0.867	0.891	0.930
Residual Std. Error	5.244 (df = 2334)	4.762 (df = 2330)	3.819 (df = 2326)
F Statistic	5,100.286*** (df = 3; 2334)	2,723.035*** (df = 7; 2330)	2,811.880*** (df = 11; 2326)
Note:			*p<0.1, **p <0.05, ***p<0.01

Table 3. Results of hierarchical multiple linear regression

Table 3 shows results from the multiple regression of Enterprise Conditions and 12 pillars of prosperity. The overall fit of the model measured by Adjusted R2 is fairly high in all three cases and close to 1. This means that the model explains much of the variance in the dependent variable.

The results from the first model suggest that Investment Environment with 0.737 strong and positively affect Enterprise Conditions ($p < 0.01$). This result is expected because investment environment is considered highly related to the enterprise conditions. Market Access and Infrastructure, 0.120 also positively affect Enterprise Conditions, but lower ($p < 0.01$). Both results are following Mervar survey results [5]. Compared to the results from Radman-Funaric, Pisker, Radman [9] for EU countries Market Access and Infrastructure influence is lower at the global level than in the EU countries ($\beta = 0.706$, $p < 0.01$). The result is expected as Market Access and Infrastructure elements valued combine Communications, Resources, Transport, Border Administration, Open Market Scale, Import Tariff Barriers, Market Distortions which favour EU countries in regard to its common (single) market policy when compared to the rest of the world [2]. Surprisingly, Economic Quality is insignificant and does not affect the Enterprise Conditions. The same result was gained in the research encompassing EU countries by Radman-Funaric, Pisker, Radman [9]. Overall, Fiscal Sustainability, Macroeconomic Stability, Productivity and Competitiveness, Dynamism and Labour Force Engagement do not affect the level of entrepreneurial activity neither in the EU countries nor in the countries of the world.

The results in Model 2 included the variables from the first model, adding the second pillar group to the analysis: Living Conditions, Health, Education, and Natural Environment. The exhibit shows all independent variables are statistically significant. By adding those variables to the regression, no significant change in impact of the first group of variables on Enterprise Conditions was observed. The impact of new variables is strong and positively affect 0.688 Enterprise Conditions ($p < 0.01$), Market Access and Infrastructure gained 0.386, and unlike the first model Economic Quality becomes statistically significant ($\beta = 0.066$, $p < 0.01$). Living Conditions is, as expected, negativly interrelated to Enterprise Conditions ($\beta = -0.213$, $p < 0.01$). As institutional environment is a societal legacy, economic growth does not ultimately reach societys sustainable targets and is not necessarily directly related to economic and societal quality in general, especially in underdeveloped countries as noted by Prasetyo and Kistanti [10].

Unexpectedly, Education variable also shows negative and statistically significant interrelation ($\beta = -0.073$, $p < 0.01$) towards Enterprise Conditions. This interconnection could be explained thou through the sociological theory of a marginal man developed by Park, R.E. (1928) claiming how profession, institutional embedded individual (nowadays also highly educated) is less likely to develop an innovative and entrepreneurial mind, therefore development of new business forms or the opening up of new market opportunities may not be as sensitive to an increasing educational requirement. Empirical research conducted by Collins and Moore in 1970 [11] and Channon in 1979 [12] also show how entrepreneurs often come from disadvantaged background, humble origins and receive an education through secondary school only. Therefore, this thesis can explain the findings from the Model 2. Health ($\beta = 0.038$, $p<0.5$) and Natural Environment ($\beta = 0.024$, $p < 0.1$) show slightly smaller interrelation towards Enterprise Conditions.

Comparing the results gained with the EU research results Radman-Funaric, Pisker, Radman from 2021 [9] significant differences are observed. EU data analysis show only two elements are statistically significant (p<0,01), Natural Environment (-0.068) and Education (0.440) with different sign then in world observed sample. In the countries of the world, the statistical significance of these variables is somewhat smaller, but while Education in the EU is positively related to Enterprise Conditions, in the world this correlation is negative. The results are at odds with the results revealed by Romer [6] and Mervar [5]. The results may suggest that despite poorer education in developing and underdeveloped countries there is a better Environment for Business Creation, higher Labour Market Flexibility and lower Burden of Regulation, which are included in the assessment of overall Enterprise Conditions. Also, the Natural Environment is negatively connected in the EU and positively in the world. This result may indicate that European countries have already reduced the quality of the Natural Environment due to greater development, while developing and underdeveloped countries still have an unspoiled natural environment.

Lastly, the third regression model included all 12 pillars of prosperity. It is the strongest model in regard to Adjusted *R* squared. It comprises additional four new variables grouped under the PI Inclusive Societies domain: Safety & Security, Personal Freedom, Governance, Social Capital. All of these independent variables are statistically significant at the significance level $p < 0.1$, the other ten variables for which $p < 0.01$. This model introduced variables novelties meant a rather significant decrease of the impact of Investment Environment on Enterprise Conditions; precisely it decreased from 0.688 in 2nd model to 0.269 in 3rd model. This brings a key note in the discussion showing how Inclusive Societies domain does not fit in mutually positively with Investment Environment in Enterprise Conditions. This is due to the fact how globally dominant sector of production is still predominantly embedded in a domain of favouring profit before people and not jet tuned in with societal frame tailored and fitted for the rise of the Inclusive Societies domain favouring people before profit. Economic Quality and Natural Environment in second previously presented model had a positive influence to Enterprise Conditions, while in this model they take on a negative sign, which indicates that by increasing the number of independent variables, which we believe have a positive impact on the level of entrepreneurship, they give a different perspective on the prerequisites for the Enterprise Conditions development.

Safety and Security shows negative relation (-0.041) to the Enterprise Conditions in accordance to Pinazo-Dallenbach, Roig-Tierno & Mas-TurAre [13]. Social Capital (0.125) is positively related to the Enterprise Conditions, which is partly consistent with Leonardi, Naneti, Putman [14], Fukuyama [15], Tortosa-Ausina & Peiró-Palomino [16], Radman-Funaric [17], Borozan & Radman-Funaric [18], Radman-Funaric [19]. Partly only, as the research is based solely on social trust or the results differ in individual elements (variables) of social capital. The same results were gained by Radman-Funaric, Pisker, Radman [9] research when analysing this impact in EU countries.

Governance is positively related to the Enterprise Conditions (0.535), as verified in the EU countries in accordance to Hardin [20] and Knack & Keefer [21] research proving more reliable and efficient state institutions lead to greater mutual trust and respect for norms which creates a more favourable entrepreneurial environment. Surprisingly, Personal Freedom is negatively interrelated to Enterprise Conditions (-0.142) although this relation is not strong, while EU countries research [9] did not reveal interconnection inbetween Personal Freedoms and Enterprise Conditions. As Baumol noted depending on the overal societal validation frame there are two types: a society favouring productivity through economic and personal freedoms that will be rewarded with higher allocation of entrepreneurial energy to innovation. On the other hand, a society with a high level of political intervention, high level of taxation and regulation, will see more entrepreneurial energy devoted to non-productive activities such as rent-seeking [22], [23]. Nyström also finds that countries with smaller government, stronger legal system and rule of law, better protection of property rights, and fewer regulations have higher level of self-employment concluding that the overall effect of freedom on entrepreneurial activity is positive and statistically significant [24], [25].

4. Conclusion

Prosperity goes beyond narrow, single sided material pleasures and transcends material concerns. It resides in the quality of our lives and the health and happiness of our society. It is present in the strength of our relationships and our trust in the community. It is evidenced by our satisfaction at work and sense of shared meaning and purpose of our existence. It hangs on our potential to participate actively in the life of society. Prosperity consists in our ability to flourish as human beings – within the ecological limits of our finite planet. The challenge for our society is to create the conditions under which this is possible. It is the most urgent task of our times and our generation. The study presented through the panel sample included 167 world countries analysis for a period between 2007 and 2020. The overall number of observations is 2338 and the analysis included two methods: bivariate correlation and hierarchical multiple linear regression (through three models) The study results show that Investment Environment, Market Access and Infrastructure strong and positively affect Enterprise Conditions. Health, Natural Environment, Governance and Social Capital pillars also show positive, but slightly weaker relation, while Education and Living Conditions as well as Safety and Security show negative relation towards entrepreneurship. Statistically significant but weak negative link between Personal Freedom towards Enterprise Conditions is also shown.

Practical implications of the study presented lead towards a need for different stakeholder action (from legal framework, board managers, employees and civil society sector) in contributing towards developmental pathway in global positioning of prosperous society as a dominant category. As it is evident from the study results presented societies cannot be prosperous if undermining any of the aspects of the holistic developmental PI approaches (opened economies, empowered people, inclusive society). The time has come for the leaders to reshape the developmental list of priorities placing people and planet before greedy economies in order to keep the ecosystem balance in a humbler vision of a sustainable future to come. It is on the business leaders to enable a sustainable future success.

Study limitations oblige us to declare how results show no interconnection between Enterprise conditions and Economic quality including GDP growth. Future research call upon research towards the relationship between Enterprise Conditions and all elements of Economic quality to discover which of the specific element is not interrelated to Enterprise conditions and which one has got the strongest connection. Furthermore, the analysis would improve due to the separate elements and indicator analysis in social capital towards Enterprise Conditions influence, a task for future research.

Since it is expected for all positive results to be closely bonded with conditions of entrepreneurship development it is necessary to select those variables that are not statistically relevant towards Enterprise conditions from the present analysis and closely check which of the variables included in the model disturb the results expected.

5. References

[1] Legatum Prosperity Index (2020). LPI Summary, Available from: https://www.prosperity.com/about/summary Accessed: 2021-3-15

[2] The Legatum Institute (2020). Sources and indicators used, and changes made since the 2019 index 2020. A tool for transformation, Available from: https://docs.prosperity.com/4416/0518/8129/2020_Index_Sources_Indicators_and_Changes.pdf Accessed: 2021-3-15

[3] Porter, M.E. (2004). Building the microeconomic foundations of prosperity: findings from the microeconomic competitiveness index, Oxford University Press, ISBN 0195173600, 9780195173604, Oxford

[4] Brakman, S.; Garretsen, H.; Van Marrewijk, C. & Van Witteloostuijn, A. (2006). Nations and firms in the global economy: an introduction to international economics and business, Cambridge University Press, ISBN 0521832985, Cambridge

[5] Mervar, A. (1999). "Pregled modela i metoda istraživanja gospodarskog rasta", Privredna kretanja i ekonomska politika, Vol. 9, No. 73, 1999, pp. 20–61., ISSN 1330-187X

[6] Romer, P. (1993). "Idea gaps and object gaps in economic development", Journal of Monetary Economics, Vol. 32, No. 3, 1993, pp. 543–573., ISSN 0304-3932

[7] The Legatum Institute (2020). The Legatum Prosperity Index, The 14th edition of the Legatum Prosperity Index Available from: https://www.prosperity.com/about/resources, Accessed: 2021-2-12

[8] Kamki, J. (2017). Digital Analytics: Data Driven Decision Making in Digital World, Notion Press, ISBN 978-1946556196, Chennai

[9] Radman-Funaric, M., Pisker, B. & Radman, M. (2021). Future at Stake: Key Pillars Aggravating Croatian Pathway to Prosperity, Proceedings of 10th International Scientific Syposium Region, Entrepreneurship, Development, Osijek, ISSN 1848-9559 Leko Simic, M. & Crnkovic, B. (Ed.), pp. 967-983., EFOS, Osijek

[10] Prasetyo, P. E. & Kistanti, N. R. (2020). "Human capital, institutional economics and entrepreneurship as a driver for quality & sustainable economic growth", Entrepreneurship and Sustainability Issues, Vol. 7, No. 4, pp. 2575-2589., DOI https://doi.org/10.9770/jesi.2020.7.4(1)

[11] Collins, O. F. & Moore, D. G. (1970). The organization makers: A behavioral study of independent entrepreneurs, Appleton-Century-Crofts, ISBN 0390203122 9780390203120, New York

[12] Channon, D. (1979). "Leadership and corporate performance in the service industries", Journal of Management Studies, Vol. 16, No. 2, pp. 185–201., DOI https://doi.org/10.1111/j.1467-6486.1979.tb00382.x

[13] Pinazo-Dallenbach, P., Roig-Tierno, N. & Mas-Tur, A. (2015). Are Problems with Violence and the Lack of Public Safety a Barrier to Entrepreneurship?, Proceedings of Annual Conference of the Global Innovation and Knowledge Academy, ISBN 9783319222042, Valencia

[14] Leonardi, R.; Nanetti R. Y. & Putnam, R. D. (2001). Making democracy work: Civic traditions in modern Italy, Princeton University Press, ISBN 9780691037387, Princeton

[15] Fukuyama, F. (1995). Social capital and the global economy, Foreign Affairs, Vol. 74, No. 4., pp. 89-103., ISSN: 0015-7120, 2327-7793

[16] Tortosa-Ausina, E. & P. J. Peiró (2012). "Social Capital, Investment and Economic Growth: Evidence for Spanish Provinces", Spatial Economic Analysis, Vol. 10, No. 1, pp. 102-126., Routledge

[17] Radman-Funaric, M. (2013). Osnaživanje socijalnog kapitala u cilju poticanja gospodarskog razvoja hrvatskih regija, Ph.D. Dissertation, Faculty of Economics in Osijek, J. J. Strossmayer University of Osijek, Osijek, Croatia

[18] Borozan, D. & Radman-Funaric, M. (2016). "Social capital in Croatia: Measurement and regional distribution," Innovation: The European Journal of Social Science Research, Vol. 29, No. 4, pp. 481–505.

[19] Radman-Funaric, M. (2018). "The Trust in Institutions and Manager Association," International Journal Vallis Aurea, Vol. 4, pp. 37 - 48.

[20] Hardin, R. (1982). Collective Action, RFF Press, ISBN 9780801828195, Washington

[21] Knack, S. & Keefer, P. (1997). "Does social capital have an economic payoff? A cross-country investigation," The Quarterly journal of economics, Vol. 112, No. 4., pp. 1251–1288., DOI https://doi.org/10.1162/003355300555475

[22] Baumol, W. J. (2010). The microtheory of innovative entrepreneurship, Princeton University Press, ISBN: 9780691145846, Princeton

[23] Hall, J.; Pulito J. & Van Metre, B. (2012). Freedom and entrepreneurship: New evidence from the 50 states, Mercatus Centre Working Paper No. 12-13., Available from: https://www.mercatus.org/system/files/Freedom-and-Entrepreneurship-Working-Paper.pdf, Acessed: 2021-2-12

[24] Nyström, K. (2008). "The institutions of economic freedom and entrepreneurship: evidence from panel data," Public choice, Vol. 136, pp. 269–282. DOI https://doi.org/10.1007/s11127-008-9295-9

[25] Nyström, K. (2008). Is Entrepreneurship the Salvation for Enhanced Economic Growth?: A review of the empirical evidence of the effect of entrepreneurship on employment, productivity and economic growth, Working Paper Series in Economics and Institutions of Inovation 143, Royal Institute of Technology, CESIS - Centre of Excellence for Science and Innovation Studies

DOI: 10.2507/32nd.daaam.proceedings.071

APPLICATION OF BOX-JENKINS METHODOLOGY IN ELECTRIC CONSUMPTION FORECASTING OF A METALLURGICAL ENTERPRISE

Gulnara Ibrayeva, Iosif Breido & Juliya Bulatbayeva

This Publication has to be referred as: Ibrayeva, G[ulnara]; Breido, I[osif] & Bulatbayeva, J[uliya] (2021). Application of Box-Jenkins Methodology in Electric Consumption Forecasting of a Metallurgical Enterprise, Proceedings of the 32nd DAAAM International Symposium, pp.0493-0499, B. Katalinic (Ed.), Published by DAAAM International, ISBN 978-3-902734-33-4, ISSN 1726-9679, Vienna, Austria
DOI: 10.2507/32nd.daaam.proceedings.071

Abstract

The article discusses the construction of a predictive model of power consumption of a large industrial facility based on the Box-Jenkins methodology. This method belongs to the tools for statistical data processing. Retrospective data on power consumption of the metallurgical enterprise JSC "ArcelorMittal Temirtau" were taken as the initial data for forecasting. The graphs of the initial time series, autocorrelation function and private autocorrelation function are presented. Based on the analysis of auxiliary graphs, alternative parameters of the predictive model are selected. Among the results of the experiment, the best parameters of the model were selected based on the assessment of adequacy and accuracy.

Keywords: short-term forecasting; time series analysis; autocorrelation function; ARIMA model; power consumption of the metallurgical plant; statistical analysis.

1. Introduction

The laws of modern wholesale electricity markets in post-Soviet countries are undergoing constant changes, which puts new demands on its participants. Such transformations include the introduction of a balancing electricity market, the integration of which is expected in 2022 in Kazakhstan. The balancing market principle can be described as follows. To connect them to the national electric grid, consumers submit hourly applications for the planned consumption of electric energy every day in advance. The claims represent a daily electrical load profile. According to the Law on the Electricity Industry of the Republic of Kazakhstan, large consumers with a load capacity of more than 1 MW are considered to be participants in the wholesale electricity market. After the end of each operational day, the System Operator of the electrical network monitors the planned and actual consumption. For each hour when the consumer has exceeded the declared capacity, it is necessary to pay the difference at the inflated tariff. Thus, in order to incur lower financial losses, the consumer must minimize the difference between the declared and actual consumption.

This can be achieved by managing the electricity consumption schedule or more accurately forecasting the declared consumption. However, managing the schedule of electricity consumption in large enterprises, especially with different types of production, is a difficult task. Therefore, consumers need to analyse their own power consumption and create working forecasting models.

There is no unambiguously suitable model among the methods for forecasting electricity consumption. For each individual enterprise, an analysis should be carried out, alternative models should be built and the optimal model should be chosen. Hourly energy consumption data is a time series. Time series forecasting methods are divided into statistical (regression, extrapolation, exponential smoothing, etc.) and structural models (neural networks, fuzzy logic models, Markov chains, etc.). In this study, the mathematical model ARIMA (Integrated Moving Average Autoregression Model) is considered. This model is based on the Box-Jenkins methodology, which is an adaptive model. Electricity consumption of the metallurgical plant of ArcelorMittal Temirtau JSC was taken as the initial data. The initial time series is the weekly profile of the electrical load, and the daily load profile is taken as the forecasting horizon.

2. Box-Jenkins methodology for building ARIMA model

Electricity consumption of the metallurgical plant of ArcelorMittal Temirtau JSC was taken as the initial data. The initial time series is a weekly electric load profile, a daily load profile is taken as the forecasting horizon, the sampling period is one hour. ARIMA's approach to time series is that the stationarity of the series is assessed first. Various tests reveal the presence of unit roots and the order of integration of the time series (usually limited to the first or second order). Further, if necessary, the series is transformed by taking the difference of the corresponding order and some ARMA model is built for the transformed model, since it is assumed that the resulting process is stationary, in contrast to the original non-stationary process [1]. The time series forecasting algorithm based on the Box-Jenkins methodology consists of the following stages:

1) checking the time series for stationarity by building graphs of the time series, the correlation function and the private autocorrelation function;
2) estimation of the parameters of the ARIMA model;
3) forecasting and assessing the accuracy of the ARIMA model [2].

The choice of the parameters of the ARIMA model is reduced to a separate consideration of the AR component and the MA component. The AR component of the ARIMA model is a separate autoregressive model, in which the values of the time series at the moment are linearly dependent on the previous values of the same series. An autoregressive process of order p for describing the time series of power consumption is defined as follows (1):

$$x(t) = c + \sum_{i=1}^{p} a_i \cdot x_{t-i} + \varepsilon_t, \tag{1}$$

where t – time series step,

$x(t)$ – power value of the electrical load at time t,

c – constant coefficient that determines the power of the electrical load at the current time without the influence of external factors (by default it is equal to zero),

i – time series delay from baseline,

p – AR model order (defined as the maximum delay value from the original series),

$a_1, \dots, a_p$ – AR model coefficients describing the degree of dependence of the current value of the power of the electrical load on the past values,

x_{t-i} – previous power values of the electrical load,

ε_t – white noise.

The MA component of the ARIMA model is a separate moving average model, in which the current value of the time series is equal to the average value over a certain previous period of time. The moving average model of order q has the form (2):

$$x(t) = \sum_{j=0}^{q} b_j \cdot \varepsilon_{t-j}, \tag{2}$$

where t – time series step,

$x(t)$ – power value of the electrical load at time t,

i – time series delay from baseline,

q – MA model order (defined as the maximum delay value from the original series),

$b_1, \dots, b_q$ – MA coefficients of the model, describing the degree of dependence of the current value of white noise on the past values,

ε_t – white noise.

The Autoregressive Moving Average Model (ARMA) is one of the mathematical models used to analyze and forecast stationary time series in statistics. The ARMA model generalizes two simpler time series models - the autoregressive model (AR) and the moving average (MA) model. The ARMA(p,q) model, where p and q are integers specifying the order of the model, is the following process of generating a time series (3):

$$x(t) = c + \sum_{i=1}^{p} a_i \cdot x_{t-i} + \sum_{i=1}^{q} b_i \cdot \varepsilon_{t-i} + \varepsilon_t, \tag{3}$$

where t – time series step,
$x(t)$ – power value of the electrical load at time t,
i – time series delay from baseline,
c – constant coefficient that determines the power of the electrical load at the current time without the influence of external factors (by default it is equal to zero),
p – AR model order (defined as the maximum delay value from the original series),
q – MA model order (defined as the maximum delay value from the original series),
$a_1, \dots, a_p$ – AR model coefficients describing the degree of dependence of the current value of the power of the electrical load on the past values,
$b_1, \dots, b_q$ – MA coefficients of the model, describing the degree of dependence of the current value of white noise on the past values,
x_{t-i} – previous power values of the electrical load,
ε_t – white noise.

The ARIMA model is an extension of ARMA models for non-stationary time series, which can be made stationary by taking differences of some order from the original time series. The ARIMA(p,d,q) model means that time series differences of order d obey the ARMA(p,q) model.

The ARIMA(p,d,q) model for a non-stationary time series has the form (4):

$$\Delta^d x(t) = c + \sum_{i=1}^{p} a_i \cdot \Delta^d x_{t-i} + \sum_{j=1}^{q} b_j \cdot \varepsilon_{t-j} + \varepsilon_t, \tag{4}$$

where t – time series step,
Δ^d – operator of the time series difference of order d, determined when converting the current time series of the power of the electrical load into a stationary form,
$x(t)$ – power value of the electrical load at time t,
i – time series delay from baseline,
c – constant coefficient that determines the power of the electrical load at the current time without the influence of external factors (by default it is equal to zero),
p – AR model order (defined as the maximum delay value from the original series),
q – MA model order (defined as the maximum delay value from the original series),
$a_1, \dots, a_p$ – AR model coefficients describing the degree of dependence of the current value of the power of the electrical load on the past values,
$b_1, \dots, b_q$ – MA coefficients of the model, describing the degree of dependence of the current value of white noise on the past values,
x_{t-i} – previous power values of the electrical load,
ε_t – white noise [3].

The theoretical autocorrelation (ACF) and partial autocorrelation function (PACF) for the AR, MA and ARMA models are known and are different for each model. These differences between models are important to keep in mind when choosing models. Table 1 provides a description of the behaviour of ACF and PACF for AR, MA and ARMA models.

Model type	**ACF behaviour**	**PACF behaviour**
AR(p)	Tails off gradually	Cuts off after p lags
MA(q)	Cuts off after q lags	Tails off gradually
ARMA(p,q)	Tails off gradually	Tails off gradually

Table 1. Comparative data on the properties of ACF and PACF

To select the type of model (AR, MA, ARMA or ARIMA), you should build graphs of the initial time series, ACF and PACF and evaluate their behaviour [4].

3. Checking the time series for stationarity

The stationarity of the time series means that the probabilistic characteristics of the time series are unchanged over time. Unlike regular sample data, time series data is ordered. The study of the time series graph allows us to draw preliminary conclusions about stationarity. Figure 1 shows the hourly electricity consumption schedule of ArcelorMittal Temirtau JSC from January 1 to January 7, 2020. The absence of a trend gives a preliminary conclusion about stationarity, with the exception of the peak component on January 6.

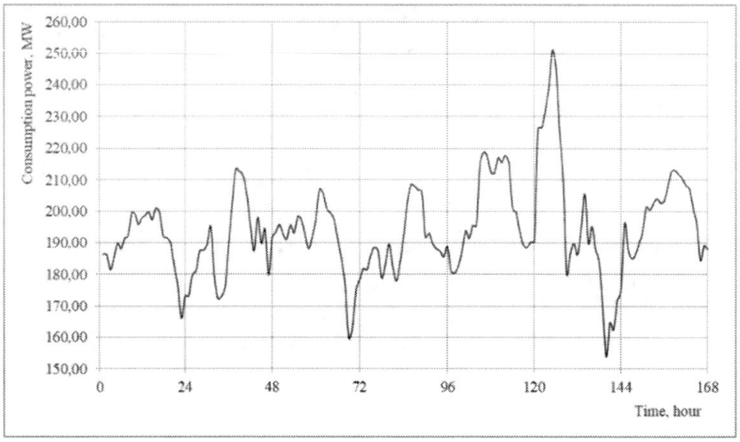

Fig. 1. Hourly power consumption chart for one week

The autocorrelation function (ACF) and the partial autocorrelation function (PACF) give more detailed characteristics of the time series, which make it possible to further evaluate the parameters of the ARIMA model. In particular, the ACF reports a correlation between points separated by different time delays, called lags [5].

In time series analysis, ACF shows the degree of linear statistical relationship between time series values. Numerically, the ACF is a sequence of correlation coefficients between the original series and its copy shifted by a given number of series intervals, called the lag L (5):

$$f(L) = \sum_{L=0}^{n} r_{t,t-L}, \tag{5}$$

where $f(L)$ – autocorrelation function,
n – number of time series levels,
L – time series delay,
$r_{t,t-L}$ – correlation coefficient for lag L.

The graph of the dependence of the correlation coefficient on the lag is called a correlogram. It is an effective tool for studying the dynamic properties of time series [6].

The ACF is maximal at $L = 0$, when the series is completely correlated with itself. Since the value of the autocorrelation coefficient varies from -1 to 1, the ACF can only take values from the same range. The autocorrelation coefficient is determined as follows (6):

$$r_{t,t-L} = \frac{cov_{y_t,y_{t-L}}}{\sigma_{y_t} \cdot \sigma_{y_{t-L}}} = \frac{\Sigma(y_t - \bar{y}_t) \cdot (y_{t-L} - \bar{y}_{t-L})}{\sqrt{\Sigma(y_t - \bar{y}_t)^2 \cdot \Sigma(y_{t-L} - \bar{y}_{t-L})^2}}, \tag{6}$$

where $r_{t,t-L}$ – correlation coefficient for lag L,
n – number of time series levels,
L – time series delay,
$cov_{y_t,y_{t-L}}$ – covariance coefficient,
σ_{y_t} – standard deviation of the original time series,
$\sigma_{y_{t-L}}$ – lagged sample standard deviation,
y_t – current value of the original time series,
y_{t-L} – current sample value with lag,
$\bar{y}_t$– mean of the original time series,
$\bar{y}_{t-L}$ - sample mean with lag L [7].

If the series is affected by any long-term external factors, then this leads to the appearance of trends in the series and a cyclical component, which allows the ACF to be detected. Based on the type of the ACF correlogram, the following conclusions can be drawn:

- if the expressed maximum of the ACF correlogram turns out to be for the lag $L = k$, then the time series contains a cyclic component with a period of k;
- if the correlogram of the ACF does not have pronounced maxima, then the series does not contain a trend and a cyclical component, and a random component dominates in it [8].

Figure 2 shows the ACF of the investigated time series, which contains a break after 4 lags.

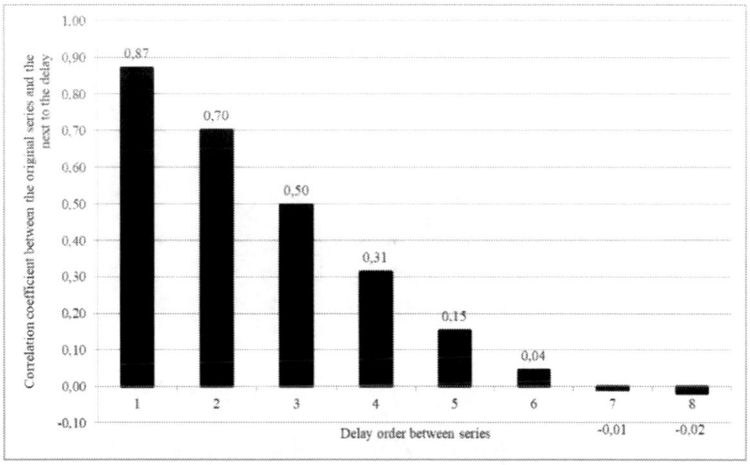

Fig. 2. Autocorrelation function of the time series of electricity consumption

Another way of preliminary estimation of the stationarity of the time series is the private autocorrelation function. Partial correlation coefficients between two random variables with a fixed third characterize the tightness of the relationship between these two values, cleared of the influence of the third variable. Therefore, if the paired correlation coefficient between the same two random variables turned out to be greater than the corresponding partial coefficient, then it is concluded that the third fixed value enhances the relationship between the studied values, i.e. the higher value of the pair coefficient is due to the presence of the third value. A lower value of the pair correlation coefficient in comparison with the corresponding partial indicates a weakening of the relationship between the studied values due to the action of the fixed value [9]. Figure 3 shows the PACF graph of the investigated time series, which contains a break after 3 lags.

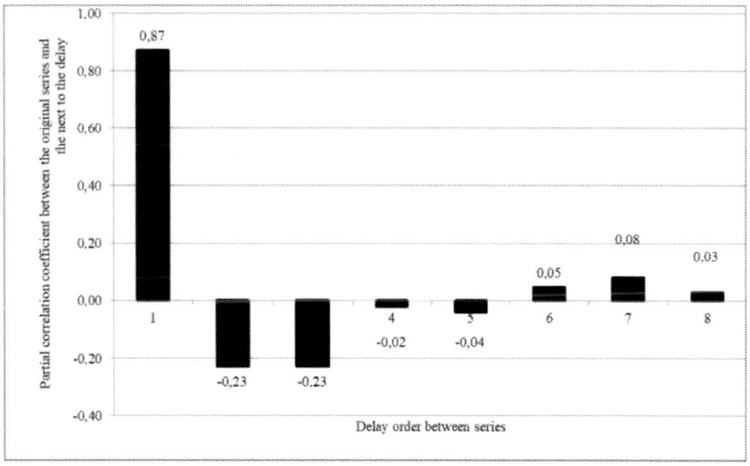

Fig. 3. Private autocorrelation function of the time series of electricity consumption

Based on the time series plots, ACF and PACF and theoretical data on the behaviour of the AR, MA and ARMA model, it can be concluded that the proposed model will have the form ARIMA (p,0,q) or ARMA (p,q). Parameters p and q are selected experimentally.

4. Predicting and evaluating the accuracy of the ARIMA model

Analysis of the time series graph indicates stationarity and the order of integration $d = 0$, the ACF gives the assumed range of coefficients of the AR-component of the model $p = 0 \ldots 4$, the analysis of the PACF gives the range of coefficients $q = 0 \ldots 3$. It is necessary to build a forecast for January 8, 2021. When assessing the predictive ability of the model, the parameter of the root-mean-square estimate RMSE was used (7):

$$RMSE = \sqrt{\sum_{i=1}^{n} \frac{(y_i - \hat{y}_i)^2}{n}}, \tag{7}$$

where y_i– actual value of the time series,
$\hat{y}_i$– time series predicted value,
n – time series length [10].

Table 2 shows an estimate of the possible combinations of p, d, q and the RMSE [11] coefficients of the constructed models.

Model type	RMSE
ARIMA(3,0,1)	0.0641
ARIMA(3,0,2)	0.0661
ARIMA(3,0,3)	0.0646
ARIMA(4,0,1)	0.0679
ARIMA(4,0,2)	0.0644
ARIMA(4,0,3)	0.0622

Table 2. Assessing the accuracy of alternative ARIMA models

The ARIMA model (4,0,3) has the best predictive power, as suggested by ACF and PACF charts.

5. Conclusion

The management of the electricity consumption schedule at large enterprises in Kazakhstan with various types of production is an urgent task. To predict the consumption of electric energy of ArcelorMittal Temirtau JSC on the basis of retrospective data, an adaptive model of the ARIMA type was developed, built according to the Box-Jenkins methodology. The model is highly flexible, and the disadvantages associated with its rather low versatility are compensated by the presence of detailed time series of power consumption of an energy-intensive metallurgical plant. Its advantages include the fact that for its construction it is sufficient to use the data of only the original time series.

It was found that for the forecast horizon of power consumption per day with a sampling step of one hour, it is optimal to consider a sample with hourly data for the week proceeding the day under study. The dynamics of changes in the power consumption of the electric load graph, the gradual attenuation of the ACF and PACF allows us to conclude about the stationary nature of the power consumption process. When analysing the weekly profile of electrical load, a functional dependence of time data with a delay of up to 4 orders of magnitude was revealed with a confidence interval of 95% (high correlation coefficient with a delay: at 1 hour - 0.8 and 2 hours - 0.7; significant correlation coefficient with a delay in: 3 hours - 0.5 and 4 hours - 0.31).

The ARIMA model in the presence of an initial time series can be applied to predict the power consumption of large energy-intensive industries of the mining and metallurgical complex, taking into account the fact that there is a detailed methodology, following which you can choose a model that is most suitable for each specific enterprise. The models have a very clear mathematical and statistical basis, which makes them one of the most promising of the whole variety of options for predicting the power consumption of large enterprises with various types of production. For this model, the weekly profile of the electrical load in winter was taken as retrospective data. In the future, it is planned to consider the monthly profile of the electrical load. It will also be possible to review data for other seasons, identify the seasonal component and obtain a SARIMA model.

6. References

[1] Kazantsev V. P.; Petrochenkov A. B.; Romodin A. V. & Khoroshev N. I. (2011). Some aspects of technology of use of electrical objects on the basis of methods of short-term forecasting of technical condition. Russian Electrical Engineering, Vol. 82, No. 11, pp. 600–606.

[2] Kuntush, Y. V.; Breido, J. V. & Zyuzev, A. M. (2020). Developing a mathematical model of a horizontal looper taking into account the features of a steel strip. EAI Endorsed Transactions on Energy Web, Vol. 7, No. 28, E11.

[3] Taylor J. W. & McSharry P. E. (2008). Short-Term Load Forecasting Methods: An Evaluation Based on European Data. IEEE Transactions on Power Systems, Vol. 22, pp. 2213-2219.

[4] Mindaugas C. (2020). Lithuanian electricity market price forecasting model based on univariate time series analysis. Energetika, Vol. 66, No. 1, pp. 39–46.

[5] Kaytez F.; Taplamacioglu M. C.; Cam E. & Hardalac F. (2015). Forecasting electricity consumption: A comparison of regression analysis, neural networks and least squares support vector machines. International Journal of Electrical Power & Energy Systems, Vol. 67, pp. 431–438.

[6] Lee N. C.; Leal V. M. S. & Dias L. C. (2018). Identification of objectives for national energy planning in developing countries. Energy Strategy Reviews, Vol. 21, pp. 218–232.

[7] Lopez M.; Valero S.; Rodriguez A.; Veiras I. & Senabre C. (2018). New online load forecasting system for the Spanish Transport System Operator. Electric Power Systems Research, Vol. 154, pp. 401–412.

[8] Jimenez J.; Donado K. & Quintero C. G. (2017). A methodology for short-term load forecasting. IEEE Latin America Transactions, Vol. 15, No. 3, pp. 400–407.

[9] Chinnathambi R. A.; Mukherjee A.; Campion M.; Salehfar H.; Hansen T. M.; Lin J. & Ranganathan P. (2019). A multi-stage price forecasting model for day-ahead electricity markets. Forecasting, Vol. 1, pp. 26–46.

[10] Breido J. V.; Kalinin A. A. & Lissitsyn D. V. (2018). Algorithms of energy efficient control of electric technological complex for autonomous heat supply. EAI Endorsed Transactions on Energy Web, Vol. 18, No. 19, e2.

[11] Kim B. S. & Kim T. G. (2019). Cooperation of simulation and data model for performance analysis of complex systems. International Journal of Simulation Modelling, Vol. 18, No. 4, pp. 608–619.

DOI: 10.2507/32nd.daaam.proceedings.072

STRATEGIC DEVELOPMENT OF MARITIME RELATED INDUSTRIES: THE ROLE OF MARITIME CLUSTER RESEARCHES

Kaidi Nommela & Kati Korbe Kaare

This Publication has to be referred as: Nommela, K[aidi] & Korbe Kaare, K[ati] (2021). Strategic Development of Maritime Related Industries: The Role of Maritime Cluster Researches, Proceedings of the 32nd DAAAM International Symposium, pp.0500-0507, B. Katalinic (Ed.), Published by DAAAM International, ISBN 978-3-902734-33-4, ISSN 1726-9679, Vienna, Austria
DOI: 10.2507/32nd.daaam.proceedings.072

Abstract

Maritime clusters have an important impact on the development of regional industry. A well-functioning maritime cluster can enhance social and economic benefits and innovation on entire surrounding region. The European Union has placed cluster concept to the centre of region's strategic policy planning. Joint implementation of the maritime cluster concept is a way to meet challenges and achieve strategic development goals to increase the competitiveness of industries in a region. At the same time, cluster concept can have different meanings and there are different ways to analyze maritime cluster and it's positive impacts. Researches of maritime cluster related studies in the same area are using different approaches and frameworks. A coherent approach to cluster research has not yet been developed for the maritime sector; such research is incomparable and cannot be used as input for integrated policymaking. The aim of this study was to examine different approaches and monitoring frameworks that describe maritime clusters in the Baltic Sea Region which maritime activities make a significant contribution to the EU's maritime economy. The results of the study emphasize the need to develop a coherent framework of indicators to research and monitor maritime cluster performance as a basis for regional strategic policymaking.

Keywords: maritime industries; cluster performance; policymaking; Baltic Sea Region.

1. Introduction

Maritime sector have major impacts on the development of maritime related industries in the Baltic Sea Region (BSR) with considerable contribution to the entire economy of the region. Traditional maritime activities and emerging new areas are determining the regional competitiveness and international trade [1]. Maritime sectors offer a significant opportunity for economic growth and the development of the regional industries that rely in particular on cluster concept and innovation driven by green revolution and digital development. The digitalization process has changed the development of the economy through digital technologies affecting the business operations [2]. Enterprises together with regional policymakers are searching for opportunities to improve their performance in order to increase the competitive advantage of the region [3]. In addition, special empathizes in the BSR has also placed on sustainability goals, it is important to plan and be prepared for the future to support maritime sectors as a whole in the European Union (EU), incl. in the BSR [4].

The EU's Integrated Maritime Policy [5] sets strategic maritime objectives for the Union, i.a. maximizing the sustainable use of seas, building knowledge and innovation, improving the quality of life in coastal regions, etc. The Policy identifies maritime cluster as one of the most important instruments for achieving these goals. Although, maritime clusters in the EU are separate clusters, which often compete with each other, joint implementation of the maritime cluster concept is a way to meet the challenges and to increase the competitiveness of the region [6]. In addition, the concept of cluster in policymaking provides an opportunity to achieve jointly the global environmental goals for greener and more sustainable shipping sector as well as EU's target to develop prosperous maritime and related activities in an environmentally sustainable way.

According to Ketels [7], there are different ways to analyze maritime cluster and to explore its advantages and relatedness. Several authors have indicated that cluster concept can have different meanings and it may vary according to context of the study, economic sector, geographical scope, dimensions of a study, etc. [8], [9], [10]. Among cluster concept researchers, there is a well-known fact that common definition of a cluster has not yet been agreed [8], [11], [12], [13] and the debate on what constitutes a cluster is an on-going process in economic research [14]. Consequently, maritime cluster related studies of the same region use different approaches and frameworks for researching and monitoring maritime clusters and these studies are therefore incomparable and cannot be used as a basis for common policymaking to shape strategies and joint goals.

The aim of this study is to examine different methods and monitoring frameworks that describe maritime clusters in the BSR. This study identifies the need for a common methodology to maritime cluster research and monitoring frameworks that could help to achieve joint strategic goals and increase the competitiveness of the region through integrated maritime clusters and policymaking. A comparative analysis was carried out to establish the most relevant aspects and characteristics, which differentiate maritime cluster related studies. Analysis was applied to the maritime cluster related studies of seven Baltic Sea countries: Denmark, Estonia, Finland, Germany, Latvia, Lithuania and Sweden. The focus was to determine the following aspects of the studies revealed from literature review: the main aim of the study, applied cluster definition, study object, methods and main focus area of the study, and to identify differences in approaches and implementations of these aspects in the selected studies.

2. Background – The concept of the maritime cluster in European Union's policymaking

In the EU's economy, the implementation of a cluster concept has been an important way of analyzing economic developments and planning future strategies. However, the implementation of the concept is multifaceted, depending on the approach of implementation and the chosen economic theory. Different authors apply different theories in understanding the cluster concept, and the multidirectional knowledge and understanding has also reached to policymaking. The following is a brief background overview of the development and implementation of the maritime cluster concept in EU's policymaking.

In economics, clusters became known in the late 20th and early 21st centuries, when globalization had peaked and economic theories introduced ways to promote geographically concentrated communities to strengthen regional development. So far, there is still no universally recognized standard definition of a cluster [8], [11], [12], [13] and the debate on what constitutes as a cluster is an on-going process [14]. One of the leading academic authorities in this area [12] Professor Michael E. Porter defined clusters as "*geographically proximate group of interconnected companies and associated institutions in a particular field, linked by commonalities and complementarities. The geographic scope of clusters ranges from a region, a state, or even a single city to span nearby or neighbouring countries.*" [15] Since Porter, several authors has built the definition of a cluster on Porters' concept with own additions, regardless of the study field, aim etc.

The implementation of a cluster concept as a basis for policy has been supported and valued by international associations: the European Union [5], [16], [17]; the World Bank [18]; the Organisation for Economic Co-operation and Development [19], etc. After the EU launched an integrated policy approach to maritime related issues with a focus on cluster concept in the early 2000s, maritime cluster related researches have been carried out throughout the EU, including in the BSR, through various programs, projects and other actions. Although several joint cluster platforms have been set up, and several researches and projects have been carried out to study maritime clusters in the region, they all take a relatively different approach to the concept of a maritime cluster. As a result, there is no common methodology for researching and monitoring maritime clusters that can be used as a basis for creating political strategies. Instead, the research process starts each time from the beginning according to a specific goal, and leads to single-purpose work.

The concept of a cluster first appeared in EU's maritime policy documents in the early 2000s. The European Commission's (EC) Green Paper on towards a future Maritime Policy for the Union (2006) identified clustering as one way to increase the competitiveness of maritime sector, given its high degree of interaction [20]. The Paper emphasized the need for intra-sectoral knowledge sharing, joint research and innovation activities, harmonization of education and training, sharing of innovative technologies and methods, and joint marketing and advertising activities, especially in sectors with a complex supply chain such as logistics and maritime sector. The Integrated Maritime Policy Document (2007) placed maritime cluster at the centre of EU's maritime policy [5]. The Policy encouraged the creation of cross-sectoral clusters and regional maritime centres of excellence and promoted the creation of European network of maritime clusters. The EC Staff Working Document on Maritime Clusters (2007) emphasized the need to integrate regional and national clusters with maritime and environmental aspects in order to contribute better quality and higher standards of European maritime products and services [16].

In 2009, the EC compiled an overview of European maritime clusters with the aim of gathering information on the size, specialization and focus of maritime sectors or clusters in the EU [21]. The paper defined cluster as a part of the maritime sector or region, which was divided into the traditional maritime sector, coastal tourism and recreational activities, and fisheries. This paper highlighted the need to analyze cluster policy and the success of the sector to provide an overview of how the implemented cluster policy has affected the development of the maritime sector. Since this project, the preparation of maritime cluster analysis became popular in the EU Member States, incl. in the BSR countries, although a number of different approaches were used.

The EU's first Smart Guide to Cluster Policy [22] was issued in 2016. The Guide explained the definition of clusters, the economic relevance of clusters and cluster policy, the process of creating a cluster policy, possibilities for monitoring and evaluation of cluster policy, etc. The Guide defined cluster as "concentration of economic activities in groups of related industries in a specific location that are connected through multiple linkages and spill-overs" and highlighted five main characteristics of a cluster: critical mass, related industries, location, linkages and the fact that vast majority of cluster cases are not 'created' [22]. Although the Guide covered all the necessary topics related to cluster policy, it did not contain precise instructions for identifying the existence and the performance of a cluster, but referred instead to the database of the European Cluster Observatory [23], where e.g. only 25 maritime related institutions from Estonia are represented (as of February 2021).

The EU's new Smart Guide to cluster policy was issued in 2020 with the focus on monitoring and evaluating a cluster policy [24]. This Guide defined cluster using Porter's [25] definition: "*geographic concentrations of interconnected companies, specialized suppliers, service providers, firms in related industries, and associated institutions ... in particular fields that compete but also cooperate*". The Guide emphasized the complex system of clusters, which rely on formal and non-formal interactions and the multiple variations of cluster policy forms, and highlighted the fact, there is no simple guide to monitor and evaluate cluster policies and each situation needs separate guidance how to deal with challenges in different scenarios [24].

The concept of cluster entered the European maritime policy documents in the early 2000s and since then, its implementation has taken place in different directions. Although the EU has encouraged Member States for the last twenty years to implement the cluster theory and concept in policymaking, different approaches and methodologies have made the process diverse and the results incomparable. The same has happened in the Baltic Sea region, where regional specificities and joint challenges need a coherent approach to regional policymaking, supported by the integrity of the implementation of the cluster concept.

2.1 The Baltic Sea Region's maritime sector in the EU

The Baltic Sea is located in the northern part of Europe and it is one of the densest shipping areas in the world, with around 2,000 ships operating at the same time [26]. The EU's countries around the Baltic Sea are Sweden, Finland, Estonia, Latvia, Lithuania, Poland, Germany and Denmark (see Figure 1, the gross weight of Germany's goods includes only the Baltic Sea part). There are more than 400 seaports on the Baltic Sea, of which more than 90 are of international importance.

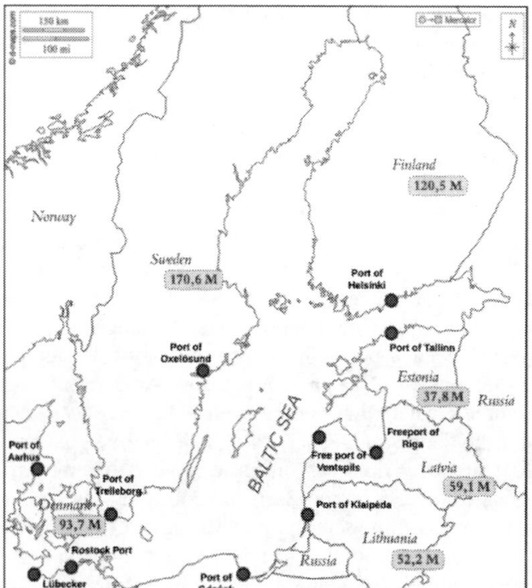

Fig. 1. Map of the Baltic Sea Region with states, major ports and gross weight of goods handled in all ports in 2019 (million tons). Source: d-maps.com [27], Eurostat [28], amended by the authors.

Shipping and other related maritime activities are essential for the BSR development and international trade [1]. The development of the maritime sector in the BSR has always been an important part of the entire EU maritime sector. The BSR is a considerable cargo transit corridor for north-south and east-west traffic. In 2019, the total gross weight of goods handled in BSR main ports was over 586 million tons, which accounted almost 15% of all European cargo traffic through seaports in the same year. Compared to previous ten years (2010–2019), the BSR share of cargo traffic through the main seaports has remained stable (see Figure 2) [29]. In 2019, the total number of passengers transported to and from BSR main ports (both national and international) was over 100 million people, which accounted for 42% of all European passenger traffic through seaports in the same year. Compared to previous ten years (2010–2019), the BSR share of passenger traffic through the main seaports has also remained stable (see Figure 2) [30].

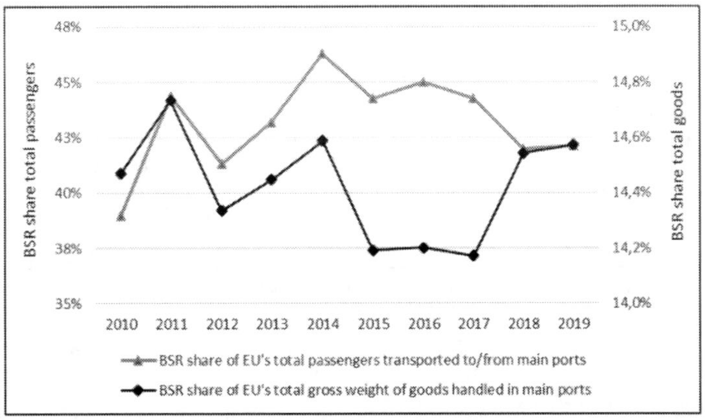

Fig. 2. BSR share of EU's total goods and passengers transported in main ports in 2010 – 2019
Source: Eurostat [29], [30], compiled by the authors

The BSR maritime sector accounts for a significant share of EU maritime affairs. Therefore, it is important that EU's policymaking process takes into account the challenges and development opportunities of the region. The application of the cluster concept in regional policymaking can be of greater value in regions where the development is affected by similar trends. Consequently, it is necessary to identify different methods of defining the existence and describing the current state and performance of maritime clusters in the BSR to increase the competitiveness through integrated maritime clusters and policymaking.

3. Methods

The research design of this work is based on the premise that there are several possibilities and different ways to approach a research of maritime cluster, depending on the research object, theme, study objective, methodology, etc. However, to achieve the goals of a coherent policy and to increase the competitiveness of the region, a common approach and monitoring framework to maritime clusters is important. In this work, we compare maritime cluster related studies conducted in the BSR in order to examine different approaches and explore the need for a common methodology to maritime cluster research, which can be a basis for policymaking in maritime sector. Nine maritime cluster related studies were included to this analysis. The maritime clusters related studies in the BSR selected for this analysis are:

- Competencies in the Danish Maritime Cluster: A benchmarking-analysis [31],
- Maritime Cluster Research of Estonia [32],
- The Finnish Maritime Cluster 2004 [33],
- The Finnish Maritime Cluster 2008 [34],
- The Finnish maritime cluster towards the 2020s [35],
- A Differentiation Framework for Maritime Clusters: Comparisons across Europe [36],
- Development of the Latvian Maritime Policy; A Maritime Cluster Approach [37],
- Lithuanian maritime sector's clustering economic impact evaluation [38],
- Maritime Clusters in Västra Götaland [39].

Based on literature review of cluster economic concept, a comparative analysis framework to distinguish main differences between selected maritime cluster related studies was created. A combination of dimension of cluster research was used for the analysis. This framework included different comparison dimensions: 1. main aim; 2. cluster definition; 3. study object; 4. main methods; 5. main focus area of the study. The last contained main characteristics of a cluster proposed by different authors in literature review: workforce (workforce structure and needs); relationships between members; competition; competitiveness; geographical scope; geographical orientation; structure of members; number of members [15], [25], [40], [41].

4. Comparative analysis of maritime cluster related studies

A comparative analysis was carried out in order to examine different approaches and monitoring frameworks used in maritime clusters related studies in the BSR. The aim was to identify the main differences and links between the used approaches and the need for a common methodology to maritime cluster research and monitoring frameworks that could be valuable input to policymaking and evaluation in the region. A comparative analysis allows to find out differences in approaches across the main dimensions of cluster analysis and thus explain the impact of the used approaches on the results of the researches. Overview of the results of the comparative analysis is shown in the following table (see Table 1).

Comparison dimensions	Differences between maritime cluster related studies
Main aim	Different in all studies, incl.: • describe the current state and extract main characteristics; • assess future prospects and growth areas, evaluate the economic impact of the cluster; • outline a strategy or plan to further developments of the cluster; • compare the cluster with international maritime clusters, define maritime cluster networks, ensure the continuity of research.
Cluster definition	• Porter's definition (1998, 1999); • authors' own definition (mostly, based on different theories and authors).
Study object	• In all studies: shipping and shipping related areas, maritime industry (incl. shipbuilding and repair, equipment industry), ports and port operators; • in most studies: maritime related services (transport services, companies related to shipping, maritime logistics, classification institutions, finance and insurance, etc.), public sector (authorities, education, science); • in few studies: fishing and aquaculture, maritime tourism and recreation, offshore technology and marine energy, maritime construction, blue (marine) biotechnology.
Main methods	• Literature review, incl. internet research, reports, books and scientific articles; • quantitative analysis of maritime related companies' data, and qualitative analysis, incl. interviews with actors, workshops, seminars, etc.; • benchmarking-analysis.
Main focus area	• In most studies: quantitative overview of the cluster, incl. contribution to national economy, number of employees, turnover, number of ships, etc.; • in few studies: current state and overview of the cluster, incl. historic overview, development prospects, future opportunities, growth prospects of the cluster.

Table 1. Overview of the results of the comparative analysis

All analysed studies had different main aims and focus in certain respects. Majority of studies main aim was to study and describe the current state of maritime cluster in the region based on an overall overview or focusing on specific characteristics. Describing the current situation was generally a common first step in examining a cluster. Multiple studies focused on identifying future possibilities and prospects and providing an analysis for the future growth areas. Other studies were compiled with the aim to outline a strategy or to initiate an action plan for further development strategy or maritime policy. Fewer studies were conducted to describe and evaluate the economic and social impact or value of a cluster. Least studies were conducted to compare maritime cluster of a region with international maritime clusters, to define maritime cluster networks or to ensure the continuity of research (i.e. were a follow-up research to previous cluster researchers).

In defining the cluster, the analysed studies used mainly two approaches: 1. the definitions issued by M. E. Porter in the late 1990s; 2. the authors formulated an appropriate definition themselves (usually, based on different theories and authors). The authors defined cluster mainly as a unit or concentrations of companies that uses maritime related know-how and that interaction are proven to produce advantages. Other authors included to the definition other actors (in addition to companies), such as public authorities, organisations and research institutes. One study defined cluster within a specific regional area, and another defined cluster as geographically concentrated economic activities that are interconnected. As a result, the interpretation of a cluster varies widely in different studies.

Traditional maritime areas have been included to the research object in all analysed studies. These traditional areas include shipping with related fields, ports and port operators and maritime industry, such as shipbuilding and ship repair and marine equipment industry. Most studies also considered as a research object maritime related services, such as maritime logistics, transport services, classification institutions, finance and insurance, etc. Half of the studies included to the research object public sector (authorities, education and science institutions), fishing and aquaculture, maritime tourism and recreation and offshore technology, incl. marine energy. The inclusion of the latter depended in particular on the development of the sector in a specific region. Only few studies included maritime construction and marine (blue) biotechnology to the research object.

The level of description of the research methodology differed in all analysed studies. Most studies referred to literature review and different quantitative and qualitative methods as main data collection methods. The sources of literature review were mainly internet based, reports and scientific literature. Economic data from companies were mainly described as the source of quantitative analysis and interviews with actors and workshops as the source of qualitative analysis. Some studies mentioned benchmarking as the main method of analysis, but the majority did not mention any specific method.

The main focus areas of the studies were also different and depended mostly on the main aim of the study. In most cases, the focus area was a descriptive overview of the current state and historic development of a cluster, or quantitative overview of a cluster's contribution to the national economy, incl. number of members, number of employees, total turnover, number of ships, etc. Mostly used indicators to analyse focus areas were geographical scope and competitiveness of a cluster, which was generally presented in a descriptive way. Workforce and geographical orientation were also used as analysis indicators, but in different level of detail. Relationships between members, structure and the concept of intra-cluster competition was the least analysed indicators in these studies.

The comparison of maritime cluster related studies of the BSR countries showed that there are different approaches to maritime cluster research and monitoring frameworks depending on the main aim of the study, applied cluster definition, research object, methods and focus areas. Majority of studies were carried out to describe the current state of maritime cluster in the region based on an overall overview. Other studies were designed to explore a specific narrow topic. In defining the concept of a cluster, the Porter's definitions or the authors' own interpretation of various explanations of the concept were mainly used. Studies that used Porter's definition took more similar approach to the cluster concept, but the studies that used the authors' own definitions had broader difference in the interpretation of the concept. The research methods varied greatly and did not depend on the purpose or focus of the study. At the same time, the focus area and analysed characteristics depended on the main topic of the study. Geographical scope and competitiveness of a cluster were the most analysed indicators, although they were examined in different ways and mainly descriptively.

All these differences in the approaches and frameworks hinder the use of cluster research in international regional policy-making. The diverse application of the aspects of cluster research described in economic theory makes the results of cluster research incomparable and, consequently, unsuitable input for joint policymaking in the region.

5. Conclusion

In the BSR, maritime clusters have an important impact on the development of regional industry. Although, clusters often compete with each other, they also face similar challenges and opportunities at the same time, which can be addressed through joint strategic policymaking. The EU has set common policy goals for the region to enhance the interaction of maritime clusters and policymaking. Although a number of maritime cluster studies have been carried out in the BSR over the years, these studies have used multiple theoretical and methodological approaches, as there is no right or wrong way to apply cluster concept. The results of this study revealed that there are different approaches and monitoring frameworks to maritime cluster research in the region and that these researches cannot be considered as comparable for policymaking. The aim of most researches was to study and describe the current state of maritime cluster in the region based on an overall overview or focusing on specific characteristics. Some studies also had the goal to identify future possibilities and growth areas of the maritime cluster.

As the main purpose of the studies differed, so did the applied cluster definitions, methods and focus areas and analysed characteristics. Many authors defined cluster with Porter's definition from the late 1990s or used their own interpretations of different definitions. Studies that used similar definition (e.g. Porter's definition) took more similar approach to the cluster concept, but the studies that used own interpretations had broader difference. The focus areas of the studies and analysed characteristics mostly depended on the purpose of the work. Although literature review highlights main characteristics describing a cluster, the analysed studies gave an overview of those characteristics that were related to the purpose of their work and did not address other characteristics of a cluster. As a result, there was a lack of a comprehensive understanding of the region's maritime clusters on the basis of cluster characteristics.

One of the dimensions that were similar in most analysed studies was the object of research that depended on the geographical location of a cluster or the national economy rather than on the research approach. As maritime sector is one of the traditional sectors in coastal nations' economy, which covers areas of activity with a long history, thus maritime cluster related studies often examine similar maritime activities as a research object. Most of the studies covered maritime activities, such as shipping, ports and port operators, maritime industry (incl. shipbuilding and ship repair) and maritime related services (incl. maritime logistics, transport services, classification institutions, finance and insurance). Less was studied public sector (incl. authorities, education and science institutions), fishing and aquaculture, maritime tourism and recreation and offshore technology (i.a. marine energy) and marine (blue) biotechnology.

Several joint cluster platforms have been set up, and multiple researches and projects have been carried out to research maritime clusters in the EU and in the BSR. However, due to the wide range of interpretations of the cluster concept, different approaches are still in use. This makes it difficult to implement the cluster concept and its potential benefits and opportunities in policymaking due to the lack of a comprehensive understanding of the current situation and future opportunities of the maritime clusters in the region. Consequently, the authors of this work emphasize the need for a coherent approach to maritime cluster research in the regions where common strategic policy objectives are set and common challenges and opportunities lie ahead.

Although the results of this work have shown that it is necessary to create a common approach and framework for conducting maritime cluster related studies to ensure their comparability, the main aim and focus of the studies are always the decision of the researchers and/or clients. A common framework would not prevent the different aims and focus themes of the studies, but would ensure the comparability of the maritime cluster dimensions and research components, their use as input for policymaking and the possibility of follow-up researches.

As a follow-up to this study, the authors plan to carry out a comparative analysis of the region's maritime clusters based on previously developed monitoring frameworks in order to compare the results of different analysis techniques. Such analysis would provide necessary and comparative overview of the BSR maritime clusters and serve as a basis for identifying competitive advantages of clusters, development trends and future scenarios affecting the region, which will be an important input for cluster policymaking. In addition, the authors also have started to develop a common framework for defining and identifying maritime clusters, which can be used as a basis for analyzing maritime related industrial clusters.

6. References

[1] Matczak, M. (2018). Exploring the future of shipping in the Baltic Sea. Baltic Lines, Quo Vadis. Coherent Linear Infrastructures in Baltic Maritime Spatial Plans. Available from: https://vasab.org/wp-content/uploads/2018/08/20180730_FutureShippingQuoVadis.pdf Accessed: 2021-05-20

[2] Kokolek, N., Jakovic, B. & Curlin, T. (2019). Digital Knowledge and Skills – Key Factors for Digital Transformation, Proceedings of the 30th DAAAM International Symposium, pp.0046-0053, B. Katalinic (Ed.), Published by DAAAM International, ISBN 978-3-902734-22-8, ISSN 1726-9679, Vienna, Austria. DOI: 10.2507/30th.daaam.proceedings.006

[3] Stojkic, Z. & Bosnjak, I. (2019). An Overview of Performance Measurement Methods in SMEs, Proceedings of the 30th DAAAM International Symposium, pp.0518-0524, B. Katalinic (Ed.), Published by DAAAM International, ISBN 978-3-902734-22-8, ISSN 1726-9679, Vienna, Austria. DOI: 10.2507/30th.daaam.proceedings.070

[4] Routa, T. (2016). Evolution of the maritime cluster in a changing world. The maritime cluster in the Baltic Sea region and beyond. BSR Policy Briefing 1/2016, Centrum Balticum Foundation. ISBN: ISSN 2342-3153

[5] European Commission. (2007a). Communication from the Commission to the European Parliament, the Council, the European Economic and Social Committee and the Committee of the Regions. An Integrated Maritime Policy for the European Union Brussels, 10.10.2007 COM(2007) 575 final

[6] SmartComp. (2012). Maritime cluster analysis on the Central Baltic region. SmartComp Research Report No 1, University of Turku, December 2012. Available from: https://www.utu.fi/sites/default/files/media/PEI_SmartComp_ResearchReport1.pdf Accessed: 2021-03-15

[7] Ketels, C. (2017). Cluster Mapping as a Tool for Development. Institute for Strategy and Competitiveness. Harvard Business School. Available from: https://www.hbs.edu/ris/Publication%20Files/Cluster%20Mapping%20as%20a%20Tool%20for%20Development%20_%20report_ISC%20WP%20version%2010-10-17_c46d2cf1-41ed-43c0-bfd8-932957a4ceda.pdf Accessed: 2021-03-21

[8] Cortright, J. (2006). Making Sense of Clusters: Regional Competitiveness and Economic Development. A Discussion Paper Prepared for the The Brookings Institution Metropolitan Policy Program. The Brookings Institution.

[9] Brett, V. & Roe, M. (2010). The potential for the clustering of the maritime transport sector in the Greater Dublin Region. Maritime Policy & Management, 37:1, 1-16. DOI: 10.1080/03088830903461126

[10] Pinto, H., Cruz, A. R., Combe, C. (2015). Cooperation and the emergence of maritime clusters in the Atlantic: Analysis and implications of innovation and human capital for blue growth. Marine Policy, Volume 57, July 2015, Pages 167-177. Available from: https://doi.org/10.1016/j.marpol.2015.03.029 Accessed: 2021-02-17

[11] Deloreux, D., Shearmur, R. (2009). Maritime clusters in diverse regional contexts: The case of Canada. Marine Policy 33 (2009) 520–527. DOI: 10.1016/j.marpol.2008.12.001

[12] Han, C-H. (2006). Comparative Analysis on World's Major Maritime Cluster. The Journal of Maritime Business, No.81, pp. 89-114 Dec 2006

[13] Lagoudis, I., Madentzoglou, E. M., Theotokas, I. N., Yip, T. L. (2019). Maritime Cluster Attractiveness Index. Maritime Business Review, vol. 4 No. 2, 2019 pp. 169-189. DOI: 10.1108/MABR-11-2018-0044

[14] Andersson, T., Schwaag-Serger, S., Sörvik, J., Hansson, E. W. (2004). The Cluster Policies Whitebook. IKED – International Organisation for Knowledge Economy and Enterprise Development, Holmbergs August 2004

[15] Porter. M. E. (2000). Location, Competition, and Economic Development: Local Clusters in a Global Economy. Economic Development Quarterly 14:15-34. DOI: 10.1177/089124240001400105

[16] European Commission. (2007b). Commission Staff Working Document, Maritime Clusters. Brussels, 17.10.2007 SEC(2007) 1406

[17] European Union. (2010). Clusters and clustering policy: a guide for regional and local policy makers. INNO Germany AG, ISBN: 978-92-895-0506-2

[18] International Trade Department. (2009). Clusters for Competitiveness. A Practical Guide & Policy Implications for Developing Cluster Initiatives. The World Bank, February 2009

[19] OECD (2005). Business Clusters: Promoting Enterprise in Central and Eastern Europe. OECD Local Economic and Employment Development (LEED) Programme, august 2005, ISBN: 9264007105
[20] European Commission. (2006). Green Paper. Towards a future Maritime Policy for the Union: A European vision for the oceans and seas. Brussels, 7.6.2006 COM(2006) 275 final Volume II – ANNEX
[21] European Communities. (2009). The role of maritime clusters to enhance the strength and development in European maritime sectors, executive summary. Office for Official Publications of the European Communities, Available from: https://op.europa.eu/en/publication-detail/-/publication/2d594b8b-10fa-441f-8d81-7afea1dfca47 Accessed: 2021-03-05
[22] European Commission. (2016). Smart Guide to Cluster Policy. Directorate-General for Internal Market, Industry, Entrepreneurship and SMEs. DOI: 10.2873/48105
[23] European Cluster Observatory. (2021). Cluster Observatory, data. Available from: http://www.clusterobservatory.eu/data Accessed: 2021-02-10
[24] European Union. (2020). Smart Guide to cluster policy monitoring and evaluation. European Commission's Directorate-General for Internal Market, Industry, Entrepreneurship and SMEs, COSME programme, Publication Office of the European Union
[25] Porter, M. E. (1998). Clusters and the New Economics of Competition. Harvard Business Review, November–December 1998
[26] Aro, E., Rytter, N. G. M., Itälinna, T. (2020). Maritime industry processes in the Baltic Sea Region, Synthesis of eco-inefficiencies and the potential of digital technologies for solving them. ECOPRODIGI Research Report 2020, Turku School of Economics. Available from: https://ecoprodigi.eu/wp-content/uploads/2020/02/ECOPRODIGI-Research-Report-1-2020-final.pdf Accessed: 2021-03-05
[27] d-map.com. (2021). Baltic Sea map. Retrieved from: https://www.d-maps.com/pays.php?num_pay=182&lang=en Accessed: 2021-05-11
[28] Eurostat. (2019a). Gross weight of goods handled in all ports by direction - annual data. Retrieved 3.07.2021, European Commission. Available from: https://ec.europa.eu/eurostat/databrowser/view/mar_go_aa/default/table?lang=en Accessed: 2021-04-16
[29] Eurostat. (2019b). Gross weight of goods handled in main ports by direction and type of cargo - quarterly data. Retrieved 25.05.2021, from European Commission. Available from: https://ec.europa.eu/eurostat/databrowser/view/MAR_GO_QMC__custom_992283/default/table?lang=en Accessed: 2021-04-16
[30] Eurostat. (2019c). Passengers transported to/from main ports by direction and type of traffic (national and international) - quarterly data. Retrieved 18.05.2021, European Commission. Available from: https://ec.europa.eu/eurostat/databrowser/view/mar_pa_qm/default/table?lang=en Accessed: 2021-04-16
[31] Gammelgaard, B., Sornn-Friese, H., Hansen, J., Jessen, M., Larsen, M. (2013). Competencies in the Danish Maritime Cluster: A benchmarking-analysis. Danish Maritime Cluster (DKMK), ISBN: 978-87-996049-1-3
[32] Portsmuth, R., Hunt, T., Terk, E., Nõmmela, K., Hartikainen, A. (2011). Maritime Cluster Research (Merenduse klasteruuring). Eesti Mereakadeemia toimetised nr 12, Estonian Maritime Academy, ISSN: 1736-207
[33] Viitanen, M., Karvonen, T., Vaitse, J., Hernesniemi, H. (2003). The Finnish Maritime Cluster. Technology Review 145/2003, National Technology Agency
[34] Karvonen, T., Vaitse, J., Hernesniemi, H. (2008). Finland Maritime Cluster 2008 (Suomen meriklusteri 2008). Tekesin katsaus 226/2008, Tekes 2008, ISSN: 1239-758X
[35] Karvonen, T., Grönlund, M., Jokinen, L., Mäkeläinen, K., Oinas, P., Pönni, V., Ranti, T., Saarni, J., Saurama, A. (2016). Finnish maritime cluster towards the 2020s (Suomen meriklusteri kohti 2020-lukua). Ministry of Economic Affairs and Employment, ISBN: 978-952-327-127-2
[36] Monteiro, P., de Noronha, T., Neto, P. (2013). A Differentiation Framework for Maritime Clusters: Comparisons across Europe. Sustainability 2013, 5, 4076-4105. DOI: 10.3390/su5094076
[37] Gailitis, R., Jansen, M. (2012). Development of the Latvian Maritime Policy; A Maritime Cluster Approach. International Journal on Marine Navigation and Safety of Sea Transportation, volume 6, number 2. DOI: 10.1201/b11347-19
[38] Viederyte, R. . (2014). Lithuanian maritime sector's clustering economic impact evaluation. Procedia – Social and Behavioral Sciences 156 (2014) 292 – 297, 19th International Scientific Conference; Economics and Management 2014, ICEM 2014, 23-25
[39] Wenblad, A., Lindegarth, S., Hanning, A. (2012). Maritime Clusters in Västra Götaland. Gothenburg and Tjärnö in October 2012, Available from: https://scanbaltbusiness.com/wp-content/uploads/2013/02/Maritime-Clusters-V%C3%A4stra-G%C3%B6taland2.pdf Accessed: 2021-03-15
[40] Koschatzky, K., Lo, V. (2007). Methodological framework for cluster analyses. Fraunhofer Institute for Systems and Innovation Research ISI. Karlsruhe 2007, ISSN: 1438-9843
[41] Boja, C. (2011). Clusters Models, Factors and Characteristics. International Journal of Economic Practices and Theories, Vol. 1, No. 1, 2011

DOI: 10.2507/32nd.daaam.proceedings.073

MINIMIZING COATING PROCESS SOURCED HYDROGEN EMBRITTLEMENT OF FASTENERS

Ugur Buyuk, Anil Akdogan, Ali Serdar Vanli & Numan Durakbasa

This Publication has to be referred as: Buyuk, U[gur]; Akdogan, A[nil]; Vanli, A[li] S[erdar] & Durakbasa, N[uman] (2021). Minimizing Coating Process Sourced Hydrogen Embrittlement of Fasteners, Proceedings of the 32nd DAAAM International Symposium, pp.0508-0513, B. Katalinic (Ed.), Published by DAAAM International, ISBN 978-3-902734-33-4, ISSN 1726-9679, Vienna, Austria
DOI: 10.2507/32nd.daaam.proceedings.073

Abstract

Fasteners are devices that mechanically fix two or more objects to each other used in many joints. The phenomenon of hydrogen embrittlement can be very common in fastener materials, causing problems in the assembly of fasteners and reducing their mechanical properties. Hydrogen embrittlement means the embrittlement of a metal by diffusible hydrogen due to the absorption of its atoms or molecules by the material. Appropriate plating or coating processes should be used in the surface coating of fasteners to minimize hydrogen embrittlement. In this paper, we examined spring type straight pins as fasteners to be coated to avoid unexpected failures due to hydrogen diffusion such as low mechanical properties. Choosing the right coating type and conditions improved assembly productivity and the performance of spring type straight pins.

Keywords: Fasteners; Coating; Hydrogen embrittlement; Mechanical properties

1. Introduction

Fasteners are hardware tools which mechanically joins two or more objects together. Fasteners usually being used to create temporary joints; that is, joints that can be removed or disassemble without damaging the joining parts. It is overcritical to choose the right fasteners for production. First, it should be suitable for parts and applications. Additionally, there are two important decisions to take at design stage.

- Selecting the most appropriate material,
- Applying the right coating operation.

Designers have a simple and efficient way when selecting materials in the design phase to prevent damage and defect in structural materials throughout the process lifetime and occasionally further on to improve to application life. They must know if a material is susceptible to cracking. In addition, quick decisions or process measures are often obligatory throughout service time, to prevent or delay this type of damage which is hard situation [1].

Sometimes it is not enough to choose a suitable material for mechanical durability and a suitable coating process against corrosion to achieve good performance. It is essential to establish a relation between material and coating process. Hydrogen embrittlement can occur if the material and coating process do not suitable to each other.

Unfortunately, it effects mechanical properties of the material in a negative way. Hydrogen embrittlement means the embrittlement of a metal by diffusible hydrogen due to the absorption of its atoms or molecules by the material. In many studies, hydrogen can affect crack distribution, such as stress corrosion cracking and corrosion fatigue. Three situations are imperative to establish crack initiation that can lead to failure. The first is the existence of hydrogen, the second is the tensile stresses and the last is the sensitivity of the material [2]. The phenomenon of hydrogen embrittlement can be very common in fasteners and cause problems during assembly. The selected proper coating type improved assembling productivity and performance of spring type straight pins.

Spring Type Straight Pin is common fastener which could be used on many applications. There is an international standard numbered ISO 8752 about this hard situ. Spring Type Straight Pin has a special flexibility according to the standard. It can easily fix to hole which has suitable dimension defined by ISO 8752. Figure 1 determines the geometry of spring type straight pin, and the Figure 2 defines proper dimension of the hole. Appropriate plating or coating processes are needed to avoid hydrogen embrittlement if the pins are surface coated. The pins shouldn't be electroplated or phosphate-coated due to the risk of hydrogen embrittlement. If electroplating or phosphate plating is required for corrosion prevention, it is obligatory that the pins be thermally treated immediately after plating to minimize the risk of hydrogen embrittlement. Nevertheless, preventing hydrogen embrittlement is not absolutely guaranteed in such processes [3].

This paper details the negative effects of HE in the materials in their usage lifetime and the importance of selecting proper coating type in improvement of assembling productivity and performance of spring type straight pins. It also consists of an example of how hydrogen embrittlement phenomena can be reduced by the right plating process for fasteners.

2. Hydrogen Embrittlement in Materials

Hydrogen is the most abundant element in the universe, is safe and clean fuel gas of high calorific value, available in infinite quantities or amounts by the electrolysis of water [4]. High strength materials are too sensitive against hydrogen embrittlement when they stand on hydrogen environment [5]. There are three main reasons which are effect directly hydrogen embrittlement. The first is the stress, the second is the material, and the third is the environmental conditions (Figure 1). When the researchers search for material, they need to consider many variables such as the microstructure, the strength of material, its residual stress, and the applied strain rate. Moreover, the surface conditions and the heat treatments are the other variables. Metallic plating and specific precipitates, applied strain ratio and surface condition of a material, processing pressure, temperature and exposure time, concentration hydrogen and amount of hydrogen trap, solution that make a reaction with metals could be considered as the rest variables.

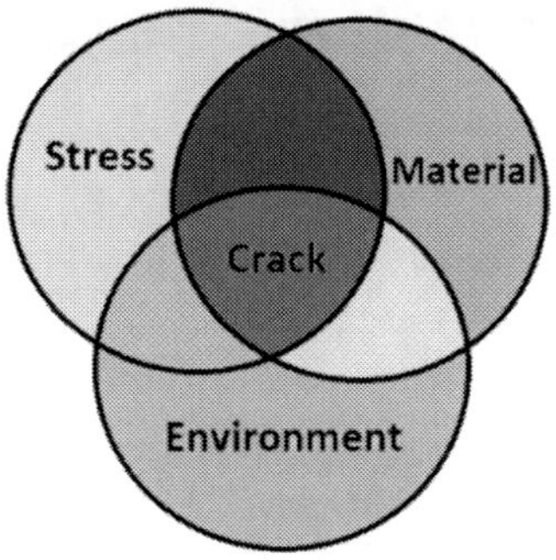

Fig. 1. Three Main Reasons of Hydrogen Embrittlement [6]

Effect of hydrogen embrittling process is the degradation of mechanical properties or delay failure which can occur due to internal residual stresses or applied external stresses. The most mechanical degradation is that normally ductile material become brittle [7]. There are three situations which increase the potential of hydrogen embrittlement. These are high tensile strength, hardness or surface hardening which have could absorbed hydrogen under tensile stress. If the hardness is higher than 320 HV, the operation examination shall be using a test to aware of hydrogen embrittlement [8]. If there is any embrittlement, revision of the production operation will be obligatory such as the inclusion of a treatment process. Of course, hydrogen embrittlement prevention cannot be %100 guaranteed. If a reduced possibility of occurrence hydrogen embrittlement is desired alternative methods should be evaluated. Steels whose strength are more than 1000 MPa is susceptible for HE [9]

To avoid from HE on high strength materials, the main reason of it and its mechanism must be defined clearly. Choosing of a proper material is crucial. The material should be optimum strength and high resistant alloy for HE. Design of part must be properly detailed. Sharp corners and critical point must be avoided and if any there are any residual stresses, they should be eliminated before the operation. The thermal treatment processes should be applied to detach hydrogen which is consumed and can cause any failure occurs.

Heat treatment operation conditions must be arranged according to the process performed on the material [10]. Additional cleaning operation should be applied to eliminate some flake and oxide parts from material with the use of acidic solvent. This solution helps to avoid hydrogen diffusion like mechanical means such as sand and grit blasting. Avoid of HE could be done by proper metallic alloys to the base material and applying protective plating, as well [5]. Some plating methods are vacuum deposited, and some are mechanical. Proper inhibitors also help to prevent from HE. Titanium may reduce the HE sensitivity of a hot stamped boron steel by forming titanium carbide in the material if titanium is suitable in a large batch [11].

When atomic hydrogen occurs in steels and other metals like aluminium alloys it can cause loss of ductility or load carrying capability resulting cracking or catastrophic brittle failures at applied stresses well below the yield strength or even the normal design strength for alloys. This situation often occurs in alloys that do not show important loss in ductility when measured by normal tensile tests. This is usually referred to as hydrogen induced postponed brittle failure, hydrogen stress cracking or hydrogen embrittlement. These can be occurred throughout heat treatment, gas carburizing, cleaning, pickling, phosphating, electroplating, autocatalytic operations and in the service environment due to cathodic plating reactions or corrosion due to the break-down of inappropriate lubricants as well as throughout welding or brazing process. There are some parts such as by machined ground cold-formed and hardening heat treatment processed ones more sensitive to hydrogen embrittlement damage [8].

The steps of the production process and its parameters must be checked to eliminate the possibility of embrittlement. Possibility of HE can be reduced during the production process with many ways as mentioned earlier; plating or coating process, baking or heat treatment process, cleaning processes and stress relief. This article examines the possible reasons and solutions of HE in literature and gives an example how hydrogen embrittlement can be reduced by right parameters of plating process for fasteners. Heat-treated or cold worked steels hardened more than 365 HV high cathodic efficiency electroplating solutions are advisable for fasteners. [8]

3. Case Study

In this work the designed pin is considered for HE occurrence. When we design our spring-loaded latches, we decide to use slotted spring pin as an examined sample. The technical drawing and photo of the slotted spring pin can be found in Figure 2. ISO 8752 international standard describes the materials for split type straight pins. Suitable slotted spring pin has been chosen by simulation according to the related standard (Table 1). The chosen pin has 5 mm diameter. The tolerances of it given in Table 2.

	Steel		Austenitic stainless steel	Martensitic stainless steel
	St		A	C
	All pin diameters	Alternative for pin diameters d_1 > 12 mm	Chemical composition limits (chemical analysis) %	
	Chemical composition limits (chemical analysis) %			
Material[a]	C ⩾ 0,64 Mn ⩾ 0,60 Si ⩾ 0,15 Cr[b] P ⩽ 0,04 S ⩽ 0,05	C ⩾ 0,38 Mn ⩾ 0,70 Si ⩾ 0,20 Cr ⩾ 0,80 V ⩾ 0,15 P ⩽ 0,035 S ⩽ 0,04	C ⩽ 0,15 Mn ⩽ 2,00 Si ⩽ 1,50 Cr 16 to 20 Ni 6 to 12 P ⩽ 0,045 S ⩽ 0,03 Mo ⩽ 0,8	C ⩾ 0,15 Mn ⩽ 1,00 Si ⩽ 1,00 Cr 11,5 to 14 Ni ⩽ 1,00 P ⩽ 0,04 S ⩽ 0,03
	Hardened and tempered to a Vickers hardness of 420 HV to 545 HV Hardness testing according to ISO 6507-1.		Cold worked	Hardened and tempered to a Vickers hardness of 460 HV to 560 HV Hardness testing according to ISO 6507-1.

Table 1. Requirements and reference International Standards [3]

Dimension of Hole		nom.	1	1,5	2	2,5	3	3,5	4	4,5	5	6	8	10
d_1	before mounting	max.	1,3	1,8	2,4	2,9	3,5	4,0	4,6	5,1	5,6	6,7	8,8	10,8
		min.	1,2	1,7	2,3	2,8	3,3	3,8	4,4	4,9	5,4	6,4	8,5	10,5

Table 1. Dimension with tolerances [3]

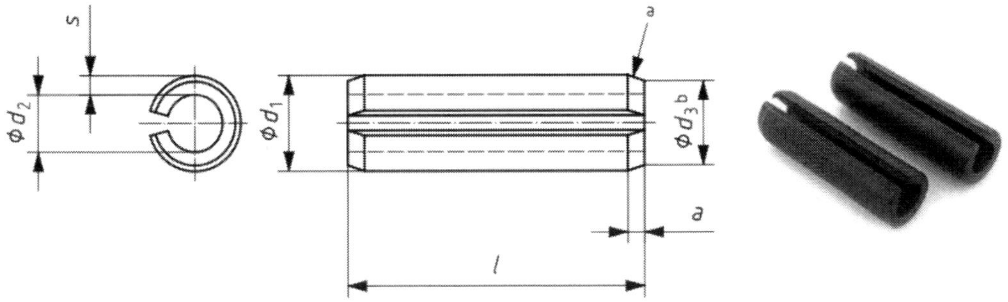

Fig. 2. Slotted Spring Type Straight Pin

The pin is being used on the spring-loaded latches. The slotted pin has been placed on the product via hand press machine as shown in Figure 3. The detailed assembled product shown in Figure 4. While the lock was working open to close position the pin exposes to instant high load. Each position changes of the lock create stress on pin. This cycling effects the lifetime of the examined product.

On the assembling operations the zinc coated pins' sudden cracks could be observed even pins don't have any sign during the process. And it can cause failure cracks during open-close position cycles. It directly effects the product quality and cause customer dissatisfaction. When we assemble the pin from the lock product under the high tensile stress the zinc plated pin getting cracked easily with some reason. When we try the uncoated pin (From same batch) on the product there is no any cracking issue. The Zinc Coating effects pin durability directly. We research the plating to find root cause and we found zinc coating process conditions of us can cause HE phenomena.

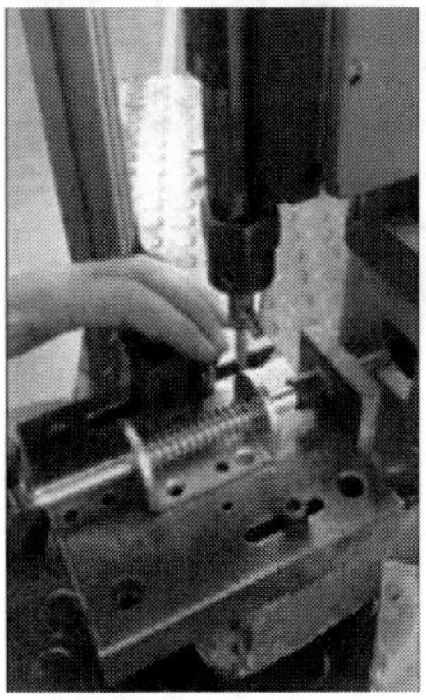

Fig. 3. Assembly operation of the pin

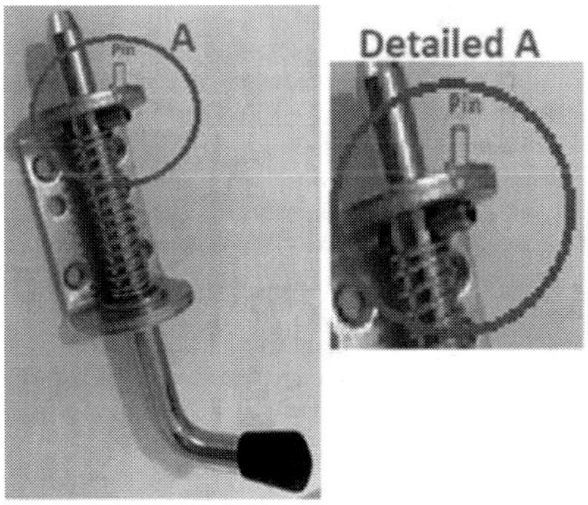

Fig. 4. Assembled Product

The process of Zn plating consists of many steps such as cleaning, electroplating, cascade rinsing, bright dip, water rinse, drying, heat treatment, bright dip, water rinse, passivation and finally cascade rinsing. Electroplating is defined as the process or procedure in which a coating is placed by electrolysis, that is, a metal is deposited on an electrode dipped in an electrolyte by passing a suitable electric current through the electrolyte [12]. Zinc plating is not suitable this type of fasteners due to the application process HE. Sometimes, a coating such as graphene coating, CrN, TiAlN coatings etc. are preferred over the other coatings to reduce the effects of the hydrogen embrittlement on high strength material.[12]

We begin by considering other types of coatings. Using a coating process that does not introduce hydrogen into the material (particularly those that do not utilize acids for cleaning) will help avoid this problem A number of dip-spin coatings are considered hydrogen embrittlement "free" because they use mechanical processes (abrasive blasting) for descaling. We found another type of coating that reduce the HE possibility on the item. The pins have been applied two different coating to test and compare. One of them is our classical zinc coating which we have been using already, the other one is a commercial one called Geomet (NOF Metal Coating Asia Pacific CO., LTD. Presented GEOMET® License). This is brand name which invented the special coating by NOF Metal Coating Asia Pacific CO., LTD. Also, similar coatings know by the industry with the name "Zinc Flake Coating".

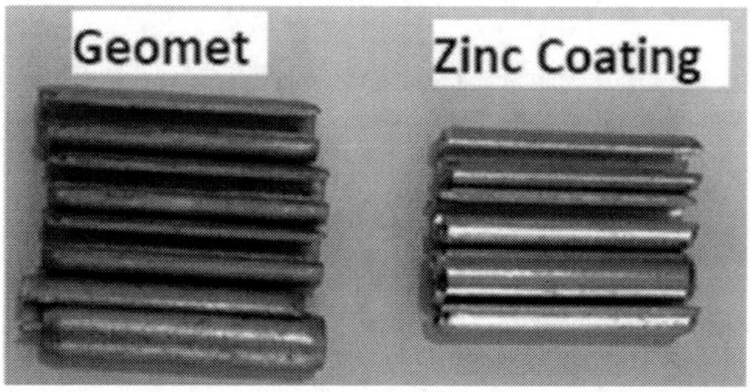

Fig. 5. Visual Check for Zinc & Geomet Coatings

Thickness	1.Point (μ)	2.Point (μ)	3.Point (μ)	4.Point (μ)	Average (μ)
Zinc Coating	9,42	9,07	8,80	8,24	8,88
Geomet Coating	11,7	11,5	9,99	14,4	11,90

Table 2. Thickness Measurement According Coating Types

After both coating operations the pins was observed at first. The zinc coating looks brighter than the brand mark as seen in Figure 6. Average zinc coating thickness is 8,88 μm measured. Average geomet coating thickness is 11,90 μm measured The measurement results are given in the Figure 6. When we applied the Geomet on the pin and produce our pilot products, there was not any issue on process of assembly. R&D started the all the lifespan test on the new plated pin. The performance seems to be satisfied. Cracking failure has been not seen on our products. In the case study show us, when different plating process applied on the same material, it increases HE according with type of coating. It is possible to avoid from HE with to choose right plating type on it. Geomet coating solved issue of case study, it never come up again.

4. Conclusions

Study issue is, it occurs HE problem on fasteners because of appliying wrong type coating. Hardness is a major contributor to hydrogen embrittlement. Harder, stronger materials are more susceptible to failure than weaker, softer ones. In general, if the hardness of the fastener is less than 35 HRC, there will probably be little difficulty with hydrogen embrittlement. However, if the fastener has hardness above 40 HRC, problems are more likely to occurIt has to been change the coating procees with different type. The other option is to use a fastener which is of a lower hardness. Unfortunately this is not possible to change our hardness because of our application. After research and sample test, we found there was another type of coating. It is named as "Zink Flake". It has coating process that does not introduce hydrogen into the material will help avoid this problem. We continued to our test with Geomet which is a kind of Zinc flake coating.

After our study for coating, we solved the cracking issue with a commercial coating for our specific application. This solution helped to reduce the minimum to occur HE and cracking issues. Also, when we compare the Geomet vs. Zinc coating, we discovered that, the Geomet has many advantages besides to avoid HE. It has better corrosion protection (around the 10 times more) and excellence resistance against some chemical abrasives than usual zinc coating. It needs to be considered HE on the design stage by engineers when they choose their material. There are many ways to prevent from HE. It is crucial to choose right material with suitable coating and correct process.

This study shows us, how significant is to prevent HE with change to coating type. Product performance effected by all process with right material. These should be made assessment all together, otherwise any disharmony will come up with problem. There are no possibilities to eliminate HE occur totally but it is possible to reduce to potential of HE occurrences via choosing the right material with the right processes. This should be considered especially for high tensile or strength application parts like fasteners. For future, we need to be apply right coating for fasteners on our design phases which their hardeness more than 35. It should be "Zinc Flake" type of coating. First choice will be Geomet for similar example after our case study.

5. References

[1] G.Gabetta, P.Cioffi, R.Burschi (2018). Engineering Thoughts On Hydrogen Embrittlement.

[2] Brahimi S. (2014). Fundamentals of Hydrogen Embrittlement in Steel Fasteners Ibeca Tech. Corp.

[3] EN ISO 8752:(2009). Spring-Type Straight Pins Slotted, Heavy Duty.

[4] Cwiek J. (2009). Hydrogen degradation of high-strength steels. Journal of achievements in materials and manufacturing. 2009;37(2):193-212

[5] Sandeep Kumar Dwivedi, Manish Vishwakarma (2018). Hydrogen Embrittlement in Different Materials: A review.

[6] Herring DH (2010). Hydrogen embrittlement wire Form TechnolInt.

[7] Davis JR. (1998). Metals handbook. ASM international.

[8] EN ISO 4042:(2019). Fasteners - Electroplated Coatings.

[9] Venezuela J, Liu Q, Zhang M, Zhou Q, Atrens A. (2016). A review of hydrogen embrittlement of martensitic advanced highstrength steels. Corrosion Rev 2016 Jun 1;34(3):153e86

[10] Grobin Jr AW (1988). Other ASTM committees and ISO committees involved in hydrogen embrittlement test methods. Hydrogen Embrittlement: Prev Contr.

[11] Kim HJ, Jeon SH, Yang WS, Yoo BG, Chung YD, Ha HY, et al. (2018). Effects of titanium content on hydrogen embrittlement susceptibility of hot-stamped boron steels. J Alloy Comp

[12] Amit Kharea, Sandeep Kumar Dwivedib, Dr.Manish Vishwakarmac, Dr.Siraj Ahmedd (2018). Experimental Investigation of Hydrogen Embrittlement during Coating Process and Effect on Mechanical Properties of High Strength Steel used for Fasteners

DOI: 10.2507/32nd.daaam.proceedings.074

KNOWLEDGE BASED DECISION MAKING IN MECHANICAL WORKSHOPS

Kaarel Kruuser, Jueri Riives & Indrek Kiolein

This Publication has to be referred as: Kruuser, K[aarel]; Riives, J[ueri] & Kiolein, I[ndrek] (2021). Knowledge Based Decision Making in Mechanical Workshops, Proceedings of the 32nd DAAAM International Symposium, pp.0514-0521, B. Katalinic (Ed.), Published by DAAAM International, ISBN 978-3-902734-33-4, ISSN 1726-9679, Vienna, Austria DOI: 10.2507/32nd.daaam.proceedings.074

Abstract

Now-days manufacturing is going more complex, dynamic and changing over time even chaotic in the extended value chain. In order to being able to satisfy the demand for high-quality products in an efficient manner, it is essential to use modern technologies based on AI. These areas, which have great developments are expert systems and machine learning. These have the new challenges to manage the manufacturing in different application areas. There are different possibilities and different approaches for solving multi criteria based engineering tasks, influencing directly to the performance in a workplace. How to combine and use them to improve and manage the performance of the production process in the workplace is a main objective of the current article. The integration of the AI models with FBM templates and libraries can also give a great advantage to standardize the way of decision making in a certain product family (like injection moulds). In the article, the models from both categories considered with tooling questions have been examined. This can help improve manufacturing consistency and reduce the tool inventory by setting shop tool standards for similar operations.

Keywords: Artificial intelligence; Feature Based Machining; Machine learning; Knowledge based manufacturing.

1. Introduction

Production automation and digitalization is a definite trend in business today. At the same time, production technologies and the equipment that supports them are becoming increasingly complex. The goal is to produce various products in small batches, with high productivity, high quality and as efficiently as possible. For this objective the complex, dynamic and changing over time solutions are needed. In order to being able to satisfy the demand for high-quality products in an efficient manner, it is essential to use modern technologies based on Artificial intelligence (AI). AI is the key driver in this change and it is critical for effective decision-making [1], [2]. These areas, which have great developments in manufacturing are expert systems and machine learning. These two components have the new challenges to manage the manufacturing in different application phases (process planning, operation planning, NC code developing, etc).

In a small or medium-sized company that does not have its own brand or is not focused on series production, has a great challenge to prepare for production for every different detail. In addition to the company, they often specialize in a market segment, which simplifies production preparation because the products have similar technologies and can be broadly divided into families. Tool manufacturers are just the type of companies where each product is different but the manufacturing technology is similar. Knowledge is be understood as the ultimate competitive advantage, the creation and capture of value in the unpredictable fast changing conditions [3]. The products can be divided into families according to the types of manufacturing technology. Machine learning can be applied in these types of situations. By analyzing products, you can compile type technologies on the basis of which you can develop rules for the production of product manufacturing technology.

The aim of this paper is describe the ontology of discrete manufacturing and look at the possibilities of AI applications in process planning decision making. A framework for intelligent manufacturing is proposed. Based on the proposed framework , the use case is solved.

2. Ontology in a discrete manufacturing

Discrete part of manufacturing integrates a design, planning and execution phase in which product information is defined to a manufacturing phase in which the processes would be fulfilled in the best possible way (cost efficient, resource utilization efficient, time efficient, etc). In the process planning typically culminates with the generation of numerical control (NC) programs for specific equipment. To perform the machining process in a workshop there is needed the following decision-making tasks such as technology selection, planning of processing trajectories, optimal selection of machining tools, optimization of operational technology, quality control solutions, use of technological possibilities of machine tool, use of machine tool resources. These tasks are going more complicated as the equipment possibilities are growing constantly. Expert systems, based on accumulated and/or gathered knowledge is a justified solution.

From the other hand in manufacturing process there are different problems influencing directly to the efficiency and depending from the big number of different parameters. These are like tool life optimization, energy saving, machine tool capability utilization, etc. Having sophisticated machining centers and producing high-quality complex workpieces, these have a high significance. Production process planning (PPP) is highly knowledge-intensive activities sequence planning, considering resources, production routes (with alternatives), machining operation planning with tool path planning, fixtures and tools choosing. The production process constantly involves a large amount of data, information and knowledge, which is either static or dynamically changing during the process. As there are different users, an important contribution to the success of application, is its common taxonomy. An approach to defining manufacturing taxonomy and axioms, based on manufacturing systems engineering (MSE) ontology was presented in [4].

Ontology is a type of semantic tool providing a source of shared and precisely defined terms that can be used to describe both knowledge and concepts (see Figure 1). Ontology can be expressed in standard formal languages like XML which would ensure the shareability and cooperation. These characteristics make ontology a notable knowledge modelling tool in many domains like medical science, digital library, manufacturing [5], etc. An ontology is the representation of knowledge based on conceptualization in a formal and explicit manner [6], in another word, explicit, formal specifications of terms in the domain and of the relations among them. The advantage of ontology is that it offers the concepts and their relations in a domain in a commonly agreed and formal expression that is machine-readable [7] and it has the reasoning capability that makes the implicit information explicit [8].

The described ontology model (see Fig.1) is the bases for development of models for the manufacturing process and the creation interaction of data, information and knowledge (see Fig.4., Fig. 9.) The ontology reflects the relationships and dependencies of the business model and the production process; production process and product; product and technology; technology and device, etc. By filling these attributes with data and analyzing the values of the output parameters, process performance can be optimized. The landscape where easy-to-understand and rational decisions and solutions for complex systems and relationships are sought is illustrated in Fig. 2.

The input is a manufacturing task: {product type = plate; parametrical model = CAD model}.

Decisions in the operation level:

- optimal sequence of machining;
- optimal selection of tools;
- optimal cutting conditions;
- tool life optimization.

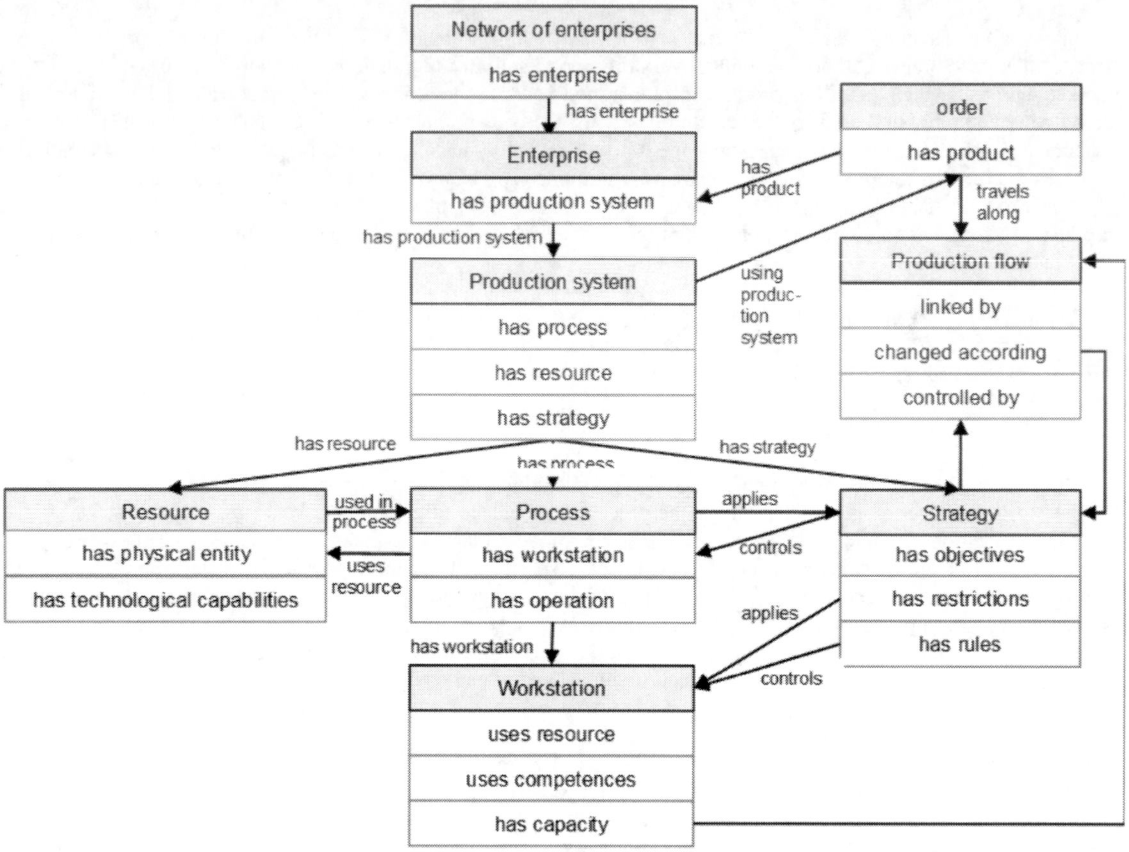

Fig. 1. Ontology of modern manufacturing

Decisions on the production system level:
- optimized total throughput time;
- optimized OEE;
- optimized SMED.

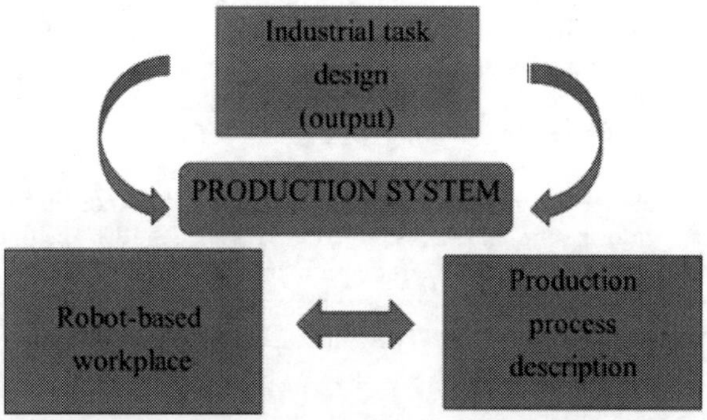

Fig. 2. Basic relationships in manufacturing

For the decisions of abovementioned parameters there are three data and knowledge flows:
- theory (theoretical models);
- recommendations from suppliers of equipment, instruments, etc.;
- knowledge, accumulated by practice.

These all are as input for the knowledge - based decision-making.

3. AI application possibilities in process planning decision making

There has been developed a feature-based machining approach for the sophisticated parts and 5-axes machining centers. The principles of using feature - based machining (FBM) methods were described in [7]. Today's production is characterized by: a reduction in batch sizes, a shorter time to market, increased quality requirements and increased price pressure. The equipment and technologies used, are increasingly complex and have a wide range of technological capabilities. All this creates certain complications for SMEs. There is no exception also the tooling industry.

A two-level approach is needed to make production more efficient. First, it is important to quickly identify the product and its features. There is need to classify and to systematize these, and to develop the processing technologies for different features shape clusters. On the other hand, workplace performance is shaped by productivity, the sustainable use of resources and the cost of meeting quality requirements. Mathematical models and decision-making plays an important role here. An example of a productivity model is given below (1), which takes into account the nature of the product and its features, the technological possibilities of the machine tool and the process planning principles.

$$Q = \frac{ns}{(ns) * \frac{l_{xx}}{s_{xx}} + (ns) * IA * t_y + (ns) * t_s + \frac{(ns) * l_{xz}}{s_{xz}}} \tag{1}$$

Where, n - spindle speed (sec) -1, s - tool feed (mm/sec), lxx - the length of the path that the tool travels with a quick placement from one feature to another (mm), sxx -Machine Rapid movement time (sec), IA - number of tools used, tv - tool exchange time (sec), ts - setup time (sec), lxz - tool rapid movement length in tools axis (mm), sxz - machine rapid movement speed in Z axis. The problem is the constant change of product nomenclature, the constant improvement of the company's strategies, as well as technological equipment and accompanying technologies. Static models do not work here and flexibility has a great importance. Here has been developed a framework for intelligent manufacturing. This framework consists of four basic frames having different objectives (see Fig. 3.). The framework is based on the ontology model (see Fig. 1). These frames are all as independent parts that have tight interactions. Content of the frames depends on the manufacturing application. Example of the realization of the framework for real manufacturing task solving is given on the bases of a use-case:

1. Parametric frame for part unification with corresponding grouping the features and realizing the possibilities for feature - based machining.
2. Process frame, using AI based rules and patterns is used for optimal decision making in process planning activities. The process frame is working in interaction with the parametric frame.
3. Data frame lays on the physical content of the manufacturing system and the data output depends from the technological capabilities of equipment used, but also how these are used in the production process. Therefore, the data models here are the important part and logic for data obtaining and analysis.
4. Analysis frame is working tightly together with the data frame and machine learning algorithms are used for process improvement.

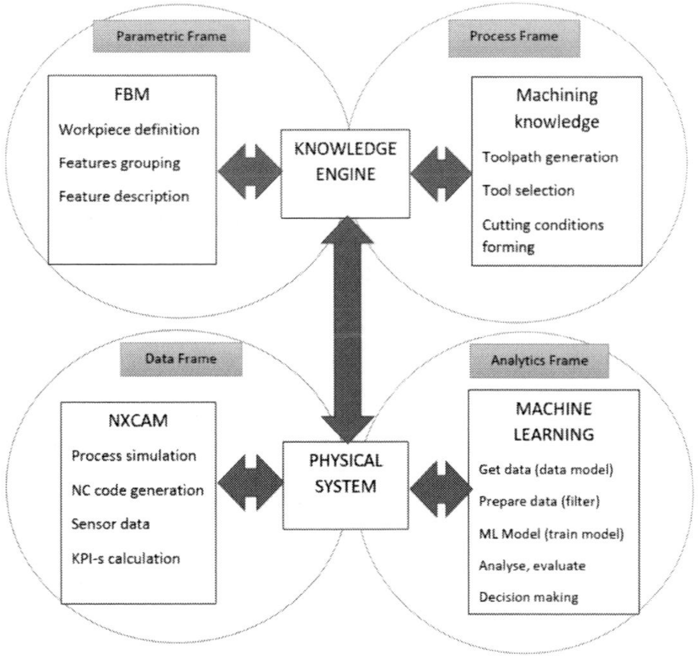

Fig. 3. Framework of intelligent manufacturing

Feature based machining is based on the principles that using designed rules, we can compile the machining of a certain workpiece (using catalogues of standard technologies and feature classification principles) according to a predicted logic. In principle, the CAD model of a product must be transformed into a language understood by the machine control software (such as NX CAM) of the machining center and the device must be "trained" to manufacture this product intelligently. Here we come across the principles of machine learning.

Machine learning is a possible tool in these situations. Machine Learning is a branch of computer science that enables computers to act without explicitly programmed. Machine learning uses statistical and pattern matching techniques to accomplish prediction and data analysis task. Moreover, machine learning is a subfield of artificial intelligence. Its goal is to enable computers to learn on their own. A machine's learning algorithm enables it to identify patterns in observed data, build models that explain the world, and predict things without having explicit pre-programmed rules and models [8].

In machine learning, the major objective is to find a model or function that maps input variables to output variables. Consider we have set of input variables (X) and set of output variables (Y), the objective of machine learning is to find a function f that maps the input variable (X) to (Y). There are different possibilities and different approaches for solving multicriteria based engineering tasks, influencing directly to the performance in a workplace. How to combine and use them to improve and manage the performance of the production process in the workplace is a main objective of the current article. The integration of the AI models with FBM templates and libraries can also give a great advantage to standardize the way of decision making in a certain product family (like injection moulds). In the article the models from both categories considered with tooling questions have been considered. This can help improve manufacturing consistency and reduce the tool inventory by setting shop tool standards for similar operations

Machine learning describes a set of techniques that are commonly used to solve a variety of real-world problems with the help of computer systems which can learn to solve a problem instead of being explicitly programmed. [9]. Here the role of big data in decision making was examined [10], [11]. Within the context (see Fig.1, Fig3 and productivity model) the production process is represented by big data attributes, considering workpiece, feature, machine tool, cutting tool, machining process, cutting conditions and machining results (cycle time, operation fulfilment time, production cost). Data classification, decision trees, decision making rules are developed using big data patterns and artificial neural networks for determining the reasonable solutions in an on-line and off-line manufacturing process. Some examples are described in the following use-cases. In the analysis phase is the interaction of input parameters {cutting conditions, tool selection, tool life} with the output parameters {productivity, machine utilization rate, main drive power utilization rate}. The objective is to minimize the cycle time (increase the productivity) with low manufacturing cost.

Additionally, there are a lot of process parameters, which have an influence to the cycle time and productivity. Here we have considered: number of tools used for machining and their allocation in tool magazine [12], tool path optimization and operation sequencing [13], sequence of machining one feature, sequence of machining the features, optimization of cutting parameters, etc. For developing basic rules there is necessary the classification of workpieces and to use the possibilities of group technology. Some examples, based on the above-mentioned logic, are described in the following use-cases.

4. Practical use cases of FBM with the AI application

4.1 Production flow and engineering tasks in mechanical workshops.

There are always certain basic rules, based on the production process. In order to start a production process, there must be production resources that allow the process to take place. Work centers, cutting instruments and fixtures can be considered as production resources. With these resources, we can add value to the raw materials. The more complex the processing process, the greater the added value it brings to the product.

In addition to production resources, production must have production processes. Describing the production process as accurately as possible and without excessive noise allows for efficient production. The more precisely and clearly understood the production process is described, the more efficiently it can be realized. In order to describe the production process more efficiently, production processes must be standardized. In an ideal world, the whole process could be standardized according to the same principles. Unfortunately, this cannot be done in real production, because the products are different and the same technological solution cannot be used. However, it is possible to divide the products into groups or families on the basis of similarities, on the basis of which similar basic principles can be used in the development of standard technologies. This allows you to optimize the production process and reduce the cost of processing the production resource.

There is no one-size-fits-all solution for optimizing manufacturing technology. The solutions are different and give different results. The choice of technology largely depends on the processing equipment and the choice of instruments used by the company. However, there are still many specific flaws in the implementation options. However, not all options are as effective and equal, although they may eventually lead to the same result. There is an alternative technology to each technology, but it may not be as effective.

As the Fig. 4. shows, how the data moves and how many different results can be obtained to produce a single feature element. When creating a work path, different inputs must be considered in order to achieve the most efficient work path. Even a small change in one parameter can drastically change the result achieved with the path.

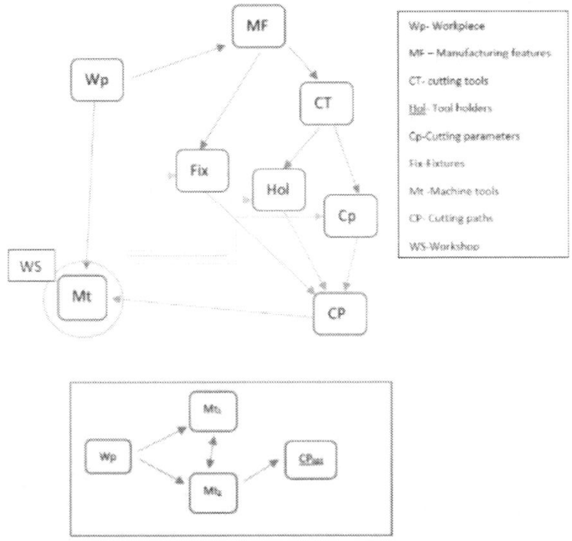

Fig. 4. Data movement in Cutting path decision making

Traditionally process parameters are determined by the operator's experience, the conservative technological data provided by the machining equipment manufacturers and trial-and-error operations. This leads to inconsistent machining performance since operator's experience is limited and subjective while the manufacturer data is based on safety-conscious principles and it only includes applications on certain machining materials.[14] Optimizing the process parameter makes it possible to record the company's knowledge and knowhow. When an employee leaves the company then such action does not leave a big knowledge gap to the company's manufacturing knowhow.

4.2 Use case

Let's take a closer look at the use case. In the production of injection molds, there is a situation where a single part produced is never identical to the previous one, but this production technology is always similar. Based on this, we can divide the products into families. Let's look at plate-type products and how to produce them, and how you integrate machine learning technology into production preparation. For plate type products, illustrated Fig. 5. We see different types of geometries that have different workmanship and quality requirements. The information required for production is attached to the 3D model with PMI information

Processing type	Remarks	Color values	Positional tolerance	Dimensional tolerance Hole	shaft
Milling of the cavity	Milling of the faces according to the CAD of the part	Olive green 175,255,175	+/- 0,05	+0.05	-0.05
Finish milling or grinding	Grinding, etc. or special fits (slides alignment bore)	Ivory 255,255,175	+/- 0,01	+0.01	-0.01
Finish milling	Super-fine finishing (slide seat)	Pink 255,175,175	+/- 0,02	+0.02	-0.02
Fit H7	Special milling		+/- 0,02	+0.02	
Clearance	Milling	[illegible]	+/- 0.1	+0.1	-0.1
Thread	Metric right-hand thread to DIN / ISO	Yellow 255,255,000	+/- 0.2		
Fine-pitch thread	Metric fine-pitch thread to DIN / ISO	Bright orange 255,175,000			
Special thread	All other special threads, for example Whitworth, tube threads (Whitworth R, Rp)	Orange 255,125,000			
Special drilled hole / stepped drilled hole	Complex holes, combination of holes, cooling system	[illegible]			
Red shows changed areas		[illegible]			

Fig. 5. Plate type product with production information applied with PMI information

We focus on one specific feature and analyze the technology needed to make it so that the image meets the set quality requirements. For analysis, we select the insert cavity, shown in Fig 6., And consider what the output variables must be. What results are expected from this product.

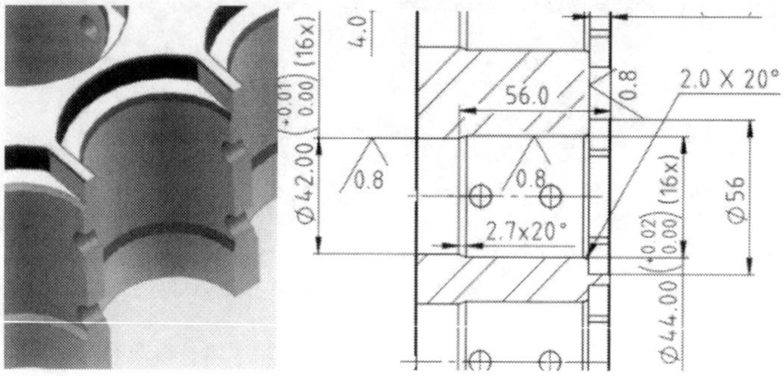

Fig. 6. Feature in plate type product, with PMI data and drawing data

Many different technologies can be used to make this feature, shown in Fig. 7. The choice of which may depend on the experience of the engineer and the work experience in the given company. Each technology has different efficiencies and easily produces different results. The development of standard manufacturing technology for this design makes it possible to keep efficiency and manufacturing costs stable, as well as to ensure stable and consistent quality.

Rough milling →Semi finish milling → Finish milling → Chamfering
Drilling→ Semi finish milling →Finish milling → Finish Boring→ Chamfering
Rough milling →Semi finish milling → Finish milling → Finish Boring→ Chamfering

Fig. 7. Different types of production technology for the feature.

When choosing needed technology, it is important to choose the appropriate instruments and cutting parameters for machining. There are different instruments in the factory that can be used to perform the same technology, see Fig 8. But different instruments do not have the same efficiency in performing the same technology. The instruments work with different parameters in different materials, and with incorrect parameter, the tool does not work effectively or can even break.

Technological operation	Type	Cutter diameter	Cutting edge radius	Nr. Of flutes
Rough milling	Indexable	18	0.8	3
	Indexable	18	1.8	3
	Endmill	18	0.5	3

Fig 8. Selection of instruments on same technological operation.

In order to use Machine learning applications, the tools, modes and technologies used in the factory must be mapped and corresponding databases must be created. This allows data to be used in the machine learning process.

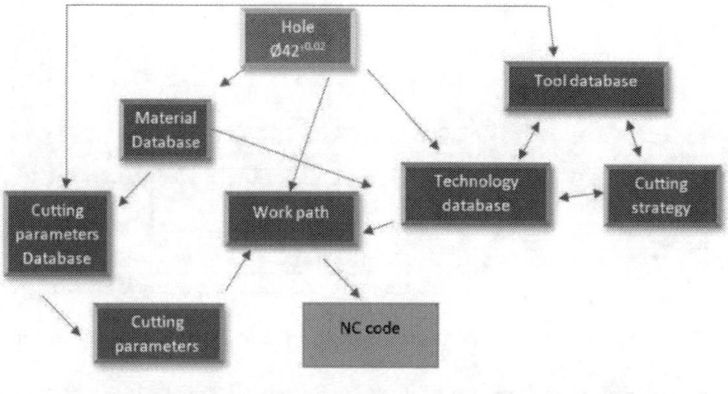

Fig. 9. Decision making in Feature machining

As described on the Fig. 9, first input comes from the features, which describes the geometry and processing requirements (tolerances and surface quality). In order to compile the technology, a suitable cutting strategy must be selected, based on the choice of material and tool. Technology is chosen first. Depending on the technology chosen, tools and cutting strategies will be selected for each operation. To do this, the tools in the tool database are compared according to which tools are available for this type of operation. The tools are selected according to the optimal diameter and the appropriate tool geometry. When the optimal tool is found, the appropriate parameters are searched in the table of cutting parameters based on the material of the workpiece. If the tool does not fit, the next tool is selected based on the geometry fit. Depending on the choice of technology and tool, it will choose a cutting strategy. As a result, a work path is generated from which the NC code can be generated using the postprocessor.

In order to make the rules for technological selection, the capabilities of the workshop must first be examined and the rules must be adapted to the possibilities. Rules that do not support the technological capability of workshop are ineffective and rather harmful.

5. Conclusions

To automate and digitize increasingly complex technologies, dynamic solutions that adapt to time is needed to be developed. These solutions must meet the needs and capabilities of the company and be able to adapt to the changes taking place in the company. This article described the ontology of modern production and relationships involved in manufacturing. The role of big data in decision making was studied. Based on the proposed framework, the use case solution was presented. So far, the research has mainly focused on developing a theoretical framework for decision making. The next step of research is to monitor the operating parameters of the equipment and to collect information on the efficiency in relation to the cutting parameters used. Based on this data, develop a decision-making smart system to select more efficient parameters.

6. References

[1] Buchmeister, B; Palcic, I.; Ojstersek, R. (2019). Artificial Intelligence in Manufacturing Companies and Broader. An Overview. DAAAM International Scientific Book (pp 81-98). DAAAM International Publishing.

[2] Plant Automation (2019). Technology: The future of artificial intelligence in manufacturing industries, Available from: https://www.plantautomationtechnology.com/articles. Accessed on 2019-03-10

[3] Henriques, E.; Peças, P. (2012). New Business Models for the Tooling Industry. In book: Advances in Business and Management - Vol.4 (pp.1-33) Publisher: Nova Science Publishers, New York Editors: W.D. Nelson.

[4] Lõun, K.; Riives, J.; Otto, T. (2011). Evaluation of the Operation Expedience of Technological Resources in a Manufacturing Network. Estonian Journal of Engineering, 2011, 17,1, pp51-65

[5] Huang1, Z; Qiao1, L; Anwer, N; Mo, Y. (2014). Ontology Model for Assembly Process Planning Knowledge. 21st International Conference on Industrial Engineering and Engineering Management 2014 (IEEM 2014)

[6] Lu, Y. ; Panetto, H. ; Ni, Y. ; Gu, X. (2013). Ontology alignment for networked enterprise information system interoperability in supply chain environment. In. J. Comp. Integ. Manufac. 26(1-2), 140-151 (2013).

[7] Kruuser, K.; Riives, J. ; Tšukrejev, P. ; Kiolein, I. (2019). Implementation of a knowledge-based manufacturing on the example of Sumar Tools OÜ. Proceedings of the Estonian Academy of Sciences, 2019, 68, 4, 407–412

[8] Wuest, T.; Weimer, D.; Irgene,C.; Thoben, K. (2016) Machine learning in manufacturing: challenges and applications. Production and Manufacturing Research, 4. 1., 23-45, 2016

[9] Kühl, N.;Goutier, M.;Hirt, R. ;Satzger, G.; (2019). Machine Learning in Artificial Intelligence: Towards a Common Understanding, Proceedings of the 52nd Hawaii International Conference on System Science, Grand Wailea, Maui, Hawai.

[10] Fan, J.; Han, F.; Liu, H. (2014). 'Challenges of big data analysis', National Science Review, 1 (2), pp. 293–314.

[11] Jeble,S., Kumari, S., Patil, Y. Role of Big Data in Decision Making. Operations and Supply Chain Management. V.11, N. 1. 2018, pp 36-44

[12] Dereli,T.; ,Filiz,I.H. (2000). Allocating optimaal index positions on tool magazines using genetic algorithms. Robotics and Autonomous Systems, 33 (2-3), pp 155-167

[13] Guo, Y.W.; Mileham, A.,R.; Owen, G.W.; Maropoulos, P.G.; Li,W.D. (2009). Operation sequencing optimization for five-axes prismatic parts using a particle swarm optimization approach. Proceedings of the Institution of Mechanical Engineers. Part B – Journal of Engineering Manufacture, 223 (5), pp 485-497

[14] Krishna Mohana Rao G.; Rangajanardhaa G.; Hanumantha Rao D.; Sreenivasa Rao M. (2009).'Development of hybrid model and optimization of surface roughness in electric discharge machining using artificial neural networks and genetic algorithm', J. Mater. Process. Technol., vol. 209, no. 3, pp. 1512–1520, 2009

DOI: 10.2507/32nd.daaam.proceedings.075

BLOCKCHAIN-BASED METROLOGICAL TRACEABILITY

Almira Softic, Nermina Zaimovic Uzunovic & Samir Lemes

This Publication has to be referred as: Softic, A[lmira]; Zaimovic Uzunovic, N[ermina] & Lemes, S[amir] (2021). Blockchain-based Metrological Traceability, Proceedings of the 32nd DAAAM International Symposium, pp.0522-0526, B. Katalinic (Ed.), Published by DAAAM International, ISBN 978-3-902734-33-4, ISSN 1726-9679, Vienna, Austria DOI: 10.2507/32nd.daaam.proceedings.075

Abstract

Calibration of measuring instruments is common requirement for quality control, and calibrated instrument has to be traceable. Metrological traceability is the property of the measurement result relating it to a stated reference standard through an unbroken chain of comparisons with stated uncertainties. The data integrity of the traceability chain is vulnerable, as digital calibration certificates can potentially be altered. This paper evaluates how blockchain technology could be used to prevent data alteration and improve data integrity in calibration certificates. This is of great importance in digital quality infrastructure for innovative products and services in Industry 4.0.

Keywords: Metrological Traceability; Blockchain; Industry 4.0; Digital Calibration Certificate.

1. Introduction

The International vocabulary of metrology (VIM) [1] defines metrological traceability as the "property of a measurement result whereby the result can be related to a reference through a documented unbroken chain of calibrations, each contributing to the measurement uncertainty". The International Laboratory Accreditation Cooperation (ILAC) considers the elements for confirming metrological traceability to be an "unbroken metrological traceability chain to an international measurement standard or a national measurement standard, a documented measurement uncertainty, a documented measurement procedure, accredited technical competence, metrological traceability to the SI, and calibration intervals" [2].

The metrological traceability chain is defined in [1] as a "sequence of measurement standards and calibrations used to relate a measurement result to a reference". The calibration hierarchy defines the metrological traceability chain used to establish the metrological traceability of a measurement result to a measurement unit. The reference is the international definition of a measurement unit through its practical realization [1]. It is necessary to establish such a chain of traceability where each step is described, and the uncertainty connected with each step is evaluated to maintain the metrological traceability for quality control. Ačko et al. in [3] analysed some novel approaches to assuring traceability of in-process measurements by using robust and thermal invariant multi-purpose material standards and a mobile simulator to emulate shop floor environmental conditions. The traceability of the measurements guarantees end users' confidence in the correctness of measurement results, ensuring customer protection and product quality. The requirements of ISO 17025:2017 standard in its latest revision [4] introduced the concept of risk-based thinking, following the ISO 9000 series of standards for quality management.

The lack of metrological traceability identifies potential risks in calibration and testing laboratories [5]. The overwhelming digitalization and upcoming Industry 4.0 automation introduce some new risks, introducing potential breaches of information integrity. The calibration and accreditation system relies on a well-established hierarchy of accredited calibration laboratories, making this process error-proof to a certain extent. For example, around 10.000 calibrations to a national standard are performed in Germany each year, 100.000 calibrations to reference standards and millions of calibrations to working standards [6]. To address the risk of data integrity breach and eliminate the risk of compromising calibration certificates, the Physikalisch-Technische Bundesanstalt (PTB) introduced the concept of a Digital Calibration Certificate (DCC), which serves for the electronic storage, the authenticated, encrypted and signed transmission and the uniform interpretation of calibration results in Germany. The information integrity is based on the official national electronic signature and time stamps.

It is estimated that there are 850 million measuring instruments on the EU market [7], which are responsible for a share of 4 % to 6 % of the European GDP, equivalent to 660 to 990 billion euros per year. Each year, almost 350 million measuring instruments are being in the European market [7]. Emerging digital technologies, such as embedded systems, cloud computing, big data, Internet of Things (IoT), and cyber-physical systems, opened some new challenges for instruments that are using distributed and cloud-based data storage and processing. There is a need to establish a digital quality infrastructure for European Union and even globally. Smart data and smart services could be connected with databases and infrastructure through a "European Metrology Cloud" [7], ensuring that all measurements are reliable and traceable to protect the customer and the end-user, as European directives within the New Approach regulate it. However, the concept presented in [7] could be centralized, distributed or hybrid, making sure that appropriate surveillance and quality control mechanisms are in place. This research evaluates how blockchain could improve data integrity and prevent data alteration in calibration certificates. To open the space for this technology, it is important to establish the link between the legal metrology, software engineering and to leverage the two previously unrelated engineering disciplines.

2. Digital metrology

A novel concept of digital metrology primarily intended for IoT devices is presented in [8], which relies on a digital calibration certificate (DCC) and a Digital Twin concept. They proposed using the X.509 certificate to cryptographically bound a measurement device with supporting authentication of the instrument's identity. X.509 is a cryptography standard frequently used in electronic signatures used to define the format of public-key certificates. This standard relies on certificate authorities, making it unsuitable for decentralized systems. The use of false certificates issued through illegal compromise of certification authorities was reported by national security services, such as Dutch certification authority DigiNotar [9].

A blockchain-based system for metrological purposes is mentioned in [10], where a vehicle speed measuring system was implemented using the Hyperledger Fabric blockchain platform. They compared three different measuring system models: the traditional measuring instruments, a cloud-based centralized and hierarchical measuring system, and distributed and decentralized blockchain-based measuring system model. The biggest advantage of the blockchain-based model is the fact that the measuring process as a cloud-based service is supported by resources owned and managed by anyone who has no interest to alter the measured quantity. The quantitative assessment showed that blockchain-based measurements are less costly but highly rely on internet connectivity.

A similar concept, evaluated on the CMM dataset, was presented in [11]. They fully implemented the dual digital traceability of the metrological data approach using X.509 certification standard and SHA-256 hashing algorithm to represent the instrument calibration and cryptographically sign measurements on record level. Dual digital traceability has been successfully implemented in a software prototype, allowing publicly stored fingerprints of a measurement record able to validate invalid or retracted calibration certificates and unwanted modification. The system was tested using IOTA cryptocurrency distributed infrastructure (https://www.iota.org/).

Aimagmbetova et al. in [12] discussed how metrological traceability is achieved in the Internet of Measurement (IoM) concept. Without going into details, they concluded that there is a need to perform remote measurements and calibration using centralized databases, but the risk of jeopardizing information security opens the space for multistage cryptographic hash functions. These functions are used to protect the data in a distributed chain of encrypted blocks. The data encryption and data checks are performed by special users of the IoM network called miners, like those in blockchain-based cryptocurrencies. Anto and Nugraha in [13] designed a system of Blockchain-based technology for Digital Multimeter calibration used to improve the data integrity and to prevent potential data manipulation (tampering-data) for Digital Multimeter calibration results. The most important advantage that motivated that research is the possibility of storing track-record of previous calibration data and losing calibration records.

3. Blockchain

Blockchain, at first used only for cryptocurrencies, is a concept emerging in many different areas, providing data integrity and reliability by utilizing distributed data processing and cryptographic techniques. It is already well established in the supply chain processes [14], but it opens its path into other disciplines of Industry 4.0, such as distributed CAD environments: ERP, BIM, PDM, PLM [15]. In order to describe how it can be leveraged to support metrological traceability, there is a need to explain the concepts behind the blockchain briefly.

The blockchain relies on three IT concepts: peer-to-peer networking, public-key cryptography, and distributed consensus based on the resolution of a random mathematical challenge [15]. This technology creates fixed-size blocks of information using cryptographic hash functions. These blocks are added to an array called blockchain in a way that every new block has the same size but contains irreversibly encrypted information about all its predecessors. The blockchain contains the encrypted version of the complete history of changes of all blocks.

Fig. 1 shows how blocks in the blockchain are signed and verified. The entire process is based on cryptography based on private and public keys. Each transaction in the chain is verified by the previous block owner's public key and signed by his private key. The hash function ensures data integrity as it is irreversible and has no inverse function. The accredited laboratories 1, 2 and 3 have different vertical levels, from working (industrial), national, regional, to international. As these laboratories also act as nodes in the network, they are also distributed horizontally. The nature of the blockchain provides the error-proof traceability of calibration certificates and procedures.

Each data transaction (calibration that needs metrological traceability) is accompanied by the timestamp, making the data also temporally identifiable and traceable. The process relies on the consensus of the nodes, and each case of altered or invalid data is easily detected and excluded from the chain. The nonce is an arbitrarily chosen string, changed multiple times until the hash value of the block header is below the predefined target value. This process is called mining, and it is performed by the nodes of the network. Only if the solution is valid, the block is included in the blockchain, and the transaction considered complete.

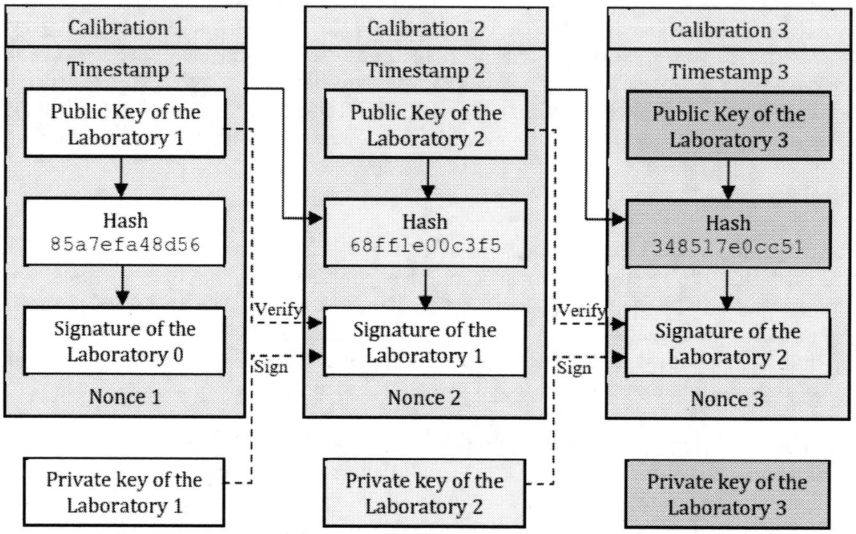

Fig. 1. The signing and verification in the blockchain, modified from [15]

The blockchain is prone to alterations, as any change in any phase of data transfer irreversibly alters the final output. This feature makes it a state-of-the-art digital tool to ensure data integrity. The blockchain relies on mathematical functions (hash functions), which create the fixed-size bit-string output (hash). It is practically impossible to guess the length of the hash by decrypting the blockchain. The hash algorithm produces a unique output as a unidirectional function having no inverse function. The blockchain uses these hash functions to encrypt information and to digitally sign the information from all previous steps. Cryptocurrencies Bitcoin and Ethereum use SHA-256 (secure hash algorithm), developed in 2001 by the National Security Agency (NSA) in the USA.

The list of currently used cryptographic hash functions includes 128–320-bit RACE Integrity Primitives Evaluation Message Digest (RIPEMD), Merkle tree-based 128–512-bit message digest algorithm (MD5, MD6), 224–512-bit secure hashing algorithm (SHA-2 also known as SHA-256 or SHA-512, SHA-3), 224–512-bit BLAKE (BLAKE2, BLAKE3), Russian 256, 512-bit Streebog, etc. These hash functions can contain more different algorithms, and they are implemented in programming languages as classes.

The top feature of the blockchain is the fact that information stored in the blockchain can never be altered or lost. The blockchain replicates into the same number of copies as there are nodes in the network. The blockchain also stores the complete history of all previous states of information stored. In that way, anyone could check the final state validity simply by using the same hashing algorithm to all information from the beginning to the end.

Bucci et al. in [16] tackled the problem of privacy if blockchain technology is used for the purpose of distributed measurement systems, especially in the European Union where GDPR (General Data Protection Regulation) is used. They also emphasized the problem of network bandwidth and connectivity, as it plays a crucial role in the overall performance of the system.

Takatsuji et al. in [17] demonstrated how blockchain technology could be used to improve visualization of metrological traceability. As only the first higher standard is shown in the calibration certificate, the end-user lacks the data about the traceability to an international standard, making space for immutability and falsification of calibration standards. As the traceability chain could be long, there is a possibility that the highest-level calibration in the traceability chain was performed a long time ago. Blockchain can overcome this by using timestamps which are included in each block and secured with hash functions. Zhu et al. in [18] explained how blockchain is superior to classic electronic signatures and digital certificates, as it disables data alteration. Peters et al. in [19] raised an issue of how could embedded devices handle the huge amount of data in a blockchain, ignoring the fact that data is stored in a cloud, and only a digital signature is stored locally.

There is no doubt that blockchain is an emerging technology, which can be used in many metrological disciplines, from distributed measuring systems to calibration certificate management.

4. Metrological traceability

From the very beginning, the concept of metrological traceability was intended to be centralized, assuring that any measurement result is traceable to the highest-level standard. On the contrary, blockchain is decentralized by its nature, and it is not obvious that this technology can be used to support metrological traceability. Table 1 summarizes what should be taken into account to utilize the blockchain as a means to make calibration certificates non-repudiable and traceable to the international standard.

Metrological traceability			Blockchain
The interlaboratory comparison relies on a network of laboratories. The higher number of comparisons, the higher confidence in measurement results.	😐	😐	Blockchain relies on a high number of network nodes. The higher number of nodes, the better information security.
Measuring instrument traceable to a highest calibration reference standard	😐	😐	Unchangeable information stored in a distributed network of nodes
An unbroken chain of comparisons of instrument's measurements	😐	😐	A continuous chain of encrypted data blocks
Calibration certificates (also digitally signed)	😐	😐	Encrypted digital certificates stored in the cloud
Strictly vertically centralized hierarchical structure	😐	😐	Still, a vertically centralized hierarchical structure, reinforced with a horizontal network of nodes
Calibration certificates could be falsified	☹	☺	Calibration certificates cannot be falsified
Fixed calibration intervals	☹	☺	Flexible calibration intervals
Slow and expensive system of checking authenticity and integrity of the calibration certificates	☹	☺	Automated and fast system of checking authenticity and integrity of the calibration certificates
Low dependence on internet connectivity and bandwidth	☺	☹	High dependence on internet connectivity and bandwidth

Table 1. Relations between the metrological traceability and the blockchain

The statements presented in Table 1 reveal that the only weakness in blockchain-based metrological traceability is high dependence on internet connectivity and high requirements for computational resources. The advantages prevail, and it is plausible to expect that blockchain will find its way into this area of use. Measurement and instrumentation will benefit if metrological traceability is supported by a distributed ledger of information about calibration certificates.

5. Conclusion

The general requirement for each measuring instrument is a calibration certificate, stating that the measurement results are reliable, accurate, and within prescribed measurement uncertainty interval. The calibrated instrument has to be traceable to the highest level of the reference standard, and this chain should be verifiable and prone to alterations. However, these certificates can be easily falsified and it is very important to provide their integrity.

Blockchain is an emerging digital technology constantly finding new applications towards Industry 4.0 and cyber-physical systems. Initially used only as a basis for cryptocurrencies, its capability to ensure data integrity and authenticity made it an unavoidable tool to secure any information which needs a high level of information security. Its nature makes this technology a reliable solution for the problem of calibration certificate data integrity.

Recent practices opened the path for blockchain to be used in distributed digital measurements, sensors in smart devices, the Internet of Things and all instruments requiring calibration. It is not enough to have surveillance, auditing and legal mechanisms against falsification of calibration certificates. These certificates can be reinforced with blockchain-based infrastructure made of accredited laboratories.

These laboratories would appoint a part of their computational resources to verify the cryptographic hash functions in the same way the cryptocurrency miners verify the financial transactions or smart contracts. International metrology infrastructure still does not support the implementation of this concept, but there are some initiatives at the national level, where authorities recognized the need to increase the level of digitization of metrology infrastructure.

In the near future, the national metrological institutions should first get familiar with the blockchain technology, then implement a pilot project with one type of measuring instrument, and finally provide a legal framework to make blockchain-based calibration certificates digitally traceable.

6. Acknowledgments

The publication of this paper is supported by the University of Sarajevo, Mechanical Engineering Faculty.

7. References

[1] JCGM 200:2012 International vocabulary of metrology - Basic and general concepts and associated terms (VIM), 3rd edition, Joint Committee for Guides in Metrology (JCGM).

[2] ILAC P10:07/2020: ILAC Policy on Metrological Traceability of Measurement Results. International Laboratory Accreditation Cooperation (ILAC). Available online from https://ilac.org/?ddownload=123220 29.5.2021

[3] Acko B, Klobucar R, Acko M (2015) Traceability of In-process Measurement of Workpiece Geometry, In: Procedia Engineering, Vol.100, pp 376-383, ISSN 1877-7058, DOI: 10.1016/j.proeng.2015.01.381

[4] Standard ISO/IEC 17025:2017 General requirements for the competence of testing and calibration laboratories. The International Organization for Standardization (ISO).

[5] Vasilnakova, A. (2018). Risk management in accredited testing laboratories. In: Proceedings of the 29th DAAAM International Symposium, pp.1071-1075, B. Katalinic (Ed.), DAAAM International, ISSN 1726-9679, pp 1071-1075. DOI: 10.2507/29th.daaam.proceedings.153

[6] Hackel, S, Härtig, F, Hornig, J, Wiedenhöfer, T. (2017). The Digital Calibration Certificate. In: PTB-Mitteilungen 127 (2017), Heft 4. pp 75-81. ISSN 0030-834X. DOI: 10.7795/310.20170499

[7] Thiel, F, Esche, M, Grasso Toro, F, Peters, D, Oppermann, A, Wetzlich, J, Dohlus, M. (2017). A Digital Quality Infrastructure for Europe:The European Metrology Cloud. In: PTB-Mitteilungen 127 (2017), Heft 4. pp 83-97

[8] Mustapää, T, Autiosalo, J, Nikander, P, Siegel, J E & Viitala, R. (2020). Digital Metrology for the Internet of Things. In 2020 Global Internet of Things Summit (GIoTS), pp 1-6. IEEE. DOI: 10.1109/GIOTS49054.2020.9119603

[9] Zetter K. (2011). DigiNotar Files for Bankruptcy in Wake of Devastating Hack, Wired online magazine, Condé Nast. https://www.wired.com/2011/09/diginotar-bankruptcy/ 29.05.2021

[10] Melo, W S, Bessani, A, Neves, N, Santin, A O & Carmo, L F R C. (2019). Using Blockchains to Implement Distributed Measuring Systems. IEEE Transactions on Instrumentation and Measurement, ISSN: 1557-9662. pp 1-12. DOI: 10.1109/TIM.2019.2898013

[11] Peterek, M, & Montavon, B. (2020). Prototype for dual digital traceability of metrology data using X.509 and IOTA. CIRP Annals, 69(1), pp 449-452. DOI: 10.1016/j.cirp.2020.04.104

[12] Aimagmbetova, R Z, Ershov, I A, & Stukach, O V. (2017). Towards the problem of measurement traceability in the Internet of measurement concept. In: Dynamics of Systems, Mechanisms and Machines (Dynamics), pp 1-4. IEEE. DOI: 10.1109/Dynamics.2017.8239425

[13] Anto, I A F & Nugraha, I G B B. (2018). Blockchain-based for calibration of digital multimeter system design. In 2018 International Conference on ICT for Smart Society (ICISS), pp 1-6. DOI: 10.1109/ICTSS.2018.8549924

[14] Madhwal, Y & Panfilov, P (2017). Blockchain And Supply Chain Management: Aircrafts' Parts' Business Case, In: Proceedings of the 28th DAAAM International Symposium, pp.1051-1056, B. Katalinic (Ed.), DAAAM International, ISSN 1726-9679, pp 1051-1056. DOI: 10.2507/28th.daaam.proceedings.146

[15] Lemeš S. (2020). Blockchain-Based Data Integrity for Collaborative CAD, In: Mixed Reality and Three Dimensional Computer Graphics (B. Sobota, D. Cvetković, Eds.), IntechOpen. DOI: 10.5772/intechopen.93539

[16] Bucci, G, Ciancetta, F, Fiorucci, E, Fioravanti, A, Prudenzi, A & Mari, S. (2019). Challenge and future trends of distributed measurement systems based on blockchain technology in the european context. In IEEE 10th International Workshop on Applied Measurements for Power Systems (AMPS), ISSN: 2475-2304. pp 1-6. DOI: 10.1109/AMPS.2019.8897782

[17] Takatsuji, T, Watanabe, H & Yamashita, Y. (2019). Blockchain technology to visualize the metrological traceability. Precision Engineering, 58, pp 1-6. DOI: 10.1016/j.precisioneng.2019.04.016

[18] Zhu, Y, He, J, Yuan, K & Yang, Y. (2019). Research on Modify Protection of Metrology Electronic Certificate Based on Blockchain Technology. In 2019 14th International Conference on Computer Science & Education (ICCSE), pp 1020-1024. DOI: 10.1109/ICCSE.2019.8845467

[19] Peters, D, Wetzlich, J, Thiel, F & Seifert, J-P. (2018). Blockchain applications for legal metrology. 2018 IEEE International Instrumentation and Measurement Technology Conference (I2MTC). ISBN:978-1-5386-2222-3. DOI: 10.1109/i2mtc.2018.8409668

DOI: 10.2507/32nd.daaam.proceedings.076

JOINING GALVANIZED STEEL SHEETS WITH PLASTICS BY THE RESISTANCE SOLDERING OF BIMETALLIC ELEMENTS

Branislav Vanko, Pavol Sejc & Alexander Schrek

This Publication has to be referred as: Vanko, B[ranislav]; Sejc, P[avol] & Schrek, A[lexander] (2021). Joining Galvanized Steel Sheets with Plastics by the Resistance Soldering of Bimetallic Elements, Proceedings of the 32nd DAAAM International Symposium, pp.0527-0535, B. Katalinic (Ed.), Published by DAAAM International, ISBN 978-3-902734-33-4, ISSN 1726-9679, Vienna, Austria
DOI: 10.2507/32nd.daaam.proceedings.076

Abstract

The proposed method of joining galvanized steel sheets with plastics by the resistance soldering of bimetallic elements provides a permanent joint of two dissimilar materials without breaking the zinc layer of the steel sheets and thermal degradation of the plastics. The method is based on the principle of the resistance element welding (REW) but uses innovative bimetallic joining elements with a hard case and low-melting core enabling joining at lower temperatures (up to 450 °C). It is a suitable alternative for joining plastics to galvanized steel sheets by adhesive bonding or using rivets and screws. This article describes the method in detail and presents the first laboratory tests. The advantage of the method is an intact zinc layer of the steel sheets, non-degraded plastics, possibility to use almost all plastics and a smooth surface of reverse side of the steel sheets without visible joints. The disadvantage of the method is a lower strength of the joints and a more exacting production of joining elements. The use of the method requires solving the problems of creating stronger joints and reliable production of bimetallic elements.

Keywords: Metal-to-Plastic Joining; Resistance Element Welding; Soldering; Bimetallic Elements.

1. Introduction

In practice, it is often necessary to join dissimilar materials, for example metals with plastics, which can be problematic due to the requirements for the properties or the look of the joints. Plastics are light and cost-effective, but generally have relatively low mechanical properties and a low softening or degradation temperature [1], [2]. When joining plastics to galvanized steel sheets, which are widely used especially in the automotive industry, emphasis is placed not only on the final quality of the joint, but also the condition of plastics and the zinc layer of the steel sheets after joining. For joining such a combination of materials, it is possible to use, for example, riveting [3], [4], [5], [6], clinching [3], [4], [5], [6] or screwing [3], [4], [5], but these methods create undesirable protrusions on the joined materials and in many cases the zinc layer of the steel sheet is disrupted at the joint. This is the cause of later corrosion. Suitable alternatives are various welding methods [3], [4], [5], [6], [7], which also disrupt the protective zinc layer, or gluing [3], [4], [5], [6], [8], which may not be applicable to every plastic and a long curing time is usually required.

As an alternative to all these methods, it was designed a new hybrid mechanical-metallurgical method of joining such a combination of materials using the principle of resistance element welding (REW) technology [3], [9], which is currently used mainly in the automotive industry for joining steel sheets with aluminum components. The purpose of the study is to validate this new hybrid method of joining galvanized steel sheets with plastics.

The principle of the REW technology (Fig. 1) comprises the use of a steel joining element (weld rivet) with a cylindrical shank and a head, by means of which an overlap joint of the upper pre-punched cover sheet with the lower base sheet is created by resistance heating. Resistance heating is used to create the weld nugget. For applying REW, a resistance welding gun primarily intended for resistance spot welding (RSW) is used. This involves the creation of a fusion joint between the base sheet and the joining element in combination with a formation of a mechanical joint between the joining element and the cover sheet.

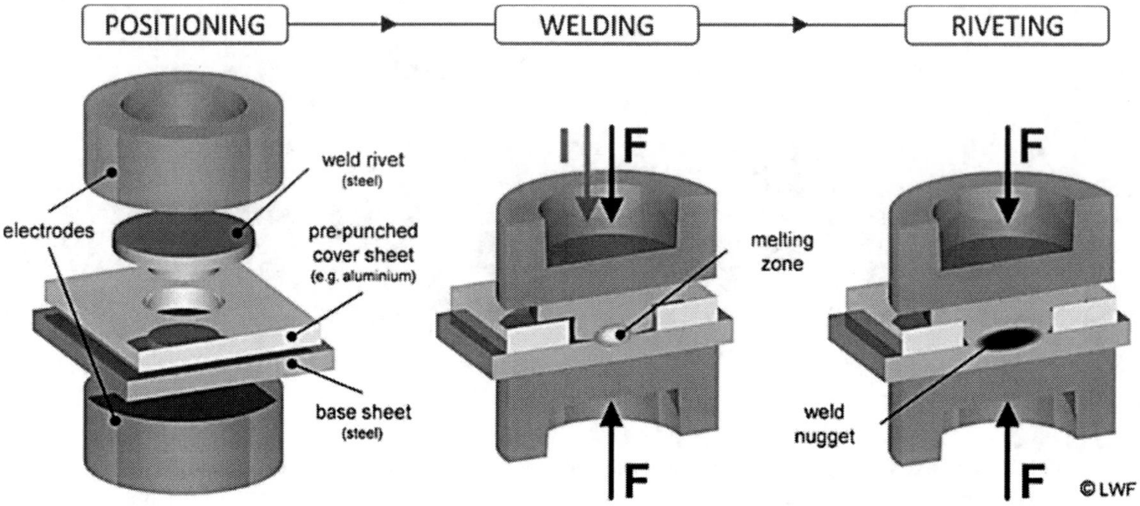

Fig. 1. The principle of welding by REW technology [9]

The main process parameters of the REW are welding time, welding current, and clamping force. Welding time is defined as the time during which the welding current flows through the welded materials. Welding time must be long enough to cause the joining element to melt, but, at the same time, as short as possible to avoid over-heating of the weld area. The clamping force influences the thermal phenomena during welding by affecting the transition resistance between the materials and ensuring that the welded materials are assembled and remain in contact. Unlike conventional RSW, the clamping force also ensures the deformation of the joining element. If the clamping force is sufficiently high, a tight force contact is formed between the joining element and the cover sheet due to deformation.

A solution for joining metal-to-plastic by REW technology can be the use of a joining element made of a material suitable for soldering the galvanized steel base sheet (e.g. tin-based solder). Thus, the joint between the joining element and the steel base sheet does not occur by fusion welding but by soldering. When joining, the steel sheet is only heated to a soldering temperature. Solders have a melting point of less than 450 °C. This is much lower than a fusion welding temperature. Lower heat input reduces heat-induced stresses and deformations that occur during heating and cooling. Plastic cover sheet with a very low softening and degradation temperature can be joined to the steel base sheet without unwanted thermal degradation. Thermal degradation of the plastic sheet significantly affects the appearance, quality, and strength of the joint.

Unfortunately, solders have a significantly lower strength than the steel used for joining elements in REW. Thus, the use of solder materials to produce a joining element for REW technology does not fulfill the condition to produce the desired tight force contact with the plastic cover sheet. Therefore, it was designed a unique bimetallic joining element consisting of a hard case and a soft core. The case was made of a Cu tube and the core was made of a Sn60Pb40 solder. The hard case can create sufficient force contact between the plastic cover sheet and the steel base sheet. The soft core is suitable for soldering of galvanized steel sheets. The melting point of the Sn60Pb40 solder is lower than the thermal degradation temperature of most thermoplastics. The mechanical properties of the Cu tube and the Sn60Pb40 solder are given in Table 1. The shape and dimensions of the bimetallic elements are shown in Fig. 2.

Material	**Yield Strength YS [MPa]**	**Tensile Strength UTS [MPa]**	**Elongation to Fracture A [%]**	**Melting Point T_M [°C]**
Cu	60 ÷ 220 (290)	200 ÷ 300 (345)	4 ÷ 30 (4)	1083
Sn60Pb40	- (40)	40 ÷ 80 (60)	40 (11)	183 ÷ 190

Table 1. Basic properties of Cu and Sn60Pb40 from the literature [10, 11] and measured given in brackets

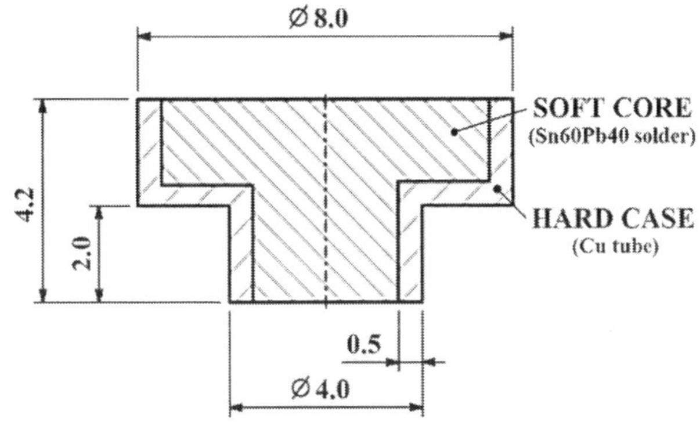

Fig. 2. Shape and dimensions of bimetallic element

2. Experiment

Bimetallic elements were made by casting molten solder into a hot Cu tube (D = 4 mm) housed in a graphite case against an unheated graphite plate as shown in Fig. 3. The casting temperature and the preheating temperature of the Cu tube in the graphite case were 210 ° C. Prior to casting, a flux was applied to the Cu tube to remove surface oxides and improve the wettability. The semi-finished products were cut into short pieces of the required length, which were cold formed by the upsetting process (Fig. 4) in the shape of joining elements (Fig. 5).

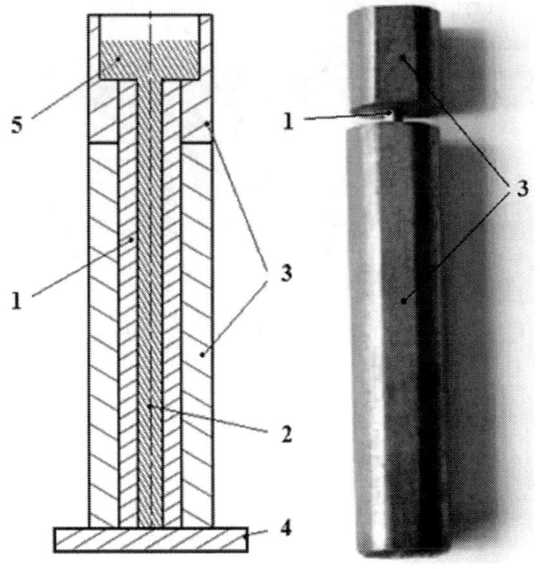

Fig. 3. Casting of semi-finished product for the forming of bimetallic elements (1 - Cu tube, 2 - Sn60Pb40 solder, 3 - graphite case, 4 - graphite plate, 5 – pouring cup)

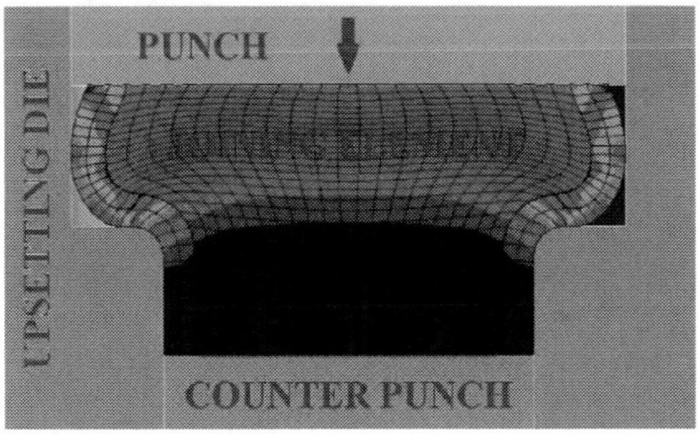

Fig. 4. Cold forming of the bimetallic element by the upsetting process

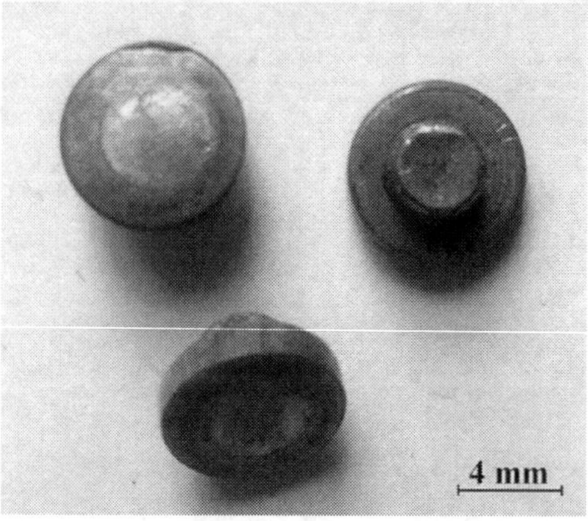

Fig. 5. Finished bimetallic elements

In these first experiments, there was a problem in achieving a good quality of the bimetallic elements. During casting, defects occurred in the form of the porosity of the solder, and during forming, wrinkles and eccentricity were formed (Fig. 6). Defects that were not evident after the production of bimetallic elements during joining caused problems and did not create quality joints (Fig. 7). These defects could also affect the resulting mechanical properties of the joints, which at first look appeared to be of good quality.

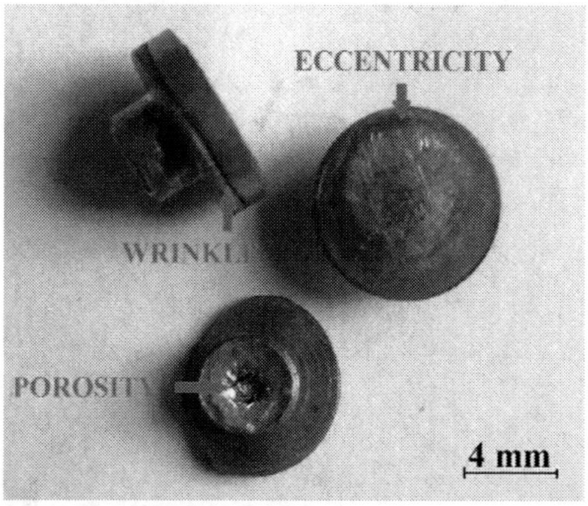

Fig. 6. Defects detected on the bimetallic elements (porosity, wrinkle, eccentricity)

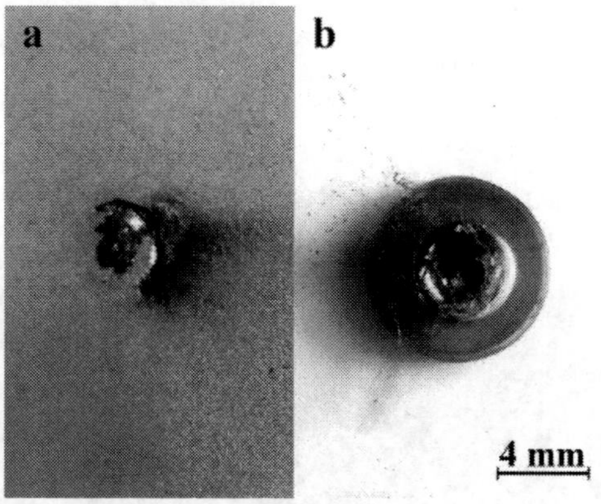

Fig. 7. Joint defect due to hidden porosity in the joining element: a) steel sheet; b) bimetallic element

To verify the proposed hybrid metal-to-plastic joining method the both side galvanized bake-hardening steel sheet HX220BD with a thickness of 0.8 mm and a 100MBO type zinc coating (approx. 15 μm of Zn) was used. Chemical composition and mechanical properties of the steel are given in Table 2 and Table 3.

C [%]	Ti [%]	Si [%]	Mn [%]	P [%]	S [%]	Nb [%]
0.1	0.12	0.5	0.7	0.01	0.03	0.09

Table 2. Chemical composition of HX220BD bake-hardening steel [12]

Yield Strength YS [MPa]	Tensile Strength UTS [MPa]	Elongation to Fracture A [%]
220 ÷ 280	320 ÷ 400	32

Table 3. Basic mechanical properties of HX220BD bake-hardening steel [12]

The steel sheet was joined to the QUINN XT PMMA (Polymethyl Methacrylate) thermoplastic, which was colorless and 2 mm thick. PMMA is primarily intended for deep hot drawing, polishing, gluing, as well as a construction element used for manufacturing safety covers and parts for the instruments. The basic properties of this thermoplastic material are shown in Table 4. The PMMA degradation temperature is 90 °C higher than the melting point of the Sn60Pb40 solder (see Table 1).

Tensile Strength UTS [MPa]	Elongation to Fracture A [%]	Vicat Softening Temperature T_S [°C]	Degradation Temperature T_D [°C]	Max. Temperature Short Term Use T_X [°C]
70	4	110	280	90

Table 4. Basic properties of QUINN XT PMMA plastic [13]

The position of the materials before joining is shown in Fig. 8. After cutting the material to a required size, a hole (d = 4.1 mm) with a recess (D = 8.1 mm, h = 1 mm) was drilled into the thermoplastic to insert the joining element. Before joining, galvanized steel sheets were cleaned of grease and dirt with acetone. Subsequently, SF-601 liquid flux was applied to the surface of the steel sheet at the connection point area.

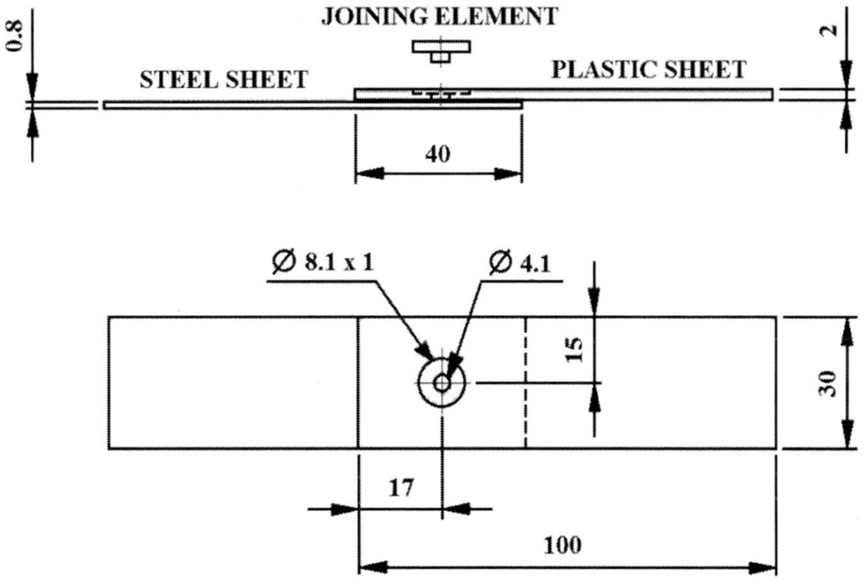

Fig. 8. Dimensions and position of joined materials

The ARO type XMA 26 kVA resistance welding gun with ULB 1.4 universal VTS control system was used for heating of materials at the joint. The parameters were set as follows: welding (soldering) current I = 3.5 kA, clamping force F = 1610 N, and heating (soldering) time t = 10, 15 and 20 cycles. For a 50 Hz power supply, one cycle last 20 milliseconds. By changing the time, the heat input (Q) to the joint was regulated. The real values of the selected process parameters were measured by using the Miyachi measuring device of the Weld Checker MM-356 B type. The set and measured process variables of current, voltage, heating time and heat input are given in Table 5.

The type II (CuCrZr) electrodes designed for RSW of galvanized steel sheets were used. On the side of the joining element, an electrode with a planar contact surface (FC geometry type) was used. On the side of the galvanized steel sheet, an electrode with a rounded surface (FB geometry type) was used. In addition to the electrical current supply, the planar surface of the electrode had the task of positioning the joining element in such a way that its axis was perpendicular to the surface of the steel sheet.

Sample No.	Set Parameters			Measured Parameters		
	Current I [kA]	Clamping Force F [N]	Heating Time t [cycles (ms)]	Current I_R [kA]	Voltage U [V]	Heat Input Q [J]
1.	3.5	1610	10 (200)	3.50	0.42	294
2.	3.5	1610	10 (200)	3.52	0.42	296
3.	3.5	1610	10 (200)	3.51	0.42	295
4.	3.5	1610	10 (200)	3.50	0.41	287
5.	3.5	1610	15 (300)	3.62	0.36	391
6.	3.5	1610	15 (300)	3.61	0.37	401
7.	3.5	1610	15 (300)	3.60	0.38	410
8.	3.5	1610	15 (300)	3.67	0.35	385
9.	3.5	1610	20 (400)	3.62	0.37	536
10.	3.5	1610	20 (400)	3.60	0.38	547
11.	3.5	1610	20 (400)	3.61	0.37	534
12.	3.5	1610	20 (400)	3.63	0.37	537

Table 5. Set and measured parameters of joining process

3. Results and discussion

After overcoming the initial complications of setting the correct welding position with a welding jig, a series of joints was created according to the parameters listed in Table 5. The variable parameter was the heating time by which the heat introduced into the joint (heat input) was regulated. The joints could be reliably made with a heating time in the range of 10 to 20 cycles. In an attempt to use heating times longer than t ≥ 25 cycles, there were persistent manufacturing problems, so these joints were included in this study only marginally, when examining the cross-sectional appearance of the joints. A typical appearance of the joints is presented in Fig. 9 and Fig. 10.

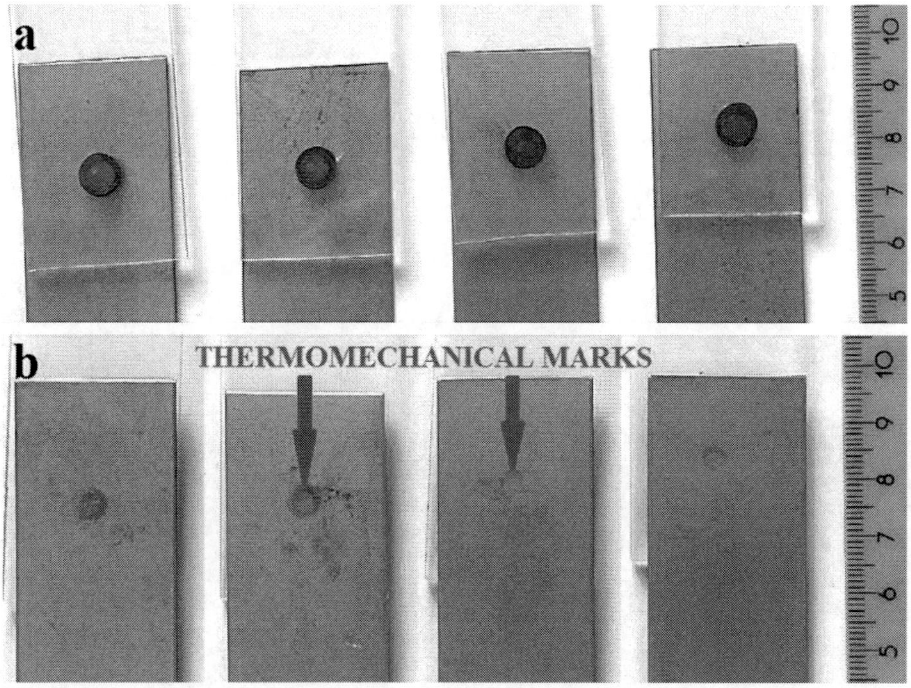

Fig. 9. Samples no. 5, 6, 7 and 8 (t = 15 cycles): a) viewed from the side of joining elements and plastic sheets; b) viewed from the side of galvanized steel sheets

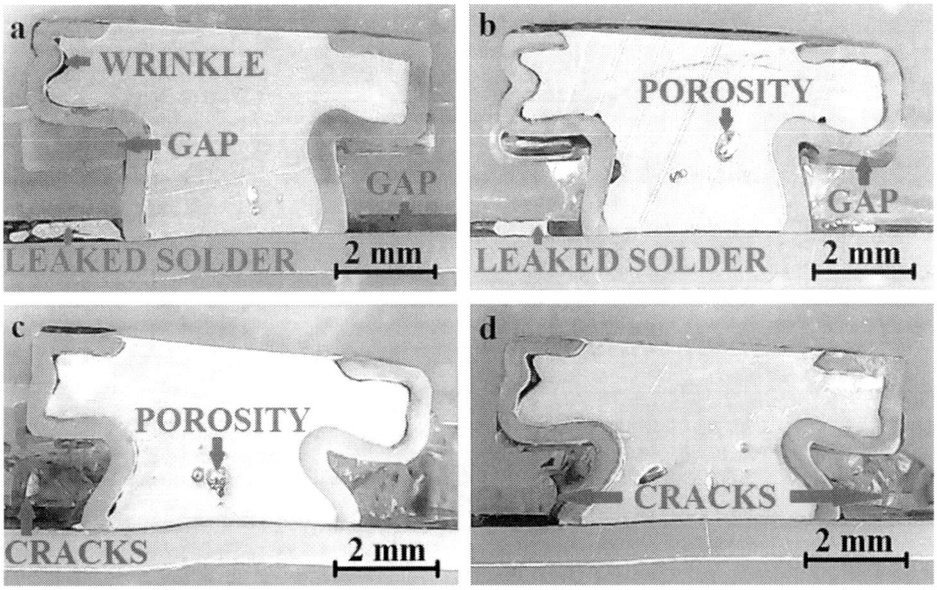

Fig. 10. Samples made with different heating times in cross section view: a) t = 10 cycles; b) t = 15 cycles; c) t = 20 cycles; d) t = 25 cycles

In Fig. 9, it can be seen that the undersides of the steel sheets have imprints from the welding electrode (thermomechanical marks), but the zinc layer has not been broken. This protective anti-corrosion layer was not broken in any joining mode. Also, the plastic did not degrade in any joining mode, but at longer heating times ($t \geq 20$ cycles) micro-cracks formed in the plastic during joining, due to excessive deformation of the shank of the joining element (Fig. 10 c and Fig. 10 d). This deformation was caused by excessive heating and softening of the joining element due to the large amount of heat introduced into the joints (heat input). Micro-cracks in the joint area occasionally caused the plastic and not the solder joint to break during the tensile test. In contrast, the small amount of heat introduced at shorter heating times ($t \leq 15$ cycles) did not cause sufficient deformation of the shank of the joining element to form a tight force connection between the plastic and the joining element (Fig. 10 a and Fig. 10 b). From this point of view, between 15 and 20 cycles seems to be the most suitable heating time. For future experiments, it is worth considering a small increase in the diameter of the pre-drilled hole in the plastic sheet or a shortening of the shank of the joining element, which should ensure less cracking in the plastic during joining.

In Fig. 10 it can be seen that the right sides of the joining elements are more compressed than the left side, i.e. the absolute perpendicularity of the joined sheets to the welding electrodes was not observed during the joining. This deviation from the perpendicularity did not have a significant effect on the quality of the joints.

The tensile test was performed using the Instron 1195 universal testing machine. A typical fracture surface after the tensile test is shown in Fig. 11 on which it can be seen that the solder wets the surface of the galvanized steel sheet very well and forms a high-quality soldered joint. It can be seen in the Fig. 10 and Fig. 11 that a small amount of solder can leak or spurt out between the plastic and the steel sheet during the process of joining. The results of tensile test of overlap joints are shown in Table 6.

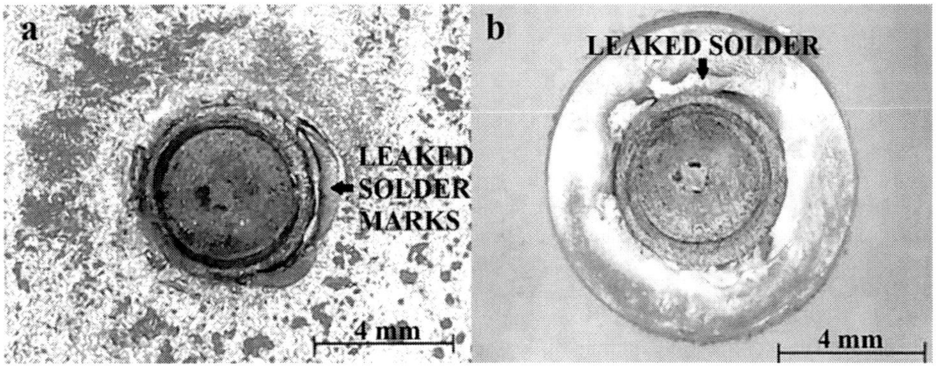

Fig. 11. Detail of the fracture surface of sample no. 6 (t = 15 cycles) after the tensile test: a) galvanized steel sheet; b) bimetallic element in the plastic sheet

Sample No.	Clamping Force F [N]	Heating Time t [cycles (ms)]	Heat Input Q [J]	Max. Tensile Force F_{Tmax} [N]	Standard Deviation
1.	1610	10 (200)	294	416	
2.	1610	10 (200)	296	308	
3.	1610	10 (200)	295	241	
4.	1610	10 (200)	287	Lost Sample	
Average value	1610	10 (200)	293	322	88
5.	1610	15 (300)	391	186	
6.	1610	15 (300)	401	425	
7.	1610	15 (300)	410	493	
8.	1610	15 (300)	385	291	
Average value	1610	15 (300)	397	349	137
9.	1610	20 (400)	536	515	
10.	1610	20 (400)	547	632	
11.	1610	20 (400)	534	472	
12.	1610	20 (400)	537	331	
Average value	1610	20 (400)	539	488	124

Table 6. Tensile test results

The average values of the joint strength (evaluated by the average values of the maximum tensile force) ranged from 322 N (t = 10 cycles) to 488 N (t = 20 cycles), but the variance of the measurements was quite large (see standard deviations in the Table 6). The results showed that the heating time (heat input) influence the mechanical properties of the joints made by a bimetallic element. As the heat input increased, the deformation of the joining element shank increased (see Fig. 10), which caused a larger contact area between the joining element and the steel sheet, therefore the strength of the joint increased.

Samples with a preheating time t ≥ 25 cycles were not tested because there were not enough of them. During production, solder escaped from the top of the joining element (contact area between the welding electrode and the joining element), which caused the higher deformation of the joining element and more extensive cracking of the plastic sheet. This phenomenon occurred because there was too much molten solder (about 90 %) in the bimetallic element due to the large amount of heat input. To create a quality joint the ideal amount of remelted solder is about 70 to 80 %, which occurred at a heating time of 15 and 20 cycles (Fig. 12).

Fig. 12. Sample no. 8 (t = 15 cycles) with approximately 70 % of remelted solder in the joint

4. Conclusion

The proposed hybrid mechanical-metallurgical method of joining galvanized steel sheets with plastics using bimetallic elements without thermal degradation of plastics and breaking of the galvanized layer has been successfully verified. After the production of the first bimetallic elements of acceptable quality, reliable joints were successfully created which could be mechanically loaded. The tensile load for breaking the joint was relatively low, the average value of the tensile force ranged from 322 to 488 N, in addition with a large variance of the values (deviation from the average value), so it is necessary to optimize this method of joining dissimilar materials.

Based on the achieved results, it was concluded that the strength of joints cannot be increased by changing process parameters such as heating time, because there is only one optimal setting of the joining process, which will bring quality and strong joints. For a given current of 3.5 kA and a clamping force of 1610 N, the optimal heating time would be somewhere between 15 and 20 cycles, which will provide the required amount of molten solder in the joining element (about 70 to 80 %), and this will allow the required deformation of the element shank.

The strength of the joints was affected by the size of the contact area between the joining element and the steel sheet and the quality of the bimetallic elements. For this reason, a larger diameter of the Cu tube (D = 6 mm), a modified technology of filling the Cu tube with solder and a new tool for forming the joining elements were chosen for further experiments.

In further experiments, we would like to verify the quality and reliability of bimetallic elements produced by a new technological process, evaluate the impact of a larger contact area between the joining element and the steel sheet on the mechanical properties, and optimize the size of the hole for the joining element. Furthermore, we would like to replace the Sn60Pb40 solder, which due to its lead content is not in accordance with the EU Directive 2002/96/EC on Waste Electrical and Electronic Equipment (WEEE Directive) of 2003, a lead-free solder (e.g. Sn91Zn9) or a solder suitable for soldering other metallic materials, e.g. aluminum alloys.

5. Acknowledgments

We thank the Slovak grant agency VEGA for financial support of the VEGA 1/0405/19 project and the University Science Park STU Bratislava for the support of the ITMS 26240220084 project.

6. References

[1] Pandzic, A[di]; Hodzic, D[amir] & Milovanovic, A[leksa] (2019). Effect of Infill Type and Density on Tensile Properties of PLA Material for FDM Process, Proceedings of the 30th DAAAM International Symposium, pp.0545-0554, B. Katalinic (Ed.), Published by DAAAM International, ISBN 978-3-902734-22-8, ISSN 1726-9679, Vienna, Austria. DOI: 10.2507/30th.daaam.proceedings.074

[2] Pandzic, A[di]; Hodzic, D[amir] & Milovanovic, A[leksa] (2019). Influence of Material Colour on Mechanical Properties of PLA Material in FDM Technology, Proceedings of the 30th DAAAM International Symposium, pp.0555-0561, B. Katalinic (Ed.), Published by DAAAM International, ISBN 978-3-902734-22-8, ISSN 1726-9679, Vienna, Austria. DOI: 10.2507/30th.daaam.proceedings.075

[3] Meschut, G.; Janzen, V. & Olfermann, T. (2014). Innovative and Highly Productive Joining Technologies for Multi-Material Lightweight Car Body Structures. Journal of Materials Engineering and Performance, Vol. 23, May 2014, pp. 1515-1523. ISSN 1059-9495. DOI: 10.1007/s11665-014-0962-3

[4] Martinsen, K.; Hu, S. J. & Carlson, B. E. (2015). Joining of Dissimilar Materials, In: CIRP Annals - Manufacturing Technology, Vol. 64, Iss. 2, pp. 679-699, Elsevier, ISSN 0007-8506. DOI: 10.1016/j.cirp.2015.05.006

[5] Kah, P.; Suoranta, R.; Martikainen, J. & Magnus, C. (2014). Techniques for Joining Dissimilar Materials: Metals and Polymers. Reviews on Advanced Materials Science, Vol. 36, No. 2, 2014, pp. 152-164, ISSN 1605-8127.

[6] Chastel, Y. & Passemard, L. (2014). Joining Technologies for Future Automobile Multi-material Modules. Proceedings of Procedia Engineering 81, 11th International Conference on Technology of Plasticity, ICTP 2014, 19-24 October 2014, Nagoya, Japan, ISSN 1877-7058, Ishikawa, T. & Mori, K. (Ed.), pp. 2104 – 2110, Elsevier. DOI: 10.1016/j.proeng.2014.10.293

[7] Katayama, S. & Kawahito, Y. (2008). Laser Direct Joining of Metal and Plastic. Scripta Materialia. Vol. 59, Iss. 12, December 2008, pp. 1247-1250, ISSN 1359-6462. DOI: 10.1016/j.scriptamat.2008.08.026

[8] Banea, M. D. (2021). Dissimilar Joining of PMCs to Metals – Adhesive Bonding, In: Reference Module in Materials Science and Materials Engineering, Elsevier, ISBN 9780128035818. DOI: 10.1016/B978-0-12-819724-0.00061-6

[9] Meschut, G.; Hahn, O., Janzen, V. & Olfermann, T. (2014). Innovative Joining Technologies for Multi-Material Structures. Welding in the World, Vol. 58, January 2014, pp. 65-75, ISSN 1878-6669. DOI: 10.1007/s40194-013-0098-3

[10] Vianco, P. T. (1999). Soldering Handbook, 3rd edition, American Welding Society, ISBN 978-0-87171-618-7.

[11] ASM Specialty Handbook: Copper and Copper Alloys (2001). Edited by J. R. Davis, Davis & Associates, ASM International, ISBN 978-0-87170-726-8.

[12] https://www.usske.sk/en/products/hot-dip-galvanized/bake-hardening-steels, (2019). U. S. Steel Košice, Bake-Hardening Steels, Accessed on: 2021-07-07.

[13] http://vascony.com/wp-content/uploads/2014/01/en-quinn-xt-10-07.pdf, (2019). Quinn Plastics, Product Information: QUINN XT, Accessed on: 2021-07-07.

DOI: 10.2507/32nd.daaam.proceedings.077

USING COLOUR-BASED OBJECT DETECTION FOR PICK AND PLACE APPLICATIONS

Moritz Abdank, Mohamed Aburaia & Wilfried Woeber

This Publication has to be referred as: Abdank, M[oritz]; Aburaia, M[ohamed] & Wöber, W[ilfried] (2021). Using Colour-based Object Detection for Pick and Place Applications, Proceedings of the 32nd DAAAM International Symposium, pp.0536-0541, B. Katalinic (Ed.), Published by DAAAM International, ISBN 978-3-902734-33-4, ISSN 1726-9679, Vienna, Austria
DOI: 10.2507/32nd.daaam.proceedings.077

Abstract

Pick and place tasks are one of the most typical applications in robotics. Conventional robots perform a fixed task which restricts the workspace in terms of flexibility. With development in mobile robotics [1] and automation it is necessary to introduce sensor systems that enable a more flexible approach. This paper demonstrates a computer vision system which uses a colour-based approach to object detection and estimation of its 3D world coordinates based on a 2D image. The system is used to enable a robot arm to pick up an object and place it on a CNC-mill. Based on the use case the vision system can tell the robot if an object has been detected and where it is relatively to the camera. Using the robot operating system (ROS) and OpenCv, image thresholding is performed for detecting the object. The Canny edge detection algorithm is deployed to mark the object with a bounding box. The retrieved information from those operations is used for the 3D pose estimation. Using a linear calibration, the error of the system can be reduced, which results in an accuracy of 0.09cm.

Keywords: Computer vision; linear regression; object detection; HSV; ROS

1. Introduction

Widely available and cheap cameras enable the opportunity for different computer vision systems. Open-source libraries such as OpenCV [2] and ROS [3] provide a solid basis for developing robotic systems. Therefore, vision systems can be used to expand typical pre-coded robotic tasks such as pick and place applications to a more flexible setup which is not restricted to a predefined pick-up location [4]. In this paper, a colour-based object detection approach is implemented using image thresholding and linear calibration for error compensation. The vision system estimates the 3D world coordinates of a red metal cylinder based on the 2D image stream from a conventional webcam. To test the system, a robot arm is used to pick up the defined object and place it in a CNC-mill. The proposed system is developed with an autonomous task in mind, where a mobile robot transfers the object to the stationary robot arm. To compensate for the inaccuracy of the mobile robot, a camera observes the delivery zone. When the object is placed in the workspace of the robot, a vision system estimates the 3D position of the object and therefore the accuracy of the mobile delivering robot does not have to be exact. The addressed problem is the calculation of the 3D pose based on the 2D stream of the image. The contribution of this paper is the accuracy analysis of a computer vision framework and the object detection without machine learning.

2. State of the art

One of the more challenging tasks with computer vision systems is object detection. Khan, et. al. [5] especially mention viewpoint changes, scale, and illumination as some of the major variables which result in challenges for object detection algorithms. Since colour can change based on the illumination and the surroundings, colour information is often discarded [5] as an option for object detection [6]. managed to use colour-based information for the estimation of the location of traffic signs. The location was then used in combination with a support vector machine (SVM) [7] to classify the sign.

Image thresholding is one of the fastest methods for object detection [6] and was successfully implemented by [8] for object tracking with autonomous unmanned aerial vehicles (UAV). Using the HSV colour spectrum instead of the RGB spectrum, [8] implemented a threshold-based colour detection for tracking their defined object. To reduce the complexity mentioned in [5], they defined certain parameters and restrictions for the system.

State of the art object detection system can also utilise convolutional neural networks. YoloV3 [9] is one of the fastest real time object detection systems, with an image processing rate of 30 FPS on a Titan X. Such implementations can be used for various applications. An UAV approach to detect forest fires with this network resulted in 6FPS [10] and another YoloV3 based detector mentioned in [11] managed to reach up to 10FPS for a real-time traffic sign recognition. As mentioned by [18] the limitations of a real-time Yolo system is highly dependent on the hardware.

3. Methods

To implement the vision system, a Logitech C270 RGB-Webcam is used. After calibration, the intrinsic matrix can be used to get the first parameters for estimating the 3D pose. With the premise of the pinhole camera model [12], the relation of similar triangles is the only restriction left, requiring that the object has a plane surface and is parallel to the camera. This setup resembles the premise from the UAV object tracking setup to reduce the complexity of the setup and therefore enable a fast and accurate detection method. The overall process flow is presented in Fig. 1.

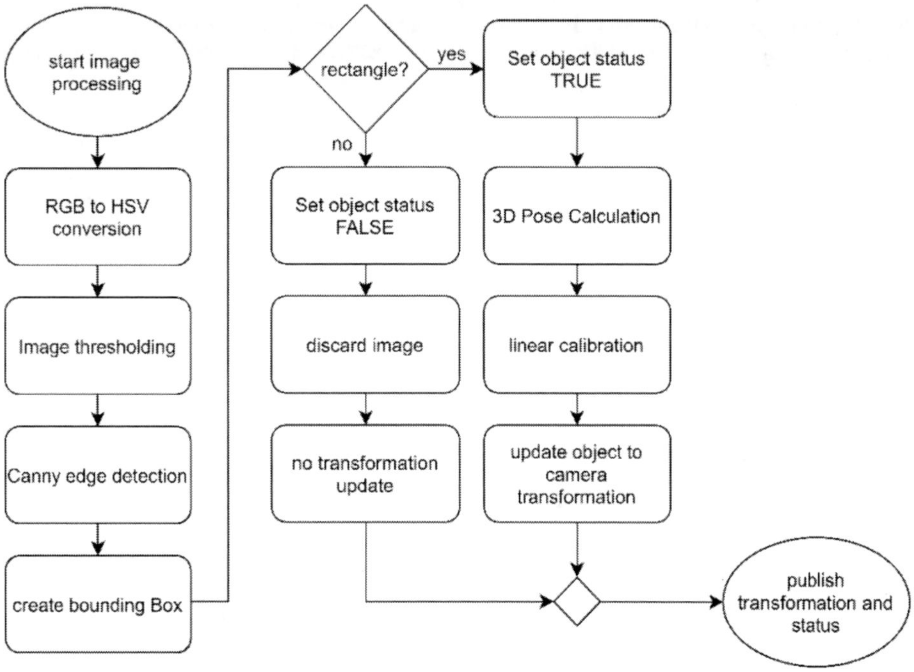

Fig. 1. Program flow diagram showing the different stages of the image processing and decision if the object is truly detected or if a false recognition is the case

For the object detection algorithm, which defines the range for the 3D pose estimation, a colour-based approach has been chosen to see if it can perform comparably to state of the art detection systems like the mentioned YoloV3 implementations [10] and [11]. The complete process of the colour-based system is shown in Fig. 1.In the first step, the rectified image is converted from the RGB-colour spectrum to the HSV-colour range. Since colour-based detection models are highly dependent on the lighting conditions, this enables the option to adapt the desired colour intensity and saturation more easily.

To detect the object, a threshold must be defined from experiments. An erode and dilate filter was implemented to eliminate noise [13] and prevent false detection to a certain degree. With the threshold image in place, the Canny algorithm [14] is deployed to find the edges of all the visible objects and the corresponding bounding boxes. The biggest bounding box is selected as the parameters of the detected object.

The bounding box provides the needed information for the pose estimation of the object. First, based on the now known image-coordinates and size of the bounding box, the centre of the object can be defined and will also be the origin of the object coordinate system. These image coordinates in pixel size will later be converted to real life references. Afterwards, the size of the bounding box can then be used to calculate the distance between the camera and the object.

Depending on the use case and the object, the third parameter is the orientation of the object. To get the orientation, the bounding box must be adapted to enclose the smallest area of the detected contours. The work piece for this paper was defined as a cylinder, resulting in the orientation being irrelevant. With this parameter, it is possible to filter false detection based on the bounding box. If the detected bounding box is not a square, the centre of the coordinate system will not be in the centre of the object and therefore must be discarded. With the now known parameters, (1) can be used to calculate the Z-coordinate of the object which represents the distance between the camera and the surface of the object.

$$Z = \frac{f \cdot width_{real}}{width_{image}} \tag{1}$$

This equation is based on the knowledge of the size of the object ($width_{real}$) and the width of the object ($width_{image}$). The proposed linear calibration is performed using (2).

$$x = \frac{(y - d)}{k} \tag{2}$$

This results in the correction of the measurement y with the linear calibrated parameters k and y. To perform no linear calibration the default values are d=0 and k=1. To calibrate the system and to adjust the parameters, a polyfit trendline calculation from the NumPy library [15] is performed based on known measurements from an experimental setup. To calculate the X and Y-coordinates, (3) and (4) are used.

$$X = \frac{u \cdot Z}{c_x \cdot f_x} \tag{3}$$

$$Y = \frac{v \cdot Z}{c_y \cdot f_y} \tag{4}$$

The same linear calibration, that was deployed for the Z-coordinate, is also implemented for the X and Y calculation. To make the detection compatible with different robot systems, the ROS tf package [16] is used as an interface between different robot systems. This allows for the possibility to define the coordinate systems for the camera and the object. Knowing that, a camera model can be implemented in a running robot operating system and the tf package will take care of the transformations between the different coordinate frames, calculating a direct transformation between the object and robot coordinate systems. For example, this is used in the experimental setup to calculate the transformation between the now detected object and the robot's end effector.

For the use case, the camera can be mounted virtually in reference to the robot system using RVIZ [17]. Different coordinate systems represent the different parts of the setup. This can be helpful to align the robot coordinate frame with the camera coordinates. The system was implemented and tested on ROS Noetic Ninjemys [3], OpenCV 4.4.0 [2] and NumPy 1.16.6 [15].

4. Experimental setup and analysis

To test the accuracy of the vision system, the camera is mounted parallel to the tabletop (see Fig. 2). The distance between the vertical camera and the use case object that must be recognized, resulting in a distance of 69cm. For the vertical (Z-Axis) measurements, the object was positioned at ten different places on the tabletop. At each position, 100 measurements were recorded periodically with a frequency of 10Hz using a tf-listener to intercept the frame transformations between object and camera. This data was then used for analysis and the proposed linear calibration. The same measurements as before were performed to evaluate the optimized system. For the X- and Y-axis, the object was placed along the desired axis. Then it was moved in 4cm intervals. Each position was recorded 100 times and evaluated.

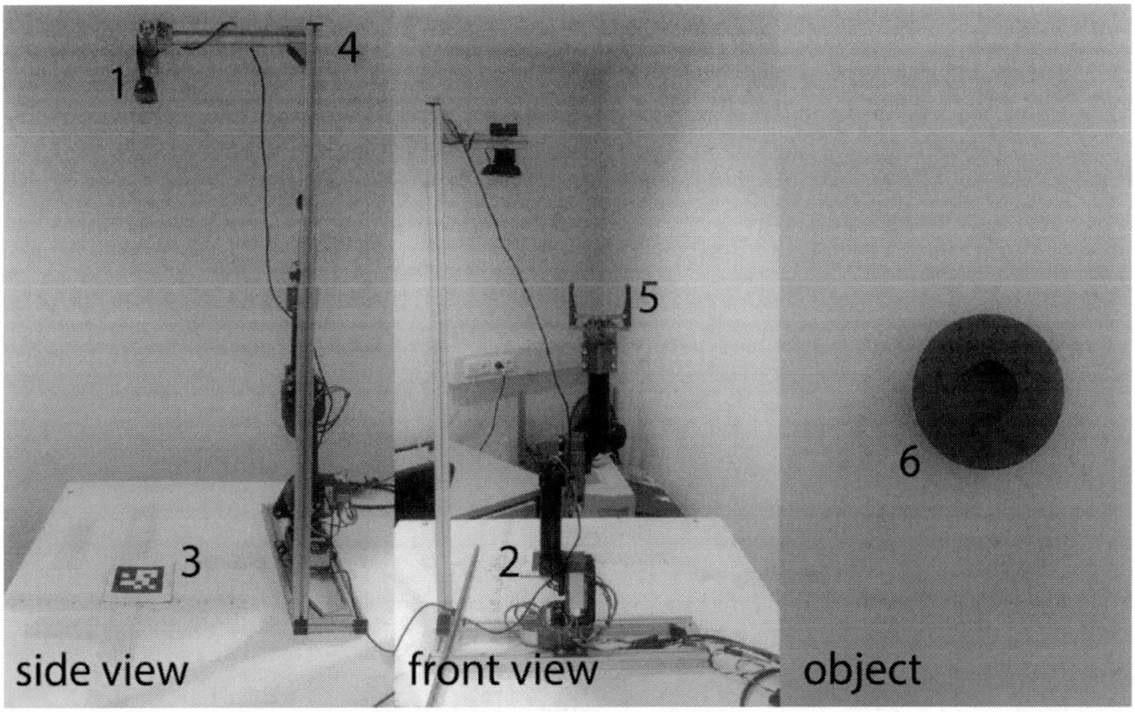

Fig. 2. Experimental Setup of the Vision System in reference to the designated Use Case. 1. webcam, 2. robot arm, 3. expected pick up location, 4. frame, 5. end effector, 6. object

These measurements were processed the same way as the Z-axis and then re-evaluated with the same measurements as before. The X-measurements were performed over a distance of 36cm and the Y-measurements were carried out over a distance of 32cm. Fig. 3 shows the calculated trendline (red) compared to a perfect system (green) that has no error.

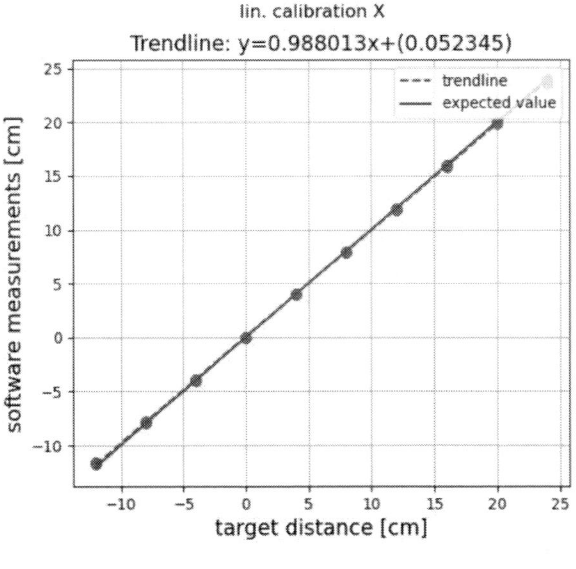

Fig. 3. Linear calibration diagram for calculating the corresponding parameters (d,k)

The difference in error is shown in Fig. 4. The trendline notation follows (3). The results are listed in table 1. The Z-axis calculation is straight forward since the measurement should always be 69cm. Therefore, the average was taken of the measurements and subtracted from the defined height. The raw calculation of the Z-distance had an average error of about 0,035mm with a scattering around 0.5565. The X and Y-coordinates had a bigger average error of 0.1194cm and 0.1531cm, but a smaller mean variation compared to the Z-axis.

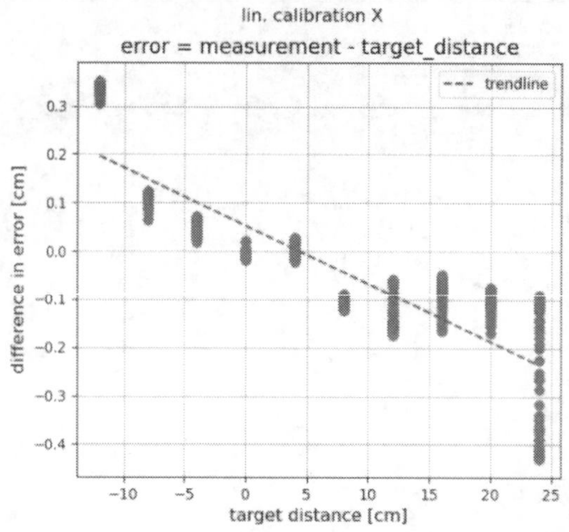

Fig. 4. Difference in error based on the linear calibration for the X-coordinates

measurement	result (mean)	difference	Standard deviation
Z	69,035cm	0,035cm	0,5565
calibrated Z	67,587	-1,413	1,9194
X	not calculated*	0,1194cm	0,08665
calibrated X	not calculated*	0,08889cm	0,06657
Y	not calculated*	0,15307cm	0,11442
calibrated Y	not calculated*	0,1465cm	0,11429

Table 1. Data Evaluation

*Mean measurements were not calculated because of the measurement type for X and Y. For the difference and scattering, the individual measurement was subtracted from the corresponding step (4cm intervals). The absolute value was used for comparison/further calculations. Corresponding Steps/Coordinates in X: [-12, -8, -4, 0, 4, 8, 12, 16, 20, 24], Steps/Coordinates in Y: [-20, -16, -12, -8, -4, 0, 4, 8, 12]

5. Discussion

This paper proposed a simple colour-based object detection for different robotic tasks, including a linear calibration method, with focus on a specific pick and place use case. A standard RGB-Webcam stream was used to detect a certain colour and obtain information which then was filtered and evaluated based on the predefined conditions of the system. ROS enables the processing of the calibrated camera stream and the interface to the pick and place robot arm. A coordinate transformation is used to pass the calculated world coordinates to the robot arm that places the detected object into a CNC-mill. The experimental setup was designed for a mobile robot to deliver the object that must be manipulated. Therefore, the vision system is needed to compensate for possible inaccuracy of the mobile robot, since the drop-off location can vary with each cycle.

During the tests an accuracy of around 0.12cm (optimized) was achieved in the X/Y-plane, which resembles the tabletop of the setup where the object rests. The average Z-distance accuracy for the system resulted in around 0.035cm without any optimization and a transformation publishing frequency of around 7Hz. The optimization for the Z-axis resulted in a false compensation and therefore distorted the measurements and had to be discarded. The overall accuracy for the system was sufficient to successfully grab the object and place it into the CNC-mill.

For more complex object shapes, the orientation might be needed. Simple orientation tests with a square object have been performed. The yaw orientation was successfully calculated but no accuracy measurements were performed, since the robot end effector was designed to grab a cylinder. Given the restricted parameters of the setup, the calculation of the orientation is promising and is worth pursuing in future work. Although the vision system and its accuracy are dependent on the surrounding lighting conditions, the advantages of a lightweight, flexible, and accurate setup should not be disregarded when designing a robot station.

6. Acknowledgments

This work was supported by the city of Vienna (MA23 – Economic Affairs, Labour and Statistics) through the research project AIAV (MA23 project 26-04)

7. References

[1] Gacovski, Z. (2011). Mobile Robots: Current Trends
[2] Bradski, G. (2000). The OpenCV Library Dr. Dobb's Journal of Software Tools
[3] Quigley, M.; Gerkey B.; Conley K.; Faust J.; Foote T.; Leibs J.; Berger E.; Wheeler R.; Ng A. (2009). ROS: an open-source Robot Operating System, ICRA workshop on open source software
[4] Andhare, P. & Rawat, S. (2016). Pick and place industrial robot controller with computer vision, International Conference on Computing Communication Control and automation (ICCUBEA)
[5] Khan, F. S.; Anwer, R. M.; Weijer, J. v. d.; Bagdanov, A. D.; Vanrell, M. & Lopez, A. M. (2012). Color attributes for object detection, IEEE Conference on Computer Vision and Pattern Recognition
[6] Soendoro, D. & Supriana, I. (2011). Traffic sign recognition with Color-based Method, shape-arc estimation and SVM, Proceedings of the 2011 International Conference on Electrical Engineering and Informatics
[7] Bishop, C. M. (2006). Pattern recognition and machine learning, New York, NY
[8] Kadouf, H. H. A. & Mustafah, Y. M. (2013). Colour-based Object Detection and Tracking for Autonomous Quadrotor {UAV}, {IOP} Conference Series: Materials Science and Engineering, Bd. 53
[9] Redmon, J. & Farhadi, A. (2018). YOLOv3: An Incremental Improvement, arXiv
[10] Jiao, Z.; Zhang, Y.; Xin, J.; Mu, L.; Yi, Y.; Liu, H. & Liu, D. (2019). A Deep Learning Based Forest Fire Detection Approach Using UAV and YOLOv3, 1st International Conference on Industrial Artificial Intelligence (IAI), pp. 1-5
[11] Rajendran, S. P.; Shine, L.; Pradeep, R. & Vijayaraghavan, S. (2019). Real-Time Traffic Sign Recognition using YOLOv3 based Detector, 10th International Conference on Computing, Communication and Networking Technologies (ICCCNT), pp. 1-7
[12] Lu, X. X. (2018). A Review of Solutions for Perspective-n-Point Problem in Camera Pose Estimation, Journal of Physics: Conference Series
[13] Jamil, N.; Sembok, T. M. T. & Bakar, Z. A. (2008). Noise removal and enhancement of binary images using morphological operations, International Symposium on Information Technology
[14] Topal, C.; Akınlar C. & Genç, Y. (2010). Edge Drawing: A Heuristic Approach to Robust Real-Time Edge Detection, 20th International Conference on Pattern Recognition
[15] Harris, C. R.; Millman, K. J. & Walt S. J. v. d. (2020). Array programming with {NumPy}, Nature, pp. 357-362
[16] Foote, T.; Marder-Eppstein E. & Meeussen, W. (2017). Available from: http://wiki.ros.org/tf, Accessed: 2021 03-16
[17] PickNikRobotics. (2020). Rviz Visual Tools: C++ API wrapper for displaying shapes and meshes in Rviz
[18] Deac, G. C.; Deac, C. N.; Popa, C. L.; Ghinea M. & Cotet, C. E. (2017). Machine Vision in Manufacturing Processes and the Digital Twin of Manufacturing Architectures, Proceedings of the 28th DAAAM International Symposium, pp.0733-0736, B. Katalinic (Ed.), Published by DAAAM International, ISBN 978-3-902734-11-2, ISSN 1726-9679, Vienna, Austria, DOI: 10.2507/28th.daaam.proceedings.103

DOI: 10.2507/32nd.daaam.proceedings.078

UNIFORM MANUFACTURER-INDEPENDENT MOBILE ROBOT PROGRAMMING

Lucas Muster[1], Wilfried Woeber[2] & Mohamed Aburaia[3]

[1]Lucas Muster is with the Department of Industrial Engineering, University of Applied Sciences Technikum Wien, 1200 Vienna, Austria lucas.muster@technikum-wien.at

[2]Wilfried Wöber is with the Department of Industrial Engineering, University of Applied Sciences Technikum Wien, 1200 Vienna, Institute for Integrative Nature Conservation Research, University of Natural Resources and Life Sciences, Vienna Austria wilfried.woeber@technikum-wien.at

[3]Mohamed Aburaia is with the Department of Industrial Engineering, University of Applied Sciences Technikum Wien, 1200 Vienna, Austria mohamed.aburaia@technikum-wien.at

This Publication has to be referred as: Muster, L[ucas]; Woeber, W[ilfried] & Aburaia, M[ohamed] (2021). Uniform Manufacturer-Independent Mobile Robot Programming, Proceedings of the 32nd DAAAM International Symposium, pp.0542-0545, B. Katalinic (Ed.), Published by DAAAM International, ISBN 978-3-902734-33-4, ISSN 1726-9679, Vienna, Austria
DOI: 10.2507/32nd.daaam.proceedings.078

Abstract

Reliability, flexibility and minimum processing time are the key production requirements for sustainably competitive manufacturing. A high degree of flexibility often requires adjustments to both the individual robot-operated systems and the mobile robots used on the process side, resulting in an automatic increase in setup times. A uniform manufacturer-independent programming method for robots would radically reduce programming time. This work describes the development and implementation of a programming concept for mobile robots which allows the use of uniform commands independent of the manufacturer. The resulting programming library breaks these commands down to a level where autonomous code can be generated for the individually used robot, eliminating the need for the user to deal with complex programming tools.

Keywords: Mobile robot; Programming library; Robot skills; Abstract programming

1. Introduction

To keep the adaptability in production lines as high as possible, a special focus is put on robotics, since it has become a decisive factor for years [4]. Due to the large number of different robot types, programming languages and applications, it is time consuming and difficult to create or adapt robot applications. Especially mobile robots are complex to program since they often have to rely on statistical algorithms [11]. If mobile robots are extended with an additional manipulator, a high degree of automation can be created [1].

However, this results in even greater complexity in the programming implementation of certain applications. Without a standard specification for robotic software the workload of programming is exceedingly high [2],[7]. Therefore, this work describes the development of an autonomous robot programming concept for a mobile robot. The contribution of this study is the development, implementation and evaluation of a generic skill-based library for mobile robot applications. This work is structured as follows. The next chapter describes the current state of the art. Then in chapter 3 the used methods are described. The results are presented in chapter 4. Finally, this paper is summarized in chapter 5.

2. State of the Art

In order to be able to design an abstract programming concept, a systematic model must be created, which lies above the actual programming of the robot. A possible implementation of this model can be done with a decision system, a state machine for the individual operations in the process, an operating system, reflecting the functions of the existing robot such as the control of a motor and the reading of a sensor and an information system, processing all forms of internal and external communication [3]. As an approach for the decision system and thus the generation of robot sequence programs can be designed with the help of Planning Domain Definition Language (PDDL) [13]. By determining so-called domain files for predicates, actions and problem files for objects, initial states and goal specifications, a certain procedure and code can be generated, providing the solution to a specific problem. In retrospect, a process can thus be changed very easily by integrating new states or constrains, for example, since the code or sequences generate themselves.

To implement all the functions of the robot, a certain sequence must also be generated in the software system. It is necessary to create a state machine for the individual commands in a way that the program can handle all kinds of functions, which can be divided into smaller commands, as shown by [10].

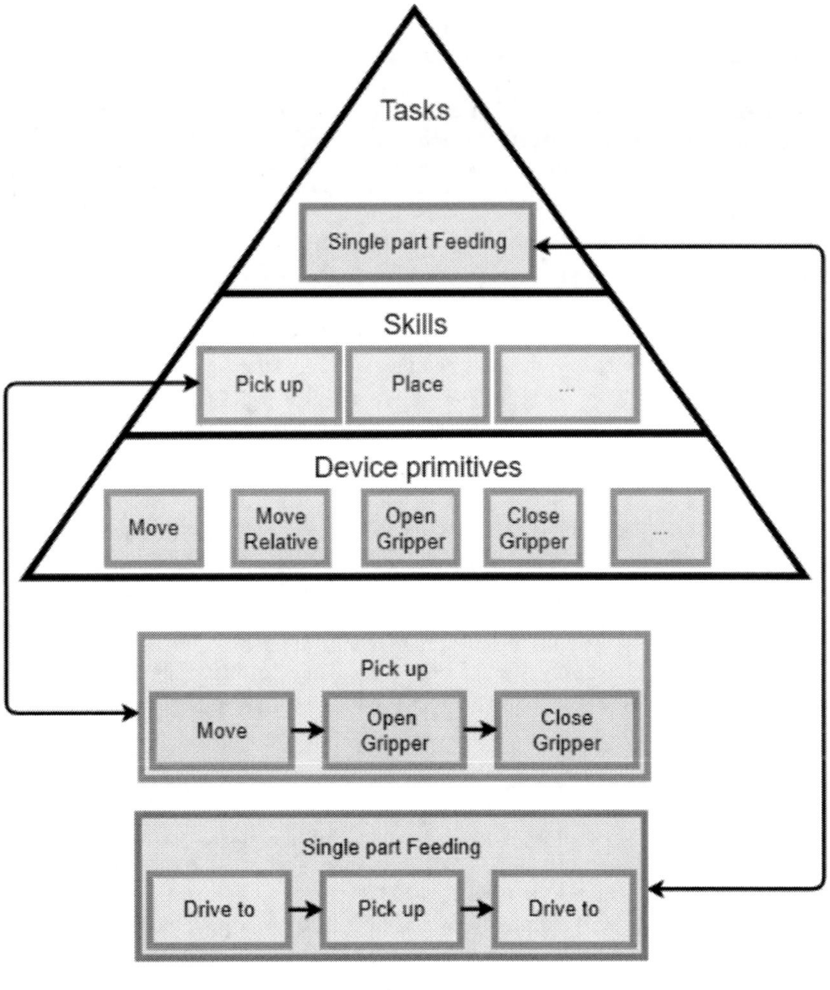

Fig. 1. Separation of individual functions into sub-functions [10]. The blue boxes, individual tasks defined by the user, are divided into smaller skills (yellow boxes), which in turn are divided into device primitives (green boxes). Thus, larger tasks can be divided into a sequence of smaller functions.

3. Materials and Methods

Basic commands must be identified and described a prerequisite for all methods. Cartesian (robot-independent) motions, object manipulation or status reports are typical functions every robot must have included, and which are required as basic functions. The Canonical Robot Command Language (CRCL) is used for this purpose [8]. It describes all robot programming commands that are necessary to execute any applications, implemented. Furthermore, internal parameters like speed or tolerances are passed and adapted if necessary. In this work the library is integrated to the MiR100 [5] mobile robot which is used as a transport robot and is connected to an industrial computer. The manufacturer of the robot provides a REST communication, where the required information is sent or requested to the robot via commands.

Parameter that are not changeable via the REST communication must be processed over the Robot Operating System (ROS) framework [9]. In order to be able to implement all the functions of the robot, a certain sequence must be generated in the software system. These functions are divided into smaller steps, so that the direct function of the robot can be used [7]. The step-by-step structure of the individual functionalities is shown in figure 1 [7]. A top-down variant is shown, starting at the top with the process task, which is specified by the user and divided into individual skills. These skills are again formed into device primitives.

4. Results and Discussion

The proposed architecture was tested in a test factory of the UAS Technikum Wien [6] for transport and pick-and place tasks in a processing procedure for the mobile robot MiR100. Primarily, functions were tested, which were already defined by the REST API, such as driving to a defined position or docking to an integrated marker. If the mission is successfully created by the program as described above, it is passed to the mobile robot and will be executed. If many missions are created and saved one after the other, the mobile robot works through them in order until there are none left. Furthermore, the user can add additional actions to an already existing mission. The actions are first stored together with their priority and handed over to the robot. Based on this priority number, the sequence of actions in the mission is automatically defined and processed accordingly during testing.

A mission requiring the robot arm UR5 [12], needs an already existing program and can be called through this library as an own action. A sequence was tested where the mobile robot docks at a position and then a UR5 program is passed to the robot arm via the MiR100 link. Further tests were performed with different integrated actions to ensure a broad evaluation. However, no direct errors occurred, resulting in the mission and action being successfully sent to the robot and processed.

5. Conclusion

This work shows a robot programming concept and how it can be implemented. Therefore, to solve the problem of various types of robots with different applications and programming languages, such an abstract way of programming can be used in a complex manufacturing factory. Consequently, employees, who have no knowledge of robot programming can still implement new programs and missions in a quick and easy way. For technicians who are directly integrated in the process flow, the concept gives a possibility to change processes without having to read the specific robot applications and data sheets. As a result, time can be saved, especially in production since complex programs are not required.

In the further course of the development of this described library, additional functions will be integrated in order to implement any possible application with a mobile robot. Further tests and evaluations of the individual tasks and missions of the robot will be carried out. In addition, safety and security aspects have to be considered, as this is a critical factor especially in mobile robots.

This work has dealt exclusively with the abstract programming of the MiR100 mobile robot. To be able to program a complete production chain, including many types of industrial and mobile robots with different interfaces in, such a library has to be designed for each individual robot, which looks complicated in the first place, but offers clear advantages in the future.

6. References

[1] R. E. Andersen, E. B. Hansen, D. Cerny, S. Madsen, B. Pulendralingam, S. Bøgh, and D. Chrysostomou (2017). "Integration of a Skill based Collaborative Mobile Robot in a Smart Cyber-physical Environment," Procedia Manufacturing, vol. 11, no. October 2019, pp.114–123.

[2] Andreev, V[iktor]; Kim, V[alerii] & Eprikov, S[tanislav] (2020). ModRob: the Hardware-Software Framework for Modular Mobile Robots Prototyping, Proceedings of the 31st DAAAM International Symposium, pp.0391-0402, B. Katalinic (Ed.), Published by DAAAM International, ISBN 978-3-902734- 29-7, ISSN 1726-9679, Vienna, Austria DOI: 10.2507/31st.daaam.proceedings.054

[3] A.-L. Courbis, K. Luu, B. Grondin, and K. Roussel (2019). "A Model Driven Architecture Framework for Robot Design and Automatic Code Generation," 15th China-Europe International Symposium on Software Engineering Education.

[4] G. Fragapane, D. Ivanov, M. Peron, F. Sgarbossa, and J. O. Strandhagen (2020). "Increasing flexibility and productivity in Industry 4.0 production networks with autonomous mobile robots and smart intralogistics," Annals of Operations Research.

[5] Mobile Industrial Robots A/S. (2019). MiR100 robot, [Online]. Available: https://www.mobile-industrial-robots.com/en/solutions/robots/mir100/.

[6] R. Otrebski, D. Pospisil, C. Engelhardt-Nowitzki, N. Kryvinska, and M. Aburaia (2019). "Flexibility enhancements in digital manufacturing by means of ontological data modeling," Procedia Computer Science, vol. 155, no. 2018, pp. 296–302.

[7] M. R. Pedersen, L. Nalpantidis, R. S. Andersen, C. Schou, S. Bøgh, V. Krüger, and O. Madsen (2020). "Robot skills for manufacturing: From concept to industrial deployment".

[8] F. Proctor, S. Balakirsky, Z. Kootbally, T. Kramer, C. Schlenoff, and W. Shackleford (2016). "The Canonical Robot Command Language (CRCL)," Industrial Robot, vol. 43, no. 5, pp. 495–502.

[9] M. Quigley, K. Conley, B. Gerkey, J. Faust, T. Foote, J. Leibs, R. Wheeler, and A. Ng (2009). "Ros: an open-source robot operating system," ICRA Workshop on Open Source Software, vol. 3.

[10] B. Ridge, T. Gaspar, and A. Ude (2017) "Rapid state machine assembly for modular robot control using meta-scripting, templating and code generation," IEEE-RAS International Conference on Humanoid Robots, pp. 661–668.

[11] S. Thrun, W. Burgard, and D. Fox (2006). Probabilistic Robotics. MIT Press: Cambridge, MA, USA.

[12] Universal Robots. (2014). Technical Specifications UR5. Available at: http://www.universal-robots.com

[13] B. Wally, J. Vyskocil, P. Novak, C. Huemer, R. Sindelar, P. Kadera, A. Mazak, and M. Wimmer (2019). "Flexible Production Systems: Automated Generation of Operations Plans Based on ISA-95 and PDDL," IEEE Robotics and Automation Letters, vol. 4, no. 4, pp. 4062–4069.

DOI: 10.2507/32nd.daaam.proceedings.079

ANALYSIS OF THE CURRENT APPLICATION OF AR IN THE CONTEXT OF TPM IN SLOVAKIA ORGANIZATIONS

Lucia Grajzova, Samuel Janik, Milos Cambal & Miroslava Mlkva

This Publication has to be referred as: Grajzova, L[ucia]; Janik, S[amuel] & Mlkva, M[iroslava] (2021). Analysis of the Current Application of AR in the Context of TPM in Slovakia Organizations, Proceedings of the 32nd DAAAM International Symposium, pp.0546-0554, B. Katalinic (Ed.), Published by DAAAM International, ISBN 978-3-902734-33-4, ISSN 1726-9679, Vienna, Austria
DOI: 10.2507/32nd.daaam.proceedings.079

Abstract

Industrial production has undergone significant changes in the last decade. Competition has increased dramatically, and customers are focusing on product quality, delivery time and product costs. For this reason, companies should focus not only on the continuous improvement of their products, but also on their processes. Total Productive Maintenance (TPM) is a methodology aimed at increasing the efficiency and longevity of equipment and machinery. Augmented reality is a direct or indirect view of the physically real environment, parts of which are enriched in digital, mostly textual, or pictorial form with additional information relevant to the object one is looking at. The aim of the paper was to analyse and describe the current situation of implementation of AR technology in industrial enterprises of Slovakia as a means to improve and explore opportunities for the future adaption of the TPM philosophy in the Industry 4.0 environment.

Keywords: augmented reality; digital technologies; Industry 4.0; total productive management; 5S;

1. Introduction

The Fourth Industrial Revolution is a topic that is currently being fully discussed in all areas. We can simply say that this is a consistent introduction of digital technologies into production. Digital technologies have developed rapidly in recent years and their use is becoming a major challenge for businesses. An industrial society that wants to succeed in the modern world must never afford to neglect the use of modern technology in production processes. The efficiency with which it is able to bring a new product to market also often reflects the profitability of companies with a competitive advantage and, ultimately, may also determine the company's future survival in the market.

A manufacturing organization should be able to provide its customers with quality products. In order to stay active in today's market, these organizations must guarantee that their processes are being monitored and constantly improved. More efficient quality systems are required when the production volumes are increased, there are a number of techniques for controlling product and process quality like Failure Mode and Effect Analysis (FMEA), Statistical Process Control (SPC), Design of Experiments (DoE) and Total Productive Maintenance (TPM) [1]. In this article, the authors focused on identifying the extent of AR application in industrial enterprises in Slovakia. The research was carried out using the questionnaire method and 230 answers were obtained from medium and large companies.

The purpose of this paper is to analyse the extent and manner of application of AR in the maintenance processes and to identify future assumptions of its use.

In the next part of the article, we will try to find answers to the following questions.
1. Is there a presumption that Slovak companies will be able to innovate their processes by using new technologies offered by the Fourth Industrial Revolution?
2. Do our respondents see potential in the use of AR?
3. Is there still room for investment in AR in order to make it more efficient given the current state of its application?
4. Do our organizations have the ability to adapt in the implementation of other future possibilities that could AR offer in relation to the global market predictions with the goal to stay competitive in the ever-changing environment?

1.1 Total productive maintenance

The origin of TPM can be traced back to 1951 when preventive maintenance was introduced in Japan. However, the concept of preventive maintenance was taken from the USA. Nippondenso was the first company which introduced plant-wide preventive maintenance in 1960 [2]. Total productive maintenance is maintenance activities that are productive and implemented by all employees. TPM involves everyone in the organization from assembly line workers to senior management in equipment improvement. It encompasses all departments including: Maintenance, Operations, Facilities, Design Engineering, Project Engineering, Construction Engineering, Inventory and Stores, Purchasing, Accounting and Finance, Plant and Site Management [3].

Total Productive Maintenance is an extremely comprehensive approach to operational efficiency and maintenance included. It is one of the basic blocks used in lean manufacturing, after all it is not possible to improve processes if we cannot rely on equipment and machines. TPM is a maintenance management system that includes the care of fixed assets and considers the impact of maintenance on productivity, quality, safety and all matters that affect the employees in the company. TPM represents the Japanese approach to maximizing the efficiency of machines and equipment and thus creating optimal conditions in the relationship "Human - Machine - Environment" [4].

TPM goals

TPM has the following five goals [3]:
1. Improving equipment effectiveness.
2. Improving maintenance efficiency and effectiveness.
3. Early equipment management and maintenance prevention.
4. Training to improve the skills of all people involved.
5. Involving operators in routine maintenance.

TPM Built on 5S

TPM has as its foundation in another management methodology, also originating from Japan, known as 5S. This is an organizational strategy, based around the Japanese concepts of seiri, seiton, seisō, seiketsu, and shitsuke. Translated, these words mean “Sort," "Set in order," "Shine," "Standardize," and "Sustain." 5S is focused on organizing the workplace environment to improve efficiency and effectiveness. This methodology establishes rules and creates a mindset of efficiency and order. As such, it helps connect the individual worker to his space and tools in a productive and efficient way and lays the groundwork for a system like TPM to be as effective as possible [5].

TPM pillars

The key areas of TPM are addressed by eight different activities, popularly termed as eight pillars of TPM. The Eight-pillar methodology of TPM results in rising labour productivity via contraction in maintenance costs and production stoppages and thereby having cut back on downtimes [6].

The eight TPM pillars include:
1. Autonomous Maintenance,
2. Focused Maintenance,
3. Planned Maintenance,
4. Quality Maintenance,
5. Early Equipment Management,
6. Education and Training,
7. Safety, Health & Environmnet,
8. Office TPM [7].

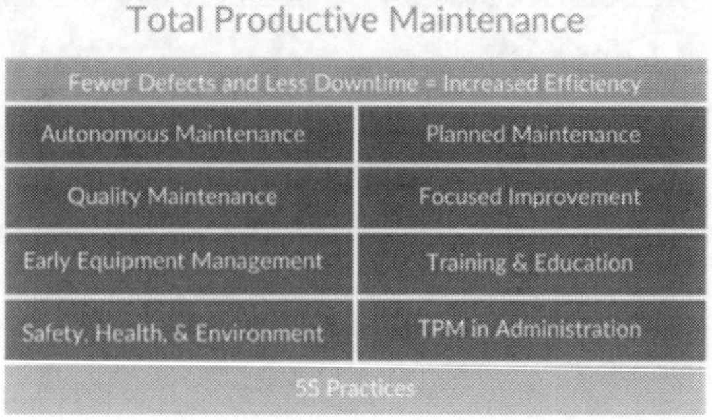

Fig. 1. Pillars of TPM [8]

The pillars shown in Figure 1 represent a house consisting of eight columns, with a logical sequence. It is necessary to build the pillars before it is possible to build a "house". The definition of the pillars depends on the philosophy and internal structure of the organization and should be adapted to the existing culture and approach of the organization [9].

1. **Autonomous Maintenance** – this pillar aims at developing operator ownership. The operator performs day-to-day tasks to be able to develop skills and in turn mastery of the equipment, thereby freeing up the skilled maintenance people. This way they can spend more time on technical repairs and other value-added activities. In this activity, operators are responsible for safeguarding their equipment to avoid operating abnormalities [10].
2. **Focused Maintenance** – Recurring problems are identified and resolved by cross-functional teams. Combines the collective talents of a company to create an engine for continuous improvement.
3. **Planned Maintenance** – Maintenance pre-empts malfunction while interventions by high-level technicians are carefully planned so that minimal downtime is required for any software updates or part replacements. Planned Maintenance follows the concept of zero failures similar to Focused Improvement. Zero failures mean zero defects, zero reworks and zero accidents [5] [10].
4. **Quality Maintenance** – It is aimed towards customer delight through the highest quality through defect-free manufacturing. Focus is on eliminating non-conformances in a systematic manner, much like a focused improvement. QM activities are to set equipment conditions that preclude quality defects, based on the basic concept of maintaining perfect equipment to maintain perfect quality of products. The condition is checked and measure in time series to very that measure values are within standard values to prevent defects [11].
5. **Early Equipment Management** – is also occasionally referred to as TPM for Design to highlight the fact that it is about the early involvement of existing operators and maintainers at the very start of a capital equipment investment, namely at the concept and definition stage [12].
6. **Education and Training** – Continuous improvement includes operator and work area education and training which improves morale, retention, and efficiency.
7. **Safety, Health and Environment Control** – Facility-wide safety is prioritized which positively impacts the sustained success of the TPM program.
8. **Office TPM** – the TPM office should be formed after activating the previous pillars. Office or, in other words, the office should be formed to monitor the improvement of productivity, efficiency in administrative work to eliminate losses, which involves analysing processes and procedures [13].

The 8 pillars remain the fundamentals of the industry. But under a large modernization movement called Industry 4.0 and on which we can argue the revolutionary nature, one can ask a simple question:

"How is this revolution going provide forever more efficient maintenance? "

The answer to this question is in Industry 4.0, Big Data or the Internet of Things, artificial intelligence, cyber-physical system, or even augmented reality appear as a digital upgrade and as many potential levers for improvement.

2. Industry 4.0

Industry 4.0 is currently one of the most popular topics in the world, both in industry and in academia. The Fourth Industrial Revolution plays an important role in defining new organizational strategies and seizing the opportunities of digitization at all levels of production and services. Its implementation is a combination of several physical and digital technologies, such as artificial intelligence, cloud computing, additive manufacturing, adaptive robotics, the Internet of Things (IoT), and augmented reality. The main goal of the implementation of industry 4.0 technologies is industrial transformation, increasing efficiency and productivity to achieve greater competitiveness of organizations) [15]. Industry 4.0 is a revolution in manufacturing that offers new opportunities based on the integration of Industry 4.0 technologies with manufacturing and its parts, which can lead to improved performance with minimal resources [16].

Like the lean manufacturing paradigm, the thinking paradigm in the context of Industry 4.0 offers promising solutions to the future challenges facing organizations. The only question remains whether and how these two philosophies can merge [17].

Augmented reality

Masood [18] speaks of the augmented reality as a key technology of the fourth industrial revolution. AR supports the implementation and development of the concepts of the fourth industrial revolution. By using AR technology, employees can overcome barriers between the physical and increasingly important digital worlds [20]. To define the operation of AR devices more specifically, Azuma [20] talks about the need to combine real and virtual objects in a real environment and time. Both real and virtual objects must be geometrically aligned in the real world. The basic idea of AR is to effectively provide the necessary information to the user virtually without restricting him at work [14]. Current developments in technology and industry are dynamic and complex, leading to a highly competitive and dynamic business environment. The manufacturing industry faces new challenges that require a holistic view of the four major classes of manufacturing attributes, i. j. cost, time, quality, and flexibility. The challenge is to implement integrated AR-supported simulation tools that could improve manufacturing operations as well as product and process development [19].

AR technology is used in a wide range of industries such as marketing, logistics, manufacturing, and maintenance. Studies that focus on identifying Industry 4.0 and maintenance relationships show positive results that support their synergy. Maintenance staff using AR can work more efficiently in performing technical maintenance tasks, and AR can also serve as support in maintenance decision-making processes [20] [15]. Complex tasks that are associated with maintenance in industrial enterprises are suitable candidates for the implementation of augmented reality. Flexible training options and access to real-time documentation meet the requirements for the appropriate application of AR technology by maintenance technicians [14].

The first attempts to implement AR into maintenance processes date back to the early 1990s. In 2013, Volkswagen introduced a tablet, smartphone application called MARTA (Mobile Augmented Reality Technical Assistance), which provided technicians with detailed instructions on car repair and maintenance [21]. At present, repair and maintenance technicians are beginning to use AR headsets and glasses in the course of their work to provide them with useful information on site, to suggest possible repairs, and to point out potential problem areas [22]. The first attempts to implement AR into maintenance processes date back to the early 1990s. In 2013, Volkswagen introduced a tablet application called MARTA (Mobile Augmented Reality Technical Assistance), which provided technicians with detailed instructions on car repair and maintenance [23]. At present, repair and maintenance technicians are beginning to use AR headsets and goggles in the course of their work to provide them with useful information on site, to suggest possible repairs, and to point out potential problem areas [24].

According to Kelley [26], we know of five ways in which augmented reality will improve industry:

- simplified installation instructions,
- increasing employee productivity,
- effective maintenance of equipment,
- reduction of operating costs,
- quality assurance.

If we want to implement AR in a way that is truly effective and transformative, we need to focus more broadly on the real basic business needs, what further business changes we can make (in the short, medium, and long term), and how AR solutions can overlap with all these changes. To provide an effective and reliable contribution to the transformation of society to achieve goals [25]. In the coming years, we will see manufacturing transformation as companies adopt augmented reality applications as a day-to-day tool for gathering real-time operational reports, improving maintenance and service, 3D design, reducing staffing costs, and developing powerful platforms for remote collaboration without the need for teams nearby.

3. Materials and Methods

This section describes the activities performed in order to analyse the current state of use of augmented reality in TPM in industrial enterprises in Slovakia. Using a questionnaire survey, which was designed for medium and large industrial enterprises, we wanted to determine the use of AR in various industries. Data collection took place during the time horizon: 1.4. 2021 - 15.5. 2021. The questionnaire was distributed via Google Form - Questionnaire to 230 industrial enterprises in Slovakia.

The first two questions were focused on the characteristics of industrial enterprises, 1. the size of the enterprise 2. the industry in which the enterprise operates. From the questionnaire survey, which was distributed among medium and large industrial enterprises in Slovakia, 135 respondents filled out the questionnaire, the number of employees ranging from 250 to 499 employees. The remaining 95 respondents work in a company with more than 500 employees.

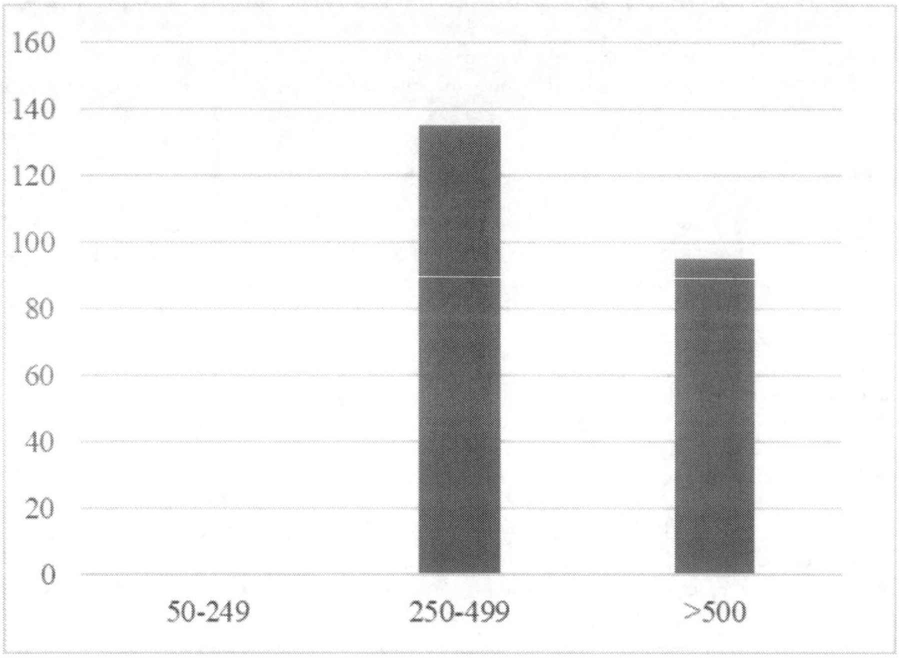

Fig. 2. The number of employees in the company

The second question, *"In what industry does your company operate?"* was compiled based on the classification of the FinStat site, which was used in creating the database of E-mail contacts. Fig. 3 shows the evaluation of the second question.

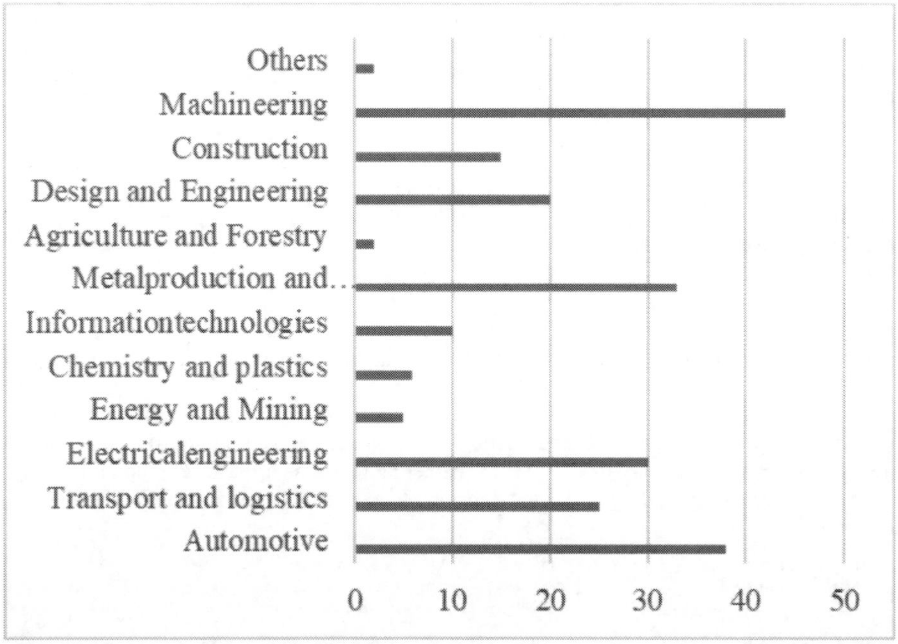

Fig. 3. Industrial sector

Fig. 3 shows that most companies were from the Machineering (44), Automotive (38), Metalworking and Metallurgy (33) sectors. The smallest were industrial enterprises from Agriculture and Forestry (2), Energy and Mining (5), and Chemicals and Plastics (6). The third question of the questionnaire survey focused on the use of AR in industries. *"In which industries does AR have the strongest application?"*

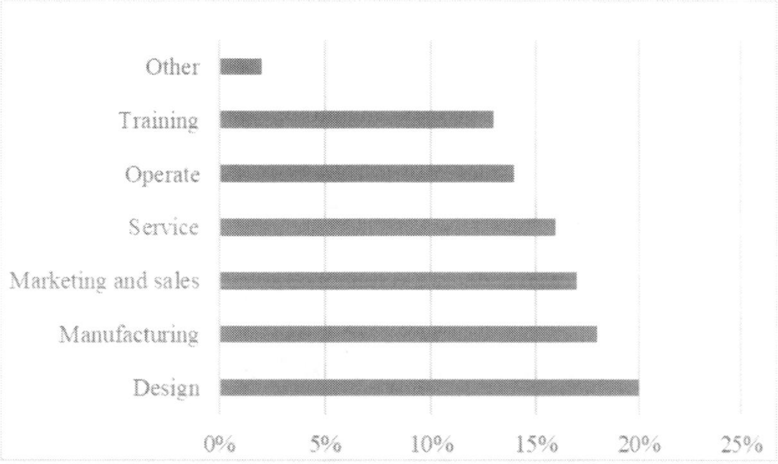

Fig. 4. Representation of AR in Industry

Fig. 4 categorically shows the strongest representation in the design sector (20%), where AR provides instruments able to improve the prototyping and visualization in the design phase. However, the second is the manufacturing (production) (18%) the use of which expands reality is the most efficient way and one of the largest industrial cases of AR use. Any operation which provides step-by-step procedures can benefit from the use of AR; installations and assemblies are just a few examples where the AR can make a difference by making the fourth-generation factory. The increased sector consists of marketing and sales (17%), service (16%), operate (14%) and has the weakest representation training (13%). In general, the complexity of the technology and the constant need for improvements in terms of time, errors, safety, and costs are the drivers for justifying the utilization of AR. Each field of application seems to have its specific needs and reasons for investing in AR.

4. Results

The Results section presents the results of the concept of maintenance support using augmented reality in industrial enterprises in Slovakia. The questionnaire survey involved 230 medium and large industrial enterprises, of which 35% of companies use AR. Respondents' answers are processed into graphs. The first question from the questionnaire was aimed at identifying the potential of individual characteristic maintenance operations that can be performed by the application of augmented reality by maintenance technicians.

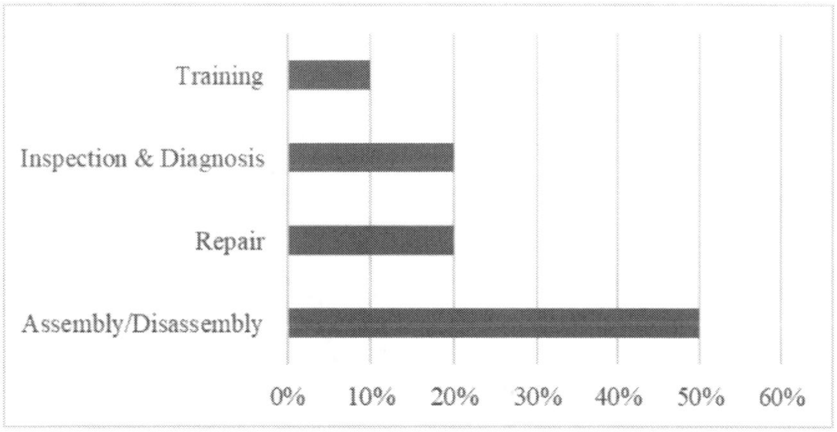

Fig. 5. Maintenance operations

Based on the results of the fig. 5, it is clear that respondents see the greatest potential for the use of AR for maintenance technicians during assembly and disassembly, these operations are the most common operations performed by maintenance technicians. Repairs of equipment and machines and the associated inspection and diagnosis have the same potential (20%) in the application of AR, based on the respondents. In the training of maintenance technicians, respondents see the least potential (10%).

In the following question, we tried to identify the type of hardware currently used by Slovakia industrial organizations, in connection with augmented reality.

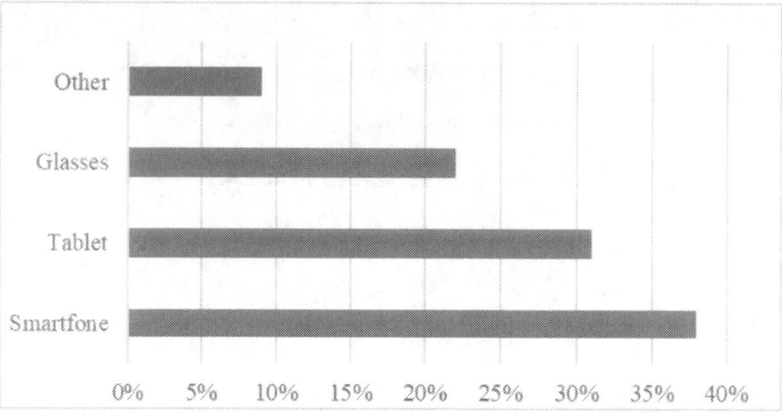

Fig. 6. Use of AR hardware

Based on fig. 6 almost 70% of respondents use the AR image via smartphones and tablets, which is not surprising given the affordability of these devices. The striking fact is that only 22% of respondents indicated the possibility of using smart glasses, which expand the abilities and skills of operators, adjusters, installers, maintenance workers, or even quality controllers. By using smart glasses compared to smartphones or tablets, the fact is that glasses are more work efficient, they ensure higher quality and safety of their user.

5. Discussion and conclusion

Based on the outputs from the materials section, the authors state that currently, the representation of AR in the industries is the strongest in the field of Design. However, this reality is probably only temporary, due to the COVID-19 pandemic, industry organizations have been forced to increase their investment towards digitization. When comparing the current representation in the sectors fig. 4 and increasing investment in the sectors fig. 7 it is highly likely that the use of AR technologies in manufacturing will be the largest in the future.

Certain research limitations can be attributed to the ongoing COVID-19 pandemic, which has changed the way employees carry out their work activities. The authors of the article assume that a certain part of the respondents to the questionnaire survey did not have to have up-to-date information on the state of implementation of selected Industry 4.0 technologies as a result of working from home. Another limitation can be a result of the individual approach of organizations to the implementation of augmented reality due to the different ways of production of finished goods, their complexity, and strategic visions for the future.

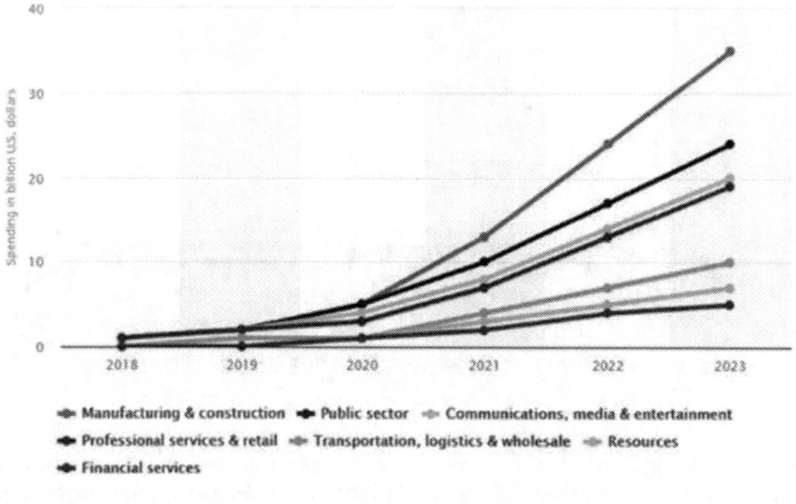

Fig. 7. Expected growth of AR market after considering the pandemic effect. [26]

Innovations in augmented reality technologies, which have enabled wider implementation in the field of production, have come in recent years and we can assume that their progress will continue in the near future. The size of the global AR market in 2021 is estimated at approximately $ 26.75 billion. The expected increase from 2021 to 2028 with a compound annual growth rate could reach up to 43.8% per year, and in 2028 the size of the AR market could be as high as 360.16 billion [24]. The authors think that a higher share of AR in manufacturing would come regardless of the pandemic, but it has accelerated this process as the priority of digitization of organizations has increased.

Based on the answers of the respondents, fig. 5, the authors found that respondents see the greatest potential in the connection of AR and maintenance of the organization during the assembly and disassembly of equipment and machines. Implementing assembly and disassembly processes using AR technology has many advantages. Workflows that are part of AR technology ensure accurate process progress, which can increase assembly/disassembly productivity, reduce errors, increase product quality, improve ergonomics, and eliminate unnecessary paper documentation and thus ensure environmental benefits. However, this does not mean that the potential that comes with the augmented reality is negligible in inspections and diagnostics or repairs and training of employees.

During repairs, it is possible with AR to use remote cooperation through communication between the technician and the engineer in the field, the engineer can navigate the technician to eliminate the error without his immediate knowledge of the problem. Inspection and diagnostics with AR technology ensure compliance with standardized procedures, thus eliminating the possibility of errors caused by human factors. In conjunction with RFID, QR technology, it is possible to unambiguously identify the inspected parts of the equipment and optimize the inspection process. The inspection record is automatically digitally recorded, so there is no need to create paper records of inspections and diagnostics.

Another category that can be made more efficient by using AR is Employee Training. Training and coaching of employees using AR can streamline the entire process of hiring new employees. AR during training shortens training time, increases its effectiveness, improves quality control during training, prevents errors and at the same time, it is possible to identify potential waste in processes.

Based on the answers of the respondent's fig. 6 to the question of what hardware they use in connection with AR technology in Slovakia, it is clear that most companies currently use tablets and smartphones (69%). AR glasses, which are becoming more available in Slovakia, are not so much used yet, and are used by only 22% of respondents. The authors of the article predict that in the future Slovak organizations will adopt more AR technologies in conjunction with AR glasses. AR glasses have many advantages over smartphones and tablets. Users are able to follow instructions without looking at a smartphone, tablet. They can use both hands at work and are not limited by holding the device. A voice-controlled system and speakers that contain AR glasses are one of the advantages offered by such glasses as they allow control of an AR device without direct contact with the device. Through the implementation of AR technology into maintenance processes, organizations can streamline processes, ensure higher quality and reduce costs associated with maintenance processes. AR technologies offer additional opportunities in improving the philosophy of lean and TPM.

Future research will lead to a more detailed understanding of the potential and possibilities that augmented reality could bring not only as support for TPM but as a change in the perception of current reality in manufacturing and other components of industrial organizations. The aim is to expand knowledge about the ongoing revolution and lay the foundations for further development in the future in terms of technology and their contribution to increasing the competitiveness of organizations in Slovakia. Slovakia is a small country but with the right resources and vision, it could become a powerhouse for industrial organizations across the world.

6. Acknowledgments

The authors would like to thank for providing financial support. The paper is a part of project VEGA No. 1/0721/20 *„Identification of priorities for sustainable human resources management with respect to disadvantaged employees in the context of Industry 4.0".*

7. References

[1] Segovia D., Mendoza, M., Mendoza E., González, E. (2015). Augmented Reality as a Tool for Production and Quality Monitoring. Procedia Computer Science. Volume 75. Pages 291-300

[2] Bhadury, B. (2000). Management of productivity through TPM. Productivity Vol.41 No. 2 pp. 240-51.

[3] Wireman, T. (2004). Total productive maintenance. Industrial press, Inc, 200 Madison Evenue, New York, 2nd ed. ISBN 0-8311-3172-1

[4] Macinnes, R., L., (2006). Štíhlý podnik Memory Jogger: vytvářejte hodnotu a eliminujte ztráty v celém vašem podniku. Praha: Česká společnost pro jakost. ISBN 80-02- 01849-4

[5] Kitain, L. (2018) Industry 4.0 Brings Total Productive Maintenance into the Digital Age. https://medium.com/@lior.kitain/industry-4-0-brings-total-productive-maintenance-into-the-digital-age-ad050dcb1c87

[6] Ahuja, I.P.S. and Khamba, J.S. (2008). Total productive maintenance: literature review and directions- International Journal of Quality & Reliability Management, Vol. 25 No. 7, pp. 709-756

[7] Ireland, F. and Dale, B.G. (2001). A study of total productive maintenance implementation. Journal of Quality in Maintenance Engineering, Vol. 7 No. 3, pp. 183-91.

[8] Diáz-Reza, J.R., Garciá-Acaraz, J.L., Martínez-Loya, V. (2019). Impact Analysis of Total Productive Maintenance. In. Springer Nature Singapore Pte Ltd., 346s. ISBN 978-3-030-01725-5
[9] Augustiady, T.K., Cudney E.A. (2016). Total Productive Maintenance – Strategies and Implementation Guide. CRC Press NW, 292s. ISBN 978-1-4822-5538-6.
[10] Parikh, Y., Mahamuni P. (2015). Total Productive Maintenance: Need & Framework. International Journal of Innovative Research in Advanced Engineering (IJIRAE) Issue 2, Volume 2, ISSN: 2349-2163
[11] Wakjira, W. M., Singh P. A., (2012). Total Productive Maintenance: A Case Study in Manufacturing Industry. Global journal of researches in engineering industrial engineering. Vol. 12, Version 1.0, ISSN: 2249-4596
[12] Brunskill A., Willmott, P., Quirke J., (2019). TPM – a foundation of operational excellence. Published by SA Partners. ISBN 978-9993748-1-5.
[13] Zlatić, M. (2019). TPM – Total productive maintenance. Quality Festival. ISSN2620-2832
[14] Schwald, B., Laval, B., Sa, T., Guynemer, R., (2003). An Augmented Reality System for Training and Assistance to Maintenance in the Industrial Context. J. WSCG 2003, 11, 1–3.
[15] Ustundag, A., Cevikcan, E., (2018). Industry 4.0: Managing The Digital Transformation. Springer Nature Singapore Pte Ltd., 285s. ISBN 978-3-319-57870-5.
[16] Pabbathi, K.,K. (2018). Quick Start Guide to Industry 4.0: One-stop reference guide for Industry 4.0. Createspace Independent Publishing Platform, 204 pages. ISBN 978-1718978614
[17] Mayr, A., Weigelt, M., Kühl, A., Grimm, S., Erill, A., Potzel, M., Franke, J., (2018). Lean 4.0 - A conceptual conjunction of lean management and Industry 4.0. 72, pp. 622-628, ISSN 2212-8271.
[18] Masood, T., Egger, J., (2019). Augmented reality in support of Industry 4.0—Implementation challenges and success factors, Robotics and Computer-Integrated Manufacturing. 58, pp.181-195, ISSN 0736-5845.
[19] Halenár, I., Juhásová, B., Juhás M., (2018). Rozšírená realita ako nástroj výučby priemyselnej automatizácie. Available from: https://www.atpjournal.sk/novetrendy/rozsirena-realita-akonastrojvyucby-priemyselnej-automatizacie-1.html?page_id=27134.
[20] Azuma, R., Baillot, Y., Behringer, R., Feiner, S., Julier, S., MacIntyre, B., (2001). "Recent advances in augmented reality", 21, pp. 34-47, ISSN 1558-1756.
[21] Williams, D., (2017). The History of Augmented Reality (Infographic). Available from: https://www.huffingtonpost.com/dennis-williams-ii/the-history-of-augmented-_b_9955048.html.
[22] Paine, J., (2017). 10 Real Use Cases for Augmented Realit).Available from: https://www.inc.com/james-paine/10-real-use-cases-for-augmented-reality.html.
[23] Grand view research, (2021). Augmented Reality Market Size, Share & Trends Analysis Report By Component, By Display (HMD & Smart Glass, HUD, Handheld Devices), By Application, By Region, And Segment Forecasts, 2021 - 2028. Available from: https://www.grandviewresearch.com/industry-analysis/augmented-reality-market.
[24] Wheelwirght, G., (2019). The key role of Augmented Reality in Industry 4.0 for Manufacturing: Part 1. Available from: https://iiot-world.com/industrial-iot/connected-industry/the-key-role-of-augmented-reality-in-industry-4-0-for-manufacturing-part-1/.
[25] Saracco, R., (2020). AR Market Gets a Boost from Pandemic. Available from: https://cmte.ieee.org/futuredirections/2020/07/19/ar-market-gets-a-boost-from-pandemic/
[26] Kelley, M., (2019). Five Ways AR Apps Will Enhance Industrial Work in Next Five Years. Available from: https://www.industryweek.com/technology- and-iiot/article/22027338/five-ways-ar-apps-will-enhance-industrial-work-in-next-five-years.

DOI: 10.2507/32nd.daaam.proceedings.080

STUDY AND ANALYSIS OF MACHINE AND TOOL PARAMETERS OF STAINLESS-STEEL TURNING USING MULTI COATED TOOLS

Viktors Gutakovskis, Ajay Varma Buddaraju, Antons Steklejns, Anita Avisane & Gatis Muiznieks

This Publication has to be referred as: Gutakovskis, V[iktors]; Buddaraju, A[jay] V[arma]; Steklejns, A[ntons]; Avisane, A[nita] & Muiznieks, G[atis] (2021). Study and Analysis of Machine and Tool Parameters of Stainless-steel Turning Using Multi Coated Tools, Proceedings of the 32nd DAAAM International Symposium, pp.0555-0565, B. Katalinic (Ed.), Published by DAAAM International, ISBN 978-3-902734-33-4, ISSN 1726-9679, Vienna, Austria
DOI: 10.2507/32nd.daaam.proceedings.080

Abstract

Metal cutting industries are confronted with the task of increasing production rate while lowering costs and improving final product quality. Manufacturing companies are under great pressure to discover and develop new technologies in order to fulfil increasing demands for productivity, closed tolerance, dimensional stability, and cost. As a result, coating an existing tool can help you enhance production rates, extend tool life, and save money. The purpose of this study is to optimize machining parameters including feed rate, depth of cut, and spindle speed when turning AISI316L using three coated tool inserts with different coatings. Using design expert software and response surface methodology (RSM), parameters in the direction of surface roughness are optimized. The condition of surface roughness was discovered to be ideal. For each solution, second-order polynomial models developed in this study were employed to attain the given ideal conditions. Derringer's desirability function technique was used to optimize several responses. The best surface roughness values found with CNMG 120404 coated with TiN (Titanium nitride), CNMG 120408 coated with TiAlN(Titanium aluminium nitride), and CNMG 120412 coated with AlCrN (Aluminum Chromium Nitride) are 1.311 m, 0.897 m, and 0.965 m, respectively. To validate the model, triplicate tests were performed under ideal conditions, and the mean values of the experimental data were compared to the expected values.

Keywords: Metal cutting; turning; lathe; stainless steel; AISI316L

1. Introduction

Because of the quick technical breakthroughs and digital advances, tool technology and production processes are particularly difficult to comprehend. It's crucial to use the correct material for production so that the product's design lasts. Cutting conditions, machining processes, tool geometry, chip formation, tool material, tool wear, and vibrations are all factors that affect the work material's quality. Cutting speed, feed rate, depth of cut, and tool noise radius are all frequent characteristics in the turning process. Austenitic stainless steel AISI 316L is used in a wide range of industries, including the chemical industry, nuclear power plants, and medical devices, due to its excellent mechanical properties and corrosion resistance. Machinability of stainless steel is an enthralling topic.

Metal-cutting businesses face the challenge of increasing output while lowering costs and improving product quality. Manufacturing companies are under great pressure to discover and develop new technologies in order to fulfil increasing demands for productivity, closed tolerance, dimensional stability, and cost. As a result, coating an existing tool can help you enhance production rates, extend tool life, and save money. Modern businesses must deal with difficulties such as high-speed machining, hard turning, cutting difficult-to-cut materials, and dry machining. Thin layer coating was developed since this involves the use of coated equipment. Coated tools are used for a variety of purposes, including wear resistance, oxidation resistance, reduced friction, fatigue resistance, and thermal shock resistance. Aside from that, machining processes are intended to minimize the environmental impact of cutting fluid usage. Coated carbide cutting tools currently account for 80% of all machining operations in order to resolve the concerns described above.

A number of previous studies have documented the effects of cutting variables on traditional objective optimization (i.e. surface roughness, cutting force, material removal rate, and tool wear). And the vast bulk of the research was carried out with a specific objective in mind. Single-objective methods are typically used to determine the ideal value of a specific response and are limited to establishing the value of the optimum cutting parameters. In addition, lubricated stainless steel is regularly machined. Furthermore, better performance is necessary for hard finish turning when the necessary parameter circumstances are selected, due to the high cost of a CBN cutting edge." The purpose of this scientific work is to optimize the performance of a CBN tool in wet environments by adjusting machining parameters (such as feed rate, depth of cut, and spindle speed) with a stainless steel 316L coated tool insert. Using design expert software and response surface methodology (RSM), parameters are optimized. In addition, optimal conditions are attained.

1.1. Aim of the work

1. To study the turning of stainless steel 316L on a CNC lathe machine with various parameter combinations using cooling.
2. To optimize the performance characteristics of various coating tool inserts using the Response Surface Methodology (RSM) approach to produce the lowest feasible surface roughness.

Tasks to achieve:

1. To receive experimental results on wet turning of stainless steel 316L on a CNC lathe machine with different parameter combinations.
2. To use response surface methodology (RSM) to optimize the performance parameters for the lowest possible surface roughness

2. Metal cutting using Turning Process

As demonstrated in fig. 1, the turning process is clearly found in the machining method, which is a material removal approach for obtaining rotating components by cutting away undesired material by reducing its diameter. This turning procedure necessitates the task, fixture, and cutting tool. A pre-shaped material is fastened to the fixture in this piece, and it uses multi-point tools to complete numerous processes. The cutting device feeds into the rotating work piece or job, and the tool cuts material and generates chips in the desired shape. Self-assembly with the rotating device, allowing for faster rotation. The cutter is often a single-point cutting instrument that is also attached to the device.

This process is most commonly used to make rotatable, mainly axi-symmetric components with a range of features like holes, grooves, threads, tapers, multiple diameter steps, and curved surfaces. Components that are completely turned frequently contain parts that can be utilised in limited quantities, potentially for prototypes, such as bespoke produced shafts and fasteners. Turning is also a common approach to improve or add quality to components that have been manufactured in another means. Turning is useful for adding precise rotating functions to a factor when a standard shape has already been developed because of the higher tolerances and surface finishes it can deliver.

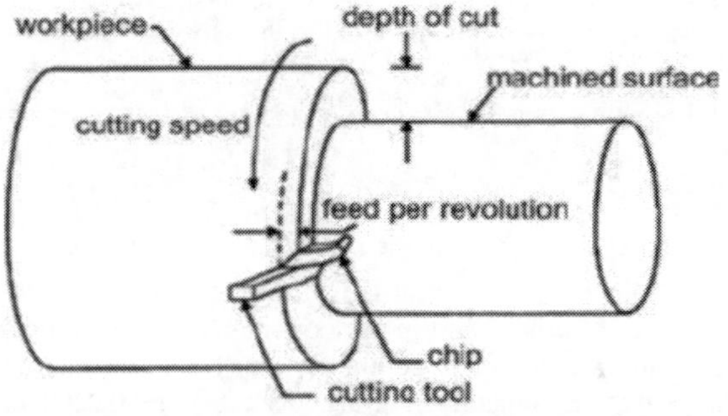

Fig. 1. Turning Process [1]

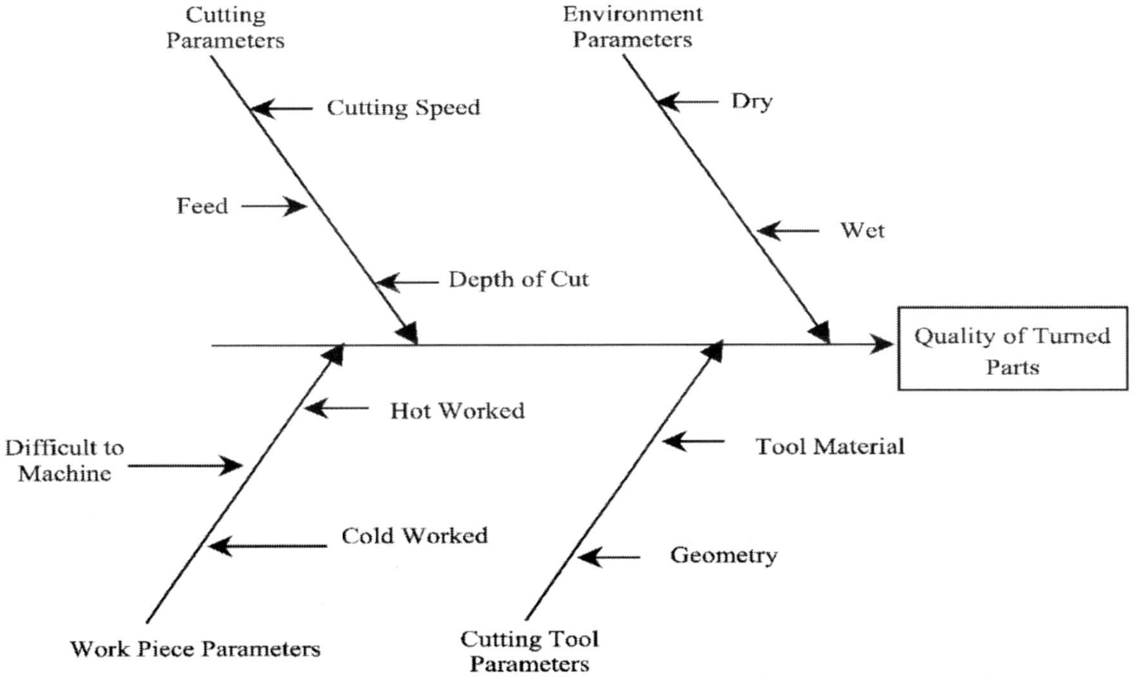

Fig. 2. Ishikawa cause- Effect diagram of a Turning process [2]

2.1 Adjustable Cutting Factors in Turning

Ishikawa cause and effect diagram has been created in an effort to recognise the process parameter that would affect the machining features of work pieces, as shown in fig. 2, Cutting tool parameters include tool shape and material, as well as cutting characteristics such as cutting velocity, feed, depth of cut, dry cutting, and wet cutting.

Spindle speed and cutting speed

The speed of the spindle and the work piece is frequently used to distinguish them. When expressed in revolutions per minute, it displays their rotating speed (rpm). However, for some turning operations, the surface speed, or the pace at which the work piece material goes past the cutting tool, is critical. Millimetres per minute are the measurement unit. Although the spinning speed remains the same, different work pieces with various diameters will always have a varying cutting speed.

Feed

It relates to how quickly the tool improves as it progresses along the cutting path. The feed of the tool affects both the tool's running speed and the surface roughness. As the feed rate increases, the running speed increases. When the feed is low, the outer surface polish is outstanding. There are two sorts of feeds: manual feeds that are manually turned and operated, and automated feeds that are automatically turned and operated. Manual mailing is recommended for beginners. Because serious accidents can occur, such as hitting the rotating chuck around the byte in all power-fed lathes' automatic feed.

Depth of Cut

"DOC" is a phrase that nearly goes without saying. “It can be defined as the thickness of the layer being removed (in a single pass) from the work piece, or the distance between the uncut and cut areas of the work, both in millimetres.”
D and d represent the job's starting and ending diameters in millimetres. The diameter of the task, however, has shrunk by twice the depth of cut because this layer has been eliminated from both sides of the work.

Cutting Tool Material selection

For successful task completion, selecting the appropriate cutting tool material for a certain application area is critical. Cutting speeds can be increased to some amount to boost productivity, but they shorten tool life, raise the expense of regrinding / replacing the tool, and cause production issues.

There is no one-size-fits-all course that will satisfy all of your requirements. The tool's qualities are a compromise; for example, hardness increases, and hardness lowers in general. The following characteristics can be found in the contents of the ideal cutting tool.

- Difficult compared to cutting work
- Maintains temperature stability
- Protects against wear and thermal stress.
- Impact resistance and chemical inertness
- Impact resistant and chemically inert

The machinist must have the following precise characteristics in order to successfully select a tool for machining:

- First, the shape and completion of the parts
- Hardness of working water
- Material tensile strength
- Material wear
- Generated chip type
- Preparing for employment

The machinist must always be involved in the performance and performance of the machine tool.

3. Other research review

Because of the growing demand for high-quality smooth surfaces, machining processes has been optimised. For decades, researchers have tried to optimise parameters that affect tool life, such as feed, depth of cut, and material hardness[3]. Tool life relates to how long a single point tool can function well before being declared failed, or how long a tool point can execute a metal cutting process before being pronounced failed. In every machining sector, the study of various types of tool inserts and the development of methods for monitoring tool life is a crucial feature that can save a lot of time and money. The tool life is greatly influenced by the material used to make it. Depending on the type of application, the current tool is coated with single or multi-layered materials such as titanium or carbide using CVD/PVD technology. Multi-layered coated carbide is cost-effective and has a superior surface polish for mass-production industries [4]. The amount of heat created during chip removal at the face and flank has a significant impact on tool performance during any metal cutting process. Furthermore, the tool undergoes physical and even chemical changes as a result of temperature and mechanical stress, causing the tool to lose its cutting characteristics and even fracture. Stainless steel is in high demand these days due to its high strength and low heat conductivity, despite its poor machinability [5].

Dabreo et al. [6] used a CNC lathe machine to conduct a turning experiment to estimate the tool life of a coated carbide insert in a wet environment. For this project, SS316L stainless steel was employed. The tool life was computed using the theoretical methodology (Taylor's tool life Equation), which took into account the number of jobs turned until it was pronounced fail. During the turning process, failure mechanisms of the tool were detected.

The compositions of Prasad et al. [7] coated inserts were identified, and their hardness was tested using X-ray diffraction analyses. "The grain structures of coatings were studied using scanning electron microscopy. Dry cutting experiments on AISI 304 stainless steel were conducted to evaluate the efficacy of various coatings in terms of wear and surface polish imparted to the workpiece. A 3D confocal laser microscope was used to examine the flank wear. A grey relation analysis was performed to optimize the machining parameters. The toughest coating was nc-AlTiN/Si3N4coating, which measured 28 GPa, according to studies."

The coated carbide tool, as opposed to the traditional cemented carbide tool, is the most important cutting tool in the metal cutting business. According to Ghani et al. [8], utilizing a coated carbide tool reduces the coefficient of friction between the tool's rake face and the cut chip, improving cutting performance and lowering costs. Depending on the use, a variety of coating materials are available commercially. Some of these materials include "TiN (titanium nitride), TiCN (titanium carbinitride), AlTiN (aluminium titanium nitride), AlCrN (aluminium chromium nitride), Al203 (aluminium oxide), new ceramic coating, Cr-Based coating, and others. According to Uhlmannet al. [9], the innovative Cr-based coatings CrTiAlN and CrMoTiAlN had the smallest wear land width of all the coatings tested." PVD and CVD are two technologies for producing thin film layers with various coating layer topologies ranging from monolayer to multilayer.

TiN coating is the most commonly used coating in industry [10] because of its high hardness, corrosion and wear resistance, and gold-like appearance. Coated carbide cutting tools are carbide inserts coated with a range of materials, including titanium carbide (TiC), titanium nitride (TiN), and aluminium oxide (Al203) [11].
The flank wear of coated and uncoated carbide tools in tool steel bar turning was examined by Haron et al. [12]. (23HRC). A 350 diamond-shaped insert with simple groove was chosen to look at tool wear based on flank wear data." Coated carbide tools beat bare carbide tools in terms of performance, and flank wear was minimal. Wet machining for coated carbide tools is also better than dry machining because it reduces friction and heat, but chipping occurs at the end of tool life due to the substantial temperature gradient at the surface. Ashok Kumar Sahoo and bidyadhar Sahoo [13] investigated the machinability of uncoated and multilayer (TiN and ZrCN) coated carbide inserts in finish hard turning of AISI 4340 steel.

According to the conclusions of this study, coated carbide inserts had a longer tool life, lower machining forces, and lower machining costs than uncoated carbide inserts. According to the quantitative data, the tool life of TiN and ZrCN coated carbide inserts was 19 and 8 minutes, respectively. Coated carbide inserts reduced machining costs by 93.4 percent, whereas ZrCN coated carbide inserts cut expenses by 40%.

Federico et al.[14] "reported that the coated carbide tool is also used for cutting hardened steel in his study of hardened steel using TiCN/Al203/TiN coated carbide and PCBN tools. Machine hardened steel with a TiCN/Al203/TiN coated carbide cutting tool was both productive and cost-effective, according to the findings. Kaladhar et al. [15] conducted an experiment to evaluate the efficiency of coated materials in machining austenite stainless steel. The results revealed that a PVD-coated TiAlN-TiN insert produces the best surface finish. Yigit et al.[16] investigated the performance of multilayer TiCN/TiC/Al203/TiN and TiCN/TiC/Al203/TiN carbide tools with an external TiN layer and found that TiN's high hardness improves wear resistance and lowers surface roughness in hard turning. Noordin et al.[17] studied AISI steel turning with two types of coated carbide, one with an Al203 outer layer and the other with a TiN layer. They discovered that a tool with a TiCN/TiN CVD/TiN and TiN PVD coating sequence generated the best results, with lower cutting forces, great surface quality, and a chip with a tiny secondary shear zone."

Aurich et al. [18] investigated the effect of four different coating methods on hot AISI4140 steel turning: TiN, TiN/TiAlN/TiN, TiN/TiCN/Al2O3, TiN/TiCN/Al2O3/TiN. The TiN/TiCN/Al2O3/TiN multilayer coating system had the smallest crater depth, while the TiN-TiCN-Al2O3-TiN coating system had the longest tool life travel. When cutting AISI1015 steel, Hasan and Muammer [19] looked at how surface roughness was affected by uncoated and coated carbide tools. They tested three-layer coatings with TiN outermost layer, AlTiN, TiAlN, and three-layer coatings with TiN outermost layer, finding that three layer coatings with TiN outermost layer improve surface roughness."

Kulkarni et al. [20] optimised cutting settings for dry turning of austenitic stainless steel AISI 304 with an ALTICRN coated tool. A fine-grained K-grade (ISO K-20) cemented carbide cutting insert received the coating via a physical vapour deposition (PVD) procedure. As turning parameters, cutting speeds of 200 and 260 m/min, feed rates of 0.20 and 0.26 mm/rev, and coating thicknesses of 3.6 and 4.6 m at two levels were examined.

To improve the surface roughness and turning of EN8 medium carbon steel, Inamdar et al. [21] employed the Taguchi technique. Feed rate and cutting depth at working speed are utilised to improve surface roughness. The Taguchi methodology is used to calculate the effect of the principal surface roughness parameters using levels and coefficients. Experiments can be designed using L9 orthogonal arrays. 95% of the important factors were also investigated (ANOVA). Cutting speed, rather than feed rate or cutting depth, appears to have a greater impact on surface sharpness in the turning process, according to the data.

Kumar & Shanmuga [22] chose the cutting speed, feed rate, and cutting depth based on the cutting parameters. Before beginning the experimental machining method, the ideal range of cutting parameters for Martentic Stainless Steel 416 machining is chosen. Assemble as many parameter combinations as possible. Change the cutting parameters and see what happens (surface roughness, cutting force). After you've optimised your machining skills, use Minitab software to implement the widely used Taguchi method. The Taguchi method is simple and effective, yielding the optimal set of machining parameters. Finally, a check test of precise machining parameters was done to ensure the reliability of the previously collected experimental results. The verified results were regarded exceptional and appropriate. Padma et al. [23] used a Taguchi and Gray Relative Analysis combination to optimise EN 9 and carbon steel turning process parameters, utilising gears as hooks) to deliver the least cutting power and the least rough surfaces for lathes. The trials with the L9 orthogonal array, as well as Taguchi's experimental design, are now complete. Analysis of variance was used to assess the principal impact of the experimental data' processing parameters (ANOVA). In addition, both the process parameter and the answer were flipped. The final result, which depicts the true cutting depth, is an important factor in the effects of cutting force and surface roughness that transfer and cutting oil consider.

Jagade et al. (2917) aimed for the highest quality in terms of partial accuracy, surface finish, and speed. As a result, selecting the appropriate machining parameters is critical for lowering production costs while also ensuring product quality. This study provides a review of the current literature for in-depth investigations on optimising process parameters in CNC turning operations. It aids understanding of the impacts and effects of CNC input process parameters on response variables in this research. To optimise the process of determining the most significant parameters, the Taguchi and ANOVA techniques were applied.

The surface of the EN47 material was studied by Sharma et al. (2017). Various cutting settings were used in the CNC lathe testing. Hard turning produces great surface roughness, according to research. The rotational performance characteristics of the EN 47 combination material applied by the composite site were investigated using the L9 orthogonal sequence, which shows sound relationships and deviation analysis (TNMG 160408-FMTN8135). The proposed surface sharpening styles for the materials currently under consideration were created, evaluated, and evaluated using new data.

Maheswara Rao and K. Venkatasubbaiah (2016) Cutting parameters (spindle, feed, and cutting depth) of the AA7075 CNC lathe on low surface roughness with tungsten carbide insert were investigated. The Taguchi orthogonal array was used to construct it in the research. Analysis of variance was used to examine the importance of parameters cut by surface roughness (ANOVA). The ideal range of surface roughness levels was then calculated after that. Finally, a collaboration between variable cutting and MINITAB-16 application inference was built, and regression analysis was conducted. As you can see, the values produced by comparing anticipated and experimental values are quite near, indicating that the model is accurate and adequate.

M Ganesh et.al (2015) While confining the EN-8 metal, he employed a genetic algorithm to optimise the processing elements and set the surface roughness. For spindle speed, feed rate, and cutting depth, this research focuses on CNC rotation with cemented carbide tools. The minimal surface area (Ra) produced by GA is 2.51. The ideal simulation for attaining the highest possible linear model adaptation is the genetic algorithm.
Using the Taguchi methodology, Krupalpawar et al. (2015) evaluated the influence of variations in tool nose radius on surface sharpness and MRR, ANOVA analysis, and inverse SSS. Nose radius on CNC lathes: 0.4, 0.8, and 1.2mm The annotation coated carbide tip tool is configured to allow high-speed river turning of SSS (M2) and research variables employing four parameters in three steps.

Rajpoot Bheem Singh et al. (2015) used the surface response methodology to assess the effects of cutting settings on average surface clarity and material removal rates when turning Al 6061 alloys. RSM was used to determine the effect of each element of surface sharpness separately. The image's sharpness was determined using three separate points. We utilised Design Expert 8.0.4.1 software to analyse the results of 20 tests to determine surface roughness and material loss rates. Surface roughness has been assessed using regression models that take into consideration actual parameters. The significance of regression models was investigated using an ANOVA with a 95% confidence level. Cutting speed, which is one of three cutting parameters, was found to effect surface roughness and MRR.

4. Methodology and experimentation

The purpose of this work has already been discussed in previous chapter. Therefore, this study was conducted through the following steps.

1. First, check and prepare the machine to perform the machining work.
2. Cut the stainless-steel rod using the electric saw and perform the first turning of the part to get the desired dimensions of the work piece (50mm in diameter and 250mm in length).
3. After cutting the work piece, turn the workpiece in a wet condition according to the Box-Behnken design.
4. The cutting length was maintained constant at 5mm.
5. After processing, the next step is to measure the surface roughness. This work is done with the help of the portable stylist Talysurf Profiler.

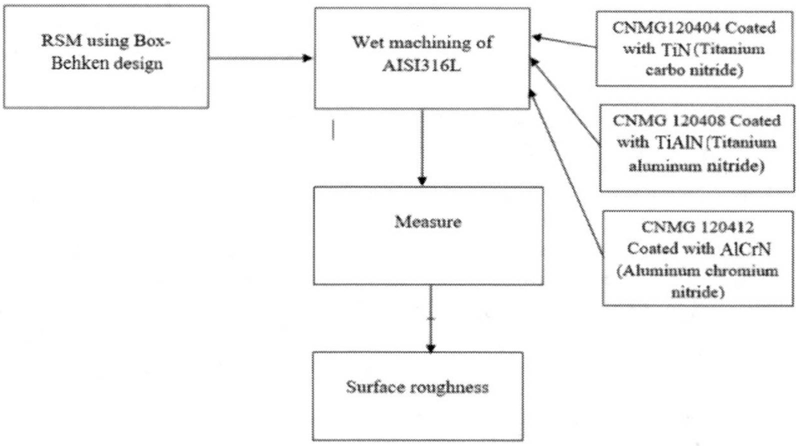

Fig. 3. Experimetal work chart.

5. Selection of Process Parameters and their Levels

After doing a literature research, the parameters and their levels are determined. According to the literature review, the experimental process parameter ranges have been finalised. The outcomes of the pilot studies were also taken into account. The parameters' stages were determined based on the outcomes of pilot tests. The process variables in this experimental investigation were the parameters chosen. Table 1 shows the parameters for machining with units (and notations).

Parameters/Factors		Level		
		-1	0	1
A	Spindle speed (rpm)	500	600	800
B	Feed rate (mm/rev)	0.15	0.17	0.2
C	Depth of cut (mm)	0.4	0.6	0.8

Table 1. Process variables and their limits

6. Material selection

Austenitic stainless steel AISI 316L was chosen for this study due to its excellent mechanical properties and corrosion resistance. Chemical plants, nuclear power plants, and medical gadgets all use it. It's fascinating to look into stainless steel machinability. [2].

Tensile strength, MPa	Yield strength, MPa	Elastic modulus, GPa	Poisson's ratio
515	205	190-210	0.265-0.275

Table 2. Mechanical property of AISI 316L

Element	Iron, Fe	Chromium, Cr	Nickel, Ni	Manganese, Mn
%	71.935	16.439	10.032	1.594

Table 3. Chemical property of AISI316L

7. Cutting tool insert

Coated carbide tools have been shown to outperform uncoated carbide tools when it comes to turning steel. As a result, a commercially available CVD coated carbide insert was used in this investigation. CNMG 120404, 120408, and 120412 nose radius of 0.4, 0.8, and 0.12 mm were used with WIDIA multi-layer coated cutting inserts. Cutting tool geometry is shown in Table 3.5.1. The manufacturer's instructions was used to choose the cutting tool. Three cutting tool inserts with different coatings were employed in the wet turning of AISI316L in this investigation, as they are often utilised in AISI316L machining.

1. CNMG 120404 coated with TiN (Titanium carbo nitride)
2. CNMG 120408 coated with TiAIN (Titanium aluminum nitride)
3. CNMG 120412AlCrN (aluminum chromium nitride)

8. Design of Experiment

A total of fourteen trials were done. Each trial experiment was repeated three times. The turning was done in a completely dry environment. To determine the effect of factors on responses, the RSM approach was utilised. The RSM experimental model is depicted in Table 3.2. RSM [13, 14] outlines a method for determining the link between causes and responses. The following is the functional relationship between the response Y, variable Xj, and the levels of the n input variables (1):

$$y = f(X_1, X_2, X_3, \dots\dots\dots Xn) \quad (1)$$

If the answer can be described by a linear function of independent variables, then the approximation function is a first-order model. A first-order model with two independent variables can be stated as follows (2):

$$y = \beta_0 + \beta_1 x_1 + \beta_2 x_2 + \varepsilon \quad (2)$$

If the response surface is curved, you should use a higher degree polynomial. An approximation function with two variables is specified as a second-order model (3):

$$y = \beta_0 + \beta_1 x_1 + \beta_2 x_2 + \beta_{11} x_{11}^2 + \beta_{22} x_{22}^2 + \beta_{12} x_1 x_2 + \varepsilon \quad (3)$$

Fourteen experiments have been performed as per above experimental plan by using one cutting tools. A total of 42 (14*3) experiments have been performed with three cutting tool inserts with different coatings under wet conditions.

9. Data Collection

Surface roughness measurements.

Turning tests with a cutting length of 220 mm were carried out on a CNC machine. A cylinder made of AISI316L with a diameter of 50 mm and a length of 250 mm was used as the work piece. Three cutting tool inserts with different coatings were employed in the wet turning of AISI316L. The machining tests were carried out on a DoE planed with a Box-Behnken design that included three variables and three levels, for a total of 14 experiments. The test was conducted with Mitutoyo's Surftest SJ 201 surface roughness tester. The probe was changed in order to find the Ra value. Three millimetres were added to the probe's length (see fig.4.)

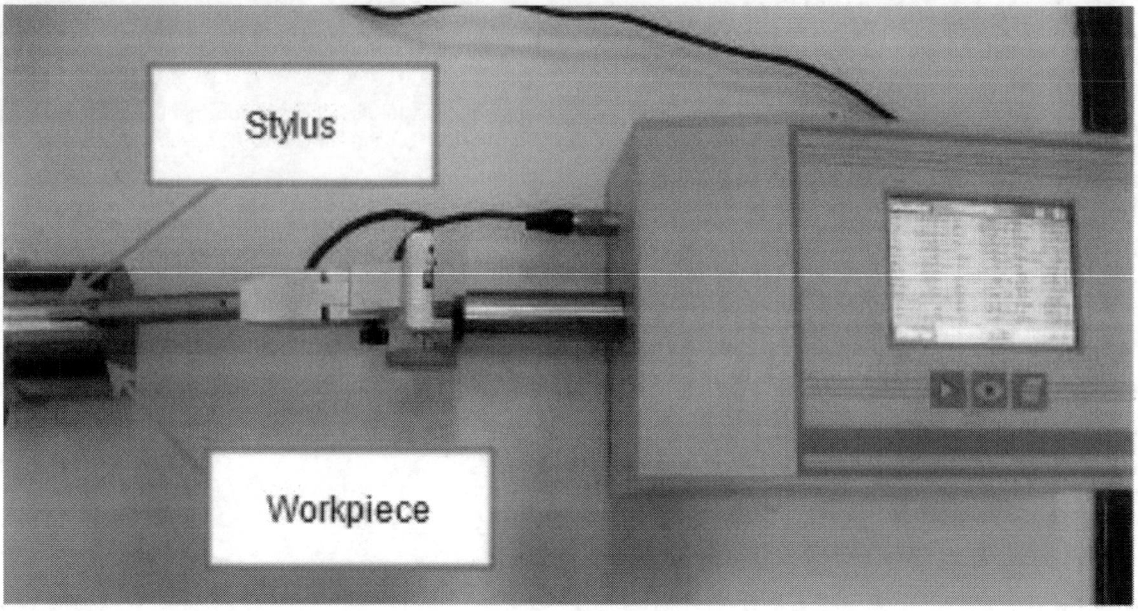

Fig. 4. Surface roughness tester.

10. Results and discussion

Adequacy of the Model

The statistical analysis was performed using Design Expert Statistical Software package (8.1 trial version) (State Ease Inc., Minneapolis, USA). The experimental data were analyzed using multiple regressions and significance of regression coefficients were evaluated by F-test and the significance of the F-values were verified at probability levels ($p<0.05$). Modeling was started with a quadratic model including linear, squared, and interaction terms. The model adequacies were checked in terms of R^2, adjusted R^2, and prediction errors sum of squares. The significant terms in the model were found by Pareto analysis of variance (ANOVA). For each response, ANOVA table was generated. The regression coefficients were used to make statistical calculations to generate response surface plots from the regression models. The responses were optimized by numerical optimization technique using Derringer's desirability function method. This function searches for a combination of factor levels that jointly optimize a set of responses by satisfying the requirements for each response in the design.

The optimization was accomplished by three steps, i.e. converting each response Yi(i=1, 2,..., m) into a dimensionless desirability scale that defines a partial desirability function (di), combining the individual desirabilities to obtain global desirability function (D), and finally identifying the optimal factor settings. The desirability function's scale ranges between0 (completely undesirable response) and 1 (fully desired response).The individual desirabilities (d) for each response were obtained by specifying the goals, i.e. minimize, maximize, or target the response, and boundaries required for each one. A weight factor was assigned which explains the shape of the desirability function for each response. The suitability of the developed model equations for predicting the optimum response values was verified using the optimal conditions. Triplicate experiments were carried out in the optimal condition and the mean experimental values were compared with the predicted values in order to determine the validity of the models.

Surface roughness vs. cutting parameters such as spindle speed (rpm), feed rate (mm/min), and cut depth are plotted in this section (mm). As illustrated in Fig.5,6,7 the visualization of residual vs. run number exhibits a similar pattern with little change across treatments. There is no proof that the procedure has been slowed down. It's useful to figure out how reliable a result is and how skilled a researcher is at running them. The graphic displays random scatter residuals that are contained within the stated domains and do not exceed them, indicating that the model is valid. The residuals are placed on the up and down sides of the straight center line at random. When it comes to answers, the residuals are usually between 2 and -2. As a result, the errors are spread in a natural fashion (see fig.5,6,7.).

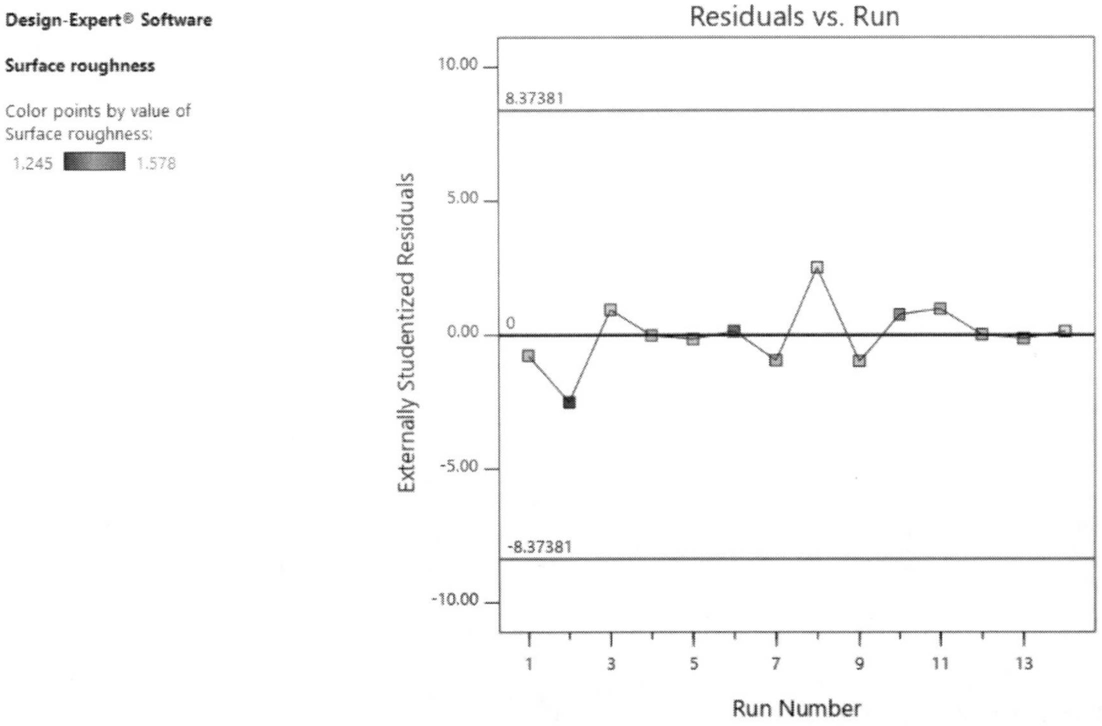

Fig. 5. Residuals vs run plot for a surface roughness achieved with CNMG 120404 coated with TiN

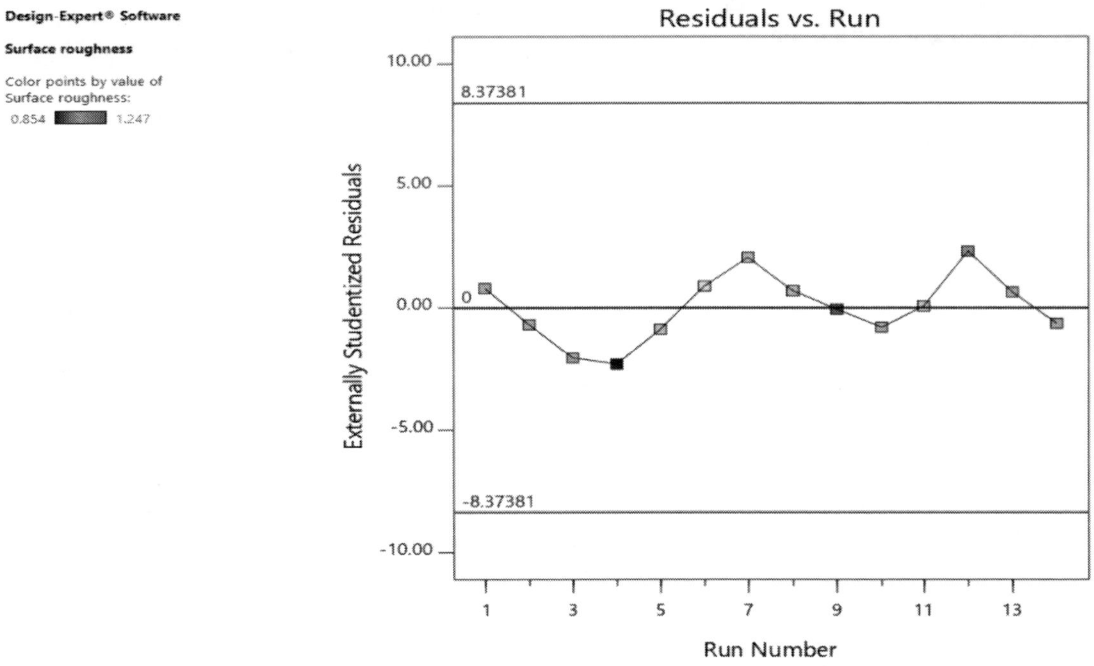

Fig. 6. Residuals vs run plot for a surface roughness achieved with CNMG 120408 coated with TiAIN (Titanium aluminum nitride)

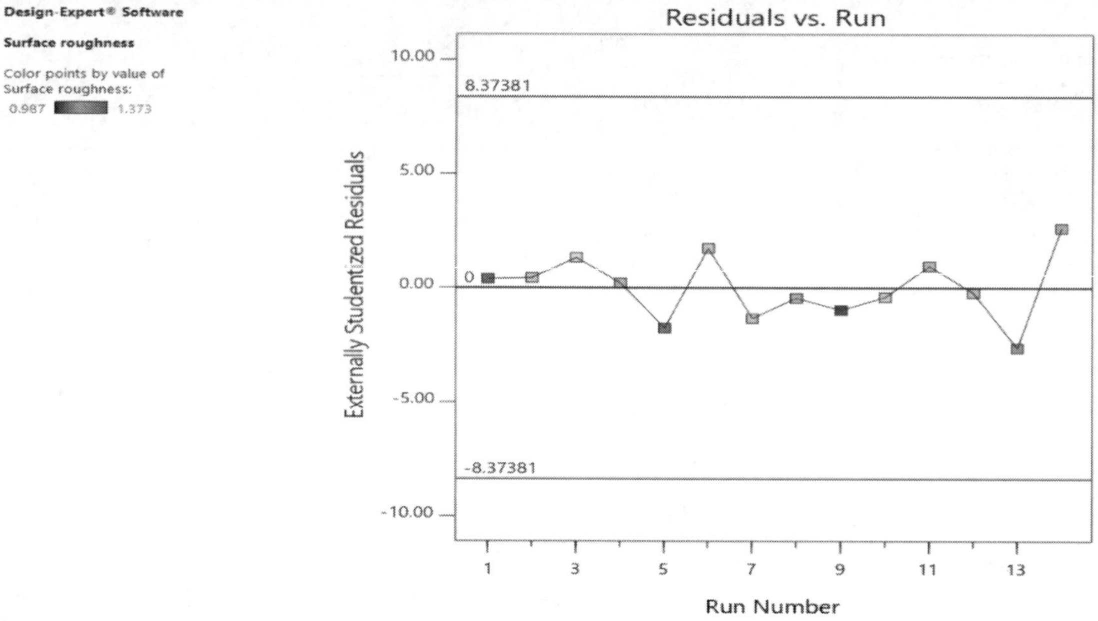

Fig. 7. Residuals vs run plot for a surface roughness achieved with CNMG 120412 AlCrN (aluminum chromium nitride)

11. Selection of Optimum Conditions

Surface roughness was found to have an optimal condition. Second-order polynomial models developed in this study were employed for each answer in order to attain the given optimum conditions. Derringer's desirability function technique was used to optimize a large number of responses. With CNMG 120404 coated with TiN, the best surface roughness was achieved, CNMG 120408 coated with TiAIN and CNMG 120412 coated with AlCrN is obtained as 1.311 μm, 0.897 μm and 0.965 μm.

12. Conclusions

Using three different cutting tool inserts, this study shows how to optimise process parameters in wet turning of AISI 316L. To optimise turning process parameters, the RSM method is employed. An ideal state of surface roughness was discovered. Second-order polynomial models developed in this work were employed for each answer in order to achieve the given optimum conditions. Using Derringer's desirability function method, many responses were optimized. As a result, the following conclusion can be derived from the current research:

1. The optimum surface roughness achieved with CNMG 120404 coated with TiN, CNMG 120408 coated with TiAINand CNMG 120412 coated with AlCrN is obtained as 1.311 μm, 0.897 μm and 0.965 μm.
2. The plot shows random scatter residuals that are contained inside and do not exceed the stated domains, showing that the model is viable. The residuals are dispersed along the up and down sides of the straight centre line at random. Residues are normally between 2 and 2 for replies. As a result, the mistakes are evenly distributed.
3. To confirm and check the normality assumption, a normal probability plot of the residual was constructed. According to the findings, the largest residuals are not as large as expected. Because there are no outliers and the linear deviations are small, the experiment produces a desirable outcome that fits the normality assumption.
4. The ANOVA approach is used to determine the model's relevance in predicting surface roughness. The response surface roughness produced by CNMG 120404 coated with TiN, CNMG 120408 coated with TiAIN, and CNMG 120412 coated with AlCrN had R2 values of 0.9657, 0.9734, and 0.9788, respectively, indicating that the model suited the experimental data reasonably well. The R2 values were changed as 0.9515, 0.9645, and 0.9634 respectively. The expected R2 values were 0.9497, 0.9551, and 0.9568, respectively. The correlations between surface roughness and process factors were found to adequately account for data variability. The adjusted R2 and anticipated R2 values showed a high degree of connection between observed and expected values. The quadratic regression model's ANOVA shows that it was very significant. This was evident from the Fisher's F-Test (F value = 19.6871 cutting with CNMG 120404 coated with TiN, F value = 114.01 cutting with CNMG 120408 coated with TiAIN and F value = 203.03 cutting with CNMG 120412 coated with AlCrN) and a very low probability value (p= 0.0001).

5. "To validate the model, triplicate tests were performed under ideal conditions, and the mean values of the experimental data were compared to the expected values. According to the findings, the percentage difference between expected and experimental surface values ranged from 3.71 percent to 11.51 percent."
6. The research plans will be targeted on using more perspective tools, for example with Duratomic- coating technology (Seco) and other Stainless Steel grades, like AISI 304 and AISI 420.

13. References

[1] Buldum, Baris & Eşme, Uğur & Kulekci, Mustafa & Tarsun-Mersin, & ŞIK, Aydın & Ankara, & Kazançoğlu, Yiğit. (2012). Use of Grey-Taguchi Method for the Optimization of Oblique Turning Process of AZ91D Magnesium Alloy. Materials Testing. 54. 779-785.

[2] Gill, Dr. Surinder & Gupta, Meenu & Satsangi, P.. (2015). Multiple-response optimization of cutting forces in turning of UD-GFRP composite using Distance-Based Pareto Genetic Algorithm approach. Engineering Science and Technology, an International Journal. 3. 10.1016/j.jestch.2015.04.010.

[3] Ghubade, A., Gupta, A., Abrol, A., & Singh, S. (2015). Study of Uncoated and Coated Carbide Insert on Tool Life, Surface Roughness and Material Removal Rate in Machining of EN27 steel. Journal for Manufacturing Science and Production, 15(2).

[4] Klocke F, Krieg T. Coated tools for metal cutting - features and applications. Ann Rhe ClRP 48:7999.3.

[5] Grzesik W. (1998), The role of coatings in controlling the cutting process when turning with coated indexable inserts. J Mater Process Technol; 179:133–43.

[6] Al-Ahmari A M A)2008(. Prediction and optimization models for turning operations, International Journal of Production Research, 464061–81.

[7] Dasgupta S and Mukherjee S)2016(. Effects of machining parameters on tool life and its optimization in turning mild steel with brazed carbide cutting tool, IOP Conf. Ser.: Mater. Sci. Eng. 149012005.

[8] Tamrakar R and Vishwakarma D U)2017(. A Review on Optimization of Cutting Parameters for Tool life and Surface Roughness in Turning Process 34.

[9] Prince Dabreo, Samhita Pashte, Larisa Dmonte, Lavin Dabre (2021). Estimation of Tool Life by Industrial Method and Taylors Method Using Coated Carbide Insert in Turning of Work- Material SS316l, IOP Conf. Series: Materials Science and Engineering, 1070.

[10] Kaushik Vijaya Prasad, Ratna Pal, and Vishnu Kuncham Rajendra Prasad (2016). Optimization of Cutting Parameters during Dry Turning of Austenitic Stainless Steel Using nc-AlTiN/Si3N4, TiAlN, and TiN Coated Inserts. Journal of Materials, Volume 2016, Article ID 3495698, 5 pages.

[11] Ghani JA, Choudhury IA, Masjuki HH. Wear mechanism of TiN coated carbide and uncoated carbide tools at high cutting speed application. J Mater Process Technol 2004; 153–154:1067–73.

[12] Uhlmann E, Wiemann E, Yang S, Krumeich J, Layyous A. New coating developments for high performance cutting tools.

[13] Stoiber M, Badisch E, Lugmair C, Mitterer C. Low-friction TiN coatings deposited by PACVD. J Surf Coat Technol 2003; 163–164:451.7.

[14] Sahin Y, Motorcu AR. Surface roughness model for machining mild steel with coated carbide tool. Mater Des 2005; 26 : 321–6.

[15] Che Heron CH, Ginting A, Goh JJ. Wear of coated and uncoatedcarbides in turning tool steels. J Mater Process Technol 2001; 116:49–54.

[16] Ashok Kumar Sahoo Bidyadhar Sahoo. (2012). Experimental investigations on machinability aspects in finish hard turning of AISI4340 steel using uncoated and multilayer coated carbide inserts. Measurement; 45: 2153–65.

[17] Federico MA, Coelho Reginaldo T, Brandao Lincoln C. Turning hardened steel using coated carbide at high cutting speed. JBraz Soc Med Sci Eng 2008;30:104–9.

[18] Kaladhar M, Venkata Subbaiah K, Srinivasa Rao Ch. Performance evaluation of coating materials and process para-meters optimization for surface quality during turning ofAISI304 austenitic stainless steel. Inter J Eng Sci Technol2011;3:89–102.

[19] Yigit R, Erdal C, Fehim F, Sakip K. Tool life performance of multilayer hard coatings produced by HTCVD for machining of nodular cast iron. Int J Refract Met Hard Mater 2008; 26: 514–24.

[20] Noordin MY, Venkatesh VC, Chan CL, Abdullah A. (2012). Performance of cemented carbide tools in turning AISI 1010 steel. J Mater Process Technol; 116: 16–21.

[21] Auricha JC, Eyrischa T, Zimmermanna M. (2012). Effect of the coating system on the tool performance when turning heat treated AISI4140. Procedia CIRP; 1:214–9.

[22] Hasan Gokkaya, Muammer Nalbant. (2006). The effects of cutting tool coating on the surface roughness of AISI 1015 steel depending on cutting parameters. Turkish J Eng Environ Sci; 30:307–16.

[23] Kulkarni, Atul & Joshi, Girish & Sargade, Vikas. (2013). Design optimization of cutting parameters for turning of AISI 304 austenitic stainless steel using Taguchi method. Indian Journal of Engineering and Materials Sciences. 20. 252-258.

DOI: 10.2507/32nd.daaam.proceedings.081

AUGMENTED REALITY IN MARKETING – STATE OF ART

Ksenija Mitrovic, Nebojsa Novakovic, Jelena Spajic & Ilija Cosic

This Publication has to be referred as: Mitrovic, K[senija]; Novakovic, N[ebojsa]; Spajic, J[elena] & Cosic, I[lija] (2021). Augmented Reality in Marketing – State of Art, Proceedings of the 32nd DAAAM International Symposium, pp.0566-0575, B. Katalinic (Ed.), Published by DAAAM International, ISBN 978-3-902734-33-4, ISSN 1726-9679, Vienna, Austria
DOI: 10.2507/32nd.daaam.proceedings.081

Abstract

Digitalization influenced significant changes in marketing activities, giving them a digital form. Implementing Augmented Reality (AR) technology aims to create a competitive advantage and better engage (potential) consumers. With growing usage of digital technologies in the field of marketing, the literature on Augmented Reality in marketing is accumulating. This paper aims to consolidate and organize the literature on the use of AR technology in marketing, providing a comprehensive state of the art review. Existing literature published from 2016 to May 2021 is reviewed and analyzed. As the result, the paper deepens an understanding of different applications of Augmented Reality in marketing. The resulting framework provides future research directions.

Keywords: Augmented Reality; Marketing; Systematic literature review; Digitalization

1. Introduction

Consumers' adoption of the practice to harness digital information resulted in a surge of new Augmented Reality (AR) applications in many different fields [1]. Marketers commonly seek opportunities to enrich the intensity of emotional responses to attract consumers. Therefore, Augmented Reality is recognized as one of the most disruptive new technologies in the field of marketing [2]. It enables greater consumer involvement [1], using interaction to promote consumer responses [3]. By scanning a brand's logo or a different image, users can access additional content or experience a digital product [4]. This paper aims to present the benefits and functions discovered in the context of using Augmented Reality in marketing. Technological progress facilitates the promotion of the products more interactively [5] putting AR to become a subject of interest among marketers.

Hence, AR marketing as a strategic concept opens the questions to which the researchers are encouraged to answer [6]. Augmented reality has become established research field, and its utilization in marketing activities has received great interest in both academic and managerial literature [7]. A systematic literature review was applied to identify the state of art and address the directions for future work. After a short overview of AR technology and its usage in marketing, the benefits, attributes, and application fields will be presented. In a result, we suggest courses for future research that would enrich the existing literature.

2. Augmented Reality (in) Marketing

AR places users in an environment that is enriched with simulated objects, views, or creatures added to the frame of their actual surroundings [8]. It provides consumers with the resources to visualize and assess products via their smart devices in real-time [9]. Covering the physical world with digital information is a key feature of this technology [1]. Superimposing virtual content into the physical environment [4] gives many opportunities to marketers. AR provides a realistic atmosphere that can be useful to online consumers. In a technology-supported setting, they can create their virtual self while enjoying an enhanced shopping experience [10].

Augmented Reality Marketing presents a digital marketing technology applied in physical settings [11]. However, Rauschnabel et al. (2019) differentiate AR marketing from other digital marketing forms, suggesting it is a long-term strategic capability [3]. Chylinski et al. (2020) define it as "creating digital affordances for consumer experiences" that engage the consumer in a contextually and experientially rich manner [11]. Considering AR as a supportive tool for communication and promotion places this technology on a tactical level. However, AR marketing should extend to strategic research also. Research conducted by Raushnabel [7] gives a radical approach to presenting AR as a potential substitute for a physical world. The results show that AR content can serve as a substitution for utilitarian and digitized products, but the ones that are less symbolic, visible, or decorative.

3. Methodology

Given the beneficial effects of AR usage in the marketing context, it is important to theoretically and practically explore how an application of AR technology changes and shapes consumer experience [12]. Systematic Literature Review was applied to examine the level of implementation of AR in the marketing context, analyze relevant studies and synthesize the key findings [13]. The review assists in the derivation of research directions in AR-based marketing [4].

Research protocol	Details description
Research online databases	The search was conducted in May 2021 in Scopus and the Web of Science databases.
Publication type	Only peer-reviewed literature was considered. The search was limited to articles, reviews, and conference proceedings.
Language	The results were limited to the English language.
Date range	Publications from 2016 to May 2021 were considered.
Search fields	Titles, abstracts and keywords.
Search terms	•Scopus: TS = (("Augmented Reality") OR ("AR") AND ("Marketing")) •Web of Science: TS = (("Augmented Reality") AND ("Marketing"))
Inclusion criteria	Only publications that present an AR application in an area related to marketing were selected.
Exclusion criteria	Papers with a purely technical focus were excluded.
Data analysis and synthesis	We analyzed and summarized the content of the papers to explore the level of AR usage in the field of marketing, present the most used attributes that are relevant for this topic, identify the main gaps and suggest avenues for a future research agenda.

Table 1. Research protocol

Research protocol (Table 1) from [13] was applied to describe the research process. This paper aims to address the existing level of research applications and outcomes related to AR that can contribute to marketing literature. Relevant articles were searched through Scopus and Web of Science, which are known as reliable databases containing quality papers [14]. After defining the search terms, we made a selection of all the papers that met the research protocol criteria. English language studies published from 2016 to May 2021 were considered for this paper. We limited the search results to articles, reviews, and conference proceedings. To explore and understand the wider context of AR usage in marketing activities, the inclusion criteria covered different aspects of consumer behavior, retail marketing, advertising, tourism marketing, sensory marketing, product marketing, etc.

Figure 1. presents the selection process of papers chosen for the review. In the first phase, 325 papers in Scopus and 402 papers in the Web of Science that appeared in the search results were reviewed based on the paper's abstract and keywords. The search results on the two databases coincided, so the duplicates were removed. In the pre-selection process, we excluded papers that are not in accordance with the research protocol. In the result, 127 papers remained for full-text reading. Excluded papers were mostly technology-oriented and therefore irrelevant for a marketing framework. After full-text reading of the chosen studies, we selected 77 papers eligible for the review. In the final phase, we synthesized the findings of the existing literature. In the result, section 4 presents different applications of researched AR in the marketing context, and also the benefits that AR technology offers to the marketing field.

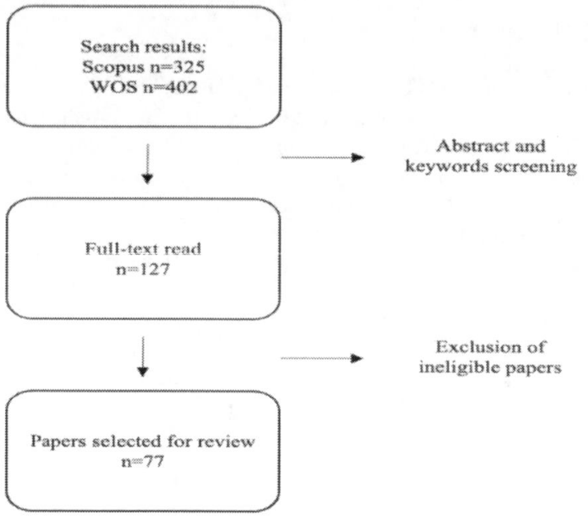

Fig. 1. Selection process

4. Review results of the AR usage in marketing

AR enhances marketing [10] by providing authentic, lifelike, and memorable brand experiences [2]. Widely researched fields imply the growing importance of identifying the advantages of using AR technology over traditional forms of media communication. In the result of the conducted systematic literature review, it is concluded that the usage of augmented reality in marketing has the ability to shape consumer behavior [2], [11]. AR marketing changes the marketing focus from attributes of a product to affordances of the situation where engagement creates value [11]. Javornik (2016b) suggests that AR functions can transform experiential activities: product trials, information search, acquisition, and product try-on [4]. Virtual garment try-on has become one of the most promising fields of AR technology usage [15], giving more knowledge and understanding of different features of a product without physically experiencing it [16]. AR is transforming brick and mortar stores into digital showrooms allowing consumers to visualize a product in a desirable virtual context. Virtual representations assist consumers in learning about the product's experiential attributes [17], especially for products that are too big or heavy, expensive, sensitive, and hard to display in every setting [18].

Another form of utilization is personalizing and optimizing the shopping experience with information popping out to locate the product, navigate the store, suggest discounts, promotions, or recommendations [19], [20]. Product attributes are enriched with immersive technology providing reliable information [9], [21]. Researchers examined the level of informativeness, as Javornik suggested [4]. It is revealed that AR enables enhancing the process of the information searching [22], [23], [24], [25], [26]. Overlaying additional information about the product enriches the physical settings and provides more knowledge [27].

Javornik's proposed function related to product trials is useful for risk reduction. Taking into consideration that one of the marketing goals is the enhancing of the likelihood of the purchase, perceived risk should be minimized [27]. AR is proposed as a tool that has the ability to reduce the risk associated with the purchase decision. The ability of testing products in a safe environment reduces the purchasing risks [18], [20], [27]. Moreover, authors suggest that this interactive technology impacts the reduction of other possible risks as well: social, financial, psychological, performance, physical, and time risk [20], [27].

Gallardo et al. (2018) show the capability of AR to improve all the consumer's activities when interacting with the product: selection, adaptation, approval, and purchase [5]. AR marketing has been found useful along a **customer journey** [1], [29], [30], [31] becoming a new form of its touchpoint [32]. It helps consumers to visualize the product in the pre-purchase stage augmenting beyond usual [9]. Cuomo et al. (2020) found that AR shows great usefulness in terms of brand trust and security [28], since it deepens user's knowledge about the product [32]. Hoyer et al. (2020) suggest that AR has great relevance in the pre-purchase stage with the ability to facilitate imagination and provide a vivid experience of the product [33]. Also, it enriches the experience within the purchase stage and improves post-purchase services by upgrading and enriching consumption [9], [33]. However, the need of future research within the post-purchase phase was proposed, since this phase has the lack of identified benefits of AR. Its opportunities consist of social and experiential benefits that engage users additionally [32].

Also, it is found that Augmented Reality technology has a significant impact on an overall **marketing mix** [6], [7]. It generates new forms of communication, offers different opportunities for distribution, changes the price of the augmented product, and can improve and extend the core product [6] or even replace it with the virtual one [7]. Carrozzi et al. (2019) suggest that holograms are arising as a new type of digital product [34].

Augmented object is not perceived just as a digital copy of a product since it is placed in the user's personal space [35]. Also, consumers are willing to buy more and pay a premium price for products chosen with the support of AR [22], [36].

In several studies an AR application was developed in order to examine the application effect on a specific attribute of consumer behavior [5], [31], [37], [38], [39], [40], [41], [42], [43]. Other studies used existing applications for the same reason, and the AR application *IKEA Place* was used mostly [1], [3], [36], [44], [45]. Ikea recognized consumer's challenges and insecurities regarding fitting the product in the existing arrangements, enabling consumers to see the virtual furniture right next to the non-augmented objects [46]. For developing new applications, authors used a tool called *Vuforia* in most cases [31], [38], [47], [48], [49]. In the following section, the variety of the usage of AR in different marketing fields will be presented.

4.1. AR usage in different marketing fields

Utilizing AR technology in the field of marketing simplifies contextual relations that are complex to consumers [11]. It is found useful in the **retail marketing** context [1], [10], [15], [17], [19], [20], [22], [28], [35], [50], [51]. The retail landscape is transforming to omnichannel [28], forcing both consumers and retailers to accept integrated shopping experiences that cover in-store, mobile, and internet shopping [26]. AR is used in this case to extend the possibilities of the physical settings [15]. Retailers have been given the opportunity to reinvent the shopping experience, making it more attractive than traditional. AR effects on consumer behavior are mostly investigated in terms of apparel retailing [21]. The fashion and beauty industry are specific for the need of trying the product before making a buying decision. Therefore, Pantano et al. (2017) explored how AR characteristics influence purchase decisions, using *Ray-Ban Virtual Mirror* in two different countries. Results show that both samples consider AR as useful and enjoying. While Italians value enjoyment, Germans give priority to the perceived usefulness [15].

AR usage within marketing is mostly covered in the **advertising** context [29]. The advertising world recognizes AR as a powerful and effective tool. AR empowered ads add life to their 2D stationary match, making them memorable [25] and interactive [47]. Yang et al. (2020) explored AR effectiveness, revealing that AR ads get positive ratings for their creativity, informativeness, and effectiveness The authors found that the effect of AR advertising is mediated by curiosity and attention, with a specified level of familiarity with AR ad technology. It is demonstrated that AR advertising is more effective compared to the traditional one when it comes to influencing user's attitudes toward the ad [52]. Pozharliev et al. (2021) suggest that AR advertising stimulates consumers' emotional responses and positive behavioral intentions [36]. Also, the authors found that it is successful for both novel and traditional products. Divya Udayan et al. (2020) investigated the power of AR to promote brand building compared to a traditional advertisement. Valves are used as an adequate example for trying to bridge the gap between marketing and engineering. Comparing traditional and AR ads, a large number of attributes go in favor of AR: memory retention, product knowledge, novelty, vividness, representational richness [25]. Analyzing different studies of AR usage in advertising [2], [14], [36], [47], [49], [52], [53], it is concluded that AR is a suitable advertising medium.

Brand-related outcomes can benefit from AR [7], since it is revealed that AR is relevant for aiding brand-interactive marketing [54]. AR enables context-aware **branding** [18], providing extended branded content [28]. Existing research suggests that AR enriches brands, driving direct contact with it [21]. Rauschnabel et al. (2019) proposed a conceptual model for exploring the relationship between consumer perceived benefits and brand attitudes [3]. The authors demonstrated the process of AR influence on a brand. The study revealed that inspiration is a mediating construct between the proposed relationship. Scholz and Duffy (2018) state that AR enables inviting brands into the consumers' personal spaces, resulting in the creation of a strong consumer-brand relationship [35]. With an imprint about the product created in the consumer's mind, brand recall is significantly enhanced [25] and brand attitudes are better communicated. Different aspects of the brand can be enhanced. AR is effective in strengthening the brand's image [5] and boosting brand awareness [20], [50].

Benefits of AR technology are widely recognized in a **tourism marketing** [31], [43], [55], [56], [57], [58]. Tourism marketing goes beyond giving information but instead immersing tourists in various destinations [43]. Cranmer et al. (2020) investigated the AR value in the tourism sector. Their results show that marketing and sales value were found the most prominent in the tourism industry, regarding AR [57]. Its great marketing potential reflects in the ability to present accurate information and promote destinations with a 'human touch'. Tourism marketing is challenged by distrust of consumers and the need to introduce a destination attractively and credibly [31]. Huertas & Gonzalo (2020) studied the factors that generate a satisfactory tourist experience in order to contribute to a destination brand. Their results show that AR apps impact satisafction and are able to communicate attributes of a destination brand [58]. While some applications enhance the tourist experience, others facilitate the knowledge related to a specific destination [55]. In this field, museum marketing is also researched [59], [60]. AR enables a vivid world of artifacts and an understanding of its origin and history [31]. Visitors are attracted to museums with the motivation to experience AR technology while exploring the intangible dimensions of a culture or heritage and immersing in the historical spaces. Enjoyment created by the enriched exhibition is expected, along with the vivid original appearance of the monument [60].

Avila (2017) suggests that AR found its usage in **library marketing** too [61]. Augmented technology can be effective in communicating information that exceeds physical signage. Therefore, library communication can become vivid and informational marketing enriched. Technology advancements enabled virtual wine-label technology to deliver unlimited content for **wine marketing** [62]. Many platforms are designed to connect consumers and wine labels, telling their stories, recommendations, or stimulations that bring labels to life. The authors suggest that the wine industry benefits from AR usage in terms of product promotion, with the great utilization of technology-based storytelling. Implementation of AR technology in the context of **sports marketing** should be considered as a complementary activity that contributes to the interactivity, but without interfering with the game action. This technology boosts the experience of sports fans which consequently leads to positive word of mouth [63]. There are several studies related to the effectiveness of AR in **housing** marketing [5], [38], [41], [42]. Traditional catalogues are now outdated and replaced by 3D visualization of houses in order to assist prospective buyers [41].

Augmenting consumer's surroundings with multiple touch-points enables greater participation and sharing of extended content [28]. Since consumers pursue participation and interaction with brands, generating engaging content comes to focus [32]. Convoy et al. (2019) examined the potential of AR technology in the context of **participatory marketing**, transforming consumers from passive actors to proactive co-creators [64]. AR technology is considered as innovative way to support **sensory marketing** [65]. Also, this immersive technology extends the experiential value and has a great significanse in the field of **experiential marketing** [66]. AR in the context of experiential marketing is recognized as relevant within a tourism sector [66], while it also has the ability to create experiential consumption [22].

4.2. Researched variables used in AR marketing studies

AR is reshaping consumer behavior, which demands new strategies to fulfill the upcoming consumer needs [67]. AR attributes: interactivity, vividness, and novelty are used often in the research. However, besides these, different attributes were examined in order to identify their effect on marketing.

Researched variables	References
Interactivity	[1], [4], [9], [15], [22], [24], [27], [44], [48], [49], [68], [69]
Engagement	[10], [20], [22], [37], [46], [49], [68], [69], [70], [71]
Usefulness	[1], [15], [21], [25], [28], [44], [63], [69]
Curiosity	[49], [52], [70]
Word of Mouth	[17], [30], [63]
Ease of use	[9], [15], [17], [21], [25], [28], [67], [69], [72]
Attitude	[1], [3], [14], [15], [17], [19], [21], [26], [49], [52], [63], [70], [73], [74]
Satisfaction	[2], [9], [10], [17], [24], [35], [49], [56], [58], [71]
Informativeness	[9], [17], [37], [44], [45], [75]
Control	[10], [17], [25], [30], [68], [76]
Flow	[10], [17], [24], [68], [77]
Purchase intention	[2], [14], [20], [28], [35], [44], [45], [56], [73], [74]
Experience	[5], [17], [18], [22], [27], [28], [40], [58], [62], [66], [67], [76]
Behavioral intentions	[14], [15], [17], [21], [23], [26], [28], [44],
Enjoyment	[10], [15], [17], [23], [24], [69], [76]
Inspiration	[3], [22], [72]
Vividness	[22], [24], [25], [69]
Novelty	[22], [24], [25], [69], [75]
Customization	[22], [34], [71]
Aesthetic	[15], [70], [78]
Entertainment	[22], [24], [59], [70], [75]
Sense of presence	[1], [10], [29], [30], [36], [58], [68], [74]
Relationship with a brand	[17], [36], [49], [51]

Table 2. Researched variables in AR marketing

Existing literature is predominantly focused on the acceptance of AR technology [9], [22], and on its technological aspects [15], [22], [32]. Several studies examined technological attributes influence on consumer behavior [15], [32].

It is found that this novel technology has a great relevance in supporting the decision making [1], [11], [15], [28], [30], [36], and sales increase [5], [20], [31], [50]. Experience is widely acknowledged by researchers [24] and it is one of the most examined component of AR marketing, as it is shown in Tabel 2. Engagement is also recognized as a beneficial characteristic that is used in many studies. McLean & Wilson (2019) established the variables relevant for influencing brand engagement. They revealed that AR attributes (interactivity, novelty, and vividness) and technology attributes (perceived ease of use, usefulness, enjoyment and subjective norms) are the drivers of engagement, which consequently results in satisfactory experience.

Mobile AR applications are recognizable for a high level of interactivity, which promises a favorable user experience [9]. Regarding interactivity, Park & Yo (2020) explored its effects on shopping behavior in mental imagery context. The authors suggest that controllability and playfulness positively influence mental imagery. AR amplifies the interactivity ladder with the brand, but also among users [32]. Greater interactivity is achieved in bringing the brand's offering in a personally relevant context, within the desired environment. Also, it can be supported by getting control over the simulated experience [30]. Qin et al. (2021) found that interactivity affects hedonic and utilitarian values. Nikhahsemi et al. (2021) examined AR attributes relationship with hedonic and utilitarian components. They found that vividness and novelty positively influence both hedonic and utilitarian values. However, interactivity does not exhibit a positive relationship with utilitarian components. On the other hand, Hilken et al. (2017) found a positive relationship between spatial presence and hedonic and utilitarian value perception of the online shopping experience [30]. Results from the research conducted by Pantano et al. (2017) also show that AR is considered a useful and enjoyable tool, thus utilitarian and hedonic values are repetitively recognized [15].

4.3. Future research agenda

Authors propose examining the influence of AR on ad format belief, ad trust, and ad message involvement. With the great potential of the usage in advertising, we believe that it is crucial to reveal the consumer responses toward proposed attributes. Also, customization abilities during AR experience should be explored, taking into consideration that it can provide relevant managerial implications. Grzegorczyk et al. (2019) indicate that creating products and services that fulfill consumer desire has great relevance [71]. However, the effects of customized experiences on consumer responses were not examined. Since most studies used students as respondents [79], it is required to broaden the experiment sample to generalize the results. Younger generations the with focus on alpha generation can be an interesting sample to discover the complete potential of using this immersive technology in marketing activities, making them the long-term strategic capability.

5. Conclusion

Marketers now have the opportunity to approach and interact with consumers in a more exciting and effective way [10], empowered by technology deployment. It is shown that there is a significant number of papers that examine the usage of AR in the marketing context, which is required given the importance of its benefits. Marketing scholars studied different variables that impact consumer behavior, brand, and acceptance of this technology by consumers. A broad overview should be used as a course in examining a specific topic that is relevant to AR in marketing. This paper summarized the relevant literature and addressed the future research agenda. In the result of reviewing current studies, it is proposed that AR should be considered as a new paradigm in marketing research and practice [6].

6. Acknowledgments

Participation in the 32nd DAAAM International Symposium on Intelligent Manufacturing and Automation was supported by The Ministry of Education, Science and Technological Development of the Republic of Serbia.

7. References

[1] Alimamy, S. & Al-Imamy, S. (2021): Customer perceived valuethrough quality augmented reality experiences in retail: The mediating effect of customer attitudes, Journal of Marketing Communications, 27, 1, ISSN 1466-4445

[2] Sung, E. C. (2021). The effects of augmented reality mobile app advertising: Viral marketing via shared social experience. Journal of Business Research, 122, 75-87, ISSN 0148-2963

[3] Rauschnabel, P. A., Felix, R., & Hinsch, C. (2019). Augmented reality marketing: How mobile AR-apps can improve brands through inspiration. Journal of Retailing and Consumer Services, 49, 43-53, ISSN 0969-6989

[4] Javornik, A. (2016b). Augmented reality: Research agenda for studying the impact of its media characteristics on consumer behaviour. Journal of Retailing and Consumer Services, 30, 252-261, ISSN 0969-6989

[5] Gallardo, C., Rodríguez, S. P., Chango, I. E., Quevedo, W. X., Santana, J., Acosta, A. G., & Andaluz, V. H. (2018). Augmented reality as a new marketing strategy. International conference on augmented reality, virtual reality and computer graphics, ISBN 978-3-319-95269-7, 351-362, july 2018, Otranto, Italy, Springer, Cham, https://doi.org/10.1007/978-3-319-95270-3_29

[6] Dwivedi, Y. K., Ismagilova, E., Hughes, D. L., Carlson, J., Filieri, R., Jacobson, J., & Wang, Y. (2021). Setting the future of digital and social media marketing research: Perspectives and research propositions. International Journal of Information Management, 59, 102168, ISSN 0268-4012

[7] Rauschnabel, P. A. (2021). Augmented reality is eating the real-world! The substitution of physical products by holograms. International Journal of Information Management, 57, 102279, ISSN 0268-4012

[8] Värno, K., Mahmood, K., Otto, T., & Kuts, V. (2019). Development of a smart workstation by using AR technology. Annals of DAAAM. Proceedings, 30, October 2019, Zadar, Craotia, EU, ISSN 1726-9679, ISBN 978-3-902734-22-8, pp.1061-1067, Katalinic, B. (Ed), DAAAM International, Vienna, Austria, EU, https://doi.org/10.2507/30th.daaam.proceedings.148

[9] Qin, H., Peak, D. A., & Prybutok, V. (2021). A virtual market in your pocket: How does mobile augmented reality (MAR) influence consumer decision making?. Journal of Retailing and Consumer Services, 58, 102337, ISSN 0969-6989

[10] Huang, T. L., & Liao, S. L. (2017). Creating e-shopping multisensory flow experience through augmented-reality interactive technology. Internet Research, 27, 2, ISSN 1066-2243

[11] Chylinski, M., Heller, J., Hilken, T., Keeling, D. I., Mahr, D., & de Ruyter, K. (2020). Augmented reality marketing: A technology-enabled approach to situated customer experience. Australasian Marketing Journal (AMJ), 28, 4, 374-384, ISSN 1441-3582

[12] Loureiro, S. M. C., Guerreiro, J., Eloy, S., Langaro, D., & Panchapakesan, P. (2019). Understanding the use of Virtual Reality in Marketing: A text mining-based review. Journal of Business Research, 100, 514-530, ISSN 0148-2963

[13] Muñoz-Leiva, F., López, M. E. R., Liebana-Cabanillas, F., & Moro, S. (2021). Past, present, and future research on self-service merchandising: a co-word and text mining approach. European Journal of Marketing, ISSN 0309-0566

[14] Yussof, F. M., Salleh, S. M., & Ahmad, A., L. (2019). Factors of Augmented Reality Technology Adoption in Influencing Attitude and Purchasing Intention: A Review on Advertising Context. International Journal of Advanced Science and Technology, 28, 18, 321 – 328, ISSN 2005-4238

[15] Penco, L., Serravalle, F., Profumo, G., & Viassone, M. (2020). Mobile augmented reality as an internationalization tool in the "Made In Italy" food and beverage industry. Journal of Management and Governance, 1-31, ISSN 1385-3457

[16] Sihi, D. (2018). Home sweet virtual home: The use of virtual and augmented reality technologies in high involvement purchase decisions. Journal of Research in Interactive Marketing, ISSN 2040-7122

[17] Perannagari, K. T., & Chakrabarti, S. (2019). Factors influencing acceptance of augmented reality in retail: insights from thematic analysis. International Journal of Retail & Distribution Management, ISSN 0959-0552

[18] Habib, A., & Hakim, A. (2016). Context aware augmentational marketing. 2016 SAI Computing Conference (SAI), july 2016, ISBN 978-1-4673-8461-2, pp.1227-1231, IEEE, London, UK, https://doi.org/10.1109/SAI.2016.7556135

[19] van Esch, P., Arli, D., Gheshlaghi, M. H., Andonopoulos, V., von der Heidt, T., & Northey, G. (2019). Anthropomorphism and augmented reality in the retail environment. Journal of Retailing and Consumer Services, 49, 35-42, ISSN 0969-6989

[20] Parekh, P., Patel, S., Patel, N. & Shah, M. (2020). Systematic review and meta-analysis of augmented reality in medicine, retail, and games. Visual Computing of Industry, Biomedicine and Art, 3, 21, ISSN 2096496X

[21] Ross, H. F., & Harrison, T. (2016). Augmented reality apparel: An appraisal of consumer knowledge, attitude and behavioral intentions. In 2016 49th Hawaii international conference on system sciences (HICSS), january 2016, ISSN 530-1605, ISBN 978-0-7695-5670-3, 3919-3927, IEEE, Koloa, HI, USA, https://doi.org/10.1109/HICSS.2016.487

[22] Nikhashemi, S.R., Knight, H.H., Nusair, K. & Lian, B. C. (2021) Augmented reaility in smart retailing: A (n) (A) Symmetric Approach to continuous intention to use retail brand's mobile AR apps. Journal of Retailing and Consumer Sevices, 60, 102464, ISSN 0969-6989

[23] Saprikis, V., Avlogiaris, G., & Kataracha, A. (2021). Determinants of the intention to adopt mobile augmented reality apps in shopping malls among university students. Journal of Theoretical and Applied Electronic Commerce Research, 16, 3, 491-512, ISSN 0718-1876

[24] Barhorst, J. B., McLean, G., Shah, E., & Mack, R. (2021). Blending the real world and the virtual world: Exploring the role of flow in augmented reality experiences. Journal of Business Research, 122, 423-436, ISSN 0148-2963

[25] Divya Udayan, J., Kataria, G., Yadav, R., & Kothari, S. (2020). Augmented Reality in Brand Building and Marketing – Valves Industry. 2020 International Conference on Emerging Trends in Information Technology and Engineering (ic-ETITE), ISSN 21570485, Vellore, doi:10.1109/ic-etite47903.2020.425

[26] Park, M., & Yoo, J. (2020). Effects of perceived interactivity of augmented reality on consumer responses: A mental imagery perspective. Journal of Retailing and Consumer Services, 52, 101912, ISSN 0969-6989

[27] Alimamy, S., Deans, K. R., & Gnoth, J. (2018). The role of augmented reality in the interactivity ofco-creation: A critical review. International Journal of Technology and Human Interaction (IJTHI), 14, 3, 88-104, ISSN 1548-3908

[28] Cuomo, M. T., Tortora, D., Festa, G, Ceruti, F & Metallo, G. (2020). Managing omni-customer brand experience via augmented reality: A qualitative investigation in the Italian fashion retailing system. Qualitative Market Research, 23, 3, 427-445, ISSN 1352-2752

[29] Wedel, M., Bigné, E., & Zhang, J. (2020). Virtual and augmented reality: Advancing research in consumer marketing. International Journal of Research in Marketing, 37, 3, 443-465, ISSN 0167-8116

[30] Hilken, T., de Ruyter, K., Chylinski, M., Mahr, D., & Keeling, D. I. (2017). Augmenting the eye of the beholder: exploring the strategic potential of augmented reality to enhance online service experiences. Journal of the Academy of Marketing Science, 45, 6, 884-905, ISSN 0092-0703

[31] Lin, C. M., Lin, T. C., Lin, Y. C., Wang, C. M., & Dow, C. R. (2018). Community interaction and marketing using 3D coloring augmented reality in Zhongxing New Village. In 2018 15th International Symposium on Pervasive Systems, Algorithms and Networks (I-SPAN), October 2018, ISBN 978-1-5386-8535-8, pp. 272-276, IEEE, Yichang, China, https://doi.org/10.1109/I-SPAN.2018.00052

[32] Penco, L., Serravalle, F., Profumo, G., & Viassone, M. (2020). Mobile augmented reality as an internationalization tool in the "Made In Italy" food and beverage industry. Journal of Management and Governance, 1, 31, ISSN 1385-3457

[33] Penco, L., Serravalle, F., Profumo, G., & Viassone, M. (2020). Mobile augmented reality as an internationalization tool in the "Made In Italy" food and beverage industry. Journal of Management and Governance, 1-31, ISSN 1385-3457

[34] Carrozzi, A., Chylinski, M., Heller, J., Hilken, T., Keeling, D. I., & de Ruyter, K. (2019). What's mine is a hologram? How shared augmented reality augments psychological ownership. Journal of interactive marketing, 48, 71-88, ISSN 1094-9968

[35] Scholz, J., & Duffy, K. (2018). We ARe at home: How augmented reality reshapes mobile marketing and consumer-brand relationships. Journal of Retailing and Consumer Services, 44, 11-23, ISSN 0969-6989

[36] Pozharliev, R., De Angelis, M., & Rossi, D. (2021). The effect of augmented reality versus traditional advertising: a comparison between neurophysiological and self-reported measures. Marketing Letters, 1, 16, ISSN 0923-0645

[37] Ludwig. T., Hoffmann, S., Jasche, F & Ruhrmann, M. (2020). VacuumCleanAR: augmented reality-based self-explanatory physical artifacts. In: Alt, F., Schneegass, S. & Hornecker, E. (Hrsg.), Mensch und Computer 2020 - Tagungsband. pp. 291–302, New York: ACM, doi: 10.1145/3404983.3405526

[38] Adrianto, D., Hidajat, M., & Yesmaya, V. (2016). Augmented reality using Vuforia for marketing residence. In 2016 1st International Conference on Game, Game Art, and Gamification (ICGGAG), december 2016, ISBN 978-1-5090-5480-0, pp. 1-5, IEEE, Jakarta, Indonesia, https://doi.org/10.1109/ICGGAG.2016.8052642

[39] Ajibola, D., Shafombabi, V., Petrus, P., Shilongo, N., Thielen, E., & Sieck, J. (2018). Using augmented reality to enhance printed magazine articles about Namibian lodges. In Proceedings of the Second African Conference for Human Computer Interaction: Thriving Communities, December 2018, Windhoek, Namibia, ISBN 9781450365581, pp. 1-4, Association for Computing Machinery, New York, NY, United States, https://doi.org/10.1145/3283458

[40] Tsai, T. H., Chang, H. T., Yu, M. C., Chen, H. T., Kuo, C. Y., & Wu, W. H. (2016). Design of a mobile augmented reality application: An example of demonstrated usability. In International Conference on Universal Access in Human-Computer Interaction, june 2016, ISBN 978-3-319-40243-7, pp. 198-205, Springer, Cham, https://doi.org/10.1007/978-3-319-40244-4_19

[41] Setyadi, R., & Ranggadara, I. (2020). Augmented reality using features accelerated segment test for property catalogue. Telkomnika, 18, 1, 140-147, ISSN 1693-6930

[42] Mufid, M. R., Tahir, M., Meiyuana, D. B., Dwileksa, A. & Kumiawaty, E. (2020), Housing design in planet green tambora using augmented reality for promotion media. IES 2020 – International Electronics Symposium: The Role of Autonomous and Intelligent Sustems for Human Life and Comfort. September 2020, ISBN 978-172819580-8, pp. 669-673, doi: 10.1109/IES50839.2020.9231546

[43] Jung, T. H., Lee, H., Chung, N., & tom Dieck, M. C. (2018). Cross-cultural differences in adopting mobile augmented reality at cultural heritage tourism sites. International Journal of Contemporary Hospitality Management, ISSN 0959-6119

[44] Kowalczuk, P., Siepmann, C., & Adler, J. (2021). Cognitive, affective, and behavioral consumer responses to augmented reality in e-commerce: A comparative study. Journal of Business Research, 124, 357-373, ISSN 0148-2963

[45] Haile, T. T., & Kang, M. (2020). Mobile augmented reality in electronic commerce: investigating user perception and purchase intent amongst educated young adults. Sustainability, 12, 21, 9185, ISSN 2071-1050

[46] Scholz, J., & Smith, A. N. (2016). Augmented reality: Designing immersive experiences that maximize consumer engagement. Business Horizons, 59, 2, 149-161, ISSN 0007-6813

[47] Wijaya, A. C., Munandar, M. W. A., & Utaminingrum, F. (2019). Usability testing of augmented reality for food advertisement based on mobile phone using system usability scale. In 2019 International Conference on Sustainable Information Engineering and Technology (SIET), september 2019, ISBN 978-1-7281-3881-7, pp. 266-269, IEEE, Lombok, Indonesia, https://doi.org/10.1109/SIET48054.2019.8986118

[48] Hing, V., & Khoo, H. K. (2017). Business card reader with augmented reality engine integration. 9th International Conference on Robotic, Vision, Signal Processing and Power Applications, september 2016, ISBN 978-981-10-1719-3, pp. 219-227, Springer, Singapore, https://doi.org/10.1007/978-981-10-1721-6_24

[49] Sravanthi, P. V. L. (2021). Litterature review on brand building through augmented reality. International Journal of Modern Agriculture, 10, 2, 595-604, ISSN 2305-7246

[50] Jäger, A. K., & Weber, A. (2020). Increasing sustainable consumption: message framing and in-store technology. International Journal of Retail & Distribution Management, ISSN 0959-0552

[51] Ramadan, Z. B., & Farah, M. F. (2017). The Pokémonisation of the first moment of truth. International Journal of Web Based Communities, 13, 2, 262-277, ISSN 1477-839

[52] Yang, S., Carlson, J. R., & Chen, S. (2020). How augmented reality affects advertising effectiveness: The mediating effects of curiosity and attention toward the ad. Journal of Retailing and Consumer Services, 54, 102020. ISSN 0969-6989

[53] de Ruyter, K., Heller, J., Hilken, T., Chylinski, M., Keeling, D. I., & Mahr, D. (2020). Seeing with the customer's eye: Exploring the challenges and opportunities of AR advertising. Journal of Advertising, 49, 2, 109-124, ISSN 0091-3367

[54] Ng, C. C., & Ramasamy, C. (2018). Augmented reality marketing in Malaysia—future scenarios. Social Sciences, 7, 11, 224, ISSN 20760760

[55] Lacka, E. (2020). Assessing the impact of full-fledged location-based augmented reality games on tourism destination visits. Current Issues in Tourism, 23, 3, 345-357, ISSN 1368-3500

[56] Ashraf, F., & Nagy, G. (2019). HBIM as a tool for branding historical Egyptian sites by utilizing augmented reality. In IOP Conference Series: Earth and Environmental Science, november 2017, New Cairo, Egypt, ISBN 17551307 IOP Publishing, https://doi.org/10.1088/1755-1315/397/1/012012

[57] Cranmer, E. E., tom Dieck, M. C., & Fountoulaki, P. (2020). Exploring the value of augmented reality for tourism. Tourism Management Perspectives, 35, 100672, ISSN 2211-9736

[58] Huertas, A., & Gonzalo, J. (2020). The role of augmented reality in destination branding. Tourism and hospitality management, 26, 2, 419-436, ISSN 1330-7533

[59] Nechita, F., & Rezeanu, C. I. (2019). Augmenting museum communication services to create young audiences. Sustainability, 11, 20, 5830, ISSN 2071-1050

[60] Recupero, A., Talamo, A., Triberti, S., & Modesti, C. (2019). Bridging museum mission to visitors' experience: activity, meanings, interactions, technology. Frontiers in psychology, 10, 2092, ISSN 1664-1078

[61] Avila, S. (2017). Implementing augmented reality in academic libraries. Public Services Quarterly, 13, 3, 190-199, ISSN 15228959

[62] Vrigkas, M., Lappas, G., Kleftodimos, A., & Triantafillidou, A. (2021). Augmented reality for wine industry: Past, Present, and Future. In SHS Web of Conferences, may 2021, ISSN 2261-2424, 04006, EDP Sciences, https://doi.org/10.1051/shsconf/202110204006

[63] Goebert, C., & Greenhalgh, G. P. (2020). A new reality: Fan perceptions of augmented reality readiness in sport marketing. Computers in Human Behavior, 106, 106231, ISSN 0747-5632

[64] Conway, N., Soro, A., Brown, R., & Turkay, S. (2019). Exploring uses of augmented reality in participatory marketing. In Extended Abstracts of the 2019 CHI Conference on Human Factors in Computing Systems, May 2019, Glasgow, Scotland Uk, Association for Computing Machinery, New York, NY, United States, pp. 1-6, ISBN 9781450359719, https://doi.org/10.1145/3290607.3313086

[65] Crofton, E. C., Botinestean, C., Fenelon, M., & Gallagher, E. (2019). Potential applications for virtual and augmented reality technologies in sensory science. Innovative Food Science & Emerging Technologies, 56, 102178, ISSN 1466-8564

[66] Kavran, A. K., Lončarić, D. & Dlacic, J. (2016). Augmented reality experiential marketing in tourism. 5th International Scientific Symposium Economz of Eastern Croatia – Vision and Growth, june 2016, Osijek, ISSN 1848-9559

[67] Wiebach, N., & Send, H. (2019). Augmented Reality Shopping Services: Key Factors Affecting Customer Evaluation and Acceptance. 14th European Conference on Innovation and Entrepreneurship, ECIE 2019, September 2019, 1111-1118, ISBN 20491050, ISBN 978-191276434-1

[68] Javornik, A. (2016a). Augmented reality: Research agenda for studying the impact of its media characteristics on consumer behaviour. Journal of Retailing and Consumer Services, 30, 252-261, ISSN 0969-6989

[69] McLean, G., & Wilson, A. (2019). Shopping in the digital world: Examining customer engagement through augmented reality mobile applications. Computers in Human Behavior, 101, 210-224, ISSN0747-5632

[70] Dodoo, N. A., & Youn, S. (2021). Snapping and chatting away: Consumer motivations for and outcomes of interacting with Snapchat AR ad lens. Telematics and Informatics, 57, 101514, ISSN 0736-5853

[71] Grzegorczyk, T., Sliwinski, R., & Kaczmarek, J. (2019). Attractiveness of augmented reality to consumers. Technology Analysis & Strategic Management, 31, 11, 1257-1269, ISSN 0953-7325

[72] Hinsch, C., Felix, R., & Rauschnabel, P. A. (2020). Nostalgia beats the wow-effect: Inspiration, awe and meaningful associations in augmented reality marketing. Journal of Retailing and Consumer Services, 53, 101987, ISSN 0969-6989

[73] Phua, J., & Kim, J. J. (2018). Starring in your own Snapchat advertisement: Influence of self-brand congruity, self-referencing and perceived humor on brand attitude and purchase intention of advertised brands. Telematics and Informatics, 35, 5, 1524-1533, ISSN 0736-5853

[74] Arlati, S., Spoladore, D., Baldassini, D., Sacco, M., & Greci, L. (2018). VirtualCruiseTour: An AR/VR application to promote shore excursions on cruise ships. In International Conference on Augmented Reality, Virtual Reality and Computer Graphics, July 2018, pp. 133-147, ISBN 978-3-319-95269-7, Springer, Cham, https://doi.org/10.1007/978-3-319-95270-3_9

[75] Feng, Y., & Xie, Q. (2018). Measuring the content characteristics of videos featuring augmented reality advertising campaigns. Journal of Research in Interactive Marketing, ISSN 2040-7122

[76] Irshad, S., & Awang, D. R. B. (2016). User perception on mobile augmented reality as a marketing tool. 2016 3rd International Conference on Computer and Information Sciences (ICCOINS), august 2016, Kuala Lumpur, Malaysia, ISBN 978-1-5090-2550-3, pp.109-113, IEEE, https://doi.org/10.1109/ICCOINS.2016.7783198

[77] Shin, M. H., Lee, Y. M., & Kim, J. H. (2019). Impact factors analysis on AR shopping services immersion. The Journal of Distribution Science, 17, 12, 13-21, ISSN 1738-3110

[78] Vongurai, R. (2021). Factors influencing experiential value toward using cosmetic AR try-on feature in Thailand. Journal of Distribution Science, 19, 1, 75-87, ISSN 1738-3110

[79] Shahab, M. H., Ghazali, E. & Mohtar, M. (2021). The role of elaboration likelihood model in consumer behavior research and its extension to new technologies: A review and future research agenda. International Journal of Consumer Studies, ISSN 14706423

DOI: 10.2507/32nd.daaam.proceedings.082

RESEARCH INTO DYNAMICS OF MOTION PERFORMED BY MODULAR POWER UNIT AS PART OF PLOUGHING TRACTOR-IMPLEMENT UNIT

Volodymyr Bulgakov, Jueri Olt, Semjons Ivanovs,
Volodymyr Kuvachov & Tormi Lillerand

This Publication has to be referred as: Bulgakov, V[olodymyr]; Olt, J[ueri]; Ivanovs, S[emjons]; Kuvachov, V[olodymyr] & Lillerand, T[ormi] (2021). Research into Dynamics of Motion Performed by Modular Power Unit as Part of Ploughing Tractor-Implement Unit, Proceedings of the 32nd DAAAM International Symposium, pp.0576-0585, B. Katalinic (Ed.), Published by DAAAM International, ISBN 978-3-902734-33-4, ISSN 1726-9679, Vienna, Austria
DOI: 10.2507/32nd.daaam.proceedings.082

Abstract

The reduction of the power input required for soil ploughing has been and remains one of the most important targets in the agricultural industry. One of the ways of solving this problem is increasing the grip and traction properties of the power unit. This aim can be achieved by introducing an additional process module with drive wheels in the composition of the ploughing tractor-implement unit. As the block-module ploughing unit travels in operation, especially on stubble field, it receives the jolts and impacts caused by the irregularities of the cultivated land, the variation of the plough's tractive resistance and other factors. The generated oscillations of the unit in the vertical plane impair the operator's working conditions, reduce the reliability and decrease the productivity of the unit's operation.

Keywords: dynamic system; modelling; modular power unit; smoothness of movement; vertical oscillation.

1. Introduction

The reduction of the power consumption in soil ploughing has been and remains one of the most important targets in the agricultural industry [1], [2], [3], [4]. One of the ways to solve this problem is to increase the grip and traction properties of the tractor [5], [6], [7], [8]. As regards the composition of the ploughing tractor-implement unit, this can be implemented by introducing an additional process module with drive wheels, thus forming a modular power unit (Fig. 1) [9], [10], [11]. The main purpose of the process module in the modular power unit is to generate additional tractive effort by using the excess power of the energy intensive tractor's engine [12], [13].

In order to ensure the adequate control and stability of the motion performed by the modular power unit, the process module has to be rigidly connected with the tractor [14], [15]. The immobility of the tractor and the process module relative to each other on the working pass is provided by the automatic locking of the latter's vertical hinge [9]. However, such a design solution provides for the stability and control of the motion performed by the modular power unit under consideration, on the one hand, but impairs the smoothness of its movement, on the other hand.

That results from the fact that, during the motion in operation of the modular power unit as part of the ploughing unit, it receives the jolts and impacts caused by the irregularities of the cultivated land, the variation of the plough's tractive resistance and other factors. The generated oscillations of the unit in the vertical plane make worse the working conditions for its operator, reduce the reliability and decrease the productivity of its operation.

Fig. 1. Modular power unit as part of ploughing unit

In order to eliminate the negative effect of the oscillatory processes in agricultural units, numerous studies have been carried out exploring the effect achieved by introducing various elastic members and damping devices for the purpose of improving the smoothness of motion, [16], [17], [18], [19]. However, the effect that the elastic damping properties of the process module have on the reduction of the vertical oscillations performed by the modular power unit remains unexplored. In view of the above, the problem of designing high-performance and energy-efficient tractor-implement units on the basis of energy intensive tractors, including the approach that suggests improving the grip and traction properties and increasing the tractive effort through the use of a process module with elastic damping properties is a topical one.

The review of the known designs using additional drive axles in tractors [9], [14], [15] has resulted in revealing the general trend in the layout solutions, which suggests making up the traction units on the basis of tractors and process modules. The most effective arrangement of units with tractors and process modules uses the serial connection of the module with the tractor by means of a three-point linkage, which allows one degree of freedom for the motion of the module in the vertical plane. In order to provide for the sufficient manoeuvrability of the tractor with the process module and the agricultural machine, the method of "unit inflection" similar to the hinged frame articulation is employed. The process module is equipped with linkage systems similar to the main linkages of higher draught class tractors for mounting agricultural implements in the field conditions.

Due to its high functional versatility, the modular power unit features a multitude of options for the process part connection arrangement. For example, the employment of the front implement-attaching linkage increases the number of process spaces (Fig. 2) and provides for ganging up wide-span units under several patterns: one implement at the front and two implements behind or two implements at the front and one implement behind with the use of a process container.

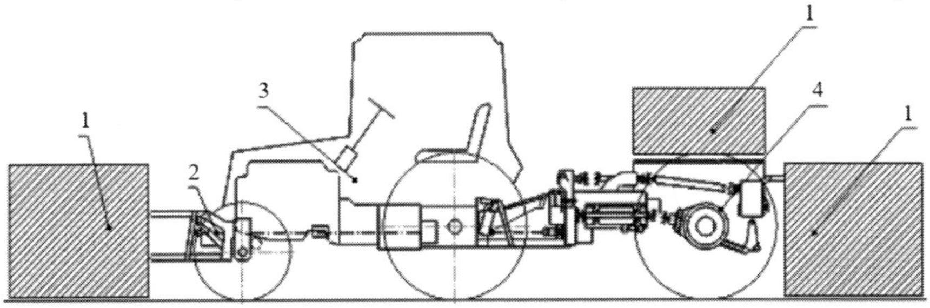

Fig. 2. Process spaces of tractor with process module: 1 – process spaces; 2 – front implement-attaching linkage; 3 – tractor; 4 – process module

The completed analysis of the design and operation characteristics of the known modular power and process equipment has proved that, in general, the layout solutions meet the requirements to the configuration of field tractor-implement units and ensure their efficient operation with tractors of draught classes 2 and 3. The review of the known modular power and process equipment has revealed the absence of any practical design solutions and research results regarding the formation of power units with the use of the modular approach. In case of making up modular power units with the use of the modular approach and on the basis of tractors of draught class 1.4, it is necessary to take into account their intended application as a multi-purpose and ploughing solution for the performance of a certain range of agricultural process operations: first of all, the ploughing ones. Such a layout solution for a multi-purpose and ploughing modular power unit, for example, in ploughing operations will have a negative effect on the smoothness of its movement.

The performance of the modular power unit as a dynamic system to a considerable extent depends on how individual members of this system are geared up to the operation under the conditions of oscillatory action. In order to eliminate the negative effect of the oscillatory processes, many studies have been carried out with the aim to investigate the effect of installing elastic members and damping devices at different points in the tractor. The modified design facilitates the reduction of the impact that the varying loads have on the tractor as a whole or on its individual assemblies.

Alongside with the positive aspects, it is worthwhile to note that the elastic damping members installed on the wheels transfer significant moments, therefore, such devices are bulky and have insufficient reliability, the drives have low energy capacity. Moreover, they are unable to protect the tractor from the vibration of the uncushioned masses. The linear response of such cushioning systems can bring about resonant modes of operation at some types of work. The pneumohydraulic cushioning systems are of specific interest, because they feature such advantages as a high energy capacity and a possibility to achieve nonlinear response. The use of pneumohydraulic elastic damping cushioning in the drive wheels of the tractor unitised with a five-bottom plough has resulted in the reduction of its vibration by 17–22%, the slipping of its running gear by 16–20% and the increase of the productivity by 7–9% [5], [10], [11], [13].

As has been shown by the studies [18], this approach also produces such a positive by-effect as the damping of the horizontal oscillations generated by the resistance force of the agricultural implement. The oscillation suppression is achieved due to the elasticity of the pneumatic tyres on the wheels and the inertia of the process module's mass. The research into the elastic damping properties of the process module is of special interest, because they can provide for reducing substantially the dynamic load on the tractor as well as improving the vibration protection of the operator.

The aim of the study is to establish the relations representing the impact that the stiffness parameters of the elastic dissipative connection between the tractor and the process module have on the characteristics of the oscillations performed by the ploughing unit in the longitudinal-and-vertical plane, as these relations provide for justifying the rational value of stiffness, i.e. the value, at which the variance of the relative vertical oscillations in the unit reaches the minimum and the amplitude-frequency response comes close to the desirable one.

2. Materials and Methods

For the purpose of designing the mathematical model of the motion performed by the modular power unit as part of the ploughing unit in the vertical plane, Lagrange's equations of the second kind have been used [15]:

$$\frac{d}{dt}\left(\frac{\partial T_t}{\partial \dot{q}_i}\right)-\frac{\partial T_t}{\partial q_i}+\frac{\partial E_t}{\partial q_i}+\frac{\partial F_t}{\partial \dot{q}_i}=Q_i, \quad (1)$$

where q_i, Q_i – i-th generalised coordinate and force, respectively;

T_t, E_t – kinetic and potential energy of the power unit, respectively;

F_t – dissipation function.

In view of the fact that the kinetic energy depends only on the velocity and is independent of the generalised coordinate, the following holds true:

$$\frac{\partial T_t}{\partial q_i}=0. \quad (2)$$

Then, with the use of Lagrange's equations of the second kind, the motion of the dynamic system in the vertical plane is described as follows:

$$\frac{d}{dt}\left(\frac{\partial T_t}{\partial \dot{q}_i}\right)+\frac{\partial E_t}{\partial q_i}+\frac{\partial F_t}{\partial \dot{q}_i}=Q_i. \quad (3)$$

The differential equations obtained in this way are solved with the use of the operator calculus methods. In the context of the operator calculus, it is known that, in case of dynamic systems, the main operators are transfer functions and frequency response characteristics. They are just the operators that provide the fullest and physically sound perception of the system's response to various perturbations as well as the transient and constant processes in its operation. Hence, in the process of solving the equations, transfer functions and amplitude-frequency response characteristics of the vertical oscillations have been used as the operators.

In order to calculate the necessary amplitude-frequency characteristics of the vertical oscillations, Lagrange's differential equations of the second kind have to be presented in the operator form. For that purpose, Laplace transformation is applied to the obtained system of differential equations. In accordance with it, the differentiation operator is replaced with the value $i \cdot \omega$ (where i – imaginary unit, $i^2 = 1$, ω– frequency of the perturbing action). After the respective transformations, the transfer functions are obtained, then, finally, the frequency response characteristics of the displacements and accelerations are arrived at:

$$\dot{A}_{Z/h}(\omega)=\left[U(\omega)^2+V(\omega)^2\right]^{\frac{1}{2}}, \tag{4}$$

where $U(\omega)$, $V(\omega)$ – expressions for the real and imaginary components of the frequency response characteristics of the dynamic system under consideration:

$$U(\omega)=M\cdot M_1+\frac{N\cdot N_1}{M_1^2+N_1^2}, \tag{5}$$

$$V(\omega)=N\cdot M_1-\frac{M\cdot N_1}{M_1^2+N_1^2}. \tag{6}$$

where

$M=b_0-b_2\cdot\omega^2+b_4\cdot\omega^4$,

$N=b_1\cdot\omega-b_3\cdot\omega^3+b_5\cdot\omega^5$,

$M_1=a_0-a_2\cdot\omega^2+a_4\cdot\omega^4-a_6\cdot\omega^6$,

$N_1=a_1\cdot\omega-a_3\cdot\omega^3+a_5\cdot\omega^5$.

Thus, the obtained amplitude-frequency response characteristics show the amplitude of the vertical accelerations (displacements) produced by the profile of the support surface at the unit amplitude.

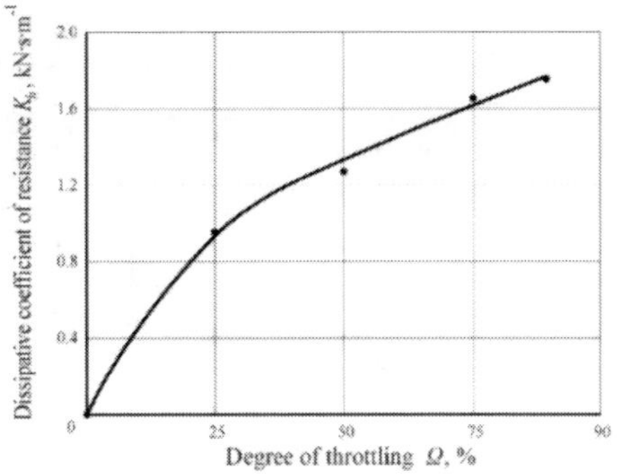

Fig. 3. Nomogram of relation between resistance coefficient K_h of elastic dissipative connection between tractor and process module, on one hand, and degree of throttling Ω in hydraulic system of tractor's rear linkage, on the other hand

In the experiments, the elastic dissipative connection between the tractor and the process module of the modular power unit was established by means of throttling the hydraulic system of the tractor's rear linkage. At the same time, the preliminary laboratory measurements had proved that the degree of throttling Ω applied to the hydraulic system of the tractor's rear linkage changed the resistance coefficient K_h of the elastic dissipative connection between the tractor and the process module in direct proportion (Fig. 2).

3. Results and Discussion

The general method of solving the dynamic problems of the interrelated vertical oscillations of the tractor and the process module in the modular power unit as part of the ploughing unit is represented by the system of rather complex differential equations obtained on the basis of Lagrange's equations of the second kind. With the aim of simplifying their generation in the process of mathematical modelling, the following provisions and assumptions have been accepted: for the purpose of generating the equivalent schematic model of the ploughing unit, it is appropriate and sufficient to analyse the oscillations of a plane figure that has the shape of the side projection only in the vertical plane that coincides with the longitudinal symmetry plane of the modular power unit and plough; the motion of the ploughing unit is assumed to be uniform and straight-line and the profile of the supporting surface is assumed to be the same under the wheels of both the right and left sides; during their translation on the field, the wheels of the tractor and the process module remain in constant contact with the surface of the agricultural background, which is represented by a random stationary ergodic function of the travelled distance; the oscillations of the plough's tractive resistance do not have a significant effect on the translational motion velocity of the unit – in view of that, the latter is assumed to be constant; the angular oscillations of the plough can be ignored; at small heading angles φ_t of the tractor, the following equalities hold true: $\tan\varphi_t \approx \varphi_t$, $\cos\varphi_t \approx 1$.

In order to simplify the process of solving the set problem, the equations that describe the vertical oscillations of the modular power unit and the plough are generated separately for each of the components. In accordance with the universally accepted principle of removing the constraints and replacing them by the reaction forces (free-body diagram), the interaction between the tractor and the process module can be expressed in terms of the force F_z applied at the instantaneous centre of turn of the tractor's rear linkage (point O, Fig. 3). The interaction between the process module and the plough can be represented by the reaction R_p applied at the instantaneous centre of turn of the process module's rear linkage (point C_p, Fig. 4). In accordance with the made assumptions, the real tractor can be represented by the following equivalent dynamic model (Fig. 3).

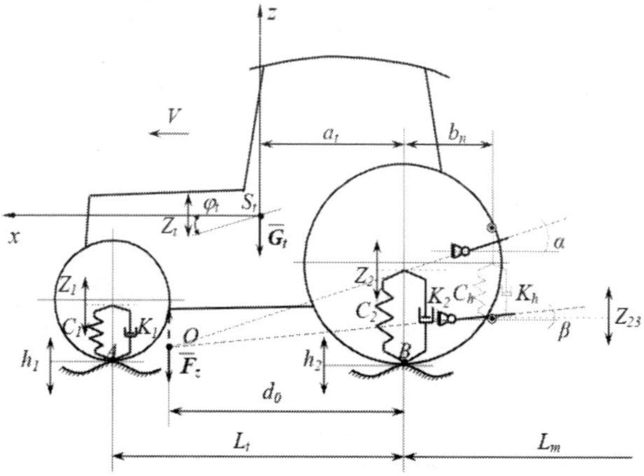

Fig. 4. Analytical model of tractor's oscillations in longitudinal-and-vertical plane

The analytical dynamic model of the two-dimensional motion performed by the tractor in the vertical plane (Fig. 3) has two degrees of freedom: the vertical displacement (Z_t) of its centre of mass (point S_t) and the angular oscillations of its frame φ_t. Similarly, the real process module can be represented by its equivalent dynamic model (Fig. 4).

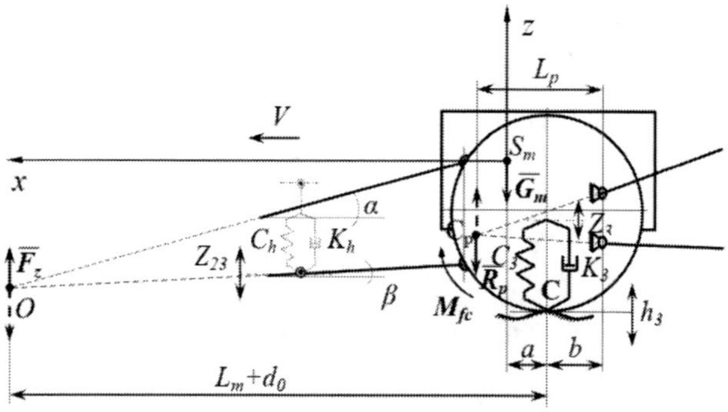

Fig. 5. Analytical model of oscillations performed by process module in longitudinal-and-vertical plane

The analytical dynamic model of the two-dimensional motion performed by the process module in the vertical plane (см. Fig. 4) exhibits only one degree of freedom – the vertical displacement ($Z_3 = Z_m$) of its centre of mass (point S_m). The majority of state-of-the-art ploughs with three to six bodies do not feature elastic members on their carrier wheels. Due to the slipping of the wheels of the tractor and the process module, the longitudinal profile of the track is smoothed out to a certain extent. As a result of that, the variance of the plough's vertical oscillations, when it is unitised with a modular power unit is significantly smaller, than in case of unitising it with a single tractor. Moreover, under the action of the force of gravity (G_p) and the vertical component of the plough's tractive resistance (R_z) (Fig. 5) its carrier wheel deforms the soil even more, which further reduces the heights of the agricultural background profile irregularities. That results in the decreased frequency and energy (variance) of the plough's vertical oscillations, which has a positive effect on the uniformity of cultivation with regard to its depth. Due to the presence of the substantial resultant action of the forces G_p and R_z, the vertical oscillations of the tractor produce little effect on the dynamics of the plough's vertical displacements. Accordingly, it is reasonable to assume that the properties of the plough's vertical oscillations are defined not so much by the dynamics of the ploughing unit motion as by the characteristics of the agricultural background's longitudinal profile. At the same time, the variations of the ploughing depth that exceed the limits set by the agronomical standards (±2 cm) take place, as is known from the practice, only in case considerable free play is present at the points of connection between the plough and the power unit's linkage. Moreover, such variations are not the same on the whole length of the plough: they are smaller after the passage of the first body, but they are greater after the last body.

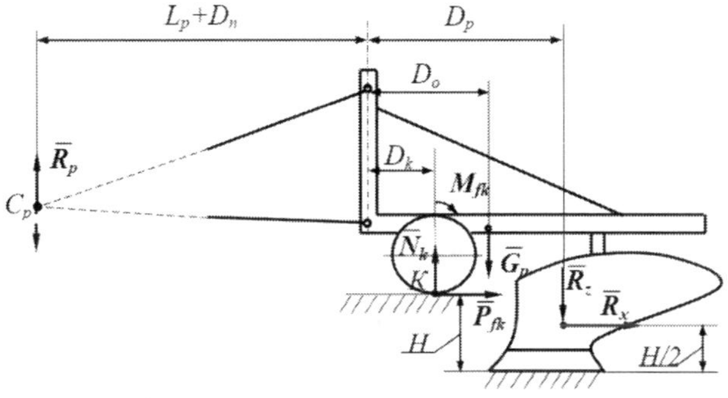

Fig. 6. Schematic model of forces acting on equivalent plough body in longitudinal-and-vertical plane

The following forces act on the equivalent plough body in the longitudinal-and-vertical plane (Fig. 5): G_p – force of gravity of the plough; R_x, R_z – horizontal and vertical components of the forces acting on the plough from the soil; N_k – normal reaction on the plough's carrier wheel. Hence, in view of the above-stated, the analytical dynamic model of the ploughing unit based on the modular power unit has three degrees of freedom: the vertical displacement Z_t of the tractor's centre of mass, the angular oscillations of its frame φ_t, the vertical displacement $Z_3 = Z_m$ of the centre of mass of the process module. Taking into consideration the following equalities:

$$Z_t = \frac{Z_1 a_t + Z_2 (L_t - a_t)}{L_t}, \tag{7}$$

$$\varphi_t = \frac{Z_2 - Z_1}{L_t}, \tag{8}$$

where Z_1, Z_2 – vertical displacements of the front and rear axles of the power unit, respectively;
L_t, a_t – base and longitudinal coordinate of the tractor's centre of mass, the vertical displacements Z_1, Z_2 and Z_3 of the axles in the modular power unit are taken as the generalised coordinates. After finding the kinetic energy T_t and the potential energy E_t as well as the respective dissipative F_t and generalized Q_i forces, the mathematical model of the vertical oscillatory motions of the modular power unit as part of a ploughing combined unit is obtained in the form of the following system of differential equations [15]:

$$\left.\begin{aligned} \grave{A}_{11} \cdot \ddot{Z}_1 + \grave{A}_{12} \cdot \dot{Z}_1 + \grave{A}_{13} \cdot Z_1 + \grave{A}_{14} \cdot \ddot{Z}_2 &= f_{11} \cdot \dot{h}_1 + f_{12} \cdot h_1 + f_{13} \cdot R_{\delta} + f_{14}, \\ \grave{A}_{21} \cdot \ddot{Z}_2 + \grave{A}_{22} \cdot \dot{Z}_2 + \grave{A}_{23} \cdot Z_2 + \grave{A}_{24} \cdot \ddot{Z}_1 &= f_{21} \cdot \dot{h}_2 + f_{22} \cdot h_2 + f_{23} \cdot R_{\delta} + f_{24}, \\ \grave{A}_{31} \cdot \ddot{Z}_3 + \grave{A}_{32} \cdot \dot{Z}_3 + \grave{A}_{33} \cdot Z_3 &= f_{31} \cdot \dot{h}_3 + f_{32} \cdot h_3 + f_{33} \cdot R_{\delta} + f_{34}, \end{aligned}\right\} \tag{9}$$

where $\grave{A}_{11} = \dfrac{M_t \cdot a_t^2 + J_t}{L_t^2}$;

$A_{12} = K_1$; $A_{13} = C_1$; $\quad \grave{A}_{14} = \dfrac{2\left[M_t \cdot a_t \cdot (L_t - a_t) - J_t\right]}{L_t^2}$;

$\grave{A}_{21} = \dfrac{M_t \cdot (L_t - a_t)^2 - J_t}{L_t^2}$; $\quad \grave{A}_{22} = \hat{E}_2$; $A_{23} = C_2$; $\quad \grave{A}_{24} = \dfrac{2\left[M_t \cdot a_t \cdot (L_t - a_t) - J_t\right]}{L_t^2}$;

$A_{31} = M_m$; $\grave{A}_{32} = \hat{E}_3$; $A_{33} = \tilde{N}_{\varsigma}$;

$f_{11} = K_1$; $f_{12} = C_1$;

$f_{13} = \dfrac{d_0 \cdot (L_p - b_m) \cdot \left[0.5H - 0.2(D_p - D_k)\right]}{L_t \cdot (L_p + D_n + D_k) \cdot (L_m + d_0)}$; $\quad f_{14} = \dfrac{d_0}{L_t \cdot (L_m + d_0)} \cdot \left[G_m \cdot (a_m - f_m \cdot r_{km}) - \dfrac{G_p \cdot (L_p - b_m) \cdot (D_0 - D_k)}{(L_p + D_n + D_k)}\right]$;

$f_{21} = K_2$; $f_{22} = \tilde{N}_2$;

$f_{23} = \dfrac{(1 - d_0 \cdot L_t^{-1}) \cdot (L_p - b_i) \cdot \left[0.5H - 0.2(D_p - D_k)\right]}{(L_p + D_n + D_k) \cdot (L_m + d_0)}$; $\quad f_{24} = \dfrac{(1 - d_0 \cdot L_t^{-1})}{(L_m + d_0)}\left[G_m \cdot (a_m - f_m \cdot r_{km}) - \dfrac{G_p \cdot (L_p - b_m) \cdot (D_0 - D_k)}{(L_p + D_n + D_k)}\right]$;

$$f_{31} = K_3;\; f_{32} = \tilde{N}_3;$$

$$f_{33} = \frac{\left[0.5H - 0.2\left(D_p - D_k\right)\right]}{\left(L_p + D_n + D_k\right)} \cdot \left[1 - \frac{L_p - b_m}{L_m + d_0}\right];\; f_{34} = \frac{G_p \cdot \left(D_0 - D_k\right)}{\left(L_p + D_n + D_k\right)} \cdot \left[\frac{L_p - b_m}{L_m + d_0} - 1\right] - \frac{G_m \cdot \left(a_m - f_m \cdot r_{km}\right)}{\left(L_m + d_0\right)}.$$

In the system of equations (9), the following parameters are denoted: M_t, J_t – mass and moment of inertia in the longitudinal and vertical plane of the power module; M_m, G_m – mass and gravity force of the process module; G_p – gravity force of the plough; f_m, r_{km} – resistance coefficient and rolling radius of the wheels of the process module; R_x – horizontal component of the plough tractive resistance; K_1, K_2, K_3 and C_1, C_2, C_3 – respectively, the reduced resistance coefficients of the dissipative and elastic members in the process of damping the oscillations of the modular power unit, which are described by the following expressions:

$$\hat{E}_1 = \frac{\hat{E}_{s1} \cdot \hat{E}_h \cdot b_n^2}{\hat{E}_h \cdot b_n^2 + 2 \cdot \hat{E}_{s1} \cdot L_t^2};\quad \hat{E}_2 = \frac{\hat{E}_{s2} \cdot \hat{E}_h \cdot (L_t + b_n)^2}{\hat{E}_h \cdot (L_t + b_n)^2 + 2 \cdot \hat{E}_{s2} \cdot L_t^2};\quad \hat{E}_3 = \frac{\hat{E}_{s3} \cdot \hat{E}_h \cdot b_n^2}{\hat{E}_h \cdot b_n^2 + 2 \cdot \hat{E}_{s3} \cdot \left(L_m - b_n\right)^2};$$

$$\tilde{N}_1 = \frac{\tilde{N}_{s1} \cdot \tilde{N}_h \cdot b_n^2}{\tilde{N}_h \cdot b_n^2 + 2 \cdot \tilde{N}_{s1} \cdot L_t^2};\quad \tilde{N}_2 = \frac{\tilde{N}_{s2} \cdot \tilde{N}_h \cdot (L_t + b_n)^2}{\tilde{N}_h \cdot (L_t + b_n)^2 + 2 \cdot \tilde{N}_{s2} \cdot L_t^2};\quad \tilde{N}_3 = \frac{\tilde{N}_{s3} \cdot \tilde{N}_h \cdot b_n^2}{\tilde{N}_h \cdot b_n^2 + 2 \cdot \tilde{N}_{s3} \cdot \left(L_m - b_n\right)^2},$$

where K_{s1}, K_{s2}, K_{s3} and C_{s1}, C_{s2}, C_{s3} – respectively, the resistance coefficients of the dissipative and elastic members in the cushioning systems of the axles of the power and process modules.

The nature of other design parameters is made obvious by Fig. 3. If the Laplace transformations are applied to the system of differential equations (9), the following mathematical model in operator notation is obtained for the dynamics of the motion performed in the longitudinal and vertical plane by the modular power unit as part of a ploughing combined unit:

$$\left.\begin{array}{l} K_{11} \cdot Z_1(s) + K_{12} \cdot Z_2(s) + K_{13} \cdot Z_3(s) = \\ = F_{11} \cdot h_1(s) + F_{12} \cdot h_2(s) + F_{13} \cdot h_3(s) + F_{14} \cdot R_x(s) + F_{15}, \\ K_{21} \cdot Z_1(s) + K_{22} \cdot Z_2(s) + K_{23} \cdot Z_3(s) = \\ = F_{21} \cdot h_1(s) + F_{22} \cdot h_2(s) + F_{23} \cdot h_3(s) + F_{24} \cdot R_x(s) + F_{25}, \\ K_{31} \cdot Z_1(s) + K_{32} \cdot Z_2(s) + K_{33} \cdot Z_3(s) = \\ = F_{31} \cdot h_1(s) + F_{32} \cdot h_2(s) + F_{33} \cdot h_3(s) + F_{34} \cdot R_x(s) + F_{35}, \end{array}\right\} \qquad (10)$$

where $K_{11} = A_{11} \cdot s^2 + A_{12} \cdot s + A_{13}$;

$$K_{12} = A_{14} \cdot s^2;\qquad K_{13} = K_{23} = K_{31} = K_{32} = 0;$$

$$K_{21} = A_{24} \cdot s^2;\qquad K_{22} = A_{21} \cdot s^2 + A_{22} \cdot s + A_{23};$$

$$K_{33} = A_{31} \cdot s^2 + A_{32} \cdot s + A_{33};$$

$$F_{11} = f_{11} \cdot s + f_{12};\qquad F_{12} = F_{13} = F_{21} = F_{23} = F_{31} = F_{32} = 0;\qquad F_{14} = f_{13};\; F_{15} = f_{14};$$

$$F_{22} = f_{21} \cdot s + f_{22};\; F_{24} = f_{23};\; F_{25} = f_{24};$$

$$F_{33} = f_{31} \cdot s + f_{32};\; F_{34} = f_{33};\; F_{35} = f_{44};$$

$s = \frac{d}{dt}$ – operator of differentiation.

In the system of equations (10), the heights of soil surface irregularities under the front (h_1) and rear (h_2) wheels of the power module and the wheels of the process module (h_3) as well as the plough tractive resistance (R_x) act as the input values. The initial parameters comprise the amplitudes of displacements, that is, the oscillations of the front (Z_1) and rear (Z_2) axles of the power module and the oscillations of the process module's frame (Z_3). In case of dynamic systems, the principal operators are transfer functions and frequency-response characteristics. Therefore, the impact that the values of the resistance coefficients of the dissipative and elastic link between the two modules have on the behaviour of the vertical oscillations of the ploughing combined unit has to be estimated with the use of the amplitude-frequency characteristics of the displacements and accelerations observed in the response of the dynamic system to the agronomic background profile irregularities and the variation of the plough tractive resistance. The analysis of the mathematical model (10) has revealed that the main spectrum of the oscillations performed by the modular power unit as part of the ploughing unit is concentrated within the frequency range of 0–16 s^{-1} (Fig. 6).

In terms of the dynamic system adequately responding to the impact of the agricultural background profile irregularities, increasing the stiffness of the viscoelastic connection between the tractor and the process module as parts of the unit is an effective measure, because it makes the amplitude-frequency response characteristics closer to the desirable ones. As the degree of throttling in the hydraulic system of the tractor's rear linkage increases to 90%, the amplitudes of the oscillations performed by its rear axle become reduced almost by a factor of 10, while the resonance peaks of the frequency response characteristics shift into the lower frequency range from 16.5 s^{-1} to 6.0 s^{-1} and simultaneously their maximum values decrease almost by a factor of 4 (see Fig. 6). That is a positive result in terms of increasing the smoothness of the motion performed by the modular power unit as part of the ploughing unit, because such a follow-up dynamic system tends to copy the agricultural background irregularities without their amplification.

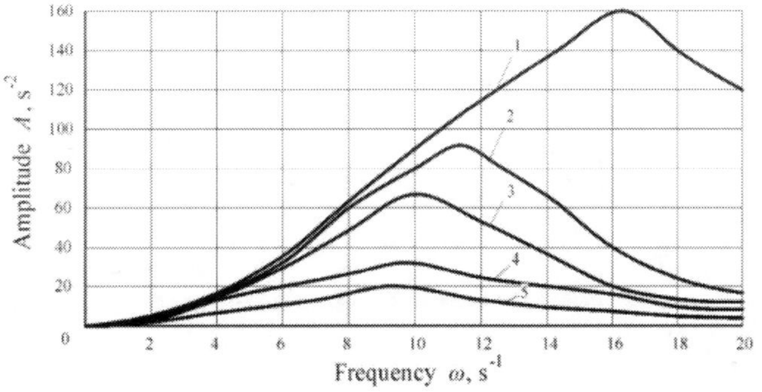

Fig. 7. Amplitude-frequency response characteristics of accelerations during vertical oscillations of tractor's rear axle copying variations in track profile at different degrees of throttling in hydraulic system of tractor's rear linkage

The comparative analysis of the theoretical and experimental amplitude-frequency response characteristics of accelerations during the vertical oscillations of the tractor's rear axle in accordance with the data in Fig. 6 has proved that the discrepancy does not exceed 8-15%, which is acceptable for practical purposes. In accordance with Fisher's ratio test, at a statistical significance level of 0.05 the null hypothesis of the equality of the compared variances is not rejected within the whole variation range of the resistance coefficient of the dissipative connection between the two modules.

According to the results of the analysis carried out with regard to the plotted amplitude-frequency response characteristics of the oscillations performed by the modular power unit as part of the ploughing unit, increasing the stiffness of the connection between the process module and the tractor shifts the frequency response characteristics into the lower frequency range, simultaneously decreasing their maximum values. That makes possible to reduce the variance of the interrelated vertical oscillations in the modular power unit by means of throttling the hydraulic system of the tractor's rear linkage.

The analysis of the plotted amplitude-frequency response characteristics of the vertical oscillations performed by the modular power unit responding to the perturbations caused by the variation of the plough's tractive resistance has proved (Fig. 7) that the oscillations of the tractive resistance offered by the agricultural implements have a significantly smaller effect on the smoothness of its motion, than the oscillations of the longitudinal profile of the track. The maximum value of the oscillation amplitude decreases with the increase of the throttling degree Ω in the hydraulic system of the tractor's rear linkage. When $\Omega = 0\%$, it is observed at a resonant frequency of $\omega = 13.0$ s^{-1} and is equal to just 1.2 mm·kN^{-1}, while at $\Omega = 90\%$ – it corresponds to a resonant frequency of $\omega = 10.0$ s^{-1} and is equal to 0.4 mm·kN^{-1}. As the value of Ω rises, the frequency response characteristics subside, their maximums simultaneously shifting into the lower frequency range, which is greatly desired.

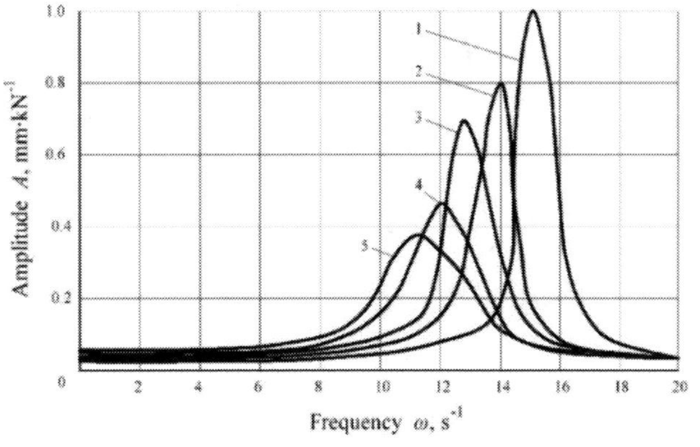

Fig. 8. Amplitude-frequency response characteristics of vertical oscillations performed by tractor's rear axle responding to variations of plough's tractive resistance at different degrees of throttling in hydraulic system of tractor's rear linkage

It has been established by means of experiments that the range of the root-mean-square deviation of the tractive resistance offered by the plough under the conditions of laboratory-and-field research is ±2.2-3.9 kN. Therefore, even at the above-mentioned resonant frequency of 13.0 s^{-1} (Fig. 7) the mean deviation of the vertical oscillations performed by the modular power unit is expected to be at a level of just ±3-5 mm.

The relation between the amplitude values of the characteristics of the vertical oscillations performed by the modular power unit and the degree of throttling in the hydraulic system of the tractor's rear linkage, when the ploughing unit responds to external perturbations, is not uniform. The intensive decrease in the amplitude values of the accelerations in the vertical oscillations performed by the tractor's rear axle copying variations in the track profile (Fig. 8) takes place, when the degree of throttling in the hydraulic system of the tractor's rear linkage rises up to 60-75%. The further rise in the throttling applied to the tractor's hydraulic system results in the less significant decrease in the vertical oscillation acceleration amplitudes.

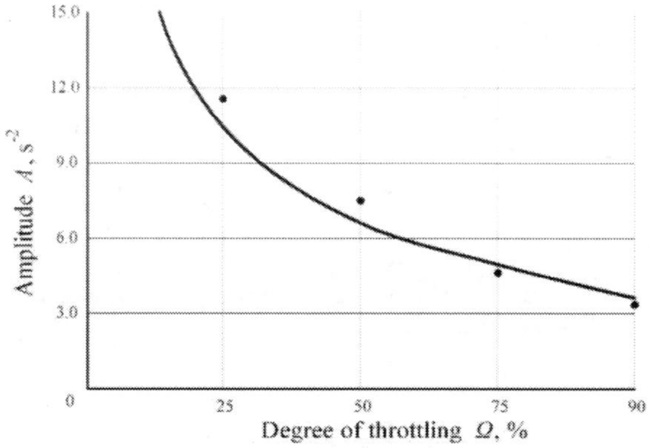

Fig. 9. Relation between amplitude A of tractor's vertical oscillations, when it copies variations in track profile, and degree of throttling Ω in hydraulic system of its rear linkage

The analysis of the amplitude-frequency response characteristics of the vertical oscillations performed by the tractor's rear axle responding to the variations in the plough's tractive resistance has produced a similar picture (Fig. 9). In the two cases (Fig. 8 and Fig. 9), the critical point is found at a throttling degree of 69.8%. The obtained value is exactly the required rational level of throttling for the hydraulic system of the tractor's rear linkage.

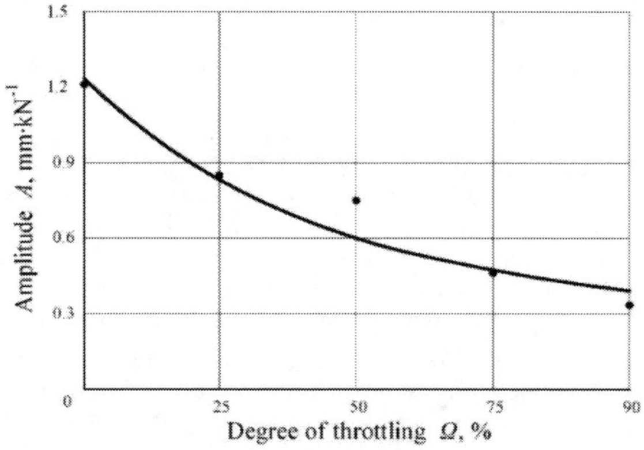

Fig. 10. Relation between amplitude A of tractor's vertical oscillations, when it responds to variations in plough's tractive resistance, and degree of throttling Ω in hydraulic system of its rear linkage

The analysis of the calculated amplitude-frequency response characteristics of the oscillations performed by the modular power unit as part of the ploughing unit has proved that increasing the stiffness of the connection between the process module and the tractor shifts the frequency response characteristics into the lower frequency range and simultaneously decreases their maximum values. That provides for reducing the variance of the interrelated vertical oscillations in the modular power unit by means of rationally throttling the hydraulic system of the tractor's rear linkage.

4. Conclusion

An appropriate mathematical model has been generated for the integrated dynamic system "tractor + process module + plough" enabling the research into the motion of the block-module ploughing unit in the longitudinal-and-vertical plane.

The analysis of the model has proved that setting the stiffness of the elastic dissipative connection between the tractor and the process module at the rational level allows to reduce the tractor oscillation amplitudes by a factor of 4 to 10, simultaneously shifting the resonance peaks of the frequency response characteristics into the lower frequency range, that is, from 16.5 s^{-1} to 6.0 s^{-1}. At the same time, the peaks of the amplitude-frequency response characteristics for the modular power unit responding to the variations in the plough's tractive resistance shift from 13.0 s^{-1} to 10.0 s^{-1}, while their maximum values become 3 times lower. The overall result is the significantly improved smoothness of the motion performed by the block-module ploughing unit. According to the results of the mathematical modelling, the variation of the tractive resistance offered by the agricultural implements has a 10 times smaller effect on the smoothness of the motion performed by the modular power unit, than the oscillations of the longitudinal profile of the track.

5. References

[1] Andreev, A.F.; Kabanau, V.I.; Vantsevich, V.V. (2010). Driveline Systems of Ground Vehicles. Theory and Design. CRC Press, Taylor & Francis Group, ISBN 978-1-4398-1728-5, 758 p.

[2] Kutzbach H.D. (2000). Trends in Power and Machinery. Journal of Agricultural Engineering Research, Vol. 76, No. 3, pp. 237-247, ISSN 0021-8634. DOI: 10.1006/jaer.2000.0574, available online at http://www.idealibrary.com

[3] Simikič, M.; Dedovič, N.; Savin, L.; Tomič, M.; Ponjičan, O. (2014). Power delivery efficiency of a wheeled traktor at oblique drawbar force. Soil and Tillage Research, Vol. 141, pp. 32-43.

[4] Steindorff, K.; Flecczoreck, T.; Kattenstroth, R.; Schattenberg, J.; Jünemann, D.; Robert, M.; Wulfmeier, K. (2010). Trends in agricultural machines and tractors: Observations on the occasion of the Agritechnica in Hannover. Olhydraulik und Pneumatik, Vol. 54, pp. 57-63.

[5] Bulgakov, V., Adamchuk, V., Kuvachov, V., Arak, M. & Olt, J. (2017). Study into movement of wide span tractors (vehicles) used in controlled traffic farming. – In: Proceedings of the 28th DAAAM International Symposium: 28th DAAAM International Symposium "Intelligent Manufacturing and Automation" (Ed. B. Katalinic). 2017 November 08–11, Zadar, Croatia, Vienna, Austria: DAAAM International Vienna, 0199–0208.10. 2507/28th.daaam.proceedings.027.

[6] Hac, A., Fulk, D. & Chen, H., 2008. Stability and Control Considerations of Vehicle-Trailer Combination. SAE Int. J. Passeng. Cars - Mech. Syst. Vol. 1, pp. 925–937. https://doi.org/10.4271/2008-01-1228.

[7] Song, P., Zong, C. & Tomizuka, M. (2014). A terminal sliding mode based torque distribution control for an individual-wheel-drive vehicle. J. Zhejiang Univ. Sci. A, Vol. 15, 681–693. https://doi.org/10.1631/jzus.A1400101

[8] Szakács, T. (2010). Developing stability control theories for agricultural transport systems. Acta Polytechnica Hungarica, Vol. 7, No 2, pp. 25–37.

[9] Bulgakov, V., Holovach, I., Nadykto, V., Parakhin, O., Kaletnik, N., Shymko, L., Olt, J. (2020a). Motion stability estimation for modular traction vehicle-based combined unit. Agronomy Research. Vol. 18, No. 4, pp. 2340-2352. doi: 10.15159/AR.20.183.

[10] Nadykto V. (2010). Modular energy units. The Ukrainian Farmer. Vol. 7, pp. 70-71. (In Ukrainian).

[11] Nadykto, V. (2010). The role of modular power units in the formation of the type of tractors in Ukraine. Tractors and agricultural machinery. Vol. 6, pp. 22-25. (In Russian).

[12] Bulgakov, V., Nadykto, V., Kyurchev, S., Nesvidomin, V., Ivanovs, S., Olt, J. (2019). Theoretical background for increasing grip properties of wheeled tractors based on their rational ballasting. Agraarteadus/Journal of Agricultural Science XXX, p-ISSN 1024-0845, e-ISSN 2228-1893, Vol. 30, No. 2, pp. 78–84. DOI: 10.15159/jas.19.07.

[13] Nadykto, V. (2012). Role of tractors' energy saturation rate in making of their range. Tractors and agricultural machinery 3, 16–21.

[14] Bulgakov, V., Kyurchev, V., Nadykto, V., Olt, J. (2015). Structure development and results of testing a novel modular power unit. Agriculture and Agricultural Science Procedia, ISSN 2210-7843, Vol. 7, pp. 40–44, DOI: 10.1016/j.aaspro.2015.12.028.

[15] Bulgakov, V., Kuvachov, V., Olt, J. (2020b). Theory of smoothness of movement of multiple-axle agricultural combined tractor-implement units, Proceedings of the 31st DAAAM International Symposium, pp.0056-0065, B. Katalinic (Ed.), Published by DAAAM International, ISBN 978-3-902734-29-7, ISSN 1726-9679, Vienna, Austria. DOI: 10.2507/31st.daaam.proceedings.008.

[16] Bulgakov, V., Kuvachov, V. & Olt, J. (2019a). Theoretical Study on Power Performance of Agricultural Gantry Systems. – In: 30th International DAAAM Symposium "Intelligent Manufacturing & Automation" (Ed. B. Katalinic). Zadar, 2019 October 23.–26, Vienna, Austria: DAAAM International: 0167–0175. 10.2507/30th.daaam.proceedings.022.

[17] Sidorov, M.V. (2014). Elastic-damping properties of the transport and technological module as part of an agricultural machine-tractor unit. International Research Journal. Vol. 41, No. 10-2, pp. 119-121. (In Russian).

[18] Sidorov M.V. (2016) Increasing efficiency of use machine-tractor unit due to the application of the technological module with leading engines for tractor class 1.4. Ph.D. Dissertation, Institute of Mechanical and Power Engineering, Russian State Agrarian University, Kaluga, 153 p. (In Russian).

[19] Sudin J.; Gritsik B. (1989). Self-propelled machines for chemicalization. Tractors and agricultural machinery. Vol. 7 pp. 40-42. (In Russian).

DOI: 10.2507/32nd.daaam.proceedings.083

FIRE SAFETY THREAT RISK ANALYSIS FOR SOFT TARGET

Adam Malatinsky, Irena Drofova,
Lucie Sousedikova & Martin Hromada

This Publication has to be referred as: Malatinsky, A[dam]; Drofova, I[rena]; Sousedikova, L[ucie] & Hromada, M[artin] (2021). Fire Safety Threat Risk Analysis for Soft Target, Proceedings of the 32nd DAAAM International Symposium, pp.0586-0592, B. Katalinic (Ed.), Published by DAAAM International, ISBN 978-3-902734-33-4, ISSN 1726-9679, Vienna, Austria
DOI: 10.2507/32nd.daaam.proceedings.083

Abstract

The level of threat risk within soft targets is constantly increasing during the development of the company. It is, therefore, necessary to pay more attention to this issue. This article deals with a qualitative risk analysis of soft target threats. When determining the level of risk, it is essential to characterize soft targets and their types. Qualitative risk analysis of soft targets threats compares and evaluates their interrelationship. Based on this correlation, it is essential to develop criteria through which the importance and significance of threat risks can be determined within the framework of soft targets. In addition to the established and evaluated analysis, it is also essential to develop proposals for measures to minimize the effects of potential threats under soft targets. The threats relate mainly to fire safety. A specific type of soft target for this threat risk analysis is a shopping centre.

Keywords: Soft target; Threat risk analysis; Fire safety; Safety

1. Introduction

Safety as a category is found in all sectors of society. One of the oldest categories concerns fire safeties. The article deals with fire safety in connection with a soft target. The need for safety and protection for soft targets is constantly increasing in the field of fire protection. Therefore, it is essential to respond promptly to the development of society and address the issue through various proposals, ideas, possible uses and their application. The first part after the introduction describes the soft goal, its degree of assessment, the primary type and its attractiveness. The following section describes the analysis, which identifies the most significant threats related to fire protection for a soft target - a shopping centre. The next part of the article focuses on assessing the significance of threats according to the established procedure. The following chapter shows the evaluated threats explicitly according to their significance for primary, secondary and tertiary. The last chapter of the article proposes measures for individual threats. Specifically, these are proposals for primary and secondary threats, as they pose a greater risk. The motivation for creating a threat risk analysis for a soft target in fire protection is to increase the protection of the population during shopping and minimize the impact of threats in the future.

2. Soft target

A soft target is an object, space, or action that is characterized by the presence of a large number of people and, at the same time, the absence or low level of security against attacks. The degree of threat to a soft target is assessed by type, specification, the density of people presents, public access, locality, security level, availability of means to attack, the occurrence of a given type of attack, the complexity of implementation of a given type of attack, impacts on lives and health impacts and impacts on the functioning of the affected community. The basic types of soft targets include transport, religious places, hospitals, nightclubs, restaurants, shops and shopping centres, government buildings, social and cultural events, sports events, schools and school facilities, tourist centres, streets, pedestrian zones, squares and places with immigrants present. The different types of soft targets are different from each other but have a common characteristic that makes them attractive to potential attackers. Common characteristics include openness to the public, presence of security personnel, number and concentration of persons, police presence, media attention and symbolism of the target. [2], [4]

3. Description of the analysis

It is appropriate to use qualitative risk analysis to evaluate the most significant risks in fire protection. Specifically, it is a qualitative analysis of competing risks. This method needs to follow the correct procedure in the following steps:

- Processing of risk inventory
- Compilation and completion of a correlation table of risks
- Creating the sum of risk correlations and calculating the coefficients of activity and passivity of individual risks
- Graphical risk assessment
- Calculation of axes of activity coefficients and passivity of individual risks
- Segment evaluation [1]

The individual segments determine how significant are the risks of the given threats for the specified object.

4. Analysis procedure

Based on the individual steps in the previous chapter, the given method is applied.

4.1 Processing of risk inventory

Risk inventory means determining the most common and most likely possible threats that may occur in a given building [3]. As part of the fire protection for the shopping centre building, the following threats are identified:

- Fire
- Smoke
- Flooding
- Explosion
- Leakage of hazardous substances
- High temperature in the building
- Escape exit blocking
- Disorientation
- Failure of the fire protection system and its components
- Undefined assembly point
- Untrained staff
- Absence of fire extinguishers
- Power outage

4.2 Compilation and completion of a correlation table of risks

Based on the individual risks of a threat in the shopping centre within the fire protection framework, a table of correlation of risks will be created, which expresses the interrelationships between the identified risks. The table will contain individual threats arranged vertically and horizontally. For the specific situation for the shopping centre, the 13 most significant threats that can occur will be used. A threat table is created in which the following three values are expressed:

x – this letter means that it cannot pose a risk of itself
1 – expresses the real possibility that the risk may give rise to another risk
0 – expresses a state where a given risk cannot give rise to another risk

Number	Threat	1	2	3	4	5	6	7	8	9	10	11	12	13
1	Fire	x	1	0	1	1	1	1	1	0	0	0	0	1
2	Smoke	0	x	0	0	0	0	0	1	0	0	0	0	0
3	Flooding	0	0	x	1	1	0	1	1	1	0	0	0	1
4	Explosion	1	1	0	x	1	1	1	1	1	0	0	0	1
5	Leakage of hazardous substances	1	1	0	1	x	1	1	1	1	0	0	0	1
6	High temperature in the building	1	1	0	1	1	x	0	0	0	0	0	0	1
7	Escape exit blocking	0	0	0	0	0	0	x	1	0	0	0	0	0
8	Disorientation	0	0	0	0	0	0	0	x	0	0	0	0	0
9	Failure of the fire protection system and its components	0	0	0	0	0	0	0	0	x	0	0	1	0
10	Undefined assembly point	0	0	0	0	0	0	0	1	0	x	0	0	0
11	Untrained staff	1	0	0	1	1	0	1	0	1	1	x	1	0
12	Absence of fire extinguishers	0	0	0	0	0	0	0	0	0	0	0	x	0
13	Power outage	0	0	0	0	0	0	0	1	1	0	0	0	x

Table 1. Interrelationships between threats

4.3 Creating the sum of risk correlations and calculating the coefficients of activity and passivity of individual risks

In this section, two coefficients are expressed: an activity coefficient, which expresses the total potential risk of causing additional threat risks, and a passivity coefficient, which expresses the number of all risks that a given risk may cause. The following equations are used to calculate the coefficients:

$$K_A R_i = \frac{\sum R_i}{x - 1} \tag{1}$$

$$K_P R_i = \frac{\sum R_i}{x - 1} \tag{2}$$

Where:
$K_A R_i$ – activity coefficient
$K_P R_i$ – passivity coefficient
$\sum R_i$ – the sum of risks (for the activity coefficient it is the horizontal axis and vertical axis for passivity coefficient
x – total number of threats [1]

After calculating the given coefficients for all threat risks, the resulting stocks are multiplied by the number 100 to the percentage.

Number	Threat	Activity coefficient [%]	Passivity coefficient [%]
1	Fire	58	33
2	Smoke	8	33
3	Flooding	50	0
4	Explosion	67	42
5	Leakage of hazardous substances	67	42
6	High temperature in the building	42	25
7	Escape exit blocking	8	42
8	Disorientation	0	67
9	Failure of the fire protection system and its components	8	42
10	Undefined assembly point	8	8
11	Untrained staff	58	0
12	Absence of fire extinguishers	0	17
13	Power outage	17	42

Table 2. Determination of activity and passivity coefficients

4.4 Graphical risk assessment

Based on Table 2, the values of the coefficients of individual threats are recorded in a graph, where the values of the coefficient of activity are located on the horizontal axis, and the values of the passivity coefficient are located on the vertical axis.

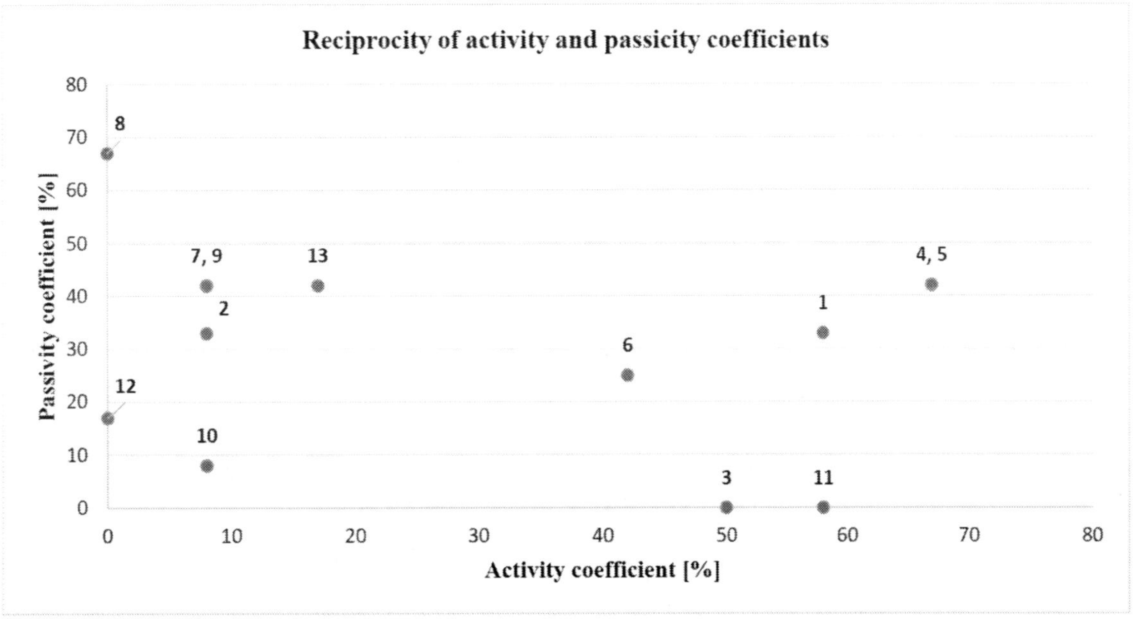

Graph 1. Threats according to the reciprocity of the activity and passivity coefficients

4.5 Calculation of axes of activity coefficients and passivity of individual risks

To determine the significance of threat risks, it is necessary to divide Chart 1 into four segments. To divide the graph, we will need two lines (horizontal and vertical), which will divide the graph into four parts. The first segment is expected to account for 80% of these most significant risks. The following mathematical relationships are used to calculate the axes.

$$S_1 = K_{Amax} - \frac{(K_{Amax} - K_{Amin})}{100} * 80 = 67 - \frac{(67 - 8)}{100} * 80 = 19{,}8 \quad (3)$$

$$S_2 = K_{Pmax} - \frac{(K_{Pmax} - K_{Pmin})}{100} * 80 = 58 - \frac{(58 - 8)}{100} * 80 = 18 \quad (4)$$

Where:
S_1 – vertical axis for activity coefficient
S_2 – horizontal axis for passivity coefficient
K_{Amax} – maximum value within the activity coefficients
K_{Amin} - minimum value within the activity coefficients;
K_{Pmax} – maximum value within the passivity coefficients
K_{Pmin} – minimum value within the passivity coefficients

For greater accuracy, values outside the value of 0 are calculated. [1]

4.6 Segment evaluation

The resulting values of the lines are plotted in the graph, and the graph is divided into four segments. According to these segments, it is possible to determine which group of significance the given threats belong to. The individual segments are:

- Segment I – primarily significant threats – high value of coefficient of activity and passivity
- Segment II and III – secondary significant threats – high value coefficient of activity or passivity
- Segment IV – tertiary significant threats – low value coefficient of activity and passivity [1]

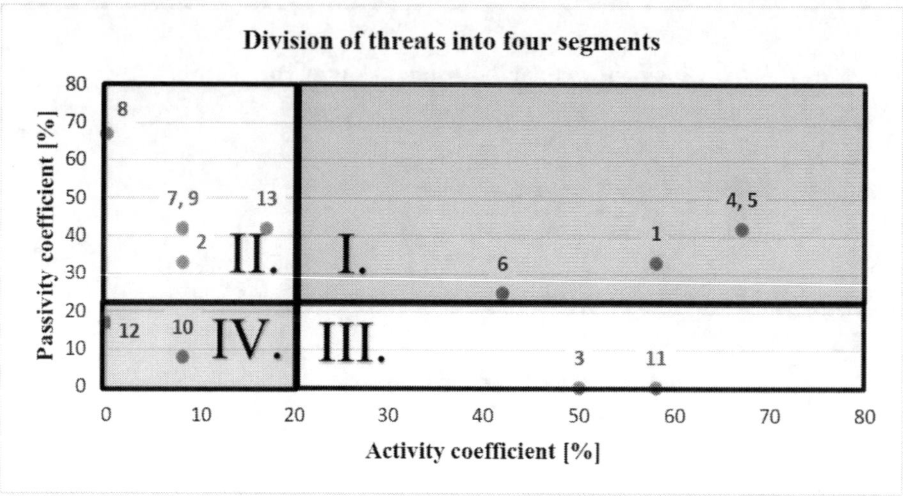

Graph 2. Division of threats into four segments

5. Threat risk assessment

Based on Graph 2, it is possible to determine the significance of the threat risk for the shopping centre within the fire protection framework.

Significance of the threat	Threat number	Threat
Primarily	1	Fire
	4	Explosion
	5	Leakage of hazardous substances
	6	High temperature in the building
Secondary	2	Smoke
	3	Flooding
	7	Escape exit blocking
	8	Disorientation
	9	Failure of the fire protection system and its components
	11	Untrained staff
	13	Power outage
Tertiary	10	Undefined assembly point
	12	Absence of fire extinguishers

Table 3. Threats division by significance

6. Proposal measures

This part of the article focuses on designing measures for the most significant threats identified according to the qualitative analysis of the concurrence of risks listed in Table 3. The design measures are intended for primary and secondary threats, according to Table 3.

6.1 Fire

For the threat of fire, it is proposed to cover the shopping centre space with elements of a fire safety system with more frequent and regular inspection of their functionality by security staff, fire protection technicians, private security staff or responsible employees in the shopping centre. Another proposed measure is fire extinguishers to operations where there is a higher chance of fire. Typical facilities with a higher chance in the shopping centre include facilities providing services, entertainment, cinemas and electronics stores. Another proposal is to create a fire patrol with at least three members, who would be located directly in the shopping centre and the event of a fire, to minimise the fire's impact in the presence of a fire protection unit. It would also be advisable to place fire curtains to prevent the spread of fire. [5]

6.2 Explosion

For the threat of explosion, it is proposed to measure the temperature in the shopping centre, especially where there is a higher chance of elevated temperature. In explosion prevention, it is ideal for carrying out a practice evacuation to a safe assembly point at least once a year. Temperature measurement is near critical objects that could cause an explosion.

6.3 Leakage of hazardous substances

If there is a risk of leakage of a dangerous substance, it is advisable to use a device for measuring the concentration of the presence of a dangerous substance in the room. Ideally, this measurement would be performed by a fire protection technician or a responsible member of the fire brigade located directly in the shopping centre.

6.4 High temperature in the building

Due to the threat of increased temperature in the building, it is ideal for reducing the increased temperature in the premises to the ideal state by opening the windows, turning on the air conditioning or cooling the outdoor areas with water. [6]

6.5 Smoke

For the threat of smoke, it is advisable to apply devices for smoke extraction and fresh air supply. Another preventive measure in case of the threat of smoke is the use of virtual reality in which the space of the shopping centre would be virtually created, and a member of the fire brigade could better orient himself in the space in case of the real presence of smoke. [7]

6.6 Flooding

Due to the threat of flooding, it is advisable to evacuate all persons present to a safe place outside the areas affected by the flood. It is ideal for creating pumps in the building to drain the water in the building to the nearest river or stream located near the shopping centre.

6.7 Escape exit blocking

For the threat of blocking the escape exit, it is ideal for checking the functionality of the escape exit regularly with a fire patrol unit every day before opening the shopping centre to the public. [8]

6.8 Disorientation

For the threat of disorientation of persons, it is ideal to create a sufficient number of information signs with the location of individual operations and with an escape exit.

6.9 Failure of the fire protection system and its components

For the threat of failure of the fire protection system and its element, it is ideal for checking the correctness of the fire control panel at least twice a day and to check the correctness of the elements at least once a week by members of the fire brigade or fire protection technician. In the event of a fault, it is essential to contact the service immediately.

6.10Untrained staff

Due to the threat of untrained personnel in fire protection, it is ideal for training every new employee before starting work and regular training on occupational safety at least once a year.

6.11Power outage

For the threat of a power outage, it is advisable for the shopping centre to procure a unit that would at least manage the fire protection system and its elements in the event of a threat.

7. Conclusion

Fire safety is one of the oldest types of safety in the development of the company. The article dealt with analysing the risk of threats to the soft target shopping centre under fire protection. In the beginning, the article described the definition of a soft target, its types, the degree of assessment and attractiveness.

The next chapter described the analysis used to evaluate the significance of threat risks for a given object type. Specifically, it is a qualitative analysis of the correlation of risks. The next chapter dealt with a specific procedure for assessing the significance of threats based on this chapter. First, a list of threats was created, based on which a table of risk correlations was created, which assessed whether the individual risks were related to each other. Two coefficients were created from the given data: the activity coefficient and the passivity coefficient, which were graphically recorded in the graph. To determine the significance of threats, it was essential to create four segments to identify them. To divide the graph into segments, it was necessary to calculate lines: horizontal and vertical, which divided the graph, where threats belonging to the first segment meant significant primary threats, to the second and third segments meant secondary threats and threats falling into the fourth segment represented tertiary (less significant threats). Based on these data, a table was created in the next chapter, where the individual threats were sorted according to their significance, where the significant primary threats include: fire, explosion, leakage of a dangerous substance and elevated temperature in the building. Secondary significant threats include smoke, flooding, escape route blocking, disorientation, fire protection system failure and its components, untrained personnel, and power outage. Tertiary significant threats include unspecified assembly points and the absence of a fire extinguisher. The last part of the article proposed design measures for primary and secondary significant threats, as they pose a greater threat risk than tertiary threats. These proposed measures are the subjective opinion of the author of the article. This process of creating a qualitative analysis to determine the risk of threats was created based on the ever-increasing risk of soft targets in fire protection. Therefore, it is essential to constantly update these risk probabilities with the development of society to identify new and primary threats. The plan to identify soft target threats in the future is to create and update new major threats as often as possible. Based on current threats, it is also essential to create and apply measures to minimise potential threats to soft targets to increase society's security continuously.

8. Acknowledgements

This research was based on the support of the Internal Grant Agency of Tomas Bata University in Zlín, IGA/CebiaTech/2021/004 and The Department of Security Engineering, Faculty of Applied Informatics.

9. References

[1] DELOITTE. (2012). Methodology for ensuring the protection of critical infrastructure in the field of production, transmission and distribution of electricity [online]. 1. Prague, 2012, 55 p. Available from: http://www.hzscr.cz/soubor/metodika-zajis-te-ni-ochrany-kriticke-infrastruktury-v-oblasti-vy-roby-pr-enosu-a-distribuce-elektricke-energie-pdf.aspx

[2] Nevrkla, J. and group. (2019). Soft targets: Identification, Hazards and Protection [online]. Zlín: Soft Targets Protection Institute, z.ú. ISBN 978-80-270-7066-4.

[3] Vichova, K.; Hromada, M.; Valasek, J. & Paulus, F. (2020). Comparison Analysis the Use of Modern Technologies by Fire Rescue Service, Proceedings of the 31st DAAAM International Symposium, pp.0535-0541, B. Katalinic (Ed.), Published by DAAAM International, ISBN 978-3-902734-29-7, ISSN 1726-9679, Vienna, Austria DOI: 10.2507/31st.daaam.proceedings.074

[4] Lapkova, D.; Kotek, L. & Kralik, L. (2018). Soft Targets – Possibilities of Their Identification, Proceedings of the 29th DAAAM International Symposium, pp.0369-0377, B. Katalinic (Ed.), Published by DAAAM International, ISBN 978-3-902734-20-4, ISSN 1726-9679, Vienna, Austria DOI: 10.2507/29th.daaam.proceedings.053

[5] Davis, J. (2019). Fire Safety in Modern Shopping Malls [online]. Available from: https://www.coopersfire.com/news/fire-safety-in-modern-shopping-malls/

[6] Fire & Safety, Security. (2014). Fire Safety Challenges in Malls & Shopping Centre [online]. Available from: https://firesafetysecurityindia.com/fire-safety-challenges-in-malls-shopping-centre/

[7] Health and Safety Authority. (2021). Fire Prevention [online]. Available from: https://www.hsa.ie/eng/Topics/Fire/Fire_Prevention/

[8] Health and Safety Executive. (2021). Fire Safety [online]. Available from: https://www.hse.gov.uk/toolbox/fire.htm

[9] Furness, A. & Muckett, M. (2007) Introduction to Fire Safety Management. Routledge. ISBN 9780750680684.

[10] Xin, J. & Huang, C. (2013). Fire risk analysis of residential buildings based on scenario clusters and its application in fire risk management. Fire Safety Journal [online]. 2013, 62. ISSN 0379-7112.

DOI: 10.2507/32nd.daaam.proceedings.084

RESEARCH AND MODELLING INDUCED ENERGY DISTRIBUTION PROCESSES IN THE GROUND WIRE CABLE

Iosif Breido, Vladimir Kaverin & Sholpan Yezhebayeva

This Publication has to be referred as: Breido, I[osif]; Kaverin, V[ladimir] & Yezhebayeva, S[holpan] (2021). Research and Modelling Induced Energy Distribution Processes in the Ground Wire Cable, Proceedings of the 32nd DAAAM International Symposium, pp.0593-0599, B. Katalinic (Ed.), Published by DAAAM International, ISBN 978-3-902734-33-4, ISSN 1726-9679, Vienna, Austria
DOI: 10.2507/32nd.daaam.proceedings.084

Abstract

Experimental studies have been carried out on high-voltage overhead power lines (HVOPL) in order to determine the energy characteristics of the induced energy in the ground wire cable, taking into account the peculiarities of its grounding. The experimentally obtained results have been used in the development of simulation models of the HVOPL-500 sections located between the anchor supports. Simulation models of the anchor span with different numbers of supports have been built. As a result of simulation experiments, the dependences of power consumption as a function of the load resistance and voltage across the ground wire have been obtained. On the basis of the obtained dependences, recommendations have been developed for various options of power supply for the telemetry system components placed on the HVOPL-500 kV supports. In the course of simulation experiments, there have been determined the maximum values of power and the corresponding voltage values on the ground wire as a function of the load resistance value. Technical solutions and recommendations have been proposed for providing power supply to the components of diagnostic systems and monitoring the state of structural elements of the supports.

Keywords: ground wire; high-voltage overhead power line; induced energy; power supply of diagnostic and monitoring systems.

1. Introduction

High-voltage overhead power lines (HVOPL) are the main facilities for transporting electricity from producers to consumers. In Kazakhstan, where the length of the HVOPL with a voltage of 110-500 kV is more than 70 thousand km, reducing losses in the course of electricity transportation is a particularly urgent scientific and technical task [1]. In world practice, research and development on remote monitoring systems designed for monitoring and diagnostics of HVOPL are known [2], [3]. The known systems of remote HVOPL monitoring contain a measuring unit, telemetric communication channels and power supply elements [4]-[7]. Due to the remarkable remoteness of the HVOPL from the settlements, there is a problem with organizing the power supply to the components of the remote monitoring system placed on the supports. Power consumption of electrical equipment located on the support does not exceed 50 W at a voltage of 12 V.

At present, the following energy sources can be used to power the monitoring system units: solar batteries, converters the energy of the electromagnetic field from the HVOPL conductive wires [5]. Solar batteries (SB) with additionally mounted accumulator batteries (AB) are the most common option in world practice. The CAT-1, DiLin, EMO, ThermalRate System, Snow Load Monitoring Systems operate on solar batteries that are directly mounted on the HVOPL supports [8]. They are used in lines ranging from 110 to 735 kV. However, with power supply from solar panels, a number of problems arise that reduce the reliability of the SB operation [9]. Due to the uneven SB operation (especially on cloudy days), the algorithm of the monitoring system is required to minimize power consumption. Implementation of a power supply from an SB with an additional AB requires high costs since electronic units must be placed on a large number of supports.

The power supply can also be performed from electromagnetic fields of the HVOPL conductive wires [6]. Works [4]-[5] describe low-energy options of the monitoring system power supply. The energy of the considered types of sources is not sufficient to power the local systems and indicator sensors located on the support. There is another unconventional method of generating energy for the diagnostic system that is based on the use of induced energy in-ground wire cables. The idea of obtaining induced energy from a ground wire was proposed by the Canadian Hydro-Québec Research Institute, but it was not brought to implementation and remained unrealized [10]. This work, there are analysed the possibilities of using this method of power supply to the blocks of the distributed protection system and diagnostics of structural elements of supports for high-voltage power lines of 220-500 kV [6]-[7].

2. Research methods

In this work, experimental studies have been carried out at the HVOPL-220 kV in order to determine the energy characteristics of the induced energy in the ground wire cable, taking into account the peculiarities of its grounding. At this, the method of simulation experiments has been used. The data obtained in the process of experimental studies, in accordance with the principle of similarity, have been used in the development of simulation models and in carrying out simulation experiments for the HVOPL-500 kV. Simulation experiments consisted of three stages: 1. Obtaining the dependences of power consumption as a function of active load resistance for different numbers of supports. 2. Determining the ground wire voltage dependences as a function of the value of the load resistance for different numbers of supports. 3. Determining a sufficient number of intermediate spans to provide power supply to the diagnostic system components located on the supports.

3. Characteristics of the object and its experimental studies

The objects of research are conductive elements of structures and ground wires of HVOPL-220 kV and HVOPL-500 kV. There have been studied the energy characteristics of the energy induced in the ground wire of a high-voltage power transmission line. The design of the ground wire for HVOPL-220 and HVOPL-500 is identical. As a rule, ground wire on supports is suspended on insulators and grounded at one end of anchor support [5]. In order to assess the energy characteristics, scientists of Karaganda Technical University, jointly with specialists of the KEGOC JSC, have carried out experimental studies of the induced electricity technical characteristics. The open-circuit voltage and short circuit current have been measured in partially de-energized conductive wires on the 220 kV power transmission line. During the measurements, it has been found that the average voltage across the ground wire did not exceed 100 V at short-circuit current lower than 2 A. The induced energy in the ground wire is sufficient to power the electronic diagnostic units located on the support. In the presence of electricity in the conductive wires, the induced energy in the ground wire is present regardless of climatic conditions and time of day.

The results obtained experimentally have been used in the development of simulation models of the HVOPL-500 sections located between the anchor supports. Technical characteristics of the induced energy substantially depend on the anchor span length. In the HVOPL-500 sections in the Karaganda region, the length of the anchor span varies from 2.5 km to 17.5 km. Each anchor span contains from 8 to 58 supports. In this work, the energy characteristics of the induced energy have been studied, taking into account the length of the inter-anchor spans under the conditions of the normal operation of the HVOPL without taking into account the effect of lightning discharges, atmospheric phenomena (ice formations) and switching overvoltage in conductive wires [11]-[12].

4. Developing a simulation model

A simulation model of the induced voltage distribution in the ground wire cable along its entire length of one anchor span has been developed in the MATLAB-Simulink software environment. The model considers sections of the anchor span of HVOPL-500 kV with the initial data presented in Table 1. The static characteristics of the energy induced in the ground wire have been studied in the normal mode of operation, without taking into account the disturbing effects (disturbing factors) on the characteristics of the induced energy. The equivalent circuit of the support is shown in Fig. 1. The equivalent circuit of the support includes the distribution of the capacitive components, taking into account the design features within one portal-type support. Ground wire cables 2 and 4 are suspended on suspension insulators 1 and 3, respectively. Conductive wires 6, 8 and 10 suspended on suspension insulators 5, 7 and 10, respectively, have a capacitive connection with ground wire cables.

Moreover, each ground wire cable is geometrically located at the same distance from the two phases of the conductive wires. Ground wire cable 2 has a capacitive connection (C1 and C2) with conductive wires 6 and 8, and ground wire cable 4 has a capacitive connection (C3 and C4) with conductive wires 8 and 10. The value of the capacitance between the conductive wires and the ground wire cable is determined by the dielectric constant of the air, the diameter of the cables and their length along the intermediate span. Additionally, the live wires have capacitance to the ground (C5, C6 and C7).

1. The line section length	264 км
2. The conductive wire brand	AC-300/39, AC-300/66
3. The wire section diameter	25 mm
4. The ground wire brand	2×TK-11
5. The diameter of the ground wire section	11 mm
6. The number of chains on the support	one
7. The number of wires per phase	three
8. The anchor span length:	
a) minimum	2500 m (8 supports)
b) maximum	17500 m (58 supports)

Table 1. Technical Specifications of the HVOPL Ek-GRES-PS Nura Section

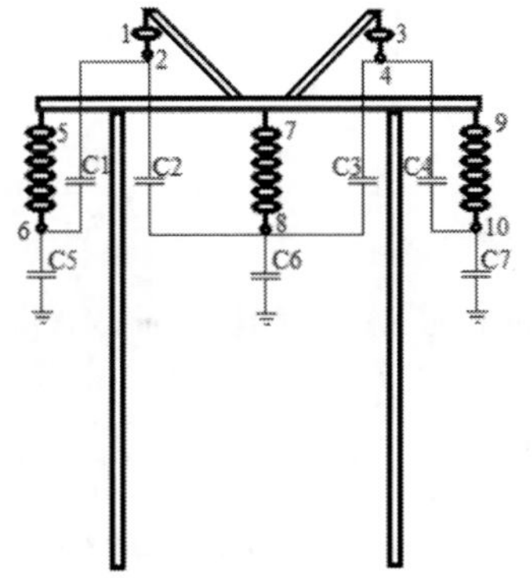

Fig. 1. The equivalent support circuit

The distance between live wires and ground is several times greater than the distance between the live wire and the ground wire. The number of intermediate spans within the anchor span in the analyzed section of HVOPL-500 varies in the range from 8 to 58, which corresponds to the number of supports at the minimum and maximum distances, respectively, within the anchor span. The equivalent circuit of an anchor span section with a ground wire cable is a sequential chain of span models describing the processes in the ground wire cable between the supports (Fig. 2).

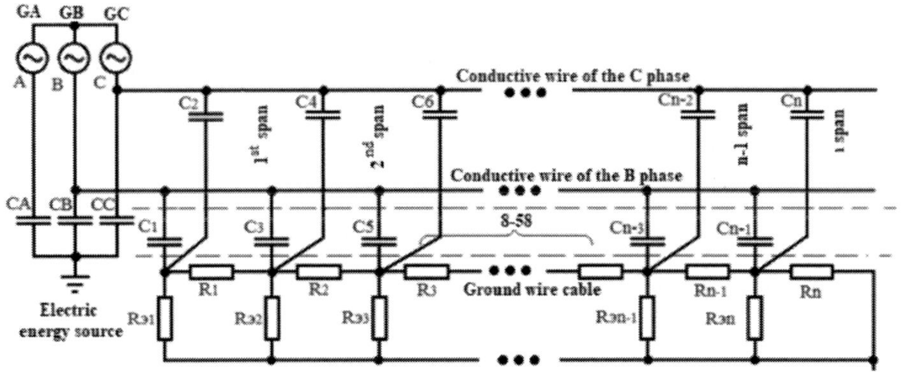

Fig. 2. The equivalent circuit of an anchor span section with a ground wire cable

One end of the ground wire cable is grounded on the anchor support, and the opposite end is structurally fixed to the suspended insulator. In the equivalent circuit, the electric capacitance between the supports relative to phase C of the conductive wire is modelled by capacitors C2, C4, C6, Cn-2 and Cn, and between the conductive wire of phase B and the ground wire it is modelled by capacitors C1, C3, C5, Cn-3 and Cn-1.

The active resistance of the ground wire between the supports is modelled by resistors R1, R2, R3, Rn-1 and Rn and the load characterizing the energy consumption of the local system for collecting, processing and transmitting the information of the state of structural elements to the control room is modelled by resistors Rэ1, Rэ2, Rэ3, Rэn-1 and Rэn. The numerical values of the parameters of the equivalent circuit elements are presented in Table 2.

Circuit elements	CA, CB and CC, pF	C1÷Cn, pF	R1÷Rn, Ohm
Value	2.27 10^{-6}	1.3 10^{-9}	0.43

Table 2. The Parameters of the Equivalent Circuit Elements

The simulation model for the equivalent circuit of the anchor span is shown in Fig. 3.

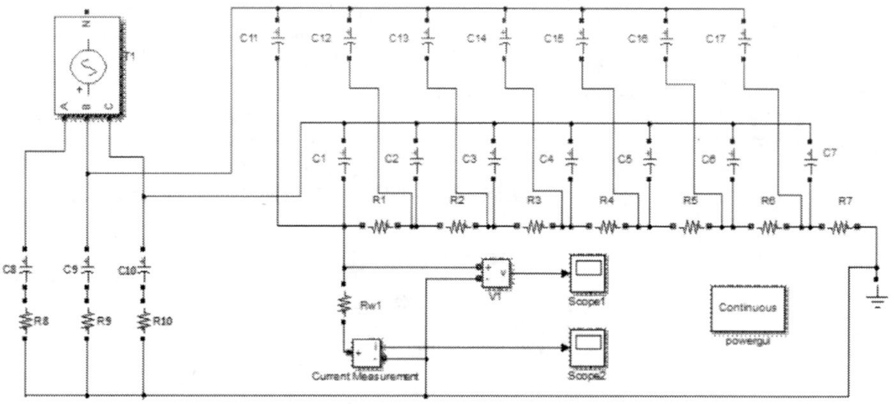

Fig. 3. The simulation model of the anchor span with seven supports

Simulation experiments have been carried out for two extreme values corresponding to the minimum (8) and maximum (58) number of supports for one anchor span. A feature of the simulation model is the need to include current-limiting resistors R8, R9 and R10 in series with capacitors C8, C9 and C10, in order to prevent infinitely large current at the moment of the simulation experiment start. The values of these resistances are selected from the condition of obtaining no more than 1% error of simulation experiments. At the same time, the load RW1 simulates power consumption of the components of the remote diagnostics system located on the supports.

5. Discussing the results of the studies

As a result of simulation experiments, the dependences of power consumption as a function of the load resistance and voltage across the ground wire as a function of the load resistance have been obtained. The dependences have been obtained for 8, 16, 24 and 32 supports. On the basis of the dependences obtained, recommendations have been developed for various options of power supply to the system components placed on the HVOPL-500 kV supports. As a result of implementing the first stage, the dependences of power consumption as a function of the load resistance have been obtained. The dependencies are shown in Fig. 4. The Figure shows that the dependences of power consumption as a function of the load resistance are of an extreme nature. Simulation experiments have been carried out for different numbers of supports in the anchor span.

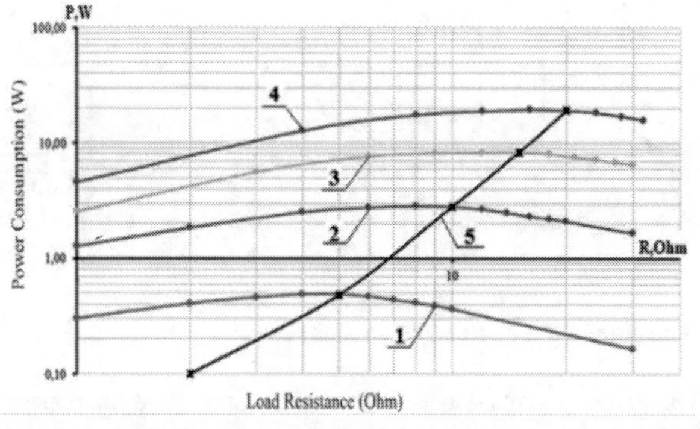

Fig. 4. The dependences of power consumption as a function of the load resistance

In the process of simulation experiments, the number of supports varied in multiples of 8. In Fig. 4, they are numbered counting from the grounded end of the lightning storm on the anchor support: - for the first eight supports - dependence 1; - for the first 16 supports - dependence 2; - for the first 24 supports - dependence 3; - for the first 32 supports - dependence 4. Dependence 5 is plotted graphically by the points of extreme values of dependences 1-4. The maximum power of the energy induced in the ground wire in the section of the first eight supports is 0.5 W with a load resistance of 5 Ohm. The power consumed by the elements of the system changes in different modes of their operation. For this reason, the power of the energy induced in the ground wire can be insufficient to power the system components.

The internal resistance of the power source is mainly determined by the resistance of the ground wire. As the length of the ground wire increases, its resistance increases. The extremum is reached when the internal resistance of the power source and the load resistance is equal. In this regard, with increasing the internal resistance of the power source, the extremum shifts towards increasing the load resistance. An insignificant effect on the law of variation of the extremum as a function of load resistance (dependence 5 in Fig 4) is exerted by increasing the distributed electrical capacitance between the conductive wires of phases B and C and the ground wire with increasing the span between the anchor supports (Fig. 3), which determines nonlinearity of dependence 5 (Fig. 4). Thus, dependence 5 corresponds to the maximum value of the power supplied to the load as a function of the load resistance value for different numbers of supports in the anchor span. The optimal power values for the corresponding number of supports are presented in Table 3.

Diagram number (Figure 4)	1	2	3	4
Number of supports in the anchor span, pcs.	8	16	24	32
Load resistance, Ohm	3	6	9	12
Power supplied to load, W	0.5	2.4	7.4	17.6
Load voltage, V	1.52	5.2	11	19

Table 3. The Optimal Power Values

An important parameter in matching the induced energy in the ground wire with the load parameters is load voltage for different numbers of supports in the anchor span. Due to the fact that the magnitude of voltage depends on the magnitude of the load resistance and the length of the anchor span, it is necessary to study the effect of the above parameters on the magnitude of the voltage across the ground wire at the point of load connection using simulation tools. In order to determine the magnitude of the amplitude value of voltage, dependencies are obtained as a function of the load resistance. The dependencies are shown in Fig. 5.

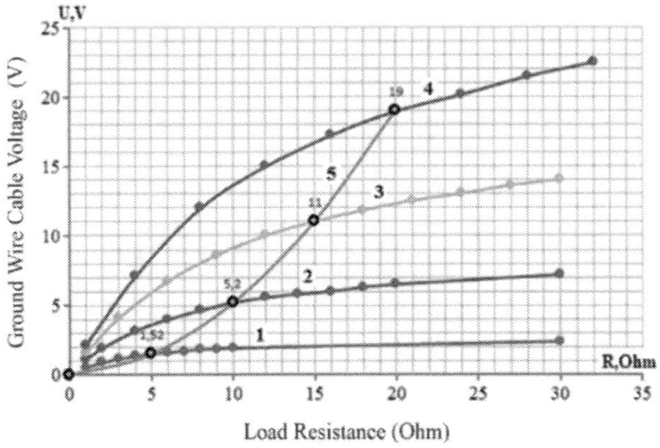

Fig. 5. The ground wire cable voltage dependences as a function of the load resistance

With increasing the load resistance in the range of 1÷30 Ohm, there is observed voltage increasing. With increasing the number of supports in the anchor span in the range of 8÷32 pcs., there is also observed increasing voltage across the load in the range of 0.5÷22 V. Fig. 5 shows dependence 5 of the voltage value for the optimal values of the load resistance. As a result of simulation experiments, it has been found that voltage across the load at the optimal value of the load resistance, with increasing the number of supports in the anchor span in the range of 8÷32 pcs., increases according to the law that is close to the parabolic law in the range of 1.5÷19 V. The voltage value on the ground wire for different numbers of supports and the optimal value of the load resistance is presented in Table 3. Thus, with increasing the number of supports more than 24 pcs., starting with the anchor support, where the grounding of the ground wire cable is implemented, the power of the energy induced in the ground wire can be considered sufficient to power the local system. With the number of supports from 16 to 24, it is desirable to use buffer energy storage in the form of a storage battery or high-capacity capacitors when developing a power supply unit.

In the section of the anchor span, where there are fewer than 16 intermediate spans of the induced energy in the ground wire, the power of the energy induced in the ground wire is not sufficient to power the elements of local systems. Therefore, it is necessary to use other technical solutions, for example, solar batteries with buffer energy storage.

6. Conclusion

The unconventional method of generating power energy for the diagnostic and monitoring systems of high-voltage overhead power lines, based on the use of induced energy in-ground wire cables, taking into account the peculiarities of its grounding, was not brought to implementation and required study. The paper analysed the possibilities of using this power supply method to the diagnostic systems for high-voltage power lines of 220-500 kV. The equivalent circuit of the anchor span section with a ground wire cable for the HVOPL-500 kV "HVOPL Ek-GRES-PS Nura" section has been developed. A simulation model of the distribution of the induced energy in the lightning protection wire of the anchor span has been developed using the MATLAB software package. In simulation experiments, the maximum power values and the corresponding voltage values across the ground wire have been determined as a function of the load resistance value. The possibilities of power supply to the components of diagnostic and monitoring systems located on the HVOPL supports have been determined. There have been proposed technical solutions and recommendations for providing power supply to the components of diagnostic systems and monitoring the state of structural elements of the supports. The results obtained can be extended to HVOPL-500 kV with portal type supports and a similar structure for placing ground wires on the supports. In the future, the power supply system requires development, where the use of solar panels with buffer energy storage, when the induced energy in the ground wire is not sufficient to power the elements of local systems.

7. Acknowledgements

This work has been supported by grant funding for scientific and (or) scientific-technical projects at the level of the Ministry of Education and Science of the Republic of Kazakhstan on topic No. IRN AR08856344 "Development of a complex for remote monitoring the state of structural elements of high-voltage overhead power lines during operation based on intelligent noise-immune telemetry systems".

8. References

[1] Atakhanova, Z. & Howie, P. (2007). "Electricity demand in Kazakhstan", Energy Policy, Vol. 35, No. 7, 2007, pp. 3729-3743, DOI 10.1016/j.enpol.2007.01.005.

[2] Yang, Y., Xie, G., Xu, X. & Jiang, Y. (2011). "A Monitoring System Design in Transmission Lines based on Wireless Sensor Networks", Energy Procedia, Vol. 12, 2011, pp. 192-199, ISSN 1876-6102, DOI 10.1016/j.egypro.2011.10.027.

[3] Morozovska, K. & Hilber, P. (2017). "Study of the Monitoring Systems for Dynamic Line Rating", Energy Procedia, Vol. 105, 2017, pp. 2557-2562, ISSN 1876-6102, DOI 10.1016/j.egypro.2017.03.735.

[4] Peabody, A. (2009). "Evolution of Real-time Monitoring and its Future Benefits", IWAIS, No. XIII, Sep. 2009.

[5] Kaverin, V. V.; Khomchenko, V. G. & Yezhebayeva, Sh. Ye. (2020, Jun.). Review and analysis of existing power supplies for monitoring systems located on the supports of overhead power lines, Presented at Proceedings (Saginov's Readings). No. 12, pp. 743-745.

[6] Breido, I., Kaverin, V. & Ivanov V. (2018). Telemetric Monitoring Insulation Condition of High Voltage Overhead Power Lines, Presented at Proceedings of the 29th DAAAM International Symposium, pp. 0319-0328, B. Katalinic (Ed.). DOI 10.2507/29th.daaam.proceedings.046.

[7] Breydo, I., Kaverin, V., Ivanov, V., Voytkevich, S. & Levin, I. (2017). "Distributed system of protection and diagnostics of support structural elements of high-voltage power lines", EAI Endorsed Transactions on Energy Web. 4. 152983. 10.4108/eai.3-8-2017.152983. 017.

[8] Uski-Joutsenvuo, S. & Pasonen, R. (2013, Feb.). Maximizing power line transmission capability by employing dynamic line ratings – technical survey and applicability in Finland, VTT Technical Research Centre of Finland, Finland.

[9] Bessel, V. V.; Kucherov, V. G. & Mingaleyeva R. D. (2016). Studying solar photovoltaic cells. (Publishing Center of the Russian State University of Oil and Gas named after I.M. Gubkin).

[10] Blais R. & Berthiaume R. (1980). "Supplying fixed and stroboscopic light beacons from the overhead ground wire on 735 kV transmission lines", IEEE Transactions on Power Apparatus and Systems, Vol. PAS-99, No. 1 Jan./Feb, 1980.

[11] Breido, I. V.; Kaverin, V. V.; & Voitkevich, S. V. The system of cathodic protection of elements of supports of high-voltage power lines sections, RK Patent No. 29977, May 19, 2015.

[12] Breido, I. V.; Kaverin, V. V.; Em, G. A.; Sichkarenko, A. V.; Ivanov, V. A. & Voitkevich, S. V. (2018). Method of telemetric measurement of ice loads of high-voltage power lines and a system for its implementation, RK Patent No. 33249, Oct. 26, 2018.

DOI: 10.2507/32nd.daaam.proceedings.085

QUANTUM COMPUTING FOR DESIGNING BEHAVIORAL MODEL AND QUANTUM MACHINE LEARNING ON A HUMANOID ROBOT

Jacob Viertel & Mohamed Aburaia

This Publication has to be referred as: Viertel, J[acob] & Aburaia, M[ohamed] (2021). Quantum Computing for Designing Behavioural Model and Quantum Machine Learning on a Humanoid Robot, Proceedings of the 32nd DAAAM International Symposium, pp.0599-0606, B. Katalinic (Ed.), Published by DAAAM International, ISBN 978-3-902734-33-4, ISSN 1726-9679, Vienna, Austria
DOI: 10.2507/32nd.daaam.proceedings.085

Abstract

Today's robots are controlled with classical computers that make decisions based on algorithms designed for a classical computing architecture. However, the amount of data increases continuously - a fact which pushes classical computers that control these robots to their limits. In comparison to classical computing, fundamentally different concepts of physics are the basis of quantum computing and thereby quantum computing has the potential to solve specific computational problems more efficiently. In addition to the aforementioned complications, the subgroup of robots that are subject to the research of this paper, namely humanoid robots, not only need more computational capacity but also require an algorithmic design based on a behavioral model which avoids causing antipathy when interacting with human beings. Using a real quantum computer with quantum phenomena such as superposition, entanglement, and quantum parallelism, new behavioral models and algorithms can be designed to improve human robot interaction and speed up machine learning processes. This paper promotes a hybrid-controlled robot using the combination of a classical computer- and quantum computer. With four quantum circuits to define a behavioral model and three different quantum machine learning algorithms we show the advantage of the hybrid-controlled robot.

Keywords: Quantum Robotics; Human-Robot-Interaction; Quantum-Machine-Learning; Grover-Algorithm, Quantum-Support-Vector-Machine.

1. Introduction

Classical computers can perfectly optimize smaller systems, but when it comes to larger systems such as transportation routes or product pricing, they only achieve incremental improvements [1]. With increasing problem size, the amount of data often grows exponentially. This faster than polynomial growth of data leads to an accumulation of computational costs that in most cases can be hard to handle by a classical computer. Not only the safety standards have to increase when robots collaborate with human beings, but also the quality of the interaction between them. Humanoid robots need to express resemblances of feelings and emotions like aggression or fear in order to increase the level of communication and create a higher acceptance of a robot participating in a conversation with humans [2]. The Uncanny-Valley theory [3] relates the human affinity to the human-like appearance and motions of a robot.

If the robots' appearance is somewhere in-between the humanlike-robot and the cartoon-like-robot its appearance will arouse human antipathy. Designing a behavioral model with classical computing is cost-intensive and further stresses the capacity of classical computers. With the deterministic nature of classical computing, it is difficult to imitate the nondeterministic behavioral model of a human being. Therefore, within this paper we use the probabilistic nature of quantum computing to create a complex behavioral model with lower computational costs than a classical behavioral model. Areas like quantum machine learning [4], [5] or mood and emotion generation with quantum computing [6], [7] have been researched more frequently in recent publications to solve problems in a way that is impossible for classical computers. One contribution of this paper is an approach to use an adaptation of the Grover algorithm in order to solve several variations of the Shortest Path problem. The Grover algorithm adaptation can analyse a graph if a path between starting node and target node exists. Despite the early state of quantum computers, the Grover adaption can keep up with various classical algorithms such as Dijkstra or Bellman Ford.

The goal of this paper is not to outperform known classical computing systems, but simply to show the raw potential of quantum humanoid robots. It shows that it is possible to overcome the problem of setting up a control system for a humanoid robot that consists of two different computers that are based on fundamentally different concepts of physics. Additionally, this paper introduces various methodologies how to transfer a classical robotic related problem in order to use quantum computing to solve them. With respect to the diversity of the research field quantum robotics, this paper focuses on motion generation with quantum computing and solving robotic related problems such as the Shortest Path problem or vision. According to the concepts developed in this paper quantum robots have classical sensors, a hybrid-computer, and classical effectors. The sensor data from the classical computer must be converted with respect to the different physics of the quantum computer before being further processed within a quantum circuit. Afterwards, the results of these quantum circuits had to be mapped again to trigger any actions of the classical effectors.

2. Fundamentals of Quantum Circuits

Every operation or gate in quantum computing can be represented as a matrix transformation of a state vector along a sphere. A quantum bit or qubit is a basic unit to store information in quantum computing and it represents the most trivial two-state quantum system [8]. The transformations in quantum computing happen in Hilbert Space, which is a specific type of vector-space. Regarding a quantum behavioral model, within the Hilbert Space the probabilistic nature of measuring a quantum circuit can be used to design cost-efficient characteristic quantum circuits that imitate different behaviours similar to the Braitenberg Vehicle concept [9]. The main idea of the Braitenberg Vehicle is to mimic complex behaviour like fear or shyness with simple methods. As the research area of quantum cognition is built on the theory that the human behaviour model is closer related to quantum computing than to classical computing [10], quantum computing is more suitable to define the artificial behavioral model of a robot that interacts with human beings.

Furthermore, quantum computing can also speed up algorithmic problems using quantum algorithms such as Shor [12] or Grover [11]. The Grover algorithm [11] is a quantum algorithm that is used for searching unstructured data and it provides a quadratic speed-up in comparison to the best classical computing algorithm. The heart of the Grover algorithm is the so-called Grover-oracle. Adapting this oracle to solve a problem that is applicable to robotics is the main challenge during the implementation of the Grover algorithm. Another potential for the use of quantum machine learning is due to the fact that quantum computing is superior in comparison to classical computing in terms of calculating the dot-product. The dot-product is used when calculating the inner product. Therefore, quantum computing has greater potential doing the Swap-Test [13] to classify data points in comparison to classical computing.

This paper proposes a hybrid-controlled robot using a classical computer for the main control system that can outsource specific tasks to a quantum computer. This setup causes an additional problem that is not of quantum nature only such as handling the noise. A collaborative problem of classical and quantum computing is that most data are stored on classical computers, therefore, they must be mapped suitable for quantum computers onto qubits before algorithms like Quantum Support-Vector-Machine can use them for classification.

3. Quantum Motions Generators

A quantum motion generator circuit uses quantum phenomena like entanglement and superposition to generate motions with behaviours such as:

- Deterministic
- Classical probabilistic
- Quantum probabilistic
- Mixed probabilistic

The motion behaviours are encoded onto the control register as follows:

- $|00\rangle$ Deterministic
- $|10\rangle$ Quantum Probabilistic
- $|01\rangle$ Quantum Probabilistic
- $|11\rangle$ Classical Probabilistic

In the four behaviour models above the operator U in the circuit below is defined as unitary matrix and has no effect during the calculation. Fig.[1] shows the circuit to realize all four behavioral motions.

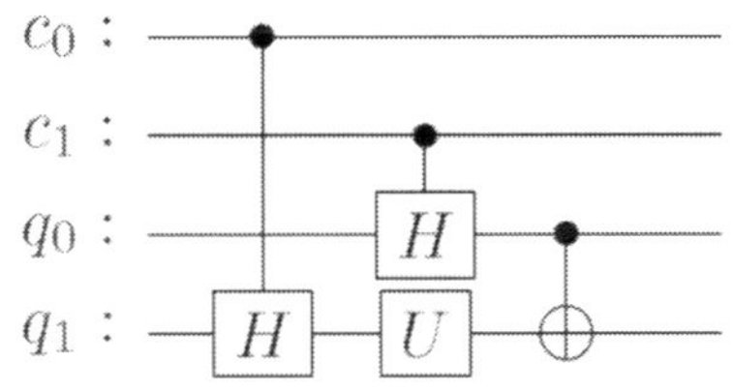

Fig. 1. The circuit can be divided in two quantum registers, target- and control-register. According to c_0 and c_1 the three behaviours deterministic, classical, and quantum will be executed.

Editing the transformation matrix U, a mixed probabilistic behaviour can be generated, which is a mixture of both classical and quantum probabilistic. Eq. [1] shows the matrix of the single transformation operator U.

$$U = \begin{bmatrix} cos\frac{\theta}{2} & -e^{-i\omega} sin\frac{\theta}{2} \\ e^{i\omega} sin\frac{\theta}{2} & e^{i\lambda + i\omega} cos\frac{\theta}{2} \end{bmatrix} \quad (1)$$

The circuit was applied to the greeting behaviour of the biped robot. With the deterministic or probabilistic behaviour, the robots greeting behaviour were not suitable for a high-quality interaction with human beings. The deterministic behaviour is too monotone and the probabilistic behaviour too random to be considered human-like. With quantum probabilistic or mixed probabilistic behaviour model, the robots greeting became more human-like.

4. Quantum Behavioral Modelling

To define a more complex behavioral model of the robot, two additional quantum circuits are needed. With the mood and handed circuit it is possible to create a behavioral model of the humanoid robot with less computational cost than other models including convolutional nets such as proposed in [14]. The mood-circuit influences the speed and range of the motion execution according to its internal state by mapping two moods, happy and exhausted, onto the circumference of a qubit. For the classical computer to adjust speed and range of the robot the mapped state vector from the quantum behavioral model must be measured. Fig. [2] shows the circuit to obtain the state vectors orientation using phases kickback within an inverse Quantum Fourier Transform.

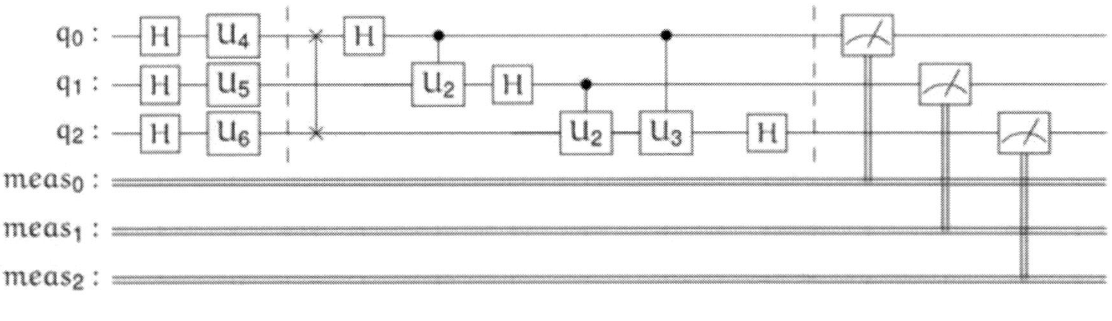

Fig. 2. The illustrated circuit obtains the quantum state vectors orientation in order to be further processed on the classical computer.

Even before a human is born, he/she/it already has the predisposition to be either left- or right-handed [15]. To add such a native characteristic to the behavioral model of the robot in order to form a more in-depth persona with little computational cost, the handed circuit defines the robots' strong hand. To define the robot's handedness, the quantum circuit empathizes left- or right-handed motions with probabilistic distribution. Fig. [3] shows the circuit for a left-handed persona, that is divided into three parts with dashed lines.

The Grover circuits for left- and right-handed characteristics are both initialized with two Hadamard gates at the beginning of the circuit. The second part of the circuit, the so-called Grover oracle, is the only part that varies for the left- and right-handed persona. The last part of the circuit, the so-called reflection around average is the same for both circuits.

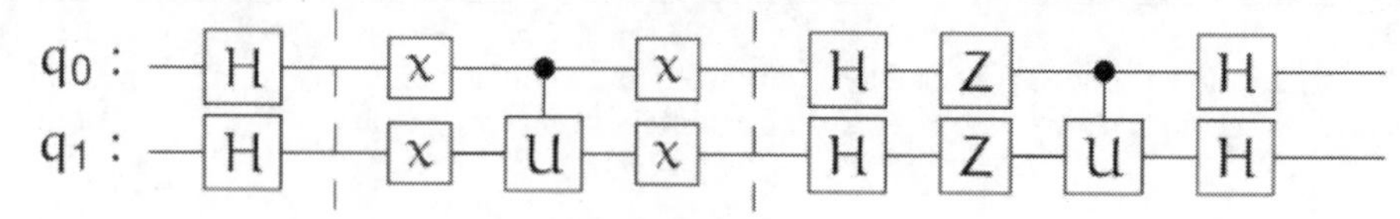

Fig. 3. The circuit for the right-handed persona varies only in the middle part of the circuit.

The last circuit for the quantum behavioral modelling realizes the first quantum random walk [16] on a humanoid biped. This circuit is used to train the robot a method to find its target using randomized patterns. The robot tries to find a person in a room by randomly changing its direction and generating a random walking or turning behaviour. When the robot finds the target person, it will approach the person and greet him/her with a quantum generated waving motion. This is the first realization of the quantum random walk on a humanoid biped robot.

5. Quantum Machine Learning

This approach uses the famous Grover Oracle to solve the algorithmic problem of the Shortest Path. To adapt the Grover oracle to the Shortest Path problem, the edges of a graph are encoded as qubits and the nodes as total symmetric Boolean functions. Fig. [4] shows the example graph with its six vertices and seven edges.

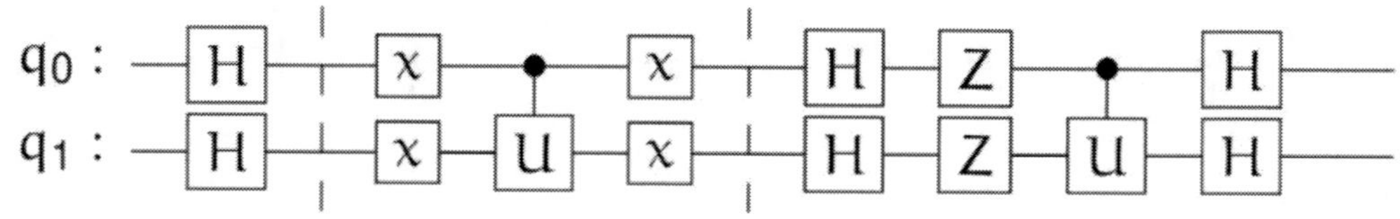

Fig. 4. A is the starting node and F is the target node.

With this approach it is possible to find a valid path in a given graph in $O\left(\frac{\pi}{4}\sqrt{\frac{2^E}{M}}\right)$ with M being the valid solutions, V being the numbers of vertices and E being the number of edges. More nodes or vertices are nowadays not possible, as the number of qubits is limited on real quantum computers and on quantum simulations. The runtime O of three classical algorithms and our Grover adaption can be seen below with V representing the number of vertices and E representing the number of edges:

1. Bellman Ford $O(V * E) = 42$
2. Dijkstra $O((V + E)\,log(V)) = 13$
3. Floyd-Warshall $O(V^3) = 216$
4. Grover oracle $O(\frac{\pi}{4}\sqrt{\left(\frac{2^E}{M}\right)}+\frac{\pi}{4}\sqrt{\left(\frac{2^E}{M-1}\right)}+\frac{\pi}{4}\sqrt{\left(\frac{2^E}{M-2}\right)} \approx 20$

In addition to the Shortest Path problem the Grover adaption can also solve problems such as:

- Does a path exist?
- What is a path?
- How many edges does the path have?

After designing the Grover oracle, the problem Does a path exist? can be solved using the quantum counting algorithm [18]. In this algorithm, the Grover oracle is used in combination with an inverse Quantum Fourier Transform. The result of this quantum circuit is needed in order to distinguish how often the Grover circuit has to be iterated to obtain a valid solution. In the runtime equation for the Grover oracle, $O\left(\frac{\pi}{4}\sqrt{\frac{2^E}{M}}\right)$, the number of solutions is represented with *M*. Using a wrong number of iterations will lead to a decreased accuracy.

After obtaining the number of valid paths the other three problems can be solved:

- What is a valid path?
- How many edges does the path have?
- What is the shortest valid path?

For the example graph, the result of the quantum counting algorithm indicated M=3. Fig. [5] shows the probabilities of measuring one of the three possible paths for the example graph.

Fig. 5. When a measurement is done after three iterations, one of the three paths is measured. The paths active edges are encoded in the qubit string at the X-axis.

The methodology of finding the Shortest Path requires, additional to the classical Grover oracle two components:

- A Quantum-Adder
- A Quantum-Comparator

After finding a valid path, the costs of the path are calculated using a quantum circuit of half- and full adder. If this path was the first one that was found, it will define the new threshold. Afterwards the Grover oracle is edited in order to no longer consider the founded path as a valid solution. In the next phase, the costs of the next founded path will be compared with this threshold using the quantum comparator circuit. Fig. [6] shows the entire circuit for solving the Shortest Path problem.

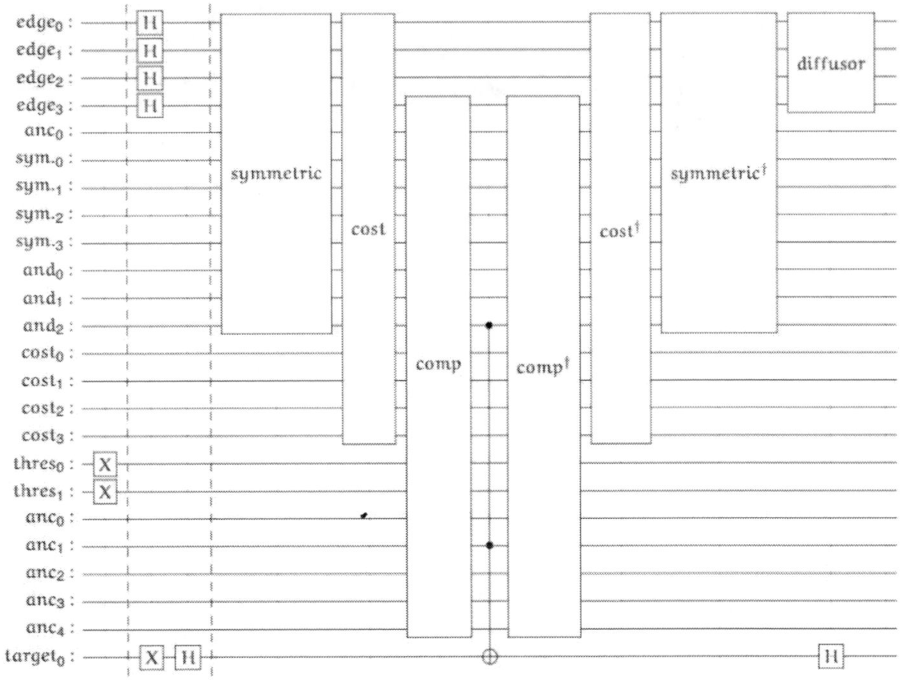

Fig. 6. The three components of the Grover oracle, symmetric, cost and comp have to be mirrored at the end of each iteration in order to restore the qubits state.

Research in quantum computing will continuously develop, and with the right technology such as the Noisy Intermediate-Scale Quantum (NISQ)-technology [17] quantum computers with limited computational power will be commercially available in the near future. As the name suggests, those NISQ-computers will be more prone to noisy environment, therefore, the circuits must be designed short in order to lower the impact of noise on the results.

Two Quantum algorithms, Quantum Support Vector Machine algorithm [4] and the Swap Test method [13], were compared by solving the same problem. Both algorithms were designed and improved to classify 200 testing samples of two different numbers with only one training sample of each number. The result was a 100% accuracy for both algorithms but with less computational time, the Swap Test is better suited for decoherence prone quantum computers. Fig. [7] shows the circuit for the Quantum Support Vector Machine algorithm and Fig. [8] shows the circuit for the Swap-Test algorithm.

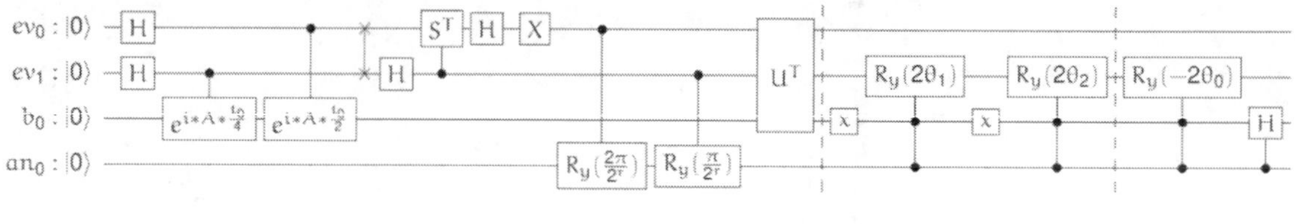

Fig. 7. A qubit in this Quantum Support Vector Machine circuit is 4.160 μs exposed to noise.

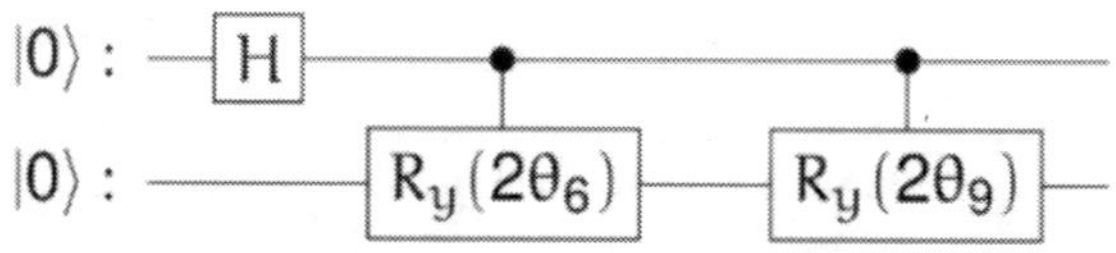

Fig. 8. A qubit in this Swap-Test circuit is 1.500 μs exposed to noise.

As the swap test is better suited for NISQ-computers, further research was conducted with this algorithm. To extend the problem of classifying two numbers, a methodology was designed in order to distinguish three different digits. The Swap Test method was able to classify those different digits of the OCR-dataset without a misclassification using an additional data mapping step. Fig. [9] illustrates the first step of this methodology in which the digits were divided into two groups according to the horizontal and vertical ratio of the pixels.

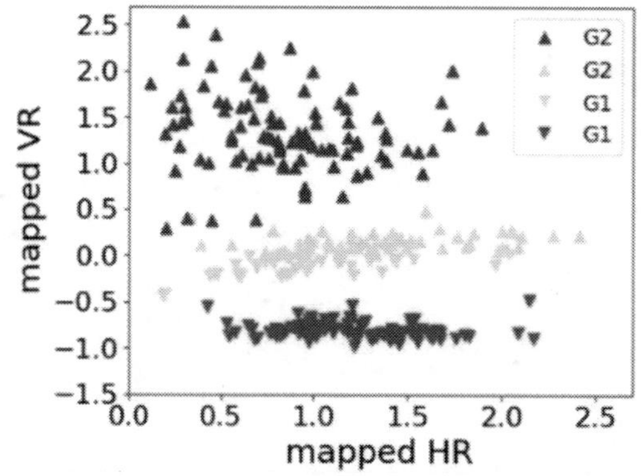

Fig. 9. The blue triangles represent the digits 8 the yellow triangles represent the digits 6 and the red triangles represent the digits 9. It is not possible to divide the digits with a straight line without any misclassification.

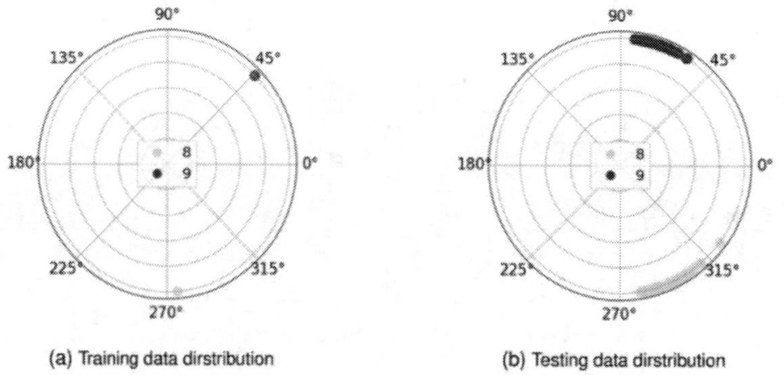

Fig. 10. The blue triangles represent the digits 8 the yellow triangles represent the digits 6 and the red triangles represent the digits 9. It is not possible to divide the digits with a straight line without any misclassification.

After the digits were grouped a linear mapping was per formed to further divide the subgroups. Fig. [10] shows how the subgroup, containing the digits 8 and 9, was mapped onto the circumference of a qubit. To extend the Swap-Test methodology, a more complex dataset, the Breast Cancer dataset [19], was mapped onto the entire Bloch Sphere and not only the circumference. After conducting a Principal Component Analysis (PCA), the four highest components were used to transform these testing samples along the Bloch sphere. Fig. [11] shows the circuit that transforms the qubits state according to the four PCA-variables and also performs the Swap-Test.

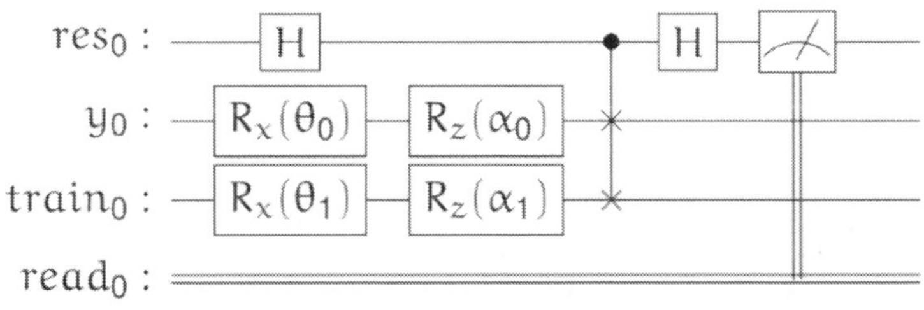

Fig. 11. The result of the Swap-Test is encoded on the qubit res_0.

The entire Breast Cancer dataset [19] was classified with this circuit with an accuracy of 92% which is only 5% below the accuracy of the best-known classical algorithm. Fig. [12] shows the entire dataset being mapped onto a Bloch sphere.

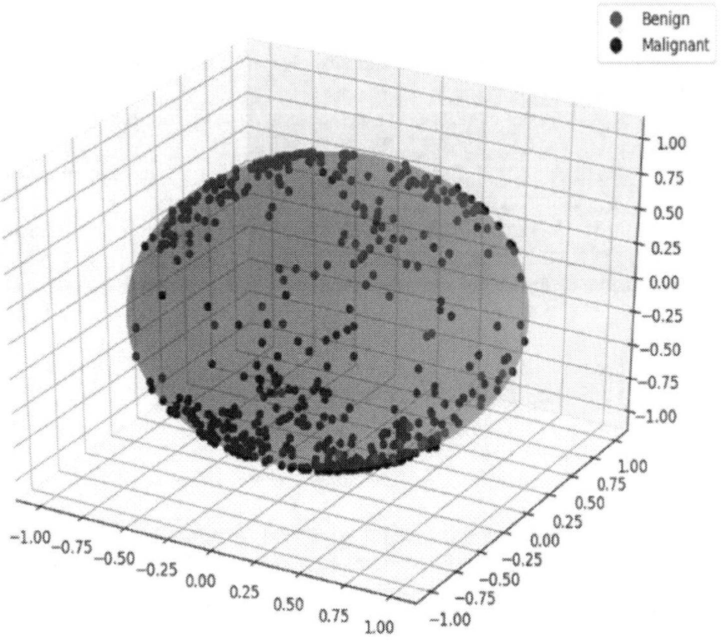

Fig. 12. All data points of the Breast Cancer Dataset mapped onto a sphere.

6. Conclusion

This research proposes potential solutions to two fields of modern robotics, namely Human-Robot-Interaction, and machine learning. More specifically, three questions were used to focus the research. For Human-Robot-Interaction we asked 1) how to control emotion and mood, while for machine learning we considered 2) how to solve the shortest path calculation and 3) how to apply a dataset to quantum computing or more precisely to a qubit. Within this research we presented two quantum circuits and showed how to set up quantum circuits on a real humanoid robot in order to express emotions. The hybrid-controlled robot was able to express certain behaviors like happiness and exhaustion. In future studies the aim will be to verify the theory of quantum cognition. This theory states that human behaviour has more parallels to quantum- than to classical-computing. In order to verify the theory of quantum cognition, more powerful quantum behavioural modelling circuits as well as quantum motion generation circuits will be required. Furthermore, it will be interesting to investigate the degree to which behaviour models should imitate humanlike behaviour. Is it enough that the humanoid robot executes motions according to left or right handedness or does the handedness also need to affect the thinking process of the robot as it is with human beings?

One possibility could be to expand behavioural circuits according to the wheel of emotions of Plutchik [20]. Mapping the internal state according to the wheel of emotions of Plutchik likely enables better scalability in comparison to the mood-circuit presented in this paper. Another contribution of this paper is in the field of quantum machine learning. A Grover-Circuit was built in order to solve the shortest path problem. This Grover-circuit could not match the best classical algorithm in existence today, but it performed better than most of the classical algorithms for the shortest path problem. Additionally, different circuits were compared in order to figure out which algorithms are best suited for NISQ computers. Finally, we were able to classify the Breast-Cancer dataset using the swap test algorithm and thereby mapping the data points onto the entire Bloch-Sphere of a qubit. Here again it was not possible to show quantum supremacy, but this was not the goal of this paper. The primary objective was to bridge the gap between robotics and quantum using a humanoid robot and a real quantum computer. In future work the quantum library must be changed in order to achieve even better performance. The Qiskit simulation has problems with more than 30 qubits. In Bravyi & Gosset [21] a new idea for simulating quantum circuits was introduced that can simulate even bigger circuits with up to a few hundreds of qubits. With more qubits a standalone Grover circuit can be built to solve the Shortest Path problem autonomously with all variations. Even with a simulator of better performance, the concept of the Shortest Path must be changed in order to get rid of the polynomial running time and be as efficient as the best classical algorithms.

The research question "How to implement quantum computing on a robot to create a behavioral model and solve algorithmic problems?" was answered with various circuits that ran on the humanoid robot. Larger circuits with more than 10 qubits were simulated while smaller ones were executed on real quantum computers. With several simulated and real quantum circuits this research not only verified a method for controlling a robot with the combination of a classical and a quantum computer, but also showed some practical robotic-relevant applications like solving the shortest path problem or classifying various datasets.

7. References

[1] Malham, S. (2015). An introduction to Lagrangian and Hamiltonian mechanics, 10.13140/RG.2.1.2914.8003.

[2] Karpova, I & Karpov, V. (2018). Some mechanisms for managing aggressive behavior in group robotics, 10.2507/29th.daaam.proceedings.082.

[3] Mori, M.; MacDorman, K. & Kageki, N. (2012). The Uncanny Valley [from the field], IEEE Robotics & Automation Magazine, vol. 19, no. 2, pp. 98-100, 10.1109/MRA.2012.2192811.

[4] Li, Z.; Liu, X.; Xu, N. & Du, J. (2015). Experimental realization of a quantum support vector machine, Physical review letters, 10.1103/physrevlett.114.140504.

[5] Rebentrost, P.; Mohseni, M. & Lloyd, S. (2013). Quantum support vector machine for big data classification, Physical Review Letters. 113. 10.1103/PhysRevLett.113.130503.

[6] Bevacqua, E. & Mancini, M. (2004). Speaking with emotions.

[7] Lukac, M. & Perkowski, M. (2007). Quantum mechanical model of emotional robot behaviors, 37th International Symposium on Multiple Valued Logic (ISMVL 07), pp. 19-19, 10.1109/ISMVL.2007.44.

[8] http://community.qiskit.org/textbook, (2020). Learn quantum computation using Qiskit, Accessed on: 2020-04-15

[9] Braitenberg, V. (1986). Vehicles: Experiments in synthetic psychology, MIT Press, 9780262521123.

[10] Bruza, P.; Wang, Z. & Busemeyer, J. (2015). Quantum cognition: a new theoretical approach to psychology, Trends in Cognitive Sciences, 10.1016/j.tics.2015.05.001

[11] Grover, L. (1996). A fast quantum mechanical algorithm for database search, STOC '96, Association for Computing Machinery, 10.1145/237814.237866.

[12] Shor, P. (1997). Polynomial-Time algorithms for prime factorization and discrete logarithms on a quantum computer, Society for Industrial & Applied Mathematics (SIAM), 10.1137/s0097539795293172.

[13] Wittek, P. (2014). Quantum machine learning: What quantum computing means to data mining, Academic Press, 9780128009536.

[14] http://roboticsbusinessreview.com, (2018). Applying machine learning to robotics, Accessed on: 2020-06-03.

[15] Corey, D.; Hurley & M.; Foundas, A. (2001). Right and left handedness defined: A multivariate approach using hand preference and hand performance measures, Neuropsychiatry, neuropsychology, and behavioral neurology, 14(3), 144–152.

[16] Aharonov, Y.; Davidovich, L. & Zagury, N. (1993), Quantum random walks, Physical review, A. 48, 1687-1690. 10.1007/BFb0083545.

[17] Preskill, J. (2018). Quantum computing in the NISQ era and beyond, Quantum Physics, 10.22331/q-2018-08-06-79.

[18] Brassard, G.; Hoyer, P. & Tapp, A. (1998). Quantum counting, Lecture Notes in Computer Science, Springer Berlin Heidelberg, 9783540686811.

[19] https://archive.ics.uci.edu/ml/datasets/Breast+Cancer+Wisconsin, (1995). Breast Cancer Wisconsin (Diagnostic) Data Set, University of Wisconsin, Accessed on: 2020-07-11.

[20] Plutchik, R. (1982). A psychoevolutionary theory of emotions. Social Science Information, 21(4-5), pp.529–553, 10.1177/053901882021004003

[21] Bravyi, S. & Gosser, D. (2016). Improved classical simulation of quantum circuits dominated by Clifford gates, Physical Review Letters, 116(25), 10.1103/PhysRevLett.116.250501.

DOI: 10.2507/32nd.daaam.proceedings.086

A DIGITAL TWIN BASED SELF-CALIBRATION TOOL FOR FAULT PREDICTION OF FDM ADDITIVE MANUFACTURING SYSTEMS

Fabio Corradini & Marco Silvestri

This Publication has to be referred as: Corradini, F[abio] & Silvestri, M[arco] (2021). A Digital Twin Based Self-Calibration Tool for Fault Prediction of FDM Additive Manufacturing Systems, Proceedings of the 32nd DAAAM International Symposium, pp.0607-0616, B. Katalinic (Ed.), Published by DAAAM International, ISBN 978-3-902734-33-4, ISSN 1726-9679, Vienna, Austria
DOI: 10.2507/32nd.daaam.proceedings.086

Abstract

Among the advantages of introducing digital twins on production systems, there is the ability to identify their eventual critical state and to enable predictive maintenance policies. The failure of a manufacturing process, in general, can be anticipated in phase of simulation, if tied up to wrong settings, or in phase of operation, if tied up to environmental variables. In both cases, knowing the conditions that could cause the failure allows to intervene in a timely and effective manner. Here a method is proposed to explore the system operating parameters in a systematic way: the system is able to process signals collected in real time by machine's sensors and to reproduce both the trajectories of the moving parts and the material deposition process. This also makes possible to predict manufacturing tolerances that will be obtained. On a FDM Cartesian 3D printer a self-calibration procedure is used to find the maximum torque that can be delivered by the drives at different speeds in an automatic and repeatable way to find the maximum speed and acceleration at which the machine can operate safely. Additional accelerometers were installed on the machine to validate the adopted procedure: tests results demonstrate the effectiveness of the system.

Keywords: self-calibration; digital twin; additive manufacturing; fault prediction

1. Introduction

In modern manufacturing plants, the use of digital twins offers significant advantages. A digital twin is a multiphysics digital simulation directly connected to its real counterpart, which can be a plant, a machine, or a product, through sensors, interfaces, and electronic control systems [1], [2]. The advantages offered using a digital twin can be of various nature [3]:

- Simplified monitoring of working parameters and automatic management of data flows.
- Real-time or offline optimization of work settings.
- Prediction and prevention of machining errors.

One of the fields where digital twins have proven to be particularly useful is in manufacturing machining, particularly additive manufacturing. In fact, since additive manufacturing is a relatively recent technology based on complex physical phenomena, it can benefit greatly from the ability to collect and organize machining data in a rational and automatic way.

Moreover, it must be considered that additive manufacturing highest costs are mainly related to raw material and machining time, so the optimization of production process becomes even more important [4]. Over time, there have been several proposals for digital twins for additive manufacturing, most often referring to metal 3D printers [5], [6], [7], [8], [9], as the cost of raw material better justifies the expense of optimizing the process. Many works were also proposed on FDM 3D printers as they are very popular and easily available [10], [11], [12], [13], [14], [15], [16].

One of the problems encountered in the realization of digital twins is the correct characterization of the physical system, in fact the digital model on which the operation of the system is based must be calibrated to correctly perform the procedures of optimization and error prediction. The need to overcome this difficulty is at the basis of the rationale for the study presented here.

With reference to the prediction of errors, it was decided to carry out a series of tests that would force a prejudicial behaviour for the processing in progress, without damaging the machine, in order to understand the operating limits of the system and, using the physical model of the digital twin, characterize some relevant aspects of the model itself. In practice, by varying the input data to the real machine, the dual purpose of identifying the precise value of the harmful input and characterizing the physical model in a more appropriate way is achieved.

2. Material and Methods

The concept exposed above is applied to the error prediction function of a digital twin of an FDM 3D printer. In laboratory Cartesian FDM printers, two of the most common reasons for failure of a machining process are layer mismatch or extruder cogging. Both problems can be traced back to the stepper motors that misses steps while driving the corresponding axes and they can be greatly reduced by carefully choosing the printing settings (maximum speeds and accelerations). The problem is that choosing conservative kinematic limits can also greatly increase the machining time, with consequent increase in costs.

The correct calibration of the kinematic parameters can be carried out by the digital twin, thanks to its physical model of the machine, but it is necessary to know both the pull-out curve of the stepper motors and the dynamic behaviour of the system, including non-linear effects. The torque curve of stepper motors is usually provided by the manufacturer, but the test conditions are almost never equivalent to the working conditions (different drivers, micro-steps settings, power supply). It is possible to obtain the curve using a test bench (that is not always available), or through analytical calculations [17] or software simulations [18], [19], solutions that require in any case the knowledge of the internal characteristics of the motor.

To overcome this problem, an algorithm has been devised that generates a series of trajectories capable of stressing the motors with increasing speed and loads. Using the sensors installed on the machine, it is possible to monitor the response of the system and in this way not only detecting critical accelerations and speeds under load conditions, but also deducing the pull-out curve of the motors.

2.1. 3D printer and sensors

The FDM 3D printer used as a test is a Cartesian type, built specifically for printing large parts (Fig. 1). The printer features an extruder block that is movable in both X and Y directions and houses two extruders with motors. The motion of the extruder block is provided by two NEMA 17 42BYGHW811 stepper motors for the X-axis and a motor of the same type for the Y-axis. Both actuators operate via toothed belt drive with a 40 mm circumference pulley. The X-axis motors are connected in parallel to the same MKS TB6600 driver powered at 12V and set to a maximum current of 3A, the Y-axis motor is connected to a driver of the same type but set to 1.5A. The printing table is mobile along the Z axis in vertical direction, has a usable area of 626 X 355 mm, and can be heated to improve adhesion.

The extruders are direct drive type, suitable for receiving a 1.75 mm filament, whose feeding is obtained through a pinion with 40 mm circumference. Brass nozzles have an inner diameter of 0.4 mm. The extruders motor is a NEMA 17 42BYGHW811B dual shaft and is powered by an A4988 driver also at 12V. All motors are controlled with 1/16 pitch micro-steps. Details about operation and sensors of Z-axis are not provided because in this type of machine its calibration has almost no impact on the printing speed. The printer controller is an Arduino Mega board with RAMPS 1.4 expansion board, on the microcontroller is installed the firmware Marlin v 1.1.9 [20].

To provide the digital twin with feed-back data, the system has been equipped with various types of sensors, the most important of these are optical encoders that measure axis position. X and Y axes are monitored by two ENS1J-B28 L00256L encoders placed on the idler pulley, the system has a resolution of 20.477 pulse/mm referred to the extruder feed. For testing purposes, only one of the extruders was kept active, the feedback is given by an encoder of the same type mounted on the rear shaft of the E-axis motor, the resolution obtained is 25.6 pulses/mm referred to the filament feed. Other sensors have also been installed, although they are not relevant in the axis calibration procedure, among them an encoder for the Z-axis, two additional thermistors and a computer-vision tracking system using a video camera and a dedicated processor.

The printer is also equipped with 6 EPOSMote III smart sensors [21] to collect data from the other sensors and to send them to the interface of the digital twin. In addition, they provide an LSM330 integrated accelerometer. The data coming from the accelerometer of the sensor placed on the extruder block will be used as reference for the tests performed on X and Y axes.

The digital twin core operates on a workstation placed next to the printer and receives the data from the smart sensors via USB serial connection.

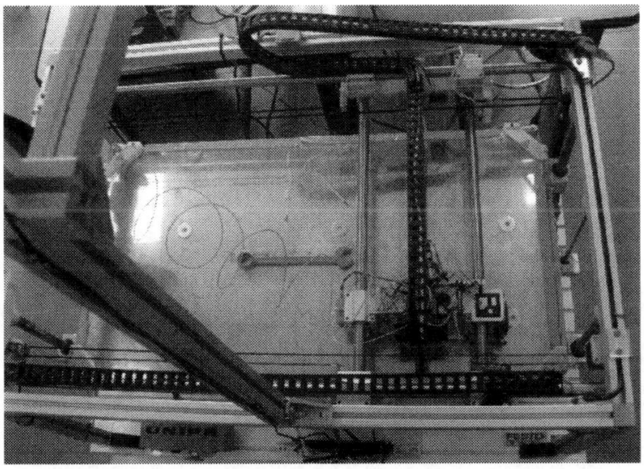

Fig. 1. Printer used for test

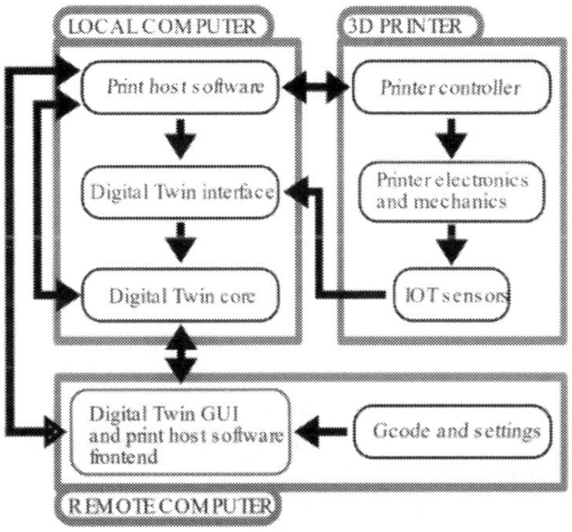

Fig. 2. Digital twin scheme

2.2. The digital twin

The digital twin is a complex system that contains various intercommunicating parts (Fig. 2). The main parts are the print host software, the interface, the core, and the GUI:

- The digital twin interface is a python script that operates locally on a computer near the printer. It is responsible for receiving data from the smart sensors and the digital vision system and sending it to the digital twin core. It is also able to receive from the print host software the information related to the printing in progress. It additionally performs the task of synchronizing the timestamps at system start-up and sending the collected data to a cloud repository from where it can be accessed or used for other projects.
- The print host software is not technically part of the digital twin but is used by it to act on processes in progress. This program performs the task of receiving commands from user (even remotely) and forward them to the printer controller via serial connection. In this system, this program no longer interacts directly with the user but receives commands and sends information both to the core and to the GUI of the digital twin. A version of Octoprint [22] was used as print host software since it offers a practical plugin system to make changes and an API to manage the printer remotely without a graphical interface.
- The digital twin core is an application programmed in python and C++ that operates on a workstation located near the printer running the model of the physical system. It performs various functions including, error detection, calculation of significant parameters for printing, offline simulation of printing processes, calculation of wear and tear of various components, real-time building of the virtual model of the printed part.
- The GUI is the part of the program dedicated to interaction with the end user, it can also run on a remote device and be used for remote control. The GUI has the basic function of allowing user to launch prints, keep the current process monitored and view the results of old jobs. It can also perform more advanced task such as allowing to initiate wizards to recover from an error situation and performing post processing of the data collected during printing to identify deviations between the model of the printed part and its CAD representation.

2.3. The calibration procedure

One of the main reasons that can lead to 3D printing failure is when one of the stepper motors misses one or more steps. This event occurs when the torque exerted by the motor at the current speed is insufficient to support the load, in such case the motor shaft cannot move forward by the commanded pitch and consequently cannot maintain the motion profile set.

To determine the conditions that cause this damaging event, a series of trajectories have been designed to gradually increase the speed and load required for the axes whose operation is more critical. During the execution of the movements, the digital twin of the machine is kept active so that, by acquiring displacements and accelerations, it is able to determine the critical values and, through its physical model, to give an estimate of the maximum loads the motors can bear and of their pull-out curve.

This operation has not been conceived to be performed normally during the operation of the printer and the digital twin but only during its set-up or after a relevant hardware modification (change of moving masses, replacement of the motor, driver, or nozzle). For this reason, the calibration procedure cannot be managed remotely from the GUI but must be launched directly at the start-up of the digital twin core of which it is an integral part.

The calibration procedure operates on the most significant axes for printing purposes, namely the X, Y, and E axes. The Z axis has been excluded because on machines of this type it plays a very marginal role, limiting itself to moving a few tenths of a millimetres at the end of each layer. The calibration procedure uses the connection of the digital twin core with the print host program to send to the printer controller the g-code and m-code lines containing the information necessary to execute the desired profile (point of arrival, maximum speed, and acceleration), then the feedback of the encoders is used to evaluate the outcome of the test. The printer controller was set to perform trapezoidal speed profiles with constant acceleration transients.

2.3.1. The procedure for X and Y axes

Based on the structure of the machine, it can be said that the main load acting on the X and Y axes is dynamic, due to the high mass of the extruder block. For this reason, the trajectories that these axes must follow will be characterized by increasing acceleration values, so as to gradually increase the load. At the same time the maximum speed is also increased since it is known that at high speeds the torque that can be delivered by the motors decreases. The trajectories for these axes have been generated to have trapezoidal speed profiles and satisfying the following criteria:

- The maximum usable space depends on the length of the axis, but a safety margin must always be maintained between the end of the movement and the limit switch so that a failure in the braking phase does not cause damage.
- The acceleration and deceleration phase cannot be longer than 8 s, as this causes problems to the printer controller firmware (overflow of a timer causes irregular movements).
- The acceleration must be maintained for a sufficient distance and overcome the backlash and elasticity of the kinematic chain; in this way it can be sure that the motor is subject to the desired load for a certain number of steps.
- The maximum speed allowed for a test at a given acceleration must be in any case lower than the one that caused a failure at a lower acceleration.
- After the acceleration ramp the axis maintains the set speed for a certain time. A maximum value relative to this time is set to reduce the duration of the test.

To satisfy these conditions for each acceleration a_i set, a series of maximum speeds must be tested between v_{inf} and v_{sup} calculated in this way:

$$v_{inf} = \sqrt{2 \cdot s_{acc\,min} \cdot a_i} \tag{1}$$

$$v_{sup} = \min\left(v_{max},\ a_i \cdot t_{max},\ a_i \sqrt{\frac{s}{a_i}} \right) \tag{2}$$

Where v_{max} is the lowest speed at which a failure was detected, a_i is the acceleration of the profile, t_{max} is the maximum possible duration for the acceleration and deceleration phase, s is the space available to perform the test, $s_{acc\,min}$ is the minimum space for which the acceleration phase must extend. For each acceleration a_i it is then possible to execute n tests at different speeds dividing the interval between v_{inf} and v_{sup} in steps of about 80 mm/s. Of course, if $v_{sup} < v_{inf}$ it is not possible to execute any test for the corresponding acceleration and a new acceleration is calculated using (3).

$$a_{new} = \frac{{v_{sup}}^2}{s_{acc\,min}} \tag{3}$$

Once the trajectory for each test is defined, it is possible to calculate the time needed to execute it. Given a certain trajectory the test is considered passed if the assigned law of motion is executed in the required time both in the outward and in the return phase (i.e., the same trajectory is executed twice, forward and backward, so that at the end of the test the axes are in the initial configuration). In practice the law of motion is considered executed if the feedback position signal is close enough to the target position for enough time. More precisely, if the difference between the actual position measured by the encoders and the target position remains less than a threshold value (settle window) for at least a certain time (settle time). If this condition does not occur within a time equal to 1.5 times the time theoretically required to make the movement, it is considered that the axis was not be able to reach the target or has exceeded it by failing to decelerate.

If a test is passed at a speed v_j it is repeated at speed v_{j+1} until the v_{sup} for the current acceleration is reached. At this point a new set of trial is started at a higher acceleration (twice the previous one). If the test is not passed, two conditions can occur, if j = 0 (i.e., $v_j = v_{inf}$) the motor has failed the movement at the minimum possible speed, therefore the current acceleration is considered the maximum bearable by the motor (corresponding to the holding torque) and the test for the current axis is concluded. If, instead, the test fails for $v_{j+1} > v_{inf}$, then v_j is set as the limit speed for the current acceleration, the variable v_{max} is updated and a new phase with higher acceleration is performed. In this way in any case the tests performed at higher acceleration cannot exceed the speed that caused the failure of tests with lower accelerations. This algorithm is shown in Fig 3. After a series of tests the parameters of the algorithm were fixed according to Table 1.

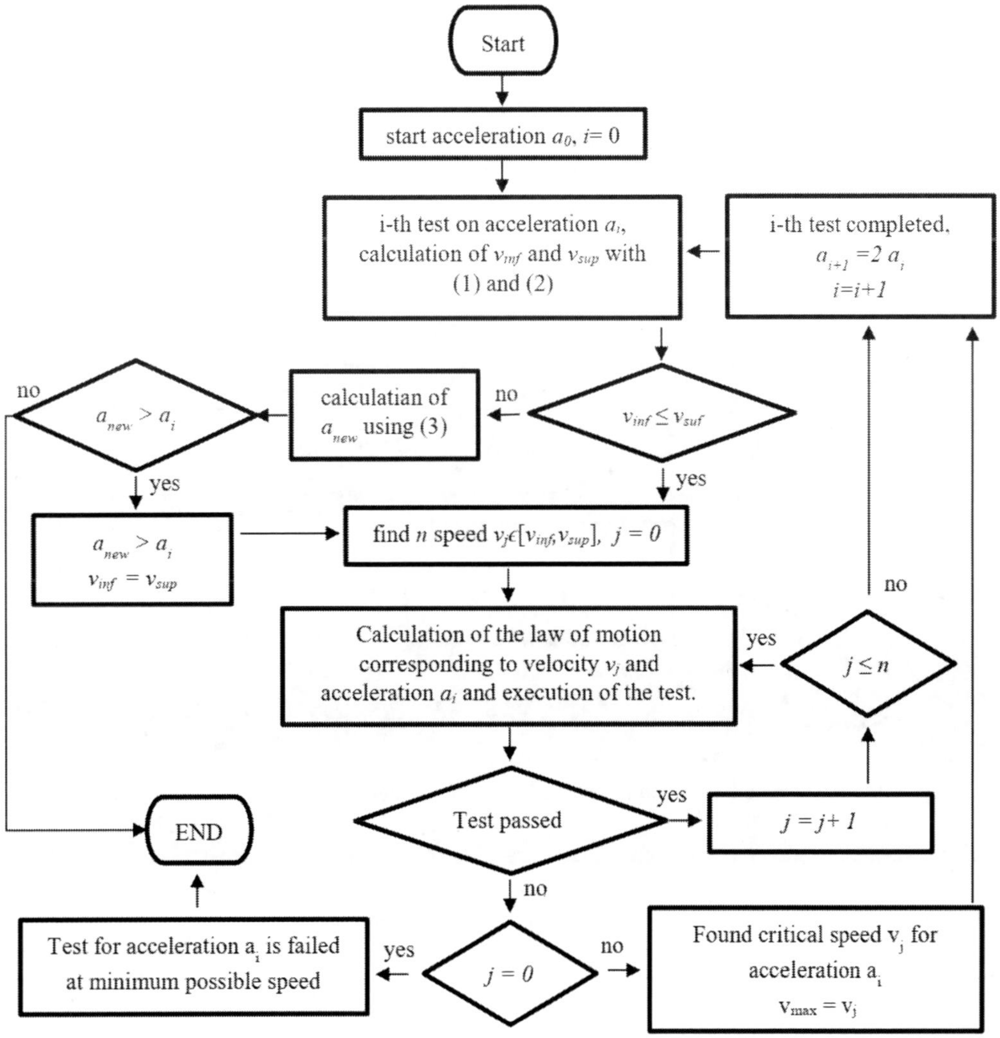

Fig. 3. Flowchart diagram explaining algorithm for calibration of X and Y axes. Every failed test is recorded.

2.3.2. The procedure for E-axis

Regarding the E-axis, the main load to which the motor is subjected is viscous and therefore dependent on extrusion speed rather than acceleration. This test is very useful since the maximum printing speed greatly affects the processing speed. To perform this test, the extruder is first brought to the most suitable working temperature for PLA to achieve dimensional accuracy (200 °C) [23], then a given amount of material is extruded at constant speed. If the extrusion succeed is retried with a greater speed (every round the speed is increased by a k factor).

When, by means of the encoder mounted on the extruder shaft, an occlusion is detected, the test is repeated with halved acceleration but maintaining the speed that caused the failure so as to exclude any influence of the acceleration. This is repeated until the acceleration fall below threshold value a_{min}. The algorithm, shown in Fig 4, is simpler than that for the X and Y axes. Table 2 shows the data used in the testing phase. For the test to be valid it is necessary to set an amount of extruded filament s such that the compressibility of the material is overcome. To ensure the reliability and repeatability of the results is necessary to tighten the screw that adjusts the force with which the pinion presses on the filament to prevent its slippage.

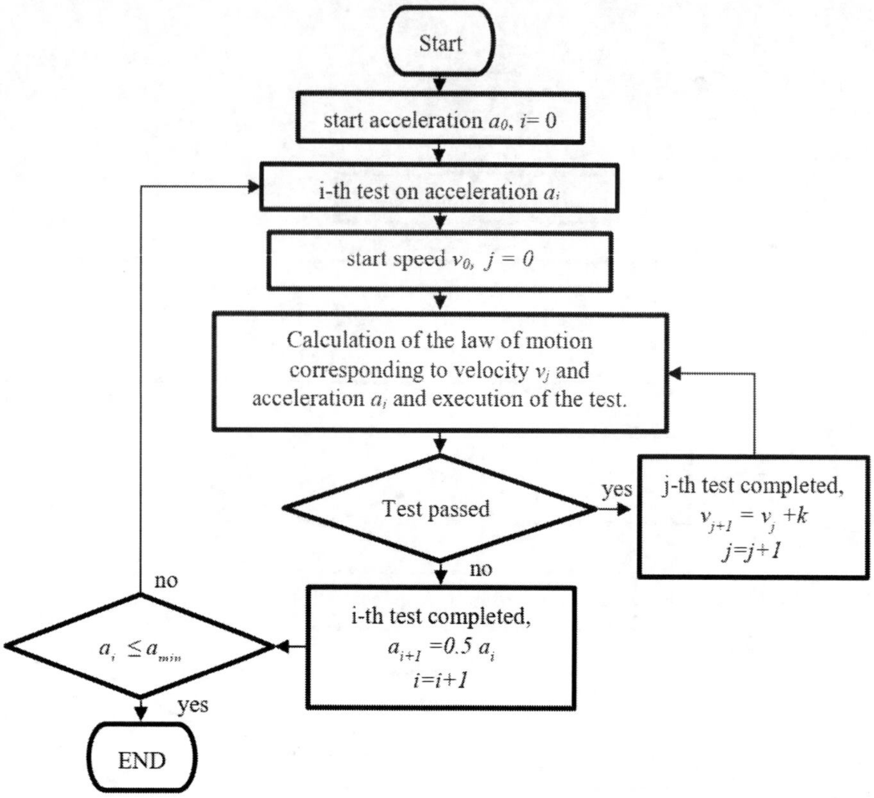

Fig. 4. Flowchart diagram explaining algorithm for calibration E-axis. Every failed test is recorded.

Parameter	X-axis	Y-axis
s	500 mm	300 mm
t_{max}	8 s	8 s
$s_{acc\ min}$	5 mm	5 mm
a_0	16 mm/s^2	16 mm/s^2
$t_{max\ const\ speed}$	1 s	1 s
Settle window	1 mm	1 mm
Settle time	0.1 s	0.1 s

Table 1. Parameters for tests on X and Y axes

Parameter	E-axis
a_0	10000 mm/s^2
v_0	2.090 mm/s
s	40 mm
a_{min}	1250 mm/s^2
k	1.045 mm/s
Settle window	0.5 mm
Settle time	0.1 s

Table 2. Parameters for tests on E axes

3. Results and discussion

Tests performed and the results obtained for the x and y axes are reported:

Acceleration [mm/s2]	max speed [mm/s]						
16	12.649	89.443					
32	17.889	72.190	126.491				
64	25.298	102.092	178.885				
128	35.777	108.179	180.581	252.982			
256	50.596	127.390	204.184	280.977	357.771		
512	71.554	143.956	216.358	288.759	361.161	433.563	**505.964**
1024	101.193	168.655	236.117	303.579	371.041	**438.503**	
2048	143.108	216.957	290.805	364.654	438.503		
4096	**202.386**						

Table 3. Test performed on X-axis, in bold are reported the failed tests.

Acceleration [mm/s2]	max speed [mm/s]						
16	12.649	69.282					
32	17.889	57.934	97.980				
64	25.298	81.931	138.564				
128	35.777	89.171	142.565	195.959			
256	50.596	126.107	201.618	277.128			
512	71.554	135.627	199.700	263.773	327.846	391.918	
1024	101.193	176.703	252.214	327.725	403.235	**478.746**	
2048	143.108	210.236	277.363	344.491	411.618	**478.746**	
4096	202.386	271.476	340.566	409.656	**478.746**		
8192	286.217	350.393	**414.569**				
16384	**404.772**						

Table 4. Test performed on Y-axis, in bold are reported the failed tests.

For reasons of space, it is not possible to report the data collected during each test, therefore it is reported only an example of successful test and one of failed test for the X axis. Figure 5 (a) shows the data taken by the encoder during the test at acceleration of 512 mm/s^2 and maximum speed of 361.161 mm/s. The speed data were obtained by deriving the position values and using a moving window filter. The acceleration data were obtained by deriving and filtering the speed data (hence also they are derived from the encoders), but, for validation purposes, accelerometers were mounted on machine and their measured values show a good agreement with the calculated ones. So, it is demonstrated that future applications could rely on the use of encoders alone. From the graphs it can be seen that the assigned joint trajectory is respected but executed with a certain delay (about 0.2 s), due to the fact that when the motion command is sent (and the timer starts to run) the digital twin core must send the instructions to the host software via TCP/IP connection. This in turn forwards the message via serial connection to the printer controller that reads and processes it. Further delays are given by the time needed to activate the drivers and magnetize the motor windings and by the effect of mechanical backlashes and belt elasticity.

Figure (b) shows the same data but related to a test in which the motor misses some steps (acceleration of 512 mm/s2 and speed of 505.964 mm/s). In the graph it can be clearly seen that at the instant 1.08 s, almost at the end of the acceleration phase, the movement suddenly stops in an uncontrolled way and then resumes later at a lower speed in the return phase. The abrupt stop and subsequent sudden restart create a considerable impulsive force that affects the structure of the printer making it oscillates on its support (this oscillation is clearly visible in the accelerometer trace).

Result of E-axis test are also reported:

Acceleration [mm/s2]	max speed [mm/s]		
10000	2.090	3.135	**4.180**
5000	**4.180**		
2500	**4.180**		

Table 5. Test performed on E-axis, in bold are reported the failed tests.

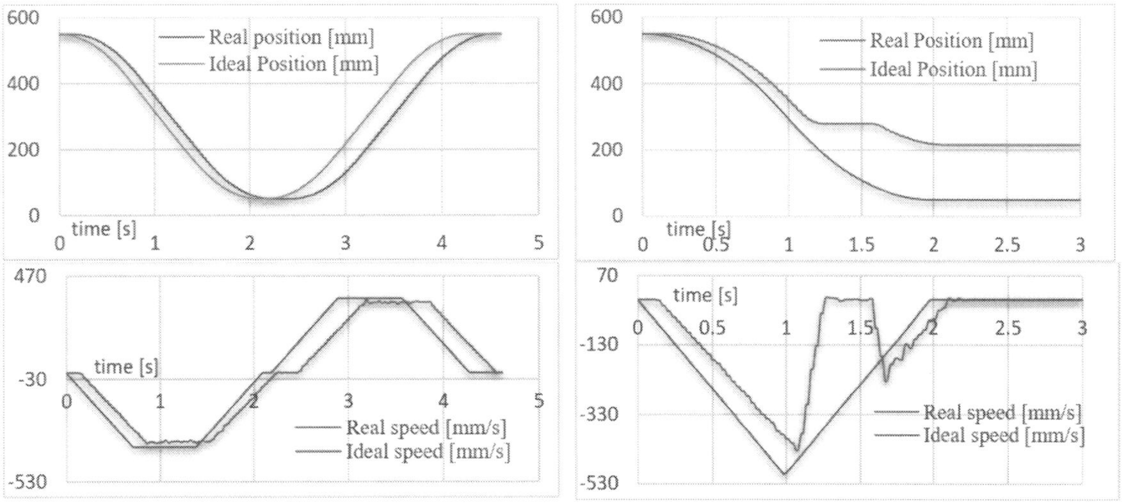

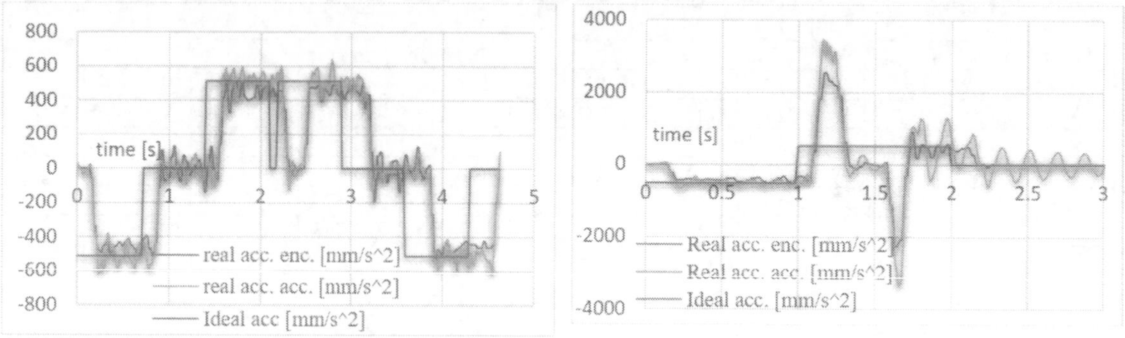

(a) (b)

Fig. 5. Figure (a) represents position, speed, and acceleration of a successful test (X-axis, acc. = 512 mm/s^2, max speed = 361.161 mm/s); Figure (b) represents the same measure for a failed test (X-axis, acc. = 512 mm/s^2, max speed = 505.964 mm/s)

In Fig 6 can be seen the E-axis motor missing steps at least 4 times, at each time is visible a backflow of material pushing back the pinion, a phenomenon due to the elasticity of the molten polymer and the nozzle pipe.

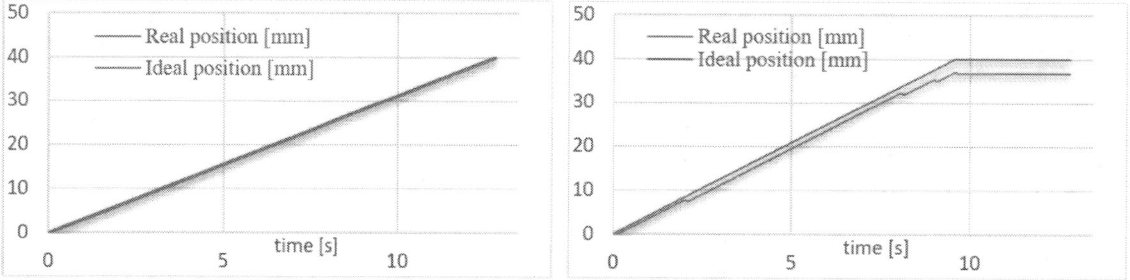

(a) (b)

Fig. 6. Figure (a) represents the position of a successful test (E-axis, acc. = 10000 mm/s^2, max speed = 3.135 mm/s); Figure (b) represents the same measure for a failed test (E-axis, acc. = 10000 mm/s^2, max speed = 4.180 mm/s)

Once the tests are finished it is possible to order the values where failure occurred at increasing speed, for this purpose the point with the highest acceleration is considered at zero speed as it is the point that best approximates the static torque of the motor. Considering that the mass of the X-axis moving carriage is 10.82 kg while that of the Y-axis is 2.43 kg, neglecting all non-inertial loads the pull-out curves for the analysed motors can be obtained. They are shown in Fig. 7 and compared with the factory curve of a motor of similar size. It can be seen that the curve obtained is always lower, although of the same order of magnitude, this may be due to differences in the power circuit and to the fact that the implemented physical model does not take into account the friction forces and therefore underestimates the actual load acting on the motor. It can also be seen that the motors that move the X-axis have much lower performance than those of the Y-axis, this may be due to the fact that X-axis is driven by two motors operating in parallel, each with its own belt, and a certain load unbalancing could occur over time.

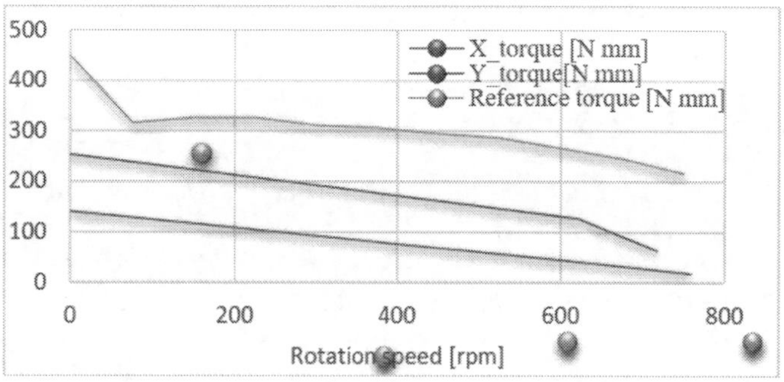

Fig. 7. X-axis and Y-axis motor pull-out curves compared to that provided by the manufacturer for a similar motor.

Axis	acc. [mm/s^2]	speed limit [mm/s]
X	2048	292.33
Y	8192	414.57
E	5000	4.180

Table 6. Proposed settings for fast and safe machine movement

As far as the load on the E axis is concerned, the speeds reached are not sufficient to draw a pull-out curve and therefore only the holding torque can be computed. The load is calculated from the physical model of the digital twin which models the nozzle as a capillary ending in an orifice. The pressure loss due to friction is calculated with the generalized Reynolds number for non-Newtonian fluids [24] in laminar regime using the properties of molten PLA [25], resulting in a holding torque of 268.94 N mm which is in line with what has been observed so far. Once the characterization of the motors has been obtained, it can be used to generate optimized working settings. The maximum acceleration that can be set for each axis will be half of the one corresponding to the maximum torque (to keep a safety margin) while the maximum speed will be that one corresponding to the torque found. The results for the series of tests just completed are shown in table 6.

4. Conclusion

An algorithm to calibrate the error prediction function of a 3D printer has been developed, implemented through a digital twin and tested in lab. Its use allows to optimize the settings of the machine and, at the same time, to characterize its operational behaviour. The procedure was applied to the prediction of layer mismatch and under-extrusion errors of a Cartesian-type FDM 3D printer. The calibration routine was defined to be carried out in a fully automatic way and provided as results the printing limit settings and the characterization of the motors under investigation. A solution of this type has the potential to be extended to different types of machinery and errors as long as they are repeatable and do not endanger the integrity of the machine or the safety of the operator. In the future is proposed to repeat the test using a model that takes into account friction and elastic parts of the machine in order to verify how a more accurate model can improve the results.

5. References

[1] Tao, F.; Zhang, H.; Liu, A. & Nee, A. Y. C. (2019). Digital Twin in Industry: State-of-the-Art, IEEE Transactions on Industrial Informatics, Vol 15, No. 4, pp. 2405–2415., ISSN: 1551-3203, DOI: 10.1109/TII.2018.2873186.

[2] Sbaglia, L.; Giberti, H. & Silvestri, M.(2019) 'The Cyber-Physical Systems within the industry 4.0 framework', Advances in Italian Mechanism Science – Mechanisms and Machine ScienceVolume 68, 2019, pp. 415-423, Springer Netherlands, DOI: 10.1007/978-3-030-03320-0_45

[3] Kostenko, D.; Kudryashov, N.; Maystrishin, M; Onufriev, V.; Potekhin, V. & Vasiliev, A. (2018). Digital Twin Applications: Diagnostics, Optimisation and Prediction, Proceedings of the 29th DAAAM International Symposium, pp.0574-0581, B. Katalinic (Ed.), Published by DAAAM International, ISBN 978-3-902734-20-4, ISSN 1726-9679, Vienna, Austria DOI: 10.2507/29th.daaam.proceedings.083

[4] Frazier, W. E. (2014). Metal Additive Manufacturing: A Review., Journal of Materials Engineering and Performance, Vol. 23, No. 6, pp. 1917–1928. doi: 10.1007/s11665-014-0958-z.

[5] Knapp, G. L.; Mukherjee, T.; Zuback, J. S.; Wei, H. L.; Palmer, T. A.; De, A. & DebRoy, T. (2017). Building blocks for a digital twin of additive manufacturing. Acta Materialia, Vol. 135, pp. 390–399. https://doi.org/10.1016/j.actamat.2017.06.039

[6] Sousa Agostino, Í. R., Morosini Frazzon, E.; Medeiros Fröhlich, A. A.; Silvestri, M.; Bornia, A. C.; Spengler, A.; Martina, J.; Fettermann, D. & Tortorella, G. (2020). Perspectives for IoT-based integration of distributed and automated manufacturing lines for mass customization, Proceedings on 25th International Joint Conference on Industrial Engineering and Operations Management – IJCIEOM, Springer International Publishing, 2020, pp. 31-41, DOI: 10.1007/978-3-030-43616-2_4

[7] DebRoy, T.; Zhang, W.; Turner, J. & Babu, S. S. (2017). Building digital twins of 3D printing machines. Scripta Materialia, Vol. 135, 119–124. https://doi.org/10.1016/j.scriptamat.2016.12.005

[8] Mandolla, C.; Petruzzelli, A. M.; Percoco, G. & Urbinati, A. (2019). Building a digital twin for additive manufacturing through the exploitation of blockchain: A case analysis of the aircraft industry. Computers in Industry, Vol. 109, pp. 134–152. https://doi.org/10.1016/j.compind.2019.04.011

[9] Liu, C.; Le Roux, L.; Körner, C.; Tabaste, O.; Lacan, F. & Bigot, S. (2020). Digital Twin-enabled Collaborative Data Management for Metal Additive Manufacturing Systems. Journal of Manufacturing Systems. https://doi.org/10.1016/j.jmsy.2020.05.010

[10] Balta, E. C.; Tilbury, D. M. & Barton, K. (2019). A Digital Twin Framework for Performance Monitoring and Anomaly Detection in Fused Deposition Modeling. 2019 IEEE 15th International Conference on Automation Science and Engineering (CASE), pp. 823–829. https://doi.org/10.1109/COASE.2019.8843166

[11] Hu, L.; Nguyen, N. T.; Tao, W.; Leu, M. C.; Liu, X. F.; Shahriar, M. R. & Al Sunny, S. M. N. (2018). Modeling of Cloud-Based Digital Twins for Smart Manufacturing with MT Connect. Procedia Manufacturing, Vol. 26, pp. 1193–1203. https://doi.org/10.1016/j.promfg.2018.07.155

[12] Chhetri, S. R.; Faezi, S.; Canedo, A. & Faruque, M. A. Al. (2019). QUILT: Quality inference from living digital twins in IoT-enabled manufacturing systems. IoTDI 2019 - Proceedings of the 2019 Internet of Things Design and Implementation, pp. 237–248. https://doi.org/10.1145/3302505.3310085

[13] Avventuroso, G.; Silvestri, M. & Morosini Frazzon, E. (2018). Additive Manufacturing Plant for Large Scale Production of Medical Devices: A Simulation Study, IFAC-PapersOnLine, Vol. 51, No. 11, June 2018, pp. 1442-1447. ISSN: 2405-8963 https://doi.org/10.1016/j.ifacol.2018.08.312

[14] Cai, Y.; Wang, Y. & Burnett, M. (2020). Using augmented reality to build digital twin for reconfigurable additive manufacturing system', Journal of Manufacturing Systems, Vol. 56, pp. 598–604. doi: 10.1016/j.jmsy.2020.04.005.

[15] Moretti, M.; Rossi, A. & Senin, N. (2021). In-process monitoring of part geometry in fused filament fabrication using computer vision and digital twins, Additive Manufacturing, Vol. 37, p. 101609. doi: 10.1016/j.addma.2020.101609.

[16] Avventuroso, G.; Foresti, R.; Silvestri, M. & Morosini Frazzon, E. (2017). Production paradigms for additive manufacturing systems: A simulation-based analysis, 2017 International Conference on Engineering, Technology and Innovation (ICE/ITMC), Madeira Island, Portugal, 2017, pp. 973-981. doi: 10.1109/ICE.2017.8279987

[17] Russell, A. P. & Pickup, I. E. D. (1982). Calculation of the pull-out torque of hybrid stepping motors in the half-step mode, Electric Machines and Power Systems, Vol. 7, No. 2, pp. 103–114. doi: 10.1080/03616968208955440.

[18] Mariolo, L.; Rubino, A.; Spelta, D. & Frosini, L. (2021). Modeling of hybrid stepper motor finalized to the optimization of the holding torque. Proceedings - 2021 IEEE Workshop on Electrical Machines Design, Control and Diagnosis, WEMDCD 2021, 107–112. https://doi.org/10.1109/WEMDCD51469.2021.9425647

[19] Bjekić, M.; Milovanović, A. & Koprivica, B. (2015). Prediction of pull-in and pull-out torque characteristics of the permanent magnet step motor, Revue Roumaine des Sciences Techniques Serie Electrotechnique et Energetique, Vol. 60, No. 1, pp. 29–38.

[20] https://marlinfw.org/, (2021), Home | Marlin Firmware (2021)., Accessed on: 2021-04-19

[21] Scheffel, R. M.; Fröhlich, A. A. & Silvestri, M. (2021). Automated fault detection for additive manufacturing using vibration sensors. International Journal of Computer Integrated Manufacturing, Vol. 34, No. 5, pp. 500–514. https://doi.org/10.1080/0951192X.2021.1901316

[22] https://octoprint.org/, (2021), OctoPrint.org, Accessed on: 2021-04-19

[23] Cekic, A.; Muhamedagic, K.; Begic-Hajdarevic, D. & Djuzo, N. (2020). Effect of Process Parameters on Dimensional Accuracy and Tensile Strength of FDM Printed Parts, Proceedings of the 31st DAAAM International Symposium, pp.0066-0071, B. Katalinic (Ed.), Published by DAAAM International, ISBN 978-3-902734-29-7, ISSN 1726-9679, Vienna, Austria DOI: 10.2507/31st.daaam.proceedings.009

[24] Madlener, K.; Frey, B. & Ciezki, H. K. (2009). Generalized reynolds number for non-newtonian fluids, Progress in Propulsion Physics, 1, pp. 237–250. doi: 10.1051/EUCASS/200901237.

[25] Fang, Q. & Hanna, M. A. (1999). Rheological properties of amorphous and semicrystalline polylactic acid polymers, Industrial Crops and Products, Vol. 10, No. 1, pp. 47–53. doi: 10.1016/S0926-6690(99)00009-6.

DOI: 10.2507/32nd.daaam.proceedings.087

AN INDUSTRY 4.0 TOOL TO ENHANCE HUMAN-ROBOT COLLABORATION

Athanasios Sidiropoulos, Vasileios Sidiropoulos,
Dimitrios Bechtsis & Dimitrios Vlachos

This Publication has to be referred as: Sidiropoulos, A[thanasios]; Sidiropoulos, V[asileios]; Bechtsis, D[imitrios] & Vlachos, D[imitrios] (2021). An Industry 4.0 Tool to Enhance Human-Robot Collaboration, Proceedings of the 32nd DAAAM International Symposium, pp.0617-0622, B. Katalinic (Ed.), Published by DAAAM International, ISBN 978-3-902734-33-4, ISSN 1726-9679, Vienna, Austria
DOI: 10.2507/32nd.daaam.proceedings.087

Abstract

The Industry 4.0 is strongly connected with the use of robotic systems and specifically the use of autonomous ground vehicles. A critical issue is the manipulation of the information flows in the industrial facilities and the integration of the robotic systems in the production line. Human Machine Interaction interfaces enable production line employees to monitor in almost real-time the information streams and handle robotic systems and robotic vehicles, but the use of handheld devices is rather limited. In most of the cases the robotic systems are controlled from a central system, and this prohibits the flexibility of the overall system and the re-initialization of the system when an error occurs. The proposed research work focuses on robotic vehicles that are working within the industrial facility and presents a user-friendly Industry 4.0 software tool for transparently monitoring the information streams in the production line and controlling the vehicle's behaviour using mobile devices (e.g., mobile phones and tablets). The developed software tool enables the transparent communication and interaction among the industrial machinery, the robotic systems, and the employees, supports the execution of everyday activities by providing information about the vehicles and assesses the operation of the mobile vehicles.

Keywords: ReactJS; web-based ROS tool; human-robot collaboration; Robot Operating System; visualization.

1. Introduction

In the era of Industry 4.0 and Industrial Internet of Things (IIoT) the Autonomous Mobile Robots (AMRs) are used in a 24/7 perspective and the human-robot collaboration is a critical issue at industrial facilities. On a recent research it was stated that the drivers of the fourth industrial revolution are automation, big data, cloud computing, Internet of Things (IoT), mass customization and production time improvement [1]. Moreover, as stated by [2] the intelligent autonomous vehicles constitute a radical innovation that could assist in the efficient management of production lines. Information technology tools are extensively used to tackle the above-mentioned issues and transform the industry to a more data-driven approach. Health and safety protocols in industrial facilities and real time location of the robotic vehicles must be managed while the complexity is increasing as new robotic systems are introduced in a daily basis.

Some researchers developed a safety Industry 4.0 environment as they created a human-centric robotic system navigation in order to ensure the workers' safety within the industrial facilities [3]. The research interest is also focusing on the collaborative tasks that need a human – machine interface for the communication with the industrial robots in the manufacturing sector and thus an online monitoring system is proposed to enhance the collaboration between them, while ensures the security of the executed data transactions [4].

The employees and the autonomous vehicles are moving at the facilities at the 24/7 perspective, and a transparent collaboration framework is necessary for executing common industrial activities. The AMRs are split into 3 categories based on their skills: i) autonomous sensing (information acquisition and data transformation) making observations and refining information, ii) autonomous planning (information interpretation and decision selection) reacting to information or deciding actions and schedule, and iii) autonomous acting (action implementation) executing a planned task or producing reflexive reactions [5]. The autonomous sensing and autonomous planning AMRs enable vehicles to execute tasks in a fully automated manner based on a predefined schedule while the autonomous acting AMRs execute user driven tasks. Robotic vehicles are used at industrial facilities for executing a variety of activities according to their skills. For example, on an assembly procedure the autonomous sensing and autonomous planning AMRs are preferred while in semi products transportation the autonomous acting AMR is the ideal type.

Moreover, in modern industrial facilities and warehouses, picking activities should be executed automatically in order to optimize space and resource utilization and minimize the delivery times. Thus, software tools based on Robot Operating System (ROS) for time and distance efficient AMR routing across multiple points in industrial facilities have been proposed [6]. Furthermore, there are several tools for the visualization of ROS activities while, RViz is the dominant built-in graphical tool [7] for visualizing data-streams from the sensors of the AMR. Indicatively, RViz presents data obtained from cameras, laser range finders, inertial measurement units, and many more [8]. However, RViz needs direct access to the AMR while it cannot work on systems without a preconfigured ROS environment.

The proposed solution is developed on top of the ROS platforms and TurtleBot2 is used for validating the developed tool in real-world scenarios. ROS is an open-source meta-operating system with a collection of sophisticated tools that can be used for enabling the scheduling and planning activities of mobile robots at industrial environments [6]. The proposed web-based software informs the end-user about: (i) the position of the AMR in the facility using the Occupancy Grid Map (OGM), (ii) the task that robot executes, while it enables the capabilities of (iii) on-line AMR programming and (iv) reinitialization of the robots' parameters, as the disorientation of the AMRs is a big issue.

The rest of the paper is structured in 4 sections. Section 2 analyses the research methodology and the background of the human-robot collaboration issues and solutions. Section 3 explains the proposed software tool that implemented for: (i) human-robot collaboration enhancement and (ii) robot on-line programming. Finally, the Section 4 ends up with conclusions and further research.

2. Literature Review

The Industry 4.0 is the main trend at manufacturing as the use of Information Systems, IoT, cloud, big data and artificial intelligence is rapidly increasing. A research concluded that IoT is an enabler for Industry 4.0 and smart manufacturing while it has privacy/security issues that need to be addressed in order to ensure the secure access of the data streams to the employees [9]. Moreover as stated by [10], enterprise information systems (EISs) will benefit from Industry 4.0. Specifically, new types of advanced manufacturing and industrial processes revolving around machine-to-human collaboration and symbiotic product development will emerge. In industries the main goal is to be more competitive by minimizing costs and maximizing their efficiency and effectiveness while for operational cost reduction they tend to use more robotic systems. Thus, information technologies can help industries by adding more robotic systems and AMRs in their production lines.

A common problem that AMRs are facing in a daily basis is the obstacle avoidance, while more robots are constantly added on the existing facilities and humans are also moving within the facilities, the space utilization of the industrial area is increasing. In this overloaded environment, human-centric robot navigation approaches have been published in order to enhance the safety of the employees in the facilities [11]. These approaches are based on Artificial Intelligence (AI) and Machine Learning (ML) and the safety cannot be fully ensured. Thus, a backup tool that allows the user to access the operation of the robot is mandatory. On the safety improvement the ISO 15066 and ISO 18646-2 standards provide guidance on human-robot collaboration and robot navigation inside the industrial facilities by dividing the industrial area into three pieces, namely: robot workspace, robot navigation space and robot collaborative space. The division of the facility area enhances the safety of the employees but adds limitations on human-robot collaboration. Additionally, as stated by [12], the localization of AMRs is a critical issue and the authors proposed cooperative localization using Robotic Swarms. Moreover, an online monitoring tool based on ROS and ReactJS has been developed and the authors concluded that it's an easy to deploy tool as it is supported by all the available web browsers [8]. Other researchers proposed a solution on optimal trajectory problem with affective algorithms where multi-robot (robot swarms) cooperation can be achieved [13].

In modern robotic systems, every established manufacturer has developed proprietary controllers and software tools where off-line programming tasks (OLP) could be executed. The term OLP refers to the technique in which the programming of the robotic tasks is defining before its activation. On the other side the On-line programming technique means the programming of the robotics tasks in real-time.

In case of AMRs where their task programming is affected by many factors the real-time human interference is obligatory. A research about assembly tasks coordination between humans and robotic systems has been conducted. They simulated a ROS based robot and they managed to enable some features of the on-line programming technique through a developed Graphical User Interface (GUI) as they were able to change the executed tasks in real-time [14]. As a result, when the collaboration between human and robots is the main objective, a tool for on-line programming of the robotic vehicles must be developed in addition to the existent OLP technique as the access to the AMR's operation is an issue that needs to be addressed.

All the above-mentioned limitations exist for quite long time and software tools like RViZ are used to facilitate these shortfalls. While the Web-based applications and services are growing rapidly a new approach for a web-based software tool is proposed and developed for the enhancement of the human-robot collaboration and human's safety within the industrial facilities.

3. Proposed Solution

This manuscript proposes a software tool for visualizing the AMR's location in the facility, the dynamic dispatching of the AMRs for executing specific tasks and the AMR's re-initialization, using a user-friendly Graphical User Interface (GUI) implementation that interconnects robotic systems with the employees in the industrial facility. The proposed software tool uses: (i) ROS to control the AMR units, (ii) ReactJS library to plot the visualization map on a webpage, (iii) rosbridge-server as a middleware between Robot units and Web-Based GUI, (iv) nodeJS to create the backbone server and the authentication mechanism and (v) MySQL to save the users' and robots' data.

At Fig. 1 the system architecture of the proposed software tools is presented. The system consists of four main modules: (i) the ROS that controls the robot and all the necessary topics, actions, and messages, (ii) the front-end ReactAPI that creates the webpages for the end-users, (iii) the nodeJS server that enables the access to the database and (iv) the database that contains the users' and robots' data. The main purpose of the React API module is the creation of the GUIs and it is based on the React library. React allows the direct creation of objects that contain and manipulate data without strictly defining how they are used. The basic building block of a React application is the React Component class. React views the UI as a state machine and renders all of its components with a particular state. Finally, the react library connects the created components and generates a multi-page working API. The npm package and yarn package managers are also used to control JavaScript packages while the "GUI data Request" and "authentication Request" routines were implemented to access the database data through the Node JS server. Moreover, on Fig. 1, the ROS JavaScript library (ROSLIBJS) is presented which is responsible for message transactions between web pages and ROS. ROSLIBJS is a library that transforms the ROS messages on a JSON format and transmits them to the ROS environment through the rosbridge module.

From the side of the ROS, the rosbridge server was used which is a middleware abstraction layer that provides robot information to application programmers who are not themselves roboticists [15]. Moreover, the rosbridge protocol as a general relies on ROS topic messaging in a client-server paradigm suitable for wide area networks, and human-robot interaction at a global scale through modern web browsers [16]. The interconnection among ROSLIBJS and rosbridge was used to provide the access of the ROS's actions and topics to the front-end react API.

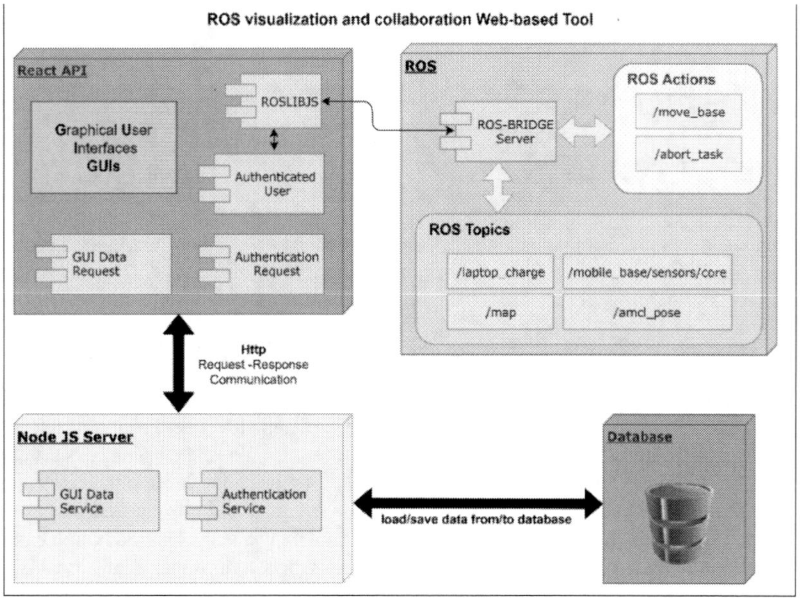

Fig. 1. System Architecture

Thus, the react library can assign new ROS messages by communicating with the ROS and getting response messages. The following ROS topics used to get the desired information from the robot: (i) "/laptop_charge" topic which provides the battery level of the laptop that we used on our implementation, (ii) "/map" topic that includes the information about the OGM, (iii) "/mobile_base/sensors/core" topic that provides information for the Kobuki® base such as its battery level and (iv) "/amcl_pose" topic that has all the information about the current location of the AMR. Also, the ROS actions "/move_base" and "/abort_task" were used for the real-time end-user informing. Specifically, the action "/move_base" used for the movement of the robot to a certain location and the "/abort_task" action was used to enable the ability of real-time task's cancelation in case of emergency. Moreover, there are the MySQL Database and the Node JS server modules which were used to ensure the privacy and security of the saved data as only an authenticated user can access the system's data. MySQL is an open-source relational database management system that organizes the data into tables while the Node JS server has direct access to the database and exports only the desired data to the external users.

From the perspective of the software web tool the ReactAPI and NodeJS are collaborating to create a full-stack web-based API. The React library is used as a client-side application while the nodeJS used as a server-side script. As mentioned before, the only way that the front-end side can access the database's data is by using http requests from the authenticated users, providing the token that the server sends on the authentication procedure. The developed system shares the information about the AMR only to authenticated users. Thus, the flow, to access the robot information, is analysed as follows: (i) the registered user must enter his username and password on the Log In page (Fig. 2), (ii) the backbone mechanism (server) retrieves user's data from the database and authenticates the user (if he has provide the correct username and password), (iii) the system provides access of the ROSLIBJS to the webpage with the data that retrieve from server and (iv) the system presents the robot information on the webpage while, the user can interact with the AMR and assign new tasks. The available tasks that an authenticated user can assign to AMR are: (i) move the robot to a certain position, (ii) abort the task that is executed in case of emergency, (iii) initialize the robot position and orientation on the map and (iv) assign new tasks. Moreover, the assignment of the tasks is implemented through ROS actionlib package, that enables users to cancel the request or to get periodic feedback about the progress of the assigned task. This package provides tools to create servers that execute long-running goals that can be pre-empted. For the demonstration of the created software tool the prerequisite is a map of the robot's working area. Furthermore, the tool can help the authenticated user to the mapping process as it gives the capability to the end-user to move the robot by clicking on the map area on the webpage. After the map is completed, the API enables the AMR's visualization, and the user monitors the AMR's executed tasks.

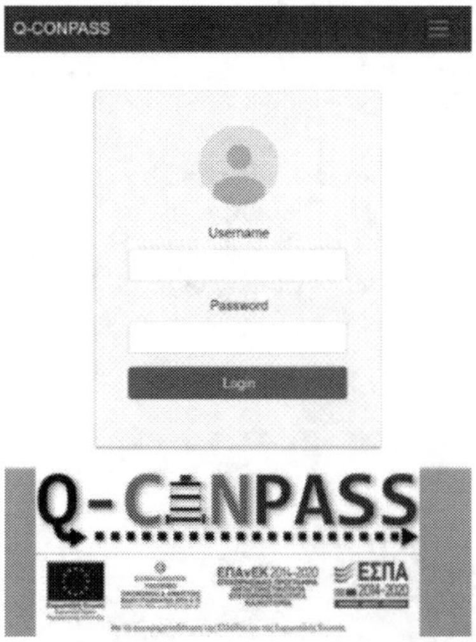

Fig. 2. Login Page of the proposed software tool

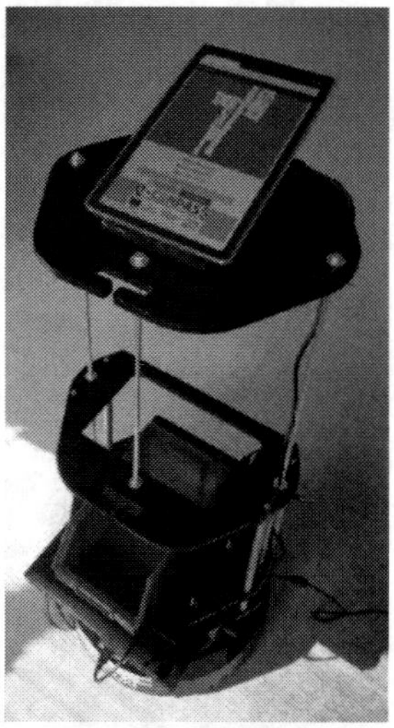

Fig. 3. Developed robot with Tablet and kobuki base

The user's webpage of the proposed software tool is presented at Fig. 4 after the user's authentication procedure. On this page the OGM of the Laboratory of Statistics and Quantitative Analysis Methods and the current position of the robot is presented on the left side of the webpage, while the system provides additional information about the robot on the right side such as robot status, robot's base battery level and laptop's battery level that runs the ROS. Moreover, with the buttons the user can assign the above-mentioned tasks to the robot units. When the user clicks a button, the green colour indicates that the next click on the OGM will assign the defined task to the robot through the action server package. While the robot is moving the application updates the robot's cursor on the map and pop-up messages inform the user for the progress of the executed task through the action server that is running on the ROS. The red button named "Abort execution task" can abort the executed tasks instantly in case of an emergency or a failure that is observed by an employee within the industrial facility.

Moreover, on Fig. 3 the main parts of the developed robot are presenting, namely: (i) the Kobuki base, (ii) the Slamtec® RPlidar which is located below the laptop's rack, (iii) the laptop with GPU, (iv) the tablet that we used for controlling the robot through the developed software tool and (v) the bin box for the placement of small items. Thus, the robot can move to user-defined points within the industrial facility and transport small parts among employees and workstations. This ability reduces employees' transportation times while enhancing human-robot collaboration as it creates a more flexible and efficient production line.

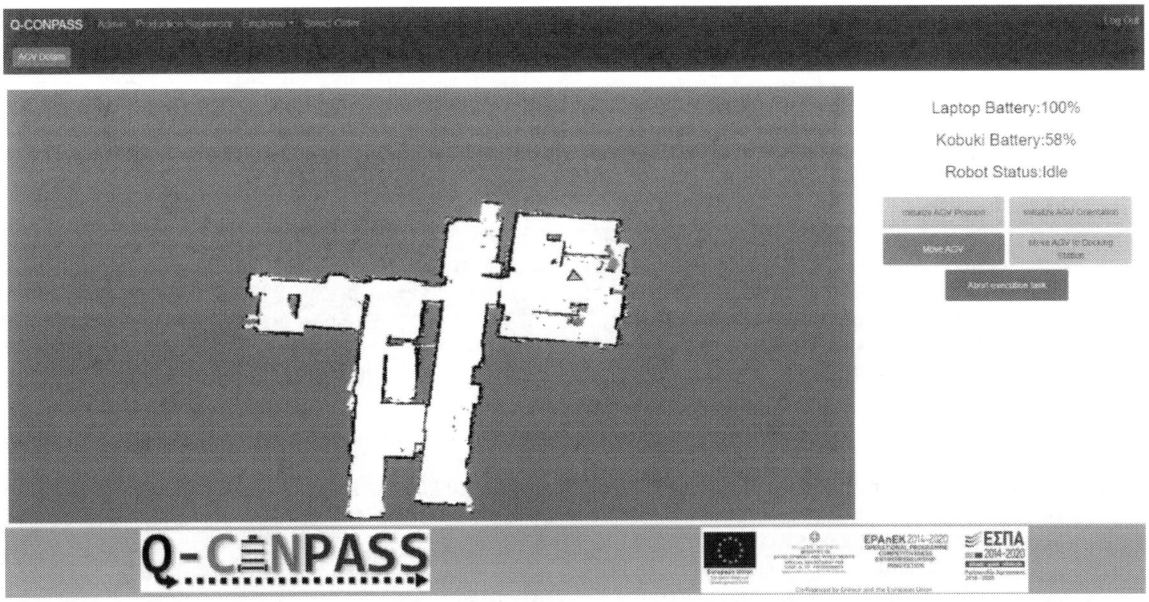

Fig. 4. Webpage with Robot OGM visualization

4. Limitations and Conclusions

The development and testing of our proposed solution took place at the laboratory of Statistics and Quantitative Analysis Methods and that set some limitations on our results, as the available area was way smaller than an industrial area and the number of the available end users was also limited. Moreover, the proposed solution tested without taking into consideration robot-user interconnection problems that can occur in a large industrial area.

While robotic systems and AMRs are vastly adopted by industries, the development of tool for user-friendly human-machine interfaces that allow employees to directly handle the robotic systems in the production line, and to be dynamically informed about the AMR's status, as well as the status of the executed activities is still in an infancy state and should be further examined. The proposed user-friendly software tool addresses some open issues in industrial environments as it can (i) move the AMR during the mapping procedure in the industrial area, (ii) visualize in real-time the industrial area, (iii) assign new tasks to AMRs such as the transportation of small items within the industrial facility and (iv) visualize the position of the AMRs on the OGM. The solution was designed with the use of Kobuki, a small size Turtlebot2 series mobile robot, but due to openness of ROS it can be used also in all ROS based robots. Also, due to the web-based nature of the proposed tool, it categorized as an easy-to-use software tool in industrial areas, through mobile phone and tablets, when access to a computer is impossible.

The proposed software tool could be further developed to improve its functionality. The next step is the prioritization of the tasks assigned by an end-user according to her authorization role. Thus, a high-level user will have priority on task assignment in comparison with an production line employee. Moreover, a semantic mapping and the separation of the industrial area into divisions would help to add speed limits on the AMRs according to the human activity on each division. Thus, on divisions with low human activity, the AMRs could be moved faster and reduce transportation times while on divisions with high human activity the lower speed limits can enhance safety and assist human-robot collaboration.

5. Acknowledgments

This paper was partially supported by the Greek Secretariat of Research and Technology, Operational Programme Competitiveness, Entrepreneurship and Innovation 2014-2020, project "Dynamic Quality CONtrol on Production lines using intelligent AutonomouS vehicleS (Q-CONPASS)".

6. References

[1] Roblek, V.; Meško, M.& Krapež, A. (2016). A Complex View of Industry 4.0, SAGE Open, Vol. 6, No. 2. doi:10.1177/2158244016653987

[2] Bechtsis, D.; Tsolakis, N.; Vlachos, D.& Srai, J. S. (2018). Intelligent Autonomous Vehicles in digital supply chains: A framework for integrating innovations towards sustainable value networks, Journal of Cleaner Production, Vol. 181, 60–71. doi:10.1016/j.jclepro.2018.01.173

[3] Papa, M.; Kaselautzke, D.; Radinger, T. & Stuja, K. (2017). Development of a Safety Industry 4.0 Production Environment, Proceedings of the 28th DAAAM International Symposium, pp.0981-0987, B. Katalinic (Ed.), Published by DAAAM International, ISBN 978-3-902734- 11-2, ISSN 1726-9679, Vienna, Austria, DOI: 10.2507/28th.daaam.proceedings.136

[4] Kuts, V.; Sarkans, M.; Otto, T. & Tähemaa, T. (2017). Collaborative Work Between Human And Industrial Robot In Manufacturing By Advanced Safety Monitoring System, Proceedings of the 28th DAAAM International Symposium, pp.0996-1001, B. Katalinic (Ed.), Published by DAAAM International, ISBN 978-3-902734-11-2, ISSN 1726-9679, Vienna, Austria, DOI: 10.2507/28th.daaam.proceedings.138

[5] Stubbs, K.; Wettergreen, D. & Hinds, P. J. (2007). Autonomy and common ground in human-robot interaction: A field study, IEEE Intelligent Systems, Vol. 22, No. 2, 42–50. doi:10.1109/MIS.2007.21

[6] Karamanos, X.; Mallioris, P.; Poulimenos, D.; Bechtsis, D. & Vlachos, D. (2019). A ROS Tool for Optimal Routing in Intralogistics, Proceedings of the 30th DAAAM International Symposium, pp.0832-0838, B. Katalinic (Ed.), Published by DAAAM International, ISBN 978-3-902734-22-8, ISSN 1726-9679, Vienna, Austria, DOI: 10.2507/30th.daaam.proceedings.115

[7] Kam, H. R.; Lee, S. H.; Park, T. & Kim, C. H. (2015). RViz: a toolkit for real domain data visualization, Telecommunication Systems, Vol. 60, No. 2, 337–345. doi:10.1007/s11235-015-0034-5

[8] Ivanov, A.; Zakiev, A.; Tsoy, T. & Hsia, K. (2021). Online Monitoring and Visualization with ROS and ReactJS, 0–3

[9] Thoben, K. D.; Wiesner, S. A. & Wuest, T. (2017). "Industrie 4.0" and smart manufacturing-a review of research issues and application examples, International Journal of Automation Technology, Vol. 11, No. 1, 4–16. doi:10.20965/ijat.2017.p0004

[10] Lu, Y. (2017). Industry 4.0: A survey on technologies, applications and open research issues, Journal of Industrial Information Integration, Vol. 6, 1–10. doi:10.1016/j.jii.2017.04.005

[11] Tan, Q.; Tong, Y.; Wu, S. & Li, D. (2019). Anthropocentric Approach for Smart Assembly: Integration and Collaboration, Journal of Robotics, Vol. 2019. doi:10.1155/2019/3146782

[12] Pires, A. G.; Rezeck, P. A. F.; Chaves, R. A.; Macharet, D. G. & Chaimowicz, L. (2021). Cooperative Localization and Mapping with Robotic Swarms, Journal of Intelligent and Robotic Systems: Theory and Applications, Vol. 102, No. 2. doi:10.1007/s10846-021-01397-z

[13] McLurkin, J. & Yamins, D. (2005). Dynamic task assignment in robot swarms, Robotics: Science and Systems, Vol. 1, 129–136. doi:10.15607/rss.2005.i.018

[14] Tsarouchi, P.; Makris, S.; Michalos, G.; Matthaiakis, A. S.; Chatzigeorgiou, X.; Athanasatos, A.; Stefos, M.; Aivaliotis, P. & Chryssolouris, G. (2015). ROS based coordination of human robot cooperative assembly tasks-An industrial case study, Procedia CIRP, Vol. 37, 254–259. doi:10.1016/j.procir.2015.08.045

[15] Crick, C.; Jay, G.; Osentoski, S. & Pitzer, B. (n.d.). Rosbridge : ROS for Non-ROS Users, 493–504. doi:10.1007/978-3-319-29363-9

[16] Toris, R.; Kammerl, J.; Lu, D. V.; Lee, J.; Jenkins, O. C.; Osentoski, S.; Wills, M. & Chernova, S. (2015). Robot Web Tools: Efficient messaging for cloud robotics, IEEE International Conference on Intelligent Robots and Systems, Vols. 2015-Decem, 4530–4537. doi:10.1109/IROS.2015.7354021

DOI: 10.2507/32nd.daaam.proceedings.088

IMPACT OF MACHINING ON THE ENVIRONMENT

Tetjana Tomaskova & Katerina Bicova

This Publication has to be referred as: Tomaskova, T[etjana] & Bicova, K[aterina] (2021). Impact of Machining on the Environment, Proceedings of the 32nd DAAAM International Symposium, pp.0623-0630, B. Katalinic (Ed.), Published by DAAAM International, ISBN 978-3-902734-33-4, ISSN 1726-9679, Vienna, Austria
DOI: 10.2507/32nd.daaam.proceedings.088

Abstract

Since about the middle of the last century, the state of the environment has been deteriorating and, in connection with this, the pressure to protect the environment has begun to increase, so industrial companies must comply with environmental legislation. Industrial companies' implementation of the environmental management system on a voluntary basis and by implementing them undertake to carry out all their activities in an environmentally friendly manner. This trend affects not only social but also economic development and, last but not least, the satisfaction and health of the population. Sustainable development makes it possible to meet the current needs of the human population without compromising the living standards of future generations and preserving the planet's ecosystem. The article describes the issue of the impact of the metalworking industry, specifically machining, on the environment. The authors aim to design a register of environmental aspects identified in machining according to ISO 14001. This register of environmental aspects will serve as a basis for the subsequent processing of the required documentation for the certification of the company according to ISO 14001.

Keywords: Environment; environmental aspects; sustainable development; metalworking industry; machining;

1. Introduction

Society has changed over the years and with it the view of certain problems that previous generations did not perceive as problems that need to be addressed. If we talk about the natural environment, then man was an integral part of human evolution at the beginning. Among the reasons for this was, first and foremost, the effort to stay alive, which was later replaced by the fulfilment of human needs, which, however, resulted in man's disruption of the environment in connection with the scientific and technological revolution and industrial development. Under all the above (i.e. the natural environment, the scientific and technological revolution, industrial development and socio-economic conditions), it is possible to imagine an intertwined system, which is referred to as the environment [1], [2], [3], [4].

According to the ISO 14 001 standard, the environment is defined as the environment of a company, including air, water, soil, natural resources, plants, animals, people and the relationships between them. Because of this, the environment cannot be referred to as a business problem, but it must be referred to as a global system problem. The aim of this research is to focus on the impact of the metalworking industry, specifically machining, on the environment in general, and to try to design a general register of environmental aspects according to the requirements of ISO 14001, which can be identified in machining. The aim is to prepare a basis for the subsequent processing of the required documentation for the certification of the company according to ISO 14001.

1.1. Machining and environment

The environment is an issue that does not only concern the corporate sphere, but it is a global sphere. The same is true of the machining environment. The machining process itself includes the energy and resource flows that are necessary for a given production. Other aspects influencing the environmental impact of the machining process include the factory itself or the company's building with a heating, ventilation and air conditioning system, and even the supply chain (see Fig. 1) [5].

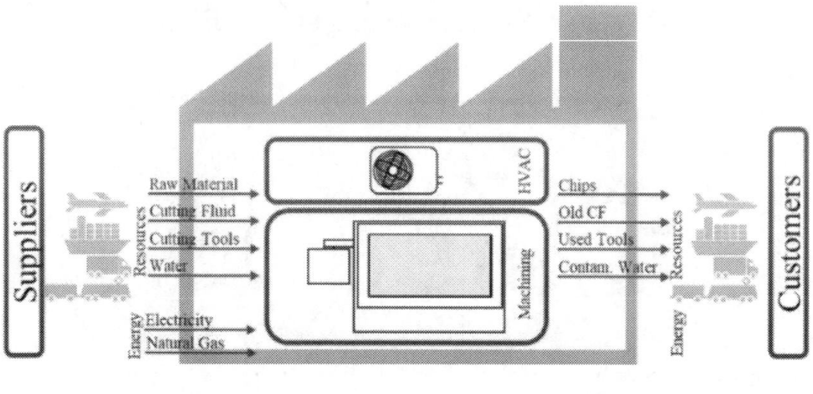

Fig. 1. Energy and Resource Flows [5]

The place where the factory is located is very important here. From a global perspective, there is a different flow of energy and resources when assessing the environmental impact of a company located in Germany or the Czech Republic and a different flow when assessing a company located in India. Especially in terms of key factors: supply chain, climate, available energy carriers, disposal and recycling (see Fig. 2) [5].

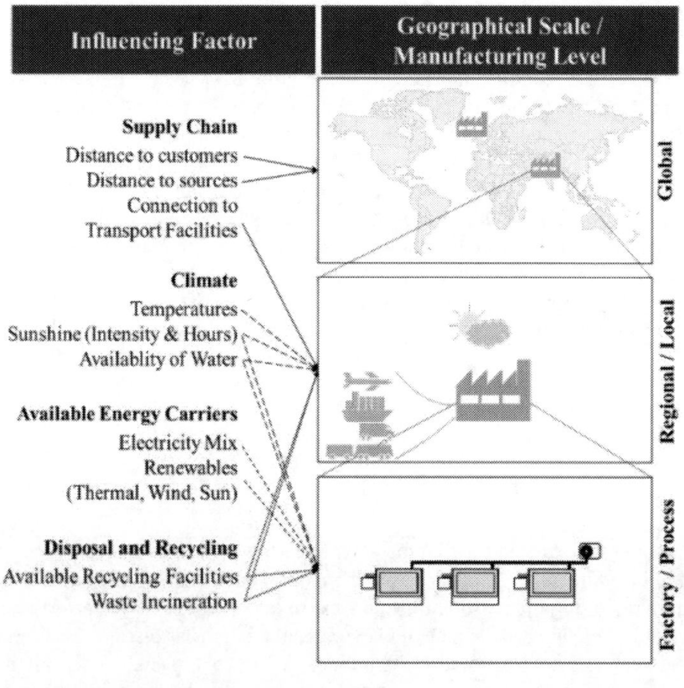

Fig. 2. Factors influencing the sustainability of a machining process [5]

The first of these is the supply chain, which includes mainly the distance of the factory from suppliers and customers, but also the mode of transport. Each country has a different infrastructure, and the possibility of using a certain transport may be one of the options where the company can reduce the burden on the environment. In addition to the availability of transport, this also includes the distance of the service, warehouse and auxiliary resources (e.g. cutting fluids, tools).

The second factor is the climate of the area where the company is located. Climatic conditions affect the energy required for the HVAC system (Heating, ventilation and air conditioning) and for machine tools. The main problem is businesses that are located in areas with extreme climatic conditions, because intensive heating or cooling is required there, which leaves an ecological footprint behind. The availability of water resources is also important here, as an important source for material processing and cooling during machining processes. The third important factor influencing the impact on the environment during the machining process is the type of energy mix available in the company. CNC machining consumes electricity in various operations and, as a result of this production, generates emissions that depend on the location of the company, the resources used and the technology used. The last key factor is the distance of sales and recycling equipment from the place of business, including recycling itself and the used recycling technology. In general, however, environmental policy strictly promotes the availability of high - level recycling and disposal facilities [5], [6], [7].

2. Impact of industry on the environment

2.1. Impact of the metalworking industry on the environment

The company, which deals with metal processing, follows the scheme of the production process. This means that there are certain inputs that will be converted into certain outputs by a given production process. The environment also influences the whole production process. Figure 3 shows that the inputs include mainly input materials, consumables, semi-finished products, but also energy. It is also shown that among the outputs we have products, residual materials, but from an environmental point of view mainly emissions, polluted materials, polluted wastewater or polluted air. These are the environmental burdens that companies interested in the environment try to minimize. [8], [9].

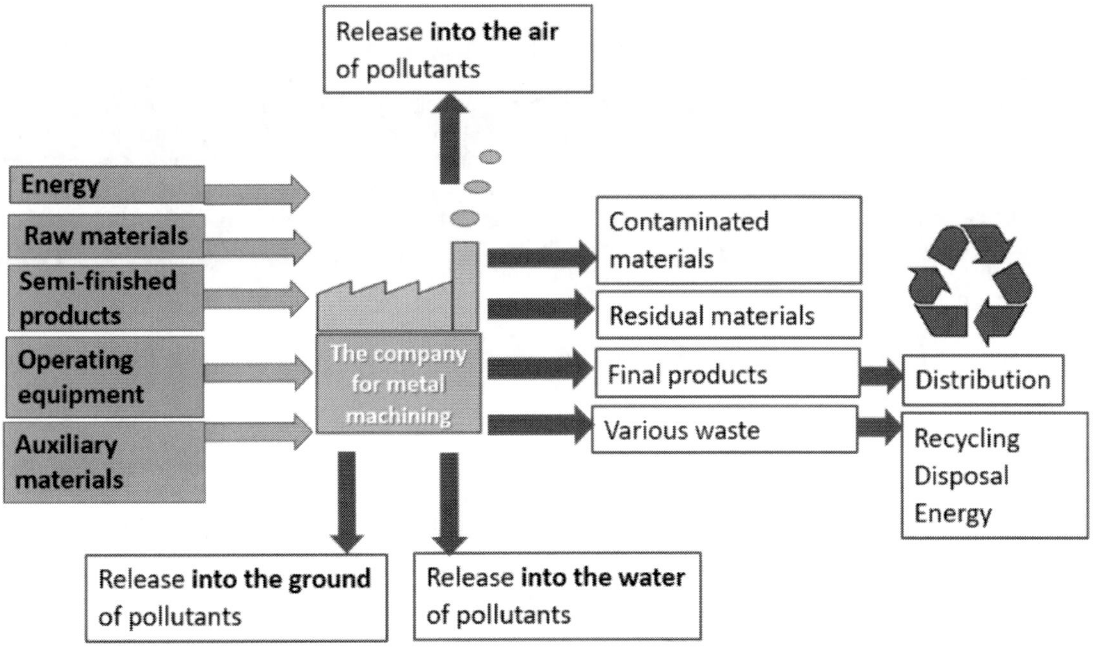

Fig. 3. Inputs and outputs of an industrial company for metal machining

2.2 Impact of machining on the environment

To determine the correct impact of machining on the environment, it is first necessary to realize what the production process of machining looks like and what it is affected from the point of view of the environment (see Fig. 4). An example can be the actual transport of material to the company or the actual extraction of this material. The energy consumed, which is needed for proper operation, also has a certain impact, but the outputs that are created together with the product are a bigger problem. The first of these outputs is chips. If chips are treated badly, they have a certain impact on the environment. In most cases, these chips are recycled and further processed, so that the impact is minimized. The second output is coolants and lubricants, which are the real problem. This can be a substance with a long disintegration time or a cancer-causing substance. At the same time, these products can pollute wastewater if used incorrectly, which can have irreversible consequences for the environment. If it is possible to use water-based products, it is important to use them, because chemical waste is dangerous from the waste generated during production.

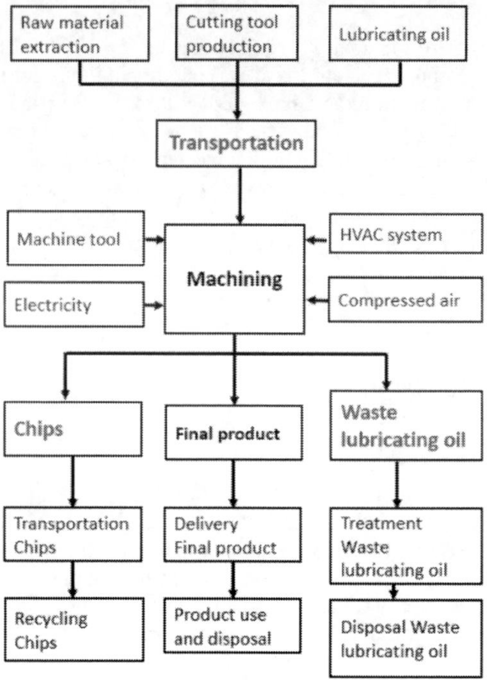

Fig. 4. Machining process

Emissions can be another impact on the environment during machining. An example of the cause of air pollution is dust. Here it depends mainly on the type of material that is ground. E.g. when grinding aluminum materials, dust is released into the work area, which can form an explosive atmosphere with the air. In this case, it is necessary to extract dust directly during the grinding process, and the most efficient way is to extract abrasive aluminum dust directly at the source. By summarizing everything that has been said, it can be stated that it is necessary to evaluate all the environmental impact of machining. Furthermore, it is necessary to know the exact production process of a particular product, because even the environmental impacts may differ from product to product, for example, because other production methods or other means of operation can be used.

3. Preparation of a register of environmental aspects

ISO 14 001 requires an organization to establish and follow a procedure to identify the environmental aspects of its activities, services and products. The result is aspects that the organization can directly manage or have some influence on, but the most important thing is to identify those aspects that have or may have a significant impact on the environment. According to the standard, it is necessary to map all aspects of activities by which the company affects the environment. However, the standard does not specify the exact procedure for carrying out this analysis, as it is assumed that this procedure will be proposed by the organization itself. However, the condition remains that the subject of interest will be all activities, products and services that are the subject of the organization's activities or over which the organization has an influence.

The company for which we are preparing environmental aspects consists only of a production hall and a warehouse for storage of inputs and outputs. Therefore, the administration, the vehicle fleet will not be taken into account, but also the dining room, which usually does not occur in the same way in similarly small companies. Nor will it be assumed that the company shares the place with other organizations (ie the company is considered as a separate entity). To facilitate the listing of individual negative effects on the environment, the following typical groups are listed:

- Use of hazardous chemicals and preparations
- Waste production (with emphasis on hazardous waste)
- Water pollution (e.g. waste water discharges, leaks)
- Soil contamination
- Emissions to air
- Odor nuisance
- Excessive noise emissions
- Exceeding hygienic limits for the working environment
- Consumption of raw materials and energy
- Impacts on human health and safety.

Once all the environmental aspects have been mapped, they need to be evaluated and the essential aspects identified. How to carry out this evaluation again is not directly determined by the standard and it is only up to the company itself what criteria and procedures it will choose. The following criteria were chosen for this work.

Frequency of occurrence or duration of the effect:

- Continuously 5
- Often (daily) 4
- Regularly (monthly) 3
- Rarely (annually) 2
- Almost never 1

Severity of influence:

- Serious 4
- Significant 3
- Normal 2
- Negligible 1

Possibility of an accident:

- It must be assumed 5
- Exceptional 4
- Not expected, but has ever happened 3
- Almost impossible 2
- Impossible 1

Costs:

- Threatening 3
- Significant 2
- Insignificant 1

Activity regulated by legislation:

- Yes Y
- No N

The criterion "Legislative activity" only indicates whether the environmental aspect is regulated or not. It has no value of significance and is therefore not taken into account in the overall assessment, but in the case of significant aspects that are regulated by law, their significance needs to be taken with greater emphasis.

Rating:

$$\text{Frequency} \cdot \text{Severity} \cdot \text{Accident} \cdot \text{Cost} = \text{Significance EA} \tag{1}$$

Significance of EA:

- Common aspect 1-50
- Significant aspect 51-150
- Alarming significant aspect 150-300

There is no need to implement any corrective measures for common aspects. Significant aspects need to be further addressed and appropriate corrective measures implemented to reduce their relevance. In the case of alarmingly important aspects, immediate action is needed, because they are so serious that ignoring them could have irreversible consequences for the environment. For the proposed register, the impact is related to the group water, soil, air, resources and man. In practice, the evaluation takes place in a working team of people who work together to implement EMS (environmental management system) in the company.

Register of environmental aspects															
Machining hall															
Proces	**Impact**					**Assessment**						**Significance of aspects**	**Control measures**	**Monitoring**
Environmental aspect	**Water**	**Soil**	**Air**	**Resources**	**Man**	**Frequency**	**Severity**	**Accident**	**Costs**	**Legislation**	**Product**			
Storage of waste, hazardous waste and hazardous substances														
Generation of general waste		•			•	4	2	2	1	A	16	Common	Training	Continuous monitoring
Hazardous waste generation	•	•	•		•	3	2	3	1	A	18	Common	Training	Continuous monitoring
Leakage of hazardous substance outside the designated containment tank	•	•	•		•	2	4	3	3	A	72	Significant	Emergency plan	Accident record
Use of resources and energy (electricity, water, natural gas)														
Use of non-renewable resources		•		•	•	5	2	3	1	N	30	Common	Responsible person	Continuous monitoring
Drinking water consumption	•			•	•	4	2	2	1	N	16	Common	Responsible person	Continuous monitoring
Domestic water consumption	•			•		4	2	2	1	N	16	Common	Responsible person	Continuous monitoring
Natural gas consumption		•		•		4	2	2	1	A	16	Common	Responsible person	Continuous monitoring
Emissions from natural gas combustion			•		•	4	2	2	1	N	16	Common	Responsible person	Continuous monitoring
Oxygen consumption during natural gas combustion			•	•	•	4	2	1	1	N	8	Common	Responsible person	Continuous monitoring
Production process														
Generation of general waste		•			•	4	2	2	1	A	16	Common	Training	Continuous monitoring
Hazardous waste generation	•	•	•		•	3	3	3	1	A	27	Common	Training	Continuous monitoring
Consumption of metallurgical material		•		•		4	2	1	1	N	8	Common	Training	Production records

Generation of waste from metallurgical material	•	•	•		•	4	3	3	2	A	72	Significant	Training	Production records
Use of cutting oil in machining	•	•	•	•		4	3	3	2	A	72	Significant	Training	Production records
Waste oil formation	•	•	•		•	4	3	4	2	A	96	Significant	Training	Production records
Use of chemicals			•	•	•	2	3	3	1	A	18	Common	Training	Continuous monitoring
Formation of inconsistent dust emissions during grinding			•		•	4	3	1	1	A	12	Common	Training	Continuous monitoring
Noise	•	•	•		•	4	2	1	1	N	8	Common	Training	Continuous monitoring

Table 1. Register of environmental aspects for the machining hall

4. Conclusion

The key part is the description of the draft register of environmental aspects in the field of machining, which be used as a basis for EMS documentation according to ISO 14001. Criteria were also used to calculate the significance of the identified environmental aspects of the machining process. The following criteria were chosen for this work: frequency of occurrence or duration of the impact, severity of the impact, the possibility of an accident, costs, and whether the activity is regulated by legislation. After calculating the significance assessment, it was found that important aspects during machining include the leakage of hazardous substances, the generation of waste from metallurgical material, the use of cutting oil in machining and the generation of waste oil. According to the created register, environmental objectives were selected for the identified significant aspects, for which possible corrective measures were written, which could help to reduce their significance.

The implementation of the environmental management system in the company can be divided into five stages of initiation, planning, implementation, control and evaluation. Opening, or preparatory stage in the framework for which the register of environmental aspects described above has been proposed. The planning stage will be the next step in which the revision of legislative requirements, the definition of environmental objectives (selection and planning of solutions to a very specific environmental problem), the elaboration of an environmental program (short-term, medium-term and long-term measures to be implemented in order to achieve defined environmental objectives), setting the organizational and process structure (it is essential to establish responsibilities and powers in order to promote environmental policy objectives and programs). In the future, we also plan to perform an analysis of the implementation stage, which will include the achievement of the set goals according to the developed environmental programs. The task of the control phase will to implement such measurement and evaluation measures that ensure that the company's environmental management system operates in accordance with the established environmental management program and that areas corrective action are identified as soon as possible. Within the evaluation stage, the setting of the environmental management system in the company will be examined. An indicator will be used to evaluate the environmental profile and its changes according to ISO 14031.

Our research has confirmed that the company's impact on the environment can be great. Better use of resources, reduction of pollutant emissions, prevention of risks associated with environmental accidents and creation of safe working conditions is the goal of every company with an established environmental management system. Should however, to become not only the goal of every company, but also every employee of the company. Therefore, in the future we plan to design a plan for training the company's employees, the content of the training, including a number of practical workshops and a check of the professional competence of the employees.

5. Acknowledgments

This article was created under the project SGS-2019-008: Research and Development for Innovation in the Field of Manufacturing Technology - Machining Technology III.

6. References

[1] Cascio, J., Woodside, G. & Mitchell, P. (1996). ISO 14000 Guide, The New International Enviromental Standars. McGraw-Hill, New York, pp. 3–217

[2] Kralikova, R.; Pinosova, M.; Koblasa, F.; Wessely, E. & Rusko, M. (2020). Environmental and Health Impact of Paint Products, Proceedings of the 31st DAAAM International Symposium, pp.0035-0043, B. Katalinic (Ed.), Published by DAAAM International, ISBN 978- 3-902734-29-7, ISSN 1726-9679, Vienna, Austria DOI: 10.2507/31st.daaam.proceedings.005

[3] Kudlacek, I. (2002). Industry ecology (in Czech), Prague ISBN 80-01-02495-4, Prague

[4] Veber, J. et al. (2010). Management of quality, environment and work safety: legislation, systems, methods, practice (in Czech), Prague ISBN 978-80-7261-210-9, Prague

[5] Sihag, N. et al. (2019). The Influence of Manufacturing Plant Site Selection on Environmental Impact of Machining Processes, Available from: https://1url.cz/OzYmQ

[6] Sobotova, L.; Badida, M. & Wessely, E. (2020). Research of Laser Cleaning of Materials and Environmental Requirements, Proceedings of the 31st DAAAM International Symposium, pp.0176-0183, B. Katalinic (Ed.), Published by DAAAM International, ISBN 978-3-902734-29-7, ISSN 1726-9679, Vienna, Austria DOI: 10.2507/31st.daaam.proceedings.023

[7] Ojstersek, R.; Palcic, I. & Buchmeister, B. (2020). Simulation Modelling of Environmental and Economic Aspects in Manufacturing Systems, Proceedings of the 31st DAAAM International Symposium, pp.0436-0443, B. Katalinic (Ed.), Published by DAAAM International, ISBN 978-3-902734- 29-7, ISSN 1726-9679, Vienna, Austria DOI: 10.2507/31st.daaam.proceedings.061

[8] HÄBERLE, G. et al. (2003). Environmental technology for school and practice (in Czech), Prague ISBN 80-86706-05-2, Prague

[9] Blazevic, A.; Bibic, D.; Trobradovic, M.; Filipovic, I. & Gebert, K. (2020). Effective Reduction of NOx Emissions of Modern Medium Speed Diesel Engines, Proceedings of the 31st DAAAM International Symposium, pp.0812-0818, B. Katalinic (Ed.), Published by DAAAM International, ISBN 978-3-902734-29-7, ISSN 1726-9679, Vienna, Austria DOI: 10.2507/31st.daaam.proceedings.113

DOI: 10.2507/32nd.daaam.proceedings.089

OPTIMIZATION OF ALUMINUM OPTICAL TABLE PRODUCTION

Jakub Cerveny, Michal Povolny & Lubos Kroft

This Publication has to be referred as: Cerveny, J[akub]; Povolny, M[ichal] & Kroft, L[ubos] (2021). Optimization of Aluminum Optical Table Production, Proceedings of the 32nd DAAAM International Symposium, pp.0631-0639, B. Katalinic (Ed.), Published by DAAAM International, ISBN 978-3-902734-33-4, ISSN 1726-9679, Vienna, Austria
DOI: 10.2507/32nd.daaam.proceedings.089

Abstract

Machining of aluminum optical tables is a current topic due to its relevance to laboratory research in the field of laser technology, but also in the pharmaceutical or chemical industries. The purpose of this article is to present proposals to produce aluminum optical tables using modern machining methods focused on the production of holes and threads with qualitative aspects for the use of optical tables in vacuum atmosphere. Firstly, the main product, the machine and the original production method will be introduced. Subsequently, modern methods and tools for potential production improvement will be introduced. The third part deals with an experiment focused on the creation of threaded holes and their qualitative evaluation. The last part evaluates the economic aspect of production. To summarize the results, it is impossible to neglect the fact that innovations in cutting tools for aluminum alloys are a big step forward in mechanical engineering. The impact these tools have on production is very much needed, especially in industries such as the laser industry, where high demands are placed on quality, which is amplified by using optical tables in a vacuum. At the same time, they have a very positive effect on the economic aspect of production, which cannot be neglected while planning a production.

Keywords: Optical table; Aluminum alloy; Drilling; Threading.

1. Introduction

Today, production requirements are increasing at a very fast pace. It is mainly about quality, quantity, and accuracy with constant correction by economic factors. This increase in requirements is due to technological development, which brings new innovative methods of production. Some sectors of the machinery industry are more affected by this trend than others. It all depends on the range of production and possible results. The vacuum equipment industry, which is the subject of this article, is one of the areas that must respond promptly to customer requirements. It is mainly about increasing the requirements for accuracy, quality, and economy of production. The production of aluminum optical tables, that are used in vacuum chambers, entails several technological tasks, and therefore production requirements. This work will focus on the appropriate choice of tools for the technology of making holes and threads. Research is made for company Streicher, spol s r o. where production of optical tables takes place. Aim of this research is to recommend to the company new strategies of production based on using modern tools that can combine more then just one machining technology. Tools must be suitable for machining of alluminium alloys and to be able to fulfil qualitative aspects of production for products that will operate in vacuum atmosphere. Therefore, tools will be most probably coated in thin layers and specifically constructed for this type of machining. Machining conditions will be also tested and recomended based on corporate machinery. [1], [2]

Optical table:
It serves as a clamping plate for various devices and equipment. Devices are connected to table by screws. There are many threads, evenly distributed over the entire table area. Optical tables for vacuum chambers have additional specific requirements due to their use. This applies to the requirement to degas the material. This means the release of particles into the vacuum space from a material while gas is pumped out of the vacuum chamber. When creating a vacuum, an effort is made to keep the number of particles of a certain size as small as possible contained in a certain volume. Degassing degrades the vacuum value due to the increase of particles in the chamber space.

Fig. 1. Optical table on supports and in use [3], [4]

Semi-finished product:
The semi-finished product to produce the optical table is cast milled plate of aluminum alloy EN AW 5083.

Machine:
Production will take place on a horizontal milling machine Soraluce SLP 8000. This is a machine suitable especially for machining larger workpieces. For this reason, the maximum possible machine speed of 5000 rpm is low and therefore not ideal for production of holes with M6 threads.

Original concept of production:
The original proposal to produce optical tables was to divide individual technological operations among specialized tools for each of them. The production process consisted of creating a hole using a DC150-05-05.000A1-WJ30RE drill, subsequent countersinking using a 90° HSSE conical countersink and Enorml-AL-GLT-8 M6 tapping. This is a procedure for creating impenetrable holes with M6 threads of a 15 mm depth.

New concept of production:
Due to the development of the machine industry, there are better production options than the use of many single-purpose tools. The improvement is mainly in the possibility of combining technological operations performed by only one multi-purpose tool. This aspect of production brings a great reduction in the machining process itself, not only the time when the tool is in contact with the workpiece, but especially at the time of transitions and other secondary times, which will significantly affect the total machining time. Among these times, tool adjustment and replacement times must be considered. If the production process is so technologically advanced that the secondary times cannot be influenced anymore, there remains a need to focus on shortening the main times during which the tool meets the workpiece. The initial modification is the choice of a machine that allows better production conditions. In this case there is a problem with the use of machine with low spindle speed. It is possible to supply the machinery equipment with an accelerating head, which will allow the spindle speed to double at least. [3], [4]

Variant 1
The first option for improving and optimizing production is to design combined tools. These tools consist of a combined drill that combines drilling and countersinking operations, and a tapping cutter that adds a threading operation to these operations. The use of a combination drill will reduce the number of tools required by one and the use of a milling cutter by two. A lower number of tools provides very substantial reduction in production times, which means an economic advantage and possibly a hiher production productivity. The reduction of production times is caused by the elimination of tool change times, which also causes the reduction of transition times between individual holes, as well as the reduction of additional times associated with the adjustment and maintenance of tools and the times of machining itself. Thus, in general, there is no reduction in fixed costs, as more modern tools tend to be more expensive and structurally complex, but this leads to a reduction in variable costs, which are specified as the cost of producing individual piece. Based on this change, there is an advantage especially in such production, where there are as many products as possible. In the eventual solution of this work, the construction of the optical table itself brings economic advantage when using combined tools with an increasing number of holes in the product.

Variant 2

The second option is to use the same combined tools that will be used in variant 1 with the addition of an accelerator head. The accelerator head is a machinery that allows you to increase the maximum speed of the machine spindle. This increase leads to improved machining productivity. At the same time, it should improve production, as higher speeds are recommended when machining aluminum alloys. Increased productivity means faster production and thus economic benefits. In the case solved in this work, it will be an expected increase from 5000 rpm to 12,000 rpm. Due to the higher purchase price of the accelerator head and the smaller time differences that its use will bring compared to the optimization with the combined tools from variant 1, it is expected to show the advantage of this variant after a larger number of produced holes.

Variants:	**Processes and tools**		**Using the accelerator head**
	Driling and countersinking	**Threading**	
1 A	Drill – DC260-03-05.000A0-WJ30ET	Tap –N205166-M6	No
1 B	Drill – DC260-03-05.000A0-WJ30ET	Tap – Enorml-AL-GLT-8 M6	No
1 C	Milling cutter BGF VHM IKZ HB M6 - 2,5xD		No
2 A	Drill – DC260-03-05.000A0-WJ30ET	Tap –N205166-M6	Yes
2 B	Drill – DC260-03-05.000A0-WJ30ET	Tap – Enorml-AL-GLT-8 M6	Yes
2 C	Milling cutter BGF VHM IKZ HB M6 - 2,5xD		Yes

Table 1. Description of different variants

2. Experiment:

Variants:	**Processes and tools**				**Using the accelerator head**
	Drilling and countersinking:	**Speed [min-1]**	**Threading:**	**Speed [min-1]**	
1 A	Drill – DC260-03-05.000A0-WJ30ET 	5 000	Tap –N205166-M6 	1 000	No
1 B	Drill – DC260-03-05.000A0-WJ30ET 	5 000	Tap – Enorml-AL-GLT-8 	1 000	No
1 C	Threading cutter BGF VHM IKZ HB M6 - 2,5xD 			5 000	No
2 A	Drill – DC260-03-05.000A0-WJ30ET 	12 000	Tap –N205166-M6 	1 500	Yes
2 B	Drill – DC260-03-05.000A0-WJ30ET 	12 000	Tap – Enorml-AL-GLT-8 	1 500	Yes
2 C	Threading cutter BGF VHM IKZ HB M6 - 2,5xD 			12 000	Yes

Table 2. Overview of tools for each variant [5], [6], [7], [8], [9]

The aim of the experiment is to determine the economic cost of individual variants of optical table production and to verify the functionality of the proposed solutions. The experiment was divided into several stages. First, aluminum plates were machined by specified methods. Subsequently, they were cut with a water jet into smaller pieces and milled in the center of predetermined holes for the possibility of subsequent measurement. The measurements of the produced holes were performed on a microscope and a roughness gauge/profilometer.

Variants:	Production time [h:min:s]		Total time [h:min:s]
	Drilling and countersinking:	Threading:	
1 A	3:20:48	3:19:48	6:40:36
1 B	3:20:48	3:19:48	6:40:36
1 C	10:53:18		10:53:18
2 A	2:25:48	3:01:48	5:27:36
2 B	2:25:48	3:01:48	5:27:36

Table 3. Production times

3. Technical evaluation:

Tool wear was measured with a KEYENCE VHX-6000 optical microscope with a maximum magnification of up to 1000 times and compared individual instruments before and after the production process, which amounted to 2304 repetitions of the technological operation, according to a certain instrument.

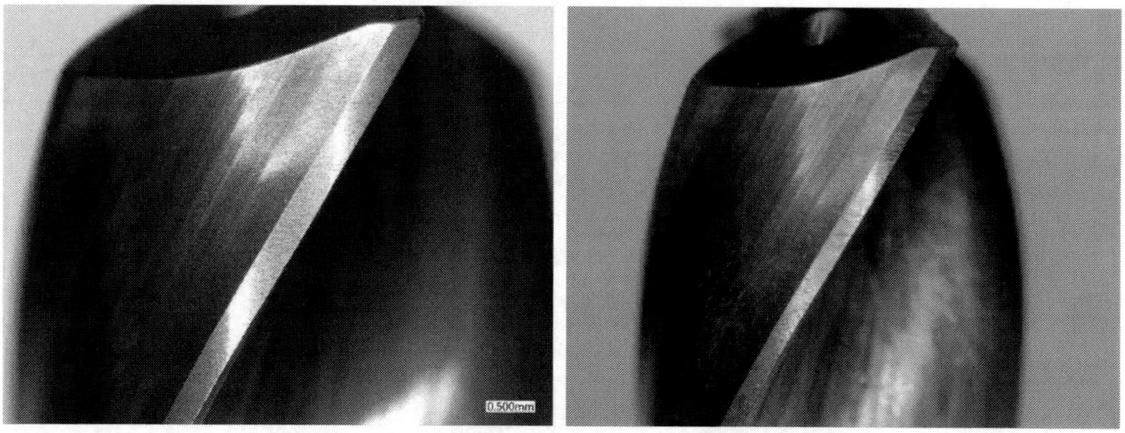

Fig. 2. Tool wear of drill observed on microscope

The drill shows no signs of wear. The only visible change is the change in color due to heat and abrasion. Apart from the visual difference, the change in color does not bring any change related to the quality of production.

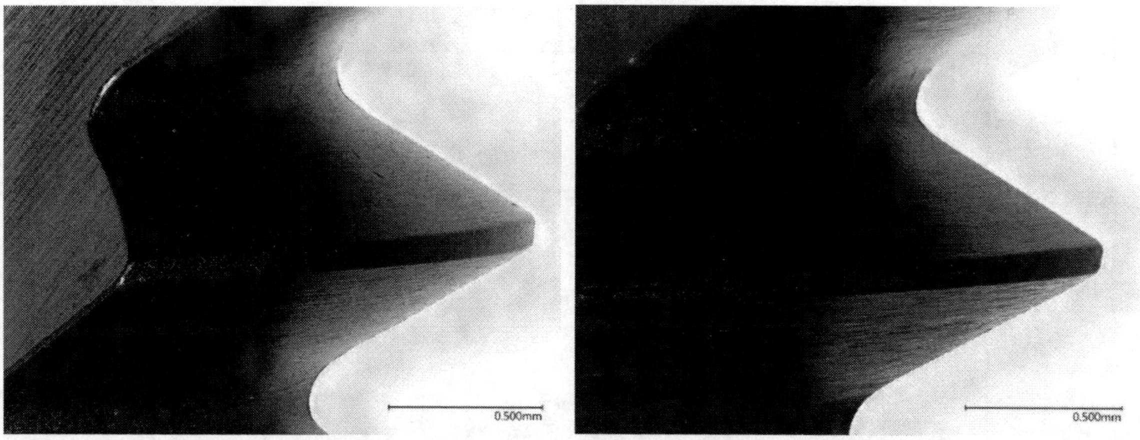

Fig. 3. Tool wear of tap observed on microscope

Walter and Emunge taps have the same characteristic when progressing in time. This is a phenomenon in which the tool geometry is smoothed after performing a certain number of operating cycles. This phenomenon has a positive effect on machining and on product quality. As with the drill, the negative effects did not appear. It can be assumed that the taps and the drill will have an extreme service life when used for machining the aluminum alloy from which the optical tables are made. From this assumption we can evaluate the wear of the tool as a minor aspect of production, which does not require much attention and does not occur as a control and decision parameter in the economic evaluation of production.

Quality of threads measured on microscope:
The quality of the threads, as well as the wear of the tools, was measured on an optical microscope. The microscope allows you to measure 3D elements of the thread and then evaluate the pitch, roughness, angle of inclination and height of the tooth of the thread. Individual samples were observed at 100 times magnification, and then at 200 times magnification. At 100 times magnification, 3 threads were photographed in a 3D image. At 200 times magnification, only one thread was photographed and similarly converted to 3D and evaluated. The comparison led to the conclusion that none of the threading methods showed better results than the others.

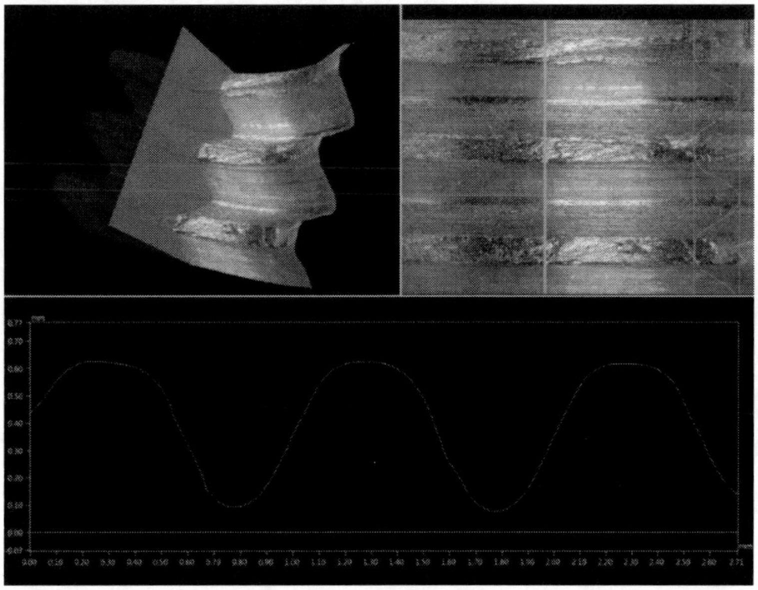

Fig. 4. 100 times magnifitcation of threads observed on microscope

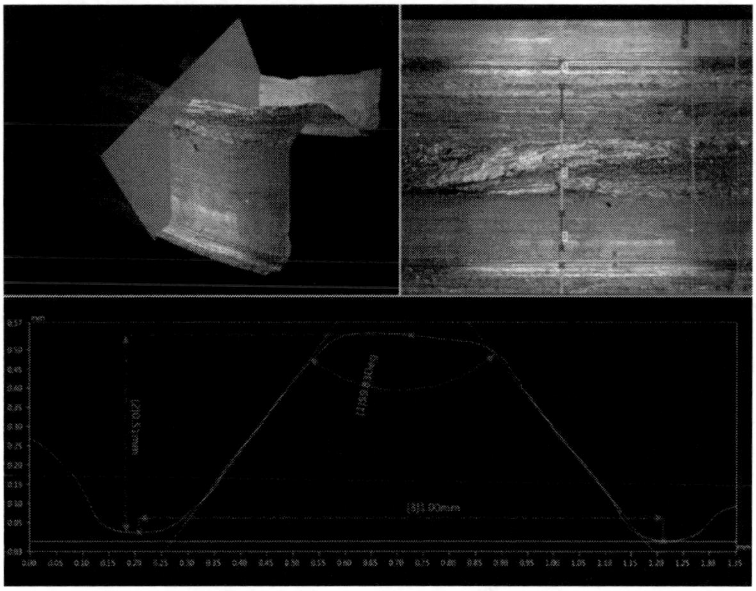

Fig. 5. 200 times magnification of thread observed on microscope

The similarity and technically different results could be caused by two reasons. One reason could be the true technical similarity with minimal deviations in the measured parameters, and thus the qualitative conformity of the threads regardless of the production method. The other reason could be an unsuitable measurement method and insufficient accuracy of the measuring device, which cannot distinguish the monitored elements and leads to misleading measurement results. For verification, the threads were also remeasured with a roughness gauge/profilometer. The differences between the 49th, 1,152th and 2,256th threads are minimal and do not indicate the need to run the tool before production or, conversely, the loss of quality due to wear.

Quality of threads measured on roughness tester/profilometer:
Measurements on a Hommel Etamic T8000 roughness tester/profilometer achieve greater accuracy than measurements on an optical microscope. The values measured by this device are therefore more authoritative than the values measured on a microscope. The disadvantage of the roughness gauge/profilometer is the inability to convert the measurement into a 3D model for visual evaluation and comparison of the thread area. The measurement was performed again for all variants on all pieces representing individual predetermined holes. Threads were measured at 10 mm. Measurements of the roughness on the roughness tester/profilometer resulted in verification of the observation on the microscope. The possibility of invalid results due to inaccuracy of microscopic measurements was ruled out. Nevertheless, the measurement with a roughness gauge/profilometer remains more authoritative precisely due to the mentioned inaccuracies of the microscope.

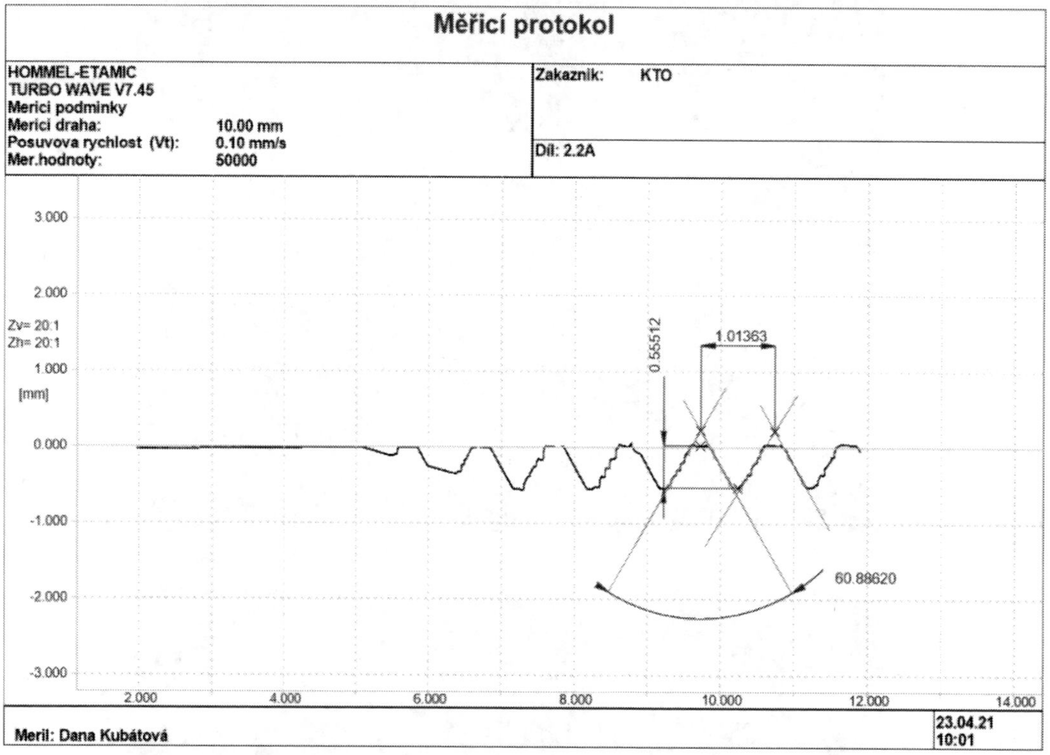

Fig. 6. Observation on roughness tester/ profilometer

4. Economical evaluation:

Variants:	Number of holes corresponding to the turning point Q: [pcs] Hourly rate of the machine 2000 Kč/h
0x1A	628
0x1B	897
0x2A	21 541
1Ax2A	107 652
1Cx2A	17 352

Table 4. Economical turning points

The economic evaluation is focused on the production price for individual variants in comparison with the original variant. For each variant, turning points were calculated to determine the advantage of the pairs of variants relative to each other. [8]

Grafical evaluation of different concepts of production:
The graphical comparison shows the economic suitability of the individual variants. Based on the turning points, the variants can be compared with each other and tey can determine from which production volume which of the variants will be advantageous. The points cannot be determined with sufficient precision from the graphs, so they were calculated in the previous table. The graphs are used to clearly verify the calculated values and to easily distinguish the advantages of the compared variants. Variants 1B and 2B were not included in the graphs due to their clearly distinguishable disadvantages compared to variants 1A and 2A.

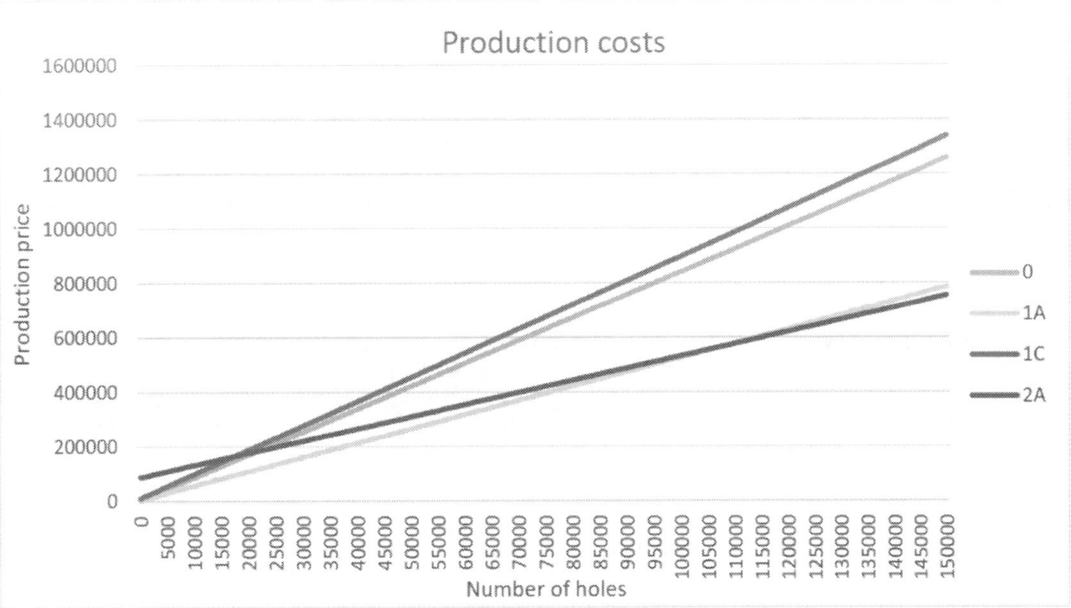

Fig. 7. Graph of production cost

Variants 1A and 2A are the most economically advantageous variants. The difference is visible in the reduced required number of holes produced for the economic advantage of the variants. This is the position of the turning point with a value of 107,652 produced holes, which again confirms the previous calculations. In conclusion, a comparison of the economic advantages of all proposed production variants, considering the hourly rate of the machine of CZK 2,000/hour, resulted in 4 variants 1A, 1B, 2A and 2B, from which variants 1A and 2A are best based due to their fixed costs. The advantage is reversed between these variants for the number of holes, which is the value of 107,652 holes produced. For a lower number of productions, variant 1A is more advantageous and for a higher volume variant 2A. Other variants were evaluated as economically disadvantageous.

5. Conclusion:

The aim of this article was to propose the optimization of the technology of production of aluminum optical tables. The optimization proposal consisted of the design of 6 new methods of holes and threads prodction, which differed from one another by production tools and machining conditions. This was followed by the experiment and its verification from a qualitative and an economic point of view. To ensure the qualitative aspect of the proposed variants, tools designed for machining non-ferrous metals, especially for aluminum and its alloys, were selected. The economic suitability of the designs was focused on reducing production times. Initially, the proposals focused on minimizing downtime, which corresponds to crossing times and tool changes. A secondary aspect was shortening of the main times, which correspond to the machining times when the tool meets the workpiece.

The original method did not show sufficient signs of modernization that today's machinery industry offers. It used only single-purpose tools for each operation separately, namely it separated the drilling, countersinking and threading operations. The newly designed variants consisted in the use of modern combined tools, which can combine these operations into one production cycle. Due to the differences of the tools, three variants were proposed. The first two used a combined drill, which drilled and countersinked the hole in one operating cycle, and then the thread was cut with two different taps.

The third variant used a thread milling cutter, which is designed to perform all production operations of the threaded hole during one cycle. For these variants, such conditions were proposed that they correspond to the maximum possible speed of the production machine on which the newly introduced production will take place. Furthermore, the extension of machine accessories with an accelerating head was proposed, which managed to increase the maximum speed of the machine spindle. The machining conditions were adapted to this increase and these were subsequently applied to the previous three variants

After the completion of the design and determination of the machining parameters, an experiment was performed, which put the individual proposed variants into practice. Test plates made of EN AW 5083 material were machined, which corresponded to the material of the manufactured optical tables. 2,304 holes with M6 threads were drilled into each plate. Testing was performed on a DMG MORI DMU 65 monoBlock machine, which was able to simulate the speed before and after using the accelerator head.

The machined plates were then cut into smaller pieces with a water jet and milled through the center of predetermined holes to allow subsequent observation on a KEYENCE VHX-6000 microscope and a Hommel Etamic T8000 roughness tester/profilometer. The results of the experiment showed a qualitative agreement of the threads produced by the individual variants. Another qualitative aspect of the evaluation was the determination of tool wear. Wear did not occur in any of the variants that used a combined drill – for this reason, a high tool life is assumed. For machining with a thread milling cutter at lower spindle speeds, the tool broke and thus the experiment did not occur under increased cutting conditions. The cutter broke during the hole drilling operation, probably due to insufficient cutting fluid pressure, which left loose chips in the cut, on which the cutter rested and broke while drilling.

The economic evaluation of the experiment, which was focused on the comparison of individual variants in terms of fixed and variable costs, led to the determination of variants with a combined drill and tap Walter as the most economically advantageous. The advantage is determined by the value of 107,652 produced holes for an hourly rate of the machine of 2,000 CZK/hour in favor of the variant without an accelerating head before reaching this production volume. In case of exceeding this volume of yield, an economically more advantageous variant with an accelerating head is issued. The original variant and variants using thread cutters have proven to be unsuitable. For this reason, variants with combined drill and tap Walter will be recommended to the firm Streicher, spol. s.r.o. Pilsen based on their production to perform machining with the best qualitative and economical aspects.

Next step of research is focused on milling operation. Milling is also taking place in machining of aluminum optical tables. Use of these products in vacuum atmosphere puts higher demands on milling operation as well as drilling, countersinking, and threading. Therefore, the research will lead to finding the best way and tools for machining of the optical tables. [10], [11]

6. Acknowledgments

I would like to thank to the co-authors of this paper for all their help and personal interest in this work. I would also like to thank to the colleagues from STREICHER spol. s.r.o. Pilsen, a company this research was done for, for their advice and all the support they have provided.

This article was created under the project SGS-2019-008: Research and Development for Innovation in the Field of Manufacturing Technology - Machining Technology III.

7. References

[1] Buranska, E[va] & Buransky, I[van] (2018). Cutting Environment Impact on the Aluminium Alloy Machining (DFA), Proceedings of the 29th DAAAM International Symposium, pp.1158-1163, B. Katalinic (Ed.), Published by DAAAM International, ISBN 978-3-902734-20-4, ISSN 1726-9679, Vienna, Austria DOI: 10.2507/29th.daaam.proceedings.166

[2] Steininger, A., Siller, A., Bleicher, F. (2014) Investigations regarding process stability aspects in thread tapping Al-Si alloys, Available from: https://www.scopus.com/record/display.uri?eid=2-s2.0-84937032739&origin=resultslist&sort=plf&src=s&sid=a15cbf7e20a92e56f8168a2535fba248&sot=b&sdt=b&sl=38&s=KEY%28machining+tools+threads+aluminium%29&relpos=4&citeCnt=19&searchTerm= Accessed: 2014-11-26

[3] www.binteractiv.cz, www cliquo cz. (2021) Laser center ELI Beamlines test the most poverfull laser on earth. Laserové centrum ELI Beamlines už testuje nejvýkonnější laser na světě a brzy zahájí plný provoz. Je to šance i pro vědce z Liberce. Available from: https://tuni.tul.cz/rubriky/veda-a-vyzkum/id:103812/laserove-centrum-eli-beamlines-uz-testuje-nejvykonnejsi-laser-na-svete-a-brzy-zahaji-plny-provoz-je-to-sance-i-pro-vedce-z-liberce Accessed on: 2021-01-05

[4] Streicher, spol s r o. (2020) Modular Vacuum Chamber. Streicher. Available from: https://www.scubechamber.com/ Accessed on: 2020-12-26

[5] Walter tools. Walter (2021) DC260-03-05.000A0-WJ30ET - Step drills - ToolsUnited. Available from: https://www.toolsunited.com/App/EN/Article/ArticleDetailsPage/181022154023869 Accessed on: 2021-05-05

[6] Emuge high performance tools (2021) Enorm 1-AL-GLT-8 | Emuge-Franken Website Available from: https://www.emuge-franken-group.com/gb/en/thread-technology/machine-tap/enorm-1-al-glt-8/p/000000000010418138 Accessed on: 2021-05-23

[7] Walter tools. Walter - N205166-M6 (2021) Taps with reinforced shank - ToolsUnited Available from: https://www.toolsunited.com/App/EN/Article/ArticleDetailsPage/24696400130374852 Accessed on: 2021-05-23

[8] DC260-03-03.300A0-WJ30ET (2021) Combined drill., Sdružený vrták obrázek Available from: https://www.boie.de/dc260-03-a0-zidPRD-17927 Accessed on: 2021-05-05

[9] Synek, Miloslav (1999) Business administration. in Prague, Podniková ekonomika. V Praze: C.H. Beck, 2006. ISBN 978-80-7179-892-7.

[10] Kroupa, T[omas]; Dana, M[ilan] & Zetek, M[iroslav] (2017). Using of Waterjet Technology for Cutting Aluminum Alloy, Proceedings of the 28th DAAAM International Symposium, pp.0486-0492, B. Katalinic (Ed.), Published by DAAAM International, ISBN 978-3-902734-11-2, ISSN 1726-9679, Vienna, Austria DOI: 10.2507/28th.daaam.proceedings.068

[11] Sogorovic, D[anijel] & Knezevic, A[nica] (2019). Testing of the Surface Roughness Parameters of Aluminium-Lithium Parts Machined by Milling, Proceedings of the 30th DAAAM International Symposium, pp.0148-0154, B. Katalinic (Ed.), Published by DAAAM International, ISBN 978-3-902734-22-8, ISSN 1726-9679, Vienna, Austria DOI: 10.2507/30th.daaam.proceedings.019

DOI: 10.2507/32nd.daaam.proceedings.090

EXPLAINABLE ARTIFICIAL INTELLIGENCE FOR ROBOT ARM CONTROL

Simon Schwaiger, Mohamed Aburaia, Ali Aburaia & Wilfried Woeber

This Publication has to be referred as: Schwaiger, S[imon]; Aburaia, M[ohamed]; Aburaia, A[li] & Woeber, W[ilfried] (2021). Explainable Artificial Intelligence For Robot Arm Control, Proceedings of the 32nd DAAAM International Symposium, pp.0640-0647, B. Katalinic (Ed.), Published by DAAAM International, ISBN 978-3-902734-33-4, ISSN 1726-9679, Vienna, Austria
DOI: 10.2507/32nd.daaam.proceedings.090

Abstract

In this paper, we investigate reinforcement learning model explainability through a pick and place task. Two robots with three degrees of freedom learned to solve the pick and place task in simulation as well as reality. To investigate the explanatory factors implicitly learned by the models, we derive robot parameters, i.e., the length of the robot segments. To overcome the black box nature of reinforcement learning models and provide a physical explanation of the results, the robot dimensions are derived from the learned reinforcement learning model and compared to the real dimensions. The hypothesis in the presented work is that converged reinforcement learning models must learn the robot parameters implicitly in order to learn a task. This transforms black box models into white box models, where each model's decisions can be interpreted. Our experiments show that robot parameters can be derived from learned models and that the chosen reinforcement learning model implicitly learns physical context. In order to create robust and trustworthy AI systems for intelligent factories, we suggest that a physical interpretation of all black box models must be done.

Keywords: Explainable Artificial Intelligence; XAI; Reinforcement Learning; Movement Control; Parameter Extraction

1. Introduction

In automation, motion planning is required for robots to perform tasks. Motion planning algorithms depend on physical models of robots as well as knowledge about the environment, such as obstacles and the goal pose. Information of the robot model, such as robot type and configuration, the physical properties and transforms between the robot coordinate frames need to be known in order to enable adequate motion planning and execution.

Accurate motion planning is made increasingly difficult by unknown robot parameters as well as a changing environment, creating new problems that a robot needs to solve [1], [2]. While artificial intelligence (AI) based approaches, such as reinforcement learning (RL) [3], can be used to automate planning procedures and partially overcome these problems, the models resulting from RL methods are so-called black box models [4], which must be turned into white box models for explainable and reliable application. Creating such explanation through physical context, however, still poses an open problem [5]. Motivation for solving this problem comes from the dependency of machine learning algorithms on data representation [6] in order to be successful.

In AI, reaching a level of so-called "usable intelligence" [7] is done in multiple stages, with explanation of the data in context of an application domain especially being a challenge. In order to meet this challenge, the field of explainable AI (XAI) engages with the concept of the interpretability of AI, aiming to enable models to be understood by a human through explanation and interpretation [8]. Understanding the functionality of a model allows for justification of its decisions, control of model behaviour, improvement of model architecture and discovery of new strategies for problem solving [4]. In the context of robotics, XAI can be used for feasibility explainability of task- and motion planning [4], enabling inspection of possible solutions to planning problems.

This paper tackles the identification of latent explanatory factors implicitly learned by RL methodologies for robot path planning problems. We trained a three degree of freedom robotic arm to perform a task using reinforcement learning and extract robot parameters from the model during application on the real robot. To this end, we initially trained a model in simulation. Afterwards, the model is adopted based on information from the real-world during deployment on the real robot. Finally, after the learning procedure, we investigate the transition models and the robot's kinematics to derive the physical characteristics of the joints. We hypothesise that physical context, i.e., the robot's kinematic structure, is learned implicitly.

This study is structured as follows. Chapter 2 depicts the current state of the art regarding RL applications and provides an overview of strategies of creating XAI solutions. Chapter 3 describes used methods, including an introduction to the concept of RL, as well as calculation of the kinematic structure of the tested robots. In chapter 4, results of the experiments performed on both robots are presented and discussed in chapter 5. A summary of the contents of the paper is given in chapter 6 and an outlook on future work is provided.

2. State of the Art

Reinforcement learning engages with the process of mapping the state of a system to an action in order to maximise a numerical reward [3]. Tasks can be formulated as an interaction between an agent and an environment, called the Markov-Decision process [3]. The goal of RL is to find a policy that determines which action the agent should take in a certain state. This concept provides a computational approach to understanding and automating goal-centric learning and decision making [3]. While RL models based on policy gradients [9], [10] and deep Q-networks (DQN) [11] have been applied to benchmarks, such as Atari games [12], continuous control based on Deep Deterministic Policy Gradients (DDPG) [13] has been used for practical RL implementations, such as process control [14], robot arm control [2] and motion planning for industrial robots [15].

XAI is part of the third wave of AI with one of the objectives of this wave being the precise creation of algorithms that explain themselves [4]. Used approaches for explaining black box models either look into the model itself or assume the model to be complex, trying to explain behaviour without knowing and altering the inner works of the model [4]. The first category is especially suitable when a model is fundamentally designed to be explainable, however, the more complex a model is, the more difficult this kind of explanation is to achieve. Since increased model accuracy often requires complex prediction methods [16], this results in a trade-off between model interpretability and accuracy [17].

The interpretation can be either done in a global manner [18], [19], explaining the entire reasoning for a model's decisions, or locally, based on explanation of a single decision [20]. Local explanations are commonly used for understanding image classification models [21], where the area most influential for the model's decision is highlighted [22], [23]. However, few work is done in the context of XAI in robotics. Explanatory factor extraction for mobile robots based on sparse Gaussian processes is done in [24]. This study examines black box models using motion data for mobile robot control. Using kinematics, vehicle parameters are derived from the black box model, transforming the black box model into a white box model.

3. Methods

In order to achieve a computationally efficient system, a robot with three degrees of freedom was used in a discrete environment. The environment is assumed to be fully observable in order to enable the use of policy and value iteration for computing a policy that moves the robot's tool centre point (TCP) to a desired location, while avoiding collisions with itself and obstacles. The environment and problem are formulated in a universal way in order to enable the use of multiple robots as well as multiple reinforcement learning methods for training. The computed model is globally explained by calculating the kinematics from application of the model on the real robot, showing that the kinematics are implicitly learned.

This allows for the black box model to be turned into a white box model by disclosing the internal design and explain the model in physical context. Similarly to the methodology applied in [24], we use robot kinematics and trained RL movement models to derive robot parameters and thus unveil explanatory factors implicitly learned by the RL agent. Joint state data as well as the resulting TCP position are measured during application of the trained model on the robot. Based on this data, robot parameters representing the real robot are calculated using kinematics. A full overview of the resulting software solution is provided in Fig. 4.

3.1 Reinforcement Learning

For application of RL, the problem is formulated as a Markov Decision process [3]. Environment and agent are created based on a simulated robot (see Fig. 1). The environment is fully observable and discrete with the state being based on the robot's joint positions. A positive reward is given to the agent whenever the robot's TCP reaches the desired position and a negative reward is given when the robot collides with itself or any obstacle. A transition function is created based on the simulation, which maps the resulting next state for each possible action-state pair to the associated reward (success, failure or neutral). Based on this function, a policy is computed using policy and value iteration [3], [25], determining what action the agent takes in each state in order to maximise reward. The experiments are based on two robots, one of which has three rotational joints and the other with two rotational and one translational axes.

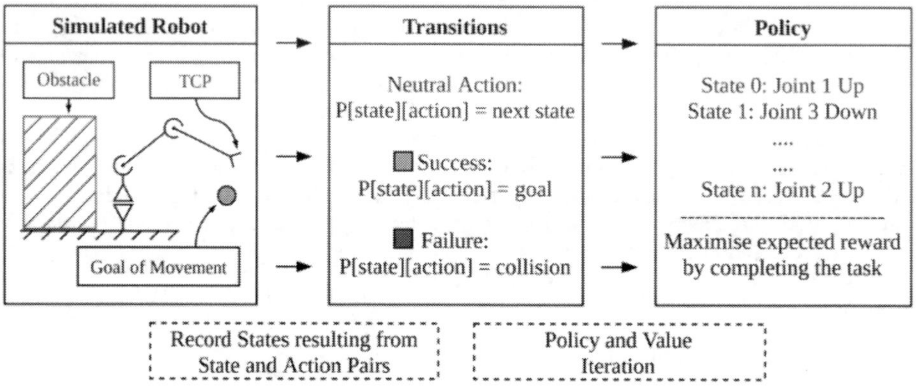

Fig. 1. The underlying structure of the RL model consists of transitions, created from a virtual version of the robot. The policy is computed based on these transitions with the goal of maximising reward. Reward is given upon the TCP reaching the desired destination, while a negative reward is given upon collisions.

3.2 Extraction of the first Kinematic Model

In order to show that the RL model implicitly learns the robot kinematics, robot parameters are calculated using TCP position and joint state data during application of the model on the real robot. Robot 1 is a three degree of freedom robotic arm with three rotatory axes based on the Hebi X-Series [26] platform (see Fig. 2). TCP displacement along the y-axis (when $\theta_1 = 0$) is combined into one metric, noted as o_1, resulting in four unknown robot parameters (l_1, l_2, l_3, o_1). The forward transform is not formulated based on the Denavit-Hartenberg notation [27], since the resulting equation system was not solvable due to the high complexity of the equations. Instead, the transform is established geometrically.

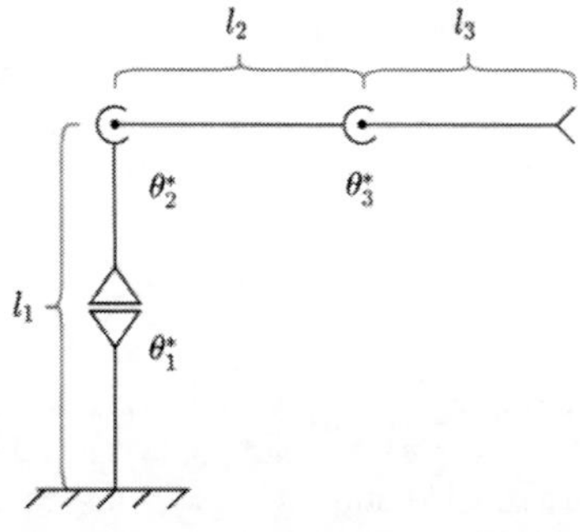

Fig. 2. Kinematic structure of the first robot.

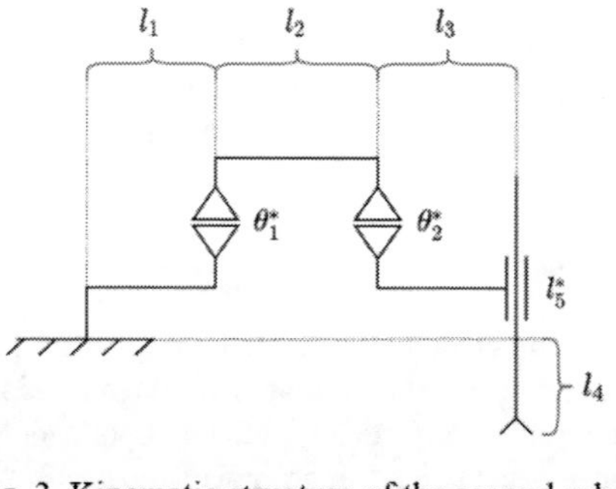

Fig. 3. Kinematic structure of the second robot.

Assuming $\theta_1 = 0$, terms for z and x-coordinates of an intermediary point P, which is offset by o_1 from the TCP position, are formulated ((2) and (1)). The resulting term for P_z is used directly, while the term for P_x is projected onto the xy-plane with the inclusion of θ_1. The resulting term (3) describes the coordinates of P. Since this offset is orthogonal to the first joint's x-axis, θ_1 can be used to project the offset onto the xy-plane and add it to the intermediary point.

$$P_x(\theta_1 = 0) = l_3 \cdot \cos(\theta_2 + \theta_3) + l_2 \cdot \cos(\theta_2) \quad (1)$$

$$P_z = l_1 + l_2 \cdot \sin(\theta_2) + l_3 \cdot (\theta_3 + \theta_2) \tag{2}$$

$$\begin{pmatrix} TCP_x \\ TCP_y \\ TCP_z \end{pmatrix} = \begin{pmatrix} P_x \\ P_y \\ P_z \end{pmatrix} + \begin{pmatrix} o_1 cos\left(\theta_1 + \frac{\pi}{2}\right) \\ o_1 sin\left(\theta_1 + \frac{\pi}{2}\right) \\ 0 \end{pmatrix} = \begin{pmatrix} cos(\theta_1)[l_3 cos(\theta_2 + \theta_3) + l_2 cos(\theta_2)] \\ sin(\theta_1)[l_3 cos(\theta_2 + \theta_3) + l_2 cos(\theta_2)] \\ l_1 + l_2 \cdot sin(\theta_2) + l_3 \cdot (\theta_3 + \theta_2) \end{pmatrix} + \begin{pmatrix} o_1 cos\left(\theta_1 + \frac{\pi}{2}\right) \\ o_1 sin\left(\theta_1 + \frac{\pi}{2}\right) \\ 0 \end{pmatrix} \tag{3}$$

When substituting the known joint states, (3) results in a linear system, allowing for a simple determination of the unknown variables, since o_1 can be calculated as the orthogonal distance between the first joint's x-axis and the TCP position. o_1, the joint states and the TCP positions can be substituted into the equation system, resulting in a system that has three unknown variables and three equations, making the system solvable.
In order to improve the reliability of the parameter calculation, the TCP position for $\theta_1 = \theta_2 = \theta_3 = 0$ (see (4)) is measured as well and substituted into one of the lines in (3), as seen in (5).

$$\theta_1, \theta_2, \theta_3 = 0 \rightarrow (TCP_x) = (l_3 + l_2) \tag{4}$$

$$\begin{pmatrix} TCP_x \\ TCP_x(\theta_1, \theta_2, \theta_3 = 0) \\ TCP_z \end{pmatrix} = \begin{pmatrix} cos(\theta_1)[l_3 \cdot cos(\theta_2 + \theta_3) + l_2 \cdot cos(\theta_2)] + o_1\left(\theta_1 + \frac{\pi}{2}\right) \\ l_3 + l_2 \\ l_1 + l_2 \cdot sin(\theta_2) + l_3 \cdot (\theta_3 + \theta_2) \end{pmatrix} \tag{5}$$

Robot parameters are calculated by first calculating o_1 as the orthogonal distance between TCP and the first joint's x-axis. The TCP positions are measured on the robot during operation, while θ_1, θ_2 and θ_3 are determined using the built-in feedback from the used servo motors. The joint states, as well as o_1 and the TCP positions are substituted into the equation system, resulting in a linear system, with three terms and three unknown variables. A linear solver is used to determine the parameters l_1 to l_3.

3.3 Extraction of the second Kinematic Model

Robot 2 is a three degree of freedom robot with two rotating and one translational axis based on the Makeblock smart servo [28] platform. In a similar manner to the first robot, the transform from base to TCP is calculated geometrically in order to obtain a linear equation system for determination of the robot parameters l_1 to l_4 (see Fig. 3). X and y-coordinates of the TCP are expressed based on positions of the first two joints. Due to the third joint being a translational axis, the TCP position along the z-axis is determined by the displacement of the robot's linear axis (l_5), as well as l_4. The transformation from base to TCP is depicted in (6).

$$\begin{pmatrix} TCP_x \\ TCP_y \\ TCP_z \end{pmatrix} = \begin{pmatrix} l_3 \cdot cos(\theta_1 + \theta_2) + l_2 \cdot cos(\theta_1) + l_1 \\ l_3 \cdot sin(\theta_1 + \theta_2) + l_2 \cdot sin(\theta_1) \\ l_4 + l_5 \end{pmatrix} \tag{6}$$

Since the robot has four unknown parameters, (6) is not solvable without additional information. Therefore, the TCP position's x-coordinate is measured for $\theta_1 = \theta_2 = l_5 = 0$ and formulated as seen in (7). In combination with the forward transform, this results in a solvable linear equation system (8) that allows for calculation of the robot parameters (l_1 to l_4).

$$\theta_1, \theta_2 = 0 \rightarrow (TCP_x) = (l_3 + l_2 + l_1) \tag{7}$$

$$\begin{pmatrix} TCP_x \\ TCP_y \\ TCP_z \\ TCP_x(\theta_1, \theta_2, l_5 = 0) \end{pmatrix} = \begin{pmatrix} l_3 \cdot cos(\theta_1 + \theta_2) + l_2 \cdot cos(\theta_1) + l_1 \\ l_3 \cdot sin(\theta_1 + \theta_2) + l_2 \cdot sin(\theta_1) \\ l_4 + l_5 \\ l_3 + l_2 + l_1 \end{pmatrix} \tag{8}$$

The robot parameters are calculated by measuring TCP positions, as well as θ_1, θ_2 and l_5 on the robot during operation and substituting them into (8). The resulting equation system is then solved for the robot parameters using a linear solver.

3.4 Used Software Packages

The robot operating system (ROS Noetic) [29] provides communication between different software components and manages coordinate frame transformations. The built-in software Rviz and the Rviz Visual Tools plugin [30] are used for visualisation of the virtual robot's training procedure.

OpenAI Gym [31] is used as the base framework for reinforcement learning since it provides a standardised way of formulating the RL problem as well as an interface between agent and environment. Even though the OpenAI Gym software offers multiple implemented environments, a custom environment for controlling the tested robots was implemented. Due to the standardised nature of the OpenAI Gym framework, any reinforcement learning algorithm written for discrete OpenAI Gym environments can be used with the implemented software solution.

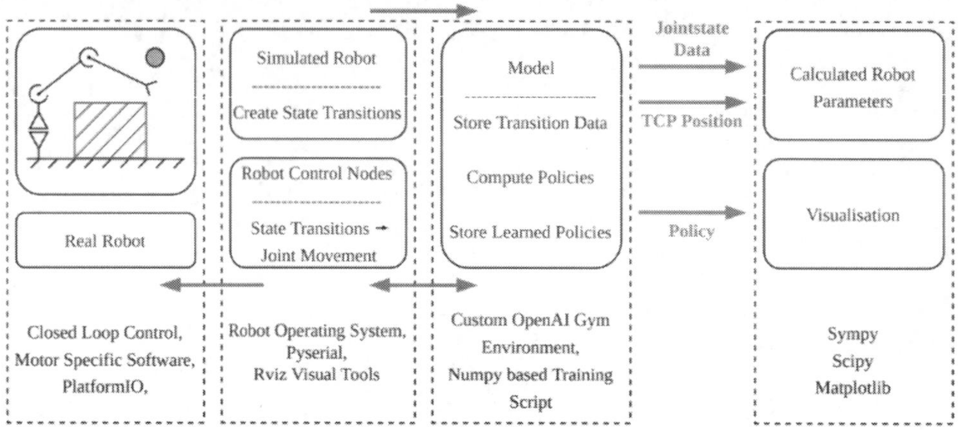

Fig 4. Multiple openly available software packages have been used for implementation of the system.

Since robot 1 is based on the HEBI-Robotics X-series platform [26], the provided Hebi ROS, C++ and Python libraries are used for robot control. Robot 2 is based on the Makeblock platform. Therefore, the Makeblock Servo Library [28], Pyserial [32] and PlatformIO [33] are used to control the real robot from the RL application. Sympy, a symbolic mathematics library for Python [34] and the linear algebra module of Scipy [35] are used for calculating the robot kinematics. Data is stored and handled using Numpy [36].

4. Experimental Results

In order to measure the error introduced by the robot parameter calculation, a pick and place task was learned in simulation on both robots. Simulated joint state and TCP position data was sampled ten times during operation at different robot poses. Based on this data, the robot parameters were calculated for each pose based on the kinematics presented in chapter 3. The calculated robot parameters are compared to the actual parameters used for simulation, in order to determine the accuracy of the implemented kinematics solution. Table 1 shows the parameters of the simulated robot, as well as average discrepancy between the calculated and real robot parameters in absolute and percent values.

	Robot 1			
	l1	l2	l3	o1
Robot len.[m]	0.086	0.175	0.252	0.043
Avg. dev.[m]	0.00014	0.00043	0.00043	1.48E-05
Avg. dev.[%]	0.16311	0.2469	0.17115	0.03481

	Robot 2			
	l1	l2	l3	l4
Robot len.[m]	0.047	0.093	0.073	0.042
Avg. dev.[m]	1.65E-16	2.62E-16	1.26E-16	1.39E-17
Avg. dev.[%]	3.51E-13	2.82E-13	1.73E-13	3.30E-14

Table 1. Data from simulations of both robots was used to measure the average error introduced by calculation of the robot parameters.

In order to show the system's ability to correct an inaccurate virtual robot model, a simulation was set up for both robots with robot parameters that intentionally deviate from the real robots. The use case depicted by this experiment intends for a model of a robot with partially unknown parameters to be trained in an inaccurate simulation. The model can then be corrected using data from the real robot, gained during deployment.

Point	Joint State			TCP pos.[m]		
P1	0.229	0.324	0.109	1.06	0.59	-0.88
P2	0.222	-0.234	0.11	-0.68	1.02	-1.52
P3	0.17	0.393	0.107	1.26	0.29	-0.4

Point	Robot Param.			
P1	0.0862	0.1747	0.2603	0.0414
P2	0.0855	0.175	0.26	0.0424
P3	0.0859	0.1739	0.2611	0.0417
Actual	0.086	0.175	0.26	0.0425

Table 2. Calculated robot parameters of robot 1 for the measured points. TCP_x (θ_1, θ_2, θ_3=0) was measured as 0.435m.

Three TCP positions (P_1, P_2, P_3) were measured by hand on each robot and used with the respective joint state data to calculate the robot parameters. The TCP coordinates and corresponding joint state data, as well es the calculated and actual robot parameters are depicted in in tables 2 and 3 for each respective robot. Based on the data in tables 2a and 3a, robot parameters were calculated using the kinematics presented in chapter 3. The resulting robot parameters as well as the actual parameters used for training an accurate representation of the real robots are depicted in tables 2b and 3b.

Point	Joint State			TCP pos.[m]		
P1	-0.5061	0.9599	-0.048	0.194	-0.011	-0.09
P2	0.6283	0.6283	-0.043	0.145	0.124	-0.0855
P3	-0.8029	-1.1694	0	0.083	-0.135	-0.0425

Point	Robot Param.			
P1	0.0469	0.0908	0.0753	-0.042
P2	0.0469	0.0936	0.0726	-0.0425
P3	0.0457	0.0947	0.0727	-0.0425
Actual	0.047	0.093	0.073	-0.0425

Table 3. Calculated robot parameters of robot 2 for the measured points. TCP_x (θ_1, θ_2=0) was measured as 0.213m.

5. Discussion

The presented system is able to train a model based on a simulated robot and use data from deployment on the real robot to correct the trained model. Table 1 shows that extraction of the robot parameters introduces an absolute error in the range of (3.3E−14m; 1.4E10−4m). This error, however, is smaller than other possible sources of inaccuracy (e.g., inaccurate manual measuring of the TCP positions or deviations resulting from closed loop motor control). The accuracy achieved in the practical implementation of the system allows for the completion of a pick and place task using robot 1. The robot trajectory is planned by the RL agent, who is able to adapt to different obstacles that need to be avoided. In practice, this allows for the system to be flexible in terms the surroundings of deployment as well as in terms of the problem that needs to be solved. Thus, we accept our hypothesis since the reinforcement learning model did learn the kinematic structure implicitly.

6. Summary and Outlook

In this paper, a system for autonomously fulfilling a pick and place task of three degree of freedom robots was implemented and tested. The system is capable of determining the robot's trajectory using reinforcement learning (RL), reaching the desired TCP position while avoiding obstacles as well as correcting the learned model by means of explainable artificial intelligence (XAI). This allows for a model to be learned in simulation and corrected afterwards, using data gained from the real world in order to achieve a model that is accurate to the real robot. The system was tested on two robots of different configurations, in simulation and reality. Future work on application of XAI on robot control is possible regarding multiple aspects of XAI due to the novelty of the topic. Explanation techniques, such as the one presented in this paper, are inherently only able to explain a certain type of model.

A generalised approach for explanation of robot behaviour still poses an open problem [4]. Furthermore, the trade-off between model accuracy and model explainability [17] still prevents models that are not designed to be explainable from being fully interpretable.

7. Acknowledgments

This work was supported by the city of Vienna (MA23 – Economic Affairs, Labour and Statistics) through the research project AIAV (MA23 project 26-04).

8. References

[1] Vecerik, M., Hester, T., Scholz, J., Wang, F., Pietquin, O., Piot, B., Heess, N., Rothörl, T., Lampe, T. and Riedmiller, M. (2017). Leveraging demonstrations for deep reinforcement learning on robotics problems with sparse rewards. arXiv preprint arXiv:1707.08817.

[2] Andrychowicz, M., Wolski, F., Ray, A., Schneider, J., Fong, R., Welinder, P., McGrew, B., Tobin, J., Abbeel, P. and Zaremba, W. (2017). Hindsight experience replay. arXiv preprint arXiv:1707.01495.

[3] Sutton, R. S., & Barto, A. G. (2018). Reinforcement learning: An introduction, A Bradford Book, 978-0-262-03924-6, Cambridge, MA, USA.

[4] Adadi, A., & Berrada, M. (2018). Peeking inside the black-box: a survey on explainable artificial intelligence (XAI). IEEE access, 6, 52138-52160.

[5] Pearl, J., & Mackenzie, D. (2018). The book of why: the new science of cause and effect. Basic books, 978-0-465-09760-9, New York, NY, USA.

[6] Bengio, Y., Courville, A., & Vincent, P. (2013). Representation learning: A review and new perspectives. IEEE transactions on pattern analysis and machine intelligence, 35(8), 1798-1828.

[7] Holzinger, A. (2018). From machine learning to explainable AI. In 2018 world symposium on digital intelligence for systems and machines (DISA) (pp. 55-66). IEEE.

[8] Anjomshoae, S., Najjar, A., Calvaresi, D., & Främling, K. (2019). Explainable agents and robots: Results from a systematic literature review. In 18th International Conference on Autonomous Agents and Multiagent Systems (AAMAS 2019), Montreal, Canada, May 13–17, 2019 (pp. 1078-1088). International Foundation for Autonomous Agents and Multiagent Systems.

[9] Sutton, R. S., McAllester, D. A., Singh, S. P., & Mansour, Y. (2000). Policy gradient methods for reinforcement learning with function approximation. In Advances in neural information processing systems (pp. 1057-1063).

[10] Silver, D., Lever, G., Heess, N., Degris, T., Wierstra, D., & Riedmiller, M. (2014, January). Deterministic policy gradient algorithms. In International conference on machine learning (pp. 387-395). PMLR.

[11] Mnih, V., Kavukcuoglu, K., Silver, D., Rusu, A.A., Veness, J., Bellemare, M.G., Graves, A., Riedmiller, M., Fidjeland, A.K., Ostrovski, G. and Petersen, S. (2015). Human-level control through deep reinforcement learning. nature, 518(7540), 529-533.

[12] Mnih, V., Kavukcuoglu, K., Silver, D., Graves, A., Antonoglou, I., Wierstra, D., & Riedmiller, M. (2013). Playing atari with deep reinforcement learning. arXiv preprint arXiv:1312.5602 .

[13] Lillicrap, T.P., Hunt, J.J., Pritzel, A., Heess, N., Erez, T., Tassa, Y., Silver, D. and Wierstra, D. (2015). Continuous control with deep reinforcement learning. arXiv preprint arXiv:1509.02971.

[14] Spielberg, S., Tulsyan, A., Lawrence, N. P., Loewen, P. D., & Gopaluni, R. B. (2020). Deep Reinforcement Learning for Process Control: A Primer for Beginners. arXiv preprint arXiv:2004.05490.

[15] Meyes, R., Tercan, H., Roggendorf, S., Thiele, T., Büscher, C., Obdenbusch, M., Brecher, C., Jeschke, S. and Meisen, T. (2017). Motion planning for industrial robots using reinforcement learning. Procedia CIRP, 63, 107-112.

[16] Breiman, L. (2001). Statistical modeling: The two cultures (with comments and a rejoinder by the author). Statistical science, 16(3), 199-231.

[17] Sarkar, S., Weyde, T., Garcez, A., Slabaugh, G. G., Dragicevic, S., & Percy, C. (2016). Accuracy and interpretability trade-offs in machine learning applied to safer gambling. In CEUR Workshop Proceedings (Vol. 1773). CEUR Workshop Proceedings.

[18] Yang, C., Rangarajan, A., & Ranka, S. (2018). Global model interpretation via recursive partitioning. In 2018 IEEE 20th International Conference on High Performance Computing and Communications; IEEE 16th International Conference on Smart City; IEEE 4th International Conference on Data Science and Systems (HPCC/SmartCity/DSS) (pp. 1563-1570). IEEE.

[19] Lundberg, S.M., Erion, G., Chen, H., DeGrave, A., Prutkin, J.M., Nair, B., Katz, R., Himmelfarb, J., Bansal, N. and Lee, S.I. (2020). From local explanations to global understanding with explainable AI for trees. Nature machine intelligence, 2(1), 56-67.

[20] Ljubobratovic, D., Guoxiang, Z., Brkic Bakaric, M., Jemric, T., & Matetic, M. (2020). Predicting Peach Fruit Ripeness Using Explainable Machine Learning, Proceedings of the 31st DAAAM International Symposium, pp.0717-0723, B. Katalinic (Ed.), Published by DAAAM International, ISBN 978-3-902734-29-7, ISSN 1726-9679, Vienna, Austria

[21] O'Shea, K., & Nash, R. (2015). An introduction to convolutional neural networks. arXiv preprint arXiv:1511.08458.

[22] Baehrens, D., Schroeter, T., Harmeling, S., Kawanabe, M., Hansen, K., & Müller, K. R. (2010). How to explain individual classification decisions. The Journal of Machine Learning Research, 11, 1803-1831.
[23] Woods, W., Chen, J., & Teuscher, C. (2019). Adversarial explanations for understanding image classification decisions and improved neural network robustness. Nature Machine Intelligence, 1(11), 508-516.
[24] Wöber, W., Novotny, G., Mehnen, L., & Olaverri-Monreal, C. (2020). Autonomous Vehicles: Vehicle Parameter Estimation Using Variational Bayes and Kinematics. Applied Sciences, 10(18), 6317.
[25] Russell, S., & Norvig, P. (2002). Artificial intelligence: a modern approach, Prentice Hall Press, 978-0-13-604259-4, One Lake Street Upper Saddle River, NJ, USA.
[26] https://www.hebirobotics.com/, (2021). HEBI Robotics, X-series actuator, Accessed on: 2021-10-15
[27] Angeles, J., & Angeles, J. (2002). Fundamentals of robotic mechanical system, Springer Verlag, 978-3-319-01850-8, NY, USA.
[28] https://github.com/Makeblock-official/Makeblock-Libraries, (2020). Makeblock-official, Makeblock-libraries: Arduino library for makeblock electronic modules, Accessed on: 2021-10-15
[29] Quigley, M., Conley, K., Gerkey, B., Faust, J., Foote, T., Leibs, J., Wheeler, R. and Ng, A.Y. (2009). ROS: an open-source Robot Operating System. In ICRA workshop on open source software (Vol. 3, No. 3.2, p. 5).
[30] https://github.com/PickNikRobotics/rviz visual tools, (2020). PickNikRobotics, Rviz visual tools: C++ api wrapper for displaying shapes and meshes in rviz, Accessed on: 2021-10-15
[31] Brockman, G., Cheung, V., Pettersson, L., Schneider, J., Schulman, J., Tang, J., & Zaremba, W. (2016). Openai gym. ArXiv Preprint ArXiv:1606.01540.
[32] https://pypi.org/project/pyserial, (2020). C. Liechti, pyserial 3.4, Accessed on 2021-10-15
[33] https://platformio.org/, (2020), PlatformIO labs, A professional collaborative platform for embedded development platformio, Accessed on: 2021-10-15
[34] Meurer A, Smith CP, Paprocki M, Čertík O, Kirpichev SB, Rocklin M, Kumar A, Ivanov S, Moore JK, Singh S, Rathnayake T, Vig S, Granger BE, Muller RP, Bonazzi F, Gupta H, Vats S, Johansson F, Pedregosa F, Curry MJ, Terrel AR, Roučka Š, Saboo A, Fernando I, Kulal S, Cimrman R, Scopatz A. (2017) SymPy: symbolic computing in Python. PeerJ Computer Science, 3, e103, https://doi.org/10.7717/peerj-cs.103
[35] Virtanen, P., Gommers, R., Oliphant, T.E., Haberland, M., Reddy, T., Cournapeau, D., Burovski, E., Peterson, P., Weckesser, W., Bright, J. and Van Der Walt, S.J. (2020). SciPy 1.0: fundamental algorithms for scientific computing in Python. Nature methods, 17(3), 261-272.
[36] Harris, C.R., Millman, K.J., van der Walt, S.J., Gommers, R., Virtanen, P., Cournapeau, D., Wieser, E., Taylor, J., Berg, S., Smith, N.J. and Kern, R. (2020). Array programming with NumPy. Nature, 585(7825), 357-362.

DOI: 10.2507/32nd.daaam.proceedings.091

CFD INVESTIGATION OF TEXTILE DRYING TEMPERATURE AND DENSITY IN A INDUSTRIAL STENTER FRAME SECTION

Mesud Ramic, Ejub Dzaferovic & Amra Hasecic

This Publication has to be referred as: Ramic, M[esud]; Dzaferovic, E[jub] & Hasecic, A[mra] (2021). CFD Investigation of Textile Drying Temperature and Density in a Industrial Stenter Frame Section, Proceedings of the 32nd DAAAM International Symposium, pp.0648-0651, B. Katalinic (Ed.), Published by DAAAM International, ISBN 978-3-902734-33-4, ISSN 1726-9679, Vienna, Austria
DOI: 10.2507/32nd.daaam.proceedings.091

Abstract

Drying as a basic scientific phenomenon has been intensively investigated but limited for specific and real running stenter. Authors approach was to apply scientific methods to investigate real machine performance. This work investigates phenomenon of textile drying under specific technological conditions. Temperature inside the machine chamber has been scientifically proven as a parameter that significantly correlates to energy consumption. Research on such large systems and machines is too expensive and the investigation must involve the digitalization of process by modern computational methods. Previous experimental investigation and 2D numerical simulation of the machine have been used as base for extended investigation with 3D numerical model and appropriate simulation. Goal is to understand temperature and density distribution inside the textile dried in a stenter chamber with simulation under real value parameters.

Keywords: temperature; stenter; CFD; drying; energy efficiency;

1. Introduction

The first energy efficiency project in Bosnia-Hercegovina (2011) with appropriate measures reflected to 43% reduced energy consumption [1]. Industrial drying and finishing of textiles is a part of the finishing processes which are energy-intensive processes and interesting from the aspects of energy consumption. Measures for energy efficiency are the focus on energy management, recovery, alternative energy, alternative processes and more [2]. This The basic principles of thermodynamic dependences in drying process of polyester textile materials can be found in the papers where the dependences of drying costs in relation to the humidity parameter [2], nozzle temperature, chamber climate [3] and material stabilization time were investigated [4]. According to the research [6] 39.7% of energy belongs to heating the air while 41% is used in the evaporation process of energy breakdown 6.2 GJ/tonne material for typical stenter. Reducing the supply of fresh air to the combustion chamber from 10 to 5 kilograms of air per kg material, energy savings of 57% are obtained. Further studies were related to the development of mathematical models [5], numerical simulations and appropriate analysis [6]. The authors investigate the effects of turbulent flows [7], flow velocities on convective heat transfer at individual droplets [8] and turbulent channel flow [9].

Base for this investigation are presented in studies with experimental part and defined indicators to understand stenter energy consumption [11], statistical correlation analysis to define relations between parameters [12], 2D and 3D model to understand turbulent flow, temperature and density distribution in area between nozzles and textile [13]. Intelligent manufacturing is one of the key roles in the Industry 4.0 [14], therefor this CFD numerical investigation has been made as further work to understand new aspects of the physical modelling of the process regarding temperature and density distribution inside the textile that is subject of drying. Mathematical modelling in this research include model of heat transfer, model of turbulent mixture flow and mass transport, material characteristics model and boundary conditions [6].

2. Numerical method and results

In this study standard Ansys Fluent Academy Package R20 was used for numerical simulation. The investigation model is based on Finite Volume Methodology and numerical solution of equations is based on methods of discretization. Discretization is the process of converting integral-differential equations into algebraic equations [15]. In numerical simulation were used mixture model of the fluid that simulates the phases of water and evaporation, model of solid used for material characteristics. Textile inside the section was approximated as porous media with its porosity characteristics calculated based on [9]. The mixture flow was simulated using RANS approach for the three-phase flow. It solves the transport equation, pressure and velocity by SIMPLE algorithm. A realizable high-Re number k-ε model was used for prediction of turbulence. The standard wall treatment was applied for providing the wall boundary conditions. Pressure-based solver was used and the results are visualized with different settings such as contours and field vector. Real geometric shape of section, numerical model and boundary conditions are presented at Figure 1. At Figure 2. are showed air temperature and phase density inside the textile for time periods; t_1=0.25s, t_2=0.5s, t_3=0.75s, t_4=1s. Temperature inside the textile for time periods t_1=0.25s, t_2=0.5s, t_3=0.75s, t_4=1s in 3D are presented at Figure 3. The geometric insensitivity test was performed by iterative network fragmentation according [13].

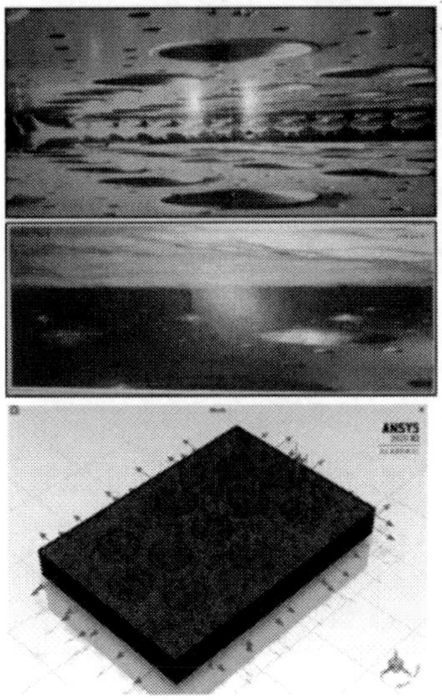

DESCRIPTION	VALUE	UNIT
Model dimension	70*12,5*100	cm
Textile position in Y	0	mm
Textile thickness	5	mm
Mesh elements	429071	pcs
Mesh element size	0,011	m³
Textile velocity	0.2	m/s
Textile weight	0,25	kg/m³
Outlet surface area of nozzles	1727	cm²
Inlet surface area of air nozzles	372,68	cm²
Nozzle diametar outlet	10	cm
Nozzle diameter inlet	14	mm
Air velocity at inlet nozzle	44	m/s
Temp of air at inlet nozzle	443,15	K
Porosity of textile	15,79	%
Viscous resistance of textile	9.13e15	1/m²
Time step duration	0,25	S
Time steps	4	pcs
Spec. heat cap. polyesther	988	J/kgK
Spec. heat cap. water	4182	J/kgK
Heat cond. coeff. polyesther	0.727	w/mK
Density polyesther	1320	kg/m³
Pressure at inlet nozzles	105	kPas
Wall temp.	300	K
Spec. heat of vapor	2014	J/kgK
Thickness of slide	1	mm
Textile temp.	293,15	K
Operational pressure	101325	Pas
Fluid zones	3	pcs

Fig. 1. Real section geometry (left top), 3D numerical model (left bottom) and calculation parameters (right) used for numerical simulation

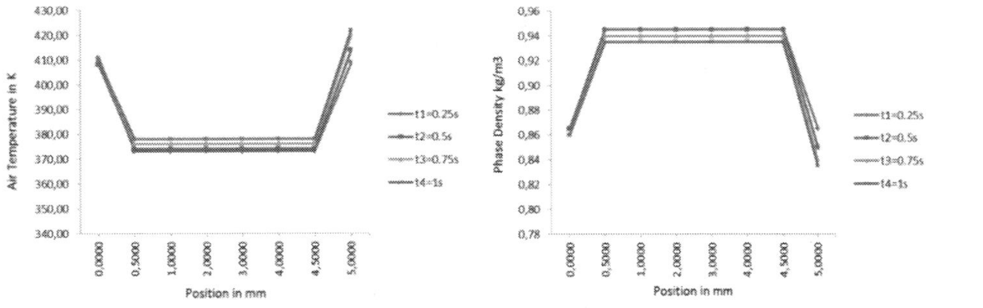

Fig. 2. Air temp. (left) and phase density (right) inside the textile for time periods; t1=0.25s, t2=0.5s, t3=0.75s, t4=1s

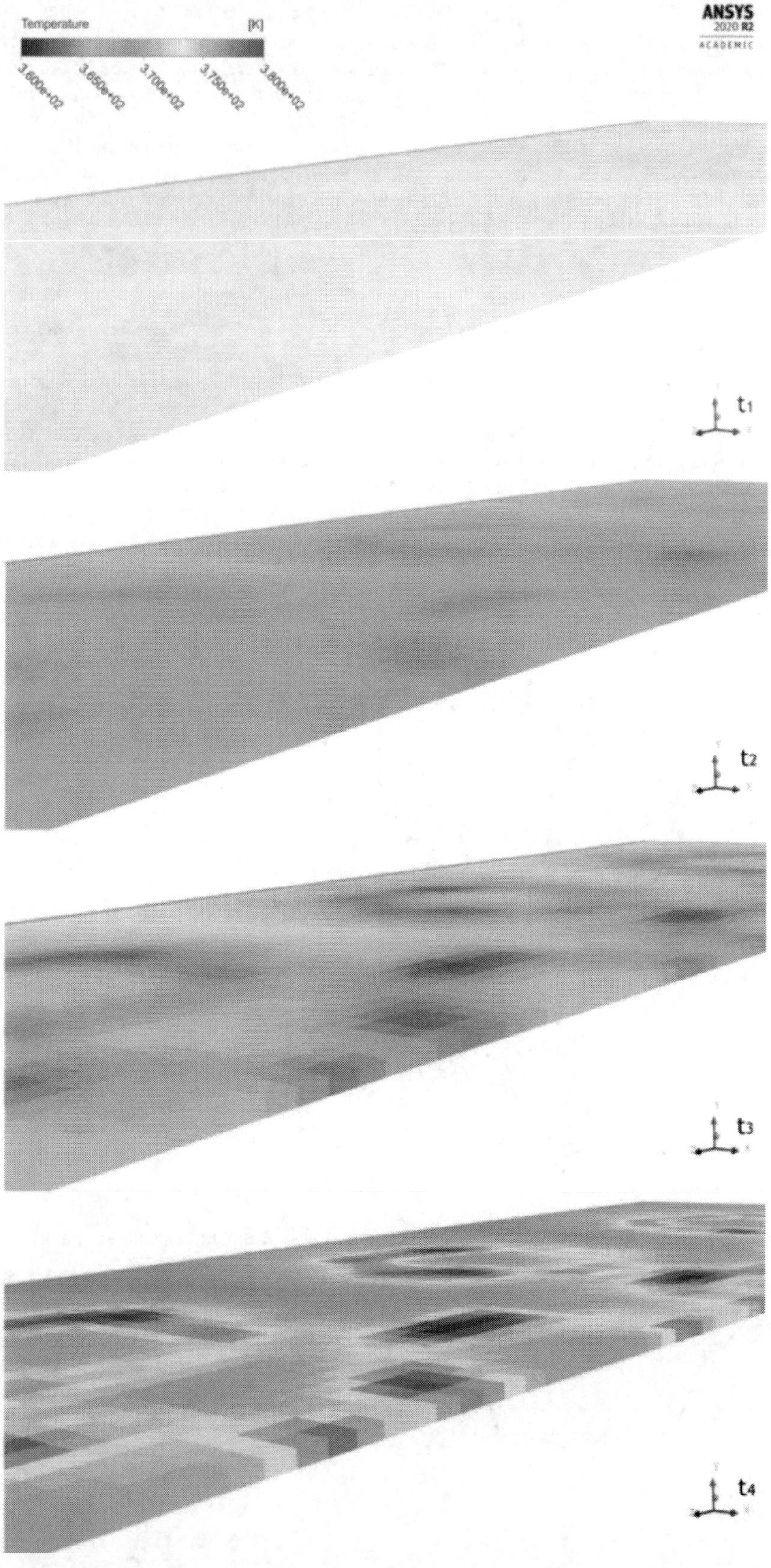

Fig. 3. 3D temperature distribution inside the textile for time periods; t_1=0.25s, t_2=0.5s, t_3=0.75s, t_4=1s

3. Conclusion

The motivation for this research was unknown temperature and density distribution inside the textile in a stenter frame section. The solution for this was to numerically simulate the drying process and analyse temperature and density distribution inside the textile. The temperature and density were calculated and presented graphically. Textile temperature decreases linearly from outside, from both sides to the central line of material. In case of density the process is inversely proportional and the calculated values increase linearly to the central segment of material. From the view of the textile middle section to surfaces, during evaporation process density is decreasing and temperature is increasing. These method and results help us to understand running phenomena in textile and chamber section and could be used as base for simulation with real material characteristic. Limitation in this research represents the model of textile, where phenomena of porous media was used and not the real textile structure with his porosity. Future plans are to investigate the same drying process and phenomena under conditions where textile is modelled as real as possible with yarn characteristics and weaving structure. This will lead to recognition of real interdependence between temperature and density for the appropriate parameters of stenter exploitation. The contributions of this and planed investigation are multiple and this research work is useful from both scientific and professional aspects. The research methodology is traceable and integrative because it includes research in real conditions, calculations and statistical analysis in combination with numerical process simulations, and as such, in addition to its scientific validity, it also has its market applicability. The collected results represent one base for the profession, plant design organizations, designers and textile materials technologists. The database is also useful for efforts to reduce energy consumption its harmful impact on the environment. The results of this study can improve the drying concepts in the textile industry, as well as in all other applications with porous material, and bring closer knowledge about potential savings.

4. Acknowledgments

This research did not receive any specific grant from funding agencies in the public, commercial, or not-for-profit Sectors.

5. References

[1] Zečević, N.; Husika, A.; Džaferović, E. (2018). Impact of Energy Efficiency Measures on Indoor Air Quality in Building of Mechanical Engineering Faculty Sarajevo, Proceedings of the 28th International DAAAM Symposium, ISBN 978-3-902734-11-2

[2] Al Musa, F.; Qahtani, H.; Owaidh, M.; Qahtani, S.; Noureldin M. (2011). GHG Emissions Reduction Via Energy Efficiency Optimization, ISBN: 978-953-51-4328-4,

[3] Hasanbeigi, A. (2010). Energy-Efficiency Improvement Opportunities for the Textile Industry, National Laboratory, Berkeley

[4] Pabst, M. (1979). Trocken und thermofixieren von Poliesterartikeln unter wirtschaftlichen Gesichtspunkten, Lenzinger Berichte, Vol 47

[5] Wang, Sh. (2020). Process analysis and optimization of open-width fabric continuous drying based on numerical simulation, Textile Research Journal 1–25

[6] Wang, Sh. (2019). A numerical model of the open-width coupling drying process for cotton fabrics based on the theory of heat and mass transfer in porous media, Textile Research Journal Vol. 90

[7] Schmidt, M. (2009). CFD Untersuchung des Auftrags hochviskoser polymerer Systeme in Textilen Verarbeitungsprozessen, Dissertation, Universitaet Duisburg

[8] Guo, N.; Finnerman O.; Strom H. (2016). The effect of turbulent velocity fluctuations on the convective heat transfer to droplets subjected to evaporation and thermolysis, AIP Publishing

[9] Wei, Y.; Hua, J.; Ding, X. (2017). A mathematical model for simulating heat and moisture transfer within porous cotton fabric drying inside the domestic air-vented drum dryer, Text Inst Proc,

[10] Fontana, E.; Donca, R.; Mancusi E.; Augusto, A.; Guelli, S. (2016). Mathematical modeling and numerical simulation of heat and moisture transfer in a porous textile medium", The Journal of the Textile Institute, Volume 107 No 5, 672-682

[11] Ramić, M.; Zrno, A.; Husika A. (2016). Energy flow analysis of stenter Artos Babcock by textile producer Prevent Fabrics, World Textile Conference AUTEX, Ljubljana SLO

[12] Ramić, M.; Hasečić, A.; Džaferović, E. (2016). Dependence analysis of parameters from thermo-fixing process based on IBM SPSS Statistical tool, IEEE Conference, Zlatibor, SRB,

[13] Ramić, M.; Hasečić, A.; Džaferović, E. (2021). Numerical analysis of drying proces in stenter frame section, ICNAAM Conference of Numerical Analysis and Applied Mathematics, Accepted paper, Greece

[14] Kržalić, A.; Hodžić L. (2019). Sustainable engineering challenges towards Industry 4.0, Acomprehensive review, Sustainable Engineering and Innovation, ISSN 2712-0562Vol.1, No.1, pp.1-23

[15] https://ansyshelp.ansys.com, ANSYS Fluent Users Guide, 2021

DOI: 10.2507/32nd.daaam.proceedings.092

THE CURRENT STATE OF THE USE OF SELECTED INDUSTRY 4.0 TECHNOLOGIES IN MANUFACTURING COMPANIES

Martin Polivka & Lilia Dvorakova

This Publication has to be referred as: Polivka, M[artin] & Dvorakova, L[ilia] (2021). The Current State of the Use of Selected Industry 4.0 Technologies in Manufacturing Companies, Proceedings of the 32nd DAAAM International Symposium, pp.0652-0659, B. Katalinic (Ed.), Published by DAAAM International, ISBN 978-3-902734-33-4, ISSN 1726-9679, Vienna, Austria
DOI: 10.2507/32nd.daaam.proceedings.092

Abstract

The technologies of the Fourth Industrial Revolution (Industry 4.0) include both technologies related to the physical aspect of production process, and those related primarily to data and information infrastructure, whose deployment in enterprises affects their IT landscape including the Enterprise Resource Planning (ERP) system. The research aims to identify the Industry 4.0 technologies that have a direct impact on ERP systems and to determine the current level of use of these technologies in medium-sized and large industrial enterprises in the Czech Republic. Based on the deductive analysis, Industry 4.0 technologies having a direct impact on ERP systems are identified, and data obtained from the questionnaire survey are statistically processed. Big data, system integration, cloud, and automatic identification and data capture have been evaluated as the Industry 4.0 technologies that have a direct impact on ERP systems. The research results document that the technology of in-house system integration is the most frequently used one. The ongoing research is directed at analysing the reasons for this finding in order to evaluate why the other technologies are not used more intensively and how enterprises make decisions on the choice of technical solution when deploying technology.

Keywords: ERP systems; Industry 4.0; Industry 4.0 technologies; Information systems

1. Introduction

The last decade has seen the emergence of a number of technologies in the field of informatics and cybernetics that have the potential to significantly influence the shape of the future economy as well as the entire society. The development and the increasingly widespread practical use of these technologies are collectively referred to in the literature as the so-called the "Fourth Industrial Revolution", which is expected to result in the emergence of "Industry 4.0" or, in the broader sense, the entire "Society 4.0" [1], and especially in Japan, they even use the term "Society 5.0" in this context (see, for example, [2], [3] or [4]). The defining feature of this new environment is, above all, to be the close connection between physical and virtual reality and the maximum possible degree of cooperation between humans and machines (for more details see, e.g., [5] or [6]). The spectrum of technologies considered to be part of the Fourth Industrial Revolution is very broad and includes technologies mainly from the fields of digitization, automation, robotics, and artificial intelligence. Some of these technologies relate especially to the physical aspect of the production process, and their implementation is thus primarily reflected in changes in the layout and operation of production lines.

However, there are also technologies of the Fourth Industrial Revolution that primarily involve data and information infrastructure. Their deployment in the enterprise will thus primarily affect company information systems and technologies and they are referred to as the "IT landscape" (see, e.g., [7]). Changes in the enterprise´s IT landscape will then logically affect its central element, which is the company ERP system.

2. Research objectives and methods

This paper presents research which, firstly, aimed to identify the technologies of Industry 4.0 that have a direct impact on ERP systems and, secondly, to determine the current state of the use of these technologies in industrial companies in the Czech Republic. The research method used for the first objective was the desk research of existing literature on Industry 4.0 technologies combined with the deductive analysis, the principles of which are presented in Chapter 3. To achieve the second objective, data were collected by means of a questionnaire survey conducted at the beginning of 2021 in medium-sized and large industrial enterprises in the Czech Republic. A total of 1,758 enterprises were successfully approached to fill in the questionnaire and 68 of them completed the questionnaire. Therefore, the return rate of the questionnaire was approximately 4%. The data were then statistically processed, and, considering the nature of the data, the non-parametric Kruskal-Wallis test combined with the Nemenyi method was chosen for more detailed analysis. The detailed description of the questionnaire survey, together with the analysis and discussion of the results, is presented in Chapter 4.

3. Relevant technologies of the industrial revolution

Since the use of the term "Industry 4.0" has recently undergone some inflation [8], and is thus used in the multitude of meanings, it is possible to encounter a variety of technologies that are described as part of it in various academic papers, but the list of these technologies varies significantly depending on the opinion of a particular author. Hence, some papers emphasize cyber-physical systems, the Internet of Things and cloud computing (e.g. [9]) or even edge computing [10], while other authors stress the importance of augmented reality [11] and digital twin [12], collaborative robots [13] or blockchain [14].

The inconsistency in the definition of technologies of the Fourth Revolution is logical on principle, since the estimation of the fact which technologies will see mass diffusion in the future and significantly influence the shape of the economy is ultimately largely arbitrary and depends on the opinion of a particular author or institution. However, for the purposes of the research presented here, it was necessary to first define a core set of technologies from which the technologies with a direct impact on a firm´s ERP system could be selected on the basis of the analysis.

Therefore, the so-called "9 Pillars of Industry 4.0" were used as a core set of technologies for further selection in our research. This term, introduced by Rüssman et al. [15] and adopted by many other authors (see, e.g., [6] or [16]), refers to nine technologies (or technological areas) that are the basic building blocks of "Industry 4.0". The core technologies are as follows:

1. Big data
2. Autonomous (collaborative) robots
3. Simulations
4. System integration
5. The Internet of Things
6. Cyber-physical systems
7. Cloud technologies
8. Additive manufacturing
9. Augmented reality

In addition to these nine pillars, some authors add the tenth pillar, collectively referred to as "Other Technologies", which contains advanced technologies with a narrower focus, always applicable only for the purposes of selected industries (cf. [17]). However, based on the analysis presented in the study of Polívka and Dvořáková [18], the authors of this article believe that it is possible to identify one more modern technology the development of which has a similar impact on the course of the "Fourth Industrial Revolution" as the aforesaid 9 pillars and which is also universally applicable across industries. This technology is Automatic Identification and Data Capture (AIDC), which has been added to the nine pillars listed above for the reasons given above.

The first objective of the research was to identify within these ten core technologies those that are directly related to the issue of ERP systems. The basic starting point of the analysis was the fact that the ERP system is the central element of the enterprise´s IT landscape, where the summary data on all business processes are concentrated. These data are subsequently reported and further evaluated within the ERP system and serve as a basis for further management of the enterprise. However, not all of these data are directly collected within the ERP system. ERP systems usually manage both administrative processes, such as invoicing or accounting, and cross-section processes using data from the entire organisation, such as central purchasing and production planning. However, the actual operational management of physical processes, such as production or logistics, is rather the domain of specialized, narrowly focused systems such as Manufacturing Execution System (MES), Warehouse Management System (WMS), etc.

Compared to these narrowly focused solutions, ERP systems do not have comparable functionality in this area, and due to their technical nature, they are not suitable or even not at all applicable for direct management of some physical processes. Typically, these are situations where the production management system has to communicate online with production machines and equipment, and the communication protocol requires a feedback speed of milliseconds. However, such response time is technically not necessarily achievable in the case of ERP systems containing large, highly interconnected databases with a high degree of business logic.

Therefore, in the first part of our research, technologies that are primarily related to the management of administrative processes and/or the evaluation of centralized data were selected from the core set of ten Industry 4.0 technologies, as these are the main domain of ERP systems. Big data, system integration, cloud, and automatic identification and data capture have been evaluated as such technologies.

As far as the technology of big data is concerned, in order to use it, an enterprise must be able to collect large volumes of primary data, and it must be able to store and evaluate them. Even for the actual collection of data on some processes, the interaction of ERP system is necessary - while data describing physical processes are collected primarily through sensors and detectors, the progress of administrative processes is usually monitored in the form of records in the ERP system, through which these processes are controlled and coordinated. In order to ensure appropriate storage and analysis of the collected big data, the ERP system must then contain suitably designed, sufficiently robust data structures, and must either have powerful tools for their evaluation itself, or at least be able to cooperate with specialized analytical tools. It is therefore clear that the technology of big data is directly related to ERP systems.

System integration consists in the interconnection of information systems and the exchange of data between them either within the company (sometimes referred to as vertical integration) or across the supply and customer chain (the so-called horizontal integration). Although individual integration solutions differ significantly in practice, both in terms of technical solution and functionality, ERP systems usually act as one of the basic elements. In the case of in-house integration, the usual scenario is the connection of partial, specifically focused subsystems (e.g., in the form of MES, WMS or PLM systems) to the central element of the company IT landscape represented by the ERP system, while horizontal integration across the supply and customer chain logically concerns mainly mutual business relations between companies, which are represented in data terms by the circulation of orders and invoices, while these entities are normally recorded and processed within ERP systems.

As far as the cloud technology is concerned, it enables the transfer of data repositories (and possibly other elements of the IT landscape, e.g., in the form of application servers) to a shared environment operated by the provider and accessible via the Internet. It is therefore closely linked to the issue of enterprise information systems, which, in the case of transfer to the cloud, are no longer operated in the internal environment of the company, but their availability is provided to the company in the form of a service. ERP systems are the part of the IT landscape that is a suitable candidate for the transfer to the cloud - on the one hand, the ERP system as a central hub of the information system contains a large amount of data, for the effective processing of which local devices may not be sufficient, and on the other hand, unlike operational systems such as MES or WMS, it is not a business critical application whose temporary failure would completely stop the operation of the company. The short-term unavailability of cloud services caused by, for example, physical interruption of data lines can thus be viewed as an acceptable risk, which affects only administrative processes, but does not entail costs such as the complete stoppage of production, inventory devaluation, etc.

Although the technology of automatic identification and data capture is seemingly associated primarily with physical devices in the form of sensors or detectors, these devices are only tools for the automated data acquisition (whether they are accounting or business data in the case of machine document mining or data on the physical movements of goods, materials and products obtained by reading barcodes or RFID transponders). Then, these data are again processed, stored, and analyzed in central information systems, typically in ERP. Therefore, this technology is also closely related to the ERP issue.

As for the other technologies of the Fourth Industrial Revolution, these are primarily related either to physical processes (in the case of autonomous robots and additive manufacturing) or to their software control (in the case of simulations and cyber-physical systems) and the acquisition of data for this control (in the case of the Internet of Things). Therefore, they are not primarily related to ERP systems and were not included in the following part of the research.

4. The questionnaire survey results and discussion

To obtain information on the current use of the four "Industry 4.0" technologies identified as relevant in relation to ERP systems in the Czech Republic, a questionnaire survey was carried out as part of a broader study examining the extent to which "Industry 4.0" influences the requirements that companies place on ERP systems and the process of selecting such systems.

Three criteria were applied when selecting the basic population. From the geographical point of view, companies based in the Czech Republic were selected, which corresponds to the defined research objectives. The size of enterprises was chosen as the second criterion, where only those falling into the category of medium-sized and large ones according to the EU categorization, i.e., enterprises with more than 50 employees and an annual turnover of more than EUR 10 million, or a balance sheet total of more than EUR 10 million, were selected for the survey. The size criterion was used to define the basic population because the implementation of Industry 4.0 technologies is, primarily due to the investment costs associated with that, still mainly a matter of large and medium-sized enterprises [19].

Similarly, ERP systems are mostly used in medium-sized and large enterprises, while micro and small ones often still manage with simple accounting and business software. Conducting such a focused survey in micro and small enterprises would thus be unlikely to yield relevant data for these reasons. The third criterion was the sector in which the enterprises operate, where enterprises in the manufacturing industry, i.e., those falling under section "C" of the Czech Statistical Office´s classification, were selected for the basic population. The reason for defining the specific sector is that although Industry 4.0 is no longer considered exclusively a matter of manufacturing companies, as evidenced by the existence of terms such as "Agriculture 4.0", "Logistics 4.0" or the general term "Society 4.0", yet it is manufacturing and especially the so-called "smart factories" that are most closely linked to this concept [20]. It is therefore logical to examine the current state of the use of Industry 4.0 technologies in this particular environment.

A total of 1,758 companies were approached to participate in the survey. Companies were asked to have employees who are either key users of the ERP system or IT managers responsible for the operation of the ERP system in the company complete the survey. The questionnaire was completed by the representatives of 68 enterprises, so the return rate was about 4%. In the questionnaire, the four selected technologies were assessed individually, with the system integration technology further divided into the horizontal and vertical ones. For each technology, excluding the cloud, the respondents were asked two questions to establish the following facts:

1. Whether the company uses the technology.
2. If the company uses the technology, what specific technical solution is implemented.

As for the cloud technology, the respondents were only asked the first question mentioned above. This is because the technical nature of the cloud is homogeneous on principle. Although there are, of course, different options of using cloud services as well as different options of their hosting, from the technical point of view it is ultimately always about providing server resources in the form of a service. The second question would thus be irrelevant in the case of cloud technologies.

The questions were closed-ended, and in the case of the first question, the answers were defined by a five-degree scale, with each degree also described verbally for better understanding by the respondents. The specific wording of each question and possible answers are available from the authors of the paper. The possible answers to the second question were then defined by a list of potentially applicable technical solutions. The answers to the first set of questions investigating to what extent firms use individual technologies are shown in Table 1:

Technology	**Response frequency on the scale of 1-5 (1 = technology is not used at all, 5 = technology is used with the maximum intensity)**					**Average**
	1	**2**	**3**	**4**	**5**	
Big data	9	12	29	11	7	**2.93**
In-house system integration	4	4	12	19	29	**3.96**
Inter-company system integration	24	17	8	16	3	**2.37**
Cloud solutions	14	19	29	4	2	**2.43**
AIDC	26	11	15	14	2	**2.34**

Table 1. The rate of the use of the selected Industry 4.0 technologies in industrial enterprises in the Czech Republic.

As can be seen from Table 1, the most used Industry 4.0 technology related to ERP systems in the conditions of Czech manufacturing companies is clearly in-house system integration, i.e., the interconnection of several information systems. Big data have been evaluated as the second most used technology; however, the rate of its use is significantly lower compared to the in-house system integration. According to the respondents, the remaining technologies examined are used to a comparable, and relatively small, extent.

Whether the observed differences between the use of different technologies are indeed a statistically significant phenomenon or merely a random variation was further examined by means of the Kruskal-Wallis test, which tests the null hypothesis of distribution functions concordance of more than two random variables. The non-parametric Kruskal-Wallis test was chosen instead of the ANOVA test, which assumes the normality of data, both because of the nature of the input data (discrete values in a limited range) and because of the limited sample size. The Kruskal-Wallis test criterion was calculated according to equation (1):

$$H = \frac{12}{N(N+1)} \sum_{i=1}^{C} \frac{R_i^2}{n_i} - 3(N+1) \tag{1}$$

Where C is the number of groups, n_i is the number of observations in a group, N is the total number of observations in all groups, and R_i is the sum of the rank of the members of the i-th group.

The calculated value of the criterion was 66.78. This value was compared with the critical value given by the χ^2 distribution with $(C - 1)$ degrees of freedom. At significance level $\alpha = 0.01$, this critical value is equal to 13.27. The value of test criterion H therefore exceeds the critical value and the concordance hypothesis of the distribution functions of the individual groups can thus be rejected. Therefore, the differences in the use of individual technologies can be declared statistically significant.

However, the Kruskal-Wallis test only shows that some of the examined random variables have a different distribution function than the others, without allowing a detailed analysis of these differences. Therefore, to compare in detail which technologies are statistically significantly more used than the others, the Nemenyi method was used, which allows the mutual comparison of the individual variables. The principle of this method is to compare the absolute value of the difference in the average ranking of each of the two groups examined with the value of the test criterion, which is calculated for the given sample range according to equation (2):

$$K = q_\alpha(C, \infty)\sqrt{\frac{1}{12}C(Cm + 1)} \qquad (2)$$

Where $q_\alpha(C, \infty)$ is the critical value of C distribution of independent random variables with a normal distribution at significance level α and m is the sample range in one group (i.e., for equal group sizes, m equals any n from equation 1). In our case, the critical value equals 54.87 at the significance level of 0.01.

The results of the Nemenyi method for the first question are shown in Table 2, with statistically significant differences indicated by an asterisk (Note that values below the diagonal are not completed as they would mirror the values above it.):

Technology	Big data	In-house system integration	Inter-company system integration	Cloud	AIDC
Big data	0	72*	40.38	36.70	42.10
In-house system integration		0	112.38*	108.70*	114.10*
Inter-company system integration			0	3.68	1.72
Cloud				0	5.40
AIDC					0

Table 2. The results of the Nemenyi method in comparison with the use of the selected "Industry 4.0" technologies

Table 2 shows that only the rate of the use of in-house system integration is statistically significant compared to the other technologies. Although the use of the big data technology appears to be higher than the other three technologies based on the results at first glance, the hypothesis of their concordance cannot be rejected at the significance level of 0.01, and it must therefore be assumed that this difference is caused only by a random deviation in the sample. In terms of the specific technical solutions used by the companies within the individual technologies, the frequency of each variant is shown in the following series of figures.

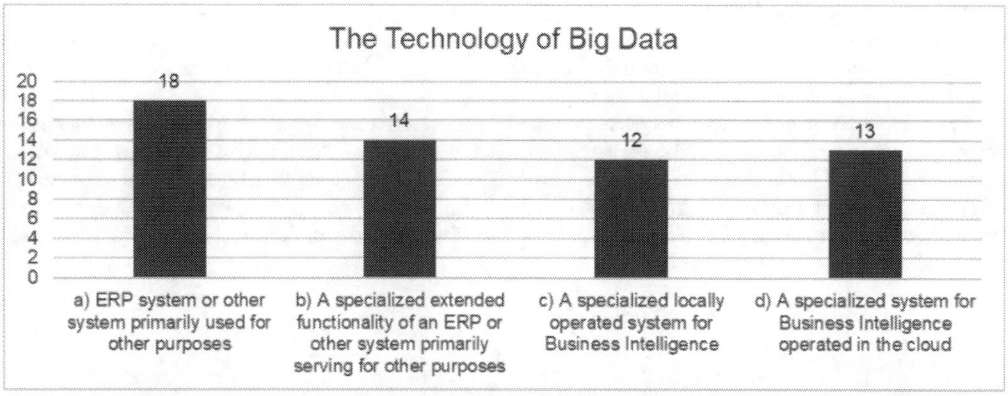

Fig. 1. The frequency of the specific technical solutions within the technology of big data

Only those technical solutions that are used by at least one responding company are shown in the figures; the complete list of technical solutions that were offered to respondents as possible answers in the questionnaire can be requested from the authors of this paper. The sum of frequencies of individual technical solutions for each technology does not have to correspond to the number of the respondents. Not all the companies of respondents use all technologies, while other companies may use several different technical solutions within one technology.

As can be seen from Figure 1, which shows the frequency of the specific technical solutions used for big data, the companies that participated in the research are still using systems primarily designed for other purposes in the slight majority of cases. This option, represented by responses a) and b), occurred in a total of 32 companies. Eighteen of these companies do not even have any specialized module for dealing with large volumes of data within their information system and thus they have to manage with the standard functionalities offered by their system. On the other hand, twenty-five surveyed companies use specialized Business Intelligence software for storing and analysing big data, with locally operated and cloud-based solutions being represented to a similar extent. Thus, although large suppliers of big data technologies, such as Microsoft or Oracle, have focused mainly on cloud platforms in recent years, it is clear that cloud solutions in the domain of big data are not yet definitely preferred in the Czech industrial companies.

Figure 2 demonstrates that the integration between the internal information systems of the responding companies is most often implemented in the form of database interconnection. Traditional solutions consisting in generating and importing files, whether structured (XML, JSON, EDI standards) or in the simple csv format, are also used relatively intensively. On the other hand, more modern integration methods in the form of web services are used for the needs of in-house system integration in far fewer cases - REST-based services were mentioned in only three cases and SOAP-based services are used by none of the respondents.

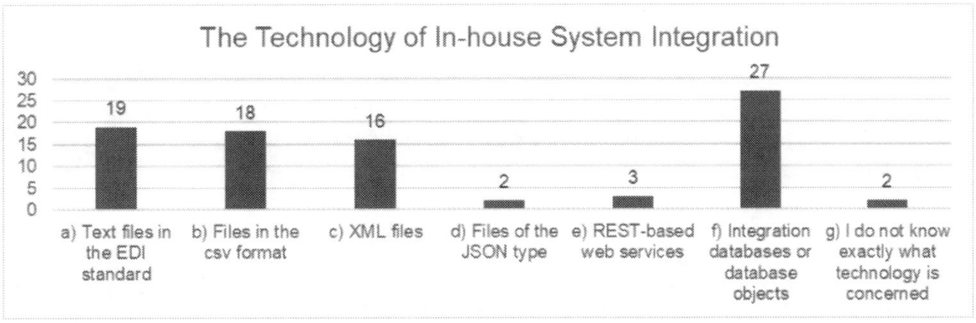

Fig. 2. The frequency of the specific technical solutions within the technology of in-house system integration

If we compare this situation with the technical solutions used for the needs of inter-company integration, the frequency of which is shown in Figure 3, it becomes apparent (of course, taking into account the fact that the overall frequency of inter-company integration solutions is lower, as this technology is significantly less used compared to in-house system integration, as evidenced by the answers to Question 1) that the big difference is mainly in the rate of the use of database integration. It is very frequent in the case of in-house integration, but it is rarely used for interconnection with business partners′ systems, despite the fact that it has a number of advantages over other solutions, for example, it can transfer large volumes of data without negative impacts on performance, or the fact that database technologies are standardized to a large extent, which brings significant simplification in the design and implementation of interconnections.

However, text files in the EDI format are also used intensively in inter-company integration, which is probably related to the fact that their use is widespread especially in the automotive sector, which plays a significant role in the manufacturing industry of the Czech Republic. A rather surprising fact is that none of the surveyed companies use web services for the inter-company system integration, which are technically suitable for this purpose and are currently a standard part of many ERP systems as their API layer.

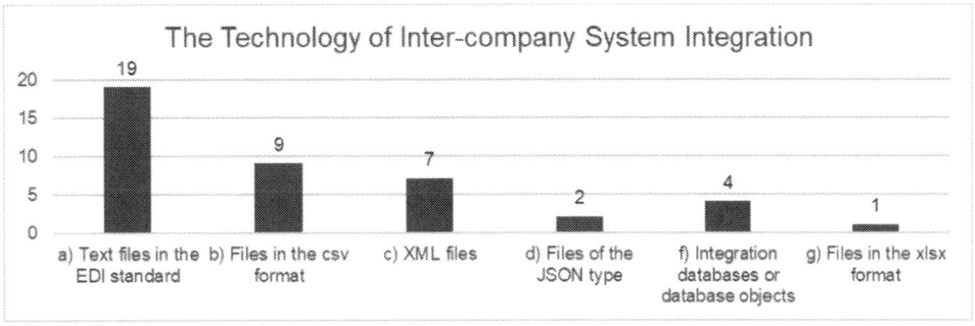

Fig. 3. The frequency of the specific technical solutions within the technology of inter-company system integration

Figure 4 shows the frequency of technical solutions used by the respondents´ companies in the domain of automatic identification and data capture. The most used solution is definitely the identification of objects and activities by barcodes or QR codes. The high rate of use corresponds to a long-term trend in WMS and MES systems, where the manipulation of individual objects or the recording of operations in production through code scanning instead of manual data entry is one of the basic functionalities used because of the higher speed and lower error rate compared to manual data entry.

In contrast, the second most common solution, image recognition for document mining, is significantly newer and its wider use in companies has only been made possible by the increase in computing power of hardware resources in recent years. The fact that even in such a short time this solution has become relatively widely used is evidence of the attractiveness of this technology, which allows significant time savings for office staff. Solutions based on other principles, with the exception of radio-frequency identification, which is used by six of the companies surveyed, occurred only sporadically among the respondents.

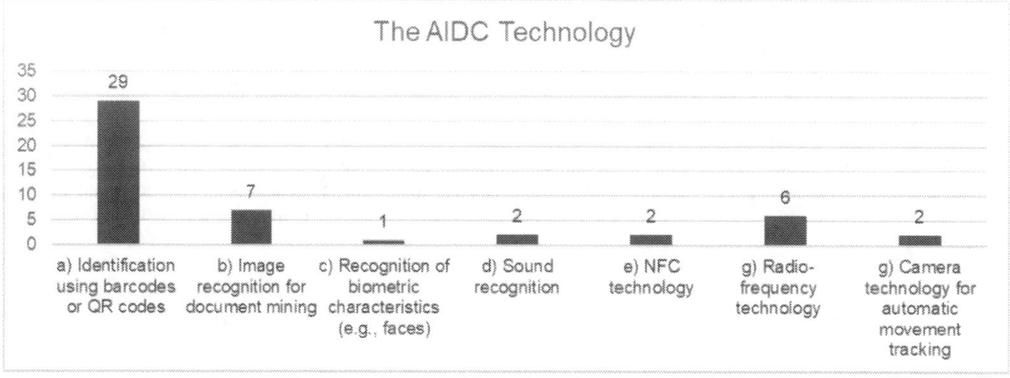

Fig. 4. The frequency of the specific technical solutions within the AIDC technology

5. Conclusion

The paper has presented the research aiming to identify Industry 4.0 technologies that are directly related to ERP systems and to determine the current state of the use of these technologies by industrial enterprises in the Czech Republic. Based on the search of available literature and the subsequent analysis, the basic Industry 4.0 technologies were identified at first, and then those related to the management of administrative processes and/or the storage and analysis of centralized data, which are the traditional tasks of ERP systems, were identified. These technologies are big data, system integration, cloud, and automatic identification and data capture. In the more comprehensive questionnaire distributed to medium-sized and large manufacturing enterprises based in the Czech Republic, enterprises were then asked questions mapping both the extent to which these technologies are used by enterprises and the specific technical solutions chosen by them in this area. The results of the questionnaire survey show that the most used technology is clearly the system integration for in-house purposes.

The research presented in this paper focused on determining the current state of the use of the Fourth Industrial Revolution technologies. Analysing the causes of this state in order to obtain answers to the question of why other technologies, besides the in-house system integration, are not used more intensively and how enterprises make decisions in choosing a particular technical solution when deploying a particular technology appears to be a possible direction for further research.

6. References

[1] Vacek, J. (2017). On the Road: from Industry 4.0 to Society 4.0, Trendy v Podnikání, Vol. 7, No. 4, pp. 43-50, ISSN 1805-0603.

[2] Fukuda, K. (2020). Science, technology and innovation ecosystem transformation toward society 5.0, International Journal of Production Economics, Vol. 220, ISSN 0925-5273, DOI: 10.1016/j.ijpe.2019.07.033

[3] Takakuwa, S.; Veza, I. & Celar, S. (2018). Industry 4.0 in Europe and Asia, Proceedings of the 29th International DAAAM Symposium "Intelligent Manufacturing & Automation", 24-27th October 2018, Zadar, Croatia, ISSN 1726-9679, ISBN 978-3-902734-20-4, Katalinic, B. (Ed.), pp. 61-69, Daam International, Vienna, DOI: 10.2507/29th.daaam.proceedings.009

[4] Sawaragi, T.; Horiguchi, Y. & Hirose, T. (2020), Design of Productive Socio-Technical Systems by Human-System Co-Creation for Super-Smart Society, IFAC PapersOnLine, Vol. 53, No. 2, pp. 10101-10108, ISSN 2405-8963, DOI: 10.1016/j.ifacol.2020.12.2734

[5] Rojko, A. (2017). Industry 4.0 Concept – Background and Overview, International Journal of Interactive Mobile Technologies, Vol. 11, No. 5, pp. 77-90, ISSN 1865-7923.

[6] Vaidya, S.; Ambad, P. & Bhosle, S. (2018). Industry 4.0 – A Glimpse, Procedia Manufacturing, Vol. 20, pp. 233-238, ISSN 2351-9789, DOI: 10.1016/j.promfg.2018.02.034

[7] Leyh, Ch.; Schäfer, T.; Bley, K. & Forstenhäusler, S. (2017). Assessing the IT and Software Landscapes of Industry 4.0-Enterprises: The Maturity Model SIMMI 4.0. In Information technology for management: new ideas and real solutions, E. Ziemba (Ed.), pp. 103-119, Springer, ISBN 978-3-319-53076-5, Cham.

[8] Inkermann, D.; Schneider, D.; Martin, N.L.; Lembeck, H.; Zhang, J. & Thiede, S. (2019). A framework to classify Industry 4.0 technologies across production and product development, Procedia CIRP, Vol. 84, pp. 973-978, ISSN 2212-8271, DOI: 10.1016/j.procir.2019.04.218

[9] Zhong, R.Y.; Xu, X.; Klotz, E. & Newman, S.T. (2017). Intelligent Manufacturing in the Context of Industry 4.0: A Review, Engineering, Vol. 3, pp. 616-630, ISSN 2095-8099, DOI: 10.1016/J.ENG.2017.05.015

[10] Bajic, B.; Cosic, I.; Katalinic, B.; Moraca, S., Lazarevic, M. & Rikalovic, A. (2019). Edge Computing vs Cloud Computing: Challenges and Opportunities in Industry 4.0, Proceedings of the 30th International DAAAM Symposium "Intelligent Manufacturing & Automation", 23-26th October 2019, Zadar, Croatia, ISSN 1726-9679, ISBN 978-3-902734-22-8, Katalinic, B. (Ed.), pp. 864-871, Daam International, Vienna, DOI: 10.2507/30th.daaam.proceedings.120.

[11] Egger, J. & Masood, T. (2020). Augmented reality in support of intelligent manufacturing – a systematic literature review, Computers & Industrial Engineering, Vol. 140, pp. 1-22, ISSN 0360-8352, DOI: 10.1016/j.cie.2019.106195

[12] Novák, P.; Vyskočil, J. & Wally, B. (2020). The Digital Twin as a Core Component for Industry 4.0 Smart Production Planning, IFAC PapersOnLine, Vol. 53, No. 2, pp. 10803-10809, ISSN 2405-8963, DOI: 10.1016/j.ifacol.2020.12.2865

[13] Schmidbauer, Ch.; Komenda, T. & Schlund, S. (2020). Teaching Cobots in Learning Factories – User and Usability-Driven Implications, Procedia Manufacturing, Vol. 45, pp. 398-404, ISSN 2351-9789, DOI: 10.1016/j.promfg.2020.04.043

[14] Esmaeilian, B.; Sarkis, J.; Lewis, K. & Behdad, S. (2020). Blockchain for the future of sustainable supply chain management in Industry 4.0, Resources, Conservation & Recycling, Vol. 163, pp. 1-15, ISSN 0921-3449, DOI: 10.1016/j.resconrec.2020.105064

[15] Rüssman, M.; Lorenz, M.; Gerbert, P.; Waldner, M.; Engel, P.; Harnisch, M. & Justus, J. (2015). Industry 4.0: The Future of Productivity and Growth in Manufacturing Industries, Available from: https://www.bcg.com/publications/2015/engineered_products_project_business_industry_4_future_productivity_growth_manufacturing_industries Accessed: 2021-09-09

[16] da Silva, E.R.; Shinohara, A.C.; Nielsen, Ch.P.; Pinheiro de Lima, E. & Angelis, J. (2020). Operating Digital Manufacturing in Industry 4.0: the role of advanced manufacturing technologies, Procedia CIRP, Vol. 93, pp. 174-179, ISSN 2212-8271, DOI: 10.1016/j.procir.2020.04.063

[17] Büchi, G.; Cugno, M. & Castagnoli, R. (2020). Smart factory performance and Industry 4.0, Technological Forecasting & Social Change, Vol. 150, pp. 1-10, ISSN 0040-1625, DOI: 10.1016/j.techfore.2019.119790

[18] Polívka, M. & Dvořáková, L. (2019). Využití automatické identifikace a sběru dat prostřednictvím radiofrekvenčních technologií v prostředí průmyslu 4.0, Trendy v podnikání, Vol. 9, No. 2, pp. 53-61, ISSN 1805-0603.

[19] Masood, T. & Sonntag, P. (2020). Industry 4.0: Adoption challenges and benefits for SMEs, Computers in Industry, Vol. 121, pp. 1-12, ISSN 0166-3615, DOI: 10.1016/j.compind.2020.103261

[20] Alcácer, V. & Cruz-Machado, V. (2019). Scanning the Industry 4.0: A Literature Review on Technologies for Manufacturing Systems, Engineering Science and Technology, an International Journal, Vol. 22, pp. 899-919, ISSN 2215-0986, DOI: 10.1016/j.jestch.2019.01.006

DOI: 10 2507/32nd daaam proceedings.093

INDUSTRY 4.0 - AUTOMATION IN THE CONDITIONS OF RECONFIGURATION OF PRODUCTION LINES

Aleksandr Aryskin, Aleksey Grigoriev, Maksim Petrakov, Valentin Pryanichnikov, Anastasia Travushkina & Aleksander Travushkin

This Publication has to be referred as: Aryskin, A[lexander]; Grigoriev, A[leksey]; Petrakov, M[axim]; Pryanichnikov, V[alentin]; Travushkina, A[nastasia] & Travushkin, A[leksandr] (2021). Industry 4 0 - Automation in the Conditions of Reconfiguration of Production Lines, Proceedings of the 32nd DAAAM International Symposium, pp.0660-0665, B. Katalinic (Ed.), Published by DAAAM International, ISBN 978-3-902734-33-4, ISSN 1726-9679, Vienna, Austria DOI: 10.2507/32nd.daaam.proceedings.093

Abstract

In the modern world, increasing digitalization of production processes is becoming more important for industry in order to increase the competitive advantages of manufactured products. Using the example of the company "Techfom", the practical implementation of the technology of "smart production" is presented, including such components as: digitalization of the main control elements; the use of remote access to the software being developed and to the production process itself; optimization of technological systems by digital means and architecture of software and hardware microprocessors and computer systems tied into a single local area network. By using various electromechanical elements in it (sensors, input devices, tachometers), a computing platform (operating system), a "network structure", a so-called cyberphysical system was created. Examples of the use of smart manufacturing components on specific complexes of the "Techfom" robotic industrial lines are shown and further tasks in their implementation are outlined to improve its productivity and product quality.

Keywords: Innovative material smart manufacturing; Industry 4.0; digitalization and cyberphysical systems; automation with digital twins; robotization.

1. Introduction

Reducing costs, improving the efficiency and competitiveness of products are the main areas of optimization of production and development of any enterprise. Currently, many companies are trying to implement the concept of Industry 4.0 at their production facilities and, thereby, are in search of new methods and approaches in the development and creation of their products.So, in China, the government has created five key national projects for the development and implementation of Industry 4.0 in the country's industry. Thanks to this policy, a large network of metallurgical enterprises Baosteel Group, engaged in technological research and development, has created a cloud business platform Ouyeel, which uses such aspects of "smart production" as the industrial Internet, big data and other modern technologies [1] Many global companies not only use innovative technologies to optimize their production, but also create such productions for customers.

For example, Siemens, using elements of Industry 4.0 – the "Digital Twin" of production, the "Internet of Things" (IoT), "Big Data" - creates a continuous integrated process for its customers: "construction - technological preparation - production" [2]. Some countries are realizing Industry 4.0 components not only in individual productions, but also creating special spaces. In Germany, the equivalent of "Silicon Valley" is a cluster called OWL (Intelligent Technical Systems OstWestfalenLippe). There are 173 companies on its territory, that implement the concept of "smart factories" in practice [3]. In Russia, basically, only large companies can realize Industry 4.0 technologies to increase their financial and economic indicators. For example, JSC Russian Railways uses a big data system, which makes it possible to predict the condition of train nodes, increase the degree of readiness of rolling stock [4].

The development and implementation of such technologies is considered in this article on the example of the creation of a robotic industrial line for the manufacture of products from innovative thermal and sound insulation material "Techfom" at one of the Moscow region enterprises. As noted, one of the approachers of Industry 4.0 is related with cyber-physical capabilities to ensure the connections of technical systems and society [5,6]. Using the example of the "Techfom" enterprise, one can see the practical implementation of the "smart production" technology, which includes such components as: digitalization of the main control elements and, as a result, the use of remote access to the software being developed and to the control process itself; optimization of technological systems by digital means, the choice of architecture of software and hardware microprocessor and computer systems united into a single local network.

The production process of innovative thermal and sound insulation material is carried out as follows: silicate block, water, chrysotile, salt, foaming agent are supplied to the input of the technological line. At the exit from it is the "Techfom" material, which has a different density (from 80 to 300 kg / cubic meter) and purpose (depending on the current dozens of formulations/recipes being executed), in addition, packed in polyethylene with fixed dimensions of 800x600x900 mm. The company "Techfom" has implemented cooking, mixing, drying and cutting complexes (production sites), which produce the corresponding technological processes shown in Figure 1. Consider the elements of "smart production" implemented in these production sites in more detail. "Digitalization" means not only the conversion of information into digital form from analog measured parameters (temperature, pressure, volume), but also the unification of data, coding, filtering to ensure the necessary sensor polling frequencies corresponding to the capabilities of receiving and processing their readings by microprocessors [7] and, most importantly, the logical analysis of hundreds of recipes, both current executable and newly created by the technologist.

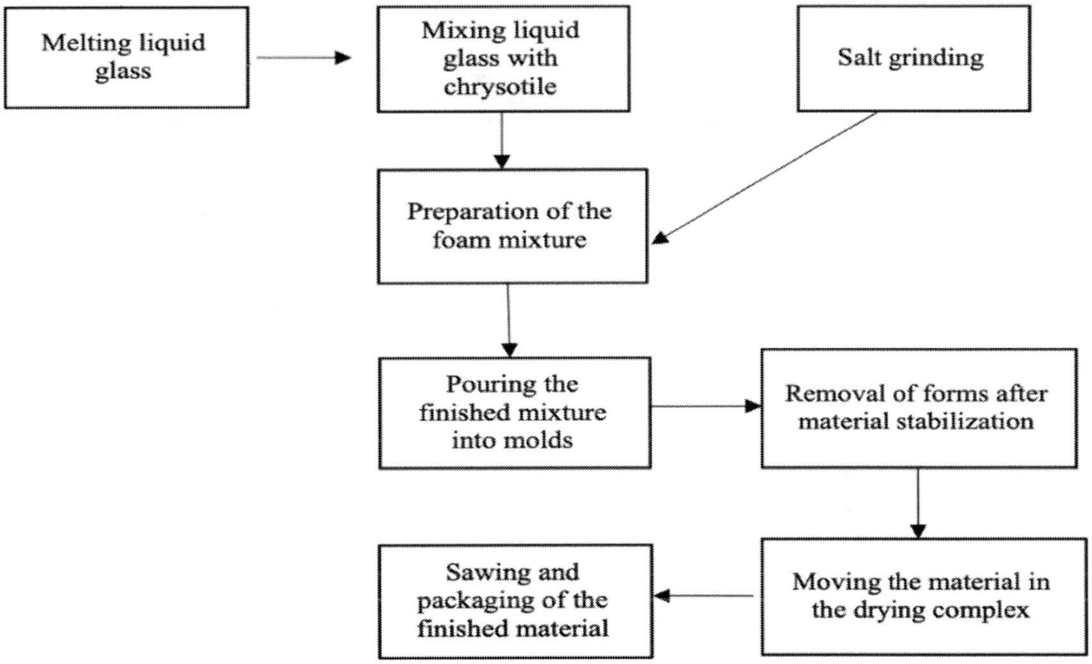

Fig. 1. The main technological processes of the flexible production line

The telemetry information received from the sensors is fed to the microprocessor control system, analyzed by it for issuing control commands and, according to the developed algorithm, transmitted to the actuators of the robotic industrial line (electric motors, relays, electric valves, gate valves). To control the trolleys (more than 60 pieces, each up to 2 tons) of the drying complex (Fig. 2), it is necessary to make their coordinated movement along the rail tracks, similar to a game of "tag". That is, it is possible to move a new trolley to one of the 4 drying lines only when the trolley at another end of the line is removed from this line to the driver line moving it in the transverse direction.

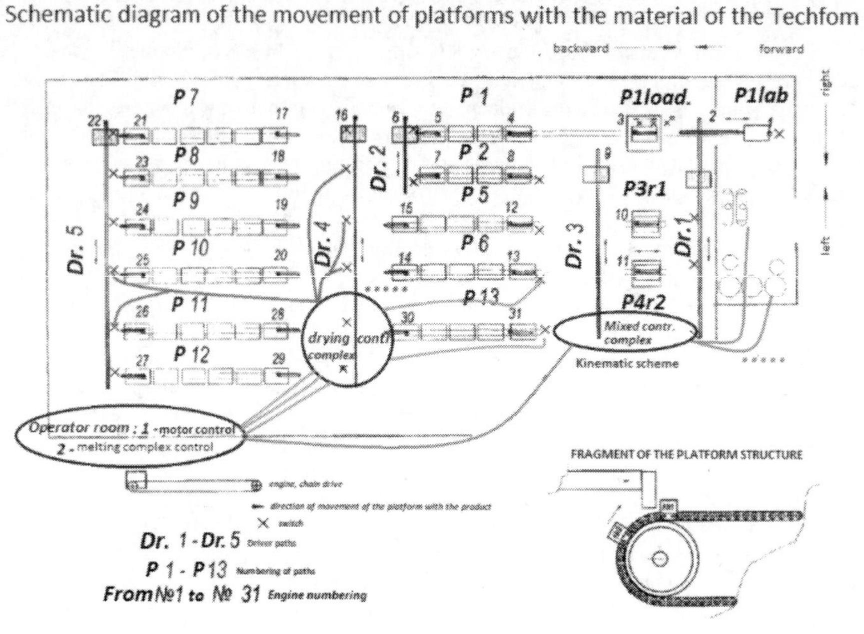

Fig. 2. The scheme of the lines of the drying complex

Therefore, any error in the control algorithm of switching on the motors, moving the trolleys, can lead to their falling off the rails, on which they move, and, as a consequence, to a technological accident at work. It is also important to obtain accurate telemetric information from the position sensors of these trolleys, which is embedded in their control algorithm. The main problem to be solved is to ensure the reliability of control due to the developed apparatus of logical analysis and comparison of the readings of large groups of sensors, which allows detecting abnormal situations even in conditions of failure of a small part of these sensors. The previously described production cycle control system has now received its expansion, modernization and additional application in the production of Techfom material, while the technology used allows modernization, improvements and changes without stopping production and in remote mode [8].

Tele-surveillance systems are also installed at the production site, which carry out tele-visual (and partially automated) control over the operation of mixing, drying and sawing complexes. This part of the "smart production" is one of the leading ones, since with the help of sensors and a tele-surveillance system, the operator monitors the production process remotely. The next criterion of smart production is end-to-end automation, which allows to improve and stabilize the quality of products, optimize the work of the entire enterprise, increase the quantity and volume of manufactured material.

Using various electromechanical elements (sensors, input devices), computing platforms (operating environment), a "network structure", a so-called cyberphysical system is created [9]. Control complexes with microcontrollers via transistor switches and intermediate relays, as well as sensors installed at all production sites, the information from which is converted by SCADA software and displayed on the operator and technologist's video monitors, are responsible for the process of "automation" in the production of innovative material "Techfom" (Fig. 3).

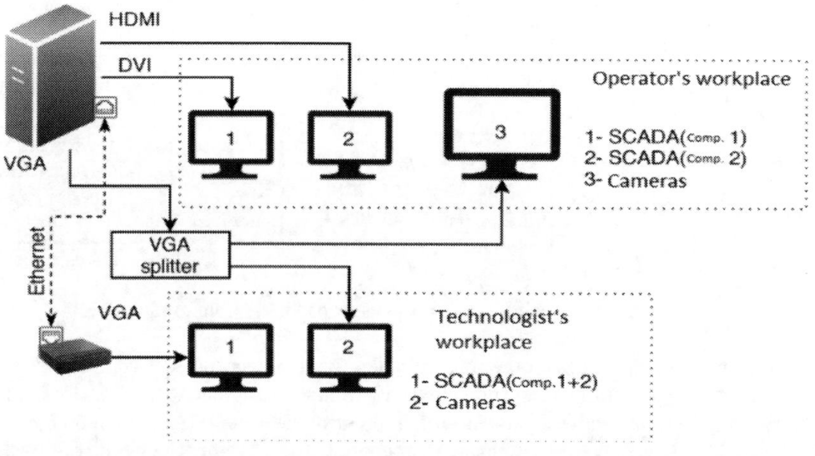

Fig. 3. Display of the received information for the operator and the technologist

Automation is the main component of "smart production", thanks to which it is possible to realize the flow of almost individual orders, the productivity of the technological line increases significantly, using less human resources, as well as increasing the quality and volume of products [10]. During the Fourth Industrial Revolution (Industry 4.0), not only the production methods of the product have to be optimized, but also the interaction of man and machine are changing. Robots began to appear in many enterprises, thanks to which "robotization" has become one of the aspects of "smart production" [11]. Creating "intelligent robots" requires efficient software. With the help of it, it is possible not only to reduce the costs of "creating control programs", but also to ensure control over "service robots" [12]. At the considered production of the innovative material "Techfom", robotization is implemented by forming qualitative and quantitative parameters of the mixture of the material "Techfom" and subsequent control of transport and processes of displacement, filling of casting molds. This is carried out according to the characteristics set by the operator/technologist on the control computer and in the microprocessor control system, which monitors the required parameters of the manufactured mixtures using feedbacks from sensors installed on the cooking and mixing complexes (Fig. 4)

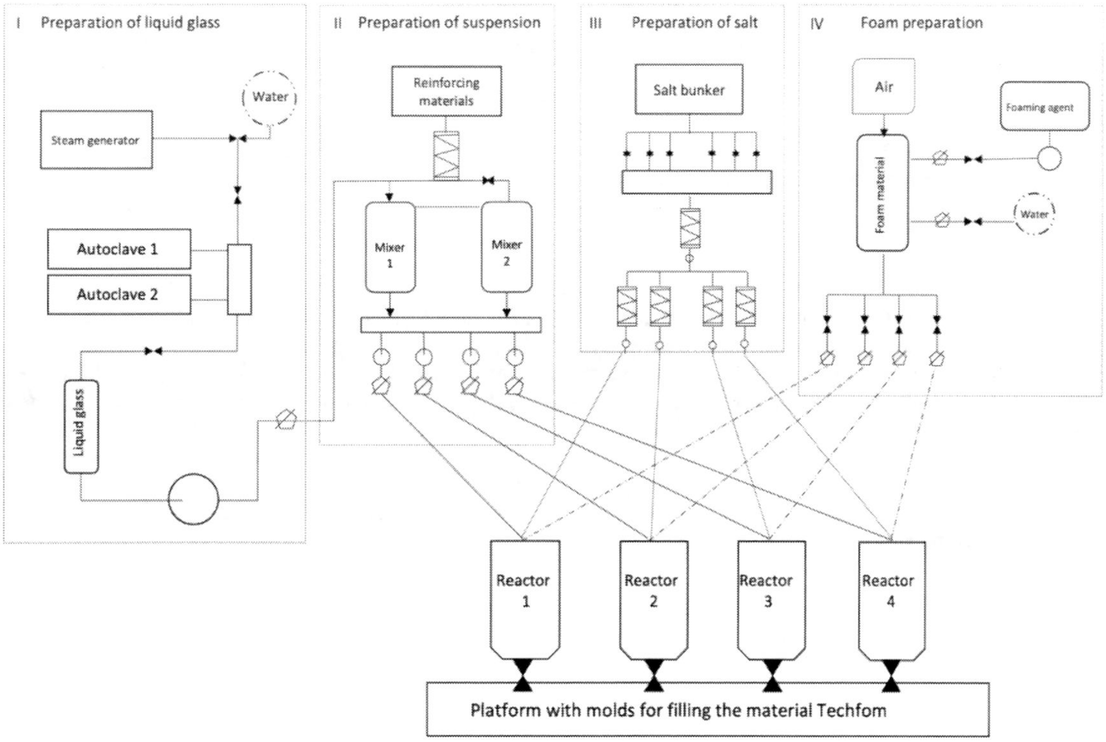

Fig. 4. Structure of technological interaction of the line complexes

In this case, robotization/ intellectualization is a subject of control based on feedbacks and ways of their logical verification due to algorithmic comparison of non-direct sensor readings. For example, when filling the container "Mixer 2" of complex II "Preparation of suspension" (Fig. 4), the liquid level is monitored in it using a conductometric level sensor placed in this container. But what happens if this sensor fails, or its contacts are oxidized, or the signal cable is broken? In these cases, in the absence of only a single simple feedback, a tank overflow may occur, which will lead to a technological accident at the enterprise. To solve this problem, it is necessary to implement processes consisting, for example, in the following program checks:

- IF a control signal is given to open the valve to fill the "Mixer 2" tank, and the readings of its level gauge do not change (in expected time), THEN it is necessary to issue a signal for additional verification and conclude, that either the valve is damaged or the level gauge does not work. Then an additional check is turned on - whether the liquid glass supply pump of complex I "Preparation of liquid glass" has worked.
- IF the pump is running, the valve is open, and the level gauge does not change its readings, THEN it is faulty and the operator must receive information about this on his video monitor for making subsequent decisions, and through the SCADA software it is necessary to stop the pump and turn off the valve.

So, a complete logical scheme of mutual influence/duplication of sensors and continuous testing of equipment directly during operation is compiled. For this purpose, a system of RULES (logical and expert conditions and appropriate cross-checks) is being created, which are implemented in the robotization of the technological line for the production of "Techfom" material.

Smart production includes several aspects, one of which is the "digital twin of production". The digital twin is based on a model, that is created as an analogue of the physical production itself [13]. At the moment, there are various data collection systems, but it is difficult to find common applications to support them and practical examples of implementing digital twins with forecasting and contral capabilities [14]. Most often, the "digital twin" technology is used in aerospace and automotive companies, however, through the introduction of this methodology, it is possible to improve the financial and economic performance of companies operating in other sectors of the economy [15]. For this implementation of such technology, an effective data exchange is necessary, for which the "Techfom" company uses SCADA software tools to work out the technology for creating a component base of material, that differs in significant diversity both in density, composition and purpose:

- "Techfom – 80…120" with a density of 80…120 kg / cubic meter can be used for insulation of residential premises;
- "Techfom – 120…150" with a density of 120…150 kg /cubic meter can be used for noise reduction in residential/non-residential premises;
- "Techfom – 150...220" with a density of 150...220 kg / cubic meter can be used as structural elements of residential/non-residential premises (interior/inter-apartment partitions, enclosing structures /hinged panels).

The development of the technology for obtaining various types of "Techfom" material is carried out by the operator / technologist on the "digital twin" of production. This aspect of "smart production" allows using proper software to virtually work out the qualitative parameters of the required composition of the mixture for any customer, to assess the feasibility of the recipes. Thus, in this article, the implementation of a robotic industrial production line of products made of innovative material "Techfom" (based on the concept of Industry 4.0) was considered, thanks to this, costs are reduced, it becomes possible to increase the production capacity of the enterprise.

2. Conclusion

In this article, using the example of the "Techfom" enterprise, the practical implementation of the "smart production" technology is presented, including the digitalization of the main control elements and the use of remote access both to the software being developed and to the control process itself; optimization of technological systems by digital means, the choice of architecture of software and hardware microprocessor-computer systems united into a single local network. The use of various electromechanical elements (sensors, input devices, tachometers), a computing platform (operating environment) tied into a network structure actually led to the implementation of the so-called cyberphysical system. The article provides examples of the use of components of "smart production" on a specific complex of the robotic industrial line "Techfom", identifies further tasks of their implementation to improve productivity and quality of products. Thus, the application of the considered technology of digitalization of production processes leads to an increase in the competitive advantages of the products

3. References

[1] Goloventchik G.G, Yanhai He. (2021). New business models in digital globalisation of industry (China experience) Journal of the Belarusian State University Economics 2021;1:86–93 Russian

[2] V. Kishorre Annanth, M. Abinash & Lokavarapu Bhaskara. (2021). Rao Intelligent manufacturing in the context of industry 4.0: A case study of siemens industry // Journal of Physics: Conference Series, 1969 (2021) 012019

[3] Novikov, S. V. (2020). Ekonomika izmerenij v usloviyah cifrovizacii / S.V. Novikov // Inzhenerno-fizicheskie problemy novoj tekhniki: Sbornik materialov XIV Vserossijskoj nauchno-tekhnicheskoj konferencii s mezhdu-narodnym uchastiem, posvyashchyonnoj 85-letiyu so dnya rozhdeniya zasluzhennogo rabotnika VSH RF, doktora fiziko-matematicheskih nauk, professora M I Kiselyova, Moskva, 17–19 marta 2020 goda / Redkollegiya: A A Kransuckaya, E V Tumakova, E V Krechetova – Moskva: Obshchestvo s ogranichennoj otvetstvennost'yu «Diona», 2020 – S 236-242 (in Russian)

[4] Kapustina Larisa Mihajlovna, Kondratenko YUliya Nikolaevna. (2020). K voprosu o ponyatii «umnogo predpriya-tiya» v cifrovoj ekonomike // Voprosy upravleniya 2020 №4 (65) Available from: https://cyberleninka ru/article/n/k-voprosu-o-ponyatii-umnogo-predpriyatiya-v-tsifrovoy-ekonomike (date of the application: 30 09 2021, in Russian)

[5] Gribanov Yu I. (2019). Key aspects of the theory and methodology of digital transformation of social and economic systems / Gribanov Yu I // Vestnik Altajskoj akademii ekonomiki i prava – February 2019 – №2 (part 1) – P 83-89

[6] Reiner A. (2014). Industrie 4.0 — Advanced Engineering of Smart Products and Smart Production // Technological Innovations in the Product Development 19th Intern Seminar on High Technology Piracicaba, Brasil, October 9th, 2014 Available from: https://cyberleninka ru/article/n/kiberfizicheskaya-sistema-kak-razvitie-avtomatizatsii-na-vseh-etapah-zhiznennogo-tsikla-deyatelnosti-predpriyatiya-na-osnove (date of the application: 02 09 2021)

[7] Khalin V. G., Chernova G. V. (2018). Digitalization and its impact on the Russian economy and society: advantages, challenges, threats and risks // Administrative Consulting 2018;(10):46-63 (In Russ) Available from: https://cyber-leninka ru/article/n/tsifrovizatsiya-i-ee-vliyanie-na-rossiyskuyu-ekonomiku-i-obschestvo-preimuschestva-vyzovy-ugrozy-i-riski (date of the application: 02 09 2021)

[8] V. E. Pryanichnikov, A. A. Aryskin, S. R. Eprikov, K. B. Kirsanov, R. V. Khelemendik, A. Ya Ksenzenko, E. A. Prysev & A. S. Travushkin (2017) Technology of Multi-Agent Control for Industrial Automation with Logical Processing of Contradictions, Proceedings of the 28th DAAAM International Symposium, pp 1202-1207, B Katalinic (Ed), Published by DAAAM International, ISBN 978-3-902734-11-2, ISSN 1726-9679, Vienna, Austria, DOI: 10 2507/28th daaam proceedings 167

[9] Kudriavtceva A. S. (2019). Cyber-physical system as the development of automation processes at all stages of the life cycle of the enterprise through the introduction of digital technologies // SAEC 2019 №1

[10] Lu Y., Xu X., & Wang L. (2020). Smart manufacturing process and system automation – A critical review of the standards and envisioned scenarios // Journal of Manufacturing Systems, 56, 312-325

[11] Dopico M., Gomez A., De la Fuente D., García N., Rosillo R., Puche J. (2016). A vision of industry 4.0 from an artificial intelligence point of view // International Conference on Artificial Intelligence (ICAI 2016 – WORLDCOMP 2016) Las Vegas, USA: CSREA Press; 2016 pp 407-413

[12] Eprikov, S.; Kim, V.; Pryanichnikov, V.; Prysev, E. & Punenkov, O. (2019) Cloud Platform for Robotariums with Wireless Coverage Networks, Ensuring Service Robots with Parallel Simulation, Proceedings of the 30th DAAAM International Symposium, pp 1160-1165, B. Katalinic (Ed), Published by DAAAM International, ISBN 978-3-902734-22-8, ISSN 1726-9679, Vienna, Austria DOI: 10 2507/30th daaam proceedings 163

[13] Travushkina A. A. (2021). Obzor tekhnologii umnogo proizvodstva / A. A. Travushkina // Studencheskaya nauchnaya vesna: Vserossijskaya studencheskaya konferenciya: sbornik tezisov dokladov, Moskva, 01–30 aprelya 2021 goda – Moskva: Izdatel'skij dom "Nauchnaya biblioteka", 2021 – S 91-92 (in Russian)

[14] Alekseev, A.; Efremov, V.; Lyadskiy, D. & Potekhin, V. (2020) Technical Implementation of Individual Elements of the Digital Twin Concept, B. Katalinic (Ed) Proceedings of the 31st DAAAM International Symposium, Vienna, Austria, pp 0254-0260

[15] Janda, P.; Hajicek, Z. & Bernardin, P. (2019) Implementation of The Digital Twin Methodology, Proceedings of the 30th DAAAM International Symposium, pp 0533-0538, B. Katalinic (Ed), Published by DAAAM International, ISBN 978-3-902734-22-8, ISSN 1726-9679, Vienna, Austria, DOI: 10 2507/30th daaam proceedings 072

DOI: 10.2507/32nd.daaam.proceedings.094

THE APPLICATION OF KNOWLEDGE MANAGEMENT ELEMENTS IN DYNAMICALLY CHANGING ORGANISATIONS

Katerina Kasova & Martin Melichar

This Publication has to be referred as: Kasova, K[aterina] & Melichar, M[artin] (2021). The Application of Knowledge Management Elements in Dynamically Changing Organisations, Proceedings of the 32nd DAAAM International Symposium, pp.0666-0672, B. Katalinic (Ed.), Published by DAAAM International, ISBN 978-3-902734-33-4, ISSN 1726-9679, Vienna, Austria
DOI: 10.2507/32nd.daaam.proceedings.094

Abstract

This paper focuses on the application of the selected knowledge management elements in a dynamically changing organisation. The main research question is how to secure the transfer of knowledge in a team with annual fluctuation of its members. For this single case study, we picked the Formula Student team by the West Bohemian University in Pilsen, which consists of 25-30 students. Part of the investigation is a description of basic terms of knowledge management such as knowledge, information, and data. Then we analyse the organisation, where we found that the current status of knowledge transfer is not satisfactory. Based on the findings, we suggested and applied several knowledge management elements in the organisation. Finally, we defined expected benefits and possible risks related to applying knowledge management in the respective organisation.

Keywords: Knowledge; Information; Management; Knowledgeable workers.

1. Introduction

If any organisation wants to be perceived differently and uniquely, it is very important to use its technologies and methods in a specific and effective way, different from its competitors. Any organisation needs to be able to master their knowledge and make it to their advantage. This article aims to apply knowledge management elements in the Formula Student team by the West Bohemian University in Pilsen. Regular Formula Student teams comprise 20-60 members, mostly university students of technical majors. The racing period covers the whole academic year and usually, 30% of the team changes annually. Most of the members stay for 2-5 years and gain enough experience which they pass further to the new members. Especially this feature defines a dynamically changing team where we need to apply knowledge management elements.

2. Knowledge management

Many publications have been written about knowledge management. [1], [2], [3] These publications focus mostly on knowledge management on its general level or describe its application in large corporations. Knowledge management development goes hand in hand with data, information, and knowledge collection.

Knowledge management helps companies to increase their ability to succeed and is successfully used by educational institutions and in the cooperation among various companies. [4] To successfully exploit knowledge management benefits, it is necessary to further develop intellectual capital and informational technologies. Our time is characterised by easily accessible information. This accessibility goes together with the need to correctly identify the necessary information, get rid of the unnecessary ones, and process them effectively, which is very difficult and time-consuming. Knowledge management ensures this information will be used effectively. [1]. The main goal of knowledge management is to provide organisations with a set of tools and methods, that make their efforts more effective. Most organisations work with knowledge but without any primary rules, which means they don't necessarily reach their desired goals. [2]

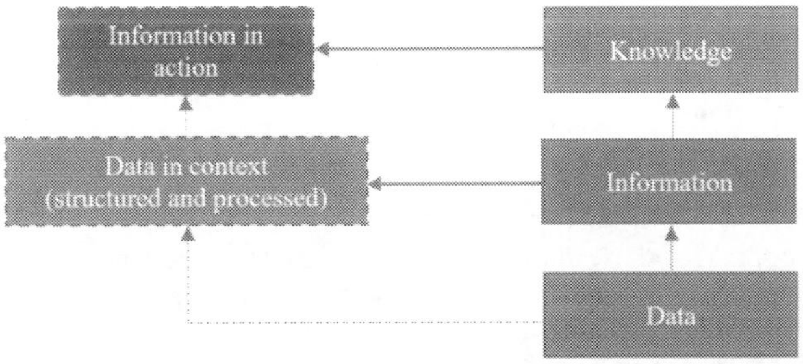

Fig. 1. Relations between data, information, and knowledge

Knowledge assumes various preconditions and experience should be used while working with knowledge. Hence, you can express knowledge with this formula:

$$knowledge = information + x \tag{1}$$

where x is the amount of familiarity, experience, skills, mental relations, values, and principles obtained in the past.[5]

3. Definition of the term organisation for the use of knowledge management

The theory is applied to the Formula Student team, which now has 20 active members and 15 new members. Formula Student is a prestigious competition organised by SAE International. There are usually 540 competing teams from technical universities. [6] It's the most famous technical competition that focuses on developing young people and motivate them to take further steps in their education by a very intensive experience in the field of design and production while having to consider economic aspects as well. [7]

3.1 Formula Student team knowledge management requirements

The Formula Student team has partly specific requirements compared to regular racing teams or organisations. The requirements are keeping a very high level of information sharing among most members, quick accessibility, timing, and validity. This is caused mostly by a high fluctuation of members as desired based on the philosophy of the whole competition. A regular Formula Student team comprise 20-60 students of various classes, while roughly 20-30% of the members change every year. Therefore, there's a huge need for swift information transfer and knowledge management in general.

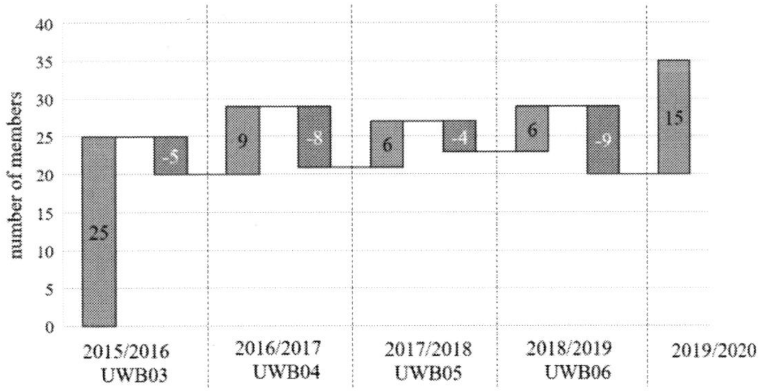

Fig. 2. Evolution of the Formula Student team member's number

Fig. no. 2 shows the number of members in the Formula Student Team by the West Bohemian University in Pilsen. (The total number of members is seen on the "Number of members" axis, the red columns show the number of leaving members at the end of every season. The green columns show newly entered members at the beginning of every new season.) Based on these data, the fluctuation is roughly 30% of all members annually. The main benefit of knowledge management application expected by the team is a more effective way of knowledge transmission among the individual members without losing the important know-how in between various racing seasons.

The Formula Student is a highly competitive and dynamic environment, where innovation and development play a significant role. It is not uncommon for leading teams to lose their position quickly when not reflecting the fast pace of necessary changes. They might fall from the top of the race to middle or low positions. Hence, if teams want to be successful, they must judge the situation quickly and avoid repeating mistakes or less functional solutions. It is important to see the opportunities, quickly use them or abandon them, based on the situation. To achieve this in a long-term horizon, it is essential to have the data, information, and knowledge from the past seasons and use them effectively.

4. Data collection and evaluation

To effectively choose and apply the knowledge management elements with the greatest benefit for the team, it is important to map the status, processes, and knowledge transfer across the team.

- **Main knowledge and skills**: With the leading workers, we defined the most important knowledge and skills necessary to designing and producing a new racing car and to ensure the everyday operations throughout the racing season and at the race itself. These essential knowledge and skills are quickly transforming into various new types of knowledge. It would be very difficult, even impossible, to realise the whole project without this knowledge. Therefore, the human and intellectual capital is essential for a successful realisation. The knowledge varies from administrative, to design, up to the construction and operations of the vehicle at the race.
- **Respective position within the team**: To keep up with the rules of the competition, it is important to perform a lot of essential tasks every season. And on top of that, if any teams aim to score on the top, it is essential to do even more than that. Therefore, we mapped the following positions (Fig 3), which are essential to ensure the operation. These positions have been defined based on historical experience

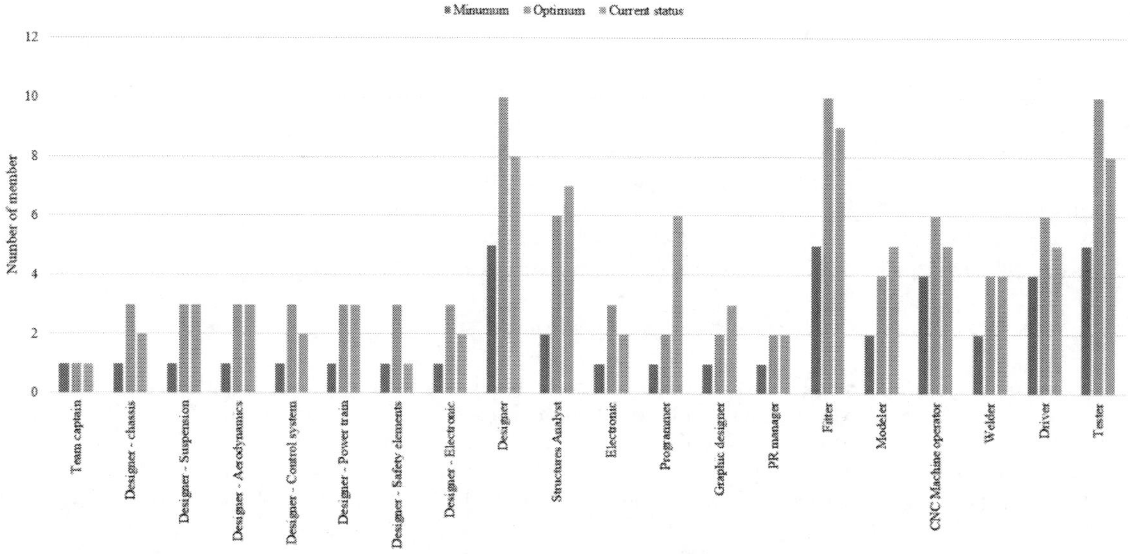

Fig. 3. Comparison of the actual status and desired status

- **Current communication and knowledge sharing process:** Every type of communication is suitable for various situations and adds another value to the knowledge transfer. Therefore, we distributed a survey where we asked more experienced members to rate various communication tools based on their effective knowledge transfer. In total, 15 members filled out the survey. As you can see in the next Fig. 4, it is evident that the most effective tool is face-to-face communication, such as meeting/brainstorming sessions, personal meetings and/or conversations with authority, and practical demonstrations. Furthermore, diaries and news have been rated as another effective way of knowledge transfer.

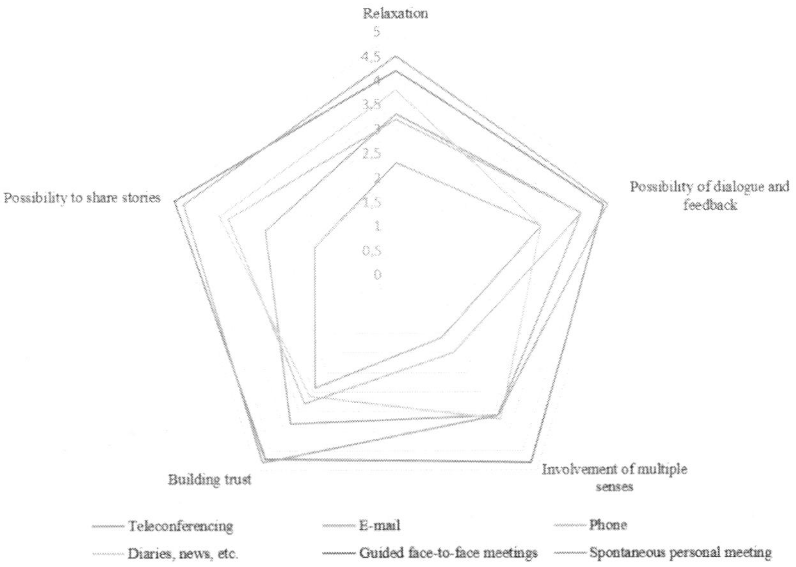

Fig. 4. Knowledge transfer survey – results

While evaluating the overall situation, we found out that the status is not suitable for easy and organized information access for the new members and the regular members as well. Information gets lost due to the big amount of data, information, and knowledge or, on the other hand, duplicity occurs very often. Generally, information transfer occurs verbally in a form of a certain apprenticeship. This system is not suitable due to the high fluctuation of the members and time consumption of the informational transfer. Therefore, much information gets lost when any person leaves the team. In terms of future team management, applying knowledge management seems like an important step. Systematic effort to save and store the knowledge in the team would decide whether the team will be successful in the future or not.

5. Knowledge management elements application

Based on the knowledge status analysis in the Formula Student team, we choose and applied the following knowledge management elements: main knowledge definition, key workers definition, process setup, knowledge sharing, and targeted education. These elements should strengthen processes in knowledge management control to make possible the creation of new ways of work and to further develop know-how, find suitable ways of knowledge presentation, and put them to use for further application, revision, preservation, and verification in Fig. 5. [8]

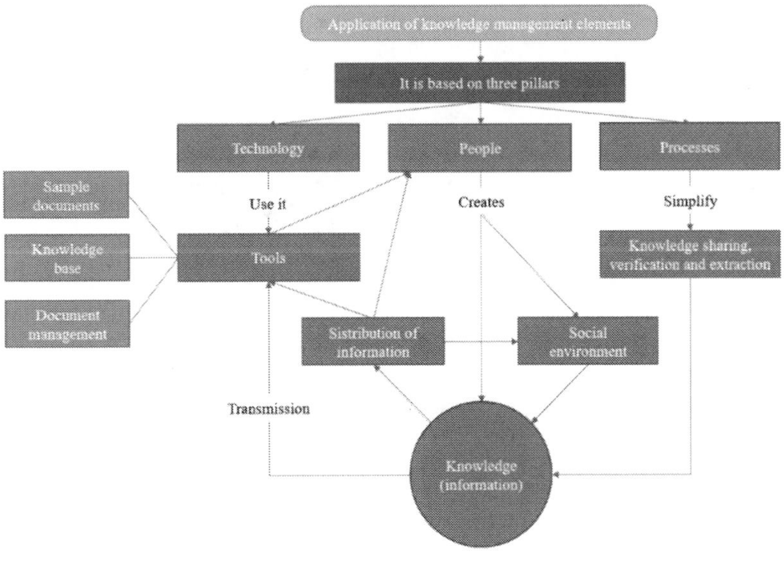

Fig. 5. Use of knowledge management elements

5.1. Main knowledge and core workers identification

To identify the main knowledge in the team, it was necessary to perform a status analysis, which was based on the communication among the team members and the breakdown of individual processes. It was also important to consider the long-term goals and visions. We had to consider not just the internal knowledge but to map the external knowledge (such as the knowledge of the competing teams) as well.

Of all the workers defined by the organisational structure, we picked those members (with the help of current members), which have unique knowledge and therefore, we can mark them as the key workers. [9] We defined five key workers with knowledge covering all the unique knowledge throughout the project and assigned key knowledge as demonstrated in Table 1.

	Knowledge	Team Captain	Head of Tech.	Head of sponsors.	Test driver	Test technician
1.	Knowledge of Formula Student rules	■	■			■
2.	Knowledge of the technical solutions of competing teams	■	■		■	
3.	Estimation of future competitive intentions		■			
4.	Knowledge of team management	■	■			
5.	Knowledge of the possibilities of technical solutions		■			
6.	Knowledge of design solutions		■			
7.	Selection of participation in races from the formula program	■	■	■		
8.	Selection of team members to participate in specific races	■				
9.	Provision of conditions at the site of the plant		■			
10.	Knowledge of possible solutions for individual structural nodes		■			
11.	Knowledge of the team's production capabilities		■			■
12.	Knowledge of possible support at the UWB	■		■		
13.	Knowledge of automotive equipment producers in the region	■	■			
14.	Knowledge of business creation	■		■		
15.	Knowledge of advertising rules and possible tax breaks			■		
16.	Knowledge of advertising and sponsorship contracts			■		
17.	Estimate of financing of other formula teams	■		■		
18.	Knowledge of the paraments affecting formula performance		■		■	■
19.	Knowing how each node's function will work on formula performance		■		■	■
20.	Knowledge of the possibility of loading individual parts of a formula		■			■
21.	Knowledge of the real driving characteristics of the formula				■	
22.	Knowledge of the theoretical driving characteristics of a formula		■			■

Table 1. Main knowledge identification

5.2. Knowledge sharing process setup

Among main goals of knowledge management, you can find knowledge identification, use of know-how, experience, and skills exploitation. Furthermore, another goal is to make sure all the above-mentioned would be shared and used further, as necessary. Doing that improves the overall quality, productivity, and the effectivity of work. It should be a task for most knowledgeable workers to provide their knowledge to other team members, who can then perform better and avoid mistakes. Another very important aspect is to be able to preserve the transferred knowledge to make sure the knowledge remains in the team even if the key worker leaves. [10]

It is also important to transfer the knowledge in a continuous way as the average time people spend in team is 2-4 years. A good example of knowledge transfer (among the drivers in Fig. 6) is when the test driver can convey the knowledge of handling and current power of the power unit to ensure the driver would use the right gear ratio in a suitable driving regimen. This significantly shortens the testing period of new drivers.

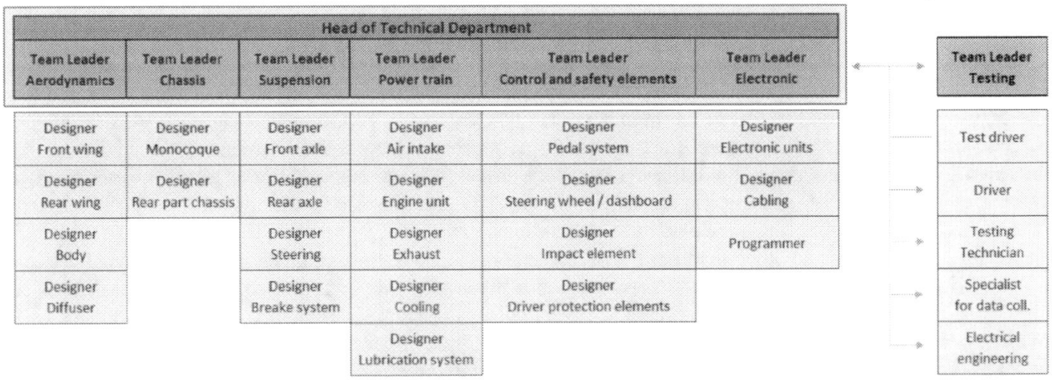

Fig. 6. Test drive knowledge transfer scheme

5.3. Team members education

A new educational system has been designed for the team members to ensure sufficient and effective knowledge and experience transfer and support the introduction and application of various knowledge management elements. Among others, important features of the system are hands-on experience training, a one-week introductory course, mentoring of the new members (every new member has a mentor to obtain the necessary experience and knowledge in the very first weeks in the team), and support courses from external partners.

6. Main expected benefits and possible risks

A comparable level of knowledge and skills of the individual members is an essential precondition for national and international success. It is important to stress that fundamental improvement might be reached only through synergy when various elements of knowledge management enhance each other. Knowledge management by itself can't guarantee 100% success. [11] Fundamental benefits that can be obtained via knowledge management are innovation capacity augmentation – better use of human and other resources, quality improvement. Faster reaction period – higher flexibility, speed and correctness of decisions, more effective team education. Improvement of inner qualities – more effective process innovation, morale, and operational effectivity. More benefits can be seen from the point of view of managers and leaders. It helps them to easily and timely define the knowledge, key performers, key skills and experience, and to improve communication channels, management tools, and strategy tools. [2]

6.1. Risks

Applying knowledge management elements in the team presents a big challenge because every project is different and bears various risks and issues related to the realisation. Successful implementation and use of suggested solutions might be hindered especially by change avoidance, fear of the unknown, unwillingness to share information, lack of support from the leadership, not clearly defined responsibilities, lack of ability to recognise knowledge and possibilities of their application, lack of practical use of knowledge, frustration because of wrong expectations and more. Some of the above-mentioned situations might be prevented by monitoring the risks and evaluating the possibilities of their occurrence. The selected issues and related solutions are presented in the following overview Table 2.

Problems	**Possible risks**	**Risk lowering/eliminating measurements**
Goal definition	Not defining milestones. Defining unmeasurable milestones.	Defining measurable and clear milestones, regular information sessions.
Factual, time, and financial range.	Clearly defined scope. Missing on the time frame. Exceeding the estimated costs.	Dividing the process into several phases. Controlling the progress. Primarily focusing on the knowledge chains, not only on the used technologies.
Implementation process	Uncontrolled transfer of information. Overly technical approach. Unclear leadership and low level of support.	Choose a responsible authority. Inform periodically the key members of the overall progress. The realisation must be supported by leading authorities, who should be able to motivate others.
Measuring the expected benefits	Lack of capability to evaluate success/failure of individual steps. Result don't match the expectations.	Inform everyone about the success/failure as soon as possible. Request feedback from the key members regularly. Regular checks of progress, goals achievement and the overall benefits.

Table 2. Possible risk related to realisation and countermeasures

7. Conclusion

This paper investigates the application of knowledge management in a dynamically changing team. The main research concern is how to secure the knowledge transfer in a team with a very high (annual) fluctuation of the members. The subject of research in this single case study is the Formula Student team of the University of West Bohemia, composed of 20-35 students. Roughly 20-30% of the members change every year. In the theoretical part, we described the term knowledge and its relation to data and information. To define appropriate applicable elements, we used the status analysis. We defined the main knowledge and skills necessary to ensure the functioning of the team, job positions included. Furthermore, we applied the member education process, which describes the possibilities of transferring the knowledge/information on the new members of the team especially. Figure 2 shows an example of the knowledge transfer between the test drive pilot and other members of the realisation team. All the applied elements of knowledge management might be effectively used as tools to ensure knowledge transfer in the long-term horizon. In the final part of this paper, we described the benefits and risks of applying the knowledge management elements. Among the biggest benefits are innovation capacity augmentation, faster reaction period, and improvement of the inside qualities. On the other hand, among the biggest risks is the unwillingness of the team members to provide and convey the information verbally. Another area for improvement, and thus the idea for further research, would be the possibility of applying an information knowledge base to the assessed team, which is now missing here.

8. Acknowledgments

The article contribution has been prepared under project SGS-2019-008.

9. References

[1] Davenport, T. H. & Prusak, L. (1998). Working knowledge: how organizations manage what they know, Harvard Business School Press, ISBN 978-0-87584-655-2, Boston

[2] Pasher, E. & Ronen, T. (2011). The complete guide to knowledge management: a strategic plan to leverage your company's intellectual capital, John Wiley, ISBN 978-0-470-88129-3, Hoboken.

[3] Collison, Ch. & Parcell G. (2005). Knowledge management: praktický management znalostí z prostředí předních světových učících se organizací, Computer Press, ISBN 80-251-0760-4, Brno

[4] Majstorovic, V. & Majstorovic, A (2020): Strategic and Cultural Dimension of the Management by Projects, Proceedings of the 31st DAAAM International Symposium, pp.0049-0055, B. Katalinic (Ed.), Published by DAAAM International, ISBN 978-3-902734-29-7, ISSN 1726-9679, Vienna, Austria

[5] Vymetal, J.; Diacikova, A. & Vachova, M. (2005). Informační a znalostní management v praxi, LexisNexis, ISBN 80-86920-01-1, Praha

[6] https://mazur-events.de/fs-world/C/, (2020). Formula Student Combustion - World Ranking List, Accessed on: 2020-02-12

[7] https://fsaeonline.com/, (2021). Formula Student General Inforamtion, Accessed on: 2021-09-15

[8] Senge, P. (2006). The fifth discipline: the art and practice of the learning organization, Doubleday, ISBN 978-0385517256, New York

[9] Mladkova, L. (2005). Management znalostí, Oeconomica, ISBN 80-245-0878-8, Prague

[10] Bettiol, M.; Di Maria, E. & Micceli, S. (2020). Knowledge management and industry 4.0: new paradigms for value creation, Springer International Publishing, ISBN 978-3-030-43588-2

[11] Schopflin, K. & Walsh, M. (2019). Practical knowledge and information management, Facet Publishing, ISBN 978-1-78330-335-9, London

DOI: 10.2507/32nd.daaam.proceedings.095

SMART MAINTENANCE FRAMEWORK BASED ON SPATIO-VISUAL COMPUTING: THE CASE OF AIRCRAFT MRO

Polina Bortsova & Peter Panfilov

This Publication has to be referred as: Bortsova, P[olina] & Panfilov, P[eter] (2021). Smart Maintenance Framework Based on Spatio-Visual Computing: The Case of Aircraft MRO, Proceedings of the 32nd DAAAM International Symposium, pp.0673-0684, B. Katalinic (Ed.), Published by DAAAM International, ISBN 978-3-902734-33-4, ISSN 1726-9679, Vienna, Austria
DOI: 10.2507/32nd.daaam.proceedings.095

Abstract

Many organizations in aviation industry are facing the problem of establishing an overall coherent picture of their business and operations, while disparate datasets from different departments and geographically dispersed locations are united only by experience and often-heterogeneous hardware and software platforms prevent from efficient enterprise application integration. This paper focuses on approaches to increasing the value proposition of management information systems of aviation organization through integrated spatial and visual computing solutions while using the domain of aircraft maintenance, repair and overhaul organization as an illustrative example. The subject of the research is the analysis of the application of new computing approaches to digital transformation of complex geographically distributed services that require timely coordination, accurate localization and efficient planning and execution. The main object of the study is a new integrated spatio-visual framework for smart MRO service of aircraft engines that includes technology and software solutions in support of operations at different levels in aviation organization from top management personnel to the aircraft technician.

Keywords: aircraft MRO; spatial computing; visual computing; augmented reality.

1. Introduction

In the last decade, visual computing is under intensive study in a wide range of application areas. In particular, augmented reality (AR) as a subarea of visual computing has received special attention as an advanced approach to improve the user interface at the level of management information system (MIS) architecture both in the industrial sector and in the sector of providing various services. The visual computing interfaces can be designed for different kinds of users, and support several objects and applications in several areas, depending on the specifics of the system being built. Spatial computing is a new computing paradigm driven by advances in location-aware smart applications and services, machine learning (ML) and the Internet of Things (IoT) that enables seamless interactions between people, machines and environments to optimize processes in real or near real time. Spatial computing acts as a mean that allows people to understand better, how machines, products and process participants move in space, and what relationships exist between them and how all the information related to their location can be used for business processes.

Location Intelligence (LI) is achieved through visualization and analysis of geospatial data, enhancing the ability to make decisions and predictions. The main goal of the presented study is to consider approaches to unlocking the potential of spatial and visual computing by suggesting an integrated spatio-visual framework that integrates existing geospatial solutions into the future intelligent transportation systems (ITS). We use an example of aviation industry and a specific sector of aircraft engine maintenance, repair and overhaul (MRO) service as a business case for the framework prototyping and demonstration.

1.1. Digital transformation of transportation sector

Intelligent Transportation Systems (ITS) provide solutions to the current transportation problems with the help of new digital technologies and computing paradigms. ITS is a "smart" system that uses innovative developments in the modelling of transport systems and the regulation of traffic flows, providing end users with greater information content and safety, as well as qualitatively increasing the level of interaction of transportation system users compared to conventional transportation systems. The overall idea behind the ITS is to improve decision making, particularly in real time, by vehicle operators, transport network controllers and other users, thereby improving the operation and safety of the entire system. ITS is an integrated system that implements a broad range of computing, communication, control, vehicle sensing and electronics technologies to help in monitoring and managing vehicle status, traffic flow, enhancing productivity of the system, and saving lives, time and money. Emergency Management System (EMS) is the newest research field in ITS domain. EMS is mainly concerned with the application of different digital technologies to develop a transport system that can provide help in the emergency conditions. It can help in reducing the fatality rate in the accidents [1].

Achieving mobility, safety and sustainability are in focus of ITS research and development. Although significant progress has been made in visual computing techniques such as computer vision in ITS, researchers from both academia and industry are still facing several major challenges that hinder further advances in ITS development that are represented with the following How questions:

- How to improve the visual perception for ITS in adverse weather conditions?
- How to take full advantage of multi-sensor perceptual data in ITS?
- How to develop the ITS-specific visual computing solutions through advanced artificial intelligence techniques?
- How to develop a visual computing-based traffic monitoring system and enhance the situational awareness and safety?
- How to guarantee the effectiveness of different autonomous vehicles using visual computing techniques?

There exists a broad spectrum of technologies intended to support applications and systems developed for ITS. The scope of modern technologies and possibilities of their fully integration in a typical smart transportation service project are going to be discussed in in this work.

1.2. Digital transformation of aviation transportation: Aircraft health monitoring example

Although avionics and control technologies have produced continuous advances in aircraft systems, there is still significant opportunity for greater efficiency, enhanced functionality, and better integration of aviation systems. This is particularly true for systems that reduce the burden on the crew of flying the aircraft and systems that allow for increased capacity of the global air traffic management (ATM) system. Using big data and data analytics together with secure communications capabilities opens up a host of possibilities to monitor an aircraft's components and on-board equipment status. Real-time data enables aircraft operators to apply predictive maintenance to monitor broken parts. Moreover, data analytics can improve fuel usage, as fuel is one of the biggest expenses for airlines. The big challenge, however, is extracting relevant data and making it useful, i.e. transforming airline operations data from maintenance to flight safety in the end-to-end manner [2]. More and more value is being extracted from aircraft data, delivering intelligence and analysis that are able to transform operations. Data can be unlocked from every part of the aircraft – cockpit, cabin, aircraft engine, and used for enhancement of performance and service. Accessing this data is rather complex and that is why effectiveness of handling data is playing a key role in digital transformation of aviation transportation.

Current state of the aviation industry requires new digital "smart" solutions [3], because the majority of reporting, diagnosis, planning and maintenance services happen on ground. Due to aircraft and engine condition issues, costly delays and cancellations of transportation services occur and, consequently, airplane serviceability is interrupted. Aviation industry experts and specialists are increasingly discussing issues of big data technologies, data ownership, access, and management in relation to the aircraft health monitoring (AHM), as well as the benefits and challenges for airlines, aircraft OEMs, maintenance, repair and overhaul (MRO) organizations, and the role of each player in the AHM market. More and more data from aircraft engines, airplanes, and on-board avionics systems is available for the measurement, storage, and transmission that require a new way of operating aircraft — a connected and integrated operation that brings increased efficiency for airlines [4].

Aircraft engines are the most expensive part of an aircraft, and the carrier is interested in keeping its power plants in good condition. In addition to the fact that the engines are important for fast and safe flight, unscheduled maintenance due to problems can become very expensive. The most expensive parts of the engine are the control systems, turbine blades and bearings. Prices for aircraft power plants range from $12 million to $45 million. Despite the fact that they are made of the most modern and durable materials, jet engines cannot avoid wear and tear. This occurs because of vibration, friction, high temperatures and corrosion from external influences or damage from foreign objects. High speed, high temperature, high load of work environment contribute to corrosion, wear and failure condition. Aircraft engine maintenance standards are the most important thing in ensuring flight safety and have a huge impact on the financial condition of airlines around the world. Answering the question, why do accidents happen, statistics provide the figures that highlight above all human error factor and situations when industrial equipment (engines in particular) is elevated or is hard to access.

Many airlines partially or fully outsource MRO services from repair shops and system suppliers. MRO software manages the maintenance process and offers solutions to improve efficiency and enhance the end-user experience. Airlines require software that will optimize productivity and keep operations up with demand. New technologies and the digitalization of services are creating benefits for airline and MRO operators to streamline maintenance, reduce operating costs, and take full advantage of the skyrocketing air travel demand. Key factors such as rising demand for air travel increase in digitalization and upgrade of the legacy MIS to digital MRO Software influence growth within the MRO Software market. Most of leading manufacturers of aircraft engines invest into digital transformation of their businesses including engine production, sells, repair, and maintenance services (Table 2).

OEM	Digitalization initiative
GE	Flight efficiency/operational management Fuel management Predictive maintenance Fleet management
Rolls-Royce	Analytics service into Total Care for analysing engine data in real-time Managing customers' engine maintenance Monitoring aircraft availability
Pratt & Whitney	Data ecosystem System for capturing thousands of engine and aircraft parameters throughout the full flight cycle Monitoring engine performance for minimization of disruptions

Table 1. Engine manufacturers' digitalization initiatives

1.3. Research goals and tasks

The complexity and difficulty of aviation transportation management has increased due to the wide range of business areas and departments involved in aviation transportation business. Within the framework of this research, the main interest is put to the information flow between the parties of the organizational process of the aviation transportation service as it is provided in the Figure 1. The distributed connected (networked) structure of the aviation transportation organizations and their operations implies that in case of digital transformation of aviation services real time exchange of data and the distributed nature of the MIS in organization are important requirements and prerequisites.

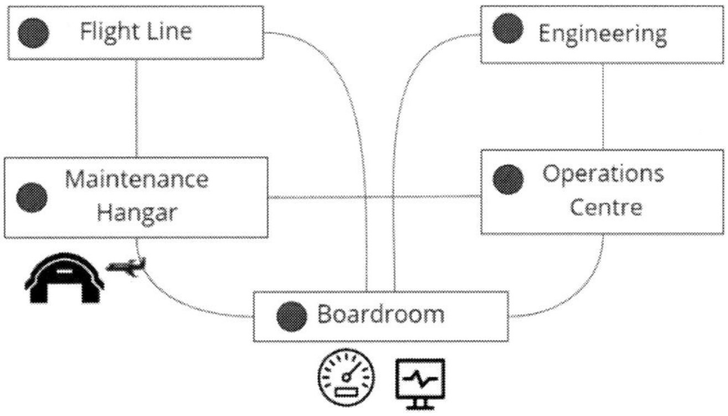

Fig. 1. Key elements of the aviation transportation service organization.

The Figure 2 provides the ICF Aviation [5, 6] described a high-level conceptual view of the AHM practice as it. Elements of advanced digital technologies of the smart MRO service application should address the issues during health checks and effective communication between OEM Tech Support, Airline Maintenance Control Centre, Aircraft and Maintenance Engineering Operations. The position of each instance relative to each other as well as full description of the common information with regular updates, parallel processing of spatial data, visualization of geolocations of the aircrafts and conditions of all systems play a decisive role in AHM performance improvement [7, 8]. Key factors such as rising demand for air travel increase in digitalization and upgrade of the legacy MIS to digital smart MRO influence growth within the MRO Software market.

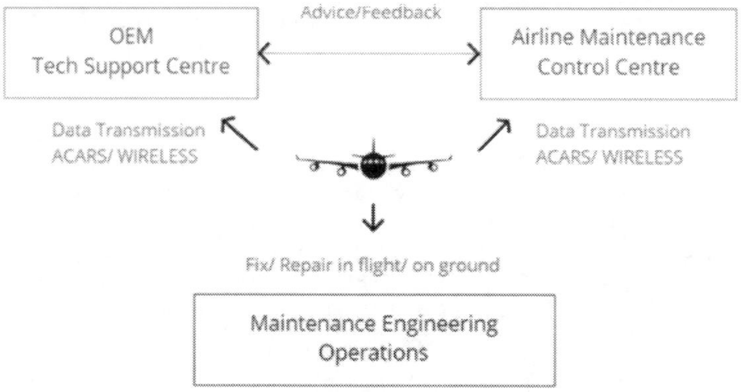

Fig. 2. Aircraft health monitoring schematic.

Based on the analysis of existing digitization initiatives [5] and the research on aircraft MRO topic the following forms of initiatives that can be supported by the suggested smart maintenance framework were formulated:

- Transformation of technical services
- Automated and secure transparent system of digital reporting on work with components of aircraft
- Engine monitoring based on the processing of an engine status information collected from sensors
- Building a model for accurate prediction of the remaining useful lifetime (RUL) and optimal scheduling of the engine replacement/repair service with exact location tag
- Profound data analytics, a work with historical data for mining and collecting insight to enhance operations and reduce maintenance delays
- Collecting fleet data for maximization of the customer's specific engine performance and time-on-wing while maintaining predictable MRO
- Spatial computing bolsters interaction of workers with the physical systems in industrial environments
- Full usage of Visual Computing potential.

Supposedly, in future, the customer will support the maintenance company or the machine manufacturer over long distances with the repair of the machine via live transmission. The customer will need a high-speed Internet connection and data glasses, and the manufacturer a PC. With the benefits of innovative Visual Computing technologies, the aviation technicians equipped with AR data glasses will have a projection surface in front of their eyes that will show different files: images, videos, and sound or text document. These displays overlay and complement the real image, i.e. the view of the equipment to be serviced or repaired. The repair and maintenance costs will therefore decrease; long journey times for the service technician, as well as machine downtime and production failure at the customer side will be outdated.

In this section, the application of the proposed architecture to the solution of aviation transportation problems is considered given the example of, particularly, aircraft engine maintenance, repair and overhaul (MRO) smart service [7, 8]. Firstly, the subject area, environment, key participants and level of current digitalization as well as the scope of up-to-date technologies used by manufacturers are analysed. Then, possible implementation of the proposed framework from the previous section, enriched with specific recommendations, is discussed.

The aircraft engines MRO was chosen in this work as a specific illustrative example for detailed analysis and implementation of the application framework. The MRO in aviation is the repair, service, or inspection of an aircraft or aircraft component(s). It is applicable to all maintenance activities that are conducted to ensure safety and airworthiness of all aircrafts by international standards. Safety of aircraft is influenced by the condition of the core component of the aircraft; it is needed for OEMs and airlines to understand permanently, whether engine could be operated normally.

The modern digital technologies such as spatial and visual computing can offer desired Smart MRO solutions that could lead to minimization of risks that threaten safety in the air, for instance, by giving an opportunity to track the location of engines in the sky, building digital twins for them and developing new approaches and tools for engine maintenance.

To achieve this goal, the authors have identified a number of tasks to be solved on the way to the design and implementation of the universal spatio-visual computing framework:

- Discussion of spatial and visual computing areas, as well as their interaction and comparison
- Analysis of the possibilities of practical application of these technologies to aviation transportation problems
- Design of universal integrated framework for supporting information systems, combining advanced related technologies: AR, digital twins, GIS, geographic positioning systems (GPS), data analysis, ML
- Considering the application of spatial and visual computing within the proposed framework for the example application of aircraft engine MRO.

2. Methodology

For the study, methods of synthesis of information on the selected topic, deduction, critical analysis of technologies, their comparison and search for complementary components were used.

2.1. Spatial computing

Digital transformation includes a network of interconnected processes and solutions. To transfer value of the network it is necessary to provide its transparent visibility of the connections, incl. in physical space. As part of the digital transformation, many organizations rely on geographic information systems (GIS) technology to generate location information. The technology has recently expanded and now it includes advanced spatial computing capabilities. Spatial computing ranks as one of the 10 most promising technologies of 2020 according to the World Economic Forum's report in terms of the value of their potential contribution to the transition to Industry 4.0 [9]. More than 90% of companies believe that location information is critical for achievement of business success. Geospatial data and spatial analysis of location are used to generate the knowledge needed to make decisions. Applications of this nature can be used to find the best places to select a marketplace and solve bottlenecks on the roads, maintain and repair vital infrastructures. In computer science, MIT Media Lab alumnus Simon Greenwold in his futuristic dissertation in 2003 firstly defied the concept [10]. Spatial computing is primarily the ability to relate the location in space of one object to its operational control or its virtual complement elsewhere. From a practical point of view, spatial computing is the digitization of the actions of machines, people, objects and the environment in which they are located in order to ensure and optimize their actions and interactions. The key difference between traditional and spatial computing is the orientation of spatial computing to the three-dimensional location of the object in question [11].

Technically speaking, spatial computing is a technology that uses software to seamlessly integrate user interface that provides human-computer interaction into the 3D physical world. In the literature devoted to the study of spatial computing the following areas of knowledge are considered for studying:

1. *The science of spatial computing* that cover among others the problems of spatio-temporal prediction, working with qualitative spatio-temporal data and development of spatial and spatio-temporal computing standards.
2. *Spatial computing systems* issues include definition and development of spatial computing infrastructure; collection, integration and storing sensed data; computational aspects of spatial Big Data, and augmented reality.
3. *Crosscutting solutions* focus on ubiquitous computing arrangements, sustained sensing and monitoring, reliable transportation systems and geo-confidentiality.

Technologies. The main approach in spatial computing is to enable the real world to interact with digital entities. Thanks to a smart combination of technologies such as AR, which provides an interface between a digital world and a physical one, the industrial Internet of Things (IoT), which is responsible for the control of products and assets, machine learning, sensor technologies, a large number of processes in logistics and production, can be optimized. Due to virtual and augmented reality, the concept of the digital twin in spatial commutation is becoming extremely valuable. Digital twin [9] refers to a digital copy or model of a physical twin: processes, people, places, systems and devices that can be used for various purposes, such as modeling, predicting behavior, monitoring performance, restoring an object for 3D printing, checking scenarios, and workouts, connecting to other objects in the digital space.

To create a digital world, software algorithms (such as methods for optical measurement of distance or speed, video recording, and other geolocation technologies) integrate the digital map with sensor data and digital representations of objects and people to purposefully influence the real world. A physical object has IoT sensors that constantly collect various forms of data and transmit information to a digital twin to analyze and perform the above functions. The results obtained with the virtual models can be applied further to the physical twin.

In terms of architecture, spatial computing provides a great opportunity for scripting, giving a possibility and access to generating plenty of ideas, especially for dynamic spaces such as rooms, buildings, neighborhoods, cities, and everything that populates these spaces. Spatial computing and augmented reality together enable architecture with an ideal and realistic view of the space available and the "virtual lens" mechanism through which new information can be viewed and applied.

Application areas. One of the most visible aspects of spatial computing are its analytical capabilities. For instance, using spatial heat maps, companies can obtain the following information: where workers are located most often during their work process, where unexpected machine downtime occurs, which routes are most frequently used. Additionally, the considered approaches can provide safety. With the help of augmented reality within a factory, safety zones and routes can be made for both people and cars. Spatial computing helps to understand the user's location, find the surrounding elements in space and give directions on the way to a right safe place using AR. During an emergency evacuation, the spatial computing can help to find the most efficient exit route for each individual. Information about the route can be sent via push notifications to a smartphone. The user clicks on the notification and launches an augmented reality scene that shows him the best route.

2.2. Visual computing

Visual computing is a field of computing that deals with the acquisition, analysis, and synthesis of visual data using computer resources. It covers several areas of science, in particular, computer science, mathematics, physics, computer vision, graphics, animation, modelling of complex systems related to visualization, as well as cognitive sciences. Visual computing is used to visualize the results of computer calculations, can help to understand the meaning contained in a huge datasets. Visual computing focuses on the management of activities (for example, business processes in an organization) and interaction through the manipulation of visual virtual images and physical objects [13]. Visual computing is concerned with the creation, processing and analysis of digital images. Images, 3D models, videos, block diagrams and even simple signs can be used as media.

Technologies. In applied research, innovative solutions in the field of data visualization and augmented reality are developed mainly through a combination of computer graphics and computer vision. Visual computing today relies on the processing power of graphics processing units (GPUs). Thanks to the parallel architecture, the GPU is best suited for solving visual computing tasks, providing efficient processing of large amounts of data. Computer vision, image processing, and artificial intelligence are necessary to create the intellect of the object with which the information system interacts - the digital twin, the principle of which is described above.

Application areas. People process information much faster when it is presented visually. Maps and other location illustrations create a powerful entrance into the world of big data and digital transformation, providing a context that can reveal hidden insights, increase efficiency, productivity and profit. As noted above, almost all of the applications of spatial computing need support from visual computing side.

2.3. Interrelation between spatial and visual computing

Spatial computing enhances the benefits of visual computing by creating true depth perception in a 360 ° user-centred space [10]. Therefore, at the intersection of spatial and visual computing, Immersive Analytics (IA) emerges, which seeks to understand how immersion technologies based on spatial computing can be used to create more intuitive and efficient visual structures supported by visual computing. Interaction provides an overview of the information behind the design, creating a mixture of rendering and spatial computing tasks. Visual computing makes complex data accessible and understandable, spatial computing makes it in its turn useful through data visualization. The synergy of these technologies will enhance the combined potential for obtaining the interpretation of insights.

Spatial computing combines distributed smart sensors with computer vision algorithms, thus mixing the physical and the digital assets to better understand and visualize/display the physical world. Location data can give enterprise access to valuable data to support its digital transformation. Let us say a retailer runs a mobile ad campaign by tracking the number of clicks. At the same time, the POS and online sales data generated during the campaign is recorded. In a traditional data analytics model, these two subsets of data — ad clicks and sales — are isolated. However, if the data is accompanied with [geo] spatial information and displayed on a map, managers can quickly see which ad clicks are associated with specific physical stores. The visual reconstruction of shopping behaviour can reveal valuable information, such as the correlation between the number of clicks and the distance to the store, whether online sales were made at home or while commuting. Combining multiple datasets with physical locations creates a rich model for data analysis and an intuitive platform for stakeholders to view and understand large amounts of data. Spatial visual analytics spreads insights into data throughout the enterprise.

	Spatial computing	Visual computing
Definition	Area of computing which is used to create user interface for sending input information and manipulating the output from the computer	Area of computing, which deals with the acquisition, analysis and synthesis of visual data using computer resources
Tools	Software for creating AR apps Software for developing digital twins	Tools for creating 2D and 3D real-time visualization NVIDIA GPU graphics processors
Functions	Using the surrounding 3D space as a canvas for the user interface Mapping digital and physical locations	Real-time creation of 3D models (3D-reconstruction) based on data received from sensors Processing and generation of digital images, creation, capture, analysis and manipulation of visual data, 2D-image processing (convolution, spectral analysis, pattern recognition, object detection)
Enables	Interface for interaction beyond physical capabilities or as support Perception of space through AR / VR headsets Matching human cognitive capabilities Seamless interaction between people, products, processes, physical spaces Understanding how physical space can be used to improve efficiency and safety	Digitizing data generating visual content Effective 3D UX design Using information from cameras and sensors for people
Benefits for digitalization	High operating speed, low training time, increased productivity, minimized malfunctions Optimization of complex work in the environment by obtaining a complete picture of movements in space in real time or over a certain period	Human ability to understand vast and complex information Interactive data analysis Realistic virtual images

Table 2. Characteristics of spatial and visual computing

This paper focuses on Spatial and Visual computing systems and their interplay in real world ITS applications. It should be noted that specific scenarios from aviation industry discussed further require high-quality visual solutions.

3. Spatio-visual framework for distributed smart service application

In this section, we describe and discuss those components of the data processing/analytics pipeline that can be included in the universal integrated spatio-visual framework. All of them serve to solve problems of spatial and visual computing within the enterprise MIS environment. To identify the major components, we use the top-down approach to the design of the framework that implies a stepwise refinement of the required functionality through several subsystem levels until the specification of the base elements (modules) that can provide it. The following subsections provide information about the constituents of our model framework, the interaction between components, and their functionality.

3.1. Conceptual view of the system

Location-based analytics are used by leading organizations to address business challenges and unlock new opportunities. Location information is an essential addition to business intelligence (BI) tools. It is important to note that spatio-temporal data requires specific (spatial computing) approaches to its handling and the use of dedicated techniques and tools.

Maps as a core of the geo-spatial data visualization provide meaningful context through which the various participants in the process (providing a service or working in a production) can view specifically the data that is required for their individual roles - and for a broader, common understanding of the entire company. Broadly speaking, maps can serve as an effective spatio-visual framework for organizational integration.

The user interface, that allows employees and customers interact with information via mobile devices, computers and the IoT devices, maps provide an intuitive picture of work in different sections and planes. Smart maps that have functionality for managing processes can also be called custom Dashboards. Layers of data such as demographics, traffic, and weather can be added to them. With the help of smart maps, organizations can get information about their assets, their condition and location.

Easy-to-use smart maps and spatial analytics provide companies with the confidence to make big, often very costly, decisions. Presenting data visually, so that stakeholders can understand not only what happened, but where and in what context, instantly makes big data more understandable and valuable. By creating and mixing physical and visual context, businesses can better evaluate and make more intelligent use of the vast amounts of information at their disposal. The visual representation of the map is supported by geographic system information (GIS) technology, which serves to manage, visualize, analyse and ultimately understand the importance of location information, obtain actionable information based on the integration of various types of data; CRM information system designed for customer relationship management; business intelligence (BI) tools, as well as systems for managing records, assets and many others.

Below, we provide more details on the components of the suggested application system framework based on spatial and visual computing technologies (Fig.3). Note, the proposed solution features exclusively Open-Source technology-based solutions for versatility.

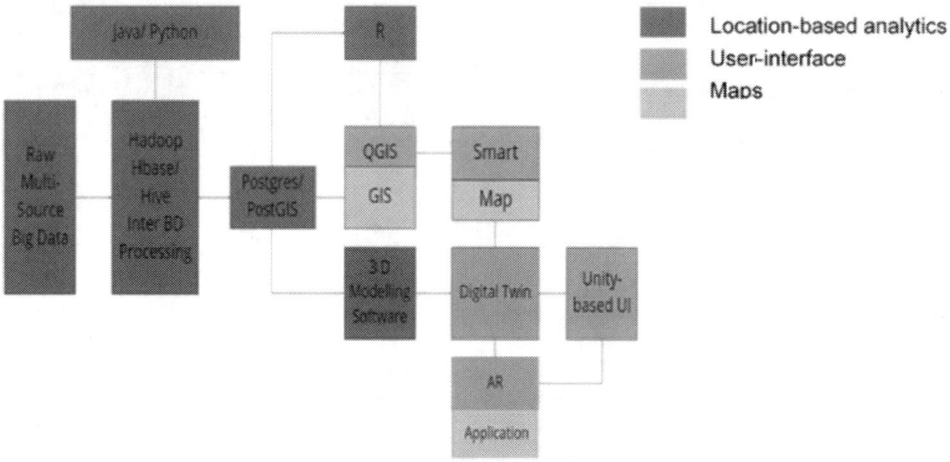

Fig. 3. Block diagram of the integrated spatio-visual application framework.

In terms of architecture, spatial computing provides a great opportunity for scripting, giving a possibility and access to generating plenty of ideas, especially for dynamic spaces such as rooms, buildings, neighbourhoods, cities, and everything that populates these spaces. Spatial computing and augmented reality together enable architecture with an ideal and realistic view of the space available and the "virtual lens" mechanism through which new information can be viewed and applied.

3.2. Architecting the system

The application system is architected in a way to provide a web-based user interface to the spatio-temporal analytical and visualization services through ad-hoc APIs and interactive maps.

An important role in the creation of such an application system is the collection of data according to certain principles. It is important to note the nature of the data, in complex objects, which are mainly implied in the writing of this work, both visual and spatial data - Big Geospatial Data. The received data is integrated into the built-in map and then transferred to build an actual 3D model. Binding to geolocation allows restoring the complete picture, objects from the environment, and their representation in three-dimensional space.

Modern open-source technologies such as Big Data technology stack from Apache.org software foundation have potential to handle massive data with high scalability and low latency. Big data computing platforms such as Spark and Hadoop do not offer native support for spatial data. There are Hadoop-based spatial frameworks for big data: Spatial Hadoop and Hadoop GIS, which can become a core part of the Hadoop ecosystem in this project and provide data pre-processing with Hbase and Hive. It is important to note that these frameworks are only capable of performing spatial operations on datasets available in text file formats. Therefore, Python and Java will be used in pre-processing support.

Spatial database [14] is a database solution for storing and accessing spatial data or data defining geometric space. Spatial databases perform a number of spatial computation operations that can be found in the specification of the standard by The Open Geospatial Consortium (OGC). These include spatial dimensions, spatial functions, spatial predicates, geometry constructors, observer functions. In addition, spatial databases are capable of handling 3D object data, which means that information about each 3D model can be stored in the same database. The diagram shows that among all the alternatives, PostGIS was chosen as an open source library and a spatial database extender for PostgreSQL that runs on top of it and adds spatial datatypes and geometry processing functions for queries.

Once spatial big data becomes available, data analytics tools can provide more information and improve the accuracy of results [15], leveraging the full potential of spatial computing. In this project, we have explored the potential of the data analysis tool provided by R, an OSS environment for statistical calculations and graphics, the programming language has more than 100 available spatial packages, as well as basic geo-visualization tools.

GIS is a tool for processing and visualizing any type of geospatial data. The technology allows users to visualize, analyse, query results, create templates, and view relationships. GIS software contains a variety of applications that combine the use of digital maps and geo-referenced data. QGIS is represented on the proposed model, since this open source software performs all the necessary functions for a project to collect, store, query, analyse, visualize and output data. It can be connected to other systems that support spatial functionality: PostGIS, R, and Python.

The data visualization capabilities of the framework should demonstrate analytical power and dynamic layouts. Applications can be developed for the dashboard generation to display the results of data analysis. The visual analytics implementation can lead to savings in operating costs by more than 20% due to the reduction of staff time and effective adjustment. GIS-based maps and interactive visualizations can then serve as a communication tool between different teams, departments and geographically distant locations. The Smart Map can act as a platform and interface in which people, space, hardware, software, content, services, IoT devices are integrated. The dashboard can display a planner, calendar, instructions, messages, tools for filtering information and submitting requests. In addition, the digital twin can be accessed through the map interface.

4. Prototype smart MRO system

The proposed framework has prospects of application and implementation in the chosen domain of aircraft engine MRO. The section is describing elements of the application project.

4.1. The smart MRO application framework

From the operational point of view, the project may contain the following application parts:

1. Dashboard for engine location tracking (users are Managers at OEM's, Airlines Representatives):

- Includes the newest information on condition of the engine
- Allows to observe and monitor the mobility of the engine from airport to airport, from airport to service
- Operative check of the operating status of an aircraft engine
- Role-based reports.

2. Spatial computing systems issues include definition and development of spatial computing infrastructure; collection, integration and storing sensed data; computational aspects of spatial Big Data, and augmented reality:

- Provides guidance for technicians, who are working with the physical object
- Includes the newest information on condition of the engine.

3. Cross-cutting solutions focus on ubiquitous computing arrangements, sustained sensing and monitoring, reliable transportation systems and geo-confidentiality:

- Help to see the engine rendering the complexity of the 3D models.

As the next step, let us take a look at all the parts of the model through the prism of the aviation MRO. The figure 4 shows how Big Spatial Data gets into database at maintenance site.

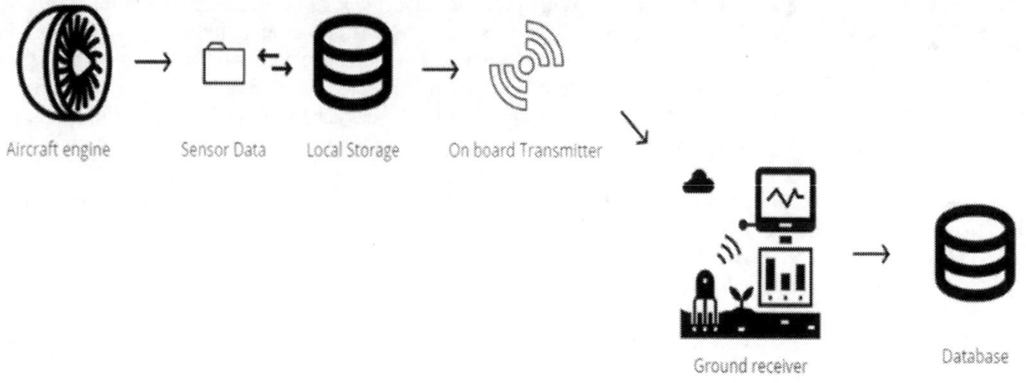

Fig. 4. Aircraft engine data path from data origins to the MRO service database

Coming back to the Figure 3, the type of the database will be essentially the spatial database. Within the scope, spatial databases are capable of handling 3D object data, so that in the same database the information on 3D model of each engine and the information from sensors of an actual engine might be stored and then displayed on the same dashboard, where the locations of engines are presented, too.

Spatial computing can play a significant role as using geo-oriented technologies serving the problem of finding an optimal decision, better orientation in terms of nodes, where are suitable technicians or spare parts, how fast can they reach this place. Analysis based on spatial computing and growing knowledge base will result in building a profound network of all players in this process. Efficient organization of technician's work, who exactly know where and when they have to manage the repair quickly.

The analysis will answer the following questions: who is available and who have right competencies for the task. Flight crewmembers will make qualified decisions; determine the required action that is needed to be done. Suppliers will be on time notified. There will be no need to replace an aircraft if all problems are solved on time.

Use cases of GIS:

- Engine manufacturers are getting knowledge about what is happening in geographical space with their products
- Enabling efficient scheduling of maintenance activities for engines, assigning work to a particular repair team in a particular location, and configuring tasks based on the visualized [geo-] spatio–temporal data.

The *Data Visualization* framework will show the analytical results with dynamic layouts. A dashboard application will be designed to depict the analytical results. Elements of a dashboard include:

1. *Visualize aircraft engine performance parameter:* Visualize Efficiency, Visualize Pressure Ratio, etc...
2. *Visualize models of aircraft engines:* Visualize nominal geometry; Visualize engine.
3. *Inspect models:* Navigation in 3D Space, Zoom-in / zoom-out / rotate the model.
4. *Associate it with particular model*: Observe and Compare differences between the nominal (semi-transparent shape) and particular (overlaying shape) geometries

As a result, implementations can lead to savings over 20% in operational expenses through reduction of staff working time and efficient adjustment. GIS-based maps and visualizations will serve in this case as a communication tool between different teams, departments and geographical far apart from each other locations. Manufacturing directors can visualize everything happening with the products. On the dashboard scheduler, calendar, repair time, needed details, checks from technicians (who checked, who made double check), logs, critical parameters will be mapped. In addition, an access to the digital twin will be realized through the map interface, VR (the whole view of computer-generated picture).

Visual inspection as a predictive/corrective maintenance technology is highly useful [16]. Monitoring for anomaly detection and prevention is an essential work during normal flight and mission execution. With this support, the technicians on site have the opportunity to check the status and position of all planned components and work much more precisely.

In addition, they can coordinate ad-hoc with component developers in their offices and workshops using collaborative tools. The transfer of the extensive CAD data to the terminal requires high bandwidths - even inside the parked aircraft. Key benefits of using Visual Analytics in maintenance are as follows:

- Visibility of a system status
- Match between the system and the real world
- User control and freedom
- Help users recognize, diagnose errors
- Closeness of mapping

AR Application to MRO is the next idea of the project when an AR application system for a technician is provided to enable him to visually explore the object details through the AR prism. By establishing a connection between the remote expert and the on-site worker, a precise visual guidance can be provided through AR pointers and annotations, displayed on a head-mounted system or a tablet. AR solution will be needed to show 3D model of the engine in real time, allow to get inside a virtual AR cockpit or explore an immersive engine model and to see the working parts from various angles. When user is gazing over an interaction object, which is internally facilitated by the ray tracing, the object highlights itself, that is, the user receives a return feedback signal in the form of a visual clue. As for specific hardware, a device in combination with glasses/head gear will be proposed, as it would be more convenient to have free hands while working. Technicians will fix the event with help of detailed information and required spare parts provided to them to the right time. Thus, maintenance activities are completed faster thus reducing the total Aircraft on Ground (AOG) time. Suitable will be those models of AR glasses that are compatible with Unity3D and transitively with Vuforia that suggest a toolbox for realization of a spatial application.

The benefits of using AR in maintenance are as follows:

- Better ramp handling – ground handling workers can work more efficiently, while they do not have to go through papers or look at tablets
- Faster ramp operations with less errors, for the AR reducing gaps where mistakes used to occur
- More efficient assembling process, because technicians equipped with AR glasses receive instructions in real time with better accuracy. AR headgear gains simple access to engine data (also during the flight)
- Enhanced communication. AR helps users to establish an instant video/audio communication. Expertise can be provided on demand from any remote location
- Convenient Data Access

5. Discussion

Spatial computing and visual analytics are used all over the world to discover new information and knowledge from raw data and make informed decisions. Organizations that rely on spatial data analysis represent a wide range of human activities - these are state and local government, national agencies, various businesses, engineering companies, colleges and universities, NGOs. Consequently, this model, which proposes to combine technologies, focused on the features of spatial and visual computing, can have a wide range of applications. For example, the field of aviation [3], namely the technical support of engines that was considered in the work. Aviation industry is facing two major challenges of safety and performance improvement. Business intelligence built within the framework of the proposed model can provide the following: as a result, the application will eliminate the need to replace the aircraft, because decisions on repair and its progress will be applied in a timely manner.

Spatial computing can play a significant role, as the use of geo-oriented technologies serves the task of finding the best solution, finding the best network node with the right technicians or spare parts. Analysis based on spatial computations and a growing knowledge base will create a deep network of all participants in this process. An optimal organization of the work of technicians can be built, who know exactly where and when they have to cope with the repair. The analysis can provide answers to the questions: who is available, and who has the necessary competencies to complete the task.

Flight crewmembers will make informed decisions, determine the actions to be taken, and part suppliers and part manufacturers will be notified in a timely manner. Moreover, cases where spatial computing is used in an industrial plant and several machines and robots need to be connected have to be considered. To work with robots, a large number of sensors are used, and it is necessary to check a huge number of time series with coordinates in the drop-down menus on the desktop, track various data structures.

The framework described in this work may potentially solve the task, by providing the aviation technician with an interface that combines all necessary knowledge based on pre-processed data and information.

6. Conclusion

In this work, the focus of research was made on leveraging technologies of spatial and visual computing for modernization and improvement of aviation MRO efficiency and on new ways of orchestrating them in integrated application framework intended for improvement of the aircraft engine MRO service. In the frames of the work, it was revealed that it is advisable to carry out digital transformation of the transportation business using spatial computing, for example, in cases where advanced personnel interacting with the physical world are involved in business processes. Spatial computing promises to be transformative in a narrower sense and within a targeted set of use cases and verticals (e.g. Smart Health, Smart Transport, and Smart Cities). Spatial analytics add essential context to everything from shopping habits to transportation patterns and industrial service needs. Spatial analysis can combine information from various independent sources and extract qualitatively new information using complex combinations of spatial operations. A comprehensive collection of spatial analysis tools helps to find answers to complex questions related to the organization of working space. Statistical analysis can help to understand whether the observed patterns are meaningful, and visual analysis can reveal changes over time and in real time.

For future research it is planned to adjust the suggested spatio-visual framework to a real industry case. The further steps of the work will be devoted to the creation of a digital twin of an aircraft engine or its core parts and development of a visual analytics tool. Thus, a large part of a future work can be defined as work with data: gathering, transformation, pre-processing, modelling and with implementation of a discussed software. Additionally, analysis of other domains that could benefit from the spatio-visual framework will be conducted.

7. Acknowledgments

The reported study was partially supported by RFBR grant No 20-07-00958 and the HSE University Project Group Competition grant in 2020 — 2022.

8. References

[1] Singh, B. and Gupta, A. (2015) "Recent trends in intelligent transportation systems: a review," Research Directory, J. Transp. Lit. 9 (2), Apr 2015

[2] Peters, J. (2018) "Big data: Unlocking the value of data in air transport", International Airport Review, June 19, 2018. URL: https://www.internationalairportreview.com/article/79124/unlocking-value-data-air-transport/. Accessed 15 July 2021.

[3] Lyu, Y., Liem, R.P. (2020) "Flight performance analysis with data-driven mission parameterization: mapping flight operational data to aircraft performance analysis," Transportation Engineering, 2020.

[4] Brown, R. (2017) "The Connected Fleet: Further Implications of Aircraft Health Monitoring for the Aviation Supply Chain", ICF White Paper, 2017.

[5] Vieira, D.R. and Loures, P.L. (2016) "Maintenance, Repair and Overhaul (MRO) Fundamentals and Strategies: An Aeronautical Industry Overview," 2016.

[6] Chang, H.M. and Abdullah, K. (2014) "The Operation Management Model of Aircraft Maintenance, Repair and Overhaul (MRO) Business," Int'l J. of Trends in Economics Management & Technology (IJTEMT), 2014.

[7] Bachan, A. (2019) "How to navigate the complex landscape of emerging MRO technology," ICF, June 27, 2019. URL: https://www.icf.com/insights/transportation/navigating-emerging-mro-technology. Accessed 15 July 2021.

[8] Bachan, A. (2020) "Using digital threads and twins to improve the MRO lifecycle," ICF, March 26, 2020. URL: https://www.icf.com/insights/transportation/improving-mro-lifecycle. Accessed 15 July 2021.

[9] World Economic Forum Top 10 Emerging Technologies of 2020 Special Report, November 2020.

[10] Greenwold, S. (2003) "Spatial Computing", Manuscript submitted to school of architecture and planning, Massachusetts Institute of Technology, Cambridge, June 2003. URL: http://acg.media.mit.edu/people/simong/. Accessed 17 July 2021.

[11] Hashplay Inc. Company Presentation Spatial Computing for Enterprises, June 20, 2017.

[12] Nakamura, T. (2020) "Digital Twin Computing Initiative", NTT Technical Review, Vol. 18, No. 9, Sept. 2020.

[13] Posada, J. et al. (2015) "Visual Computing as a Key Enabling Technology for Industrie 4.0 and Industrial Internet," in IEEE Computer Graphics and Applications, vol. 35, no. 2, pp. 26-40, Mar.-Apr. 2015, doi: 10.1109/MCG.2015.45.

[14] Advances in Spatial and Temporal Databases: 12th International Symposium, SSTD 2011. Challenge and Vision Track. Springer, 2011.

[15] Gelfand, E. et al. (2010) "Handbook of Spatial Statistics," Chapman and Hall/CRC Handbooks of Modern Statistical Methods (ISBN: 1420072870), 2010.

[16] Katona, A.and Panfilov, P. (2018) "Building Predictive Maintenance Framework for Smart Environment Application Systems," in Proceedings of the 29th DAAAM International Symposium, Vienna, Austria, 2018, pp.0460-0470. DOI: 10.2507/29th.daaam.proceedings.068

[17] Beverungen, D., Müller, O., Matzner, M. et al. Conceptualizing smart service systems. Electron Markets 29, 7–18 (2019). https://doi.org/10.1007/s12525-017-0270-5

DOI: 10.2507/32nd.daaam.proceedings.096

TANGENTIAL LASER MACHINING USING FS-PULSED LASER

Simon Syrovatka, Adam Cermak, Pavel Kozmin,
Ondrej Marsalek & Tomas Zatloukal

This Publication has to be referred as: Syrovatka, S[imon]; Cermak, A[dam]; Kozmin, P[avel]; Marsalek, O[ndrej] & Zatloukal, T[omas] (2021). Tangential Laser Machining Using Fs-pulsed Laser, Proceedings of the 32nd DAAAM International Symposium, pp.0685-0691, B. Katalinic (Ed.), Published by DAAAM International, ISBN 978-3-902734-33-4, ISSN 1726-9679, Vienna, Austria
DOI: 10.2507/32nd.daaam.proceedings.096

Abstract

For the production of very small and precise rotating parts, it is often necessary to use the best of conventional and unconventional machining methods. Laser machining is an unconventional machining method that is increasingly used because with its use it is possible to produce very precise and miniature shapes that are not thermally affected. The aim of this article is to explore the technology of laser tangential machining (which is inspired by the classical machining turning process) on a used femtosecond laser system. And find out the ideal setting for the laser tangential machining process to obtain the best surface roughness and the highest process productivity.

Keywords: laser; laser turning; tangential machining; surface roughness

1. Introduction

These days the laser machining technology represents one of the most advanced technologies in the mechanical industry. There are a number of different laser systems that are used during laser machining. These devices can be divided according to their average laser power P_{av}, size of the wavelength λ, length of the laser pulses τ_p, type of the laser source, etc. [1]. However, this article is focusing on the lasers with the ultrashort duration of the laser pulses namely pulses with a duration of the order of femtoseconds. The advantage of these lasers is that due to the ultra-short lengths of the laser pulses and the large energy in one pulse, the material does not melt, but the ablation of material is happening. This is desirable because machining with these lasers does not significantly thermally affect the surrounding material [2]. The femtosecond lasers are usually used for producing small shapes or parts because of the diameter of the laser spot which value is commonly in the tens of microns.

Nowadays is it more usually to reduce the dimensions of products to such an extent that sometimes it is not even enough to use the process of classical radial laser ablation for production. In such cases, a tangential laser process can be used. This is a process where the laser beam does not impact perpendicular to the workpiece surface, but tangentially[3]. This type of process can be also used in the production of rotating products where the laser beam moves around the axis of the workpiece and impinges tangentially on the workpiece while the workpiece is rotating around its axis [3], [4]. This process is closely linked to the conventional method of turning.

It is also possible that the laser beam is not tangential to the workpiece but directly perpendicular to the workpiece surface, which can result in greater material removal and a shorter process time. The output qualities which are important for customers can be the surface roughness and shape accuracy of the product. These qualities depend on several laser parameters which are investigated in scientific articles. For example, Kibra et al. [6] made an experiment with a nanosecond pulse laser which goal was to observe the relations between overlaps of laser spots and surface roughness during different pulses frequencies and velocities of rotation of the workpiece. They also found out that with increasing percentage overlap of laser spots the roughness decreased with each tested frequency of laser pulses. It can be also said that the surfaces with lower values of roughness can be produced by using bigger laser power [6].

Another experiment was done by Ackerl et al. [7] where the aim of this experiment was to observe the influence of laser spot overlap (in the axis of the workpiece (*lp*) and vertically to the axis of the workpiece(*lb*)) on the roughness of the final surface. The scheme of both types of overlaps is shown below.

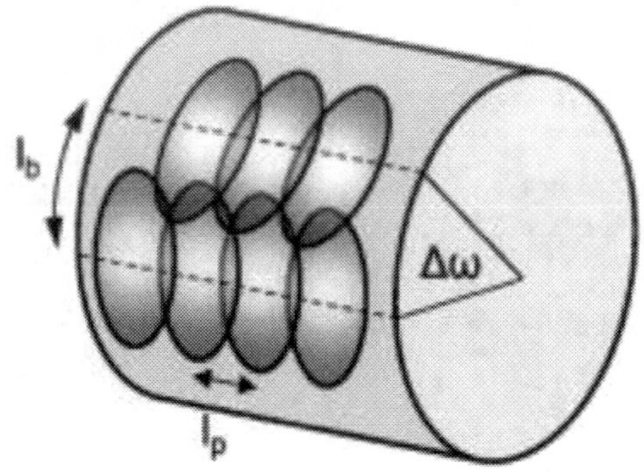

Fig. 1. Two types of overlaps (lb - line overlap; lp - pulse overlap) [7]

They found out that with the increasing value of the laser spot overlap in the axis of the workpiece (*lp* – pulse overlap) the roughness of the final surface was better (the value of R_a was smaller). For the line overlap applies that with a smaller value of the line overlap (*lb*) the surface roughness was better (the value of R_a was smaller) [7]. Other parameters that affect tangential laser machining are for example the depth of the cut, the feed rate and the rotational speed (cutting speed) [8]. These parameters are also used in conventional turning [9]. Their influence on the final state of the products is investigated in the experiment of this article.

2. Experiment

The aim of this experiment was to find out the relations between the parameter cutting speed and final state of surface roughness Ra, time of the machining process (process time) and the accuracy of the final shape according to the required shape. This experiment was done in the laboratory of the company HOFMEISTER s.r.o. where the company's laser system was used. This device was a femtosecond laser with a wavelength of 515 nm, a laser pulse length of 240 fs, a laser spot size of 16 µm, a frequency of 200 kHz and an average maximum power of 3.2 W. The semi-finished products for the experiment were sintered carbide rods (Grade F06D [10] - Table *1*) with a diameter of 2 µm and a length of 200 mm.

Grade of sintered carbide	grain size [µm]	C_o [%]	hardness (HV30)	hardness (HRA)	density [g/cm³]
F06D	≤ 0.6	6.0	1840	93.4	14.80

Table 1. Specification of the grade of sintered carbide [10]

This experiment consists of two parts. The first part concerns the preparation of a good initial state of the workpiece. During this first part of the experiment, the semi-finished product is machined via laser to reduce the diameter of the semifinished product from 2 mm to 1.87 mm in the length of 6 mm. This diameter reduction is performed due to compensation of the workpiece's radial run-out after its clamping into the laser system. After that, the second part of the experiment follows.

This is the main part of the experiment which aim is to produce material removal with different values of the parameter cutting speed (v_c) (five different values) with a constant value of feed rate per revolution (f_{rev}), depth of cut (a_p) and all other laser parameters (diameter of the laser spot, frequency, laser power) to observe the influence of the parameter v_c on the roughness of the final surface machined diameter of the workpiece, the accuracy of the final diameter and process time. The removal of the material is from diameter 1.87 mm to 1.5 mm to a range of 0.5 mm which is shown in Fig. 2.

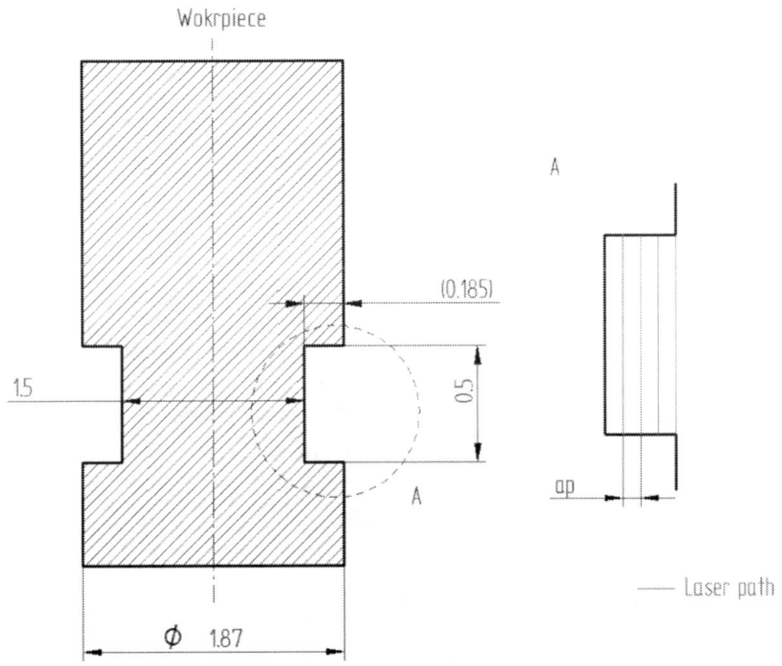

Fig. 2. Scheme of the experiment

This material removal was done by using the laser technology called tangential machining which was described in the previous chapter. The setup of the experiment is shown in Fig. 3.

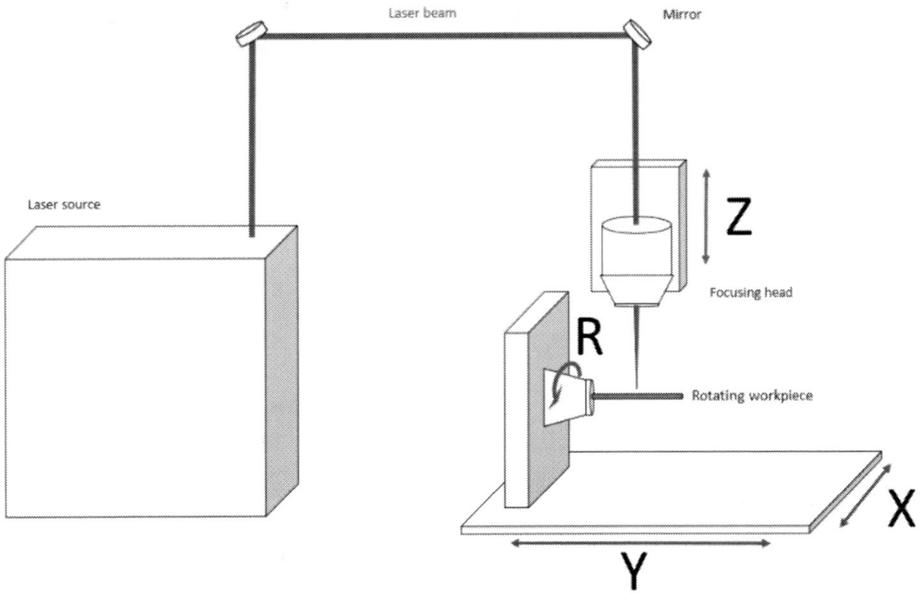

Fig. 3. Experiment setup

The main machining parameters which were used for this experiment are shown in the table below. As for other laser parameters the pulse energy E_p was 15 μJ, its value was constant so as not to affect the results.

v_c [m/min]	f_{rev} [mm]	a_p [mm]
0.06; 0.08; 0.1; 0.5; 0.7	0.025	0.005

Table 2. Parameters used for the experiment

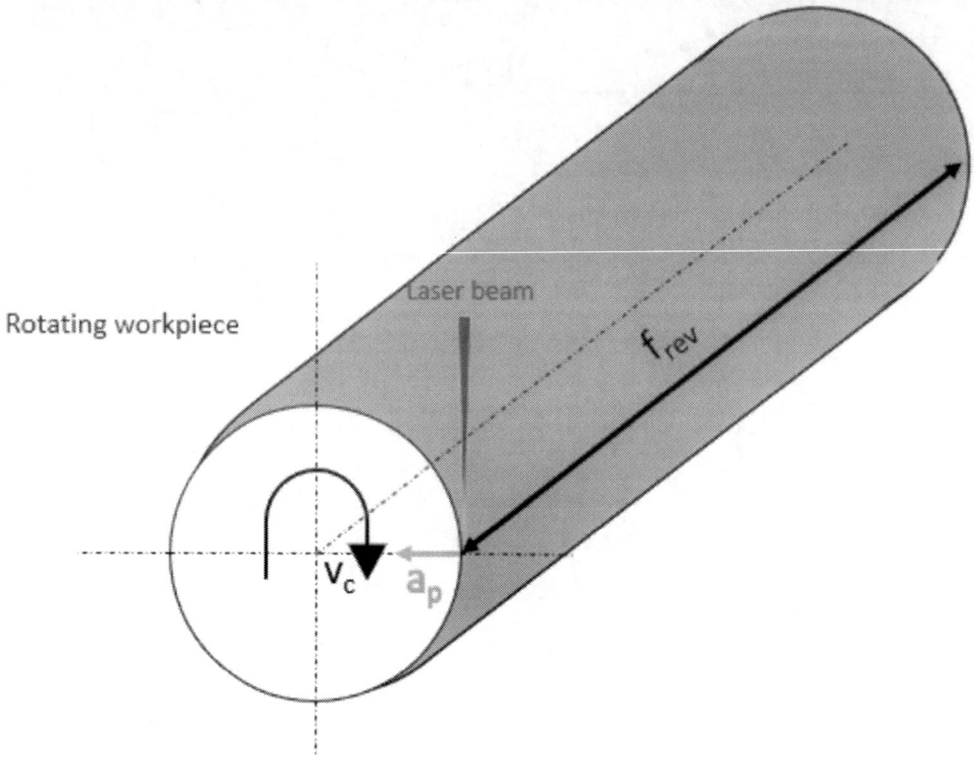

Fig. 4. Scheme of the parameters which were used for the experiment

3. Results

The final workpiece was analysed using an Olympus Lext confocal microscope (Fig. 5), where the roughness R_a of the resulting surfaces and the accuracy of the manufactured dimensions were measured.

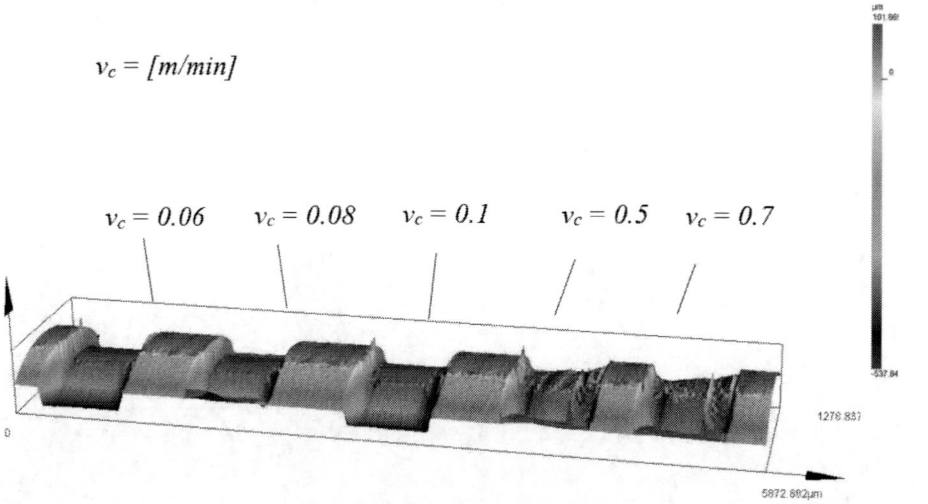

Fig. 5. Scan of final workpiece

The results obtained from the performed experiment via confocal microscope were analysed and inserted into the following graphs.

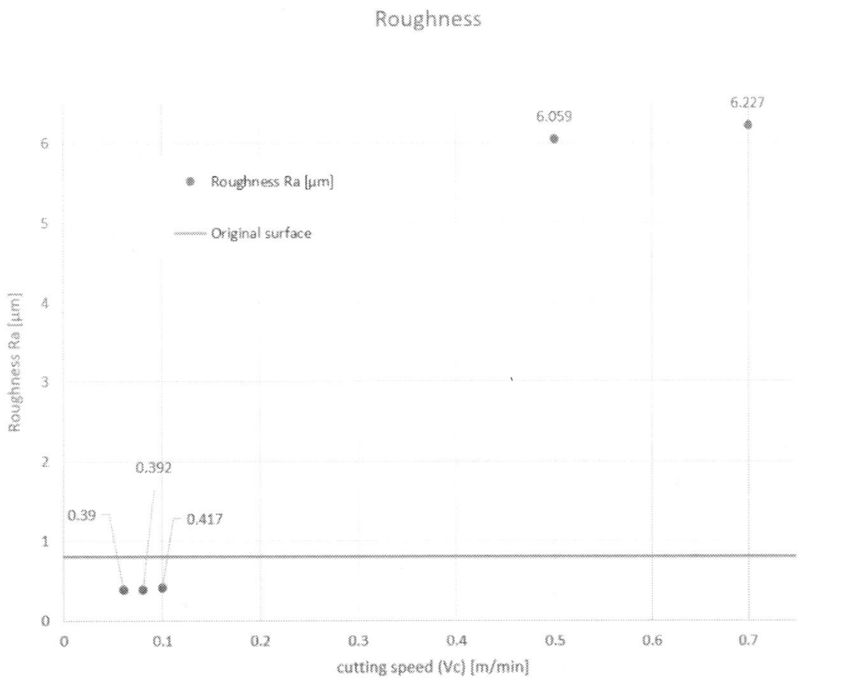

Graph 1. The dependence of the parameter cutting speed (v_c) on the surface roughness

From Graph 1 can be seen that the value of the parameter cutting speed (v_c) has a significant influence on the resulting value of the surface roughness R_a. It is evident that with the increasing value of v_c the value of the resulting surface roughness (R_a) also increased. Furthermore, it can be seen from the graph above that the roughness R_a of the resulting surface was lower than the original in the cases where the values of cutting speeds were 0.06; 0.08; and 0.1 m/min. The lowest achieved surface roughness R_a was 0.39 µm (v_c = 0.06 m/min) and the highest value of surface roughness R_a was 6.227 µm (v_c = 0.7 m/min).

Another quantity that was examined was the size of the resulting workpiece diameter. This quantity is important for the evaluation of the performed experiment in order to determine whether the required dimensions have been achieved by tangential laser micromachining or not. Measurement of the resulting values of workpiece diameters was also done using a confocal microscope and the obtained results were processed in the graph below.

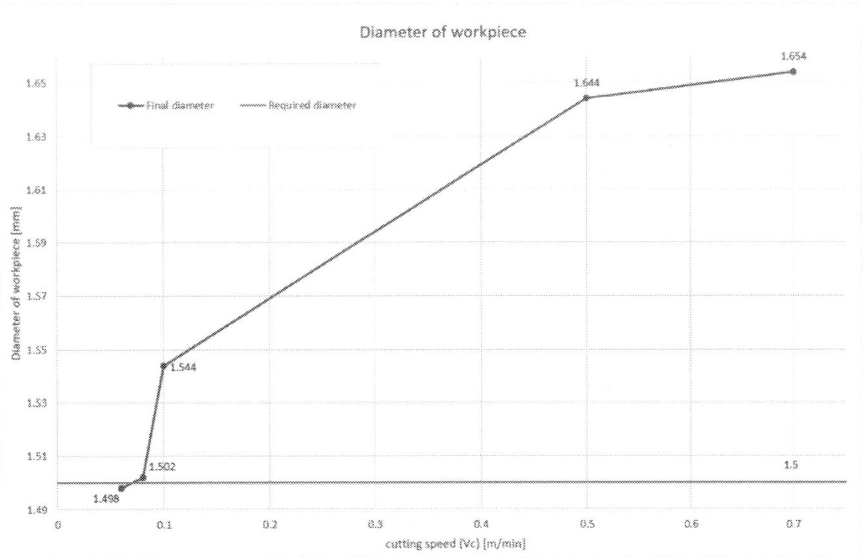

Graph 2. The dependence of the parameter cutting speed (vc) on the value of the diameter of the workpiece

It can be seen from the graph (Graph 2) that the size of material removal also depends on the value of the cutting speed. Also, it can be observed that with decreasing value of v_c, the size of material removal increases. The results that most closely resembled the desired shapes were obtained using a low cutting speed. If the value of v_c was equal to 0.06 m/min, the resulting workpiece diameter was measured as 1.498 mm and for the v_c = 0.08 m/min the final workpiece diameter was measured as 1.502 mm, which means that the deviation of these produced diameters is only 0.002 mm from the desired size of the final diameter of the workpiece which was 1.5 mm.

The last quantity for which the influence of the value of the cutting speed v_c on its size was investigated was the length of the process time. The process time is a very important variable in laser machining but also in other types of machining due to the fact that nowadays the hourly rates of machines are high and every minute of process time can have a big impact on the final price of the product. The values of the process time were determined from the used laser femtosecond device which after each machining process shows its exact duration. The obtained values were put in the following graph.

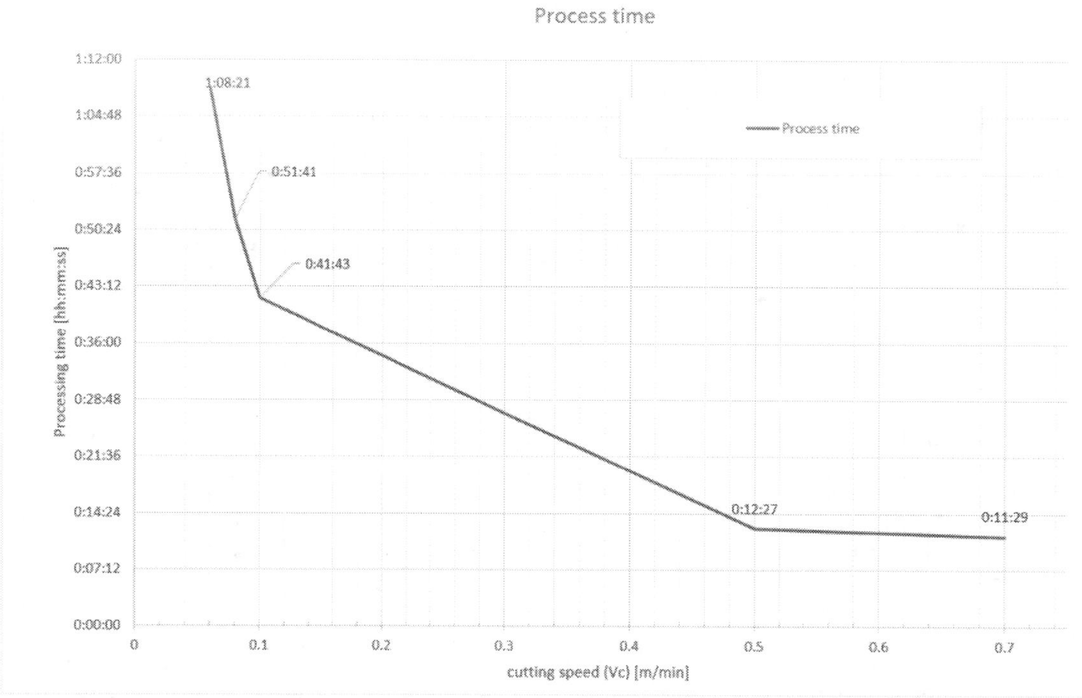

Graph 3. The dependence of the parameter cutting speed (vc) on the length of the process time

Graph 3 showing the relationship between parameter v_c and the length of the process time. It can be said that the process parameter v_c has an influence on the length of the process time and also it can be seen that the length of the process time increases with decreasing cutting speed. It was found that the fastest production time (00:11:29) was achieved using the cutting speed of 0.7 m/min.

4. Conclusion

Based on the performed experiment, which contained the tangential laser micromachining of sintered carbide rods, the effects of cutting speed on the final parameters of the workpiece were investigated therefore other process parameters were constant. The investigated parameters were the surface roughness R_a, the accuracy of the final shape and the length of the process time. It was found that all of these parameters are affected by the setting of the cutting speed. Regarding the dependence of the length of the process time on the cutting speed, it was found that the higher values of the cutting speed shorten the process time (the fastest process time was 11m 29s for used cutting speed equal to 0.7 m/min). However, with this high cutting speed, a surface with a high roughness R_a (6.227 μm) was created and the final diameter of the workpiece did not correspond to the required (the deviation from the required shape was 0.154 mm). It is possible, despite the very short process time, high cutting speeds are not suitable for the production of microscopic entities. When using a low cutting speed (v_c = 0.06 m/min), a surface with a small roughness R_a (0.39 μm) was produced and the deviation of the accuracy of the produced workpiece diameter from the required one was only by 0.002 mm. But the length of the process time was very long (1h 08m 21s). It can therefore be said that when using low cutting speeds, the result is a very precise shape with good surface roughness Ra but the process is not effective due to long process times.

The value of the executed experiment is that it was successfully put into operation the tangential laser machining on the laser system from HOFMEISTER s.r.o. and that limits on cutting speed values have been tested. Another possible continuation of this experiment could be to try a wider range of cutting speed values or to investigate the influence of other laser parameters on the tangential laser micromachining process.

5. Acknowledgments

This article was created under the project SGS-2019-008: Research and Development for Innovation in the Field of Manufacturing Technology - Machining Technology III.

6. References

[1] L. Orazi, L. Romoli, M. Schmidt, and L. Li. (2021). 'Ultrafast laser manufacturing: from physics to industrial applications', CIRP Ann., vol. 70, no. 2, pp. 543–566, 2021, doi: 10.1016/j.cirp.2021.05.007.

[2] G. Raciukaitis. (2021). 'Ultra-Short Pulse Lasers for Microfabrication: A Review', IEEE J. Sel. Top. Quantum Electron., vol. 27, no. 6, pp. 1–12, Nov. 2021, doi: 10.1109/JSTQE.2021.3097009.

[3] M. Hajri, J. Pfaff, H. Büttner, M. Voegtlin, R. Kaufmann, and K. Wegener. (2018). 'Fabrication of a ball end nose micro milling tool by tangential laser ablation', Procedia CIRP, vol. 77, pp. 654–657, 2018, doi: 10.1016/j.procir.2018.08.184.

[4] J. Zettl, C. Esen, M. Klar, and R. Hellmann. (2020). 'Generation of Rotationally Symmetric Micro Tools using Ultrashort Laser Pulses', J. Laser MicroNanoengineering, Sep. 2020, doi: 10.2961/jlmn.2020.02.2007.

[5] M. Warhanek, C. Walter, M. Hirschi, J. Boos, J. F. Bucourt, and K. Wegener. (2016). 'Comparative analysis of tangentially laser-processed fluted polycrystalline diamond drilling tools', J. Manuf. Process., vol. 23, pp. 157–164, Aug. 2016, doi: 10.1016/j.jmapro.2016.06.023.

[6] G. Kibria, B. Doloi, and B. Bhattacharyya. (2017). 'Laser Micro-turning Process of Aluminium Oxide Ceramic Using Pulsed Nd:YAG Laser', in Non-traditional Micromachining Processes, G. Kibria, B. Bhattacharyya, and J. P. Davim, Eds. Cham: Springer International Publishing, 2017, pp. 179–226. doi: 10.1007/978-3-319-52009-4_5.

[7] N. Ackerl, M. Warhanek, J. Gysel, and K. Wegener. (2019). 'Ultrashort-pulsed laser machining of dental ceramic implants', J. Eur. Ceram. Soc., vol. 39, no. 4, pp. 1635–1641, Apr. 2019, doi: 10.1016/j.jeurceramsoc.2018.11.007.

[8] D. Sriram, G. Jayaprakash, D. Arulkirubakaran, M. Prabu, and A. Ajithkumar. (2021). 'Laser turning of alumina (Al2O3) ceramic by Nd:YAG laser technique', Mater. Today Proc., vol. 39, pp. 731–735, 2021, doi: 10.1016/j.matpr.2020.09.405.

[9] W. Zębala, R. Kowalczyk, and A. Matras. (2015). 'Analysis and Optimization of Sintered Carbides Turning with PCD Tools', Procedia Eng., vol. 100, pp. 283–290, 2015, doi: 10.1016/j.proeng.2015.01.369.

[10] D. F. Ilgen. (2020). 'Catalogue for Carbide Rods', p. 24, 2020, [Online]. Available: https://pcg-gmbh.de/default.asp?kat=193&ukat=262&lang=eng

DOI: 10.2507/32nd.daaam.proceedings.097

OPTICAL CHARACTER RECOGNITION METHODS FOR NUMBER PLATE RECOGNITION: A SYSTEMATIC LITERATURE REVIEW

Andrea Gutai, Sara Havzi*, Darko Stefanovic, Andras Anderla & Srdjan Sladojevic

This Publication has to be referred as: Gutai, A[ndrea]; Havzi, S[ara]; Stefanovic, D[arko]; Anderla, A[ndras] & Sladojevic, S[rdjan] (2021). Optical Character Recognition Methods For Number Plate Recognition: A Systematic Literature Review, Proceedings of the 32nd DAAAM International Symposium, pp.0692-0700, B. Katalinic (Ed.), Published by DAAAM International, ISBN 978-3-902734-33-4, ISSN 1726-9679, Vienna, Austria
DOI: 10.2507/32nd.daaam.proceedings.097

Abstract

In the last decade, data processing of vehicle license plates became a common subject of research with numerous methodologies proposed to address the problem. The license plate recognition technology converts image data from a camera into a character format that can be further processed and manipulated in a database in a specific application area. The characters on license plates are often unambiguous in identifying vehicles that are facilitated by optical character recognition methods. The aim of this systematic literature review is to provide an overview of the most common methods used for optical character recognition in automated number plate recognition systems, as well as their level of recognition accuracy.

Keywords: Optical Character Recognition; Methods; Automated Number Plate Recognition; Systematic Literature Review;

1. Introduction

The need to extract and interpret certain information from images has been a key to the development of digital image processing in recent decades. As a result, a system that can provide the user with appropriate information based on a digital image, without the need for human intervention, is required. Adequate identification of vehicles based on image strongly depends on the number plate recognition that benefits security systems. In the field of security, the license number can be utilized as a unique identifier to aid the process. Not only can vehicles be recognized in security applications, but there are also a variety of methods for facial detection and recognition [1]. Object recognition in systems based on image processing methods is influenced by numerous factors including camera resolution, low visibility, and weather conditions [2]. The characters on license plates, which are easy to read by humans but not by machines, are used to identify vehicles in the majority of cases. Due to the variety of plate formats and no uniform outside illumination circumstances during image acquisition, this task is rather demanding [3]. Additionally, the plate recognition process is affected by a variety of colors, languages, and fonts that can be found on license plates [4]. The aim of this study is to present available methods for optical character recognition in automated number plate recognition systems by conducting a systematic literature review.

The rest of the paper is organized as follows: Section 2 contains a theoretical foundation about number plate recognition and optical character recognition, Section 3 represents a methodology for systematic literature review with defined research questions and inclusion and exclusion criteria. Section 4 describes related discussion for achieved results that are presented in Section 3, subsection 3.3 through graphs and tables. Finally, Section 5 represents conclusions and limitations of the study as well as research directions.

2. Background and Related Work

The widespread integration of information technologies in the majority aspects of life, contributes to the development of needs for information systems and data processing of license plates, as well as focus of researchers to propose different methodologies to address the problem. The three primary phases of an automated number plate recognition (ANPR) system are license plate detection, segmentation and character recognition which is shown in Figure 1. The main purpose of number plate recognition systems is to extract the license plate completely automatically and recognize the characters, which are then written to a text file for data manipulation and storage. Converting image data from a camera into a character format that may be processed in a database for further manipulation in a specific application area is another objective of such systems. The input parameter is an image, while the output of the same process is a text file.

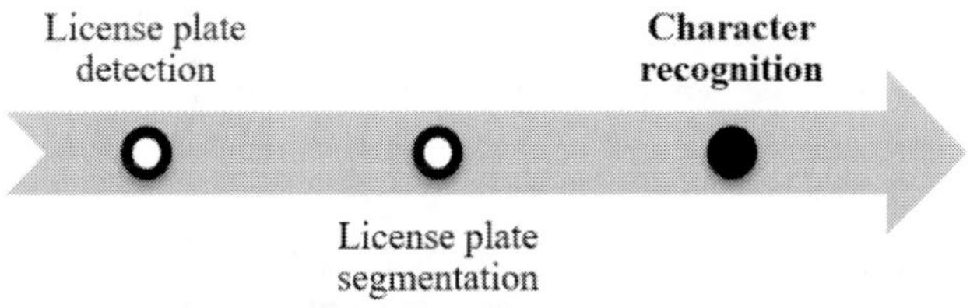

Fig. 1. Number plate recognition methodology

In order to obtain a text file numerous methods for character recognition step can be employed. Four general approaches of pattern recognition are: template matching, statistical techniques, structural techniques and artificial neural networks [5]. According to the specificity of licenses plates and field where these methods are supposed to be implemented, different modifications are required. Each license plate possesses a set of characters, letters and numbers, where an error in a single character leads to inaccurate recognition of the vehicle [6].

In order to achieve character recognition, three steps are required: segmentation, feature extraction and classification [5]. The character is recognized when it is assigned to a predetermined class or category. Different languages have an impact on the recognition process, because every methodology requires adjustments in order to recognize alphanumeric adequately, necessitating the adoption of methods for character recognition depending on the type of plate.
Each country has its own license plates with different font, letters, numbers and various other markings as well as shape and background colour of license plate. Creating a universal system for recognizing license plates would require enormous time resources, consequently researchers propose different methodologies focused on specific type of license plate in accordance with the regulations of the area.

Recognition accuracy represents a crucial parameter for the evaluation of proposed methods, which is calculated by the ratio of successfully recognized license plates to a total number of images used for method testing. Consequently, number of images used for the testing method is another parameter that should be considered while evaluating the method. Equation (1) is proposed by Yang et al. [7]

$$\text{rate} = \left(1 - \frac{error}{total}\right) \times 100\% \qquad (1)$$

3. Methodology

This research is intended to represent the methods that are widely used for character recognition phase that is a vital part of automated number plate recognition process. Kitchenham B. established a guideline for systematic reviews [7], which was used to conduct the review.

3.1. Planning the Review

In order to gather necessary information about the need for a systematic literature review, a thorough review of previous work is required [8]. Consequently, no explicit systematic literature review summarizing optical character recognition methods in automated number plate recognition systems was found. An inevitable part of review is formulation of research questions, which are formulated as follows:

RQ1: Which optical character recognition methods are most frequently used in ANPR?
RQ2: Which optical character recognition algorithms in ANPR have the highest rate of accuracy?
RQ3: Does the type of method used for optical character recognition affects recognition accuracy?

For the literature review, the following databases were used:

- *Scopus* and
- *Web of Science.*

Words, terms, and phrases used for automatic search in the aforementioned databases are stated as follows:

("optical character recognition" OR "ocr")
AND
("artifical neural networks" OR "ann" OR "kernel" OR "statistical method" OR "template matching" OR "structural pattern recognition")
AND
("automatic number plate recognition" OR "automated number plate recognition" OR "anpr" OR "alpr" OR "number plate recognition" OR "npr" OR "vehicle plate " OR "number plate" OR "license plate" OR "lpr" OR "car plate"))
AND
PUBYEAR > 2009

Definition of inclusion and exclusion criteria facilitate the selection process of primary studies [8]. Only primary studies that are relevant to research questions will be taken into consideration.

Inclusion criteria is defined as follows:
IC1: Paper has to be written in English.
IC2: Paper has to present a study on ANPR with the rate of character recognition accuracy.
IC3: Paper has to include detailed information about the optical character recognition step in automated number plate recognition for the proposed method.
IC4: License plates used for the proposed method should include alphabet letters and numbers.

Exclusion criteria is defined as follows:
EC1: If a paper is focused only on license plate recognition without character recognition, the paper should be removed.
EC2: Duplicate papers found in different databases should be removed.
EC3: Non-availability of full text.

Data extraction is the next step, where we should state which information will be obtained from studies [8]. Extraction of required information for all primary studies will include:

1. Publication year and source type;
2. Method used for optical character recognition;
3. The rate of recognition accuracy; and
4. Number of images used for method testing.

3.2. Conducting the Review

The first activity of the review phase is to identify studies that will be included in the systematic literature review. The review methodology, based on the inclusion and exclusion criteria is utilized to identify publications that will be taken into consideration. Figure 2 illustrates the flow diagram of the systematic literature review procedure, that resulted in 42 selected primary studies.

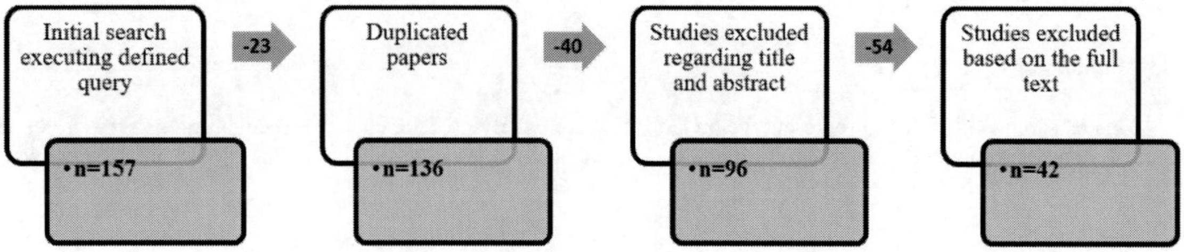

Fig. 2. Overview of the selection process

The initial search in the databases for a previously determined query resulted in 157 publications. Due to the usage of two databases on the same query, several papers were duplicated, resulting in a total of 23 duplicated papers that were excluded. Based on the title and abstract of publications, 40 studies were excluded from further process. Inclusion and exclusion criteria were applied, resulting in the exclusion of 54 primary studies. Finally, a systematic literature review was conducted using a total of 42 publications.

3.3. Data extraction and summarization

In this chapter, previously selected primary studies (42) will be presented in form of tables and graphs that provide visual representation of extracted data. Depending on the source of publication, the majority of the publications, 33 publications (79%) of a total of 42 were conference papers, while the remaining 9 publications (21%) were articles, as presented in Table 1.

Source Type	Primary studies	Number	%
Conference paper	[9], [10], [11], [12], [13], [14], [15], [16], [17], [18], [19], [20], [21], [22], [23], [24], [25], [26], [27], [28], [7], [29], [30], [31], [32], [33], [34], [35], [36], [37], [38], [39], [40]	33	79
Journal article	[41], [42], [43], [44], [45], [46], [47], [48], [49]	9	21

Table 1. Primary studies according to source type

The distribution of number of studies over the period from 2010 to 2020 is presented in Table 2. In the period of the previous ten years, the number of papers was similar per year, while in 2019 there was an increase in the number of papers, which is shown in the line diagram. Line diagram shown in Figure 3 illustrates that there was a peak in the number of published papers in 2019. In 2019, nine papers were published, compared to three to four papers each year on average per previous years.

Year	Primary studies	Number	%
2010	[39], [38], [37]	3	7.14
2011	[48], [36], [40]	3	7.14
2012	[35], [34], [33], [32]	4	9.52
2013	[31], [30]	2	4.76
2014	[47], [29], [7]	3	7.14
2015	[28], [27], [26]	3	7.14
2016	[25], [46], [45], [24]	4	9.52
2017	[44], [23], [22], [21]	4	9.52
2018	[20], [19], [18]	3	7.14
2019	[17], [16], [15], [14], [13], [43], [42], [12], [49]	9	21.43
2020	[11], [10], [41], [9]	4	9.52

Table 2. Distribution of primary studies in the last 10 years

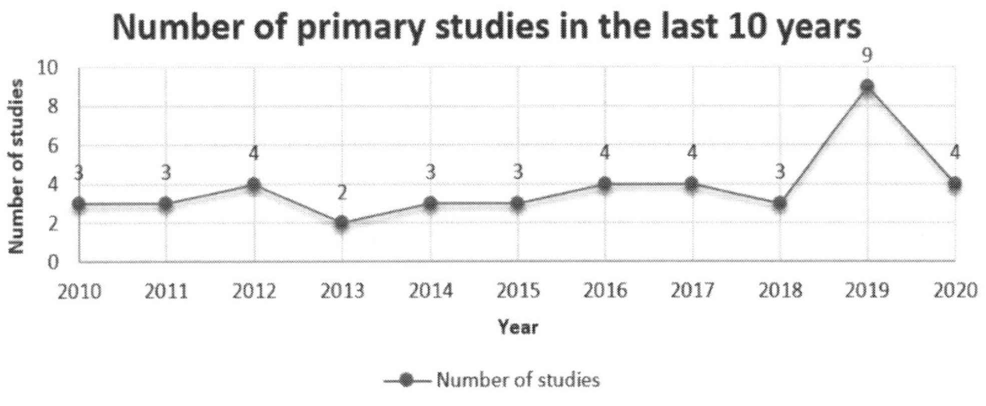

Fig. 3. Number of primary studies per year

The papers were distributed according to the type of method used for the optical character recognition step in number plate recognition systems. Table 3 provides the response to the first research question. The template matching approach was employed in up to 55% of the articles, followed by a set of neural network methods that represents 38%, Support Vector Machine (SVM) method in two studies (5%), and finally K-Nearest Neighbours Algorithm (k-NN) method in only one study (2%).

Method		Studies	Number	%
Template Matching (TM)		[41], [10], [42], [14], [15], [16], [19], [20], [23], [24], [46], [25], [27], [28], [7], [30], [31], [33], [35], [48], [37], [38], [40]	23	55
Neural Networks	Convolutional Neural Network (CNN)	[9], [13], [17], [49]	4	38
	Deep Convolutional Network (Deep CN)	[43]	1	
	Artificial Neural Network (ANN)	[18], [22], [44], [26], [32], [34], [36], [39]	8	
	Extreme Learning Machine (ELM)	[29]	1	
	Back Propagation (BP) Neural Network	[47], [45]	2	
Support Vector Machine (SVM)		[11], [12]	2	5
K-Nearest Neighbors Algorithm (k-NN)		[21]	1	2

Table 3. Frequency of usage methods for OCR

Set of neural networks methods consists of five methods that are used for optical character recognition and their percentage share is shown in Figure 4. Artificial neural networks (ANNs) are the most widely used method in the group of neural networks, 50% of the total number of papers are related to neural networks (8 out of 16), followed by Convolutional neural networks (CNN) which occupy 25% that is 4 publications out of 16. Back Propagation Neural Network is employed in two publications (13%), while the remaining percentages are evenly distributed because the remaining methods, Deep Convolutional Network and Extreme Learning Machine had been proposed in one paper each.

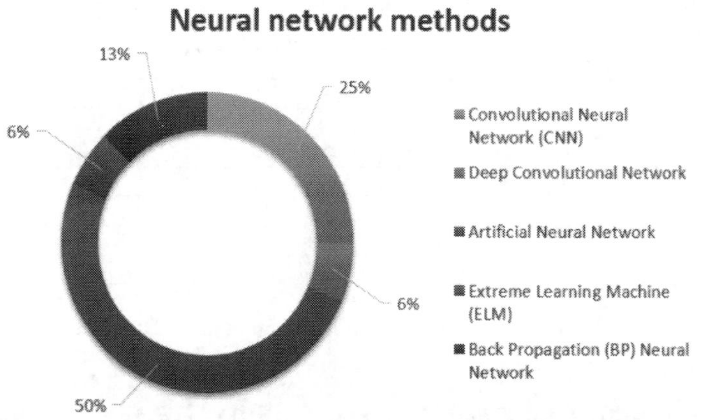

Fig. 4. Percentage share of five identified group of Neural network methods

The rate of recognition is another important aspect of evaluating proposed approaches, as well as number of images used for testing the method, which are shown in Table 4. The overall recognition rate is reported in the majority of articles (81%). Researchers have provided rates for each step of localization, segmentation, and recognition in 11 articles (26%). In 6 publications (14%) researchers have stated two out of three rates. Overall recognition rate was the highest while using set of neural network methods. The highest overall recognition rate is shown in [17], where CNN method was used. Approaches that used deep convolutional networks [43] stated high recognition accuracy as well as in [9] where CNN method was used. Subsequently, two approaches that stated similar overall recognition rate were approaches [31] and [30] that used template matching technique. The highest rates for localization step is shown in template matching approach suggested by Pun et al. [48], for segmentation step the highest rate was achieved with template matching approach in [40], and finally the highest character recognition step had the highest percentage in again template matching approach suggested by [31].

The number of images used to test the method also contributes to a better understanding of the success rate of the method. In 83% of publications, researchers have stated the number of images used for testing their methodology. The highest number of image dataset used for method test was, undoubtedly, in convolutional neural network methods [13], [17] and [49] where number of images range from 4800 to 9000 images.

Primary study	Method	Localization Rate	Segmentation Rate	Recognition Rate	Overall Recognition Rate (%)	Image dataset
[41]	TM	96.5	95.6	94.4		400
[10]	TM		96	98		30
[42]	TM	95	95	95	95	22
[14]	TM				82.5	40
[15]	TM	97	97	98	92	100
[16]	TM	94.66	96.45	95.41		100
[19]	TM				82.6	3
[20]	TM				95	357
[23]	TM	93.33	86.67	93.33		45
[24]	TM	78.1	96	95	78.1	500
[46]	TM				85.87	80
[25]	TM				88.67	165
[27]	TM				92.78	100
[28]	TM	89.13	95.12	97.94		230
[7]	TM				96.08	1462
[30]	TM				97.46	30
[31]	TM	97.5	97.71	100	98.73	80
[33]	TM				87.96	
[35]	TM	94.11		90.62	89.7	102
[48]	TM	100		95	95	100
[37]	TM	85	90	91.66		60
[38]	TM				92.8	500
[40]	TM	98.2	98.5	96.8		288
[9]	CNN				98.73	
[13]	CNN				90	5000
[17]	CNN				99.98	9997
[49]	CNN	91		93	92	4800
[43]	Deep CN				99.32	
[18]	ANN	92	92	87	89.5	100
[22]	ANN				94.32	13
[44]	ANN				61	30
[26]	ANN				97	200
[32]	ANN				97.3	
[34]	ANN		83.5	92	88	
[36]	ANN				91.59	150
[39]	ANN				94.1	
[29]	ELM	94.85		97.9		286
[47]	BP NN				91.2	1000
[45]	BP NN				70	80
[11]	SMV				95	
[12]	SMV				92.6	130
[21]	k-NN				87.43	30

Table 4. Recognition accuracy rate and number of images used for testing

A one-way ANOVA was conducted to compare the effect of different methods on overall recognition accuracy stated in Table 4. The dependent variable was overall recognition accuracy, while the independent variable represented different methods used for optical character recognition. According to the null hypothesis, all population means are equal. The p-value (0.383) is greater than the significance level (0.05), which leads to the conclusion that there is no enough evidence to reject the null hypothesis and that there are no statistically significant differences between the means.

4. Discussion

The findings of the literature review reported in the previous data extraction and summarization will be discussed in this chapter. The number of publications over the years was approximate, except in 2019, there was an increase in the number of articles published, and it is expected that the number of publications will continue to increase. Looking at previous years, methodologies regarding automated number plate recognition have gained attention due to employment in a variety of fields, such as parking surveillance, toll collecting, different access control systems where license plates are stored in databases for further manipulation.

The most frequently used method for optical character recognition step in automated number plate recognition systems, in 23 publications out of 42 included in review are based on template matching technique. Followed by subcategory of neural network methods, artificial neural networks that are employed in 8 publications. The first research question (RQ1) is answered in this way. Overall recognition rate was highest while using set of neural network methods, convolution neural networks (CNNs). The highest overall recognition rate is shown in [17], where CNN method was used, additionally highest number of images was used to test the method. This provides an answer to the second research question (RQ2). The results of the ANOVA reveal that there is no statistically significant difference between the methods. ANOVA results gives a response to the third research question (RQ3).

The main limitation of this research is that every researcher has its own way of stating rate of accuracy, where majority of researchers have stated overall recognition accuracy. Hidayatullah et al. [46] used Likert scale to classify rate of recognition ranging from 0% to 100%. Regarding Likert scale majority of publications included in systematic literature review could be classified as excellent, considering that they have a percentage range higher than 81. Only three publications [24], [22] and [45] can be stated as good due to the belonging to the range from 60 to 81%.

Another limitation is that methodologies were not tested on the same images, researchers used their own dataset and based on that they presented the success rates of the methodology. The use of an identical dataset would make it easier to compare all these methods. Thakur et al. [26] state that their methodology cannot be compared to other approaches regarding the same issue due to unavailability of set of images used to test the approach. Arafat et al. [41] used public datasets that consists of 126 additional images, besides their own dataset (400 images), to assess their own performance against other methodologies. This methodology could be taken in consideration as provable, while other methodologies could not be verified due to the unavailability of their datasets.

5. Conclusion

This study presented a review of optical character recognition methods in ANPR systems that have been proposed in numerous publications. The importance of the license plate recognition system, especially optical character recognition methods is affected by the need to reduce the human factor in this process, achieving a higher level of vehicle access control in a specific area, parking billing solutions, and automatic calculation of the time and money required to use the certain services. The most frequently used method for optical character recognition step is template matching technique. In state-of-the-art techniques, Convolutional Neural Network based recognition algorithms provide higher recognition rates comparing to traditional Template Matching technique, due to publications that we took into consideration. Due to the limitations stated in the discussion, if the image dataset is not available, it is hard to compare and evaluate proposed methods. Public datasets could be optimal solution for aforementioned limitation, and if future researchers would use public images, it would make methods suitable for comparison and verification. Further research should include stating accuracy rate for every step as well as, stating overall recognition accuracy should increase traceability and verifiability of specific methodology.

6. References

[1] Kotkova, B[arbora]; Hromada, M[artin]; Mach, V[aclav] & Blahova, M[arta] (2019). Detection and Face Recognition of People by Camera for Security Applications, Proceedings of the 30th DAAAM International Symposium, pp.0610-0614, B. Katalinic (Ed.), Published by DAAAM International, ISBN 978-3-902734-22-8, ISSN 1726-9679, Vienna, Austria DOI: 10.2507/30th.daaam.proceedings.083

[2] Mach, V[aclav]; Kotkova, B[arbora] & Hromada, M[artin] (2019). Detection and Recognition of People by Camera – Reliability and Use, Chapter 18 in DAAAM International Scientific Book 2019, pp.233-240, B. Katalinic (Ed.), Published by DAAAM International, ISBN 978-3-902734-24-2, ISSN 1726-9687, Vienna, Austria DOI: 10.2507/daaam.scibook.2019.18

[3] Anagnostopoulos, C. N. E.; Anagnostopoulos, I. E.; Psoroulas, I. D.; Loumos, V. & Kayafas, E. (2008) "License plate recognition from still images and video sequences: A survey," IEEE Transactions on Intelligent Transportation Systems, vol. 9, no. 3, pp. 377–391, doi: 10.1109/TITS.2008.922938.

[4] Du, S.; Ibrahim, M.; Shehata, M. & Badawy, W. (2013) "Automatic License Plate Recognition (ALPR): A State-of-the-Art Review," IEEE Transactions on Circuits and Systems for Video Technology, vol. 23, no. 2, pp. 311–325, doi: 10.1109/TCSVT.2012.2203741.

[5] Chaudhuri, A.; Mandaviya, K.; Badelia, P. & Ghosh, S. K. (2017) "Optical Character Recognition Systems for Different Languages with Soft Computing," Springer, ISBN 978-3-319-50251-9

[6] A. Farhat et al. (2018) "Optical character recognition on heterogeneous SoC for HD automatic number plate recognition system," Eurasip Journal on Image and Video Processing, vol. 2018, no. 1, doi: 10.1186/s13640-018-0298-2.
[7] Yang, X.; Zhao, Y.; Fang, J.; Lu, Y.; Zhang, Y. & Yuan, Y. (2014) "A license plate segmentation algorithm based on mser and template matching," 12th International Conference on Signal Processing (ICSP), pp. 1195-1199, doi: 10.1109/ICOSP.2014.7015189.
[8] Kitchenham, B. (2004) Procedures for Performing Systematic Reviews, Keele University.
[9] Cai, Y.; Lin, X.; Qian, H. & Lu, P. (2020) "FPGA Accelerator Design for License Plate Recognition Based on 1BIT Convolutional Neural Network," Journal of Physics: Conference Series, vol. 1621, doi: 10.1088/1742-6596/1621/1/012022.
[10] Vaishnav, A. & Mandot, M. (2020) "Template Matching for Automatic Number Plate Recognition System with Optical Character Recognition," Advances in Intelligent Systems and Computing, vol. 933, Springer Verlag, pp. 683–694. doi: 10.1007/978-981-13-7166-0_69.
[11] Ding, H. & Shen, J. (2020) "Research on license plate recognition of Chinese vehicle based on GWO-SVM algorithm," in WCSE 2020: 2020 10th International Workshop on Computer Science and Engineering, pp. 132–137. doi: 10.18178/wcse.2020.06.022.
[12] Pingping, L. & Wenjing, X. (2019) "Research on Recognition Technology of License Plate Image," Proceedings - 2nd International Conference on Computer Network, Electronic and Automation, ICCNEA, pp. 52–56. doi: 10.1109/ICCNEA.2019.00020.
[13] Suvarnam, B. & Ch, V. S. (2019) "Combination of CNN-GRU Model to Recognize Characters of a License Plate number without Segmentation," 2019 5th International Conference on Advanced Computing & Communication Systems (ICACCS), pp. 317–322. doi: 10.1109/ICACCS.2019.8728509.
[14] Brillantes, A. K. M.; Bandala, A. A.; Dadios, E. P. & Jose, J. A. (2019) "Detection of Fonts and Characters with Hybrid Graphic-Text Plate Numbers," IEEE Region 10 Annual International Conference, Proceedings/TENCON, Feb. 2019, vol. 2018-October, pp. 629–633. doi: 10.1109/TENCON.2018.8650097.
[15] Chai, D. & Zuo, Y. (2019) "Extraction, segmentation and recognition of vehicle's license plate numbers," Advances in Intelligent Systems and Computing, vol. 887, pp. 724–732. doi: 10.1007/978-3-030-03405-4_52.
[16] Indira, K.; Mohan, K. V. & Nikhilashwary, T. (2019) "Automatic license plate recognition," Advances in Intelligent Systems and Computing, vol. 727, pp. 67–77. doi: 10.1007/978-981-10-8863-6_8.
[17] Fu, M.; Chen, N.; Hou, X.; Sun, H.; Abdussalam, A. & Sun, S. (2019) "Real-time vehicle license plate recognition using deep learning," Lecture Notes in Electrical Engineering, vol. 494, pp. 35–41. doi: 10.1007/978-981-13-1733-0_5.
[18] Jagtap, J. & Holambe, S. (2018) "Multi-Style License Plate Recognition usingArtificial Neural Network for Indian Vehicles," 2018 International Conference on Information, Communication, Engineering and Technology (ICICET), pp. 1–4. doi: 10.1109/ICICET.2018.8533707.
[19] Kashyap, A.; Suresh, B.; Patil, A.; Sharma, S. & Jaiswal, A. (2018) "Automatic Number Plate Recognition," Proceedings - IEEE 2018 International Conference on Advances in Computing, Communication Control and Networking, ICACCCN 2018, Oct. 2018, pp. 838–843. doi: 10.1109/ICACCCN.2018.8748287.
[20] Jiang, Z.; Lin, Z.; Tang, J.; Li, H. & Menglu, Y. (2018) "The fast recognition of vehicle license plate based on the improved template matching," MATEC Web of Conferences, vol. 176. doi: 10.1051/matecconf/201817601029.
[21] Quiros, A. R. F.; Bedruz, R. A.; Uy, A. C.; Abad, A.; Bandala, A.; Dadios, E. P. & Fernando, A. (2017) "A kNN-based Approach for the Machine Vision of Character Recognition of License Plate Numbers," TENCON 2017-2017 IEEE Region 10 Conference. IEEE, p. 1081-1086. doi: 10.1109/TENCON.2017.8228018.
[22] Alyahya, H. M.; Alharthi, M. K.; Alattas, A. M. & Thayananthan, V. (2017) "Saudi License Plate Recognition System Using Artificial Neural Network Classifier," 2017 International Conference on Computer and Applications, ICCA 2017, Oct. 2017, pp. 220–226. doi: 10.1109/COMAPP.2017.8079759.
[23] Babu, K. M. & Raghunadh, M. V. (2016) "Vehicle Number Plate Detection and Recognition using Bounding Box Method," International Conference on Advanced Communication Control and Computing Technologies (ICACCCT), pp. 106–110. doi: 10.1109/ICACCCT.2016.7831610.
[24] Saleem, N.; Muazzam, H.; Tahir, H. M. & Farooq, U. (2016) "Automatic License Plate Recognition Using Extracted Features," 2016 4th international symposium on computational and business intelligence (ISCBI). IEEE, 2016. p. 221-225. doi: 10.1109/ISCBI.2016.7743288
[25] Choudhury, A. & Negi, A. (2016) "A new zone based algorithm for detection of license plate from Indian vehicle," 2016 Fourth International Conference on Parallel, Distributed and Grid Computing (PDGC). IEEE, 2016. p. 370-374. doi: 10.1109/PDGC.2016.7913178
[26] Thakur, M.; Raj, I. & Ganesan, P. (2015) "The cooperative approach of genetic algorithm and neural network for the identification of vehicle license plate number," 2015 International Conference on Innovations in Information, Embedded and Communication Systems (ICIIECS). IEEE, 2015. p. 1-6. doi: 10.1109/ICIIECS.2015.7193090
[27] Jalil, N. A.; Basari, A. S. H.; Salam, S.; Ibrahim N. K. & Norasikin, M. A. (2015) "The utilization of template matching method for license plate recognition: A case study in malaysia," Lecture Notes in Electrical Engineering, 2015, vol. 315, pp. 1081–1090. doi: 10.1007/978-3-319-07674-4_100.

[28] Palai, C. & Jena, P. K. (2015) "Automatic vehicle identification: LPR with enhanced noise removal technique," Smart Innovation, Systems and Technologies, 2015, vol. 31, pp. 143–153. doi: 10.1007/978-81-322-2205-7_14.
[29] Gou, C.; Wang, K.; Yu, Z. & Xie, H. (2014) "License plate recognition using MSER and HOG based on ELM," Proceedings of 2014 IEEE International Conference on Service Operations and Logistics, and Informatics. IEEE, 2014. p. 217-221. doi: 10.1109/SOLI.2014.6960724
[30] Mutholib, A.; Gunawan, T. S.; Chebil, J. & Kartiwi, M. (2013) "Optimization of ANPR algorithm on android mobile phone," 2013 IEEE International Conference on Smart Instrumentation, Measurement and Applications (ICSIMA). IEEE, 2013. p. 1-5. doi: 10.1109/ICSIMA.2013.6717950.
[31] Shaikh, S.; Lahiri, B.; Bhatt, G. & Raja, N. (2013) "A novel approach for Automatic Number Plate Recognition," 2013 International Conference on Intelligent Systems and Signal Processing (ISSP). IEEE, p. 375-380. doi: 10.1109/ISSP.2013.6526938
[32] Zhai, X.; Bensaali, F. & Sotudeh, R. (2012) "OCR-Based Neural Network for ANPR," 2012 IEEE International Conference on Imaging Systems and Techniques Proceedings. IEEE, p. 393-397. doi: 10.1109/IST.2012.6295581
[33] Hidayatullah, P.; Syakrani, N.; Suhartini, I. & Muhlis, W. (2012) "Optical character recognition improvement for license plate recognition in Indonesia," Proceedings - UKSim-AMSS 6th European Modelling Symposium, EMS 2012, pp. 249–254. doi: 10.1109/EMS.2012.95.
[34] Mutholib, A.; Gunawan, T. S. & Kartiwi, M. (2012) "Design and Implementation of Automatic Number Plate Recognition on Android Platform," 2012 International Conference on Computer and Communication Engineering (ICCCE). IEEE, 2012. p. 540-543. doi: 10.1109/ICCCE.2012.6271245
[35] Rasheed, S.; Naeem, A. & Ishaq, O. (2012) "Automated Number Plate Recognition Using Hough Lines and Template Matching," Proceedings of the world congress on engineering and computer science. 2012. p. 24-26.
[36] Roy, A., & Ghoshal, D. P. (2011) "Number Plate Recognition for Use in Different Countries Using an Improved Segmentation," 2011 2nd National Conference on Emerging Trends and Applications in Computer Science. IEEE, 2011. p. 1-5. doi: 10.1109/NCETACS.2011.5751407
[37] Ahmad, W.; Hassan, S. A.; Irfan, M. A.; Bais, A.; Hassan, G. M.; Shah, S. A. A. & Yahya, K. M. (2010) "Design and Implementation of Real Time LPRSystem on a Fixed Point DSP." 2010 6th International Conference on Emerging Technologies (ICET). IEEE, 2010. p. 159-163. doi: 10.1109/ICET.2010.5638496
[38] Wang, J. X.; Zhou, W. Z.; Xue, J. F. & Liu, X. X. (2010) "The research and realization of vehicle license plate character segmentation and recognition technology," 2010 International Conference on Wavelet Analysis and Pattern Recognition. IEEE, 2010. p. 101-104. doi: 10.1109/ICWAPR.2010.5576426
[39] Feng, J.; Li, Y. & Chen, M. (2010) "The Research of Vehicle License Plate Character Recognition Method Based on Artificial Neural Network," 2010 2nd International Asia Conference on Informatics in Control, Automation and Robotics (CAR 2010). IEEE, 2010. p. 317-320. doi: 10.1109/CAR.2010.5456536
[40] Anishiya, P. & Joans, S. M. (2011) "Number Plate Recognition for Indian Cars Using Morphological Dilation and Erosion with the Aid Of Ocrs." International Conference on Information and Network Technology. 2011.
[41] Arafat, M. Y.; Khairuddin, A. S. M. & Paramesran, R. (2020). "Connected component analysis integrated edge based technique for automatic vehicular license plate recognition framework," IET Intelligent Transport Systems, vol. 14, no. 7, pp. 712–723, Jul. 2020, doi: 10.1049/iet-its.2019.0006.
[42] Dabholkar, R. S. & Phakade, S. V. (2019) "Revolution by using optical character recognition technique to identify registered number plate," International Journal of Innovative Technology and Exploring Engineering, vol. 8, no. 10, pp. 1185–1188, Aug. 2019, doi: 10.35940/ijitee.I8643.0881019.
[43] Liu, Z.; Cai, Y.; Chen, L.; Wang, H. & He, Y. (2019) "Vehicle license plate recognition method based on deep convolution network in complex road scene," Proceedings of the Institution of Mechanical Engineers, Part D: Journal of Automobile Engineering, vol. 233, no. 9, pp. 2284–2292, doi: 10.1177/0954407019851339.
[44] Gunawan, T. S.; Mutholib, A. & Kartiwi, M. (2017) "Performance evaluation of automatic number plate recognition on android smartphone platform," International Journal of Electrical and Computer Engineering, vol. 7, no. 4, pp. 1973–1982, doi: 10.11591/ijece.v7i4.pp1973-1982.
[45] Wang, J., & Yan, W. Q. (2016) "BP-neural network for plate number recognition," International Journal of Digital Crime and Forensics, vol. 8, no. 3, pp. 34–45, doi: 10.4018/IJDCF.2016070103.
[46] Hidayatullah, P.; Feirizal, F.; Permana, H.; Mauluddiah, Q. & Dwitama, A. (2016) "License plate detection and recognition for Indonesian cars," International Journal on Electrical Engineering and Informatics, vol. 8, no. 2, pp. 331–346, doi: 10.15676/ijeei.2016.8.2.7.
[47] Qu, Z.; Chang, Q. L.; Chen, C. Z. & Lin, L. D. (2014) "An Improved Character Recognition Algorithm for License Plate Based on BP Neural Network," The Open Electrical & Electronic Engineering Journal, vol. 8, pp. 202–207, doi: 10.2174/1874129001408010202
[48] Pun, C. M. & Ho, W. Y. (2011) "An Edge-Based Macao License Plate Recognition System." International Journal of Computational Intelligence Systems, 4.2: 244-254.
[49] Hashmi, S. N.; Kumar, K.; Khandelwal, S.; Lochan, D. & Mittal, S. (2019) "Real Time License Plate Recognition from Video Streams using Deep Learning," International Journal of Information Retrieval Research, vol. 9, no. 1, pp. 65–87, doi: 10.4018/ijirr.2019010105.

DOI: 10.2507/32nd.daaam.proceedings.098

INFORMATION SUPPORT FOR CRISIS MANAGEMENT IN THE ZLÍN REGION WITH A FOCUS ON METEOROLOGICAL INFORMATION

Miroslav Tomsu & Robert Pekaj

This Publication has to be referred as: Tomsu, M[iroslav] & Pekaj, R[obert] (2021). Information Support for Crisis Management in the Zlin Region with a Focus on Meteorological Information, Proceedings of the 32nd DAAAM International Symposium, pp.0701-0711, B. Katalinic (Ed.), Published by DAAAM International, ISBN 978-3-902734-33-4, ISSN 1726-9679, Vienna, Austria
DOI: 10.2507/32nd.daaam.proceedings.098

Abstract

The article deals with information in crisis management, its importance, significance, and effectiveness. Information plays a crucial role in crisis management, as it is a warning and protection of the population. It defines the main requirements for the crisis management information support system and its elements and maps most of the previously used systems, including current and current ones. One of the critical situations in the use of information is the distribution of meteorological information. The article's content is also a presentation of a unique information and communication system in the Zlín region. Finally, the practical use of information in population protection and crisis management will be presented.

Keywords: information; information support; crisis management; meteorological information.

1. Introduction

All events and phenomena that arise around us, develop and disappear over time, are inextricably linked to a concrete, real and unique space. The space in which this process takes place is called the information space. All the consequences that have occurred in the crisis area have a retroactive effect on that information space and usually change it. information conditions significantly influence every crisis, both in the phase of development of the crisis itself and its solution.

The advent of new technologies also brought new opportunities in the field of information about the territory. The entry of information technology into the field of geodesy, cartography, and geography caused a sharp increase in the possibilities of collecting and processing information about the territory and at the same time opened up a vast space and unprecedented opportunities in their interpretation, modeling, and general use. The advent of digital technologies currently supports the development of spatial simulations that can be used very effectively to model potential phenomena that could occur in crisis situations [1], [2].

The aim is to streamline work in the event of emergencies and crisis meteorological situations. The meteor radar will have a significant impact on the early warning of the population in the event of a threat, a faster response in the event of a crisis, which will ultimately reduce the extent of damage to life, health, and property in the event of a crisis.

2. The importance of information in crisis management

One of the decisive factors that directly affect the level of crisis management is information. In the conditions of various destructive activities, fast, concise, exhaustive, purposeful, and at the same time, accurate information is invaluable. The resolution of specific crises requires timely notification and warning of the population and continuous information of crisis management bodies about the development of the crisis and the progress of its resolution. Last but not least, it is not possible to solve crises without several various pieces of information necessary to secure people evacuated from crisis areas and units performing rescue and liquidation work.

Information (from the Latin in-formation, formation, rendering) is most often defined as a notification that reduces the degree of uncertainty of the system at the recipient (for example, the recipient/user of the information). Such notification is of practical importance to the recipient, especially its usefulness and importance for solving a particular problem.

There are basically two groups of information requirements:

- General:
    - need - unnecessary information unnecessarily overwhelms and makes decision-making more complex,
    - truthfulness - the use of false information could cause irreparable damage and reduce confidence in the information provided and its source,
    - completeness - incomplete information is wrong information. Based on it, we can make a completely wrong judgment if we do not have enough relevant information,
    - timeliness - outdated information becomes incorrect, untrue information. Ensuring that information is up-to-date is, therefore, a condition for ensuring its integrity.
- Specific - they consist mainly of content, confidentiality, processing, and transmission methods and are usually determined by the contracting authority requesting this information.

Information is a message that one of the possible phenomena from the existing phenomena has occurred, reducing the recipient's ignorance of this phenomenon. But there are several levels of awareness. The following terms characterize information levels:

- **Data** - are specific images of the properties of the monitored objects, regardless of their more profound meaning and other contexts. These are simple facts expressed in symbols.
- **Information** - arises from data containing a specific context and purpose. They represent new knowledge for recipients with a specific meaning.
- **Knowledge** is created by combining information and creative thinking of a person - it is a state of intimate knowledge gained through learning, experience, or association.

The importance of information in crisis management stems primarily from the fact that it is indispensable for all managerial functions and because working with information represents a substantial part of managerial activities in terms of content and time. Information can be perceived as an input assumption and part of crisis management activities [3].

Effective and efficient work with information in crisis management must respect the following principles:

- information processing must be based on an analysis of information needs,
- effective crisis management must have at its disposal essentially the information necessary for the quality solution of the tasks performed,
- the information must be provided in a differentiated and aggregated manner, according to the levels of management,
- information base at the top level should be more oriented towards more comprehensive information on the course and development of the crisis, which is monitored on an ongoing basis, it is necessary to increase the share of the information supplied in real-time,
- the information must be used mainly to support the decision-making process.

The success of executive elements of crisis management depends on the information available to each level of management. Therefore, each manager must formulate, with regard to the conditions of his activity, the information needed for management and constantly update it. This presupposes the creation of a flexible system that would provide the necessary information on the course and development of the situation in the entire external and internal environment of crisis management so that they can be used in preventive activities and in resolving a specific crisis [5].

3. Crisis management information support

Crisis management information support is a process (set of information activities) supporting information management, decision-making, and cognitive functions. The information need aims to satisfy, through information activities, the information needs necessary for the performance of activities related to crisis management [4].

3.1. Crisis management information system

An information system is generally any system that is capable of processing and providing information. In the narrower sense, it is a set of procedures, activities, people, and technologies to collect relevant data, store them until needed, process them to provide answers to a specific set of questions, and finally communicate information to their users. They will take place. The Crisis Act addresses the crisis management information systems in particular, which states that crisis management bodies use crisis management information systems when planning crisis measures and resolving crises. It is further stipulated that the implemented and used crisis management information systems must meet the standards of public administration information systems and the rules:

- transmission of information to superior, subordinate, and cooperating crisis management bodies,
- technical and program adaptation for operation under challenging conditions,
- security of the information stored laid down for the information with the highest degree of classification contained in the processed documentation.

Therefore, the information system is an integral part of crisis management and must be built at the same time as its development, management system, and organizational structure.

The basic mission of a quality information system is to enable easier orientation in complex crisis management processes and to enable the acceleration and improvement of decisions. The role of the information system is to facilitate and improve the decision-making process and to help turn information into decisions on the process of risk analysis, organization of preventive measures, and ensuring the process of resolving specific crises.

The crisis management information system comprises information components (partial information activities) and information elements (individual information means), including their relations, internal and external links, which form an organized unit supporting ensuring protection, defense, internal security, and order.

It consists of the following elements:
- information files,
- information media,
- staff involved in information activities,
- information,
- links between those elements.

The crisis management information system aims to achieve a safe connection of risky objects and localities with monitoring centers, control points, and possible centers of assistance and rescue and mediation of objective reports in resolving crises. The main task of the crisis management information system is the processing, transmission, and use of the necessary information for the effective resolution of crises, including the interconnection of management and executive elements [5], [6].

4. Software tools to support crisis management

Each crisis is specific and requires different resolution procedures. For this reason, there is a relatively wide range of software products in our market that deal with this issue. These software tools can be divided as follows:

- accounting, planning, management and combined,
- monitoring,
- computational and modeling,
- analytical,
- special.

Modeling and simulation is an integral part of crisis management, especially in the areas of prevention or prediction of the devastating effects of natural disasters and technological accidents and forecast of the spread of the cause of an emergency (e.g., chemical, biological, and radiation substances, flood wave progression, explosion). Individual model situations are based on a mathematical calculation of the propagation of effects depending on external conditions, such as air temperature, wind direction and strength, surface structure, the size of the source of destructive effects. There are also software products that provide information flows to support communication in resolving a crisis [7].

5. Specific crisis management activities related to SW tools

5.1. Analyzes in crisis management of the organization

Risk analysis is a critical stage in solving the safety and emergency plans of the organization. It should also be part of the project and planning of changes (mergers, change of the company's registered office, development of a new direction of the business, development of a new product, implementation of a new information system, etc.). The entry into this process is the organization's security policy, i.e., the basic strategy of the organization in the field of security, legislative standards, and, among other things, the experience of processors in the field of crisis management. The output is a threat catalog, which serves as input data to the action planning process. A threat is understood as a phenomenon which, provided that at least one threat is realized, can lead to an emergency.

5.2. Modeling

Modeling is an integral part of crisis management, especially in prevention, prediction of the devastating effects of natural and technological accidents, and prediction of the spread of the cause of other emergencies (chemical, biological and radiation substances, flood wave progression, explosion, etc.). The products are based on a mathematical calculation of the propagation effects depending on external conditions such as air temperature, wind direction, strength, surface structure, size of the source of destructive effects, etc. The more accurate the required calculation, the more complex and precise the input information required. There are several products whose focus is different.

5.3. Monitoring

Regular monitoring is also an integral part of crisis management. Up-to-date knowledge of the situation is essential input information when dealing with an emergency and can be used to detect exceeding the health of a safe condition. Another function that monitoring plays in informing the public.

5.4. Planning

Plan - is a binding schedule of procedures for a specific activity (set of measures), which sets out the tasks, forces, places, resources, responsibilities, and timing of the planned move to meet the set objectives [8], [9].

6. ICWS in the Zlín Region

ICWS is unique in connecting clients using optical fibers to each crisis staff in the Zlín municipality. The advantage is that the system can also be used in case of a power failure and can communicate with other crisis staff or flood ommissions. ICWS is also part of the meteoradar that monitors weather and precipitation in the municipality.

6.1. ICWS as a unified platform for crisis management

The Information, Communication and Warning System (ICWS) of the Zlín Region is a unified communication platform integrating all relevant crisis management workplaces (at the level of the Zlín Region, the Fire and Rescue Service of the Zlín Region and relevant municipalities) into a unified platform at the regional level. Thanks to this, there is a significant strengthening of crisis management tools in the Zlín Region.

The implementation of the project created a modern independent system that allows:
- better crisis prevention and preparedness
- faster response in the event of a crisis
- early warning of the population in the event of a threat
- streamlining the work of the components of the integrated rescue system and crisis management
- reduction of the extent of possible damage to the lives and property of the inhabitants and visitors of the city
- minimizing the duration of the crisis and reducing the costs of intervention
- coordinated elimination of the consequences of the action of harmful factors of crises
- addition of new elements to the system [10]

6.2. Technical specification of the ICWS central workplace

ICWS software will control external applications running on the same computer or remote computers on an IP network. To this end, the ICWS software must set up a TCP port (the specific number determined by the configuration) on which it will receive and process commands. Acceptance of the connection may be limited on the server-side to the addresses (or their range) specified during installation.

After accepting the link, command messages can be sent using the ICWS software. The IS will also receive confirmation and diagnostics from the technology server, both in direct response to the command sent and spontaneously. The VoIP techno-legion will be used as a communication standard to ensure direct voice reporting. The voice compression algorithm currently used in VoIP transmits voice only in a part of the waveband and significantly increases efficiency [11].

6.3. ICWS server

The basis of the system will be the server of the central control room ICWS, which enables the control of local systems and data collection through individual access points at the local level (individual municipalities). SW ICWS Server will be created with the help of Microsoft .Net Framework 4.5. The basic architecture of the application will be a three-layer composed of: a presentation layer, an application (domain) layer and a data layer. The data layer will cover access to the Microsoft SQL Server database. The Hibernation Framework will be used for object-relational mapping (ORM).

WPF (Windows Presentation Foundation) technology will be used for the presentation layer. The graphical user interface will be server-side with no interactivity. It will display only the server status, the list of currently connected clients, and presently related access points (individual municipalities). The application (domain) layer will contain the essential functions and logic of the application. Its implementation will not directly link to the presentation or data layer, so a possible future change or modification of the technology at the level of the presentation or data layer will simply be possible. Communication between the ICWS server and client workstations is implemented in networks using the TCP / IP protocol family. The transport layer protocol uses a connection-oriented, reliable TCP (Transmission Control Protocol), which ensures reliability in the event of packet loss by retransmissions, including the rearrangement of received packets (confirming the correct order). The TCP protocol uses port numbers to distinguish between communicating applications. These, together with the IP address, form a so-called socket - unambiguous identification of the endpoint of communication within the network.

An integral part of the server is also data synchronization across connected clients. I.e., in the event of a data change that one client triggers, this change is automatically propagated to other connected clients in the network (without the need for manual refresh, etc.). The same applies to asynchronous events triggered within access points (municipalities) [11].

6.4. ICWS client

The client workplace creates an interface for users to control the entire system, monitor the status, and manage it. The ICWS client will be operated at all workplaces of individual municipalities and designated stations within the Zlín Region and the Fire and Rescue Service of the Zlín Region. Communication between the ICWS server and client workstations is implemented in networks using the TCP / IP protocol family. The connection layer protocol is a connection-oriented, reliable protocol TCP (Transmission Control Protocol). In case of packet loss, ensure reliability by repeated sending, including the rearrangement of received parquets (ensuring the correct order). SW ICWS Client will be created with the help of Microsoft .Net Framework 4.5. The basic architecture of the application will be a three-layer composed of: a presentation layer, an application (domain) layer, and a data layer.

The data layer will primarily cover the synchronization of data between the client and the server. Furthermore, the management of locally stored map data and layers, including work with other local files (documents, etc.) WPF (Windows Presentation Foundation) technology will be used for the presentation layer. The graphical user interface is further discussed in the text in individual points. The application (domain) layer will contain the essential functions and logic of the application. Its implementation will not directly link to the presentation or data layer, so a possible future change or modification of the technology within the presentation or data layer will simply be possible.

One of the main functions of the client SW is the presentation of information and the status of the system on the GIS map data of the Zlín Region. "Tiles" of map data and individual layers will be obtained through the WMS service and stored in two-layer caches to achieve maximum rendering speed and the best possible user impression. Firstly, the part in the client's operational memory and the rest on the disk (possibly in the local database) of the client's workplace. Thanks to this system, the map data will be functional even in a connection failure between the client workplace and the GIS server of the Zlín Region. A WMS query for an individual tile will be performed only if it is not in any caching layer. I.e. at the initial request to draw the given "tile" from the client workplace at all. After downloading, the "tile" is permanently stored at the client's workplace. It follows that the data flow between the client and the GIS server of the Zlín Region will be minimal in the long run. At the same time, users will be able to delete locally stored layers (cache) via the user interface and thus force the automatic re-download of these layers from the GIS server [11].

6.5. ICWS services

From the user's point of view, the software itself will be based on the application of selected layers from the map base with the possibility of any display and further from the button interface for the transition of individual menus and their subgroups. The display and processing of individual operating and alarm messages will be provided.

Another functional part of the system will be its communication module to control local exchanges of individual ORPs through their servers for the control and monitoring of integrated systems.

Display of the map layer from the GIS of the Zlín Region:

- Background maps
  - Basic map
  - Orthophoto map with roads, watercourses, descriptions
- Floodplains
- Emergency planning zones

Map layers from web services on the Internet:

- Czech Hydrometeorological Institute - reporting profiles - import of water levels
- traffic information, deadlines, accidents, etc.
- Czech Geological Survey - slope instabilities (landslides)
- traffic detours

Map layers from I"CWS:

- ICWS elements
- Point layers with the possibility of editing:
  - Extraordinary events
  - Activated / non-activated flood commissions
  - Medical facility (capacity)
  - Educational facility for evacuation (capacity including meals)
  - Gas station
  - Objects of possible attack
  - Large poultry farms - the threat of bird flu

All functions listed in the basic requirements for ICWS software, including the static location of units and points of interest, will be displayed in individual layers with a choice. A dynamic change of the status of units in a separate layer of branches, electronic sirens, and control stations or level sensors will be ensured if the VIS system in a specific ORP will support this function.
The SW architecture also consists of a large number of modules:

a) Central control room of the ICWS of the Zlín Region.
b) Videoconferencing systems (interconnection of crisis management participants).
c) Information boards (monitors) for visualization.
d) City camera and surveillance systems.
e) Elements of environmental and chemical detection of hazardous substances.
f) Traffic information.
g) Level sensors on watercourses.
h) Meteorological radar of the Zlín region.
i) Warning information systems.
j) Crisis management documents.
k) Alternative energy sources [11]

7. Meteoinformation in ICWS

7.1. Meteorological radars

At present, meteorological radars belong to the important and standard equipment of all advanced meteorological services. These devices are known under the general name radar, which originated from the English "RAdio Detection And Ranging". Meteorological radar is a device that is used to detect significant precipitation clouds up to distances of about 100-250 kilometers. In practice, we can encounter the terms large and small meteorological radar. These radars differ not only in size, but mainly in wavelength. Each wavelength size has its letter designation.

- European operational radars (large radars) with a C-band (wavelength 5 cm, frequency 6 GHz) are suitable for the detection of precipitation clouds in temperate latitudes with a more extended usable range (150-300 km).
- Mobile radars (small radars) with X-band (wavelength 3 cm, frequency 10 GHz) are suitable for the detection of precipitation clouds in temperate and high latitudes with a smaller usable range (50-150 km) [10], [11].

The MMR50 meteorological radar and the CZRAD network radars have different parameters and properties, which are listed in Table 1:

Location	Holešov industrial zone	Skalky in Protivanov	Brdy-Prague
Region	Zlín Region	South-Moravian region	Central Bohemian Region
Latitude		49,501 N	49,658 N
Longitude		16,790 E	13,818 E
Altitude	200 m	730 m	860 m
In operation since	2014	2016	2016
Measurement interval	2 minutes	5 minutes	5 minutes
Producer	Microstep-MIS	Vaisala	Vaisala
Radar type	MMR-50	Vaisala WRM200	Vaisala WRM200
Zone	X-band	C-band	C-band
ANTENNA			
Diameter of parabolic antenna	0.55 m	4.5 m	4.5 m
Range of motion in elevation	1 to + 90 °	2 to + 92 °	2 to + 92 °
Number of elevation angles	13	16	16
Antenna gain	32 dB	46 dB	46 dB
Polarization	Linearly horizontal	Horizontal / horizontal + vertical	Horizontal / horizontal + vertical
TRANSMITTER			
Frequency	9.41 GHz	5.645 GHz	5.630 GHz
Wavelength	3.0 cm	5.3 cm	5.3 cm
Transmitter power	50 kW	250 kW	305 kW
Repetition frequency	1500 Hz	200-2400 Hz	200-2400 Hz
RECEIVER	Linear	Polarization (dual)	Polarization (dual)
Receiver sensitivity	- 111 dBm	-117 dBm	-117 dBm
Dynamic range	85 dB	101 dB	101 dB
Maximum radar range	100 km	260 km	256 km

Table 1. Comparison of parameters of mobile radar MMR50 and radars of CZRAD network [13]

In general, it can be said that large radars of the CZRAD network have a higher transmission power, a higher maximum range (up to 250 km), and therefore also lower attenuation in precipitation than in the case of small radars. The research of precipitation measurements of the CZRAD radar network and the MMR50 meteorological radar revealed the following findings:

- In both types of meteorological radars (MMR50 and Radar Skalka), the same results were achieved when monitoring precipitation with higher radar reflectivity (from 35 dBZ and higher). This type of precipitation is characteristic of the occurrence of intense showers and storm cores.
- During the passage of precipitation directly above the radar, there was a large attenuation of electromagnetic waves in the precipitation. It has been experimentally proven that the most significant attenuation occurs when detecting snowfall (but also precipitation with very high radar reflectivity above 50 dBZ) occurring directly above the radar. Due to the significant attenuation, it was then not possible to detect rain at greater distances.
- The low transmission power of 50 kW for radars operating in the X-band is insufficient to detect collisions with higher radar reflectivity (above 35 dBZ) arranged in successive bands. These radars can usually detect high collisions with high attenuation with high accuracy so that the rays do not penetrate beyond this collision field. Further precipitation behind this field may no longer be detected, a severe problem in predicting rainfall [13].

7.2. Meteoradar in the Zlín Region

In 2015, in addition to the end elements for information, warnings and notifications, a meteorological radar was put into operation within the ICWS project. A meteorological station consisting of its meteor radar located on a mast structure and a control data server is installed in the administrative building of the industrial zone in Holešov.

The meteor radar is used to obtain information on dangerous meteorological phenomena detectable in the X-zone within a radius of 160 km and to predict hazardous situations. For the Zlín Region, radio signal coverage means the detection of impending dangerous storm precipitation and torrential rains that directly threaten cities and municipalities in the Zlín Region. The meteoradar transmits the obtained data to the control server to the crisis management workplace (CR). The data are presented using the ICWS software at the CR control workplace in Zlín but also at all ORP workplaces in the Zlín Region.

Meteoradar MMR50 consists of the following main parts:

- support part, with upper rotating part, including cover and random
- antenna
- waveguide part with rotating waveguide coupler
- radar unit
- elevation drive, including sensors
- azimuthal drive, including sensors
- main computer
- complete heating, cooling, and ventilation systems
- power supply and the source part

Fig. 1. Meteorological radar of Zlín region [10]

A part of the control computer is the control SW, which enables a fully automatic function of the device without the operator's intervention. It ensures the successful execution of a meteorological measurement, within which it sets up the radar unit and provides the control of the drives. During the measurement, it receives and continuously forwards the measured meteorological data.

The meteorological radar can provide the following functions:

- scanning of backscatter on meteorological objects from the volume of the radar horizon
- editing and assigning data to a spatial matrix
- processing of input data in databases for analyzed and prognostic radar meteorological maps (storms, showers, rain, snowfall, hail, etc.) and distribution of bases towards other systems.

The MMR 50 meteorological radar is a compact device designed to monitor the current state of the earth's atmosphere and obtain information on dangerous meteorological phenomena within a radius of 150 kilometers from the installation site.

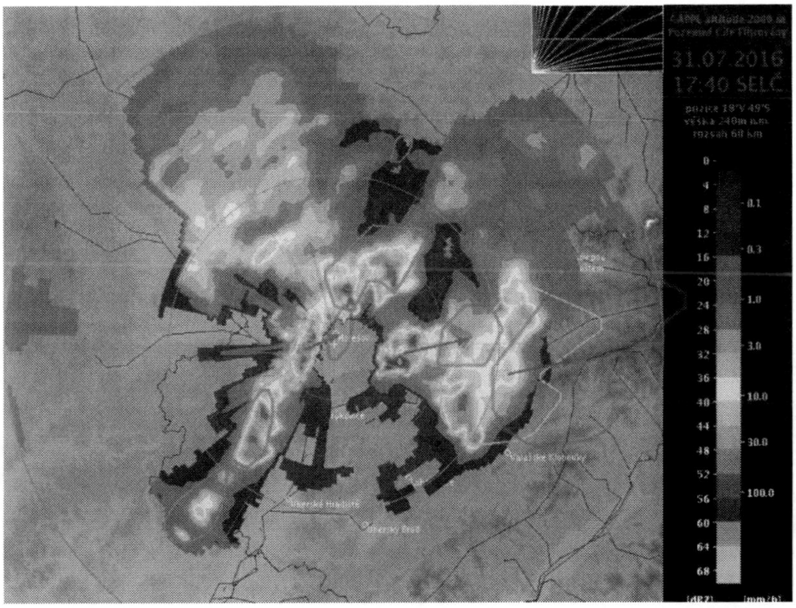

Fig. 2. Measurement of precipitation intensity using the CAPPI 2 km product from 31.7.2016 [13]

CAPPI module 2 km

We will use this product to detect and predict intense precipitation and hail, which will most likely fall on the earth's surface and thus can significantly support the occurrence of flash floods.

Color expression of precipitation intensity (radar reflectivity) is the main output information about how intense precipitation occurs at the current time over a given place. To be able to use the information on the current intensity of rainfall to estimate the risk of flash floods, we must monitor the total duration of precipitation over a given locality. In general, the more extended precipitation occurs above a site, the greater the likelihood of flash floods.

VIL module

The second important product determining the probability of torrential rains and floods is VIL (Vertically Integrated Luquid). The VIL product differs in the kg / m2 scale, which is currently intended to estimate the occurrence of torrential rains that can cause flash floods.

In practice, the assessment of the probability of a flash flood will be carried out in the following steps:

- We assign the risk of a flash flood to a given precipitation field according to color.
- Finally, we determine the resulting risk of flash floods at the Radar Information by averaging the probabilities of CAPPI 2km and VIL outputs [13].

7.3. Critical places of implementation

The software can be implemented in two ways:

- Direct integration into ICWS
- Indirect integration into ICWS

As part of direct integration into ICWS, implementation into the intranet is planned. The application will run on the servers of the Zlín Region. The software is designed to be usable for web and mobile applications and thus meets current rules in software architecture design, including compatibility with existing solutions. One of the critical points could only be an incompatibility with an older version of, for example, the ICWS ZK intranet web interface. The essential point of usability of outputs in practice for the direct variant will then be the login to the intranet, and thus the overall approach to the application. As not all representatives of crisis management and flood planning of municipalities with extended powers in the Zlín Region currently have access, this may mean a specific problem that would need to be solved by another form of access to the ICWS intranet. Indirect integration of the meteor radar into ICWS would only mean that the application would run on servers in the Zlín Region outside the intranet. A particular advantage would be a simplified approach compared to the direct variant both for flood and crisis authorities of municipalities in the Zlín Region and for the public. However, this option would need to be further analyzed in terms of applicability for practical work.

7.4. Practical activities when using meteorological radar MMR 50

The outputs from the MMR 50 meteor radar are used to send warning SMS messages, which inform us about precipitation with an intensity of 20 mm in 30 minutes and higher within a radius of 5 km. Alert SMS messages are sent only to selected people according to the settings. This setting is still in trial operation, and the "sensitivity" of the setting can be changed.

Here is the general procedure of the crisis management staff using the MMR 50 meteorological radar:

In everyday life:
- Ensure the sending of alerts to the Czech Hydrometeorological Institute.
- Ensure the sending of warning SMS from the MMR 50 radar.
- Ensure a permanent connection to the ICWS and regularly check the operation of the meteor radar.
- Set MMR 50 to "CAPPI" or "VIL" depending on the situation.

When warned by the Czech Hydrometeorological Institute:

- Inform the crisis management authorities according to the system set up in the relevant municipality.
- Check the outputs from the ALADIN model.
- Check meteor radar outputs
    - precipitation intensity estimate (CAPPI),
    - estimation of the amount of liquid water inside the clouds (VIL),
    - precipitation speed forecast.

When receiving a warning SMS from the MMR 50 meteoradar:
- Study meteoradar data immediately.
- Evaluate meteor radar outputs.
- Evaluate soil saturation from the CHMI Flash Flood Indicator for areas where intense precipitation occurs according to the radar or experience.
- According to the development of the situation, inform the crisis management authorities, check the situation on the spot, organize activities according to the elaborated plans and procedures usual for the relevant ORP with emphasis on the functionality of notifications and warnings [13].

8. Conclusion

Information systems systematize the storage of information, their quick retrieval, and many of them have tools that analyze available information and thus provide new information. However, no information system offers a comprehensive proposal for resolving a crisis. The design of the solution is always the subject of a correct and high-quality interpretation of the obtained data and results, which must be performed by an expert, not an information system. ICWS is a unique information and notification system in its scope and concept. Its development is still expected in the future, and I assume that there will be a connection to a similar nationwide system one day.

Through this communication infrastructure, it is possible to use selected services provided by the region. These are expected to expand in the future. The infrastructure also enables connection, via optical fibers, of contributory organizations of the Zlín Region and, last but not least, in cooperation with the Zlín Region, brings opportunities to obtain funds for the construction or development of metropolitan networks in individual municipalities in the region. In this way, information can be disseminated that is valuable primarily to warn the population and protect lives from natural disasters, especially unstable weather.

9. Acknowledgments

This work was supported by Grant Project IGA/FAI/2021/003 "SW Population Protection and Soft Target Support" - resolved in 2021.

10. References

[1] Comfort, L. K.; Sungu, Y.; Johnson, D.; Dunn, M. (2001). Complex systems in crisis: Anticipation and resilience in dynamic environments. Journal of Contingencies and Crisis Management, pp. 144-158.

[2] Hu, Q.; Kapucu, N. (2016). Information communication technology utilization for effective emergency management networks. Public Management Review, pp. 323-348.

[3] Ošmera, P. (1991). Information systems, VUT, ISBN 80-214-0286-5, Brno.

[4] Lukáš, L.; Hrůza, P.; Kný, M. (2008). Information Management in Security Components. Ministry of Defence of the Czech Republic, ISBN 9788072784608, Prague.
[5] Drozdek, M.; Jelšovská, K. (2013). Information Support of Crisis Management. Silesian University in Opava, Opava
[6] http://www.giswebsite.com/lkc/refs/er/sld001.htm (1995) Lowe WJ. GIS Application Design for an Emergency Management Information System
[7] Adamec, V.; Štolba, L. (2004). Information Technology for Integrated Rescue System and Crisis Management. Public administration online, supplement to the magazine Municipality and finance
[8] Bláhová, M.; Mach, V.; Pavlík, L.; Hromada, M.; Ficek, M. (2019). The Information Security to Software of Crisis Management, Proceedings of the 30th DAAAM International Symposium, pp.1019-1025, B. Katalinic (Ed.), Published by DAAAM International, ISBN 978-3-902734-22-8, ISSN 1726-9679, Vienna, Austria
[9] Víchová, K.; Hromada, M. (2018). Information Support of Crisis Management. Crisis Management - Theory and Practice, pp. 38-45, DOI: 10.5772/intechopen.74094
[10] http://www.colsys.cz, (2017). COLSYS, ICWS of Zlín municipality, Accessed on 2021-09-10
[11] Information, communication and warning system of the Zlín Region (2015). Regional Office of the Zlín Region, Internal document.
[12] Řezáčová, D. et. al. (2007). Physics of clouds and precipitation. 1 ed., Academia ISBN 978-80-200-1505-1, Prague
[13] Šaur D. (2016). Methodology of using the meteorological radar of the Zlín Region for the needs of crisis management. Internal document.

DOI: 10.2507/32nd.daaam.proceedings.099

FINITE ELEMENT MODEL UPDATING METHODS FOR STRUCTURAL APPLICATION

Suzana Ereiz, Ivan Duvnjak & Javier Fernando Jimenez-Alonso

This Publication has to be referred as: Ereiz, S[uzana]; Duvnjak, I[van] & Jimenez-Alonso, J[avier] F[ernando] (2021). Finite Element Model Updating Methods for Structural Application, Proceedings of the 32nd DAAAM International Symposium, pp.0712-0718, B. Katalinic (Ed.), Published by DAAAM International, ISBN 978-3-902734-33-4, ISSN 1726-9679, Vienna, Austria
DOI: 10.2507/32nd.daaam.proceedings.099

Abstract

Structural finite element models are a modern tool for continuous monitoring of structures, damage detection, prediction of service life and determination of optimal strategy for maintenance. It is well known that we get uncertain numerical models due to the input parameters such as stiffness, mass, and boundary conditions. To solve this problem, finite element model updating methods are used in combination with the results of structural dynamic tests or structural health monitoring data. In this paper, a brief overview of the procedure for updating finite element models using experimentally determined structural dynamic parameters - natural frequencies and natural modes - is given. Some of the most commonly used methods are listed and their advantages and disadvantages are highlighted.

Keywords: Finite element model updating; Structural health monitoring; Dynamic analysis of structures; experimental investigation

1. Introduction

Finite element model updating (FEMU) emerged in the 1990s as a topic of the great importance for the design [1], [2], [3], construction [4], [5], [6] and maintenance of mechanical systems [7], [8], [9] and structures [10] [11] [12] [13]. As for civil engineering structures, it refers to the updating of its models to assume its dynamics behaviour more precise and accurate. To minimize the differences and maximize the correlation between the numerical model and the real structure, the model updating (MU) of the civil engineering structure is most often performed by applying the structural dynamic test results.

In addition to dynamic, the static test results [14], [15], [16], [17] or its combination with dynamic test results is also apply [18], [19], [20], [21]. The conducting of FEMU is mainly focused on the determination of most relevant structural parameter values such as geometry, material properties, and boundary conditions. Those parameters minimize the differences between the structural behaviours predicted by the numerical model from the actual behavior of the structure. The reasons why there are differences between the predicted behavior of a numerical model of the structure and its actual behavior include model structure errors, model order errors, model parameter errors, and errors in measurements [22].

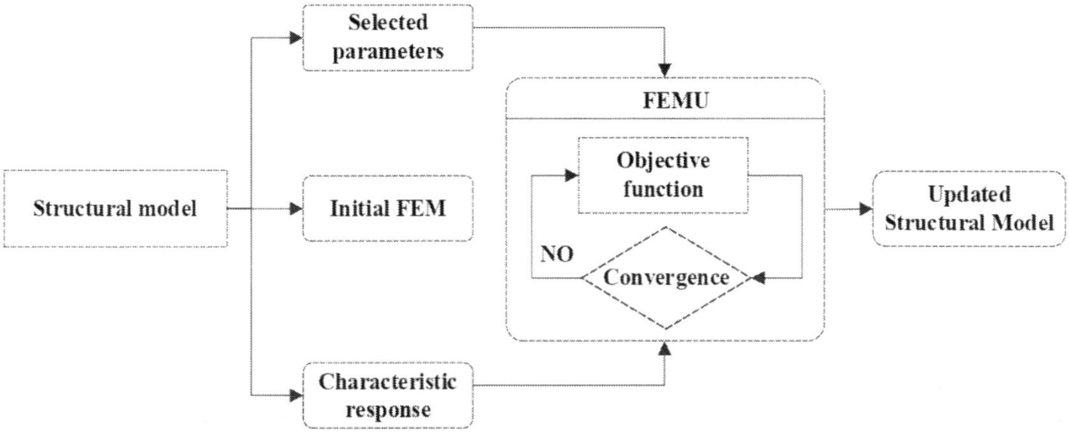

Fig. 1. Flowchart of FEMU procedure

Model structure errors arise from incorrect assumptions of the mechanical properties and physical behaviour of structures [23], [24]. Second one, model order errors, most often arise from the difficulty in the modelling of the nonlinearity [14]. Model parameter errors arise from the incorrect assumptions of model parameters- material and geometrical properties. Measurement errors occur in the measurements, where due to the influence of various environmental effects and other different factors an error may occur. Many of investigation is mainly focused on the finite element model updating using structural dynamic parameters - natural frequency, mode shapes and damping ratio. The advantages of using these parameters are that they are easy to apply for damage detection. Also, they are most appropriate when there is a large error in the numerical modelling, and they are directly related to the topology of the structure. In this type of data, most of the structural parameters that best describe the global behavior are include [25]. On the other hand, the determination of the structural dynamic properties for identification is computationally intensive and prone to additional noise. Moreover, it is not able to extract the modes of the frequency bandwidth and is only suitable for highly damped and linear structures [25].

The FEMU method can generally be classified as the automated and manual methods [26], and iterative(non-direct) and non-iterative (direct) methods [27]. The main difference between the manual and automated method lies in the number of selected updating parameters and in the way the model updating is performed - using trial and error methods (manual FEMU) or automated process. There is also a difference in the obtained results, where in manual methods the difference between predicted and actual behavior of structure is less than 5% while in automated methods it is less than 1% [28]. The division of MU methods into iterative and non-iterative methods is somewhat more concrete. As their name say the differences between them is in the way the model updating is implemented, using iterative process or not. The iterative methods are further divided into deterministic methods and Bayesian method. Based on the targeted responses used to update the numerical model, deterministic methods can be further divided into two groups: (1) methods that use eigenvalues and eigenvectors (eigenfrequencies, mode shapes, damping coefficient) and (2) methods that use frequency response data. The Bayesian method treats numerical models as statistical problems and aims to develop a set of numerical models that will represent the actual behavior of the structure [29]. They are aimed at quantifying the uncertainties of the selected updating parameters [30]. The aim of this paper is to present the procedure of model updating by selecting the parameters for numerical model updating, defining the difference function between the predicted and the actual behavior of the structure, and by presenting the most commonly used methods for minimizing the difference. In this way, an review as well as the capacity of the approaches proposed so far is given, on the basis of which some new approaches can be proposed. In the following (second section) the procedure for selecting the parameters of the numerical model is described, through which iteration the numerical model is improved to match the real behavior of the structure. The third section describes how to define the difference between the real and the predicted behavior of the structure, depending on whether the approach to model updating is deterministic or stochastic. Section 4 gives a review of the methods most commonly used in model updating and their advantages and disadvantages.

2. Selection of the updating parameters

The success of the finite element model update implementation depends on the selection of the updating parameters. It is very important to properly define the model parameterization and estimate the unknown parameters from ill-conditioned equations. In addition to parameterization, the uniqueness of the updating parameters, computational efficiency, ill-conditioned equations, and the use of incomplete data is also important. The selected parameters of the numerical model that are being updated should be able to explain the uncertainties or inaccuracies of the model, provided that the output data are sensitive to the input parameters of the model [31]. There are several methods that are used for selection of updating parameters.

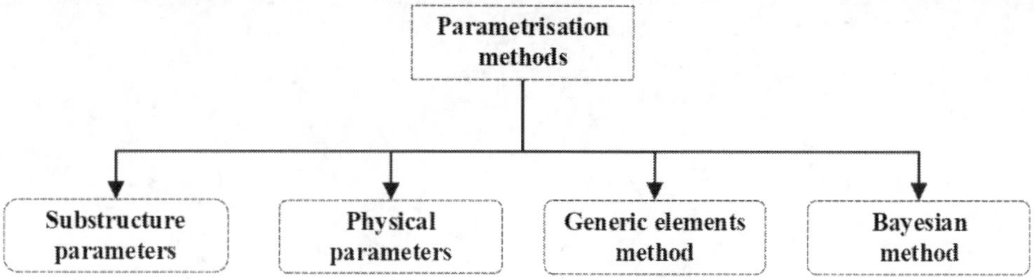

Fig. 2. Parametrisation methods for FEMU

One of them is to define the scalar multipliers associated with the mass, stiffness and damping matrices [32]. Second parametrization method employs the direct material and geometrical properties of structures [33]. Third one, generic method [34] by adjusting the eigenvectors and eigenvalues of individual elements or sub-structure update the stiffness and mass matrices. Regardless of the selected parameterization method, the set of selected updating parameters should be as small as possible to eliminate unnecessary parameters and reduce computational cost. Their number is also limited by the amount of information available from measurements in a limited frequency range.

3. Definition of the objective and probability density function

Formulation of the objective function for the FEMU problem must take into account the residuals that are sensitive to the selected updating parameters [35]. Objective function in problem of FEMU is the residuals which defined the differences between structural behaviour predicted by numerical model and its actual behaviour. Usually, it is most defined by considering two types of residuals. The first one is associated with the natural frequencies while the second is associated with the corresponding mode shapes. Also, the objective function can be defined by considering the residual associated with the damping ratio [36]. The influence of those residuals on the objective function can be evaluated via two approaches: the single objective function (1) and multi objective function (2) approach.

$$F(x) = \sum_{i=1}^{NF} w_i F_i(x) \tag{1}$$

$$\left(F_i(x)\right) = \left(F_1(x), F_2(x), .., F_{NF}(x)\right) \tag{2}$$

Single objective function (1) approach is characterized by only one objective function. This function is defined in terms of the sum of weighted residuals. The values of weighted residual are usually obtained using the trial-and-error method or a probabilistic method in terms of the standard deviation of the experimental data. On the other hand, in the multi objective function approach (2) each component of the objective function is defined in terms of unweighted residuals. Several non-dominated solutions are obtained which form a Pareto front. As the solutions is non-dominate a decision-making problem must be solved. The best solution is selected from the Pareto front by determining the knee point. Knee point is a solution from the Pareto front that has a maximum marginal utility. It represent the locally the best trade - off solutions among the considered objective values [36]. In the stochastic approach, a numerical model is updated to the multiple sets of predicted results using multivariate multiple regression. In this way, a parameter sensitivity matrix is created that accounts for the entire population of randomized values for all parameters together. Differences between the experimentally and numerically obtained data are minimized using a gradient and regression approach to obtain improved estimates of the mean value of the randomized parameters at each step [37].

In Bayesian method [38] instead an objective function probability density function (PDF) is defined in terms of the posterior PDF-$P(\theta|\tilde{M},\tilde{M}_m)$, likelihood function-$P(\tilde{M}|\theta,\tilde{M}_m)$, normalisation function-$P(\tilde{M}|\tilde{M}_m)$ as follows (3)

$$P(\theta|\tilde{M},\tilde{M}_m) = \frac{P(\tilde{M}|\theta,\tilde{M}_m)P(\theta|\tilde{M}_m)}{P(\tilde{M}|\tilde{M}_m)} \tag{3}$$

4. Finite element model updating methods

4.1. Matrix update methods

The matrix update method is mainly based on updating the structural mass, stiffness or damping matrices. They are implemented to minimize the differences between the analytical and measured matrices. The main concept is based on changing the global stiffness, mass or damping matrices to an improved numerical model that accurately reproduces the experimental results [39]. Their application directly minimizes the differences between experimentally and numerically obtained structural dynamic parameters. These methods are computationally expensive and it is difficult to find a global minimum through the optimization techniques since there are several stationary points [22]. Besides, it is difficult to achieve a solution with the physical meaning.

4.2. Sensitivity based methods

Sensitivity-based methods are the most successful approach with many applications to the MU problems. In these methods, the experimentally obtained structural dynamics parameters – natural frequencies and mode shapes are adopted for updating the initial numerical model. These methods allow a wide choice of the parameters for model updating and measured outputs to be weighted. With the respect to parameters, determination of the sensitivity of the measurements is computationally intensive and the approximation of the sensitivity may only be obtained. The sensitivity is most often the non-linear function of the updating parameters, an iterative procedure is necessary [40]. This iterative procedure is associated with the convergence problem to solve for the selected parameters [41]. Sensitivity based method has some limitations related to request of existence of a sensitivity matrix with the respect to all updating parameters. This led to the expensive computation. On the other hand this method may not be applicable for structure which contain a considerable amount of damage [22].

4.3. Dynamic perturbation and regularization method

As the sensitivity based method works well when the differences between the predicted and actual structural behaviour is relatively small, in the case where such difference is sufficiently large, the linear or first order approximations may be inappropriate [41]. An exact relationship between the perturbation of structural parameters and the perturbation of the associated dynamic parameters is used for the FEMU. This method directly adopts the measure incomplete structural dynamic parameters and in the FEMU process it doesn't require mode shape expansion or reduction [42]. This method needs much less computational effort to estimate updating parameters. It provides optimised solutions for MU in the least squares sense without requiring optimisation techniques [43]. Also, this method offers reliable estimates of structural updating parameters, even in the cases where relatively large modifications in structural parameters and/or in modal properties exist between the finite element model and the tested structure [44].

4.4. Iterative optimization procedure

Iterative optimization methods transform the FEMU into the optimization problem. On this way the objective function is defined as (here there is a typo) the relative differences, i.e., residuals, between the behaviour of structure predicted by numerical model and its actual behaviour. There are two different ways in which these residuals can be defined. The first one is as single objective optimization problem, while the second one is the multi-objective optimization problem [45]. As the name says, the first optimization problem- single objective is characterized with only one objective function define in terms of the sum of the weighted residuals. The final values of the weighted residuals are usually derived from a trial-and-error process.

4.5. Bayesian method and Fuzzy methods

The Bayesian FEMU method approaches on the development of the numerical model as a statistical problem focusing on the estimating the probabilistic density function of the numerical model physical parameters. This method uses Bayesian probability theory [46] for estimating the posterior probability density function of the physical parameters of the model. Density function is estimated in terms of the likelihood and prior probability density functions [29]. The likelihood function measures the discrepancies between the numerical and experimental structural dynamic properties. The prior probability density function reflects a previous assumption about the statistical behavior of the physical parameters of the FE model. Including its advantages, this method has a disadvantage related to a complex numerical model, which is often time consuming and computationally limited, limiting its applicability to the large real structures.

4.6. Computational intelligence algorithms

Finite element model updating is an optimization method in which the design variables are the uncertain parameters of the model [47]. The objective is to minimize the distance between the structural behavior predicted by numerical model and its actual behavior obtained by conducting the experimental investigation. To solve the optimization problem of FEMU the computational intelligence algorithms are used (Figure 3).

This techniques mostly include the Nelder mead simplex method [48], sequential programming technique [49], genetic algorithm [50], particle swarm optimization [51], simulated annealing [44], response surface method [52], hybrid optimization [53] and multi objective optimization [54] and all the other algorithms

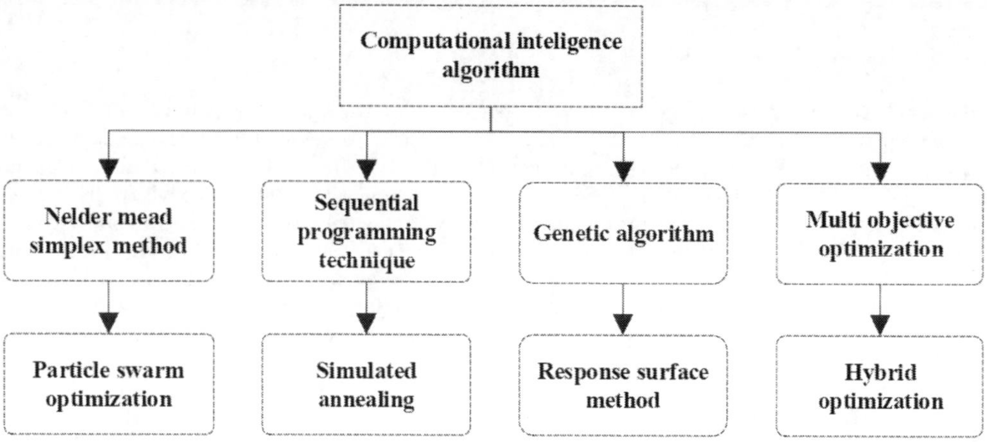

Fig. 3. Computation intelligence algorithm for FEMU

5. Conclusion

With the ever-advancing development of computer software and structural health monitoring techniques, ever-increasing demands are being placed on the numerical modelling of structures. This has created a need for the development of methods that combine computer software and SHM techniques into numerical models to describe real structure as accurately and precisely as possible. The numerical models obtained can be used for various purposes, from monitoring the construction phases, to the maintenance of the structure and its restoration, to important related decisions. This paper gives a brief overview of the process of implementing model updating and the most used methods for this purpose. Moreover, a way to select the updating parameters and to define the differences between the behavior of the structure predicted by the numerical model and the actual one is given. Based on the review performed, it can be concluded that the selection of the model updating method depends on the test settings of the structure and the numerical model, on the complexity of the structure and on the way of defining the difference between the behavior of the structure predicted by the numerical model and its actual behavior. It turned out to be very important to correctly define the function and the appropriate algorithm that ensures its solutions. Due to the complexity of numerical models for certain types of structures, updating their model is a very time-consuming and expensive process and requires a large amount of computation. Therefore, for further research, it is proposed to develop an algorithm that reduces the time and computational cost required to improve the numerical model without compromising the accuracy of the updated solution (optimal value of the selected parameters).

6. Acknowledgments

This research was funded by the European Union through the European Regional Development Fund's Competitiveness and Cohesion Operational Program, grant number KK.01.1.1.04.0041, project "Autonomous System for Assessment and Prediction of infrastructure integrity (ASAP)."

7. References

[1] Zapico, J.L., González, M.P., Friswell, M.I., Taylor, C.A., Crewe, A.J. (2003). Finite element model updating of a small scale bridge, Journal of Sound and Vibration, 268, (993–1012), https://doi.org/10.1016/S0022-460X(03)00409-7.

[2] Modak, S. V., Kundra, T.K., Nakra, B.C. (2002) Use of an updated finite element model for dynamic design, Mechanical Systems and Signal Processing, 16, (303–322), https://doi.org/10.1006/mssp.2001.1431.

[3] Structures, T., Technical, C., Republic, C., Arora, V., Ilcik, J., Republic, C. (2014). Updating finite element model of façade scaffolds anchor considering experimental data, (1–11),.

[4] Li, Z., Feng, D., Feng, M.Q., Xu, X. (2017). System identification of the suspension tower of Runyang Bridge based on ambient vibration tests, Smart Structures and Systems, 19, (523–538), https://doi.org/10.12989/sss.2017.19.5.523.

[5] Zong, Z., Lin, X., Niu, J. (2015). Finite element model validation of bridge based on structural health monitoring—Part I: Response surface-based finite element model updating, Journal of Traffic and Transportation Engineering (English Edition), 2, (258–278), https://doi.org/10.1016/j.jtte.2015.06.001.

[6] Sequera Gutierrez, D., Solano Rodríguez, L.F., Muñoz Díaz, E.E., Alvarado Vargas, Y.A., Villalba Morales, J.D., Gasch Molina, I. (2020). Updating the finite element model of a Colombian Bridge with Ansys, Dyna, 87, (209–218), https://doi.org/10.15446/dyna.v87n212.79785.

[7] Wang, W., Mottershead, J.E., Ihle, A., Siebert, T., Reinhard Schubach, H. (2011). Finite element model updating from full-field vibration measurement using digital image correlation, Journal of Sound and Vibration, 330, (1599–1620), https://doi.org/10.1016/j.jsv.2010.10.036.

[8] Butkewitsch, S., Steffen, V. (2002). Shape optimization, model updating and empirical modeling applied to the design synthesis of a heavy truck side guard, International Journal of Solids and Structures, 39, (4747–4771), https://doi.org/10.1016/S0020-7683(02)00372-4.

[9] Kwon, K.S., Lin, R.M. (2004). Frequency selection method for FRF-based model updating, Journal of Sound and Vibration, 278, (285–306), https://doi.org/10.1016/j.jsv.2003.10.003.

[10] Schommer, S., Nguyen, V.H., Maas, S., Zürbes, A. (2017). Model updating for structural health monitoring using static and dynamic measurements, Procedia Engineering, 199, (2146–2153), https://doi.org/10.1016/j.proeng.2017.09.156.

[11] Altunişik, A.C., Karahasan, O.Ş., Genç, A.F., Okur, F.Y., Günaydin, M., Kalkan, E., Adanur, S. (2018). Modal parameter identification of RC frame under undamaged, damaged, repaired and strengthened conditions, Measurement: Journal of the International Measurement Confederation, 124, (260–276), https://doi.org/10.1016/j.measurement.2018.04.037.

[12] Nguyen, A., Kodikara, K.T.L., Chan, T.H.T., Thambiratnam, D.P. (2019). Deterioration assessment of buildings using an improved hybrid model updating approach and long-term health monitoring data, Structural Health Monitoring, 18, (5–19), https://doi.org/10.1177/1475921718799984.

[13] Bayraktar, A., Sevim, B., Can Altunişik, A. (2011). Finite element model updating effects on nonlinear seismic response of arch dam-reservoir-foundation systems, Finite Elements in Analysis and Design, 47, (85–97), https://doi.org/10.1016/j.finel.2010.09.005.

[14] Sun, L., Li, Y., Zhu, W., Zhang, W. (2020). Structural response reconstruction in physical coordinate from deficient measurements, Engineering Structures, 212, (110484), https://doi.org/10.1016/j.engstruct.2020.110484.

[15] Kim, S., Young, K., Lee, J. (2016). Bridge Finite Model Updating Approach By Static Load Input / Deflection Output Measurements : Field Experiment, in: The 2016 Structures Congress (Structures 16), Jeju Island, Korea,

[16] Schlune, H., Plos, M., Gylltoft, K. (2009). Improved bridge evaluation through finite element model updating using static and dynamic measurements, Engineering Structures, 31, (1477–1485), https://doi.org/10.1016/j.engstruct.2009.02.011.

[17] Mylonas, C., Abdallah, I., Chatzi, E.N. (2010). Conference Proceedings of the Society for Experimental Mechanics Series, in: T. Proulx (Ed.), Bayesian Finite Element Model Updating Using Static and Dynamic Data, Springer, Bethel, USA, : pp. 189–196.

[18] Ismail, Z., Kuan, K.K., Yee, K.S., Chao, O.Z. (2014). Examining the trend in loss of flexural stiffness of simply supported RC beams with various crack severity using model updating, Measurement: Journal of the International Measurement Confederation, 50, (43–49), https://doi.org/10.1016/j.measurement.2013.12.036.

[19] Oh, B.K., Kim, D., Park, H.S. (2017). Modal Response-Based Visual System Identification and Model Updating Methods for Building Structures, Computer-Aided Civil and Infrastructure Engineering, 32, (34–56), https://doi.org/10.1111/mice.12229.

[20] Zhou, L., Wang, L., Chen, L., Ou, J. (2016). Structural finite element model updating by using response surfaces and radial basis functions, Advances in Structural Engineering, 19, (1446–1462), https://doi.org/10.1177/1369433216643876.

[21] Seon Park, H., Kim, J.H., Oh, B.K. (2019). Model updating method for damage detection of building structures under ambient excitation using modal participation ratio, Measurement: Journal of the International Measurement Confederation, 133, (251–261), https://doi.org/10.1016/j.measurement.2018.10.023.

[22] Marwala, T. (2010). Finite-element model Updating Using Computational Intelligence Techniques, 1st ed., Springer-Verlag London, London, England, https://doi.org/10.1007/978-1-84996-323-7.

[23] Sun, L., Li, Y., Zhang, W. (2020). Experimental Study on Continuous Bridge-Deflection Estimation through Inclination and Strain, Journal of Bridge Engineering, 25, (04020020), https://doi.org/10.1061/(asce)be.1943-5592.0001543.

[24] Farshadi, M., Esfandiari, A., Vahedi, M. (2017). Structural model updating using incomplete transfer function and modal data, Structural Control and Health Monitoring, 24, (1–13), https://doi.org/10.1002/stc.1932.

[25] Ewins, D.J. (2000). Modal Testing: Theory, Practice and Application, Second Edi, Research Studies Press Ltd, Hertfordshire, .

[26] Altunişik, A.C., Karahasan, O.Ş., Okur, F.Y., Kalkan, E., Ozgan, K. (2019). Finite element model updating and dynamic analysis of a restored historical timber mosque based on ambient vibration tests, Journal of Testing and Evaluation, 47, https://doi.org/10.1520/JTE20180122.

[27] Marwala, T. (2010). Finite-element-model updating using computional intelligence techniques: Applications to structural dynamics, 1st ed., Springer-Verlag London, London, England, https://doi.org/10.1007/978-1-84996-323-7.

[28] Altunişik, A.C., Kalkan, E., Okur, F.Y., Karahasan, O.Ş., Ozgan, K. (2020). Finite-Element Model Updating and Dynamic Responses of Reconstructed Historical Timber Bridges using Ambient Vibration Test Results, Journal of Performance of Constructed Facilities, 34, https://doi.org/10.1061/(ASCE)CF.1943-5509.0001344.

[29] Goller, B., Schuëller, G.I. (2011). Investigation of model uncertainties in Bayesian structural model updating, Journal of Sound and Vibration, 330, (6122–6136), https://doi.org/10.1016/j.jsv.2011.07.036.

[30] Simoen, E., De Roeck, G., Lombaert, G. (2015). Dealing with uncertainty in model updating for damage assessment: A review, Mechanical Systems and Signal Processing, 56, (123–149), https://doi.org/10.1016/j.ymssp.2014.11.001.

[31] Naranjo-Pérez, J., Jiménez-Alonso, J.F., Pavic, A., Sáez, A. (2021). Finite-element-model updating of civil engineering structures using a hybrid UKF-HS algorithm, Structure and Infrastructure Engineering, 17, (620–637), https://doi.org/10.1080/15732479.2020.1760317.

[32] Stubbs, N., Kim, J.T. (1996). Damage localization in structures without baseline modal parameters, AIAA Journal, 34, (1644–1649), https://doi.org/10.2514/3.13284.

[33] Friswell, M.I., Mottershead, J.E., Ahmadian, H. (2001). Finite-element model updating using experimental test data: Parametrization and regularization, Philosophical Transactions of the Royal Society A: Mathematical, Physical and Engineering Sciences, 359, (169–186), https://doi.org/10.1098/rsta.2000.0719.

[34] Ahmadian, H., Gladwell, G.M.L., Ismail, F. (1997). Parameter selection strategies in finite element model updating, Journal of Vibration and Acoustics, Transactions of the ASME, 119, (37–45), https://doi.org/10.1115/1.2889685.

[35] Alkayem, N.F., Cao, M., Zhang, Y., Bayat, M., Su, Z. (2018). Structural damage detection using finite element model updating with evolutionary algorithms: a survey, Neural Computing and Applications, 30, (389–411), https://doi.org/10.1007/s00521-017-3284-1.

[36] Grafe, H. (1998). Model Updating of Large Structural Dynamics Models Using Measured Response Functions, University of London https://doi.org/10.1002/eqe.2925.

[37] Mares, C., Mottershead, J.E., Friswell, M.I. (2006). Stochastic model updating: Part 1-theory and simulated example, Mechanical Systems and Signal Processing, 20, (1674–1695), https://doi.org/10.1016/j.ymssp.2005.06.006.

[38] Marwala, T., Boulkaibet, I., Adhikari, S. (2017). Probabilistic Finite Element Model Updating using Bayesian Statistics. Application to aeronautical and mechanical engineering, First, John Wiley & Sons, Ltd Registered, Chichester, UK, .

[39] Girardi, M., Padovani, C., Pellegrini, D., Porcelli, M., Robol, L. (2020). Finite element model updating for structural applications, Journal of Computational and Applied Mathematics, 370, (112675), https://doi.org/10.1016/j.cam.2019.112675.

[40] Jung, D.S., Kim, C.Y. (2013). Finite element model updating on small-scale bridge model using the hybrid genetic algorithm, Structure and Infrastructure Engineering, 9, (481–495), https://doi.org/10.1080/15732479.2011.564635.

[41] Chen, H.-P. (2018). Structural health monitoring of civil engineering structures, John Wiley and Sons Inc., https://doi.org/10.1002/pse.2260010313.

[42] Zhou, X.Q., Xia, Y., Weng, S. (2015). L1 regularization approach to structural damage detection using frequency data, Structural Health Monitoring, 14, (571–582), https://doi.org/10.1177/1475921715604386.

[43] Hua, X.G., Ni, Y.Q., Ko, J.M. (2009). Adaptive regularization parameter optimization in output-error-based finite element model updating, Mechanical Systems and Signal Processing, 23, (563–579), https://doi.org/10.1016/j.ymssp.2008.05.002.

[44] Chen, H.P., Tee, K.F. (2014). Structural finite element model updating using incomplete ambient vibration modal data, Science China Technological Sciences, 57, (1677–1688), https://doi.org/10.1007/s11431-014-5619-9.

[45] Naranjo-Pérez, J., Infantes, M., Fernando Jiménez-Alonso, J., Sáez, A. (2020). A collaborative machine learning-optimization algorithm to improve the finite element model updating of civil engineering structures, Engineering Structures, 225, https://doi.org/10.1016/j.engstruct.2020.111327.

[46] Bayes, M., Price, M. (1763). An essay towards solving a problem in the doctrine of chances, by the late rev. mr. bayes, frs communicated by mr. price., in a letter to john canton, amfrs.

[47] Jiménez-Alonso, J.F., Naranjo-Perez, J., Pavic, A., Sáez, A. (2020). Maximum Likelihood Finite-Element Model Updating of Civil Engineering Structures Using Nature-Inspired Computational Algorithms, Structural Engineering International, (1–13), https://doi.org/10.1080/10168664.2020.1768812.

[48] Olsson, D.M., Nelson, L.S. (1975). The nelder-mead simplex procedure for function minimization, Technometrics, 17, (45–51), https://doi.org/10.1080/00401706.1975.10489269.

[49] Jin, S.S., Jung, H.J. (2016). Sequential surrogate modeling for efficient finite element model updating, Computers and Structures, 168, (30–45), https://doi.org/10.1016/j.compstruc.2016.02.005.

[50] Levin, R.I., Lieven, N.A.J. (1998). Dynamic finite element model updating using simulated annealing and genetic algorithms, Mechanical Systems and Signal Processing, 12, (91–120), https://doi.org/10.1006/mssp.1996.0136.

[51] Kennedey, J., Eberhart, R. (1995). Particle Swarm Optimization, in: Proceedings of ICNN'95 - International Conference on Neural Networks, : pp. 1942–1948 https://doi.org/10.1002/9780470612163.

[52] Ren, W.-X., Fang, S.-E., Deng, M.-Y. (2011). Response Surface–Based Finite-Element-Model Updating Using Structural Static Responses, Journal of Engineering Mechanics, 137, (248–257), https://doi.org/10.1061/(asce)em.1943-7889.0000223.

[53] Hao, X., Liu, Y. (2011). Updating the finite element model of a bridge model using a hybrid optimization method, Key Engineering Materials, 456, (37–50), https://doi.org/10.4028/www.scientific.net/KEM.456.37.

[54] Jin, S.S., Cho, S., Jung, H.J., Lee, J.J., Yun, C.B. (2014). A new multi-objective approach to finite element model updating, Journal of Sound and Vibration, 333, (2323–2338), https://doi.org/10.1016/j.jsv.2014.01.015.

DOI: 10.2507/32nd.daaam.proceedings.100

ANALYSIS OF VEHICLE PATH RADII ON HORIZONTAL CURVES FOR TWO-LANE RURAL ROADS

Biljana Maljkovic & Drazen Cvitanic

This Publication has to be referred as: Maljkovic, B[iljana] & Cvitanic, D[razen] (2021). Analysis of Vehicle Path Radii on Horizontal Curves for Two-Lane Rural Roads, Proceedings of the 32nd DAAAM International Symposium, pp.0719-0723, B. Katalinic (Ed.), Published by DAAAM International, ISBN 978-3-902734-33-4, ISSN 1726-9679, Vienna, Austria
DOI: 10.2507/32nd.daaam.proceedings.100

Abstract

In order to obtain locations with reduced level of consistency, vehicle path radii on horizontal curves for two-lane rural roads were analysed. The data collected in an experimental survey on a segment of the two-lane state road DC1 in Croatia were used, in which the drives of a representative sample of drivers were recorded with a high frequency GPS device. The stepwise linear regression was used to investigate the possible influence of geometric characteristics of the road on the driver's choice of path radius and no statistically significant correlation was found. The analysis showed that the vehicle path radii were mainly smaller than curve radius (on average for 12%). The largest deviations of vehicle path radii from designed values were found on sharper curves (R < 150 m), indicating an increased risk of skidding which may be related to the fact that sharp curves are associated with high accident rates.

Keywords: vehicle path radius; horizontal curve; design consistency; global positioning system (GPS); vehicle stability.

1. Introduction

Four factors mainly affecting road safety, sorted in ascending order are: equipment failure, road design, poor roadway maintenance and driver behaviour [1]. Road accident "black spots" indicate the fact that, in addition to human factor, road characteristics can often be the cause of accidents. Horizontal curves are high-risk locations, i.e. locations where traffic accidents occur more often. More detailed research has shown that there is a relationship between the number of traffic accidents and curvature. Specifically, the accident frequency increases with decreasing curve radius, especially in curves with radii less than 200 m (Fig. 1.).

Traffic accidents frequently occur due to lack of geometric design consistency, i.e., when road characteristics are not in line with expectations of drivers. A lot of research has been done to explore this concept, resulting in various quantitative measures for evaluation of design consistency. Specifically, when talking about consistency in horizontal curves, vehicle stability should certainly be highlighted as an important consistency measure. Quantitative consistency measure based on vehicle stability is provided by checking the margin of safety (the difference between the limiting value of side friction and side friction demand) [3]. Ensuring vehicle stability is important because insufficient side friction at horizontal curves may result in vehicle skidding or rollover.

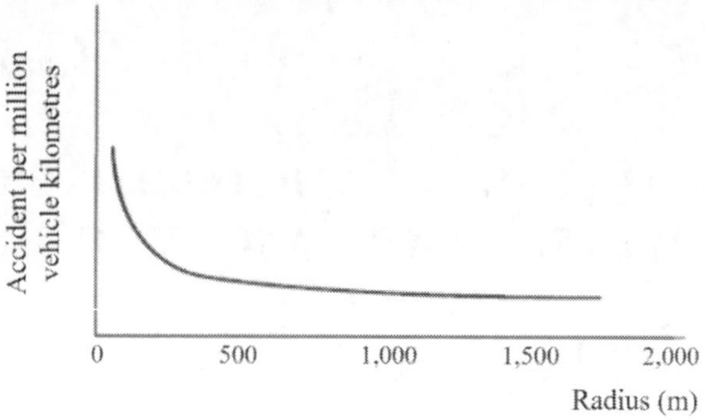

Fig. 1. Relationship between horizontal curve radius and accident risk [2]

Furthermore, most of previous road safety and design consistency studies focused on operating speed, whereas only a few authors were analysing driver behaviour in terms of actual vehicle path radii [4], [5], [6], [7], [8], [9]. Therefore, the aim of this paper is to analyse design consistency in horizontal curves, by comparing design curve radii and actual vehicle path radii, as well as to determine the impact of speed, speed differential and geometric features of the road on driver's choice of vehicle path radius in curve. The path radii were collected in an experimental investigation on a segment of the two-lane state road, conducted as part of [10].

2. Data collection methodology

An experimental survey was conducted on a 24 km long road segment of the state road DC1 (Fig. 2.) to collect the driver behaviour data. High performance GPS device was used to record the test drives of 20 drivers (of different ages and with different driving experiences), performed with their own personal cars. The selected road segment is a two-lane rural road with relatively low traffic volume (the average annual traffic is about 1400 veh/day) and with no main intersections what allowed to collect the data in free-flow conditions. The selected road segment consists of alignment elements with a wide range of geometric characteristics: 97 horizontal curves with radii varying from 80 to 1000 m and 92 tangents with lengths up to 700 m. The survey was conducted under optimal weather conditions (daylight and dry pavement). Based on precisely recorded positions and headings (ten times a second), GPS associated software automatically calculates vehicle path radii:

$$R = \frac{57.3 \cdot L}{D_C} \tag{1}$$

where L (m) is curve length and D_c (degree) is the change in heading. Due to significantly higher accident risk in small-radius curves (as shown in Fig. 1.), as well as the fact that the precision of radius estimating is decreasing with the increase in curve radius [11], only sharper curves (radii from 80 to 315 m) were considered. As relevant vehicle path radius for each curve and individual test run, the value of the smallest constant radius was selected, i.e., relatively constant value of the radius (with up to 5% deviations in curvature) with a driving length of at least 0.5 seconds. As recommended by some authors ([4], [12]), the 15th percentile path radius (R_{15}) was selected for the critical curve path radius (path radii of only 15% of drivers are smaller than R_{15}).

Fig. 2. Segment of the state road DC1 [10]

3. Analysis and results

To improve road design consistency and safety, road characteristics should be consistent with driver's expectations. The road design consistency can easily be checked by comparing the values of driver's behaviour parameters (e.g., vehicle path radius) with the corresponding actual design geometric alignment (e.g., design curve radius), and detecting their differences. So, the collected naturalistic data were used to compare design curve radii and critical 15th percentile path radii. In addition, stepwise linear regression was used to analyse the possible influence of speed, speed differential and geometric characteristics of the road on the driver's choice of vehicle trajectory. The percentage difference between design curve radius and critical 15th percentile vehicle path radius was selected for dependent variable, thus enabling analysis of horizontal curves with different curvatures:

$$\Delta R = \frac{R-R_{15}}{R} \cdot 100 \tag{2}$$

The following independent variables were considered in the regression procedure: 85th percentile curve speed $V_{85,c}$ (km/h), 85th percentile speed differential $\Delta_{85}V$ (km/h), curve radius R (m), deflection angle α (°), slope s (%), curve length L_c (m), previous and following tangent length L_{t1}, L_{t2} (m), previous and following spiral length L_{s1}, L_{s2} (m). Except the above geometric characteristics of the observed curve, speed differential was also considered as an independent variable, representing driver behaviour in the transition between adjacent alignment elements.

At first, the effect of each independent variable was examined by their individual regressions and this preliminary analysis showed that the influence of geometric features of adjacent elements on ΔR is negligible. Therefore, the regression plots for the following variables are shown in Fig. 3.: curve radius (Fig. 3. a), curve length (Fig. 3. b), deflection angle (Fig. 3. c), curve operating speed (Fig. 3. d) and speed differential – north-south direction (Fig. 3. e). As can be seen from Fig. 3., the selected independent variables do not explain much of the ΔR variability, while L_c has the greatest influence. However, the conducted analysis could provide some regularities about drivers' behaviour. In general, drivers tend to perform path radius sharper than the curve radius, regardless of the driving direction, with critical vehicle path radius on average 12% smaller than curve radius. Similar results were found in some previous research [8], [9].

The largest range of vehicle path radii values was observed on sharp curves (Fig. 3. a). With the intention of more comfortable ride, in sharp and short curves with small deflection angles, many drivers "cut the curve" performing a path radius greater than the curve radius (Fig. 3. a, b and c). Such manoeuvres could partially be explained by limited sight distance that is characteristic of sharp curves. At the same time, the largest positive values of ΔR (of even 25%) were observed in curves with radii less than about 150 m. Since these are small-radii curves, they are associated with lower operating speeds, regardless of the curve length and deflection angle. In order to achieve more comfortably ride in sharp curves, drivers often underestimate actual curvature. Somewhere along the curve the mistake is realized and is usually corrected by abrupt turn of the steering wheel resulting in path radius much smaller than design curve radius. Negotiating curve with smaller path radius results in an increase of centripetal acceleration and thus the side friction demand. In poor pavement conditions and with inappropriate speed, such sudden turns of the steering wheel can result in vehicle instability and can potentially lead to single-vehicle accident. The greatest path radii inconsistencies (highest values of ΔR) were observed in sharp curves (R < 150 m), which indicates an increased risk of skidding, and can be associated with an increased accident rate in small-radii curves (Fig. 1.).

After individual regressions, the stepwise linear regression was performed. Since several independent variables have been analysed, it is necessary to determine which of them are the most significant. Individual regressions (Fig. 3. a, b and c) indicate that for some independent variables (R, L_c and α), the approximation withe logarithmic form would be better than the linear one. Therefore, the logarithmic values of these variables were used when performing multiple regression analysis. As expected with respect to the individual regression diagrams, the stepwise regression resulted with only the curve length as independent variable:

$$\Delta R = 8.41 \cdot \ln L_c - 27.09 \tag{3}$$

Other selected independent variables were excluded from the model because they were found not to contribute significantly to the overall model. The developed model shows that the percentage difference between the curve radius and the critical vehicle path radius increases with increasing curve length. However, although there is a certain dependence between the percentage of the radii difference and curve length, with adjusted coefficient of determination of $\bar{R}^2$ = 0.348, this model could be considered weak and unreliable.

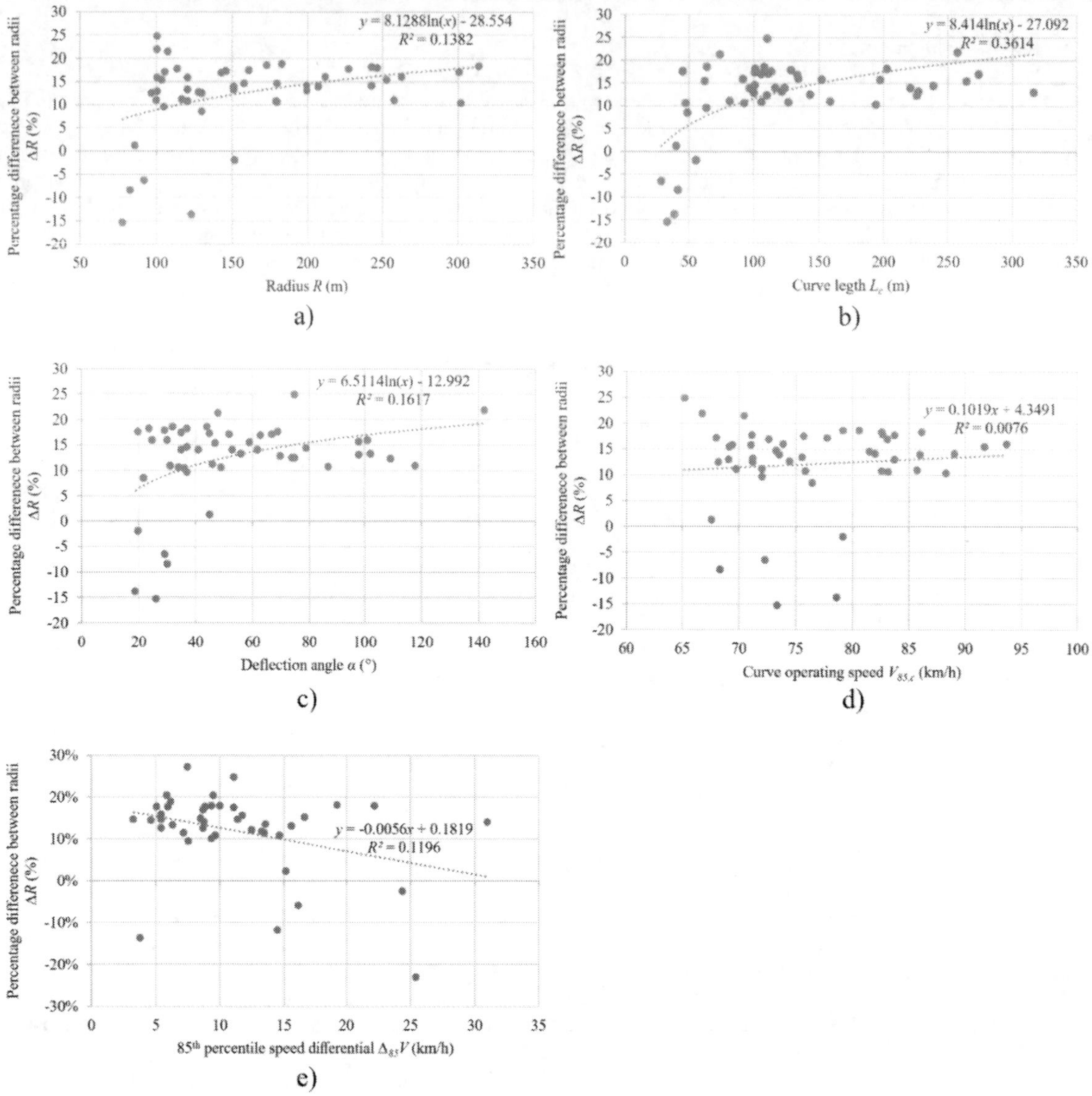

Fig. 3. Individual scatter plots of percentage difference between radii versus: a) curve radius; b) curve length; c) deflection angle; d) curve operating speed; e) speed differential (north-south direction) [10]

4. Conclusions

Traffic accident statistics, according to which horizontal curves (especially sharp curves) are high-risk locations, present a fundamental motivation for this research. Accidents usually occur due to lack of design consistency. Most of previous design consistency and road safety studies were based on operating speed whereas only a few of them were analysing the consistency in curves in terms of vehicle path radii. Under this scope, the main objective of this paper was to analyse the conformance of the road geometry with the drivers' behaviour, by comparing curve radii and critical 15th percentile vehicle path radii. Continuous data recording methodology, based on GPS device, was used to collect the driver behaviour data on a segment of the two-lane state road. Differences between radii were observed and it was attempted to find the possible causes of these deviations. Using stepwise regression analysis, it was analysed whether the speed, speed differential and geometric characteristics of the road affect the driver's choice of path radius. Although the performed analysis did not result in a statistically significant model, it was found that drivers generally track vehicle path in sharper curvature than that of road alignment, regardless of driving direction. The vehicle path radii were on average 12% smaller than curve radius. Additionally, higher differences in curvature were observed on sharper curves, whereas smaller radius of vehicle path causes greater centripetal acceleration, i.e., increased side friction demand. Therefore, in poor road conditions and speed not adjusted to the curve geometry, the loss of lateral vehicle stability can easily occur in sharp curves and potentially lead to traffic accident.

Finally, the most important conclusion of the present paper is the fact that the cause of single-vehicle accidents on curves, which are common and usually occur due to loss of vehicle control or stability as a result of exceeding the available values of side friction, cannot be attributed only to inappropriate speed. The cause can also be found in the drivers' behaviour in terms of vehicle trajectory or drivers' inability to maintain lane position. The results of this paper can be used to improve design consistency measure regarding to the vehicle stability. Since the existing vehicle stability procedure is based on the assumption that the driver's path radii are equal to the curve radius, it is logical to use the value of vehicle path radius when calculating side friction demand, to more realistically predict lateral vehicle inconsistency. Considering that the critical vehicle path radius 12% lower than curve radius was observed in this research, it is proposed to use this value when calculating the margin of safety. Smaller values of curve radii would result in lower margins of safety and thus inconsistencies would be much easier detected and eliminated. Therefore, horizontal curves of the final road design would represent the satisfactory level of consistency even in real conditions for the majority of drivers. Designing horizontal curves that are consistent with driver behaviour is a base to achieving safer roads.

Limitations of the present research relate to the size of the driver sample that cannot be considered sufficient for reaching wider applicable results. In addition, it should be pointed out that the analysed section was designed several decades ago when, according to the former guidelines, the use of spirals was not obligatory. Therefore, in most of the cases, there is no spiral transition between tangent and circular curve or there is a short one (mostly lengths from 20 to 30 m). This absence of spirals or the application of very short ones might have had the negative impact on the driver's choice of path radius because the driver failed to adjust both, speed and path radius. So, it would be interesting to analyse the driver's behaviour on sections with spiral transitions of sufficient lengths. In addition, future work may focus on collecting more data, covering a larger sample of drivers and test drives from several different road sections (i.e., from different areas). More representative sample of drivers and larger number of road sections would result in a more detailed and comprehensive analysis of the design consistency in horizontal curves.

5. Acknowledgments

This research was partially supported through project KK.01.1.1.02.0027, co-financed by the Croatian Government and the European Union through the European Regional Development Fund – the Competitiveness and Cohesion Operational Programme.

6. References

[1] Bartunek, M., Moravcik, O. & Schreiber, P. (2011). Safety distance for vehicles through simulation. In Annals of DAAAM for 2011 & Proceedings of the 22nd International DAAAM Symposium, Volume 22, No. 1, ISSN 1726-9679, ISBN 978-3-901509-83-4, Editor B. Katalinic, Published by DAAAM International, Vienna, Austria.

[2] PIARC World Road Association (2003). Road Safety Manual.

[3] Lamm, R., Psarianos, B. & Mailaender, T. (1999). Highway design and traffic safety engineering handbook, McGraw-Hill, ISBN 0-07-038295-6, New York, United States.

[4] Said, D., Abd El Halim, A. O., Hassan, Y. (2009). Methodology for driver behaviour data collection and analysis for integration in geometric design of highways, Proceedings of the 4th International Symposium on Highway Geometric Design, June 2.-5. 2009; Valencia, Spain.

[5] Fitzsimmons, E. J. et al. (2013). Determining vehicle operating speed and lateral position along horizontal curves using linear mixed-effects models, Traffic Injury Prevention, 14(3): 309–321. DOI: 10.1080/15389588.2012.701356.

[6] Wong, Y. D., Nicholson, A. (1993). Speed and lateral placement on horizontal curves, Road and Transport Research, 2(1): 74–87.

[7] Wong, Y. D., Nicholson, A. (1992). Driver behaviour at horizontal curves: risk compensation and the margin of safety, Accident Analysis and Prevention, 24(4): 425–436. DOI: 10.1016/0001-4575(92)90053-L.

[8] Glennon, J. C., Weaver, G. D. (1972). Highway curve design for safe vehicle operations, Highway Research Record, 390; 15-26.

[9] Glennon, J. C., Neuman, T. R., Leisch, J. E. (1983). Safety and operational considerations for design of rural highway curves, Report FHWA-RD-86-035, Federal Highway Administration, U. S. Department of Transportation, Washington, DC, 1983.

[10] Maljković, B. (2021). Unapređenje kriterija konzistencije u horizontalnim krivinama vangradskih dvotračnih cesta uz primjenu bicikl-modela vozila te realnih parametara ponašanja vozača, dokstorska disertacija (Evaluation of design consistency on horizontal curves for two-lane state roads using bicycle-model and realistic parameters of driver behaviour), Ph.D. Dissertation, Department of Transportation Engineering, University of Split, Faculty of Civil Engineering, Architecture and Geodesy, Split, Croatia.

[11] Carlson, P. J., Burris, M., Black, K., Rose, E. R. (2005). Comparison of radius-estimating techniques for horizontal curves, Transportation Research Record, 1918(1): 76–83. DOI: 10.3141/1918-10.

[12] Nicholson, A. (1998). Superelevation, side friction, and roadway consistency, Journal of Transportation Engineering, 124(5): 411–418. DOI: 10.1061/(ASCE)0733-947X(1998)124:5(411).

DOI: 10.2507/32nd.daaam.proceedings.101

COMPARING HDFS – GREENPLUM DATA LOADING OPTIONS

Alexander Suleykin, Anna Bobkova, Peter Panfilov & Ilya Chumakov

This Publication has to be referred as: Suleykin, A[lexander]; Bobkova, A[nna]; Panfilov, P[eter] & Chumakov, I[lya] (2021). Comparing HDFS – Greenplum Data Loading Options, Proceedings of the 32nd DAAAM International Symposium, pp.0724-0731, B. Katalinic (Ed.), Published by DAAAM International, ISBN 978-3-902734-33-4, ISSN 1726-9679, Vienna, Austria
DOI: 10.2507/32nd.daaam.proceedings.101

Abstract

In the last five years, many companies around the world have been successfully implemented Apache Hadoop as a main Data Lake storage for all data presented in the organization. At the same time, the adoption of other Open-Source technologies has been also increasing over years, such as classical MPP-based systems for Analytical workloads. Thus, the issue of efficient and fast data integration of Apache Hadoop and other organizational data storage systems becomes highly important for enterprises, where business and decision makers require the minimum delay of heterogeneous data exchange between Hadoop and other storages. In this paper, we compare different options for loading data from Apache Hadoop, representing the Data Lake of organization, into Open-Source MPP Greenplum database with the role of classical data warehouse for analytical workloads, and choose the best one. Also, we identify potential risks of using different data loading methods.

Keywords: Big Data; Hadoop Distributed File System; Greenplum; Massively Parallel Processing

1. Introduction

1.1. Hadoop HDFS as a main Data Lake storage

As it is defined in the IBM Redbook, Data Lake represents a set of centralized repositories, that contains vast amounts of structured and unstructured data, described in metadata, and organized into identifiable datasets [1]. One of the well-known distributed file systems for projects with a large amount of data is Hadoop Distributed File System (HDFS) [2], [3]. It is needed for storing large files and allows streaming access to data that are distributed in blocks of computing cluster that may consist of various hardware. HDFS is written in Java for Hadoop framework. There are factors that make HDFS a number one choice for data storage of an organization. HDFS can improve data processing and integrate different systems into a single Data Lake. Another factor is the fact that Hadoop is 10 to 100 times less expensive to deploy than conventional data warehousing. Today, companies in all industries starting to extract and place data for analytics into a single Hadoop-based repository, e.g., Google, Yahoo, Facebook, Netflix [4]. As is known from a study conducted by BARC (Business Application Research Center) in 2016, the main benefits of Hadoop are better heterogeneous data analysis – 59%, increased customer understanding - 53% and greater flexibility – 47% [5].

HDFS as the main Data Lake storage is the choice for most organizations with large amounts of unstructured data. Now, most of the medical data about patients, their anamnesis, the results of clinical trials are presented in an unstructured form, and their transformation and reduction to a structure takes a lot of time from medical organizations. There are several studies on this topic that support the point that Hadoop is a suitable solution for analytics of medical data [6], [7], [8]. As for other industries, namely in the Russian market, most often companies in the IT market also give their preference to building Data Lake with data storage in Hadoop, e.g., Kaspersky Lab, Aliexpress.ru, OZON, X5 Retail Group.

1.2. MPP databases

Massively parallel processing (MPP) databases in comparison to traditional database solutions make data reading from nodes and data processing on average much faster. The distinctive feature of MPP database architecture is physical separation of the memory of compute nodes combined into a cluster. Each node of MPP database of a cluster works only with its own hard disks by parallelizing data reading and writing operations. After each node finishes its computations and sort the results, it sends its portion of data to other nodes for getting data from them as well. During operation of receiving data each node selects only necessary records that are needed for computations. Other data is filtered out and does not take up space in the RAM. Commercial data warehouses are also in demand among MPP databases, such as Teradata, Oracle Exadata, Vertica, Netezza, etc. Since the comparison of databases is not in the scope of this paper, the Greenplum MPP-database will be considered in more detail. Greenplum is an MPP analytics database that adopts a 'shared-nothing' architecture with multiple cooperating processors. Greenplum manages the storage and processing of large amounts of data by balancing the load across multiple servers to create an array of individual databases that work together to present a single image of the database [9].

Database management system Greenplum is one of MPP-databases which is based on PostgreSQL for managing large-scale analytical data warehouses. It is considered that Greenplum cluster is a number of instances of PostgreSQL object-relational database with modification that work together as one Database management system. The programming code of PostgreSQL for Greenplum was modified for working with parallel processing, especially for parallelization of SQL queries. Greenplum uses ideas of the concept 'Shared nothing' when cluster nodes that interacts with each other for performing computing operations do not share resources. In this case, each of them has its own memory, operating system, and hard disk. It allows efficiency to parallel the load on performing analytical queries to multi-terabyte data stores.

Thus, having an idea of what Hadoop HDFS and Greenplum are, we can outline the purpose of this study. Now, there are a lot of tools for loading data. The paper presents the results of experiments with three instruments, such as data loading with python, data loading with Greenplum-Spark Connector, and data loading with PXF HDFS Connector. The result of this work is considered the most optimal tool for loading data from HDFS to Greenplum.

2. Integrating Hadoop HDFS and Greenplum

Greenplum flexibly complements Hadoop. It does not only allow using HDFS for data storage but also provides extensive opportunities for separate storage of partitions. For instance, some partitions can be stored on SSD for quick access. Other partitions with rarely used data on HDD because it is cheaper. Third partitions can be stored on HDFS. One of the features of the Greenplum DBMS is the usage of HDFS for creating external tables which avoids significant data loading time [13], [14]. The database includes the ability to provide gphdfs protocol privileges to the owner of external tables that access files on HDFS. Moreover, there is a software-based Hadoop connector that provides high-performance parallel import and export of compressed and uncompressed data from Hadoop clusters.

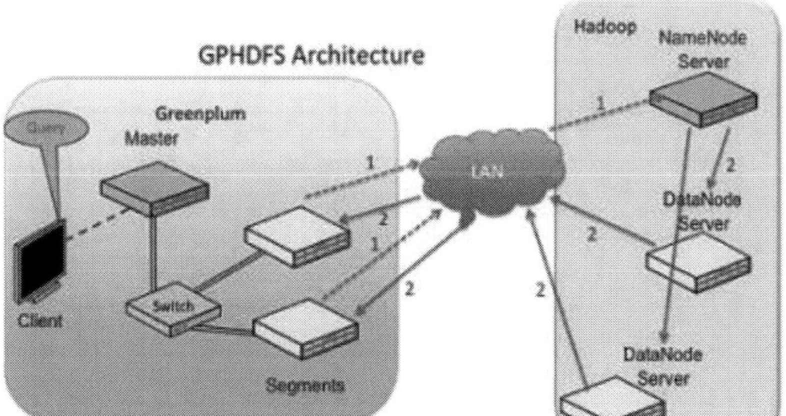

Fig. 1. Process of transferring data files from a Hadoop cluster to the Greenplum database

Custom format data in Hadoop can be converted to GPDB format using MapReduce and then imported into it for additional optimization of resource consumption during loading. This provides a much more efficient and flexible data exchange between Hadoop and the Greenplum DBMS [10], [11]. Greenplum supports parallel processing of SQL and MapReduce with data volumes ranging from hundreds of gigabytes, tens to hundreds of terabytes, to several petabytes. Fig. 1 illustrates the process of transferring data files from a Hadoop cluster to the Greenplum database upon reading request initiated by it. Reading request is sent to NameNode that controls access to the requested file data blocks. NameNode sends DataNode instructions to provide blocks of files to the Greenplum database. Greenplum has the framework PXF that allows exchanging data with third-party heterogeneous systems, including HDFS. The interaction of PXF with HDFS is explained in Fig. 2. Greenplum database is represented by a master server (MasterHost) and several segment servers (Segment Hosts). PXF framework consists of the Greenplum database protocol, client library for integrated data stores and Java service.

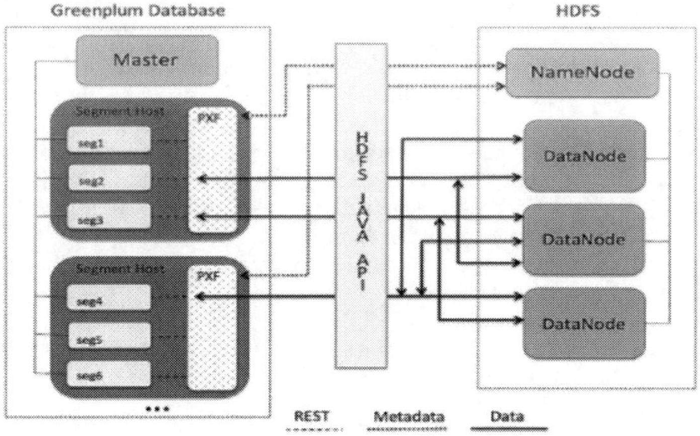

Fig. 2. The interaction of PXF with HDFS

PXF framework provides connectors for Hadoop, Hive, and HBase datastores that define profiles that support Binary and TEXT, Avro, JSON, RCFile, Parquet, SequenceFile and ORC. PXF connectors include the clients of the datastores in use on each segment host of the Greenplum database [6]. PXF extension implements a protocol called pxf is used to create an external table that references data in the integrated storage. The command to create an external table is used after the PXF extension is registered in the databases to access external data and privileges for the pxf protocol are assigned to those roles that need access. When executing a request for data in HDFS, the Master Host sends it to all segment servers. Each segment instance connects with a PXF running on its segment host and when PXF receives a request from segment instance it calls the Java HDFS API to retrieve the metadata information from the NameNode for the HDFS file and provides it to the segment instance. PXF is effective for integration with Greenplum since it excludes a network component during data exchange, has great flexibility and the ability to connect third-party systems with programming its own connector.

3. Experiments with Greenplum data loading

Name	Role	CPU, amount	RAM
demo-1	segment	6	32 Gb
demo-2	segment	6	32 Gb
demo-3	segment	6	32 Gb
demo-4	segment	6	32 Gb
demo-5	segment	6	32 Gb
demo-6	segment	6	32 Gb
demo-m	master	2	12 Gb
demo-s	master standby	2	12 Gb

Table 1. Experimental setup: Greenplum database configuration

In our experiments, we have tested three different data loading methods into Greenplum - using pure python, Apache Spark and PXF. Greenplum configuration in the test environment is shown in table 1: All hosts used SAS disks from shared storage. Machines did not change it configuration during all stages of testing.

3.1. Greenplum data loading with python

Load testing was carried out using a self-written python script. The dataset consists of an array of random strings 18 * 100,000 in size. The script was run with insert Single_insert (row insertion), Execute_many, Execute_batch, Execute values methods. To communicate with the database, the psycopg2 python library was used. The dataset was handled by Pandas and NumPy. Below the results are presented of different insert types with python libraries into Greenplum system (Fig. 3, table 2):

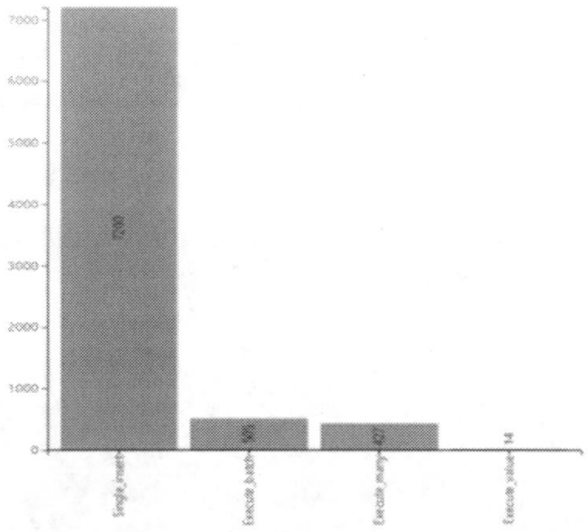

Fig. 3. Different Insert types in Greenplum using python, figure

Insert Type	Time, sec	Speed, rows/sec
Single Insert	7200	17
Execute Batch	505	198
Execute Many	427	234
Execute Value	14	7145

Table 2. Different Insert types in Greenplum using python, table

3.2. Greenplum data loading with Greenplum-Spark Connector

The source file in Apache Parquet format (with Snappy compression) includes 18 columns with numeric and text arbitrary data, all records are unique. The average row size of a test dataset in Greenplum is 933 bytes. The number of rows in the test dataset is presented in the corresponding column in the totals table. Based on the source file, a Spark DataFrame is formed with the number of sections corresponding to the number of Greenplum segments to achieve maximum parallelism (distribution keys also match). Filling of internal Greenplum tables is performed using the Greenplum-Spark Connector from the generated DataFrame. ZSTD compression is used for internal Greenplum tables.

Dataset Size, rows	Data Loading Time, sec	Deserialization Time, sec	Data Loading Speed, rows/sec
10 000 000	324	168	20325
20 000 000	900	396	15432
50 000 000	2100	1080	15723

Table 3. Apache Spark data loads, performance comparison

The test evaluation of cluster load was performed using Grafana – an open-source tool for metrics visualization. We measured the most important metrics when performed load tests which are the CPU load and RAM usage. Every node in Greenplum cluster is shown in different color. The following Figures show the CPU data load (Fig. 4 – 6) and RAM memory load (Fig. 7 – 9) visualizations with 10 mln., 20 mln., and 50 mln. rows inserts with Apache Spark Connector, respectively.

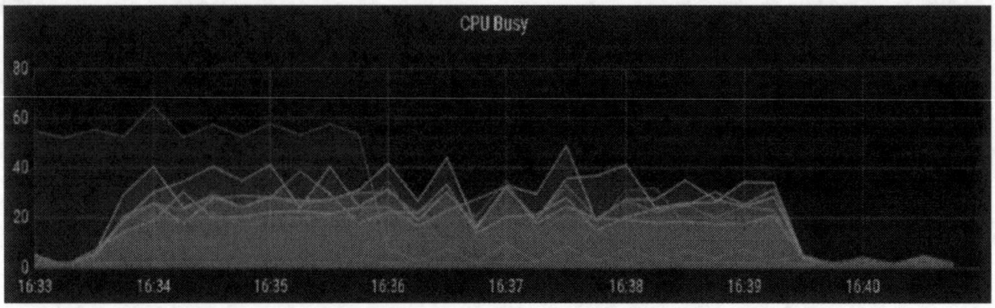

Fig. 4. CPU load with Apache Spark Connector and 10 000 000 records loading

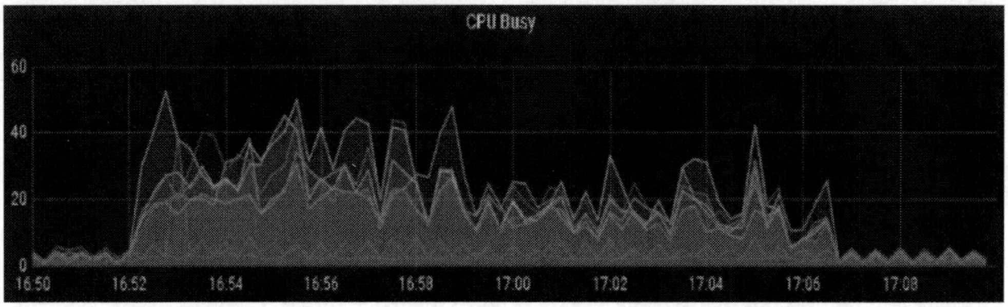

Fig. 5. CPU load with Apache Spark Connector and 20 000 000 records loading

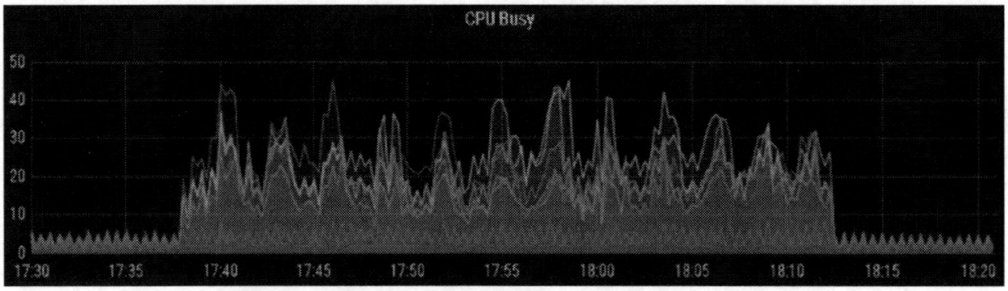

Fig. 6. CPU load with Apache Spark Connector and 50 000 000 records loading

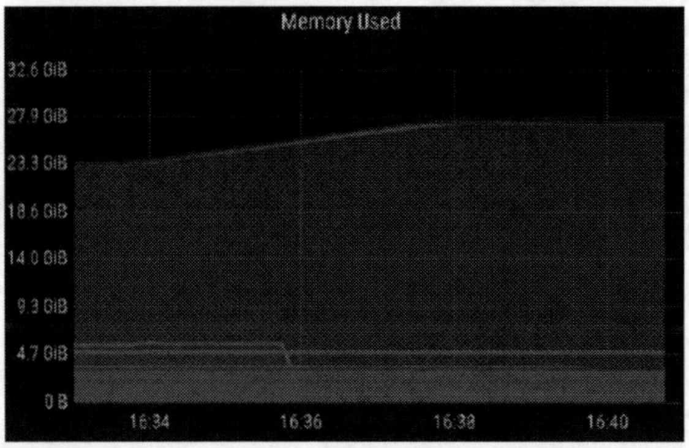

Fig. 7. RAM used with Apache Spark Connector and 10 000 000 records loading

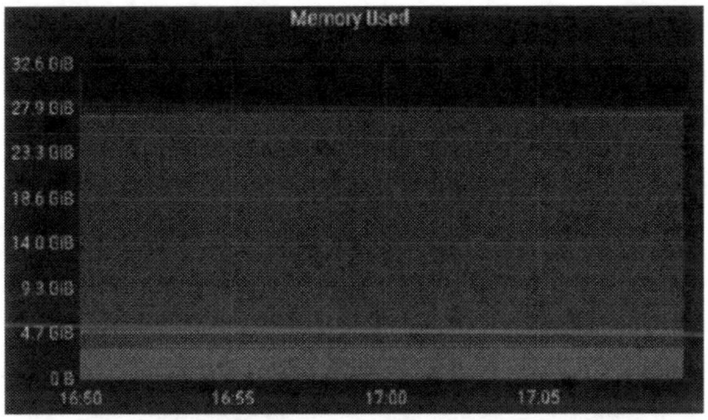

Fig. 8. RAM used with Apache Spark Connector and 20 000 000 records loading

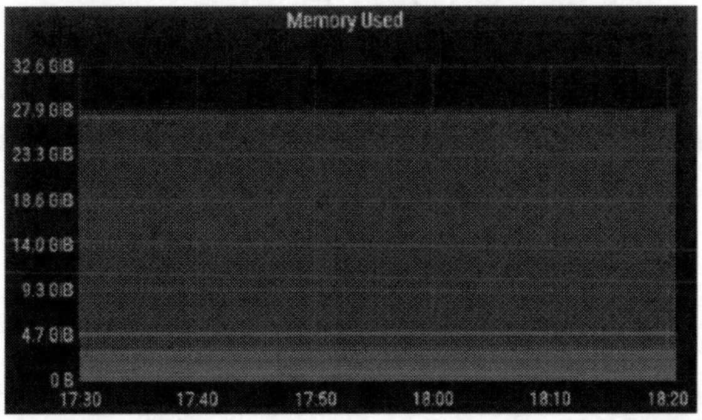

Fig. 9. RAM used with Apache Spark Connector and 50 000 000 records loading

3.3. Greenplum data loading with PXF HDFS Connector

The structure of the dataset is the same as using Greenplum-Spark Connector test set. Based on the source file, the external Greenplum tables are populated using PXF HDFS Connector, and the data is transferred from them to the internal Greenplum tables. For internal tables, ZSTD compression is used. Table 4 depicts the results for different number of records with PXF Connector:

Number of Rows in Dataset, rows	**Loading Time, sec**	**Loading Speed, rows/sec**
10 000 000	144	69444
20 000 000	279	71685
50 000 000	759	65876

Table 4. PXF data loadings, performance comparison

The following Figures show the CPU data load (Fig. 10 and 11) and RAM memory load (Fig. 12 and 13) visualizations with 10 mln. and 20 mln. rows inserts with PXF Connector, respectively.

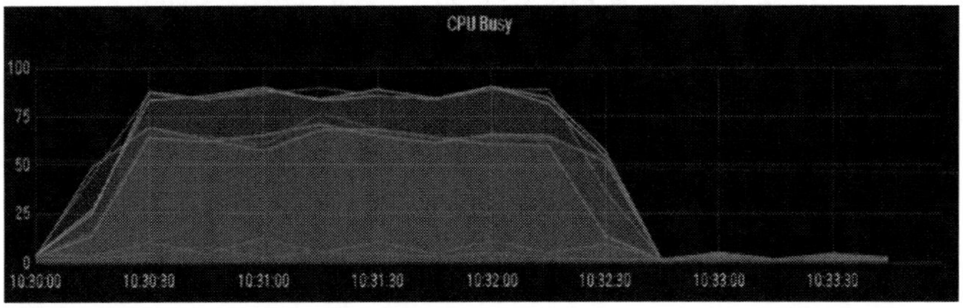

Fig. 10. CPU load with PXF Connector and 10 000 000 records loading

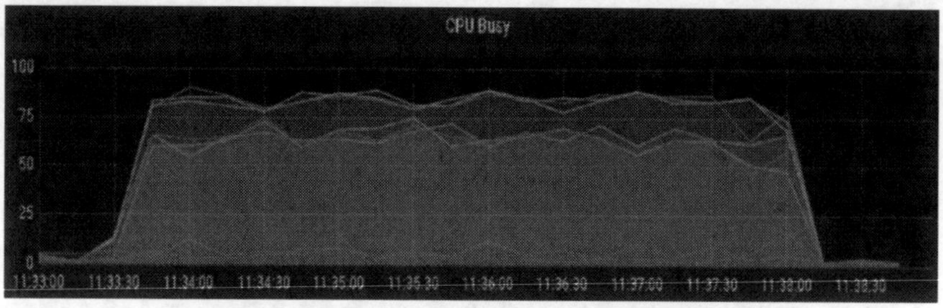

Fig. 11. CPU load with PXF Connector and 20 000 000 records loading

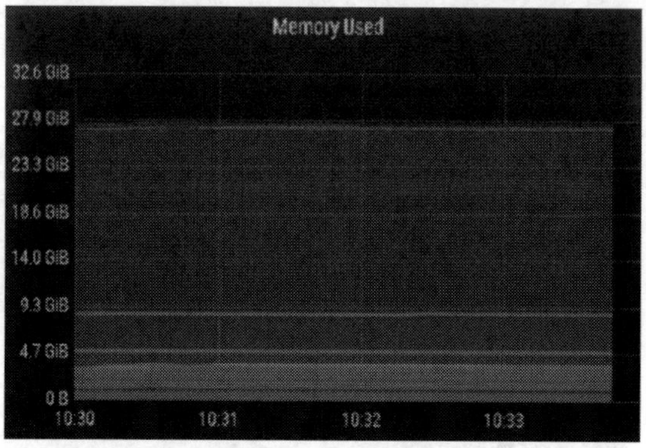

Fig. 12. RAM used with PXF Connector and 10 000 000 records loading

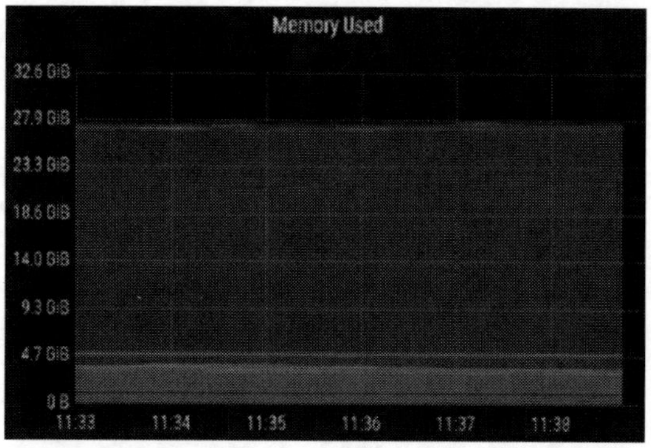

Fig. 13. RAM used with PXF Connector and 20 000 000 records loading

With similar settings, loading data through PXF is faster, including due to higher equipment utilization. No degradation of productivity was observed with increasing volumes. When using the PXF protocol, the load is evenly distributed among the Greenplum segment nodes. Master node usage is minimal. Combined with the high speed of the PXF protocol and the absence of degradation, this option is suitable for loading large amounts of data into Greenplum.

4. Discussion

The use of direct inserts (SQL inserts) through the Master node is immediately excluded, since has an obvious problem loading the master, which is used to parse requests and check permissions. Greenplum-Spark Connector provides loading that bypass the master and possibly has the potential for horizontal scaling. Formally, this is enough to complete the task, but there are nuances. Among the obvious drawbacks, it has a mandatory step of loading data into a Spark DataFrame, which creates additional overhead. To download data through this connector, the Spark competency (Java / Scala) is required, which increases the TCO of the solution and entails additional difficulties in replicating updates. PXF HDFS Connector also provides data loading that bypass the master node, that allows for the solution to scale horizontally. When installed, it seamlessly integrates with the Hadoop cluster. During data access, automatic load balancing occurs on the Greenplum side, which allows for abstracting from the Hadoop cluster configuration. To work with data through this connector, it is enough to know only the declarative part of SQL, and no knowledge of other languages is required.

5. Conclusion

In this paper, we have presented the practical study of the usage, architectural features, and main benefits of Apache Hadoop HDFS-based data infrastructure in many organizations. As a popular alternative solution and enhancement to data infrastructure, we have also discussed architecture, implementations, and advantages of the Open-Source MPP-based database Greenplum usage in organizations, as well as different ways of how data exchanges between HDFS and Greenplum can be arranged. Finally, we have tested different HDFS-Greenplum integration options based on Master node row insert, Spark-Greenplum connector and PXF usage. As a result of test implementation, the best option for Hadoop-Greenplum data exchange has been proposed and different risks for the development and production exploitation were considered. Returning to the question of which tool is the most optimal for loading data from HDFS to Greenplum, our experiments have shown that it is PXF HDFS Connector. But it is important to note that with the correct application of any of these tools, the desired result can be achieved. Therefore, as one of the ways for further research, the settings of these tools for solving various kinds of tasks can be analyzed.

6. Acknowledgements

The reported experimental research was partially supported by RFBR grant No 20-07-00958 and the HSE University's Project Group Competition grant.

7. References

[1] Chessell, M., Scheepers, F., Strelchuk, M., van der Starre, R., Dobrin, S., Hernandez, D. (2018). The Journey Continues: From Data Lake to Data-Driven Organization. IBM Corp. Available from: https://www.redbooks.ibm.com/redpapers/pdfs/redp5486.pdf. Accessed: 2021-10-12

[2] Beakta, R. (2015). Big Data and Hadoop: A Review Paper. International Journal of Computer Science & Information te.2. Vol. 2, Issue 2. Available from: https://www.researchgate.net/publication/281403776_Big_Data_And_Hadoop_A_Review_Paper/, Accessed: 2021-10-12

[3] Depardon, B., Le Mahec, G., Séguin, C. (2013). Analysis of Six Distributed File Systems. [Research Report] 2013, pp.44. ffhal-00789086. Available from: https://hal.inria.fr/hal-00789086/file/a_survey_of_dfs.pdf, Accessed: 2021-10-12

[4] Stein, B., Morrison, A. (2014). The enterprise data lake: Better integration and deeper analytics, Available from: http://www.smallake.kr/wp-content/uploads/2017/03/20170313_074222.pdf, Accessed: 2021-10-12

[5] Grosser, T., Bloemen, J., Mack, M. & Vitsenko, J. (2016). Hadoop and Data Lakes: Use Cases, Benefits and Limitations. BARC – Business Application Research Center, a CXP Group Company. Available from: http://barc-research.com/research/hadoop-and-data-lakes/, Accessed: 2021-10-12

[6] Rangarajan, S., Liu, H., Wang, H., Wang, Ch.-L. (2018). Scalable Architecture for Personalized Healthcare Service Recommendation Using Big Data Lake, Available from: https://link.springer.com/chapter/10.1007/978-3-319-76587-7_5, Accessed: 2021-10-12

[7] McPadden, J., Durant, T.J., Bunch, D.R., Coppi, A., Price, N., Rodgerson, K., Torre, Jr C.J., Byron, W., Hsiao, A.L., Krumholz, H.M., Schulz, W.L. (2019). Health Care and Precision Medicine Research: Analysis of a Scalable Data Science Platform, Available from: https://www.jmir.org/2019/4/e13043/, Accessed: 2021-10-12

[8] Soroushmehr, S.M.Reza, Najarian, K. (2016). Transforming big data into computational models for personalized medicine and health care, Available from: https://www.ncbi.nlm.nih.gov/pmc/articles/PMC5067150/, Accessed: 2021-10-12

[9] Albertini, O., Bhargov, D., Denissov, A., Guerrero, F., Jayaram, N., Kak, N., Khanna, E., Kislal, O., Kumar, A., McQuillan, F., Owen, L., Raghavan, V., Valdano, D., Zhang, Y. (2020). Image Classification in Greenplum Database Using Deep Learning, Available from: http://greenplum.org.s3.amazonaws.com/wp-content/uploads/2020/05/12170349/Image-Classification-in-Greenplum.pdf, Accessed: 2021-10-12

[10] Bani, F.C.D., Suharjito, Diana, Girsang, A.S. (2018). Implementation of Database Massively Parallel Processing System to Build Scalability on Process Data Warehouse. Available from: https://www.sciencedirect.com/science/article/pii/S1877050918314376, Accessed: 2021-10-12

[11] Greenplum® Database 4.1 Administrator Guide, [online] Available from: media.gpadmin.me/wp-content/uploads/2011/ 05/GP-4100-AdminGuide.pdf, Accessed: 2021-10-12

[12] Kasibhotla, D. (2012). Greenplum and Hadoop HDFS integration, Available from: https://dwarehouse.wordpress.com/2012/ 10/10/greenplum-and-hadoop-hdfs-integration/, Accessed: 2021-10-12

[13] New Functionality in Greenplum Database 4.2. Welcome to Greenplum Database 4.2. Updated: Nov. 23. 2011. Available from: http:// media.gpadmin.me/wp-content/uploads/2012/11/ GPDB_4200_README.pdf, Accessed: 2021-10-12

[14] Using PXF to Read and Write External Data. Greenplum Platform Extension Framework (PXF). Available from: https://gpdb.docs. pivotal. io/550/pxf/using_pxf.html, Accessed: 2021-10-12

DOI: 10.2507/32nd.daaam.proceedings.102

DIGITAL TRANSFORMATION, SHARING ECONOMY AND EFFECTS ON SOCIETY

Cvijeta Djevojic, Mirta Brajak & Iris Mihajlovic

This Publication has to be referred as: Djevojic, C[vijeta]; Brajak, M[irta] & Mihajlovic, I[ris] (2021). Digital Transformation, Sharing Economy and Effects on Society, Proceedings of the 32nd DAAAM International Symposium, pp.0732-0739, B. Katalinic (Ed.), Published by DAAAM International, ISBN 978-3-902734-33-4, ISSN 1726-9679, Vienna, Austria

DOI: 10.2507/32nd.daaam.proceedings.102

Abstract

This paper examines the importance of innovative tools that help companies overcoming the uncertainties. Assuring an optimal platform designed under a single architecture should meet the interests of both market poles as common, based on technology that creates value and care society. First part of paper has an emphasis on terminology definition of digital transformation as model, business process and prerequisites that define its implementation as necessary as it is. Second part of the paper speaks in favour of sharing economy and its effects on society. SE is based on 'technology' and how it facilitates new social and economic configurations. In some discussions this technology is an 'algorithm', while in others it is a 'platform' or simply 'technology'. Furthermore, there is not always agreement on these terms, as researchers have different definitions of an 'algorithm' defining the technology as a platform, but only concern themselves with one algorithmic process of that platform. In other cases, technological changes are reduced to quantifiable trends, such as increasing computational power, speed of matchmaking, or the omnipresence of personal devices. Results speak in favour of business framework that should meet "soft socially fulfilling responsibilities" as market insists today. In the long run, they will certainly guarantee success and profitability.

Keywords: Digital transformation; Sharing economy; Tourism; Communication; Society.

1. Introduction

Finding a way out of the uncertainty that determines the business environment through the practice, past, present and future conditions, still seems being as an ambitious task. Historically, looking at technological progress based on creative destruction that is in synergy with scientific knowledge from other fields at the same time, the great efforts has been made to discover a model that could guarantee an existence and progress. Evolution and transformation are the only constants in the natural environment, society and economy, driven by the search for efficiency, not necessarily an excellence. Consequently, today's new measurable indicators that put pressure on companies should be found in the challenges of dynamic market requests. Innovative tools that help companies overcoming the uncertainties are in the central research interest of scientists, and the authorities mentioned in this paper.

Analysing efficiencies of algorithms, we found out that the most energies are paid for an idea of ensuring accessibility, flexibility, personalization, and convergence. Convincing that new platforms designed under a single architecture meet the interests of both market poles as common, the technology creates value. Efforts are being made, and are going to be made to search for semiotic values of antonyms "as physical, analogous, mappings" that will transform a standardized operating model into a flexible one with cyber, digital, sharing capabilities that will be focused or redirected on understanding of business to "soft social" fulfilling responsibilities "as the market demands today, which in the long run will surely guarantee success and profitability by economic standards.

2. Digital transformation

Digital business involves the use of digital technologies to conduct business processes. Digital technologies include the use of computers, mobile devices and other technology that serves businesses in the modernization of business processes in order to achieve better business results. The main feature of digital technologies is their accessibility [1], which greatly facilitates the transition to new platforms. With the development of internet marketing, new opportunities have opened up for customers, who were able to receive and view information on their smartphones and tablets in real time. Social networks play a significant role in transforming the structure of social relationships through digital technologies, both in the consumer and business worlds [2].

Digital transformation is the adaptation of the way of doing business to a new, digital environment, using digital technologies, with all the benefits that e-business brings. Digital transformation should affect the improvement of quality of life, encourage economic growth and environmental protection, which are social values that business models must have, in addition to commercial values. Prerequisites for achieving digital transformation and spreading innovation in services are strengthening cooperation between government, companies and society as a whole, as well as harmonizing policies in their interaction [3].

Recently, the digital transformation has been the subject of research by many experts, both theorists and practitioners, who have published a large number of papers and articles in which they express their, often different, views on the subject. Thus, it is difficult to draw a general definition of digital transformation from the existing literature. To better describe what digital transformation is, what the drivers are, what the characteristics are, and what areas it affects, Morakanyane, Grace, and O'Reilly [4] put existing concepts into a matrix. Each of the authors of the concept of digital transformation wants to give their contribution in defining this phenomenon. Therefore, there is not just one definition of digital transformation. To obtain a general definition of digital transformation, it is necessary to look for keywords and different models [4]. Generally speaking, digital transformation is the use of digital technologies with the aim of increasing value and achieving a competitive advantage in business processes.

IT strategy is important at all levels [2] and needs to be constantly reviewed. Businesses, using IT technology, gain a competitive advantage by reducing costs. As a result, a new hierarchy of market priorities emerges, technologies, platforms and applications gain in importance. The new time brings with it the competition of companies in which the main factors of competition are time and information used for profit. Through these new customer-customer models using new technologies facilitates economic and social configurations.

The foundation of the digital transformation is digital technology. In recent years, digital technologies have become more accessible to a wider circle of people and increasingly easy to use, so mobile phones and social networks have become the main means of communication between private and business entities. Wider availability prevents companies from using digital technology to gain a competitive advantage, but companies can take advantage of the use of digital technologies in business to improve their image and create added value through faster and better communication with customers and other businesses [5]. Rapid changes in the way products are distributed to the market result in a new experience for all market participants, which conditions the transformation of the way of doing business by applying new technologies [4]. The definition and analysis of the process leads us into a broad field of problematics, giving us an insight into the development of organizational structure, organizational resources and organizational culture, using new concept and an innovative vision [6]. Priorities should be re-defined .

The implementation of digitalisation can take a long time due to the fact that it is a big and radical change, and those companies that accept the change faster and easier have a competitive advantage. The new industrial revolution is thought to have begun in the early 21st century and we call it Industry 4.0 [7], [8], [9], for which digitalization is the main driving force. The change in living conditions has been reflected in the use of new technologies that have facilitated business operations and contributed to trends in product placement. Such a new, innovative business creates a positive user experience that changes the production paradigm in the new industrial revolution [10], [11], [12]. A key factor in the fourth industrial revolution is digitalization, which will change the way we produce, market, and provide services, and the improvement resulting from these changes is reflected in lower costs and higher quality. By using digital technologies, communication between buyers and sellers will become more personalized, as well as marketing tools: pricing, distribution, customer management, etc. [13].

The result of digital transformation through the use of digital technologies are new business models, and together with new products, services and new markets, they form the digital economy [14]. The basic features of the digital economy is the use of information and communication technology in business. Existing physical resources are gaining new, digital resources in this way.

Informatization lies at the core of the digital economy, and all economic aspects are adapted to it, the way of thinking, organizational structure, organizational culture, production processes and analysis of business results. The rapid changes in the external environment caused by global digitalization are forcing companies to react quickly in adapting. The biggest transformations are happening in the areas of communication, interaction with partners, consumers, and production. As a result of these changes, there is great interest in researching the benefits that digitalisation has on economies and the ways in which it affects the functioning of economic operators. Therefore, it has been proven that the introduction of the latest information and computer technologies in all spheres of the economy is one of the priority directions of the state's internal policy [15], and research in this area is extremely important.

Thanks to digitalization, we measure success by the level of flexibility to respond to changes and adapt business to the new environment, and not as before by the size of capital and the length of existence of the company. According to a group of scientists [16], a modern enterprise represents a complex production and organizational system. Its components are constantly changing. Increased competition leads to increased complexity of production analysis, management, planning, production processes and internal and external relations with suppliers and intermediaries. [17], [18].

2.1. Digital Economy and Society Index (DESI)

The Digital Economy and Society Index (DESI) [19] has been published annually by the European Commission since 2014, and its task is to measure the progress of EU members in the digital economy and society, which collects relevant indicators and consists of 5 main policy areas with a total of 37 indicators:

Policy area	**Indicators**
1 Connectivity	Fixed broadband take-up, fixed broadband coverage, mobile broadband and broadband prices
2 Human capital	Internet user skills and advanced skills
3 Use of internet	Citizens' use of internet services and online transactions
4 Integration of digital technology	Business digitisation and e-commerce
5 Digital public services	e-Government

Table 1. DESI's five principal policy areas

The following chart shows the progress of EU member states looking at the overall level of digitalisation of the economy and society observed in the last 5 years. DESI progression over that time period is measured.

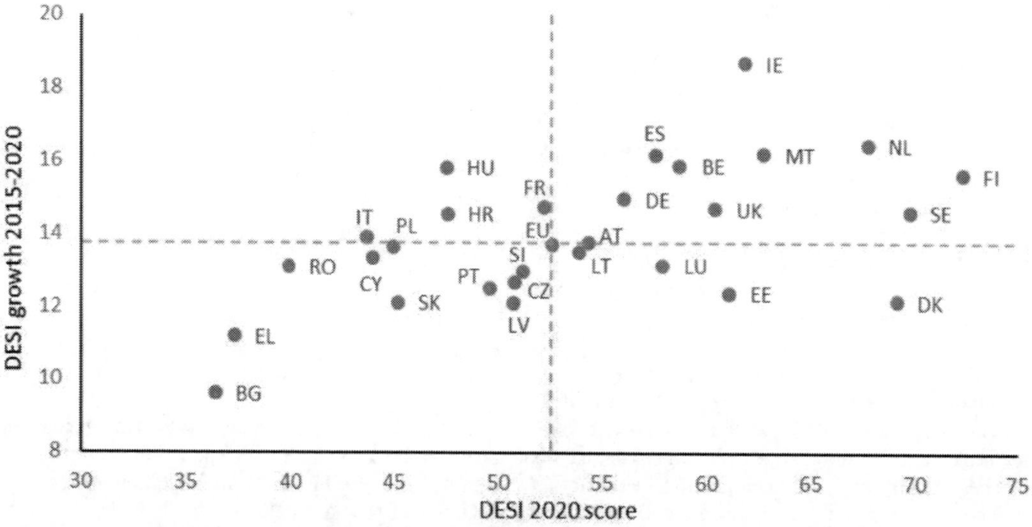

Fig. 1. Digital Economy and Society Index – Member States' progress, 2015-2020

Significant progress has been made in Ireland, the Netherlands, Malta and Spain, which are well above the EU average measured by the DESI score. Finland and Sweden are slightly above average in terms of the progress made in the last five years, along with Belgium and Germany, although they are among the leaders in overall performance in the digital world. Among the high-performing Member States in the overall DESI scale are Denmark, Estonia and Luxembourg, although they have shown relatively low progress in digitization in the last five years. Significantly, most countries that are below the EU average in terms of digitization have not made significant progress in the observed period.

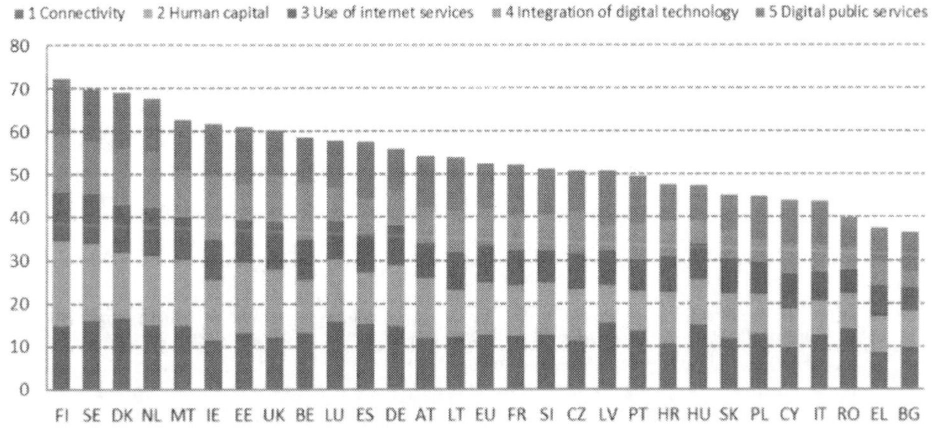

Fig. 2. Digital Economy and Society Index, 2020

Scandinavian countries, Finland, Sweden, Denmark, along with the Netherlands have the most advanced digital economies in the. Eastern European countries Bulgaria, Greece, Romania, and Italy have the lowest scores on the index.

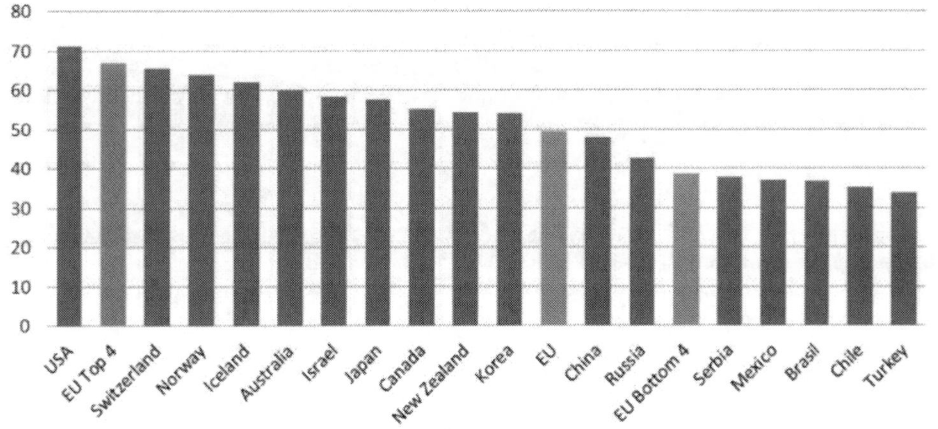

Fig. 3. International Digital Economy and Society Index, based on 2018 data

The digital performance of EU countries is compared with 17 non-EU countries. The International DESI (I-DESI) evaluates the performance of both the individual EU countries and the EU as a whole in comparison to some non-EU countries. The top four EU countries (Finland, Sweden, the Netherlands and Denmark) are just behind the US and considered the global leaders, with scores even higher than Korea and Japan. EU's average in digital performance is, however, lower than those three countries.

2.2. Main findings of DESI 2020 in the 5 digital areas

Broadband connectivity

Access to a fast and reliable broadband connection (including fixed and mobile connections) is crucial in light of the ongoing coronavirus pandemic in which key social and economic services are delivered online. Modern and robust digital infrastructure provides the necessary coverage for these services. During the crisis, networks faced significantly increased demand, while at the same time having to ensure the use of broadband-enabled services. Overall connectivity has improved, both in terms of supply and demand. In 2019, new generation access (NGA) coverage increased to 86% of households compared to 83% a year earlier, while very high capacity fixed networks (VHCN) are available to 44% of households.

Over a period of 5 years, more and more people are using broadband services of at least 100 Mbps, now with the current level of 26% of households, five times higher than five years ago. 4G networks cover almost the entire European population, but little progress has been made in allocating 5G spectrum. Only 17 Member States have already allocated spectrum in the pioneering 5G bands.

Human capital - digital skills

Digital skills are the backbone of the digital society. They allow people to use digital services and engage in online activities, especially when mobility is limited. Basic and advanced digital skills need to be strengthened in school curricula through academic offerings in EU Member States. Similarly, digital skills are essential for the effective use of distance learning solutions, including support for schools and families, with special attention to those at risk of social exclusion (e.g. by making hardware and training available). In 2019, the percentage of people who have at least basic digital skills reached 58% (compared to 55% in 2015). However, much of the EU population still lacks basic digital skills, although most jobs require such skills.

Internet use of citizens

Internet use by individuals increased during the pandemic. The general restriction has led to repetitive access to social media and entertainment platforms, as well as distance work, e-commerce, e-government services and distance learning. This trend already existed before the pandemic, as internet use continued to increase from year to year, with 85% of Europeans surfing the internet at least once a week (compared to 75% in 2014). The use of video calls has increased the most, from 49% of Internet users in 2018 to 60% in 2019. Internet banking and shopping are also more popular, used by 66% and 71% of Internet users, respectively. In contrast, only 11% of Europeans completed the 2019 online course.

Integration of digital technology by enterprises

Small and medium-sized enterprises (SMEs) that have not embedded digital solutions are a challenge to provide their staff with the opportunity to work from home. One of the many obstacles to the digitalization of SMEs is the digital knowledge gap caused by the low level of digital literacy among owners, managers and employees. Addressing these shortcomings will be vital to ensure a strong recovery. Businesses became more and more digital, and leading companies took over. 38.5% of large companies already relied on advanced cloud services, and 32.7% used big data analytics. However, the vast majority of SMEs reported not yet using these technologies, with only 17% using cloud services and only 12% using big data analytics.

Digital public services

There has been an upward trend in digital public services. In 2019, the quality and use of digital public services also increased. 67% of people who normally use the Internet who have submitted forms to their public administration reported that they now use Internet channels (compared to 57% in 2014), which shows the growing convenience of online procedures compared to paper ones.

3. Sharing econoomy in terms of Social Challenges

There is no generally accepted term or name for the sharing economy. Many different terms are still used in the scientific and professional literature for the term sharing economy [20]. The term sharing economies first appeared in 2008. [21] and can be said to be the most commonly used term today [22]. There are other terms that are being used, such as "peer-to-peer economy", "on-demand economy", "gig economy" [23] (Belk, 2014), "access-based consumption" [24] and "collaborative economy" [25]. The traditional business model of companies, in which production factors are used to offer products or services on the market, is increasingly being replaced by a new business model. The new business model relies on the use of digital and interconnected platforms to share physical resources among a large number of people - the focus is no longer on ownership, but on direct experience and mutual trust of service providers and users. The sharing economy in newer history is a term for a set of business models, exchanges and platforms [26]. "Sharing is an alternative to the private ownership that is emphasized in both marketplace exchange and gift giving. In sharing, two or more people may enjoy the benefits (or costs) that flow from possessing a thing" [23].

The development of the sharing economy was driven by the financial crisis that caused the economic collapse of 2008, as well as the rapid development of technology. Connecting and networking became easier, and information was shared more easily. The sharing economy evolves still, and is not a complete concept. It is oriented exclusively to the present, which means that it does not require a start from scratch or drastic change and revolution, but seeks what is possible and feasible in the current situation. It does not find its reliance on the support and support of the state and its institutions, which are expensive and slow in resolving the most important issues of today, such as unemployment. While using the latest technology, it allows people to create the content of a product or service themselves. The very important fact is that technology today is available to almost everyone. It thus creates new forms of demand and new forms of economy that did not exist before. This is why we are not looking at the sharing economy as a transformation of the existing, but something completely new [27]. The development of technology, ways of communication through social networks, as well as changes in consumer attitudes and behaviour, have led to the need for more efficient use of products and services. The benefits are lower prices, availability to the average consumer or user, with resource savings and better effects on the environment and society [28].

The main goal of this new business model – Shared economy and "share goods" which implies redistribution, mutualisation and shared mobility.[29]. The sharing economy connects service providers directly or through intermediaries. Users in the sharing economy can also be suppliers. Users search for information about different products and services and compare them according to various criteria such as price and rating added by other users. To use a certain service payment must be made. After using a certain service users can evaluate it themselves. The goal of sharing economy platforms in the role of intermediaries is to connect the user and the service provider and maintain the electronic market, enable the transactions and billing through the application, for which they get a commission. [21].

The sharing economy has flourished in tourism and travel, which are an indispensable part of human lives. The local population was given the opportunity to provide services to tourists. There are many examples, and we can single out the most popular, such as:

- Transfer services (Uber)
- Restaurant services (Eatwith),
- Accommodation services (Airbnb).
- Services for tour guides (Vayable) etc. [30]

The Airbnb was created in 2008 and soon became worldwide famous. It encompasses three main categories of participants: service providers, service users and intermediaries. To become what it is today and to make this work, Airbnb needed to cope with few key issues: getting hosts and visitors on board avoid negotiation and build a trust between users for transactions to take place [31].

Technological innovations, sociological and philosophical perspectives, and economic perspectives are the main drivers of social change. Airbnb followed those main drivers. It followed the growth of technological innovations as we move into the time frame of the Internet of Things and later of energy and distribution. People's mind-sets have moved from industrial economy with focus on product ownership to an experience economy. Airbnb is moving from a time frame of possession to the age of access which means they are even eliminating concepts of "property" and "ownership" [32].

Society is greatly affected by the growth of the sharing economy. The benefits are various, and we can extract some of them:

- Lower prices of goods and services caused by the expansion of the sharing economy
- Flexible working hours
- Use of unused capacity and encouragement of additional consumption
- Improving consumer experience and development of trust mechanisms between the user and the service provider [33]

In addition to the many positive effects, we need to consider the consequences of such business models. It is necessary to consider the main shortcomings or challenges posed by the sharing economy, such as lower personal income instability, the level of protection of workers and benefits, professional development and education, the issue of privacy and security of user data and access to sharing economy platforms [33]. The rise of the sharing economy may affect not only consumers' interest in purchasing products but also their preferences for product features and quality. Future researchers should consider hypothesis that bear in mind the consumer's decision of becoming providers and buyers of higher-quality products— luxury rather than budget brand products – to make themselves more attractive to users of the sharing economy. There are also few insights related to dynamics of such changes in behaviour as the sharing economy evolves of an optimal manner the efficient producers should accordingly adapt their product mixes.

4. Conclusion

The modern way of life is now in digital spaces and the platforms are the gateway. This research phenomenon has drawn scholars from a large number of different fields and disciplines into an emerging research area. Given the variety of perspectives represented, there is a great need to collect and connect what has been done, and to identify some common themes and threats, which will serve as a basis for future discussions on the crucial roles played by digital platforms in the sharing economy. Academic studies on sharing economy and related terms, supported by digital transformation results identify some trends and the ways in which the new sharing economy is disrupting the traditional way of doing business in high intensive labour sector as service sector is.

Finally, this paper presents advantages of digitalization that stimulated the emergence new technology as appropriate model, algorithms as sharing economy. Observations which are circumscribe often missing the deep insight on technology's role in terms of impacts in an area of social also time and space convergence. In that manner the advantages in time savings are found out in efficiencies of usage of digital tools. The enterprises will evaluate the complementary resources that they need and conduct strategic alliances (as DESI indicators support) to ensure the availability of resources for survival and to reduce environmental uncertainties. As many scholars of the sharing economy have explained, the act of sharing has roots in the ancient bartering systems and communal ways of life. However, it is only in the last two decades that an intense discussion of sharing and economic collaboration has risen up around the term 'sharing economy'.

This is in part due to the fact that, although sharing has been around for a long time, digital platforms and other large-scale mediating technologies have not. Indeed, much of the excitement over the sharing economy and collaborative consumption surrounds digitally-supported businesses and communities which have enjoyed commercial success in recent years, upsetting established businesses. As a result, the presence of such technologies has been put forward as a defining characteristic, separating those businesses and communities under the term "sharing economy" from traditional sharing contexts. Thus, we can describe the sharing economy as an economic model defined as a peer-to-peer based activity of acquiring, providing, or sharing access to goods and services that is often facilitated by an online platform. The sharing economy has grown out of the open-source community to refer to peer to peer-based sharing of access to goods and services, but the term is now often more broadly used to describe any sales transactions conducted online. At the base of the increasing concept of the SE is the role of digital technologies. In many conceptions, the SE system is predicated on some kind of efficient, scalable technology, which brings large networks of people together and matches them to the goods or services they need with a lot of comfort at the same time. The market successes of SE businesses, as well as the social futures of collaborative networks, are often tightly associated with the technologies on which they run. More broadly, the SE (Sharing Economy) presents some new contexts for the use of technology, and for the types of social relations which are carried out.

Dynamics of demand have implications as stimuli and challenge for definition of model or algorithm to support the sharing values in accordance with desirable experiences, addressed on info users. In that manner the dynamics of platforms' capacity should be considered through the appropriate models as well. Future research should have an emphasis on concept that respect a compromise between added values in time savings, simultaneously communicating with potential market niches, while sharing true values that multiply positive experience and shape future demand. While appearing as virtue of supplying services through the platforms, providers are effectively distributors of the platforms' service capacity. Sharing assets through sharing economy market channels is alternative, ecologically friendly mode of consumption, which brings more value to people from ownership and reduces costs of usage of goods and services [34]. However, current research into the sharing economy has some directions for additional research into important areas such as offering providers of intellectual assets to rent those assets. It should consider premise if non-commercial platforms sharing of intellectual assets would could be an additional promising avenue.

Future research should also clarify the possible implications, predict appropriate (reasonable and responsible) model of selection of users by providers and that is taken from experiences have been affecting on behaviour of users in practice.

5. References

[1] Ross, J. W., Sebastian, I. M., Beath, C., Scantelbury, S., Mocker, M., Fonstad, N., Kagan, M., Moloney, K., Krussel, S. G. (2016). Designing Digital Organizations, CISR Working Paper No. 406, Sloan School of Management, Cambridge, Massachusetts

[2] Bharadwaj, A., El Sawy, O.A., Pavlou, P.A., Venkatraman, N. (2013). Digital business strategy: Toward a next generation of insights. MIS Quarterly, Vol. 37 No. 2, 471-482, Available from: https://misq.org/misq/downloads/download/editorial/581/, Accessed on: 2021-10-15

[3] Haberli Junior, C., Oliveira, T., Yanaze, M., Spers, E. E. (2019). Performance, farmer perception, and the routinisation (RO) moderation on ERP post-implementation. Heliyon Vol.5, No. 6. Available from: http://www.scopus.com/inward/record.url?scp=85067337657&partnerID=8YFLogxK, Accessed on: 2021-03-23

[4] Morakanyane, R., Grace, A. A., O'Reilly, P. (2017). Conceptualizing Digital Transformation in Business Organizations: A Systematic Review of Literature, Proceedings of BLED 2017, 21, Available from: https://aisel.aisnet.org/bled2017/21

[5] Spremić. M. (2017). Governing digital technology – how mature IT governance can help in digital transformation?, International Journal of Economics and Management Systems, 2, 1; 214-223, Available from: https://www.bib.irb.hr/934731, Accessed on: 2021-10-15

[6] Biočina Lobaš, I., Račić, M. (2002). The importance of process reenngineering in the Hotel Industry // Annals of DAAAM for 2002. & Annals of DAAAM for 2002 & Proceedings of the 13th DAAAM International Symposium on Intelligent Manufacturing and Automation / Katalinic, Branko (Ed.).Vienna: DAAAM International, pp. 35-36, p.35.

[7] Schwab, K. (2017). The Fourth Industrial Revolution, Penguin Random House, New York

[8] Liao, Y., Deschamps, F., de Freitas Rocha Loures, E., Pierin Ramos, L. F. (2017). Past, present and future of Industry 4.0 - a systematic literature review and research agenda proposal, International Journal of Production Research, Vol. 55, No. 12, pp.3609-3629

[9] Vaidya, S., Ambad, P., Bhosle, S. (2018). Industry 4.0 - A Glimpse, Procedia Manufacturing, Vol. 20, pp.233-238

[10] Moavenzadeh, J. (2015). The 4th Industrial Revolution: Reshaping the Future of Production, World Economic Forum

[11] Geissbauer, R., Vedsø, J., Schrauf, S. (2016). A Strategist's Guide to Industry 4.0, Strategy + Business, Vol. 83.

[12] Baur, C., Wee, D. (2015). Manufactu ring's next act, McKinsey Quarterly.

[13] Rekettye, G., Rekettye, G. Jr. (2019). The Effects of Digitalization on Customer Experience, In Proceedings of the ENTRENOVA – ENTerprise REsearch InNOVAtion Conference (Online) 5, no. 1, 340-346. Available from: https://hrcak.srce.hr/ojs/index.php/entrenova/ article/view/13774/6921

[14] Spremić, M. (2020). Digital business transformation (Digitalna transformacija poslovanja), Ekonomski fakultet – Zagreb
[15] Bilovodska, O., Gaidabrus, N., Ruban, D. (2017). An analytical study on logistics outsourcing impact on logistical service quality in supply chains (case study: industrial enterprises of Sumy region). Problems and Perspectives in Management, 15(3), pp. 201-211. Available from: http://10.21511/ppm.15(3-1).2017.04.
[16] Babenko, V., Romanenkov, Yu., Yakymova, L., Nakisko A. (2017). Development of the model of minimax adaptive management of innovative processes at an enterprise with consideration of risks. Eastern-European Journal of Enterprise Technologies, Vol 5, No 4 (89), pp. 49-56.
[17] Bilovodska, O., Gryshchenko, O., Syhyda, L. (2016). Specific features of regional logistics potential consideration in the process of the enterprise distribution system development. Economic Annals-XXI, 160(7-8), pp. 105-110. https://doi.org/10.21003/ea.V160-21.
[18] Chmutova, I. & Vovk, V. & Bezrodna, O. (2017). Analytical tools to implement integrated bank financial management technologies. Economic Annals-XXI, Vol. 163(1-2(1)), pp. 95-99. Available from: http://soskin.info/ea/2017/163-1-2-1/201726.html
[19] Digital Economy and Society Index (DESI) (2020. Questions and Answers, 2020. Available from: https://ec.europa.eu/commission/presscorner/detail/en/qanda_20_102
[20] Frenken, K., Schor, J. (2017). Putting the sharing economy into perspective, Environmental Innovation and Societal Transitions, 23, ppt. 3-10.
[21] Puschmann, T., Alt, R. (2016). Sharing Economy, Business & Information Systems Engineering, 58 (1), ppt. 93-99.
[22] Gobble, M., M. (2017). Defining the Sharing Economy, Research-Technology Management, 60 (2), ppt. 59-63.
[23] Belk, R. (2007). Why not share rather than own?. The Annals of the American Academy of Political and Social Science, 611, 126-140
[24] Frechette, M. (2016). Peer-To-Peer Collaborative Consumption, Social Distance, ppt.1-116
[25] Botsman, R. (2013). Fact Company, The sharing economy lacks a shared definition. Available from: https://www.fastcompany.com/3022028/the-sharing-economy-lacks-a-shared-definition
[26] Allenm, D. & Berg, C. (2014). The sharing economy: How over-regulation could destroy. Available from: https://ipa.org.au/publications/2312/the-sharing-economy-how-over-regulation-could-destroy-an-economic-revolution
[27] Sajter, D., (2014). Sharing economy (Ekonomija dijeljenja), Available from: https://bib.irb.hr/datoteka/722791.Ekonomija_dijeljenja.pdf
[28] Dodig I., Rađenović Kozić B. (2018). Sharing economy and marketing aspects of its development (Ekonomija dijeljenja (sharing economy) i marketinški aspekti njenog razvoja)
[29] Demailly, D., Novel, A.-S. (2014). The sharing economy: make it sustainable, Studies N°03/14, IDDRI, Paris, France, 30 p.
[30] Botsman, R., & Rogers, R. (2011). What's mine is yours: how collaborative consumption is changing the way we live. Collins London. Avaible from http:// appli6.hec.fr/amo/Public/Files/Docs/241_fr.pdf
[31] Guttentag, D. (2013), Airbnb: disruptive innovation and the rise of an informal tourism accommodation sector, Current Issues in Tourism, pp. 1-26
[32] Jeroen Oskam, Albert Boswijk, (2016). Airbnb: the future of networked hospitality businesses, Journal of Tourism Futures,Vol. 2 Issue: 1, pp.22-42
[33] Brozović V., Hrstić F., Meštrović L., Bilić H., Naletina D., Petljak K, (2019.), Proceedings of the Faculty of Economics iz Zagreb (Zbornik Ekonomskog fakulteta u Zagrebu), year 17, nr. 1
[34] Brkljač, M, Sudarević, T. (2018). Sharing Economy and „Industry 4.0" as the Business, Environment of Millennial Generation - a Marketing Perspective, Annals of DAAAM for 2018. & Proceedings of the 29th DAAAM International Symposium on Intelligent Manufacturing and Automation / Katalinic, Branko (Ed.).Vienna: DAAAM International, pp. pp. 1092-1101

DOI: 10.2507/32nd.daaam.proceedings.103

A REVIEW OF EXISTING MASONRY ARCH BRIDGES IN CROATIA

Domagoj Trajber, Davorin Penava, Fulvio Rinaudo,
Vasilis Sarhosis & Lars Abrahamczyk

This Publication has to be referred as: Trajber, D[omagoj]; Penava, D[avorin]; Rinaudo, F[ulvio]; Sarhosis, V[asilis] & Abrahamczyk, L[ars] (2021). A Review of Existing Masonry Arch Bridges in Croatia, Proceedings of the 32nd DAAAM International Symposium, pp.0740-0744, B. Katalinic (Ed.), Published by DAAAM International, ISBN 978-3-902734-33-4, ISSN 1726-9679, Vienna, Austria
DOI: 10.2507/32nd.daaam.proceedings.103

Abstract

This study offers a review of masonry arch bridges in Croatia, it presents their most common typologies, geometric characteristics, and damages. Data about bridges have been collected from official institutions, state archives, and conservation institutes. Statistical analysis considering materials, arch geometry, arch thickness, number of spans, and superstructure type was conducted using collected data. The most common typologies are stone bridges with constant arch thickness, closed spandrel walls, and circular arch for the single-span and segmental arch for multi-span bridges. Additional data about geometry and damage of road brickwork bridges has been obtained with a rapid visual in-situ inspection. Observed damages indicate the need for further investigation for purpose of better understanding the capacity of these bridges, to secure their safe use through adequate maintenance and retrofitting.

Keywords: masonry arch bridges; construction material; typology; geometry; damage.

1. Introduction

There are approximately 300000 masonry arch bridges on railway and national road systems networks across Europe [1]. Papers e.g. [2], [3], [4] are just some of the papers published that offer a review of masonry arch bridges in some European countries. The majority of these bridges are over 100 years old. During their lifetime they have been exposed to natural hazards, extreme weather conditions, and ever-increasing traffic loads. These effects cause material degradation, deformation, and damage development which decreases the load-bearing capacity of bridges and brings their safety to question. This highlights the need for a reliable assessment method of masonry bridges. [5] gives a review of experimental investigations and assessment methods for masonry bridges. Currently, there is no unified database of masonry arch bridges in Croatia, nor are they classified in any manner. This hinders research that looks to assess their serviceability state and load-bearing capabilities. This work presents data collected on masonry arch bridges and results of statistical processing.

Statistical processing of data on masonry arch bridges in Croatia, obtained from authorized institutions: Croatian Railways, Croatian Roads and the County Road Administration, State Archives and Restoration Institutes, determined that there are approximately 178 masonry arch bridges. The majority of them are located on roadways (85%), while the rest are on railways (15%).

They were mostly built during the second half of the 19th and the first half of the 20th century out of locally available materials, clay bricks in continental regions and natural stone in mountain and coastal regions. Two of the oldest bridges (Figure 1.) that are still in use today were built at the beginning of the 18th century.

Data were collected and processed for railway and road bridges made out of both stone and clay bricks. However, further research will be limited only to brickwork road bridges. For that reason, more detailed data about geometry and damage of brickwork bridges was obtained by fast visual *in-situ* inspection. The need for a more detailed inspection to assess the condition of these bridges was identified based on observed damage, to continue their intended use and preserve durability with appropriate maintenance and repair measures.

Fig. 1. Old bridge Valpovo (top) (image source: author)
Bridge on river Dobra (bottom) (image source: Google)

2. Typological characteristics

Based on typological characteristics, masonry arch bridges have been classified in accordance with [4], with an adjustment that included segmental arches and excluded irregular arches (as they were not registered). Characteristics considered were: materials, arch geometry, arch thickness, number of spans and superstructure type (Table 1.).

Material	→	Brick
		Stone
		Combination of both
Arch geometry	→	Half-circle
		Parabolic
		Elliptical
		Segmental
		Inclined
Arch thickness	→	Constant
		Variable
Number of spans	→	Single-span
		Multi-span
Spandrel walls	→	Closed
		Open

Table 1. Classification of masonry arch bridges according to typological characteristics

Statistical analysis of collected data gives the following results:

- 86% of bridges are made out of natural stone, 13% out of brick, and the remaining 1% is a combination of both materials
- most represented types of arch geometry are segmental and circular arches, presenting 28% and 25% of bridge stock respectively.
- 69% of bridges have a constant arch thickness
- around 66% of bridges are single-span, while the rest (33%) are multi-span bridges
- the most common typologies are stone bridges with constant arch thickness, and closed spandrel walls, with circular arch for single-span (15%) and segmental arch for multi-span bridges (10%)

3. Geometrical characteristics

To compare masonry arch bridges according to their geometrical characteristics, data about the number of spans, span length, rise, arch geometry, and thickness of arch in crown and springing was. Bridges were then classified into six groups in accordance with [6], three of which are based on rise to span length ratio, and the remaining three on span length.

Groups based on rise to span length ratio are:

- shallow arches $0.00 < r/s \leq 0.25$,
- semi-shallow arches $0.25 < r/s \leq 0.40$,
- deep arches $0.40 < r/s$.

Groups based on span length are:

- short span $0.0 < s \leq 7.5$,
- medium span $7.5 < s \leq 15.0$,
- large span $15.0 < s$.

Figure 2 shows the relationship between the rise to span length ratio and span length, as well as corresponding areas of bridge groups. The frequency of bridges certain groups is shown in Table 2.

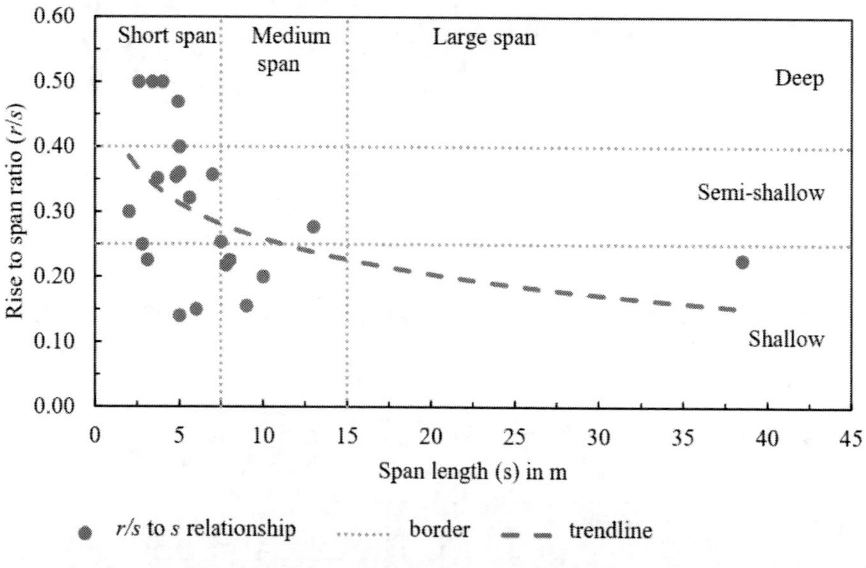

Fig. 2. Relation between rise to span length ratio (r/s) and span length (s)

	Short span	Medium span	Large span	Σ
Deep arch	**4 (18%)**	-	-	4 (18%)
Semi-shallow arch	**7 (30%)**	**1 (4%)**	-	8 (34%)
Shallow arch	**4 (18%)**	**6 (26%)**	**1 (4%)**	11 (48%)
Σ	15 (66%)	7 (30%)	1 (4%)	

Table 2. Classification of masonry arch bridges according to relation between rise to span length ratio (r/s) and span length (s)

Based on geometrical features, three reference brickwork spans of road bridges have been selected. Large-span bridges and bridges with deep arches were not considered here because of their low frequency. Geometrical characteristics of reference brickwork spans are shown in Table 3.

	Semi-shallow arch	Shallow arch
Short span	s = 4,70 m	s = 4,20 m
	r/s = 0,35	r/s = 0,20
	d_c = 0,60 m	d_c = 0,60 m
	(Bridge A)	(Bridge B)
Medium span	-	s = 8,40 m
		r/s = 0,20
		d_c = 0,80 m
		(Bridge C)

Table 3. Geometrical characteristics of reference brickwork spans

4. Damage and in-situ inspection of bridges

During their life span arch bridges are exposed to damages which affect their load-bearing capacities, and serviceability. Damages are mostly caused by deterioration of material due to environmental effects, age and inadequate maintenance. In accordance with [7] damages are classified into six categories:

- contamination – appearance of any type of plant vegetation or dirtiness
- deformation – geometry changes (deflection of backfill, protection or masonry, and slipping and swelling of masonry)
- destruction – the deterioration of physical and chemical structural features (such effects can be recognized for example by spalling and/or chipping of bricks)
- discontinuity – cracks
- displacement – displacement of the whole structure or its structural components (rotation or translation of spandrel walls)
- loss of material

In-situ visual inspection and classification of damages has been conducted for 22 brickwork road arch bridges (Fig. 3). Most commonly occurring damages are contamination, destruction and loss of material. Salt attack and moss growth are the most common types of contamination, and in some cases water penetration and heavy vegetation were noticed. There were also a few cases of significant discontinuities, like ring separation, and possible detachment of spandrel walls from arch. These kinds of damages cause significant loss of load-bearing capacity and should be repaired as soon as possible.

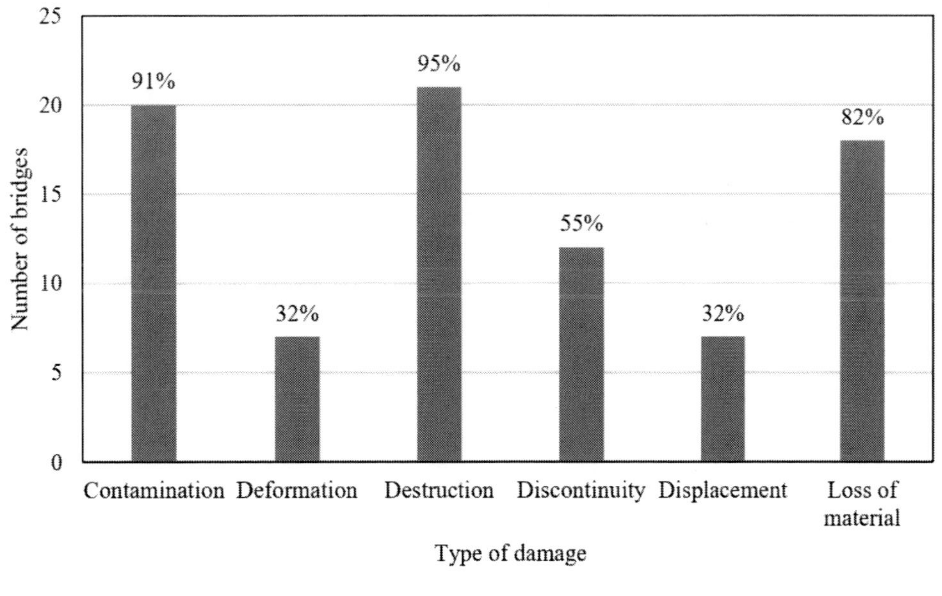

Fig. 3. Relation between rise to span length ratio (r/s) and span length (s)

5. Conclusion

Currently, there is no unified database of masonry arch bridges in Croatia, nor are they classified in any manner. This hinders research that looks to assess the serviceability state and load-bearing capabilities of those bridges. Data about bridges have been collected from official institutions, state archives, and conservation institutes, as well as by visual in-situ inspection. Classification of bridges by their typological features was conducted, based on which most common typologies were found. Since further work will be focused on brickwork roadway bridges additional classification of them by their geometrical features gave three reference brickwork spans. Reference brickwork span will be used to develop experimental and computer models, to provide a reliable assessment of serviceability state and load-bearing capacity of bridges. Damages observed during visual in-situ inspection imply the need for further detailed inspection to assess the state of bridges, to secure their safe use through adequate maintenance and retrofitting. The number of masonry arch bridges in Croatia is modest when compared to other European countries. However, they are an essential part of transportation infrastructure and construction heritage.

6. Acknowledgments

This research has been conducted with the support of the Erasmus+ programme of the European Union project "Environmental Risk Assessment and Mitigation on Cultural Heritage Assets in Central Asia - ERAMCA" (Project Reference: 609574-EPP-1-2019-1-IT-EPPKA2-CBHE-JP) and Croatian Science Foundation (HRZZ) "Young Researchers' Career Development Project – Training of Doctoral Students" programme. The authors are also grateful to Croatian Railways, Croatian Roads, County Road Administrations, and Croatian State Archives for providing the data that contributed to this research.

7. References

[1] Brencich, A.; Morbiducci, R. (2007). "Masonry Arches: Historical Rules and Modern Mechanics", International journal architectural heritage, Vol. 1, No. 2, 2007, pp. 165-189., DOI: 10.1080/15583050701312926.

[2] De Santis, S.; De Felice, G. (2014). "Overview of railway masonry bridges with a safety factor estimate", International journal architectural heritage, Vol. 8, No. 3, 2014, pp. 452-474., DOI: 10.1080/15583058.2013.826298.

[3] Page, J. (1993). Masonry arch bridges. State of the art review, The Stationery Office, ISBN: 0115511903, United Kingdom

[4] Bień, J.; Kamiński, T. (2004). Masonry arch bridges in Poland, Proceedings of the 4th International Conference on Arch Bridges, Roca, P. & Oñate, E. (Ed.), pp. 10, Barcelona

[5] Sarhosis, V.; De Santis, S. & De Felice, G. (2016). "A review of experimental investigations and assessment methods for masonry arch bridges", Structure and Infrastructure Engineering, Vol. 12, No. 11, 2016, pp. 1439-1464., ISSN: 1573 – 2479, DOI: 10.1080/15732479.2015.1136655.

[6] Olivera, D. V.; Lourenco, P. B. & Lemos, C. (2010). "Geometric issues and ultimate load capacity of masonry arch bridges from the northwest Iberian Peninsula", Engineering Structures, Vol. 32, No. 12, 2010, pp. 3955-3965, DOI: 10.1016/j.engstruct.2010.09.006.

[7] Bień, J.; Kamiński, T. (2007). Damages to masonry arch bridges - proposal for treminology unification, Proceedings of the 5th International Conference on Arch Bridges, ISBN: 9789728692315, Lourenço, Paulo, B. (Ed.), pp. 341-348, Madeira

DAAAM International Vienna
Bahnallee 10
A-2120 Wolkersdorf
Austria, EU

ISBN 978-1-7138-4139-5

CURRAN ASSOCIATES INC.
proceedings
.com